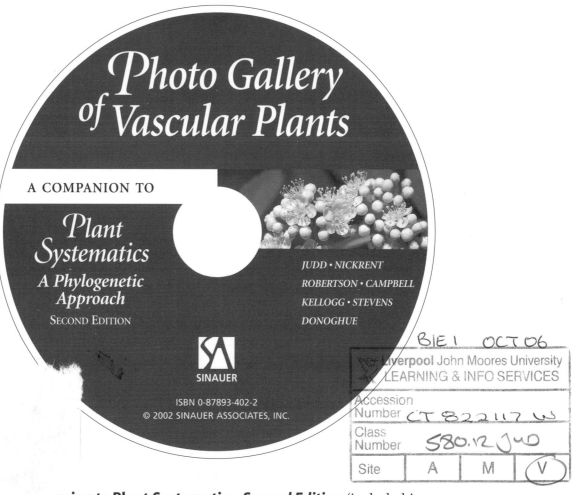

Photo Gallery of Vascular Plants

A COMPANION TO

Plant Systematics
A Phylogenetic Approach
SECOND EDITION

JUDD • NICKRENT
ROBERTSON • CAMPBELL
KELLOGG • STEVENS
DONOGHUE

SINAUER

ISBN 0-87893-402-2
© 2002 SINAUER ASSOCIATES, INC.

...*anion to* **Plant Systematics,** *Second Edition* (included in
... of the book) contains over 2,200 color photographs
... the variability of the vascular plant families covered in the
... 1,300 species from 175 families are represented, most with
... photos, including habit, flower, fruit and sections of flowers
...nd longitudinal sections). The photographs are organized by
...ame, by scientific name, and by phylogenetic arrangement.
...⊃ also includes a glossary of plant terminology, in which each
... is linked to photographs that illustrate the term. Three supple-
...al appendices (PDF files) list the families covered in the book
...rding to the taxonomic systems of (1) the Angiosperm Phylogeny
...p, (2) Arthur Cronquist, and (3) Robert Thorne.

Plant Systematics

A PHYLOGENETIC APPROACH

Second Edition

Plant Systematics

A PHYLOGENETIC APPROACH

Second Edition

Walter S. Judd
University of Florida

Christopher S. Campbell
University of Maine

Elizabeth A. Kellogg
University of Missouri, St. Louis

Peter F. Stevens
*University of Missouri, St. Louis,
and Missouri Botanical Garden*

Michael J. Donoghue
Yale University

Sinauer Associates, Inc. • Publishers
Sunderland, Massachusetts U.S.A.

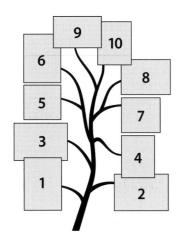

The Cover

(1) Club moss, *Lycopodium annotinum* (Lycopodiaceae). (2) Fern, *Blechnum tuerckheimii* (Polypodiaceae). (3) Conifer, *Pinus elliottii* (Pinaceae). (4) Magnoliid, *Magnolia virginiana* (Magnoliaceae). (5, 6) Monocots: orchid, *Eriopsis biloba* (Orchidaceae); grass, *Andropogon glomeratus* (Poaceae). (7, 8) Rosids: *Merania brevipedunculata* (Melastomataceae); *Sorbus aucuparia* (Rosaceae). (9, 10) Asterids: *Leucothoe racemosa* (Ericaceae); *Cubanola domingensis* (Rubiaceae). Photographs of *E. biloba* by W. Mark Whitten, of *M. brevipedunculata* by J. Dan Skean, and of *C. domingensis* by Scott Zona; all others by Walter S. Judd.

Plant Systematics: A Phylogenetic Approach, Second Edition

Sinauer Associates, Inc.
23 Plumtree Road
Sunderland, MA 01375-0407   USA
FAX: 413-549-1118
EMAIL: orders@sinauer.com, publish@sinauer.com
www.sinauer.com

Library of Congress Cataloging-in-Publication Data

Plant systematics: a phylogenetic approach / Walter S. Judd . . . [et al.].--2nd ed.
        p.   cm.
    Includes bibliographical references.
    ISBN 0-87893-403-0 (cloth)
    Botany--Classification.   I. Judd, Walter S.
QK95.P548   2002
580'.1'2--dc21

2002010248
CIP

Printed in U.S.A.

5 4 3

With appreciation, we dedicate this textbook to our mentor and friend,
Carroll E. Wood, Jr.,
whose kindness and knowledge of plant systematics
have helped many students and colleagues.

Contents in Brief

Contents

6 The Evolution of Plant Diversity 121

7 An Overview of Green Plant Phylogeny 153

Lycophytes, Ferns and Their Allies, and Extant Gymnosperms 185

Phylogenetic Relationships of Angiosperms 217

Preface

We wrote this book because we sensed a need for a treatment of plant systematics (or taxonomy) that can be used in undergraduate courses and that incorporates recent exciting developments in the field. In the past two decades abundant new systematic data, especially from DNA sequences, have come forth, and a rigorous phylogenetic approach has established novel ways of analyzing data and practicing systematics.

The first edition of this book was well received by many students and other botanists. It has been translated into French and Italian, and the authors of the first edition (Walter S. Judd, Christopher S. Campbell, Elizabeth A. Kellogg, and Peter F. Stevens) received the 1999 Engler Silver Award, an annual award of the International Association of Plant Systematists for the most outstanding book or monograph in plant systematics. Encouraged by this positive reception, we prepared this second edition.

The basis for this book is the tree of life—the idea that all life is interrelated, like branches on a tree. We deal with the part of this tree occupied by green plants and focus on the vascular plants or tracheophytes, an important group that dominates most of Earth's terrestrial ecosystems and provides us with medicines, beautiful ornamentals, fiber for paper and clothing, and most of our food. Historically, information about these and other attributes of vascular plants was essential to the development of human civilization. Survival rested upon knowing which plants were good to eat; which were poisonous to people or to potential food animals; which were good for tools and weapons; which might cure ailments; and which might be useful in many other ways. Systematics as a science grew out of such interests in plants and animals and evolved from systems of information about the uses of different organisms to a broader understanding of biodiversity. To this end, plant systematists ask questions such as, How should the tremendous diversity within the orchid family be partitioned or classified into genera and then into species? What are the relationships among the major groups of grasses? How do we know whether or not a plant is a member of the oak family? This text deals with how these and thousands of similar questions are answered; the final two chapters present some current answers to questions about relationships among the major groups of tracheophytes.

Charles Darwin led the way in establishing the evolutionary perspective of modern biology. One impact of this view of life was to make it a major goal of systematics to uncover the evolutionary history—or phylogeny—of groups of organisms. Willi Hennig and W. H. Wagner, along with others, developed explicit methods of hypothesizing and testing phylogenetic relationships and reflecting these in classifications. We follow a phylogenetic approach throughout this book.

Chapter 1 introduces and explains the concept of phylogeny, and Chapter 2 explains how phylogenies are reconstructed from systematic evidence. The process of transforming evidence into a representation of phylogenetic relationships in an evolutionary tree is first presented using simple hypothetical data and then extended to complex real situations. This "primer of phylogeny" also incorporates the conversion of phylogenetic hypotheses into a classification.

The central position of phylogeny in classification has emerged only recently, but many groups of plants, such as orchids, grasses, and oaks, have long been recognized. Chapter 3 outlines how systematists perceived groups in the past and the historical background of phylogeny. Over several centuries of study, a tremendous store of information about vascular plant relationships has accumulated. This information, which is the basis for the formulation of phylogenies and classifications, comes from a wide array of sources, including structural aspects of organisms, especially morphology (Chapter 4), and molecular features, especially DNA (Chapter 5). These chapters on phylogeny, classification, and taxonomic evidence provide essential information for understanding the vascular plant groups we cover.

Diversification and the evolution of plant species—how plant species form, how they interact, and how we define them—are the fascinating and challenging topics of Chapter 6. Chapter 7, which is new in this second edition, presents an overview of the evolutionary history of plants and explains the origin of many of the important features used to identify plants. Chapter 7 lays the foundation for the final two chapters, which focus on plant diversity.

Chapter 8 covers tracheophytes that do not have flowers, including ferns and related plants plus conifers. Although these plants are certainly important, they are far surpassed in terms of number of species and ecological dominance by the flowering plants or angiosperms, the subject of Chapter 9. Chapters 8 and 9 present the diversity of these groups, with descriptions, illustrations, and discussions of phylogenetic relationships. Exceptionally high-quality illustrations from the Generic Flora of the Southeastern United States, which show taxonomically useful information for many families of plants, are a special asset of Chapters 8 and 9.

Chapters 8 and 9 also emphasize characterization of major groups of plants, especially plant families. Two appendices provide additional information related to plant identification, including botanical nomenclature (the application of scientific names), collection of scientific plant specimens, an overview of plant identification, information about the plant systematics literature, and some useful sites on the World Wide Web.

An important tool for plant identification is the CD that accompanies this book. It contains over 2200 images of flowers, fruits, detailed dissections of flowers and fruits, and other parts of plants. Students can access these images through an alphabetical list of species, an alphabetical list of families (with the species listed alphabetically within each), or a list of orders and families following the arrangement adopted in this text. The CD was an effective study aid for students using the first edition of this text, and it is greatly expanded in number of images, in floral and fruit dissections, and in an illustrated glossary, a list of terms with links to particular images on the CD illustrating the term.

The science of plant systematics has a large number of specialized terms. Such terms used in this book are not only included in the glossary, but they are also boldfaced and defined in the first six chapters and included in the Subject Index, where they are also boldfaced.

Scientific names (and common names) are included in the Taxonomic Index.

Our knowledge of plant systematics is growing very rapidly, making this an exciting time to study plant systematics. We encourage students to get caught up in this excitement, to appreciate the beauty and importance of plants, and to follow the advice of poet and botanist Johann Goethe (1749–1832): "Begin it now!"

Acknowledgments

This book is a group effort. Although each of us wrote certain chapters, we all benefited from numerous helpful comments from our coauthors. Walter S. Judd had primary responsibility for Chapters 4 and 9, and the two appendices. Christopher S. Campbell was responsible for Chapters 1, 6, and 8; he also contributed the treatment of Rosaceae in Chapter 9, the sections on pollination biology, embryology, chromosomes, and palynology in Chapter 4, and the section on the World Wide Web in Appendix 2. Elizabeth A. Kellogg wrote Chapter 5, as well as the treatments for Poaceae, Juncaceae, Cyperaceae, and Santalales in Chapter 9, and the section on arguments against ranks in classifications in Appendix 1. Peter F. Stevens wrote Chapter 3 and Michael J. Donoghue wrote Chapter 7. Chapter 2 was co-written by Elizabeth and Walter.

The CD was the responsibility of Daniel L. Nickrent, Kenneth R. Robertson, Walter S. Judd, and Christopher S. Campbell; this group also provided most of the photographs. Walter compiled the three appendices on the CD.

Carroll E. Wood, Jr. generously gave us permission to use numerous plates prepared for the Generic Flora of the Southeastern United States and assisted in their organization and electronic digitalization. These beautiful illustrations, which greatly enhance the value of this text, were drawn by a series of exceptionally talented artists:

IAS Ihsan Al-Shehbaz

YS Virginia Savage

Irene Brady

KSV KS Karen Stoutsenberger Velmure

ADC Arnold D. Clapman

LT. LaVerne Trautz

SD Sydney B. DeVore

mvm Margaret van Montfrans

DCJ Diane C. Johnston

LA Vorobik 1989 Linda A. Vorobik

Dorothy H. Marsh

Rachel A. Wheeler

Sue Sargent (not initialed)

Wendy B. Zomlefer

Carroll also provided editorial assistance with the figure captions. Edward O. Wilson gave permission to use the previously unpublished plate of *Yucca filamentosa* and its pollinator; we thank Michael D. Frohlich for supplying materials for this plate and Kathy Horton for assistance in locating and transmitting it. Robert Dressler kindly allowed the use of his original illustration of *Encyclia cordigera*. H.-Dietmar Behnke generously prepared the informative plate illustrating various sieve-element plastid conditions, and Y. Renea Taylor prepared the photo of epicuticular waxes. We are grateful for the numerous colleagues who contributed other illustrative materials; they are acknowledged in the figure captions.

Allison R. Minott assisted in the preparation of the captions for Chapter 9 and Reuben E. Judd helped take the photographs in Figure 1 of Appendix 1. Brian Moore and Mary Walsh helped with the figures for Chapter 7; Susan Donoghue edited the manuscript of this chapter. J. Richard Abbott, Alison E. Colwell, Steven P. Darwin, Peter K. Endress, Miguel A. García, Litton J. Musselman, Roger W. Sanders, David S. Seigler, J. Dan Skean, Douglas E. Soltis, Margaret H. Stone, W. Mark Whitten, and Scott Zona provided some very useful slides for the CD that accompanies this text, and Jason Dirks and Christopher Small handled technical aspects relating to its production. We are deeply grateful for all the above-listed contributions.

We express our sincere appreciation to the following individuals, who read and commented on various sections of the book and/or provided useful reprints or unpublished manuscripts: Pedro Acevedo, Victor A. Albert, Lawrence A. Alice, Ihsan Al-Shebaz, Arne A. Anderberg, William R. Anderson, George W. Argus, Daniel F. Austin, David S. Barrington, David A. Baum, Paul E. Berry, Camilla P. Campbell, Lisa M. Campbell, Philip D. Cantino, Heather R. Carlisle, Mark W. Chase, Lynn G. Clark, David S. Conant, Garrett E. Crow, Steven P. Darwin, Claude W. de Pamphilis, Alison C. Dibble, James A. Doyle, Robert Dressler, Mary E. Endress, Peter Goldblatt, Shirley A. Graham, Michael H. Grayum, Arthur D. Haines, Peter C. Hoch, Sara B. Hoot, Joachim W. Kadereit, Christine M. Kampny, Robert Kral, Kathleen A. Kron, Matthew Lavin, Steven R. Manchester, Paul S. Manos, Lucinda A. McDade, Alan W. Meerow, Laura C. Merrick, David R. Morgan, Cynthia M. Morton, Daniel L. Nickrent, Eliane M. Norman, Richard G. Olmstead, Clifford R. Parks, Gregory M. Plunkett, Robert A. Price, John F. Pruski, Kathleen M. Pryer, Susanne S. Renner, Karen S. Renzaglia, James L. Reveal, Kenneth R. Robertson, Edward E. Schilling, Alan R. Smith, Douglas E. Soltis, William L. Stern, Henk van der Werff, Thomas F. Vining, Terrence Walters, Grady L. Webster, W. Mark Whitten, John H. Wiersema, Norris H. Williams, Wesley A. Wright, George Yatskievych, Wendy B. Zomlefer, and Scott A. Zona. We also thank the numerous individuals who sent corrections and/or suggestions relating to the first edition.

Walter thanks the University of Florida, College of Liberal Arts and Sciences, for providing a one-term sabbatical during which he was able to initiate work on this text.

We thank Andy Sinauer and all the staff at Sinauer Associates for their excellent advice and guidance, and we especially thank Carol Wigg, Jason Dirks, Christopher Small, Jefferson Johnson, and Marie Scavotto. Without their valuable work this project would never have reached completion.

Finally, Christopher, Michael, and Walter thank their wives, Margaret, Susan, and Beverly, respectively, for their emotional support and for their forbearance with their husband's preoccupation with plant systematics and this book. Elizabeth and Peter thank their son, Harry Stevens, for continuing to be patient with his parents.

We, the authors, assume all editorial responsibility for this book. Comments and corrections on the first edition have been helpful in preparing the second edition, and we would greatly appreciate any comments or corrections on this edition as well.

WALTER S. JUDD, *Gainesville, Florida*

CHRISTOPHER S. CAMPBELL, *Orono, Maine*

ELIZABETH A. KELLOGG, *St. Louis, Missouri*

PETER F. STEVENS, *St. Louis, Missouri*

MICHAEL J. DONOGHUE, *New Haven, Connecticut*

1

The Science of Plant Systematics

[T]he characters which naturalists consider as showing true affinity between any two or more species, are those which have been inherited from a common parent, all true classification being genealogical.

Charles Darwin 1859: 391

What exactly is plant systematics? The question turns out to be more difficult than you may have imagined because both *plant* and *systematics* are rather hard to define. Considering these concepts in some detail will help us to define the science and clarify our aims.

What Do We Mean by Plant?

Most people have a commonsense notion of what a plant is: A plant is something green that doesn't move around very much. For some it includes fungi, which are not green, and botany and plant biology departments in many colleges and universities include people who study fungi (mycologists). For other people, the word *plant* is restricted to green organisms living on land, and the word *algae* is used for most aquatic plants. But people who study algae also usually reside in botany departments. Obviously, there are conflicting views.

1

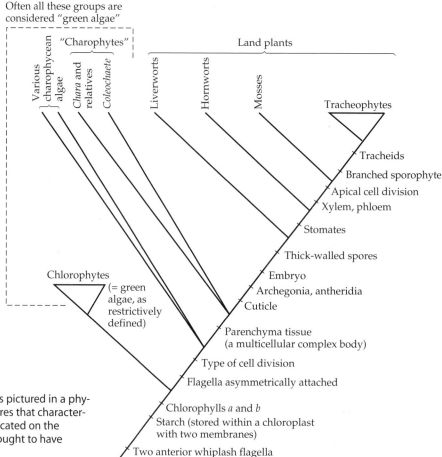

Figure 1.1 Phylogeny of green plants (as pictured in a phylogenetic tree). Distinctive structural features that characterize various groups of green plants are indicated on the branches where the characteristics are thought to have evolved.

How do we understand *plant* for the purposes of this book? In keeping with the phylogenetic approach we take in this book, we wish to focus on a single branch of the evolutionary tree of life. But which one? The fungi do not appear to be related to any other group that has been called a plant, but instead are probably more closely related to animals. This still leaves us with quite a diversity of photosynthetic organisms. Several of these lineages—including red algae, brown algae, and related groups—probably originated separately from one another and are independent endosymbiotic associations between eukaryotic organisms and bacteria. We will concentrate on the major branch known as the green plants. This lineage includes the so-called green algae and the land plants (Figure 1.1), which share a number of features, including (1) the presence of the photosynthetic pigments chlorophylls *a* and *b*, (2) storage of carbohydrates, usually in the form of starch, and (3) the presence of two anterior whiplash flagella (often modified or sometimes lost).

Within the green plants we will concentrate on the land plants—that is, the **embryophytes** (a few of which actually live in water), whose closest extant relatives are members of the "charophytes," a green algal group. The land plants have a life history involving alternation of two morphologically distinct bodies (diploid sporophyte and haploid gametophyte), thick-walled spores, an embryonic stage in the life cycle, specialized structures that protect the gametes (archegonium for eggs and antheridium for sperm), and a cuticle (a waxy protective layer over the epidermal cells). Numerous DNA characters also show that this group represents a single branch on the tree of life.

The land plants consist of three groups of rather small plants—liverworts, hornworts, and mosses—and the **tracheophytes**. Tracheophytes, sometimes referred to as **vascular plants**, are by far the largest group of green plants, including about 260,000 species. They form the dominant vegetation over much of Earth's land surface and are the focus of this book.

The great majority of tracheophytes, all but about 12,000 species, are **flowering plants**, or **angiosperms**. In importance to the world's ecosystems and to human nutrition, medicine, and overall welfare, flowering plants far surpass the other tracheophytes. Most of this book is devoted to angiosperms. However, the basic principles of systematics used here apply to all organisms.

What Do We Mean by Systematics?

Systematics is the science of organismal diversity. It entails the discovery, description, and interpretation of biological diversity, as well as the synthesis of information on diversity in the form of predictive classification systems. According to paleontologist George Gaylord Simpson (1961: 7), "Systematics is the scientific study of the kinds and diversity of organisms and of any and all relationships among them." This view is so broad that it may be seen as encompassing what we normally think of as ecology, and perhaps other disciplines, so it is necessary to consider in more detail the kinds of relationships that have specifically concerned systematists.

In our view, the fundamental aim of systematics is to discover all the branches of the evolutionary tree of life, to document all the changes that have occurred during the evolution of these branches, and to describe all species—the tips of these branches. Systematics is therefore the study of the biological diversity that exists on Earth today and its evolutionary history.

Systematists endeavor to reconstruct the entire chronicle of evolutionary events, including the splitting of populations into separate lineages and any and all evolutionary modifications in the characteristics of organisms associated with these branching events, as well as with periods between branching. A secondary but critical aim of systematists is to convey their knowledge of the tree of life—of the terminal branches and all their phylogenetic relationships to one another—in an unambiguous system of classification, which can then orient our understanding of life and the rest of the world around us. This is the phylogenetic approach to systematics.

We explicitly take the view that systematics is not just a descriptive science, but also aims to discover evolutionary relationships and real evolutionary entities that have resulted from the process of evolution. We assume that evolution has occurred and is occurring, so we take as a starting point the separation of a lineage into two or more lineages. We also assume that evolutionary modifications within these lineages have happened and will continue to happen.

The central problem, then, is to reconstruct the history of the separation of lineages and the history of their modifications as accurately as possible by bringing as much relevant information as possible to bear on the problem. One of the things systematists continually do is to put forward hypotheses about the existence of branches of the tree of life and test them with evidence derived from a wide variety of sources. Alternative hypotheses are evaluated, and some are provisionally chosen over others.

Some systematists see their work rather differently. They think of themselves as attempting to achieve a theory-neutral description of the similarities and differences evident in the organisms around us. They view branching diagrams (such as Figures 1.1 and 1.2) or classifications (such as Figure 1.3) only as efficient depictions of these similarities and differences. According to this approach, entities recognized by systematists are summaries of observed information and no more, whereas in our view these entities are hypothesized branches of the evolutionary tree. Thus the approach we have adopted extends beyond summation of the data at hand to statements about entities that we do not directly observe, but that we have reason to believe came into being through the evolutionary process.

This tension between theory-neutral and theory-grounded approaches pervades the history of science. There have always been those who consider theory-neutral observations to be both possible and desirable and who wish to define the basic terms of scientific discipline in terms of particular operations performed on the data. And there are those, like us, whose concepts, definitions, and inferential procedures are explicitly grounded in theory, and who wish to go beyond summarizing the evidence at hand to making claims about the world at large. Clearly, this is not an issue we can settle here.

To some, the distinction we have just made may seem small, and in practice, it is true that systematists with rather different views of their activities conduct their research in much the same way. We highlight the difference here because it may help interested readers to comprehend some of the literature of systematics and because it helps to explain the orientation of our own treatment of plant systematics. Most importantly, throughout this book we will focus on how we interpret evidence of all kinds in relation to the fundamental aim of systematics stated here.

A Phylogenetic Approach

Our view of the science of plant systematics also explains an outlook that we hope will emerge again and again throughout this book—namely, that systematics is linked directly and centrally to the study of evolution in general, from the study of fossils to the study of genetic changes in local populations. The basic connection is extraordinarily simple: Studies of the process of evolution benefit (usually enormously!) from knowledge of what we think has actually taken place during the evolution of life on Earth. For example, when one sets out a hypothesis about the evolution of a particular characteristic of an organism, it is assumed that the trait of interest did in fact originate within the group being studied. Furthermore, such hypotheses usually rely on some knowledge of the precursor condition from which the trait in question arose.

This kind of information on the sequence of evolutionary events is obtained from the systematist who has reconstructed the **phylogeny**, the evolutionary chronicle, of the organisms. Similarly, studies of the rate of evolutionary change and of the ages and diversification patterns of lineages depend directly on knowledge of phylogenetic relationships.

Whereas there has sometimes been a tendency to think of systematics as a discipline completely separate from evolutionary biology, we think instead that it occupies a central position in that field. Systematics will play an increasingly important role in many other disciplines, including ecology, molecular biology, developmental biology, and even linguistics and philosophy.

A phylogeny consists of simple sets of statements of the following nature: Groups A and B are more closely related to each other than either is to group C. Consider the simple example of three members of the rose family (Rosaceae): blackberry, cherry, and raspberry. Following the injunction "by their fruits ye shall know them," we can infer evolutionary relationships using only fruits as evidence. Blackberries and raspberries both have numerous small and fleshy fruits (drupes or stone fruits) that are clustered together (see Chapter 4 for a description of fruit types). Cherry fruits are also drupes, but they are borne singly and are larger than those of blackberries and raspberries.

With this information about fruits, we would judge that blackberries and raspberries are more closely related to each other than either is to cherries. This is equivalent to saying that blackberries and raspberries share a more recent **common ancestor** than either does with cherries. Blackberries and raspberries are said to be **sister groups**, or closest relatives. An abundance of other evidence, from other structural features, chemical constituents, and DNA sequences, leads to the same conclusions about these relationships. We can present these phylogenetic relationships as a phylogenetic tree (Figure 1.2). This is one of the simplest trees in this book, which has scores of other trees.

How does an interest in phylogeny affect the groups of organisms, or **taxa** (singular **taxon**), recognized by systematists? A phylogenetic approach demands that taxa be **monophyletic.** To see how one identifies monophyletic taxa, consider the example in Figure 1.2. Assume for the sake of this example that the rose family contains only these three groups (it is, of course, much larger; see Chapter 9); that all three groups are themselves monophyletic; and that the figure shows the true phylogenetic relationships among the three groups.

Which subsets of the three groups are monophyletic? There are three possible subsets of two of the three taxa: (1) blackberries and raspberries, (2) blackberries and cherries, and (3) cherries and raspberries. Only subset 1 includes all the entities on a single branch of the tree and is therefore monophyletic. The two components of the definition of monophyly—*entities* and *single branch*—may be restated as *all the descendants of a common ancestor.* Monophyletic groups are also referred to as **clades.**

Another way to understand monophyly, which may be useful for larger phylogenetic trees, is with the simple rule that a monophyletic group is one that can be removed from the tree by one "cut" of the tree. Try this with Figure 1.2; removal of a nonmonophyletic group, such as blackberries and cherries, requires two cuts. Chapter 2 discusses monophyly in more detail, as well as how we interpret evidence for or against it.

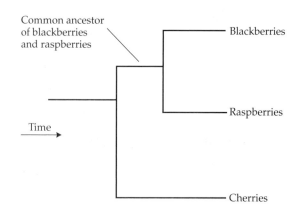

Figure 1.2 Simple phylogeny of three groups in the rose family.

This particular definition of monophyly has been adopted only recently (see Chapter 3), and many traditionally recognized plant taxa are not monophyletic by this definition. A familiar example of a commonly recognized group that is not monophyletic is the "dicots." These angiosperms have characteristics, such as two cotyledons and often flowers with multiples of four or five parts, that make them easily recognized. Nevertheless, they do not form a monophyletic group. The monocots, which apparently are monophyletic, are also descendants of the common ancestor of the "dicots," and the monocots are nested within the "dicots." Thus the "dicots" do not contain all the descendants of their common ancestor, and more than one cut is required to remove them from the tree of life.

In keeping with our phylogenetic approach, we recognize only monophyletic groups, and we reject groups that are not monophyletic. We reject the "dicots" as a formal group, and place the names of this and similar groups in quotation marks.

In some cases, the evidence is not unambiguously for or against monophyly. For example, the extant gymnosperms, which include familiar plants such as pines and redwoods, do not form a monophyletic group according to some data, while other data support their monophyly. Recent molecular data generally support extant gymnosperm monophyly, but further study may well show that the group is not monophyletic. We provisionally recognize extant gymnosperms in our classification. Further discussion of this difficult issue can be found in Chapters 7 and 8.

An important exception to the rule of monophyly in the recognition of taxa occurs at the level of species. The problem with monophyly at the species level has to do with the nature of relationships above and below the level of species. Above the level of species, the tree of life generally splits into separate branches, as in Figures 1.1 and 1.2. This is so because blackberries and cherries, for example, do not cross or hybridize with one another. Within species, in contrast, branches join through mating between members of a species. Thus, during the separation of one species into two, matings may occur between members of the nascent lineages such that one cannot identify a common ancestor that is unique to either or both species. This problem and others concerning species are taken up in Chapter 6.

The Practice of Plant Systematics

Two important activities of plant systematists are classification and identification. **Classification** is the placement of an entity in a logically organized scheme of relationships. For organisms, this scheme is usually hierarchical, consisting of large, inclusive groupings such as the green plant kingdom (all green plants), and less inclusive, progressively nested groups such as families, genera, and species. The largest, most inclusive groups are the three great domains of life: the Bacteria, the Archaea, and the Eukarya. Within the Eukarya are several kingdoms, including the animals, the fungi, and the green plants. Phylogenetic studies deal with groups ranging from domains to species. About 1.5 million species of organisms have been named, but the Earth supports perhaps 10 to 20 times that number of species.

Figure 1.3 presents an example of the place of one plant species, *Aster tenuifolius* (the perennial salt marsh aster), in a hierarchical classification scheme (see also Appendix 1).

In addition to storing information about organisms, classifications may be useful as predictors. For example, the discovery of biochemical precursors of the drug cortisone in certain species of yams (the genus *Dioscorea* in the family Dioscoreaceae; see Chapter 9) prompted a search for, and subsequent discovery of, higher concentrations of the drug in other *Dioscorea* species (Jeffrey 1982). The fact that these species are relatives of yams made it likely that they would share genetically controlled features, such as many chemical constituents, with the yams. The more a classification reflects phylogenetic relationships, the more predictive it will be.

Although classification has always focused on describing and grouping organisms, it has only relatively recently been concerned with phylogenetic relationships. Publication of Charles Darwin's *On the origin of species* in 1859 stimulated the incorporation of general evolutionary relationships into classification, an ongoing process that has yet to be fully realized (de Queiroz and Gauthier 1992). A critical step in this process has been the development of a phylogenetic perspective, to which Willi Hennig (a German entomologist, 1913–1976), Walter Zimmermann (a German botanist, 1892–1980), Warren H. Wagner, Jr. (a North American botanist, 1920–2000), and many others have contributed.

Evolution has long been recognized as an integral part of plant systematics (Lawrence 1951). Phylogenetic systematists focus on the outcome of the processes of descent with modification and separation of lineages

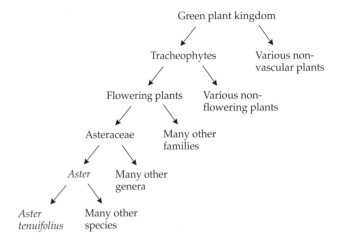

Figure 1.3 Portion of a hierarchical classification, showing the placement of the species *Aster tenuifolius*, the perennial salt marsh aster. Downward-pointing arrows indicate groups that are nested within the group above. In each case there are two arrows, one leading to the group containing the perennial salt marsh aster and the other to all other groups at the same level in the hierarchy.

(de Queiroz and Gauthier 1992), and their goal is to discover the products of these processes and to chronicle the evolutionary events that have resulted in present diversity.

Systematics encompasses the discipline of **taxonomy**, a term tied to the word *taxon*. In taxonomy, groups of organisms are described and assigned scientific names. Names of taxa give us access to information about them, and thus it is valuable to have one name by which all can refer to a group of plants. This is especially true for species, which hold a special place in terms of usefulness and overall importance to humanity. The application of scientific names is the field of **biological nomenclature** (see Appendix 1).

Identification involves determining whether an unknown plant belongs to a known, named group of plants. In temperate regions, where the flora is generally well known, one can usually match a plant with a known species. An environmental consultant conducting an inventory of the plants of a salt marsh in the temperate zone may find a plant, an aster, that he/she cannot identify to species. There are numerous species of asters (a genus with the scientific name *Aster*), and they are sometimes difficult to distinguish from one another. The consultant needs to record information about the plant that will make its identification possible, but does not want to damage a plant of an unknown species because it might be rare or endangered. Therefore, he/she takes careful notes and perhaps a photograph to document the appearance of the plant. It may be helpful in some circumstances to collect a specimen, which can be preserved by pressing and drying, in order to accurately identify a plant (see Appendix 2).

There are three primary ways to identify an unknown plant. The fastest way is to ask someone, such as a professional botanist or well-trained naturalist, who knows the plants of a region. If our consultant does not know such a person, he/she could use the literature about the plants of the region. For most temperate regions there are books devoted to naming the plants, some of which cover all the plants, others of which may focus on a portion of the flora. Perhaps our consultant owns pertinent books and determines that the unknown plant is the perennial salt marsh aster, *Aster tenuifolius*. A third path to identification is to visit an **herbarium** (plural **herbaria**), a facility for the storage of scientific plant collections that is a standard component of universities and botanical institutions, and match the specimen with a named specimen there.

Plant identification in the tropics is far more challenging than in the temperate zone because tropical floras contain many more species than temperate floras, tropical species are far less well studied, and many tropical species have not yet been recognized, described, named, and collected for herbaria. Here the role of the specialist is critically important, yet there are fewer specialists in this area every year.

Why Is Systematics Important?

Systematics is essential to our understanding of and communication about the natural world. The basic activities of systematics—classification and naming—are ancient human methods of dealing with information about the natural world, and early in human cultural evolution they led to remarkably sophisticated classifications of important organisms. We depend on many species for food, shelter, fiber for clothing and paper, medicines, tools, dyes, and myriad other uses, and we can use these species in part because of our systematic understanding of the biota.

Systematics plays an essential role in many endeavors. Knowledge of systematics, for example, guides the search for plants of potential commercial importance. In the 1960s, while studying plants in the Peruvian Andes, botanist Hugh Iltis collected a species of the tomato or potato genus (*Solanum*). Iltis knew that wild relatives of the cultivated tomato could be important for the breeding of improved tomato crops. He sent some seeds to tomato geneticist Charles Rick in California, who identified the plant as a new species, *Solanum chmielewskii*, named in honor of the late Tadeusz Chmielewski, a Polish tomato geneticist.

Rick crossed this wild relative with a cultivated tomato and thus introduced genes that markedly improved the taste of tomatoes. Similar advances, by the hundreds, have improved yield, disease resistance, and other desirable traits in crops, commercial timber species, and horticultural varieties. Systematics is also critical in biological sciences involving diversity, such as conservation biology, ecology, and ethnobotany.

Systematics advances our knowledge of evolution

because it establishes a historical context for understanding a wide variety of biological phenomena, such as ecological diversification and specialization, coevolutionary relationships of hosts and parasites and of plants and their pollinators, biogeography, adaptation, speciation, and rates of evolution. Three examples will be presented to show the importance of systematic approaches to evolutionary biology.

The Hawaiian silverswords of the daisy family (Asteraceae) are like many other groups on the Hawaiian Islands in exemplifying evolutionary radiations. The silversword alliance is a monophyletic group of 28 species in three genera (*Argyroxiphium*, *Dubautia*, and *Wilkesia*) that is endemic to the Hawaiian Islands and that evolved from a single founding individual of a Californian ancestral species (Baldwin and Robichaux 1995). This group includes some of the most visually spectacular members of the Hawaiian flora. *Argyroxiphium* species, for example, have a basal rosette of green- or silvery-haired, swordlike leaves from which emerges a terminal inflorescence up to 2 m tall and with as many as 600 large heads (Figure 1.4).

In addition to rosette plants, members of this group have diversified into many other growth forms, including trees, shrubs, subshrubs, cushion plants, and lianas. Silverswords occupy an exceptionally wide range of habitats, from 75 to 3750 m in elevation, and from less than 400 to more than 12,300 mm in annual precipitation, and a common pathway in the evolutionary radiation of the group apparently involves inter-island dispersal followed by numerous, major ecological shifts along moisture gradients and into bogs.

We can see this pattern of radiation in a phylogeny of the silversword alliance. Figure 1.5 shows a molecular

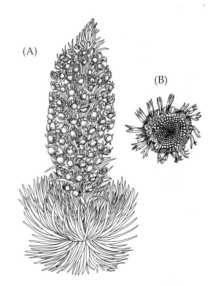

(A)

(B)

Figure 1.4   Habit of one of the Hawaiian silverswords, *Argyroxiphium sandwicense*, showing (A) the basal cluster of sword-shaped leaves and the massive inflorescence, which can be up to 2 m tall and (B) one head of flowers. (From Baldwin and Robichaux 1995.)

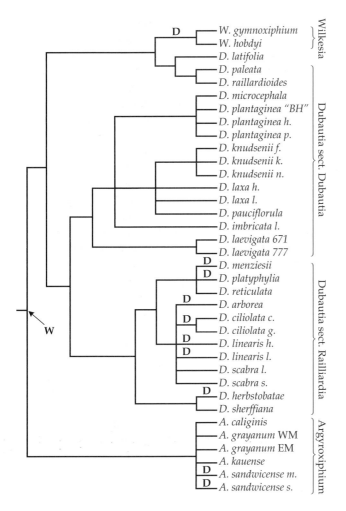

Figure 1.5 Phylogenetic tree of the three genera (*Argyro-xiphium, Dubautia,* and *Wilkesia*) and 25 species of Hawaiian silverswords, based on sequences from nuclear ribosomal DNA internal transcribed spacers. The letters above the branches of the tree indicate the habitat (dry, D, or wet, W) of the lineage defined by the branch. (From Baldwin and Robichaux 1995.)

nated and had many of the reproductive features of the traveler's tree.

The family may have evolved in areas, such as Madagascar, where there are few alternative pollinators, such as large pollinators that fly. The rest of the family consists of two genera: one in South Africa that has adapted to pollination by sunbirds, the other in South America that is specialized for pollination by bats. The sunbird-pollinated plants have brightly colored flowers to attract sunbirds, and the bat-pollinated plants bear flowers that are held well away from the canopy so that they are easily accessible to the flying bats and that open at night when bats are active. The floral features associated with pollination by vertebrates and other aspects of pollination biology are dealt with in Chapter 4.

The southern beeches (the genus *Nothofagus* of the southern beech family, Nothofagaceae) exemplify the value of phylogeny to biogeography, the study of the geographic distribution of organisms. There are about 35 species of southern beeches, and they now grow in southern South America, New Zealand, Australia, New Caledonia, and New Guinea (Figure 1.7). They are represented by fossil pollen, flowers, and leaves in these areas and Antarctica but nowhere else in the world.

These areas are parts of the former supercontinent East Gondwana that the forces of continental drift separated starting about 80 million years ago. **Continental drift** consists of a series of separations, like the splitting of lineages that make up a phylogeny. The geological history of the breakup of East Gondwana as it is currently understood is presented in Figure 1.8. Geological evidence, for example, shows that New Caledonia plus New Zealand separated as one landmass from the remainder of the continent about 80 million years ago and separated from one another sometime between 70 and 40 million years ago.

Phylogenetic relationships of the species of *Nothofagus* are well understood from studies of nuclear DNA sequences, morphology, and chloroplast DNA sequences (Figure 1.9) (Manos 1997). The genus is divided into four clades, which correspond to subgenera. Subgenus *Brassospora* contains two clades: one that grows only on New Caledonia, and one that is confined to New Guinea. Subgenus *Nothofagus* is limited to South America, and the final two subgenera—*Fuscospora* and *Lophozonia*—both have species in South America, Australia, and New Zealand. *Nothofagus* was an important part of the flora of East Gondwana, and all four subgenera had evolved before the breakup of the supercontinent.

tree of the group, which has been inferred from analyses of DNA sequences. When the habitat preferences of silversword species are mapped onto the molecular tree, it is clear that there have been numerous shifts from wet to dry habitats accompanying the evolution of the species. Some of the genetic aspects of silversword diversification are presented in Chapter 6.

The second example concerns evolutionary adaptations for pollination. The traveler's tree (*Ravenala madagascariensis*, Strelitziaceae) is an endemic tree of Madagascar up to 30 m tall with a crown of large leaves that resemble those of banana plants (Figure 1.6A). The flowers of traveler's trees (Figure 1.6C) are pollinated by the roughed lemur (Figure 1.6B), which is highly dependent on the floral nectar of the traveler's tree at certain times of the year.

Traveler's tree flowers are adapted for lemur pollination in being easily accessible to these climbing animals, enclosed in protective structures that require a strong pollinator to open, tough and able to withstand rough handling, and a large source of sucrose-dominant nectar for a sizable animal. Studies of the phylogeny and pollination biology of the Strelitziaceae (Kress et al. 1994) suggest that the ancestor of this family was lemur-polli-

Figure 1.6   (A) The traveler's tree (*Ravenala madagascariensis*, Strelitziaceae) of Madagascar and (B) its pollinator, the black and white ruffed lemur (*Varecia variegata variegata*). The flowers (C) are in groups of 10 to 16 that are surrounded by tough protective structures (bracts). Abundant nectar is released into the base of the flower, and while drinking nectar, a lemur brushes against the pollen-bearing parts of the flower. Pollen stuck to hairs on the lemur is transported to other flowers of the traveler's tree and deposited on the stigma (the receptive surface) to complete pollination. (From Kress et al. 1994.)

The splitting of the once continuous geographic distribution of a group of organisms through continental drift, followed by the evolution of the group on separated continents, is well known in many groups of organisms. How much of this process, which is called **vicariance**, explains the current distribution of southern beeches? An alternative explanation, long-distance fruit dispersal, is possible but seems unlikely over great expanses of ocean for the relatively large fruits (nuts) of these plants.

Analyses of *Nothofagus* phylogeny, the *Nothofagus* fossil record, and the geological history of East Gondwana give most of the credit for the pattern of geographic distributions in the genus to continental drift. In subgenus *Fuscospora*, a postulated vicariance event occurred with the initial breakup of East Gondwana about 80 million years ago (see Figures 1.8 and 1.9). Similarly, the division of subgenus *Brassospora* into clades in New Caledonia and New Guinea represents a vicariance event corresponding to the same first splitting of the supercontinent. The separation of Antarctica/South America and the Australian areas at least 35 million years ago was the basis of a vicariance event in subgenus *Lophozonia*.

Aims and Organization of This Book

This book takes a phylogenetic approach to plant systematics. Chapter 2 establishes the basic concepts and

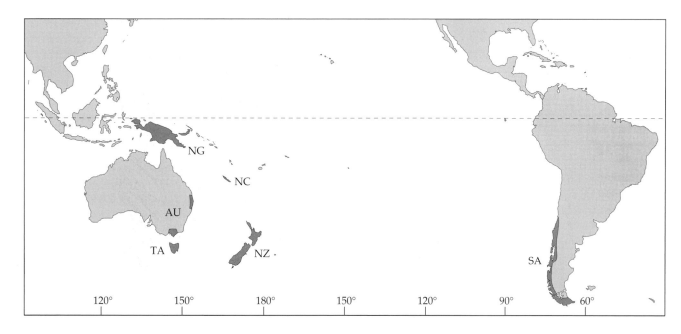

Figure 1.7 Current geographic distribution of the southern beeches (*Nothofagus*). These areas, along with Antarctica, made up the supercontinent of East Gondwana. The southern beeches were widespread on East Gondwana, and groups of southern beeches (see Figure 1.9) were separated by continental drift into their current distribution. AU, Australia; NC, New Caledonia; NG, New Guinea; NZ, New Zealand; SA, South America; TA, Tasmania. (From Swenson et al. 2001.)

practices of phylogenetic systematics. To appreciate this approach, it is important to understand some of the history of plant systematics, the topic of Chapter 3.

We do not know the actual phylogeny of any nontrivial group of organisms; instead we must infer phylogenies from the data available to us. Systematics is sometimes considered a science without its own data, instead taking data from a wide range of other disciplines. These data include both structural aspects of organisms (anatomy and morphology, or the external features of organisms) and molecular features (biochemical constituents, such as proteins and flavonoids, and DNA). Chapter 4 is devoted to structural and non-DNA molecular systematic evidence; Chapter 5 focuses on evidence from DNA sequences.

Chapter 6 is about plant diversification and the evolution of plant species. How are species formed and maintained? How do we determine that two individuals belong to the same or different species? How do hybridization, polyploidy, and breeding systems shape diversification?

Chapter 7 presents an overview of the evolutionary history of plants, explaining the origin of many of the important features used to identify plants, and creating a foundation for the final two chapters.

This book approaches identification by focusing on plant families, a good entry level for coverage of plant diversity. There are far too many plant species to learn effectively in an undergraduate course. Moreover, many plant families—such as the oak, pine, rose, grass, mustard, pea, and orchid families—are already familiar to many people and are relatively easy to recognize. Learning to recognize important families provides a classification within which one can then learn genera and species.

Common methods of identifying organisms are illustrations, verbal descriptions, and **keys**. Keys organize

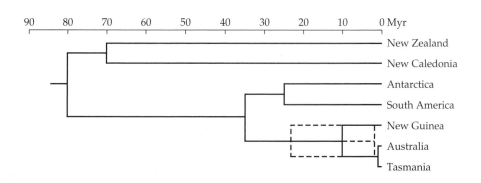

Figure 1.8 Tree showing the hypothesized breakup of East Gondwana, with a time scale in millions of years. The timing of the split between New Guinea and Australia is uncertain. (From Swenson et al. 2001.)

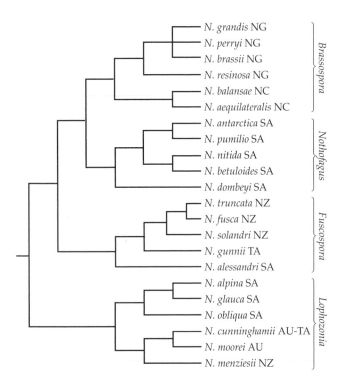

Figure 1.9 Phylogeny of 22 species of southern beeches (the genus *Nothofagus*), showing the four subgenera: *Brassospora*, *Nothofagus*, *Fuscospora*, and *Lophozonia*. Letters following the species are their current locations; abbreviations of geographic areas as in Figure 1.7. (From Manos 1997.)

The CD contains over 2200 images of flowers, fruits, and other parts of plants that are useful for learning about the groups covered in this book. These features are useful for identification, as are images that show detailed dissections of flowers and fruits. The CD also contains an illustrated glossary and a list of terms with links to particular images that illustrate the terms.

The CD was an effective study aid for students using the first edition of this text, and now, greatly expanded in number of images and with floral and fruit dissections, it is an even better resource. Images in the CD show many of the synapomorphies and field identification characters in the treatments of families in this text. The images not only complement the family descriptions, but also facilitate recognition of living representatives of important plant families.

The CD contains three appendices, one each for the families covered in this text arranged in the major plant classification systems of Cronquist (1981), Thorne (1992), and the Angiosperm Phylogeny Group (1998, 2002). The latter is largely followed in this text.

This edition of *Plant Systematics* has a detailed glossary in the printed text as well as a glossary on the CD. Entries in the CD glossary are linked to one or more images that illustrate the defined feature. In addition, the book provides both taxonomic and subject indices.

It is important to appreciate that our knowledge of plant systematics is growing very rapidly. New phylogenetic hypotheses are emerging at a rapid pace, and we can expect major changes even within the next few years. Under these circumstances it is impossible for a textbook to remain current, and it will undoubtedly be necessary to augment the material presented here with additional information, perhaps obtained through computer databases (see Appendix 2 for a discussion of plant systematics on the World Wide Web).

In this book we present our current understanding of phylogenetic relationships, but it will be obvious that in many cases great uncertainties remain. We fully expect that some (perhaps many) of the hypotheses of phylogenetic relationships described here will be modified or even completely abandoned in favor of others. Students may find it frustrating that, even in a discipline as old as plant systematics, our knowledge still needs frequent revision. We hope instead that this rapid change will be viewed positively, as a measure of the excitement and vitality of the field.

As in any science, knowledge in plant systematics is forever provisional and must change to reflect new discoveries. Fortunately, recent developments in the logic and methods of inferring phylogenetic relationships, along with the availability of new sources of evidence, give us hope of achieving an increasingly accurate picture of evolutionary history. We will be satisfied if others join in the fun and help us achieve a better understanding of plant systematics.

information about a group, such as the families of conifers, in ways that facilitate identification. Chapters 8 and 9, which focus on plant diversity, contain numerous illustrations, descriptions, and keys. Chapter 8 covers the tracheophytes that are not angiosperms—specifically, 18 families from eight major groups: lycophytes (including club mosses and related plants), Psilotaceae, Equisetaceae (the horsetails), ferns, cycads, Ginkgoaceae, conifers (pines, spruces, and others), and gnetophytes.

Angiosperm diversity, covered in Chapter 9, is very great; we describe more than 140 angiosperm families. The angiosperms include all grain and major vegetable crops, the broad-leaved trees such as oaks and maples, and plants adapted for growth in almost every kind of habitat the world offers. As noted earlier in the chapter, because of our phylogenetic approach we have endeavored to define families that are consistent with what is known about their phylogeny.

Two appendices deal with important components of the practice of systematics. In Appendix 1 we explain botanical nomenclature, the application of scientific names to plants. Appendix 2 covers the collection of scientific plant specimens, gives an overview of plant identification, and provides a brief guide to keeping abreast of future advances in plant systematics, using both the plant systematics literature and the World Wide Web.

Literature Cited and Suggested Readings

Angiosperm Phylogeny Group. 1998. An ordinal classification for the families of flowering plants. *Ann. Missouri Bot. Gard.* 85: 531–553.

Angiosperm Phylogeny Group. 2002. An update of the Phylogeny Group classification for the orders and families of flowering plants: APGII. *Bot. J. Linnaean Soc.* In press.

Baldwin, B. G. and R. H. Robichaux. 1995. Historical biogeography and ecology of the Hawaiian silversword alliance (Asteraceae). New molecular phylogenetic perspectives. In *Hawaiian biogeography. Evolution on a hot spot archipelago*, W. L. Wagner and V. A. Funk (eds.), 259–287. Smithsonian Institution Press, Washington, DC.

Bremer, K. and H. Wanntrop. 1978. Phylogenetic systematics in botany. *Taxon* 27: 317–329.

Briggs, D. and S. M. Walters. 1997. *Plant variation and evolution*, 3rd ed. Cambridge University Press, Cambridge, UK.

Cronquist, A. 1981. *An integrated system of classification of flowering plants*. Columbia University Press, New York.

Darwin, C. 1859. *On the origin of species*. Mentor edition, 1958. New American Library, New York.

Davis, P. H. and V. H. Heywood. 1963. *Principles of angiosperm taxonomy*. Oliver & Boyd, Edinburgh, Scotland.

de Queiroz, K. and J. Gauthier. 1992. Phylogenetic taxonomy. *Annu. Rev. Ecol. Syst.* 23: 449–480.

Donoghue, M. J. and J. W. Kadereit. 1992. Walter Zimmerman and the growth of phylogenetic theory. *Syst. Biol.* 41: 74–85.

Graham, L. E., M. E. Cook and J. S. Busse. 2000. The origin of plants: Body plan changes contributing to a major evolutionary radiation. *Proc. Natl. Acad. Sci. USA* 97: 4535–4540.

Hennig, W. 1966. *Phylogenetic systematics*. University of Illinois Press, Urbana.

Hoch, P. C. and A. G. Stephenson. 1995. *Experimental and molecular approaches to plant biosystematics*. Missouri Botanical Garden, St. Louis.

Iltis, H. H. 1988. Serendipity in the exploration of biodiversity: What good are weedy tomatoes? In *Biodiversity*, E. O. Wilson and F. M. Peter (eds.), 98–105. National Academy Press, Washington, DC.

Jeffrey, C. 1982. *An introduction to plant taxonomy*. Cambridge University Press, Cambridge, UK.

Kress, W. J., G. E. Schatz, M. Andrianifahanana and H. S. Morland. 1994. Pollination of *Ravenala madagascariensis* (Strelitziaceae) by lemurs in Madagascar: Evidence for an archaic coevolutionary system. *Am. J. Bot.* 81: 542–551.

Lawrence, G. H. M. 1951. *The taxonomy of vascular plants*. Macmillan, New York.

Manos, P. S. 1997. Systematics of *Nothofagus* (Nothofagaceae) based on rDNA spacer sequences (ITS): Taxonomic congruence with morphology and plastid sequences. *Am. J. Bot.* 84: 1137–1155.

Niklas, K. J. 1997. *The evolutionary biology of plants*. University of Chicago Press, Chicago.

Radford, A. E., W. C. Dickison, J. R. Massey and C. R. Bell. 1974. *Vascular plant systematics*. Harper & Row, New York.

Renzaglia, K. S., R. J. Duff, D. L. Nickrent and D. J. Garbary. 2000. Vegetative and reproductive innovations of early land plants: Implications for a unified phylogeny. *Philos. Trans. R. Soc. London [Biol]* 355: 769–793.

Simpson, G. G. 1961. *Principles of animal taxonomy*. Columbia University Press, New York.

Stace, C. A. 1980. *Plant taxonomy and biosystematics*. University Park Press, Baltimore.

Stuessy, T. F. 1990. *Plant taxonomy: The systematic evaluation of comparative data*. Columbia University Press, New York.

Swenson, U., R. S. Hill and S. McGloughlin. 2001. Biogeography of *Nothofagus* supports the sequence of Gondwana breakup. *Taxon* 50: 1025–1041.

Thorne, R. F. 1992. The classification and geography of the flowering plants. *Bot. Rev.* 58: 225–348.

2 Methods and Principles of Biological Systematics

Biological systematics (or taxonomy) is the theory and practice of grouping individuals into species, arranging those species into larger groups, and giving those groups names, thus producing a **classification**. Classifications are used to organize information about plants, and keys can be constructed to identify plants.

There are many ways to construct a classification. For example, plants could be classified on the basis of their medicinal properties (as they are in some systems of herbal medicine) or on the basis of their preferred habitats (as they may be in some ecological classifications). A phylogeny-based classification, such as the one followed in this book, attempts to arrange organisms into groups on the basis of their evolutionary relationships.

There are two main steps in producing such a classification. The first is determining the **phylogeny**, or evolutionary history. The second is basing the classification on this history. These two steps can be, and often are, separated, such that every new theory of relationships does not lead automatically to a new classification. This chapter will outline how one goes about determining the history of a group, and then will discuss briefly how one might construct a classification, given that history.

Figure 2.1 Evolution of two hypothetical species. Each circle (or square) represents a plant. A mutation in the lineage on the left causes a change to woody stems, which is then transmitted to descendant plants. Woody-stemmed plants gradually replace all the herbaceous ones in the population. A similar mutation in the lineage on the right leads to a group with red petals.

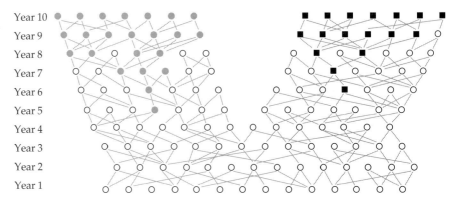

○ Petals white, stems herbaceous, leaves non-hairy, stamens five, fruit dry, seed coat smooth

● Petals white, stems woody, leaves non-hairy, stamens five, fruit dry, seed coat smooth

■ Petals red, stems herbaceous, leaves non-hairy, stamens five, fruit dry, seed coat smooth

What Is a Phylogeny?

As described in Chapter 1, evolution is not simply descent with modification, but also involves the separation of lineages. Imagine for a moment a population of organisms that all look similar to each other. By some process the population divides into two populations, and these two populations go on to evolve independently. In other words, two **lineages** (ancestor–descendant sequences of populations) are established. We know this has happened because the members of the two new populations acquire, by the process of mutation, new characteristics in their genes, and possibly changes in their overall form, making the members of one population look more similar to each other than to members of the other population or to the ancestral population. These characteristics are the evidence for evolution.

For example, a set of plants will produce offspring that are genetically related to their parents, as indicated by the lines in Figure 2.1. The offspring will produce more offspring, so we can view the population over several generations, with genetic connections indicated by lines.

If the population divides into two separate populations, each will have its own set of genetic connections, and eventually will acquire distinctive characteristics. For example, the population on the right could develop red flowers, and the stems of the population on the left could become woody. Red flowers and woodiness are evidence that each of the two populations constitutes a single lineage.

The same process can repeat, and each of the new populations can divide (Figure 2.2). Again, we know this has happened because of a new set of characteristics

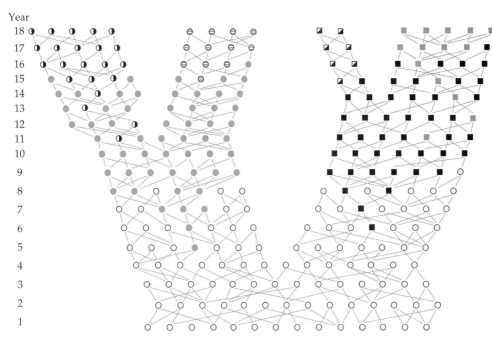

○ Petals white, stems herbaceous, leaves non-hairy, stamens five, fruit dry, seed coat smooth

● Petals white, stems woody, leaves non-hairy, stamens five, fruit dry, seed coat smooth

⊖ Petals white, stems woody, leaves non-hairy, stamens five, fruit dry, seed coat spiny

◐ Petals white, stems woody, leaves non-hairy, stamens five, fruit fleshy, seed coat smooth

■ Petals red, stems herbaceous, leaves non-hairy, stamens five, fruit dry, seed coat smooth

◼ Petals red, stems herbaceous, leaves hairy, stamens five, fruit dry, seed coat smooth

◪ Petals red, stems herbaceous, leaves non-hairy, stamens four, fruit dry, seed coat smooth

Figure 2.2 The same hypothetical set of plants as in Figure 2.1 after 8 years and two more speciation events.

(A)

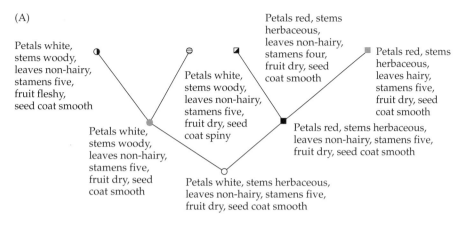

Petals white,
stems woody,
leaves non-hairy,
stamens five,
fruit fleshy,
seed coat smooth

Petals white,
stems woody,
leaves non-hairy,
stamens five,
fruit dry, seed
coat spiny

Petals white,
stems woody,
leaves non-hairy,
stamens five,
fruit dry, seed
coat smooth

Petals red, stems
herbaceous,
leaves non-hairy,
stamens four,
fruit dry, seed
coat smooth

Petals red, stems
herbaceous,
leaves hairy,
stamens five,
fruit dry, seed
coat smooth

Petals red, stems herbaceous,
leaves non-hairy, stamens five,
fruit dry, seed coat smooth

Petals white, stems herbaceous,
leaves non-hairy, stamens five,
fruit dry, seed coat smooth

(B)

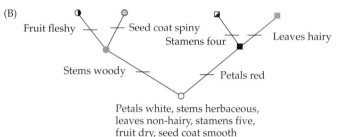

Fruit fleshy — Seed coat spiny
Stamens four — Leaves hairy
Stems woody — Petals red

Petals white, stems herbaceous,
leaves non-hairy, stamens five,
fruit dry, seed coat smooth

Figure 2.3 (A) A simple way to redraw the pattern of change shown in Figure 2.2. Full descriptions are provided for each of the ancestors and their descendants. (B) A simpler way to redraw Figure 2.3A, showing only the mutations that occurred in the various lineages.

acquired by the newly formed populations. Some of the woody plants have fleshy fruits, and another group has a spiny seed coat. Meanwhile, some of the red-flowered plants now have only four stamens, and another set of red-flowered plants have hairy leaves.

The characteristics of plants, such as flower color or stem structure, are generally referred to as **characters**. Each character can have different values, or **character states**. In this case the character "flower color" has two states: white and red. The character "stem structure" also has two states—woody and herbaceous—and so forth. All else being equal, plants with the same state are more likely to be related than those with different states.

The critical point in this example, however, is that characteristics such as red petals and woody stems are new, and they are **derived** relative to the ancestral population. Only such new characters tell us that a new lineage has been established; retention of the old characteristic (white flowers, herbaceous stems, non-hairy leaves, five stamens, dry fruit, smooth seed coat) does not tell us anything about what has happened.

A character state that is derived at one point in time will become ancestral later. In Figure 2.2, woody stems are derived relative to the original population, but they are ancestral relative to the groups with fleshy fruits or spiny seed coats.

A group composed of an ancestor and all of its descendants is known as a **monophyletic group** (*mono*, single; *phylum*, lineage). We can recognize it because of the shared derived characters of the group (synapomorphies). **Synapomorphies** are character states that have arisen in the ancestor of the group and are present in all of its members (albeit sometimes in modified form). This

concept was first formalized by Hennig (1966) and Wagner (1980).

The diagrams of Figures 2.1 and 2.2 are cumbersome to draw, but each of them can be summarized as a branching tree (Figure 2.3A). It is also inconvenient to repeat the ancestral character states retained in every group, so systematists commonly note only the characters that have changed, placing tick marks on the appropriate branches to indicate the relative order in which the character states originated (Figure 2.3B).

The shared derived characters in Figure 2.3B can be arranged in a hierarchy from more inclusive (e.g., stems woody or petals red) to less inclusive (e.g., leaves hairy, seed coat spiny). These then lead to the obvious conclusion that the plants themselves can be arranged in a hierarchical classification that is a reflection of their evolutionary history. The plants could be divided into two groups: one with herbaceous stems and red petals, the other with woody stems and white petals. Each of these groups could also be divided into two groups. Thus the classification could be derived directly from the phylogeny.

Note that the hierarchy is not changed by the order in which the branch tips are drawn. The shape, or **topology**, of the tree is determined only by the connections between the branches. We can tell the evolutionary "story" by starting at any point in the tree and working up or down. This means that the terms *higher* and *lower* are not really meaningful, but simply reflect how we have chosen to draw the evolutionary tree.

From this point of view, a plant systematics course could as well begin by covering the Asteraceae, which some textbooks consider an "advanced" family, and then working out to other members of the asterid clade, as by starting with the so-called "primitive" families, such as Magnoliaceae and Nymphaeaceae. The latter simply share a set of characters thought to be ancestral, but these are combined with a large set of derived characters as well.

(A)

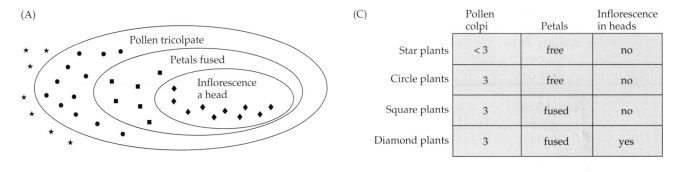

(C)

	Pollen colpi	Petals	Inflorescence in heads
Star plants	< 3	free	no
Circle plants	3	free	no
Square plants	3	fused	no
Diamond plants	3	fused	yes

(B)

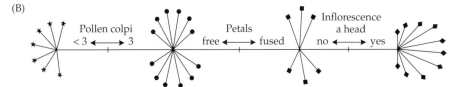

Figure 2.4   (A) Venn diagram of a set of plants. A large group has tricolpate pollen. Of those plants, a smaller group has fused petals, and of the plants with tricolpate pollen and fused petals, a subset has flowers arranged in a head. (B) The pattern of part A redrawn as an unrooted network. (C) The pattern of parts A and B redrawn as a matrix.

Determining Evolutionary History

CHARACTERS, CHARACTER STATES, AND NETWORKS

In the example in Figures 2.1, 2.2, and 2.3, we have described evolution as though we were there watching it happen. This is rarely possible, of course, so part of the challenge of systematics is to determine what went on in the past.

Extant species must first be examined closely for characteristics that are believed to be heritable. These can be any aspect of the plant that can be passed down genetically through evolutionary time and still be recognizable. For example, petal color, inflorescence structure, and plant habit are all known to be under genetic control, and they are generally stably inherited from one generation to the next. They thus would be good taxonomic characters. Many examples of such heritable characters are described in Chapter 4. Characters of DNA and RNA (described in Chapter 5) can also be used.

It may be harder than you think to determine which structures in one plant can be compared to structures in another plant. Two structures may be similar because they are (1) in a similar position in the different organisms, (2) similar in their cellular and histological structure, or (3) linked by intermediate forms (either intermediates at different developmental stages of the same organism or intermediates in different organisms). These numbered criteria are **Remane's criteria** of similarity. (Remane called them criteria of homology, but we are using the term *homology* in a more restricted sense.) The assessment of similarity is the basis of all of comparative biology and of systematics in particular.

Systematics entails the precise observation of organisms. Without accurate comparative morphology, classification of any sort is impossible. Without careful description of characters and their states, phylogeny reconstruction and the description of history are meaningless.

Determining similarity is the first step in determining **homology**, or identity by descent. Be aware, however, that the word *homology* has many different meanings, and it is impossible to summarize the literature here. Many phylogenetic systematists argue that homology can be determined only by construction of an evolutionary tree—a viewpoint that will be followed in this text. When reading the literature, it is worth checking what particular authors mean when they use the term.

From observations of plants, groups that share particular characteristics can be identified. For example, a large group of plant species has pollen with three grooves, or germination furrows, called colpi (singular *colpus*); the pollen is thus described as tricolpate. Within this large group is a smaller group that has fused petals, and within the fused-petal group is a still smaller group with flowers arranged in a head. These nested groups can be diagrammed as a set of ovals (a **Venn diagram**) as in Figure 2.4A, with the individual shapes (stars, circles, squares, diamonds) representing different plant species.

The same information can be drawn as a **network** (Figure 2.4B). In this case the number of colpi in pollen, fusion of petals, and type of inflorescence are shown as vertical lines. Whereas all shapes (species) to the left of the pollen line have fewer than three colpi, everything to the right of the line has tricolpate pollen, as indicated by the numeral 3. Likewise, the line for petals indicates a shift between free petals and fused petals, and the inflorescence line indicates a shift between flowers clustered in a head and flowers borne separately.

All the plants indicated by the same shape are drawn as though they arose at the same point in time. The reason is that for the purposes of this simplified example, we have not provided any information on their order of origin.

We can count the number of changes along the network, which is defined as the length of the network. For example, proceeding from right to left, there is one change each in inflorescences, petals, and pollen grooves, so the network can be described as having a length of 3.

The same information can also be presented as a **matrix** (plural **matrices**) (Figure 2.4C). This time the rows correspond to plants, and the columns correspond to characters of the plants. The character states are then used to fill in the matrix. These are, or are hypothesized to be, genetic changes that potentially distinguish the groups of plants in the matrix. Thus the three changes in the network of Figure 2.4B represent three changes in character states or genes of plants.

In the example here we have implied that determination of the character states is perfectly obvious. This is often not the case, however, particularly with morphological characters. The variation among similar structures must be described by dividing the character into character states. This is a hypothesis of underlying genetic control, although it is rarely framed this way. For example, if two species differ in the color of their flowers, we may score the character petal color as having two states: red and blue. By scoring it this way, we are hypothesizing that underlying genes have switched, over evolutionary time, to produce red flowers from a blue-flowered ancestor, or blue flowers from a red-flowered ancestor.

In this case we know that there are genes (e.g., components of the anthocyanin pathway) that do in fact control flower color, and thus the inference of two states controlled by a genetic switch is probably a reasonable guess. In most cases, however, we have no idea of the genetic control of the structural characters observed. In making hypotheses about the nature of the underlying switches, then, often all we can do is be sure that the character states really are distinct. For quantitative characters such as leaf length or corolla tube width, this means drawing a graph to be sure that the species we are studying have measurements that do not overlap.

For many characters, the measurements do overlap, such that any guess as to the underlying switches, and therefore division into character states, is unsupported by any evidence. In these cases the characters should be omitted from the phylogenetic analysis (unless the overlap is caused by only a few plants, in which case the character could be scored as polymorphic and retained in the analysis). Even though such characters probably reflect genetic changes over evolutionary time, with our current knowledge it is difficult to extract from them information on the underlying changes, although methods of dealing with plants with variable characters have been developed.

Variability and overlap in morphological characters are, of course, good reasons why many systematists have turned to molecular data in constructing phylogenies. The recognition of molecular character states (i.e.,

nucleotides) is often easier and more precise, although even this can be difficult if gene sequences are hard to align, or if restriction fragments are similar in size (see Chapter 5).

EVOLUTIONARY TREES AND ROOTING

Figure 2.4 shows three different ways of recording and organizing observations about plants. Even though the network (Figure 2.4B) looks somewhat like a time line, it is not. It could be read from left to right, from right to left, or perhaps from the middle outward. To turn the network into an **evolutionary tree**, we must determine which changes are relatively more recent and which occurred further in the past. In other words, the tree must be **rooted**, which causes all character changes to be **polarized**, or given direction. (Some workers make distinctions between an evolutionary tree, a phylogeny, and a branching diagram or cladogram, but in this text the terms are used interchangeably.)

If you imagine that the network is a piece of string, you can keep the connections exactly the same, even when you pull down a root in many different places. The network from Figure 2.4B is redrawn in Figure 2.5, but rooted in three different places. Notice that the length of each tree (or **cladogram**) is the same as the length of the original network—3—and that all the connections are the same, but that the order of events differs considerably.

For example, in the rooting shown in Figure 2.5A, the ancestral plants had pollen with fewer than three colpi, petals not fused, and flowers not in heads, whereas in Figure 2.5B, we would conclude that the ancestral plants had exactly the opposite. In Figure 2.5C, the tree is rooted in such a way that the ancestor had tricolpate pollen. The pollen later changed to having fewer than three colpi in one lineage, whereas the other lineage kept the pollen character state of three colpi and later acquired fused petals and flowers in heads.

As is obvious from inspection of Figure 2.5, rooting the tree is critical for interpreting how plants evolved. Different roots suggest different patterns of changes (character polarizations). There has been much discussion among systematists of how the position of the root should be determined. One frequent suggestion is that one should use fossils. But just because a plant has been fossilized does not mean that its lineage originated earlier than plants now living; we know only that it died out earlier.

In determining evolutionary history, we are interested in determining when lineages—that is, **taxa** (singular **taxon**), or taxonomic groups—diverged from one another (when taxa originated). When taxa die out is interesting to know, but that fact by itself does not help in determining origins. (Fossils are, of course, extremely useful when included as additional taxa in a phylogeny. They often have combinations of characteristics that no longer occur in extant taxa and can affect the overall structure of the tree, sometimes in surprising and informative ways.)

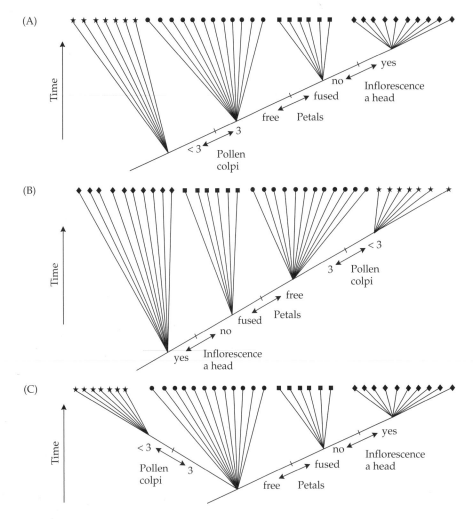

Figure 2.5   Three possible rootings of the network in Figure 2.4B. Note that in each case the number of evolutionary steps (character state changes) is the same as in the unrooted network.

In general, evolutionary trees are rooted using a relative of the group under study: an **outgroup**. When selecting an outgroup, one must assume only that the ingroup members (members of the group under study) are more closely related to each other than to the outgroup; in other words, the outgroup must have separated from the ingroup lineage before the ingroup diversified. Often several outgroups are used. If an outgroup is added to the network, the point at which it attaches is determined as the root of the tree.

In the case of Figures 2.4 and 2.5, the plants shown are all flowering plants (angiosperms), and their closest living relatives are the conifers, cycads, gnetophytes, or ginkgos or a set of these (see Chapter 7). In Figure 2.6A, a conifer is added to the matrix from Figure 2.4C. (We could have used all "gymnosperms" as outgroups, but to keep the example simple we have chosen only one.)

Because conifers do not have petals or flowers, two of the characters must be scored as not applicable, but we do know that conifer pollen does not have three colpi. With this information the conifer can be added to the

network as an outgroup, as in Figure 2.6B. Because it attaches among the star species, the tree can be rooted and redrawn as in Figure 2.6C. This tree corresponds to the rooted tree in Figure 2.5A and strengthens the hypothesis that Figure 2.5A accurately reflects evolutionary history.

Note that the tree can be drawn in different ways and still reflect the same evolutionary history. Comparing Figures 2.7A and B with Figure 2.6C shows that we can rotate the branches of the tree around any one of the branch points (**nodes**) without affecting the inferred order of events.

With a rooted tree (and only with a rooted tree), we can determine which groups are monophyletic. Therefore, in the example laid out in Figure 2.6C, the diamond plants are monophyletic (i.e., they form a **clade**). In fact, the flowering plants with fused petals and flowers arranged in a head are the family Asteraceae, which are known to form a monophyletic group. Thus, having flowers in heads is a synapomorphy for (i.e., is a shared derived character for, or indicates the monophyly of) the Asteraceae, having fused petals is a shared derived character (synapomorphy) uniting the square species with the diamond species, and having tricolpate pollen indicates the monophyly of the circle plus square plus diamond species.

Notice how important rooting is for determining monophyly. If Figure 2.5B were the correct rooting of the flowering plant phylogeny, then fused petals and flowers in heads would be ancestral character states (usually called **symplesiomorphies**) rather than derived synapomorphies. In this case, the species indicated by diamonds and squares would not share any *derived* character.

Notice also that a group including the diamond plus square plants in Figure 2.5B does not include *all* the descendants of their common ancestor; some of those descendants went on to become the circle and the star plants. Therefore, if Figure 2.5B were correct, the diamond plus square species would not be a monophyletic group. Such a group is called **paraphyletic**; a paraphyletic group includes a common ancestor and some but not all of its descendants.

Some taxonomic groups are cladistically unresolved; they cannot be determined to be either positively para-

(A)

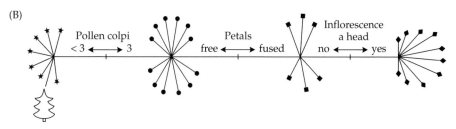

	Pollen colpi	Petals	Inflorescence in heads
Star plants	< 3	free	no
Circle plants	3	free	no
Square plants	3	fused	no
Diamond plants	3	fused	yes
Conifer	< 3	not applicable	not applicable

Figure 2.6 (A) The matrix from Figure 2.4C, but with character states added for a conifer. (B) The unrooted network from Figure 2.4B, but with the conifer attached according to the character states in part A. (C) The network of part B rooted with the conifer. Note that the evolutionary history is now the same as in Figure 2.5A.

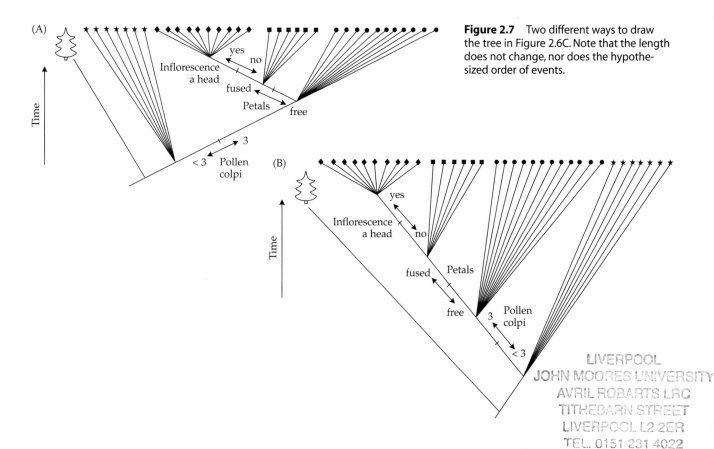

Figure 2.7 Two different ways to draw the tree in Figure 2.6C. Note that the length does not change, nor does the hypothesized order of events.

phyletic or positively monophyletic, and they are referred to as **metaphyletic**. The way we have drawn the circle plants in Figure 2.5, for instance, indicates that we do not know whether they have a synapomorphy, so they form a metaphyletic group.

As mentioned earlier, a character state that is derived (synapomorphic) at one point in time will become ancestral later. In our example, tricolpate pollen is a shared derived character of a large group of flowering plants. It is a synapomorphy and indicates monophyly of the group sometimes called the eudicots.

For the group with fused petals, however, tricolpate pollen is an ancestral, or **plesiomorphic**, character. It is something that all of the species in the group inherited from their common ancestor and thus does not indicate relationship. Plesiomorphic similarities cannot show genealogical relationships in the group being studied because they evolved earlier than any of the taxa being compared, and merely have been retained in various lineages (taxa).

Sometimes monophyly of a group is indicated by the fact that the character states do not occur in any other organism. For example, all members of the grass family

BOX 2A   *Hennigian Argumentation*

Figure 2.8   Three imaginary species (I, II, and III) and an outgroup.

In the examples presented thus far, a network is constructed and is then polarized by determination of where the outgroup attaches. Some systematists, however, prefer to polarize the characters first, using one or several outgroups, and then construct the phylogeny.

Consider, for example, the hypothetical organisms presented in Figure 2.8. In this case, the character states of the outgroup are assumed to be ancestral (plesiomorphic) and represented by 0; derived states are represented by 1 or higher numerals (Table 2.1). These states are then used to produce a character × taxon matrix (Table 2.2).

Next a phylogenetic tree (or cladogram) is constructed in which taxa are grouped (placed on the same branch) according to evidence provided by shared derived charac-

TABLE 2.1   *States of morphological characters used in cladistic analysis of the three imaginary species in Figure 2.8.*

	Character state[a]	
Morphological character	**Plesiomorphic**	**Apomorphic**
1. Roots	Less than 1 mm thick (0)	Greater than 5 mm thick (1)
2. Stems	Glabrous (0)	Pubescent (1)
3. Leaves	Alternate (0)	Opposite (1)
4. Venation	Pinnate (0)	Palmate (1)
5. Petiole	Lacking (0)	Present (1)
6. Base of blade	Acute (0)	Cordate (1)
7. Perianth parts	4 (0)	3 (1)
8. Perianth parts	Separate (0)	Fused (1)
9. Flowers[b]	In a group of 3 (0)	Solitary (1)

[a]Character state codings are given in parentheses.

[b]Note that inflorescence condition (flowers solitary versus flowers in groups of 3) cannot be polarized unless additional outgroups are employed.

(Poaceae) have an embryo that is unlike the embryo of any other flowering plant. We can thus hypothesize that the grass embryo is uniquely derived in (is a synapomorphy for) the family and indicates that the family is monophyletic. This is the same as saying that any reasonable rooting of the phylogenetic tree will lead to the same conclusion.

It is often possible to find evidence that a group is monophyletic even without a large computer-assisted phylogenetic analysis. Indeed, most cladistic analyses were done by hand until the mid-1980s. Characters are divided into character states, as with any cladistic analysis. The character state in the outgroup (or outgroups) is then assumed to be ancestral (Stevens 1980; Watrous and Wheeler 1981; Maddison et al. 1984). In other words, the character is polarized, or given direction. The shared derived, or synapomorphic, state can then be used as evidence of monophyly, and cladograms can be constructed on the basis of shared derived character states (see Box 2A). This kind of thinking is often useful in providing a first guess as to whether taxonomic groups might be monophyletic and thus named appropriately.

TABLE 2.2 *Character × taxon matrix for the three hypothetical species of Figure 2.8, based on characters in Table 2.1.*

Taxa	Characters								
	1	2	3	4	5	6	7	8	9
Species I	0	1	0	0	0	0	0	0	1
Species II	0	0	1	0	1	0	0	0	1
Species III	1	1	1	1	1	1	1	1	1
Outgroup(s)	0	0	0	0	0	0	0	0	0

ter states (synapomorphies). The presence of a derived character state (apomorphy) in two taxa suggests that they share a unique common ancestor in which the apomorphy first evolved; the two taxa have inherited the apomorphy (or evolutionary novelty) from this ancestor. The cladogram then represents the simplest hypothesis that explains the pattern of derived characters, following the principle of parsimony.

A hypothesis of relationships for species I, II, and III of Figure 2.8 is presented in Figure 2.9. Species II and III are hypothesized to share a unique common ancestor because they share the derived character states of characters 3 and 5 (see Table 2.1). They both have opposite, petiolate leaves, which are hypothesized to have evolved in their common ancestor. Similarly, the shared possession of solitary flowers supports the recognition of a more inclusive monophyletic group containing species I, II, and III.

The presence of hairy stems in species I and III is homoplasious; that is, hairy stems are hypothesized to have evolved in parallel in these two species, so their similarity is not based on common ancestry. Note, however, that hairy stems could have evolved in the most recent common ancestor of all three species, and then been lost (a reversal) in species II.

The sharing of pinnate leaves with narrowed bases and flowers with four separate perianth parts by species I and II is symplesiomorphic; these are shared ancestral characters. Such characters do not indicate relationship. In contrast, palmate venation, heart-shaped leaf bases, and flowers with three fused perianth parts are derived characters unique to species III. These unique derived character states (autapomorphies) also tell us nothing about the phylogenetic relationships of species III.

Figure 2.9 Cladogram illustrating hypothesized phylogenetic relationships of the three imaginary species of Figure 2.8. Synapomorphies are indicated on the tree.

(A)

	Pollen colpi	Petals	Inflorescence in heads	Cotyledon number
Black star plants	< 3	free	no	2
Gray star plants	< 3	free	no	1
White star plants	< 3	fused	no	1
Circle plants	3	free	no	2
Square plants	3	fused	no	2
Diamond plants	3	fused	yes	2
Conifer	< 3	not applicable	not applicable	> 2

Figure 2.10 (A) A character × plant matrix. (B) Unrooted network based on the matrix in A. Note that petal fusion appears to change twice. Network length is 5. (C) Another possible unrooted network based on the matrix in A. Unlike the network in B, petal fusion changes only once, but cotyledon number and pollen colpi change twice. Network length is 6.

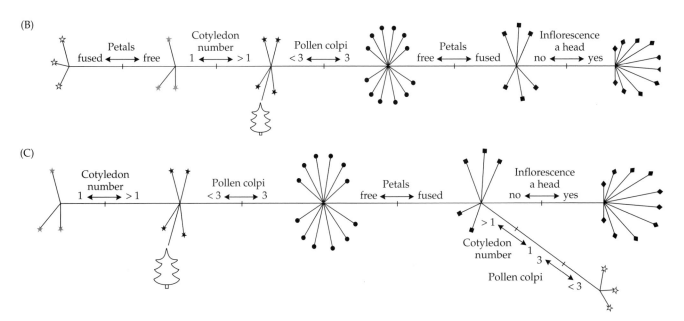

(B)

(C)

CHOOSING TREES

As the preceding sections have shown, determining the evolutionary history of a group of organisms is conceptually quite simple. First, characters are observed and divided into character states. Second, from the character states a Venn diagram (see Figure 2.4A), character × taxon matrix (see Figure 2.4C), and branching network (see Figure 2.4B) can be constructed. Third, by inclusion of an outgroup the network can be rooted to produce an evolutionary tree, cladogram, or phylogeny.

Two phenomena, however, make it much harder in practice to determine evolutionary history: **parallelism** and **reversal**, which sometimes are referred to together as **homoplasy**. Parallelism is the appearance of similar character states in unrelated organisms. (Many authors make a distinction between parallelism and convergence, but for this discussion we will treat them as though they were the same.) A reversal occurs when a derived character state changes back to the ancestral state.

To provide a clear example, divide the group that we have called "star plants" into black star plants, gray star

plants, and white star plants. Let us assume that the gray star and white star plants have only one cotyledon, whereas all the rest of the organisms have more than one (including the conifer). Let us further assume that the white star plants have fused petals. Add the character cotyledon number to the matrix in Figure 2.6A to create the matrix in Figure 2.10A, which gives the same information as the network in Figure 2.10B.

Now we see that, according to this network, there have been *two* changes in petal fusion. Counting the number of changes on this network (its length), we find five: one each in pollen colpi, flowers in heads, and cotyledon number, and two in petal fusion.

A group based on fused petals would be considered **polyphyletic**. Polyphyletic groups have two or more ancestral sources in which the parallel similarities evolved. (Although we distinguish here between paraphyletic and polyphyletic, many systematists have observed that the difference is slight and simply call any para- or polyphyletic group nonmonophyletic.) Petal fusion in this case is nonhomologous because it fails the ultimate test of homology: congruence with other characters in a phylogenetic analysis.

Why not draw the network in such a way that petal fusion arose only once? Such a network is shown in Figure 2.10C. Now we have one change in petal fusion, but that requires two changes in cotyledon number and also two changes in number of pollen colpi, making the network six steps long.

Each of the networks (Figure 2.10B and C) can be converted to a phylogeny by rooting at the conifer, but they make different suggestions about how plants have evolved. In one case, cotyledon number and number of pollen colpi have been stable over evolutionary time, whereas petal fusion has appeared twice, independently. In the other case, we postulate that cotyledon number and number of pollen colpi have changed twice over evolutionary time, while petal fusion has evolved only once. By drawing either of these networks, we are making a hypothesis about how evolution has happened—about which genetic changes have occurred, at what frequency, and in which order.

As the two networks show, the hypotheses differ. How do we determine which one is correct? There is no way to be certain. No one was there to watch the evolution of these plants. We can, however, make an educated guess, and some guesses seem more likely than others to be correct. One way to proceed is to ask, What is the simplest explanation of the observations? This rule, used throughout science, is known as **Occam's razor**: Do not generate a hypothesis any more complex than is demanded by the data. Applying this principle of simplicity, or **parsimony**, leads us to prefer the shorter network. The fact that it is shorter does not make it correct, but it is the simplest explanation of the data.

There are other ways to construct and to choose among evolutionary networks and trees. We have presented a simple step-counting (parsimony) method here because it is the most widely used, the most easily applicable to morphological changes, and possibly also the most intuitive method. Parsimony works well when evolutionary rates are slow enough that chance similarities (due to the evolution of identical derived characters independently in two or more lineages) do not overwhelm characters shared by the common ancestor.

Other methods use other criteria for choosing the preferred (optimal) tree. Instead of choosing the tree with the fewest evolutionary changes, one could convert the character matrix to a measure of similarity or dissimilarity among the plants, and then build a network that minimizes the dissimilarity; this is known as the **minimum-distance method**. Alternatively, one could develop theories about the probability of change from one character state to another and then use those probabilities to calculate the likelihood that a given branching diagram would lead to the particular set of data observed. The tree with the highest likelihood is preferred, so this approach is known as the **maximum likelihood method** (Felsenstein 1981; Hillis et al. 1993; Huelsenbeck 1995; Swofford et al. 1996).

Maximum likelihood methods are particularly suited to molecular data (see Chapter 5), for which it is easier to model the probability of genetic changes (mutations). For a more comprehensive description of methods of phylogeny reconstruction, see Swofford et al. 1996.

In the examples we have presented, in which there are few characters and little homoplasy, it is easy to construct the shortest network to link the organisms. In most real cases, however, there are many possible networks, and it is not immediately obvious which one is the shortest. Fortunately, computer algorithms have been devised that compare trees and calculate their lengths. Some of the most widely used programs are PHYLIP (Felsenstein 1989), Hennig86 (Farris 1989), Nona (Goloboff 1993), and PAUP 3.1.1/PAUP* 4.0 (Swofford 1993, 2000). These programs either evaluate data over all possible trees (an exhaustive search) or make reasonable guesses about the topology of the shortest trees (branch-and-bound searches or heuristic searches).

With numerous taxa, only heuristic algorithms can be used. These algorithms may not succeed in finding the shortest tree or trees because of the large number of possible dichotomous trees. For example, the possible interrelationships of three taxa can be expressed by only three rooted trees, [A(B,C)], [B(A,C)], and [C(A,B)]. But with larger numbers of taxa the number of potential trees expands rapidly; for example, four taxa yield 15 trees, five yield 105 trees, six yield 945 trees, and ten yield 34,459,425 trees!

SUMMARIZING EVOLUTIONARY TREES

Often parsimony analyses will find multiple trees, all with the same length but with different linkages among the taxa. Sometimes, too, different methods of analysis will find trees showing different topologies and therefore different histories for the same taxa. In addition, studies using different kinds of characters (e.g., gene sequences, morphology) may find still other trees. Rather than choosing among the trees in these cases, often systematists simply want to see what groups are found in all the shortest trees, or by all methods of analysis, or among different kinds of character matrices. The information in common in these trees can be summarized in a **consensus tree**.

Strict consensus trees contain only monophyletic groups that are common to all trees. For example, analyses of different sets of data have produced different results regarding the early evolution of the angiosperms. A study of morphological characters and 18S rRNA sequences led to the evolutionary tree shown in Figure 2.11A (we have omitted some taxa for the purposes of this example) (Doyle et al. 1994). A study of *rbcL* sequences led to the tree in Figure 2.11B (Albert et al.

Figure 2.11 (A) Phylogeny of angiosperms based on data from morphology and 18S rRNA sequences. (B) Phylogeny of angiosperms based on data from *rbcL* sequences. (C) Strict consensus of the trees in A and B. (D) Semi-strict consensus of trees in A and B. (Part A based on Doyle et al. 1994; part D modified from Albert et al. 1994.)

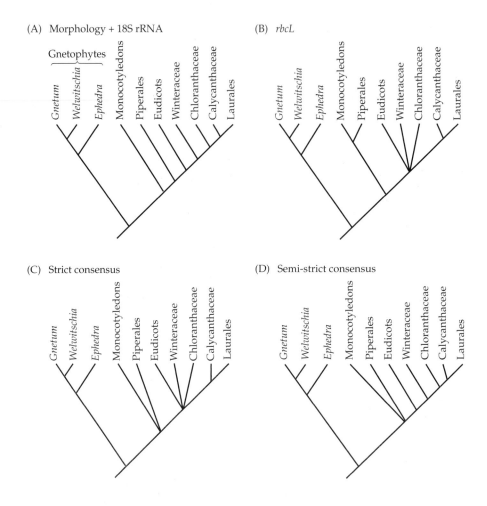

(A) Morphology + 18S rRNA

(B) *rbcL*

(C) Strict consensus

(D) Semi-strict consensus

1994). (For descriptions of 18S rRNA and *rbcL* data, see Chapter 5.) The trees both show *Gnetum* as sister to *Welwitschia*, and those two as sister to *Ephedra*. (These three genera together make up the gnetophytes; see Chapter 8.) Both trees also suggest that the Calycanthaceae and Laurales are closely related (see Chapter 9). The strict consensus of the two cladograms (Figure 2.11C) also shows the gnetophyte and the Calycanthaceae/Laurales clades.

There are differences between the two evolutionary hypotheses, however. The *rbcL* tree suggests that the Piperales are sister to the monocots, but the morphology/rRNA tree tells us that the Piperales arose after the monocot lineage diverged, such that the Piperales are more closely related to the rest of the dicots. In Figure 2.11B, the eudicots, Winteraceae, and Chloranthaceae appear as though they arose at the same time. This means that *rbcL* data cannot tell us whether they arose together or one after the other, nor can we determine the order.

Having multiple lineages arising at the same apparent point in the diagram is usually an expression of ambiguity. The difference in the position of the Piperales combined with the ambiguity in the *rbcL* tree leads us to conclude that we really do not know which early angiosperm lineages appeared first. This uncertainty is reflected in the strict consensus tree by the fact that all those lineages are drawn as though they arose at the same time.

When many trees are compared, it is sometimes interesting to know whether a clade appears in most of the trees, even if it doesn't occur in all of them. A **majority-rule consensus tree** can show all groups that appear in 50% or more of the trees. If a particular clade is present in the majority of the most-parsimonious trees, then this clade will be represented on the majority-rule tree (along with an indication of the percentage of most-parsimonious trees showing that clade). The majority-rule consensus tree will be inconsistent with some of the original trees, and thus provides only a partial summary of the phylogenetic analyses.

A **semi-strict**, or **combinable-component, consensus tree** is often useful, particularly in comparisons of phylogenies with slightly different terminal taxa, or from different sources of characters. It is common, for example, to construct trees from two different sets of characters (e.g., a gene sequence and morphology) and to find that both sets of characters indicate monophyly of a particular group of species. Only one set of characters, however, may resolve relationships among the species. The semi-strict consensus then indicates all relationships supported by one tree or both trees and not contradicted by either.

BOX 2B *Long Branch Attraction*

If there are great differences in the rates of character evolution between lineages such that some lineages are evolving very rapidly, and if the pattern of variation is sufficiently constrained (i.e., only a limited number of character states exist), then unusually long branches are often connected to each other whether or not they are actually closely related (Figure 2.12) (Felsenstein 1978).

This phenomenon occurs because numerous random changes, some of which appear in parallel in the two lineages, outnumber the information that shows common ancestry. The problem cannot be circumvented by acquisition of more characters; these merely add to the number of parallelisms linking the two rapidly evolving lineages.

This situation, often called long branch attraction or the "Felsenstein zone," can affect all methods of tree construction. With the correct model of evolution, however, maximum likelihood methods will not have this

problem (although determining the correct model may be difficult). The situation basically is a sampling

problem, and may be alleviated by including the taxa that are related to those terminating the long branches.

Figure 2.12 Long branch attraction, a situation in which strongly unequal evolutionary rates cause parsimony to fail. (A) True phylogeny. Dashed lines show character states that have arisen in parallel in the lineages leading to a and b. (B) Phylogeny as reconstructed by parsimony. The number of parallelisms shared by a and b is greater than the number of characters linking a and c, so a and b appear to be sister taxa, with parallelisms (in the true phylogeny) treated as shared derived characters of a and b.

(A) True phylogeny:
28 steps

(B) Tree generated by parsimony analysis: 26 steps

For example, although the *rbcL* tree (Figure 2.11B) does not give us information about the order in which eudicots, Winteraceae, and Chloranthaceae originated, the tree in Figure 2.11A does. The two trees are not really conflicting; the morphology/rRNA tree just provides more precise information. The semi-strict consensus thus follows the morphology/rRNA arrangement of those three groups (Figure 2.11D).

THE PROBABILITY OF EVOLUTIONARY CHANGE IN CHARACTERS

In trying to infer the evolutionary history of a group, we depend on an implicit or explicit description (model) of the evolutionary process. The more accurately the description reflects the underlying process, the more accurately we will be able to estimate the evolutionary history. This is particularly important for very divergent species in molecular phylogenies, for which parsimony methods often produce misleading results (Box 2B). For nucleotides in a DNA sequence, mutation is usually assumed to be random, although this assumption is often then modified to reflect hypothesized mechanisms of molecular evolution.

A model is much more difficult for morphological characters because we usually have no idea how many genes are involved, nor do we know what kinds of changes in the genes lead to different character states. Nonetheless, certain assumptions must be made if one is to proceed at all. (And, we note, there are *no* methods

that are entirely free of assumptions!) The major assumptions have to do with the likelihood of particular changes of character states, and the likelihood of reversals and parallelisms.

Ordering character states   The characters in Figure 2.10A have only two states. Such two-state characters are interpreted as representing a single genetic switch— "on" producing one state (e.g., pollen is tricolpate), "off" resulting in the other state (e.g., pollen is one-grooved, or monosulcate). Over evolutionary time, of course, such characteristics can continue to change. For example, tricolpate pollen is modified in some Caryophyllales so that it is spherical, with many pores evenly spaced around it (looking rather like a golf ball); this pollen is pantoporate.

If we were to include the character "pollen grooves" in a matrix, it would now have three states: monosulcate, tricolpate, and pantoporate. This is now a multistate character, in contrast to the two-state, or binary, characters discussed previously. Multistate characters create a dilemma: How many genetic switches are involved?

It is possible that monosulcate pollen changed to tricolpate, which then changed to pantoporate pollen, and this progression matches what we think happened in evolutionary time (Figure 2.13A). (Recall that the outgroup does not have tricolpate pollen.) This scenario implies two genetic switches. It also implies that they

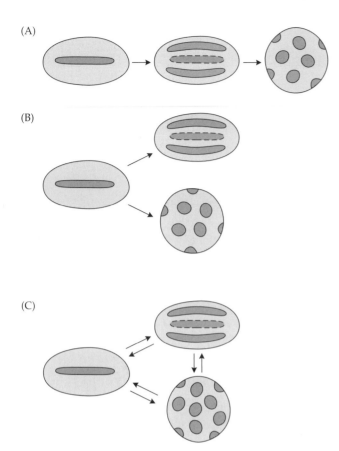

(A)

(B)

(C)

Figure 2.13 Three alternative hypotheses about the evolution of pollen morphology. (A) Monosulcate changed to tricolpate, which then changed to pantoporate. As drawn, the character is ordered and irreversible. (B) Monosulcate changed to tricolpate, and then independently changed to pantoporate. Again, the character is ordered and irreversible. If the arrows were drawn as double headed, the character would be interpreted as reversible. (C) Any pollen type can change to any other pollen type. The character is unordered and reversible.

must have occurred in order; that is, pantoporate pollen could arise only after tricolpate pollen. If we accept this series of events, the multistate character is considered to be **ordered**.

If we decide to allow for reversals of character states—that is, consider the possibility that pantoporate pollen might switch back to tricolpate and tricolpate to monosulcate pollen—the character is still ordered. It requires two evolutionary (genetic) steps to go from monosulcate to pantoporate pollen, or two to go from pantoporate to monosulcate pollen. A phylogenetic analysis in which all characters are treated as ordered is sometimes referred to in the literature as **Wagner parsimony**.

If we didn't know anything about the plants involved, we might want to consider the possibility that monosulcate pollen changed to tricolpate pollen, and in an independent event, monosulcate pollen changed to pantoporate pollen (Figure 2.13B). This sequence would suggest that there is a genetic switch from monosulcate to tricolpate pollen, as well as a switch that allows

change from monosulcate to pantoporate pollen, but a change from tricolpate to pantoporate pollen is impossible. The character in this case is still ordered, but in a different way from what Figure 2.13A shows. If reversals are possible, then two steps are required to get from tricolpate to pantoporate, and two from pantoporate to tricolpate pollen.

With morphological characters and character states, we are usually unsure of which switches are possible, so it is common to treat multistate characters as unordered (Figure 2.13C); this is sometimes called **Fitch parsimony**. In the case of an unordered character, we postulate only one switch between any two states. DNA sequence characters are multistate characters with four states (adenine, thymine, guanine, cytosine). To treat these as ordered would be nonsensical; adenine does not need to change to cytosine before changing to guanine. DNA characters are therefore always treated as unordered and fully reversible.

Reversals, parallelisms, and character weighting In the network in Figure 2.10B, we hypothesized that petal fusion arose twice, independently. To make the slightly longer network in Figure 2.10C, we had to let cotyledon number change from one to more than one and back to one again—that is, to reverse. In comparing the trees in Figure 2.10B and C, therefore, we are comparing the hypotheses that (1) mutations in the genes leading to petal fusion have happened more than once versus (2) mutations in the genes controlling cotyledon number have happened and then their effects have been reversed. In deciding that the network in Figure 2.10B was shorter than the one in Figure 2.10C, we counted all the steps equally, whether they were parallelisms, reversals, or unique origins.

This approach may or may not be reasonable. Dollo's law, for example, suggests that for very complex characters, parallel origin is highly unlikely, whereas reversal may be quite easy (Mayr and Ashlock 1991). The assumption is that many genes must change in order for a morphological structure to be created, but only one of those genes needs to be modified for the structure to be lost.

We can build Dollo's law into the process of choosing a tree by making gains of structures count for more than losses; the process is then known as **Dollo parsimony**. (To define the terms *gain* and *loss*, of course, requires a rooted tree; hence Dollo parsimony cannot be applied to an unrooted network.)

Certain characters are sometimes **weighted** in cladistic analyses. This weighting reflects the assumption that certain characters should be harder to modify in evolutionary time than others. One might hypothesize, for example, that leaf anatomy is less likely to change than leaf hairiness (pubescence), and therefore a change in a leaf anatomical character could be counted as equivalent to two changes in pubescence for the purposes of counting steps in the tree.

Such weighting decisions can easily become subjective or arbitrary, and they risk biasing the outcome of the study toward finding particular groupings. (For example, the investigator might theorize, "My favorite species group has interesting leaf anatomy; therefore I think that leaf anatomy is phylogenetically important; therefore I will give it extra weight in the phylogenetic analysis." In this case, it is no surprise when the favorite species group is shown to be monophyletic.)

Because of the possibility of bias, systematists generally attempt to base weighting decisions on an objective criterion. One approach is to do a preliminary phylogenetic analysis with all characters assigned equal weights. The results of this analysis will identify which characters have the least homoplasy on the shortest tree(s); the characters with less homoplasy can then be given more weight in subsequent analyses, a process known as **successive weighting**.

Another approach is to base weights on knowledge of the underlying genetic basis of characters. For example, transversions (purine → pyrimidine or pyrimidine → purine changes) are weighted over transitions (purine → purine or pyrimidine → pyrimidine changes) because transitions are known to occur more frequently and to be easier to reverse. Restriction site gains may be weighted over site losses because there are fewer ways to gain a restriction site than to lose one (see Chapter 5). And complex characters (presumably controlled by many genes) may be weighted over simple characters (presumably controlled by fewer genes), again because the latter are thought to be easier for selection to modify over evolutionary time.

The most common approach, used in most preliminary analyses, is to consider all characters of equal weight. Although this approach sometimes is described as "unweighted," in fact it assumes that all characters are equally likely to change and weights them accordingly.

Underlying all discussion of weights is the assumption that characters of organisms evolve independently. This assumption requires that change in one character does not increase the probability of change in another character. As with the previous assumption, this one may be violated frequently. For example, a change in flower color might well lead to a shift in pollinators, which would then increase the probability that corolla shape would change. Violating this assumption obviously affects character weighting, in that the likelihood of change of two characters is not the same.

DO WE BELIEVE THE EVOLUTIONARY TREE?

An evolutionary tree is simply a model or hypothesis, a best guess about the history of a group of plants. It follows that some guesses might be better, or at least more convincing, than others. Much of the current literature on phylogeny reconstruction involves ways of determining how much credence we should give to a particular evolutionary tree.

Use of an optimality criterion is one way to evaluate the evolutionary tree; of all possible descriptions of history, we prefer the one that requires the fewest steps, or gives the maximum likelihood, or the minimum distance. Trees can be evaluated more precisely, however. For the purposes of this discussion, we will continue to focus on phylogenies generated according to maximum parsimony (i.e., the fewest evolutionary steps).

Measuring support for the whole tree: Assessing homoplasy

Parsimony analyses minimize the number of characters that change in parallel or reverse. If there are many such homoplasious characters, then the phylogenetic tree may be an artifact of the characters we have chosen, and a slight change in characters will lead to a different tree. The simplest, and most common, measure of homoplasy is the **consistency index** (**CI**), which equals the minimum amount of possible evolutionary change (the number of genetic switches) divided by the actual tree length (the number of actual genetic changes on the tree).

In the network shown in Figure 2.4B, each of the three characters represents a single genetic switch, and each one changes only once, so the consistency index is $3/3 = 1.0$. In the network in Figure 2.10B, there are four binary (one-switch) characters, but one of them (petal fusion) changes twice on the tree, so the consistency index is $4/5 = 0.80$.

Consistency indices may also be calculated for individual characters and equal the minimum number of possible changes (one, for a binary character) divided by the actual number of changes on the tree. For example, the CI of petal fusion (see Figure 2.10B) is $1/2 = 0.50$. For a given matrix (set of characters and taxa), the shortest network or tree will also have the highest consistency index. Lower consistency indices indicate the presence of many characters that contradict the evolutionary tree.

Comparing consistency indices across data sets is hazardous because the CI has some undesirable properties. For one thing, a character that changes once in only one taxon will have a consistency index of 1.0, but such a character says nothing about relationships. Such a uniquely derived character is sometimes called an **autapomorphy**. For example, if one of the black star plants in Figure 2.10B had hairy leaves while all other plants studied had hairless ones, leaf hairiness would not be of any help in indicating the relationship of the hairy-leaved plant. The character is **uninformative**.

Because the uninformative character changes only once, however, it has a CI of 1.0. If we added many uninformative characters into the analysis, the overall CI would be inflated accordingly and would give a misleading impression that many characters supported the tree. Uninformative characters, therefore, are often omitted before the consistency index is calculated.

The consistency index is also sensitive to the number of taxa in an analysis (Sanderson and Donoghue 1989):

Analyses with many taxa tend to have lower CIs than analyses with fewer taxa. This relationship is true of both molecular and morphological data, and of analyses of species, genera, or families.

Other measures are used to describe how characters vary over the tree. One of these, the **retention index (RI)**, is designed to circumvent another limitation of the CI (Wiley et al. 1991; Forey et al. 1992). The CI varies between near 0 (a character that changes many times on the tree) and 1.0 (a character that changes only once). But consider the plants in Figure 2.10 once more. For those described by the matrix in Figure 2.10A, only two groups—the white star plants and the gray star plants—have a single cotyledon. If the single-cotyledon plants are all on a single branch of the network, as in Figure 2.10B, then the CI for cotyledon number is 1.0. If they are unrelated, as in Figure 2.10C, then the CI is 0.5 (1/2), which is the lowest possible value on the tree. Thus, instead of varying between 0 and 1, the CI in this case varies between 0.5 and 1.0. The RI corrects for this narrower range of the CI by comparing the actual number of changes in the character to the maximum possible number of changes.

The RI is computed by calculation of the maximum possible tree length, which is the length that would occur if the derived character state originated independently in every taxon in which it appears (i.e., if all taxa with the derived character state were unrelated). The minimum tree length and actual tree length are computed the same way they are for calculating the CI. The RI then equals the maximum length minus the actual length, divided by the maximum length minus the minimum length: $(L_{max} - L_{actual})/(L_{max} - L_{min})$. In Figure 2.10B, then, the RI is $(9 - 5)/(9 - 4) = 4/5 = 0.80$.

Measuring support for parts of trees

With parsimony methods, the shortest available tree is preferred over one that is longer. However, some parts of the tree may be more reliable than others. This will be the case if reversals and parallelisms (or simple misinterpretation of characters) affect some groups of plants more than others, or if there were very few evolutionary changes in the history of a particular group.

One simple way to evaluate support for a particular part of a tree is to note the number of genetic changes that occur on the branch leading to a particular group, along with the consistency indices of the characters. For

example, one of the morphological trees produced by Doyle et al. (1994) found 18 changes on the branch leading to the angiosperms (Figure 2.14), and of these, 11 were in characters that had a CI of 1.0. In other words, over half of the genetic changes that occurred during the origin of the angiosperms produced novel characteristics, found nowhere else. Groups like the angiosperms that share numerous characters that do not change elsewhere on the cladogram are more believable than a group that shares only a few highly homoplasious characters.

Another way to assess how well the data support the tree is to determine whether a group of interest occurs in other trees that are almost equally short. Suppose, in other words, we ask if there are other ways to analyze the homoplasious characters that lead to trees that are one, two, or three steps longer.

For example, in the tree shown in Figure 2.14, the shortest tree indicates that the earliest diverging lineages in the angiosperms were the monocots and the water

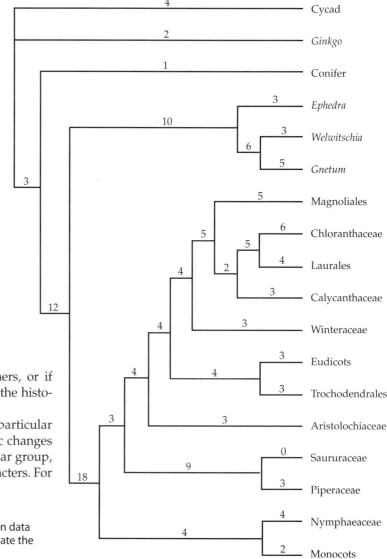

Figure 2.14   Phylogeny of the angiosperms, based on data from Doyle et al. 1994. Numbers above branches indicate the number of characters changing along that branch.

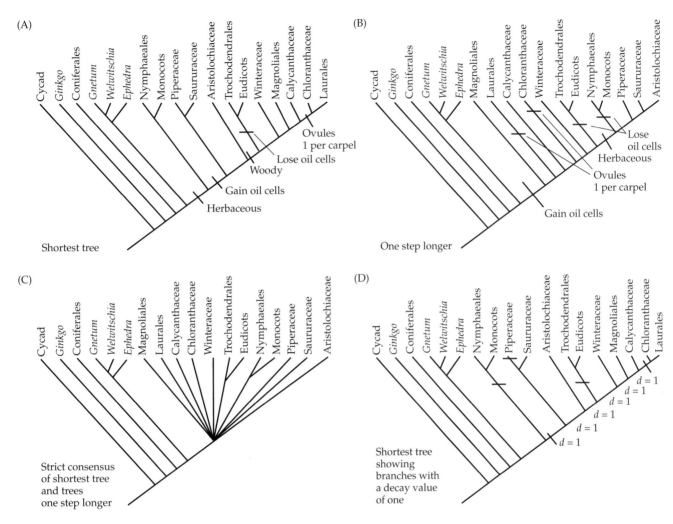

Figure 2.15 (A) The same tree as in Figure 2.14, indicating patterns of change in the presence or absence of oil cells, ovule number per carpel, and plant habit. (B) An alternative tree, only one step longer than the tree in A, showing patterns of change in the same characters. Note that herbaceousness now is hypothesized to have evolved only once, but loss of oil cells and reduction of ovule number occur twice. (C) Strict consensus of the shortest tree and trees one step longer (Figures 2.15A and B). (D) The same tree as in A and Figure 2.14, showing branches with a decay value of 1. (Data from Doyle et al. 1994.)

lilies (Nymphaeaceae; see Chapter 9). This implies that the character of herbaceous stems is gained once and then lost, whereas reducing the number of ovules per carpel to one occurs only once, and oil cells are gained once and lost once (Figure 2.15A). On the other hand, trees one step longer, in which the earliest angiosperm lineages led to the magnolias, suggest that herbaceous stems evolved once, but reduction in ovule number occurred twice, and there were three changes in oil cells (gained once and lost twice or vice versa) (Figure 2.15B).

Thus, by looking at trees one step longer, we can hypothesize that some characters are less homoplasious, but some are more so. If we now take the strict consensus of all the trees, including the shortest ones and those

one step longer, all early angiosperm lineages are drawn as though they radiate from a single point, indicating uncertainty about the order in which they evolved (Figure 2.15C).

In other words, many of the branches in the shortest trees do not appear in trees one step longer. Thus all those branches are not drawn in the strict consensus; they "collapse." We can indicate this by placing a 1 next to each of the collapsing branches of the shortest tree (Figure 2.15D). The number is the **decay index**, sometimes called Bremer support, which represents how many extra steps are required to find trees that do not contain a particular group. It provides a relative measure of how much the homoplasy in the data affects support for a particular group.

The decay index is not statistical, which, depending on one's point of view, is either a virtue or a drawback. Because history, and therefore the phylogeny, happened only once and cannot be repeated, it is impossible to replicate the evolutionary experiment. It is certainly possible, however, to test whether character data are different from random, although there are many possible ways to randomize systematic data. Many tests have been

	Pollen colpi	Petals	Pollen colpi	Cotyledon number
Black star plants	< 3	free	< 3	2
Gray star plants	< 3	free	< 3	1
White star plants	< 3	fused	< 3	1
Circle plants	3	free	3	2
Square plants	3	fused	3	2
Diamond plants	3	fused	3	2
Conifer	< 3	not applicable	< 3	> 2

Figure 2.16 The matrix from Figure 2.10A, sampled with replacement, as it would be for the first step of a bootstrap analysis. Note that in the sampling process, the character "pollen colpi" has been sampled twice, whereas the character "inflorescence a head" has been omitted.

devised that use some sort of randomization technique. Probably the most widely used is bootstrap analysis.

Bootstrap analysis randomizes characters with respect to taxa. As an example, begin with the matrix in Figure 2.10A and randomize the columns while leaving the rows in place. Choose a column at random to become the first column of the new matrix. Then choose another column from the original matrix to become the second column, and so on until a new matrix is created with the same number of columns as the original. Because one returns to the original matrix each time to choose a new column, some characters may be represented several times in each new matrix, while others are omitted. This is usually described as random sampling with replacement.

Figure 2.16 shows the matrix in Figure 2.10A sampled with replacement; note that the first character ("pollen colpi") has been selected twice, whereas the third character ("inflorescence a head") was missed

Figure 2.17 The single most parsimonious tree found in branch-and-bound analysis of the *Lyonia* group (taxa in boldface type) using *matK* data. Branch lengths appear above lines; bootstrap values are in parentheses; decay indices (*d*) appear below lines. Length = 425, consistency index = 0.60. (From Kron and Judd 1997.)

by the random selection process. Multiple randomized matrices are constructed, and the most-parsimonious tree(s) are found for each new matrix. This process leads to a set of at least 100 trees, which can be summarized by a consensus tree (see pages 23–25). In the bootstrap consensus tree, a clade with a bootstrap value of, say, 95% was present in 95% of the cladograms generated in the bootstrap analyses.

An example of a cladistic analysis giving both bootstrap and decay values (along with branch lengths) is shown in Figure 2.17. We see that bootstrap and decay values are high for the genus *Lyonia*, indicating that the data support monophyly of the genus, whereas the linkage of *Agarista* and *Pieris* is supported by only 51% of the bootstrap trees, and in trees only one step longer the two genera are not sisters, indicated by the notation *d* (decay) = 1.

Another excellent way to gain confidence in the groupings present in a tree is to compare phylogenies that have been based on different sets of characters. For example, phylogenies based on morphology, chloroplast DNA nucleotide sequences (cpDNA), and nuclear DNA nucleotide sequences could be (and often are) compared. If these phylogenies show similar groups, we can be more confident that they reflect the true order of events. For example, the monophyly of such families as the Poaceae, Onagraceae, Ericaceae, Asteraceae, and Orchidaceae has been supported by phylogenetic analysis of many kinds of data, including information from morphology, chloroplast gene sequences, and nuclear gene sequences.

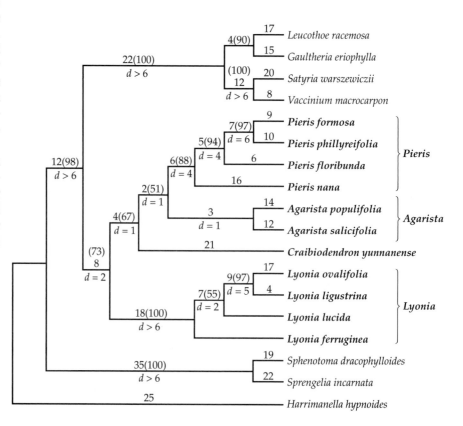

Comparing trees is often particularly intriguing when the data come from different genes, as will be discussed in more detail in Chapter 5. It is also common to combine morphological and DNA characters in a single phylogenetic analysis, which often leads to more strongly supported phylogenies than either kind of data can produce alone. Morphological phylogenies assume that hybridization has not occurred, or at least is rare (Box 2C); multiple molecular trees can often test this assumption.

Describing Evolution: Mapping Characters on Trees

Once created, phylogenies may be used as the basis of classification. This is one major goal of systematics and is described in detail in the next section. Phylogenies can also be used to describe the evolutionary process and to develop hypotheses about adaptation, morphological and physiological change, or biogeography, among many other uses. If a phylogeny is to be used to describe history, however, careful attention must be given to the characters and character states used in the description. In the discussion that follows we will focus on morphological characters, but many of the points apply to any kind of character.

Consider a group of plants for which the tree is known; a good example is the Ericaceae, for which much information is available (Figure 2.18). Assume for the purposes of this discussion that this tree is an accurate reflection of history, and that each of the terminal genera really is monophyletic, as demonstrated by studies of multiple species of each. Then consider a study that is concerned with the gain or loss of fused petals, which are intimately connected with the evolution of pollination systems. This is the kind of study that systematists frequently engage in, because the details of character

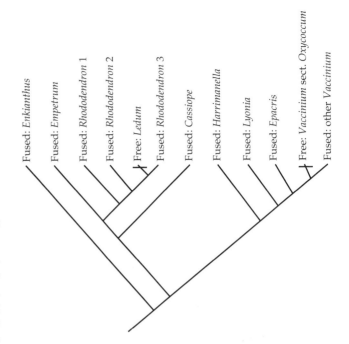

Figure 2.18 Phylogeny of a portion of the Ericaceae, based on data summarized in Stevens 1998. The genus *Rhododendron* is paraphyletic and is represented by three separate lineages, numbered 1 to 3. Two changes to free petals are hypothesized.

evolution lead to hypotheses about how natural selection has worked. In addition, when constructing classifications, one frequently wants to know what morphological characters can be attributed to and distinguish a particular monophyletic group.

Figure 2.18 shows the observed character states for the genera. It seems trivially obvious from looking at the distribution of characters and character states that free petals must have evolved in the lineage leading to *Ledum*

BOX 2C *Phylogenetic Analysis Assumes That Evolution Can Be Diagrammed as a Branching Tree*

Phylogenetic studies assume that after two lineages diverge, they never exchange genetic information again. This assumption may in fact be violated frequently. If hybridization is common, a plant may share the derived characters of two unrelated parent plants, and the history will look more like a piece of macramé than like a tree. Phylogenetic analysis will always produce a treelike diagram, whether appropriate or not. Phylogenetic methods presuppose divergent evolution and cannot give the correct phylogeny

for hybrids, which have reticulating evolutionary histories.

Interspecific hybridization is known to be common in plants, and the proper treatment of hybrids in cladistic analyses has been much discussed (Bremer and Wanntorp 1979; Wagner 1980, 1983; Bremer 1983; Funk 1985; Kellogg 1989; Kellogg et al. 1996). Most systematists have suggested that hybrids be identified and removed from analyses because their inclusion could lead to increased homoplasy, an increased number of most-parsimonious trees,

and a distortion of the patterns of relationships among nonhybrid taxa.

However, recent studies by McDade (1990, 1992, 1997) indicate that hybrids are unlikely to create problems in phylogenetic analysis unless they are between distantly related parental species. When hybrids are recognized and their ancestry determined (see Chapter 6), they can be manually inserted into the cladogram, which then indicates not only cladogenetic events (brought about through speciation) but also reticulating histories (developed through interspecific hybridization).

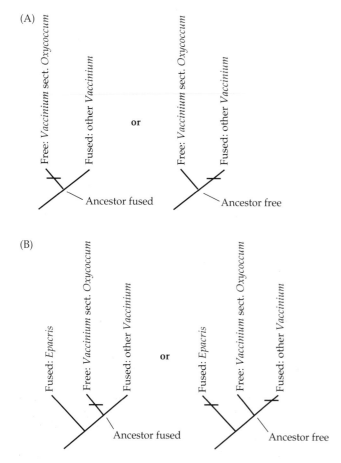

Figure 2.19 (A) Two taxa differ in character states. It is impossible to determine from this information alone what the character state of the ancestor was because either assumption will involve one change in one descendant lineage. (B) The addition of an outgroup determines the character state of the ancestor. In this case, it is simpler (requires fewer steps) to assume that the ancestor had fused petals.

(Labrador tea) and again in the lineage leading to *Vaccinum* sect. *Oxycoccum* (cranberries). Phrased another way, the ancestor of *Vaccinium* sect. *Oxycoccum* and all other vacciniums (blueberries) had fused petals, as did the ancestor of *Ledum* plus *Rhododendron* sect. 3.

Examine this "obvious" conclusion more closely. If we were studying only species of *Vaccinium*, we would have no way of knowing whether fused petals were ancestral or derived (Figure 2.19A). There must have been one genetic change, but it could have happened as easily in the lineage leading to the cranberries (sect. *Oxycoccum*) as in the lineage leading to the blueberries (other *Vaccinium*).

Only by reference to the outgroup *Epacris* can we determine when petal fusion was lost. Because *Epacris* has fused petals, free petals must have originated within *Vaccinium*; it is simplest (most parsimonious) to assume just one genetic change, from fused to free (Figure 2.19B). This is the same as saying that the ancestor of blueberries plus cranberries had fused petals. If we were to postulate that the ancestor had free petals, we would need

two changes to fused petals: one in *Epacris* and one in the blueberries. The same argument applies in the case of *Rhododendron* and *Ledum*.

Now suppose we were studying only species of *Vaccinium*, but this time, instead of using *Epacris* or other Ericaceae as outgroups, we used only *Ledum*. This could easily be the case if material of the other genera were hard to obtain, or if those genera were extinct and we didn't even know they had existed. Now we would conclude that the ancestor of all vacciniums had free petals, and that in response to an unknown selective pressure there was a change to fused petals (Figure 2.20A). *This is exactly the opposite conclusion from the one reached in the previous example*, and the only difference is the genera included in the analysis.

One might try to improve the situation by using additional outgroups. For example, consider the same study of *Vaccinium*, but now use both *Ledum* and *Rhododendron* as outgroups. In this case the direction of change is completely ambiguous (Figure 2.20B). It is as simple to pos-

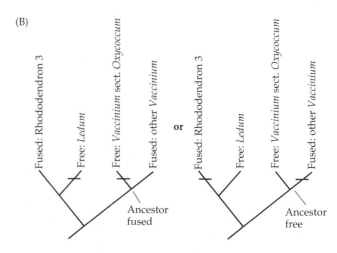

Figure 2.20 (A) Analysis of character state change in *Vaccinium* using a different outgroup. Note that the inference of the ancestral state is exactly the opposite of that reached when *Epacris* is used as an outgroup. (B) Analysis of character state change in *Vaccinium* using two outgroups that differ in state. It is now impossible to determine the character state of the ancestor.

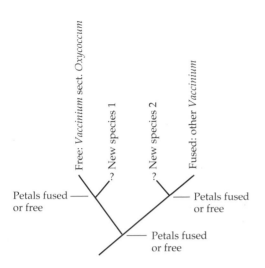

Figure 2.21 Addition of species for which the character state is unknown can prevent any inference about the ancestral state.

tulate that the ancestor of the group had fused petals and there were two changes to free petals as it is to postulate that the ancestor had free petals and there were two changes to fused.

These two choices are known as **equally parsimonious reconstructions**. For many characters on many trees, there are multiple equally parsimonious reconstructions. In other words, there are multiple equally good hypotheses about the direction and timing of character state change. If you return to the example in Figure 2.15, you should be able to find equally parsimonious reconstructions that differ from the ones shown.

Ambiguity can also come from including taxa for which the character state is not known. Suppose, for example, two new taxa are discovered such that, on the basis of other characters, one is clearly sister to *Vaccinium* sect. *Oxycoccum*, and the other is sister to the rest of *Vaccinium* (Figure 2.21). In addition, suppose that it is unclear whether the petals are fused or free. (This type of ambiguity is more common than you might think; it can occur when the original description is vague and/or illustrations are unclear, or when the original plant is known only from fruiting material.) In this example we do not know what the ancestral state was for *Vaccinium*, so we cannot make any hypothesis about the direction of evolutionary change. Also we cannot be sure that the character "petals fused" is a synapomorphy for the genus.

Various algorithms have been developed to assign character state changes to particular portions of trees (see Chapters 3 and 4 of Maddison and Maddison 2000 for a lucid and comprehensive discussion of these). Depending on the algorithm used, the character changes can be biased in favor of parallelisms (the so-called "delayed transformation," or DELTRAN, algorithm) or in favor of reversals ("accelerated transformation," or ACCTRAN). The results can have implications, some-

times major, for hypotheses about the evolutionary process, and they may also affect how organisms are described in a classification.

Constructing a Classification

The theory of classification is a topic with which systematists have been wrestling for centuries, leading to a broad and frequently contentious literature (see Chapter 3). The principles of phylogenetic classification outlined here are commonly but not universally held. In general, however, there are several goals of classification. A classification is a common vocabulary designed to aid communication. A classification should be stable; names that are frequently changed become useless for communication. A classification should be predictive; that is, the name of a plant should help you to learn more about that plant and guide you to its literature.

Systematists generally agree about the goals of classification, but they may disagree profoundly about how to reach those goals. In this text we take a particular point of view, using phylogenetic classifications throughout. Thus, as far as possible, we recognize monophyletic and avoid paraphyletic or polyphyletic groups. In the few cases in which a nonmonophyletic family or order has not yet been divided into monophyletic units, we have placed the taxon name in quotation marks. Monophyly of many genera of angiosperms is questionable, but so few phylogenetic analyses are available at this level that possible or probable paraphyly or polyphyly of genera is not indicated.

The biological diversity on Earth is the result of genealogical descent with modification, and monophyletic groups owe their existence to this process. It is appropriate, therefore, to use monophyletic groups in biological classifications so that we may most accurately reflect this genealogical history. Classifications based on monophyletic groups are more predictive and of greater heuristic value then those based on overall similarity or idiosyncratic weighting of particular characters (Farris 1979; Donoghue and Cantino 1988).

Phylogenetic classifications, because they reflect genealogy, will be the most useful in biological fields, such as the study of plant distributions (phytogeography), host–parasite or plant–herbivore interactions, pollination biology, and fruit dispersal, or in answering questions related to the origin of adaptive characters (Nelson and Platnick 1981; Humphries and Parenti 1986; Brooks and McLennan 1991; Forey et al. 1992). Because of its predictive framework, a phylogenetic classification can direct the search for genes, biological products, biocontrol agents, and potential crop species. Phylogenetic information is also useful in conservation issues. Finally, phylogenetic classifications provide a framework for biological knowledge and the basis for comparative studies linking all fields of biology (Funk and Brooks 1990).

Constructing a classification involves two steps. The first step is delimitation and naming of groups. In a phylogenetic classification this is uncontroversial: Named groups must be monophyletic. The second step involves ranking the groups and placing them in a hierarchy. This remains problematic.

GROUPING: NAMED GROUPS ARE MONOPHYLETIC

A phylogenetic classification reflects evolutionary history and attempts to give names *only* to groups that are monophyletic—that is, an ancestor and all its descendants. In the example in Figure 2.6C, we infer that the Asteraceae (diamond plants) are monophyletic because they have flowers in heads. The square plants plus Asteraceae are also monophyletic because they share the derived character state of fused petals; this group has a name, the Asteridae (or the asterid clade). Similarly, the entire group of plants with tricolpate pollen (circle plants plus Asteridae) is monophyletic and is known as the eudicots (or the tricolpate clade). This group could be given a formal Latin name, but it does not have one at the moment and may not actually need one.

In cladistic classification, paraphyletic groups are not named. In Figure 2.6C, a group made up of square plants plus circle plants would be paraphyletic. The most recent common ancestor shared by any square plant and a circle plant is also the most recent common ancestor of any circle plant and a diamond plant. In other words, the circle plants are as distantly related to square plants as any one of them is to diamond plants. Naming a group that included the square plus the circle plants would imply that the two plants are closely related even though they are not.

There are many examples in this book of named groups of plants that we now believe to be paraphyletic. One well-known example is "bryophytes," a group that traditionally includes the nonvascular land plants (liverworts, hornworts, and mosses; see Figure 1.1). But the liverworts, hornworts, and mosses are more distantly related to each other than the mosses are to the vascular plants (tracheophytes). Without quotation marks the name *bryophytes* implies a closer relationship than actually exists.

Several traditionally recognized plant families, such as Apocynaceae and Capparaceae, are paraphyletic. In this text these families have been recircumscribed so as to recognize monophyletic groups: Apocynaceae have been combined with Asclepiadaceae, and Capparaceae with Brassicaceae.

NAMING: NOT ALL GROUPS ARE NAMED

A phylogenetic classification attempts to name only monophyletic groups, but the fact that a group is monophyletic does not mean it needs to have a name. The reasons for this are practical. We could put every pair of species into its own genus, every pair of genera into its own family, every pair of families into its own superfam-

ily, and so on. But such a classification would be cumbersome; it also would not be stable because our view of sister species would change each time a new species was described, and our view of the entire classification would have to shift accordingly.

In practice, many monophyletic groups are not named. For example, the genus *Liquidambar* (sweet gum) is monophyletic and contains four species. Although the relationships among the four species are quite clear, the pairs of species are not named, and few systematists would consider doing so. In another example, over half of the genera of the grass family fall into a single large clade that contains four large, traditionally recognized subfamilies, plus two smaller ones. Although agrostologists refer to this clade as the PACCAD clade (an acronym for Panicoideae-Arundinoideae-Centothecoideae-Chloridoideae-Aristidoideae-Danthonioideae), it has no formal Latin name.

How do systematists decide which monophyletic groups to name? There is no codified set of rules, but several criteria have been suggested by various authors, and some criteria are in common use despite not being fully articulated. A major criterion—perhaps *the* major criterion—is the strength of the evidence supporting a group. Ideally, only clades linked by many shared derived characters should be formally recognized and named in classifications. This makes sense if a classification is to function as a common vocabulary.

Names are most useful if they can be defined, and the more precise the definition the better. In other words, if a clade is to be named, it should have a set of characters by which it can be distinguished from other clades, or **diagnosed**. This restriction also relates to nomenclatural stability. If the meaning of a name shifts every time a new phylogeny is produced or a new character is examined, the name becomes effectively meaningless.

A second criterion is the presence of an obvious morphological character. Although systematists are not likely to agree on the importance of this criterion, it is an important extension of the idea of a well-supported group and is also relevant to the use of classifications by nonsystematists for identification purposes. If, for example, the only way a field biologist can identify an organism is by knowing whether it has an alanine or a serine at position 281 in its ribulose-1,5-bisphosphate carboxylase/oxygenase molecule, she may not find the classification much help in making predictions about the organism.

If, on the other hand, she knows that the organism is a grass with a particular spikelet structure, she can easily and reliably infer many aspects of its biology. (Lack of an obvious morphological synapomorphy is one of several reasons that the PACCAD clade of the grasses is not given a name.) The characters used for classification do not have to be those used for identification, but many systematists prefer to name clades that are easily recognized morphologically.

Another criterion is size of the group. Human memory is easily able to keep track of small numbers of items (in the range of 3 to 7) (Stevens 1998), but to organize and remember larger numbers of items requires additional mnemonic devices. (As an example, consider how many 9-digit zip codes you can remember compared to the 5-digit variety, or to 7-digit telephone numbers.) Dividing a large group into smaller groups is a way to organize one's thinking about large numbers of taxa.

In the words of Davis and Heywood (1963: 83), "We must be able to place taxa in higher taxa so that we can find them again." The genus *Liquidambar* could be redefined to include only *Liquidambar styraciflua* and *L. orientalis*, and a new genus could be described to include *L. acalycina* and *L. formosana*. There seems little reason to do this, however, because four species is not a difficult number to keep track of. That said, there seems little reason to divide a large group if well-supported clades cannot be identified within it.

A fourth criterion is nomenclatural stability. A classification is ultimately a vocabulary, a means of communication. It cannot function this way if the meanings of the names continually change. Thus, given a set of well-supported, diagnosable, monophyletic groups, groups that have been named in the past can—and we would argue should—continue to be named. This is yet another argument against formally naming the PACCAD clade of the grasses, in that it would entail an unnecessary set of changes affecting long-standing taxonomic usage. It is also an argument against dividing *Liquidambar* into two genera, even though both would be monophyletic and well-supported; both size of group and nomenclatural stability argue against the division (Backlund and Bremer 1998; Stevens 1998).

RANKING: RANKS ARE ARBITRARY

Having decided which monophyletic groups to name, we still have the question of exactly how to name them. The groups could, for example, be numbered, and a central index could list what is encompassed by each numbered group. This approach is similar to the system used by the telephone company to organize telephone numbers. The difficulty, of course, is that without a telephone book (a central index) and/or an excellent memory, the system is inaccessible.

Biological classification attempts to provide a working vocabulary that conveys phylogenetic information, yet can be learned by biologists who are not themselves primarily systematists. Because a phylogeny is similar in structure to a hierarchy, in which small groups are included in larger groups, which themselves are included in still larger groups, it makes sense to reflect it as a hierarchy.

Botanical classification uses a system developed in the eighteenth century, in which taxa are assigned particular ranks, such as kingdom, phylum, class, order, family, genus, and species (i.e., Linnaean ranks) (see Chapter 3 and Appendix 1). A classification of named monophyletic groups should be logically consistent with the phylogenetic relationships hypothesized for the organisms being classified (as expressed in the sequence of branching points in the cladogram). That is, the categorical ranks of a Linnaean classification can be used to express sister-group relationships.

It is important to realize that even though monophyletic taxa are considered to represent real groups that exist in nature as a result of the historical process of evolution, the categorical ranks themselves are only mental constructs. They have only relative (not absolute) meaning (Stevens 1998). In other words, the familial level is less inclusive than the ordinal level and more inclusive than the generic level, but there are no criteria available indicating that a particular taxon, such as the angiosperms, should be recognized at the level of phylum, class, or order.

In Figure 2.22, a cladogram of imaginary taxa A through E is first converted into a hierarchical classification according to Linnaean ranks. Note that subgenus DE is nested within genus CDE, which is in turn nested within family ABCDE. (But we could have treated clade ABCDE as an order, clade CDE as a family, and clade DE as a genus.) These procedures often lead to difficulties because in order to fully express the sister-group relationships (in the cladogram), one needs more ranks than are available (in the taxonomic hierarchy).

Although additional ranks can be created by use of the prefixes *super-* and *sub-*, the number of ranks may still be insufficient. Therefore, modifications to the method of

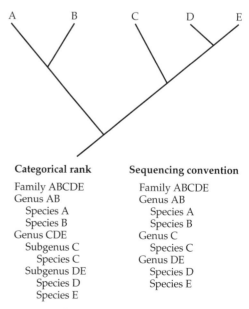

Categorical rank	Sequencing convention
Family ABCDE	Family ABCDE
Genus AB	Genus AB
Species A	Species A
Species B	Species B
Genus CDE	Genus C
Subgenus C	Species C
Species C	Genus DE
Subgenus DE	Species D
Species D	Species E
Species E	

Figure 2.22 Alternative classifications based on the phylogeny of a hypothetical group of taxa A, B, C, D, and E. The classification on the right uses only three ranks (family, genus, and species) plus a sequencing convention; the one on the left uses four ranks (family, genus, subgenus, and species).

classification outlined here have been proposed (Wiley 1979, 1981). For example, the **sequencing convention** states that taxa forming an asymmetrical part of a clado-gram may be placed at the same rank and arranged in their order of branching. The sequence of names in the classification denotes the sequence of branching in the cladogram. Note that this is the same as saying that not all monophyletic groups are given names.

Even though ranking is arbitrary, the criteria described here for deciding which groups to name can also be applied to deciding the level at which to rank a group (see Stevens 1998 for full discussion). Nomenclatural sta-bility again becomes important. Often one of the mono-phyletic groups that could be given the name of family already has a commonly used family name, so it makes sense to continue to use the term *family* for these taxa.

For example, it has recently been shown that the earli-est-diverging lineage in the Poaceae includes only two extant genera—*Anomochloa* and *Streptochaeta*—so the phylogeny looks like that in Figure 2.23. One could, in principle, create a new family for *Anomochloa* and *Strep-tochaeta*; after all, it would be monophyletic and would leave the Poaceae as also monophyletic. For the purpos-es of stability, however, it makes sense to leave the two genera in Poaceae, where they have been given a subfa-milial name: Anomochlooideae.

Some systematists have proposed abandoning the Linnaean system altogether and replacing it with a "phylogenetic taxonomy." Full exploration of this possi-bility is beyond the scope of this text, but the arguments against the use of Linnaean ranks are addressed briefly in the discussion that follows.

Because rank is arbitrary, a genus (group of species) in one family may not be the same age, encompass the same amount of variation, or indeed have anything in common at all with a genus in another family—other than the fact that they are both monophyletic groups. Trained systematists are generally aware of this (Darwin was, for example) and realize that genera, families, and so on are not comparable units (Stevens 1997). Some sci-entists, however, frequently use such categories as though they were real. For example, it is common to measure plant diversity by listing the number of families represented by a local flora, even though the unit "fami-ly" does not mean anything in particular.

If rank is arbitrary, then one logical step would be to eliminate ranks altogether. Taxa would be placed in named groups, but the groups would not be designated as genus, family, order, or other rank. Such categoriza-tion already exists informally, particularly among groups above the level of order. For example, the eudi-cots are widely recognized as monophyletic, but are not given a particular Linnaean rank. Similarly, few system-atists worry about whether the angiosperms should be recognized as a division, class, subclass, superorder, or other rank; they are clearly monophyletic and designat-ed by the non-Linnaean name *angiosperm*.

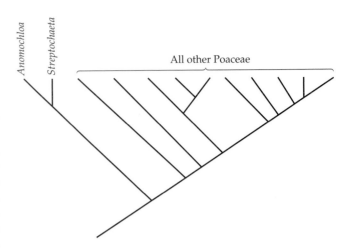

Figure 2.23 Phylogeny of Poaceae showing the position of the genera *Anomochloa* and *Streptochaeta*.

Eliminating ranks becomes more problematic among orders, families, and genera. Groups assigned to those ranks are familiar and in common use, and an entirely new sort of nomenclature is unlikely to be accepted rapidly or without protest. Nonetheless, an alternate sys-tem of phylogenetic nomenclature, known as the *Phy-loCode*, is being developed. The PhyloCode is designed entirely outside the rules of the International Code of Botanical Nomenclature (ICBN), which governs the use of Linnaean ranks and has long been used by all plant taxonomists (see Appendix 1). It is an alternative nomen-clatural system rather than a revision of the existing sys-tem (see the PhyloCode Web site at *www.ohiou.edu/ phylocode*).

Another result of phylogenetic studies is the observa-tion that many phylogenies are only partially resolved, so precise placement of taxa is impossible, given the avail-able data. This means that some species cannot be placed for certain in a genus, and some genera cannot be reliably assigned to a family. The current system allows uncertain placements above the rank of species to be reflected by the category of *incertae sedis*—literally, "of uncertain posi-tion." An alternative would be a rank-free system, in which neither placement in a larger group nor naming all branches of a dichotomy or polychotomy is necessary.

The authors of this text have been involved in reclas-sification of genera, families, and orders on the basis of phylogenetic data and have found that—as long as the phylogeny is clear—use of the standard Linnaean hierar-chy is quite easy (especially when supplemented by unranked informal names). When the phylogeny is unclear, it is usually reasonable to wait for more data before modifying the classification.

For more discussion of the problems encountered in using the Linnaean system in phylogenetic classification, consult Wiley 1981; de Queiroz and Gauthier 1990, 1992; Wiley et al. 1991; Forey et al. 1992; and Hibbett and Donoghue 1998.

COMPARING PHYLOGENETIC CLASSIFICATIONS WITH THOSE DERIVED USING OTHER TAXONOMIC METHODS

Not all taxonomists use phylogenetic methods, although this is the majority approach. Some systematists have held the view that although evolution has occurred, parallelism and reversal are so common that the details of evolutionary history can never be deciphered. This point of view led to a school of systematics known as **phenetics**. Pheneticists argued that because evolutionary history could never be unequivocally detected, organisms might best be classified according to overall similarity. Thus, similar organisms were placed together in a group, while very different organisms were placed in different groups (Sneath and Sokal 1973).

One serious difficulty with the phenetic point of view was that many systematists produced treelike diagrams that grouped organisms by overall similarity, but these diagrams were then interpreted as though they reflected evolutionary history. Sometimes this approach led to results similar to those produced by a phylogenetic analysis, but sometimes it led to the production of "groups" made up of organisms that shared only the fact that they were different from everything else, including each other. Such groups have since proven to be paraphyletic or polyphyletic.

The development of phenetic methods was an important prelude to the acceptance and use of phylogenetic approaches. A taxonomist constructing a phenetic classification first carefully observed as many characters as possible. These characters were divided into states, or the quantitative value of the character was recorded (e.g., a series of measurements of leaf length, with the mean recorded for each taxon). This information was arranged in a character × taxon matrix similar to that in Figure 2.10A.

The matrix was converted to a similarity (taxon × taxon) matrix by the use of any of several mathematical measures of similarity (or dissimilarity; see Sneath and Sokal 1973; Abbot et al. 1985). The systematist then grouped the taxa that were most similar, with the similarity relationships illustrated on either a maplike or a treelike diagram (a phenogram) (Figure 2.24). Phenograms were constructed using clustering algorithms, while maplike diagrams resulted from ordination studies employing multivariate statistical procedures (see Abbot et al. 1985).

Phenetic methods were used to produce classifications, many of which are useful for identification and information retrieval. These classifications were not designed to retrieve evolutionary history, however, and are thus not appropriate for asking evolutionary questions. Phenetic systems do not distinguish between synapomorphy and convergent or parallel evolution.

Evolutionary taxonomy differed from phylogenetic taxonomy in its approach to classification. The morphological similarity of a group was of utmost importance,

(A) Map

(B) Phenogram

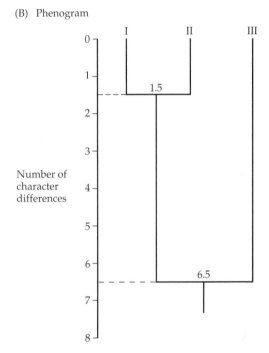

Figure 2.24 Two graphical means of expressing phenetic relationships. (A) Maplike diagram. (B) Phenogram.

and monophyly and paraphyly (in the strict cladistic senses of those words) were secondary. Thus a group could be recognized on the basis of some combination of derived and ancestral, unique and shared characters (Figure 2.25).

Importance was given to the recognition of "gaps" in the pattern of variation among phylogenetically adjacent groups (Simpson 1961; Ashlock 1979; Cronquist 1987; Mayr and Ashlock 1991). Characters considered to be evolutionarily (or ecologically) significant were stressed, and the expertise, authority, and intuition of individual systematists were central. Finally, although evolutionary classifications usually referred to evolution, and the groups recognized in such classifications were often called monophyletic, the taxa were expected to be morphologically homogeneous, and to be separated from each other by discrete gaps (Ashlock 1979; Stuessy 1983, 1990; Stevens 1986; Mayr and Ashlock 1991).

It has been said that systematics is as much an art as a science (although this statement begs the question of how

Figure 2.25 Phylogeny and a resulting non-phylogenetic classification produced according to the evolutionary school. The classification includes a mix of monophyletic and paraphyletic groups, separated from each other by morphological "gaps."

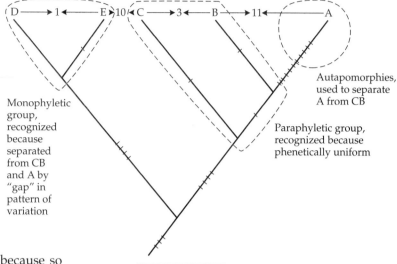

Monophyletic group, recognized because separated from CB and A by "gap" in pattern of variation

Autapomorphies, used to separate A from CB

Paraphyletic group, recognized because phenetically uniform

Classification

Family ABCDE
  Genus DE
    Species D
    Species E
  Genus CB
    Species C
    Species B
  Genus A
    Species A

one might define art and science), in part because so many aspects of the discipline seemed to have no objective basis. One fortunate result of phylogenetic systematics is that at least one major aspect—the delimitation of groups—has become formalized such that there is general agreement on how it should be done. Whereas phenetic and evolutionary classifications were ambiguous about grouping criteria, phylogenetic classifications are precise. A named group can be taken as monophyletic, including all descendants of a single common ancestor.

Literature Cited and Suggested Readings

Items marked with an asterisk (*) are especially recommended to those readers who are interested in further information on the topics discussed in this chapter.

Abbot, L. A., F. A. Bisby, and D. A. Rogers. 1985. *Taxonomic analysis in biology.* Columbia University Press, New York. [An introduction to various phenetic methods.]

Albert, V. A., A. Backlund, K. Bremer, M. W. Chase, J. R. Manhart, B. D. Mishler and K. C. Nixon. 1994. Functional constraints and *rbcL* evidence for land plant phylogeny. *Ann. Missouri Bot. Gard.* 81: 534–567.

Ashlock, P. D. 1979. An evolutionary systematist's view of classification. *Syst. Zool.* 28: 441–450. [Presentation of an easy-to-follow explicit method to construct an evolutionary taxonomic classification.]

*Backlund, A. and K. Bremer. 1998. To be or not to be: Principles of classification and monotypic plant families. *Taxon* 47: 391–400.

Bremer, K. 1983. Angiosperms and phylogenetic systematics. *Verh. Naturwiss. Verein Hamburg* 26: 343–354. [Discussion of reticulate evolution and cladistics.]

Bremer, K. 1988. The limits of amino acid sequence data in angiosperm phylogenetic reconstruction. *Evolution* 42: 795–803. [Decay indices.]

Bremer, K. and H.-E. Wanntorp. 1979. Hierarchy and reticulations in systematics. *Syst. Zool.* 28: 624–627.

*Brooks, D. R. and D. A. McLennan. 1991. *Phylogeny, ecology, and behavior.* University of Chicago Press, Chicago. [Excellent presentation of biological uses of phylogenetic hypotheses.]

Cantino, P. D., H. N. Bryant, K. de Queiroz, M. J. Donoghue, T. Eriksson, D. M. Hillis, and M. S. Y. Lee. 1999. Species names in phylogenetic nomenclature. *Syst. Biol.* 48: 790–807.

Cronquist, A. 1987. A botanical critique of cladism. *Bot. Rev.* 53: 1–52.

*Dahlgren, R. and F. N. Rasmussen. 1983. Monocotyledon evolution: Characters and phylogenetic estimate. In *Evolutionary biology*, Vol. 16, M. K. Hecht, B. Wallace and G. T. Prance (eds.), 255–395. Plenum Press, New York. [Contains a simple introduction to cladistic methods.]

Davis, P. D. and V. H. Heywood. 1963. *Principles of angiosperm taxonomy.* Krieger, New York.

*de Queiroz, K. and J. Gauthier. 1990. Phylogeny as a central principle in taxonomy: Phylogenetic definitions of taxon names. *Syst. Zool.* 39: 307–322. [Proposal to abandon the Linnaean system.]

de Queiroz, K. and J. Gauthier. 1992. Phylogenetic taxonomy. *Annu. Rev. Ecol. Syst.* 23: 449–480.

de Queiroz, A., M. J. Donoghue, and J. Kim. 1995. Separate versus combined analyses of phylogenetic evidence. *Annu. Rev. Ecol. Syst.* 26: 567–581.

*Donoghue, M. J. and P. D. Cantino. 1988. Paraphyly, ancestors, and the goals of taxonomy: A botanical defense of cladism. *Bot. Rev.* 54: 107–128.

Doyle, J. A., M. J. Donoghue and E. A. Zimmer. 1994. Integration of morphological and ribosomal RNA data on the origin of angiosperms. *Ann. Missouri Bot. Gard.* 81: 419–450.

*Eldredge, N. and J. Cracraft. 1980. *Phylogenetic patterns and the evolutionary process: Methods and theory in comparative biology.* Columbia University Press, New York.

*Farris, J. S. 1974. Formal definitions of paraphyly and polyphyly. *Syst. Zool.* 23: 548–554. [A polyphyletic group is defined as "a group in which the most recent common ancestor is assigned to some other group and not the group itself."]

Farris, J. S. 1979. The information content of the phylogenetic system. *Syst. Zool.* 28: 458–519.

Farris, J. S. 1989. Hennig86, Version 1.5. Port Jefferson Station, NY.

Felsenstein, J. 1978. Cases in which parsimony and compatibility methods will be positively misleading. *Syst. Zool.* 27: 401–410.

Felsenstein, J. 1981. Evolutionary trees from DNA sequences: A maximum likelihood approach. *J. Mol. Evol.* 17: 368–376.

Felsenstein, J. 1989. *PHYLIP 3.2 manual.* University of California Herbarium, Berkeley.

*Forey, P. L., C. J. Humphries, I. L. Kitching, R. W. Scotland, D. J. Siebert and D. M. Williams. 1992. *Cladistics: A practical course in systematics.* Oxford University

Press, Oxford, UK. [Summary of then-current cladistic methods.]

*Frohlich, M. W. 1987. Common-is-primitive: A partial validation by tree counting. *Syst. Bot.* 12: 217–237. [The principle that says that given an ingroup and an outgroup, both with states A and B of a homologous character, if state A is much more common than state B *in the outgroup*, then state A is likely to be ancestral *within the ingroup*.]

Funk, V. A. 1985. Phylogenetic patterns and hybridization. *Ann. Missouri Bot. Gard.* 72: 681–715.

*Funk, V. A. and D. R. Brooks. 1990. *Phylogenetic systematics as the basis of comparative biology* (Smithsonian Contributions to Botany, No. 73). Smithsonian Institution Press, Washington, DC. [Examples of biological uses of cladograms.]

Gift, N. and P. F. Stevens. 1997. Vagaries in the delimitation of character states in quantitative variation: An experimental study. *Syst. Biol.* 46: 112–125.

*Givnish, T. J. and K. J. Sytsma. 1997. Homoplasy in molecular vs. morphological data: The likelihood of correct phylogenetic inference. In *Molecular evolution and adaptive radiation*, T. J. Givnish and K. J. Sytsma (eds.), 55–101. Cambridge University Press, Cambridge, UK.

Goloboff, P. A. 1993. *NONA*, Version 1.5.1. P. A. Goloboff, Tucumán, Argentina.

*Hennig, W. 1966. *Phylogenetic systematics*. University of Illinois Press, Urbana.

Hibbett, D. and M. J. Donoghue. 1998. Integrating phylogenetic analysis and classification in fungi. *Mycologia* 90: 347–356.

Hillis, D. M., M. W. Allard and M. M. Miyamoto. 1993. Analysis of DNA sequence data: Phylogenetic inference. *Methods Enzymol.* 224: 456–487.

*Huelsenbeck, J. P. 1995. Performance of phylogenetic methods in simulation. *Syst. Biol.* 44: 17–48. [Comparison of performance in computer simulations of parsimony, maximum likelihood, and overall similarity methods of tree construction.]

*Huelsenbeck, J. P. and D. M. Hillis. 1993. Success of phylogenetic methods in the four-taxon case. *Syst. Biol.* 42: 247–264. [Several sorts of parsimony methods are relatively successful in reconstructing phylogeny.]

Humphries, C. J. and V. A. Funk. 1982. Cladistic methodology. In *Current concepts in plant taxonomy* (Systematics Association Special Volume, No. 25), V. H. Heywood and D. M. Moore (eds.), 323–362. Academic Press, London. [Introduction to basic cladistic methods.]

Humphries, C. J. and L. R. Parenti. 1986. *Cladistic biogeography*. Clarendon, Oxford, UK.

Kellogg, E. A. 1989. Comments on genomic genera in the Triticeae. *Am. J. Bot.* 76: 796–805.

Kellogg, E. A., R. Appels and R. J. Mason-Gamer. 1996. When genes tell different stories: The diploid genera of Triticeae (Gramineae). *Syst. Bot.* 21: 321–347.

Kitching, I. J., P. L. Forey, C. J. Humphries and D. M. Williams. 1998. *Cladistics: The theory and practice of parsimony analysis*, 2nd ed. Oxford University Press, Oxford, UK. [Summary of parsimony methods for constructing phylogenetic hypotheses, including issues dealing with characters.]

Kron, K. A. and W. S. Judd. 1997. Systematics of the *Lyonia* group (Andromedeae, Ericaceae) and the use of species as terminals in higher-level cladistic analyses. *Syst. Bot.* 22: 479–492.

*Maddison, D. R. and W. P. Maddison. 2000. *MacClade: Analysis of phylogeny and character evolution*, Version 4.0. Sinauer Associates, Sunderland, MA. [A useful program for exploring patterns of character change on a cladogram.]

Maddison, W. P., M. J. Donoghue and D. R. Maddison. 1984. Outgroup analysis and parsimony. *Syst. Zool.* 33: 83–103.

Mayr, E. and P. D. Ashlock. 1991. *Principles of systematic zoology*, 2nd ed. McGraw-Hill, New York.

*McDade, L. A. 1990. Hybrids and phylogenetic systematics. I. Patterns of character expression in hybrids and their implications for cladistic analyses. *Evolution* 44: 1685–1700.

*McDade, L. A. 1992. Hybrids and phylogenetic systematics. II. The impact of hybrids on cladistic analyses. *Evolution* 46: 1329–1346.

McDade, L. A. 1997. Hybrids and phylogenetic systematics. III. Comparison with distance methods. *Syst. Bot.* 22: 669–683.

Nelson, G. and N. Platnick. 1981. *Systematics and biogeography*. Columbia University Press, New York.

Pankhurst, R. J. 1991. *Practical taxonomic computing*. Cambridge University Press, Cambridge, UK. [Use of computers in databases, identification, and phenetic and cladistic analyses.]

*Quicke, D. L. J. 1993. *Principles and techniques of contemporary taxonomy*. Blackwell, London.

*Sanderson, M. J. and M. J. Donoghue. 1989. Patterns of variation in levels of homoplasy. *Evolution* 43: 1781–1795.

Schuh, R. T. 2000. *Biological systematics: Principles and applications*. Cornell University Press, Ithaca, NY. [Detailed coverage of cladistic methods and application of cladistic results in classification, biogeography, ecological investigations, and conservation.]

Simpson, G. G. 1961. *Principles of animal taxonomy*. Columbia University Press, New York.

Sneath, P. H. A. and R. R. Sokal. 1973. *Numerical taxonomy*. Freeman, San Francisco.

Sokal, R. R. 1986. Phenetic taxonomy: Theory and methods. *Annu. Rev. Ecol. Syst.* 17: 423–442.

*Sokal, R. R. and F. J. Rohlf. 1980. An experiment in taxonomic judgement. *Syst. Bot.* 5: 341–365.

Stevens, P. F. 1980. Evolutionary polarity of character states. *Annu. Rev. Ecol. Syst.* 11: 333–358.

*Stevens, P. F. 1984. Homology and phylogeny: Morphology and systematics. *Syst. Bot.* 9: 395–409.

Stevens, P. F. 1986. Evolutionary classification in botany, 1960–1985. *J. Arnold Arbor.* 67: 313–339.

*Stevens, P. F. 1991. Character states, morphological variation, and phylogenetic analysis: A review. *Syst. Bot.* 16: 553–583.

Stevens, P. F. 1997. How to interpret botanical classifications: Suggestions from history. *BioScience* 47: 243–250.

Stevens, P. F. 1998. What kind of classification should the practising taxonomist use to be saved? In *Plant diversity in Malesia III: Proceedings of the 3rd International Flora Malesiana Symposium 1995*, J. Dransfield, M. J. E. Coode, and D. A. Simpson (eds.), 295–319. Royal Botanical Gardens, Kew, London.

Stuessy, T. F. 1983. Phylogenetic trees in plant systematics. *Sida* 10: 1–13.

Stuessy, T. F. 1990. *Plant taxonomy*. Columbia University Press, New York.

*Swofford, D. L. 1993. *PAUP: Phylogenetic analysis using parsimony*, Version 3.1.1. Distributed by the Illinois Natural History Survey, Champaign.

*Swofford, D. L. 2000. *PAUP*: Phylogenetic analysis using parsimony and other methods*, Version 4.0. [Beta-test edition, distributed by Sinauer Associates, Sunderland, MA.]

*Swofford, D. L., G. J. Olsen, P. J. Waddell and D. M. Hillis. 1996. Phylogenetic inference. In *Molecular systematics*, 2nd ed., D. M. Hillis, C. Moritz and B. K. Mable (eds.), 407–514. Sinauer Associates, Sunderland, MA. [Excellent summary of methods of tree construction.]

Sytsma, K. J. and J. C. Pires. 2001. Plant systematics in the next 50 years—Re-mapping the new frontier. *Taxon* 50: 713–732.

*Wagner, W. H., Jr. 1980. Origin and philosophy of the groundplan-divergence method of cladistics. *Syst. Bot.* 5: 173–193.

Wagner, W. H., Jr. 1983. Reticulistics: The recognition of hybrids and their role in cladistics and classification. In *Advances in cladistics: Proceedings of the second meeting of the Willi Hennig Society*, N. I. Platnick and V. A. Funk (eds.), 63–79. Columbia University Press, New York.

Watrous, L. E. and Q. D. Wheeler. 1981. The outgroup comparison method of character analysis. *Syst. Zool.* 30: 1–11.

Weins, J. J., ed. 2000. *Phylogenetic analysis of morphological data*. Smithsonian Institution Press, Washington, DC.

Wiley, E. O. 1979. An annotated Linnaean hierarchy, with comments on natural taxa and competing systems. *Syst. Zool.* 28: 308–337.

*Wiley, E. O. 1981. *Phylogenetics*. Wiley, New York. [Detailed discussion of cladistic principles.]

*Wiley, E. O., D. Siegel-Causey, D. R. Brooks, and V. A. Funk. 1991. *The compleat cladist: A primer of phylogenetic procedures* (University of Kansas, Museum of Natural History, Special Publication, No. 19). University of Kansas, Museum of Natural History, Lawrence. [Summary of then-current cladistic methods.]

3 *Classification and System in Flowering Plants: Historical Background*

Throughout history, scientists have tried to determine the best way to classify living things. Their ideas on how to do this have changed considerably over time. In this chapter we discuss the main ways that botanists have classified plants, and some of the reasoning behind those classifications. This is only part of the history of systematics; discussion of the complex relationships between professional and amateur botanists and the general public is largely ignored. These relationships are an integral part of the historical background of our discipline, even if we know all too little about them (but see, for example, Allen 1976).

If you look at the phylogenetic trees in Chapters 7 through 9, you will see that it is possible to base classifications on them that capture precisely the clades in those phylogenies (see Chapter 2). Both phylogenies and classifications are hierarchical and are made up of groups nested within groups. However, some classifications in use today—in particular, evolutionary classifications—do not try to represent phylogenies in this way. Indeed, they are not strictly hierarchical. What they are trying to represent can be understood only in the context of a long history stretching back before anyone had any idea about evolution. Understanding classifications thus means that we need to understand their history.

In the past, both users and makers of classifications had quite different ideas about nature and about the role of classification than they do today, yet we tend to assume that our and their ideas are the same. The problem is made worse because many terms have changed meaning over time. The term system is a prime example. It now refers to sets of relationships in genealogies (de Queiroz 1988), but in the late eighteenth century, system was used as a term to belittle classifications based on a single character—except in England, where it referred to what European botanists would have called method, classifications based on many characters!

Finally, plant systematists, perhaps even more than other systematists, have long distrusted theory, and have considered classification a theory-free, "empirical" operation (see Stevens 1986, 1990, 1994, 1998a, and Kornet 1991 for discussion). Theories should not affect a systematist's observations or classifications. For this reason, plant systematists have often been unwilling or unable to explain the reasons for their decisions regarding classification.

This chapter describes some of the history of botanical classification in order to show how earlier, often non-phylogenetic, ideas about nature were incorporated into current classifications. First we discuss the long-standing and continuing tension between the makers of classifications, most of whom want to understand relationships (although note that the term relationship has meant different things to different people), and many users of classifications, who simply want names to be stable. Then we discuss how relationships are understood and how nature is visualized, how higher taxa are delimited, and how the circumscription of some of the major groups has changed. (Here we discuss only higher taxa—genera and above; for a discussion of species concepts, see Chapter 6 and Stevens 1992, 1997b.)

Classification, Nature, and Stability

For hundreds of years, botanists have tried to develop a classification that was "natural." Until recently we have assumed that history of systematics is in part the history of a single "natural" system that has been gradually developing over the centuries. Its principles were first outlined by Caesalpinus. Tournefort and Linnaeus described "natural" genera, and Linnaeus suggested a number of "natural" families, although he did not describe them. The "natural" method then received a major boost from A.-L. de Jussieu in his great *Genera plantarum* of 1789, in which he described both genera and families and placed the latter in classes. This Jussieuan foundation is the basis of our current classification, and although new families have been added, the limits of existing families have been modified, and higher taxa such as orders have been added, nothing fundamental has changed.

Unfortunately, the word natural has no fixed meaning; rather, authors have used it to mean something that agrees with their own ideas about nature, or about constructing classifications or systems. Eighteenth-century systematists had ideas about nature that were very different from ours—they were certainly not evolutionary—and their systematic practice and classifications are best interpreted in terms of how they understood nature. Nineteenth-century systematists built on the work of their predecessors. Although they generally did not describe clearly their understanding of nature, which was changing, the way they discussed and depicted relationships was not much different from that of the preceding century. And many aspects of nineteenth-century classification persisted through the twentieth century.

Some historians of classification see a trend from analytic ways of grouping that were particularly common in the eighteenth century toward more synthetic procedures in the nineteenth century. In analytic grouping procedures, one or a few characters are used successively to define groups, so that organisms are divided up into smaller and smaller groups. The whole process is rather like using a key. In synthetic grouping, many characters are used and groups are built up ("synthesized") (Mayr 1982).

The distinction between the two procedures is not always clear. Even in the twentieth century some botanists used single-character, analytic (divisive) systems (such as John Hutchinson, who divided dicotyledons into woody and herbaceous groups, a classification that even Linnaeus had dismissed as "lubricious"; see Hutchinson 1973). On the other hand, Jussieu's method for recognizing relationships, developed in the later part of the eighteenth century, is synthetic.

Classifications have been expected to do much more than reflect nature, however. They have also been expected to be (1) easy to use, (2) stable, (3) an aid to memory, (4) predictive, and (5) concise—a set of goals that are sometimes in conflict. These goals of classification were spelled out by Andreas Caesalpinus in 1583 (see Greene 1983, Vol. 2: 815–817). Thus the systematist must not only describe nature (whatever he/she thinks nature is), but also serve a community of users, many of whom may have no interest in the systematist's ideas of nature. Before the twentieth century, those users were largely medical personnel, but they now include a broad array of biologists and non-biologists (even interior designers).

Stability of names has been a perennial problem. Systematists have often wanted to leave the names of taxa unchanged—even if those names conflicted with their views of relationship—lest the users of their classifications become upset

Andreas Caesalpinus
(1519–1603, Italian)

(Stevens 1994: Chapter 10; 1997a). Even George Bentham and J. D. Hooker, authors of the great three-volume *Genera plantarum* (1862–1883), circumscribed some taxa to reflect custom and convention; that is, they constructed some taxa that were not natural, even by their own definition of the word. Bentham, at least, ignored some of his own taxon circumscriptions when it came to discussing distributional patterns (Stevens 1997a). In this book we have attempted to change our classification so that it better reflects relationships, but it is highly likely that some readers will wish we had not changed so much. This point of view is, if nothing else, traditional!

J. S. L. Gilmour (1940) promoted the idea that the best classification had maximum general utility, although this leaves open the definition of "utility." The needs of groups of people who use classifications may change over the years, and needs may conflict (Stevens 1998b); the result may even be classifications that conflict. Certainly Gilmour's ideas sharpened the conflict between those who wanted classifications to reflect history and those who were not at all interested in relationships.

As early as 1778, however, Lamarck had suggested a solution to the problem: The characters used in the formal classification did not have to be those used for identification. The keys (see Appendix 2) that he promoted link users and experts. Easily visible characters could be used in keys, and these would not necessarily be the same as the sometimes inconspicuous characters used to distinguish the groups recognized in the formal classification. Keys made it easy to give plants their correct names, so groups in classifications did not—and do not—have to be easily recognizable.

A general respect for authority has also affected classifications. Some plant groups have been recognized for a long time, for example, the Labiatae, Liliaceae, Cruciferae, and Compositae. Many of these do not end in the conventional -*aceae*, which indicates that they are not based on particular genera, and the groups may even pre-date "scientific" classification. The fact that such groups have always been recognized is sometimes used as evidence that they are "natural" groups. If they have been historically recognized by the acknowledged masters of the discipline, so the argument goes, they must be correctly delimited. Systematists have been generally reluctant to modify such groups.

These paradigmatic groups are generally ones that are obvious in the European flora, a fact that reflects the European origin of botanical systematics. Not only did the discipline originate in Europe, but it was dominated by Europeans for centuries. Not until Asa Gray (1810–1888) was there a North American botanist considered by Europeans to be fully their equal, and only with Charles Bessey (1845–1915) did North American botany become fully independent (Dupree 1959; Cuerrier et al. 1996). In other parts of the world, especially those that were then colonies of European nations, this domination persisted longer, and it was especially evident in the floras of countries that were written by European botanists and based on material held in European institutions.

Understanding Relationships

Asa Gray
(1810–1888, American)

We mentioned in the preceding section that eighteenth- and nineteenth-century systematists saw nature quite differently from the way we do now. How can we know what they were thinking? The analogies they used when describing patterns of relationships that they saw in nature and the diagrams they prepared to show these patterns are particularly good sources of information.

Although it is obvious to us that relationships can often be represented as treelike diagrams, the reason is that we share a common set of assumptions about how organisms came to be the way they are. Thus, throughout this book, we diagram relationships somewhat like pedigrees, with extant organisms linked by extinct ancestors (this sort of diagram is known as a **Steiner tree**).

Many of our predecessors did not share this view of nature. Put yourself in the world before Darwin and before ideas of evolution. How would you imagine plants are related to each other? What would the word *relationship* mean? In fact, we find that many eighteenth- and nineteenth-century botanists thought of relationships as being like the relationships between countries on a map or stars in a constellations; that is, they were reticulating.

Antoine-Laurent de Jussieu described many of the families whose evolution we now attempt to study. His genera and families have been interpreted for two centuries as though they were more or less distinct groups, but this is not how he saw them. For Jussieu, relationships in nature formed continuous series that lacked any clear breaking points. Any divisions in these series were the work of man, not of nature. Jussieu emphasized that groups were linked, and his natural families, such as the Compositae, were natural precisely because they were examples of this continuous nature; not surprisingly, genera in such families were difficult to recognize.

For Jussieu, as for his colleague Lamarck, the taxonomic hierarchy was simply a set of words, each of which individually referred to a part of the

Antoine-Laurent de Jussieu
(1748–1836, French)

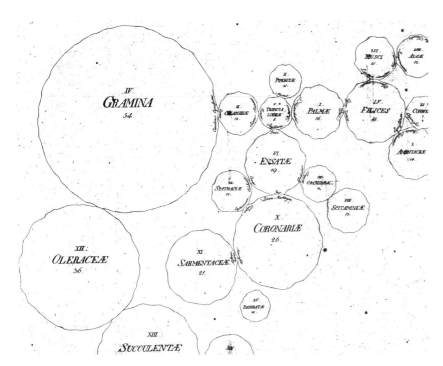

Figure 3.1 A portion of P. D. Giseke's "genealogical-geographical" map (1792).

continuum and together allowed the whole to be recalled to mind. Complications arose from Jussieu's descriptions of the groups he recognized. These groups were rarely wholly characterized by the features he listed for them, and a family description often referred only to characters of genera in the middle of the sequence in which he placed them (i.e., in the middle of the continuum).

A particularly interesting diagram from about this time is P. D. Giseke's "genealogical-geographical" map of 1792 (Figure 3.1). In this diagram, circles of various sizes, representing families, are placed at varying distances from one another. Giseke took pains to say that the relationships he showed were not those between "grandfather and grandson," but rather between "cousins or relatives by marriage." He noted whether or not there were intermediates between the families shown in the diagram, and he carefully distinguished between different kinds of relationships as he described the complex, two-dimensional spatial relationships between groups.

Much later, Bentham and Hooker's *Genera plantarum* (1862–1883) reflected the principles first outlined by Bentham in 1857. Both Bentham and Hooker thought that groups showed reticulating relationships, and that their boundaries were sometimes, or even often, indistinct. All in all, their ideas were not too different from those of Giseke.

Through much of the nineteenth century, and even much of the twentieth, botanical relationships have been portrayed as highly complex and reticulating. Even when tree diagrams were used to show evolution-

ary relationships, extant groups were linked directly to other extant groups. (These diagrams are known as **minimum spanning trees**.) Such diagrams imply that groups that exist in the world today are the ancestors of other groups that also currently exist, which doesn't make much sense in terms of evolutionary processes.

A more evolutionary view would be to say that two extant groups are descended from a single extinct ancestor. The direct links between extant groups were drawn in part because many botanists, from at least as early as 1786 (Johann Georg Forster), have disliked talking about ancestors, whether because ancestor-descendant relationships could not be seen directly or, later, because the fossil record was simply too poor to detect them. There are a few examples of early trees with extinct ancestors, but these are much less common (for illustrations, see Lam 1936 and Voss 1952).

In addition to showing that groups were linked simultaneously to many other groups, the goal of these complex diagrams was often to indicate the relative "highness" or "lowness" of groups. This aim can be seen in the work of Charles Edwin Bessey, a major figure in North American botany at the end of the nineteenth century. He produced numerous diagrams showing relationships, and in the latter part of his career these diagrams showed extant groups joined directly (Figure 3.2) (Cuerrier et al. 1996).

Bessey drew these "trees" to show major trends in advancement (and sometimes also reversals), and his classifications are to be read as sequences that in part reflect these diagrams. It is interesting that Bessey was attempting to make systematics more philosophical, and he had specifically dismissed maplike representations of nature. In fact, his freestanding "trees" are more like archipelagos or maps, even if they have been given an axis. They are conceptually similar to the representations of nature he dismissed, being quite like those of Linnaeus and Jussieu. Furthermore, although Bessey repeatedly emphasized that classifications should reflect phylogeny, the way he produced classifications made that goal very difficult to achieve.

Charles Edwin Bessey
(1845–1915, American)

Figure 3.2 One of Charles Bessey's freestanding "trees." (From Bessey 1915.)

Other major classification systems, such as that of Arthur Cronquist (1981), are also minimum spanning trees (Figure 3.3) that sometimes even allow reticulations between groups. They are certainly not readily interpretable in phylogenetic terms. An unwillingness to specify only historical connections between groups is associated with a tendency to emphasize parallel evolution, parallel tendencies, or even ideas of **orthogenesis** (directed evolution), as is particularly evident in Cronquist's earlier work. If two groups are not related directly, the argument goes, the occurrence of the same characters in these groups must be explained by independent evolution.

H. F. Wernham, in an influential series of papers (1911–1912), asserted that the Sympetalae were polyphyletic—as were monocots, dicots, and even angiosperms as a whole. He felt that all important characters had evolved in parallel several times in closely related but independent lineages. In fact, suggestions that there were large-scale parallelisms in patterns of relationships are quite common from the late eighteenth century onward, and some researchers have even seen quite close parallelisms between the series into which animals and plants could be placed. The existence of such parallelisms was taken as evidence that the "real" pattern of relationships in nature had been discovered!

Rolf Dahlgren's name has become associated with diagrams representing a cross section of a phylogenetic tree ("Dahlgrenograms") (Figure 3.4). Groups are represented by bubbles of different sizes between which relationships are implied—though not clearly shown—by the way the diagram is drawn. These diagrams have

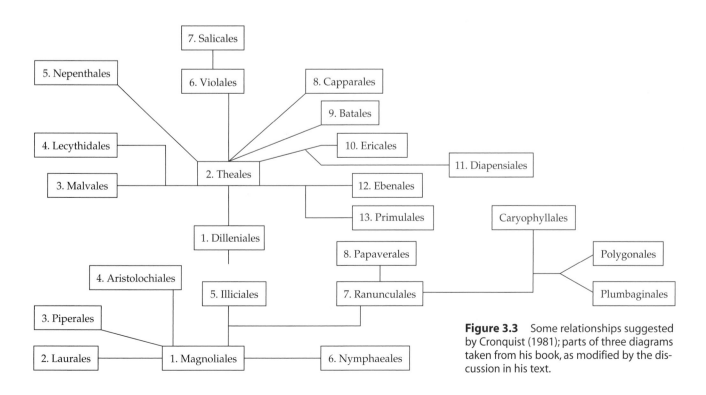

Figure 3.3 Some relationships suggested by Cronquist (1981); parts of three diagrams taken from his book, as modified by the discussion in his text.

Figure 3.4   An example of a "Dahlgrenogram," a diagram representing a cross section of a phylogenetic tree. The diagram on the right is a three-dimensional representation of the diagram on the left. (From Dahlgren et al. 1985.)

been much used to display results of broad surveys of variation of characters, such as the distribution of iridoids or the types of plastids in sieve tubes. Conceptually, Dahlgrenograms are more closely related to Bessey's cactuslike diagram (see Figure 3.2) or Giseke's "genealogical-geographical" map (see Figure 3.1) than to the phylogenies used in this book. (Note, however, that Dahlgren himself was much interested in phylogeny.)

Indeed, for some authors the fact that such diagrams did not have evolutionary implications was a virtue. They permitted one to think about general relationships, without worrying about evolution (Heywood 1978). Note that in this context, again, the term *relationship* can't mean evolutionary or, still less, phylogenetic relationships.

Classifications and Memory

The use of classification as an aid to memory was critical in a time before computers existed and when even books were not common. A classification had to have a moderate number of families, and these had to be divided into subgroups that were neither too small nor too large. Linnaeus emphasized the value of system, grouping organisms in tens; this approach allowed him to place all of the fewer than 10,000 plant species that he thought existed into groups at just four hierarchical levels.

Jussieu, whose classification was imposed on what he saw as continuous natural variation, recognized only taxa that he thought were neither too small (there had to be at least two included members) nor too big (100 members may have been the uppermost limit). Thus, he recognized no monogeneric families, and the Compositae,

which he thought were very "natural" but had well over 100 genera, were divided into three families.

Similarly, Bentham and Hooker, and some of their colleagues, such as Asa Gray, had agreed before the monumental *Genera plantarum* was written that 200 was the upper limit of families to be recognized; otherwise there would be too many to memorize (201 was the final figure). Bentham and Hooker also agreed that the best size for taxa was 2 to 6, rarely up to 12, included members. Yet some of the families they accepted had hundreds of genera.

Bentham and Hooker reconciled their intention of having a fixed, low number of often quite large families and the need to have small taxa at all hierarchical levels by interpolating formal or informal groupings wherever needed. As a result, all groups in the *Genera* above the level of genus have fewer than 14 included members. Such small groups are best suited for storage in and recall from memory, and the emphasis on recognizing large genera and families also minimized the number of names in general used and reduced the burden on the memory (Stevens 1997a, 2002).

The Formation of Higher Taxa

The idea that almost all plants belong in genera with two or more species was suggested by Conrad Gessner around the middle of the sixteenth century (Morton 1981). However, similar groupings of plants (and animals) are evident in classifications of local peoples worldwide (Atran 1990; Berlin 1992). The recognition of such groupings is based on their **salience**, or obviousness and importance to the observer. This idea of salience is based on things like the degree of similarity among members of a group, their commonness, and their utility for humans.

Joseph Pitton de Tournefort
(1656–1708, French)

The basic units in such classifications often are named with binomials, or, more generally, a noun + adjective construction.

In herbals and other early botanical literature, plants are grouped in various ways (alphabetical arrangements are common), although why these groupings were recognized is often unclear. In 1694, Joseph Pitton de Tournefort provided clear guidelines for describing genera (see Dughi 1957). Generic characters should be recognizable in all members of the genus, he argued, and should be visible without the use of a microscope. When possible, these characters should be taken from features of the flower and fruit.

Tournefort called groups based on these features primary genera. However, if these genera were too big, features from other parts of the plant could be used to characterize smaller genera (see also Walters 1986 and references). Groups characterized by nonreproductive features he called secondary genera. He also suggested that it was important to keep the total number of genera to about 600 (Stevens 1998a). This number is in line with the basic units in folk taxonomies worldwide (see Berlin 1992). Tournefort's classification indeed has much in common with folk taxonomies.

Carolus Linnaeus focused on genera, and his descriptions were much more detailed than Tournefort's. Linnaeus believed that individual genera and species existed in nature, and the ranks of genus and species were distinct ranks in the organization of nature, whereas larger groupings were matters of human convenience. He emphasized that features of the flower and fruit should be used to distinguish genera, and hence he combined most of Tournefort's secondary genera with the primary genera. Vegetative features distinguished species. But, as Linnaeus observed, "Characterem non constituere Genus, sed Genus Characterem" (Linnaeus 1751: 119)—loosely interpreted, genera exist in nature independently of the features used to characterize them, certainly independently of any rigid application of "generic" characters.

This and similar dicta, when coupled with the way Linnaeus went about recognizing and describing genera (for instance, he did not always change generic descriptions as he added new species to them), make his actual practice seem almost unprincipled at times (Stafleu 1971; Stevens 2002). Like Tournefort, Linnaeus con-

Carolus Linnaeus
(1707–1778, Swedish)

sidered large groups unwieldy, and he preferred small taxa at all hierarchical levels. Indeed, it has recently become common to think of Linnaeus as being some kind of essentialist, a person who believes that taxa have certain kinds of fixed characters without which they cannot be formally recognized (again, a loose interpretation). However, this is at best a great oversimplification (Winsor 2001; Stevens 2002).

Today's genera are ultimately built on this Linnaean foundation. Although many taxonomists, at least in theory, tolerate genera distinguished by characters other than those of flowers and fruits, in practice reproductive characters have served as a major source of generic-level differences. Nevertheless, many important nineteenth-century taxonomists certainly did not believe that there was a rank of genus (or family, for example) in nature, and by the middle of the century this was true of the rank of species too (Stevens 2002).

By the 1870s it even seemed that most genera were known, but this state of affairs did not last long. Genera were groups of species separated by morphological gaps of adequate size (see below). Hall and Clements (1923: 6) called for "experimental and statistical studies of the generic criteria in use" in a paper whose title indicated that they wanted to clarify how systematists detected phylogenies, but their aim was really to maintain the status quo, the conventional (broad) delimitation of genera (and species) because of "the significance of system, and of the mechanism of memory" (Hall and Clements 1923: 7). Moreover, they provided no new way of detecting phylogenies.

The idea that there were families of plants was specifically suggested by Pierre Magnol in 1689 (see Adanson 1763–1764, Vol. 1: xxii–xxvii). Magnol used characters taken from all parts of the plant, or sometimes, an "affinité sensible" that could not be expressed in words. He did not recognize all the families he might have because he wanted to keep their number small. He listed 76 families.

Linnaeus described classes and orders (= families) in his sexual system. Plants were assigned to groups primarily on the basis of stamen number and arrangement, and secondarily on the basis of ovary number (more exactly, on the number of stigmas or styles). For example, *Datura* and *Verbascum*, both of which have two carpels but only a single style, were placed in the Pentandria Monogyna.

Linnaeus also outlined a natural method, the goal of botany, in which genera were grouped into natural families (he recognized 67 in the year 1751, with a substantial residue of unplaced genera). He stressed the need for characters unique to a family and found in all its members, since without such characters the natural method would be like a bell without a clapper, as he graphically described the problem. However, Linnaeus was unable to provide such characters for even the most natural of families, such as the Umbelliferae (Apiaceae).

Michel Adanson
(1727–1806, French)

Although his largely artificial sexual system and his natural genera both rely almost exclusively on characters taken from the flower and fruit, when it came to natural families, Linnaeus (1751: 117) observed, "Habitus occulte consulendus est"—that is, "the habit should secretly be consulted." Habit, for Linnaeus, comprised all other parts of the plant, including features such as leaf vernation, and could also be used to distinguish families. He regretfully noted that his natural method was incomplete because plants showed relationships in several directions, like territories on a map; some plants were not yet discovered; and the habits of plants were poorly known (Linnaeus 1751: 26–36, 137).

The need, then, was to find features that indicated higher-level relationships. Between 1763 and 1789 three authors outlined the issue in ways that defined the debate over the ensuing two centuries. In a series of tabulations (Adanson 1763–1764), Michel Adanson showed that every characteristic of plants *varied within* natural groups; he concluded that no character was essential for defining a group and that groups could be defined only by combinations of characters. A classification could be produced only by an exhaustive comparison of all parts and properties of plants.

Adanson did not state clearly how this classification was to be carried out, but his contemporaries, such as Marie Jean Antoine Nicolas Caritat de Condorcet (1743–1793), noted that this might entail mechanization in the recording of characters (no easy task two centuries before the personal computer!). Other naturalists, presumably not wanting to have to look at every character in every plant, needed clear guidelines for deciding whether some characters were more important than others. Jean-Baptiste de Lamarck (1778) came up with a numerical weighting scheme (the first one known in botany) that assigned similarity values to features that depended on how widely they were distributed in plants (e.g., how common a calyx was), although he took into account not simply presence or absence, but also the nature of the feature.

Jussieu (1789) built up "groups" by synthesis, successively forming species, genera, and families; ideas of general similarity seem to have guided this synthesis. He then showed

Jean-Baptiste-Pierre-Antoine de Monet de Lamarck
(1744–1829, French)

how different features characterized groups of successively smaller circumscription. He described these characters as if they were invariable at the level they characterized, and he strongly disagreed with Adanson's contention that there were no invariable, essential characters.

Augustin-Pyramus de Candolle (1813) saw a proper subordination of characters as being the third and final stage in detecting relationships, following the stages of "blind groping" and general comparison. This subordination of characters was similar to the way in which Jussieu had described the distributions of characters. (Candolle tended to give the same character equal weight, at least in related taxa.)

Note that Jussieu's emphasis on synthesis was compatible with his belief that there were no groups sharply separated from other such groups in nature; gradual synthesis produced the continuity in relationships that was a feature of continuous nature. Indeed, Jussieu confidently expected plants yet to be discovered to be intermediates that filled in the apparent gaps between groups. Candolle, on the other

Augustin-Pyramus de Candolle
(1778–1841, Swiss)

hand, tended to emphasize analysis. He asserted that there were distinct groups in nature, noting that botanical discoveries were not filling in the morphological gaps between groups, and he looked for features that characterized these groups.

During the nineteenth and twentieth centuries, the fundamental differences between Jussieu's and Candolle's understanding of nature were almost never discussed, and no accepted rationale for weighting characters was developed. Whether or how to weight remained a bone of contention for the next century and a half. Arguments between self-styled Jussieuans and Adansonians over weighting in the latter part of the nineteenth century centered more on what characters should be used, and how, and less on whether or not to use all characters.

A number of systematists, especially those in France and Germany, adopted concepts of **types**. These might be the common form in a group, or represent a "perfect" flower—a radially symmetrical form such as a peloria mutation in a bilaterally symmetrical group, or a bisexual flower in a monoecious or dioecious group. (The original peloria mutation, which excited Linnaeus so much when he found it that he initially wanted to name it as a new genus, *Peloria*, was a mutant of *Linaria* (Plantaginaceae) with five spurs radiating from the center of the flower, not the normal one in the abaxial position).

Such types could provide a way of understanding the diversity of form in a group and of relating one group to another. They were also in some ways an alternative to conventional weighting schemes, but typological thought, although widespread, never became systematized. Not only did the word type reflect a variety of very different ideas, but typological thought in general was equated by some (mostly English-speaking individuals) with speculation. Ideas of essences (i.e., essential characters, as described earlier) and types are often linked.

Acceptance of evolution did not inspire any new way of detecting relationships. Furthermore, Charles Darwin (1809–1882) provided no indication of how to rank taxa; he did, however, emphasize that relationships could be represented as groups subordinate to other groups (Figure 3.5). Systematists like George Bentham (1875) understood this—when coupled with the idea of evolution—to mean that the only difference between taxa occupying the highest and the lowest ranks of the hierarchy was one of degree. Taxa might be distinct, yet ranks were not fundamentally different; neither ranks nor individual taxa had essences. This only compounded the problem of deciding at what rank to recognize a particular group, unless reference was made to previous taxonomic practice—that is, an appeal was made to established practice or convention.

George Bentham
(1800–1884, British)

Up until the middle of the twentieth century, systematists continued to delimit groups very much in the same way that they had at the beginning of the nineteenth century, although of course they knew much more about the basic morphology and anatomy of plants. Similarity in general morphology and anatomy came to indicate closeness in evolutionary or phylogenetic relationships, but there was no way of deciding which particular characters indicated such relationships and which did not.

Systematists sometimes tried to distinguish between characters that were adaptive, and therefore less valuable in assessing relationships, and those that were not

adaptive and thus more valuable. Groups were still circumscribed by morphological gaps, but there was no agreement as to when a gap was large enough for a group to be recognized at a particular rank. Indeed, gap size has tended to be inversely proportional to the size of the groups involved (Davis and Heywood 1963).

For about 170 years it has been recognized that the application of criteria to evaluate closeness of relationship and to rank taxa has been inconsistent. So, for example, families in the Malvales have often been considered equivalent to tribes in the Rosaceae. It is not surprising that arguments over how broadly or narrowly taxa should be circumscribed have remained unresolved.

John Gilmour (1940; see also Winsor 1995, 2000) reopened discussion about the general issue of how to group organisms when he observed that the use of characters to establish evolutionary relationships tended to be circular: Characters important in establishing evolutionary relationships were those important in delimiting groups, and vice versa. He suggested that groups in natural classifications were simply those that had many characters in common.

John Scott Lennox Gilmour
(1906–1986, British)

Such groups had the useful property that they could be used for a wide range of purposes; they were general-purpose classifications.

Gilmour believed that attributes (characters) of plants were sensory data, which, he thought, were facts. Classifications were "clips" that held these data together, so if different principles of classification were used, the clips—and so the groups recognized—would be different. Evolutionary classifications, by their very definition, would always be special-purpose and so not of general use or interest.

What became known as phenetics or numerical taxonomy owes much to Gilmour's ideas. Pheneticists produced groups on the basis of overall similarity, in hopes that this process would produce an objective, stable, and repeatable classification (Sokal and Sneath 1963; for more on this topic, see Chapter 2 and Vernon 1988). (Interestingly, Gilmour himself was not sympathetic to the use of computers that his approach encouraged.) The assumption that characters were observable facts soon proved to be a considerable oversimplification; what appeared to be basic characters could be subdivided. Furthermore, different numerical algorithms produced different phenograms (see Chapter 2), and hence could be the basis of different classifications, but it was often unclear why one algorithm should be preferred over another.

Phenetic theory and, in particular, practice had little effect on higher-level systematics in North America. It

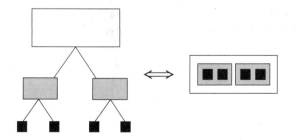

Figure 3.5 Although Charles Darwin's work did not provide any insights on how to rank the various taxa, he understood that some groups were subordinate to others.

had rather more influence in England, where botanists were perhaps particularly distrustful of what they saw as being evolutionary speculation (Vernon 1993; Winsor 1995); good systematists, it was claimed, always had been more or less Gilmourian.

It has generally been conceded that genera, and particularly families, are less "natural" than species. However, what such comparisons might mean is unclear, because systematists have not been clear as to what is meant by "natural." Is it that the rank of genus (for example) is a rank in nature, or that genera are discrete groups, or that members of a genus are more closely related to one another than they are to members of another genus?

In any event, there always have been dissenting voices. Linnaeus claimed that genera and species were equally natural. A.-L. de Candolle thought genera were more natural than species because genera were named as such by the common man (a similar idea was expressed by H. H. Bartlett). In a much-cited paper, Edgar Anderson (1940) reported on a survey that he had carried out to find out whether systematists thought genera or species were more natural. Some systematists who monographed groups, at least, were inclined to think that genera were more natural than species, despite the often expressed views to the contrary. So were biogeographers such as Ronald Good, and it should not be forgotten that the genus is the basic unit of much of the biogeographic work that has focused on global patterns of diversity and relationships.

Early in the nineteenth century, Charles-François Brisseau de Mirbel (1776–1854) suggested that there were two main kinds of families (and genera). Whereas *familles en groupes* ("families in groups") were very natural and clearly circumscribed, *familles par enchaînement* ("families by chaining") were less natural and had less clear limits. All genera in the first kind of family tended to be united by one or more characters, but they were often difficult to distinguish from one another.

In the second kind of family, individual genera were linked individually, forming a chain of similarities; these genera were often easily distinguished, even if the family was not. Similar distinctions, as between definable and indefinable families, have persisted (Davis and Heywood 1963: 107). It is perhaps ironic that some indefinable families, like Rosaceae and Ranunculaceae, have turned out to be largely monophyletic, while definable families like Lamiaceae or Liliaceae are strongly paraphyletic or polyphyletic.

Today most systematists realize that they need phylogenies. Before his untimely death in a car accident in 1989, Dahlgren had begun to elaborate relationships following more thoroughly phylogenetic principles. Such principles—especially the use of synapomorphies to diagnose monophyletic groups—were conceptualized by Willi Hennig (1950, 1966) and Warren H. Wagner (1969, 1980); they are outlined in Chapter 2. These principles provide criteria for deciding which particular features indicate phylogenetic relationships.

For almost a quarter of a century, students of Wagner and others produced Wagner trees, although most studies using them involved only a few taxa. In the late 1970s papers by Bremer, Wanntorp, and others popularized morphological Hennigian studies. A decade later, papers by Bremer (1987) and Jansen and Palmer (1987) suggested that both molecular and morphological data divided Asteraceae in a novel and exciting way. It was to be the combination of massive amounts of molecular and morphological data with methods of data analysis inspired ultimately by both Hennig and Sokal and Sneath that would transform both our ideas of higher-level relationships in land plants, and indeed how systematists work (Stevens 2000a).

In phylogenetic studies, at least, collaboration is the order of the day. No longer are systems as it were "owned" by individuals, as phrases such as "Engler's system" or "Cronquist's system" might suggest. Collaboration is common in producing and analyzing the data, as well as in suggesting possible classificatory interpretations (Angiosperm Phylogeny Group 1998; Grass Phylogeny Working Group 2001; Angiosperm Phylogeny Group 2002; see also Endersby 2001).

The arguments in systematics—often rather confused—now focus on the use of statistical methods for evaluating the support of hypotheses of phylogenetic relationships, and the proper use of evolutionary models, as in maximum likelihood methods. Perhaps surprisingly, little advance has been made in understanding the relationship between morphological observations and hypotheses of phylogeny (Stevens 2000b).

There is still disagreement over the relationship between phylogenies and the classifications that are based on them. Some systematists think that an important element shaping classifications should be how different one group looks from another; others think that classifications should be based strictly on phylogeny—that is, that all taxa should be monophyletic. The latter is the approach taken here (see Chapter 2).

This division is independent of the arguments for and against the PhyloCode, a discussion in which history has been invoked by all sides. However, not only is there little evidence for some of the particular historical arguments advanced, but history seems largely irrelevant (Stevens 2002).

A final point bears on the weight or importance of particular kinds of characters in the detection of relationships. Whole suites of characters deemed to be important come into and go out of fashion over the years. Thus in 1883 Ludwig Radlkofer proclaimed the following century to be that of the anatomical method in systematics, and in 1924 Hermann Ziegenspeck produced the "Königsberger Stammbaum," a tree showing the serological relationships of all plants (with fossils placed in their appropriate positions).

Other features, such as plant chemistry, chromosome number, and the morphology of sieve tube plastids, have all had their moment of glory. Because it was realized that single-character classifications were suspect, botanists often have restricted themselves to providing extensive surveys showing patterns of variation in individual characters. Unfortunately, despite claims that classifications synthesize all available data (Lawrence 1951; Constance 1964), prior to the advent of the computer there was no way of really integrating all the data that systematists produced. Furthermore, the coverage of systematic data has often been sadly inadequate.

Generalized anatomical surveys such as those that Radlkofer favored became unfashionable well before the end of his "century of the anatomical method." Developmental studies were not popular in systematics, despite remarkable work by J.-B. Payer in the middle of the nineteenth century. Although this work was dismissed by systematists such as J. D. Hooker, it is still widely cited in the comparative developmental studies that are adding so much to systematics today.

There has been some tension between floras, with their proper emphasis on geographically circumscribed treatments focusing on characters that help in the identification of the plants (see Frodin 2000 and Appendix 2), and monographs that deal with taxa wherever they are found and emphasize characters indicative of relationships, whether or not they can be used in identification. Major floras still often take more than 50 years to complete and are similar in their goals (and in the time it takes to finish them) to the Colonial Floras promoted by the Directors of the Botanic Gardens at Kew, England, in the later part of the nineteenth century. Accessible to a wide variety of scientists, they have helped stabilize usage of taxon names, although too little attention has been paid to just how the contributors to these floras reached the taxonomic conclusions that they did (Stevens 1997a).

Plant Groupings over the Years

It is impossible to do more than mention a few of the main changes in ideas about relationships in the years prior to the advent of phylogenetic methodology (see Lawrence 1951 for summaries). Some of the differences between what are now called monocotyledons and dicotyledons were already evident to Theophrastus in 300 B.C., but John Ray (1627–1705) was the first to make a major distinction between the two, although he subordinated it to his primary division into trees and herbs. Cotyledon number was the main character used by Jussieu in 1789 to divide plants, and it has almost always retained its prime position.

Jussieu placed the monocotyledons before the dicotyledons because they were simpler (monocots appeared to lack a corolla and had only a single cotyledon), and he began the dicots with such plants as Aristolochiaceae and many Caryophyllales (his nomenclature is modernized here); all, he thought, lacked a corolla (petals) and so were the simplest in the series of increasing complexity that his arrangement represented.

Many plants with catkins have flowers of different sexes, often on different individuals, and so they seemed to Jussieu to be by his criteria the most complex; these he placed near the end of the dicots, and thus near the end of his whole sequence. He placed the conifers (but not the cycads, which he included with the ferns) at the very end, probably in part because some genera have many cotyledons and so represented the culmination of the cotyledonary series. The distinctive nature of gymnospermy was demonstrated by Robert Brown in 1826, but "gymnosperms" (conifers and cycads) were not finally excluded from the angiosperms until much later in the century.

Two major arrangements adopted subsequently are associated with the names of A.-P. de Candolle and Adolf Engler. However, there has always been a plethora of alternative systems: In the twentieth century, Lam, Melville, Meeuse, and Hayata, to name just a few, all proposed their own sometimes very different systems. Candolle (1813) began his system—which he said was not to be interpreted as being linear—with the Ranunculaceae on the grounds that one should place well-known organisms first. Simple plants tended to be least well known; the Ranunculaceae had, he thought, the most complex flowers and were well known.

Candolle's series followed the sequence (1) Thalamiflorae (superior, sepals and petals distinct), (2) Calyciflorae (often a hypanthium, sepals and petals distinct), (3) Corolliflorae (sympetalous plants), (4) Monochlamydeae (only a single perianth series) (see Chapter 4 for discussion of these floral terms). Monocots followed the dicots, and gymnosperms, not named as such, straddled the end of the dicot series and the beginning of the monocots.

Although Bentham and Hooker largely followed the Candollean sequence, their classification delimits taxa that are quite often substantially different, and hence implies different relationships. Furthermore, they noted that they adopted the dicot sequence Thalamiflorae–Gamopetalae–Monochlamydeae for convenience only, and that many Monochlamydeae, in particular, were probably related separately to Polypetalae (i.e., plants having distinct petals); they were not much happier with Gamopetalae. Their Gymnospermae was a fourth dicot group placed just before the monocots.

Engler's system is basically a modification of that of Adrien de Jussieu (1843). Jussieu, like his father, Antoine-Laurent, allowed that the basic sequence should be simple to complex. However, he thought that monocots and dicots should be placed in parallel, not in series. Hence his dicot sequence began with plants that were absolutely simple, rather than those that Antoine-Laurent thought were most similar to monocots. Within the dicots, Adrien

Heinrich Gustav Adolf Engler
(1844–1930, German)

placed dioecious groups, divided into angiosperms and gymnosperms, first, and Amentiferae (i.e., species with reduced, wind-pollinated flowers borne in catkins or aments) were placed first within the angiosperms. The other three major groups of dicots that he recognized followed the morphological sequence (1) apetaly (petals lacking), (2) polypetaly (petals distinct), (3) monopetaly (petals fused).

Engler excluded gymnosperms from the angiosperms and divided dicots into Archichlamydeae and Sympetalae. The angiosperms began with groups such as Piperaceae and Chloranthaceae before proceeding to Amentiferae and polypetalous plants. The basic arrangement is unchanged in recent editions of this system (Engler 1964), although Piperaceae have been moved. There is some debate whether Engler thought that the Amentiferae really were primitive, but some who used the Englerian sequence (or its precursors) certainly did think this to be the case.

Bessey's system combines features of both main ways of arranging plants. Bessey's dicta—guidelines for the production of phylogenies (Bessey 1915)—have been particularly influential. Many of these dicta are specific evolutionary trends, and the identification of such trends long remained a major component of evolutionary thought. Recent systems, of which perhaps the most notable are those of Dahlgren (1983; Dahlgren et al. 1985), Thorne (1999, 2000), Takhtajan (1997), and Cronquist (1981), are largely variants of Besseyan and Englerian ideas combined (Cuerrier et al. 1996), although Thorne and particularly Dahlgren (as discussed earlier) paid more attention to phylogenetic principles.

Despite all these varying approaches, by the early 1980s something of a consensus about ideas of relationships seemed to be developing (Stevens 1986), and in North America Cronquist's system was much in use. Well documented and with descriptions that incorporated both anatomical and chemical information, it starts with families that are still considered to be members of basal angiosperm lineages, although they are now often placed in groups that are differently circumscribed. Elsewhere in Cronquist's system, particularly in groups like the Rosidae, Dilleniidae, and Liliidae, there is little in common between the groups that he recognized and those that are recognized here.

Indeed, the consensus of the late twentieth century ignored those who still followed Englerian ideas, and it has not survived the effects of cladistic theory (see Chapter 2) and the recent spate of major molecular and morphological studies that utilize it (see Chapters 8 and 9). The changes are caused by the existence of clear goals, large amounts of new data, and new methods of data analysis, as well as the realization that the systematist is no longer describing relationships in nature, but proposing hypotheses of phylogeny. Although many important aspects of phylogenies remain unclear—for instance, the relationship of monocots to other angiosperms—the broad outlines of a new arrangement are evident (Angiosperm Phylogeny Group 1998, Angiosperm Phylogeny Group 2002) and are reflected in the relationships discussed in Chapters 8 and 9 of this book and the sequence followed there.

Literature Cited and Suggested Readings

Items marked with an asterisk (*) are especially recommended to those readers who are interested in further information on the topics discussed in this chapter.

Adanson, M. 1763–1764. *Familles des plantes*, 2 vols. Vincent, Paris.

Allen, D. E. 1976. *The naturalist in Britain: A social history*. A. Lane, London.

Anderson, E. 1940. The concept of the genus. II. A survey of modern opinion. *Bull. Torrey Bot. Club* 67: 363–369.

Angiosperm Phylogeny Group. 1998. An ordinal classification for the families of flowering plants. *Ann. Missouri Bot. Gard.* 85: 531–553.

Angiosperm Phylogeny Group. 2002. An update of the Angiosperm Phylogeny Group classification for the orders and families of flowering plants: APGII. *Bot. J. Linnean Soc.*

*Atran, S. 1990. *Cognitive foundations of natural history*. Cambridge University Press, Cambridge, UK. [A challenging reinterpretation of early classifications.]

Bentham, G. 1857. Memorandum on the principles of generic nomenclature in botany as referred to in the previous paper. *J. Proc. Linnean Soc., Bot.* 2: 30–33.

Bentham, G. 1875. On the recent progress and present state of systematic botany. *Rep. Br. Assn. Adv. Sci.* (1874): 27–54.

Bentham, G. and J. D. Hooker. 1862–1883. *Genera plantarum*, 3 vols. Reeve & Co., London.

*Berlin, B. 1992. *Ethnobiological classification: Principles of categorization of plants and animals in traditional societies*. Princeton University Press, Princeton, NJ. [An excellent summary.]

Bessey, C. E. 1915. The phylogenetic taxonomy of flowering plants. *Ann. Missouri Bot. Gard.* 2: 109–164.

Bremer, K. 1987. Tribal interrelationships of Asteraceae. *Cladistics* 2: 210–253.

Brown, R. 1826. Character and description of *Kingia* . . . with observations . . . on the female flower of Cycadaceae and Coniferae. In *Narrative of a survey of the intertropical coasts of Western Australia . . .*, P. P. King (ed.), Vol. 2, 538–565. Murray, London.

Candolle, A.-P. de. 1813. *Théorie élémentaire de la botanique*. Déterville, Paris.

*Constance, L. 1964. Systematic botany—An unending synthesis. *Taxon* 13: 257–273. [A clear statement of the goals of evolutionary systematics.]

Cronquist, A. 1981. *An integrated system of classification of flowering plants*. Columbia University Press, New York.

*Cuerrier, A., R. Kiger and P. F. Stevens. 1996. Charles Bessey, evolution, classification, and the New Botany. *Huntia* 9: 179–213. [A study of the work of perhaps the most influential American systematist at the beginning of the twentieth century.]

Dahlgren, R. 1983. General aspects of angiosperm evolution and macrosystematics. *Nordic J. Bot.* 3: 119–149.

Dahlgren, R., H. T. Clifford and P. F. Yeo. 1985. *The families of monocotyledons.* Springer, Berlin.

Davis, P. H. and V. H. Heywood. 1963. *Principles of angiosperm taxonomy.* Edinburgh University Press, Edinburgh.

de Queiroz, K. 1988. Systematics and the Darwinian revolution. *Philos. Sci.* 55: 238–259.

Dughi, R. 1957. Tournefort dans l'histoire de la botanique. In *Tournefort*, R. Heim (ed.), 131–185. Muséum National d'Histoire Naturelle, Paris.

Dupree, H. 1959. *Asa Gray 1810–1888.* Belknap Press of Harvard University Press, Cambridge, MA.

Endersby, J. 2001. "The realm of hard evidence": Novelty, persuasion and collaboration in botanical cladistics. *Stud. Hist. Philos. Biol. Biomed. Sci.* 32: 343–360.

Engler, A. 1964. *Syllabus der Pflanzenfamilien*, H. Melchior (ed.), 12th ed., Vol. 2. Borntraeger, Berlin.

Frodin, D. G. 2001. *Guide to standard floras of the world*, 2nd ed. Cambridge University Press, Cambridge, UK.

*Gilmour, J. S. L. 1940. Taxonomy and philosophy. In T*he new systematics*, J. Huxley (ed.), 461–474. Oxford University Press, Oxford, UK. [A fascinating interpretation of the whys and wherefores of classification that remains worth reading.]

Giseke, P. D. 1792. *Praelectiones in ordines naturales plantarum.* Hoffmann, Hamburg, Germany.

Grass Phylogeny Working Group. 2001. Phylogeny and subfamilies of the grasses (Poaceae). *Ann. Missouri Bot. Gard.* 88: 373–457.

*Greene, E. L. 1983. *Landmarks of botanical history*, 2 vols., F. N. Egerton (ed.). Stanford University Press, Stanford, CA. [Written almost 100 years ago, but half of it published for the first time only in 1981 (but carefully edited then), this work covers botany up to the sixteenth century, although authors as late as Tournefort are also included.]

Hall, H. M. and F. E. Clements. 1923. *The phylogenetic method in taxonomy. The North American species of Artemisia, Chrysothamnus and Atriplex* (Carnegie Institute of Washington Publication No. 326). Carnegie Institution of Washington, Washington, DC.

Hennig, W. 1950. *Grundzüge einer Theorie der phylogenetischen Systematik.* Deutsche Zentralverlag, Berlin.

Hennig, W. 1966. *Phylogenetic systematics.* University of Illinois Press, Urbana.

Heywood, V. H. (ed.). 1978. *Flowering plants of the world.* Mayflower, New York.

Hutchinson, J. 1973. *The families of flowering plants; arranged according to a new system based on their probable phylogeny.* Clarendon Press, Oxford.

Jansen, R. K. and J. D. Palmer. 1987. A chloroplast DNA inversion marks an ancient evolutionary split in the sunflower family (Asteraceae). *Proc. Natl. Acad. Sci.* USA 84: 5818–5822.

Jussieu, A.-L. de. 1789. *Genera plantarum.* Hérissant and Barrois, Paris.

Jussieu, A.-L. de. 1843. *Cours élémentaire de histoire naturelle: Botanique.* Fortin Masson, Langlois and Leclerc, Paris.

Kornet, D. J. 1991. On specific and interspecific delimitation. In *The plant diversity of Malesia*, P. Baas, K. Kalkman and R. Geesink (eds.), 359–379. Kluwer, Dordrecht, Netherlands.

Lam, H. J. 1936. Phylogenetic symbols, past and present. *Acta Biotheor.* 2: 153–194.

Lamarck, J.-B.-P.-A. de M. de. 1778. *Flore françoise*, 3 vols. Imprimerie Royale, Paris.

Lawrence, G. H. L. 1951. *Taxonomy of vascular plants.* Macmillan, New York.

Linnaeus, C. 1751. *Philosophia botanica.* Kiesewetter, Stockholm.

*Mayr, E. 1982. *The growth of biological thought: Diversity, evolution and inheritance.* Belknap Press of Harvard University Press, Cambridge, MA. [History with a broad sweep, emphasizing animals and species.]

*Morton, A. G. 1981. *Outlines of botanical history.* Academic Press, London. [An invaluable account of all aspects of botanical knowledge up to the end of the nineteenth century.]

Sokal, R. R. and P. H. A. Sneath. 1963. *Principles of numerical taxonomy.* Freeman, San Francisco.

*Stafleu, F. 1971. Linnaeus and the Linnaeans. *Oosthoeck, Utrecht, Netherlands.* [The classic treatment of the work of Linnaeus and his immediate successors.]

Stevens, P. F. 1986. Evolutionary classification in botany, 1860–1985. *J. Arnold Arbor.* 67: 313–339.

Stevens, P. F. 1990. Nomenclatural stability, taxonomic instinct, and flora writing—A recipe for disaster? In *The plant diversity of Malesia*, P. Baas, K. Kalkman and R. Geesink (eds.), 387–410. Kluwer, Dordrecht, Netherlands.

Stevens, P. F. 1992. Species: Historical perspectives. In *Keywords in evolutionary biology*, E. F. Keller and E. A. Lloyd (eds.), 302–311. Harvard University Press, Cambridge, MA.

*Stevens, P. F. 1994. *The development of biological systematics.* Columbia University Press, New York. [A discussion emphasizing that authors of the period 1780–1860 often did not produce classifications in the currently accepted sense of the word, and evaluating the history of systematics accordingly.]

*Stevens, P. F. 1997a. How to interpret botanical classifications—Suggestions from history. *BioScience* 47: 243–250. [A reinterpretation of Bentham and Hooker's Genera Plantarum, a major reference source for over a century.]

Stevens, P. F. 1997b. J. D. Hooker, George Bentham, Asa Gray and Ferdinand Mueller on species limits in theory and practice: A mid-nineteenth century debate and its repercussions. *Hist. Rec. Aust. Sci.* 11: 345–370.

Stevens, P. F. 1998a. Mind, memory and history: How classifications are shaped by and through time, and some consequences. *Zoologica Scripta* 26: 293–301.

Stevens, P. F. 1998b. What kind of classification should the practicing taxonomist use to be saved? In *Plant diversity in Malesia III*, J. Dransfield, M. J. E. Coode

and D. A. Simpson (eds.), 295–319. Royal Botanical Gardens, Kew, UK.

Stevens, P. F. 2000a. Botanical systematics 1950–2000: Change, progress, or both? *Taxon* 49: 635–659.

Stevens, P. F. 2000b. On characters and character states: Do overlapping and non-overlapping variation, molecules and morphology all yield data of the same value? In *Homology and systematics*, R. Scotland and R. T. Pennington (eds.), 81–104. Taylor & Francis, London.

*Stevens, P. F. 2002. Why do we name organisms? Some reminders from the past. *Taxon* 51: 11–26.

Takhtajan, A. 1997. *Diversity and classification of flowering plants.* Columbia University Press, New York.

Thorne, R. F. 1999. The classification and geography of the monocotyledon subclasses Alismatidae, Liliidae and Commeliniidae. In *Plant systematics for the 21st century*, B. Nordenstam, G. El-Ghazaly and M. Kassas (eds.), 75–124. Portland Press, London.

Thorne, R. F. 2000. The classification and geography of the flowering plants: Dicotyledons of the class Angiospermae. *Bot. Rev.* 66: 441–647.

Tournefort, J. P. de. 1694. *Élémens de botanique*, 3 vols. Imprimerie Royale, Paris.

Vernon, K. 1988. The founding of numerical taxonomy. *Br. J. Hist. Sci.* 21: 143–159.

Vernon, K. 1993. Desperately seeking status: Evolutionary systematics and the taxonomists' search for respectability 1940–1960. *Br. J. Hist. Sci.* 26: 207–227.

*Voss, E. 1952. The history of keys and phylogenetic trees in systematic biology. *J. Sci. Lab. Denison Univ.* 43: 1–25. [A useful and well-illustrated survey.]

Wagner, W. H., Jr. 1969. The construction of a classification. In *Systematic biology: Proceedings of an International Conference* (National Research Council Publication 1692), 67–103. National Academy of Sciences, Washington, DC.

Wagner, W. H., Jr. 1980. Origin and philosophy of the groundplan-divergence method of cladistics. *Syst. Bot.* 5: 173–193.

*Walters, S. M. 1986. The name of the rose: A review of ideas on the European bias in angiosperm classification. *New Phytol.* 104: 527–546. [A valuable review with references to Walters's earlier publications.]

Wernham, H. F. 1911–1912. Floral evolution with particular regard to the sympetalous dicotyledons. *New Phytol.* 10: 73–83, 109–120, 145–159, 217–226, 203–235; 11: 145–166, 217–235, 290–305, 373–397.

*Winsor, M. P. 1995. The English debate on taxonomy and phylogeny, 1937–1940. *Hist. Philos. Life Sci.* 17: 105–130. [An illuminating account of a crucial period in systematics.]

Winsor, M. P. 2000. Species, demes and the omega taxonomy: Gilmour and the New Systematics. *Biol. Philos.* 15: 349–388.

Winsor, M. P. 2001. Cain on Linnaeus: The scientist-historian as an unanalysed entity. *Stud. Hist. Philos. Biol. Biomed. Sci.* 32C: 239–254.

4

Taxonomic Evidence: Structural and Biochemical Characters

Taxonomic evidence consists of the characters used in the phylogenetic analyses upon which plant classifications are based, and it includes characters used in describing patterns of variation at or below the species level (see Chapter 6). Taxonomic evidence can be gathered from a wide variety of sources, from all parts of a plant during all stages of its development. In this chapter we summarize the use of characters from morphology, anatomy, embryology, chromosomes, palynology, secondary plant compounds, and proteins. Nucleic acids (DNA and RNA) provide an increasingly important source of taxonomic characters; their use in plant taxonomy and the rapidly developing field of molecular systematics is discussed in detail in Chapter 5.

The practical discussion of plant characters in this chapter and the next provides a useful counterpart to the more theoretical discussion of characters in Chapter 2.

Morphology

Morphological characters are features of external form or appearance. They currently provide the characters used for practical plant identification and some of those used for hypothesizing phylogenetic relationships. These features have been used for a longer time than anatomical or molecular evidence, and they were the only source of taxonomic evidence in the beginnings of plant systematics. Morphological characters are easily observed and find practical use in keys and descriptions; the characters used in phylogeny reconstruction may not be so easily observed. Characters of both sorts are found in all parts of the plant, both vegetative and reproductive.

The vegetative parts of angiosperms are **roots**, **stems**, and **leaves**, and the reproductive parts are **flowers**, **fruits**, and **seeds**. The terms used in tracheophytes (vascular plants) to describe variation in these parts are outlined in the discussion that follows, although the reproductive terms strongly emphasize the angiosperms because they are the dominant group of vascular plants. (Specialized vegetative and especially reproductive terms relating to other groups of vascular plants are covered in Chapter 8.)

Many, if not all, of the terms outlined here should be considered merely convenient points along a continuum of variation in form. Although they are useful in communication, intermediate conditions will be encountered.

DURATION AND HABIT

Duration is the life span of an individual plant. An **annual** plant lives for a single growing season. A **biennial** plant lives for two seasons, growing vegetatively during the first and flowering in the second. A **perennial** plant lives for three or more years and usually flowers and fruits repeatedly. Perennials may be herbaceous (lacking woody tissue), with only the underground portions living for several years, or woody, with a persistent aerial stem.

The general appearance, or **habit**, of plants varies greatly. Woody tissue is present in **trees** and **shrubs** but is lacking in **herbs**. Trees produce one main **trunk** (or **bole**); shrubs are usually shorter and produce several trunks. Climbing plants may be woody (**lianas**), or herbaceous (**vines**). **Suffrutescent** plants are intermediate between woody and herbaceous.

The characteristic shape of a tree or shrub often relates to its pattern of growth or architecture, which is often of systematic value. Stems form the major plant axes and may be erect (**orthotropic**) or horizontal (**plagiotropic**). **Monopodial shoots** grow through the action of a single **apical meristem**—a single region of dividing, elongating, and differentiating cells at the tip of the shoot. In other plants, axillary branches take over the role of the main axis and provide for continuing growth, while the main axis slows growth or dies; a series of such axillary branches constitutes a **sympodial shoot**.

Stems are provided with **buds**, which are small embryonic shoots, often protected by modified leaves (bud scales) or hairs. Buds may show a period of dormancy and, when they eventually grow out, may leave scars at the base of the new shoot. Such shoots are called **proleptic**. On the other hand, buds may develop and elongate at the same time as the shoot on which they are borne, in which case the new shoot usually lacks basal scars and has an elongated first internode. These shoots are called **sylleptic**.

Proleptic shoots are characteristic of temperate species, while sylleptic shoots are common in tropical plants. Some taxa, such as many Lauraceae, have both sylleptic and proleptic shoots. All shoots on a plant may be the same, or there may be two or more kinds of shoots. Thus *Ginkgo* and *Cercidiphyllum* both have some shoots with elongated internodes (**long shoots**), and others that produce only a few leaves and scales each year, all with only very short internodes (**short shoots**). These and other criteria are combined in various ways to produce an array of distinctive architectural growth patterns in trees and shrubs (see Hallé et al. 1978). Three examples are shown in Figure 4.1.

Axillary branches are initiated externally, in tissues immediately under the epidermis. Xylem and phloem are often arranged in a ring surrounding the **pith**, a central region with more or less isodiametric cells.

ROOTS

Roots usually branch irregularly. Lateral roots are initiated internally, in the endodermis and pericycle (the cell layers surrounding the conducting tissues) and erupt

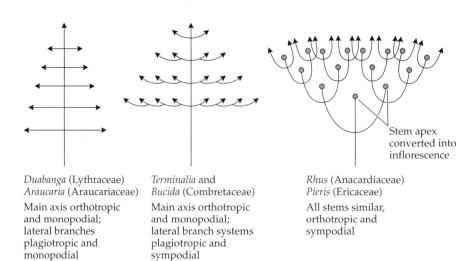

Duabanga (Lythraceae)
Araucaria (Araucariaceae)

Main axis orthotropic and monopodial; lateral branches plagiotropic and monopodial

Terminalia and *Bucida* (Combretaceae)

Main axis orthotropic and monopodial; lateral branch systems plagiotropic and sympodial

Rhus (Anacardiaceae)
Pieris (Ericaceae)

All stems similar, orthotropic and sympodial

Stem apex converted into inflorescence

Figure 4.1 Three architectural patterns of plant growth; branching is rythmic in all three.

through the cortex, although lycophyte roots branch by a forking of the apical meristem. The xylem and phloem are situated in the central portion of the root, usually resulting in a lack of pith. Roots also lack the nodes and internodes that characterize stems (see the next section), and they are usually found underground.

The primary functions of roots are to hold the plant in place, to absorb water and minerals, and to store water and carbohydrates. Some roots are specialized for other functions, such as photosynthesis (as in some reduced epiphytic Orchidaceae), penetration of the tissues of a host species (as in parasitic species such as mistletoes, Viscaceae), constriction of the trunks of supporting trees (as in strangler figs, Moraceae), or aboveground support for the trunk or branches (as in banyan figs, Moraceae, and some mangroves, Rhizophoraceae).

Some plants, such as epiphytic aroids (Araceae), have dimorphic roots, with some functioning in water and mineral uptake and others providing attachment. Most roots grow downward, but exceptions occur, as in **pneumatophores**, which are specialized roots involved in gas exchange in some mangrove or swamp species.

Roots are quite uniform in appearance, and a plant usually cannot be identified without its aboveground parts. Roots are useful, however, in determining whether a plant is an annual or a perennial, and variation in the root system is sometimes taxonomically significant. Here are a few important terms relating to roots:

adventitious developing from any plant part other than the embryonic root (radicle) or another root

aerial growing above ground or water

fibrous with all portions of the root system being of more or less equal thickness, often well branched, the primary root (taproot) absent or not obvious

fleshy thick with water or carbohydrate storage tissue

haustorial specialized for penetrating other plants and absorbing water and nutrients from them (as in parasites)

taproot the major root, usually enlarged and growing downward

STEMS

Stems—the axes of plants—consist of **nodes** (where leaves and axillary buds are produced) separated by **internodes** (Figure 4.2). They are frequently useful in identification and provide numerous systematically important characters.

Stems are usually elongated and function in exposing leaves to sunlight, flowers to pollinating agents, and fruits to dispersal agents. Some, however, are the primary photosynthetic organ of the plant (as in asparagus, Asparagaceae, and many cacti, Cactaceae), store water or carbohydrates (as in many cacti and other succulents), climb (as in hooked or twining stems of vines and lianas), or protect the plant (as in plants with thorns).

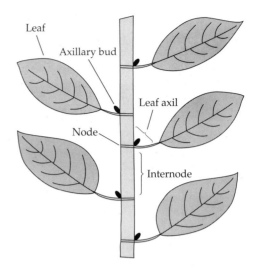

Figure 4.2 Generalized angiosperm stem showing nodes and internodes, leaves, leaf axils, and axillary buds.

Some important stem-related vocabulary is listed here.

acaulescent having an inconspicuous stem

bulb a short, erect, underground stem surrounded by thick, fleshy leaves or leaf bases

caulescent having a distinct stem

corm a short, erect, underground, more or less fleshy stem covered with thin, dry leaves or leaf bases

herbaceous not woody; dying down at the end of the growing season

internode the part of the stem between two adjacent nodes

lenticel a wartlike protuberance on the stem surface involved in gas exchange

long shoot a stem with long internodes; this term is applied only in plants in which internode length is clearly bimodal and both long and short shoots are present

node region of the stem where the leaf and bud are borne

pith soft tissue in the center of the stem, usually consisting of more or less isodiametric cells

rhizome a horizontal stem, more or less underground, bearing scale-like leaves; often called a stolon (or runner) if above ground and having an elongated internode

scape an erect, leafless stem bearing an inflorescence or flower at its apex; usually composed of a single elongated internode

scar the remains of a point of attachment, as in leaf scar, stipule scar, bud-scale scar

short shoot a stem with short internodes, other shoots having distinctly longer internodes; see long shoot

thorn a reduced, sharp-pointed stem [In contrast, a reduced, sharp-pointed leaf or stipule, or sharp-pointed marginal tooth is called a spine, and a sharp-pointed hair (involving epidermal tissue) or emergence (involving both epidermal and subepidermal tissues) is called a prickle.]

tuber   a swollen, fleshy portion of a rhizome involved in water or carbohydrate storage

twining   spiraling around a support in order to climb

woody   hard in texture, containing secondary xylem, and persisting for more than one growing season

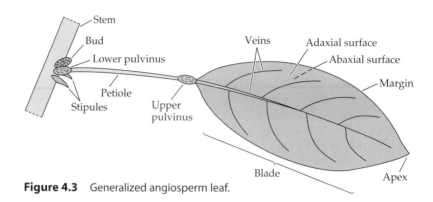

Figure 4.3   Generalized angiosperm leaf.

BUDS

Buds are short embryonic stems. They may be protected by **bud scales** (modified leaves that are sometimes represented only by stipules), a dense covering of hairs, and/or a sticky secretion. In angiosperms buds are found at the nodes, in the **leaf axil** (the angle formed by the stem and the petiole of the leaf; see Figure 4.2), or at the end of the stem. They are especially useful for identifying twigs in winter condition. Some common terms pertaining to buds are listed here.

accessory bud   an extra bud (or buds) produced on either side of, or above or below, the main axillary bud

axillary bud   a bud located in the leaf axil

flower bud   a bud containing embryonic flowers

leaf bud   a bud containing embryonic leaves

mixed bud   a bud containing both embryonic flowers and leaves

naked   not covered by bud scales

pseudoterminal bud   an axillary bud that takes over the function of a terminal bud in sympodial shoots

superposed bud   a bud located above or below the axillary bud

terminal bud   a bud at the apex of a stem (monopodial shoot)

LEAVES

Leaves are the major photosynthetic parts of most plants. They are borne at the nodes of a stem, usually below a bud (see Figures 4.2 and 4.3). In contrast to stems, leaves usually do not continue to grow year after year. They are usually flat, and have one surface facing toward the stem axis (the **adaxial**, or upper, surface) and another surface facing away from the stem axis (the **abaxial**, or lower, surface). Most leaves are **bifacial**, having definite adaxial and abaxial surfaces, but sometimes they are **unifacial**, lacking such differentiation.

Leaves are homologous structures among the angiosperms, but not among vascular plants as a whole (see Chapters 7, 8 and 9). In addition to their obvious function in photosynthesis, leaves may be modified for protection, forming sharp-pointed **spines**; for water storage, as in many succulents; for climbing, as in vines or lianas with tendril leaves; for capturing insects, as in carnivorous plants; or for providing homes for ants or mites (domatia, described later).

The major parts of a leaf are shown in Figure 4.3. The base of the **petiole** may have a narrow to broad point of attachment and may obscure the axillary bud. In monocots the leaf is almost always broadly sheathing at the base, with the edges either fused or overlapping. In taxa such as grasses (Poaceae) and gingers (Zingiberaceae) there is an adaxial flap or **ligule** at the junction of the sheath and blade. A leaf that lacks a petiole is said to be **sessile**.

Pulvini (singular **pulvinus**), somewhat swollen and morphologically distinct parts of the petiole, are often present and involved in leaf movement. These may be at the leaf base; at the apex of the petiole, as in prayer plants (Marantaceae); in the middle of the petiole, as in a few Araceae; or on the petioles of leaflets (of compound leaves, as described in the section on "Leaf Structure").

Stipules are usually paired appendages located on either side of (or on) the petiole base. Stipules are some-

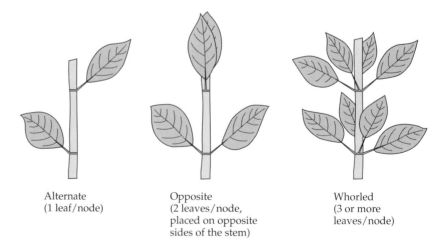

| Alternate
(1 leaf/node) | Opposite
(2 leaves/node,
placed on opposite
sides of the stem) | Whorled
(3 or more
leaves/node) |

Figure 4.4   Three major patterns of leaf arrangement.

times single, in which case they are borne between the petiole and stem. They may be leaflike, scale-like, tendril-like, spinelike, glandular, very reduced, or completely lacking. They have various functions, but most often they help in protecting young leaves. Stipules are not always homologous.

Leaf arrangement Leaves may be arranged in one of three major patterns (Figure 4.4). **Alternate** leaves are borne singly and are usually arranged in a spiral pattern along the stem. Various kinds of spirals occur; they can be evaluated by determining the angle around the stem between the points of insertion of any two successive leaves or by following the spiral around the stem from any older, lower leaf to the first younger leaf directly in line above it. Alternate leaves are sometimes placed

along just two sides of the stem (2-ranked, or distichous), or only three sides of the stem (3-ranked, or tristichous). (Two-ranked leaves that are flattened in the same plane with both surfaces identical, as in irises (Iridaceae), are called **equitant**.)

In contrast, **opposite** leaves are borne in pairs, the members of which are positioned on opposing sides of the stem. Opposite leaves may be spiraled, as in red mangroves (*Rhizophora*, Rhizoporaceae); 2-ranked, as in many Zygophyllaceae; or **decussate** (the leaves of adjacent nodes rotated 90°). The decussate arrangement is the most common condition among vascular plant species.

Finally, when three or more leaves are positioned at a node, they are considered to be **whorled**.

Leaf structure A leaf with a single blade is termed **simple**; a leaf with two or more blades, or **leaflets**, is said to be **compound**. The distinction between simple and compound leaves can be made by locating an axillary bud: An axillary bud is subtended by the entire leaf and never by individual leaflets. Leaflets may be arranged in various ways, as shown in Figure 4.5.

Leaf duration Leaves may function from a few days to many years, but most leaves function for only one or two growing seasons. **Deciduous** leaves fall (are abscised) at the end of the growing season; **evergreen** plants are leafy throughout the year. Some leaves, such as those of many members of Fagaceae, are **marcescent**; that is, they wither but do not fall off during the winter or dry season.

Venation types If there is one most prominent vein in a leaf, it is called the midvein or primary vein; branches from this vein are called secondary veins. Tertiary veins usually link the secondaries, forming a ladderlike (**percurrent** or **scalariform**) or netlike (**reticulate**) pattern (Figure 4.6).

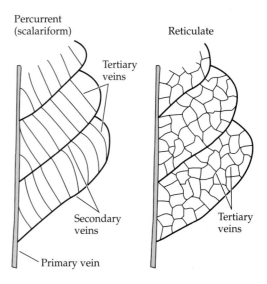

Figure 4.5 Arrangements of leaflets in compound leaves.

Figure 4.6 Two patterns of tertiary veins.

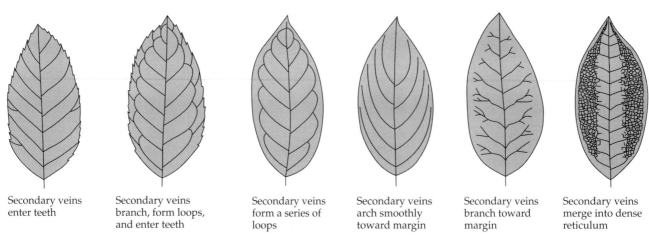

Secondary veins enter teeth

Secondary veins branch, form loops, and enter teeth

Secondary veins form a series of loops

Secondary veins arch smoothly toward margin

Secondary veins branch toward margin

Secondary veins merge into dense reticulum

Figure 4.7   Some kinds of pinnate venation. (After Hickey 1973.)

There are three major patterns of organization of the major veins. The leaf may have a single primary vein with the secondary veins arising along its length like the teeth of a comb; this pattern is termed **pinnate**. Or the leaf may have several major veins radiating from the base (or near the base) of the blade, like fingers from a palm; this pattern is called **palmate**. Many different kinds of pinnate (Figure 4.7) and palmate (Figure 4.8A) venation have been characterized (they are discussed in more detail in Hickey 1973 and Dilcher 1974). Finally, the leaf may have many parallel veins, a pattern termed **parallel** venation (Figure 4.8B).

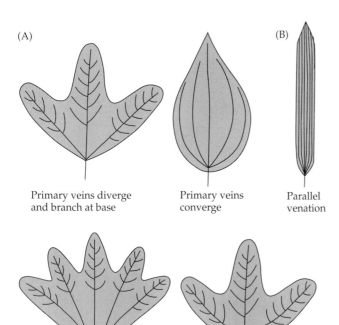

(A)

(B)

Primary veins diverge and branch at base

Primary veins converge

Parallel venation

Primary veins diverge and branch at various points

Primary veins diverge and branch above base

Figure 4.8   (A) Four kinds of palmate venation. (B) Parallel venation. (After Hickey 1973.)

Leaf shapes   A leaf may be considered to have one of four major shapes (**ovate**, **obovate**, **elliptic**, **oblong**) depending on where the blade is the widest (Figure 4.9) (Hickey 1973). The meanings of these shape terms may be adjusted by the use of modifiers such as *broadly* or *narrowly*. If the petiole is attached away from the leaf margin, such that the leaf and its stalk form an "umbrella," the leaf is termed **peltate**, and such leaves may be any of a number of different shapes. A **linear** leaf, on the other hand, is very long and narrow. Various other specialized shape terms are sometimes employed, but their use is avoided as much as possible here. The blade of a leaf may be symmetrical or asymmetrical when viewed from above.

Very different leaf shapes may occur on the same plant, a condition known as **heterophylly**. Juvenile leaves may be quite different from adult leaves, but sometimes even an adult plant will bear several different kinds of leaves (as in *Sassafras*, Lauraceae).

Leaf apex and base   Various terms relating to the shape of the leaf apex are illustrated in Figure 4.10. Terms relating to the shape of the leaf base are illustrated in Figure 4.11.

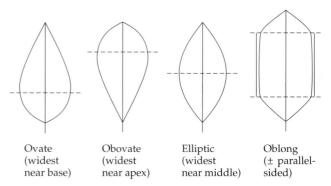

Ovate (widest near base)

Obovate (widest near apex)

Elliptic (widest near middle)

Oblong (± parallel-sided)

Figure 4.9   Leaf shapes.

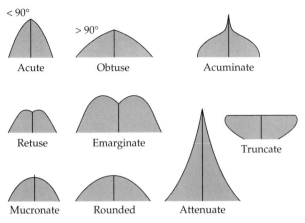

Figure 4.10 Leaf apex shapes.

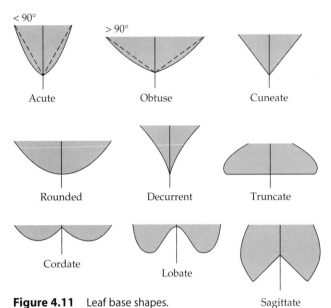

Figure 4.11 Leaf base shapes.

Leaf margin The leaf blade may have **lobed** or **unlobed** margins. These and other margin types are illustrated in Figure 4.12.

Various kinds of teeth may be defined according to anatomical features such as the pattern of the vein or veins entering the tooth, the shape of the tooth, and characters of the tooth apex such as glandularity. The more common tooth types are illustrated in Figure 4.13; others are defined where first encountered in Chapter 9 (see also Hickey and Wolfe 1975).

Leaf texture The leaf blade may be very thin (**membranous**), papery in texture (**chartaceous**), or very thick (**coriaceous**).

Ptyxis and vernation Ptyxis is the way in which an individual leaf is folded in the bud. **Vernation** is the way in which leaves are folded in the bud in relation to one another. Leaves that overlap in the bud are termed **imbricate**, while those with margins merely touching are called **valvate**. These are vernation terms; a few others are defined in Chapter 9 (in the discussion of particular families). A few ptyxis terms are illustrated in Figure 4.14 (see also Cullen 1978).

Indumentum An **indumentum** (plural **indumenta**), or covering of hairs (trichomes), on the surface of an angiosperm gives that surface a particular texture. Most indumentum terms are ambiguous, and we will use only three here: **glabrous** (lacking hairs), **pubescent** (with various hairs), and **glaucous** (with a waxy covering, and thus often blue or white in appearance). A few terms describing the indumentum are listed here; we will not use them in this text, but you may encounter them, as well as many others, in botanical keys and descriptions:

arachnoid having a cobwebby appearance
canescent gray, with dense short hairs
hirsute having long, often stiff, hairs

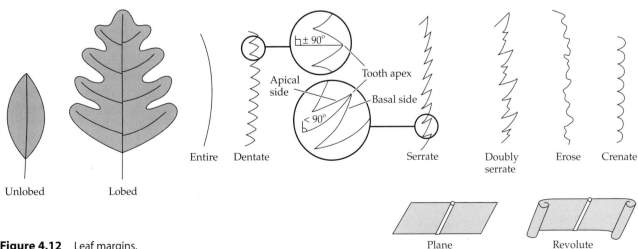

Figure 4.12 Leaf margins.

Figure 4.13 Some major tooth types.

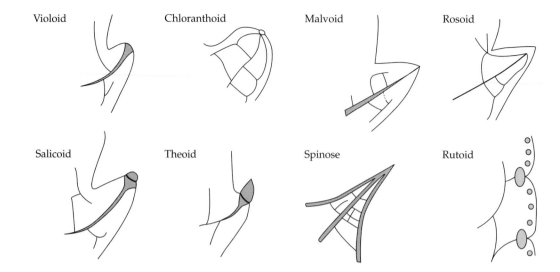

Violoid Chloranthoid Malvoid Rosoid

Salicoid Theoid Spinose Rutoid

hispid having stiff or rough hairs; bristly

lanate woolly, with long intertwined, somewhat matted hairs

pilose having scattered, long, slender, soft hairs

puberulent having minute, short hairs

scabrous rough

sericeous silky, with usually long, thin, appressed hairs

strigose having stiff hairs, all pointing in one direction

tomentose having densely matted soft hairs

velutinous velvety

villous covered with long, fine, soft hairs

We strongly recommend that the kinds of hairs occurring on a plant, along with their distribution and density, be carefully observed under a dissecting (or compound) microscope. Characters derived from such observations usually will be more useful (and consistently applicable) than the indumentum terms listed here.

Hairs may be **unicellular** or **multicellular**, **nonglandular** or **glandular**, and borne **singly** or in **tufts**, with surrounding cells of the epidermis modified or not. The shape of the individual hairs can be described in detail: Are they **branched** or **simple**? How are they branched; that is, are they **dendritic**, **stellate**, or **T-shaped**? Do they have a **flattened** or **globose** head, and is the stalk **uniseriate** (with one row of cells), **biseriate** (with two rows of cells), or **multiseriate** (with several rows of cells)? These terms are illustrated in Figure 4.15.

Some taxa have two or more kinds of hairs mixed together on their leaves or stems. For example, many species have nonglandular, unicellular hairs intermixed with gland-headed, multicellular hairs. The types of hairs, along with their density and distribution on the plant, are often of taxonomic value.

Domatia and glands Domatia (singular **domatium**)—literally, "tiny homes"—may contain organisms, usually mites or ants. Domatia occur on the leaves of many angiosperms (Pemberton and Turner 1989; Brouwer and Clifford 1990). Arthropod inhabitants of domatia assist the plant by deterring herbivory; in return, the plant provides not only a home but sometimes food as well.

Ant domatia usually are pouchlike and are typically found at the base of the leaf blade. Mite domatia are smaller and are usually at vein junctions. They may be bowl-, volcano-, or pocket-shaped, formed by axillary hair tufts or by a revolute margin. Various glandular structures may also occur on leaves. These structures usually secrete nectar and attract ants, which protect against herbivory (Bentley 1977).

Conduplicate

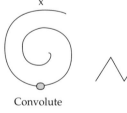

Reduplicate Convolute Plicate

Figure 4.14 A few ptyxis terms. All patterns are shown in cross-section except circinate. An "X" indicates the position of the branch bearing the leaf.

Revolute

Involute

Circinate

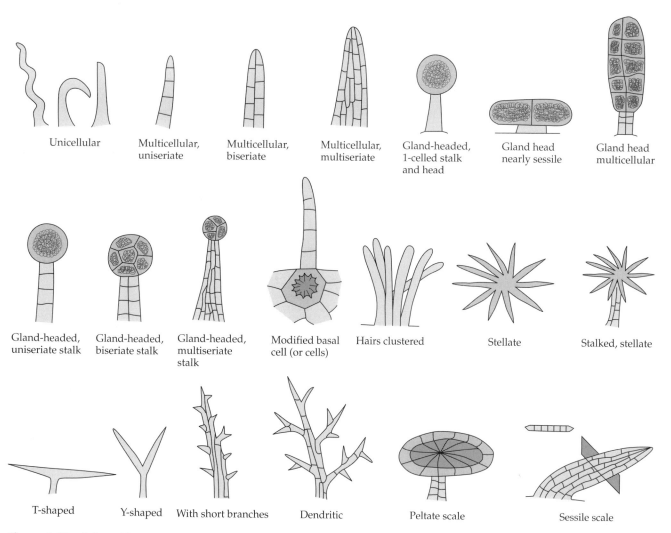

Unicellular

Multicellular, uniseriate

Multicellular, biseriate

Multicellular, multiseriate

Gland-headed, 1-celled stalk and head

Gland head nearly sessile

Gland head multicellular

Gland-headed, uniseriate stalk

Gland-headed, biseriate stalk

Gland-headed, multiseriate stalk

Modified basal cell (or cells)

Hairs clustered

Stellate

Stalked, stellate

T-shaped

Y-shaped

With short branches

Dendritic

Peltate scale

Sessile scale

Figure 4.15 Selected features of hairs.

FLORAL MORPHOLOGY

The reproductive structures of angiosperms are called flowers. In our discussion of plant reproductive parts, we will focus on angiosperms; the specialized reproductive structures of the lycophytes, ferns and their allies, conifers, cycads, ginkgos, and gnetophytes are described in Chapter 8.

A **flower** is a highly modified shoot bearing specialized appendages (modified leaves) (Figure 4.16). The modified shoot (or floral axis) is called the **receptacle**; the floral stalk is referred to as the **pedicel**. Flowers are usually borne in the axil of a more or less modified leaf, or **bract**; smaller, leaflike structures, the **bracteoles**, are often borne along the pedicel.

Flowers have up to three major parts: **perianth** (outer protective and/or colorful structures), **androecium** (plural **androecia**; pollen-producing structures), and **gynoecium** (plural **gynoecia**; ovule-producing structures). Flowers that have all three of these parts are said to be **complete**. If one or more of the three is lacking, the

flower is **incomplete**. If at least the androecium and gynoecium are present, the flower is termed **bisexual** (or **perfect**). If either is lacking, the flower is **unisexual** (or **imperfect**). It may be either **staminate**, if only the androecium is present, or **carpellate**, if only the gynoecium is present.

In **monoecious** species, both staminate and carpellate flowers are borne on a single individual; in **dioecious** species, the staminate and carpellate flowers are borne on separate individual plants. Various intermediate conditions exist, of course. **Polygamous** species have both bisexual and unisexual flowers (staminate and/or carpellate) on the same plant.

The perianth is always outermost in the flower, followed in nearly all flowers by the androecium, with the gynoecium in the center of the flower. The perianth parts may be undifferentiated, and the perianth composed merely of **tepals**. Alternatively, the perianth may be differentiated into two main parts, in which case it is composed of an outer whorl (or whorls or spirals) of **sepals**, collectively called the **calyx** (plural **calyces**), and an inner

Figure 4.16 Parts of a generalized flower. Collective terms are in boldface.

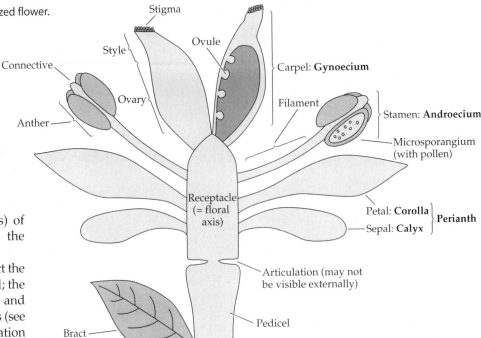

whorl (or whorls or spirals) of **petals**, collectively called the **corolla**.

The sepals typically protect the inner flower parts in the bud; the petals are usually colorful and assist in attracting pollinators (see also the section on pollination biology later in this chapter). Corollas have evolved independently in various groups of angiosperms; in some families it is clear that the petals are showy, sterile stamens; in others the petals are modified sepals.

It is important to remember that although these perianth terms are useful in descriptions and identification, they need to be used with caution in phylogenetic studies. Homology should not be assumed merely on the basis of a general similarity of form and function.

The androecium comprises all the **stamens** of the flower. Stamens are usually differentiated into an **anther** and a **filament**, although some are petal-like and are not differentiated into these two parts. Anthers usually contain four pollen sacs [or microsporangia (singular *microsporangium*], and these are often confluent in two pairs. The pollen sacs are joined to each other and to the filament by a **connective**, which is occasionally expanded, forming various appendages or a conspicuous sterile tissue separating the pollen sacs.

Meiosis occurs within the pollen sacs, leading to the production of pollen grains (male gametophytes, or microgametophytes). The androecium is therefore often referred to as the "male part" of the flower. Of course, flowers, as part of the diploid plant (or sporophyte), cannot properly be said to be male (or female) because the sporophyte is involved only in spore production (associated with meiosis). Only the haploid plant (or gametophyte) is involved in gamete production (Figure 4.17).

Anthers open by various mechanisms, and pollen usually is released through longitudinal slits, although transverse slits, pores, and valves also occur. Anthers that open toward the center of the flower are said to be **introrse**; those that shed pollen toward the periphery are **extrorse**.

The gynoecium comprises all the **carpels** of the flower. The carpel is the site of pollination and fertilization. Carpels are typically composed of a **stigma**, which collects and facilitates the germination of the pollen carried to it by wind, water, or various animals; a **style**, a usually slender region specialized for pollen tube growth; and an **ovary**, an enlarged basal portion that surrounds and protects the **ovules**. The stigmatic surface may be variably papillate and either wet or dry.

Each ovule contains the megagametophyte (the female gametophyte, or embryo sac), which produces an egg and is usually provided with two protective layers called **integuments**. The ovule is attached to the ovary wall by a stalk called the **funiculus** (plural **funiculi**) or funicle. The gynoecium is often called the "female part" of the flower, although, as noted already, this is technically incorrect. As the ovule develops into a seed, the surrounding ovary develops into a fruit.

Various floral parts may be modified for the production of nectar or other pollinator attractants, such as oils or fragrances. **Nectaries** (nectar-producing glands) often form projections, lobes, or disklike structures. Nectaries are often produced near the base of the androecium and gynoecium, or in nectar spurs formed by floral parts such as petals. Some flowers have an "extra" series of floral parts, often showy, called a **corona**. Coronal structures may be outgrowths of the perianth parts, stamens, or receptacle, and they are extremely diverse in form and function. (For a detailed discussion of the diversity of these and other floral structures, see Weberling 1989.)

The variation in floral features can be efficiently summarized by the use of floral formulas and diagrams (see Box 4A).

Floral asymmetry The parts of some flowers are arranged so that two or more planes bisecting the flower through the center will produce symmetrical halves. Such flowers have **radial** symmetry, and they are also called **actinomorphic** or **regular** (Figure 4.18A). (A few radial flowers have only two planes of symmetry; these are sometimes called **biradial**.)

The parts of other flowers are arranged so that they can be divided into symmetrical halves on only one plane. These flowers have **bilateral** symmetry, and are also called **zygomorphic** or **irregular** (Figure 4.18B).

A few flowers have no plane of symmetry and are **asymmetrical** (Figure 4.18C). In determining the symmetry of a flower, the position of the more conspicuous

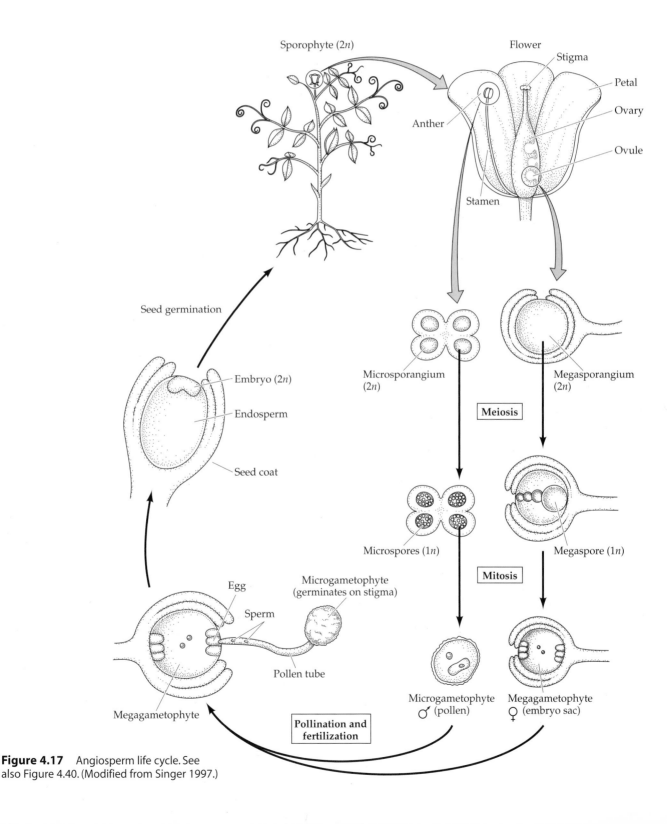

Figure 4.17 Angiosperm life cycle. See also Figure 4.40. (Modified from Singer 1997.)

BOX 4A *Floral Formulas and Diagrams*

A **floral formula** is a convenient shorthand method of recording floral symmetry, number of parts, connation and adnation, insertion, and ovary position. The formula consists of five symbols, as in the following example:

$$*, 5, 5, \infty, \underline{10}$$

The first symbol indicates either radial symmetry (*), bilateral symmetry (X), or asymmetry ($). The second item is the number of sepals (here 5); the third item the number of petals (also 5); the fourth, the androecial item, is the number of stamens (here numerous—the symbol for infinity is generally used when the number of stamens is more than about 12), and the last item is the number of carpels (here 10). The line below the carpel number indicates the position of the ovary with respect to other floral parts (here superior, and the flower hypogynous). If the ovary were inferior (and the flower epigynous), the line would have been drawn above the carpel number.

Connation is indicated by a circle around the number representing the parts involved. For example, in a flower with five stamens that are monadelphous (i.e., connate by their filaments), the androecial item of the floral formula would be indicated as

$$⑤$$

The plus symbol (+) may be used to indicate differentiation among the members of any floral part. For example, a flower with five large stamens alternating with five small ones would have the androecial item recorded as

$$5 + 5$$

Adnation is indicated by a line connecting the numbers representing different floral parts. Thus, a flower that has a sympetalous corolla with epipetalous stamens—for example, two stamens adnate to the four connate petals—would have the numbers representing the corolla and androecium parts indicated as

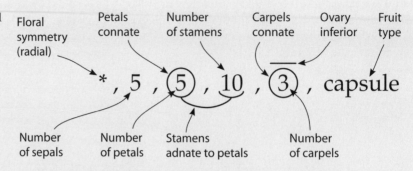

A sample floral formula.

The presence of a hypanthium (as in perigynous flowers) is indicated in the same fashion as adnation.

$$X, ⑤⑤, 10, \underline{5}$$

Sterile stamens (staminodes) or sterile carpels (carpellodes or pistillodes) can be indicated by placing a dot next to the number of these sterile structures. Thus a flower with a syncarpous gynoecium composed of five fertile carpels and five sterile ones would be represented in the formula as

Variation in the number of floral parts within a taxon is indicated by using a dash (–) to separate the minimum and maximum numbers. For example, the formula

$$*, 4 - 5, 4 - 5, 8 - 10, ③$$

would be representative of a taxon that has flowers with either 4 or 5 sepals and petals and from 8 to 10 stamens. Variation within a taxon in either connation or adnation is indicated by using a dashed (instead of a continuous) line:

$$*, 3, ③, 6, \underline{1}$$

The lack of a particular floral part is indicated by placing a zero (0) in the appropriate position in the floral formula. For example, the floral formula

$$*, 3, 3, 0, ②$$

represents a carpellate flower.

Flowers in which the perianth parts are not differentiated into a calyx and corolla, (that is, flowers with a perianth of tepals) have for-

mulas in which the second and third items (those representing sepals and petals) are combined into a single item (representing tepals). A hyphen (-) is placed before and after this item to indicate that the calyx and corolla categories have been combined. For example, an actinomorphic flower with 5 tepals, 10 stamens, and 3 connate carpels, with a superior ovary, would be represented as

$$*, -5-, 10, ③$$

The fruit type is often listed at the end of the floral formula:

$$*, -5-, 10, ③, \text{capsule}$$

A floral formula is by no means an end in itself; it is merely a convenient means of recording the information needed to identify a plant. Floral formulas also can be useful tools for remembering characteristics of the various angiosperm families. They are used extensively in this text (see Chapter 9). Their construction requires careful observation of individual flowers and of variation among the flowers of the same or different individuals.

Floral diagrams are stylized cross-sections of flowers that represent the floral whorls as viewed from above. Rather like floral formulas, floral diagrams are used to show symmetry, number of parts, their relationships of the parts to each other, and degree of connation and/or adnation. Such diagrams cannot easily show ovary position. For more information of floral diagrams, see Rendle (1925), Porter (1967), Correll and Correll (1982), Zomlefer (1994), and Walters and Keil (1995).

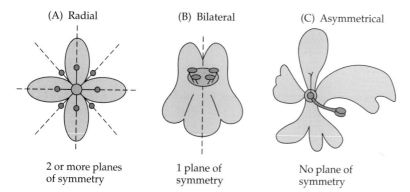

(A) Radial

2 or more planes of symmetry

(B) Bilateral

1 plane of symmetry

(C) Asymmetrical

No plane of symmetry

Figure 4.18 Patterns of floral symmetry.

structures—that is, the perianth and/or androecium—is considered.

Fusion of floral parts   Floral parts may be fused together in various ways. Fusion of like parts (e.g., petals united to petals) is called **connation**. When like parts are not fused, they are said to be **distinct**. Fusion of unlike parts (e.g., stamens united to petals) is called **adnation**; the contrasting condition is called **free** (e.g., stamens free from petals). Fused structures may be united from the moment of origin onward, or they may grow together later in development.

Various other specialized terms are used for various types of connation and adnation; some of these terms are listed here.

apocarpous   carpels distinct
apopetalous   petals distinct
aposepalous   sepals distinct
apotepalous   tepals distinct
diadelphous   stamens connate by their filaments in two groups
epipetalous   stamens adnate to corolla
monadelphous   stamens connate by their filaments in a single group
sympetalous   petals connate
synandrous   stamens connate
syncarpous   carpels connate
syngenesious   stamens connate by their anthers
synsepalous   sepals connate
syntepalous   tepals connate

The corolla shape, especially when sympetalous, may be valuable in identification, and specialized terms are applied to distinctive corolla shapes, including rotate (wheel- or disk-shaped), campanulate (bell-shaped), urceolate (urn-shaped), salverform (with slender tube and abruptly flaring limb), funnelform (funnel-shaped), tubular, and bilabiate (2-lipped).

Carpel versus pistil   The term **pistil** is sometimes used for the structure(s) in the center of the flower that contain(s) the ovules. How does this term differ from *carpel*, the term introduced earlier and used throughout this book? Carpels are the basic units of the gynoecium; they may, of course, be distinct or connate. If they are distinct, then the term *pistil* is equivalent in meaning to the term *carpel*. If, however, the carpels are connate, then the terms are not equivalent because each carpel constitutes only one unit within a pistil, which is then considered to be **compound** (Figure 4.19).

Number of parts   Flowers differ in numbers of sepals, petals, stamens, and carpels. The number of parts is usually easily determined by counting, but extreme connation, especially of the carpels, may cause difficulties, and variation between different flowers of the same plant or closely related species is common. Often it is possible to count fused carpels by counting the number of styles, stigmas, or stigmatic lobes (Figure 4.20). Placentation

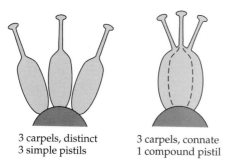

3 carpels, distinct
3 simple pistils

3 carpels, connate
1 compound pistil

Figure 4.19   The difference between the terms *carpel* and *pistil*.

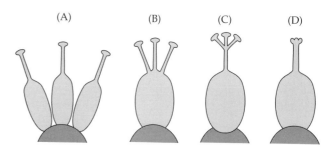

(A)   (B)   (C)   (D)

Figure 4.20   Three carpels, variously connate. (A) Three ovaries, styles, and stigmas. (B) One ovary, three styles and stigmas. (C) One ovary and style, the latter apically branched, and three stigmas. (D) One ovary and style, and three stigmas (or stigmatic lobes).

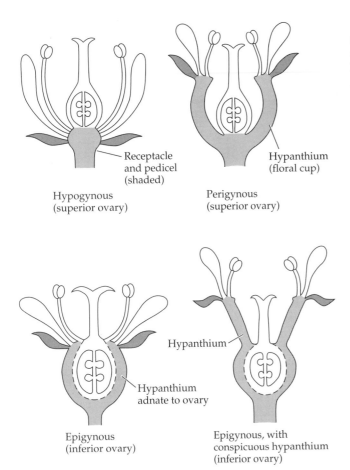

Figure 4.21   Insertion types.

(which will be discussed shortly) may also be useful in determining carpel numbers.

Most flowers are based on a particular numerical plan—that is, on patterns of three, four, five, or various multiples of those numbers. For example, a flower may have four sepals, four petals, eight stamens, and four carpels. Such a flower would be described as 4-merous; the ending *-merous*, along with a numerical prefix, is used to indicate a flower's numerical plan.

Insertion   Attachment of floral parts is called **insertion**. Floral parts may be attached to the receptacle (or floral axis) in various ways. Three major insertion types are recognized: hypogynous, perigynous, and epigynous. The position of the ovary in relation to the attachment of floral parts also varies, from **superior** to **inferior** (Figure 4.21).

Flowers in which the perianth and androecium are inserted below the gynoecium are called **hypogynous**; the ovary of such flowers is said to be superior. Flowers in which a cuplike or tubular structure surrounds the gynoecium, but without being adnate to it, are called **perigynous**. In such flowers the perianth and androecium are attached to the rim of this structure, which is called the **hypanthium** (plural **hypanthia**; or **floral cup** or **floral tube**). The ovary of such flowers is also superior.

Hypanthia have evolved from various structures, such as from the fused basal portion of the perianth parts and stamens or from the receptacle. Flowers in which the perianth and stamens appear to be attached to the upper part of the ovary due to fusion of the hypanthium (or bases of floral and androecial parts) to the ovary are called **epigynous**. The ovary of such flowers is said to be inferior.

In some epigynous flowers the hypanthium may extend beyond the top of the ovary, forming a cup or tube around the style. If the hypanthium is fused only to the lower portion of the ovary, the latter is considered **half-inferior**. Insertion type and ovary position are best determined in a longitudinal section of the flower.

Floral parts making up adjacent whorls normally alternate with each other, so one would expect to find a petal, for example, inserted at the point between two adjacent sepals. An understanding of this common pattern can assist in interpreting the number of floral parts, especially when they are obscured by connation or adnation.

The gynoecium, or the androecium and gynoecium, occasionally are borne on stalks (the **androgynophore** and **gynophore**, respectively).

Placentation   Ovules are arranged in various patterns within an ovary, allowing the recognition of various **placentation** types. Ovaries may contain from one to several chambers, or **locules**. The wall separating adjacent locules is called a **septum** (plural **septa**). The **placenta** (plural **placentae**) is the part of the ovary to which the ovules are attached. Major placentation types are illustrated in Figure 4.22. The number of ovules has no necessary correlation to the number of carpels, number of placentae, or placentation type.

Placentation type can be quite useful in determining the number of fused carpels in a flower. If the placentation is **axile**, the number of locules usually is indicative of the number of carpels. In **parietal** placentation, the number of placentae usually equals the carpel number.

Miscellaneous floral terms   The following list defines a few other floral terms commonly encountered in plant descriptions.

basifixed   referring to a structure, such as an anther, that is attached at its base

carpellode   a sterile carpel

centrifugal   developing first at the center and then gradually toward the periphery

centripetal   developing first at the periphery and then gradually toward the center

didynamous   having two long and two short stamens

exserted   sticking out, as in stamens extending beyond the corolla

included   hidden within, as in stamens not protruding from the corolla

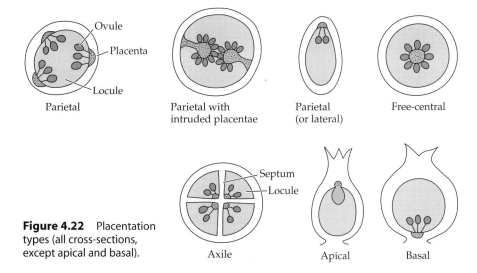

Figure 4.22 Placentation types (all cross-sections, except apical and basal).

[Parietal] [Parietal with intruded placentae] [Parietal (or lateral)] [Free-central] [Axile] [Apical] [Basal]

Ovule
Placenta
Locule
Septum
Locule

pistillode a sterile pistil

staminode a sterile stamen

tetradynamous having four long and two short stamens

versatile referring to a structure, such as an anther, that is attached at its midpoint

Pollination Biology

Plants are stationary and thus depend on external forces to bring their gametes together. The sperm of ferns swim through water to reach the egg; the sperm of seed plants are packaged in pollen grains for transport—a process referred to as **pollination**. Conifer pollination occurs largely via wind. Most flowering plants are pollinated by animals, although wind pollination predominates in some large and ecologically dominant families and occurs sporadically in many others. Water pollination is rare.

Pollination has interested people since at least 1500 B.C., which is when Babylonians discovered that date palm (*Phoenix dactylifera*, Arecaceae) flowers produce a yellow powder (pollen). Through careful observation, they found that this powder must be applied to the flowers of fruit-bearing trees in order for the trees to produce fruit. This important discovery made it possible for them to increase the production of dates simply by spreading the yellow powder on date flowers by hand.

Pollination remains essential to human welfare today. Most human food comes from cereal grains (Poaceae) and beans (Fabaceae), all of which are the result of pollination. Nearly all the edible fruits, including apples, coconuts, figs, strawberries, and tomatoes, would not exist without pollination.

Flowers are adaptations for pollination. One can often infer the pollen vector from the morphology of a flower, as Darwin did when he predicted that there had to be a moth in Madagascar with a tongue long enough to reach

the nectary in the 30-cm-long spur of the orchid *Angraecum sesquipedale* (see Chapter 6). The link between floral color, scent, time of flowering, structure, and rewards on the one hand and animal pollinator sensory capacity, behavior, and diet on the other is the basis of floral **pollination syndromes**.

This link may be strong enough that a plant and a pollinator adapt to each other. We will examine an example of such **coevolution** involving the plant genus *Yucca* and its pollinator, the yucca moth.

POLLINATION SYNDROMES

Wind and water pollination Wind-pollinated flowers are characterized by the production of a large amount of pollen that is readily transported by wind currents and by efficient means of trapping airborne pollen. Such flowers are small and lack much of a corolla (see, for example, illustrations of Betulaceae, Cyperaceae, Fagaceae, Juglandaceae, and Poaceae in Chapter 9).

If you walk in a pine forest during pollen shed and pass by a pond, its surface may be covered by a yellow film of pine pollen. The pollen grains of wind-pollinated plants are often small, light, and have a smooth surface, but some species have larger grains with air spaces, which lower the density of the pollen grain and make it float better. The pollen-receptive surface of conifers and wind-pollinated angiosperms is either a sticky or a netlike trap for airborne pollen. Commonly in wind-pollinated plants each flower has only a few ovules, often only one of which produces a seed.

Wind pollination is unusual in the tropics, especially in lowland rainforests. Temperate forests, in contrast, are dominated by wind-pollinated trees—oaks (*Quercus*, Fagaceae), beeches (*Fagus*, Fagaceae), hickories (*Carya*, Juglandaceae), walnuts (*Juglans*, Juglandaceae), and birches (*Betula*, Betulaceae) in the Northern Hemisphere, and southern beeches (*Nothofagus*, Nothofagaceae) in the Southern Hemisphere.

Pollen shed in wind-pollinated temperate species occurs at the beginning of the growing season, before or as leaves develop. A leafless forest has fewer obstacles to interrupt pollen flow. Pollen shed is also timed to avoid high humidity (which prevents drying out of the pollen and therefore reduces its buoyancy) and rain (which carries pollen down out of the air currents).

Corn (*Zea mays*, Poaceae) illustrates another common feature of wind pollination: unisexual flowers. The pollen is borne in staminate flowers in the corn "tassel," at the top of the plant, and the stigmas are the familiar "silk" that emerges, plumelike, from the top of the ear.

Many people are unhappily aware of wind pollination because of hay fever, an allergic reaction to proteins in the exterior pollen wall of some wind-pollinated species, such as ragweed (*Ambrosia*, Asteraceae).

Water pollination is limited to about 150 species in 31 genera and 11 families (Cox 1988). Almost half of these species are marine or grow in brackish water, and 9 of the families are monocots. Pollen may be transported above, on, or below the water surface. Plants that pollinate underwater often have filamentous, or eel-shaped, pollen borne in mucilaginous strands.

One of the longest-known and most fascinating examples of water pollination is *Vallisneria* (Hydrocharitaceae). In this genus, plants grow submersed. Their staminate flowers are released from the plant and float to the water surface, where they open and float about. At the same time, carpellate flowers rise to the surface on long peduncles and create a slight depression in the water surface, into which staminate flowers fall and are captured. Pollination follows this capture.

Animal pollination Animal pollination is thought to be an important factor in the evolutionary success of angiosperms. Animals are often more efficient transporters of pollen than wind, and they can be found where there is little wind (such as within a dense tropical forest). They promote cross-pollination by moving between plants. Animal pollination has driven diversification in many groups, and evolution associated with pollination is nowhere more evident than in the Orchidaceae, as we shall see shortly. Many species in this large family are apparently separated by floral reproductive isolation.

We are all familiar with bees buzzing about brightly colored flowers on warm summer days. In an apple orchard, bees fly from flower to flower, collecting pollen in special pollen baskets on their legs and drinking the flowers' sugary nectar. Back at their hive, the bees convert the nectar into honey and feed their young the protein-rich pollen. In return for these rich rewards, the bees pollinate the apple flowers. As a bee drinks nectar and collects pollen, some pollen becomes stuck to its body hairs. That pollen is removed from its body by the stigmas of the next flower it visits. Without bees, there would be no or few apples. The same can be said for many other fruits.

In essence, bees and apples have a contract: Apples are pollinated, and bees are rewarded with nectar and pollen. A plant's part of this contract consists of adaptations to attract pollinators, exploit their morphology and behavior to effect pollination, and ensure that they will return to its flowers by rewarding them. The animal's part of the contract is efficient pollination. Sometimes this contract is broken by the plant or by the animal. The lack of reward in some cases of orchid pollination (as will be discussed a little later in the chapter) represents a "breach of contract" on the part of the plant. On the side of the pollinator, some insects, called nectar robbers, chew into a flower to take the nectar without pollinating the flower.

Generally animals visiting flowers are not necessarily pollinators. For example, four species of *Marcgravia* (Marcgraviaceae) in Costa Rica are visited by small bats that take nectar from floral bracts modified into nectaries. The bats are also pollinators of these species, except for *M. nepenthoides*, which has much larger nectaries than the other three species. The distance between nectaries and flower in this species suggests that two species of opossums are more efficient than the bats and are the primary pollinators (Tschapka and von Helversen 1999).

Many, though not all, flowers pollinated by one of the major kinds of animal pollinators have characteristic suites of floral adaptations (Table 4.1). Flowers attract their pollinators with color and scent. Insects do not see the same spectrum of light that humans do; they are less sensitive to red but are able to see shorter wavelengths, down into the ultraviolet. Bee-pollinated flowers often have ultraviolet-reflecting or -absorbing patterns that attract the insects to the flower and then direct them to the nectar.

Flowers open and emit scents when their pollinators are active. Moths and bats, for example, are active at night. The flowers they visit are white and emit their scents at night, so they are easier to locate in the darkness. Many floral fragrances are pleasant to humans; others are powerful and unpleasant. For example, flowers of *Weberocereus tunilla* (Cactaceae), an epiphytic rain forest cactus, give off a fragrance that has been described as similar to that of a corpse (Tschapka et al. 1999).

The floral syndrome of nonflying mammals (see Table 4.1) encompasses a broad range of floral characteristics because of differences in morphology and foraging behavior of the three groups of pollinating mammals. Flowers pollinated by primates (such as lemurs and monkeys) tend to be unscented and very large to accommodate the large size of the animals; flowers pollinated by marsupials are usually located in the forest canopy; and rodent-pollinated flowers tend to be close to the ground and have a yeasty odor (Johnson et al. 2001). About 60 species of nonflying mammals have been documented as pollinators of about 100 plant species, especially in the Southern Hemisphere (Carthew and Goldingay 1997).

Flowers are built to fit their pollinators physically and to provide an appropriate reward. The corollas of many bee-pollinated flowers have nectar guides—lines or marks that direct bees to the nectar source. The corolla may form a landing platform that orients pollinators toward the nectar and/or pollen and forces them to perform the movements required for pollination. The pollen of many animal-pollinated flowers is covered with minute projections that stick to animal hair or feathers (see Figure 4.47E–H). Whereas bees consume both nectar and pollen, butterflies, moths, and birds are rewarded by nectar alone. In some cases pollinators are not rewarded,

TABLE 4.1 *Floral pollination syndromes.*

Pollinator	Color	Scent	Time of flowering	Corolla	Reward
Bee	Blue, yellow, purple	Fresh, strong	Day	Bilateral landing platform	Nectar and/or pollen
Butterfly	Bright; often red	Fresh, weak	Day	Landing platform; sometimes nectar spurs	Nectar only
Moth	White or pale	Sweet, strong	Night or dusk	Dissected; sometimes nectar spurs	Nectar only
Fly (reward)	Light	Faint	Day	Radial, shallow	Nectar and/or pollen
Fly (carrion)	Brownish, purplish	Rotten, strong	Day or night	Enclosed or open	None
Beetle	Often green or white	Various, strong	Day or night	Enclosed or open	Nectar and/or pollen
Bird	Bright; often red	None	Day	Tubular or pendant; ovary often inferior	Nectar only
Bat	Whitish	Musky, strong	Night	Showy flower or inflorescence	Nectar and/or pollen
Nonflying mammals	Dull-colored	Unscented to variously strong	Night	Robust, exserted styles and stamens	Copious nectar and/or pollen

The table header spans: Floral characteristics covers Color, Scent, Time of flowering, Corolla, Reward.

but rather are deceived into pollinating a flower, as we shall see shortly.

Plant species may specialize for one pollinator or may be served by a wide range of pollinators; the latter species are called **generalists**. *Acer saccharum* (sugar maple, Sapindaceae), for example, is both wind- and animal-pollinated. *Putoria calabrica* (Rubiaceae), a dwarf shrub of the Mediterranean region, is another example of a generalist, being pollinated by various flies and butterflies (Ortiz et al. 2000). The syndromes we have listed in Table 4.1 are certainly not to be thought of as complexes of characters that are highly linked to the pollination of a particular flower type by a single pollinator, although this is sometimes the case.

COEVOLUTION BETWEEN PLANT AND POLLINATOR
One of the most fascinating cases of pollination is that of the yucca (*Yucca*, Agavaceae) and the yucca moth (*Tegiticula*). The white yucca flowers open at night and attract the female yucca moth. Once inside the flower, the moth pollinates it. She carries pollen from another yucca plant in special tentacles under her head, and she stuffs this pollen into the stigmatic cavity (Figure 4.23). She then uses her ovipositor to penetrate the ovary wall and lay eggs among the ovules. Finally, just prior to leaving the flower, she climbs the filament, stabilizes herself by laying her tongue over the top of the filament, and collects pollen with her tentacles.

The moth takes no food from the flower because adult yucca moths do not eat; all consumption in the species is by the larvae. The moth eggs hatch inside the fruit at about the time the seeds ripen, and the larvae consume some, but not all, of the developing seeds. When they have eaten enough, the larvae bore out of the fruit, fall to the ground, and burrow into the soil to overwinter as pupae.

Coevolution is evident in the adaptations of yucca and moth that facilitate pollination. The plant produces stout, club-shaped filaments that the moth can easily climb to collect pollen. The anthers are positioned on top of the filament, where they are readily accessible to the moth. The moth's tentacles are an adaptation for collecting and transporting pollen and placing it in stigmatic cavities. The moth's pollination behavior ensures a food supply for her young, as well as reproduction by the plant that her species absolutely depends on.

Another tight coevolutionary relationship exists between figs (*Ficus*, Moraceae) and their fig wasp pollinators. Like the *Yucca* and its moth, figs and wasps are completely dependent on each other. There are about 750 species of figs, each with its own species of fig wasp pollinator.

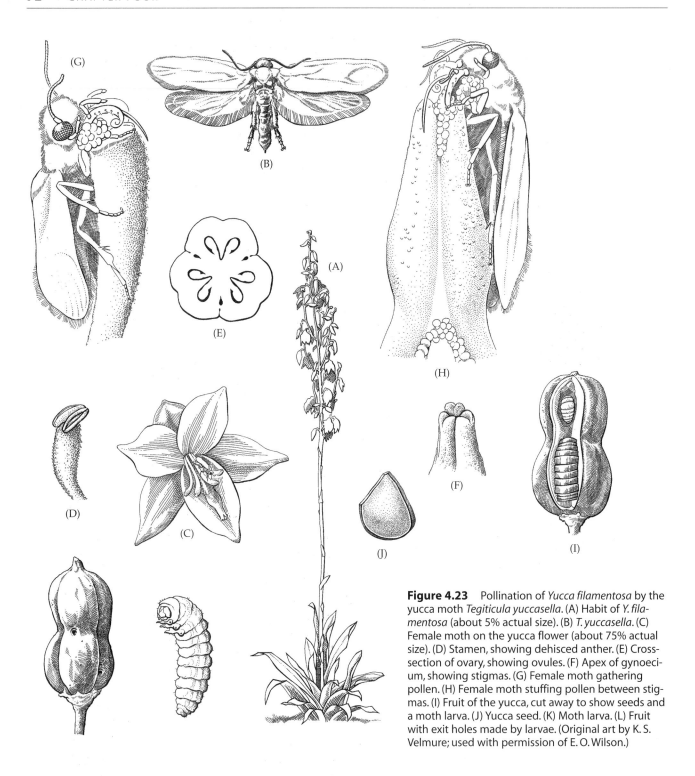

Figure 4.23   Pollination of *Yucca filamentosa* by the yucca moth *Tegiticula yuccasella*. (A) Habit of *Y. filamentosa* (about 5% actual size). (B) *T. yuccasella*. (C) Female moth on the yucca flower (about 75% actual size). (D) Stamen, showing dehisced anther. (E) Cross-section of ovary, showing ovules. (F) Apex of gynoecium, showing stigmas. (G) Female moth gathering pollen. (H) Female moth stuffing pollen between stigmas. (I) Fruit of the yucca, cut away to show seeds and a moth larva. (J) Yucca seed. (K) Moth larva. (L) Fruit with exit holes made by larvae. (Original art by K. S. Velmure; used with permission of E. O. Wilson.)

DECEPTION AND NON-NUTRITIVE AWARDS IN ORCHID POLLINATION

Of the 19,500 species of orchids, about 8000 offer no food reward to their pollinators. Instead, they deceive their pollinators or provide them with non-nutritive rewards, such as fragrant chemicals. Some orchids, like some species in other families, bear flowers that resemble rotting flesh in their scent and purplish color. These flowers, which are called carrion flowers (see Table 4.1), attract flies that lay their eggs in them because they are tricked by the flowers' odor and color into perceiving the flowers as a source of food for their offspring. In the process of laying eggs, the fly moves within the flower and pollinates it.

Far more bizarre methods of ensuring pollination have evolved in some tropical orchids. We see another example of orchid pollination deception in Chapter 6, in the flowers of *Ophrys* (Orchidaceae), which trigger mating behavior in male bees and wasps and provide no food reward.

Figure 4.24 Male *Euglossa hemichlora* falling and sliding through the flower of *Gongora grossa*. In this posed photograph, the pollinia are attached to the bee as it slides past the apex of the column. (From van der Pijl and Dodson 1966.)

Many New World tropical orchids are pollinated by male euglossine bees. After emerging from the nest, male euglossines fly through the forest, feeding on the nectar from various flowers and sleeping on the undersides of leaves or in tubular flowers. They are strongly attracted by the fragrant chemicals of certain orchids that do not produce nectar. They scratch at the source of the odor, and in the process of transferring the chemicals to their hind tibiae, involuntarily follow a path in the flower that leads to pollination.

In some species of the orchid genera *Gongora* and *Stanhopea*, the bee is turned upside down and slides down the flower in such a way that it picks up the pollinium (Figure 4.24) (see the discussion of Orchidaceae in Chapter 9 for a description of the pollinium). The bee flies to another flower of the same species and eventually slides past the stigma, where the pollinium is deposited. Male euglossines use the floral fragrances they gather to attract females.

The Orchidaceae are an encyclopedia of pollination syndromes. In addition to pseudocopulating bees and wasps, carrion flies, and euglossine bees, pollinators of orchid species include beetles, bugs, moths, butterflies, mosquitoes, and birds. There are no records of nonflying mammal pollinators of orchids.

Modes of pollination are noted for nearly all of the families treated in Chapters 8 and 9.

SELF-INCOMPATIBILITY

Pollination may occur within an individual (**self-pollination** or **selfing**) or between individuals (**cross-pollination**). It has long been known (Darwin 1876) that many plants avoid self-pollination and thus the possible harm-

ful consequences of inbreeding depression. Such avoidance may be achieved by separating female and male gametes in space or time.

Spatial separation may be accomplished through dioecism or monoecism. Temporal separation is achieved when the stigmas are receptive before the pollen is shed in a bisexual flower or monoecious individual (**protogyny**), or the anthers shed pollen before the stigmas are receptive (**protandry**). The arrangement of stamens and stigmas and the movement of pollinators may also limit selfing.

Another mechanism for avoiding self-pollination is **self-incompatibility**, the inability of a bisexual plant to produce zygotes with its own pollen. Self-incompatibility is genetically controlled in many species by multiple alleles of a gene labeled simply S. Even though a diploid individual has only two of these alleles, there may be hundreds of other alleles in other individuals of the species.

If a pollen grain has the same S allele as the carpellate plant, then mating will not be successful. Self-fertilization therefore will not occur, but with many different S genotypes possible, most matings with other individuals will be at least partially successful. Note that this mechanism will prevent not only selfing, but also matings between individuals that happen to have the same S alleles, regardless of how genetically different they are otherwise.

Self-incompatibility In **monomorphic self-incompatibility**, there are no morphological differences between the flowers of incompatible individuals. Monomorphic self-incompatibility may be gametophytic or sporophytic. In **gametophytic self-incompatibility**, the genotype of the pollen determines what matings will be successful. If two individuals have the same two self-incompatibility alleles, then no pollinations between them will be successful (Figure 4.25). If one or both self-incompatibility

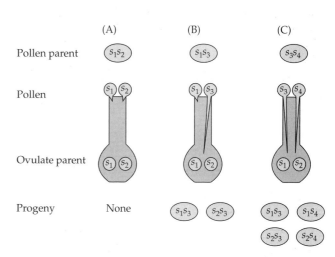

Figure 4.25 Consequences of a gametophytic self-incompatibility system on the production of progeny and the genetic variability of that progeny. (After Lewis 1949.)

(A)

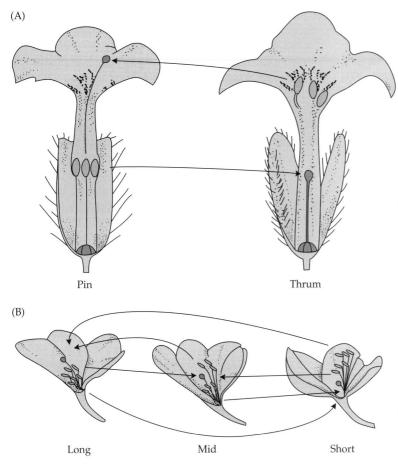

Pin   Thrum

(B)

Long   Mid   Short

Figure 4.26   Heterostyly. (A) Reciprocal anther and stigma positions in pin and thrum morphs of a distylous plant. (B) The three forms of a tristylous plant. Anthers and stigma are positioned at three levels. The arrows indicate the directions of compatible pollinations. (After Ganders 1979.)

Heterostyly   Heterostyly, or **heteromorphic self-incompatibility**, is a sporophytic self-incompatibility system that also involves floral morphological differences. The existence of two kinds of flowers in *Primula* (Primulaceae) was recognized long ago. One kind of flower has a long style and short stamens and is called a **pin** because of the resemblance of the long style to a pin. The other kind has a short style and long stamens that project out of the top of the corolla tube. This morph is called a **thrum**, referring to the resemblance of the stamens to pieces of thread coming out the end of a shirtsleeve (Figure 4.26A).

Darwin (1877) demonstrated that this heteromorphy is associated with self-incompatibility: Only pollinations between different morphs are successful. The heteromorphism is governed by a supergene, a series of genes closely spaced or linked on a chromosome. Thrum individuals are heterozygous (*Ss*), and pins are homozygous recessive (*ss*), so the cross pin × thrum gives equal numbers of thrum and pin offspring.

Heterostyly is relatively infrequent. It is known from 24 families and is especially common in the Rubiaceae, where it has been documented in about 90 genera. It may be expressed as either two (distyly; Figure 4.26A) or three (tristyly; Figure 4.26B) floral morphs. Distyly, as in *Primula* and many Rubiaceae, is far more common than tristyly, which is known from only three families (Lythraceae, Oxalidaceae, and Pontederiaceae).

alleles differ, however, then some or all matings will produce zygotes. The genetic difference between pollen parent and potential ovulate parent is most often recognized within the style, where the pollen tube with a matching self-incompatibility allele simply stops growing.

In **sporophytic self-incompatibility**, the sporophyte genotype determines the fate of pollen grains. This form of self-incompatibility occurs in the Asteraceae and the Brassicaceae and a small number of other families, and it is not as widespread as gametophytic self-incompatibility.

Figure 4.27   Inflorescences (in this case, racemes) with variouS-types of subtending leaves or bracts. Ebracteate and bracteate forms are frequently called terminal racemes because of the arbitrary delimitation of inflorescences from the vegetative region at 1. Leafy racemes (D) are often considered to have solitary, axillary flowers because of the arbitrary delimitation of inflorescences from the vegetative region at 2.

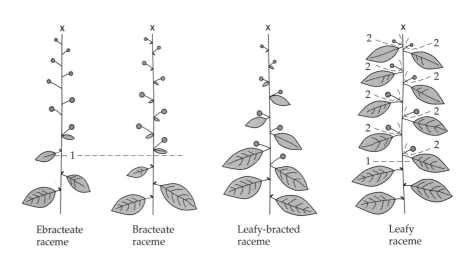

Ebracteate raceme   Bracteate raceme   Leafy-bracted raceme   Leafy raceme

Inflorescences, Fruits, and Seeds

An **inflorescence** can be defined as "the shoot system which serves for the formation of flowers and which is modified accordingly" (Troll 1964, trans. Weberling 1989: 201). The arrangement of flowers on a plant (the inflorescence form and position) is important in routine identification, as well as in the determination of phylogenetic relationships. Inflorescence categories have often been confused, however, because of the arbitrary separation of flower-bearing and vegetative regions of the plant (Figure 4.27).

Two quite different inflorescence types occur in angiosperms. In **determinate** (or **monotelic**) inflorescences, the main axis of the inflorescence ends in a flower; in **indeterminate** (or **polytelic**) inflorescences, the growing point produces only lateral flowers or partial inflorescences (groups of flowers). Typical determinate and indeterminate inflorescences are shown in Figure 4.28 (see also Weberling 1989).

The flowering sequence of determinate inflorescences usually begins with the terminal flower at the top (or center) of the cluster of flowers. In indeterminate inflorescences, the flowering sequence usually starts at the base (or outside) of the cluster. Determinate inflorescences are generally ancestral to indeterminate ones, and transitional forms are known. Various kinds of determinate and indeterminate inflorescences, based on pattern of branching, have been described.

One of the more common types of determinate inflorescences is the **cyme** (or determinate thyrse), the lateral branches of which are composed of usually numerous three-flowered units, usually showing opposite branching. Cymes can be of many different shapes because of differences in their branching patterns. If the inflorescence branches are initially monopodial—that is, producing several internodes before ending in a terminal flower—a panicle-like cyme results, and through reduction a racemelike cyme is formed.

The lateral branches (paraclades) of typical cymes or panicle-like cymes may be either alternately or oppositely arranged. **Scorpioid** and **helicoid** cymes are especially distinctive because of their coiled form, resulting from the abortion of one of the flowers within each three-flowered inflorescence unit (Figure 4.29).

The most common types of indeterminate inflorescences are racemes, spikes, corymbs, and panicles. A **raceme** is an inflorescence with a single axis bearing pedicellate flowers; a **spike** is similar, but the flowers are sessile (lacking a pedicel or stalk). In contrast, a **corymb** is a raceme with the pedicels of the lowermost flowers elongated, bringing all flowers to approximately the same level. A **panicle** is merely a compound raceme—that is, an indeterminate inflorescence that has two or more orders of branching, and each axis bears flowers or higher-order axes (Figure 4.30). Axillary racemes or cymes can become reduced in length, resulting in a fascicle.

A **head**, or **capitulum**, is a dense terminal cluster of sessile flowers. This inflorescence type can result through aggregation of the flowers of either an indeterminate or determinate inflorescence. In an indeterminate head, the peripheral flowers open first; in a determinate head the central flowers open first (compare Figures 4.29 and 4.30). An umbel is an inflorescence in which all the

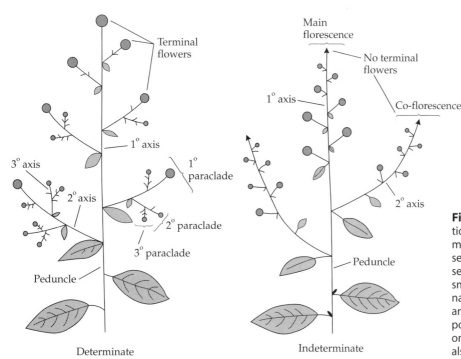

Figure 4.28 Diagrammatic representation of typical determinate and indeterminate inflorescences. The circles represent flowers; their size indicates the sequence of opening (from large to small). The individual units of determinate inflorescences are called paraclades; an indeterminate inflorescence is composed of the main florescence and secondary florescenses (coflorescences). (See also Weberling 1989.)

Figure 4.29   Some common kinds of determinate inflorescences. The circles represent flowers; their size indicates the sequence of opening (from large to small).

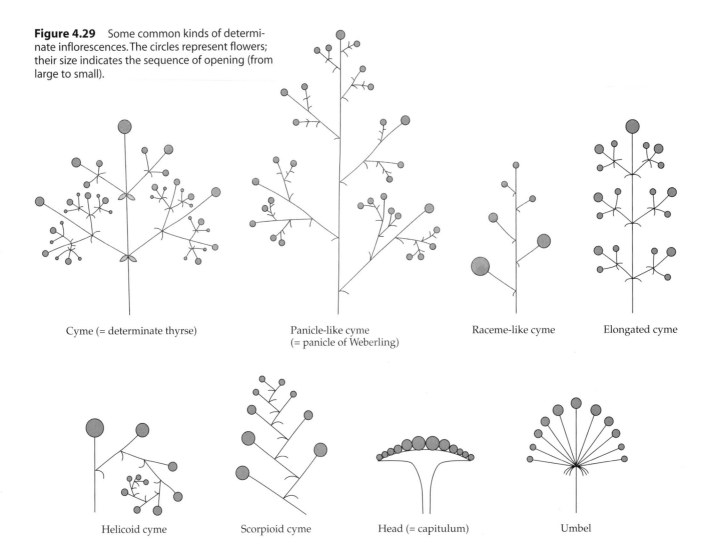

Cyme (= determinate thyrse)

Panicle-like cyme
(= panicle of Weberling)

Raceme-like cyme

Elongated cyme

Helicoid cyme

Scorpioid cyme

Head (= capitulum)

Umbel

flowers often have pedicels of approximately equal length that arise from a single region at the apex of the inflorescence axis. Umbels are often indeterminate but may also be determinate (see Figures 4.29 and 4.30).

Simple inflorescences, such as racemes, spikes, umbels, and heads, have only a single axis (i.e., one branch order). Compound inflorescences (e.g., compound racemes, compound umbels, cymes, thyrses, and panicles) have two or more orders of branches.

The term **catkin** or **ament** is used for any elongated inflorescence composed of numerous inconspicuous, usually wind-pollinated flowers. These terms are non-specific with regard to order of branching and floral arrangement: Aments may be either simple or compound, and they may be determinate or indeterminate structures.

Most inflorescences and solitary flowers are borne on young shoots, but some are borne on leaves (producing **epiphyllous** flowers or inflorescences) or on older stems and/or trunks (producing **cauliflorous** flowers or inflorescences). Epiphylly often (but not always) results from ontogenetic displacement of the bud; in early stages of growth the cells below the young bud primordium and

adjacent leaf primordium divide actively, and the bud and leaf grow out as a single unit. In contrast, cauliflory is due to the delayed development of the inflorescences, which break out through old wood.

Inflorescences may sometimes be modified for climbing by becoming elongated and twining or developing adhesive pads, thus forming tendrils. (Tendrils, of course, may also evolve from leaves, and twining stems may be tendril-like.)

FRUIT TYPES

A **fruit** is a matured ovary along with fused accessory structures (hypanthium or perianth parts). The great diversity of size, form, texture, means of opening, and anatomy of fruits has long confounded plant systematists, and many different fruit types have been proposed. All systems of fruit classification must deal with several difficulties.

Foremost is the problem of the bewildering and often continuous variation in fruit structure; van der Pijl (1972: 17) concluded that "the fruit is too versatile and has too many aspects to be divided into strict categories." Second, additional complexities come from the extensive

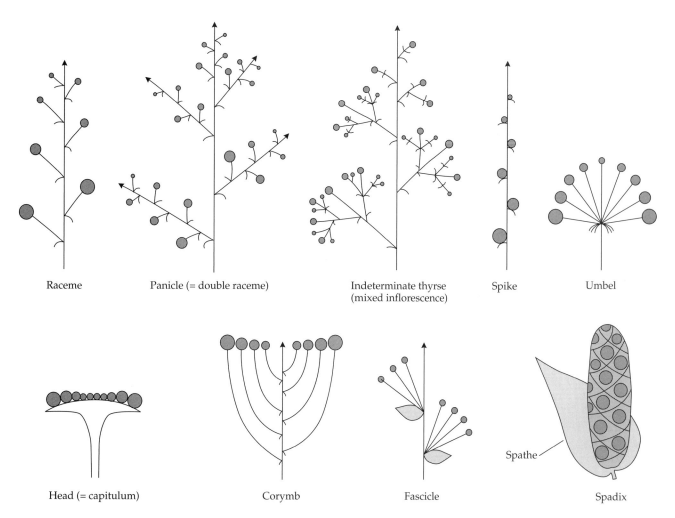

Raceme · Panicle (= double raceme) · Indeterminate thyrse (mixed inflorescence) · Spike · Umbel

Head (= capitulum) · Corymb · Fascicle · Spathe · Spadix

Figure 4.30 Some common kinds of indeterminate inflorescences. The circles represent flowers; their size indicates the sequence of opening (from large to small).

convergent evolution of fruiting structures; functionally similar fruits have arisen independently in different lineages of angiosperms from similar and different gynoecial conditions. Third, other parts of the flower (and even associated vegetative structures), in addition to the matured gynoecium, may form a functional part of the fruit. Examples of such **accessory structures** include the expanded fleshy receptacle of strawberries (*Fragaria*, Rosaceae), the fleshy perianth of seagrape fruits (*Coccoloba*, Polygonaceae), the winglike calyx of dipterocarp fruits (*Dipterocarpus*, Dipterocarpaceae), and the fleshy inflorescence axis of figs (*Ficus*, Moraceae). Finally, tropical fruits have been neglected in many traditional fruit classifications.

In this text we employ an artificial system of descriptive fruit terms, based on the traditional fruit classification of Gray (1877). This system has been widely employed. It is based on the texture of the **pericarp**, or fruit wall (fleshy, dry, or hard), the pattern of **dehiscence** or **indehiscence** (type of fruit opening, or lack thereof), the

shape and size of the fruit, and the carpel and ovule number.

Simple fruits (those resulting from a single flower) are divided into two categories: (1) those formed from a single carpel or several fused carpels, and (2) those that develop from several separate carpels of a single gynoecium (**aggregate fruits**). The individual units of an aggregate fruit may be any of the basic fruit types given in the list following this discussion. For example, the fruit of *Magnolia* (Magnoliaceae) is an aggregate of follicles, that of *Annona* (pawpaw, Annonaceae) is an aggregate of berries, and that of *Rubus* (blackberry, Rosaceae) an aggregate of drupes.

If a fruit is the product of the gynoecia of several closely clustered flowers, it is termed a **multiple fruit**. As with aggregate fruits, the individual fruits composing the cluster may consist of any of the basic fruit types outlined in the list that follows this discussion. For example, the fruit of *Ananas* (pineapple, Bromeliaceae) is a multiple of berries, that of *Morus* (mulberry, Moraceae) a multiple of drupes, and that of *Platanus* (sycamore, Platanaceae) a multiple of achenes.

The use of modifying terms (e.g., drupaceous schizocarp, winged or samaroid schizocarp, one-seeded fleshy

A Key to Fruit Types

1. Fruit the product of a single flower 2
1. Fruit the product of several flowers clustered in one mass .... **Multiple fruit**
   [go to 3 and key based on an individual unit]
2. Single fruit (carpel solitary or several and fused) .... **Simple fruit** [go to 3]
2. Several distinct fruits (carpels several and distinct) .... **Aggregate fruit**
   [go to 3 and key based on individual units]
3. Fruit not opening (indehiscent) 4
3. Fruit opening or breaking apart (dehiscent) 13
4. Fruit fleshy (at least in part) 5
4. Fruit dry 8
5. Texture of fruit ± homogeneous (except for seeds), fleshy throughout .... **Berry**
5. Texture of fruit heterogeneous 6
6. Outer part of fruit firm, hard, or leathery; inner part softer .... **Berry**
6. Outer part of fruit ± soft; inner part papery, cartilaginous, or hard 7
7. Center of fruit with 1 or more hard pits (pyrenes) enclosing seeds; ovary inferior or superior ..... **Drupe**
7. Center of fruit with papery or cartilaginous structures enclosing seeds; ovary inferior .... **Pome**
8. Fruit with several to many seeds .... **Indehiscent pod**
8. Fruit usually 1-seeded 9
9. Fruit winged .... **Samara**
9. Fruit wingless 10
10. Pericarp thick and bony; fruit generally large .... **Nut**
10. Pericarp thin; fruit smaller 11
11. Pericarp loose and free from seed .... **Utricle**
11. Pericarp firm, close-fitting or fused to seed 12
12. Pericarp firm, close-fitting, but free from seed .... **Achene**
12. Pericarp adnate (fused) to seed .... **Caryopsis (grain)**

capsule) is encouraged. This classification is presented by means of a key to a series of definitions. Although this system is admittedly arbitrary, the key and descriptions have proved useful in teaching and floristics. (For more information on fruit types, see Judd 1985, Weberling 1989, and Spjut 1994.)

achene a fairly small, indehiscent, dry fruit with a thin and close-fitting wall surrounding a single seed; includes the **cypsela**. Examples: *Bidens, Carex, Ceratophyllum, Clematis, Cyperus, Ficus, Fragaria, Helianthus, Medicago* (some), *Ostrya, Petiveria, Polygonium, Ranunculus, Rhynchospora, Rosa* (achenes enclosed in a fleshy hypanthium), *Rumex, Sagittaria, Taraxacum, Trifolium* (some), *Vernonia*. (See Figures 9.17, 9.58, 9.80, and 9.146.)

berry an indehiscent, fleshy fruit with one or a few to many seeds. The flesh may be ± homogeneous throughout, or the outer part may be hard, firm, or leathery; septa are present in some, and the seeds may be arillate (having a fleshy outgrowth of the funiculus) or with a fleshy testa (seed coat). Examples: *Actinidia, Aegle, Annona, Averrhoa, Cananga, Citrus, Cucurbita, Eugenia, Litchi, Miconia, Musa* (some), *Opuntia, Passiflora, Phoenix, Punica, Sideroxylon, Smilax, Solanum, Tamarindus, Vaccinium, Vitis*. (See Figures 9.12, 9.48, 9.63, 9.52, 9.100, 9.108, and 9.121.)

capsule a dry to (rarely) fleshy fruit from a two- to many-carpellate gynoecium that opens in various ways to release the seeds. Such fruits may have from one to many locules; if 2-locular, then the partition is not persistent. Examples: *Aesculus, Allium, Antirrhinum, Argemone, Aristolochia, Begonia, Blighia, Campsis, Clusia, Echinocystis, Epidendrum, Eucalyptus, Euonymus, Hibiscus, Hypericum, Ipomoea, Justicia, Lachnanthes, Lagerstroemia, Lecythis, Lyonia, Momordica, Oxalis, Papaver, Portulaca, Rhododendron, Swietenia, Tri-*

13. Fruit from a single carpel 14

13. Fruit from a 2- to many-carpellate gynoecium 16

14. Fruit dehiscing along a single suture (slit) .... **Follicle**

14. Fruit dehiscing by two longitudinal sutures, or breaking up by transverse sutures 15

15. Sutures longitudinal .... **Legume**

15. Sutures transverse, the fruit breaking into 1-seeded segments .... **Loment**

16. Fruit with a dry/fibrous to leathery or fleshy outer husk that early to tardily breaks
    apart; center of fruit with hard pit(s) enclosing seed(s) .... **Dehiscent drupe**

16. Fruit lacking hard pit(s) enclosing seed(s); splitting open or into 1-seeded segments 17

17. Fruit splitting into 1- or few-seeded segments (mericarps) .... **Schizocarp**

17. Fruit splitting open and releasing seeds 18

18. Fruit 2-locular, the two valves splitting away from a persistent thin partition around
    the rim of which the seeds are attached .... **Silique**

18. Fruit 1- to several-locular, the partition not persistent if the fruit 2-locular .... **Capsule** [Go to 19]

19. Dehiscence circumscissile (splitting transversely), the top coming off like a lid ....
    .... **Circumscissile capsule (pyxis)**

19. Dehiscence not circumscissile 20

20. Fruit opening by pores, flaps, or teeth 21

20. Fruit opening longitudinally or irregularly 22

21. Fruit opening by a series of apical teeth .... **Denticidal capsule**

21. Fruit opening by pores or flaps (often near the top) .... **Poricidal capsule**

22. Fruit opening irregularly .... **Anomalicidal capsule**

22. Fruit opening longitudinally 23

23. Valves breaking away from the septa (partitions between the locules) .... **Septifragal capsule**

23. Valves remaining attached to the septa (at least in part) 24

24. Fruit splitting at the septa .... **Septicidal capsule**

24. Fruit splitting between the septa and into the locules of the ovary, or
    fruit 1-locular .... **Loculicidal capsule**

odanis, Viola. (See Figures 9.32, 9.51, 9.55, 9.62, 9.71, 9.73, 9.109, 9.115, 9.116, 9.131, and 9.134.)

caryopsis (pl. caryopses) **(grain)** a small, indehiscent, dry fruit with a thin wall surrounding and more or less fused to a single seed. Examples: most Poaceae. (See Figures 9.39 and 9.40.)

dehiscent drupe a fruit with a dry or fibrous to fleshy or leathery outer husk that early to tardily breaks apart (or opens), exposing one or more nutlike pits enclosing the seed(s). Examples: *Carya, Rhamnus* (some), *Sageretia*. (See Figure 9.96.)

drupe an indehiscent, fleshy fruit in which the outer part is more or less soft (to occasionally leathery or fibrous) and the center contains one or more hard pits (**pyrenes**) enclosing seeds. Examples: *Arctostaphylos, Celtis, Clerodendrum, Cocos, Cordia, Cornus, Ilex, Juglans, Licania, Melia, Myrsine, Nectandra, Prunus, Psychotria, Roystonea, Rubus, Sabal, Scaevola, Syagrus, Ter-*

minalia, Toxicodendron. (See Figures 9.13, 9.82, 9.95, 9.113, 9.130, and 9.137.)

follicle a dry to (rarely) fleshy fruit derived from a single carpel that opens along a single longitudinal suture; the seeds may be arillate or with a fleshy testa. Examples: *Akebia, Alstonia, Aquilegia, Asclepias, Caltha, Grevillea, Magnolia, Nerium, Paeonia, Sterculia, Zanthoxylum.* (See Figures 9.47, 9.61, 9.126, and 9.127.)

indehiscent pod an indehiscent, fairly dry fruit with few to many seeds. Examples: *Adansonia, Arachis, Bertholletia, Cassia* (some), *Crescentia, Kigelia, Medicago* (some), *Thespesia* (some).

legume a dry fruit derived from a single carpel that opens along ± two longitudinal sutures. Examples: many Fabaceae. (See Figures 9.74, 9.75, and 9.76.)

loment a dry fruit derived from a single carpel that breaks transversely into one-seeded segments. Examples: *Aeschynomene, Desmodium, Sophora.*

nut   a fairly large, indehiscent, dry fruit with a thick and bony wall surrounding a single seed. Examples: *Brasenia, Castanea, Corylus, Dipterocarpus, Fagus, Nelumbo, Quercus, Shorea*. (See Figure 9.91.)

pome   an indehiscent, fleshy fruit in which the outer part is soft and the center contains papery or cartilaginous structures enclosing the seeds. Examples: most Rosaceae, subfamily Maloideae. (See Figure 9.81.)

samara   a winged, indehiscent, dry fruit containing a single (rarely two) seed(s). Examples: *Ailanthus, Betula, Casuarina, Fraxinus, Liriodendron, Myroxylon, Ptelea, Stigmaphyllon, Ulmus*. (See Figures 9.85 and 9.93.)

schizocarp   a dry to rarely fleshy fruit derived from a two- to many-carpellate gynoecium that splits into one-seeded (or few-seeded) segments (**mericarps**). If desired, the mericarps may be designated as samara-like, achenelike, drupelike, and so on. Examples: *Acer, Apium, Cephalanthus, Croton, Daucus, Diodia, Erodium, Euphorbia, Glandularia, Gouania, Heliconia, Heliotropium, Lamium, Lycopus, Malva, Ochna, Oxypolis, Salvia, Sida, Verbena*. (See Figures 9.67, 9.68, 9.112, 9.125, 9.135, and 9.139.) Fruits that show late-developmental fusion of their apical parts are not considered schizocarps—for example, *Asclepias* (follicles), *Sterculia* (follicles), *Ailanthus* (samaras), *Simarouba* (drupes), *Pterygota* (samaras).

silique   a fruit derived from a two-carpellate gynoecium in which the two halves of the fruit split away from a persistent partition (around the rim of which the seeds are attached); includes the **silicle**. Examples: many Brassicaceae. (See Figure 9.102.)

utricle   a small, indehiscent, dry fruit with a thin wall (bladderlike) that is loose and free from a single seed. Examples: *Amaranthus* (some), *Chenopodium, Lemna, Limonium*. (See Figure 9.54.)

SEEDS

A **seed** is a matured ovule that contains an embryo and often its nutritive tissues (endosperm, perisperm). The **endosperm** is usually triploid tissue derived from the union of two cells of the female gametophyte (the polar nuclei) and one sperm nucleus (see the section on embryology later in this chapter). Endosperm may be **homogeneous** (uniform in texture) or **ruminate** (dissected by partitions that grow inward from the seed coat). It may contain starch, oils, proteins, oligosaccharides, and/or hemicellulose, and it may be hard to soft and fleshy. The **perisperm** is a specialized, diploid nutritive tissue derived from the megasporangium.

The seed is surrounded by a **seed coat**, which develops from the integument(s). The details of seed coat anatomy are quite variable. The **testa** (plural **testae**) develops from the outer integument and the **tegmen** (plural **tegmina**) from the inner integument. The prefixes *exo-, meso-,* and *endo-* refer to tissues developing from the outer epidermis, the middle portion, and the inner epidermis, respectively, of each of the two integuments. The seed may be variously sized and shaped, and it may be associated with a wing or a tuft of hairs. The testa varies in surface texture due to the pattern and outgrowths of the individual cells composing its surface, and it is sometimes colorful and fleshy.

Some seeds are associated with a hard to soft, oily to fleshy, and often brightly colored structure called an **aril**. The aril is usually an outgrowth of the funiculus or the outer integument, although sometimes this term is restricted to structures derived from the funiculus, with those derived from the outer integument called **caruncles**. The seed bears a scar, called the **hilum** (plural **hila**), at the point where it was attached to the funiculus.

The embryo consists of an **epicotyl**, which will develop into the shoot; a **radicle**, which will develop into the primary root and usually gives rise to the root system; a **hypocotyl**, which connects the epicotyl and radicle; and usually one or two **cotyledons** (seedling leaves), which may be leaflike, fleshy, or modified as nutrient-absorptive structures.

FRUIT AND SEED DISPERSAL

Most fruit types may be dispersed by a variety of agents. Different parts of the fruit, seed, or associated structures (pedicel, perianth) may be modified for similar dispersal-related functions. For example, wind dispersal may be accomplished by (1) a tuft of hairs on the seeds, as in *Asclepias* (Apocynaceae), which has follicles that open to release hairy seeds; (2) wings on the fruits, as in *Fraxinus* (Oleaceae), which has samaras; (3) hair tufts on the fruits, as in *Anemone* (Ranunculaceae), which has an achene with a persistent style bearing elongated hairs; (4) a winglike perianth, as in *Dipterocarpus* (Dipterocarpaceae), which has nuts associated with elongated, winglike sepals; (5) association of the infructescence (mature inflorescence, with fruits) with an expanded, winglike bract, as in *Tilia* (Malvaceae), in which the fruits are nuts; or (6) a tumbleweed habit, as in *Cycloloma* (Amaranthaceae), in which the entire plant is blown across the landscape, dispersing its small fruits as it rolls.

Bird dispersal may be enhanced by (1) a colorful, fleshy seed coat, as in *Magnolia* (Magnoliaceae), which has an aggregate of follicles that open to reveal the fleshy seeds; (2) fleshy, indehiscent fruits, as in *Solanum* (Solanaceae), which has berries, *Prunus* (Rosaceae), which has drupes, or *Amelanchier* (Rosaceae), which has pomes; or (3) association of the fruit (or fruits) with fleshy accessory structures, as in *Coccoloba* (Polygonaceae), which has achenes surrounded by fleshy perianth, *Fragaria* (Rosaceae), which has achenes borne on an expanded, colorful, fleshy receptacle, or *Hovenia* (Rhamnaceae), which has drupes associated with fleshy pedicels and inflorescence axes.

It is easy to see that similarly functioning fruiting structures may be derived from very different floral parts. Convergence is common in fruits, and similar fruits have evolved independently in many different angiosperm families. (For additional information on fruit dispersal, see van der Pijl 1972 and Weberling 1989.)

Some heavy fruits or seeds simply drop from the plant, land on the ground, and stay there. This is not very common, however, and may be characteristic only of species that have lost their primary dispersal agent, as possibly in osage orange (*Maclura pomifera*, Moraceae).

Dispersal by the plant itself usually occurs through some kind of explosive discharge of seeds, fruits, or portions of fruits by means of swelling of seed mucilage, turgor pressure changes, or hygroscopic tissues. This category also includes passive movement of seed containers by wind, rain, or animals, and creeping dispersal units whereby the fruit or seed moves itself by hygroscopic movements of bristles.

Adaptations for transport by wind include small, dustlike seeds; seeds with a balloonlike, loose testa; inflated utricles, calyces, or bracts; or a pericarp with air spaces. Wind dispersal may be facilitated by a plume formed from a persistent style, long hairy awns, a modified perianth (such as a pappus), placental outgrowths, outgrowths of the funiculus, elongation of the integument, a wing that splits apart, or hair tufts. Wings for wind dispersal may be present on fruits or seeds, or developed from accessory parts (perianth, bracts). In tumbleweed dispersal, a large part of the plant or inflorescence breaks off and is blown around.

Dispersal by water occurs when seeds or fruits are washed away by rainfall or carried in water currents. Such seeds or fruits are often small, dry, and hard, and they may have spines or projections as anchoring structures (water burs), a slimy covering, an unwettable surface layer, or low density and thus the ability to float.

Adaptations for transport on the outside surface of animals include small seeds or fruits with spines, hooks, or sticky hairs that easily detach from the plant and are placed near ground level. This category also includes small and hard fruits or seeds that stick with mud to the feet of waterfowl, as well as trample burs that become caught in the feet of large grazing mammals. Many sticky fruits attach to the feathers of birds.

Fruits and seeds may also be transported either within an animal (after ingestion) or in its mouth. These cases can be divided into subtypes by the kind of animal carrying the fruit or seed:

- *Fish* disperse some fleshy fruits or seeds of plants of riversides or inundated areas.
- Transport by *turtles* or *lizards* characterizes some fleshy fruits that have an odor. Such fruits are sometimes colored and often are borne near the ground or dropped from the plant at maturity. Some have a hard skin; others are hard but contain arillate seeds or seeds with fleshy testae.
- *Birds* may disperse nuts or seeds by carrying them in their bills or by hiding and burying them. Some viscid

seeds stick to birds' bills. Bird-dispersed fruits or seeds often have an attractive edible part. The seeds of some fleshy fruits are protected from ingestion by a bony wall, bitter taste, or toxic compounds, and when mature, these fruits have signaling colors that attract birds (often red contrasting with black, blue, or white). These fruits have no odor and no closed hard rind (or in hard fruits, the seeds are exposed or dangling), and they remain attached to the plant. Some have colorful hard seeds that mimic the colorful fleshy fruits of other bird-dispersed species.

- Transport by *mammals* is often associated with their stockpiling of fruits (especially nuts) or seeds. Mammal-dispersed fruits often have a high oil content, and are frequently fleshy, with hard centers or leathery to hard skins that can be opened to reveal fleshy inner tissues, arillate seeds, or seeds with fleshy testae. The seeds may be toxic, bitter, or thick walled. Odor is very important in attracting mammals, but color is not essential. The fruits often drop from the plant.
- Fruits transported by *bats* share many of the characters listed in the previous entry, for mammals, but usually they are borne in an exposed position (e.g., outside the dense crown of a tree). They have drab colors, a musty, sourish, or rancid odor; they are often large, fleshy, easily digested; and they remain attached to the plant.
- Some seeds contain small, nutritious arils (or elaiosomes) and are dispersed by *ants*.

Modes of fruit and/or seed dispersal are noted for most of the families treated in Chapters 8 and 9.

Anatomy

Characteristics related to the internal structure of plants have been employed for systematic purposes for over 150 years, and they are useful in both practical identification and determination of phylogenetic relationships. **Anatomical characters** are investigated by light microscope study, and many can be seen in the laboratory by use of the simplest of techniques. Characters observable with the transmission electron microscope (TEM) are frequently referred to as **ultrastructural**; those observable with the scanning electron microscope (SEM) are often called **micromorphological**. Some important characters—anatomical, ultrastructural, and micromorphological—are discussed briefly in this section.

Various distinctive cell types may be found in a variety of tissues. Sclereids and fibers are thick-walled and lignified cells. **Fibers** are elongated cells that frequently surround and protect the vascular tissue in the stem and the veins of the leaf. Such fibers vary in their pattern of arrangement and sometimes form girders connecting the vein to the adaxial and/or abaxial epidermis. **Sclereids** are thick-walled cells of various shapes. Secretory canals or cells, and cells containing various kinds of crystals (discussed later in this section), are often diagnostic for particular taxa. In collenchyma, the angles of the cell walls, in particular, are very thickened with cellulose.

SECONDARY XYLEM AND PHLOEM

Wood (often called **secondary xylem**) is produced by a **vascular cambium** (plural **cambia**), a cylinder of actively dividing cells just inside the bark of a woody plant. Wood is a complex mixture of water-conducting cells (**tracheids** and/or **vessel elements**), cells with a support function (**fibers**), and living cells that run from the outside to the inside of the stem (**rays**) (Figure 4.31).

The water-conducting and support cells are both dead at maturity, which makes sense because cytoplasm would interfere with water transport. These cells have thick walls made up of both **cellulose** and **lignin**—the former a long chain of glucose molecules, the latter a complex polymeric substance composed of phenolic subunits. Whereas conifers and cycads have only tracheids for water conduction, gnetophytes and angiosperms usually have both tracheids and vessel elements. Angiosperms tend to have short, broad vessel elements with completely open ends that fit together like sewer pipes and conduct large amounts of water rapidly.

Many anatomists (e.g., Bailey 1944, 1951, 1957) have considered xylem types in terms of a priori defined trends in vessel element evolution; that is, the progression from tracheids to long, narrow vessel elements with slanted, scalariform perforation plates, to short, broad, vessel elements with simple perforation plates (Figure 4.31E–K). Carlquist (1988) has emphasized the linkage between xylem structure and its ecological function.

Other important aspects of wood anatomy include growth rings (bands or layers in the wood produced by seasonal variation in cambial activity) and the presence of various specialized cells containing crystals, resins, mucilage, or latex.

Secondary phloem provides fewer taxonomic characters than secondary xylem. There are two major types of carbohydrate-conducting cells in vascular plants: **sieve cells** and the more specialized **sieve tube elements**. The latter possess a distinctive sieve plate and companion cells. Stratified phloem consists of alternating bands of sieve tubes and companion cells and fibers; plants with such phloem often have fibrous bark.

Much emphasis has been placed on the structure of the sieve-element plastids of angiosperms (Behnke 1972, 1975, 1977, 1981, 1991, 1994, 2000) because differences in their structure have been correlated with major clades, such as monocots, Caryophyllales, and Fabaceae. Behnke recognized two major categories of plastids in sieve tube elements: the S-type, which accumulates starch, and the P-type, which accumulates proteins (or proteins and starch) (Figure 4.32).

NODAL ANATOMY

Nodal anatomy describes the various patterns in vascular connections between stem and leaf. The anatomy of the node is quite variable and often of systematic significance. A key feature is the number of **leaf gaps**, or **parenchymatous interruptions**, left in the secondary vascular system of angiosperms by the departure of vascular bundles (**leaf traces**) to the leaves. The configuration of these leaf gaps is used as the basis of nodal types (Figure 4.33).

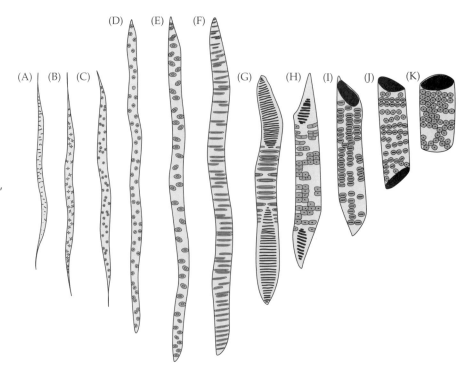

Figure 4.31 Some cell types of the secondary xylem. (A–D) Wood fibers. (E, F) Tracheids. (G–K) Vessel elements. (G) This vessel element type has many-barred scalariform perforation plates (the ladderlike pattern of openings at each end of the cell. (H) This vessel element also has scalariform perforation plates, but the bars are reduced in number. (I) This vessel element is an intermediate between H and J. (J, K) These short, broad, vessel elements have simple perforation plates. (Modified from Radford et al. 1974 and Bailey and Tupper 1918.)

(A)

(B)

(C)

(D)

(E)

Figure 4.32 A few examples of sieve-element plastids. (A) S-type, with many starch grains (common in many angiosperms). (B) *Dracaena* (Convallariaceae); P-type, with numerous cuneate crystalloids. (C) *Laurus* (Lauraceae); plastid with both proteins and starch. (D) *Petiveria* (Petiveriaceae); P-type, with central protein crystalloid surrounded by protein filaments. (E) *Alternanthera* (Amaranthaceae); P-type, with protein filaments only. C, protein crystalloid; F, protein filaments; S, starch. (From Behnke 1975.)

The nodal pattern can be expressed in terms of the number of traces (strands of vascular tissue) and gaps, or **lacunae** (singular **lacuna**) (spaces between the vascular strands). For example, a unilacunar node with a single trace would be described as 1:1; a unilacunar node with two traces would be recorded as 2:1; and a trilacunar node with three traces would be 3:3.

LEAF ANATOMY

Leaves are extremely varied anatomically and provide numerous systematically significant characters (Carlquist 1961; Dickison 1975; Stuessy 1990).

The **epidermis** (outer layer of the leaf) varies in the number of cell layers, the size and shape of individual cells, the thickness of cell walls, and the occurrence of **papillae** (singular **papilla**) (rounded bumps or projections of individual epidermal cells) or various kinds of

hairs (trichomes; see the section titled "Indumentum" on page 61). Some leaves have a **hypodermis**, which is formed from one or more differentiated layers of cells beneath the epidermis. The **cuticle** is a waxy coating over the epidermis, and it varies in thickness and surface texture. Various wax deposits (**epicuticular waxes**) may also be deposited on top of the cuticle (Figure 4.34) (see also Behnke and Barthlott 1983; Barthlott 1990; Barthlott et al. 1998).

The epidermis contains pores, or **stomates** (also *stomata*, singular *stoma*), each surrounded by specialized **guard cells** that open or close the pore by means of changes in their internal water pressure. A variety of stomatal forms occur in vascular plants. Stomates are usually classified by the relationships of their **subsidiary cells** (epidermal cells associated with the stomate and morphologically distinguishable from the surrounding

Figure 4.33   Some types of nodal anatomy. (Modified from Radford et al. 1974.)

Leaf trace
Cauline xylem

Complete vascular cylinder

Branch gap
Branch trace
Leaf gap
Leaf trace

Leaf trace
Leaf gap

Unilacunar, 1-trace node (1:1)

Median leaf trace

Lateral leaf traces

Trilacunar, 3-trace node (3:3)

Unilacunar, 2-trace   Unilacunar, 5-trace   Multilacunar   Pentalacunar

Figure 4.34   *Strelitzia*-type epicuticular waxes—large, massive secretions composed of rodlet-like subunits—in a leaf of *Latania* (Arecaceae). (Photograph by Y. Renea Taylor, used with permission.)

10 µm

epidermal cells) to one another and to the guard cells. Note that the same configuration of subsidiary cells can result from different developmental pathways; studies of these pathways often aid in the interpretation of the phylogenetic significance of stomatal form.

Anomocytic stomata are surrounded by a limited number of cells that are indistinguishable in size and shape from those of the remainder of the epidermis; other stomatal types have recognizable subsidiary cells in various arrangements (Figure 4.35). Stomates may be surrounded by a cuticular ridge or various cuticular projections, or they may be sunken into crypts or grooves.

Characters relating to the internal tissues of leaves are also important. The **mesophyll** may be differentiated into palisade and spongy layers, and the number of cell layers in each may vary. The distribution and shape of mesophyll cells and the presence or absence of intercellular spaces may also be diagnostic.

Internal leaf structure is correlated with the biochemistry of photosynthesis. Leaves associated with C_3 photosynthesis, the most common photosynthetic pathway in green plants, in which compounds with three carbon

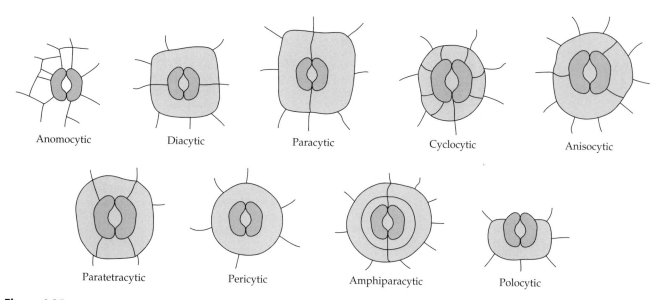

Anomocytic   Diacytic   Paracytic   Cyclocytic   Anisocytic

Paratetracytic   Pericytic   Amphiparacytic   Polocytic

Figure 4.35 Some important stomatal types.

atoms are the immediate products of carbon dioxide fixation, typically have a distinct chlorenchymatous layer of palisade cells in one to several cell layers below the leaf epidermis. In contrast, leaves associated with the C₄ photosynthetic pathway, in which compounds with four car-

bon atoms are the immediate products of carbon dioxide fixation, have prominent chlorenchymatous vascular bundle sheaths (**Kranz anatomy**) (Rathnam et al. 1976).

Xylem and phloem may be arranged in various ways in the petiole and midvein of the leaf (Figure 4.36) (see

Figure 4.36 Common petiole vascular patterns (diagrammatic transverse sections at base of blade). (Modified from Radford et al. 1994.)

Rib bundles

Flat arc with rib traces   Medullated vascular cylinder   3 free traces   Flat arc with dorsal traces

Many free traces in ring   Medullated cylinder with dorsal plate   Arc with invaginated ends   Medullary plate / Medullated cylinder with medullary plate

Free traces in concentric rings   U-shaped free bundles   Medullary bundles / Many free traces in ring with medullary bundles   Open medullated cylinder with lateral rib traces

Howard 1974 for more detail). Such patterns are best studied by making a series of cross-sections through the petiole. Usually several sections are required because the pattern often changes as one moves from the petiole base to its apex and into the leaf midvein.

SECRETORY STRUCTURES

Many plant species contain specialized cells or groups of cells that produce latex, resins, mucilage, or essential oils (Metcalfe 1966; Metcalfe and Chalk 1950). **Latex** is a more or less opaque and milky or colored (usually yellow, orange, or red, but sometimes green or blue) fluid produced by specialized cells called **laticifers**. Laticifers are located in parenchymatous tissues of any plant part, but especially stems and leaves, and they may be solitary or in series, forming tubes, which may be branched or unbranched.

Latex contains a wide variety of secondary compounds in solution and suspension and is important in deterring herbivory. Some species produce clear **resins** (aromatic hydrocarbons that harden when oxidized) or **mucilage** (slimy fluids) in scattered cells, specialized cavities, or canals. **Essential oils** are highly volatile, aromatic organic compounds that are produced by specialized spherical cells scattered in the mesophyll or in cavities created by cellular breakdown or the separation of adjacent cells. Such cells or cavities in leaves cause them to appear **pellucid dotted** when viewed with transmitted light. The presence or absence of latex, resins, mucilage, and essential oils and the form and distribution of laticifers and secretory canals or cavities are often taxonomically significant.

CRYSTALS

Crystals are common in vascular plants, usually located in cells, and variously shaped (Figure 4.37), and they are usually composed of calcium oxalate, calcium carbonate, or silica. **Druses** (spherical groups of crystals), **raphides** (needlelike crystals), and **crystal sand** are the most common. Silica bodies are often taxonomically important in monocots. Calcified bodies called **cystoliths** sometimes occur in specialized cells (called **lithocysts**); cystoliths often have systematic significance. The calcareous material in cystoliths is in the form of small amorphous particles.

ARRANGEMENT OF XYLEM AND PHLOEM IN THE STEM

The stems of most seed plants contain a ring of primary xylem and phloem (as seen in cross-section; Figure 4.38). Primary vascular tissues result from the differentiation of cells produced by the **apical meristem**, the zone of dividing cells at the apex of branches. The ring of bundles produced is called a **eustele**. In woody species, a meristematic layer of cells, the **vascular cambium** (plural **cambia**), develops between the xylem and phloem

Figure 4.37 Crystal types. (A) Druse. (B) Prismatic form. (C) Raphides. (D) Cystolith.

and adds to the thickness of the stem by producing secondary xylem toward its inner side and secondary phloem on the side toward the periphery of the stem. This is the normal pattern of secondary growth in angiosperms and seed plants in general.

Monocots usually lack a vascular cambium and secondary growth; their stems have scattered bundles, each containing xylem and phloem. Some angiosperms have various so-called anomalous patterns of secondary growth, some of which are listed next (and illustrated in Figure 4.38). Anomalous growth patterns are frequently encountered in succulents, where they allow for rapid increases in girth, and in lianas, where they resist damage due to twisting and bending. The following terms describe different anatomical patterns, most relating to secondary growth:

axes more or less flattened or furrowed   the stem is flattened or furrowed due to unequally active cambium.

cortical bundles   leaf trace bundles that run longitudinally in the cortex of the stem before joining the vascular system of the stem.

fissured xylem   xylem broken up by the development of phloem or parenchyma tissue.

included phloem   strands of phloem embedded in the secondary xylem.

internal phloem   primary phloem in the form of strands or a continuous ring (as seen in a cross-section of the stem) at the inner boundary of the xylem; the xylem is bounded by phloem on both its inner and outer surfaces. In species lacking secondary growth, development of internal phloem results in the presence of **bicollateral bundles**; that is, the vascular bundles of the stem have phloem on the inner as well as the normal outer side of the xylem.

medullary bundles   vascular bundles occurring in the pith of a stem.

xylem and phloem concentrically alternating   layers of phloem and xylem alternate in the stem because of the action of a series of vascular cambia.

Figure 4.38 Various arrangements of xylem and phloem in stems.

FLORAL ANATOMY AND DEVELOPMENT

The pattern of vascular traces in flowers is often useful for understanding vestigial structures and homologies of parts in highly modified flowers. Extreme modification of parts is a particular problem in flowers that are greatly reduced and often densely clustered together in association with the evolution of wind pollination, such as the Betulaceae (see Chapter 9). The pattern of vascular traces in the gynoecium usually can be used to indicate the number of carpels, especially in gynoecia that have completely fused carpels with free-central, basal, or apical placentation.

The positioning of floral **primordia** (singular **primordium**) (floral parts or organs in their earliest condition) and their sequence of initiation are also of taxonomic significance (Evans and Dickinson 1996). Developmental studies are important for understanding homologies of various floral parts. For example, some flowers with many stamens have stamen primordia positioned in a spiral, while in other flowers they are clustered into five or ten groups. In addition, stamen primordia may be initiated **centripetally** (from periphery to center) or **centrifugally** (from center to periphery). Such

variation suggests that numerous stamens have evolved many times among angiosperms.

Studies of corolla development have also been especially informative. In some sympetalous flowers the corolla develops from a ring primordium, which then develops lobes, while in others the corolla lobes are initiated first. Developmental studies have indicated that some flowers, such as Apiaceae and many Ericaceae, that appear to have separate petals are sympetalous early in development; they are therefore considered to have evolved from sympetalous ancestors.

Embryology

Spores, gametophytes, and gametangia show marked evolutionary trends in tracheophytes (Table 4.2). Sporophytes produce spores in **sporangia** (singular **sporangium**). Most lycophytes and ferns (see Chapter 8) have only one kind of spore; these plants are **homosporous**. Their gametophytes are completely independent of the sporophyte (see Figure 8.1), bear both kinds of gametes (egg and sperm), and are either large (about 1 cm in diameter), green, and photosynthetic or curious

TABLE 4.2   *General features of spores, gametophytes, and gametangia in major groups of tracheophytes.*

Group[a]	Heterospory	Gametophyte dependence on sporophyte	Gametophyte size (in cells)	Gametangia (antheridia and archegonia)
Lycopodiaceae, Equisetaceae, Psilotaceae, and most leptosporangiate ferns	No	None	Millions (macroscopic)	Present
Selaginellaceae, Isoetaceae, and aquatic leptosporangiate ferns (Marsileaceae and Salviniaceae)	Yes	Almost complete	Thousands (female); sometimes under 100 (male)	Present
Conifers, cycads, ginkgos, gnetopsids	Yes	Complete	1000s (female); a few (male)	Archegonia only
Angiosperms	Yes	Complete	About 7 (female); 3 or fewer (male)	Absent

[a] See Chapter 8 for more information about groups other than angiosperms and Chapter 9 for angiosperms.

subterranean structures that are saprophytic, acquiring nutrition from dead organisms with the aid of a fungus. Gametes are housed and protected in specialized structures within gametophytes called **gametangia** (singular **gametangium**): eggs in **archegonia** (singular **archegonium**) and sperm in **antheridia** (singular **antheridium**).

Heterosporous plants, which include a few genera of lycophytes and ferns and all seed plants, have two kinds of sporangia. **Megasporangia** contain **megaspores** that develop into **megagametophytes** (female gametophytes, or embryo sacs), and **microsporangia** contain **microspores** that develop into **microgametophytes** (male gametophytes). Heterospory is correlated with three gametophytic adaptations. First, gametophytes are small and develop inside the spore, unlike the relatively large, independent gametophytes of most tracheophytes that release their spores. Second, the two kinds of gametophytes are specialized: The megagametophyte is larger and is invested with a supply of nutrients, and the microgametophyte is small and dispersible. Third, the gametophytes are nutritionally dependent on the sporophyte. Reduction in gametophyte size continues in seed plants and is associated with the loss of recognizable antheridia in conifers and related plants and of antheridia and archegonia in angiosperms.

Embryological features of lycophytes, ferns and allies, conifers, cycads, ginkgos, and gnetophytes are discussed further in Chapter 8. In this section we focus on angiosperm ovules, megagametophytes, embryos, and endosperm, as well as on agamospermy.

OVULES AND MEGAGAMETOPHYTES

The **ovule** is a megasporangium surrounded by one or two protective layers (the integuments) and attached to the ovary wall by a stalk (the funiculus). The integuments nearly enclose the ovule, leaving only a small opening, the **micropyle**, through which pollen tubes usually enter the megasporangium. Ovules have been classified by their curvature. In an **orthotropous** ovule the axis of the ovule and funicle are in a straight line; in an **anatropous** ovule the ovule is inverted almost 180°. In **campylotropous** ovules, the axis of the ovule is curved, and the ovule is held at about 90° (Figure 4.39). The ovule develops into a seed, with the integuments becoming the seed coat.

The megasporangium is the site of the meiosis that generates the megaspore. In at least 70% of angiosperms, meiosis yields four haploid megaspores, three of which degenerate. The fourth megaspore undergoes three mitotic divisions to produce a megagametophyte with eight nuclei in seven cells (Figure 4.40). Typically the egg and two other cells (the *synergids*) cluster near the micropyle, and a cell with two nuclei (the *polar nuclei*) ends up near the center. Three antipodal cells (whose functions are not clearly known) lie near the end of the megagametophyte away from the micropyle.

The pollen tube enters the megagametophyte through the micropyle and releases two sperm. One sperm fertilizes the egg to form the diploid **zygote**, the first cell of the next sporophytic generation. The other sperm fuses with one (in at least some basal angiosperms), or usually both, nuclei of the central cell to form the triploid primary endosperm nucleus. This double-fertilization formation of endosperm is unique to angiosperms. The function of the other cells of the megagametophyte is unclear, although the synergids facilitate the fusion of sperm and

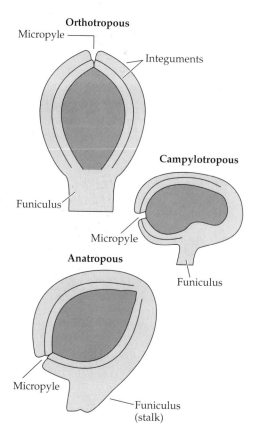

Orthotropous
Micropyle
Integuments

Campylotropous
Funiculus
Micropyle
Funiculus

Anatropous
Micropyle
Funiculus (stalk)

Figure 4.39 Three common types of ovule axes.

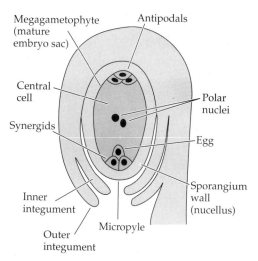

Megagametophyte (mature embryo sac)
Antipodals
Central cell
Polar nuclei
Synergids
Egg
Inner integument
Sporangium wall (nucellus)
Outer integument
Micropyle

Figure 4.40 A mature ovule.

egg in at least some species. Other, less common types of megagametophytes are restricted to certain genera or families.

EMBRYO AND ENDOSPERM

Embryo and endosperm are the two major components of angiosperm seeds. The development of embryos varies considerably within angiosperms. Mature embryos consist of an axis, one end of which is the root (radicle) and the other end the shoot (epicotyl) (Figure 4.41).

When there are two cotyledons, the shoot apex lies between them, as in Figure 4.41. Flowering plants with such embryos are called **dicotyledons** and have long been considered a taxon, but they are actually a paraphyletic group. In **monocotyledons** the shoot apex is lateral, lying next to the solitary cotyledon (Figure 4.42). When the seed germinates and the embryo begins growing into a seedling, the cotyledon or cotyledons may enlarge and become green and photosynthetic. Alternatively, they may remain below ground or not develop beyond their embryonic state.

Endosperm is the specialized tissue of angiosperms that supplies nutrients to the developing embryo and in many cases to the seedling as well. The primary endosperm nucleus may divide many times without cell walls being formed. In most species with this **nuclear type** of endosperm development, cell walls do eventual-

ly form. The **cellular type** of endosperm development involves cell wall formation from the start. In the **helobial type** of endosperm development, which characterizes most monocots, the primary endosperm nucleus produces two cells; one divides little while cells derived from the other make up most of the endosperm.

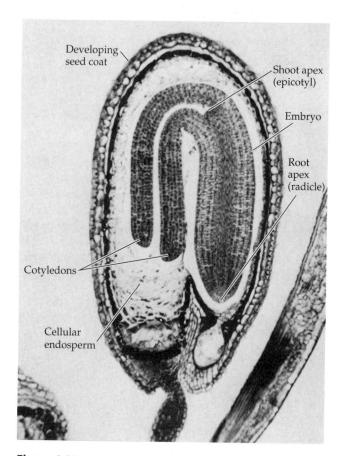

Developing seed coat
Shoot apex (epicotyl)
Embryo
Root apex (radicle)
Cotyledons
Cellular endosperm

Figure 4.41 Longitudinal section of the seed of *Capsella* (Brassicaceae), showing embryo and endosperm. (From Gifford and Foster 1988. Copyright W. H. Freeman; used with permission.)

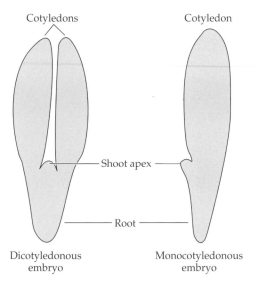

Figure 4.42 A comparison of "dicot" and monocot embryos. (Modified from Gifford and Foster 1988.)

Endosperm may be completely absorbed by the cotyledons of the embryo before it matures, as in peas, beans, and walnuts (something you can easily confirm for yourself). The cotyledons of these **exalbuminous** seeds take over the role of the endosperm in nourishing the seedling. Examples of mature seeds with conspicuous endosperm are cereal grains—such as rice, wheat, corn, and oats—and palms. Corn endosperm is the source of popcorn, and endosperm of coconut (*Cocos nucifera*) is made up of the familiar coconut "milk" (nuclear endosperm before cell walls have formed) and "meat" (endosperm after cell walls have formed).

AGAMOSPERMY

Some plant species produce embryos without producing haploid gametes and without fertilization. This phenomenon is referred to as **agamospermy** (*a* = without; *gamo* = gametes; *sperm* = seed).

There are two major forms of agamospermy. In the first, a diploid cell functions as a megaspore to produce a gametophyte with the somatic chromosome number. Such megagametophytes are often similar in appearance to sexually derived gametophytes. The egg develops into an embryo without fertilization, a process called **parthenogenesis**. The second form of agamospermy, **adventitious embryony**, is best known in citrus plants (*Citrus*, Rutaceae) and is otherwise rare. Here the embryo develops directly from a somatic cell within the ovule without the formation of a gametophyte.

The best method of detecting agamospermy is to study megaspore development to determine whether the megaspores are mitotically or meiotically derived. The taxonomic distribution and systematic significance of agamospermy are discussed in Chapter 6.

Chromosomes

Chromosome number by itself may be a useful systematic character. Similar chromosome numbers may indicate close relationship; different chromosome numbers often create some reproductive isolation through reduced fertility of hybrids. Chromosome size, the position of the centromere, special banding patterns, and other features may also be systematically informative.

CHROMOSOME NUMBER

The lowest chromosome number in the somatic cells of a plant is its **diploid** number, often designated $2n$. Rice, for example, has a diploid number of 24. The diploid genome consists of two full chromosome sets or complements, one from the ovulate (maternal) parent and one from the pollen (paternal) parent. One full set, the **haploid** chromosome number, is borne by both spores and gametes—eggs and sperm. The haploid number, n, of rice is 12. The hypothesized ancestral haploid number (often the lowest number) in a group of plants, such as a genus or family, is referred to as the **base number** and designated x. In the birches (*Betula*), for example, $x = 14$, and the hypothesized ancestral diploid number in the genus is therefore $2n = 28$ (Table 4.3).

The chromosome numbers of angiosperms range from $2n = 4$ (*Haplopappus gracilis*, Asteraceae; and some other species) to $2n \cong 250$ [*Kalanchoe* (Crassulaceae)]. Ophioglossaceae and Polypodiaceae (see Chapter 8) show unusually high chromosome numbers, the highest being $2n = 1440$ in *Ophioglossum reticulatum*. Knowledge of chromosome numbers is limited in many groups to a minority of species and often comes from a small sample of individuals within species.

The addition or loss of one or two whole chromosomes is referred to as **aneuploidy**. Examples of aneuploidy in *Clarkia* and *Lantana* are discussed later in this section. The presence of three or more whole sets of chromosomes in somatic cells is called **polyploidy**. A familiar polyploid is bread wheat (*Triticum aestivum*), which has six full sets of chromosomes. In *Triticum*, $x = 7$, and the chromosome number of bread wheat is therefore $2n = 6x = 42$. Polyploids are differentiated by the number of chromosome complements they contain. Triploids contain three chromosome complements, tetraploids four (hence *tetra-*), pentaploids five, and hexaploids six.

There are two major forms of polyploidy. **Autopolyploidy** results from the union of three or more chromosome complements from the same (hence *auto-*) species; **allopolyploidy** results from the union of two or more different (hence *allo-*) genomes. Because they have diverged, chromosomes of different genomes often do not pair with each other. In diploids and allopolyploids there are two copies of each chromosome, and in meiosis the two homologous chromosomes pair to form a **bivalent**. Autopolyploids contain three or more homologues

TABLE 4.3 *Some chromosome number variation in tracheophytes.*

Taxon	Family	2n
Asplenium trichomanes	Polypodiaceae	
ssp. *trichomanes*		72
ssp. *quadrivalens*		144
Lycopodium and related genera	Lycopodiaceae	34–36, 44–48, 68, 136, 260, 272
Betula	Betulaceae	28, 42, 56, 70, 84
Lantana	Verbenaceae	18, 22, 24, 27, 33, 44, 48, 72
Vicia	Fabaceae	10, 12, 14, 24, 28
Maloideae	Rosaceae	34, 51, 68
All genera	Pinaceae	24

of each chromosome, and pairing of more than two chromosomes to form **multivalents** is thus a possibility. Multivalents lead to gametes with unbalanced chromosome numbers and sterility problems.

Chromosomal pairing between genomes from different sources ranges from none (allopolyploids) to complete (autopolyploids), with a full range of intermediate levels corresponding to intermediate levels of genetic divergence. Hence it is best to consider autopolyploidy and allopolyploidy as extremes of a continuum. Some examples of polyploid plants are given in Table 4.4. Autopolyploidy is apparently not as common as allopolyploidy.

The addition of whole chromosome complements can occur in either somatic tissue or gametes. For example, if the nucleus of a cell that gives rise to a flowering branch fails to divide mitotically, then the chromosome number is automatically doubled. More commonly, production of chromosomally unreduced gametes leads to polyploidy. If, for example, an unreduced (2n) egg is fertilized by a reduced (1n) sperm, the zygote will be triploid (3n). If the plant that develops from this triploid zygote produces a triploid egg that is fertilized by a haploid

sperm, the offspring will be tetraploid. Evolutionary and systematic aspects of polyploidy are discussed in Chapter 6.

Chromosome number is generally constant within a species, although exceptions to this generality are fairly frequent. Chromosome number may be constant within large groups. In Andropogoneae, the large grass tribe that includes *Zea*, *Sorghum*, and many important range grasses, $x = 10$ consistently; in the great majority of the approximately 1000 species of the subfamily Maloideae, $x = 17$; and almost all members of Pinaceae are diploids ($2n = 24$).

In some species, chromosome number varies without correlated morphological variation. Autopolyploids, for example, may not differ morphologically from their diploid progenitors and are therefore often placed in the same species. Diploid ($2n = 14$) *Tolmiea menziesii* (Saxifragaceae), which grows in northern California and southern Oregon, and tetraploid ($2n = 28$) *T. menziesii*, which grows from central Oregon to southern Alaska, are morphologically indistinguishable. In spring beauty (*Claytonia virginica*, Portulacaceae) of eastern North America, aneuploidy is extensive: There are 50 diploid numbers reported for this species, ranging from 12 to about 191.

Differences in chromosome number, when associated with morphological differences, may be recognized taxonomically, as in subspecies of *Asplenium trichomanes* (see Table 4.3). Species within many genera differ in ploidy level. The white-barked birches (*Betula*, Betulaceae) of North America, for example, include diploids [$2n = 28$: gray birch (*B. populifolia*) and mountain paper birch (*B. cordifolia*)], tetraploids [$2n = 56$: *B. cordifolia* and paper birch (*B. papyrifera*)], pentaploids ($2n = 70$: *B. papyrifera*), and hexaploids ($2n = 84$: *B. papyrifera*).

Differences in chromosome number between species often lead to reduced fertility in hybrids and the creation of a species boundary. *Clarkia biloba* (Onagraceae) is a widespread and variable Californian endemic with $n = 8$. *Clarkia lingulata* ($n = 9$) differs from *C. biloba* in that its additional chromosome is derived from chromosomes of *C. biloba*. The two species differ only in petal shape, but hybrids between them show very low fertility. Such aneuploid changes, however, do not always lead to speciation, as *Claytonia virginica*

TABLE 4.4 *Examples of plant polyploids and their 2n chromosome numbers.*

Species	Type of polyploid	2n
Taraxacum officinale, common dandelion	Allotriploid	21
Nicotiana tabacum, tobacco	Allotetraploid	48
Gossypium barbadense, cotton	Allotetraploid	52
Vaccinium corymbosum, highbush blueberry[a]	Allotetraploid	48
Betula papyrifera, white birch[a]	Allopentaploid	70
Triticum aestivum, bread wheat	Allohexaploid	42
Lythrum salicaria, purple loosestrife[a]	Autotetraploid	60
Phleum pratense, timothy grass[a]	Autohexaploid	42

[a]Other ploidy levels are known for these species.

and other species with more than one chromosome number exemplify.

As an example of the taxonomic utility of chromosome number, consider *Lantana* (see Table 4.3). These tropical shrubs are taxonomically confusing due to hybridization, polyploidy, and poorly resolved generic limits. There are two base numbers (11 and 12) and perhaps a third (9) in the genus. Diploid species, with $2n = 22$ or 24, are the foundations for polyploids, especially tetraploids ($2n = 44$ or 48). Triploids ($2n = 33$ or 36) result from crosses between diploids and tetraploids.

The base number 12 characterizes *Lantana* sect. *Callioreas*, and $x = 11$, found in *Lantana* sect. *Camara*, may have evolved via aneuploidy from $x = 12$. Several apparent synapomorphies support section *Camara* as monophyletic, but section *Callioreas* strongly resembles the related large genus *Lippia*. Additional studies are required to unravel the relationships among these groups.

At the species level, chromosome studies have augmented morphological studies in *Lantana*. Most triploids have some chromosomes that do not pair as bivalents in meiosis. Instead, they occur as **univalents** (solitary chromosomes for which there is no homologue) and multivalents. The presence of univalents corroborates the morphologically based hypothesis that many of the triploids are hybrids.

Cytology has played a key role in resolving relationships among Florida *Lantana* sect. *Camara* species. Morphologically based studies initially identified two species in the state: native *L. depressa* and the introduced tetraploid *L. camara*. Chromosome studies, coupled with other data, identified three diploid varieties of *L. depressa* in Florida (Figure 4.43), each of which hybridizes with *L. camara*. Chromosome numbers were critical data in resolving this systematic problem.

CHROMOSOME STRUCTURE

Chromosome number, size, and structural features make up what is called the **karyotype**, which may be useful in discriminating taxa. Chromosomes differ not only in

Figure 4.43 Relationships among *Lantana* species in Florida. (From Sanders 1987a.)

L. depressa
▲ var. *depressa*
● var. *floridana*
★ var. *sanibelensis*
◆ intermed.?, vars. *floridana* & *sanibelensis*

△ Hybrids with *L. camara*
○ Hybrids with *L. camara*
☆ Hybrids with *L. camara*

Miami

0   km   5

0   50   100
km

Figure 4.44 Chromosomes from *Callisia fragrans* (Commelinaceae), in which 2*n* = 12. Some of the centromeres are indicated with arrows. (From Jones and Jopling 1972.)

overall length, but also in the length of the two chromosomal arms (Figure 4.44). The location of the **centromere**, the point on the chromosome where it is attached to the mechanism that separates chromosomes in cell division, determines whether the arms are more or less equal or unequal in length.

The combination of overall chromosome length and centromere location may allow discrimination of many of the chromosomes in a genome. Further distinctions are provided by specialized techniques for staining chromosomal bands (Nogueira et al. 1995). Genome mapping, an exciting approach that may soon have a major impact on systematic studies, is discussed in Chapter 5.

METHODS OF CHROMOSOME STUDY

Determination of chromosome number and other karyotypic features is a routine component of plant systemat-

ics. Chromosome number may be studied from mitotic or meiotic cell divisions (Figure 4.45). Mitosis is commonly examined in cells from actively growing root tips, but other tissues, such as expanding petals, may be used. Meiosis is studied most often because it contains more information than mitosis about relationships of genomes. **Microsporocytes**, the cells that give rise to pollen, are the cells of choice for meiotic study because they are easier to work with than **megasporocytes** and because they are easily removed from the anthers; there are also just more of them.

Protocols for chromosome study involve staining the cells with a chromosome-specific stain, such as carmine. Softening of the tissue facilitates squashing the cells so that the chromosomes separate from one another and are distinguishable for counting. Successful chromosome study may require considerable patience and skill because the chromosomes may be numerous and small, and because it may be difficult to collect material during the correct stage of meiosis. (Methods are outlined in Darlington and La Cour 1975 and Sessions 1990.)

There are now rapid procedures for estimating nuclear DNA content, which often strongly correlates with ploidy. One may therefore survey a large number of individuals for nuclear DNA content and, on the basis of direct knowledge of chromosome number in a smaller sample of individuals, at least infer ploidy (see, for example, Dickson et al. 1992, Cox et al. 1998, and Obermayer et al. 1999).

Figure 4.45 (A) Mitosis in *Rheo spathacea* (Commelinaceae), in which 2*n* = 12. Each structure consists of two duplicate DNA molecules (sister chromatids). (From Jones and Jopling 1972.) (B) Meiosis in *Andropogon gyrans* (Poaceae), in which *n* = 10 (× 940); each object is a pair (bivalent) of homologous chromosomes. The partial sphere just above the center is the nucleolus, where ribosomes are made. (From Campbell 1983.)

(A)

5μm

(B)

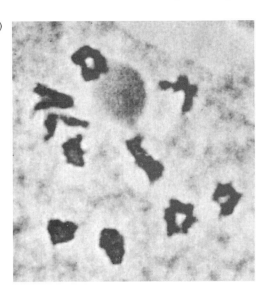

Palynology

Palynology is the study of pollen and spores. Pollen and spores are similar in size, but spores are the beginning of the gametophyte generation, whereas pollen grains are mature microgametophytes. The outer layers of pollen and spores are, however, equivalent. The outer layer often contains a special compound, sporopollenin, that resists degradation by various chemicals, bacteria, and fungi and contributes to the long persistence of pollen in sediments. Pollen, therefore, has been important in paleobotanical studies.

Pollen also interests people as a cause of the allergic reaction called hay fever. We will examine the spores of lycopodiophytes, equisetophytes, psilotophytes, and ferns in Chapter 8; here we will focus on pollen. But first we will briefly consider the development of the anther.

DEVELOPMENT OF THE ANTHER

Most anthers consist of four microsporangia arranged in pairs. The anther wall is made up of several layers, and the innermost, the **tapetum** (plural **tapeta**), plays a key role in the development of the microspores and pollen. When the pollen matures and environmental conditions are appropriate, the anther opens to release the pollen.

The opening, or **dehiscence**, of most angiosperm anthers is accomplished by a longitudinal slit on each side of the anther between the paired microsporangia (**longitudinal**, or **slit**, **dehiscence**). In a few families, such as Ericaceae and Melastomataceae, pollen is shed through a small opening or pore at one end of the anther (**poricidal dehiscence**), while in others, such as Lauraceae, pollen is shed as the anther opens by flaps (**valvate dehiscence**).

POLLEN STRUCTURE, VIABILITY, AND METHODS OF STUDY

Pollen grains may be released from the anthers singly or in clusters of two, four, or many. In many Apocynaceae (e.g., *Asclepias*) and Orchidaceae, pollen is aggregated into clusters called **pollinia** (singular **pollinium**). The smallest known pollen grains are about 10 μm in diameter, and the largest (in Annonaceae) are 350 μm in diameter. Pollen shapes range from spherical to rod-shaped (19 × 520 μm in some Acanthaceae).

The two most important structural features of pollen grains are the apertures and the outer wall. **Apertures** are areas in the pollen wall through which pollen tubes emerge during germination. Pollen grains are often described according to the shape of their aperture(s): **Colpate** pollen grains (also referred to as **sulcate** if positioned at the pole) have long and grooved apertures (Figures 4.46A,C,E and 4.47A), **porate** pollen grains have round and porelike openings (Figures 4.46B and 4.47C,D,E,G,H), and **zonate** pollen grains have ring-shaped or band-shaped apertures. **Colporate** apertures combine the groove of colpate and the pore of porate

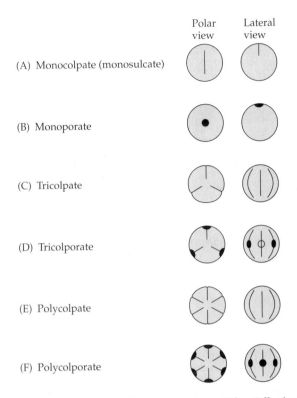

Figure 4.46 Some pollen aperture types. (After Gifford and Foster 1988, and Faegri and Iverson 1950.)

apertures (Figures 4.46D,F and 4.47B). Apertures may be located at the pole or equator of the pollen grain (Figure 4.47A,D), or they may be more or less uniformly distributed over the grain surface (Figure 4.47E,G,H).

The nature and number of apertures is constant in many plant taxa. **Monosulcate** pollen grains (see Figures 4.46A and 4.47A) characterize many putatively basal, woody angiosperms of the Magnoliales. Monocots also are basically a monosulcate group. In contrast, members of one large clade of angiosperms—the eudicots—bear **tricolpate** or tricolpate-derived pollen types (see Figures 4.46C,D and 4.47B,D).

The surface of the **pollen wall**, or **exine**, may be more or less smooth, as in many wind-pollinated species (see Figure 4.47C,D); or variously sculptured with spines, striations, reticulating ridges, knobs, and other features, as in most animal-pollinated species (see Figure 4.47E–H). These surface projections, which attach the pollen grain to animal pollinators, are a rich source of systematic characters. Systematists have also made use of internal exine features as characters at many taxonomic levels (Figure 4.48).

During the development of pollen, the microspore nucleus divides into a small **generative cell** and a much larger **vegetative cell**. The vegetative cell directs the growth of the pollen tube, and the generative cell usually divides into two sperm within the growing tube. In a minority of angiosperms, including both some tricolpates and some monocots, the generative cell divides

(A) *Magnolia grandiflora* (Magnoliaceae) Monosulcate (× 500)

(B) *Scaevola glabra* (Goodeniaceae) Tricolporate (× 1050)

(E) *Chenopodium oahuense* (Amaranthaceae) Polyporate (× 2800)

(F) *Pereskia grandifolia* (Cactaceae) 12 slit-like apertures (× 1200)

(C) *Oryza sativa* (Poaceae) Monoporate (× 1400)

(D) *Cucumis sativus* (Cucurbitaceae) Triporate (× 700)

(G) *Ipomaea wolcottiana* (Convolvulaceae) Polyporate (× 550)

(H) *I. wolcottiana* surface: spines, germinal pores, perforations in tectum (× 1500)

Figure 4.47 Scanning electron micrographs of representative angiosperm pollen grains, showing aperture types and surface features. (From Gifford and Foster 1988; original photos by J. Ward and D. Sunnell.)

into two sperm prior to anther dehiscence, and the pollen is shed in the three-celled stage.

Pollen varies greatly in its ability to function (**viability**) after being shed from the anther. Viability is strongly affected by temperature and humidity, but these effects depend on the taxonomic group. For example, whereas grass pollen is short-lived, sometimes being viable for only minutes or hours, the pollen of many other species remains viable for up to several years if properly stored. One can evaluate viability by testing pollen for its capacity to germinate, for metabolic (enzymatic) activity, or for the presence of cytoplasm.

Features of the exterior pollen wall are obvious when it is viewed by scanning electron microscopy (see Figure

4.47). In this procedure the image is formed from electron beams. The internal structure of pollen, especially the nature of the exine (see Figure 4.48), is commonly examined with transmission electron microscopy.

Secondary Plant Compounds

The biochemical characters of plants have been employed taxonomically for some 100 years, and indirectly—through the use of odors, tastes, and medicinal characteristics—for much longer. Chemical compounds have been used extensively in plant systematics, from analyses of infraspecific variation (see Adams 1977; Harborne and Turner 1984) to determination of phylogenetic rela-

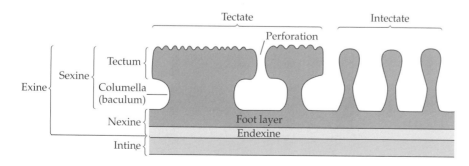

Figure 4.48 Cross-section of a typical angiosperm pollen wall (exine and intine). (After Gifford and Foster 1988.)

Figure 4.49 Representative alkaloids. (A) A secologanin-type alkaloid. (B) A tropane alkaloid. (C, D) Benzylisoquinoline alkaloids.

Corynantheine
(*Corynanthe*, Rubiaceae)

Hyoscyamine
(*Datura*, Solanaceae)

Ochotensimine
(*Corydalis*, Papaveraceae)

Thalicarpine
(*Thalictrum*, Ranunculaceae)

tionships of families and other high-level taxonomic groups (see Dahlgren 1975, 1983; Gershenzon and Mabry 1983).

Generally emphasis was placed on whether a particular compound was present or absent, but because many compounds can be formed by different synthetic pathways, much attention is now placed on the elucidation of these pathways. Two major categories of systematically useful chemical compounds can be recognized: **secondary metabolites**, chemicals that perform nonessential functions in the plant; and the information-containing proteins, DNA, and RNA. Proteins are discussed at the end of this chapter, and the taxonomic use of DNA and RNA is covered in detail in Chapter 5.

Most secondary compounds function in defense against predators and pathogens, as allelopathic agents, or as attractants in pollination or fruit dispersal (Swain 1973; Levin 1976; Cronquist 1977). The major categories of secondary compounds are discussed briefly in this section, and some aspects of their distribution among angiosperms are outlined. (For more information on the categories of secondary compounds and their taxonomic use, see Gibbs 1974; Young and Seigler 1981; Gershenzon and Mabry 1983; Goodwin and Mercer 1983; Harborne

1984; Harborne and Turner 1984; Kubitzki 1984; Giannasi and Crawford 1986; and Stuessy 1990.)

ALKALOIDS

Alkaloids are structurally diverse (Robinson 1981) and are derived from different amino acids or from mevalonic acid by various biosynthetic pathways. They are physiologically active in animals, usually even at very low concentrations, and many are widely used in medicine (e.g., cocaine, morphine, atropine, colchicine, quinine, and strychnine). A few structural classes of alkaloids are shown in Figure 4.49 (see also Li and Willaman 1976).

Betanin

Indicaxanthin

Delphinidin

Apigeninidin

Figure 4.50 Structure of betalains (A, B) and anthocyanins (C, D).

Secologanin-type indole alkaloids (Figure 4.49A) are limited to the Apocynaceae, Gelsemiaceae, Loganiaceae, and Rubiaceae of the Gentianales. **Tropane alkaloids** (Figure 4.49B) occur in a wide array of families, but similar ones are characteristic of Solanaceae and Convolvulaceae (of the Solanales). **Benzylisoquinoline alkaloids** (Figure 4.49C,D) occur in many members of Magnoliales, Laurales, and Ranunculales, as well as the Nelumbonaceae. Other groups, such as **isoprenoid alkaloids** and **pyrrolizidine alkaloids**, show a more scattered distribution among angiosperms and therefore are of less systematic interest.

BETALAINS AND ANTHOCYANINS

Betalains are nitrogenous red and yellow pigments (Figure 4.50A,B) that are restricted to Caryophyllales except for the Caryophyllaceae and Molluginaceae (Clement et al. 1994). In contrast, the pigments of most other plants are **anthocyanins** (Figure 4.50C,D).

Betalains and anthocyanins are mutually exclusive; they have never been found together in the same species. Such pigments, occurring in the perianth parts, are of course important in attracting pollinators, but they also occur in young shoots, stems, leaves, fruits, and even roots, and they probably have additional functions, such as UV absorption and deterrence of herbivory.

GLUCOSINOLATES

The **glucosinolates** (Figure 4.51), also called mustard oil glucosides, are hydrolyzed by enzymes known as myrosinases to yield pungent hot mustard oils (Rodman 1981). Glucosinolates are synapomorphic for the Brassicales. The glucosinolates of the four core families of the Brassicales (Brassicaceae, Resedaceae, Tovariaceae, and Moringaceae) are biosynthetically more complex than those of the other families of the order. Current evidence suggests that these compounds evolved only twice, in the common ancestor of Brassicales and in the common ancestor of the species of *Drypetes* (Putranjivaceae; see Euphorbiaceae) (Rodman et al. 1998).

Taxiphyllin
(*Liriodendron*, Magnoliaceae)

Gynocardin
(*Gynocardia*, Flacourtiaceae)

Figure 4.52 Representative cyanogenic glycosides.

CYANOGENIC GLYCOSIDES

Cyanogenic glycosides (Figure 4.52) are defensive compounds that are hydrolyzed by various enzymes to release hydrogen cyanide (Hegnauer 1977). This process is called **cyanogenesis**. Cyanogenesis is widespread in angiosperms, and five different biosynthetic groups of cyanogenic glycosides are known.

Some biosynthetic types probably have evolved numerous times, but others, such as the **cyclopentenoid cyanogenic glycosides**, are more restricted in distribution (in this case to Achariaceae, Passifloraceae, Turneraceae, Malesherbiaceae, and some Thymeliaceae). Cyanogenic glycosides synthesized from leucine are common in subfamilies Amygdaloideae and Maloideae of the Rosaceae. Similar cyanogenic compounds are found in Fabaceae and Sapindaceae. Cyanogenic glycosides derived from tyrosine are common in several families of Magnoliales and Laurales.

POLYACETYLENES

Polyacetylenes (Figure 4.53) are a large group of nonnitrogenous secondary metabolites formed from the linking of acetate units via fatty acids. These compounds characterize a related group of asterid families, including Asteraceae, Apiaceae, Pittosporaceae, Campanulaceae, Goodeniaceae, and Caprifoliaceae. **Falcarinone polyacetylenes** are restricted to the Apiaceae (including Araliaceae) and Pittosporaceae. These two families are also similar in their essential oils, oleanene and ursene-type saponins, caffeic acid esters, furanocoumarins, and flavonoid profiles.

TERPENOIDS

Terpenoids are a large and structurally diverse group of secondary compounds that are important in numerous biotic interactions (Goodwin 1971). They are formed by

Isopropylglucosinolate
(*Tropaeolum*, Tropaeolaceae)

2-Phenylethylglucosinolate
(*Rorippa*, Brassicaceae)

Figure 4.51 Representative glucosinolates.

Dehydrofalcarinone
(*Artemisia*, Asteraceae)

Figure 4.53 A representative polyacetylene.

Figure 4.54 Representative monoterpenoids.

Figure 4.55 Representative sesquiterpene lactones.

the union of 5-carbon isopentenyl diphosphate units formed in the mevalonic acid pathway. Terpenoids are very widely distributed, and many have primary physiological functions, being a part of membrane-bound steroids, carotenoid pigments, the phytyl side chain of chlorophyll, and the hormones gibberellic acid and abscisic acid.

The distribution of a few terpenoid types, however, is of taxonomic interest. Volatile **monoterpenoids** (Figure 4.54) and **sesquiterpenoids** (10-carbon and 15-carbon compounds, respectively) are the major components of **essential** (or **ethereal**) **oils**, which are characteristic of Magnoliales, Laurales, Austrobaileyales, and Piperales, as well as only distantly related clades, such as Myrtaceae, Rutaceae, Apiales, Lamiaceae, Verbenaceae, and Asteraceae. These compounds occur not only in vegetative tissues in spherical cells or various cavities or canals in parenchymatous tissues, but also in floral odor glands, where they are released and often function as floral attractants.

Sesquiterpene lactones (Figure 4.55), another terpenoid type, are known primarily from the Asteraceae (where they are diverse and taxonomically useful; Seaman 1982), but they also occur in a few other families, such as Apiaceae, Magnoliaceae, and Lauraceae.

Various **diterpenoids** (20-carbon), **triterpenoids** (30-carbon), and **steroids** (triterpenes based on the cyclopentane perhydro-phenanthrene ring system) are widely distributed and also have some systematic significance (Young and Seigler 1981). The triterpenoid **betulin** occurs in the bark of the white birches (*Betula papyrifera* and relatives); it is waterproof, highly flammable, and virtually unknown outside this group. Betulin is taxonomically useful at the species level (O'Connell et al. 1988). **Triterpene saponins** occur in both Apiaceae and Pittosporaceae, supporting the hypothesized close phylogenetic relationship of these two families.

The triterpenoid derivatives **limonoids** and **quassinoids** (Figure 4.56), which are biosynthetically related, are limited to the Rutaceae, Meliaceae, and Simarou-

baceae of the Sapindales; the bitter quassinoids constitute a distinctive synapomorphy of the Simaroubaceae (including *Leitneria*). **Cardenolides** are highly poisonous glycosides of a type of 23-carbon steroid that occur in Ranunculaceae, Euphorbiaceae, Apocynaceae, Liliaceae, and Plantaginaceae.

Iridoids are 9- or 10-carbon monoterpenoid derivatives that usually occur as *O*-linked glycosides (Figure 4.57). Iridoid compounds are found in many families of the asterid clade, and iridoid types have been used to support relationships within this group (Jensen et al. 1975; Jensen 1992). For example, **seco-iridoids**, a chemically derived type of iridoid compound that lacks a carbocyclic ring, occur in Gentianales, Dipsacales, and many families of Cornales and Asterales. In contrast, **carbocyclic iridoids**, which have two rings, one composed entirely of carbon, are characteristic of the Lamiales, except for the Oleaceae, Tetrachondraceae, and Gesneriaceae. The presence of iridoids in the Ericales and Cornales provides evidence that these taxa actually belong to the asterid clade, even though they frequently have been excluded from that group (see Cronquist 1981).

Figure 4.56 Representative triterpenoid derivatives.

Figure 4.57 Representative iridoids.

FLAVONOIDS

Flavonoids (Figure 4.58) are phenolic compounds that usually occur in a ring system derived through cyclization of an intermediate from a cinnamic acid derivative and three malonyl CoA molecules. They probably function in defense against herbivores and in regulation of auxin transport. Flavonoids are employed extensively in plant systematics, probably because they can be fairly easily extracted and identified. They are found throughout the embryophytes (and are also known from the charophyte algae), and they have a diverse array of side groups attached to a common system of rings.

Although useful primarily in assessing relationships among closely related species (or in studies of infraspecific variation), flavonoids are occasionally useful in assessing phylogenetic relationships at higher levels (Bate-Smith 1968; Crawford 1978; Gornall et al. 1979; Harborne and Turner 1984). For example, the presence of certain 5-deoxyflavonoids in *Amphipterygium* (which has often been placed in its own family, the Julianiaceae; see Cronquist 1981) support its placement in Anacardiaceae.

Finally, flavonoid profiles have been shown to be quite useful in studies of interspecific hybridization (see

Alston and Turner 1963; Smith and Levin 1963; Crawford and Giannasi 1982).

Proteins

Proteins are an extremely diverse class of molecules made up of amino acids linked into a chain by peptide bonds. This chain of amino acids—a **polypeptide chain**—is three-dimensionally folded, resulting in a diversity of molecular shapes. Proteins function as enzymes, storage molecules, transport molecules, pigments, and structural materials. Proteins have been used systematically in several different ways; the major methods—amino acid sequencing, systematic serology, and electrophoresis—are discussed briefly here.

AMINO ACID SEQUENCING

Phylogenetic information is contained in the **amino acid sequence** (the **primary structure**) of proteins. These sequences can be used as taxonomic characters, just as the nucleotide sequences of DNA and RNA are used to reconstruct phylogenetic relationships. Proteins such as cytochrome *c*, plastocyanin, ferredoxin, and ribulose-1,5-bisphosphate carboxylase (small subunit) were used early to determine phylogenetic relationships among diverse groups of living organisms and within angiosperms (Fairbrothers et al. 1975; Boulter et al. 1978; Fairbrothers 1983; Martin et al. 1985), albeit with mixed success.

At present, however, most taxonomic interest has shifted away from amino acid sequencing in favor of nucleotide sequencing because of technological advances in the extraction, amplification, and sequencing of DNA and RNA (see Chapter 5).

SYSTEMATIC SEROLOGY

Systematic serology first developed in the early 1900s, following the discovery of serological reactions and the discipline of immunology (Fairbrothers et al. 1975; Fairbrothers 1983). In this method, proteins—antigens—are extracted from a particular species and injected into rabbits to produce antibodies. The antibodies are then cross-reacted to antigens from a series of taxa related to the original species.

The degree of precipitation observed is an indication of how similar the taxa are: A strong precipitation reaction indicates that the antigen proteins of a particular species are very similar to the proteins of the species that originally stimulated antibody production. Whereas a high level of protein similarity is taken as evidence that taxa are closely related, a low level (as assessed by low antibody/antigen reactivity) indicates that the taxa are only distantly related.

Crude protein extracts from seeds or pollen grains are often used to produce antibodies; however, more elaborate methods have been developed, allowing single proteins from different taxa to be compared. Serological tech-

Figure 4.58 Representative flavonoids.

niques have proven to be quite useful in assessing evolutionary relationships. For example, such studies have supported the placement of *Hydrastis* in the Ranunculaceae (and not the Berberidaceae), the placement of *Mahonia* within *Berberis* (Berberidaceae), the existence of a close relationship of *Typha* and *Sparganium* (Typhaceae), and the removal of *Nelumbo* (Nelumbonaceae) from the Nymphaeaceae (Fairbrothers et al. 1975; Fairbrothers 1983).

ELECTROPHORESIS

Electrophoresis is a technique for measuring the rate and direction of movement of organic molecules (in this case proteins) in response to an electric field. The rate and direction of protein movement in a starch or an agar gel will depend on the protein's net surface charge, size, and shape. The proteins can be stained, resulting in a series of bands in the gel. Those proteins that migrate to the same place in a slab of agar gel and yield similar banding patterns when stained are considered to be homologous. The banding patterns resulting after electrophoresis of seed or pollen proteins (or of specific types of enzymes) can be compared, and the presence or absence of particular bands can be used as taxonomic characters.

Electrophoresis is most useful at the level of populations within a species or among closely related species, and the resulting taxonomic characters (banding patterns) are usually analyzed phenetically (Crawford and Julian 1976; Crawford 1979, 1983). In recent years DNA (see Chapter 5) has replaced proteins as a source of data about variation at and below the species level.

Summary

Taxonomic characters are gathered from a wide variety of sources, from all parts of a plant, during all stages of its development. This chapter summarizes the terminology associated with characters drawn from morphology, pollination biology, anatomy, embryology, chromosomes, pollen grains, secondary plant compounds, and proteins.

Morphological characters are based on variation in the form of the major vegetative (roots, stems, and leaves) and reproductive (flowers, fruits, and seeds) parts of plants. The importance of these features in practical identification and phylogeny reconstruction has lead to the development of a complex array of specialized terms, many of which should be considered merely as representing convenient points along a continuum of variation. The most important of these terms are outlined in this chapter. The functions of flowers (in pollination) and fruits (in dispersal) are also described.

Anatomical characters—features of the internal structure of plants—are important in taxonomic studies, and they are observed by use of the light microscope, transmission electron microscope, and/or scanning electron microscope. Anatomical characters include features of the primary and secondary conducting tissues (xylem and phloem), stem anatomy, nodal anatomy, and leaf anatomy, including the presence of secretory structures or crystals.

Developmental studies of plant parts—whether reproductive or vegetative—are also useful. The series of events involved in the development of the ovules and anthers, mega- and microgametophytes, endosperm, and embryonic sporophyte are significant, and frequently of phylogenetic interest. Pollen forms (e.g., shape, size, number and form of the apertures; structure and surface ornamentation of the exine) are variable, providing many useful taxonomic characters.

Chromosome number, size, and structural features make up what is called the karyotype. Although not used in routine identification, variation in the karyotype is frequently of phylogenetic significance. The lowest chromosome number in the somatic cells of a plant is its diploid number, and the diploid genome consists of two full chromosome sets. Chromosome number varies greatly in vascular plants. The addition or loss of one or a few whole chromosomes is referred to as aneuploidy; the presence of three or more whole sets of chromosomes in somatic cells is called polyploidy. Both aneuploidy and polyploidy are common evolutionary phenomena in vascular plants.

Finally, biochemical characters have been used extensively in plant systematics. Two major categories of systematically useful chemical compounds can be recognized: secondary metabolites (chemicals that do not perform essential metabolic functions in the plant) and the information-containing molecules (proteins, RNA, and DNA). Secondary compounds considered in this chapter include various alkaloids, glucosinolates, cyanogenic glycosides, polyacetylenes, terpenoids, betalains, and flavonoids (including anthocyanins). These compounds are structurally very diverse and function in defense against predators and pathogens, as allelopathic agents, or as attractants in pollination or fruit dispersal.

Proteins, a diverse class of molecules made up of polypeptide chains (i.e., chains of amino acids), are discussed briefly. They have been used in electrophoretic studies, especially in assessing variation within species, and in serological studies, assessing relationships through strength of antibody/antigen reactivity. DNA and RNA are discussed in Chapter 5.

Literature Cited and Suggested Readings

Items marked with an asterisk (*) are especially recommended to those readers who are interested in further information on the topics discussed in this chapter.

Morphology

Baumann-Bodenheim, M. G. 1954. Prinzipien eines Fruchtsystems der Angiospermen. *Bull. Soc. Bot. Suisse* 64: 94–112.

Bell, A. D. 1991. *Plant form: An illustrated guide to flowering plant morphology.* Oxford University Press, Oxford, UK.

Bentley, B. L. 1977. Extrafloral nectaries and protection by pugnacious bodygards. *Annu. Rev. Ecol. Syst.* 8: 407–427.

Brouwer, Y. M. and H. T. Clifford. 1990. An annotated list of domatia-bearing species. *Notes Jodrell Lab.* 12: 1–50.

Corner, E. J. H. 1976. *The seeds of dicotyledons,* 2 vols. Cambridge University Press, Cambridge, UK.

Correll, D. S. and H. B. Correll. 1982. *Flora of the Bahama Archipelago.* J. Cramer, Vaduz, Germany. [Contains numerous floral diagrams.]

Cullen, J. 1978. A preliminary survey of ptyxis (vernation) in the angiosperms. *Notes R. Bot. Gard. Edinburgh* 37: 161–214.

Davis, P. H. and V. H. Heywood. 1963. *Principles of angiosperm taxonomy.* Van Nostrand, Princeton, NJ. [Contains a detailed chapter on the use of morphological and anatomical characters in plant systematics.]

*Dilcher, D. 1974. Approaches to the identification of angiosperm leaf remains. *Bot. Rev.* 40: 1–157. [Very detailed summary of terms relating to angiosperm leaves.]

*Endress, P. K. 1994. *Diversity and evolution of tropical flowers.* Cambridge University Press, Cambridge, UK. [Detailed presentation of the structure and function of tropical flowers.]

Endress, P. K., P. Baas and M. Gregory. 2000. Systematic plant morphology and anatomy—50 years of progress. *Taxon* 49: 401–434.

*Hallé, F., R. A. A. Oldeman and P. B. Tomlinson. 1978. *Tropical trees and forests: An architectural analysis.* Springer, Berlin.

*Harris, J. G. and M. W. Harris. 1994. *Plant identification terminology: An illustrated glossary.* Spring Lake Publishing, Spring Lake, UT.

Heslop-Harrison, Y. 1981. Stigma characteristics and angiosperm taxonomy. *Nordic J. Bot.* 1: 401–420.

*Hickey, L. J. 1973. Classification of the architecture of dicotyledonous leaves. *Am. J. Bot.* 60: 17–33. [Detailed summary of terms related to dicot leaves.]

Hickey, L. J. and J. A. Wolfe. 1975. The bases of angiosperm phylogeny: Vegetative morphology. *Ann. Missouri Bot. Gard.* 62: 538–589. [Note especially the careful treatment of tooth types.]

Jackson, B. D. 1928. *A glossary of botanic terms, with their derivation and accent,* 4th ed. Duckworth, London.

Keller, R. 1996. *Identification of tropical woody plants in the absence of flowers and fruits: A field guide.* Birkhauser, Basel, Switzerland.

*Kiger, R. W. and D. M. Porter. 2001. *Categorical glossary for the Flora of North America project.* Hunt Institute for Botanical Documentation, Carnegie Mellon University, Pittsburgh, PA.

Lawrence, G. H. M. 1951. *Taxonomy of vascular plants.* Macmillan, New York. [Appendix II is an illustrated glossary of taxonomic terms.]

Nettancourt, D. de. 1984. Incompatibility. In *Cellular interactions encyclopedia of plant physiology,* H. F. Linskens and J. Heslop-Harrison (eds.), New Series, Vol. 17, 624–639. Springer, Berlin.

O'Dowd, D. J. and R. W. Pemberton. 1998. Leaf domatia and foliar mite abundance in broadleaf deciduous forest of North Asia. *Am. J. Bot.* 85: 70–78.

Payne, W. W. 1978. A glossary of plant hair terminology. *Brittonia* 30: 239–255.

Pemberton, R. W. and C. E. Turner. 1989. Occurrence of predatory and fungivorous mites in leaf domatia. *Am. J. Bot.* 76: 105–112.

Philipson, W. R. 1977. Ovular morphology and the classification of dicotyledons. *Plant Syst. Evol.,* Suppl. 1: 123–140.

Porter, C. L. 1967. *Taxonomy of flowering plants,* 2nd ed. Freeman, San Francisco. [Contains numerous floral diagrams.]

Proctor, M. and P. Yeo. 1972. *The pollination of flowers.* Taplinger, New York.

Rendle, A. B. 1925. *The classification of plants,* 2 vols. Cambridge University Press, Cambridge, UK.

Richards, A. J. 1986. *Plant breeding systems.* Allen and Unwin, London.

Singer, S. R. 1997. Plant life cycles and angiosperm development. In *Embryology: Constructing the organism,* S. F. Gilbert and A. M. Raunio (eds.), 493–514. Sinauer, Sunderland, MA.

Stearn, W. T. 1992. *Botanical Latin,* 4th ed. David and Charles, London.

Stuessy, T. S. 1990. *Plant taxonomy: The systematic evaluation of comparative data.* Columbia University Press, New York. [Contains a chapter on the use of morphological characters in plant systematics.]

van Balgooy, M. M. J. 1997. *Malesian seed plants,* Vol. 1: *Spot-characters. An aid for identification of families and genera.* Rijksherbarium/Hortus Botanicus, Leiden, Netherlands.

Walker-Larsen, J. and L. D. Harder. 2000. The evolution of staminodes in angiosperms: Patterns of stamen reduction, loss, and functional re-invention. *Am. J. Bot.* 87: 1367–1384.

Walters, D. R. and D. J. Keil. 1995. *Vascular plant taxonomy,* 4th ed. Kendall/Hunt, Dubuque, IA.

*Weberling, F. 1989. *Morphology of flowers and inflorescences* (R. J. Pankhurst, trans.). Cambridge University Press, Cambridge, UK. [Includes a detailed presentation of variation in floral morphology and inflorescences, and associated terms.]

Willson, M. F. 1983. *Plant reproductive ecology.* Wiley, New York.

*Zomlefer, W. B. 1994. *Guide to flowering plant families.* University of North Carolina Press, Chapel Hill. [Contains a beautifully illustrated glossary of commonly used morphological terms.]

Pollination Biology

Arditti, J. 1992. *Fundamentals of orchid biology.* Wiley, New York.

Carthew, S. M. and R. L. Goldingay. 1997. Non-flying mammals as pollinators. *Trends Ecol. Evol.* 12: 104–108.

Cox, P. A. 1988. Hydrophyllous pollination. *Annu. Rev. Ecol. Syst.* 19: 261–280.

Darwin, C. 1862. *The various contrivances by which British and foreign orchids are fertilized by insects.* Murray, London.

Darwin, C. 1876. *The effect of cross- and self-fertilization in the vegetable kingdom.* Murray, London.

Darwin, C. 1877. *The different forms of flowers on plants of the same species.* Murray, London.

*Faegri, K. and L. van der Pijl. 1979. *The principles of pollination ecology,* 3rd ed. Pergamon, Oxford, UK.

Ganders, F. R. 1979. The biology of heterostyly. *New Zealand J. Bot.* 17: 607–635. [Many other papers in this issue deal with reproduction in flowering plants.]

Gibernau, M., D. Barabé and D. Labat. 2000. Flowering and pollination of *Philodendron melinonii* (Araceae) in French Guiana. *Plant Biol.* 2: 331–334.

Johnson, S. D., A. Pauw and J. Midgley. 2001. Rodent pollination in the African lily *Massonia depressa* (Hyacinthaceae). *Am. J. Bot.* 88: 1768–1773.

Lewis, D. 1949. Incompatibility in flowering plants. *Biol. Rev.* 24: 472–496.

Nilsson, L. A. 1992. Orchid pollination biology. *Trends Ecol. Evol.* 7: 255–259.

Ortiz, P. L., M. Arista and S. Talavera. 2000. Pollination and breeding system of *Putoria calabrica* (Rubiaceae), a Mediterranean dwarf shrub. *Plant Biol.* 2: 325–330.

Pellmyr, O., J. N. Thompson, J. M. Brown and R. G. Harrison. 1996. Evolution of pollination and mutualism in the yucca moth lineage. *Am. Nat.* 148: 827–847.

Real, L. (ed.) 1983. *Pollination biology.* Academic Press, Orlando, FL.

Tschapka, M. and O. von Helversen. 1999. Pollinators of syntopic *Marcgravia* species in Costa Rican lowland rain forest: Bats and opossums. *Plant Biol.* 1: 382–388.

Tschapka, M., O. von Helversen and W. Barthlott. 1999. Bat pollination of *Weberocereus tunilla,* an epiphytic rain forest cactus with functional flagelliflory. *Plant Biol.* 1: 554–559.

van der Pijl, L. and C. H. Dodson. 1966. *Orchid flowers: Their pollination and evolution.* University of Miami Press, Coral Gables, FL.

Inflorescences, Fruits, and Seeds

Gray, A. 1877. *Gray's lessons in botany and vegetable physiology.* Ivison, Blackman, Taylor, New York. [Contains an artificial classification of fruit types.]

Judd, W. S. 1985. A revised traditional/descriptive classification of fruits for use in floristics and teaching. *Phytologia* 58: 233–242.

Spjut, R. W. 1994. A systematic treatment of fruit types. *Mem. N.Y. Bot. Gard.* 70: 1–182.

Troll, W. 1964/1969. *Die Infloreszenzen, Typologie und Stellung im Aufbau des Vegetationskörpers*, 2 vols. Gustav Fischer, Jena, Germany.

*van der Pijl, L. 1972. *Principles of dispersal in higher plants*. McGraw-Hill, New York.

Weberling, F. 1965. Typology of inflorescences. *J. Linnean Soc. Bot.* 59: 215–221.

*Weberling, F. 1989. *Morphology of flowers and inflorescences* (R. J. Pankhurst, trans.). Cambridge University Press, Cambridge, UK.

Anatomy

Ayensu, E. S. 1972. *Anatomy of the monocotyledons*, Vol. 6: *Dioscoreales*. Clarendon, Oxford, UK.

Baas, P. 1982. *New perspectives in wood anatomy*. Nijhoff/Junk, The Hague, Netherlands.

Baas, P., E. Wheeler and M. W. Chase. 2000. Dicotyledonous wood anatomy and the APG system of angiosperm classification. *Bot. J. Linnean Soc.* 134: 3–17.

Bailey, I. W. 1933. The cambium and its derivative tissues. VIII. Structure, distribution and diagnostic significance of vestured pits in dicotyledons. *J. Arnold Arbor.* 14: 259–273.

Bailey, I. W. 1944. The development of vessels in angiosperms and its significance in morphological research. *Am. J. Bot.* 31: 421–428.

Bailey, I. W. 1951. The use and abuse of anatomical data in the study of phylogeny and classification. *Phytomorphology* 1: 67–69.

Bailey, I. W. 1957. The potentialities and limitations of wood anatomy in the phylogeny and classification of angiosperms. *J. Arnold Arbor.* 38: 243–254.

Bailey, I. W. and W. W. Tupper. 1918. Size variations in tracheary cells. I. A comparison between the secondary xylems of vascular cryptogams, gymnosperms, and angiosperms. *Proc. Am. Acad. Arts Sci.* 54: 149–204.

Barthlott, W. 1981. Epidermal and seed surface characters of plants: Systematic applicability and some evolutionary aspects. *Nordic J. Bot.* 1: 345–355.

Barthlott, W. 1990. Scanning electron microscopy of the epidermal surface in plants. In *Applications of the scanning EM in taxonomy and functional morphology*, D. Claugher (ed.), 69–94. Clarendon, Oxford, UK.

*Barthlott, W., C. Neinhuis, D. Cutler, F. Ditsch, I. Meusel, I. Theisen and H. Wilhelmi. 1998. Classification and terminology of plant epicuticular waxes. *Bot. J. Linnean Soc.* 126: 237–260.

*Behnke, H.-D. 1972. Sieve-element plastids in relation to angiosperm systematics: An attempt towards a classification by ultrastructural analysis. *Bot. Rev.* 38: 155–197.

*Behnke, H.-D. 1975. The bases of angiosperm phylogeny: Ultrastructure. *Ann. Missouri Bot. Gard.* 62: 647–663.

Behnke, H.-D. 1977. Transmission electron microscopy and systematics of flowering plants. *Plant Syst. Evol.*, Suppl. 1: 155–178.

Behnke, H.-D. 1981. Sieve-element characters. *Nordic J. Bot.* 1: 381–400.

Behnke, H.-D. 1991. Distribution and evolution of forms and types of sieve-element plastids in the dicotyledons. *Aliso* 13: 167–182.

Behnke, H.-D. 1994. Sieve-element plastids: Their significance for the evolution and systematics of the order. In *Caryophyllales*, H.-D. Behnke and T. J. Mabry (eds.), 87–121. Springer, Berlin.

Behnke, H.-D. 2000. Forms and sizes of sieve-element plastids and evolution of the monocotyledons. In *Monocots: Systematics and evolution*, K. L. Wilson and D. A. Morrison (eds.), 163–188. CSIRO, Collingwood, Australia.

Behnke, H.-D. and W. Barthlott. 1983. New evidence from ultrastructural and micromorphological fields in angiosperm classification. *Nordic J. Bot.* 3: 43–66.

*Carlquist, S. 1961. *Comparative plant anatomy: A guide to taxonomic and evolutionary applications of anatomical data in angiosperms*. Holt, Rinehart & Winston, New York.

Carlquist, S. 1988. *Comparative wood anatomy*. Springer, Berlin.

Cutler, D. F. 1969. *Anatomy of the monocotyledons*, Vol. 4: *Juncales*. Clarendon, Oxford, UK.

Davis, P. H. and V. H. Heywood. 1973. *Principles of angiosperm taxonomy*. Krieger, New York. [Chapter on morphology, anatomy, palynology, and embryology.]

*Dickison, W. C. 1975. The bases of angiosperm phylogeny: Vegetative anatomy. *Ann. Missouri Bot. Gard.* 62: 590–620.

*Dickison, W. C. 2000. *Integrative plant anatomy*. Harcourt/Academic Press, San Diego, CA. [Includes chapters dealing with the usefulness of anatomical data in morphological and systematic studies.]

Esau, K. 1965. *Plant anatomy*, 2nd ed. Wiley, New York.

*Esau, K. 1977. *Anatomy of seed plants*, 2nd ed. Wiley, New York.

Evans, R. C. and T. A. Dickinson. 1996. North American black-fruited hawthorns. II. Floral development of 10- and 20-stamen morphotypes in *Crataegus* section *Douglasii* (Rosaceae: Maloidese). *Am. J. Bot.* 83: 961–978.

*Eyde, R. H. 1975. The bases of angiosperm phylogeny: Floral anatomy. *Ann. Missouri Bot. Gard.* 62: 521–537.

*Hickey, L. J. and J. A. Wolfe. 1975. The bases of angiosperm phylogeny: Vegetative morphology. *Ann. Missouri Bot. Gard.* 62: 538–589.

Howard, R. A. 1974. The stem-node-leaf continuum of the Dicotyledoneae. *J. Arnold Arbor.* 55: 125–181.

Jansen, S., P. Baas and E. Smets. 2001. Vestured pits: Their occurrence and systematic importance in eudicots. *Taxon* 50: 135–167.

Keating, R. C. 1984 [1985]. Leaf anatomy and its contribution to relationships in Myrtales. *Ann. Missouri Bot. Gard.* 71: 801–823.

Leins, P. 1964. Das zentripetale und zentrifugale Androeceum. *Berichte Deutsch. Bot. Gesellsch.* 77: 22–26.

Leins, P. and C. Erbar. 1997. Floral developmental studies: Some old and new questions. *Int. J. Plant Sci.* 158: 3–12.

Metcalfe, C. R. 1960. *Anatomy of the monocotyledons*, Vol. 1: *Gramineae*. Clarendon, Oxford, UK.

Metcalfe, C. R. 1966. Distribution of latex in the plant kingdom. *Notes Jodrell Lab.* 3: 1–18.

Metcalfe, C. R. 1971. *Anatomy of the monocotyledons*, Vol. 5: *Cyperaceae*. Clarendon, Oxford, UK.

Metcalfe, C. R. and L. Chalk. 1950. *Anatomy of the dicotyledons*, 2 vols. Clarendon, Oxford, UK.

*Metcalfe, C. R. and L. Chalk. 1979. *Anatomy of the dicotyledons*, 2nd ed., Vol. 1: *Systematic anatomy of leaf and stem, with a brief history of the subject*. Clarendon, Oxford, UK.

*Metcalfe, C. R. and L. Chalk. 1983. *Anatomy of the dicotyledons*, 2nd ed., Vol. 2: *Wood structure and conclusion of the general introduction*. Clarendon, Oxford, UK.

Owens, S. A. 2000. Secondary and tertiary pulvini in the unifoliate leaf of *Cercis canadensis* L. (Fabaceae) with comparison to *Bauhinia purpurea* L. *Int. J. Plant Sci.* 161: 583–597.

Payne, W. W. 1979. Stomatal patterns in embryophytes: Their evolution, ontogeny and classification. *Taxon* 28: 117–132.

*Radford, A. E., W. C. Dickison, J. R. Massey and C. R. Bell. 1974. *Vascular plant systematics*. Harper and Row, New York. [Includes listings of numerous morphological and anatomical terms, with illustrations.]

Raghavendra, A. S. and V. S. Rama Das. 1978. The occurrence of C4 photosynthesis: A supplementary list of C4 plants reported during late 1944–mid 1977. *Photosynthetica* 12: 200–208.

Rasmussen, H. 1981. Terminology and classification of stomata and stomatal development—A critical survey. *Bot. J. Linnean Soc.* 83: 199–212.

Rathnam, C. K. M., A. S. Raghavendra and V. S. Rama Das. 1976. Diversity in the arrangements of mesophyll cells among leaves of certain C4 dicotyledons in relation to C4 physiology. *Z. Pflanzenphysiol.* 77: 283–291.

*Stace, C. A. 1965. Cuticular studies as an aid to plant taxonomy. *Bull. Br. Mus. (Nat. Hist.). Bot.* 4: 1–78.

Stace, C. A. 1966. The use of epidermal characters in phylogenetic considerations. *New Phytol.* 65: 304–318.

Stace, C. A. 1989. *Plant taxonomy and biosystematics*, 2nd ed. Edward Arnold, London. [Chapter on structural information, including morphology and anatomy.]

Stuessy, T. F. 1990. *Plant taxonomy*. Columbia University Press, New York. [Includes a chapter on anatomy.]

Tomlinson, P. B. 1961. *Anatomy of the monocotyledons*, Vol. 2: *Palmae*. Clarendon, Oxford, UK.

Tomlinson, P. B. 1969. *Anatomy of the monocotyledons*, Vol. 3: *Commelinales-Zingiberales*. Clarendon, Oxford, UK.

Van Cotthem, W. R. J. 1970. A classification of stomatal types. *Bot. J. Linnean Soc.* 63: 235–246.

Wheeler, E. A., P. Baas and P. E. Gasson. 1989. I.A.W.A. list of microscopic features for hardwood identification. *IAWA Bull.* 10: 219–332.

Embryology

Asker, S. E. and L. Jerling. 1992. *Apomixis in plants*. CRC Press, Boca Raton, FL.

Dahlgren, G. 1991. Steps toward a rational system of the dicotyledons: Embryological characters. *Aliso* 13: 107–165.

Davis, G. L. 1966. *Systematic embryology of the angiosperms*. Wiley, New York.

Floyd, S. K. and W. E. Friedman. 2000. Evolution of endosperm developmental patterns among basal flowering plants. *Int. J. Plant Sci.* 161 (Suppl. 6): 557–581.

Friedman, W. E. and S. K. Floyd. 2001. Perspective: The origin of flowering plants and their reproductive biology—A tail of two phylogenies. *Evolution* 55: 217–231.

Gifford, E. M. and A. S. Foster. 1988. *Morphology and evolution of vascular plants*, 3rd ed. Freeman, New York.

Herr, J. M., Jr. 1984. Embryology and taxonomy. In *Embryology of angiosperms*, B. M. Johri (ed.), 647–696. Springer, Berlin.

*Johri, B. M., K. B. Ambegaokar and P. S. Srivastra. 1992. *Comparative embryology of angiosperms*, 2 vols. Springer, New York.

Chromosomes

Campbell, C. S. 1983. Systematics of the *Andropogon virginicus* complex (Gramineae). *J. Arnold Arbor.* 64: 171–254).

Cox, A. V., G. J. Abdelnour, M. D. Bennett and I. J. Leitch. 1998. Genome size and karyotypic evolution in the slipper orchids (Cypripedioideae: Orchidaceae). *Am. J. Bot.* 85: 681–687.

Darlington, C. D. and L. F. La Cour. 1975. *The handling of chromosomes*, 6th ed. Wiley, New York.

Dickson, E. E., K. Arumuganthan, S. Kresovich and J. J. Doyle. 1992. Nuclear DNA content variation within Rosaceae. *Am. J. Bot.* 79: 1081–1086.

Flora of North America Editorial Committee. 1993. *Flora of North America north of Mexico*, Vol. 2: *Pteridophytes and gymnosperms*. Oxford University Press, New York.

Grant, W. F. (ed.). 1984. *Plant biosystematics*. Academic Press, Toronto, Ontario.

Jones, K. and C. Jopling. 1972. Chromosomes and the classification of the Commelinaceae. *Bot. J. Linnean Soc.* 65: 129–162.

Moore, D. M. 1976. *Plant cytogenetics* (Outline Series in Biology). Chapman and Hall, London.

Nogueira, C. Z., P. M. Ruas, C. F. Ruas and M. S. Ferrucci. 1995. Karyotype study of some species of *Serjania* and *Urvilliea* (Sapindaceae; Tribe Paullinieae). *Am. J. Bot.* 82: 646–654.

Obermayer, R. W., K. Üwiocicki and J. Greilhuber. 1999. Flow cyrometric determination of genome size in some Old World *Lupinus* species (Fabaceae). *Plant Biol.* 1: 403–407.

Qu, L., J. F. Hancock and J. H. Whallon. 1998. Evolution in an autopolyploid group displaying predominantly bivalent pairing at meiosis: Genomic similarity of diploid *Vaccinium darrowi* and autotetraploid *V. corymbosum. Am. J. Bot.* 85: 698–703.

*Raven, P. H. 1975. The bases of angiosperm phylogeny: Cytology. *Ann. Missouri Bot. Gard.* 62: 724–764.

Rieseberg, L. H., H. Choi, R. Chan and C Spore. 1993. Genomic map of a diploid hybrid species. *Heredity* 70: 285–293.

Sanders, R. W. 1987a. Identity of *Lantana depressa* and *L. ovatifolia* (Verbenaceae) of Florida and the Bahamas. *Syst. Bot.* 12: 44–60.

Sanders, R. W. 1987b. Taxonomic significance of chromosome observations in Caribbean species of *Lantana* (Verbenaceae). *Am. J. Bot.* 74: 914–920.

*Sessions, S. K. 1990. Chromosomes: Molecular cytogenetics. In *Molecular systematics*, D. M. Hillis and C. Moritz (eds.), 156–203. Sinauer, Sunderland, MA.

Soltis, D. E. and L. R. Rieseberg. 1986. Autopolyploidy in *Tolmiea menziesii* (Saxifragaceae): Genetic insights from enzyme electrophoresis. *Am. J. Bot.* 73: 310–318.

Soltis, D. E. and P. S. Soltis. 1988. Are lycopods with high chromosome numbers ancient polyploids? *Am. J. Bot.* 75: 238–247.

Stace, C. A. 1984. *Plant taxonomy and biosystematics*. Edward Arnold, London.

Stace, C. A. 2000. Cytology and cytogenetics as a fundamental taxonomic resource for the 20th and 21st centuries. *Taxon* 49: 451–477.

Stebbins, G. L. 1971. *Chromosomal evolution in higher plants*. Addison-Wesley, Reading, MA.

Stuessy, T. F. 1990. *Plant taxonomy: The systematic evaluation of comparative data.* Columbia University Press, New York.

Palynology

Blackmore, S. and I. K. Ferguson (eds.). 1986. *Pollen and spores: Form and function.* Academic Press, London.

*Erdtman, G. 1966. *Pollen morphology and plant taxonomy.* Hafner, New York.

Faegri, K. and J. Iversen. 1950. *Textbook of modern pollen analysis.* Munksgaard, Copenhagen.

Gifford, E. M. and A. S. Foster. 1988. *Morphology and evolution of vascular plants*, 3rd ed. Freeman, New York.

Graham, A., S. A. Graham, J. W. Nowicke, V. Patel and S. Lee. 1990. Palynology and systematics of the Lythraceae. III. Genera *Physocalymma* through *Woodfordia*, addenda, and conclusions. *Am. J. Bot.* 77: 159–177.

Nowicke, J. W. 1994. Pollen morphology and exine ultrastructure. In *Caryophyllales: Evolution and systematics*, H.-D. Behnke and T. J. Mabry (eds.), 167–221. Springer, Berlin.

Nowicke, J. W. and J. J. Skvarla. 1979. Pollen morphology: The potential influence in higher order systematics. *Ann. Missouri Bot. Gard.* 66: 633–700.

Stone, J. L., J. D. Thomson and S. J. Dent-Acosta. 1995. Assessment of pollen viability in hand-pollination experiments: A review. *Am. J. Bot.* 82: 1186–1197.

*Walker, J. W. and J. A. Doyle. 1975. The bases of angiosperm phylogeny: Palynology. *Ann. Missouri Bot. Gard.* 62: 664–723.

Secondary Plant Compounds

Adams, R. P. 1977. Chemosystematics: Analyses of populational differentiation and variability of ancestral and recent populations of *Juniperus ashei. Ann. Missouri Bot. Gard.* 64: 184–209.

Alston, R. E. and B. L. Turner. 1963. Natural hybridization among four species of *Baptisia* (Leguminosae). *Am. J. Bot.* 50: 159–173.

Bate-Smith, E. C. 1968. The phenolic constituents of plants and their taxonomic significance. *J. Linnean Soc. Bot.* 60: 325–383.

Bohlmann, F. 1971. Acetylenic compounds in the Umbelliferae. *Bot. J. Linnean Soc.* 64 (Suppl. 1): 279–291.

Clement, J. S., T. J. Mabry, H. Wyler and A. S. Dreiding. 1994. Chemical review and evolutionary significance of the betalains. In *Caryophyllales*, H.-D. Behnke and T. J. Mabry (eds.), 247–261. Springer, Berlin.

Crawford, D. J. 1978. Flavonoid chemistry and angiosperm evolution. *Bot. Rev.* 44: 431–456.

Crawford, D. J. and D. E. Giannasi. 1982. Plant chemosystematics. *BioScience* 32: 114–118, 123–124.

Cronquist, A. 1977. On the taxonomic significance of secondary metabolites in angiosperms. *Plant Syst. Evol.*, Suppl. 1: 179–189.

Cronquist, A. 1981. *An integrated system of classification of flowering plants.* Columbia University Press, New York.

Dahlgren, R. 1975. A system of classification of the angiosperms to be used to demonstrate the distribution of characters. *Bot. Notiser* 128: 119–147.

Dahlgren, R. 1983. General aspects of angiosperm evolution and macrosystematics. *Nordic J. Bot.* 3: 119–149.

*Gershenzon, J. and T. J. Mabry. 1983. Secondary metabolites and the higher classification of angiosperms. *Nordic J. Bot.* 3: 5–34.

Giannasi, D. E. and D. J. Crawford. 1986. Biochemical systematics II. A reprise. *Evol. Biol.* 20: 25–248.

*Gibbs, R. D. 1974. *Chemotaxonomy of flowering plants*, 4 vols. McGill-Queen's University, Montreal, Quebec. [A detailed summary of secondary compounds occurring in various plant families, and a discussion of their systematic significance.]

Goodwin, T. W. 1971. *Aspects of terpenoid chemistry and biochemistry.* Academic Press, London.

Goodwin, T. W. and E. I. Mercer. 1983. *Introduction to plant biochemistry*, 2nd ed. Pergamon, Oxford, UK.

Gornall, R. J., B. A. Bohm and R. Dahlgren. 1979. The distribution of flavonoids in the angiosperms. *Bot. Notiser* 132: 1–30. [Includes discussions of the distribution of the variouS-types of flavonoids and of their relative evolutionary advancement.]

Harborne, J. B. 1984. Chemical data in practical taxonomy. In *Current concepts in plant taxonomy*, V. H. Heywood and D. M. Moore (eds.), 237–261. Academic Press, London.

Harborne, J. B. and B. L. Turner. 1984. *Plant chemosystematics*. Academic Press, London.

Hegnauer, R. 1962–1996. *Chemotaxonomie der Pflanzen*, 11 vols. Birkhauser, Basel, Switzerland.

Hegnauer, R. 1977. Cyanogenic compounds as systematic markers in Tracheophyta. *Plant Syst. Evol.*, Suppl. 1: 191–209.

Jensen, S. R. 1992. Systematic implications of the distribution of iridoids and other chemical compounds in the Loganiaceae and other families of the Asteridae. *Ann. Missouri Bot. Gard.* 79: 284–302.

Jensen, S. R., B. J. Nielsen and R. Dahlgren. 1975. Iridoid compounds, their occurrence and systematic importance in the angiosperms. *Bot. Notiser* 128: 148–180.

Kite, G. C., R. J. Grayer, P. J. Rudall and M. S. J. Simmonds. 2000. The potential for chemical characters in monocotyledon systematics. In *Monocots: Systematics and evolution*, K. L. Wilson and D. A. Morrison (eds.), 101–113. CSIRO, Collingwood, Australia.

Kubitzki, K. 1984. Phytochemistry in plant systematics and evolution. In *Current concepts in plant taxonomy*, V. H. Heywood and D. M. Moore (eds.), 263–277. Academic Press, London.

Levin, D. A. 1976. The chemical defenses of plants to pathogens and herbivores. *Annu. Rev. Ecol. Syst.* 7: 121–159.

Li, H. L. and J. J. Willaman. 1976. Distribution of alkaloids in angiosperm phylogeny. *Econ. Bot.* 22: 240–251.

O'Connell, M. M., M. D. Bentley, C. S. Campbell and B. J. W. Cole. 1988. Betulin and lupeol in bark from four white-barked birches. *Phytochemistry* 27: 2175–2176.

Robinson, T. 1981. *The biochemistry of alkaloids*, 2nd ed. Springer, New York.

Rodman, J. E. 1981. Divergence, convergence, and parallelism in phytochemical characters: The glucosinolate-myrosinase system. In *Phytochemistry and angiosperm phylogeny*, D. A. Young and D. S. Seigler (eds.), 43–79. Praeger, New York.

Rodman, J. E., P. S. Soltis, D. E. Soltis, K. J. Sytsma and K. G. Karol. 1998. Parallel evolution of glucosinolate biosynthesis inferred from congruent nuclear and plastid gene phylogenies. *Am. J. Bot.* 85: 997–1006.

Seaman, F. C. 1982. Sesquiterpene lactones as taxonomic characters in the Asteraceae. *Bot. Rev.* 48: 121–595.

Seigler, D. S. 1998. *Plant secondary metabolism*. Kluwer, Boston.

Smith, D. M. and D. A. Levin. 1963. A chromatographic study of reticulate evolution in the Appalachian *Asplenium* complex. *Am. J. Bot.* 50: 952–958.

Stace, C. A. 1989. *Plant taxonomy and biosystematics*. Edward Arnold, London. [Includes a chapter on chemical characters.]

Stuessy, T. F. 1990. *Plant taxonomy*. Columbia University Press, New York. [Includes a chapter on chemical characters.]

Swain, T. (ed.). 1973. *Chemistry in evolution and systematics*. Butterworth, London.

Swain, T. 1977. Secondary compounds as protective agents. *Annu. Rev. Plant Physiol.* 28: 479–501.

Young, D. A. and D. S. Seigler (eds.). 1981. *Phytochemistry and angiosperm phylogeny*. Praeger, New York.

Proteins

Boulter, D., J. T. Gleaves, B. G. Haslett and D. Peacock. 1978. The relationships of eight tribes of the Compositae. *Phytochemistry* 17: 1585–1589.

Crawford, D. J. 1979. Allozyme variation in several closely related diploid species of *Chenopodium* of the western United States. *Am. J. Bot.* 66: 237–244.

Crawford, D. J. 1983. Phylogenetic and systemic inferences from electrophoretic studies. In *Isozymes in plant genetics and breeding*, Part A, S. D. Tanksley and T. J. Orton (eds.), 257–287. Elsevier, Amsterdam.

Crawford, D. J. and E. A. Julian. 1976. Seed protein profiles in the narrow-leaved species of *Chenopodium* of the western United States: Taxonomic value and comparison with distribution of flavonoid compounds. *Am. J. Bot.* 63: 302–308.

Fairbrothers, D. E. 1983. Evidence from nucleic acid and protein chemistry, in particular serology. *Nordic J. Bot.* 3: 35–41.

*Fairbrothers, D. E., T. J. Mabry, R. L. Scogin and B. L. Turner. 1975. The bases of angiosperm phylogeny: Chemotaxonomy. *Ann. Missouri Bot. Gard.* 62: 765–800.

Martin, P. G., D. Boulter and D. Penny. 1985. Angiosperm phylogeny using sequences of five macromolecules. *Taxon* 34: 393–400.

5 *Molecular Systematics*

One of the most exciting developments in the past decade has been the application of nucleic acid data to problems in systematics. The term **molecular systematics** is used to mean macromolecular systematics: the use of DNA and RNA to infer relationships among organisms. Although technically isozyme methods and flavonoid data are also molecular, they are usually discussed separately. In this chapter we outline the major types of nucleic acid data currently available, the molecules and genomes that have been most commonly used, and some aspects of data analysis that are unique to molecular data.

This chapter cannot provide a truly comprehensive introduction to molecular systematics, for which several excellent textbooks are available. Serious students of the field should certainly take a course in plant molecular biology. As we emphasize here, good systematics requires careful evaluation of systematic characters; if they are molecular characters, then their molecular and cellular context becomes important.

Molecular data have revolutionized our view of phylogenetic relationships, although not for the reasons initially suggested. Early proponents of molecular systematics claimed that molecular data were more likely than morphological data to reflect the true phylogeny, ostensibly because molecular data reflect gene-level changes, which were thought to be less subject to convergence and parallelism than were morphological traits.

This early assurance now appears to be wrong. Molecular data are in fact subject to most of the same problems that morphological data are. The big difference is that there are simply many more molecular characters available, and their interpretation is generally easier: An adenine is an adenine, but compound leaves, for example, can form in quite different ways in different plants. As a result, molecular data are now widely used for generating phylogenetic hypotheses.

In many cases, molecular data have supported the monophyly of groups that were recognized on morphological grounds (e.g., Poaceae, Fabaceae, Rosaceae). More importantly, molecular data often have allowed systematists to choose among competing hypotheses of relationships (e.g., to decide what group is the sister group of the Asteraceae or the Poaceae). In other cases, molecular data have allowed the placement of taxa whose relationships were known to be problematic. For example, although the Hydrangeaceae were traditionally placed in or near the Saxifragaceae, it was clear that the two were unrelated. Only with molecular data, however, was there a strong alternative hypothesis for the placement of the Hydrangeaceae: in the order Cornales.

It has been fairly rare for molecular data to suggest something completely novel, although there have been a few dramatic cases, such as the monophyly of the glucosinolate clade and the placement therein of the Limnanthaceae, the inclusion of the Vochysiaceae in the Myrtales, and the documentation of introgression between species that were apparently intersterile.

Plant Genomes

The plant cell contains three different genomes: those of the chloroplast, the mitochondrion, and the nucleus (Table 5.1). Systematists have used data from all three. The two organelles are generally inherited uniparentally (usually maternally in angiosperms); the nuclear genome is biparental. The three genomes differ dramatically in size, with the nuclear being by far the largest—measured in megabases. The mitochondrial genome includes several hundred kilobase pairs (kbp) of DNA (200–2500 kbp), which is small relative to the nuclear genome, but quite large relative to the mitochondrial genomes of animals (which tend to be about 16 kbp). The chloroplast genome is the smallest of the three plant genomes, in most plants ranging from 135 to 160 kbp.

Like the eubacteria from which they are derived, mitochondria and chloroplasts have circular genomes. The order of genes in the mitochondrion is variable, and they are separated by large regions of noncoding DNA. The mitochondrial genome rearranges itself frequently, so

TABLE 5.1   *Comparison of the three genomes in a plant cell.*

	Genome size (kbp)	Inheritance
Chloroplast	135–160	Generally maternal (from the seed parent)
Mitochondrion	200–2500	Generally maternal (from the seed parent)
Nucleus	1.1×10^6 to 110×10^9	Biparental

many rearranged forms can occur in the same cell. This means that rearrangements of the genome occur so often within individual plants that they do not characterize or differentiate species or groups of species and are thus not especially useful for inferring relationships.

The chloroplast, in contrast, is stable, both within cells and within species. The most obvious feature of the chloroplast genome is the presence of two regions that encode the same genes, but in opposite directions; these are known as **inverted repeats**. Between them are a small single-copy region and a large single-copy region (Figure 5.1). Rearrangements of the chloroplast genome are rare enough in evolution that they can be used to demarcate major groups. In addition, gains and losses of genes, or their introns, are common enough to be worth looking for, but rare enough to be a stable marker of evolutionary change.

The order of genes in the nuclear genome is also presumed to be stable, at least within species, and it may be stable across groups of species as well. Some information on gene order has been revealed by classic techniques of cytogenetics, but much more detailed information is now being provided by genome mapping (see the next section). In the coming years this could become an important source of systematic information.

DNA sequences change at a rate different from the rate of genomic rearrangement. Chloroplast genes tend to accumulate mutations more rapidly than do mitochondrial genes in plants. It is more difficult to generalize about nuclear genes, which is hardly surprising because there are so many of them.

Generating Molecular Data

Molecular systematics has been and remains technique driven; as new molecular methods become available, they expand the kinds and amounts of systematic data that can be extracted from nucleic acids. A number of good textbooks describing the techniques for generating and analyzing molecular data are available (Hillis et al. 1996; Soltis et al. 1997, 1998; Page and Holmes 1998; Graur and Li 2000).

If useful comparisons are to be made across many taxa, the technique applied has to be fast and easy. This is why molecular systematics was barely possible until

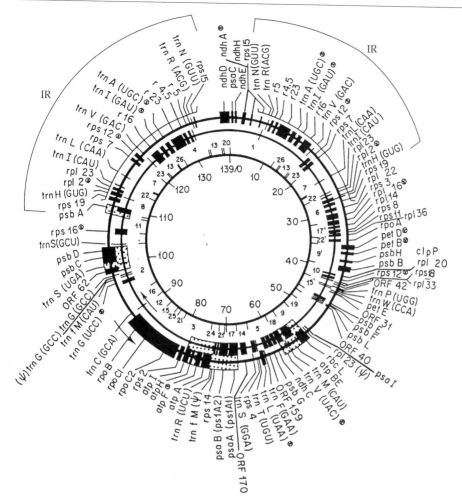

Figure 5.1 Diagram of the chloroplast genome of maize, showing the locations of some major genes and the inverted-repeat (IR) regions. (From Rodermel 1993.)

fragment sizes, then cutting it with a second enzyme, and finally cutting it with both enzymes together. This process creates a sort of puzzle from which the order of the restriction sites can be constructed.

Such studies were initially done by the laborious process of separating organelles from plant tissue, isolating their DNA, cutting it with a particular restriction enzyme, and then measuring the fragment sizes on an ethidium-stained gel. Once a couple of cloned organellar genomes became widely available, it proved easier to use them as probes.

In the probing technique the DNA of interest is cut and then transferred to a nylon membrane. A cloned piece of chloroplast DNA (the probe) is then labeled with radioactive phosphorus and denatured to produce single-stranded DNA. This single-stranded DNA is then allowed to bind to the DNA on the membrane; it will bind only to matching (chloroplast) sequences. The membrane is then placed next to a piece of X-ray film. The bands where the probe has bound appear as dark lines on the film. (This technique is known as Southern blotting, named after E. M. Southern, who invented it.)

Restriction site analysis is used less widely today than it was initially, but it remains common for studying variation in the chloroplast genome and in ribosomal RNA spacers, particularly among congeneric species and sometimes within species as well. Restriction site studies are also sometimes used for assessing variation among PCR fragments, although with recent advances in gene sequencing techniques, this is less of a shortcut than it once was.

Nuclear genome mapping continues to require a large commitment of time and effort. Two plants must be crossed and their F_1 offspring self-pollinated to produce a large number (100 or more) of F_2 plants. Then the genotypes of both parents and offspring are determined through the use of restriction site, RAPD, or AFLP markers (the RAPD and AFLP methods are described later in the chapter); these markers must be polymorphic between the parents. This technique generally requires the use of sophisticated statistical programs to infer linkage relationships.

the invention of recombinant DNA, became easier as sequencing techniques were improved, and took another leap forward with the invention of the polymerase chain reaction (PCR) technique. Current advances in automated sequencing will continue to add speed to the mechanical task of data collection, which is now being done by robots in large sequencing facilities.

RESTRICTION SITE MAPPING

Most molecular systematic studies were initially done with **restriction site analysis**. This technique can be used to generate maps of individual genes or entire genomes. Much of what we now know about chloroplast and mitochondrial genome structure comes from such studies.

In restriction site analysis, DNA is extracted from a plant and is then cut with restriction enzymes—enzymes that cut DNA at a particular sequence. The enzyme known as *Bam*HI, for example, cuts DNA everywhere it finds the sequence GGATCC, and *Eco*RI cuts at GAATTC.* A map is constructed by first cutting the DNA with one enzyme and examining the resulting pattern of

---

*The names of restriction enzymes are acronyms based on the first letter of the genus and first two letters of the species of the bacterium from which the enzyme was isolated. *Bam*HI is from *Bacillus amyloliquefaciens*, *Eco*RI from *Escherichia coli*.

GENE SEQUENCING

DNA sequencing of genes, parts of genes, or noncoding regions is now widely used in systematics. Sequencing determines the precise order of nucleotides—adenine (A), cytosine (C), guanine (G), or thymine (T)—in a stretch of DNA. The central difficulty of sequencing has always been obtaining enough DNA to work with.

The initial approach was to clone genes into bacteria and allow the bacteria to replicate the genes along with their own genomes. Genes were taken from genomic libraries, which researchers create by cutting all the DNA of an organism with a restriction enzyme and then cloning all the resulting fragments into an appropriate plasmid, bacteriophage, or other vector. This method is quite slow, but it is reliable and avoids some of the possible artifacts of more efficient methods. It is also the only method available if only a few sequences of the gene are known.

This laborious approach was later replaced by the **polymerase chain reaction (PCR)** technique, in which DNA is replicated enzymatically, allowing omission of the cloning step (Figure 5.2). PCR requires some knowledge of the sequence to be studied. Small pieces of single-stranded DNA (primers) are produced to match the DNA sequences at either end of the region of interest. These primers are placed in a tube with DNA from the organism, a DNA polymerase, and free nucleotides. The mix is then subjected to repeated heating and cooling.

As it heats, the double-stranded organismal DNA denatures and becomes single-stranded. Then as it cools, the primers bind to their complementary sequences at either end of the target region. The temperature is then raised to the point at which the polymerase becomes active. It binds to the DNA + primer complex and begins synthesizing a complementary strand using the free nucleotides in the solution. Then the temperature is raised to denature the DNA, and the cycle repeats. The DNA in the region between the primers is thus copied, and the amount increases exponentially. The PCR product can be sequenced directly, or it can be cloned and then sequenced.

This rapid method has allowed systematists to study the same region in many species of a particular group. One disadvantage of PCR is that the polymerase itself introduces occasional errors, which could affect an estimate of phylogeny, particularly if the sequences being compared are extremely similar. One way to reduce potential error in sequencing is to sequence both strands of the molecule. Some journals require this step before they will publish results, but it is not universal practice. The decision of how accurately to sequence depends on the relative costs of an error versus the costs of repeatedly sequencing the same region. Systematists must often make a choice between highly accurate sequences from fewer taxa or less accurate sequences from more taxa.

Direct sequencing of the PCR product will not generally reveal minor variants of the sequence if they are present. This is sometimes a problem with highly repetitive genes such as ribosomal genes, for which the many copies often are not identical. Direct sequencing also cannot distinguish between alleles of the same gene. Imagine that two alleles differ from each other at two positions, such that one allele has an A at the first position and a T at the second, whereas the other has a T at the first position and an A at the second. Both positions will appear as

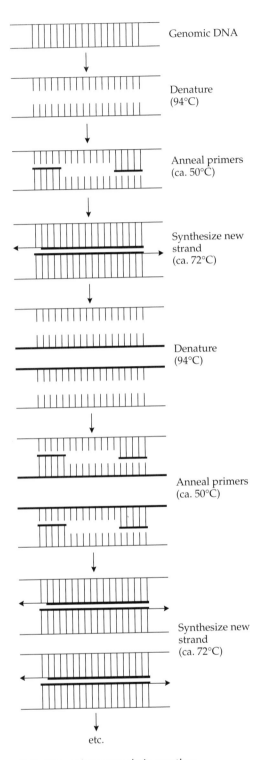

Figure 5.2   The polymerase chain reaction.

A/T polymorphisms on a sequencing gel, and it is impossible to tell which allele has which base at which position. One can avoid this problem by cloning the PCR products (which remains easier than screening genomic libraries, but PCR can create its own sampling bias).

As the major genome sequencing projects progress, the technology for gene sequencing is improving and becoming increasingly automated. It seems likely that systematics laboratories will make increasing use of commercial sequencing facilities in the future so that the attention of systematists can be focused on the more intellectually demanding work of analyzing sequences.

Types of Molecular Data

Virtually all molecular phylogenetics is now done with either genome rearrangements or sequences of DNA as characters. The former are studied through the use of restriction sites as genomic markers, and the latter are usually sequenced completely.

GENOME REARRANGEMENTS

Investigators study chloroplast and mitochondrial genomes by constructing restriction site maps, which reveal the order of genes in the genome. One of the early successes of molecular systematics was the identification of the earliest-diverging members of Compositae by Jansen and Palmer (1987). Using restriction site mapping, they found that almost all members of the family have a unique order of genes in the large single-copy region of the chloroplast genome. This order could be explained by a single inversion of the DNA. All other angiosperms lack the inversion.

The few composites that have the ancestral arrangement of the genome are members of the subtribe Barnedisiinae, a South American group with bilabiate corollas. This finding strongly suggests that the Barnedisiinae (now treated as a subfamily, the Barnedisioideae) is the sister group to the rest of the enormous sunflower family, and that the latter is monophyletic.

The context of this discovery is important. Many previous researchers had speculated on what the "ancestral" composite might have looked like, and they had suggested several extant groups that might represent the earliest lineages. The Barnedisiinae was one of several possibilities and had been supported by cladistic analyses of morphological data. In this case, then, the molecular data did not completely change our view of evolution, but they did resolve the ambiguity of the morphological data.

Other mapping data have supported a number of major groups. The grass family (Poaceae), for example, has three inversions in the chloroplast genome relative to most other angiosperms. One of these is unique to the Poaceae, one is shared with the Joinvilleaceae, and a third is shared with the Joinvilleaceae and Restionaceae. The inversion that is unique to the Poaceae is hardly a surprise; the family is unquestionably monophyletic, a result that can be confirmed by almost any sort of data. The inversions shared with the Joinvilleaceae and Restionaceae again helped clarify the morphological data, which suggested that either one might be the sister group to the grasses.

Loss of genes from the chloroplast is also common. Some groups, for instance, have lost one of the inverted repeats. This has occurred among the angiosperms (e.g., in a group of papilionoid legumes), in all conifers, and in *Euglena* (a flagellated photosynthetic eukaryote unrelated to green plants). In general, however, losses of a smaller piece of DNA, such as an intron or an entire gene, are more common than major rearrangements, and such losses may occur multiple times in evolution. For example, most angiosperms have an intron in the chloroplast gene *rpoC1*, but this intron has been lost in grasses, one subfamily of cacti (Cactoideae), at least two members of Goodeniaceae, some Aizoaceae, and some, but not all, members of the genera *Passiflora* (Passifloraceae) and *Medicago* (Fabaceae) (Downie et al. 1996).

Mapping of the nuclear genome is in its infancy, and is only rarely applied outside the model systems for molecular biology and major crop plants. Techniques for mapping are developing rapidly, however, and comparative studies using the nuclear genome have been done in the grasses, the Solanaceae, and the Brassicaceae. In addition, nuclear mapping studies are just beginning to address questions of speciation.

In a study of *Mimulus*, Bradshaw et al. (1995) found that the shift from bee pollination to bird pollination involved eight genes, which they were able to localize to linkage groups (see Chapter 6). In a similar study of *Helianthus*, Rieseberg and colleagues (1995, 1996) found that two species (*H. annuus* and *H. petiolaris*) differed by at least ten genomic rearrangements (three inversions and at least seven translocations), which affected recombination and possibilities for introgression. The genome of their hybrid derivative, *H. anomalus*, is rearranged relative to that of both parents, so the species is partially reproductively isolated from both.

Rieseberg's group then created new hybrids of *H. annuus* and *H. petiolaris*, and found that the chromosomal rearrangements in the experimental hybrids were similar to those in the naturally occurring hybrid species, *H. anomalus*. They concluded that certain combinations of genes and gene rearrangements were selectively favored in the hybrid.

It is not clear whether this approach will ever become simple enough to apply to the multiple species usually covered by a systematic study. It may, however, be extremely useful for systematists interested in the mechanics of the speciation process.

SEQUENCE DATA

By far the majority of molecular systematic studies have used DNA (or RNA) sequences. Initially these sequences were obtained indirectly by the use of restriction enzymes to sample the underlying sequence, generally as part of the same studies that generated gene or genome mapping data. Because restriction enzymes cut DNA at a particular sequence, any time a particular restriction site is found, the sequence can be inferred at that position. Conversely, if the site is not found, a mutation in one of the four or six bases in the restriction site can be inferred.

This relatively simple inference has been turned into a powerful tool for systematics. It has been used most notably in studies of the chloroplast genome. Sequences (restriction sites) are scored as present or absent, and these scores are then used as characters in phylogenetic analysis. Methods for this sort of study are the same as those used for genome mapping, which are now most commonly done with Southern blots. Yet another method, devised after PCR became widely available, is to amplify a particular piece of DNA and then cut it with restriction enzymes.

The advantage of using any restriction site approach is that it covers potentially a large stretch of DNA and thus is thought to be less sensitive to local vagaries of selection or differences in mutation rate. This is also a disadvantage, of course; generally one does not know exactly where the restriction sites are (e.g., inside a gene or outside it, in the third position of a codon or not). Therefore it is impossible to be absolutely certain that a restriction site gain or loss is in exactly the same place across several taxa.

With standard methods, estimates of the size of restriction fragments are accurate to only 50 or 100 base pairs (bp), which means that two sites very close to each other could easily be confused. And there is another problem: A restriction site is a four- or six-base sequence of DNA, and it can be lost by mutations in any one of its four or six bases. This means that different mutations will all look the same and cannot be distinguished. With a complete DNA sequence, some of these ambiguities are no longer a problem. Nonetheless, sequence data have their own complexities, which are described briefly in the next section.

Analysis of Molecular Data

There is a huge literature on the use of DNA sequences in phylogeny reconstruction. Here we will discuss some of these uses and some examples that have affected our current view of phylogenetic relationships. The major issues to be addressed are mutation rate, alignment, analytic technique, and the relationship between the history of genes and the history of organisms (gene trees versus species trees).

Genes accumulate mutations at different rates, in part because the gene products (RNA or protein) differ in how many changes they can tolerate and still function. Histones, for example, generally cease to work if many of their amino acids are replaced with different ones, whereas the internal transcribed spacer (ITS) of ribosomal RNA can still fold properly even if many of the nucleotides are changed. Thus, genes for histones do not accumulate mutations rapidly, whereas genes for the ITS do, reflecting the different functional constraints on their gene products.

This simple observation has implications for the use of particular genes in phylogenetic reconstruction. If a gene is changing slowly, a lot of data will have to be generated to find mutations from which a phylogeny can be constructed. At a very low mutation rate, the level of variation will approach the expected level of sequencing error (often estimated at about 3 in 10,000 bp for a double-stranded sequence), and inferences will become unreliable. Conversely, if a gene is changing too fast, parallelisms and reversals will accumulate to the point that all phylogenetic information is lost; the history of the sequence will be obliterated. The latter problem is particularly acute in work with noncoding sequences or remotely related taxa.

Many of the methods used to analyze molecular data, and the limitations that apply to them, are similar to those for morphological data. Some methods, however, were developed specifically for use with molecular data (e.g., neighbor joining, maximum likelihood), and some problems, although present in all data sets, become more acute with molecular data.

ALIGNMENT OF SEQUENCES

Once sequences have been generated, they must be aligned. This is a critical step in that it determines which bases will be compared. It is the stage at which the scientist makes the initial assessment of similarity of nucleotide sites. Alignment is by far the most difficult part of using sequence data, and at the moment there is no good analytic solution for it.

Many computer programs will produce alignments, although in practice most systematists rely heavily on alignment "by eye." For many molecules currently used in plant systematics (e.g., *rbcL*, *ndhF*, and other chloroplast genes), alignment is not a serious problem. For other molecules, such as genes encoding RNAs, alignment can be guided by models of secondary structure (the way the molecule folds). In this case the secondary structure is used as a template, and the sequence is mapped on it. This ensures that the proposed alignments maintain the structure of the molecule. (Methods for inferring secondary structure, however, have their own limitations.)

In protein-coding sequences, alignments generally need to maintain the reading frame, so that any postulated insertions or deletions occur in sets of three, corresponding to the gain or loss of an amino acid. Addition or subtraction of a single base pair will change the entire structure of the protein encoded by the sequence. For example, the sequence ATGTCTCCTGAA codes for the four amino acids Met-Ser-Pro-Glu (the first four amino acids of the large subunit of RuBisCO). If a single base is deleted from the second codon—leaving, for example, ATGCTCCTGAA—the protein changes and now consists of Met-Leu-Leu followed by Asn or Lys.

HOMOPLASY AND LONG BRANCHES

A reversal or convergence at a particular nucleotide is undetectable except via phylogenetic analysis; more detailed study of the character won't help. Multiple mutations (sometimes called "multiple hits") at a site are another aspect of this problem. Such multiple changes may well be invisible and can be corrected for only by the assumption of particular models of evolutionary change. Discussion of this problem is beyond the scope of this text; it is described in Swofford et al. 1996, Page and Holmes 1998, and Graur and Li 2000, among others.

With high rates of mutation, or long evolutionary times between speciation events, parallelisms and reversals will increase purely because of random changes. For example, an adenine at a particular position in a gene may have changed to guanine and then back to adenine. Thus the actual amount of evolutionary change will be underrepresented by the observed differences. If a high rate of mutation appears in only a few taxa, it creates what has become known as **long branch attraction** (see Box B in Chapter 2). If one or more of the species studied have accumulated many mutations since diverging from other taxa, they will appear on long branches in phylogenetic trees (where length equals number of mutations).

Because there are only four nucleotides, some mutations will make the sequence of the divergent taxon look more similar to the sequence(s) of other taxa than it really is, purely by chance. Random mutations tend to make those taxa (species) look alike, and they may appear in the analysis to be closely related even if they are not; the long branches "attract" each other. This can be a particular problem if there is more than one long branch, which could occur in principle with morphological data, but it is more likely with molecular data because the potential number of characters is so large and the available character states (A, C, G, T) are so few.

METHODS OF PHYLOGENY RECONSTRUCTION

Some of the methods of phylogeny reconstruction that were described in Chapter 2 are particularly appropriate for use with DNA sequence data. These methods generally rely on statistical models of how DNA has changed over time. As noted in Chapter 2, for data with little homoplasy, virtually all methods will produce the same phylogenetic tree. In some cases, however, the

choice of method will affect the result, and this is particularly true in the case of unequal rates of evolution described earlier.

Because of the problem of multiple mutations at the same site, the observed number of mutations, and thus the apparent divergence between sequences, can easily be less than the actual number of changes. In this case a correction factor can be applied to estimate the actual evolutionary distance. Which correction factor to use depends on estimates of the probability of particular types of mutations. (A full discussion of this sort of analysis can be found in Swofford et al. 1996, Page and Holmes 1998, or Graur and Li 2000.)

GENE TREES VERSUS SPECIES TREES

If a species has a single history, then we expect all parts of the plant to reflect that history. We also then expect any phylogeny based on any gene to reflect the history of the organisms bearing the genes, but we now know that this is not necessarily true. Nuclear genes may or may not track the history of the nucleus, and chloroplasts and mitochondria may or may not have a history different from that of the nucleus. There are three main reasons for these differences:

1. Mutation is a random process; therefore the phylogeny reconstructed for a particular gene may differ from other genes by chance alone.

2. Hybridization or introgression may transfer some DNA into a different lineage. This is particularly true for organelles, which are not linked to particular nuclear genomes.

3. Polymorphisms in an ancestral species can be lost in descendant species. By chance, this can result in a history of the genes that is actually different from the history of the organisms (Figure 5.3).

For many groups of organisms we now have multiple gene trees, and it is possible that none of those gene trees

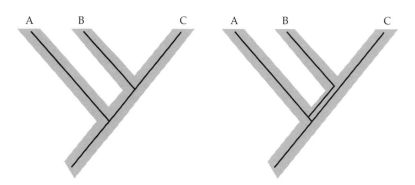

Figure 5.3 A comparison of gene trees and species trees. The gene tree is shown by a dark line, the species tree by gray shading. In the left-hand tree, the genes have the same history as the species that bear them. In the right-hand tree, a polymorphism appears in the lineage leading to species B and C. One of the two gene copies is more closely related to A. Sampling of this gene will lead to incorrect inferences about the species tree. (After Avise 1994.)

BOX 5A *History of genes may not match history of organisms*

The grass family has been the subject of many molecular systematic studies. Consider the subfamily Pooideae of the family Poaceae (Figure 5.4).

This group was identified as monophyletic by cladistic studies of morphology, but several genera or small tribes, including the genus

Brachyelytrum and the Stipeae, were sometimes placed with the pooids and sometimes placed in other subfamilies.

Figure 5.4 Phylogeny of the subfamily Pooideae. Semi-strict consensus tree, showing clades supported by particular data sets and not strongly contradicted by any other data set.

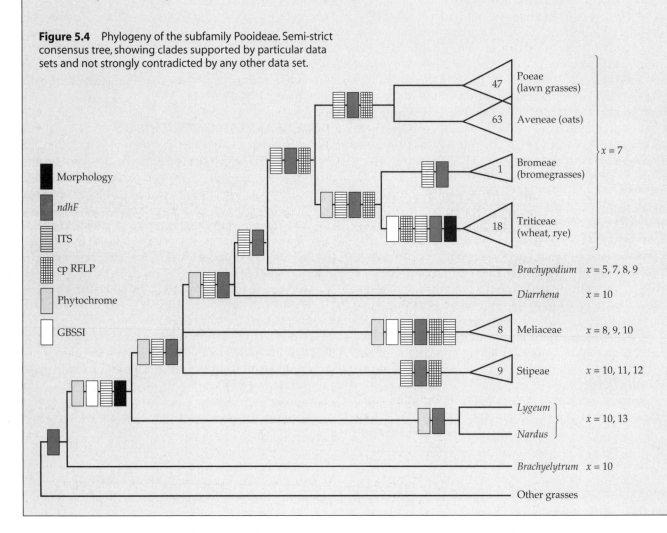

will necessarily be exactly the same as the species tree (see Box 5A). For example, many plant taxa have now been shown to have the "wrong" chloroplast, presumably because of introgression.

Molecular Characters

Good systematic work requires detailed knowledge of characters, their underlying biology, and the nature of variation. For morphological characters, this research leads naturally into studies of developmental morphology. For molecular characters, it directs our attention to molecular biology and the structure and function of particular molecules. Each molecule has its own role in the cell, and its sequence is constrained according to that role. Each molecule, like each set of morphological char-

acters, has its own natural history, reflecting historical contingency, developmental constraints, past and current adaptations (to both intra- and extracellular factors), and stochastic changes, whether fixed or transient.

This means that molecular systematists need to become as familiar with the structure and function of the molecules they study as they are with the plants themselves. (At the same time, of course, they must be careful not to overlook the plants for the molecules!) Molecular genetics and biochemistry are becoming increasingly important as tools for understanding evolution, and any aspiring systematist should consider formal course work in plant molecular biology, molecular genetics, and/or biochemistry.

In the following discussion we will describe some of the major molecules used in systematic studies and what

We now have five molecular phylogenies of the pooid clade, all of which show the Stipeae as an early-diverging lineage. The morphological characters of the Stipeae are thus a mixture of synapomorphies linking them with the pooids and symplesiomorphies, which they share with many other grasses. Two of the studies that resulted in these phylogenies were based on chloroplast DNA, using restriction site polymorphisms (cp RFLP; Davis and Soreng 1993) and sequences of *ndhF* (Catalán et al. 1997). We would expect these methods to give the same phylogeny because the chloroplast does not recombine and thus has a single history.

The other three studies were based on nuclear genes—those for the ITS (Hsiao et al. 1994), phytochrome *B* (Mathews and Sharrock 1996), and granule-bound starch synthase I (GBBSI) (Mason-Gamer et al. 1998). These studies support the same placement of the Stipeae. The fact that all data from both nuclear and chloroplast genomes suggest the same relationships indicates that the gene trees are probably good estimates of the organismal phylogeny. These data are also congruent with information on chromosome number.

A different result appears when we investigate relationships within the tribe Triticeae (Figure 5.5). For this group we have five molecular phylogenies, all dealing with the diploid genera. The two chloroplast phylogenies, based on restriction site polymorphisms (Mason-Gamer and Kellogg 1996) and *rpoA* sequences (Petersen and Seberg 1997), suggest

Figure 5.5 Phylogeny of the tribe Triticeae. Data for the two chloroplast phylogenies are from *rpoA* sequences and restriction site polymorphisms. Chromosome 1 and 5 histories are based on sequences of independent sets of 5S DNA spacers. Chromosome 7 history is based on sequences of granule-bound starch synthase I.

the same groupings, as expected. However, the three nuclear gene trees (based on three different chromosomes) are significantly different (Kellogg et al. 1996).

The explanation for this difference is not clear but may involve a history of limited gene flow among the genera. The important point is that not all genes have identical histories. This means that one gene tree needs to be compared with a second one, preferably from a different genome, if we are to begin to infer organismal histories.

they do in the cell. The literature on biochemistry and molecular biology, however, should be explored for each molecule used.

For historical reasons the most data are available on chloroplast DNA and on nuclear genes for ribosomal RNA. These pieces of DNA are abundant in the cell, making their detection on Southern blots easy. Ribosomal RNA was initially sequenced directly, without reference to the genes; this was possible because the high copy number of rRNA made it easy to extract and sequence. The strengths and limitations of each of these molecules, as well as those of other molecules, are discussed in this section. The choice of molecule for phylogenetic analysis is a difficult one and will continue to be a topic of much discussion for the foreseeable future.

CHLOROPLAST DNA STRUCTURE

Many phylogenetic studies of plants have used chloroplast DNA (see reviews by Olmstead and Palmer 1994 and Sytsma and Hahn 1997, and various chapters in Soltis et al. 1998). Analysis of restriction site variation was the method of choice in the 1980s and remains a good way to assess relationships among species that have diverged recently. For taxa that are more distantly related, restriction site variation often becomes hard to interpret—once there have been too many mutations or rearrangements, it is impossible to infer which mutations occurred in which order.

Because restriction site variation is generally lower within than between species, systematists have been somewhat slower to realize the potential of restriction site studies for assessing population histories. In one extensively studied example, Soltis et al. (1991) found that individuals of *Tellima grandiflora* (Saxifragaceae) have two distinct chloroplast genomes, with distinct geographic ranges. The two differ by at least 18 restriction sites.

The "northern" genome type occurs in plants from northern Oregon to Alaska, and the "southern" type mostly in plants from northern Oregon south into California; there are also a few southern plants on Prince of Wales Island in the Alaskan panhandle and on the Olympic Peninsula in Washington. Because the southern chloroplast genome is most closely related to that of the genus *Mitella*, it is likely that there was some ancestral introgression from *Mitella* into *T. grandiflora*.

The situation in *Tellima* is not an isolated example. The chloroplast genome of *Coreopsis grandiflora* (Asteraceae) is highly polymorphic and can be divided into two types that differ by at least 19 restriction sites (Mason-Gamer et al. 1995). Both types can be found co-occurring in populations and within different varieties of the species. Other species of the genus *Coreopsis* have one chloroplast type or the other, but it is not known how many are polymorphic like *C. grandiflora*.

The chloroplast polymorphism suggests either that *C. grandiflora* is the result of hybridization between two morphologically distinct species, or that both chloroplast types are shared by many members of the genus. The latter explanation suggests that polymorphism can be retained for many years, even through multiple speciation events.

Many other examples of chloroplast DNA variation are cited by Soltis et al. (1998). Such studies illustrate how dynamic plant populations are, and how much gene exchange can occur over evolutionary time.

RBCL

Many plant systematists were involved in a community-wide effort to generate a large database of sequences of the chloroplast gene *rbcL*. This gene encodes the large subunit of the photosynthetic enzyme ribulose-1,5-bisphosphate carboxylase/oxygenase (RuBisCO), which is the major carbon acceptor in all photosynthetic eukaryotes and cyanobacteria. The secondary structure of the gene is known (Figure 5.6), and amino acids can be assigned to particular structural components of the gene.

This gene was chosen because it is almost universal among plants (excepting only the parasites), it is fairly long (1428 bp), it presents no problems of alignment, and as part of the chloroplast it is present in many copies in the cell. The enthusiasm for sequencing the *rbcL* gene was aided by the generosity of Gerard Zurawski (then at

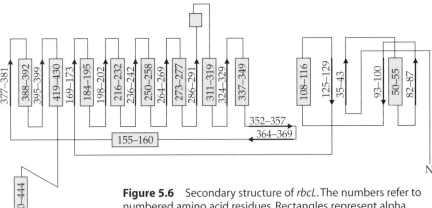

Figure 5.6 Secondary structure of *rbcL*. The numbers refer to numbered amino acid residues. Rectangles represent alpha helices; arrows represent beta sheets. (From Kellogg and Juliano 1997.)

the University of Georgia), who designed a set of near-universal PCR primers that he distributed freely to anyone who wanted them. The availability of these primers has encouraged many plant systematists to generate *rbcL* sequences and has resulted in well over 2000 sequences, primarily of seed plants. The power of this broadly collaborative approach should not be underestimated.

The gene trees generated from these *rbcL* sequences had an enormous influence on our view of relationships among angiosperm families, and they are referred to throughout this book. In particular, several studies presented in a single issue of the *Annals of the Missouri Botanical Garden* in 1993 generated many hypotheses of relationship, which have been tested with other molecular and morphological data. These initial studies were remarkable for their tremendous heuristic value.

Like any heuristic study, however, these studies have limitations. Much attention has focused on the work of Chase et al. (1993), who attempted to generate a phylogeny for all seed plants using 499 *rbcL* sequences. The published trees turned out not to be the shortest available for that data set, a few of the sequences proved to be pseudogenes, and whole families were represented by single sequences, among other problems. Reanalyses of the 499-taxa data set have found numerous equally parsimonious trees, many of which are quite different from the ones presented in the original study (Rice et al. 1997).

The results, in other words, should be interpreted with caution—a point that the authors of the paper recognized. Nonetheless, *rbcL* data, and the Chase et al. (1993) tree in particular, are widely cited and are taken as the starting point for many current research projects. We will therefore discuss the results briefly as a point of entry into the current literature.

The trees based on *rbcL* data support many ideas that were accepted on the basis of morphology. The Caryophyllidae, for example, are monophyletic. The groups that Cronquist called Rosidae and Dilleniidae are largely intermingled, as had been suspected by anyone who had tried to teach according to Cronquist's system.

Several well-known family pairs (e.g., Asclepiadaceae/ Apocynaceae, Araliaceae/Apiaceae, Brassicaceae/Capparaceae, here united to preserve monophyly) are supported by *rbcL* data, although the exact relationships between members of the pairs differ. Families long suspected of being polyphyletic (e.g., Saxifragaceae, Caprifoliaceae) appear polyphyletic on the basis of *rbcL* data. The eudicot (tricolpate) clade is supported, albeit weakly, by *rbcL* data.

In other cases, *rbcL* data helped resolve relationships that were previously ambiguous. The Ericaceae, for example, were included in Adolf Engler and Karl Prantl's Sympetalae, but recent thought had placed them well outside it. Trees based on *rbcL* data support the placement of an Ericalean clade in a larger clade with the Asteridae, reuniting much (but not all) of the Englerian Sympetalae. These data also support a monophyletic Rosaceae, within which three of the four subfamilies are monophyletic; some genera will need to be realigned, however (Morgan et al. 1994).

Finally, in a few cases *rbcL* data suggest something quite surprising. We have mentioned the placement of Vochysiaceae and Limnanthaceae in the Myrtalean and glucosinolate clades, respectively. Another striking example is the finding that the nine families with nitrogen-fixing members fall into a single large clade, along with a restricted set of families that do not fix nitrogen (Soltis et al. 1995). Because previously the N-fixing families had appeared completely unrelated, this finding suggests that these families may have more in common than previously believed.

OTHER CHLOROPLAST GENES

One of the limitations of *rbcL* as a phylogenetic marker is its slow rate of change. The protein it encodes is fundamentally a conservative molecule that is highly constrained at the amino acid level. The *rbcL* gene is therefore not particularly useful for inferring relationships within or between closely related genera. Instead, other chloroplast genes have been used for such purposes, notably those that encode subunit F of NADP dehydrogenase (*ndhF*, in the small single-copy region), those that encode the α and β″ subunits of RNA polymerase II (*rpoA* and *rpoC2*, in the large single-copy region), and a maturase gene in the intron that separates the coding region of *trnK* (*matK*, previously known as ORF, for *o*pen *r*eading *f*rame, K). Because these are all part of the same nonrecombining genome as *rbcL*, they all track the same (generally maternal) history.

The gene that encodes the β subunit of ATP synthase—*atpB*—has been used to address the same problems as *rbcL*. It appears to evolve at about the same rate and thus provides additional phylogenetically informative characters. Data from *atpB* have been combined with data from *rbcL* to refine the picture of angiosperm relationships (as will be discussed shortly) (Qiu et al. 1999; Soltis et al. 1999). Among more closely related species, a maturase gene, *matK*, has provided phylogenetically useful variation.

MITOCHONDRIAL GENES

Relatively few studies have used plant mitochondrial genes for studying phylogeny. In general these genes evolve slowly and are most useful for assessing ancient events such as the origin of the angiosperms (Qiu et al. 1999) or the evolution of the monocots (Davis et al. 1998). The commonly used genes encode subunits of ATP synthase (*atp1*, *atpA*) and a maturase (*matR*).

NUCLEAR GENES

Ribosomal RNA
Historically, the only nuclear genes with a copy number high enough for easy study were the ribosomal genes. These are arranged in tandem arrays of several hundred to several thousand copies. The general arrangement of these genes is shown in Figure 5.7. The small subunit (18S) and large subunit (26S) genes are separated by a smaller (5.8S) gene, and the whole set of genes is transcribed as a single unit.

There are short transcribed spacers (ITS) between the three genes. Each set of three genes is separated from the following set by a large spacer, variously referred to as the intergenic spacer (IGS), extragenic spacer (EGS), or nontranscribed spacer (NTS). The last is something of a misnomer, in that some of the sequence immediately upstream of the gene that encodes the 18S subunit and downstream of the one that encodes the 26S subunit is actually transcribed; these regions are sometimes called external transcribed spacers (ETS). The middle portion of the spacer is not transcribed and is made up of variable numbers of short repeated sequences (about

Figure 5.7 Structure of the ribosomal array. Coding regions of the 18S small subunit (SSU), 5.8S unit, and 26S large subunit (LSU) are shown as shaded rectangles; the transcription unit is bracketed. Spacers are indicated by black lines, and the short repeats in the intergenic spacer (IGS) are indicated by small boxes. ETS, external transcribed spacer; ITS, internal transcribed spacer.

BOX 5B *Molecular Data Reveal Ancient Hybridization*

Wendel et al. (1995) studied the evolution of the genus *Gossypium*, which includes all the species that produce cotton. They used isozymes, nuclear ITS sequences, and chloroplast restriction site analysis to study the history of both diploid and tetraploid species. Most of their data indicate that the New World diploids (with a genome designated D; Figure 5.8) are monophyletic, as are the Old World diploids (genome groups A, B, and F).

The surprise came in analyzing the New World tetraploids, including *Gossypium hirsutum*, the source of most of the world's commercial cotton. These species were formed by allopolyploidization of A and D genomes. Wendel and his colleagues found that *G. hirsutum* has a chloroplast derived from one of the African species, and that it must have acquired that chloroplast only about 1–2 million years ago, well after the formation of the Atlantic Ocean. One other New World species has elements of an Old World ITS sequence as well. We do not know how such long-distance gene flow might have occurred.

Figure 5.8 History of *Gossypium* species. The New World AD genome allopolyploids were formed by the joining of an Old World ancestor similar to *G. herbaceum* to a New World ancestor similar to *G. raimondii*. (Redrawn from Wendel et al. 1995.)

100–300 bp each). These sequences are thought to play a role in gene regulation.

A completely separate rRNA array encodes only 5S RNA (not to be confused with 5.8S rRNA), a molecule that joins with the 26S RNA and a set of proteins to form the large ribosomal subunit. The 5S rRNA genes are in tandem arrays of several thousand copies and are separated by nontranscribed spacers.

Such highly repetitive sequences undergo homogenization processes known as **concerted evolution**. If a mutation occurs in one copy of a sequence, it is generally corrected to match the other copies, but sometimes the nonmutated copies are "corrected" to match the mutated one, causing nucleotide changes to propagate throughout the array. This means that the many copies of the sequence are generally more similar to one another than they are to copies in other species. Within-species variation does occur, however, because concerted evolution is slower than the mutation rate. This means that

some highly repetitive sequences can be used to assess variation within and among populations of the same species.

Ribosomal genes were studied initially by restriction site mapping of the intergenic region. This region, particularly the short repeated sequences, is quite variable even within populations. Its major value has therefore been in studies of closely related plants, where it has been useful in determining population structure and in assessing patterns of hybridization (see Box 5B).

King (1993), for example, studied North American and European species of dandelions. All known North American dandelions are asexual, but European dandelions are both sexual and asexual. All ribosomal DNA and chloroplast DNA genotypes found in asexual plants were also found in sexual plants, but in different combinations. This finding is evidence that asexual plants are produced frequently by hybridization between sexual plants.

Methods for direct sequencing of RNA (without recourse to the DNA that encodes it) were developed earlier than rapid DNA sequencing methods, and well before PCR. The development of such methods led to some early hope that the 5S rRNA gene sequences would be the key to evolutionary history. Researchers soon recognized, however, that the 5S rRNA genes were too short (at 120 bp) and too conservative to illuminate relationships. But in closely related groups the nontranscribed spacer region has been used with some success.

Sequences of the 18S and 26S genes are more promising. These genes are large (about 1800 and 3300 bp, respectively). They have some regions that are highly conserved, which helps in alignment, and others that are quite variable, which helps in distinguishing phylogenetic groups. A large cooperative effort, analogous to the *rbcL* study, generated a database of 18S sequences (Soltis et al. 1997), which were then combined with *rbcL* and *atpB* sequences to provide a clear picture of angiosperm evolution (Soltis et al. 1999).

Use of the ITS region is becoming increasingly common for determining relationships among species. In general, the ITS region has supported relationships inferred from the chloroplast or from morphology. In other cases, however, it has proved to be polymorphic within species, or even within individual plants, suggesting that concerted evolution has not completely homogenized the repeats.

Although this polymorphism may be a problem in some cases, it also provides tools for understanding gene flow and population-level variation. Because of the problems of incomplete homogenization of ITS genes, many systematists use PCR to amplify the ITS region, but then clone and sequence ITS copies individually. This approach provides an assessment of ITS variation within a plant and allows correction for ancestral polymorphisms.

Low copy number genes Nuclear genes with low copy numbers are being used more and more commonly as phylogenetic markers. In order for a nuclear gene to be a useful phylogenetic indicator, it must not be easily confused with any other gene. This requirement can be a problem because many nuclear genes are duplicated or exist as part of a small set of genes (a **gene family**). If there are multiple genes in the gene family, they must be distinguishable and not undergo any sort of concerted evolution.

Population genetic theory suggests that allelic variation should not be misleading in studies of closely related species because alleles in one species should be more closely related to each other than they are to alleles in other species. Data on the genes for alcohol dehydrogenase in *Arabidopsis* and granule-bound starch synthase I (*waxy*) in grasses and Rosaceae support this expectation.

The gene that encodes phosphoglucoisomerase (PGI) has been studied in some detail in the genus *Clarkia*, by Gottlieb and Ford (1996). There are actually two PGI genes in angiosperms: one that functions in the cytosol and one in the plastid. The two can be distinguished easily. Isozyme studies indicated that some species of *Clarkia* had two copies of the cytosolic gene rather than just one. Studies of DNA sequences showed that both copies were present in *all* species of the genus, but that in some species the extra copy had accumulated so many mutations that it could no longer be translated properly, and thus would not appear on an isozyme gel. Such nonfunctional genes are known as **pseudogenes**. In the example described here, however, the sequences themselves provided robust and identical gene trees, which clarified relationships among the species.

The phytochrome genes, which respond to light signals and control plant growth, development, and photosynthesis, have proven to be particularly illuminating. There are four major phytochrome genes in angiosperms, given the letter names *A*, *B*, *C*, and *E*. (Phytochrome *D* occurs in Brassicales and is a relatively recent duplicate of *B*.)

All of these genes are descendants of an ancestral phytochrome gene that was duplicated once in the ancestor of the seed plants. One copy became the ancestor of phytochromes *A* and *C*, and the other the ancestor of *B* and *E*. The *A/C* duplication occurred before the origin of the angiosperms. Phytochromes *A* and *C* each have been used to produce a phylogeny of the early-diverging angiosperms (Mathews and Donoghue 1999), with results that are similar to those produced by chloroplast genes.

Introns of nuclear genes are often more variable than the ITS or the most variable chloroplast regions, and thus they are useful tools for assessing relationships among similar species or even populations. For example, Olsen and Schaal (1999) used introns in the metabolic gene glucose-6-phosphate dehydrogenase to determine where in Brazil cultivated cassava (*Manihot esculenta*) was domesticated from wild populations.

Malcomber (2002) used introns of triose phosphate isomerase, and two different genes for phosphenolpyruvate carboxylase to assess relationships among species of *Gaertnera*, a member of the Rubiaceae. Introns of the gene that encodes granule-bound starch synthase I (*GBSSI*, orthologous to the gene known in maize as *waxy*) have been used to determine relationships among species of grasses, maloid Rosaceae, potatoes, and morning glories (Mason-Gamer et al. 1998; Evans et al. 2000; Manos et al. 2001; Peralta and Spooner 2001).

When performing studies using nuclear introns as phylogenetic markers, it is often most productive to try several genes for a subset of the taxa to be studied. Because the choice of nuclear gene is not particularly critical (assuming that orthology can be demonstrated), it is most efficient to choose one that can be easily amplified through PCR, that can be readily aligned among the study taxa, and that provides sufficient variation for resolving relationships.

High copy number noncoding nuclear sequences

Unlike chloroplast, ribosomal, and nuclear protein-coding genes, high copy number noncoding sequences appear to evolve rapidly and are thus useful for addressing questions among conspecific populations. These high copy number sequences are generally short sequences that are repeated many times, often at many locations in the genome.

So-called **minisatellites** or **variable-number tandem repeats** (**VNTR**) are made up of repeated sequences that are generally tens of base pairs in length. There are also similar regions, known as **microsatellites**, in which the repeats are much shorter, consisting of only two or three nucleotides. Such repeated sequences are unstable and prone to errors in replication, usually deriving from **replication slippage** (although unequal crossing-over is also a possibility in some cases).

Replication slippage occurs as the DNA is being copied. The strands separate for replication but reanneal out of register, leading to a loop in the DNA. Mismatch repair mechanisms then either remove the loop (leading to a loss of the repeat unit) or insert extra bases on the opposite strand (leading to a duplication). Because of this instability, individual organisms often vary in the number of repeats at a particular satellite locus. This variation can be used to determine a DNA "fingerprint" unique to a particular plant or closely related group of plants. Studies of population structure generally depend on accurate assessment of relationships among individual plants, and these markers are useful for such assessment.

Another method often used in studies at the population level is the **random amplified polymorphic DNA** (**RAPD**) method. In this technique, short (10 bp) PCR primers are designed with arbitrary sequences. These short random sequences will generally match one or more sequences somewhere in the genome of the plant, and the primers will bind to and amplify a fragment of DNA.

By performing many such PCRs with random primers, one can generally find fragments that distinguish individual plants or populations. These fragments allow a rapid assessment of how many genotypes are present in a population and a rough estimate of how different those genotypes are. The technique is limited, however, because the identity of the fragments is not known. In other words, a fragment of 150 bp in one plant may not actually represent the same part of the genome as a 150 bp fragment in another plant.

Verifying the identity of the fragments requires Southern blotting or restriction site analysis, at which point the technique may become as laborious as restriction site studies or sequencing. Other techniques, such as amplified fragment length polymorphisms (AFLP), have been developed to circumvent the problems of RAPD, but a full discussion of these is beyond the scope of this book. Many studies employing high copy number genes are published in the journal *Molecular Ecology*, which is a rich source of information on systematics of closely related species.

Summary

Molecular techniques provide powerful tools for the study of evolution and phylogeny. Most data on relationships at the species level and above have so far come from the chloroplast genome and the highly repeated sequences of ribosomal RNA. Future data are likely to come also from low copy number nuclear genes. New tools are continually being developed for the study of variation within and among conspecific populations, including methods of genome mapping. As these tools come into more widespread use, they will provide new insights into the processes of population-level differentiation.

No matter how powerful the molecular data, however, morphological data will remain critical for phylogenetic studies. The major questions in plant systematics are still morphological. Questions about the origin of species, the mechanisms of diversification, and the best way to classify that diversity all require understanding of morphology as well as phylogeny. We can now envision a time when robust phylogenies will have been constructed for all groups of plants, and the question of systematics will shift from "What is the phylogeny of my group?" to "How did the morphological diversity arise?"

Literature Cited and Suggested Reading

Items marked with an asterisk (*) are especially recommended to those readers who are interested in further information on the topics discussed in this chapter.

*Avise, J. C. 1994. *Molecular markers, natural history and evolution*. Chapman & Hall, New York.

Bradshaw, H. D., S. M. Wilbert, K. B. Otto and D. W. Schemske. 1995. Genetic mapping of floral traits associated with reproductive isolation in monkeyflowers (*Mimulus*). *Nature* 376: 762–765.

Catalán, P., E. A. Kellogg and R. G. Olmstead. 1997. Phylogeny of Poaceae subfamily Pooideae based on chloroplast *ndhF* gene sequences. *Mol. Phylogenet. Evol.* 8: 150–166.

*Chase, M. W. and 41 others. 1993. Phylogenetics of seed plants: An analysis of nucleotide sequences from the plastid gene *rbcL*. *Ann. Missouri Bot. Gard.* 80: 528–580.

Davis, J. I. and R. J. Soreng. 1993. Phylogenetic structure in the grass family (Poaceae) as inferred from chloroplast DNA restriction site variation. *Am. J. Bot.* 80: 1444–1454.

Davis, J. I., M. P. Simmons, D. W. Stevenson and J. F. Wendel. 1998. Data decisiveness, data quality, and incongruence in phylogenetic analysis: An example from the monocotyledons using mitochondrial atpA sequences. *Syst. Biol.* 47: 282–310.

Downie, S. R., E. Llanas and D. S. Katz-Downie. 1996. Multiple independent losses of the *rpoC1* intron in angiosperm chloroplast DNAs. *Syst. Bot.* 21: 135–151.

Evans, R. C., L. A. Alice, C. S. Campbell, E. A. Kellogg and T. A. Dickinson. 2000. Multiple putative GBSSI loci in the Rosaceae: Characterization and potential phylogenetic utility. *Mol. Phylogenet. Evol.* 11: 388–400.

Gottlieb, L. D. and V. S. Ford. 1996. Phylogenetic relationships among the sections of *Clarkia* (Onagraceae) inferred from the nucleotide sequences of *PgiC*. *Syst. Bot.* 21: 45–62.

*Graur, D. and W.-H. Li. 2000. *Fundamentals of molecular evolution*, 2nd ed. Sinauer, Sunderland, MA.

*Hillis, D. M., C. Moritz and B. K. Mable. 1996. *Molecular systematics*, 2nd ed. Sinauer, Sunderland, MA.

Hsiao, C., N. J. Chatterton, K. H. Asay and K. B. Jensen. 1994. Molecular phylogeny of the Pooideae (Poaceae) based on nuclear rDNA (ITS) sequences. *Theor. Appl. Genet.* 90: 389–398.

Jansen, R. K. and J. D. Palmer. 1987. A chloroplast DNA inversion marks an ancient evolutionary split in the sunflower family (Asteraceae). *Proc. Natl. Acad. Sci. USA* 84: 5818–5822.

Kellogg, E. A. and N. D. Juliano. 1997. The structure and function of RuBisCO and their implications for systematic studies. *Am. J. Bot.* 84: 413–428.

Kellogg, E. A., R. Appels and R. J. Mason-Gamer. 1996. When genes tell different stories: The diploid genera of Triticeae (Gramineae). *Syst. Bot.* 21: 321–347.

King, L. M. 1993. Origins of genotypic variation in North American dandelions inferred from ribosomal DNA and chloroplast DNA restriction enzyme analysis. *Evolution* 47: 136–151.

Malcomber, S. T. 2002. Phylogeny of *Gaertnera* Lam. (Rubiaceae) based on multiple DNA markers: Evidence of rapid radiation in a widespread, morphologically diverse genus. *Evolution* 56: 42–57.

Manos P. S., R. E. Miller, and P. Wilkin. 2001. Phylogenetic analysis of *Ipomoea*, *Argyreia*, *Stictocardia*, and *Turbina* suggests a generalized model of morphological evolution in morning glories. *Syst. Bot.* 26: 585–602.

Mason-Gamer, R. J. and E. A. Kellogg. 1996. Chloroplast DNA analysis of the monogenomic Triticeae: Phylogenetic implications and genome-specific markers. In *Methods of genome analysis in plants: Their merits and pitfalls*, P. Jauhar (ed.), 301–325. CRC Press, Boca Raton, FL.

Mason-Gamer, R. J., K. E. Holsinger and R. K. Jansen. 1995. Chloroplast DNA haplotype variation within and among populations of *Coreopsis grandiflora* (Asteraceae). *Mol. Biol. Evol.* 12: 371–381.

Mason-Gamer, R. J., C. F. Weil and E. A. Kellogg. 1998. Granule-bound starch synthase: Structure, function, and phylogenetic utility. *Mol. Biol. Evol.* 15: 1658–1673.

Mathews, S. and M. J. Donoghue. 1999. The root of angiosperm phylogeny inferred from duplicate phyochrome genes. *Science* 286: 947–950.

Mathews, S. and R. A. Sharrock. 1996. The phytochrome gene family in grasses (Poaceae): A phylogeny and evidence that grasses have a subset of the loci found in dicot angiosperms. *Mol. Biol. Evol.* 13: 1141–1150.

Morgan, D. R., D. E. Soltis and K. R. Robertson. 1994. Systematic and evolutionary implications of *rbcL* sequence variation in Rosaceae. *Am. J. Bot.* 81: 890–903.

*Olmstead, R. G. and J. D. Palmer. 1994. Chloroplast DNA systematics: A review of methods and data analysis. *Am. J. Bot.* 81: 1205–1224.

Olsen, K. M. and B. A. Schaal. 1999. Evidence on the origin of cassava: Phylogeography of *Manihot esculenta*. *Proc. Natl. Acad. Sci. USA* 96: 5586–5591.

*Page, R. D. M. and E. C. Holmes. 1998. *Molecular evolution: A phylogenetic approach*. Blackwell, Oxford, UK.

Peralta, I. E. and D. M. Spooner. 2001. Granule-bound starch synthase (GBSSI) gene phylogeny of wild tomatoes (*Solanum* L. section *Lycopersicon* [Mill.] Wettst. subsection *Lycopersicon*). *Am. J. Bot.* 88: 1888–1902.

Petersen, G. and O. Seberg. 1997. Phylogenetic analysis of the Triticeae (Poaceae) based on *rpoA* sequence data. *Mol. Phylogenet. Evol.* 7: 217–230.

Qiu, Y.-L., J. H. Lee, F. Bernasconi-Quadroni, D. E. Soltis, P. S. Soltis, M. Zanis, E. A. Zimmer, Z. D. Chen, V. Savolainen and M. W. Chase. 1999. The earliest angiosperms: Evidence from mitochondrial, plastid and nuclear genomes. *Nature* 402: 404–407.

Rice, K. A., M. J. Donoghue and R. G. Olmstead. 1997. Analyzing large data sets: *rbcL* 500 revisited. *Syst. Biol.* 46: 554–563.

Rieseberg, L. H., B. Sinervo, C. R. Linder, M. Ungerer and D. M. Arias. 1996. Role of gene interactions in hybrid speciation: Evidence from ancient and experimental hybrids. *Science* 272: 741–745.

Rieseberg, L. H., C. VanFossen and A. Desrochers. 1995. Genomic reorganization accompanies hybrid speciation in wild sunflowers. *Nature* 375: 313–316.

Rodermel, S. 1993. Genetic map of *Zea mays* plastid chromosome. *Maize Newslett.* 67: 167–168.

Soltis, D. E., M. Mayer, P. S. Soltis and M. Edgerton. 1991. Chloroplast DNA variation in *Tellima grandiflora* (Saxifragaceae). *Am. J. Bot.* 78: 1379–1390.

Soltis, D. E., P. S. Soltis, and B. G. Milligan. 1992. Intraspecific chloroplast DNA variation: Systematic and phylogenetic implications. In *Molecular systematics of plants*, P. S. Soltis, D. E. Soltis and J. J. Doyle (eds.), 117–150. Chapman and Hall, New York.

Soltis, D. E., P. S. Soltis, D. R. Morgan, S. M. Swensen, B. C. Mullin, J. M. Dowd and P. G. Martin. 1995. Chloroplast gene sequence data suggest a single origin of the predisposition for symbiotic nitrogen fixation in angiosperms. *Proc. Natl. Acad. Sci. USA* 92: 2647–2651.

Soltis, D. E. and 15 others. 1997. Angiosperm phylogeny inferred from 18S ribosomal DNA sequences. *Ann. Missouri Bot. Gard.* 84: 1–49.

*Soltis, D. E., P. S. Soltis, and J. J. Doyle (eds.). 1998. *Molecular systematics of plants II: DNA sequencing*. Kluwer, Boston.

Soltis, P. S., D. E. Soltis and M. W. Chase. 1999. Angiosperm phyolgeny inferred from multiple genes as a research tool for comparative biology. *Nature* 402: 402–404.

*Swofford, D. L., G. J. Olsen, P. J. Waddell and D. M. Hillis. 1996. Phylogenetic inference. In *Molecular systematics*, 2nd ed., D. M. Hillis, C. Moritz and B. K. Mable (eds.), 407–514. Sinauer, Sunderland, MA.

*Sytsma, K. J. and W. J. Hahn. 1997. Molecular systematics: 1994–1995. *Prog. Bot.* 58: 470–499.

Wendel, J. F., A. Schnabel and T. Seelanan. 1995. An unusual ribosomal DNA sequence from *Gossypium gossypioides* reveals ancient, cryptic, intergenomic introgression. *Mol. Phylogenet. Evol.* 4: 298–313.

6 The Evolution of Plant Diversity

Planet Earth supports about 260,000 species of tracheophytes. They range from minute plants such as the floating aquatic duckweeds (*Lemna* and *Spirodela* of the Araceae), only millimeters in size and without leaves, to massive trees such as redwoods (*Sequoia*, Cupressaceae), over 100 meters tall and thousands of years old. As different as these extreme examples are, they are related through a common ancestor on the tree of life. **Evolution**—genetic change through time—is the source of this tremendous diversity and of the tree of life itself (Box 6A). The systematist's main concern is identifying and understanding groups of organisms produced by evolution.

An essential ingredient of evolution is natural variation among individuals. Hence it is important to understand how this variation arises and is distributed geographically. The processes that create discrete units of variation—**taxa** (singular **taxon**)—especially the formation of species—**speciation**—are of central interest to systematists. The role of gene flow through interbreeding in creating and maintaining plant groups is not clear. In order for one species to split into two new species, interruption of gene flow between the forming species is apparently necessary. Mating between individuals of different plant species may blur the boundary between species, but in many cases it has surprisingly little effect on morphological variation within a species.

BOX 6A   *Theories of Organic Evolution*

Evolution has most often been defined in two ways: (1) change in gene frequency and (2) descent with modification. The latter definition is associated with Charles Robert Darwin (1809–1882), (Figure 6.1), who conceived the theory now called **Darwinism**. One of the most significant ideas in history, Darwinism has profoundly influenced not only the biological sciences, but also our perspective on our position in the world.

Others, most notably Jean-Baptiste Lamarck (1744–1829), also suggested that organisms change through time, an idea encouraged by the discovery of numerous fossils and of the great age of Earth, but Lamarck believed in the inheritance of acquired characteristics. When Darwin began his work, the general view was that all organisms were the product of divine creation. Darwin's observations of natural history—both in England and during his travels as the naturalist aboard the ship HMS *Beagle* on its round-the-world voyage from 1831 to 1836—and years of thought led him to his theory, which he presented in 1859 in his book *On the Origin of Species*. Darwin perceived that all organisms are related to one another in a branching tree of

Figure 6.1   Charles Darwin in his prime. (Courtesy of the American Philosophical Society.)

life, and he conceived of a process that could generate that pattern. He called that process **natural selection**. Its logic is based on a set of facts that Darwin brought together and two major inferences based on those facts (Figure 6.2).

Thomas Malthus (1766–1834), an influential English economist, observed that human populations are superfecund—that is, that they are capable of tremendous (geomet-

ric) growth in numbers. Darwin extended this idea to the natural world. That natural populations do not experience explosive increases in their numbers, but instead generally persist in a steady state, could be explained by the constraint of limited resources. These first three facts—superfecundity, the steady state of populations, and limited resources—led to Darwin's first inference, that there is a struggle for existence. Only the best-adapted, or fittest, organisms, such as the plant individuals with flowers that most efficiently attract pollinators, will compete successfully against other individuals to survive and reproduce.

Darwin linked to this inference the fact that there is variation among individuals of a species. This concept was well understood by animal and plant breeders in Darwin's time, but it was otherwise not generally accepted. Since the ancient Greeks, a species, such as a particular species of oak, was considered the manifestation of a unique, unchangeable essence. Any deviations from this essence were regarded as unimportant. Second, from breeders and from his own experiments and observations, Darwin knew that offspring

Interspecific gene flow (**hybridization**, sometimes referred to as reticulation) plays a dual role in speciation. On the one hand, it may reduce diversity by merging species. On the other hand, it can be a powerful force leading to speciation, especially when coupled with polyploidy, an important source of genetic variation within plant species.

Plant diversity is strongly shaped by breeding systems. Uniparental reproduction, by self-fertilization or asexuality, tends to package variation in smaller, more numerous units than are found in groups with strictly biparental reproduction.

The final section of this chapter deals with species concepts, an issue that has been intensely debated by systematists and evolutionary biologists. Although many biologists define a species as a group of populations that does not exchange genes with other populations, gene flow fails as a criterion for species definition in plants because of interspecific hybridization and uniparental reproduction.

What are commonly recognized as plant species frequently hybridize, and uniparental reproduction reduces a reproductive community to one individual or

the progeny of one individual. Hence plant systematists by and large do not insist that species are equivalent to reproductive communities. Instead, they define species using a wide array of evidence that a population, or more usually a group of populations, forms an independent evolutionary lineage.

The dominant type of data is morphology: Members of a given plant species are morphologically more similar to each other than they are to members of other species. Molecular data, increasingly common, challenge and complement the delimitation of species on morphological grounds.

Plant Diversity Is the Result of Evolution

Support for evolution as the source of biodiversity comes from fossils, from the distribution of features common to groups of organisms (and some to all living organisms), from observations of the geographic variation of organisms, and from environmental studies. Fossils of flowering plants (Figure 6.3) record the evolution of structures—flowers, fruits, leaves, and other parts—

tend to resemble their parents. Although he did not understand the mechanism of heredity, he did see that it was essential to his second inference: natural selection.

When variation affects the outcome of the struggle for existence (as in the attractiveness of a flower to a pollinator), individuals with the most advantageous characteristics will survive and reproduce more often than others. The progeny of these successful individuals will be more frequent in future generations than those of the less successful individuals. Without heredity, by which successful characteristics are passed on to individuals of future generations, the population would not change through time. To summarize Darwinism, heritable variation in fitness leads to natural selection and evolution.

Although natural selection is responsible for much evolutionary change, an alternative must operate some of the time. When the force of natural selection is weak, chance will govern the course of evolutionary change. A gene may be present in a population or species not because it confers higher fitness than another gene, but merely by chance. This process, known as random genetic drift, is described elsewhere in this chapter and is the basis of the **neu-**tral theory of molecular evolution (see Li 1997). The "neutral" part of this name refers to the notion that some molecular variation does not affect (is neutral to) the survival of organisms.

Shortly after the publication of *On the Origin of Species*, an Austrian monk, Gregor Mendel (1822–1884), performed elegant experiments demonstrating heredity in the common garden pea. These experiments were not widely appreciated until the beginning of the twentieth century, when they formed the basis of the field of genetics. Later, two British scientists (R. A. Fisher and J. B. S. Haldane) and an American (Sewall Wright) developed the field of population genetics and provided a quantitative and theoretical framework for studies of evolutionary change in populations. Transformed by population genetics, Darwinism was renamed **neo-Darwinism**.

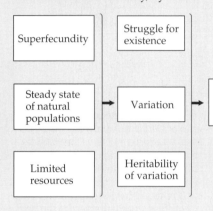

Figure 6.2 Darwin's logic. Three observations—that (1) organisms are capable of rapid increase in number of individuals (superfecundity), but (2) population size is generally stable and (3) resources are limited—led to the inference of a struggle for existence among individuals. Two additional observations, of natural variation among individuals and the heritability of this variation, led to the inference of differential reproduction (i.e., natural selection) and of evolution over many generations. (Modified from Mayr 1977.)

and the origin and extinction of species, genera, and families (Stewart and Rothwell 1993). Although there are many gaps in the fossil record, it overwhelmingly verifies morphological change in organisms (and therefore in their genes) over the millions of years of life on Earth.

The fact that living organisms function through common genetic and biochemical processes strongly supports a unique origin and common history of life on Earth. All organisms perform certain essential functions via the same metabolic pathways. Glycolysis, for example, is a series of coordinated biochemical reactions that involves the partial breakdown of glucose to drive the synthesis of adenosine triphosphate (ATP) molecules and that can occur in the cytoplasm of all living cells. The Krebs cycle is a continuation of glucose breakdown to generate more ATP, and it takes place in the mitochon-

Figure 6.3 Reconstruction of leafy branch and flower of *Archaeanthus linnenbergeri*, an early angiosperm from the mid-Cretaceous. Fossils of such plants help us understand when angiosperms evolved and what some of the first members of this lineage may have looked like. (From Dilcher and Crane 1984; original art by M. Rohn.)

(A) *Cypripedium calceolus*

(B) *Orchis maculata*

(C) *Cattleya* sp.

(D) *Nigritella nigra*

Figure 6.4   Variations in the orchid floral lip. Despite differences in size, shape, and color, the orchid floral lip is thought to have derived from one ancestral structure in the common ancestor of all orchids. Differences in orchid lips (arrows) are related to adaptations for different pollinators. In parts A–C the orchid flower is shown in front view (left) and side cross section (right). (A) The pouch-shaped lip of *Cypripedium calceolus* is designed to momentarily trap small bees of the genus *Andrena*, which are forced to pollinate the flower because the only possible way out of the lip takes them past the pollen and stigmatic surface. (B) The three-lobed lip of *Orchis maculata* attracts bees (and sometimes flies) and orients them toward the nectar in the spur of the lip; as pollinators take nectar, they pollinate the flower. (C) The large lip of this *Cattleya* species functions in much the same fashion as the lip of *Orchis*, but it attracts different pollinators—large social bees. In *Cattleya*, nectar is provided by a nectary embedded in the ovary, and the lip does not have a nectar spur. (D) The lip of *Nigritella nigra* (inflorescence on left; solitary flower in front view on right) is nearly identical to the other perianth parts because it is not a primary attractant to pollinators; pollinators (butterflies) are instead attracted by the dense clustering of the flowers. In addition, the lip is not in the lower position, as it is in the flowers in A–C; the lip of this orchid is specialized for butterflies because the opening to the small nectar spur is too small for the proboscis (tongue-like organ) of other insects. (From Faegri and van der Pijl 1979.)

drion (plural *mitochondria*), an organelle found only in eukaryotic organisms (all organisms except bacteria).

DNA makes up the **genotype**, registering genetic information in all organisms. This information is encoded in the sequence of the components of DNA (the nucleotides), forming a code that is nearly universal in living organisms and that is transmitted (often with some variations) to coordinate development of the phenotype. The **phenotype** is the product of interactions between the genotype and the environment.

Major genetic change is often recorded in phenotypes. The specialized lowermost petal [the lip, or labellum] of orchid flowers, for example, varies in size, color, and shape among species (Figure 6.4). Despite their differences, these structures trace back to a common ancestral lip; they are all modifications wrought by evolution over millions of years. Orchid lips have a fundamental similarity—known as **homology**—because of this common evolutionary history.

Many of the changes that make organisms different affect their ability to survive and reproduce. This ability is termed evolutionary **fitness** and is related to the concept of evolutionary **adaptation**, the close correspondence between organisms and their environment. Variations in the color, size, and shape of flowers, for example, are adaptations that attract, manipulate, and reward dozens of species of birds and bats, as well as thousands of species of bees, moths, and other insect pollinators (see Chapter 4). Flowers pollinated by bats, for example, are readily accessible to flying bats, are white and easily located at night, and provide a plentiful nectar reward (Figure 6.5; see also Table 4.1).

Another remarkable example of adaptation is the nectar spur of the orchid *Angraecum sesquipedale* of Madagascar. The specific epithet refers to the great length of this spur: *Sesqui* means "one-and-a-half" and *pedale* refers to "foot"; therefore, the literal meaning is "a foot and a half" (actually, the spur is only 30 cm long, not 45 cm). Pollinators of this species were not known in Darwin's time, but he hypothesized that there must be a moth with a tongue long enough to reach the nectary near the bottom of the spur. Later such a moth, a hawkmoth, was indeed discovered.

A final line of evidence for evolution is the intricacy of geographic variation within groups of organisms. Numerous studies document patterns of plant variation—morphological, ecological, phenological, and genetic—that are coincident with geographically varying physical or biotic features. Jeffrey pine (*Pinus jeffreyi*), a forest tree of montane regions from southwestern Oregon to Mexico, illustrates such a pattern (Furnier and Adams 1986).

The species maintains high genetic diversity, like most conifers, except in populations on serpentine soils (Figure 6.6). These soils, which are quite infertile, differ sufficiently from other, more fertile soils to induce the formation of edaphic (soil) **ecotypes** (populations specialized for certain ecological circumstances, such as soil conditions). Different ecotypes have different genotypes. Serpentine ecotypes of Jeffrey pine harbor less genetic diversity than populations on more fertile soils, probably because of intense natural selection for genotypes tolerant of the rigorous soil conditions.

Evolution is frequently considered on two levels. Our discussion so far has focused on change within species. For example, an individual might bear a mutation (which is heritable) for enhanced chemical defense against herbivores. That individual might therefore produce more offspring than other individuals in the population, thereby increasing the frequency of the mutation within the population and perhaps within other populations. The accumulation of such changes within a lineage is sometimes referred to as **anagenesis**. Changes

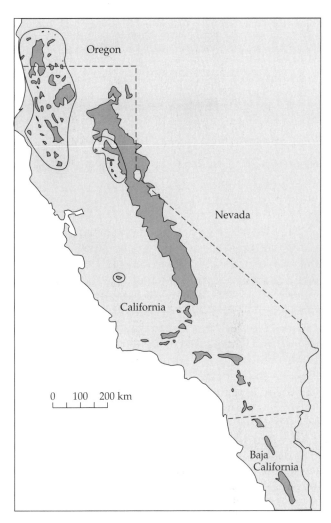

Figure 6.6 The global distribution of *Pinus jeffreyi*, with the distribution of serpentine soil ecotypes shown by the outlines in northern California and southwestern Oregon. (After Furnier and Adams 1986.)

Figure 6.5 The long-nosed bat (*Leptonycteris nivalis*; Arizona to Texas and south to Guatemala), using its highly extensile tongue to lick nectar from the cup-shaped corolla of the Saguaro cactus (*Carnegia gigantea*), which it pollinates in the process. The size, horizontal orientation, white color, and nocturnal functioning are all floral adaptations for bat pollination. Bats are not capable of hovering, like hummingbirds, but instead drink briefly at a flower and then land, returning to the flower until the nectar is gone. Rarely, bats fatally impale themselves on the long cactus spines. (From Faegri and van der Pijl 1979.)

that lead to the formation of new species by the splitting of one species into two—**cladogenesis**—are of primary interest to systematics.

Variation in Plant Populations and Species

SOURCES OF VARIATION

Mutation and genetic recombination are the primary sources of variation within plant populations and species, and they form the raw material for natural selection and random genetic drift. **Mutation** refers to alterations of DNA, ranging from changes in single bases (point mutations; Figure 6.7) to insertions, duplications, deletions, and inversions of parts of a chromosome, to gain or loss of whole chromosomes, and finally to changes in whole genomes (the full complement of chromosomes in the nucleus).

A portion of the DNA molecule, showing linear array of bases (adenine, cytosine, guanine, and thymine)

A point mution (C → T) has occurred at the third position

Figure 6.7 Point mutations in DNA are a source of genetic variation.

The extent to which a mutation spreads through a population or species varies greatly, depending on the importance of the affected region of DNA to the organism. The effect of a mutation may be lethal (when production of essential gene products is disrupted), neutral to the survival of the organism, or selectively advantageous (as when new chromosomal arrangements of genes create beneficial, coordinated gene expression). DNA coding for essential components of cellular metabolism, such as ribosomal subunits, is highly **conserved**, varying little between major groups of organisms, such as conifers and angiosperms. DNA with no clear role in the functioning of the cell, on the other hand, is free to mutate and regularly does so, creating variation that may distinguish closely related groups of organisms.

One of the most important mutation types is the duplication of genes, which creates extra copies of genes that are free to mutate into new genes (or into nonfunctional pseudogenes). A large proportion of genes are apparently the product of gene duplication. Globin genes, the products of which are essential for oxygen transport in our blood, have diversified following several duplications, and even in plants have duplicated and diverged. Leghemoglobin binds oxygen, playing an essential role in nitrogen fixation by legumes and other plants, and plastocyanin, another globin, is a photosynthetic pigment.

The largest mutations are aneuploidy and polyploidy (see Chapter 4). Aneuploidy is pronounced in some groups, such as the large genus of sedges (*Carex*, Cyperaceae), in which the haploid chromosome number ranges from 6 to 56 and every number between 12 and 43 is found in at least one species. Bluegrasses (*Poa*, Poaceae) and willows (*Salix*, Salicaceae) also contain extensive aneuploid series, although the loss of essential genetic material or excess of certain genes attending aneuploidy can be quite harmful. Polyploidy leads to gene duplication and greater genetic diversity upon which natural selection can act. Polyploidy is frequent and evolutionari-

ly important in plants; it is discussed in more detail later in this chapter.

Genetic recombination refers to the shuffling of genes that occurs primarily during meiosis, the special cell division that produces spores. Chromosomes occur in duplicates, one derived from the father and the other from the mother (Figure 6.8A)—each with the same genes in the same order. In meiosis, these duplicate, or homologous, chromosomes pair up and routinely exchange chromosomal segments so that genes are rearranged. Rather than each of the homologues having only maternal or paternal genes, this reciprocal exchange, or **crossing-over**, gives them a mixture of parental genes (Figure 6.8B).

Genes are also recombined during meiosis by **independent assortment** of the chromosomes. Homologues separate after meiotic pairing into spores (Figure 6.8C), and the maternal (or paternal) homologue of one pair may end up in a spore with the maternal or the paternal homologue of any other pair. An organism with 10 pairs of chromosomes, such as corn, can produce 2^{10} (or 1024) genetically different spores and 1024^2 (or 1,048,576) different zygotes just by independent assortment.

Recombination is a rich source of variation. It increases as the frequency of crossing-over and the number of chromosomes increase, and it is affected by population size, pollination system, seed and pollen dispersal, and other factors. Recombination is largely responsible for the great variation observed within natural populations and even among the progeny of a single mating.

Variability is also affected by gene flow, which may introduce new genetic material into a population, and by **random genetic drift**, the chance fixation of genes in small populations. To see how random genetic drift operates, consider a hypothetical population of 10 rose plants in which flower color is controlled by one gene that has two forms (**alleles**). One allele codes for red flowers and is designated by the letter *R*; the other allele codes for white flowers and is designated by the letter *W*.

Figure 6.8 Genetic recombination between parental chromosomes is a primary source of genetic variation.

Each plant bears two alleles (one from each parent). Individuals with two *R* alleles (*RR* genotype) have red flowers; individuals with *WW* genotypes have white flowers; and individuals with *RW* genotypes have pink flowers. The fitness of the three flower colors is the same. Of the 10 plants in this population, 9 have red flowers and 1 has pink flowers. There are 20 total alleles in this group of ten plants; 19 of them are *R* alleles, and 1 is a *W* allele. The frequency of the *R* allele is therefore 95%, and that of the *W* allele only 5%.

Consider the possible outcomes of mating. The red roses produce gametophytes bearing gametes with the *R* allele; the one pink rose produces gametophytes bearing gametes with an *R* allele or a *W* allele. The *W* allele might not participate in matings during one generation and therefore might be lost from the population. The *R* allele would then be the only allele in this population. The probability of the loss of the *W* allele is 95%, and it represents a genetic change that is strictly the result of chance, not of natural selection.

Individual plant variation, which is extensive in almost all plant groups, is discussed in detail by Briggs and Walters (1997). Next we will look at how this individual variation is distributed spatially and partitioned into taxonomic groups.

LOCAL AND GEOGRAPHIC PATTERNS OF VARIATION

The starting point in studies of spatial variation is the **population**. A group of individuals of a species occupying a more or less well defined geographic region and interacting with one another reproductively forms a population. Populations are difficult to define. They range in size from a single individual to millions of individuals, and they may persist for less than a year or thousands of years. They may consist of the offspring of one individual or be regularly stocked by immigrants. This variation in their history, coupled with different levels of genetic diversity, means that populations vary greatly at the genetic level.

Spatial variation may be more or less continuous, forming a **cline**, such as a continuous reduction in the height of trees from the bottom to the top of a mountain. Plants more commonly vary discontinuously in space. White fir (*Abies concolor*, Pinaceae), a conifer of mountainous regions in the western United States and northern Mexico, shows marked geographic discontinuities: Northern California populations are unique in their hairy twigs and notched leaves; Baja California populations have uniquely short, thick leaves; a cluster of populations in Utah has shorter leaves and a slightly different chemical pattern than an otherwise similar set of populations in Colorado and northern New Mexico; and populations in southern New Mexico and Arizona are chemically like Colorado populations but morphologically like southern California populations.

Patterns of variation covering large areas of the globe occur in many groups of plants. One source of some of these patterns, **continental drift**, breaks up continents into pieces and separates taxa into geographically isolated groups that have the potential to diversify. A well-known example of this process, which is called **vicariance**, is provided by the southern beeches (the genus *Nothofagus* of the southern beech family, Nothofagaceae); it is discussed in Chapter 1.

Careful studies of geographic variation often uncover patterns like that of white fir (or of Jeffrey pine, described earlier). Documenting patterns of geographic variation is an important step in understanding systematic diversity.

Speciation

Natural variation at and above the population level is usually not continuous, but occurs in discrete units, or taxa. Easily the most important taxon is the species because it is often the smallest clearly recognizable and discrete set of populations. Understanding how species form and how to recognize them have been major challenges to systematists.

Speciation can be defined as the permanent severing of population systems so that migrants from one system would be at a disadvantage when entering another. This disadvantage could stem from a lack of mates for the migrants if the two systems were reproductively isolated. Alternatively, a migrant might not compete well with residents in withstanding pathogens, pests, and predators or in attracting pollinators and animals to disperse its fruits. Speciation may result either from adaptive changes or from changes that are due to chance.

One of the problems in studying speciation is that it may be a slow process. We see only moments in time and must infer a process from a pattern. For example, in *Gilia*, a genus of small herbs in the Polemoniaceae, some species in southwestern North America contain groups of populations that have been called races. These races differ morphologically, and they often grow together, interbreed, and intergrade. Subspecies show less geographic overlap than races, but some intergradation still occurs. Species of *Gilia* generally are more strongly differentiated from one another and show less tendency to mate. From such patterns of races, subspecies, and species, the inference has been made that speciation is a gradual process of divergence and shutting off of gene flow.

Such gradual divergence through the intermediate steps of geographic races, subspecies, and finally fully separate species is the traditional view of speciation. According to this view, geographic isolation is required to prevent gene flow and allow divergence of the isolated sets of populations. This form of speciation is referred to as **allopatric**, or **geographic**, **speciation**.

Note that divergence in the capacity to interbreed does not automatically result from geographic separation. For example, the sycamore (*Platanus occidentalis*,

Figure 6.9   Distribution of pollen directly downward from its source; most pollen grains fall near the plant. (Modified from Grant 1981.)

Small populations at the edge of a species range are subject to random genetic drift, and environmental conditions at the margin may elicit adaptive change. These random and/or adaptive changes may be significant enough to create a new lineage (a **neospecies**); such neospecies may fail to spread geographically and thus go extinct. If they do expand, their success may depend on their having unique ecological adaptations that allow them to avoid competition with their progenitor. Dispersal from the founding local population binds populations of the neospecies by common ancestry.

Platanaceae) of eastern North America and *P. orientalis* of the eastern Mediterranean region have been geographically separated for millions of years and have diverged from each other morphologically, and yet the artificial hybrid is vigorous, fertile, and frequently planted as a stately ornamental. Widely and long-separated populations may also be morphologically very similar.

A strong argument against the allopatric speciation model is the lack of effective mechanisms for the transformation of geographic races into coherent species (Levin 1993). The two most likely transforming mechanisms are gene flow and natural selection that is uniform in its effect over the range of a race. Genes are transported by pollen grains and fruits, most of which fall near the parent plant and rapidly diminish in numbers away from the source (Figure 6.9).

Gene flow is usually measured in meters (Figure 6.10), and it rarely extends to a kilometer. Overall, the diffusion of genes among populations of a geographic race spread over a large region could require thousands of generations. On the other hand, occasionally pollen or fruit may travel a considerable distance, and gene flow in plants is apparently sufficient to allow the spread of strongly favorable alleles (Rieseberg and Burke 2001).

Natural selection could direct different populations of a geographic race toward a similar end point, but only if important elements of the environment are very similar in different parts of the range of the race. Given the complexity of biotic, physical, and climatic elements of the environment, such uniform change seems unlikely over large geographic regions. The likelihood that chance plays a part in speciation further reduces the plausibility of geographic speciation.

An alternative view of speciation does not require that gene flow be extensive or selection be uniform over long distances. Instead, it postulates that speciation most often occurs in local populations or metapopulations (clusters of several local populations connected by occasional gene flow). This alternative is called **local speciation**, or the peripheral isolation model of speciation (Levin 1993).

Both ecology and genomic changes can play major roles in speciation (Levin 2000). Ecology has been more predominant in some cases, such as the Hawaiian silverswords. As discussed in Chapter 1, evolutionary diversification of this group apparently involves inter-island dispersal followed by numerous, major ecological shifts along moisture gradients. Despite the morphological and ecological diversity in the silversword alliance,

Figure 6.10   Gene flow in animal-pollinated *Phlox* (Polemoniaceae). (A) The distances (in meters) that pollinators traveled between plants. (B) The distances of gene flow. The mean distance ($\bar{x}$JH: place a bar above the "*x*") is greater for gene flow than for pollinators because of carryover; that is, some pollen deposited by a pollinator is picked up by subsequent pollinators and "carried over" to other plants. (From Levin 1981.)

genomes have not diverged in ways that prevent gene flow between species and even genera (Carr 1995; Caraway et al. 2001). More generally, the Hawaiian Islands and other oceanic islands, such as the Canary Archipelago, provide many examples of evolutionary diversification illustrating the interplay of dispersal and ecological diversification (Baldwin et al. 1998).

Under special circumstances, speciation may occur without geographic separation, although some form of strong barrier to gene flow is still necessary. An example of this form of speciation, called **sympatric speciation**, is provided by the genus *Stephanomeria* (Asteraceae). A new species of annual plant, *S. malheurensis*, is postulated to have arisen recently within a population of *S. exigua* ssp. *coronaria* in Oregon (Brauner and Gottlieb 1987). There were fewer than 250 individuals of this new species when it was discovered in the 1960s and early 1970s.

The new species and its progenitor are similar genetically but differ morphologically in quantitative characters (such as achene length). The progenitor (*S. exigua* ssp. *coronaria*) cannot self-pollinate. The derivative (*S. malheurensis*) is predominantly self-pollinated, which reduces gene flow between it and the progenitor. This shift to self-pollination is governed by one gene, making the initiation of this speciation a relatively minor genetic event. Chromosomal structural differences between progenitor and derivative, however, lead to intersterility, a second barrier to gene flow.

Two species in the monkeyflower genus *Mimulus* (Phrymaceae) may illustrate rapid speciation brought about by a barrier to gene flow caused by genetic changes (Bradshaw et al. 1995). *Mimulus lewisii* and *M. cardinalis* of the western United States differ strikingly in floral morphology. Flowers of *M. lewisii* are pollinated by bumblebees; their petals are pink with yellow nectar guides and landing platforms, and they contain low quantities of concentrated nectar. Flowers of *M. cardinalis* are adapted for hummingbird pollination; they are bright red, with a narrow corolla tube and copious nectar.

Despite these major floral differences, experimental hybridizations between the two species yield vigorous and fertile offspring. Natural hybridization between the two species, however, has never been reported, even though they flower at the same time. The floral differences between the species are controlled by genes with relatively large effect. Hence reproductive isolation, and therefore speciation, could have been achieved quite rapidly. Rapid speciation is common in plants via allopolyploid speciation, which is described in the section on polyploidy later in this chapter.

PRESERVATION OF DIVERSITY AGAINST GENE FLOW

Reproductive isolating mechanisms prevent gene flow between many closely related plant species. It is important to understand that the nature of gene flow between and within plant species is not here used as a criterion for their recognition. In the section on species concepts later in this chapter, we reject the general equivalence of plant species and reproductive communities, between which gene flow does not occur and within which gene flow could act as a unifying force.

Gene flow does occur between individuals of many different plant species, and it does not occur between geographically distant populations of widely distributed species, such as the giant reed (*Phragmites australis*), which grows on several continents. Gene flow between plants from different continents must be very weak (or nonexistent), and yet the plants are globally very similar morphologically and are therefore considered members of the same species.

A CLASSIFICATION OF REPRODUCTIVE ISOLATING MECHANISMS

Reproductive isolation largely involves mechanisms that prevent successful mating. Geographic separation is often considered a reproductive isolating mechanism, but it differs from the mechanisms in this classification scheme, all of which act when two species are in contact. Physical separation does isolate species, but one must be careful not to give it too much importance. Remote populations within a wide-ranging species, such as American beech (*Fagus grandifolia*, Fagaceae), which has a geographic range from New Brunswick to northern Florida, could be considered separate species if geographic isolation were the sole criterion for the definition of species. But beeches in Canada and Florida do not differ markedly, and they are recognized universally as members of the same species.

Often sets of populations within species are adapted to different climatic or soil conditions. Eastern red cedar (*Juniperus virginiana*, Cupressaceae) commonly grows in old fields and dry uplands throughout much of eastern North America. From North Carolina to central and northwestern Florida, however, this species grows on sand dunes and coastal river sandbanks. This ecotype is not recognized as a distinct species, just as the Jeffrey pine soil ecotypes described earlier are not separate species because the ecological differences are not accompanied by morphological differences.

Some species, however, are ecologically isolated from close relatives. White ladyslipper orchids (*Cypripedium*, Orchidaceae), for example, often grow on prairies, while yellow ladyslippers inhabit forests. In this case, the ecological difference is strong enough to reduce the fitness of hybrids so it qualifies as a reproductive isolating mechanism.

Reproductive isolating mechanisms can be classified according to the time of their effect during reproduction (Table 6.1): premating (before pollination), postmating but prezygotic (after pollination and before fertilization), and postzygotic (after fertilization). Levin (2000) divides isolating mechanisms into an ecological group, corre-

TABLE 6.1   *A classification of reproductive isolating mechanisms in plants.*

Premating	Postmating, prezygotic
1. Temporal	5. Incompatibility
a. Seasonal	a. Pollen-pistil
b. Diurnal	**Postmating, postzygotic**
2. Ecological	b. Seed
3. Floral	6. Hybrid inviability, including ecological isolation
a. Behavioral	7. Hybrid floral isolation
b. Structural	8. Hybrid sterility
4. Self-fertilization	9. Hybrid breakdown

Source: After Levin 1971.

sponding to the premating group in Table 6.1 and a genomic group, corresponding to the postmating mechanisms of Table 6.1. The ecological group emphasizes that species interact with the environment (including resources like pollinators) in different ways that restrict their capacity to interbreed. Genomic divergence is required for postmating isolation.

Temporal isolation   Flowering at different times of the year (Table 6.1: 1a) is common among many closely related species, such as several *Phlox* species of east-central Illinois and many species of oaks that grow together. Seasonal isolation is also effective among some species of willows (*Salix*). Among seven species from Ontario there is an early-flowering group and a late-flowering group (Figure 6.11).

Experimental hybridizations between the two groups are easy, but natural hybridization appears to be rare, even though the species frequently grow together and are visited by the same kinds of pollinators. This suggests that seasonal isolation is effective, although molecular markers, which may be more sensitive to gene flow than morphological markers, have not been used to test this hypothesis of lack of hybridization. Hybridization also appears to be rare among the species within each group, suggesting that postmating isolating mechanisms must exist as well.

Flowering at different times of day (Table 6.1: 1b) may also effectively isolate species that would otherwise hybridize. Creeping bentgrass (*Agrostis stolonifera*, Poaceae) flowers in the morning, and another bentgrass species (*A. tenuis*) flowers in the afternoon. Unusual weather conditions may lead to simultaneous flowering and hybridization of these two species.

Floral isolation   Floral adaptations to attract different pollinators limit or prevent gene exchange between many species. These adaptations may work through floral structure or through effects on pollinator behavior (also called ethology). **Behavioral isolation** (Table 6.1: 3a) reflects the capacity of pollinators to distinguish floral signals, such as color, shape, and fragrance (see the section on pollination biology in Chapter 4). Bees have long been known for their ability to recognize floral features; their sensitivity to the fragrances of orchids, for example, is the basis of many remarkable cases of reproductive isolation.

Closely related orchids in the genus *Ophrys*, growing primarily in the Mediterranean region, produce different floral fragrances that attract males of different species of bees and wasps. Once the bee or wasp lands on the flower, continued stimulation by floral scents, along with the shape and texture of the flower, trigger the insect's mating instincts (Figure 6.12). Deceived into pseudocopulation with the flower, the insect pollinates the flower.

Another aspect of behavioral isolation is floral constancy, the restriction of a pollinator to one species even though other flowers are available. For example, bees

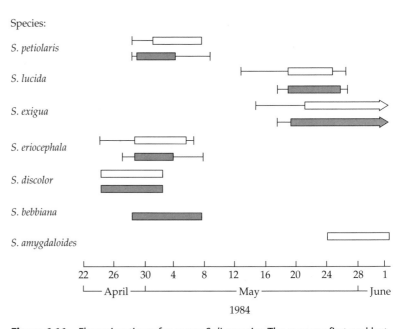

Figure 6.11   Flowering times for seven *Salix* species. The average first and last date of flowering is given by the thick bar; the thin line shows the range of flowering dates. For these dioecious species, darkened bars represent staminate plants, and open bars represent carpellate plants. (From Mosseler and Papadopol 1989.)

(A)   (B)

Figure 6.12   Pollination of orchids in the genus *Ophrys* by pseudocopulation. (A) A male *Andrena* bee attempting to mate with the lip of an *O. fusca* flower gets pollinia on the tip of its abdomen. (B) A male *Eucera* bee attempting to mate with the lip of an *O. scolopax* flower gets pollinia on its head. (From Kullenberg and Bergstrom 1976.)

attracted by the scent of one particular *Ophrys* species tend to visit the flowers of only that species during their forays, thus avoiding interspecific pollinations. Another example comes from the snapdragon species *Antirrhinum majus* and *A. glutinosum* (Plantaginaceae). Bees confine their visits to one species during a foray in mixed plantings, and seeds taken from these mixed plantings yield very few hybrids, even though the species are interfertile.

Two species of *Fuchsia* (Onagraceae) display floral isolation through floral structure (Table 6.1: 3b). Where they are allopatric, *F. parviflora* and *F. enciliandra* ssp. *enciliandra* have flowers that are similar in size and are both pollinated by hummingbirds and bees. The petals of the former species are red, and those of the latter white or pink. Where the species' geographic ranges overlap, however, their flowers diverge dramatically: *F. parviflora* petals are white, and the hypanthium is shorter and broader, thus favoring bee pollination. Plants of *F. enciliandra* in the area of sympatry have red flowers with a longer, narrower hypanthium, making them more attractive to hummingbirds. Because these species are not known to hybridize, these floral changes are thought to be adaptations to reduce competition for pollinators. Such cases of divergence are called **character displacement**.

Hummingbirds are also involved in a well-studied case of floral isolation in *Aquilegia* (Ranunculaceae), the columbines. *Aquilegia formosa* flowers are red, nodding, with short nectar spurs (Figure 6.13), and they are pollinated by hummingbirds. *Aquilegia pubescens* flowers are pale yellow to white, upright in orientation, with long nectar spurs, and they are pollinated by hawkmoths.

The two species grow mostly at different elevations, and when they do come together, some hybridization occurs, possibly because a pollinator other than hummingbirds or hawkmoths pollinates the flowers of both species. However, even though genes are exchanged between the species, the floral differences largely persist. These observations suggest that it is important for the species to maintain their floral syndromes and that floral isolation is the primary barrier to gene flow between them.

Milkweeds (*Asclepias*, Apocynaceae) exhibit another form of structural floral isolation. In this genus, pollen is bulked in packages called pollinia (singular *pollinium*)

that are delivered as a unit to the stigma. The stigmas lie within a slit, so precise orientation of the pollinia is required to effect pollination. The shapes of the pollinia of different species differ sufficiently that pollinia of one species are unlikely to fit the stigmatic slit of another species. Another example of floral structural isolation, involving the genus *Mimulus*, was discussed earlier.

Self-fertilization   The shift from outcrossing to self-fertilization, or selfing, has occurred in many plants. It requires that **self-incompatibility**, the failure of an individual's own pollen to effect self-pollination and self-fertilization, be replaced by **self-compatibility**. Self-pollination is associated with changes in floral morphology, such as reduced amounts of pollen, small corollas and corolla lobes, small stigma lobes, and spatial proximity of stigmas and anthers. Self-compatible plants do not need to invest as much in pollen and corollas to attract pollinators because reproduction can occur through selfing, which is facilitated by the close positioning of stigmas and anthers.

Character	*A. formosa*	*A. pubescens*
Flower position	Nodding	Erect
Spur and sepal color	Red	Red
Petal blade color	Deep yellow	Pale yellow or white
Spur length	10–17 mm	29–37 mm
Petal blade length	2–4 mm	9–12 mm

Figure 6.13   Flowers of *Aquilegia formosa* (left) and *A. pubescens*. Whereas the nodding flowers with short spurs (10–17 mm) of *A. formosa* are adapted for hummingbird pollination, the erect flowers with long spurs (29–37 mm) of *A. pubescens* are suited for hawkmoth pollination. (From Arnold 1997.)

Self-compatibility has arisen multiple times within the same lineage, as exemplified by a study in the genus *Linanthus* (Polemoniaceae). In one part of the genus, a molecular phylogeny based on ITS DNA sequences reveals that what has been called *L. bicolor* is represented by one clade occurring in southern California, a second clade in northern California and Oregon, and a third in Washington and British Columbia (Goodwillie 1999). Plants in these three clades are morphologically similar in floral features that are common in self-compatible plants and lead to their inclusion in one species.

Variation in less conspicuous features, such as calyx trichomes, is consistent with molecular relationships of these three clades to other species, confirms the molecular perspective that *L. bicolor* is polyphyletic, and explains the similarity between the clades as the result of convergent evolution. One hypothesis for the evolution of self-compatibility is that it makes reproduction more likely when cross-pollination is unlikely. Plants in the three clades of *L. bicolor* grow in dry habitats that contain relatively few pollinators when the plants flower. The ancestors of these clades may have responded to a lack of pollinators when they migrated into these habitats by shifting to self-compatibility.

A consequence of a shift to selfing is reduction or elimination of gene flow. Loss of attractiveness of flowers to pollinators and early deposition of pollen onto the stigma of the same flower limit cross-pollination. A well-documented case of such a shift involves grasses that have developed tolerance for heavy metals, such as copper, in the soil around mines (Antonovics et al. 1971). A bentgrass species (*Agrostis tenuis*) and sweet vernal grass (*Anthoxanthum odoratum*) have subpopulations that are tolerant of metals and grow on soil contaminated by mining. Selfing in these subpopulations limits gene flow from nearby, outcrossing but nontolerant subpopulations and the consequent loss of tolerance.

Sympatric speciation of *Stephanomeria malheurensis* (discussed earlier) was mediated in part by a shift to self-compatibility and self-pollination. Many species eliminate cross-pollination altogether with self-pollination in unopened flowers (**cleistogamy**).

Incompatibility   If pollen of one species reaches the stigma of another species, the stigma and style commonly will not permit growth of the foreign pollen tube to the ovule (Table 6.1: 5a). The angiosperm stigma and style serve as effective sieves, allowing the maternal plant to accept or reject pollen. Often the pollen of one species, such as corn (*Zea mays*), will not germinate on the stigmas of a close relative (in this case *Tripsacum*) or grow in its styles (although the reciprocal cross is successful).

A well-documented case of the greater effectiveness of intraspecific pollen over interspecific pollen comes from *Haplopappus* (Asteraceae). If pollen of *H. torreyi* is mixed with pollen of *H. graniticus* and applied to the

stigma of the first species, predominantly nonhybrid progeny are produced. If *H. graniticus* pollen is applied first to *H. torreyi* stigmas and *H. torreyi* pollen is applied up to 24 minutes later, nonhybrid production is still favored. If the interval between application of pollen from a different species and pollen from the same species exceeds 24 minutes, the offspring are mostly hybrids, but a delay of 1 to 2 hours is required to stop production of nonhybrid progeny altogether.

The success of intraspecific pollen over interspecific pollen is also striking in experimental pollinations of *Iris* (Iridaceae). Carney et al. (1994) applied a full range of mixtures of *I. fulva* and *I. hexagona* pollen to the stigmas of these species and then measured the proportion of hybrid and nonhybrid seeds. The results showed that intraspecific pollen contributes to far more seeds than interspecific pollen in all mixtures of pollen except for exclusively interspecific pollen (Figure 6.14). Intraspecific pollen may outcompete interspecific pollen because the pollen tubes of intraspecific pollen grow more quickly in styles of the same species. Hybridization does occur between these two *Iris* species and has been studied intensively (see Box 6D).

Even if a hybrid embryo forms, it may not develop into a viable seed because of incompatibility between the parental genomes within the embryo or between the hybrid embryo and the maternal endosperm (Table 6.1: 5b). The hybrid embryo resulting from the cross of *Primula elatior* and *P. veris* (Primulaceae) exemplifies this failure to mature.

Hybrid inviability   If hybrids are inviable, they will not reach reproductive maturity (Table 6.1: 6), as in *Papaver dubium* × *P. rhoeas* (Papaveraceae). In addition, the habi-

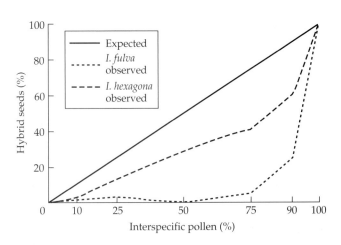

Figure 6.14   Percentage of hybrid seeds produced by different ratios of intraspecific and interspecific pollen on *Iris fulva* and *I. hexagona* flowers. The solid, straight line connecting the corners of the graph represents the expected ratio of hybrid seeds, assuming no difference in the success of intraspecific and interspecific pollen. (From Carney et al. 1994.)

tat requirements of a hybrid are likely to be different from those of either parent. Thus, hybrid inviability may be due to the lack of a suitable ecological niche, a possibility sometimes referred to as ecological isolation.

Hybrid floral isolation The term *hybrid floral isolation* (Table 6.1: 7) refers to the absence of effective pollinators for a hybrid of two species that are adapted to very different pollinators.

Hybrid sterility Hybrids may be perfectly vigorous but sterile (Table 6.1: 8) because of a failure of successful pairing of the chromosomes during meiosis if chromosomes differ in number between the parents or have diverged sufficiently to hinder pairing. Nonfunctional gametes are the result. The cross of broccoli (*Brassica oleracea*; this species of the Brassicaceae includes cabbage and related crops) with radish (*Raphanus sativus*) produces *Raphanobrassica*. First-generation hybrids are vigorous, but the chromosomes from the two parental species do not pair with one another in meiosis, so functional gametes do not form.

Mooring (2001) found extensive hybrid sterility in some F_1 individuals of species in the *Eriophyllum lanatum* (Asteraceae) species complex. In sterile F_1 individuals, chromosomes do not pair with one another successfully in meiosis, and pollen has reduced viability. F_1 offspring of the Hawaiian silverswords similarly show irregular meiosis and low pollen viability (Carr 1995). In the silverswords, however, F_1 individuals retain enough fertility to mate with one of the parents, and the offspring of this mating may have much greater pollen fertility.

Hybrid breakdown Hybrid breakdown refers to problems in later-generation hybrids (Table 6.1: 9). First-generation hybrids may be viable and fertile, but backcrosses or later-generation individuals may be inviable or sterile. The hybrid of two grasses—*Festuca rubra* and *Vulpia fasciculata*—produces few offspring, and these are weak and do not flower.

Combinations of isolating mechanisms may operate together and reinforce one another. In jimsonweeds (*Datura*, Solanaceae), for example, both interspecific pollen-stigma incompatibility and hybrid inviability prevent gene flow between species. Gene flow between two species of avens (*Geum rivale* and *G. urbanum*, Rosaceae) is prevented by seasonal isolation, behavioral isolation, hybrid inviability, and hybrid breakdown (Table 6.1: 1a, 3a, 6, and 9).

Origins of Reproductive Isolating Mechanisms

There are three possible origins of reproductive isolation. First, as two species diverge from a common ancestor, accumulated genetic changes often diminish the likelihood of successful reproduction between them. Recall that isolation may not always be the by-product of speciation (see the sycamore example earlier in this chapter).

Second, premating isolation may be the result of selection against wastage of gametes on matings with other species. To understand how natural selection could select against hybridization, consider two individuals. The pollen and eggs of the first individual are committed exclusively to successful intraspecific matings and are never used for less successful interspecific matings, and the second individual uses some gametes in unsuccessful hybridizations. The first individual will have more offspring, and its mating behavior will become more common within the species.

Two observations demonstrate that selection against hybrids can increase reproductive isolation. Removal of hybrids found in mixed plantings of two corn cultivars—white sweet and yellow flint—markedly decreased overlap in flowering time of the two cultivars (temporal isolation, described earlier). Two species of *Phlox*—*P. cuspidata* and *P. drummondii*—illustrate a shift in floral traits contributing to reproductive isolation when in contact with one another. The two species are fully cross-compatible, and their hybrids are vigorous but mostly sterile. Where the two species grow separately, both have pink flowers. Where they grow together, *P. drummondii* has undergone character displacement and has red flowers.

When both pink- and red-flowered individuals of *P. drummondii* are experimentally introduced into a population of *P. cuspidata*, butterfly pollinators tend to visit flowers of one color on their forays (floral constancy). The result is that the offspring of red-flowered *P. drummondii* include fewer hybrids than those of pink-flowered plants (Table 6.2). Reproductive isolation is further promoted by increased self-compatibility in sympatric populations of these two species.

A third possible origin of reproductive isolation involves selection for prefertilization isolation to reduce competition for animal pollinators. Closely related species often have similar flowers designed to attract the same or similar pollinators. If there are not enough pollinators to service two similarly adapted species, then seed set could be reduced. Flowering at different times and attracting different pollinators may increase the efficiency of the use of a limiting resource and thus elevate fitness.

TABLE 6.2 *Progeny of red- and pink-flowered P. drummondii grown with P. cuspidata.*

| Flower color | Progeny | |
	P. drummondii	Hybrid
Red	181 (87%)	27 (13%)
Pink	86 (62%)	53 (38%)

Source: From Levin 1985.

HYBRIDIZATION AND INTROGRESSION

Reproductive isolating mechanisms are not always effective, and gene flow between plant taxa is common (Rieseberg and Morefield 1995; Ellstrand et al. 1996; Arnold 1997). Strictly speaking, hybridization is mating between unrelated individuals, but the term is most often used for matings between species. Such interspecific hybridization is critically important in plant evolution as a source of novel gene combinations and as a mechanism of speciation. Hybridization has also been very important in plant breeding, in which desirable traits can be moved from one cultivated or even wild species into another cultivated species. The end result may be a new cultivar of agronomic (e.g., tomatoes and strawberries) or horticultural (e.g., roses) importance.

Hybridization is often associated with habitat disturbance (see Box 6B). The ecological adaptations that isolate two species may be broken down by natural disturbances (such as new pests or predators, blowdowns in a forest, fire, erosion, floods, or volcanic activity) that create a habitat suitable for hybrids. Reduced competition in the wake of disturbance also favors the growth of hybrids.

Human disturbance in Europe and North America is thought to have promoted extensive hybridization and subsequent morphological complexity in shadbush [also called serviceberry (*Amelanchier*, Rosaceae)] (Figure 6.15), hawthorn (*Crataegus*, Rosaceae), blueberries and related plants (*Vaccinium*, Ericaceae), and blackberries and raspberries (*Rubus*, Rosaceae). Natural disturbance from volcanism and erosion favors hybridization in many Hawaiian groups (Carr 1995).

Hybridization between individuals belonging to different genera is unusual. It is important to bear in mind, however, that genera are human constructs; there is no clear "mark" for genera. Intergeneric hybridization does occur naturally between genera of the Hawaiian silverswords (Carr 1995) and between several genera of the Maloideae (Rosaceae), such as *Amelanchier*, *Crataegus*, apple (*Malus*), pear (*Pyrus*), *Cotoneaster*, and *Sorbus*. In North America, for example, *Amelanchier* (with simple leaves) and *Sorbus* (with pinnately compound leaves) hybridize at times. The offspring of these morphologically divergent genera, named × *Amelasorbus*, amount to little more than curiosities. The intergeneric hybrid of wheat and rye has produced the commercially successful *Triticale*.

Frequency of hybridization Hybridization is prevalent in plants. There are about 70,000 naturally occurring interspecific plant hybrids in the world according to one estimate (Stace 1984), but not all plant groups hybridize. For example, locoweed (*Astragalus*, Fabaceae), a genus of flowering plants with about 2500 species, is not known for extensive hybridization. In contrast, hybridization is frequent between species of certain genera, some examples of which are presented in Table 6.3 (and see, for example, Figure 6.15).

TABLE 6.3 *Examples of plant groups showing frequent hybridization.*

Family	Genus
Asteraceae	Pussytoes (*Antennaria*), beggar's-tick (*Bidens*), sunflower (*Helianthus*)
Betulaceae	Birch (*Betula*)
Cupressaceae	Junipers (*Juniperus*)
Cyperaceae	Sedge (*Carex*), bulrush (*Scirpus, Schoenoplectus*)
Fagaceae	Oaks (*Quercus*)
Paeoniaceae	Peony (*Paeonia*)
Pinaceae	Pine (*Pinus*), spruce (*Picea*)
Poaceae	Bluegrass (*Poa*)
Polypodiaceae	Spleenwort (*Asplenium*)
Ranunculaceae	Columbine (*Aquilegia*)
Rosaceae	Shadbush (*Amelanchier*; Box 6B, Figure 6.13), hawthorn (*Crataegus*), avens (*Geum*), cinquefoils (*Potentilla*), roses (*Rosa*), blackberries and raspberries (*Rubus*), mountain ash (*Sorbus*)
Salicaeae	Poplars and aspens (*Populus*), willows (*Salix*)

Some large families, such as Apiaceae and Solanaceae, are not known to contain hybrids in certain well-studied parts of the world (Ellstrand et al. 1996). Genera within which hybridization is common tend to be perennials, outcrossers, pollinated by insects, and possessing some means of asexual reproduction. Hybridization is common enough to be an important creative force in plant diversity, to create complex patterns of variation, and in some cases to weaken the morphological distinctness of species.

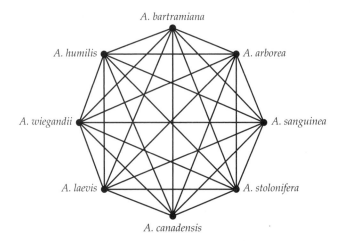

Figure 6.15 Reported natural hybridizations between members of eight species of *Amelanchier* in northeastern North America. (From Campbell and Wright 1996.)

Evolutionary consequences of hybridization Hybridization has five potential consequences (Rieseberg and Ellstrand 1993; Rieseberg and Wendel 1993; Arnold 1994, 1997): (1) reinforcement of reproductive isolating mechanisms, (2) formation of a hybrid swarm through reproduction by hybrids at one site, (3) fusion of two species through interspecific gene flow (introgression), (4) creation of genetic diversity and adaptation, and (5) evolution of new species. Hybridization can therefore maintain biodiversity (consequences 1 and 2), destroy it (consequence 3), or create it (consequences 4 and 5).

The observations on corn and *Phlox* described earlier show that selection against hybridization may strengthen reproductive isolating mechanisms. Species diversity in areas of species overlap may be preserved in this way. Numerous cases of apparent shifts in prefertilization isolating mechanisms are also likely responses to selection against hybridization.

Crossing between two species followed by backcrossing of hybrids with one or both parental species may produce a **hybrid swarm** (Figure 6.16A) when the gene flow is limited to one or a few sites where the species grow together (see Boxes 6B and 6C). The formation of hybrid swarms is sometimes referred to as local introgression.

Hybrids may have higher fitness than the parental species—a well-known phenomenon called **hybrid vigor** (**heterosis**)—and may be adapted for new habitats. Heterosis is familiar in crop varieties; its importance to agriculture is exemplified by the high-yielding varieties of corn, tomatoes, cabbage, and many other species now widely available. It is not clear how important heterosis is in wild plants.

Introgression, the permanent incorporation of genes of one species into another, has three potential consequences. First, it may lead to the merging of different species. *Gilia capitata* represents such a possible reversal of speciation by extensive gene flow (Grant 1963). This species consists of eight geographic races on the Pacific slope of North America. Three of these races are quite distinct and might be recognized as distinct taxa, were it not for the existence of intermediate races produced by hybridization among them. It is hypothesized that the three distinct races attained a maximum level of divergence in the Pliocene (roughly 2 to 5 million years ago). Later, gene flow among them created one species with continuously intergrading races.

Second, introgression may transfer genetic material from one species to another without merging the species (see Figure 6.16B and Box 6D) and thus increase genetic diversity in the introgressed species.

Third, stabilized introgressants may form new species (Figure 6.16C). Hybrid speciation may occur at the diploid level, but it is far more commonly associated with an increase in ploidy. Diploid speciation is discussed later in this section, and speciation at the polyploid level is discussed in the section on polyploidy later in the chapter.

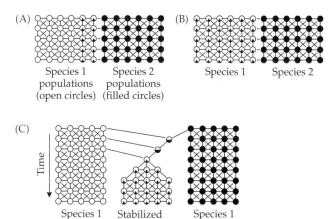

Figure 6.16 Hybrid swarm, dispersed introgression, and the origin of a stabilized introgressant. Black lines represent crosses between populations. (A) In a hybrid swarm (localized introgression), only some of the populations of either of the participating species have genes of the other species, as indicated here by the presence of partial darkening (designating species 2 genes) in some populations of species 1. (B) Dispersed introgression from species 2 into species 1 is indicated by the presence of species 2 genes (some darkening) in all species 1 populations. (C) The origin of a stabilized introgressant from hybridization is followed first by backcrossing to one parental species, and then by reproduction strictly among hybrid populations. (From Rieseberg and Wendel 1993, by permission of Oxford University Press.)

Evidence for hybridization In terms of quantitatively inherited traits, F_1 (first filial generation; i.e., first-generation hybrids) individuals tend to be intermediate between the parents, although there are many exceptions to this general rule (Hardig et al. 2000; Schwarzbach et al. 2001). The dimensions of many plant parts—such as stem height, leaf length, petal length, and fruit diameter—are quantitatively (polygenically) inherited, controlled by several genes, each of which makes a partial contribution to phenotypic expression.

Consider a hypothetical example of hybridization between a shrub with stems under 2 m tall and a tree whose stems commonly exceed 15 m. The hypothetical hybrid would receive many genes coding for short stems from the shrubby parent and many genes for tall stems from the arboreal parent. The F_1 offspring would thus have stems of intermediate height. Box 6B describes an example of F_1 intermediacy in *Amelanchier*.

F_1 intermediacy is commonly expressed in morphology, chemistry, ecology, and in other ways, depending on which characters are examined. Characters governed by just one or a few genes do not show F_1 intermediacy (Rieseberg 1995). Instead, hybrids will have parental, novel, or extreme character states. For example, *Amelanchier bartramiana* is sexual, *A. laevis* produces seed mostly asexually, and their hybrid (*A. × neglecta*), is strongly asexual. The form of asexual seed production (agamospermy) expressed in *Amelanchier* is thought to be controlled by one or a few genes.

BOX 6B *Hybridization in Amelanchier*

Amelanchier bartramiana (mountain shadbush) and *A. laevis* (smooth shadbush) commonly grow together and hybridize over much of northeastern North America. Mountain shadbush usually grows in cool, undisturbed forests or bogs, and smooth shadbush prefers early successional habitats, such as roadsides, recently burned areas, cutover timberlands, and fields. The creation of a road or timber harvesting in the vicinity of a mountain shadbush population establishes conditions for invasion by smooth shadbush and subsequent hybridization with mountain shadbush. The hybrid between these two species is named *A. × neglecta*; the multiplication symbol indicates that it is a hybrid.

Mountain shadbush and smooth shadbush differ conspicuously in their leaves and flowers. Mountain shadbush has short petioles, a tapering leaf base, and short petals. Smooth shadbush has longer petioles, a rounded or heart-shaped leaf base, and longer petals. *A. × neglecta* is intermediate between the two parental species in these characters (Figure 6.17).

A study of hybridization at one site (Weber and Campbell 1989) showed that most individuals of the hybrid are intermediate between the parental species for seven morphological characters that distinguish the two species (see Figure 6.18A in Box 6C; Box 6C provides an explanation of the quantitative methods). These

(A)

(B)

Figure 6.17 From left to right: *Amelanchier laevis*, *A. × neglecta*, and *A. bartramiana* flowers (A) and leaves (B). Note the intermediacy of *A. × neglecta* in petal length, petiole length, and the shape of the base of the leaf. Scale is in millimeters. (Photos by C. S. Campbell.)

characters include dimensions (e.g., petiole length) and quantified shapes (e.g., the angle between the base of the leaf blade and the petiole) that are likely to be quantitatively inherited.

Additional evidence that *A. × neglecta* is the hybrid of *A. bartramiana* and *A. laevis* comes from nuclear DNA content, nuclear ribosomal DNA sequences, and chloroplast DNA restriction sites (see Chapter 5). Nuclei of the hybrid contain amounts of DNA intermediate between those of the two parental species, as one would expect when two genetic constitutions are combined. The two species differ at two nuclear ribosomal DNA (nrDNA) internal transcribed spacer (ITS) nucleotide sites, and the hybrid shows the nrDNA of both parents at both sites.

Hybrids contain chloroplast DNA of one or the other parent because chloroplast DNA is inherited from the maternal parent only. Further-

more, hybridization is possible apparently because the two species flower at about the same time, are visited by the same pollinators, and produce viable, germinable seed when they are experimentally cross-pollinated.

At another site, where *A. laevis* grows with another species—*A.* "erecta" (the name is placed in quotation marks because this species has not been formally named)—morphology suggests that hybridization has proceeded well beyond creation of the F_1 generation, and a hybrid swarm has formed (see Figure 6.19 in Box 6C) (Campbell and Wright 1996). Many plants at this site contain various combinations of parental character states and span the difference between the parental species. Hybridization between *A. laevis* and *A.* "erecta" is also interesting because both species are themselves probably of hybrid origin.

Chemical and molecular characters are more likely than morphological characters to yield predictable expression in hybrids. In the case of nuclear DNA sequences, F_1 individuals show both parental sequences in what is called an additive pattern. In contrast, molecular markers that are inherited from just one parent, such as chloroplast DNA, do not show additive patterns.

Morphological data have provided primary evidence for plant hybridization in hundreds of studies. Several factors other than hybridization can result in single characters showing patterns that resemble F_1 intermedi-

acy, as will be described shortly. In addition, if there is considerable natural variation in a character, then the F_1 may not be intermediate between the parents.

The use of numerous morphological characters to study hybridization avoids these difficulties and increases the precision of analysis. With numerous characters, however, it becomes difficult to interpret all the variation simultaneously; indeed, an area of statistics—multivariate analysis—was developed by the geneticist R. A. Fisher for the purpose of analyzing hybridization in *Iris*. Quantitative morphological analysis of large data sets is described in Box 6C.

BOX 6C *Quantitative Morphological Analysis of Hybridization*

Suppose we are testing the hypothesis that species A and B have hybridized to yield putative hybrid C. If we have identified one quantitatively inherited character that clearly separates A and B, we can ask whether C is intermediate between the parental species by arranging our samples of A, B, and C on a scale for our character.

For two or three characters, we can look at our sample in two or three dimensions. Take, for example, petal length and petiole length in *Amelanchier bartramiana* and *A. laevis*. It is easy to see the intermediacy of *A. × neglecta* when petal length is plotted against petiole length (Figure 6.18A). For more than three characters, however, we cannot readily visualize relationships among A, B, and C, and we must use quantitative morphological techniques.

Figures 6.18B and 6.19 are based on **principal components analysis (PCA)** of seven and six characters, respectively. PCA is a statistical technique that reduces a data set of many characters to one, two, or three new characters. The first principal component is a composite character, the combination of all characters

Figure 6.19 Principal components analysis showing a hybrid swarm involving *Amelanchier* "erecta" (purple shadbush), *A. laevis* (smooth shadbush), and the putative hybrid of these two species. (After Campbell and Wright 1996.)

varying in a largely consistent way that captures maximal variation in the data set. Normally some variation is not explained by the first component because all characters are not perfectly consistent (correlated) with one another.

In Figure 6.19, for example, the first principal component explains 73% of the total variation in the data set. The second principal component explains the largest part of the variation not explained by the first principal component (17% of the total variation in Figure 6.19). The third and subsequent principal components explain the maximum variation not

explained by the preceding principal components.

The components summarize the data and, when plotted against one another as in Figures 6.18B and 6.19, give a visual indication of the relationships of the taxa. In Figure 6.19, 90% of all the variation in the data set of six variables is expressed by the first two principal components. We have not lost much information, but we have improved our ability to see what the data say. For another example of quantitative morphological analysis of hybridization, see Hardig et al. 2000.

Figure 6.18 Quantitative morphological study of hybridization. (A) Biplot of petiole length and petal length for individuals of *Amelanchier bartramiana*, *A. laevis*, and their putative hybrid, *A. × neglecta*. (B) Principal components analysis showing the first two principal components (PC1 and PC2) for a study of *Amelanchier* hybridization. (After Weber and Campbell 1989.)

Careful documentation of the occurrence of F_1 individuals may include a determination of whether there is a statistically significant proportion of intermediate character states (Wilson 1992) for numerous characters and should use independent sets of characters (such as morphology, chemistry, and molecular data; see Boxes 6B and 6D). Care should be taken to discount characters that are correlated or genetically linked because they contain less information than those that are uncorrelated and unlinked.

Processes other than hybridization can generate a pattern resembling F_1 intermediacy. A character state could be intermediate because of mutation, phenotypic plasticity, or an evolutionary trend. For example, taxon B in Figure 6.20 is intermediate between taxa A and C in leaf length not because it is the hybrid of A and C, but because it is an intermediate step in an evolutionary trend toward smaller leaves. One distinguishes such a case of intermediacy from those generated by hybridization by considering other characters, at least some of which are likely to show intermediacy if hybridization has occurred.

A species may also become more or less intermediate between two other species through evolutionary convergence, thus simulating hybridization. The splitting of one species into two derivatives also simulates hybridization (Figure 6.21).

A combination of multiple data sets allows one to distinguish among hybridization, convergence, and diversification. Observations of the presence of both parental species at or near F_1 individuals in nature, of simultaneous flowering by the parents, and of pollinators that transfer pollen between the parents also support hypotheses of hybridization (see Box 6B). One can be sure that a certain plant is a hybrid, however, only when it has been experimentally produced. It is therefore best to refer to potential hybrids as putative hybrids.

Local reproduction by hybrids may produce a hybrid swarm (see Figure 6.16A and Box 6C). In the first *Amelanchier* case described in Box 6C, almost all individuals of *A.* × *neglecta* appear to be F_1 offspring (see Figure

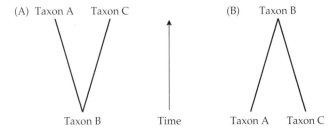

Figure 6.21 Evolutionary divergence versus hybridization. Similar patterns of intermediacy of taxon B between taxa A and C may result from (A) evolutionary divergence of taxa A and C from taxon B and (B) hybridization of taxa A and C to produce taxon B.

6.18B); matings between F_1 individuals and either parental species (**backcrossing**) or among the F_1 offspring to produce the F_2 generation are apparently very uncommon. In the second case, however, a hybrid swarm has formed, and morphological features are being transferred from one species to another, primarily from purple shadbush to smooth shadbush (see Figure 6.19).

Hybrid swarms may confound identification; not only are the two parental species present, but the intermediates potentially span the gap between the parents. Pedicel length in purple shadbush averages 1.7 cm and ranges from 1.4 to 1.9 cm; smooth shadbush fruit pedicels average 2.9 cm in length and range from 2.7 to 3.5 cm. Putative hybrid individuals have pedicels intermediate between the parental species, averaging 2.4 cm, and their range from 1.5 to 3.1 cm spans nearly the full range of the two parental species.

Introgression   Irises of southern Louisiana have been models for understanding introgression (see Box 6D and Arnold 1994). Southern Louisiana supports considerable iris diversity. In the 1930s the numerous forms there were classified into over 80 species. Later, morphological, ecological, and genetic study documented hybridization, and iris diversity was recast into four basic species plus numerous hybrids among them.

Two *Iris* species that grow together in southern Louisiana—*I. fulva* and *I. hexagona*—were central to Edgar Anderson's thesis of introgression, which he presented in his influential 1949 book *Introgressive Hybridization*. After the publication of this book, many plant systematists believed that introgression was relatively common and an important force in plant evolution. But *Iris* introgression was questioned in the 1960s, and few, if any, convincing cases of plant introgression were documented.

Recent molecular studies of Louisiana irises make clear that the first evidence for iris introgression was correctly interpreted. Introgression is common in plants, and there are now numerous well-documented cases

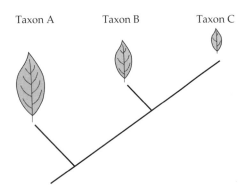

Figure 6.20 A phyletic trend, such as the decrease in leaf size depicted here, may simulate hybridization.

BOX 6D *Introgression in Louisiana Irises*

Iris fulva and *I. hexagona* hybridize in southern Louisiana where the Mississippi River bayou habitat of *I. fulva* mixes with the freshwater swamp and marsh habitat of *I. hexagona*, especially where the habitat has experienced human disturbance. Early studies used flower color—flowers are brick-red in *I. fulva* and blue in *I. hexagona*—and six other floral characters to differentiate the species and hybrids and support hypotheses of hybrid swarms and introgression.

Nuclear (ribosomal DNA, isozymes, and random amplified polymorphic DNA) and chlorplast (cpDNA) data confirm these hypotheses on the basis of morphology (reviewed in Arnold 1992, 1994). Species-specific markers of one species for all four of these types of molecular data have been detected in the other species, both where the two species grow together and where they are allopatric. The evidence therefore clearly indicates that introgression has occurred in both directions between the species.

The two species differ in the presence of an insertion in the intergenic spacer of nrDNA. Whereas allopatric populations of *I. hexagona* in Louisiana and Florida have this insertion in their nrDNA repeats, allopatric populations of *I. fulva* lack the insertion. Two populations are exceptional and exemplify dispersed introgression. One allopatric population is predominantly *I. fulva*, but 20

individuals were found to contain the insertion, although four also contained a minority of repeats lacking the insertion. Of the 12 individuals sampled from the second allopatric *I. fulva* population, four showed only the *I. fulva* type of nrDNA, and the other eight were predominantly *I. fulva*.

Patterns of gene flow in one area of overlap between *I. fulva* and *I.*

Figure 6.22 Nuclear and chloroplast DNA data for individual *Iris* plants from an area of overlap between *I. fulva* and *I. hexagona* near a road and bayou in Louisiana. The circles represent genetic data derived from single plants. (A) The relative proportion of *I. fulva* (shaded portion of circles) and *I. hexagona* (unshaded portion) nuclear markers in a sample of 37 plants. (B) The distribution of *I. fulva* (shaded circles) and *I. hexagona* (unshaded circles) chloroplast DNA markers for the same 37 plants and three additional plants. (After Arnold et al. 1992.)

hexagona differ strikingly for nrDNA and cpDNA. Whereas nuclear genotypes indicate repeated backcrossing into *I. hexagona* (Figure 6.22A), cpDNA genotypes show considerably less gene flow (Figure 6.22B). This difference may be explained by a predominance of gene flow through pollen, which carries nuclear DNA but not maternally inherited cpDNA markers.

---

and many additional possible examples (Rieseberg and Wendel 1993). The major reason for this recognition of the prevalence of introgression is the availability of molecular data. Low levels of introgression are difficult to detect with morphology, and it is difficult in some cases to refute alternatives to introgression.

Introgression may be bidirectional, in which case genes are exchanged between two species, or it may be unidirectional, with genes going exclusively from one species to another. The Hawaiian silverswords present a case of unidirectional introgression. At one site where two old lava flows are near one another, *Dubautia ciliolata* is nearly restricted to a lava flow of 1855, and *D. scabra* is limited to a lava flow of 1935. RAPD (random amplified polymorphic DNA) markers show the occurrence of F_1 individuals, F_2 individuals, and backcrosses in which *D. ciliolata* is the recurrent parent (Caraway et al. 2001).

This unidirectional introgression may be providing *D. ciliolata* with the genetic material to move from the older to the newer lava flow.

Hybrid speciation   Hybridization frequently produces new plant species. Although hybrid speciation is most often associated with polyploidy, there are a few well-documented cases of **diploid hybridization**, wherein a cross between two diploid species produces a diploid hybrid species. Considerable molecular evidence shows that two diploid sunflower species, *Helianthus annuus* and *H. petiolaris* (Asteraceae) hybridized in the past to produce three other diploid species: *H. anomalus*, *H. deserticola*, and *H. paradoxus* (reviewed in Rieseberg and Wendel 1993). These speciation events were mediated by chromosomal differences between the parental species that make the hybrids partially sterile.

Recombination in hybrids restores some fertility, but they remain at least partially sterile with the parents. Another example, *Iris nelsonii*, is a diploid hybrid of three species: the two species discussed in Box 6D (*I. fulva* and *I. hexagona*) and *I. brevicaulis* (see Arnold 1992, 1994). The hybrid of two of these three species hybridized with the third; it is not known which two species formed the initial hybrid.

Past introgression has left its trace in many groups in the form of a conflict between phylogenies based on the uniparentally inherited chloroplast genome (cpDNA) and phylogenies based on nuclear markers (Rieseberg and Soltis 1991). *Helianthus annuus* is a weedy introduced species that has hybridized with several other sunflower species in the southwestern United States. The fate of chloroplast and nuclear genomes after these hybridizations has differed. In almost all cases, *H. annuus* nuclear DNA has been eliminated through backcrossing to the other parental species. *Helianthus annuus* cpDNA has been retained, however, and is now coupled with the nuclear genome of the other species (Figure 6.23).

This phenomenon, called **chloroplast capture** or differential introgression, has been documented in many groups (Rieseberg and Wendel 1993). Chloroplast DNA is more likely than nuclear DNA to cross species boundaries and leave a trace of past introgression (Rieseberg and Soltis 1991). The combination of cpDNA from one individual and nuclear DNA from another creates obvious conflicts in phylogeny reconstruction.

Hybridization and phylogeny reconstruction   Hybrids pose problems for phylogeny because the reticulate patterns of hybridization are fundamentally incompatible with the hierarchical patterns imposed by most current methods of phylogeny reconstruction (McDade 2000). Some systematists argue that hybrids might be detectable by the disruption they cause to phylogenies. Others recommend removing hybrids from data sets prior to phylogenetic study to avoid the potential chaos and poor resolution that hybrids are thought to bring to phylogeny.

McDade (1990, 1992a) produced 17 experimental hybrids in the Central American genus *Aphelandra* (Acanthaceae) and assessed their effect on resolution (number of trees), homoplasy (as measured by the consistency index, CI; see Chapter 2), and hypotheses of phylogenetic relationships among species. Hybrids were added singly or in groups of two to five, randomly selected, to a data set of only parental species. The hybrids did not reduce resolution, although they did significantly lower the CI. They disrupted phylogeny only if they were very common in the ingroup and/or the parents were phylogenetically remote within the ingroup (Figure 6.24).

Problems in phylogeny reconstruction from frequent hybridization may be difficult in groups such as those in Table 6.3, but there are few, if any, well-documented cases of hybridization between phylogenetically remote ancestors.

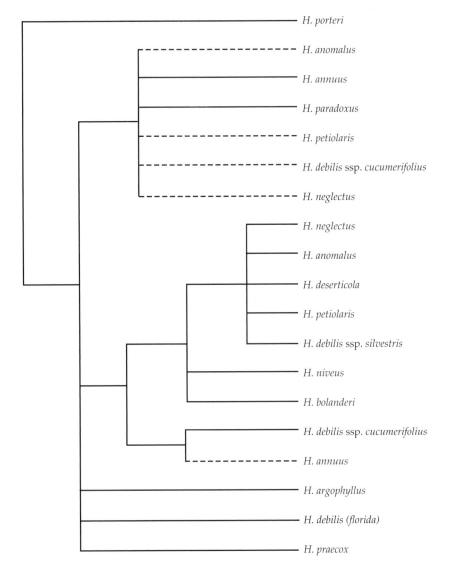

Figure 6.23   Single most-parsimonious tree for *Helianthus* sect. *Helianthus* based on chloroplast DNA data. Dashed lines indicate discrepancies between morphological classification and cpDNA type that are thought to result from cytoplasmic introgression (chloroplast capture). (After Rieseberg et al. 1991.)

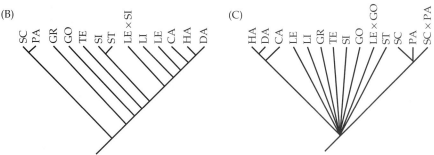

Figure 6.24 Hybridization and changes in topology and loss of resolution in phylogeny. (A) Single most-parsimonious tree for 12 species (each designated by the first two letters of the specific epithet) of Central American *Aphelandra* (Acanthaceae). (B) Effect of addition of hybrids on topology: Addition of the hybrid LE × SI causes rearrangement of the entire tree from part A, with the loss of monophyly of clade SC-PA-GR-GO-TE-SI-ST and a shift in position of clade DA-HA-CA-LE-LI. (C) Consensus tree of three equally most-parsimonious trees with inclusion of hybrids LE × GO and SC × PA, showing loss of resolution. (From McDade 1992a.)

One long-supported example of remote hybridization involved the Maloideae, the apple subfamily of the Rosaceae (Stebbins 1950). A leading hypothesis for the origin of this group was hybridization between ancestors of two other groups of the Rosaceae: subfamily Amygdaloideae (cherries and relatives) and tribe Spiraeeae (bridal wreaths). Molecular data, however, strongly refute this hypothesis and instead indicate that the Maloideae evolved from within one genus, *Gillenia*, of the southeastern United States (Evans and Campbell in press).

Similar studies of *Helianthus* species and three well-documented hybrids showed that inclusion of hybrids did not reduce resolution but did increase homoplasy for molecular data (summarized in Rieseberg and Ellstrand 1993). Homoplasy in morphological and chemical data did not change with the addition of sunflower hybrids. Finally, sunflower hybrids did not affect consensus tree topology.

Note that *Aphelandra* F_1 individuals do not behave differently from nonhybrid taxa in cladistic analyses; therefore, cladistics cannot be used to detect them. If no F_1 individuals can be detected, hybridizations that occurred in the somewhat distant past certainly cannot be detected because evidence for hybridization has been subsequently more or less obscured.

Unless an investigator knows that hybrids are numerous in the ingroup or that there is great evolutionary distance between parents, it is generally acceptable practice to include potential hybrids in phylogenetic analysis. If a hypothesis of hybrid origin is strongly supported for a taxon, then it is not necessary to include it in phylogenetic analysis. Putative hybrids may be added to trees later (e.g., Sang et al. 1995; see also Chapter 2).

POLYPLOIDY

Judging by its high incidence in many plant groups, polyploidy has played a major role in plant evolution. The addition of chromosome sets (see the "Chromosomes" section in Chapter 4) provides redundant genetic material that may be free to mutate into new, adaptive genes. Polyploids thus often exhibit biochemical diversity, producing many kinds of gene products upon which natural selection can operate.

As a result, polyploids tend to be more widely distributed and found in more extreme habitats than their diploid relatives. The shift from diploidy to polyploidy is sometimes accompanied by loss of self-incompatibility and sexuality. Polyploidy is also a prime facilitator of the rapid or quantum speciation process called **allopolyploidization** or **allopolyploid speciation**.

Frequency of polyploidy in plants
Polyploidy is rare in animals because sex determination is based on sex chromosomes and is upset by increased numbers of them. Sex chromosomes are absent in most plants, freeing them to add genomes. Polyploidy varies widely in its frequency among major groups of vascular plants. Chromosome numbers in hundreds of fern species have led to an estimate of 95% polyploidy in that group, although there is good evidence that some relatives of ferns with high chromosome numbers are actually diploid (see the discussion of Lycopodiaceae in Chapter 8).

Only a few conifers, mostly in the Cupressaceae, are polyploid. Masterson (1994) attempted to estimate the frequency of polyploidy in flowering plants by determining primitive chromosome numbers. She used the correlation between cell size and DNA content, and thus

BOX 6E *Allopolyploid Speciation in Tragopogon*

Three diploid *Tragopogon* species were introduced into North America in about 1900: *T. dubius*, *T. porrifolius*, and *T. pratensis*. Hybrids among them were first observed in 1949 in eastern Washington State. *T. dubius* has hybridized with *T. porrifolius* to produce the allotetraploid *T. mirus*, and with *T. pratensis* to produce the allotetraploid *T. miscellus* (Figure 6.25). *T. mirus* and *T. miscellus* are morphologically distinct and reproductively isolated from the parental species because backcrosses are triploid and mostly sterile.

These hybridizations have been analyzed in great detail, using mor-

phology, meiotic chromosome studies, genetic analysis of flower color, a gene controlling ligule (corolla) length, plant pigment chemistry, isozymes, and restriction site analysis of cpDNA and nrDNA (Soltis et al. 1995). Isozymes, cpDNA, and nrDNA indicate that these hybridizations have occurred 2 to 21 times for *T. miscellus* and 5 to 9 times for *T. mirus*.

Chloroplast DNA is maternally inherited in this genus, so the direction of gene flow between the diploid parents can be determined by looking for parental cpDNA in the hybrid offspring. The maternal parent of *T. mirus* is *T. dubius* in all known origins. Both parental species have served as maternal parents for *T. miscellus*. Allopolyploids have arisen numerous times in other groups, such as *Dactylorhiza* (Orchidaceae; Hedren et al. 2001) and possibly *Spartina* (Ayers and Strong 2001), and it appears that single origins of polyploids are unusual (Soltis and Soltis 1999).

What makes these allopolyploid speciation events especially interesting is that they are recent, and when and where they happened are known. They give us the opportunity to follow the ecology and dispersal of the new species, as well as their molecular and morphological evolution. Both of these allopolyploids have increased in numbers and in geographic range in their region of origin in eastern Washington and

Figure 6.26 Distribution map of allotetraploid *Tragopogon* species within five counties of eastern Washington and adjacent Idaho. Populations of *T. mirus* are indicated by stars, populations of *T. miscellus* by open squares. Localities with both species have both symbols. The gray area indicates the continuous range of *T. miscellus* in the vicinity of Spokane. (From Novak et al. 1991.)

northern Idaho (Figure 6.26). In fact, *T. miscellus* has become one of the most common weeds in the vicinity of Spokane, Washington.

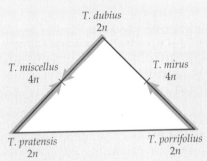

Figure 6.25 Allopolyploid speciations in *Tragopogon*. Diploid parents (2n) are at the points of the triangle, tetraploids (4n) at the midpoints of lines between parents. Arrows start at the maternal parent. Both *T. pratensis* and *T. dubius* have been maternal parents in different originations of *T. miscellus*. (From Soltis and Soltis 1993.)

chromosome number, and based estimates of ancient flowering plant chromosome number on the size of guard cells (leaf epidermal cells that control the opening and closing of stomata). The inferred chromosome number of primitive angiosperms is seven to nine, which means that approximately 70% of angiosperms have polyploidy in their ancestry.

Allopolyploid speciation A well-studied case of allopolyploid speciation involves three Eurasian goatsbeard species (*Tragopogon*, Asteraceae) that were introduced into North America in about 1900. Hybrids among these diploids were first observed in 1949 in eastern Washington. Structural differences in the chromosomes of the three parental species prevent meiotic pairing in their diploid hybrids, making their gametes nonviable. Fertility is restored, however, by chromosome doubling. In the resulting allopolyploid *Tragopogon* hybrids, there is an

exact duplicate with which each chromosome can pair so that meiosis is successful. A more detailed discussion of *Tragopogon* allopolyploidization is provided in Box 6E.

An allopolyploidization in *Spartina* (Poaceae) resembles the *Tragopogon* example in being of recent occurrence following human introduction of a species into a new part of the world. *Spartina alterniflora* (with a diploid chromosome number, 2n, of 62)—salt marsh cordgrass—is an important component of salt marshes on the coasts of eastern North America. Individuals of this species from somewhere between Boston, Massachusetts, and Newfoundland, Canada, were introduced into Southampton, United Kingdom, sometime prior to 1829 and were the seed parent in a hybridization with the native *S. maritima* (2n = 60) that resulted in the sterile F$_1$ *S.* × *townsendii* (2n = 62) by 1870.

The fertile allopolyploid *S. anglica* (2n = 120, 122, 124)—English cordgrass—appeared by the end of the

1880s in Southampton and then spread rapidly along the coast, replacing the native species. A study of RAPD and ISSR markers (Ayers and Strong 2001) showed widespread genetic variation in *S. anglica*, which could be interpreted as the result of multiple origins. Alternatively, the variation could have developed from genetic recombination or loss of entire chromosomes in *S. anglica*.

The genetic data indicate that the *S. alterniflora* component in all but one of the sampled *S. anglica* plants is identical to one plant of *S. alterniflora* in Marchwood, the United Kingdom. The Marchwood plant of *S. alterniflora* may have been the seed parent progenitor of *S. anglica*. The common name "cordgrass" indicates that these plants produce rhizomes, which allows them to persist for potentially long periods of time through vegetative reproduction.

Allopolyploid speciation has played a major role in the evolutionary history of many crops, such as corn (maize), sugarcane, cotton, and wheat, and has presumably facilitated the evolution of the traits that have made these plants useful to humans. One of the most interesting cases of hybridization involving crop species comes from *Brassica*. In 1935 the scientist U proposed that three *Brassica* species—*B. nigra* (black mustard), *B. oleracea* (the species from which brussels sprouts, broccoli, cabbage, cauliflower, kale, and kohlrabi have been developed), and *B. rapa* (= *B. campestris*, turnip)—hybridized to produce three other species.

These relationships, which became known as the "triangle of U" (Figure 6.27), have been confirmed by numerous studies (reviewed by Song et al. 1988). *Brassica* allopolyploids share with other allopolyploids, such as *Tragopogon* (see Box 6E), a history of multiple origins and introgression (Soltis and Soltis 1993).

Brassica is involved in an even earlier and better-known case of allopolyploid speciation: the intergeneric hybrid *Raphanobrassica* (discussed earlier, in the section on reproductive isolating mechanisms). In the 1920s the Russian geneticist Georgi Karpechenko attempted to create a crop combining the edible leaves of cabbage and the root of radish. He found fertile individuals of this hybrid resulting from the union of chromosomally unreduced gametes. These new fertile individuals were allotetraploids with $2n = 36$. Although this result was scientifically interesting, the new crop did not meet expectations; its roots were more like cabbage than radish, and its leaves resembled those of radish more than those of cabbage.

Recent studies show that polyploids are surprisingly dynamic (Soltis and Soltis 1999). In a brief period of time after their formation, polyploid genomes may undergo major restructuring, including exchanges of portions of chromosomes from the genomes of the parental species. The more divergent the parents of a polyploid are, the more drastic the genomic restructuring will be. Numerous studies have demonstrated that the same allopolyploid arises multiple times (see Box 6E). If the parental

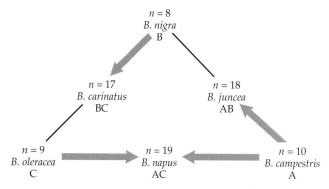

Figure 6.27 The so-called triangle of U. Diploid parents are at the points of the triangle, tetraploids at the midpoints of lines between parents. Haploid chromosome numbers (*n*) and genome consititution (A, B, and C) are given for each species. Arrows originate at the maternal parent. Both *Brassica oleracea* and *B. campestris* have served as the maternal parent in different originations of *B. napus*. (After Soltis and Soltis 1993.)

species are genetically different, then repeated hybridizations between them will produce genetically distinct allopolyploids. Matings between these allopolyploids can, in turn, further expand genetic diversity.

PLANT BREEDING SYSTEMS

The primary means of reproduction in plants is biparental (Richards 1997). Uniparental reproduction via self-fertilization or asexuality is relatively common, however, and it is one of the most frequent evolutionary transitions in plants (Takebayashi and Morrell 2001). About 20 to 25% of plant taxa are predominantly selfing (Barrett and Eckert 1990). Because uniparental reproduction markedly restricts gene flow, it is often associated with complex patterns of morphological variation and with challenges in defining species.

As a starting point, consider a species, such as orchard grass (*Dactylis glomerata*), that forms large populations of plants that always outcross. Such populations will contain considerable genetic diversity and will not be markedly differentiated from one another. In contrast, populations of plants derived from exclusively uniparental reproduction will tend to be genetically invariant and more or less differentiated from other populations of the species (Richards 1996).

Self-pollination and self-fertilization occur as a result of pollen transfer within a flower or among flowers of the same individual. Selfing affects genetic diversity within individuals. Genetic diversity is measured in terms of heterozygosity and homozygosity. **Heterozygosity** is the presence of two alleles (or potentially more if there has been gene duplication) or other genetic elements within an individual. **Homozygosity** is the presence of only one allele within an individual. Selfing reduces heterozygosity and increases homozygosity.

Consider a gene that encodes an enzyme, such as malate dehydrogenase (MDH). A diploid heterozygous

		Pollen	
		MDH-1	MDH-2
Eggs	MDH-1	MDH-1/MDH-1	MDH-1/MDH-2
	MDH-2	MDH-1/MDH-2	MDH-2/MDH-2

Figure 6.28 Self-fertilization in a heterozygote for the malate dehydrogenase gene (*MDH*). Possible genotypes (based on alleles *MDH-1* and *MDH-2*) are shown for eggs and pollen. The offspring are $\frac{1}{4}$ homozygous for *MDH-1*, $\frac{1}{4}$ homozygous for *MDH-2*, and $\frac{1}{2}$ heterozygous.

individual bears two forms of the gene, and thus two forms of the enzyme. These two forms may differ in function, giving the heterozygote some biochemical flexibility that a homozygote would lack. We represent the genotype of such a heterozygote as *MDH-1/MDH-2*, indicating two alleles (1 and 2) of the *MDH* gene. Diploid heterozygotes generate two kinds of gametes: one with the *MDH-1* allele and the other with *MDH-2*.

Self-fertilization results in equal numbers of heterozygotes and homozygotes among the offspring (Figure 6.28). Each selfing thus halves the level of heterozygosity. In our example, we started with a heterozygote and a frequency of heterozygosity of 1. After one generation of selfing, heterozygote frequency is ×. If selfing continues in the four progeny of Figure 6.28, the two homozygotes will produce only homozygous offspring, and the heterozygotes will again produce equal numbers of heterozygotes and homozygotes. After this second generation of selfing, heterozygote frequency is $(\frac{1}{2})(\frac{1}{2}) = \frac{1}{4}$.

Repeated selfing reduces heterozygosity, as well as morphological variation among offspring, to a very low level. As a result, populations of selfers may be morphologically and genetically uniform. Selfing is the most extreme form of **inbreeding**, or mating with relatives. Inbreeders generally have higher levels of homozygosity and more uniform populations than outcrossers.

Many plants rely on a variety of forms of asexual vegetative reproduction (including stolons, rhizomes, buds, and fragments of plants) to colonize an appropriate habitat and for limited dispersal. Asexual seed production, called **agamospermy** (an equivalent term is *apomixis*), differs from vegetative reproduction in retaining seed dispersal and some aspects of sexuality. Normal meiosis and full genetic recombination do not occur in the formation of female gametes by agamosperms. These gametes instead develop through an interrupted meiotic division or through mitotic division, and the egg becomes an embryo without fertilization—a process referred to as **parthenogenesis**. Nevertheless, some genetic variation does develop in agamosperms, even though all sexuality has been lost (Mogie 1992).

Agamospermy occurs in about 34 families and 130 genera of flowering plants and is especially common in the Asteraceae (35 genera contain apomicts), Poaceae (37 genera), and Rosaceae (11) (Asker and Jerling 1992; Carman 1997). Genera with agamospermous species, such as *Amelanchier, Antennaria, Calamagrostis, Cotoneaster, Crataegus, Hieracium, Malus, Poa, Potentilla, Rubus, Sorbus,* and *Taraxacum,* have been as taxonomically difficult as any. If all sexuality is lost, agamospermy is obligate and is the only means of seed production. Usually, however, agamospermy is facultative: It occurs with sexuality in the same individual.

Species Concepts

How do we determine whether two plants are members of the same or different species? This question has been, and continues to be, intensely debated among systematists and evolutionary biologists. Despite the title of his major work, *On the Origin of Species*, Darwin said that species did not matter much and that they were what competent systematists said they were. Several recent books, however, focus on species concepts and speciation (King 1993; Lambert and Spencer 1995; Claridge et al. 1997; Howard and Berlocher 1998; Wilson 1999; Levin 2000; Wheeler and Meier 2000; Schilthuizen 2001). Systematists, evolutionary biologists, population biologists, conservation biologists, ecologists, agronomists, horticulturists, biogeographers, and many other scientists are more interested in species than in any other taxon.

As a starting point, consider the situation in animals, especially large vertebrates, in which the criterion of the capacity to interbreed is usually used to distinguish species. Among most vertebrates, groups of interfertile individuals coincide with morphological, ecological, and geographic groups and units based on other criteria. Hence not only are species relatively easy to define, but also one can test them by testing interfertility.

For example, horses and donkeys are morphologically distinct, and the hybrid of a male donkey and a female horse—the mule—is sterile. This definition of a species as a group of interfertile individuals is usually called the **biological species concept**, or **BSC** (Mayr 1963), although the other species concepts are "biological" as well. The BSC, which is also referred to as the isolation species concept (ISC; Templeton 1989), has dominated the zoological literature (Coyne 1992; Mayr 1992) and, until recently, the botanical literature as well.

The appeal of the BSC lies in its simplicity, its agreement with the neo-Darwinian emphasis on gene flow and allopatric speciation, and its testability. A test of interfertility cannot be applied unambiguously in plants, however, because interfertility varies greatly among plants. The success of matings between members of different groups ranges from 0 to 100% (Figure 6.29), and species assignment in the case of intermediate levels of interfertility is ambiguous (Davis and Heywood 1963).

Figure 6.29 Variation in interfertility in the genus *Nigella* (Ranunculaceae) from the Aegean region. The thickness of lines connecting taxa indicates the percentage of fertility of hybrids formed between the taxa, as indicated in the key. (From Strid 1970.)

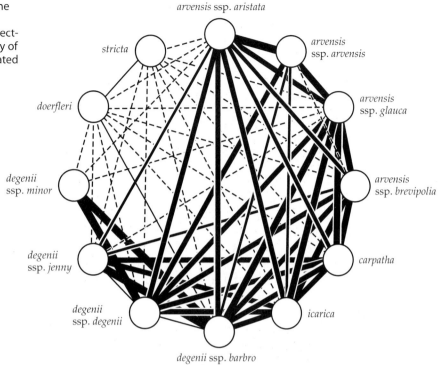

- - - - Failed
——— < 30
—— 30–44
—— 45–59
—— 60–74
—— 75–89
—— 90–100

There are other major problems with the use of mating success or gene flow to distinguish species in plants. As we have seen, gene flow varies greatly among plant groups, creating reproductive communities that range from one or a few individuals, as in selfing individuals and asexual clones, to morphologically heterogeneous assemblages when hybridization occurs between morphologically divergent groups.

Strict application of the BSC would lead to the naming of potentially vast numbers of selfing individuals and asexual clones as new species, many of which would differ little from other such species and would therefore be difficult to identify. Applying the BSC in the presence of frequent hybridization would create species that are broadly inclusive. This approach was actually advocated for the almost 1000 species included in the Maloideae, some genera of which (i.e., *Amelanchier*, *Crataegus*, *Malus*, *Cotoneaster*, and *Sorbus*) contain individuals that occasionally interbreed.

Plant systematists have largely abandoned the BSC (P. H. Davis and Heywood 1963; Ehrlich and Raven 1969; Raven 1976; Mishler and Donoghue 1982; Donoghue 1985; Mishler and Brandon 1987; Nixon and Wheeler 1990; J. I. Davis and Nixon 1992; Kornet 1993; Baum and Shaw 1995; McDade 1995). The occurrence of gene flow between plant species does not mean that they are not distinct lineages unless the hybridization is so pervasive that the species merge. Furthermore, forces other than gene flow must be responsible for the similarity of populations of cosmopolitan species, such as the giant reed (*Phragmites australis*), on different continents. For example, developmental constraints might have dictated a more or less uniform morphology throughout the geographic range.

Other species concepts have been put forth, including recognition, phenetic, evolutionary, and phylogenetic concepts (Table 6.4) (these and others are reviewed in Grant 1981, de Queiroz and Donoghue 1988, Templeton 1989, Baum 1992, Kornet 1993, Rieseberg 1994, Baum and Donoghue 1995, Baum and Shaw 1995, J. I. Davis 1997, Ghiselin 1997, Hull

---

TABLE 6.4 *Seven species concepts.*

Concept	Criterion for species definition	Reference
Biological	Gap in interfertility between species	Mayr 1963
Recognition	Common fertilization system	Paterson 1985
Phenetic	Gap in the variation between species	Sokal and Crovello 1970
Evolutionary	Common evolutionary fate through time	Wiley 1978
Autapomorphy	Monophyly	Donoghue 1985; Mishler 1985
Diagnosability	Unique combination of character states	Nixon and Wheeler 1990
Genealogical	Basal exclusivity	Baum and Shaw 1995

1997, de Queiroz 1998, and Levin 2000; they are also described briefly here). The BSC and the **recognition species concept** (RSC) both focus on the role of gene flow, either as a diversifying force when there is a gap in gene flow (BSC) or as a cohesive force maintaining the similarity of individuals within a species (RSC). The **phenetic species concept** rests on the overall similarity of members of a species, which are separated from other species by a gap in variation. The **evolutionary species concept** centers on the recognition of evolutionary lineages, although it does not clearly prescribe how those lineages are to be identified.

The ascendancy of phylogeny as an organizing principle in systematics motivated a phylogenetic species concept (PSC). A common feature of a PSC is that species are applied to taxa that mark the boundary between reticulate relationships, as one finds among interbreeding members of a species, and divergent relationships of separate lineages between which there is no exchange of genes. Unfortunately, this boundary is often blurred. Because at least three quite different criteria have been advanced for a PSC, this term is ambiguous and therefore is not used in this book.

The criterion of monophyly stipulates that a species contains all the descendants of one ancestral population and is identifiable by autapomorphies (Donoghue 1985; Mishler 1985; Mishler and Brandon 1987; de Queiroz and Donoghue 1988). This concept can be referred to as the **autapomorphy species concept**. Monophyly is an inappropriate criterion below the species level, however, because reticulate, nonhierarchical, and nondivergent relationships are incompatible with grouping by monophyly.

An alternative criterion for a PSC stresses diagnosability and defines a phylogenetic species as the "smallest aggregation of populations (sexual) or lineages (asexual) diagnosable by a unique combination of character states in comparable individuals" (Nixon and Wheeler 1990: 211; see also J. I. Davis and Nixon 1992). The character states must be fixed (invariant) within species.

Although this approach, which can be referred to as the **diagnosability species concept**, is simple and easily understood, there is uncertainty about what constitutes diagnosability. One small genetic characteristic unique to a series of populations would make that series diagnosable, but without other differences, most practicing systematists would not be willing to recognize it as a species. Insistence upon fixation is also problematic. Changes in a local population that initiate speciation and might at one time be fixed in a species can later evolve. This population would therefore not show fixation and thus would not be a species.

Diagnosability is a character-based concept, as opposed to a historically based concept (Baum and Donoghue 1995). A third criterion for a PSC—basal exclusivity—tracks history. Exclusivity stipulates that members of a group must be more closely related to one another than to any organisms outside the group. Exclusive groups that are basal (i.e., contain no less-inclusive groups) form the basis of the **genealogical species concept**.

Exclusivity may be determined by gene **coalescence**; individuals of a species are deemed more closely related to one another than to organisms from another species if their genes share more recent common ancestral genes (coalesce) than they do with individuals in the other species. To see how coalescence applies, consider that, overall, genes in siblings (brothers and sisters) share a more recent common ancestral gene than do genes in first cousins. Use of coalescence is potentially flawed because different genes often give different patterns of coalescence (Doyle 1995); some genes link some sets of individuals, and others link different sets.

From the preceding discussion it should be clear that there is no consensus about species concepts in plants. Part of the reason for this lack of consensus is that diversity is idiosyncratic. Each lineage has a unique history of genetic, morphological, and ecological changes; of interactions with other species and the physical world; and of migration, dispersal, and chance. Species differ to varying degrees in morphology, genes, ecology, geographic distribution, pollinators, breeding system, phenotypic plasticity, fruit dispersal, disease resistance, competitive ability, genome size, and numerous other aspects. We see in the world everything from the very beginning of the formation of new species and all subsequent stages of the speciation process up to species that have been established for a long time and are clearly differentiated from other species. Hence each species is unique.

Another difficulty with species concepts is that, although they should ideally be general, applicable, and theoretically significant, these three attributes often conflict (Hull 1997). For example, the BSC has theoretical significance but, as noted already, there are problems with its application. Moreover, it is not general (universal) because its requirement of sexuality excludes asexually reproducing organisms. Asexuality and hybridization have posed large challenges to species concepts. Since both phenomena are relatively frequent in plants (and sometimes are both common in the same group of plants), it is not surprising that the species problem in plants can be daunting. For the most intractable plant species situations, all species concepts are difficult to apply.

The goal of systematics is to infer the products of the unique evolutionary histories of organisms and the relationships of those organisms. To this end, all information may contribute. Because humans detect other organisms with our senses, especially sight, morphological markers are the primary evidence we use in species delimitation. Morphology is also the most accessible source of data about evolutionary relationships, and it remains the common way to recognize most species. Data from other sources, such as molecular markers, ecology, interfertility, geography, and pollination biology, are important as independent measures of the evolutionary reality of species.

Taxa are hypotheses, open to repeated testing as new data or methods of analysis become available. The basis for a taxonomic hypothesis is the systematist's knowledge of a plant group. This knowledge rests on field and laboratory study of morphology, ecology, breeding system, gene flow between closely related species, geographic distribution, and as many data as possible about structural and molecular variation.

Species are important units in disciplines other than systematics—notably agronomy, biogeography, conservation biology, ecology, genetics, horticulture, and physiology—and ideally should be recognizable by nonspecialists, people who have not dedicated months or years to the understanding of a group of plants. The use of morphologically significant variation is thus essential in practice. If species in a particular genus really are inherently difficult to distinguish, then systematists can assign those species to groups that the nonspecialist will be able to perceive without great difficulty. Some options are outlined in the following section.

CASE STUDIES IN PLANT SPECIES

Easily recognized species   An easily recognized species is characterized by outcrossing, unrestricted fertility among its members, and strong reproductive isolating mechanisms between its members and those of other species. Such species will not be internally fragmented and will not merge with other species through hybridization. Most vertebrate and vascular plant species fit this pattern. *Dactylis glomerata*, orchard grass, is such a species; it is easily recognized by field biologists and does not pose serious systematic problems. For these groups of plants, application of the different species concepts discussed in this chapter is likely to identify the same set of species.

Microspecies   Microspecies are minimally differentiated series of populations derived from uniparental reproduction. Members of the *Andropogon virginicus* species complex are abundant weeds of early succession in the eastern United States. Many of the taxa in this group differ from one another in minute, subtle ways and are difficult to identify. These taxa frequently grow together, and they flower at the same time, but they rarely hybridize (Campbell 1983). They have diverged little morphologically and yet are reproductively isolated. Some of the species in this complex are apparently outcrossing, and others are not only self-compatible but regularly self-pollinate because of cleistogamy.

Shifts from outcrossing to selfing have at least partially isolated some taxa in this group. In one case the shift from outcrossing to predominant selfing appears to be the result of precocious maturation of the flowers so that they self-pollinate before being exposed for outcrossing.

Taxa that are reproductively isolated without much apparent phenotypic divergence are also referred to as **sibling species** or **cryptic species**.

Another example of cryptic species comes from *Asplenium nidus* (Polypodiaceae), an epiphytic fern of the Old World tropics. In a study of this species in Indonesia, three types of plants could be distinguished on the basis of *rbcL* DNA sequences (Yatabe et al. 2001). These three types all have the same chromosome number ($2n = 144$) and are indistinguishable morphologically, but they have different habitat preferences.

Types A and B grow at the same elevations, but type A grows on the deeply shaded lower portions of the trunks of trees, whereas type B is found in partial shade on the upper parts of trees. Type C grows at mostly higher elevations than types A and B and not with type B and seldom with type A. Controlled experimental hybridizations between types A and B and between types A and C failed to produce any hybrids between *rbcL* types. Crosses between types B and C were not tested. The authors concluded that the three types are cryptic species.

A solution to the microspecies problem is to reserve the species category for more strongly marked taxa, which can be broadly defined to encompass closely related but more obscure taxa. In the *Andropogon virginicus* complex, two categories below the species level have been used: varieties and variants. Varieties represent a level of divergence lower than that among species. Varieties are formal taxa represented by trinomials, such as *A. virginicus* var. *glaucus*. Variants are less distinct than varieties, corresponding in some cases to microspecies. They are informal taxa, with English names, such as *A. virginicus* var. *glaucus*, the wetlands variant.

Agamospecies   Agamospecies are agamospermous microspecies. Many genera—for example, *Alchemilla*, *Amelanchier*, *Antennaria*, *Calamagrostis*, *Crataegus*, *Hieracium*, *Poa*, *Potentilla*, *Rubus*, *Sorbus*, and *Taraxacum*—contain one or more **agamic complexes**. Such groups include sexual taxa that hybridize; the hybrids acquire agamospermy. Interaction between hybridization and agamospermy may create particularly intricate patterns of variation.

Agamospermy in these genera is often facultative, and some functional pollen is regularly produced. Hence agamosperms may serve occasionally as ovulate parents and regularly as pollen parents in matings with other facultative agamosperms or sexual species. The sterility problem that plagues sexual hybrids is not present in agamosperms. Species names by the hundreds, especially in *Crataegus* and *Rubus*, have been given to introgressants stabilized by agamospermy. A key element in understanding agamic complexes is identification of sexual taxa because these are the foundations of the complex (Bayer 1987).

Taraxacum, the dandelions, is probably the most extreme example of proliferation of agamospecies. Specialists in this genus recognize over 2000 species, most of them agamospecies (Kirschner and Stepánek 1994). Up to 100 microspecies have been documented from a 1-hectare area, creating for the nonspecialist significant difficulties in identifying the bewildering multitude of weakly differentiated taxa. Use of the taxonomic rank of section, a group of species that are phenetically similar, has been advocated to simplify *Taraxacum* taxonomy.

Species that hybridize extensively Species of oak (*Quercus*, Fagaceae) readily hybridize. This genus contains 500 to 600 species, many of which are common, dominant forest trees. Hybridization between closely related species makes the task of identifying oaks—already a challenge because of the large number of species, phenotypic plasticity, and juvenile versus adult character differences—especially difficult.

The consequences of hybridization in the genus range from infrequent interspecific gene flow, to hybrid swarm formation, to sets of hybridizing species referred to as syngameons, to introgression. A **syngameon** is the most inclusive unit of interbreeding in a hybridizing species group (Figure 6.30). Adherents of the BSC use a term such as *semispecies* for each member of a syngameon.

Species that are geographically but not reproductively isolated Morphological divergence does not necessarily ensure reproductive isolation, as we saw in the case of *Platanus occidentalis* and *P. orientalis*. A spectacular case of speciation without reproductive isolation is Hawaiian *Bidens* (Asteraceae). The Hawaiian Islands have provided many insights into speciation because they are young (5.7 million to 700,000 years old) and are the most isolated land surfaces on Earth.

Colonization of these islands by a group often began with one or a small number of immigrants, which subsequently experienced spectacular diversification. Hawaiian *Bidens* follow this pattern (Ganders and Nagata 1984). Descended from one presumed ancestral immigrant, the genus now ranges from sea level to 2200 m in elevation and from semidesert with less than 0.3 m rainfall per year to rainforests and montane bogs with more than 7 m of annual rainfall.

Taxonomists recognize 19 species and 8 subspecies in Hawaii, which together exhibit a greater range of morphological and ecological diversity than the rest of the genus on five continents (Figure 6.31). The great morphological diversity within Hawaiian *Bidens* is not matched by genetic diversity, however; the different *Bidens* species are as genetically similar as populations of most other species. Morphological evolution has progressed much more rapidly than genetic evolution.

Almost all artificial crosses between Hawaiian *Bidens* species are successful; F$_1$ individuals produce good pollen and seed. Natural hybridization, however, is relatively rare; only about 5% of all possible hybridizations have been documented in nature. Hybridization does

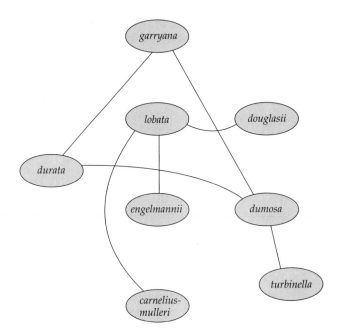

Figure 6.30 Semispecies in the California white oak syngameon. Lines connect species that produce fertile hybrids. (After Grant 1981.)

not occur in 85% of the possible cases because the species grow on different islands or different mountain ranges on the same island, and in an additional 8% because of ecological or seasonal isolation. Divergence without reproductive isolation is common in Hawaii (Carr 1995).

GUIDELINES FOR RECOGNIZING PLANT SPECIES
The following list provides a few basic guidelines aimed at making it easier to recognize plant species:

1. *Get to know the plants in the field.* Extensive field studies throughout the growing season and throughout the geographic range provide invaluable information about morphology, local patterns of variation and discontinuities, breeding system, flowering time, pollination, ecology, distribution, and the nature of any gene flow. Field studies are essential for rigorous sampling of structural, breeding system, and molecular variation.

2. *Collect data on morphological variation, molecular variation, breeding system, flowering time, pollination, ecology, distribution, and gene flow.* Combinations of data, such as molecular and morphological, may complement one another. In studies of hybridization in *Salix* (Salicaceae) (Hardig et al. 2000) and altitudinal variation in *Eucalyptus* (Myrtaceae) (McGowen et al. 2001), for example, both molecular and morphological data provided useful insights. Taxonomic hypotheses are only as good as the data on which they are based, so it is important to establish rigorous sampling plans to adequately represent the variation. In temperate regions, considerable data about some plant groups

may already exist. These extant data should not be accepted uncritically, but evaluated carefully and skeptically. Sampling by previous researchers may have been incomplete, data may have been misinterpreted or incorrectly scored, or important data may have been overlooked. A search for new systematic evidence may be very rewarding. In less-studied regions, such as the tropics, extant data will often be minimal, necessitating more intensive field study and sampling.

3. *Analyze systematic data rigorously and display the results graphically to facilitate their interpretation.* Large amounts of data are not readily synthesized and understood simply by visual inspection. Commercially available software (e.g., NTSYS-pc; Rohlf 1993) includes a wide range of tools for statistical analyses of systematic data and graphic display of results. One of the most appropriate analytic tools is ordination, such as principal components analysis (see Box 6C). Ordination and other statistical approaches have been used to assess the species issue by, for example, Freudenstein and Doyle (1994), Kellogg (1985), Hansen et al. (2000), and Spooner et al. (2000).

4. *Hypothesize speciation scenarios and test them by observation and experimentation.*

Summary

Systematists study the results of evolution. We view the organic world as a tree of life with species at the branch tips that are connected to one another by common ancestry. We seek to understand how the raw material of evolution—heritable variation—is generated, broken up into discrete units such as species, and affected by gene flow. Our greatest challenge is selecting appropriate units to call species.

Darwinian natural selection and random genetic drift have driven the evolution of tremendous plant diversity. Inherited variation leads to differential reproduction via natural selection. Mutation and genetic recombination provide the variation required for evolutionary change. Mutation includes changes in DNA, from single-base changes to the addition of whole genomes. Recombination continually presents new arrangements of genes to be "tested" for their evolutionary fitness.

Speciation usually requires some ecological or geographic separation of population systems to prevent gene flow. Allopatric speciation involves the splitting of one species into two or more population systems that diverge in physical isolation. However, gene flow is not likely to be effective in genetically integrating geographically widespread races. Local speciation best explains the origin of divergence in small, marginal populations, and the coherence of the species that result through common ancestry.

(A) *B. hillebrandiana* ssp. *polycephala* (Maui)

(B) *B. mauiensis* (Maui)

(C) *B. cosmoides* (Kauai)

(D) *B. menziesii* ssp. *filiformis* (Hawaii)

Figure 6.31   Examples of four Hawaiian *Bidens* species, most of which are interfertile. (From Ganders and Nagata 1984.)

Some closely related species might well merge when they contacted one another if gene flow were unchecked. Many mechanisms of reproductive isolation retard interspecific gene flow. Natural selection may favor premating isolating mechanisms because they prevent hybridization or reduce competition for resources. Isolating mechanisms may also have evolved simply as a by-product of genetic divergence during speciation.

Hybridization, which is common in plants, can destroy, maintain, or create species diversity. Sometimes only the F_1 generation is formed because of factors such as hybrid sterility. If hybrids reproduce, they may backcross with parents at the site of hybridization and create a hybrid swarm. If seeds or pollen of hybrid offspring disperse from the original site of hybridization, introgression may occur.

The basic incompatibility between the reticulating phylogenetic patterns of hybridization and the hierarchical patterns of most methods of phylogenetic analysis is a concern when hybrids are very common in the ingroup and/or the parents come from remote lineages in the ingroup. Hybridization is closely linked to polyploidy through allopolyploid speciation, which involves hybridization and subsequent doubling of the chromosome complement.

Polyploidy is estimated to occur in the ancestry of a majority of plant species and is therefore an important pathway in their diversification. Allopolyploids are highly dynamic, with multiple origins in the majority of studied cases, as well as the capacity for major and rapid genetic change through genome reorganization.

Plants differ from animals markedly in the diversity of their breeding systems. The prevalence of uniparental reproduction by selfing and agamospermy in plants fosters patterns of variation that do not readily fit into a taxonomic framework.

The species problem is heightened in plants by interspecific gene flow and by the compartmentalization of variation in uniparental reproducers. Consequently, plant systematists have largely abandoned the biological species concept. In defining species, they look for morphological, molecular, ecological, reproductive, geographic, and other evidence of independent and well-marked lineages. It is difficult to generalize about plant species because each one has a unique history.

Literature Cited and Suggested Reading

Items marked with an asterisk (*) are especially recommended to those readers who are interested in further information on the topics discussed in this chapter.

Anderson, E. 1949. *Introgressive hybridization.* Wiley, New York.

Antonovics, J., A. D. Bradshaw and R. G. Turner. 1971. Heavy metal tolerance in plants. *Adv. Ecol. Res.* 7: 1–85.

Arnold, M. L. 1992. Natural hybridization as an evolutionary process. *Annu. Rev. Ecol. Syst.* 23: 237–261.

Arnold, M. L. 1994. Natural hybridization and Louisiana irises. *BioScience* 44: 141–147.

*Arnold, M. L. 1997. *Natural hybridization and evolution.* Oxford University Press, New York.

Arnold, M. L., J. J. Robinson, C. M. Buckner and B. D. Bennett. 1992. Pollen dispersal and interspecific gene flow in Louisiana irises. *Heredity* 68: 399–404.

*Asker, S. E. and L. Jerling. 1992. *Apomixis in plants.* CRC Press, Boca Raton, FL.

Ayers, D. R. and S. R. Strong. 2001. Origin and genetic diversity of *Spartina anglica* (Poaceae) using nuclear DNA markers. *Am. J. Bot.* 88: 1863–1867.

Baldwin, B. G. and R. H. Robichaux. 1995. Historical biogeography and ecology of the Hawaiian silversword alliance (Asteraceae). New molecular phylogenetic perspectives. In *Hawaiian biogeography. Evolution on a hot spot archipelago.* W. L. Wagner and V. A. Funk (eds.), 259–287.

Baldwin, B. G., D. J. Crawford, J. Francisco-Ortega, S.-C. Kim, T. Sang and T. F. Stuessy. 1998. Molecular phylogenetic insights on the origin and evolution of oceanic island plants. In *Molecular systematics of plants II: DNA sequencing,* D. E. Soltis, P. E. Soltis and J. J. Doyle (eds.), 410–441. Kluwer, Boston.

Barrett, S. C. H. and C. G. Eckert. 1990. Variation and evolution of mating systems in seed plants. In *Biological approaches and evolutionary trends in plants,* S. Kawano (ed.), 229–254. Academic Press, New York.

Baum, D. 1992. Phylogenetic species concepts. *Trends Ecol. Evol.* 7: 1–2.

*Baum, D. and M. J. Donoghue. 1995. Choosing among "phylogenetic" species concepts. *Syst. Bot.* 20: 560–573.

Baum, D. and K. L. Shaw. 1995. Genealogical perspectives on the species problem. In *Experimental approaches to plant systematics* (Monographs in Systematic Botany from the Missouri Botanical Garden, Vol. 53), P. C. Hoch and A. G. Stephenson (eds.), 289–303. Missouri Botanical Garden, St. Louis.

Bayer, R. J. 1987. Evolution and phylogenetic relationships of the *Antennaria* (Asteraceae: Inuleae) polyploid agamic complex. *Biol. Zentralblatt* 106: 683–698.

Bradshaw, H. D., Jr., S. M. Wilbert, K. G. Otto and D. W. Schemske. 1995. Genetic mapping of floral traits associated with reproductive isolation in monkeyflowers (*Mimulus*). *Nature* 376: 762–765.

Brauner, S. and L. D. Gottlieb. 1987. A self-compatible plant of *Stephanomeria exigua* var. *coronaria* (Asteraceae) and its relevance to the origin of its self-pollinating derivative *S. malheurensis. Syst. Bot.* 12: 299–304.

*Briggs, D. and S. M. Walters. 1997. *Plant variation and evolution,* 3rd ed. Cambridge University Press, Cambridge, UK.

Campbell, C. S. 1983. Systematics of the *Andropogon virginicus* complex (Gramineae). *J. Arnold Arbor.* 64: 171–254.

Campbell, C. S. and W. A. Wright. 1996. Agamospermy, hybridization, and taxonomic complexity in northeastern North American *Amelanchier* (Rosaceae). *Folia Geobot. Phytotax.* 31: 345–354.

Caraway, V., G. D. Carr and C. W. Morden. 2001. Assessment of hybridization and introgression in lava-colonizing Hawaiian *Dubautia* (Asteraceae: Madiinae) using RAPD markers. *Am. J. Bot.* 88: 1688–1694.

Carman, J. G. 1997. Asynchronous expression of duplicate genes in angiosperms may cause apomixis, bispory, tetraspory, and polyembryony. *Biol. J. Linnean Soc.* 61: 51–94.

Carney, S. E., M. B. Cruzan and M. L. Arnold. 1994. Reproductive interactions between hybridizing irises: Analyses of pollen tube growth and fertilization success. *Am. J. Bot.* 81: 1169–1175.

Carr, G. D. 1995. A fully fertile intergeneric hybrid derivative from *Argyroxiphium sandwicense* ssp. *macrocephalum × Dubautia menziesii* (Asteraceae) and its relevance to plant evolution in the Hawaiian islands. *Am. J. Bot.* 82: 1574–1581.

Claridge, M. F., H. A. Dawah and M. R. Wilson (eds.). 1997. *Species: The units of diversity.* Chapman & Hall, London.

Cockrum, E. L. and B. J. Hayward. 1962. Hummingbird bats. *Nat. Hist.* 71(8): 38–43.

Coyne, J. A. 1992. Genetics and speciation. *Nature* 355: 511–515.

*Darwin, C. 1859. *On the origin of species.* Murray, London.

Darwin, C. 1876. *The effects of cross- and self-fertilization in the vegetable kingdom.* Murray, London.

Darwin, C. 1877. *The various contrivances by which orchids are fertilised by insects,* 2nd ed. Appleton, New York.

Davis, J. I. 1997. Evolution, evidence, and the role of species concepts in phylogenetics. *Syst. Bot.* 22: 373–403.

Davis, J. I. and K. C. Nixon. 1992. Populations, genetic variation, and the delimitation of phylogenetic species. *Syst. Biol.* 41: 421–435.

*Davis, P. H. and V. H. Heywood. 1963. *Principles of angiosperm taxonomy.* Van Nostrang, Princeton, NJ.

de Queiroz, K. 1998. The general lineage concept of species, species criteria, and the process of speciation. In *Endless forms: Species and speciation,* D. J. Howard and S. H. Berlocher (eds.), 57–75. Oxford University Press, New York.

de Queiroz, K. and M. J. Donoghue. 1988. Phylogenetic systematics and the species problem. *Cladistics* 4: 317–338.

de Queiroz, K. and M. J. Donoghue. 1990. Phylogenetic systematics and species revisited. *Cladistics* 6: 83–90.

Dilcher, D. L. and P. R. Crane. 1984. *Archaeanthus*: An early angiosperm from the Cenomanian of the western interior of North America. *Ann. Missouri Bot. Gard.* 71: 351–383.

Donoghue, M. J. 1985. A critique of the biological species concept and recommendations for a phylogenetic alternative. *Bryologist* 88: 172–181.

*Donoghue, M. J. 1992. Homology. In *Keywords in evolutionary biology*, E. Fox Keller and E. Lloyd (eds.), 170–179. Harvard University Press, Cambridge, MA.

Doyle, J. J. 1995. The irrelevance of allele tree topologies for species delimitation and a non-topological alternative. *Syst. Bot.* 20: 574–588.

Ehrlich, P. R. and P. H. Raven. 1969. Differentiation of populations. *Science* 165: 1228–1232.

Ellstrand, N. C., R. Whitkus and L. H. Rieseberg. 1996. Distribution of spontaneous plant hybrids. *Proc. Natl. Acad. Sci. USA* 93: 5090–5093.

Evans, R. C. and C. S. Campbell. In press. DNA sequences from the granule-bound starch synthase gene clarify the origin of the Maloideae (Rosaceae). *Am. J. Bot.*

Faegri, K. and L. van der Pijl. 1979. *The principles of pollination ecology*, 3rd ed. Pergamon, Oxford, UK.

Freudenstein, J. V. and J. J. Doyle. 1994. Plastid DNA, morphological variation, and the phylogenetic species concept: The *Corallorhiza maculata* (Orchidaceae) complex. *Syst. Bot.* 19: 273–290.

Furnier, G. R. and W. T. Adams. 1986. Geographic patterns of allozyme variation in Jeffrey pine. *Am. J. Bot.* 73: 1009–1015.

Ganders, F. R. and K. M. Nagata. 1984. The role of hybridization in the evolution of *Bidens* on the Hawaiian islands. In *Plant biosystematics*, W. F. Grant (ed.), 179–194. Academic Press, Toronto, Ontario.

Ghiselin, M. T. 1997. *Metaphysics and the origin of species*. State University of New York Press, Albany.

Goodwillie, C. 1999. Multiple origins of self-compatibility in *Linanthus* section *Leptosiphon* (Polemnoniaceae): Phylogenetic evidence from internal transcribed spacer sequence DNA. *Evolution* 53: 1387–1395.

Grant, V. 1963. *The origin of adaptations*. Columbia University Press, New York.

*Grant, V. 1981. *Plant speciation*, 2nd ed. Columbia University Press, New York.

Hansen, K. T., R. E. Elven and C. Brochmann. 2000. Molecules and morphology in concert: Test of some hypotheses in arctic *Potentilla* (Rosaceae). *Am. J. Bot.* 87: 1466–1479.

Hardig, T. M., S. J. Brunsfeld, R. S. Fritz, M. Morgan and C. M. Orians. 2000. Morphological and molecular evidence for hybridization and introgression in a willow (*Salix*) hybrid zone. *Mol. Ecol.* 9: 9–25.

Hedren, M., M. F. Fay and M. W. Chase. 2001. Amplified fragment length polymorphisms (AFLP) reveal details of polyploid evolution in *Dactylorhiza* (Orchidaceae). *Am. J. Bot.* 88: 1868–1880.

Howard, D. J. and S. H. Berlocher (eds.). 1998. *Endless forms: Species and speciation*. Oxford University Press, New York.

Hull, D. L. 1997. The ideal species concept— And why we can't get it. In *Species: The units of diversity*, M. F. Claridge, H. A. Dawah and M. R. Wilson (eds.), 357–380. Chapman & Hall, London.

Kellogg, E. A. 1985. A biosystematic study of the *Poa secunda* complex. *J. Arnold Arbor.* 66: 201–242.

King, M. 1993. *Species evolution: The role of chromosome change*. Cambridge University Press, Cambridge, UK.

Kirschner, J. and J. Stepánek. 1994. Clonality as a part of the evolutionary process in *Taraxacum*. *Folia Geobot. Phytotax.* 29: 265–275.

Kornet, D. J. 1993. *Reconstructing species: Demarcations in genealogical networks*. Unpublished doctoral dissertation, Institut voor Theoretische Biologie, Rijksuniversiteit, Leiden, Netherlands.

Kullenberg, B. and G. Bergstrom. 1976. The pollination of *Ophrys* orchids. *Bot. Notiser* 129: 11–19.

Lambert, D. M. and H. G. Spencer (eds.). 1995. *Speciation and the recognition concept: Theory and application*. Johns Hopkins University Press, Baltimore.

*Levin, D. A. 1971. The origin of reproductive isolating mechanisms in flowering plants. *Taxon* 20: 91–113.

*Levin, D. A. 1981. Dispersal versus gene flow in plants. *Ann. Missouri Bot. Gard.* 68: 233–253.

Levin, D. A. 1985. Reproductive character displacement in *Phlox*. *Evolution* 39: 1275–1281.

*Levin, D. A. 1993. Local speciation in plants: The rule not the exception. *Syst. Bot.* 18: 197–208.

*Levin, D. A. 2000. *The origin, expansion, and demise of plant species*. Oxford University Press, New York.

Levin, D. A. 2001. The recurrent origin of plant races and species. *Syst. Bot.* 26: 197–204.

*Li, W.-H. 1997. *Molecular evolution*. Sinauer, Sunderland, MA.

Masterson, J. 1994. Stomatal size in fossil plants: Evidence for polyploidy in majority of angiosperms. *Science* 264: 421–424.

Mayr, E. 1963. *Animal species and evolution*. Harvard University Press/Belknap Press, Cambridge, MA.

*Mayr, E. 1977. Darwin and natural selection. *Am. Sci.* 65: 321–327.

Mayr, E. 1992. A local flora and the biological species concept. *Am. J. Bot.* 79: 222–238.

McDade, L. 1990. Hybrids and phylogenetic systematics. I. Patterns of character expression in hybrids and their implications for cladistic analysis. *Evolution* 44: 1685–1700.

McDade, L. 1992a. Hybrids and phylogenetic systematics. II. The impact of hybrids on cladistic analysis. *Evolution* 46: 1329–1346.

McDade, L. 1992b. Species concepts and problems in practice: Insight from botanical monographs. *Syst. Bot.* 20: 606–622.

McDade, L. 1995. Hybridization and phylogenetics. In *Experimental approaches to plant systematics* (Monographs in Systematic Botany from the Missouri Botanical Garden, Vol. 53), P. C. Hoch and A. G. Stephenson (eds.), 305–331. Missouri Botanical Garden, St. Louis.

*McDade, L. 2000. Hybridization and phylogenetics. Special insights from morphology. In *Phylogenetic analysis of morphological data*, J. J. Wiens (ed.), 146–164. Smithsonian Institution Press, Washington, DC.

McGowen, M. H., R. J. E. Wiltshire, B. M. Potts and R. E. Vaillancourt. 2001. The origin of *Eucalyptus vernicosa*, a unique shrub eucalypt. *Biol. J. Linnean Soc.* 74: 397–405.

Mishler, B. D. 1985 The morphological, developmental and Phylogenetic basis of species concepts in bryophytes. *Bryologist* 88: 207–214.

Mishler, B. D. and R. N. Brandon. 1987. Individuality, pluralism, and the phylogenetic species concept. *Biol. Philos.* 2: 397–414.

Mishler, B. D. and M. J. Donoghue. 1982. Species concepts: A case for pluralism. *Syst. Zool.* 31: 491–503.

Mogie, M. 1992. *The evolution of asexual reproduction in plants*. Chapman and Hall, London.

Mooring, J. S. 2001. Barriers to interbreeding in the *Eriophyllum loanatum* (Asteraceae, Helenieae) species complex. *Am. J. Bot.* 88: 285–312.

Mosseler, A. and C. S. Papadopol. 1989. Seasonal isolation as a reproductive barrier among sympatric *Salix* species. *Can. J. Bot.* 67: 2563–2570.

Nixon, K. C. and Q. D. Wheeler. 1990. An amplification of the phylogenetic species concept. *Cladistics* 6: 211–223.

Novak, S. J., D. E. Soltis and P. S. Soltis. 1991. Owenby's *Tragopogons*: 40 years later. *Am. J. Bot.* 78: 1586–1600.

Paterson, H. E. H. 1985. The recognition species concept. In *Species and speciation* (Transvaal Museum Monograph, 4), E. E. Vrba (ed.), 21–39. Transvaal Museum, Pretoria, South Africa.

Raven, P. H. 1976. Systematics and plant population biology. *Syst. Bot.* 1: 284–316.

Richards, A. J. 1996. Breeding systems in flowering plants and the control of variability. *Folia Geobot. Phytotax.* 36: 283–293.

*Richards, A. J. 1997. *Plant breeding systems*, 2nd ed. Chapman and Hall, London.

Rieseberg, L. R. 1994. Are many plant species paraphyletic? *Taxon* 43: 21–32.

Rieseberg, L. R. 1995. The role of hybridization in evolution: Old wine in new skins. *Am. J. Bot.* 82: 944–953.

Rieseberg, L. R. and J. M. Burke. 2001. The biological reality of species: Gene flow, selection and collective evolution. *Taxon* 50: 47–67.

Rieseberg, L. R. and N. C. Ellstrand. 1993. What can molecular and morphological markers tell us about plant hybridization? *Crit. Rev. Plant Sci.* 12: 213–241.

*Rieseberg, L. R. and J. D. Morefield. 1995. Character expression, phylogenetic reconstruction, and the detection of reticulate evolution. In *Experimental approaches to plant systematics* (Monographs in Systematic Botany from the Missouri Botanical Garden, Vol. 53), P. C. Hoch and A. G. Stephenson (eds.), 333–353. Missouri Botanical Garden, St. Louis.

Rieseberg, L. R. and D. E. Soltis. 1991. Phylogenetic consequences of cytoplasmic gene flow in plants. *Evol. Trends Plants* 5: 65–84.

Rieseberg, L. R. and J. F. Wendel. 1993. Introgression and its consequences in plants. In *Hybrid zones and the evolutionary process*, R. G. Harrison (ed.), 70–109. Oxford University Press, New York.

Rieseberg, L. R., S. M. Beckstrom-Sternberg, A. Liston and D. M. Arias. 1991. Phylogenetic and systematic inferences from chloroplast DNA and isozyme variation in *Helianthus* sect. *Helianthus*. *Syst. Bot.* 16: 50–76.

Rohlf, F. J. 1993. *NTSYS-pc: Numerical taxonomy and multivariate analysis system.* Exeter Software, Setauket, NY.

Sang, T., D. J. Crawford and T. F. Stuessy. 1995. Documentation of reticulate evolution in peonies (*Paeonia*) using internal transcribed spacer sequences of nuclear ribosomal DNA: Implications for biogeography and concerted evolution. *Proc. Natl. Acad. Sci. USA* 92: 6813–6817.

Schilthuizen, M. 2001. *Frogs, flies, and dandelions: Speciation—The evolution of new species.* Oxford University Press, Oxford, UK.

Schwarzbach, A. E., L. A. Donovan and L. R. Rieseberg. 2001. Transgressive character expression in a hybrid sunflower species. *Am. J. Bot.* 88: 270–277.

Shaw, K. L. 1998. Species and the diversity of natural groups. In *Endless forms: Species and speciation*, D. J. Howard and S. H. Berlocher (eds.), 44–56. Oxford University Press, New York.

Sokal, R. R. and T. Crovello. 1970. The biological species concept: A critical evaluation. *Am. Nat.* 104: 127–153.

Soltis, D. E. and P. S. Soltis. 1993. Molecular data and the dynamic nature of polyploidy. *Crit. Rev. Plant Sci.* 12: 243–273.

Soltis, D. E. and P. S. Soltis. 1995. The dynamic nature of polyploid genomes. *Proc. Natl. Acad. Sci. USA* 92: 8089–8091.

*Soltis, D. E. and P. S. Soltis. 1999. Polyploidy: Recurrent formation and genome evolution. *Trends Ecol. Evol.* 14: 348–352.

Soltis, P. S., G. M. Pluncket, S. J. Novacek and D. E. Soltis. 1995. Genetic variation in *Tragopogon* species: Additional origins of the allotetraploids *T. mirus* and *T. miscellus*. *Am. J. Bot.* 82: 1329–1341.

Song, K. M., T. C. Osborn and P. H. Williams. 1988. *Brassica* taxonomy based on nuclear restriction fragment length polymorphisms. *Theor. Appl. Genet.* 5: 784–794.

Spooner, D. M., R. G. van den Berg and J. T. Miller. 2000. Species and species boundaries of *Solanum* series *longpedicellata* (Solanaceae) and phenetically similar species in ser. *Demissa* and ser. *Tuberosa*: Implications for a practical taxonomy of section *Petota*. *Am. J. Bot.* 87: 113–130.

Stace, C. A. 1984. *Plant taxonomy and biosystematics.* Edward Arnold, London.

*Stebbins, G. L. 1950. *Variation and evolution in plants.* Columbia University Press, New York.

Stevens, P. F. 1997. J. D. Hooker, George Bentham, Asa Gray and Ferdinand Mueller on species limits in theory and practice: A mid-nineteenth-century debate and its repercussions. *Hist. Rec. Aust. Sci.* 11: 345–370.

Stewart, W. N. and G. W. Rothwell. 1993. *Paleobotany and the evolution of plants*, 2nd ed. Cambridge University Press, Cambridge, UK.

Strid, A. 1970. Studies in the Aegean flora, XVI. Biosystematics of the *Nigella arvensis* complex with special reference to nonadaptive radiation. *Opera Bot. Soc. Bot., Lund.* 28: 1–169.

Takebayashi, N. and P. L. Morrell. 2001. Is self-fertilization an evolutionary dead end? Revisiting an old hypothesis with genetic theories and a macroevolutionary approach. *Am. J. Bot.* 88: 1143–1150.

Templeton, A. R. 1989. The meaning of species and speciation: A genetic perspective. In *Speciation and its consequences*, D. Otte and J. A. Endler (eds.), 3–27. Sinauer, Sunderland, MA.

Weber, J. E. and C. S. Campbell. 1989. Breeding system of a hybrid between a sexual and an apomictic species of *Amelanchier*, shadbush (Rosaceae, Maloideae). *Am. J. Bot.* 76: 341–347.

Wheeler, Q. D. and R. Meier (eds.). 2000. *Species concepts and phylogenetic theory: A debate.* Columbia University Press, New York.

Wiley, E. O. 1978. The evolutionary species concept reconsidered. *Syst. Zool.* 27: 17–26.

Wilson, P. 1992. On inferring hybridity from morphological intermediacy. *Taxon* 41: 11–23.

Wilson, R. A. (ed.) 1999. *Species: New interdisciplinary essays.* MIT Press, Camdridge, MA.

Yatabe, Y., S. Masyama, D. Darnaedi and N. Murakami. 2001. Molecular systematics of the *Asplenium nidus* complex from Mt. Halimun National Park, Indonesia: Evidence for reproductive isolation among three sympatric *rbcL* sequence types. *Am. J. Bot.* 88: 1517–1522.

7 An Overview of Green Plant Phylogeny

The word *plant* is commonly used to refer to any autotrophic eukaryotic organism capable of converting light energy into chemical energy via the process of photosynthesis. More specifically, these organisms produce carbohydrates from carbon dioxide and water in the presence of chlorophyll inside of organelles called chloroplasts. Sometimes the term *plant* is extended to include autotrophic prokaryotic forms, especially the (eu)bacterial lineage known as the cyanobacteria (or blue-green algae). Many traditional botany textbooks even include the fungi, which differ dramatically in being heterotrophic eukaryotic organisms that enzymatically break down living or dead organic material and then absorb the simpler products. Fungi appear to be more closely related to animals, another lineage of heterotrophs characterized by eating other organisms and digesting them internally.

In this chapter we first briefly discuss the origin and evolution of several separately evolved plant lineages, both to acquaint you with these important branches of the tree of life and to help put the green plant lineage in broad phylogenetic perspective. We then focus attention on the evolution of green plants, emphasizing several critical transitions. Specifically, we concentrate on the origins of land plants (embryophytes), of vascular plants (tracheophytes), of

153

seed plants (spermatophytes), and of flowering plants (angiosperms).

Although knowledge of fossil plants is critical to a deep understanding of each of these shifts and some key fossils are mentioned, much of our discussion focuses on extant groups. In Chapter 8 you will find detailed descriptions of the major extant groups of vascular plants and of seed plants, along with much more information on the biology of these plants. Likewise, Chapter 9 focuses on the attributes of flowering plant lineages, and their phylogenetic relationships.

Our main aim is to chronicle the evolutionary events leading up to angiosperms. We therefore pay rather little attention to major branches such as the chlorophytes, the mosses, the lycophytes, and the ferns and their allies. From a phylogenetic standpoint we could just as well "tell the story" of green plant evolution as leading up to the evolution of the mosses, the horsetails, or any other group (O'Hara 1992), but we follow the path leading to angiosperms simply because they are the focus of this book.

Before we proceed, it is important comment on the taxonomic names we will use in this chapter. Our knowledge of phylogenetic relationships among the major plant lineages has long been uncertain, and this is reflected in the existence of many contrasting classification systems. Sometimes the same name has been used to refer to different groups. For example, the name Chlorophyta is sometimes applied to the entire green plant clade, and sometimes to a branch within the green plants that includes all or most of the "green algae." In other cases different names have been used for the same group: The green plants have been called Chlorophyta by some authors and Viridiplantae by others.

To a large extent these differences reflect the attempts of different authors to assign taxonomic ranks to groups in what they believe to be an internally consistent manner. However, as we have stressed elsewhere (see Chapter 2), the assignment of taxonomic ranks is basically arbitrary, and typically it reflects only the traditions of the relevant taxonomic community. Thus, taxa assigned to a particular taxonomic rank (such as class, order, or family) are not necessarily equivalent with respect to age, species diversity, or ecological breadth.

Other problems relate to changes in our knowledge of phylogeny. Progress in discerning relationships has quite often resulted in the realization that traditionally recognized groups are not, in fact, clades. For example, the name Bryophyta has long been applied to a group that includes the liverworts, hornworts, and mosses. In recent years, however, it has become clear that these groups do not form a clade; instead, "bryophytes" refers to a grade, or paraphyletic group, at the base of the embryophytes (land plants).

As we will emphasize, the same is true of several other traditional groups, including "green algae," "seedless vascular plants," "gymnosperms," and "dicotyle-dons." In some cases it is possible to abandon such names entirely, but in others it is tempting to retain them, either as common names for certain forms of organization (e.g., the "bryophytic" life cycle), or to refer to a clade (e.g., applying "gymnosperms" to a hypothesized clade including just the extant "naked seed plants").

In this chapter we simply do not refer to taxonomic ranks. Elsewhere in the text, major clades within vascular plants are referred to orders and families, and we use the same names here. Likewise, standard genus and species names are used. However, whether a taxon is considered to be a class or an order by a particular author is not important in our discussion of green plant phylogeny.

Our choice of names reflects our sense of which ones are most commonly used in the literature and will therefore create the least confusion. Where possible, we have chosen names with rank-neutral endings, especially the ending *-phytes*, which means "plants." In addition, we avoid using names that refer to non-monophyletic groups, but when we do use such names (e.g., to clarify historical usage) we put them in quotation marks.

Endosymbiotic Events

The chloroplasts found in eukaryotes are endosymbiotic organelles derived ultimately from cyanobacteria. This view of the origin of plastids is now firmly established on the basis of structural evidence (e.g., the form and number of membranes) and molecular studies establishing that the DNA in plastids is more closely related to free-living cyanobacterial DNA than it is to DNA in the nucleus of the same cell.

Endosymbiosis entailed massive reduction in the size and gene content of the plastid genome relative to free-living cyanobacteria (see Chapter 5) (Palmer and Delwiche 1998; Delwiche 1999; Palmer 2000). An average cyanobacterium has a genome size of about 3600 kilobases and some 3200 genes. By contrast, a red algal chloroplast has on the order of 190 kilobases and only about 250 genes. Green algal chloroplasts are even smaller in most cases: about 120 kilobases and 120 genes. This reduction has involved the complete loss of some genes and the transfer of others from the chloroplast to the nucleus (e.g., Baldauf and Palmer 1990). There are many more proteins active within plastids (from 500 to 5000) than there are genes, meaning that some of these are products of genes that reside outside of the chloroplast, which then need to be imported into the plastid.

How many endosymbiotic events have there been? Recent phylogenetic evidence is consistent with just a single primary endosymbiotic event. For example, a recent analysis of eukaryote phylogeny (Baldauf et al. 2000) recovered a clade containing viridophytes (green plants), rhodophytes (red algae), and glaucophytes (Figure 7.1). This result, combined with evidence on the number of membranes and other morphological charac-

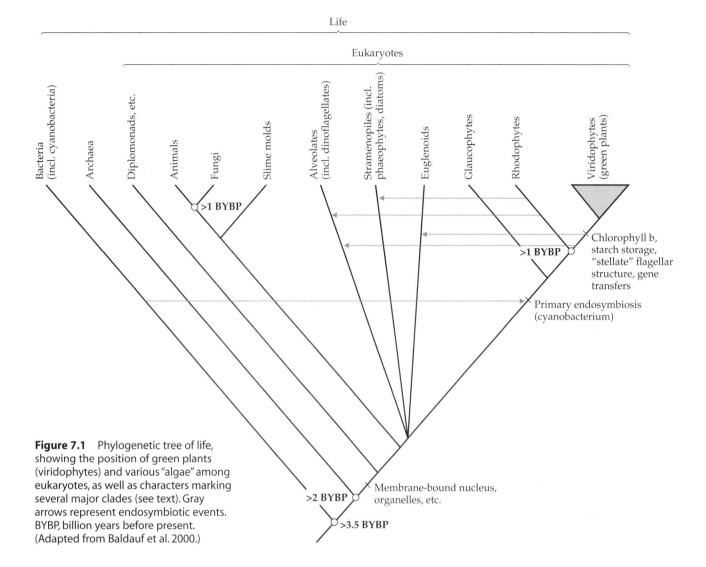

Figure 7.1 Phylogenetic tree of life, showing the position of green plants (viridophytes) and various "algae" among eukaryotes, as well as characters marking several major clades (see text). Gray arrows represent endosymbiotic events. BYBP, billion years before present. (Adapted from Baldauf et al. 2000.)

ters, suggests that a primary endosymbiotic event occurred in the common ancestor of this clade. In the glaucophytes the cyanobacterial cell wall is still present surrounding the plastid, but the wall was lost in the lineage that includes red algae and green plants.

Plastids in red algae and in green plants differ significantly from one another, which makes it possible to distinguish with considerable confidence between a red plastid lineage and a green plastid lineage (Delwiche 1999). This is important because it helps us identify situations in which plastids have been acquired by permanent incorporation of either red or green eukaryotes. It appears that red algal chloroplasts were acquired via such secondary endosymbiosis in stramenopiles (including brown algae, golden algae, and diatoms, which are discussed in the next section) and several other lineages. Endosymbiotic events involving the uptake of green algal eukaryotes appear to account for euglenas and several other groups. Dinoflagellates include a mixture of different types and may even (in one lineage) involve a tertiary symbiotic event. The origin of the remnant plastids found in apicomplexans (including *Plasmodium*, the

malaria parasite) is controversial, though recent analyses suggest a secondary endosymbiotic event involving a member of the red algal line.

Miscellaneous "Algae"

The term *algae* is applied to a wide variety of aquatic photosynthetic organisms belonging to several lineages that are not directly related to one another (see Figure 7.1). Before we provide brief descriptions of several of the major groups of "algae," it is first important to briefly review life cycle diversity. In humans and other animals, the diploid phase of the life cycle is the **dominant phase**, and the only haploid cells are the gametes (produced by meiosis). This kind of life cycle occurs among plants but is very rare. Some plants have life cycles that are basically the opposite of ours: A multicellular haploid organism is the dominant phase and gives rise to gametes by mitosis; syngamy (fusion of gametes) yields a diploid zygote that undergoes meiosis to yield haploid spores. Most autotrophic life cycles lie somewhere between these two extremes and exhibit what is known as **alternation of generations**—that is, alternation between a multicellular haploid phase (the gametophyte) and a multicellular diploid phase (the sporophyte).

The **red algae** (rhodophytes) include about 6000 species, most of which live in marine environments and in tropical waters (e.g., coral reefs). In addition to chlorophyll *a*, they have accessory pigments called phycobilins, which make it possible for them to live in dark waters well below the surface. A few red algae are unicellular, but most are filamentous and attach to rocks or to other algae (some are even parasites). The cells in these filaments are cytoplasmically connected to one another by distinctive pit connections. Red algae have no motile cells at any stage, and they often show exceptionally complex life cycles in which there may be two morphologically and ecologically distinct diploid phases. As noted already, red algae may be the sister group of the green plants.

The brown algae, diatoms, and some other groups of algae, along with the water molds (Oomycetes), make up the **stramenopiles** (see Figure 7.1). This clade is characterized by reproductive cells with two different kinds of flagellae—a smooth "whiplash" flagellum, and a "tinsel" flagellum with numerous fine hairs along its length.

The **brown algae** (phaeophytes) form a clade of some 2000 described species of mostly marine organisms, many of which are conspicuous in cooler regions. In addition to chlorophylls *a* and *c*, they have carotenoid pigments that account for the brown color. All brown algae are multicellular, but this condition presumably evolved within stramenopiles from a unicellular condition. Many brown algae are filamentous, but some are very large and show complex differentiation of the body into a holdfast, a stipe, a float, and one or more flat blades. Some of the larger forms show considerable anatomical differentiation, and some cells are even specialized for nutrient transport. Brown algal life cycles run the gamut from alternation between similar looking diploid and haploid phases to extreme differentiation (usually with a dominant diploid phase). In *Fucus* and some related kelp-like organisms, the multicellular haploid phase has been eliminated completely; in such cases the products of meiosis function directly as gametes, as they do in animal life cycles.

There are about 6000 living species of **diatoms** (bacillariophytes), and many more (perhaps as many as 40,000) that are known only as fossils. Diatoms are unicellular organisms (though sometimes they form loose filaments or clusters) found in both marine and freshwater environments. Like the closely related brown algae, diatoms produce chlorophylls *a* and *c* and carotenoids. Their most distinctive feature is the formation of cell walls made of two often elaborately sculptured silicon valves that together form a tiny box. Flagellae are lacking, except in some male gametes.

The dinoflagellates, ciliates, and apicomplexans form the **alveolates** (see Figure 7.1), a group characterized by small membrane-bound cavities (alveoli) under their cell surfaces. **Dinoflagellates** include about 3000 described species, found in both freshwater and salt water. They have two flagellae located in characteristic grooves between cellulose plates embedded in the cell walls, which together make the cell spin as it moves. Many dinoflagellates are symbiotic with other organisms, including corals, sponges, squids, and giant clams. The symbiotic forms typically lack cellulose plates and are referred to as zooxanthellae. These organisms are of great ecological importance in maintaining coral reefs; for example, the phenomenon known as coral "bleaching" involves the death of the zooxanthellae. Dinoflagellates are also responsible for periodic "red tides" or "algal blooms" that can have dramatic negative impacts owing to the production of highly toxic substances.

Viridophytes (Green Plants)

As shown in Figure 7.2, the traditional "green algae" are related to the land plants, and together these constitute a clade called the **green plants** (Viridiplantae, or what we will call viridophytes). This clade includes more than 300,000 described species, or over one-sixth of all known extant species on Earth. Molecular evidence strongly supports the monophyly of green plants, including DNA sequence data (from the nucleus and the organelles) and structural features (such as the transfer of particular genes from the chloroplast to the nucleus). This clade is also supported by numerous chemical and morphological features, including probably the loss of phycobilins (found in cyanobacteria and red algae) and the production of chlorophyll *b* (in addition to chlorophyll *a*). Green plants also store carbohydrates in the form of starch granules in their cells, and their motile cells have a characteristic stellate structure at the base of each of the usually two anterior whiplash flagellae.

It now appears that several lineages of unicellular organisms with distinctive scaly cell walls (so-called micromonads, or prasinophytes) are situated around the base of the green plant phylogeny. However, the most comprehensive analyses to date (Karol et al. 2001) support a basal split of green plants into a **chlorophyte** clade, containing most of the traditional "green algae," and a **streptophyte** clade, which includes the land plants and several other lineages formerly placed in "green algae" (see Figure 7.2). In these recent studies one of the micromonads, *Mesostigma*, appears to be the sister group of the rest of the streptophyte line.

CHLOROPHYTES

Within the chlorophytes are several well-supported clades (see Figure 7.2): Chlorophyceae, Ulvophyceae, and Trebouxiophyceae (McCourt 1995; Chapman et al. 1998).

The **Chlorophyceae** are marked by somewhat obscure ultrastructural features (such as clockwise rotation of the

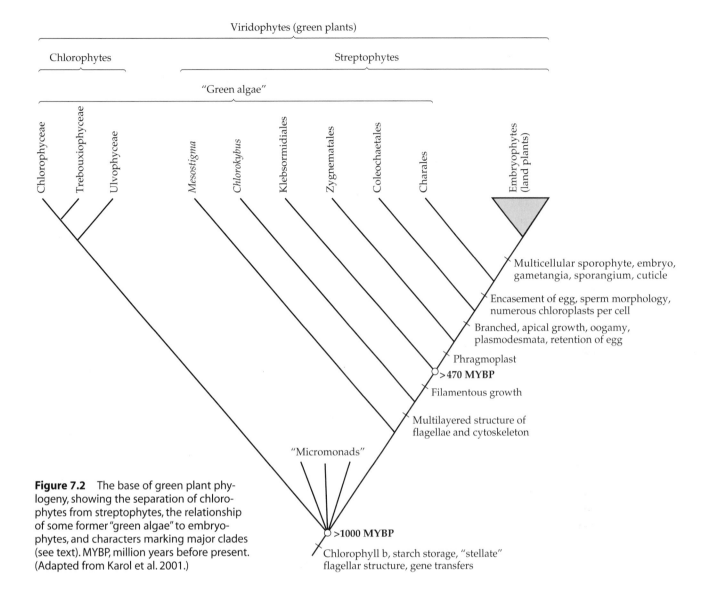

Figure 7.2 The base of green plant phylogeny, showing the separation of chlorophytes from streptophytes, the relationship of some former "green algae" to embryophytes, and characters marking major clades (see text). MYBP, million years before present. (Adapted from Karol et al. 2001.)

basal bodies), but they have been supported as a clade in most molecular studies as well. Included within this line is the so-called volvocine lineage, within which progressively more complex colonies (from 4 cells in *Gonium* to as many as 500–50,000 cells in the hollow spherical colonies of *Volvox*) are presumed to have been derived from unicells not unlike the model organism *Chlamydomonas* (Figure 7.3A, B). Recent studies indicate that the story is more complex, with several colonial lines derived independently, perhaps from within *Chlamydomonas* itself, which has hundreds of species (Larsen et al. 1991; Buchheim et al. 1994; Chapman et al. 1998).

The **Ulvophyceae** includes many marine forms and is marked by the production of multinucleate cells (Figure 7.3D–F). In some, the entire body is a coenocytic thallus, lacking walls between the nuclei except in the case of reproductive cells. Included in this group is the model organism *Acetabularia* (Figure 7.3F).

Finally, the **Trebouxiophyceae** contain forms with flagellate spores, but most are small round forms (appar-

ently derived several times independently) that completely lack motile cells at any stage. Many of the nonmotile forms live in terrestrial habitats, often in association with lichen-forming fungi or invertebrate animals. Lichen associations appear to have originated and to have been lost multiple times (Lutzoni et al. 2001).

STREPTOPHYTES

The discovery of the streptophyte lineage began in the late 1960s, when detailed ultrastructural studies of cell division first revealed a major difference in the orientation of the spindle microtubules among the organisms that had traditionally been classified as "green algae" (Pickett-Heaps 1979; Mattox and Stewart 1984; McCourt 1995). Some of these were found to have the phragmoplast orientation found in all land plants, in which the spindle is oriented perpendicular to the formation of the cell wall. A thorough survey showed that the phragmoplast condition occurred in the so-called charophycean algae (Coleochaetales and Charales). These plants show

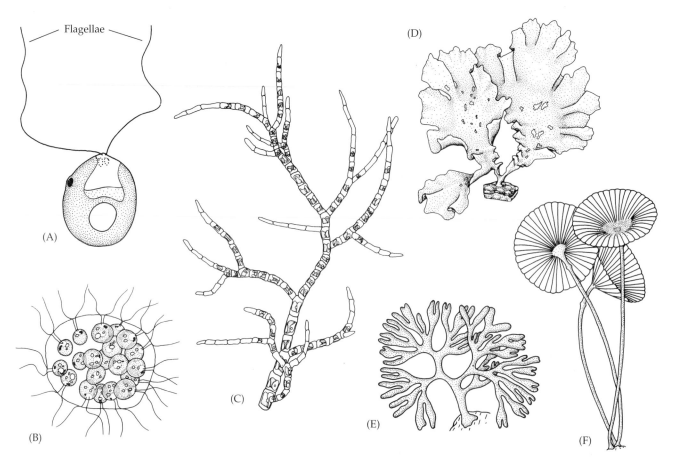

Flagellae

(A)

(B)

(C)

(D)

(E)

(F)

Figure 7.3 Morphology of chlorophytes. (A–C) Chlorophyceae: (A) *Chlamydomonas*, showing flagellae (f); (B) *Eudorina*, a colonial "volvocine" form; (C) *Stigeoclonium*, a branched filamentous form. (D–F) Ulvophyceae: (D) *Ulva*; (E) *Codium*, showing a coenocytic diploid thallus; (F) *Acetabularia*. (From Scagel et al. 1969.)

a range of different life forms (including upright, branching forms, as in *Chara* and *Nitella*, and flattened forms, as in *Coleochaete*) and live in near-shore, freshwater habitats (Figure 7.4A–C) As these organisms were studied in more detail, the idea emerged that they were actually more closely related to land plants than they were to other "green algae." It has since become clear that several other former green algal ineages belong in the streptophyte clade, including Klebsormidiales and Zygnematales (McCourt 1995; Chapman et al. 1998). Of these, the Zygnematales may be familiar as the group that includes *Spirogyra* and its relatives (Figure 7.4D–E). These are the so-called conjugating "green algae," in reference to a form of sexual reproduction that involves the formation of a tubular connection between cells of adjacent filaments, passage of the protoplast from one cell to another, and the eventual fusion of nuclei to form a zygote.

The relationships among the streptophyte groups shown in Figure 7.2 have been confirmed by molecular studies (Karol et al. 2001), including some structural

molecular data. For example, the protein elongation factor *tufA* appears to have moved from the chloroplast to the nucleus in the ancestor of the "charophyte" lineages plus the land plants (Baldauf and Palmer 1990). Coleochaetales and Charales possess some functionally important traits that are found otherwise only in land plants, such as flavonoids and the chemical precursors of cuticle. Most important from the standpoint of the evolution of the land plant life cycle is the fact that they retain the egg and sometimes even the zygote (after fertilization) on the body of the haploid plant (Graham 1993).

These phylogenetic results have many important implications for our understanding of green plant evolution. For instance, they imply that there were several independent originations of multicellularity. As we have noted, the volvocine forms explored a lifestyle in which the cells became aggregated into colonies. The larger of these colonies also show cytoplasmic interconnections and a division of labor, with some cells specialized for reproduction. Other chlorophytes formed filaments or membranous parenchymatous bodies of much larger size (such as the sea lettuce, *Ulva*, and its relatives), which show a more complex morphological integration and differentiation of cell functions. The Ulvophyceae followed a separate path involving multinucleate cells, sometimes forming filaments, and sometimes (as in

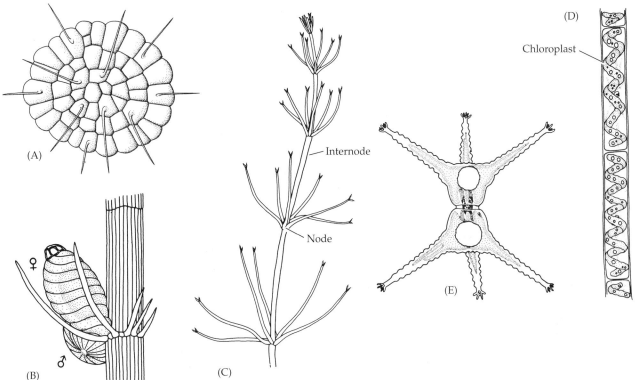

Figure 7.4 Morphology of basal streptophytes. (A) *Coleochaete*, showing a haploid discoidal thallus, with setae. (B, C) Charales: (B) *Chara*, showing a node with an egg-bearing structure (above) and a sperm-producing structure (below). (C) *Nitella* habit, showing node and internode construction. (D, E) Zygnematales: (D) *Spirogyra*, a filamentous form, showing helical chloroplasts. (E) *Staurastrum*, a unicellular desmid, forming two mirror-image semicells. (A from Taylor and Taylor 1993; B–E from Scagel et al. 1969.)

Codium) forming a thallus by densely intertwining the filaments. Finally, multicellularity evolved separately in the streptophyte line. Many Zygnematales are filamentous, and parenchymatous forms (with plasmodesmata connecting adjacent cells) are found in the charophyte lineages plus the land plants.

Among the early-diverging lineages of green plants, we also encounter a wide variety of life cycles. Alternation of similar haploid gametophyte and diploid sporophyte generations (as in *Ulva*) is quite common. In contrast, *Codium* (Ulvophyceae) evolved a life cycle like that of humans, in which gametes are the only haploid cells. In stark contrast, in charophytes the plants are haploid, and the only diploid cell in the life cycle is the zygote, which results from fertilization of a large nonmotile egg by a swimming sperm (oogamy).

Embryophytes (Land Plants)

The land plants are depicted as stemming from a single common ancestor in Figure 7.2, a finding that is strongly supported by both molecular and morphological evidence (Kenrick and Crane 1997a, 1997b; Karol et al. 2001). Land plants are also called embryophytes because they have a resting embryo stage early in the life of the sporophyte. *Embryophyte* is the preferable term in this case because several algal lineages have independently (though less conspicuously) also made the transition to life on land (e.g., in Trebouxiophyceae). In addition to the embryo, embryophytes are characterized by production of a multicellular sporophyte, multicellular reproductive structures (antheridia, archegonia, and sporangia), cuticle, and thick-walled spores with characteristic trilete marks (see Figure 7.7C).

Traditionally, embryophytes have been classified as either bryophytes or vascular plants. There are three major lineages of bryophytes—mosses, hornworts, and liverworts—which we will characterize briefly in the next few paragraphs (see also Shaw and Goffinet 2000). As we shall discuss, however, it has become increasingly clear that "bryophytes" are paraphyletic with respect to vascular plants (see Figure 7.6).

MOSSES

Mosses are probably the most familiar bryophytic plants and, with some 15,000 species, they are also the most diverse. The upright and leafy gametophyte is the dominant phase in the moss life cycle (Figure 7.5A–C). The sporophyte forms a single unbranched stalk terminated by a sporangium (or capsule). Haploid spores, produced via meiosis, are released from the sporangium; typically, dehiscence of the sporangium occurs by the detachment of a lid or operculum.

When a spore germinates, it forms a protonemal stage, which resembles a green algal filament. The pro-

Figure 7.5 Morphology of basal embryophytes ("bryophytes"). (A–C) Mosses: (A) *Dawsonia superba* habit, showing leafy gametophyte and unbranched sporophyte with terminal sporangium; (B) sporangium (capsule) of a moss prior to dehiscence; (C) apex of the dehiscing sporangium of a moss, *Fontinalis antipyretica*, showing the peristome teeth. (D, E) Liverworts: (D) A leafy liverwort, *Lepidozia reptans*, showing dehiscence of the sporangium by four valves; (E) portion of a thalloid liverwort, *Monoclea forsteri*, showing sporangia with longitudinal dehiscence. (F, G) Hornworts: (F) *Phaeoceros laevis* habit, showing the thalloid gametophyte and dehiscing sporangia of the sporophyte; (G) stomate, with guard cells, from the sporangium wall of *Anthoceros*. (A from Barnes 1998; B–G from Scagel et al. 1969.)

tonema produces one or more upright leafy gametophytes, which ultimately produce sperm and eggs in antheridia and archegonia, respectively. Fusion of the gametes yields the zygote, which develops through a series of mitotic divisions into the embryo and eventually into the mature sporophyte.

Analyses of relationships within mosses have generally supported the idea that *Sphagnum* (peat moss) is situated near the base of the tree and that *Andreaea* and a few close relatives also form an early branch (see Kenrick and Crane 1997a). The enigmatic *Takakia*, which was considered to be a liverwort until the recent discovery of

the sporophyte phase, is also probably situated near the base of the moss tree. The sporangium in *Andreaea* opens by four vertical slits, and in *Takakia* by a single helical slit, as opposed to the lid-like operculum found in the vast majority of mosses. The operculum of most mosses is also characterized by a distinctive row of tooth-like structures, which together make up the peristome.

LIVERWORTS

There are about 9000 species of **liverworts**, which come in a thalloid form or, more commonly, a leafy form (Figure 7.5D,E). Unlike mosses and hornworts, liverworts lack stomates, although some have epidermal pores without true guard cells. They also lack a characteristic columnar mass of sterile tissue (the columella) in the sporangium, which is present in mosses, hornworts, and early vascular plant lineages. These liverwort features have sometimes been interpreted as ancestral within land plants, but this is no longer clear.

Sex in liverworts involves the production of sperm-producing antheridia and egg-containing archegonia. The sporophyte phase, with its terminal sporangium, is rather small and inconspicuous. The capsule typically opens through four valves, and sterile hygroscopic cells (elaters) among the spores may aid in dispersal.

HORNWORTS

There are only about 100 species of **hornworts** (Figure 7.5A,B), which are encountered much more rarely than either mosses or liverworts. One presumably derived feature of this entirely thalloid group is the presence of an intercalary meristem in the sporophyte located at the base of the capsule. The activity of this meristem accounts for the continued upward growth of the capsule, which is quite extensive in some groups (e.g., *Anthoceros*).

PHYLOGENETIC RELATIONSHIPS WITHIN EMBRYOPHYTES

All phylogenetic analyses of land plants have concluded that "bryophytes" are paraphyletic. However, the exact relationships are still controversial (Figure 7.6). Initially, morphological analyses supported a basal split between the liverwort lineage and everything else (Figure 7.6A), and placed the mosses as the sister group to the vascular plants (Mishler and Churchill 1984, 1985). Under this view, stomates are considered to be an innovation linking hornworts, mosses, and vascular plants, to the exclusion of liverworts. Likewise, specialized cells in the stems of mosses (in both the gametophyte and sporophyte of some species), called hydroids and leptoids, have been interpreted as precursors of the water- and nutrient-conducting cells found in vascular plants. Mosses and vascular plants have sporophytes that

increase in height through cell divisions in an apical meristem, and the first vascular plants also had upright gametophytes, as do mosses. These interpretations have been upheld by some molecular studies, including an analysis showing the gain of three mitochondrial introns in all land plants except liverworts (Qiu et al. 1998).

Several recent studies of molecular data, however—alone and in combination with a variety of morphological and ultrastructural characters (especially sperm ultrastructure)—support an alternative hypothesis (Figure 7.6B). In these trees, hornworts are the sister group of all other extant land plants, and a clade containing mosses and liverworts is the sister group of the vascular plants (Nickrent et al. 2000; Renzaglia et al. 2000).

This hypothesis is consistent with detailed studies indicating that the hydroids and leptoids in mosses probably are not homologous with tracheids in the xylem and sieve cells in the phloem of vascular plants (Ligrone et al. 2000). However, such relationships would imply either the loss or the independent evolution of stomates and apical meristems. Although this issue remains

Figure 7.6 Phylogenetic relationships at the base of the embryophytes (land plants), showing characters that mark major clades under two hypotheses of how the "bryophyte" groups (hornworts, liverworts, and mosses) are related to vascular plants (see text). MYBP, million years before present. (A adapted from Mishler and Churchill 1984; B adapted from Nickrent et al. 2000 and Renzaglia et al. 2000.)

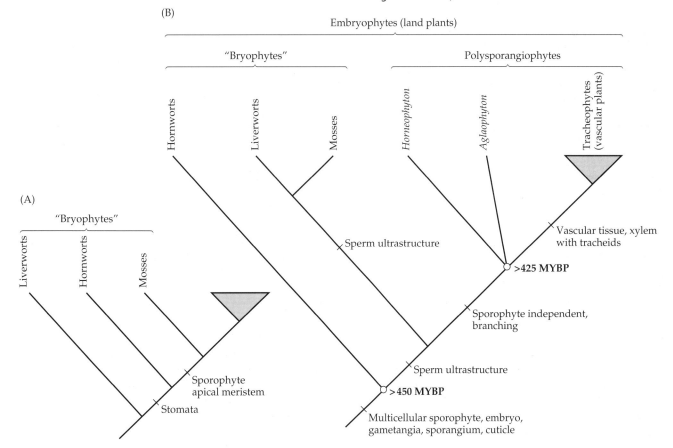

unsettled, note that all recent analyses support the view that "bryophytes" are paraphyletic with respect to vascular plants.

TRANSITION TO LAND

This phylogenetic knowledge illuminates the origin of several key adaptations to life on land (Graham 1993). Cuticle and sporopollenin (present in the thick spore wall) appear to be responses to desiccation. Gas exchange is facilitated by small pores in the epidermis or by genuine stomates with guard cells that can open or close the stomate depending on environmental conditions, thereby regulating water loss. Flavonoids help absorb damaging long-wavelength UV radiation. A glycolate oxidase system helps ameliorate the fact that carbon dioxide fixation is inhibited in the presence of oxygen, which is in much higher concentrations in the air than in water. The first land plants probably depended on symbiotic relationships with fungi to obtain nutrients from the soil, and such relationships have been documented in the major bryophytic lineages, as well as in vascular plants (where they are ubiquitous). The precursors of many of these adaptations can be found among the closely related Coleochaetales and Charales, which therefore appear to have been preadapted to make the transition to land (Delwiche et al. 1989; Graham et al. 1991; Graham 1993).

Appreciation that both the traditional "green algae" and the "bryophytes" are paraphyletic has also helped us understand the origin of the characteristic land plant life cycle, involving an alternation of multicellular gametophyte and sporophyte phases (Mishler and Churchill 1985; Graham 1993). As noted earlier, in *Coleochaete* and *Chara* the egg is retained on the haploid parent plant. In *Coleochaete* the zygote (the only diploid stage) also remains on the parent plant until it undergoes meiosis to give rise to haploid spores. A key innovation in the line including the charophyte lineages and the embryophytes was the establishment of nutrient transport between haploid and diploid phases through a placental transfer tissue (Graham and Wilcox 2000). The land plant life cycle was probably derived from a charophyte-like ancestral condition by simple delay of meiosis and interpolation of a multicellular diploid phase via a series of mitotic divisions of the zygote.

In the embryophytes, the egg—and after fertilization, the embryo—is protected by a multicellular structure called an **archegonium** (plural *archegonia*). Sperm are produced and protected by a multicellular structure called an **antheridium** (plural *antheridia*). Initially, the gametophyte phase was dominant, as it is today in hornworts, liverworts, and mosses, and the sporophyte remained attached to, and was nutritionally dependent on, the gametophyte. In vascular plants, the sporophyte became dominant and nutritionally independent, and there was progressive reduction in gametophyte size (Kenrick and Crane 1997a, 1997b).

These findings also help us interpret the absolute timing of events (see Figures 7.1, 7.2, and 7.6). Green plants may be a billion or more years old, and it is possible that some major green plant lineages existed in the Precambrian (Heckman et al. 2001). From the Cambrian (about 550 million years ago), a variety of chlorophyte fossils have been found, including well-preserved lime-secreting Ulvophyceae, such as relatives of *Acetabularia*. Charophytes (in the form of calcified Charales) do not appear in the fossil record until the mid-Silurian, but the wholesale occupation of land by green plants probably took place beginning in the mid-Ordovician, about 450 million years ago. Starting at that time (and possibly even earlier, in the Cambrian), dispersed spores are found, sometimes in envelope-enclosed tetrads or diads (sets of four or two, respectively) resembling those seen today in some liverworts. Tiny bits of cuticle and tubular structures of plant origin also appear in the Ordovician, and individual spores with the characteristic trilete marks of land plants (see Figure 7.7C) have been recovered from the early Silurian.

It is probable, therefore, that hornworts, liverworts, mosses, and vascular plants were all in existence by the late Ordovician. Somewhat later, beginning in the mid-Silurian, there are well-preserved macrofossils representing the vascular plant lineage. The occupation of land was certainly in full swing by then.

Tracheophytes (Vascular Plants)

All indications are that the first land plants were small and very simple in structure. In the case of the vascular plant lineage, the sporophyte was basically a dichotomously branching stem, about the height of a matchstick at first, with the sporangia (the site of meiosis yielding haploid spores) produced at the tips of the branches (Figure 7.7A). These plants had no leaves or roots. In some cases (e.g., *Aglaophyton*, formerly known as *Rhynia*, from the Rhynie chert in Scotland), the preservation of these plants is spectacular, and it is possible to discern many anatomical details, including stomates, spores, and the vascular tissue inside the stem.

On the basis of such fossils, it was recently discovered that the first **polysporangiophytes**—plants with branching sporophytes—did not actually produce bona fide water-conducting cells (**tracheids**) in the xylem tissue and must therefore have depended on turgor pressure to remain upright. True water-conducting cells evolved somewhat later and characterize a clade known as the true vascular plants, or the **tracheophytes** (Kenrick and Crane 1997a, 1997b).

Tracheids are elongate cells with thickened walls that are dead at maturity. Where one tracheid connects to the next, there are characteristic openings or pits, but a pit membrane (primary cell wall) remains intact, and water must pass through it as it moves from one cell to the next. In the first tracheophytes (represented by *Rhynia*),

Figure 7.7 Fossils of early tracheophytes (vascular plants). (A) Reconstruction of *Aglaophyton major*, showing the dichotomously branching stem (without leaves or roots) and terminal sporangia. (B) Enlarged terminal sporangium of *A. major*, with spores inside. (C) *A. major* spores in tetrad, and a single thick-walled spore with the trilete mark characteristic of land plants. (D) Reconstruction of *Uskiella spargens*, showing the dichotomously branched stem and terminal sporangia with distal dehiscence. (E) Reconstruction of a Devonian gametophyte, *Sciadophyton* sp.; gametangia are present on the terminal disk-shaped structures. (A–C from Stewart 1983; D and E from Kenrick and Crane 1997a.)

the tracheids were of a distinctive type, in which decay resistance (conferred by lignification of cellulose fibers) was present only as a very thin layer. Cell walls that are much more decay-resistant characterize a eutracheophyte clade, which includes all extant vascular plants (Kenrick and Crane 1997a). In these species, the strongly lignified tracheids allow more efficient water conduction and provide internal support, allowing the plants to grow much taller.

In recent years, careful paleobotanical studies have revealed that some early land plant fossils are actually haploid gametophytes, bearing antheridia and archegonia (Remy 1982; Remy et al. 1993). These fossils are remarkable because they are rather large, upright, and branched, and in general they resemble the sporophyte phase of the life cycle (see Figure 7.7E). This finding has led to the view that the first members of the vascular plant lineage exhibited alternation of more or less similar generations. Thus, in relation to the bryophytic groups, it seems that both the gametophyte and the sporophyte phases were elaborated.

This knowledge allows us to piece together a sequence of events leading to the life cycle that we see in vascular plants today. This life cycle includes a dramatic reduction in the gametophyte phase and an equally impressive elaboration of the sporophyte phase. In the first vascular plants, the gametophyte was nutritionally independent of the sporophyte, and this condition is retained today in the "free-sporing" lineages such as

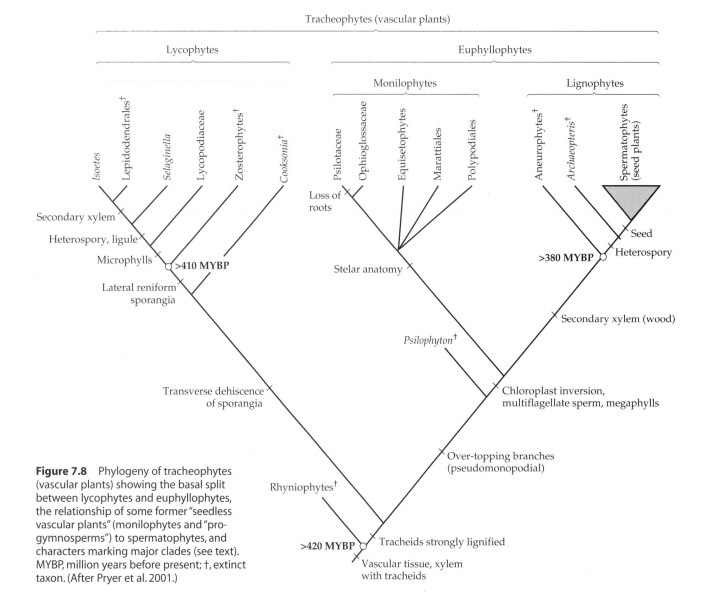

Figure 7.8 Phylogeny of tracheophytes (vascular plants) showing the basal split between lycophytes and euphyllophytes, the relationship of some former "seedless vascular plants" (monilophytes and "pro-gymnosperms") to spermatophytes, and characters marking major clades (see text). MYBP, million years before present; †, extinct taxon. (After Pryer et al. 2001.)

ferns and lycophytes. With the evolution of seed plants, however, the gametophyte became much further reduced and eventually became completely dependent on the sporophyte.

Viewed in this context, the bryophytic groups (especially the mosses) and the vascular plants appear to have explored two different mechanisms to increase the number of spores produced per fertilization event (Mishler and Churchill 1985). In mosses this increase in spore production was accomplished by intercalation of a filamentous protonemal stage that could produce numerous unbranched leafy gametophytes, each bearing a single unbranched sporophyte terminated by a single sporangium. In contrast, in the vascular plant lineage the number of sporangia was increased by branching of the sporophyte so that each branch tip could bear a sporangium.

What factors might have favored the elaboration of the sporophyte phase as opposed to the gametophyte phase (which became increasingly specialized for sexual reproduction)? One hypothesis holds that diploid organisms are buffered against deleterious mutations. But an alternative is that the sporophyte was free to become larger (which was advantageous in competing for light and may also have enhanced spore dispersal), whereas the gametophyte was dependent on water for fertilization as long as the sperm needed to swim to the egg.

Phylogenetic relationships among the major lines of extant **vascular plants** are shown in Figure 7.8. These conclusions are based on morphological and molecular evidence and are now quite strongly supported (Doyle 1998; Pryer et al. 2001). The basal split, which occurred in the early to mid-Devonian (before 400 million years ago), separated a clade that included the modern lycophyte lineage from a clade, known as the euphyllophytes, that contains all of the other extant vascular plant lineages. This split is marked by a variety of morphological features. One noteworthy feature is the presence of multiflagellate sperm in the euphyllophytes, as opposed to biflagellate sperm in the bryophytic lineages and in lycophytes (except in *Isoetes* and *Phylloglossum*, where multiflagellate sperm evolved independently). One compelling bit of molecular evidence is the presence in the euphyllophytes of a 30-kilobase inversion in the chloroplast DNA (Raubeson and Jansen 1992); lycophytes and the bryophytic plants lack this inversion.

LYCOPHYTES

The **lycophyte** lineage (see Figures 7.8 and 7.9; see also Figures 8.2 and 8.3) appeared in the fossil record very soon after the first appearance of vascular plants and is marked by the lateral position, reniform shape, and transverse dehiscence of the sporangia. Small, microphyllous leaves with a single vascular strand evolved within this lineage (possibly through modification of lateral sporangia), as did distinctive dichotomously branching roots. During the Carboniferous period lycophytes were especially diverse and abundant, and the remains of these plants account for our major coal deposits. In particular, they dominated coastal swamps of tropical lowlands (DiMichele and Skog 1992; Bateman et al. 1998).

Some lycophytes became large trees, with secondary growth allowing an increase in girth (Figure 7.9). The stems of these plants were covered by leaves, which left the distinctive leaf bases seen in fossils. These plants also evolved so-called stigmarian root systems; these are presumed to have been derived from rhizomes, in which case the spirally arranged rootlets may be modified leaves. Patterns of growth in these large plants are still poorly understood (Bateman 1994; Bateman et al. 1998), but they may have grown very slowly in height at first (while the root system became established) and later elongated rapidly. They may have died after simultaneously producing strobili at the tips of all the branches.

Today there are over 1200 species of lycophytes, belonging to several major lines (see Figures 7.8 and 7.9). Of these, rhizomatous species of *Huperzia*, *Lycopodiella*, and *Lycopodium* (club mosses) are commonly encountered in forests of the Northern Hemisphere. These plants and their tropical relatives are homosporous, meaning that they produce just a single kind of spore, which gives rise to a bisexual gametophyte, producing both sperm and eggs.

The other living lycophytes (*Selaginella*, *Isoetes*) are heterosporous, producing microspores, which give rise to male gametophytes, and megaspores, which give rise to female gametophytes. The heterosporous taxa form a clade, which is also united by the association of a leaf-like flap of tissue, the ligule, with the adaxial side of the leaf base.

Selaginella (spike mosses) (see Figure 7.9F–I), with over 700 species, is most diverse in the tropics, where many species grow as epiphytes. *Isoetes* (quillworts, with perhaps 150 species) is the only living remnant of the

clade that included the giant lycopods of the Carboniferous, though it may have been derived from plants in this lineage that never attained the size of *Lepidodendron* (see Figure 7.9D) and the other very large lycophyte trees. *Isoetes* has retained the cambium and some secondary growth, and it has rootlets that resemble those of the extinct trees (see Figure 7.9J).

EUPHYLLOPHYTES

An advance of the **euphyllophytes** (see Figure 7.8) was differentiation between a main axis and side branches (pseudomonopodial growth), an arrangement found in a variety of Devonian fossils known as trimerophytes (Figure 7.10A). According to the "telome theory" (Zimmermann 1952, 1965), large megaphyllous leaves characteristic of the euphyllophytes were derived from flattened lateral branch systems. This derivation involved planation of the branch system and then webbing to form the leaf blade. It seems clear that leaves evolved independently, and by very different pathways, in the lycophyte line versus the euphyllophyte line.

Living euphyllophytes appear to belong to two major clades (see Figure 7.8): the seed plants (spermatophytes) and a clade that includes several "fern" lineages, the horsetails, and the whisk ferns (the monilophytes; Moniliformopses of Kenrick and Crane 1997a). This new view of relationships is well supported in combined analyses of morphological characters and both chloroplast and nuclear genes (Pryer et al. 2001). Within the **monilophytes** (ferns and their allies), there are five major lineages, each discussed only briefly here: (1) leptosporangiate ferns (Polypodiales), (2) Ophioglossaceae, (3) Psilotaceae, (4) Marattiales, and (5) equisetophytes (see also Chapter 8).

The common name *fern* is applied to the members of three of these major lineages: Polypodiales, Marattiales, and Ophioglossaceae. These plants are superficially similar in usually having large (often highly dissected) frond-like leaves that unfold from a "fiddlehead" (so-called circinate vernation). These three lineages are usually divided into two groups on the basis of the structure and development of the sporangia. The Marattiales and the Ophioglossaceae are so-called eusporangiate ferns (see Figure 7.10). These appear to have retained the ancestral condition, in which the sporangium develops from several initial cells and the mature wall of the sporangium is more than one cell layer thick. Eusporangia also tend to contain large numbers of haploid spores at maturity.

In contrast, the Polypodiales are characterized by a derived development in which the sporangium arises from a single cell and the mature wall is only one cell thick. These leptosporangia are borne on a distinct stalk and have a characteristic annulus consisting of a row of cells with thickened inner walls and thin outer walls (see Figure 8.12). The leptosporangia of most species contain a relatively small and definite number of haploid spores

Figure 7.9   Morphology of lycophytes. (A) Reconstruction of the extinct *Zosterophyllum deciduum*, showing prostrate rhizome bearing leafless upright axes with lateral reniform sporangia. (B) Reconstruction of the extinct *Asteroxylon mackiei*, showing upright dichotomizing stems covered by microphyllous leaves, and rootlike axes. (C) *A. mackiei*, showing part of a fertile axis with reniform sporangia and tranverse dehiscence. (D) Reconstruction of an extinct *Lepidodendron* sp., showing the dichotomously branching root system, the massive trunk with dichotomous branching above, and terminal strobili. (E) Portion of the surface of a stem of a *Lepidodendron* sp., showing three attached leaves and the scars left by the abscission of five others. (F) Tip of a branch of *Selaginella*, showing microphyllous leaves and a terminal strobilus. (G) Microsporangium of *Selaginella* in the axil of a microsporophyll. (H) Megasporangium of *Selaginella* in the axil of a megasporophyll. (I) Longitudinal section through a strobilus of *Selaginella harrisiana*, showing megasporangia (me) with four large megaspores, microsporangia (mi) with many tiny microspores, and ligules. (J) *Isoetes bolanderi* habit, showing leaves and roots. (A and J from Kenrick and Crane 1997a; B, C, and I from Stewart 1983; D and E from Gifford and Foster 1989; F–H from Barnes 1998.)

Figure 7.10 Morphology of various euphyllophytes. (A) The extinct trimerophyte *Psilophyton forbesii*, showing pseudomonopodial growth (differentiation between a main trunk and side branches). (B) Schematic representation of an extinct treelike equisetophyte, *Calamites*, showing the stout rhizome and tall, upright, branching shoot. (C) Large, arching leaves of *Angiopteris* (Marattiales). (D) The lower (abaxial) surface of a fertile leaflet of *Angiopteris*, showing a cluster of eusporangia. (A from Stewart 1983; B and D from Gifford and Foster 1989; C from Barnes 1998.)

(e.g., 16, 32, 64), which are ejected from the sporangia by a mechanism driven by changes in moisture content in the annulus cells.

Probably the most familiar monilophytes are the **Polypodiales**, of which there are more than 12,000 living species (see Figures 8.1, and 8.8 through 8.14). Many of these plants have highly dissected pinnate leaves, of the type we commonly associate with ferns, but leaf form is actually extremely variable, and some even have simple, undissected leaves. The sporangia are typically produced in small clusters (each cluster is called a sorus, collectively the *sori*) on the undersides of the leaves. The sori are often covered by a flap of tissue called an indusium, though some are "naked." The structure and position of the sori and the indusium vary enormously from one fern group to another, and this variation has been emphasized in taxonomic treatments (see Chapter 8). Fern gametophytes are often small, heart-shaped structures, with the archegonia present near the notch and

antheridia situated among the characteristic rhizoids. There is considerable variation, however, and in some ferns the gametophyte is even filamentous.

Within the leptosporangiate line, recent morphological and molecular studies have identified several noteworthy clades (Pryer et al. 1995; Wolf et al. 1998). As has been long suggested on the basis of sporangium development (sporangia not in sori, rudimentary annulus, large number of spores), Osmundaceae (cinnamon ferns) are seen to be the sister group of the rest. One distinctive leptosporangiate clade includes the large tree ferns (Cyatheaceae), and another contains all of the heterosporous aquatic fern groups (placed in Marsileaceae and Salviniaceae). Although the aquatic ferns are morphologically quite different from one another (e.g., *Salvinia* and *Azolla* with small floating leaves, versus *Marsilea* with leaves resembling those of a four-leaf clover; see Figure 8.11), the existence of fossil intermediates also supports the monophyly of the group (Rothwell 1999; Lupia et al. 2000). Another interesting result concerns the Polypodiaceae. Despite earlier views that this diverse group was polyphyletic, recent studies indicate that its members form a clade.

The **Marattiales** are mainly plants of the wet tropics that tend to have very large pinnate fronds with thick-walled eusporangia in distinctive clusters (sometimes fused) on the lower surfaces (see Figure 7.10C, D). There are perhaps 300 living species in this clade, most of which belong to *Angiopteris* (over 100 species) or *Marattia* (about 60 species), but they have a long fossil record, and extinct relatives (especially *Psaronius*) were important components of Carboniferous swamps. Consistent with their relative morphological stasis, these plants may have a decelerated rate of molecular evolution (Soltis et al. 2002).

The **Ophioglossaceae** (with perhaps a total of 80 species) are characterized by fronds that are divided into a flattened vegetative portion (or sterile segment) and a sporangium-bearing fertile segment (see Figure 8.6). This peculiar arrangement may have been derived from a dichotomous branch system. The gametophytes are subterranean, achlorophyllous, tuber-like structures that are associated with an endophytic fungus.

The **Psilotaceae** includes about 15 species placed in *Psilotum* (the widespread whisk ferns) and *Tmesipteris* (from Australia and the South Pacific) (see Figure 8.5). Because the plant body consists of dichotomously branching stems, psilophytes have often been viewed as the last remnants of the first vascular plants. An alternative theory, based mainly on their subterranean gametophytes, which are associated with fungi, has been that they are reduced leptosporangiate ferns (possibly related to Gleicheniaceae) (Bierhorst 1977). Recent molecular phylogenetic studies have established with considerable certainty that neither of these ideas is correct (Pryer et al. 2001). Instead, it appears that Psilotaceae are most closely related to the Ophioglossaceae, with which they share some similarities in gametophytes and in the development and position of the sporangia. Under this view, the tiny leaves and the absence of true roots in the Psilotaceae are considered to be derived conditions.

Today there are only about 15 species of **equisetophytes**, or horsetails, all placed in *Equisetum* (Equisetaceae; see Figure 8.7). Equisetophytes have jointed, hollow stems, with distinct ridges where the epidermal cells deposit silica on their surfaces. The leaves are generally reduced to small scales and are borne in a whorl at each node. The haploid spores are produced in sporangia that are attached on the undersides of unusual peltate sporangiophores and clustered in strobili at the tips of the stems. Although the modern equisetophytes are homosporous, there is controversy over whether the gametophytes have separate sexes. Some gametophytes start out producing just antheridia and some only archegonia, but at least the female forms later become bisexual.

Equisetophytes are well known as fossils, which can easily be identified by the characteristic stem architecture. Like the lycophytes, these plants were present in the Devonian but became much more abundant and diverse in the Carboniferous, when some of them also had much larger leaves, evolved heterospory, and became impressive trees. The position of equisetophytes within monilophytes is uncertain (see Figure 7.8), but there is some indication of a connection to Marattiales (Pryer et al. 2001).

Spermatophytes (Seed Plants)

Spermatophytes, or seed plants, are by far the most diverse lineage within the vascular plants, with about 270,000 living species. Most of this diversity is accounted for by just one subclade: the flowering plants, or angiosperms. Morphological evidence for the monophyly of seed plants includes the seed habit itself, but also the fact that the major extant seed plant lineages all share (at least ancestrally) the production of wood (secondary xylem) through the activity of a secondary meristem called the cambium. Another noteworthy vegetative characteristic is axillary branching, as compared with the unequal dichotomous branching that preceded it within euphyllophytes.

MAJOR CHARACTERISTICS OF SEED PLANTS
To understand the seed, it helps to think about how it evolved (Figure 7.11). Seed plants are nested well within a lineage characterized by **homospory** (one kind of spore, bisexual gametophytes). A critical step in the evolution of the seed was the evolution of **heterospory**—the production of two kinds of spores (microspores and megaspores), which produce two kinds of gametophytes (male or microgametophytes, which ultimately produce sperm; and female or megagametophytes, which produce one or more eggs).

Heterospory evolved several times within separate vascular plant lineages, including the lycophytes, the

Figure 7.11 *Archaeopteris* and early seed plants. (A) Reconstruction of the habit of *Archaeopteris*, an extinct "progymnosperm" with a large trunk and flattened lateral branch systems. (B) Reconstruction of an extinct "seed fern," *Medullosa noei* (3.5–4.5 m high), showing the large compound leaves. (C) Probable steps in the evolution of the seed: (i) homospory in a distant ancestor; (ii) heterospory, with differentiation between sporangia that produce microspores and megaspores; (iii) reduction of the number of functional megaspores to one, and its development inside of the sporangium (endospory); (iv) envelopment of the megasporangium by integument tissue, leaving a micropyle at the apex (f gam, female gametophyte; int, integument; mega, megaspores; micro, microspores; nuc, nucellus or megasporangium wall; spor, sporangium). (D) Stages in the evolution of the integument in early seeds (all extinct): (i) *Genomosperma kidstoni*, (ii) *G. latens*, (iii) *Eurystoma angulare*, (iv) *Stamnostoma huttonense*. (E) Pollen-receiving structures at the apex of the ovule in early seeds (all extinct): (i) *Physostoma elegans*; (ii) *P. elegans*, longitudinal section showing pollen chamber within; (iii) *Eurystoma angulare*, showing cup-shaped opening. (F) Portion of long shoot and spur shoot of the extant ginkgophyte, *Ginkgo biloba*, showing axillary microsporangiate strobili; detail of axis and four microsporangium-bearing structures at right. (G) Portion similar to that in F of an ovule-bearing plant of *G. biloba*, showing axillary stalks, each bearing a pair of ovules; detail of the tip of a stalk at right. (H) Longitudinal section of the seed of *G. biloba* with young embryo (ii, inner layer of integument; mi, middle layer of integument; oi, outer layer of integument). (A, F, and G from Bold et al. 1967; B and D from Gifford and Foster 1989; C and H from Scagel et al. 1969; E from Stewart 1983.)

(A)

(B)

(C) Probable steps of seed evolution

Spor

Micro

Mega

Mega

f gam
nuc
int

(i)   (ii)   (ii)   (iv)

(D) Evolution of the integument

(i)   (ii)   (iii)   (iv)

(E) Pollen-receiving structures

(i)   (ii)   (iii)

(F)

(G)

(H)

oi
mi
ii
Embryo

leptosporangiate ferns, the equisetophytes, and the line including the seed plants (Bateman and DiMichele 1994). In several of these cases, the evolution of heterospory was followed by a reduction in the number of functional megaspores. In the line leading to seed plants the number was reduced to just one by abortion of all but one of the four haploid products of a single meiotic division. The single remaining megaspore was retained within the megasporangium and went on to produce a female gametophyte within the spore (endosporic development). Finally, the megasporangium became enveloped by sterile sporophyte tissue known as integument (see Figure 7.11D), but leaving open a little hole at the apex called the micropyle. In seed plants except angiosperms, the micropyle serves as the entrance for one or more pollen grains, which are microspores within which the male gametophyte has developed.

It is also helpful to consider the developmental events leading to a mature seed in a plant like a cycad or a pine tree. A single meiotic division occurs within the ovule (young seed), three of the resulting haploid products disintegrate, and within the remaining spore the female gametophyte develops. Eventually the female gametophyte may contain thousands of cells, with one or more egg cells differentiated near the micropylar end of the seed. Microspores are produced in microsporangia, which are borne elsewhere on the same plant (monoecy) or on separate plants (dioecy).

One or more pollen grains are transported to the vicinity of the micropyle—presumably by wind in the first seed plants. In many cases a drop of liquid (a pollen droplet) is exuded from the micropyle, which pulls adhering pollen grains inside when it retracts. A pollen grain germinates and sends out a tubular male gametophyte, which eventually delivers sperm to the vicinity of the egg. In modern cycads and ginkgos (discussed later in this section), the pollen tube is haustorial, ramifying slowly through the megasporangium wall, and two very large multiflagellate sperm are eventually produced. In contrast, in the remaining modern seed plant lineages a pair of nonmotile sperm are delivered directly to the female gametophyte by the pollen tube. Following fertilization, the diploid zygote develops into a new sporophyte embryo, and the female gametophyte serves as the nutritive tissue.

The second major characteristic of seed plants is the production of wood, or secondary xylem, which (along with a mechanism to regenerate the outer covering of the stem—the periderm) allows the development of a substantial trunk. Understanding how wood is produced requires some basic knowledge of how vascular plants develop. They grow in length through the activity of primary apical meristems at the tip of each shoot and of each root. These apical meristems are populated by undifferentiated cells that undergo mitotic cell divisions, leaving behind derivative cells that go on to differentiate into all of the different cell types and tissues in the plant

body. Shoot apical meristems are also the site of initiation of new buds and leaves.

Some of the cells produced by the apical meristem differentiate within the stem into distinct strands of tissue that ultimately will function as vascular tissue. Within these strands, or vascular bundles, one sees the differentiation of the first (primary) xylem, situated toward the inside of the plant axis, and of phloem, situated toward the outside. Between the xylem and the phloem there remains an undifferentiated layer of cells called the cambium. The cambium acts as a secondary meristem, giving rise to new cells both toward the inside and the outside of the stem, which then go on to differentiate into new xylem cells (such as tracheids) and new phloem cells (such as sieve cells).

The tissues that are produced through this process are referred to as secondary xylem and secondary phloem, respectively. Secondary xylem builds up over the years, forming wood, which is made up of dead, thick-walled cells that are quite sturdy and resistant to decay. Secondary phloem does not build up because phloem cells are not as thick-walled as xylem cells. Additionally, phloem cells have to be alive to carry out their function of transporting carbohydrates and nutrients up and down in the plant body.

It is interesting to note that in contrast to the bifacial cambium of seed plants, the giant lycophytes and equisetophytes of the Carboniferous seem to have had unifacial cambia, producing secondary xylem internally but not phloem. They also lacked the ability to substantially increase the size of the cambial ring, so wood production in these plants was actually quite limited (Cichan and Taylor 1990; Bateman et al. 1998).

EARLY EVOLUTION

With this background on the seed and on wood, let us briefly consider the origin and early evolution of seed plants (see Figures 7.8, 7.11, and 7.12). Our knowledge of the relevant events relies heavily on well-preserved fossils from the late Devonian and early Carboniferous, which have been called "progymnosperms" and "seed ferns" (see Figure 7.11).

Recall that the differentiation of a main trunk and side branches had already evolved in the euphyllophyte lineage. One first sees the appearance of very large trunks, with wood rather similar in structural detail to that of modern conifers, in the late Devonian. These trunks were connected to large frond-like branch systems bearing many small leaves (see Figure 7.11A). *Archaeopteris*, as this plant is now called, was found to be heterosporous, yet without seeds.

The accurate reconstruction and phylogenetic placement of *Archaeopteris* and other "progymnosperms," such as *Aneurophyton* (Beck 1981, 1988; Beck and Wight 1988), was fundamental in establishing that both heterospory and the production of wood pre-dated the evolution of the seed. The clade containing the seed plants

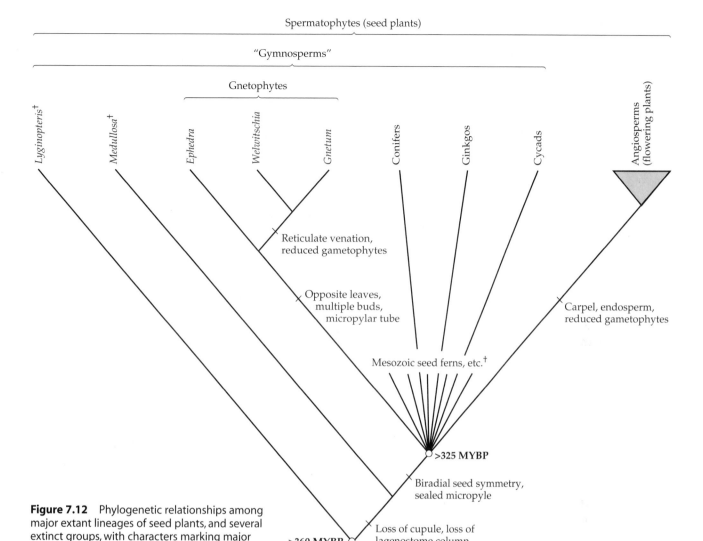

Figure 7.12 Phylogenetic relationships among major extant lineages of seed plants, and several extinct groups, with characters marking major clades (see text). MYBP, million years before present; †, extinct taxon.

plus "progymnosperms" has been called the lignophytes (Doyle and Donoghue 1986), in reference to the production of wood (see Figure 7.8).

The term *seed fern* is applied to a wide variety of early seed plants with large, frond-like leaves, resembling those seen in ferns today but bearing bona fide seeds (Stewart and Rothwell 1993; Taylor and Taylor 1993). It is clear that these are not all most closely related to one another and that a series of Paleozoic seed fern groups form a paraphyletic grade at the base of the seed plant radiation.

Careful analyses (e.g., Serbet and Rothwell 1992) have revealed that the first seeds were situated in "cupules" and that each seed had an elaborate outgrowth of the sporangium wall that formed a specialized pollen chamber. This structure presumably functioned in pollen grain capture. Integument tissue may have been derived from a series of sterilized sporangia, which initially formed lobes at the apex as opposed to a distinct micropyle (see Figure 7.11D).

Through much of the last century, extant and extinct seed plant lineages were commonly divided into two major groups: the cycadophytes and the coniferophytes. The **cycadophytes**, including modern cycads, were distinguished by rather limited production of wood with wide rays (manoxylic wood) and by large frond-like leaves and radially symmetrical seeds. In contrast, in **coniferophytes**, including the ginkgos and the conifers, the wood is well developed and dense (pycnoxylic), the leaves are simple and often needle-like, and the seeds are biradially symmetrical (platyspermic, or flattened). This distinction suggested to some workers that seed plants actually originated twice. Under this view, the cycadophyte line was derived from a progymnospermous ancestor by the modification of flattened lateral branch systems into large, frond-like leaves. In coniferophytes, on the other hand, the individual leaves of a precursor like *Archaeopteris* might have been modified into needle-like leaves. This scenario implies that the seed itself evolved twice, corresponding to the two different symmetries.

However, phylogenetic analyses that have included the extant lineages along with representative fossils have generally supported the relationships shown in Figures 7.8 and 7.12 (e.g., Crane 1985; Doyle and Donoghue 1986; Nixon et al. 1994; Rothwell and Serbet 1994). These studies imply that the seed evolved just once, and that the first seed plants were more cycadophytic, at least in having large dissected leaves and radially symmetrical seeds. Specifically, it appears that a series of Devonian-Carboniferous "seed ferns" (*Lyginopteris* and medullosans) are situated at the base of the seed plant phylogeny and that coniferophytes are variously nested well within the tree, in a platyspermic clade. This arrangement implies a later shift to small, needle-like leaves and to smaller flattened seeds—both perhaps as adaptations to arid environments.

EXTANT LINEAGES OF SPERMATOPHYTES

Today there are five major lineages of seed plants: cycads (Cycadales), ginkgos (Ginkgoales), conifers (Coniferales), gnetophytes (Gnetales), and flowering plants (angiosperms). The first four groups are often called gymnosperms, in reference to having naked seeds, as opposed to angiosperms, in which the seeds are enclosed inside of a carpel. Despite many efforts to resolve the phylogenetic relationships among these lines using morphological and molecular data, they remain quite uncertain (see Figure 7.12).

Some recent molecular analyses have indicated that the extant groups of "naked seed plants" form a clade, which is sister to the angiosperms (see the next section). However, note that even if this were true, the "gymnosperms" as a whole would not be monophyletic. They are paraphyletic when one takes into account the early-diverging fossil lineages already mentioned (see Figure 7.12), as well as several other "seed fern" lineages from the later Permian and Mesozoic, at least some of which are probably on the line leading to modern angiosperms. We will return to a discussion of these relationships following a brief introduction to each of the major groups (see also Chapter 8).

Cycads

Cycads (Cycadales) were most abundant and diverse during the Mesozoic. Today there are perhaps 130 species left. Cycads generally produce squat trunks, with limited secondary xylem, and large compound leaves resembling those of ferns or palms (see Figure 8.15). They are dioecious, meaning that some plants bear strobili producing only seeds, whereas others bear only pollen strobili. Both types of strobili are typically very large, and in some cases brightly colored. Likewise the seeds are generally large and usually have a fleshy and colorful seed coat, presumably to attract vertebrate dispersal agents.

Several cycad features may be ancestral within seed plants, such as haustorial male gametophytes and gigantic multiflagellate sperm. However, cycads are united by several apparently derived morphological characters, including the loss of axillary branching, the presence of "girdling" leaf traces, and the production of coralloid roots that house nitrogen-fixing cyanobacteria.

Within cycads, phylogenetic analyses indicate that the first split divides *Cycas* from the remaining groups. *Cycas* has retained the presumed ancestral condition (seen in some fossil relatives, such as *Taeniopteris*) of having several ovules borne on the rather leaf-like megasporophylls, which are not clustered into strobili. The derived condition, seen in the other line, is a reduction to two ovules borne on a peltate megasporophyll, with the ovules pointing in toward the axis of the strobilus.

Ginkgos

There is just one surviving species of **ginkgo** (Ginkgoales): *Ginkgo biloba* (see Figure 7.11F–H). This species is hardly known in the wild, but it has been maintained for centuries around temples in China, and in modern times it has been spread by humans as a street tree. Perhaps the most characteristic feature of the modern ginkgo is the production of deciduous, fan-shaped leaves with dichotomous venation. Ginkgos are well known in the fossil record, where a greater diversity of leaf shapes is seen.

Like cycads, ginkgos are dioecious (see Figure 7.11F, G). The ovules are borne in pairs on axillary stalks, thought to be reduced strobili. The integument tissue differentiates into a fleshy (and smelly) outer layer and a hard inner layer that encloses the female gametophyte (Figure 7.11H). Again like cycads, ginkgos retain several ancestral characteristics, including haustorial male gametophytes and swimming sperm.

Conifers

There are approximately 600 living species of **conifers** (Coniferales) (see Figures 8.17–8.20). These are shrubs or small trees with well-developed wood and often needle-like leaves. Normally the leaves are borne singly along the stem, but in pines (*Pinus*) they are clustered in short shoots. The needles often display additional adaptations to drought, such as sunken stomates. In some of the Southern Hemisphere conifers (e.g., *Podocarpus*, *Agathis*), however, the leaves are rather broad and flat, and in *Phyllocladus* the flattened branches resemble leaves.

Many conifers are monoecious, with both pollen-producing and seed-producing strobili borne on the same plant. Dioecy is found in other groups, such as in the junipers (*Juniperus*), yews (*Taxus*), and podocarps (*Podocarpus*). In the pollen cones, microsporophylls bear microsporangia on the abaxial surface. The pollen grains often have a pair of sac-like appendages, but these seem to have been lost in several lineages.

Receptive ovules are situated on the upper side of each cone scale. Meiosis occurs inside each ovule, and the one remaining haploid cell gives rise to the female gametophyte, which eventually produces one or more eggs at the micropylar end. A pollen tube grows down through the wall of the megasporangium to deliver two sperm. The phenomenon of "polyembryony" is fairly common in conifers, with multiple embryos produced either through separate fertilization events (depending on the number of eggs and pollen tubes) or, more commonly, by a characteristic subdivision of a single embryo into several genetically identical embryos early in development.

In modern conifers the pollen strobili are said to be simple, whereas the seed cones are compound. The pollen strobilus is interpreted as a modified branch, and the microsporophylls as modified leaves. The seed cone, in contrast, was derived through modification of a branch that bore lateral branches in the axils of a series of leaves. This view is supported by fossils showing a series of steps in the reduction of a lateral branch bearing a number of seeds to the highly modified cone scale that we see in the modern groups (Figure 7.13A–E) (Florin 1951, 1954). It also follows from the observation that each cone scale is subtended by a bract, which represents the modified leaf. In a few conifers the subtending bract is noticeable, sticking out from between the cone scales. This is the case, for example, in the Douglas fir (*Pseudotsuga*), where the cone scale is produced in the axil of a prominent three-pronged bract (Figure 7.14C). In many conifers, however, the bract is quite reduced. In Cupressaceae, such as *Taxodium* or *Cryptomeria*, the bract is fused to the cone scale, which still shows evidence of "leaves" (visible as small teeth or bumps).

Phylogenetic studies have yielded some important insights into the evolution of conifers (e.g., Stefanovic et al. 1998). Molecular data imply a basal split between the Pinaceae and a clade including all of the other conifers. The Pinaceae are distinguished by several features, including inversion of the ovules (micropyle facing the axis of the cone) and the wing of the seed being derived during development from the cone scale.

Within the other clade of conifers, the two major Southern Hemisphere groups—Podocarpaceae and Araucariaceae—form a clade, perhaps united by a shift to one ovule per cone scale. The Cupressaceae are marked by several potential apomorphies, such as fusion of the cone scale and the subtending bract. In turn, this group may be linked with the Taxaceae (the yews), which have highly reduced cones bearing just one terminal seed surrounded by a colorful fleshy aril.

Gnetophytes

The fourth major extant lineage of seed plants is the **gnetophytes** (Gnetales) (Figure 7.13F–I; see also Figure 8.21). This group contains only about 80 living species, which belong to three quite distinct lineages (see Doyle 1996; Friedman 1996; Price 1996). *Ephedra* (with about 40 species in deserts around the world) has very reduced scale-like leaves (see Figure 8.21). *Gnetum* (with about 35

species in tropical forests of the Old and New Worlds) has broad leaves (Figure 7.13G–I), like those seen in most flowering plants. Finally, *Welwitschia* (with only one species, *W. mirabilis*, in southwestern Africa) produces just two (rarely four) functional leaves during its lifetime, which grow from the base and gradually fray out at the tips (Figure 7.13F).

Although these plants look very different from one another, they share some unusual features, such as opposite leaves, multiple axillary buds, vessels with circular openings between adjoining cells, compound pollen *and* seed strobili, and ancestrally ellipsoid pollen with characteristic striations running from tip to tip. The seeds also have two integumentary layers—the inner one forming a micropylar tube that exudes the pollen droplet, and the outer one derived from a fused pair of bracts. Molecular studies also strongly support the monophyly of this group.

Within the gnetophytes, *Gnetum* and *Welwitschia* form a well-supported clade. Morphological synapomorphies include reticulate leaf venation, further reduction of the male gametophyte, and aspects of female gametophyte structure (tetrasporic development, loss of archegonia, free nuclei functioning as eggs). The characteristic striate pollen found in *Ephedra* and *Welwitschia* was apparently lost along the line leading to *Gnetum* (which has spiny grains with no apertures).

Aside from fossil pollen, the fossil record of this clade is rather poor; only a few macrofossils have been described (Crane 1996). Although gnetophyte pollen grains are found as far back as the Triassic, it appears that the clade containing the modern groups diversified most significantly during the mid-Cretaceous, along with the angiosperms (see the next section).

Like the angiosperms, the gnetophytes shortened the life cycle (and probably became herbaceous) and evolved insect pollination (found in some living species). In marked contrast to flowering plants, however, gnetophytes never became significant components of the vegetation at mid and high paleolatitudes, and they underwent a dramatic decline during the Late Cretaceous (Crane et al. 1995; Crane 1996).

Angiosperms (Flowering Plants)

With over 257,000 extant species, flowering plants account for most of green plant, land plant, and seed plant diversity. Strong evidence for the monophyly of angiosperms comes from molecular studies and from many shared derived morphological characters. Of these, some of the more obvious and important reproductive features are (1) seeds produced within a carpel with a stigmatic surface for pollen germination; (2) a very reduced female gametophyte, consisting in most cases of just eight nuclei in seven cells; and (3) double fertilization, leading to the formation of a typically triploid nutritive tissue called endosperm.

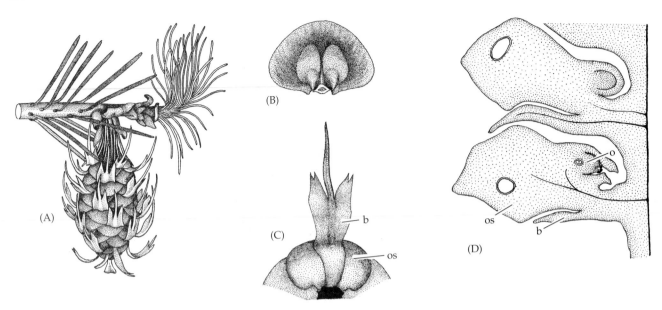

(A)

(B)

(C)
b
os

(D)
o
os
b

(E) Steps in ovuliferous cone scale evolution

o
ss
br
ds
(i)

o   ss
br
ds
(ii)

o   os
br
(iii)

(F)

(G)

(H)

(I)
mt
ob
ib
int

Figure 7.13 Morphology of conifers and gnetophytes. (A) *Pseudotsuga*, showing a branch with a first-year seed cone. (B) Single ovuliferous cone scale of *Pseudotsuga*, showing two ovules on the upper surface. (C) A single bract-scale complex of *Pseudotsuga*, showing the excerted three-pronged bract (b) subtending the ovuliferous scale (os). (D) Longitudinal section through two bract-scale complexes in an ovulate cone of *Pinus strobus*, showing an ovule (o) with micropyle directed toward the cone axis, the ovuliferous scale (os), and the subtending bract (b). (E) Probable evolutionary steps in the origin of the ovuliferous cone scale of conifers: (i) the extinct *Cordaites*, with several ovules (o) and sterile scales (ss) attached to a dwarf shoot (ds) in the axil of a bract (br); (ii) the extinct plant *Lebachia*, in which the number of ovules is reduced; (iii) extant *Pinus*, with two ovules attached to the upper surface of the ovuliferous scale. (F) General habit of the gnetophyte *Welwitschia mirabilis*, showing the short woody stem with two large leaves, axillary position of the multiple strobili, and taproot. (G) Leaves and compound microsporangiate strobili of *Gnetum*. (H) *Gnetum* with mature seeds. (I) Longitudinal section through a young seed of *Gnetum*, showing the inner integument (int) extended into a micropylar tube (mt), surrounded by inner and outer bracteoles (ib, ob). (A–D from Stewart 1983; E, G, and I from Scagel et al. 1969; F from Barnes 1998.)

Several derived vegetative characteristics are also noteworthy. Almost all angiosperms produce vessels in the xylem tissue, though this feature probably evolved within the group. Vessels differ from tracheids in that water can flow from one vessel element (an individual cell, evolutionarily derived from a tracheid) to the next without traversing a pit membrane (see Figure 4.31). Vessels are extremely efficient with respect to water transport but may be more prone to damage (especially through air embolisms) when subjected to drought stress. Angiosperm phloem differs from that of all other plants in that the sieve tube elements (living but enucleate cells functioning in the transport of carbohydrates) are accompanied by one or more "companion cells" that are derived from the same mother cell.

FLOWERS AND THE ANGIOSPERM LIFE CYCLE

The production of flowers is commonly considered the diagnostic feature of angiosperms, but the term *flower* is actually a bit nebulous. If flowers are short reproductive axes with closely aggregated sporophylls, then gnetophytes and Bennettitales might also be said to have flowers. It is the particular construction and arrangement of the flower parts that sets the angiosperms apart from all other seed plants (see Figure 4.16). Most angiosperm stamens have a stalk (filament) and a tip portion (anther) bearing usually two pairs of microsporangia (pollen sacs). The angiosperm carpel is typically differentiated into a lower portion (ovary) that encloses the ovules and an elongate portion (style) that elevates a surface receptive to pollen (stigma). The angiosperm ovule is unusual in several ways (see Figures 4.39 and 4.40). It generally becomes curved over (anatropous) during development, so that the micropyle lies near the stalk of the ovule (in

contrast to the orthotropous condition in other seed plants, where the micropyle faces away from the stalk). In addition, whereas non-angiosperm seeds have one layer of integument tissue (sometimes differentiated into fleshy and hard layers), angiosperms typically have two distinct integuments (bitegmic ovules).

The angiosperm life cycle is also remarkably derived (see Figure 4.17). The male gametophyte has just three nuclei, or sometimes just two at the time the pollen is shed. A pollen grain that lands on the right stigma sends out a pollen tube that delivers the sperm directly to the female gametophyte inside the ovule. In the development of a typical angiosperm female gametophyte, meiosis is followed by the abortion of three products, and the remaining haploid nucleus undergoes a very small series of mitotic divisions (see Figure 4.40).

Ultimately the egg is situated toward the micropylar end, along with two other cells (synergids), which appear to play a critical role in orienting the pollen tube and delivering the sperm nuclei. There are usually three cells (antipodals) at the opposite end of the female gametophyte, and two nuclei (polar nuclei) situated in a large cell in the middle. One of the two sperm nuclei fuses with the egg to give rise to the diploid zygote, and the other fuses with the two polar nuclei. This process is called double fertilization. The resulting triploid product undergoes a series of mitotic divisions to produce endosperm, which serves as the nutritive tissue in the seed (as opposed to the female gametophyte).

TIME OF ORIGIN

When did the flowering plants originate and radiate? It appears from the fossil record (pollen, leaves, flowers, and fruits) that angiosperms underwent a major radiation starting in the Early Cretaceous (Doyle and Hickey 1976; Friis et al. 1987; Doyle and Donoghue 1993; Crane et al. 1995). The oldest unequivocal angiosperm fossils are pollen grains from about 135 million years ago. Extraordinarily complete fossils from China were first described as being from the Late Jurassic (Sun et al. 1998, 2002), but now they are interpreted as Early Cretaceous in age. Many major lineages can be recognized by the mid-Cretaceous (water lilies, Chloranthaceae, Winteraceae, and eudicots are present by 125 million years ago), and by the end of the Cretaceous they had diversified extensively and were the dominant plants in many terrestrial environments (see Magallon and Sanderson 2001).

In discussing the age of the angiosperms (or any other group), it is important to distinguish clearly between the origin of the line leading to the modern group (i.e., when this lineage split from its sister lineage) and the origin of the least inclusive clade that contains all of the extant members (the crown clade). The clade that includes the "stem" lineage has been referred to as the "angiophytes," to help distinguish it from crown-clade angiosperms (Doyle and Donoghue 1993).

Figure 7.14 Reconstructions of Mesozoic fossils that may be closely related to angiosperms. (A, B) Bennettitales: (A) Habit of *Williamsonia sewardiana*, showing cycad-like trunk and compound leaves; (B) longitudinal section of a flower-like strobilus of *Williamsoniella*. B, bracts; M, microsporophyll with microsporangia; OS, stalked ovules and sterile scales borne on a central axis. (C) Caytoniales: (i) palmate leaf, *Sagenopteris phillipsi*; (ii) portion of a microsporophyll, *Caytonanthus kochi*; (iii) megasporophyll of *Caytonia nathorsti*, showing two rows of cupules; (iv) longitudinal section of a cupule of *Caytonia thomasi*, showing ovules within. (D) Glossopteridales: (i) Ovulte portion of *Denkania indica*, showing six cupule-like structures attached to a leaf; (ii) *Lidettonia mucronata*, showing seeds attached on the lower surfaces of stalked disks borne on a leaf. (A from Taylor and Taylor 1993; B, C:ii–iv, and D from Gifford and Foster 1989; C:i from Stewart 1983.)

It is possible that the angiophytes are quite ancient, whereas the crown angiosperms originated much more recently, perhaps not long before the radiation seen in the fossil record. That the angiophyte line may be quite old is suggested by the fact that all of the likely close relatives of angiosperms have fossil records going back at least to the Triassic. We might, therefore, expect to find them as fossils before the Cretaceous, though perhaps without the full complement of characters found in modern angiosperms. So far, however, putative angiosperm fossils from the Triassic and Jurassic have either turned out not to be related or are equivocal on the basis of available material.

Estimates based on molecular data are faced with the problem of shifts in the rate of molecular evolution, possibly independently in different lineages. Early molecular clock studies (e.g., Martin et al. 1989) yielded implausibly early ages for crown angiosperms. However, progress has been made in "relaxing" the molecular-clock assumption, and recent estimates place the origin of the angiosperm crown clade between 140 and 190 million years ago (Sanderson and Doyle 2001).

RELATIONSHIP OF ANGIOSPERMS TO OTHER GROUPS

Botanists have long puzzled over the relationship of angiosperms to other seed plants. As noted earlier, this problem is complicated because, in addition to the other extant clades of seed plants (cycads, ginkgos, conifers, and gnetophytes), several extinct groups bear directly on the problem (see Stebbins 1974; Beck 1988; Stewart and Rothwell 1993; Taylor and Taylor 1993). In particular, it has long been hypothesized that flowering plants are most closely related to some group of Mesozoic "seed ferns" (e.g., *Caytonia*, glossopterids), or perhaps to the Bennettitales (also known as "cycadeoids" because of their resemblance to cycads; Figure 7.14A). Bennettitales have been attractive candidates because some of them produced large, flower-like reproductive structures with pollen-producing organs surrounding a central stalk that bears the naked seeds (Figure 7.14B).

Regarding the five extant lineages, ideas on relationships have shifted over the years. In the early 1900s (e.g., Arber and Parkin 1907), Gnetales (along with the extinct Bennettitales) were widely believed to be related to angiosperms on the basis of several morphological similarities, such as vessels in the wood, net-veined leaves in *Gnetum*, and flower-like reproductive organs. These views changed by the middle of the twentieth century with the reinterpretation of these characters. For exam-

ple, vessels were interpreted as being derived independently in Gnetales (from tracheids with circular bordered pits) and in angiosperms (from tracheids with scalariform pits). This character, and several others, suggested instead that gnetophytes were related to conifers.

In the mid-1980s, several phylogenetic studies of seed plants were carried out using morphological characteristics (Crane 1985; Doyle and Donoghue 1986). These analyses concluded that angiosperms formed a clade with Bennettitales and Gnetales—a clade referred to as the "anthophytes" to highlight the flower-like reproductive structures (Figure 7.15A). Since then, a number of independent morphological analyses have yielded the same basic result. In some of these studies, the gnetophytes have even emerged as paraphyletic with respect to angiosperms (Taylor and Hickey 1992; Nixon et al. 1994). The characters that have served to unite the anthophytes vary among analyses, but they are mostly rather obscure and in some cases unknown in fossil groups—for example, lignin chemistry, the layering of cells in the apical meristems, and pollen and megaspore features (Donoghue and Doyle 2000).

In any case, the repeated recovery of the anthophyte clade favored a return to the view that gnetophytes and angiosperms are closely related. In turn, this conclusion influenced the interpretation of morphological evolution. Perhaps most notably, double fertilization (reported

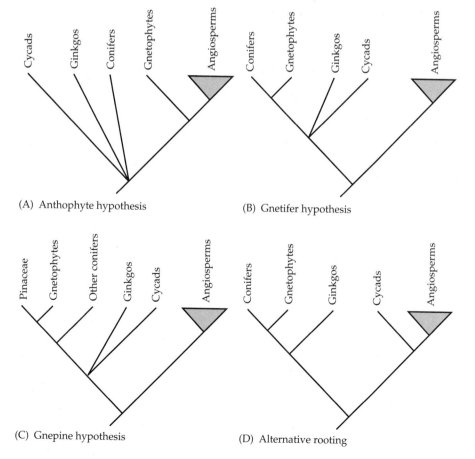

(A) Anthophyte hypothesis

(B) Gnetifer hypothesis

(C) Gnepine hypothesis

(D) Alternative rooting

Figure 7.15 Alternative hypotheses of relationships among the five major extant lineages of seed plants. (A) In the anthophyte hypothesis, gnetophytes are most closely related to angiosperms. (B) In the gnetifer hypothesis, gnetophytes are most closely related to conifers. (C) In the gnepine hypothesis, gnetophytes are most closely related to Pinaceae within the conifers. (D) An example of a tree that is difficult to reject with current data (see text).

for *Ephedra* in the early and mid-1900s, but recently clearly documented in *Ephedra* and *Gnetum*; Friedman 1990; Carmichael and Friedman 1996) was interpreted as having evolved in the common ancestor of gnetophytes and angiosperms, with polyploid endosperm evolving later in the angiosperm line (see Donoghue and Scheiner 1992; Friedman and Floyd 2001).

The first molecular phylogenetic studies of the problem yielded a variety of results and were viewed as possibly being consistent with the anthophyte hypothesis (see Donoghue and Doyle 2000). Within the last few years, however, several major studies (especially those based on mitochondrial genes or on a combination of genes from different genomes) have cast serious doubt on the existence of an anthophyte clade (e.g., Goremykin et al. 1996; Bowe et al. 2000; Chaw et al. 2000; Frohlich and Parker 2000; but see Rydin et al. 2002). These analyses suggest instead that extant gymnosperm groups form a clade that is sister to the angiosperms, and that Gnetales are related more directly to conifers (the gnetifer hypothesis; see Figure 7.13B) or may even be nested within the conifers as the sister group of the Pinaceae (the gnepine hypothesis; see Figure 7.13C). Detailed analyses of the molecular data sets (e.g., Graham and Olmstead 2000; Sanderson et al. 2000) have revealed several different signals; some partitions of the data even favor placing the gnetophytes as sister to all other extant seed plant groups.

Unfortunately, at the present time these issues remain unresolved. It has become clear, however, that there are several potentially separate issues at stake. One important question is whether an anthophyte clade exists, or whether, instead, gnetophytes may be directly related to conifers. A second issue is the rooting of the seed plant tree, especially of the portion that includes the extant lineages. One possibility is a basal split into the angiosperms on the one hand and the extant gymnosperms on the other. Other possibilities are difficult to rule out on the basis of presently available data, such as placement of the root in the vicinity of cycads and ginkgos (see Figure 7.15D).

There is a distinct possibility that no living group of seed plants is very closely related to angiosperms. Recent results therefore accentuate the importance of fitting fossils into the picture, which will depend on more and better fossils and more attention to the phylogenetic analysis of morphological characters (Donoghue and Doyle 2000; Frohlich and Parker 2000).

RELATIONSHIPS WITHIN ANGIOSPERMS

Enormous progress has recently been made in understanding phylogenetic relationships at the base of the angiosperms themselves (Figure 7.16). Until very recently, the problem of identifying the root of the angiosperms and relationships among the basal branches looked intractable (Donoghue 1994). Over the last few years, however, several different lines of evidence have converged on the same answer. These new findings are having a major impact on our interpretation of early angiosperm evolution and the factors that account for the enormous success of flowering plants.

Most students of angiosperm evolution have held that the first flowering plants were among the "Magnoliidae"—a paraphyletic group including magnolias, avocados, water lilies, and black peppers, among others (Cronquist 1988; Thorne 1992; Takhtajan 1997). Even if true, however, this conclusion is not very helpful in deriving an image of the first flowering plants because these plants display an impressive range of morphological forms. Some are woody plants and some are small herbs. Moreover, some, like magnolias, have large flowers, with many flower parts (stamens, carpels) spirally arranged on an elongate axis, while others, such as black peppers, have tiny flowers with few parts in distinct whorls. In fact, some phylogenetic results have suggested that the first flowering plants were woody with large flowers; others have implied that they were herbaceous with tiny flowers (see Doyle and Donoghue 1993).

Since 1999, several independent molecular studies have concluded that the first split within modern angiosperms was between a lineage that now includes just the single species *Amborella trichopoda* (and possibly also the water lilies, Nymphaeales) and all of the rest of the approximately 257,000 species (Mathews and Donoghue 1999; Qiu et al. 1999; Soltis et al. 1999; Barkman et al. 2000; Zanis et al. 2002; see also Zimmer et al. 2000). *Amborella trichopoda* is a shrubby plant from the island of New Caledonia with rather small flowers and a limited number of spirally arranged parts (Endress and Igersheim 2000b). Pollen-producing flowers are borne on some plants and seed-producing flowers on others. The presence of staminodes in the carpellate flowers, however, implies that this species evolved from ancestors with bisexual (perfect) flowers. Unlike almost all other angiosperms, the water-conducting cells in the xylem of *Amborella* are tracheids (Feild et al. 2000), supporting the view that the first angiosperms lacked vessels (see Figure 7.16).

The water lilies (Nymphaeales) form another very early branch of the angiosperm tree (Friis et al. 2001), as do Austrobaileyales (including Illiciaceae). Interestingly, the female gametophytes of plants in these lineages have just four cells and form diploid endosperm tissue, which may prove to be ancestral for angiosperms (Williams and Friedman 2002). Along with *Amborella*, these two lineages subtend a clade including all of the rest of the flowering plants, which we refer to as the **core angiosperms**.

Whereas in the basal-most lineages the carpels are typically sealed by a secretion, in members of the core clade the carpels usually are sealed by postgenital fusion of epidermal layers (Endress and Igersheim 2000a). In the three basal lineages, and also in Chloranthaceae (which may be at the base of the core angiosperms; Doyle and Endress 2000), the carpels are ascidiate, meaning that the primordium is U-shaped at first, and grows

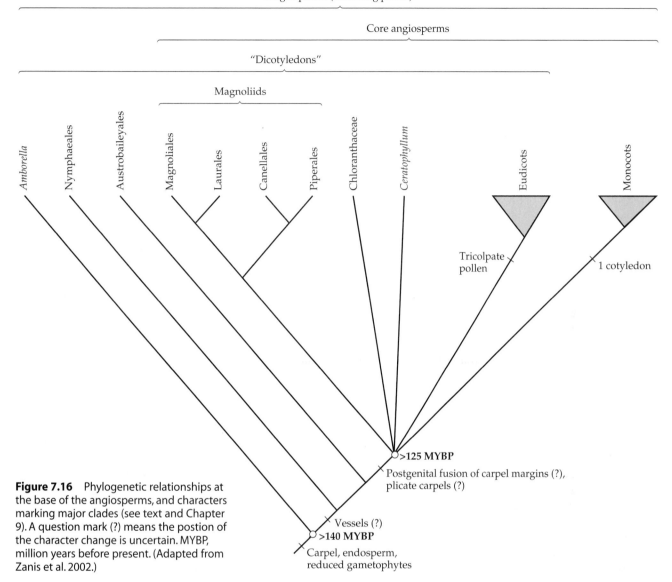

Figure 7.16 Phylogenetic relationships at the base of the angiosperms, and characters marking major clades (see text and Chapter 9). A question mark (?) means the postion of the character change is uncertain. MYBP, million years before present. (Adapted from Zanis et al. 2002.)

up like a tube, whereas in core angiosperms the carpels are plicate, like a leaf folded down the middle. Although these observations clarify the basal carpel condition in angiosperms, they leave open the controversial issue of whether the carpel was derived from a leaf or instead is a compound structure derived from a reduced branch and its subtending leaf (see Doyle 1994).

Within the core angiosperm clade, relationships are still poorly resolved, with the placement of several enigmatic groups still uncertain, especially the Chloranthaceae and *Ceratophyllum*. However, several major clades are rather well supported. First, a restricted **magnoliid clade** includes the Magnoliales plus Laurales, and the Canellales plus Piperales (including the Aristolochiaceae and Lactoridaceae). Winteraceae, a vessel-less group, is in the Canellales, implying that vessels may actually have been lost in some cases (see also Trochodendraceae in Chapter 9).

A second major lineage of core angiosperms, containing the remainder of the former dicotyledons, has been called the **eudicots**. This lineage was first recognized in morphological analyses and was initially called the tricolpate clade (Donoghue and Doyle 1989), in reference to the main morphological character marking the group—namely, the production of pollen grains with three colpi, or germinal furrows (and a variety of derivative forms; see Figure 4.46). Many eudicots also have flowers with parts in fours or fives, or in multiples of these numbers.

Altogether there are perhaps 160,000 species of eudicots. This huge group contains a number of very speciose lineages, including legumes (about 16,000 species) and composites (about 20,000 species), as well as buttercups, roses, oaks, mustards, tomatoes, mints, and snapdragons, to list only a few familiar groups among those discussed in detail in Chapter 9.

A third major clade, with some 65,000 species, corresponds to the traditional **monocotyledons**. Almost half of the species of monocots are either orchids (about 20,000 species) or grasses (about 9000 species), but this group also includes palms, bromeliads, bananas, aroids, lilies, irises, and many other familiar and important plants (see Chapter 9).

Many of the features traditionally cited in support of the monocot clade may be ancestral angiosperm conditions, including flower parts in threes and monosulcate pollen. Other features may unite the monocots, such as scattered vascular bundles and loss of vascular cambium, parallel leaf venation, and development of the leaf blade from the basal part of the leaf primordium, but this depends on exactly what their relatives turn out to be and on relationships within the monocots. In the end, the presence of one seed leaf, or cotyledon, may be the morphological character that best distinguishes the monocotyledons (see Figure 4.42).

Note that the view of relationships we have just outlined is at odds with standard classifications in which flowering plants are divided into two major groups: the monocotyledons and the dicotyledons. Instead, the monocots make up a clade that is nested within the paraphyletic "dicots."

POLLINATION, DISPERSAL, AND HABIT

Much of the diversity of flowers is related to pollination biology (see Chapter 4). Insect pollination is known from several other seed plant lineages: the modern cycads and the gnetophytes, as well as the fossil Bennettitales and possibly some Mesozoic "seed ferns." Insect pollination was apparently established by the origin of crown-group angiosperms and was probably first carried out by pollen-eating or pollen-collecting insects, especially beetles and flies. Flowers pollinated by nectar-collecting insects evolved later. This conclusion is supported by the morphology of early angiosperm fossils, as well as by knowledge of pollination mechanisms in extant members of basal lineages (Friis et al. 1987; Thien et al. 2000).

It is unclear how much pollination by insects stimulated the early diversification of angiosperms, but the evolution of flowering plants apparently did not have a major impact on the origin of the major insect lineages, which evolved much earlier. It is abundantly clear, however, that diversification within some angiosperm and insect lineages has been causally linked.

Variation in fruit morphology is largely related to the use of different dispersal agents (see Chapter 4). Cretaceous fossil fruits and seeds are generally quite small, and there is no direct evidence of specialized dispersal by mammals or birds (see Friis et al. 1987). Adaptations for dispersal by frugivorous and granivorous animals apparently did not appear until later in the Cretaceous, and in most lineages probably originated in the Tertiary. Not until the Early Tertiary do we first see evidence of complex rain forests, and the major radiation of birds

and mammals occurred during the same time. The evolution of large colorful fruits and seeds was linked to the evolution of these groups.

Finally, it is interesting to contemplate the evolution of growth form within angiosperms and what effects this might have had on diversification. Most recent studies position woody plant lineages near the base of the tree. *Amborella* and Austrobaileyales are mostly shrubs or small trees, though they show a tendency toward vine-like growth. Modern representatives, at least, live in forest understory environments, and they show various adaptations to rather low-light environments (Feild et al. 2001). A major exception among the early lineages is the water lily clade, in which the plants are herbaceous and live in high-light aquatic environments. The extinct *Archaefructus*, whose relationships are poorly resolved, was also probably an aquatic plant (Sun et al. 2002).

The herbaceous habit evolved early in angiosperm evolution, probably several times independently—for example, in Nymphaeales, Chloranthaceae, Piperales, and monocotyledons. In several cases this development appears to be correlated with movement into aquatic habitats. Larger woody forms have also re-evolved from herbaceous plants on some occasions, though the evolution of "normal" wood was precluded in the monocots by the loss of the cambium. Within the monocots, large stature has instead been attained in several other ways—for example, through a specialized thickening mechanism in the apical meristem of palms; enlarged, stiffened leaf bases in the bananas and their relatives; and an anomalous form of cambial activity in the Ruscaceae, Agavaceae, and a few of their relatives (see Chapter 9).

Within the eudicots we see enormous variation in habit, but again there have been many shifts from woody to herbaceous growth form, some of these quite early in the evolution of the group. For example, herbaceous poppies (Papaveraceae) and buttercups (Ranunculaceae) may have evolved early, and independently, within one of the first major eudicot branches, the Ranunculales. *Nelumbo*, the water lotus, presents another early example involving a shift to the aquatic environment.

An important trend within eudicots has been the derivation of herbaceous lineages of temperate climate zones from within tropical woody plant lineages (Judd et al. 1994). These transitions often appear to be correlated with upward shifts in the rate of diversification (Judd et al. 1994; Magallon and Sanderson 2001), and taken together they have had a profound impact on angiosperm diversity.

Summary

The tremendous progress made over the last few decades in establishing phylogenetic relationships is now having a major impact on our understanding of green plant evolution. Recent phylogenetic analyses have shown that some traditionally recognized groups are not monophyletic. For example, we appreciate that "plants" (autotrophic eukaryotes) originated independently through several separate endosymbiotic events. Within the green plant clade, traditional "green algae" are paraphyletic with respect to land plants, as are "bryophytes" with respect to vascular plants, "seedless vascular plants" with respect to seed plants, "gymnosperms" with respect to flowering plants, and "dicotyledons" with respect to monocots. As such groups are dismantled, major new clades are being identified, such as the streptophytes (some "green algae" plus embryophytes) and euphyllophytes (some "seedless vascular plants" plus spermatophytes).

A variety of long-standing phylogenetic questions have also recently been answered with considerable confidence. For example, the whisk ferns (Psilotaceae) are not remnants of the first vascular plants but instead are part of the monilophyte clade. Perhaps most surprisingly, the very base of the angiosperm tree is finally being resolved, with the *Amborella* and water lily branches diverging before a core angiosperm clade that includes the eudicots and the monocots.

Although phylogenetic progress has been rapid, many key questions remain unresolved. For example, we are more uncertain today than we were a decade ago about relationships at the base of the embryophytes, as well as among the major seed plant lineages. Where do the gnetophytes fit, and what really are the closest relatives of the flowering plants? And within core angiosperms, what are the closest relatives of the monocots and the eudicots?

These important questions are proving to be very difficult to resolve, but the successes of the last few decades suggest that answers will eventually be forthcoming. Experience also implies that analyses integrating evidence from a wide variety of sources—molecular data, morphology, development, and the fossil record—stand the best chance of lasting success.

Literature Cited and Suggested Readings

References marked with an asterisk (*) are other reviews focused on phylogenetic relationships and are especially recommended for additional information and viewpoints. References marked with a number symbol (#) are standard texts on comparative plant anatomy, morphology, physiology, and paleobotany that can be consulted for background information on specific characters and groups of organisms. These sources are not specifically cited in the text of this chapter, but they provide the basis for discussions of life cycles and morphological evolution. The symbol @ identifies symposia on lycophyte, gnetophyte, and angiosperm phylogeny.

Arber, E. A. N. and J. Parkin. 1907. On the origin of angiosperms. *J. Linnean Soc. Bot.* 38: 29–80.

Baldauf, S. L. and J. D. Palmer. 1990. Evolutionary transfer of the chloroplast *tuf*A gene to the nucleus. *Nature* 344: 262–265.

Baldauf, S. L, A. J. Roger, I. Wenk-Siefert and W. F. Doolittle. 2000. A kingdom-level phylogeny of eukaryotes based on combined protein data. *Science* 290: 972–977.

Barkman, T. J., G. Chenery, J. R. McNeal, J. Lyons-Weiler and C. W. de Pamphilis. 2000. Independent and combined analyses of sequences from all three genomic compartments converge on the root of flowering plant phylogeny. *Proc. Natl. Acad. Sci. USA* 97: 13166–13171.

Barnes, R. S. K. (ed.) 1998. *The diversity of living organisms.* Blackwell Science, Oxford.

Bateman, R. M. 1994. Evolutionary-developmental change in the growth architecture of fossil rhizomorphic lycopsids: Scenarios constructed on cladistic foundations. *Biol. Rev.* 69: 527–597.

Bateman, R. M. and W. A. DiMichele. 1994. Heterospory: The most iterative key innovation in the evolutionary history of the plant kingdom. *Biol. Rev.* 69: 345–417.

*Bateman, R. M., P. R. Crane, W. A. DiMichele, P. R. Kenrick, N. P. Rowe, T. Speck and W. E. Stein. 1998. Early evolution of land plants: Phylogeny, physiology and ecology of the primary terrestrial radiation. *Annu. Rev. Ecol. Syst.* 29: 263–292.

Beck, C. B. 1981. *Archaeopteris* and its role in vascular plant evolution. In *Paleobotany, paleoecology, and evolution*, K. J. Niklas (ed.), Vol. 1, 193–230. Praeger, New York.

Beck, C. B. (ed.). 1988. *Origin and evolution of gymnosperms.* Columbia University Press, New York.

Beck, C. B. and D. C. Wight. 1988. Progymnosperms. In *Origin and evolution of gymnosperms*, C. Beck (ed.), 1–84. Columbia University Press, New York.

Bierhorst, D. W. 1977. The systematic position of *Psilotum* and *Tmesipteris. Brittonia* 29: 3–13.

#Bold, H. C. 1967. *Morphology of plants*, 2nd ed. Harper & Row, New York.

#Bold, H. C. and M. J. Wynne. 1985. *Introduction to the algae: Structure and reproduction.* Prentice Hall, Englewood Cliffs, NJ.

Bowe, L. M., G. Coat and C. W. dePamphilis. 2000. Phylogeny of seed plants based on all three genomic compartments: Extant gymnosperms are monophyletic and Gnetales' closest relatives are conifers. *Proc. Natl. Acad. Sci. USA* 97: 4092–4097.

Buchheim, M. A., M. A. McAuley, E. A. Zimmer, E. C. Theriot and R. L. Chapman. 1994. Multiple origins of colonial green flagellates from unicells: Evidence from molecular and organismal characters. *Mol. Phylogenet. Evol.* 3: 322–343.

Carmichael, J. S. and W. E. Friedman. 1996. Double fertilization in *Gnetum gnemon* (Gnetaceae): Its bearing on the evolution of sexual reproduction within the Gnetales and the anthophyte clade. *Am. J. Bot.* 83: 767–780.

*Chapman, R. L., M. L. Buchheim, C. F. Delwiche, T. Friedl, V. A. R. Huss, K. G. Karol, L. A. Lewis, J. Manhart, R. M. McCourt, J. L. Olsen and D. A. Waters. 1998. Molecular systematics of green algae. In *Systematics of plants II*, D. E. Soltis, P. S. Soltis and J. J. Doyle (eds.), 508–540. Kluwer, Norwell, MA.

Chaw, S. M., C. L. Parkinson, Y. C. Cheng, T. M. Vincent and J. D. Palmer. 2000. Seed plant phylogeny inferred from all three plant genomes: Monophyly of extant gymnosperms and origin of Gnetales from conifers. *Proc. Natl. Acad. Sci. USA* 97: 4086–4091.

Cichan, M. A. and T. N. Taylor. 1990. Evolution of cambium in geological time—A reappraisal. In *The vascular cambium*, M. Iqbal (ed.), 213–228. Wiley, Somerset, UK.

Crane, P. R. 1985. Phylogenetic analysis of seed plants and the origin of angiosperms. *Ann. Missouri Bot. Gard.* 72: 716–793.

Crane, P. R. 1996. The fossil history of the Gnetales. *Int. J. Plant Sci.* 157: S50–S57.

Crane, P. R. and S. Lidgard. 1989. Angiosperm diversification and paleolatitudinal gradients in Cretaceous floristic diversity. *Science* 246: 675–678.

*Crane, P. R., E. M. Friis and K. R. Pedersen. 1995. The origin and early diversification of angiosperms. *Nature* 374: 27–33.

Cronquist, A. 1988. *The evolution and classification of flowering plants.* New York Botanical Garden, New York.

#Dahlgren, R. M. T., H. T. Clifford and P. F. Yeo. 1985. *The families of the monocotyledons: Structure, evolution, and taxonomy.* Springer, Berlin.

@Delwiche, C. F. 1999. Tracing the thread of plastid diversity through the tapestry of life. *Am. Nat.* 154: S164–S177.

Delwiche, C. F., L. E. Graham, and N. Thomson. 1989. Lignin-like compounds and sporopollenin in *Coleochaete*, an algal model for land plant ancestry. *Science* 245: 399–401.

*DiMichele, W. A. and J. E. Skog. 1992. The Lycopsida: A symposium. *Ann. Missouri Bot. Gard.* 79: 447–449.

Donoghue, M. J. 1994. Progress and prospects in reconstructing plant phylogeny. *Ann. Missouri Bot. Gard.* 81: 405–418.

*Donoghue, M. J. 2002. Plants. In *Encyclopedia of evolution*, Vol. 2, M. Pagel (ed.), 911–918. Oxford University Press, Oxford, UK.

Donoghue, M. J. and J. A. Doyle. 1989. Phylogenetic analysis of angiosperms and the relationships of Hamamelidae. In *Evolution, systematics and fossil history of the Hamamelidae*, Vol. 1, P. Crane and S. Blackmore (eds.), 17–45. Clarendon, Oxford, UK.

Donoghue, M. J. and J. A. Doyle. 2000. Demise of the anthophyte hypothesis? *Curr. Biol.* 10: R106–R109.

Donoghue, M. J. and S. M. Scheiner. 1992. The evolution of endosperm: A phylogenetic account. In *Ecology and evolution of plant reproduction: New approaches*, R. Wyatt (ed.), 356–389. Chapman and Hall, New York.

Doyle, J. A. 1994. Origin of the angiosperm flower: A phylogenetic perspective. *Plant Syst. Evol.* 8: 7–29.

Doyle, J. A. 1996. Seed plant phylogeny and the relationships of Gnetales. *Int. J. Plant Sci.* 157: S3–S39.

*Doyle, J. A. 1998. Phylogeny of the vascular plants. *Annu. Rev. Ecol. Syst.* 29: 567–599.

Doyle, J. A. and M. J. Donoghue. 1986. Seed plant phylogeny and the origin of angiosperms: An experimental cladistic approach. *Bot. Rev.* 52: 321–431.

Doyle, J. A. and M. J. Donoghue. 1993. Phylogenies and angiosperm diversification. *Paleobiology* 19: 141–167.

Doyle, J. A. and P. K. Endress. 2000. Morphological phylogenetic analysis of basal angiosperms: Comparison and combination with molecular data. *Int. J. Plant Sci.* 161: S121–S153.

Doyle, J. A. and L. J. Hickey. 1976. Pollen and leaves from the mid-Cretaceous Potomac Group and their bearing on angiosperm evolution. In *Origin and early evolution of angiosperms*, C. Beck (ed.), Columbia University Press, New York.

Endress, P. K. and A. Igersheim. 2000a. Gynoecium structure and evolution in basal angiosperms. *Int. J. Plant Sci.* 161: S211–S223.

Endress, P. K. and A. Igersheim. 2000b. The reproductive structures of the basal angiosperm *Amborella trichopoda* (Amborellaceae). *Int. J. Plant Sci.* 161: S237–S248.

#Esau, K. 1977. *Anatomy of seed plants.* Wiley, New York.

Feild, T. S., T. Brodribb, T. Jaffre and N. M. Holbrook. 2001. Acclimation of leaf anatomy, photosynthetic light use, and xylem hydraulics to light in *Amborella trichopoda* (Amborellaceae). *Int. J. Plant Sci.* 162: 999–1008.

Feild, T. S., M. A. Zweiniecki, T. Brodribb, T. Jaffre, M. J. Donoghue and N. M. Holbrook. 2000. Structure and function of tracheary elements in *Amborella trichopoda*. *Int. J. Plant Sci.* 161: 705–712.

Florin, R. 1951. Evolution of cordaites and conifers. *Acta Hort. Berg.* 15: 285–388.

Florin, R. 1954. The female reproductive organs of conifers and taxads. *Biol. Rev.* 29: 367–389.

Friedman, W. E. 1990. Double fertilization in *Ephedra*, a non-flowering seed plant: Its bearing on the origin of angiosperms. *Science* 247: 951–954.

@Friedman, W. E. 1996. Introduction to biology and evolution of the Gnetales. *Int. J. Plant. Sci.* 157: S1–S2.

Friedman, W. E. and S. K. Floyd. 2001. Perspective: The origin of flowering plants and their reproductive biology—A tale of two phylogenies. *Evolution* 55: 217–231.

Friis, E. M., W. G. Chaloner, and P. R. Crane. 1987. *The origins of angiosperms and their biological consequences.* Cambridge University Press, Cambridge, UK.

Friis, E. M., K. R. Pederson and P. R. Crane. 2001. Fossil evidence of water lilies (Nymphaeales) in the early Cretaceous. *Nature* 410: 356–360.

Frohlich, M. W. and D. S. Parker. 2000. The mostly male theory of flower evolutionary origins: From genes to fossils. *Syst. Bot.* 25: 155–170.

#Gifford, E. M. and A. S. Foster. 1989. *Morphology and evolution of vascular plants*, 3rd ed. Freeman, New York.

Goremykin, V., V. Bobrova, J. Pahnke, A. Troitsky, A. Antonov and W. Martin. 1996. Noncoding sequences from the slowly evolving chloroplast inverted repeat in addition to *rbc*L data do not support gnetalian affinities of angiosperms. *Mol. Biol. Evol.* 13: 383–396.

Graham, L. E. 1993. *Origin of the land plants.* Wiley, New York.

*Graham, L. E. and L. W. Wilcox. 2000. The origin of alternation of generations in land plants: A focus on matrotrophy and hexose transport. *Philos. Trans. R. Soc. London [Biol.]* 355: 757–767.

Graham, L. E., C. F. Delwiche and B. D. Mishler. 1991. Phylogenetic connections between 'green algae' and the 'bryophytes.' *Adv. Bryol.* 4: 213–244.

Graham, S. W. and R. G. Olmstead. 2000. Utility of 17 chloroplast genes for inferring the phylogeny of the basal angiosperms. *Am. J. Bot.* 87: 1712–1730.

Heckman, D. S., D. M. Geiser, B. R. Eidell, R. L. Stauffer, N. L. Kardos and S. B. Hedges. 2001. Molecular evidence for the early colonization of land by fungi and plants. *Science* 293: 1129–1133.

Hoot, S. B., S. Magallon and P. R. Crane 1999. Phylogeny of basal eudicots based on three molecular data sets: *atp*B, *rbc*L, and 18S nuclear ribosomal DNA sequences. *Ann. Missouri Bot. Gard.* 86: 1–32.

Judd, W. S., R. W. Sanders and M. J. Donoghue. 1994. Angiosperm family pairs—Preliminary phylogenetic analyses. *Harvard Pap. Bot.* 5: 1–51.

Karol, K. G., R. M. McCourt, M. T. Cimino and C. F. Delwiche. 2001. The closest living relatives of land plants. *Science* 294: 2351–2353.

Kenrick, P. 2000. The relationships of vascular plants. *Philos. Trans. R. Soc. London [Biol.]* 355: 847–855.

*Kenrick, P. and P. R. Crane. 1997a. *The origin and early diversification of land plants: A cladistic study.* Smithsonian Institution Press, Washington, DC.

*Kenrick, P. and P. R. Crane. 1997b. The origin and early evolution of plants on land. *Nature* 389: 33–39.

Larsen, A., M. M. Kirk and D. C. Kirk. 1991. Molecular phylogeny of the volvocine flagellates. *Mol. Biol. Evol.* 9: 85–102.

Ligrone, R., J. G. Duckett and J. S. Renzaglia. 2000. Conducting tissues and phyletic relationships of bryophytes. *Philos. Trans. R. Soc. London [Biol.]* 355: 795–813.

Lupia, R., H. Schneider, G. M. Moeser, K. M. Pryer and P. R. Crane. 2000. Marsileaceae sporocarps and spores from the late Cretaceous of Georgia, U.S.A. *Int. J. Plant Sci.* 161: 975–988.

Lutzoni, F., M. Pagel and V. Reeb. 2001. Major fungal lineages are derived from lichen symbiotic ancestors. *Nature* 411: 937–940.

Magallon, S., P. R. Crane and P. S. Herendeen. 1999. Phylogenetic pattern, diversity, and diversification of eudicots. *Ann. Missouri Bot. Gard.* 86: 297–372.

Magallon, S. and M. J. Sanderson. 2001. Absolute diversification rates in angiosperm clades. *Evolution* 55: 1762–1780.

Manhart, J. R. and J. D. Palmer. 1990. The gain of two chloroplast tRNA introns marks the green algal ancestor of land plants. *Nature* 345: 268–270.

Martin, W., A. Gierl and H. Saedler. 1989. Molecular evidence for pre-Cretaceous angiosperm origins. *Nature* 339: 46–48.

Mathews, S. and M. J. Donoghue. 1999. The root of angiosperm phylogeny inferred from duplicate phytochrome genes. *Science* 286: 947–950.

Mattox, K. R. and K. D. Stewart. 1984. The classification of green algae: A concept based on comparative cytology. In *The systematics of the green algae*, D. Irvine and D. Johns (eds.), 29–72. Academic Press, London.

*McCourt, R. M. 1995. Green algal phylogeny. *Trends Ecol. Evol.* 10: 159–163.

Mishler, B. D. and S. P. Churchill. 1984. A cladistic approach to the phylogeny of the "bryophytes." *Brittonia* 36: 406–424.

Mishler, B. D. and S. P. Churchill. 1985. Transition to a land flora: Phylogenetic relationships of the green algae and bryophytes. *Cladistics* 1: 305–328.

Mishler, B. D., L. A. Lewis, M. A. Buchheim, K. S. Renzaglia, D. L. Garbary, et al. 1994. Phylogenetic relationships of the "green algae" and "bryophytes." *Ann. Missouri Bot. Gard.* 81: 451–483.

Nickrent, D., C. L. Parkinson, J. D. Palmer and R. J. Duff. 2000. Multigene phylogeny of land plants with special reference to bryophytes and the earliest land plants. *Mol. Biol. Evol.* 17: 1885–1895.

Niklas, K. J. 1997. *The evolutionary biology of plants.* University of Chicago Press, Chicago.

Niklas, K. J., B. H. Tiffney, and A. H. Knoll. 1983. Patterns in vascular plant diversification. *Nature* 303: 614–616.

Nixon, K. C., W. L. Crepet, D. Stevenson and E. M. Friis. 1994. A reevaluation of seed plant phylogeny. *Ann. Missouri Bot. Gard.* 81: 484–533.

O'Hara, R. J. 1992. Telling the tree: Narrative representation and the study of evolutionary history. *Biol. Philos.* 7: 135–160.

Palmer, J. D. 2000. A single birth of all plastids? *Nature* 405: 32–39.

Palmer, J. D. and C. F. Delwiche. 1998. The origin and evolution of plastids and their genomes. In *Systematics of plants II*, D. E. Soltis, P. S. Soltis and J. J. Doyle (eds.), 375–409. Kluwer, Norwell, MA.

Parkinson, C. L., K. L. Adams and J. D. Palmer. 1999. Multigene analyses identify the three earliest lineages of extant flowering plants. *Curr. Biol.* 9: 1485–1488.

Pickett-Heaps, J. D. 1979. Electron microscopy and the phylogeny of green algae and land plants. *Am. Zool.* 19: 545–554.

Price, R. A. 1996. Systematics of Gnetales: A review of morphological and molecular evidence. *Int. J. Plant Sci.* 157: S40–S49.

Pryer, K. M., H. Schneider, A. R. Smith, R. Cranfill, P. G. Wolf, J. S. Hunt and S. D. Sipes. 2001. Horsetails and ferns are a monophyletic group and the closest living relatives to seed plants. *Nature* 409: 618–622.

Pryer, K. M., A. R. Smith and J. E. Skog. 1995. Phylogenetic relationships of extant ferns based on evidence from morphology and *rbc*L sequences. *Am. Fern J.* 85: 205–282.

Qiu, Y.-L., Y. Cho, J. C. Cox and J. D. Palmer. 1998. The gain of three mitochondrial introns identifies liverworts as the earliest land plants. *Nature* 394: 671–674.

Qiu, Y.-L., J. H. Lee, F. Bernasconi-Quadroni, D. E. Soltis, P. S. Soltis, M. Zanis, E. A. Zimmer, Z. D. Chen, V. Savolainen and M. W. Chase. 1999. The earliest angiosperms: Evidence from mitochondrial, plastid and nuclear genomes. *Nature* 402: 404–407.

Raubeson, L. A. and R. K. Jansen. 1992. Chloroplast DNA evidence on the ancient evolutionary split in vascular plants. *Science* 255: 1697–1699.

#Raven, P. H., R. F. Evert, and S. E. Eichhorn. 1992. *Biology of plants*, 5th ed. Worth, New York.

Remy, W. 1982. Lower Devonian gametophytes: Relation to the phylogeny of land plants. *Science* 215: 1625–1627.

Remy, W., P. G. Gensel, and H. Hass. 1993. The gametophyte generation of some early Devonian land plants. *Int. J. Plant Sci.* 154: 35–58.

Renzaglia, K. S., R. J. Duff, D. L. Nickrent and D. J. Garbary. 2000. Vegetative and reproductive innovations of early land plants: Implications for a unified phylogeny. *Philos. Trans. R. Soc. London [Biol.]* 355: 769–793.

Rothwell, G. W. 1999. Fossils and ferns in the resolution of land plant phylogeny. *Bot. Rev.* 65: 188–218.

Rothwell, G. W. and R. Serbet. 1994. Lignophyte phylogeny and the evolution of spermatophytes: A numerical cladistic analysis. *Syst. Bot.* 19: 443–482.

Rydin, C., M. Kallersjo and E. M. Friis. 2002. Seed plant relationships and the systematic position of Gnetales based on nuclear and chloroplast DNA: Conflicting data, rooting problems, and the monophyly of conifers. *Int. J. Plant Sci.* 163: 197–214.

Sanderson, M. J. and M. J. Donoghue. 1994. Shifts in diversification rate with the origin of angiosperms. *Science* 264: 1590–1593.

Sanderson, M. J. and J. A. Doyle. 2001. Sources of error and confidence intervals in estimating the age of angiosperms from *rcb*L and 18S rDNA data. *Am. J. Bot.* 88: 1499–1516.

Sanderson, M. J., M. F. Wojciechowski, J.-M. Hu, T. Sher Khan and S. G. Brady. 2000. Error, bias, long-branch attraction in data for two chloroplast photosystem genes in seed plants. *Mol. Biol. Evol.* 17: 782–797.

Savolainen, V., M. W. Chase, S. B. Hoot, C. M. Morton, D. E. Soltis, C. Bayer, M. F. Fay, A. Y. De Bruijn, S. Sullivan and Y. L. Qiu. 2000. Phylogenetics of flowering plants based on combined analysis of plastid *atp*B and *rbc*L gene sequences. *Syst. Biol.* 49: 306–362.

#Scagel, R. F., R. J. Bandoni, G. E. Rouse, W. B. Schofield, J. R. Stein and T. M. C. Taylor. 1969. *Plant diversity: An evolutionary approach*. Wadsworth Publishing Co., Belmont, CA.

Serbet, R. and G. W. Rothwell. 1992. Characterizing the most primitive seed ferns. I. A reconstruction of *Elkinsia polymorpha*. *Int. J. Plant Sci.* 153: 602–621.

#Shaw, A. J. and B. Goffinet (eds.). 2000. *The biology of bryophytes*. Cambridge University Press, New York.

Soltis, P. S., D. E. Soltis and M. W. Chase. 1999. Angiosperm phylogeny inferred from multiple genes as a tool for comparative biology. *Nature* 402: 402–404.

Soltis, P. S., D. E. Soltis, V. Savolainen, P. R. Crane and T. G. Barraclough. 2002. Rate heterogeneity among lineages of tracheophytes: Integration of molecular and fossil data and evidence for molecular living fossils. *Proc. Natl. Acad. Sci. USA* 99: 4430–4435.

Stebbins, G. L. 1974. *Flowering plants: Evolution above the species level.* Harvard University Press, Cambridge, MA.

Stefanovic, S., M. Jager, J. Deutsch, J. Broutin and M. Masselot. 1998. Phylogenetic relationships of conifers inferred from partial 28S rRNA gene sequences. *Am. J. Bot.* 85: 688–697.

Stewart, W. N. 1983. *Paleobotany and the evolution of plants.* Cambridge University Press, Cambridge.

#Stewart, W. N. and G. W. Rothwell. 1993. *Paleobotany and the evolution of plants*, 2nd ed. Cambridge University Press, New York.

Sun, G., D. L. Dichler, S. Zheng and Z. Zhou. 1998. In search of the first flower: A Jurassic angiosperm, *Archaefrutus*, from Northeast China. *Science* 282: 1692–1695.

Sun, G., Q. Ji, D. L. Dilcher, S. Zheng, K. C. Nixon and X. Wang. 2002. Archaefructaceae, a new basal angiosperm family. *Science* 296: 899–904.

Takhtajan, A. L. 1997. *Diversity and classification of flowering plants.* Columbia University Press, New York.

Taylor, D. W. and L. J. Hickey. 1992. Phylogenetic evidence for the herbaceous origin of angiosperms. *Plant Syst. Evol.* 180: 137–156.

#Taylor, T. N. and E. L. Taylor. 1993. *The biology and evolution of fossil plants.* Prentice Hall, Englewood Cliffs, NJ.

Thien, L. B., H. Azuma and S. Kawano. 2000. New perspectives on the pollination biology of basal angiosperms. *Int. J. Plant Sci.* 161: S225–S235.

Thorne, R. F. 1992. An updated phylogenetic classification of the flowering plants. *Aliso* 13: 365–389.

Williams, J. H. and W. E. Friedman. 2002. Identification of diploid endosperm in an early angiosperm lineage. *Nature* 415: 522–526.

*Wolf, P. G., K. M. Pryer, A. R. Smith and M. Hasebe. 1998. Phylogenetic studies of extant pteridophytes. In *Systematics of plants II*, D. E. Soltis, P. S. Soltis and J. J. Doyle (eds.), 541–556. Kluwer, Norwell, MA.

Zanis, M. J., D. E. Soltis, P. S. Soltis, S. Mathews and M. J. Donoghue. 2002. The root of the angiosperms revisited. *Proc. Natl. Acad. Sci. USA.* 99: 6848–6853.

@Zimmer, E. A., Y. L. Qiu, P. K. Endress and E. M. Friis. 2000. Current perspectives on basal angiosperms: Introduction. *Int. J. Plant Sci.* 161: S1–S2.

Zimmermann, W. 1952. Main results of the "telome theory." *Paleobotanist* 1: 456–470.

Zimmermann, W. 1965. *Die Telomtheorie.* Fischer, Stuttgart, Germany.

8 Lycophytes, Ferns and Their Allies, and Extant Gymnosperms

This chapter and Chapter 9 survey the diversity of tracheophytes. *Tracheo* refers to the presence of tracheids—cells specialized for transport of liquids—and the Greek root *phyte* means plant. Tracheophytes form a well-supported monophyletic group of generally large plants with branched sporophyte axes (Figure 8.1) and well-developed tissues (with tracheids in the xylem and sieve cells in the phloem) for the transport of water and carbohydrates within the plant. As described in Chapter 7, they form a major clade within the embryophytes (or land plants), nested within the paraphyletic "bryophytes" (see Figure 7.6). This implies the derivation of tracheophyte characteristics from those found in the bryophyte lineages, in which the small, unbranched sporophyte is nutritionally dependent on the dominant gametophyte phase of the life cycle.

There are two major lineages within the tracheophytes: the lycophytes and the euphyllophytes (see Figure 7.8). The euphyllophytes in turn comprise two major lineages, the monilophytes (ferns and their allies, including Psilotaceae and Equisitaceae) and the spermatophytes (seed plants). Finally, within the spermatophytes there are five major extant lineages: Cycadales, Ginkgoales, Coniferales, Gnetales, and the angiosperms (flowering plants). The first four of these groups have "naked" seeds, and are thus

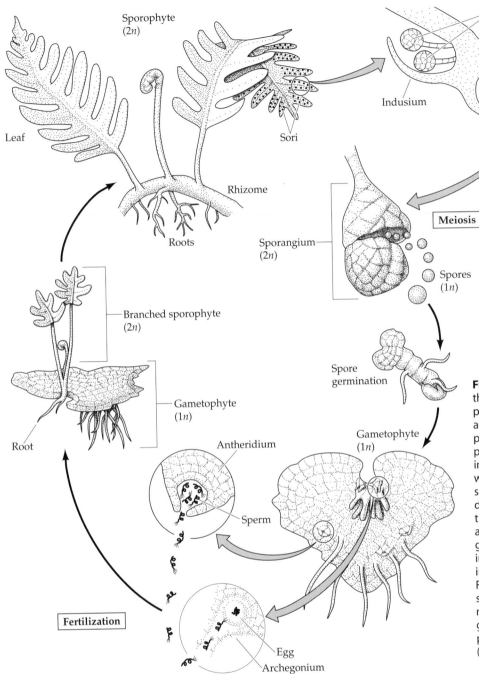

Figure 8.1 Life cycle of a fern of the genus *Polypodium*. The sporophyte generation is photosynthetic and is independent of the gametophyte. The cluster of sporangia plus a protective layer of cells called the indusium make up a sorus. Meiosis within the sporangia yields haploid spores that divide mitotically to produce a heart-shaped gametophyte that differentiates both archegonia and antheridia on one individual. The gametophyte is photosynthetic and independent, although it is reduced in size relative to the sporophyte. Fertilization takes place when sperm swim through water to the archegonia and fertilize the eggs. Unlike the gametophyte, the developing sporophyte has vascular tissue and roots. (From Singer 1997.)

commonly referred to as gymnosperms. In the angiosperms, seeds are enclosed inside one or more carpels.

Phylogenetic relationships among the seed plant lineages are not yet well resolved, although studies, including fossils, do indicate that seed plant lineages with naked seeds are paraphyletic with respect to angiosperms (see Figure 7.12). One possibility is that Gnetales are more closely related to angiosperms than they are to the other extant gymnosperms (the "anthophyte hypothesis"). However, a number of recent studies support the view that the extant gymnosperm lineages form a clade that is sister to the angiosperms.

This chapter covers the major extant lineages of tracheophytes except for the angiosperms, which are cov-

ered in detail in Chapter 9. All together there are about 12,000 species of lycophytes, monilophytes, and extant gymnosperms, or about 5% of the total number of flowering plant species. In this chapter and the next, each family treatment includes a description in which useful identifying characters are indicated in *italic print* and synapomorphies (which may also be useful for identification) in **boldface**; a brief summary of distribution and ecology; the estimated number of genera and species (including a listing of major genera); a list of major economic plants and products; and a discussion. The family discussion includes information regarding characters supporting the group's monophyly, a brief overview of phylogenetic relationships within the family, information

Key to Major Groups of Tracheophytes

1. Sperm biflagellate; .... **Lycophytes**

1. Sperm multiflagellate (or sperm without flagella)...................................... 2 (**Euphyllophytes**)

2. Dispersal by spores; seeds absent.. 3 (**Ferns and their allies**)

2. Dispersal by seeds... 6 (**Seed plants**)

3. Leaves generally less than 2 cm in length, with an unbranched vein or veinless,
   not divided into leaflets 4

3. Leaves generally more than 2 cm in length, with a branched vein, divided into leaflets or not 5

4. Leaves whorled, fused at base; branches, if present, whorled; internodes with
   conspicuous vertical ridges; sporangia clustered on peltate sporangiophores and
   aggregated in strobili; roots present and irregularly branching.............................. **Equisetaceae**

4. Leaves spiral, not fused at base; stem without vertical ridges; sporangia sometimes
   associated with forked or lobed leaves; roots lacking .... **Psilotaceae**

5. Leaves not coiled and unfolding lengthwise when enmerging; sporangia walls more
   than one cell in thickness, without an annulus, borne on or embedded in a special
   portion of the leaf (the *sporophore*).. **Ophioglossaceae**

5. Leaves coiled and unfolding lengthwise when emerging; sporangia walls
   one-cell thick, mostly with an annulus, mostly clustered in sori on the abaxial
   leaf surface, less often scattered on the abaxial leaf surface or in sporocarps ..... **Leptosporangiate ferns**

6. Carpels and endosperm present... **Angiosperms**

6. Carpels and endosperm absent .... 7 (**Gymnosperms**)

7. Vessels present; leaves opposite, joined at base.. **Gnetales**

7. Vessels absent; leaves usually spiral, rarely opposite or joined at base 8

8. Sperm not flagellate, transported to ovule by pollen tube; leaves simple, small,
   and scale-like to larger and linear, with resin canals in most species............................. **Conifers**

8. Sperm motile, swimming by means of flagella; leaves either pinnately compound or simple,
   large, and as broad or broader than long, with resin, if present, not in canals 9

9. Leaves pinnately or twice pinnately compound, persistent; stem mostly short and unbranched
   or dichotomously branching, sometimes subterranean; ovules borne on the margin of often
   peltate megasporophylls that are either loosely clustered at the shoot apex or grouped in
   strobili; outer layer of seeds mostly fleshy but not unpleasant-smelling; microsporangia
   clustered in sori on the abaxial surface of microsporophylls; mucilage canals present.............. **Cycads**

9. Leaves simple, fan-shaped, deciduous; large, freely branching trees; ovules 2 at the end
   of a long stalk, often only one maturing; outer layer of seeds fleshy, unpleasant-smelling
   at maturity; microsporangia paired at the end of slender stalk; mucilage canals absent .... **Ginkgoaceae**

regarding pollination biology and seed dispersal (where relevant), and notes on other matters of biological interest. Finally, each family treatment includes a list of references that are useful sources of additional information.

References: Bowe and de Pamphilis 1997; Bowe et al. 2001; Chaw et al. 2000; Donoghue 1994; Doyle et al. 1994; Kenrick 2000; Kenrick and Crane 1997; Mishler et al. 1994; Nickrent et al. 2000; Nixon et al. 1994; Pryer et al. 2001; Renzaglia et al. 2000; Renzaglia et al. 2001; Rothwell 1999; Rydin et al. 2002; Stefanovic et al. 1998; Stewart and Rothwell 1993.

LYCOPHYTES

Arising at least 400 MYA, lycophytes were largest in size in the Carboniferous (345–290 MYA), when arborescent members of this group, some of them over 40 m tall and 2 m in diameter at the base, dominated forests. Almost half of the plant fossils from this time are from this group. There are three families, about five genera, and about 1250 species of extant lycophytes, and they are placed in the Lycopodiales. The remains of these ancient trees are the chief components of commercially important coal

TABLE 8.1 *Families of tracheophytes as classified in this book.*[a]

LYCOPHYTES
Lycopodiales
    Lycopodiaceae (p. 187)
    Selaginellaceae (p. 189)
    Isoetaceae (p. 187)

FERNS AND THEIR ALLIES (MONILIFORMS)
Psilotales
    Psilotaceae (p. 191)
Ophioglossales
    Ophioglossaceae (p. 192)
    Marattiaceae (p. 190)
Equisetales
    Equisetaceae (p. 193)
Polypodiales (leptosporangiate ferns)
    Dipteridaceae (p. 195)

Osmundaceae (p. 196)
Cyatheaceae (p. 196)
"Dicksoniaceae" (p. 197)
Gleicheniaceae (p. 195)
Hymenophyllaceae (p. 195)
Hymenophyllopsidaceae (p. 197)
Metaxyaceae (p. 197)
Marsileaceae (p. 197)
Matoniaceae (p. 195)
Plagiogyriaceae (p. 195)
Salviniaceae (p. 197)
Polypodiaceae (including numerous segregates) (p. 198)
Schizaeaceae (p. 194)

SEED PLANTS

Gymnosperms
Cycadales (Cycads)

Cycadaceae (p. 201)
Zamiaceae (p. 201)
Ginkgoales
    Ginkgoaceae (Ginkgoales; p. 203)
Coniferales (Conifers)
    Pinaceae (p. 205)
    Cupressaceae (including "Taxodiaceae") (p. 207)
    Sciadopityaceae (p. 209)
    Podocarpaceae (p. 209)
    Araucariaceae (p. 210)
    Taxaceae (p. 211)
    Cephalotaxaceae (p. 212)
Gnetales
    Ephedraceae (p. 213)
    Gnetaceae (p. 213)
    Welwitschiaceae (p. 213)

Angiosperms: See Table 9.1

[a]Families receiving full coverage in the text are indicated in **boldface**, while those only briefly characterized are in *italics*. Page numbers (in parentheses) indicate discussion of the family in this chapter.

deposits of Europe and North America. The closest living relatives of the clade to which these ancient trees belonged are thought to be the small grass-like plants of the Isoetaceae, one of the three extant families of lycophytes. Isoetaceae and Selaginellaceae are sister groups.

Lycopodiaceae Mirbel
(Club Moss Family)

Terrestrial or epiphytic plants usually about 5–20 cm tall, with *dichotomously branching* stems. Roots dichotomously branching. Leaves simple (microphylls), 0.2–2 cm long, with 1 unbranched vein, often densely covering the stem; linear and more or less spreading away from the stem or scale-like and appressed to the stem; spiral or opposite. **Sporangia ± kidney-shaped**, *opening by a transverse slit that divides the sporangium in two; solitary in leaf axils or borne on leaf bases*, sporophylls unmodified or modified and sometimes clustered into strobili. Homosporous; spores subglobose to tetrahedral, with a 3-branched scar. Gemmae (small structures of various sizes and shapes that detach from the plant for vegetative reproduction) present in some species of *Huperzia*. Gametophytes mycorrhizal and wholly subterranean, partly green and on or near the surface of the ground, or initially mycorrhizal and later photosynthetic (Figure 8.2).

Distribution and ecology: Cosmopolitan, rare in arid areas; most diverse in tropical montane and alpine habitats.

Genera/species: ca. 3/380. ***Major genera:*** *Huperzia* (300 spp.), *Lycopodiella* (40), and *Lycopodium* (40).

Economic plants and products: The family is not significant economically. Oily, highly flammable compounds in the spore wall ignite rapidly into a flash of light and were used by magicians and sorcerers in the Middle Ages, as a flash early in photography, and in the first (experimental) photocopying machines. The spores have been used as industrial lubricants and formerly were used to prevent rubber cohesion in condoms and surgical gloves.

Key to Families of Lycophytes

1. Homosporous; leaves nonligulate ....**Lycopodiaceae**

1. Heterosporous; leaves ligulate, with a small projection on the adaxial surface 2

2. Sporangia borne singly in the axils of unmodified or modified leaves; leaves usually under 2 cm long..**Selaginellaceae**

2. Sporangia at least initially embedded in adaxial face of leaf base; leaves 2–100 cm long........Isoetaceae

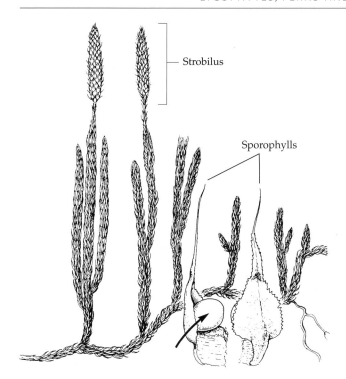

Figure 8.2 Lycopodiaceae. *Lycopodium clavatum*: habit (shown about two-thirds actual size) and sporophylls (× 7); the arrow points to a sporangium. (From Øllgaard 1990.)

Discussion: Taxonomic problems exist at generic and specific levels in the Lycopodiaceae. Traditionally treated as including just one large genus, *Lycopodium*, the family is here considered to contain four genera. Some of these genera contain monophyletic sections that some workers have recognized as genera. DNA sequences from *rbcL* divide the family into a *Huperzia* clade and a *Lycopodiella-Lycopodium* clade. A peculiar species, usually placed in a distinct genus (*Phylloglossum*), consists of plants only about 5 cm tall and highly reduced for arid environments of Australia, New Zealand, and Tasmania. Gametophyte morphology and *rbcL* sequences support inclusion of this species in *Huperzia*. Most members of *Huperzia* are tropical epiphytes with sporangia borne in the axil of the leaves and not collected into strobili. *Huperzia* is divided into to a neotropical and a paleotropical clade, which are thought to evolved after continental drift separated South America and Africa about 80 MYA. The terrestrial habit has apparently evolved at least two times in neotropical mountains. Species delimitation in *Huperzia* is poorly understood, at least in part due to frequent interspecific hybridization.

The *Lycopodiella-Lycopodium* clade is primarily temperate in distribution, and its sporangia are usually borne in conspicuous strobili. Some clades within these two genera have wider geographic distributions than the major neotropical and paleotropical clades in *Huperzia*, and some species, such as *Lycopodium clavatum*, are more or less cosmopolitan.

Lycopodiaceae have an extensive geological history extending back to about 380 MYA, but most of the mod-

ern diversity developed after 80–90 MYA, when the formation of forests of broadleaved species shaded the forest floor and favored the evolution of epiphytism in the ancestors of *Huperzia*. One of the reversions to the terrestrial habit is thought to have been in response to the formation of the Andes about 15 MYA.

Members of this family may have somatic chromosome numbers up to about 275. These high numbers may be the result of repeated episodes of polyploidy, or these plants may simply be diploids with numerous chromosomes.

References: DiMichele and Skog 1992; Kenrick and Crane 1997; Lellinger 1985; Øllgaard 1990, 1992; Soltis and Soltis 1988; Tryon and Tryon 1982; Wagner and Beitel 1992, 1993; Wikstrom 1999; Wikstrom and Kenrick 1997; Wikstrom and Kenrick 2000; Whittier and Braggins 2000.

Selaginellaceae Willk.
(Spike Moss Family)

Mostly terrestrial, herbaceous, and perennial plants under 2 cm tall. Roots dichotomously branching; rhizophores usually produced from the stem, dichotomously branching. Stems erect or creeping. *Leaves about 0.5–1 cm long, spirally arranged and often 4-ranked on the secondary and ultimate branches*; ligulate; often dimorphic, those on upper side of stem smaller than those on lower side; with a single, unbranched vein; with a ligule. Heterosporous; **sporangia borne in or near the axils of well-differentiated sporophylls, usually on 4-sided (cylindrical in a few species) strobili**; strobili usually terminating branches, with either or both megasporangia and microsporangia; megasporangia usually with 4 megaspores 200–600 μm in diameter, with distinct ridges and conspicuously patterned with variously shaped projections; microsporangia with more than 100 microspores 20–60 μm in diameter. Megagametophytes partly protruding from the megaspore wall; microgametophytes developing wholly within the microspore, the wall rupturing to release the sperm. (Figure 8.3).

Distribution and ecology: Mainly tropical, with a few species extending into arctic regions of both hemispheres; occupying a wide range of habitats.

Genera/species: 1 (*Selaginella*)/750.

Economic plants and products: The family is not significant economically, although a few species are ornamentals.

Discussion: Fossils of this family are known from tropical wetlands almost 350 MYA. Early in their history, members of Selaginellaceae evolved leaf dimorphism, with smaller leaves on the upper surface of the stem. At about the same time, it appears that closed-canopy forests were established by tree ferns. Leaf dimorphism, which has

Figure 8.3   Selaginellaceae. (A, B) *Selaginella apoda*: (A) habit (× 1.5); (B) sterile portion of plant (× 12). (C) *S. myosurus*: strobilus and vegetative branch bearing it (× 6); the arrow points to a sporangium. (A, B from Billington 1952; C from Alston 1959.)

been hypothesized to be an adaptation to poor light, may have been important for these early plants of Selaginellaceae and for modern forms as well, which mostly live on forest floors.

The majority of the species of the family occurs in tropical rainforests, but many species are capable of surviving long periods of drought with small leaves covered by a thick cuticle and by curling their branches into a ball. These plants can revive rapidly from apparent death by uncurling their branches, and they are therefore commonly called resurrection plants. The resurrection adaptation has evolved at least three times in the genus.

Although the small leaves and habit make many species of Lycopodiaceae and Selaginellaceae resemble one another, heterospory and ligulate leaves are potential synapomorphies linking Selaginellaceae and Isoetaceae. Isoetaceae are terrestrial or aquatic plants with erect or short cormlike stems and usually long leaves. Sporangia embedded in the adaxial face of the leaf bases are distinctive. The family has a nearly cosmopolitan distribution; the one genus (*Isoetes*) has about 150 species.

References: Jermy 1990a,b; Kenrick and Crane 1997; Korall and Kenrick 2002 (their Figure 3 shows important morphological features); Lellinger 1985; Manhart 1995; Taylor et al. 1993; Tryon and Tryon 1982; Valdespino 1993; Webster 1992.

FERNS AND THEIR ALLIES

The ferns and their allies span a broad range of morphological diversity. The great majority of the members of this clade are plants commonly referred to as ferns. Many ferns have a subterranean stem bearing a clump of leaves that are compound or deeply lobed, many-veined, and comparatively large (up to 7 m or more in length). Fern leaves are sometimes referred to as **fronds** and their petioles as **stipes**. Ferns also include plants with leaves that are simple or resemble those of clovers, tree ferns with stems up to 20 m tall, and aquatics with filiform leaves. Two groups, long referred to as "fern allies"—Psilotaceae and Equisetaceae—have, with few exceptions, rather small leaves with a single or no vein and superficially do not resemble ferns. For over 30 years, however, some workers have suggested that Psilotaceae and Equisetaceae are highly modified ferns, and recent DNA evidence clearly supports this view (see Chapter 7). Ferns and their allies, which are referred to in some works as moniliforms, include about 260 genera and 12,000 species ranging in size from a few cm to 20 m tall.

There are two kinds of sporangia within ferns and their allies: **eusporangiate**, in which the sporangium wall has two or more cell layers, and **leptosporangiate**, in which the sporangium wall is composed of just one cell layer. The eusporangiate condition characterizes Psilotaceae, Ophioglossaceae, Equisetaceae, and Marattiaceae (and lycophytes). The Marattiaceae contain 4–7 genera and about 300 species, most found on the shaded floor of wet tropical forests. This family is not treated in detail in this book.

The great majority of ferns and their allies belong to the leptosporangiate ferns. In addition to sporangium wall thickness, the evolution of leptosporangiate ferns resulted in four other unique structural characteristics: the **annulus** (described later); a sporangial stalk with 4–6 cells in cross-section; a vertical first zygotic division; and primary xylem with scalariform bordered pits. Rearrangements of the chloroplast genome and DNA sequence data also mark the origin of leptosporangiate ferns.

While there is strong morphological and molecular support for the monophyly of leptosporangiate ferns, morphology has not been very helpful in assigning relationships of other fern and allied groups. DNA sequences from three chloroplast genes (*rbcL*, *atpB*, and *rps4*) and the nuclear small subunit rDNA strongly link Ophioglossaceae and Psilotaceae in a clade (Figure 8.4). Relationships of Ophioglossaceae had long been a puzzle. Based on morphology, they appeared to be distantly

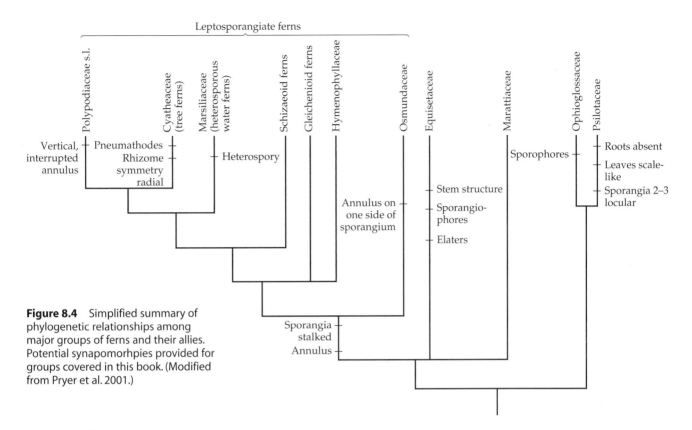

Figure 8.4 Simplified summary of phylogenetic relationships among major groups of ferns and their allies. Potential synapomorhpies provided for groups covered in this book. (Modified from Pryer et al. 2001.)

related to other tracheophytes, or perhaps they had affinities with a long-extinct group of plants called progymnosperms. Relationships of Marattiaceae are also confused. The same DNA evidence that resolved a close relationship of Ophioglossaceae and Psilotaceae place Equisetaceae and Marattiaceae in a clade that is sister to the leptosporangiate ferns (see also Chapter 7). Because support for the Equisetaceae-Marattiaceae is weak, we have not included it in Figure 8.4.

Spore shape, size, symmetry, the shape of a prominent scar—whether straight or three-branched—and the nature of the surface and layers of the spore wall have been taxonomically useful at all levels within the ferns and their allies. Spores of these plants are very small and are produced in great numbers; one estimate is that an individual tree fern in the genus *Cyathea* generates 1,250,000,000,000 spores in its lifetime. Spore dispersal is facilitated by adaptations such as elaters in *Equisetum* and the annulus in most ferns (see below). Because prothalli produced by individual spores of many homosporous ferns can produce zygotes by themselves, they can start a new population. Consequently, long-distance dispersal is probably more common in ferns than in flowering plants.

The gametophytes of ferns and their allies are generally small (less than 1 cm) and variously shaped. Many leptosporangiate ferns have heart-shaped gametophytes, although they may also be elongate, filamentous, or take various other forms. Gametophytes of most ferns are photosynthetic and live above ground. Gametophytes of Psilotaceae are subterranean, lack chlorophyll, and are mycotrophic, depending upon fungal partners for nutri-

tion. Most are homosporous, and their gametophytes are at least potentially bisexual, producing both gametes—eggs (in archegonia) and sperm, also called antherozoids (in antheridia; see Figure 8.1). However, gametophytes producing archegonia may release compounds called antheridiogens that stimulate nearby gametophytes to produce antheridia, thus promoting mating between gametophytes. Marsileaceae (see below) are heterosporous, and their gametophytes are endosporic, being wholly or mostly retained within the spore wall.

Fertilization in ferns and their allies usually requires water through which the flagellate sperm can swim from the antheridium to the archegonium. This requirement for water excludes these plants from some dry habitats. Many ferns and their allies can reproduce asexually (see Figure 8.1) by means of rhizomes, vegetative propagules on the gametophyte called **gemmae** in *Psilotum* and several leptosporangiate ferns, vegetative propagules on the sporophyte, direct outgrowths from gametophytic tissue (called apogamy), and spores that are produced via modified meiotic divisions.

Psilotales

Psilotaceae Kanitz
(Whisk Fern Family)

Herbs, terrestrial but more *often epiphytic*. **Roots absent**, the plant anchored by subterranean stems that may bear gemmae. Aerial stems erect or hanging, glabrous, simple or dichotomously branched. Leaves spirally arranged, 2-

ranked in some species; **scale-like or awl-shaped to lanceolate; simple or once-forked**, one-veined or veinless. **Sporangia 2- or 3-locular and -lobed**, sessile on or above the base of forked sporophylls. Homosporous; **spores bean-shaped**, pale in color. Gametophytes subterranean (Figure 8.5).

Distribution and ecology: Pantropical and warm temperate except in dry areas, with the greatest number of species in Southeast Asia and the South Pacific; primarily at lower elevations, especially as epiphytes on tree fern trunks.

Genera/species: 2/17. ***Major genera:*** *Tmesipteris* (15 spp.) and *Psilotum* (2).

Economic plants and products: The family is not significant economically.

Discussion: Although there is no extensive fossil record of Psilotaceae, they have long been thought to be among the most primitive extant vascular plants due to similarities to ancient and simple fossil plants. Morphological, chemical, DNA sequence data, and detailed features of the architecture of sperm cells, however, support a relationship to eusporangiate "ferns." The distinctive simplicity of these plants—part of the reason they do not resemble "ferns"—is probably reduction associated with mycotrophy. Growth of spores depends upon the presence of endophytic mycorrhizae.

References: Kramer 1990d; Kenrick and Crane 1997; Lellinger 1985; Manhart 1995; Pryer et al. 1995, 2001; Renzaglia et al. 2001; Thieret 1993b; Tryon and Tryon 1982; Wolf 1997; Wolf et al. 1998.

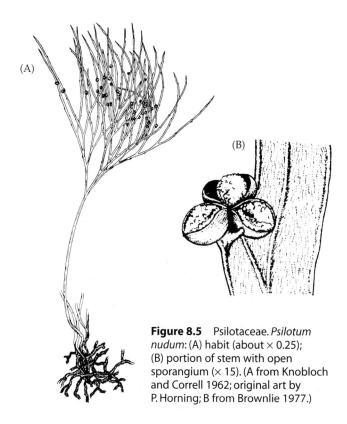

Figure 8.5 Psilotaceae. *Psilotum nudum:* (A) habit (about × 0.25); (B) portion of stem with open sporangium (× 15). (A from Knobloch and Correll 1962; original art by P. Horning; B from Brownlie 1977.)

may be associated with disturbance in pastures, old fields, and young second-growth forests, although there is considerable diversity in relatively undisturbed habitats.

Genera/species: 3/70–90. ***Major genera:*** *Botrychium* (45–60 spp.) and *Ophioglossum* (25–30).

Ophioglossales

Ophioglossaceae C. Agardh
(Adder's-tongue Family)

Terrestrial or epiphytic. Stems unbranched, short, mostly subterranean. Leaves usually one per stem, with or without a stalk that is expanded at the base into a sheath, usually divided into a photosynthetic, sterile blade and a spore-bearing portion (the ***sporophore***); the blade simple to more or less palmately compound to many times pinnately compound, from a few cm to about 50 cm long, rarely absent, conduplicate before unfolding (not or rarely somewhat circinate). Sporangia thick-walled (eusporangiate), not clustered in sori, separate or joined in synangia, **exposed on branches of the sporophore or embedded in a spike-like sporophore**; annulus absent. Homosporous; spores not green, thousands per sporangium. Gametophytes subterranean, mycorrhizal, not green (Figure 8.6).

Distribution and ecology: Worldwide, especially in temperate regions and the tropics. Members of this family

Figure 8.6 Ophioglossaceae. *Botrychium virginianum:* (A) base of plant; (B) portion of a sterile (photosynthetic) leaf; (C) fertile leaf (all × about 0.3). (From Wagner and Wagner 1993.)

Economic plants and products: None.

Discussion: The sporophore, which is assumed to be two basal leaf segments that have fused together, is a likely synapomorphy and the most unusual feature of the family. The family is morphologically distinct relative to other members of the ferns and their allies, and the DNA sequence support for its relationship with Psilotaceae has no obvious support from morphology. Chromosome numbers in *Ophioglossum* of more than $2n = 1400$ have been reported, the highest of all tracheophytes.

References: Hasebe et al. 1995; Pryer et al. 1995; Pryer et al. 2001; Tryon and Tryon 1982; Wagner 1990; Wagner and Wagner 1993.

Equisetales

Equisetaceae Michx. ex DC.
(Horsetail Family)

Terrestrial to aquatic, rhizomatous perennials. Stems to 8 m, but under 1 m in height in most species; swollen (joint-like) nodes; **internodes with alternating vertical ridges and grooves externally and usually hollow, with a central canal and smaller canals under the ridges and valleys internally**; branches none or whorled and structurally similar to the main stem. **Leaves whorled, fused into sheaths at base**, usually much less than 2 cm long, the sheaths sometimes more or less swollen. Sporangia elongate, pendant beneath the expanded apex of sporangiophores; **sporangiophores** peltate, whorled in strobili terminating stems that are green or, in a few species, not green, unbranched, and either ephemeral or becoming green and branched after the spores are shed. Homosporous; spores spherical, green, and with **4–6 straplike elaters** wrapped around the spore that rapidly straighten and assist in moving the spore out of the sporangium. Gametophytes green, growing on the surface of the ground (Figure 8.7).

Distribution and ecology: Cosmopolitan, except for Australia, New Zealand, and Antarctica, with the greatest number of species between 40° and 60° north latitude. These plants are primarily colonizers of unforested areas, lake margins, and wetlands.

Genera/species: 1 (*Equisetum*) / 15.

Economic plants and products: The family is not significant economically. Silica in the stems made them useful to early European settlers in North America for scouring cookware, a practice that is apparently the source of another common name, scouring rush.

Discussion: This monogeneric family is morphologically distinct in its grooved and hollowed stem, whorled leaves, and sporangiophores. The genus is divided into two subgenera -- subgen. *Equisetum* (8

Figure 8.7 Equisetaceae. (A–D) *Equisetum ramosissimum*: (A) habit (× 0.5); (B) node with branch (× 4); (C) strobilius (× 3); (D) cross-section through stem, showing hollow center and more or less circular canals under valley (× 6.5). (E–G) *E. arvense*: (E) habit of sterile plant (× 0.4); (F) cross-section through stem (× 9.5); (G) apex of fertile stem with strobilus (× 0.8). (A–F from Madalski 1954; G from Hauke 1990.)

species; superficial stomates; stems branched) and sub-gen. *Hippochaete* (7 species; sunken stomates; stems unbranched) -- which have been recognized as distinct genera by some. Hybridization occurs frequently between species of the same subgenus but not between species of the two subgenera. A phylogenetic study of all 15 species, based on a combined analysis of two chloroplast markers, *trnL-F* and *rbcL*, demonstrates robust support for monophyly of the two subgenera. The South American species, *E. bogotense*, however, is not a member of either clade but instead is sister to the other 14 species.

Fossil relatives of Equisetaceae are known since the Devonian (408–360 MYA), and they became most abundant as relatively small (under 1 m tall) plants of the understory of Carboniferous forests. Some members of the clade to which Equisetaceae belongs that grew in the Carboniferous has stems as tall as 20 m, and, like large Carboniferous lycophytes of that time, they went extinct. The first fossils clearly assignable to *Equisetum* are from the Eocene (54–38 MYA), but the genus may extend back into the Permian, more than 300 MYA.

References: Des Marais et al. 2001; Hauke 1990, 1993; Lellinger 1985; Pryer et al. 2001; Stewart and Rothwell 1993; Tryon and Tryon 1982.

Leptosporangiate Ferns (Polypodiales)

Terrestrial, epiphytic, or aquatic. **Primary xylem with scalariform bordered pits.** Stems rhizomatous to treelike and up to 20 m tall. Leaves megaphylls (see Figures 8.8, 8.13, 8.14), often much more than 2 cm long and with numerous and forking veins; mostly divided into separate lobes or separate leaflets, but sometimes simple and unlobed; spirally arranged or sometimes 2-ranked, their bases often persistent and more or less covering the stem or sometimes cleanly abscising; with *circinate vernation* (Figure 8.8)—forming a crozier or "fiddlehead" (i.e., coiled and unfolding lengthwise when emerging); monomorphic or dimorphic (sterile and fertile leaves different). **Sporangia with stalk 4-6 cells in cross-section,** *mostly clustered in sori* on the abaxial leaf surface of normal leaves, or in specialized portions of leaves, or on completely separate fertile leaves (see Figure 8.14), less often scattered on the abaxial leaf surface or in sporocarps; usually with an **annulus,** a cluster or row of cells with thick walls that open the sporangium and catapult the spores into the air; the sporangium wall one cell layer thick at maturity (except in Osmundaceae). Homosporous, sometimes heterosporous; spores mostly not green. Gametophytes exosporic and growing on the surface of the ground or endosporic. **First cell division of the zygote vertical.**

Discussion: Leptosporangiate ferns are distinguished by their usually large leaves that form croziers as they unfold and sporangia that often cluster in sori. The many species in this group display tremendous leaf diversity, from the large, pinnately compound forms of the stereotypic fern (see Figures 8.8, 8.9, and 8.14), to the cloverlike leaves of the water-clover ferns (see Figure 8.11), to the needlelike leaves of curly-grass ferns (Schizaeaceae, not treated here). Habit varies from aquatics (Marsileaceae), to epiphytes with ribbon-shaped leaves, to trees (Cyatheaceae).

Currently between 25 and 35 families of leptosporangiate ferns are recognized. Here we treat four families, chosen to span the evolutionary diversity of this group as currently understood, from one of the first branches within the leptosporangiate fern clade (Osmundaceae) to the derived Polypodiaceae. These four families also display tremendous morphological diversity.

Polypodiaceae are treated broadly here to include many groups that have traditionally been considered separate families. Polypodiaceae plus the tree fern clade (with Cyatheaceae), heterosporous aquatic ferns (including Marsileaceae), and schizaeoid ferns (not treated here) form a tentative clade called by some workers the higher

Figure 8.8 Polypodiaceae. *Athyrium filix-femina*: habit (about one-third to one-quarter actual size). Note circinate vernation of developing leaves. (From Taylor 1984; Milwaukee Public Museum, original art by P. Nelson.)

Key to Selected Families of Leptosporangiate Ferns

1. Plants floating in water or sometimes rooted in mud bordering water leaves divided into 2 or 4 segments or threadlike and undivided; heterosporous, the sporangia lacking an annulus and of two kinds that are borne in the same specialized, hardened and bean- or pea-shaped sporocarp ....**Marsileacae**

1. Plants mostly terrestrial; leaves mostly pinnately compound with numerous segments, at least pinnatifid, infrequently simple; homosporous, the sporangia annulate and usually borne in sori that are exposed on the leaf surface 2

2. Annulus merely a patch of cells on the side of the sporangium; sori absent; spores over 100 to several thousand per sporangium...**Osmundaceae**

2. Annulus usually a much larger vertical or obliqueband of cells; sori usually present; spores usually 64 or fewer per sporangium 3

3. Stems usually massive and often arborescent, reaching 20 m in height; leaves (0.5–) 2–3 (–5) m long; annulus oblique, not interrupted by stalk of sporangium ....**Cyatheaceae**

3. Stems creeping along the ground or subterranean, usually only tip of stem and leaves evident, or epiphytic; leaves generally less than 2 m long; annulus vertical, interrupted by the sporangium stalk..**Polypodiaceae s.l.**[a]

[a] The abbreviation "s.l." stands for *sensu lato*, meaning "in the broad sense," or broadly defined; its converse "s.s." stands for *sensu stricto*, "in the narrow sense," or narrowly defined. Both terms are used throughout Chapters 8 and 9.

(A)

(B)

leptosporangiate ferns (see Figure 8.4). Several families of basal leptosporangiate ferns, such as Dipteridaceae, Gleicheniaceae (the gleichenoid ferns), Hymenophyllaceae, Matoniaceae, Osmundaceae, and Plagiogyriaceae, are also supported, but relationships are in some cases unclear (Figure 8.4). Osmundaceae are probably sister to the remaining families. Recent papers are clarifying phylogenetic relationships of leptosporangiate ferns (Hasebe et al. 1995; Pryer et al. 1995, 2001; Smith 1995; Wolf 1997; Wolf et al. 1998, 1999; Pryer 1999; Vangerow et al. 1999; Smith and Cranfill in press).

The nature of the annulus is critical to the phylogeny of leptosporangiate ferns. In Osmundaceae it is merely a patch of cells on one side of the sporangium (Figure 8.9). Marsileaceae have no annulus, as the sporangia are encased and dispersed in the sporocarp. The annulus of Cyatheaceae (see Figure 8.10) consists of a series of cells that are not vertical in orientation—it does not lie in the same plane as the sporangium stalk—and therefore does

Figure 8.9 Osmundaceae. (A) *Osmunda regalis*: habit (about × 0.1), with an enlarged apical, partially fertile portion of the leaf to the left (about × 0.3). (B) *O. lancea*: sporangia, closed on left and open on right (× 315). Note thickened cell walls of annulus. (A from Taylor 1984; Milwaukee Public Museum, original art by P. Nelson; B from Hewitson 1962.)

not contact the stalk. In Polypodiaceae, the annulus is a line of cells whose vertical orientation puts it in the plane of the stalk, but the annulus stops where it meets the stalk, and hence it is said to be "interrupted" by the stalk (see Figure 8.12).

References: Flora of North America Editorial Committee 1993; Hasebe et al. 1995; Kramer and Green 1990; Lellinger 1985; Pryer et al. 1995, 2001; Smith 1995; Stein et al. 1992; Tryon and Tryon 1982; Vangerow et al. 1999; Wagner and Smith 1993; Wolf 1997; Wolf et al. 1998; Wolf et al. 1999.

Osmundaceae Berchtold and J. Presl
(Royal Fern Family)

Terrestrial. Stems branched, often covered by persistent leaf bases, horizontal to erect and arborescent. Leaves about 0.5–2 m long, once- to thrice-pinnately compound or once-pinnate-pinnatifid (deeply pinnately lobed), with an expanded petiolar base, circinate before unfolding; in some species wholly or partially dimorphic, with separate sporangium-bearing segments or whole leaves; often forming a vase-shaped clump. *Sporangia separate or in loose clusters, not clustered in sori, borne on wholly fertile portions of the leaf or on the abaxial surface of relatively unmodified portions of the leaf, with a short stalk of many rows of cells and a poorly differentiated* **annulus that consists of a group of thickened cells on the side of the sporangium.** Homosporous; spores green, over 100 to several thousand per sporangium. Gametophytes growing on the surface of the ground, heart-shaped to elongate, green (see Figure 8.9).

Distribution and ecology: Worldwide except for very cold and dry climates and Pacific islands. *Osmunda*, the only genus in the Northern Hemisphere, is common in wetlands, such as swamps, swales, and lowland forests.

Genera/species: 3/18. **Genera:** *Osmunda* (10 spp.), *Leptopteris* (6), and *Todea* (2).

Economic plants and products: Some species, such as cinnamon fern (*Osmunda cinnamomea*) and royal fern (*O. regalis*), are grown as ornamentals.

Discussion: The numerous spores, rudimentary annulus, sporangium wall more than one-cell thick, lack of a sorus, and *rbcL* DNA sequence data all support Osmundaceae as sister to all other leptosporangiate ferns. This conclusion is consistent with the long fossil record dating to the Permian (286–245 MYA). *Osmunda claytoniana* has apparently lived since the late Triassic, about 220 MYA.

References: Hasebe et al. 1994, 1995; Kramer 1990c; Lellinger 1985; Phipps et al. 1998; Pryer et al. 1995; Tryon and Tryon 1982; Whetstone and Atkinson 1993.

Cyatheaceae Kaulf.
(Scaly Tree Fern Family)

Stem usually a single, erect, arborescent trunk to 20 m tall, usually massive and unbranched, sometimes creeping along the ground. **Leaves distinctly scaly,** *(0.5–) 2–3 (–5) m long,* pinnately or bipinnately compound, with leaflets deeply pinnately lobed, circinate before unfolding. Sporangia borne in sori on the abaxial leaf surface; the annulus continuous, not interrupted by the sporangium stalk. Indusium (a thin, platelike structure that covers a sorus) below the sorus or absent. Homosporous; spores not green, usually 64 (sometimes only 16) per sporangium. Gametophyte heart-shaped or occasionally elongate, growing on the surface of the ground; green (Figure 8.10).

Distribution and ecology: New and Old World tropical wet montane forests and cloud forests. Some species extend into south temperate areas (New Zealand and southern South America) and into north temperate India, China, and Japan. Some species are widely distributed, but local endemics are numerous on oceanic islands and tropical mountains. Many species are early successional on landslides.

Genera/species: ca. 3–6/500. **Major genera:** *Alsophila* (230 spp.), *Sphaeropteris* (120), and *Cyathea* (110).

Economic plants and products: Members of this family are sometimes grown as ornamentals, and the fibrous rhizomes are used as a base for epiphytes in greenhouses. Human exploitation has led to extensive destruction of tree fern stands, but most species are now covered by international agreements barring trade in endangered species.

Discussion: Chloroplast DNA restriction site data establish three well-defined lineages: the *Alsophila* clade (including *Nephelea*), the *Cyathea* clade (including

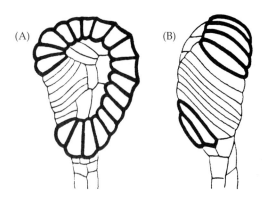

Figure 8.10 Cyatheaceae. *Cyathea capensis*: (A) sporangium, showing interrupted annulus with cells with thicked walls; (B) the same, different view (both × about 100). (From Holttum 1963.)

Trichipteris and *Cnemidaria*), and the *Sphaeropteris* clade. Relationships among these clades are unresolved.

The phylogenetic relationships of this family also are not clear, although it is phylogenetically close to two other tree fern families: the "Dicksoniaceae," which recently has been shown not to be monophyletic, and the Metaxyaceae (two species). Another relative, according to DNA sequence evidence, is Hymenophyllopsidaceae, which is surprising in that members of this family have short-creeping to suberect or erect stems and leaves only 10–30 cm long. The name of the family reflects the similarity of its delicate nature to that of members of the Hymenophyllaceae, the filmy ferns, a family of about 600 species. Hymenophyllopsidaceae, which contain a single genus and about eight species, are the most narrowly distributed fern family, restricted to the Roraima geological formation in Venezuela, Guyana, and northernmost Brazil.

Two potential synapomorphies of the broadly defined tree fern clade are pneumathodes (ventilation lines or patches on the leaf axis or petiole or on the rachis for gas exchange) and radial symmetry of the rhizome. Molecular data tentatively support a sister-group relationship between this tree fern clade and the Polypodiaceae s.l.

Groves of tree ferns, with the crown of large leaves atop the unbranched stems, make some of the most attractive displays in the natural world.

References: Conant et al. 1995; Conant and Stein 2001; Kramer 1990a; Pryer et al. 1995; Tryon and Tryon 1982; Wolfe et al. 1999.

Marsileaceae Mirbel
(Water-Clover Family)

Plants aquatic or rooted in the mud bordering bodies of water. Stems slender, glabrous, and creeping, growing on soil surface or subterranean. *Leaves long-petioled, the blade divided into 2 or 4 leaflets or filiform and not expanded;* circinate before unfolding. Sori **with no indusial opening**, enclosed in bean- or pea-shaped *sporocarps* that are borne on short stalks near or at base of petioles; each sporocarp with at least 2 sori. Heterosporous. Sporangia without an annulus; megasporangia with one megaspore; microsporangia with 16–64 microspores. Gametophytes minute, remaining within spores; megagametophytes **with one archegonium**, protruding from spores; microspores bursting as male gametophytes release sperm (Figure 8.11).

Distribution and ecology: Nearly cosmopolitan in warm temperate and tropical areas; amphibious, growing in water or near water in very wet soil.

Genera/species: 3/76. Major genera: *Marsilea* (50–70 spp.) and *Pilularia* (6).

Economic plants and products: Occasionally planted as a curiosity.

Discussion: The common name refers to the similarity between water-clover (*Marsilea*) leaves and those of clovers (*Trifolium*, Fabaceae). *Pilularia*, which is widely distributed, has filiform leaves. Leaves of *Regnellidium*, a monotypic genus of southern Brazil and adjacent Argentina, have two leaflets. Recent studies show that *Pilularia* and *Regnellidium* are sister taxa and *Marsilea* is the sister of that clade.

Morphological and DNA sequence data show that this family is the sister to the Salviniaceae, which consists of two genera of aquatic, heterosporous ferns, *Azolla* and *Salvinia*. Heterospory, therefore, is hypothesized to have arisen a single time within the leptosporangiate ferns. Salviniaceae are cosmopolitan, with *Azolla* and *Salvinia* each having fewer than 10 species, and contain plants that are floating aquatics bearing flattened sporocarps. The clade of heterosporous ferns is sister to Cyatheaceae plus Polypodiaceae.

The sporocarps of Marsileaceae are highly unusual. They protect spore viability for over 100 years in some species and may be an adaptation for growth in arid regions where rainfall is infrequent. Waterfowl consume

Figure 8.11 Marsileaceae. *Marsilea vestita*: habit (about × 0.7). Note mature, cloverlike leaves; numerous young, circinate leaves and single sporocarp (arrow). (From Taylor 1984; Milwaukee Public Museum, original art by P. Nelson.)

the sporocarps, which pass through the digestive tract and thus disperse the spores.

The fossil record indicates that the genera of this family diversified by the mid-Cretaceous.

References: Johnson 1993a; Kramer 1990b; Lellinger 1985; Lupia et al. 2000; Pryer 1999; Pryer et al. 1995.

Polypodiaceae Berchtold and J. C. Presl s.l.
(Higher Fern or Polypody Family)

Terrestrial, sometimes growing on rocks, sometimes epiphytic or climbing, infrequently aquatic; perennial, rarely annual. Stems subterranean or barely evident at soil surface (except in epiphytes, climbers, and aquatics), sometimes horizontal (rhizomatous) and creeping, reproducing vegetatively; often with epidermal scales of various sizes, shapes, and textures. Leaves mostly pinnately lobed to once, twice, or several times pinnately compound (Figures 8.8, 8.13, 8.14); less often palmately compound, simple, or ribbonlike; usually *circinate*, clustered or separated from one another on an elongate stem; petiole usually well developed, rarely absent; sterile-fertile dimorphism well developed (see Figure 8.14) to absent; with hydathodes; variously hairy, scaly, or glabrous. Sporangia with a well-developed **vertical annulus, interrupted by the sporangium stalk** (Figure 8.12); *usually borne in sori* that may be covered by an indusium; sometimes intermixed with special branched or unbranched hairs (paraphyses). Sori on the abaxial surface or near the margin of the leaf; round to elongate (Figures 8.12, 8.13). Indusium above, beside, or less often below the sorus; round to elongate; sometimes absent, replaced by the recurving leaf margin, or combining with a recurving leaf margin. Homosporous; spores 64, 32, 16, or 8 per sporangium, usually not green, monolete or trilete, the surface smooth or variously sculptured with ridges and spines. Gametophytes growing on the surface of the ground; green, heart-, kidney-, or ribbon-shaped; sometimes with small buds (gemmae) that separate from the gametophyte for asexual reproduction.

Distribution and ecology: Worldwide and in a great range of habitats, from saline and freshwater wetlands to deserts, from tropical lowlands to montane and alpine areas, from the understory of mature forests to early successional sites.

Genera/species: 193–223/7500. ***Major genera:*** *Asplenium* (700 spp.), *Cyclosorus* (600), *Elaphoglossum* (500), *Diplazium* (400), *Grammitis* (400), *Thelypteris* (280), *Dryopteris* (225), *Polystichum* (200), *Pteris* (200), *Blechnum* (175), *Adiantum* (150), *Athyrium* (150), *Cheilanthes* (150), *Ctenitis* (150), *Lindsaea* (150), *Polypodium* (150), *Tectaria* (150), and *Dennstaedtia* (45).

Economic plants and products: Many species, including maidenhair ferns (*Adiantum*), Boston fern (*Nephro-*

Figure 8.12 Polypodiaceae. (A, B) *Asplenium tripteropus*: (A) portion of underside of leaf, showing sori (× 4); (B) cross-section of fertile leaf, showing two sori and sporangia. (C) *A. nidus*: sporangium, showing interrupted annulus with cells with thickened walls (× 140). (D) *Asplenium* sp.: margin of leaf, with sori (× 8) (A, B from *Flora Tsinlingensis* 1974; C from Haider 1954; D from Pérez Arbeláez 1928.)

lepis), and brakes (*Pteris*), are grown as ornamentals, either in gardens or as houseplants. *Matteuccia struthiopteris* (Figure 8.14), a robust fern of alluvial woods in

Figure 8.13 Polypodiaceae. *Adiantum capillus-veneris*: portion of leaf; enlarged portion of a leaf segment, showing marginal sporangia. (From Paris 1993.)

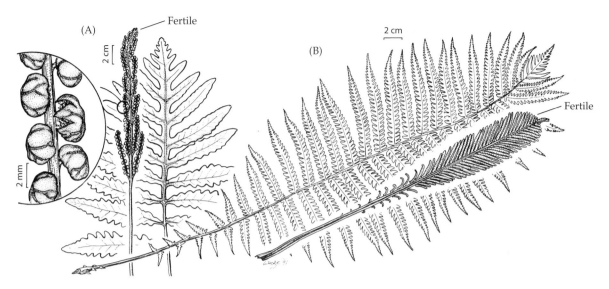

Figure 8.14 Polypodiaceae. (A) *Onoclea sensibilis*: fertile and sterile leaves; enlargement of portion of fertile leaf. (B) *Matteucia struthiopteris*: fertile and sterile leaves. (From Johnson 1993b.)

much of the Northern Hemisphere, is collected in the spring and its young leaves—often referred to as fiddle-heads—are consumed fresh or canned. This species is often planted near houses. Some species have been used to dye cloth. *Pteridium aquilinum* is an invasive weed in many parts of the world.

Discussion: The Polypodiaceae are broadly defined here* and include at least 11 families that are currently recognized by some systematists. DNA sequence and morphological data together support monophyly of some of these families, such as the Aspleniaceae (1 genus, 700 species) and Thelypteridaceae (5–25, about 1000), whereas Dennstaedtiaceae (20, 400), Dryopteridaceae (60, 3000), and Polypodiaceae s.s. (40, 500) are clearly shown to be para- or polyphyletic. More intensive taxonomic sampling and new structural and molecular data have greatly improved our understanding of relationships within Polypodiaceae s.l. The broad definition is used here because monophyly of Polypodiaceae s.l. is corroborated by morphological and DNA sequence data.

Features common to many leptosporangiate ferns—large, clumped, pinnately compound leaves and mesic forest habitat—hold for many species in this family. Nevertheless, as its description indicates, the family encompasses tremendous variation in habit, leaf morphology, and reproductive features. For example, aquatics, such as *Ceratopteris*, may have inflated petioles for flotation. Epiphytes, such as *Vittaria*, the shoestring ferns, have long, linear leaves with indistinct petioles.

Characters of sori and indusia vary greatly as well in Polypodiaceae s.l. In some, the sori and indusia are lin-

ear, and the indusium is attached beside and covers part or all the sorus. Other species bear round sori that are more or less covered by an indusium. Sori of yet other species are marginal and partially covered by the recurving leaf margin. Sori of still other species are not associated with indusia. In addition to leaf morphology and these reproductive features, scales on the stem and leaves, leaf architecture, habit, number and arrangement of vascular bundles in the petiole, hairs, venation, spore morphology, and chromosome number are taxonomically informative.

Hybridization and polyploidy play prominent roles in the evolution and systematics of *Asplenium, Athyrium, Ceratopteris, Cheilanthes, Cystopteris, Diplazium, Dryopteris, Gymnocarpium, Pellaea, Polypodium, Polystichum, Pteris, Woodsia*, and other genera.

References: Hasebe et al. 1995; Kramer and Green 1990; Lellinger 1985; Pryer et al. 1995; Smith and Cranfill 2002; Tryon and Tryon 1982; Wagner and Smith 1993; Wolf 1997.

GYMNOSPERMS

Living seed plants form a monophyletic group with five major clades: cycads, Ginkgoaceae, conifers, Gnetales, and angiosperms. The first four groups have commonly been called gymnosperms, which means "naked" (*gymno*) "seed" (*sperm*). The seeds are not enclosed in a carpel, as they are in angiosperms (*angio* means "vessel," referring to the carpel), although they may sometimes be enclosed at maturity by fused cone scales or bracts, as in juniper "berries."

---

*The broad definition of a group is indicated by s.l. and the narrow definition by s.s.; see the footnote to the key on p. 195.

Monophyly of gymnosperms has been highly controversial. For many years they were not considered to form a monophyletic group (Crane 1988; Doyle 1998; Doyle et al. 1994; Nixon et al. 1994; Price 1996; Stefanovic et al. 1998). When fossils such as the extinct seed ferns (which had naked seeds) are considered, groups with naked seeds do not represent a clade. The status of gymnosperms also depends upon the location of the root of the seed plants. If, for example, the root places cycads with angiosperms, then gymnosperms are paraphyletic. More recent, molecular data sets conflict about monophyly of extant gymnosperms (Bowe and de Pamphilis 1997, Bowe et al. 2000; Chaw et al. 2000; Goremykin et al. 1996; Rydin et al. 2002). Analyses of Bowe et al. (2000) and Chaw et al. (2000) combine sequences from all three plant genomes. Bowe et al. (2000) used two mitochondrial genes, *cox*1 and *atp*A, plastid *rbcL* and nuclear 18S rDNA, and Chaw et al. (2000) used mitochondrial small subunit rRNA, plastid *rbcL*, and nuclear small subunit rRNA. Both studies strongly support gymnosperm monophyly and, surprisingly, a close relationship of Gnetales and conifers. The nuclear data in both studies have Gnetales and conifers as sister taxa, whereas the mitochondrial and plastid genes resolve a sister group relationship between Gnetales and Pinaceae. Under this so-called "gnepine" hypothesis, conifers are not monophyletic and Gnetales become extremely divergent conifers. A relationship between Gnetales and Pinaceae is unexpected given the many morphological differences between the two groups and a major genomic difference. Pinaceae and other conifers, but not Gnetales, have lost the inverted repeat from the plastid genome.

Rydin et al.'s (2002) analyses of plastid *atpB* plus *rbcL* and nuclear 18S rDNA plus 26S rDNA confirmed monophyly of conifers and rejected the gnepine hypothesis. Resolution of relationships of seed plants awaits further studies, and here we treat conifers as monophyletic and sister to Gnetales.

Together cycads, Ginkgoaceae, conifers, and Gnetales represent only about 15 families, 75–80 genera, and about 820 species. The evolutionary success of angiosperms relative to other seed plants may be attributed to vessels and reproductive features. All gymnosperms except Gnetales have only tracheids in their xylem. Angiosperms and Gnetales also have vessels, which are more efficient than tracheids in water transport, although more likely to suffer irreparable damage during drought. Angiosperm carpels make possible stigmatic germination of pollen and are variously adapted for protection of the young ovule and seed dispersal. Gymnosperms are slow to reproduce; up to a year may pass between pollination and fertilization, and seed maturation may require 3 years. Angiosperms, in contrast, usually reproduce far more rapidly, going from a seed to a seed in weeks in some annuals. With the exception of the cycads and some Gnetales, gymnosperms are pollinated by wind. Angiosperms have adapted in numerous ways to animal pollination, and they are thus able to repro-

duce in habitats where there is little wind, such as the forest floor. The highly specific nature of animal pollination may promote speciation (see Chapter 6). Furthermore, gymnosperms are rarely polyploid and have not undergone extensive allopolyploid speciation.

Gymnosperms are all woody—being trees, shrubs, or lianas—and include no true aquatics and few epiphytes. These plants grow throughout most of the world, from 72 north to 55 south, and they are the dominant vegetation in many colder and arctic regions. Pines, spruces, hemlocks, firs, yews, cedars, and related groups are familiar as ornamentals and supply high-quality wood. They include the tallest, the most massive, and the longest-living individual plants. We will consider 8 of the 15 families of gymnosperms, which together account for the great majority of species. Phylogenetic relationships among the extant groups of seed plants are not well resolved.

References: Beck 1988; Bowe and de Pamphilis 1997; Bowe et al. 2000; Chaw et al. 2000; Crane 1988; Doyle et al. 1994; Friis et al. 1987; Gifford and Foster 1988; Goremykin et al. 1996; Kubitzki 1990; Nimsch 1995; Nixon et al. 1994; Price 1996; Rydin et al. 2002; Singh 1978; Sporne 1974; Stefanovic et al. 1998; Stewart and Rothwell 1993; Taylor and Taylor 1993.

Cycadales (Cycads)

The cycads are an ancient group that has retained clearly primitive features, such as motile sperm. Cycads evolved in the Carboniferous or early Permian, about 280 million years ago and reached their peak abundance and diversity in the Mesozoic era. Now cycads are mostly Southern Hemisphere relicts, and many of the species are endangered or threatened by extinction.

The group is monophyletic, as judged by synapomorphies in structural features, such as girdling leaf traces, a specialized (omega) pattern of vascular bundles in the petiole, the presence of mucilage canals, and distinctive meristems, as well as poisonous glycosides called **cycasins**. Cycasins and other toxic compounds may have been important in the evolution of cycads as defenses against bacteria and fungi. Cycad toxins have been responsible for partial or total paralysis of the hind limbs of livestock.

Cycads are often palmlike, with an unbranched upright stem up to 18-20 m tall and large compound leaves crowded toward the stem apex, or fernlike, with a subterranean stem and compound leaves. Most cycads bear **cataphylls**, scale-like leaves that occur among the normal leaves and often serve a protective function. Cycads are slow-growing, with stems reaching as little as 1 m in height in 500 years.

Cycad reproductive structures occur in strobili that consist of an axis and megasporophylls (ovule-bearing leaves) or microsporophylls (pollen-bearing leaves). These simple structures contrast with the complex seed cones of conifers (see below). All cycads have pollen

---

Key to Families of Cycads

1. Leaflets circinate when young, with a midvein and no lateral veins; megasporophylls leaflike, loosely clustered near stem apex and not forming a strobilus, pinnately lobed or toothed above the ovules, with 2–8 ovules attached laterally to the basal portion . . . . **Cycadaceae**

1. Leaflets flat or conduplicate when young, with or without a midvein but with numerous, ± parallel veins or midvein present, with numerous, dichotomously branched or simple lateral veins; megasporophylls greatly reduced, valvate or imbricate, forming a strobilus with 2 reflexed ovules . . . . **Zamiaceae**

---

strobili, and all except *Cycas* have ovulate strobili. Although cycads produce abundant, powdery pollen that suggests wind pollination, insects are the primary vectors of pollen. Beetles and bees are either the sole pollen vector or move pollen to the ovule after wind has carried it to the ovulate strobilus from another plant. Pollination and fertilization may be separated by up to 7 months, during which time a pollen tube forms.

Cycad seeds often have a brightly colored (pink, orange, or red) and fleshy outer layer and are commonly dispersed by birds, bats, opossums, turtles, and many other animals. Some cycads attract dispersers with brightly colored megasporophylls. Ocean currents transport cycads whose seeds are buoyant due to a spongy outer layer, while gravity disperses the seeds of others.

Dioecy, which characterizes all cycads, may be governed by sex chromosomes. Chromosome number varies considerably among, but usually not within, cycad genera, and is taxonomically useful.

All cycad species bear special roots, called **coralloid roots** because of their similarity in appearance to marine coral. These roots host cyanobacteria that carry out nitrogen fixation, like that of bacteria in legumes. The cyanobacteria convert gaseous atmospheric nitrogen, which cycads cannot use, into a form they can use, providing a source of nitrogen that promotes growth in the nutrient-poor soils cycads frequently occupy.

The cycad clade consists of 2 families, 10–11 genera, and about 130 species. Cycadaceae contains only *Cycas*; Zamiaceae comprises the remaining 9–10 genera, including *Stangeria* (which has often been placed in the monogeneric Stangeriaceae).

References: Bogler et al. 2001; Crane 1988; Johnson and Wilson 1990; Jones 1993; Norstog and Nicholls 1997; Landry 1993; Norstog and Fawcett 1989; Rai et al. 2001; Stevenson 1990, 1991, 1992.

Cycadaceae Pers.
(Cycad Family)

Palmlike plants, with an unbranched to sparsely branched, woody stem covered with old remnants of leaf bases and living foliage near the apex of the stem; or *fernlike* with a subterranean stem. Leaves persistent, spiral, *pinnately compound;* **leaflets circinate when young and unfolding, with one midvein and no lateral veins**, entire, lower leaflets often spinelike. Microsporophylls aggregated into compact strobili; pollen nonsaccate, with a single furrow. *Megasporophylls grouped at the stem apex, somewhat leaflike and not clustered into a strobilus;* ovules 2–8 on the megasporophyll margin. Seeds large, *slightly flattened,* and covered by a *brightly colored, fleshy outer layer.*

Distribution and ecology: Madagascar, possibly Africa, Southeast Asia, Malesia, Australia, and Polynesia. In forests and savannas. Many species tolerate fire because the apical meristem is protected underground or by persistent leaf bases.

Genera/species: 1 (*Cycas*)/ca. 20.

Economic plants and products: Several species are popular as ornamentals in warm climates and as houseplants. The stem is a source of starch (called sago), especially in times of food shortage. Seeds may contain 20–30% starch, which is edible after removal of toxins.

Discussion: This family is distinct because its megasporophylls have a well-developed and toothed to pinnately dissected blade and are not compactly clustered into a strobilus. *Cycas* is probably the sister genus to the remainder of the cycads, and its fossil record extends back to the Permian, at least 250 million years ago. It is currently the most widely distributed genus of cycads.

Reference: Bogler and Francisco-Ortega, in press; Johnson and Wilson 1990.

Zamiaceae Horianow
(Coontie Family)

Fernlike with subterranean stem or *palmlike* with aerial, unbranched stem to 18 m tall and large, pinnately compound leaves clustered near stem apex. Stem covered with persistent dead leaf bases or naked. Leaves pinnately (rarely twice pinnately) compound, spiral, persistent, leathery, with or without stout spines on the petiole and rachis; *leaflets flat when young and unfolding, with numerous ± parallel veins* (with a midvein and dichotmously branching secondary veins in *Stangeria*), entire, dentate, or with sharp spines. Microsporophylls aggregated into compact strobili, with numerous small microsporangia that are often clustered; pollen nonsac-

cate, with a single furrow. **Megasporangiate strobili** 1 to several per plant, more or less globose to ovoid or cylindrical, disintegrating at maturity; **megasporophylls** densely crowded, symmetrically to asymmetrically peltate, valvate or imbricate, **each with 2 ovules**. Seeds large (1–2 or more cm long), **± round in cross-section**, with an *often brightly colored and fleshy outer layer* and a hard inner layer; cotyledons 2 (Figure 8.15).

Figure 8.15 Zamiaceae. *Zamia floridana*: (A) habit of ovulate plant (some leaves removed) with strobilus at time of pollination; note large fleshy taproot with small lateral roots and coralloid roots (at right, near stem–root juncture) (× 0.75); (B) microsporangiate strobilus during pollen shedding (× 0.5); (C) adaxial surface of microsporophyll (× 4.5); (D) abaxial surface of microsporophyll, with microsporangia (× 4.5); (E) ovulate strobilus at time of pollination; note separation of megasporophylls in lower portion of strobilus, allowing entry of pollinators and pollen (× 0.75); (F) adaxial view of megasporophyll with two ovules, each with micropyle directed toward axis at bottom (× 4.5). (From Stevenson 1991, *J. Arnold Arbor.* Suppl. Series 1: pp. 367–384.)

Distribution and ecology: Tropical to warm temperate regions of the New World, Africa, and Australia. From poor, dry soils of grasslands and woodlands to dense, tropical forests. The only native cycad of the United States is a *Zamia* in Florida and southern Georgia.

Genera/species: 9–10/111. *Major genera:* *Encephalartos* (35 spp.), *Zamia* (35), *Macrozamia* (14), *Ceratozamia* (10), and *Dioon* (10).

Economic plants and products: Many species are grown as ornamentals in warm climates and as houseplants. The stem and seeds have been used as a source of starch (called sago). Removal of the toxic glycosides cycasin and macrozamin by boiling and washing is required before consumption.

Discussion: The seed strobili of this family are among the heaviest and largest of seed plant reproductive structures, weighing up to 40 kg and measuring 60 cm in length and 30 cm in diameter. The seeds are also quite large, up to 4 cm long. *Zamia* is unusual in having species with several different chromosome numbers, and species delimitation is very difficult in this genus. Stevenson (1992) divided the family into two subfamilies, each with two tribes. *Dioon*, which is distinct in having strongly asymmetrically peltate megasporophylls with stalked ovules, is strongly supported as the sister to the remaining genera. Like other cycad genera, *Dioon* was formerly much more widespread and is known as fossils from the Eocene in Alaska.

Bowenia, which has two or three species confined to tropical regions of northeastern Australia, is unique among cycads in its bipinnately compound leaves. This genus has been placed in its own family or united with *Stangeria*, which grows naturally only in southern Africa and contains only one species, in the Stangeriaceae. *Stangeria* is exceptional in the Zamiaceae in its leaf venation (see family description). Molecular data analyzed to date clearly place these two genera within the Zamiaceae, do not show a close relationship between them, but otherwise do not resolve their affinities with other members of the Zamiaceae.

Encephalartos (Africa), *Lepidozamia* (Australia), and *Macrozamia* (Australia) form a clade based on molecular data. *Lepidozamia* was in the past merged into *Macrozamia*, but molecular data actually show that *Lepidozamia* and *Encephalartos* are sister taxa.

Zamia is wide-ranging, morphologically diverse, and has numerous species, although there are differences in the recognition of species by different workers. Pollen-eating beetles are responsible for pollinating *Zamia* in Florida. Release of oil of wintergreen apparently draws the beetles to microsporangiate strobili, which provide an egg-laying site and food for adult and larval beetles. Adult beetles eat pollen and are covered by it. Pollination occurs when pollen-covered adults go to megasporangiate strobili where the beetles neither feed for lay eggs.

References: Bogler and Francisco-Ortega, in press; Johnson and Wilson 1990; Landry 1993; Norstog and Fawcett 1989; Norstog and Nicholls 1997; Stevenson 1991, 1992.

Ginkgoales

Ginkgoaceae Engler
(Ginkgo or Maidenhair Tree Family)

Trees to 30 m, with a more or less asymmetrical crown and furrowed, gray bark. Resin canals absent. Leaves simple, spiral, and widely spaced on long shoots near the tips of branches, often *closely packed on stubby short shoots on older growth;* **fan-shaped**, apically notched (the leaf then bilobed) or entire; **deciduous** and bright yellow in the fall; **dichotomously veined**. Dioecious. Pollen strobili borne on spur shoots, long and pendant; pollen not winged. **Ovules paired on a long stalk from spur shoots**; seeds frequently 1 per stalk, ca. 2.5 cm in diameter, with a juicy, unpleasant-smelling outer coat and hard inner coat; cotyledons 2–3.

Distribution and ecology: Limited to remote mountain valleys of China; possibly extinct in the wild. Little is known about the ecology of this species.

Genera/species: 1 (*Ginkgo*)/1 (*G. biloba*).

Economic plants and products: The maidenhair tree has been grown as an ornamental near religious institutions in eastern Asia for centuries. Individuals live for over a thousand years, and these old trees are probably the source of plantings in many parts of the world, where ginkgo is commonly used as ornamentals for shade and for their interesting and attractive foliage. Staminate individuals, which do not produce the unpleasant-smelling seeds, are commonly planted. The gametophyte and embryo, when boiled, fried, or roasted, are a delicacy in some Chinese dishes.

Discussion: The first representatives of this group appear in the Permian, and plants nearly identical to *Ginkgo biloba* extend back nearly 200 million years in the fossil record. During the early Jurassic, extinct relatives of *Ginkgo* were widespread and diverse, consisting perhaps of three families. Now, ironically, although rare or possibly extinct in the wild, *Ginkgo* does well as a shade tree in urban situations.

The broad, deciduous leaves of *Ginkgo* are unlike those of nearly all other gymnosperms. Sperm motility, known elsewhere in seed plants only in the cycads, is clearly a primitive feature, as is the lack of pollen tubes. Ginkgoaceae are not closely related to any extant groups.

Ginkgo is one of the few plants with sex chromosomes. Ovulate plants bear two X chromosomes, and staminate individuals are XY. Pollination occurs by wind in the spring, but fertilization is delayed for 4–7 months, after the ovules have fallen to the ground. The juiciness and odor of the seed suggest animal dispersal, but the dispersing taxa are unknown and may now be extinct.

References: Page 1990b; Whetstone 1993.

Coniferales (Conifers)

Conifers are the largest and most ecologically and economically important group of gymnosperms. Pines, spruces, firs, hemlocks, cedars, cypresses, redwoods, and giant sequoias are familiar, valued, and revered trees. Members of this group are called conifers because most bear their seeds in specialized structures called **cones** (see Figure 8.17P). Cones protect ovules and seeds and also facilitate pollination and dispersal. These structures consist of an axis bearing highly modified short shoots, the ovuliferous scales. These scales are subtended by bracts, which are either large and conspicuous (as in some Pinaceae), very small (as in other Pinaceae), or small to large and more or less fused to the scale (as in Cupressaceae). Seeds are associated with the scales. Cone scales of most members of Pinaceae and Cupressaceae are woody or leathery. The junipers (*Juniperus*) have cone scales that are more or less juicy and brightly colored, making the cones berrylike (see Figure 8.18 H,P,Q) and animal-dispersed. In Podocarpaceae cones are often reduced, with highly modified, juicy, brightly colored scales and just one ovule. Taxaceae bear solitary seeds partially or completely surrounded by a juicy aril.

Conifers date back to the Carboniferous, some 300 million years ago. Many current families developed by the late Triassic or early Jurassic, and some contemporary genera appeared in the middle Jurassic. Today conifers remain important in colder regions, such as the boreal forests of North America and Asia, where species of pine, spruce, and fir make up the dominant vegetation over large regions. Other conifers—particularly Araucariaceae, Cupressaceae, and Podocarpaceae—are prominent in cooler regions of the Southern Hemisphere. Conifers are valuable ornamentals, and their wood is used for paper production, building construction, and many other purposes. They are often referred to as "evergreens" because of the persistent foliage in most species, or as "softwoods" because their wood is softer than that of many angiosperm trees.

Pollination is by wind. Most conifers, like most nonangiosperm seed plants, use a pollination droplet, a sticky fluid extruded from the ovule at pollination, to catch airborne pollen. Pollen grains of most Pinaceae bear two **saccae**: small, winglike appendages that may serve to float the pollen grains upward in the pollination droplet toward the ovule or to orient the grains properly for germination. Alternatively, the pollen may be trapped on more or less sticky structures in the vicinity of the ovule. The pollen then germinates and grows via a pollen tube to the ovule (the sperm lack flagella).

Key to Selected Families of Conifers

1. Seeds in woody (fleshy only in *Juniperus*) cones, mostly hidden by cone scales, a few to many per cone; plants highly resinous...2

1. Seeds partially to wholly enclosed or subtended by fleshy, often brightly colored structures, usually solitary: plants slightly resinous...4

2. Seeds 1 per cone scale ....**Araucariaceae**

2. Seeds usually more than 1 per cone scale...3

3. Leaves scale-like or needle-like, spiral, opposite, or whorled, persistent on branches after dying (but most branches shed with age); pollen nonsaccate; cone scales valvate or imbricate (and then leaves scalelike and opposite), flat or peltate, fused to bract, juicy in *Juniperus*; seeds with 2 or 3 lateral wings (infrequently wingless), 1–20 per scale........**Cupressaceae**

3. Leaves linear to needle-like, spiral, shed from branches (or as short-shoots in *Pinus*); pollen often saccate; cone scales imbricate, flat, distinct from bract; seeds terminally winged (or rarely wingless), 2 per scale..**Pinaceae**

4. Seeds more or less surrounded by a specialized cone scale (the epimatium), not an aril, and usually associated with colored, juicy bracts; pollen mostly saccate; pollen strobili with 2 sporangia per microsporophyll ....**Podocarpaceae**

4. Seeds more or less surrounded by an aril, an outgrowth of the axis below the ovule; pollen nonsaccate; pollen strobili with 2–9 sporangia per microsporophyll.....................**Taxaceae**

Conifer trees are often monopodial with a dominant central shoot or trunk. With age, the crown may branch irregularly. Branches are often whorled, at least when the plant is young.

The conifers comprise 7 families, 60–65 genera, and over 600 species. Relationships among the major groups of conifers are shown in Figure 8.16. The five families treated here include all but a few conifer species.

References: Brunsfeld et al. 1994; Eckenwalder 1976; Farjon 1990; Hart 1987; Kelch 1997; Page 1990a,c–f; Price et al. 1987; Price and Lowenstein 1989; Richardson 1998; Singh 1978; Sporne 1974; Stefanovic et al. 1998; Thieret 1993a; Watson and Eckenwalder 1993.

Pinaceae Adanson
(Pine Family)

Trees (occasionally shrubs), often emitting strong fragrances from bark and/or leaves; *resin canals present in wood and leaves*. Branches whorled or opposite (rarely alternate). Leaves simple, *linear to needlelike* (rarely narrowly ovate), spiral but often appearing 2-ranked by twisting of leaf base to bring most of the leaves into one plane, clustered or fascicled in groups of 2 to 5 in *Pinus*, sessile or short-petioled, on long shoots or tightly clustered on short shoots, persistent (deciduous in *Larix* and *Pseudolarix*). *Monoecious*. Microsporangiate strobili with spirally arranged, bilaterally symmetrical microsporophylls; microsporangia 2 on the abaxial microsporophyll surface; pollen grains with 2 saccae (saccae absent in *Larix*, *Pseudotsuga*, and all but two species of *Tsuga*) and 2 prothallial cells. *Cones with spirally arranged, flattened bract-scale complexes*; scales persistent (deciduous in *Abies*, *Cedrus*, and *Pseudolarix*), *bracts free from the scale*, longer than the cone scale to much shorter than the cone scale; maturing in 2 (3) years; *ovules 2*, **inverted** (micropyle directed toward the cone axis), on the adaxial cone scale surface; archegonia few per ovule, not clustered. **Seeds with a long, terminal wing** derived from tissue of the cone scale (wing reduced or absent in some species of *Pinus*); embryo straight, cotyledons 2–18 (Figure 8.17).

Distribution and ecology: Pinaceae are almost entirely limited to the Northern Hemisphere. Three or four genera grow only in eastern Asia; one (*Cedrus*) is confined to North Africa, the Near East, Cyprus, and the Himalayas; and the remaining six genera (the major genera) all occur widely in the Northern Hemisphere. The family ranges from warm temperate climates to the limit of tree growth above the Arctic Circle, from permanently water-saturated soils to well-drained soils, and from sea level to alpine habitats up to 4800 m above sea level in eastern Tibet. The seeds of pines are primary components of the diets of many species of birds, squirrels, chipmunks, and other rodents. Members of the family provide cover for many wildlife species and are important in watershed protection.

Genera/species: 10/220. ***Major genera:*** *Pinus* (100 spp.), *Abies* (40), *Picea* (40), *Larix* (10), *Tsuga* (10), and *Pseudotsuga* (ca. 5).

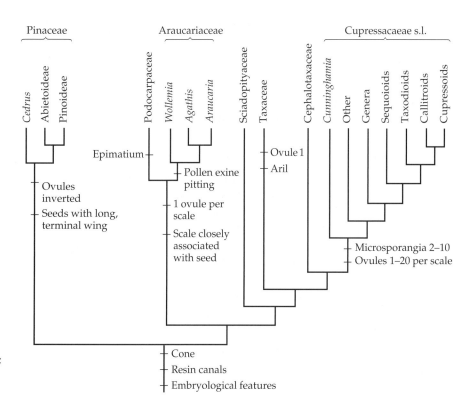

Figure 8.16 Relationships among major groups of conifers, with potential synapomorphies for the families covered in this book. (Based on Kelch 1998; Setogouchi et al. 1998; Stefanovic et al. 1998; and Wang et al. 2000.)

Figure 8.17 Pinaceae. *Pinus* subgen. *Strobus*, *P. strobus*: (A) tip of shoot with last season's leaves, new growth, and two ovulate cones at time of pollination (× 0.7); (B) short-shoots, each showing subtending bract, scale leaves, and five developing needle leaves (× 2.7); (C) detail of bases of mature short-shoots; the bracts and scale leaves have abscised (× 2.7); (D) diagrammatic cross-section of needle leaf, showing single fibrovascular bundle and two resin ducts (× 20); (E) microsporangiate strobilus just before dehiscence (× 5.3); (F,G) lateral and abaxial views of microsporophyll, showing the two abaxial sporangia and their dehiscence (× 13); (H) pollen grain, showing the two saccae (× 330); (I) ovulate cone at time of pollination (× 4); (J) detail of ovulate cone, showing scales with ovules and subtending bracts at time of pollination (× 13); (K) detail of lateral view of single cone scale and subtending bract at time of pollination; one ovule with two micropylar appendages visible (× 16); (L,M) abaxial and adaxial views of cone scale and bract at time of pollination, showing micropylar appendages on the two ovules (× 16); (N) four cones at the end of the first growing season, with all surrounding fascicles of leaves removed; lateral and terminal buds visible (× 0.7); (O) cone scale with two developing seeds at time of pollination of the next year's cones, showing remnants of the micropylar appendages and development of wings (× 4); (P) mature cone (× 0.7); (Q) mature seed, after wing has separated from cone scale (× 1.3); (R) longitudinal section of seed, wing removed, showing embryo surrounded by tissue of megagametophyte (stippled), micropyle facing base (× 7); (S) embryo, showing numerous cotyledons (× 8). (From Price 1989, *J. Arnold Arbor.* 70: pp. 247–305.)

Economic plants and products: Pinaceae are one of the leading sources of timber in the world. The wood of pines (*Pinus*), Douglas firs (*Pseudotsuga*), spruces (*Picea*), hemlocks (*Tsuga*), larches (*Larix*), and firs (*Abies*) is used extensively for construction, pulp for paper production, fenceposts, telephone poles, furniture, interior trim for houses, musical instruments, woodenware, and numerous other purposes. Pines, spruces, hemlocks, cedars (*Cedrus*), Douglas firs, and firs are used extensively as ornamentals, and hundreds of cultivars have been developed in many of the species of these genera. Pine "nuts," the more or less wingless seeds of pinon pines of southwestern North America, were a staple of native North Americans. These seeds and those of some Old World groups of pines are now a gourmet food. Rosin and turpentine are extracted from various species of pines.

Discussion: Pinaceae are the largest and both economically and ecologically the most important family of conifers. Species in the three largest genera—*Abies*, *Picea*, and *Pinus*—are the primary components of many forests in cooler and colder regions of the Northern Hemisphere. *Pinus* often dominates fire-maintained forests in warmer regions, such as the southeastern United States.

Numerous features—ovule inversion, prominent terminal seed wing, pattern of proembryogeny, protein-type sieve cell plastids, and the absence of biflavonoid compounds—establish the monophyly of Pinaceae. The family is not phylogenetically close to other extant conifer groups and is probably the sister group to the remaining conifers (see Figure 8.16).

Strongly congruent structural and seed protein immunological data divide Pinaceae into two subfamilies, Abietoideae and Pinoideae. Abietoideae include *Abies*, *Cedrus*, *Keteleeria*, *Pseudolarix*, and *Tsuga*, while *Cathaya*, *Larix*, *Picea*, *Pinus*, and *Pseudotsuga* make up Pinoideae. Phylogenetic analysis of three genes—*matK* from the chloroplast, mitochondrial *nad5*, and nuclear *4CL*—agree with this fundamental partition of the family except that *Cedrus* is sister to the remainder of the family. The Pinoideae are supported by several synapomorphies (absence of resin canals in the seed coat, absence of a narrowed, pedicillate base of the cone scales, and presence of two resin canals in the vascular cylinder of the young taproot) and contain two clades, *Pseudotsuga* plus *Larix*, and *Cathaya*, *Picea*, plus *Pinus*. Abietoideae in the molecular phylogeny has two sets of sister taxa, *Abies* plus *Keteleeria* and *Pseudolarix* plus *Tsuga*. The monotypic genus *Nothotsuga* is very closely related to *Tsuga* and is best treated as a species of *Tsuga*.

Pinus is highly distinct in its leaves, which are clustered in groups of usually two to five, and its cone scales, which are apically thickened and often armed with a prickle. This genus also has the longest fossil record of extant Pinaceae, extending back to the Jurassic or early Cretaceous. By the late Cretaceous, two monophyletic subgenera—*Pinus* and *Strobus*—had differentiated.

Members of subgenus *Pinus* are called hard pines because their wood is harder than that of subgenus *Strobus* species, which are commonly called soft pines. The two subgenera differ in other wood anatomical features and in the number of vascular bundles in each leaf, two in hard pines and one in soft pines. The leaf clusters are surrounded at the base by a sheath, which is persistent in hard pines and deciduous in soft pines. About 65 *Pinus* species are native to North America, with high concentrations of species in Mexico, California, and the southeastern United States.

Pinaceae include the longest-lived trees: intermountain bristlecone pine (*Pinus longaeva*), an alpine species of the southwestern United States, lives over 5000 years. While not matching the Cupressaceae in height and massiveness, the pine family does have some huge trees. Douglas fir (*Pseudotsuga menziesii*), for instance, grows to over 80 m in height, and many pines and spruces exceed 60 m.

References: Farjon 1990; Farjon and Styles 1997; Page 1990c; Price 1989; Richardson 1998; Stefanovic et al. 1998; Thieret 1993a; Vining 1999. Wang et al. 2000.

Cupressaceae Bartlett
(Cypress or Redwood Family)

Trees or shrubs; wood and foliage often aromatic. Bark of trunks often fibrous, shredding in long strings on mature trees or forming blocks. Leaves persistent (deciduous in three genera), simple, spiral or basally twisted to appear 2-ranked, opposite, or whorled, *scale-like, tightly appressed and as short as 1 mm to linear and up to about 3 cm long*, with resin canals, shed with the lateral branches; adult leaves appressed or spreading, sometimes spreading and linear on leading branches and appressed and scale-like on lateral branches; scale-like leaves often dimorphic, the lateral leaves keeled and folded around the branch and the leaves on the top and bottom of the branch flat. *Monoecious* (dioecious in *Juniperus*). Microsporangiate strobili with spirally arranged or opposite microsporophylls; **microsporangia 2–10 on the abaxial microsporophyll surface**; pollen nonsaccate, **without prothallial cells.** Cone maturing in 1–3 years; **scales peltate or basally attached and flattened,** juicy in *Juniperus*, fused to bracts, persistent (deciduous in *Taxodium*); **ovules 1–20, on adaxial scale surface,** erect (micropyle facing away from the cone axis; in some the ovules may eventually be inverted); archegonia quite variable in number per ovule, clustered. Seeds with 2 (3) short lateral wings (wings absent in some genera); embryo straight, cotyledons 2–15 (Figure 8.18).

Distribution and ecology: This is a cosmopolitan family of warm to cold temperate climates. About three-quarters of the species occur in the Northern Hemisphere. About 16 genera contain only one species, and many of these have narrow distributions. Members of this family

Figure 8.18   Cupressaceae. (A–J) *Juniperus virginiana*: (A) branchlets with only scale leaves, bearing mature ovulate cones (× 0.9); (B) branchlet with scale and needle leaves (× 0.9); (C) detail of branchlet with needle leaves, showing decurrent leaf bases (× 6.2); (D) microsporangiate strobilus before shedding of pollen, subtended by numerous scale leaves (× 6.2); (E) microsporophyll (abaxial view), showing dehisced sporangia (× 12); (F) branchlet with ovulate cone near time of pollination (× 9); (G) cone scale (adaxial view) with 2 erect ovules near time of pollination (× 12); (H) mature ovulate cone with fused cone scales (× 3.7); (I) cross-section of mature cone, only two seeds maturing (note resin vesicles outside seeds) (× 3.7); (J) seed, showing pits and ridges (× 6.2). (K–Q) *J. communis*: (K) branch, showing ternate leaves and axillary ovulate cones (× 0.9); (L) details of abscised portion of leaf in adaxial view, showing broad, white stomatal band (× 6.2); (M) microsporangiate strobilus after shedding of pollen (× 6.2); (N) microsporophyll, abaxial view; (O) axillary shoot with young ovulate cones at apex, showing three ovules near time of pollination (× 12); (P) portion of branchlet with mature ovulate cone; note remnant leaf bases fused to larger stem (× 3.7); (Q) apical views of ovulate cone, showing suture lines between three fused cone scales (× 3.7). (From Hart and Price 1990, *J. Arnold Arbor.* 71: pp. 275–322.)

grow in diverse habitats, from wetlands to dry soils, and from sea level to high elevations in mountainous regions. The two species of *Taxodium* in the southeastern United States often grow in standing water.

Genera/species: 29–32 / 110–130. Major genera: *Juniperus* (ca. 68 spp.), *Callitris* (15), *Cupressus* (14), *Chamaecyparis* (7), *Thuja* (5), *Taxodium* (3), *Sequoia* (1), and *Sequoiadendron* (1).

Economic plants and products: The family produces highly valuable wood. *Cryptomeria, Chamaecyparis, Juniperus, Sequoia, Taxodium, Thuja,* and several other genera are suited for house construction, siding, decking, caskets, shingles, boat construction, paneling, wooden pencils, and many other purposes. Many woods from this family are naturally fragrant and have been used as a natural moth-proofing for closets and chests and in the manufacture of perfumes. *Juniperus communis* cones are used to flavor gin. *Juniperus* pollen contains one of the most potent airborne allergens, and the tremendous amount of pollen produced by *Juniperus* species is highly correlated with nasal, sinus, and pulmonary allergies in humans and domestic animals. *Chamaecyparis, Cupressus, Juniperus, Platycladus, Thuja,* and other genera are grown extensively as ornamentals.

Discussion: This family was long split into Cupressaceae s.s. and Taxodiaceae on the basis of differences in the leaves. Cupressaceae s.s. leaves are either opposite and scale-like or whorled and linear, whereas those of Taxodiaceae are mostly spiral and linear. Leaves of *Metasequoia* (Taxodiaceae), however, are opposite, and those of *Athrotaxis* (Taxodiaceae) may be scale-like. There are numerous similarities (and potential synapomorphies) uniting these families: fusion of cone scale and bract; lateral wings on the seeds derived from the seed coat; microsporangia two or more per microsporophyll; more than two seeds per cone scale; shedding of small branches; clustered archegonia; wingless pollen grains that lack prothallial cells; peltate cone scales in many genera, and DNA sequence characters (Brunsfeld et al. 1994; Eckenwalder 1976; Hart 1987; Stefanovic et al. 1998; Tsumura et al. 1995; Watson and Eckenwalder 1993). An immunological study has shown high similarity among genera of Cupressaceae s.s. and Taxodiaceae, and much greater distance to taxa outside this group. Results of phylogenetic analysis of *rbcL* sequences are fully consistent with the immunological study. Finally, Cupressaceae s.s. are monophyletic and probably arose out of a paraphyletic assemblage, the "Taxodiaceae." Hence the evidence comes down decisively on the side of merging the two families.

Both immunology and DNA sequences from five chloroplast genes or regions identify similar groups of genera within the Cupressaceae s.l. Cupressaceae s.s. divide into two well-supported clades, the cupressoid clade with all Northern Hemisphere groups and the callitroid clade with all taxa from the Southern Hemisphere (see Figure 8.16). Cupressaceae s.s. are the sister group of the well supported taxodioid clade of three genera: *Taxodium, Glyptostrobus,* and *Cryptomeria. Taxodium* grows in the eastern United States and Mexico and, like its sister genus *Glyptostrobus* (southern and central China), is deciduous. *Cryptomeria* is wide-ranging in China and Japan. Another well supported trio of genera that diverged early in the evolution of Cupressaceae is the sequoioid clade, with *Metasequoia, Sequoia,* and *Sequoiadendron. Metasequoia* was widely distributed and one of the most common genera of Cupressaceae in the Northern Hemisphere from the late Cretaceous to the Miocene. Its native range is now restricted to an isolated region of west central China, and it was known outside its native range only as a fossil until 1944. Its deciduous habit must have evolved in parallel to that of *Taxodium* and *Glyptostrobus. Sequoia* and *Sequoiadendron,* like *Metasequoia* and several other genera, contain

just one species and are very geographically limited. *Sequoia* is restricted to coastal regions of northern California and southern Oregon and *Sequoiadendron* to the mountains of central California. *Cunninghamia,* a genus of about three species of southeast Asia, is the sister to the remainder of the family.

Juniperus is a member of the cupressoid clade and the second largest genus of conifers after *Pinus.* Junipers are mostly confined to the Northern Hemisphere, with centers of diversity in the deserts of Mexico and the southwestern United States, the Mediterranean, and central Asia-China. *Juniperus* species range from sea level to above treeline and from deserts to marches. Some species in this genus are weedy and have invaded millions of acres of rangeland and farmland.

Chamaecyparis nootkatensis, the Nootka cypress (also called the Alaska cedar and yellow cedar), has been difficult to place taxonomically because of its distinctive morphology. This important timber species of northwestern North America had been placed in *Cupressus,* and it does hybridize in cultivation with two species of *Cupressus,* an observation that has led some workers to unite these two genera. Molecular data find a better solution: return *Chamaecyparis nootkatensis* to *Cupressus.*

Sciadopitys, which is commonly called umbrella pine, was traditionally placed in "Taxodiaceae." The leaves in this genus are unusual in that they are fused in pairs so that what is actually a shoot appears to be a leaf. Numerous morphological, molecular, and other differences argue for separate family status (Stefanovic et al. 1998), i.e., Sciadopityaceae.

Cupressaceae include the tallest (*Sequoia sempervirens,* redwood, almost 112 m tall and 6.7 m in diameter) and most massive (*Sequoiadendron giganteum,* giant sequoia, 106 m tall and 11.4 m in diameter) plants on earth. Some species live 2000–3500 years or more.

The juicy cones of *Juniperus* are consumed by birds and small mammals, and long-distance dispersal by birds is suspected to have transported the genus to Atlantic islands such as the Azores, Bermuda, and the Canary Islands.

References: Adams 1993; Brunsfeld et al. 1994; Eckenwalder 1976; Gadek et al. 2000; Hart 1987; Kusumi et al. 2000; Page 1990a; Price and Lowenstein 1989; Stefanovic et al. 1998; Tsumura et al. 1995; Watson and Eckenwalder 1993.

Podocarpaceae Endlicher
(Podocarp Family)

Shrubs or trees to 60 m tall, slightly resinous. Leaves simple, entire, varying greatly in shape (broadly linear and up to 30 cm long and 5 cm wide to scale-like), persistent, alternate. *Dioecious* (rarely monoecious). Microsporangiate strobili cylindrical, with numerous spirally arranged microsporophylls each with 2 microsporangia;

pollen with 2 (0 or 3) saccae. Cones with 1 to many ovulate scales, each with 1 ovule and more or less reduced and fused to ovule, modified into a juicy structure (**epimatium**), and therefore *drupelike*, rarely resembling a cone. Cotyledons 2.

Distribution and ecology: Podocarpaceae are tropical and subtropical (less often cool temperate), especially in the Southern Hemisphere in the Old World. The family extends northward to Japan, Central America, and the Caribbean. Podocarps grow primarily in mesic forests.

Genera/species: 17/170 or more. ***Major genera:*** *Podocarpus* and *Dacrydium;* (the number of species recognized in these genus, neither of which is monophyletic, depends upon how they are subdivided).

Economic plants and products: *Dacrydium, Podocarpus,* and other members of the family have valuable timber. *Podocarpus macrophyllus* is widely planted as an ornamental in mild climates.

Discussion: Podocarpaceae are a distinct, Southern Hemisphere family. 28S rRNA gene sequences strongly support a sister-group relationship of Podocarpaceae and Araucariaceae (see Figure 8.16), sharing the following synapomorphies: one ovule per ovulate scale, ovulate scale closely associated with the seed, and possibly bract fused to scale. All but two of genera of Podocarpaceae bear an epimatium, which is generally interpreted as a modified cone scale that partially folds around the ovule and is juicy at maturity. Podocarps also have an unusual (possibly synapomorphic) binucleate cellular stage early in embryogeny, and are unusual among conifer families in their diversity of cone structure and chromosome number. The family may have been long isolated from other conifers in the Southern Hemisphere. Morphology and 18S rDNA sequence data show that neither *Dacrydium* nor *Podocarpus* is monophyletic, and some workers recognize numerous segregate genera.

References: Axsmith et al. 1998; Kelch 1997, 1998; Page 1990d; Stefanovic et al. 1998; Tomlinson 1992.

Araucariaceae Henkel et Hochstetter
(Norfolk Island Pine Family)

Long-lived trees to 65 m tall and 6 m in diameter at the base, highly resinous, usually very symmetrical and conical in growth form. Leaves simple, entire, varying in shape (awl-like, scale-like, linear, oblong, or elliptic) sometimes on the same individual, persistent, sharp-pointed in some species of *Araucaria*, spiral or opposite. Dioecious or monoecious. Microsporangiate strobili cylindrical, with numerous spirally arranged microsporophylls each with 4–20 microsporangia; pollen without saccae, the **exine pitted**. *Cones solitary, more or less erect, heavy,* maturing in 2–3

years and eventually disintegrating on the tree; *ovulate scales each with 1 ovule*, numerous, spirally arranged, flattened, linear to peltate, the bract more or less longer than and fused to the scale; *seeds large*, with or without marginal wings,. Cotyledons 2, sometimes deeply divided and appearing like 4 (Figure 8.19).

Distribution and ecology: Araucariaceae are nearly restricted to the Southern Hemisphere, ranging from southeast Asia to Australia, New Zealand, and southern South America. Members of this family usually grow in tropical and subtropical rainforests as well as more temperate areas. The family is most diverse in New Caledonia, where 5 species of *Agathis* and 13 species of *Araucaria* are endemic. The New Caledonia species of each genus form a monophyletic subgroup according to *rbcL* analyses, and there is little genetic diversification of these species within each genus. This suggests a relatively recent radiation of the genera on the unusual (ultramafic) soils of New Caledonia. In some species of *Araucaria*, sharp-pointed leaves, ability to regenerate branches, and protection of the growing apex by surrounding branches in young plants suggest adaptation against now-extinct herbivores.

Genera/species: 3/32. ***Major genera:*** *Agathis* (13 spp.) and *Araucaria* (18).

Economic plants and products: Both of the major genera produce valuable timber. Larger individuals, such as those of *Agathis australis* of New Zealand, that reach 65 m in height and over 6 m in diameter, contain large amounts of high-quality wood. The Norfolk Island pine (*Araucaria hetrerophylla*) and the monkey puzzle tree (*A. araucana*) of Chile are prized ornamentals, either as landscape plants or house plants (Norfolk Island pine).

Discussion: Araucariaceae, like Podocarpaceae, are a distinct, almost exclusively Southern Hemisphere family. The fossil record of the Araucariaceae, and *Araucaria*, extends back into the Jurassic.

Agathis and *Araucaria* differ from one another strongly in foliar and reproductive structures. The leaves of *Agathis* are opposite and broad, whereas *Araucaria* leaves are spiral and linear to broad. *Agathis* is monoecious with ovules free from the cone scale and winged seeds. *Araucaria* is dioecious, with ovules that are fused to the cone scale and usually wingless seeds. DNA sequences from *rbcL* also support monophyly of the two genera. The *rbcL* data also agree with the division of the genus into four sections based on nonmolecular characters such as the number of cotyledons, position of microsporangiate cones, and cellular features of the leaf epidermis.

Wollemia nobilis, the Wollemi pine, was found in 1994 in the Wolllemi National Park north of Sydney, Australia by National Park and Wildlife Service officer David Noble. Known to science before 1994 only as a fossil

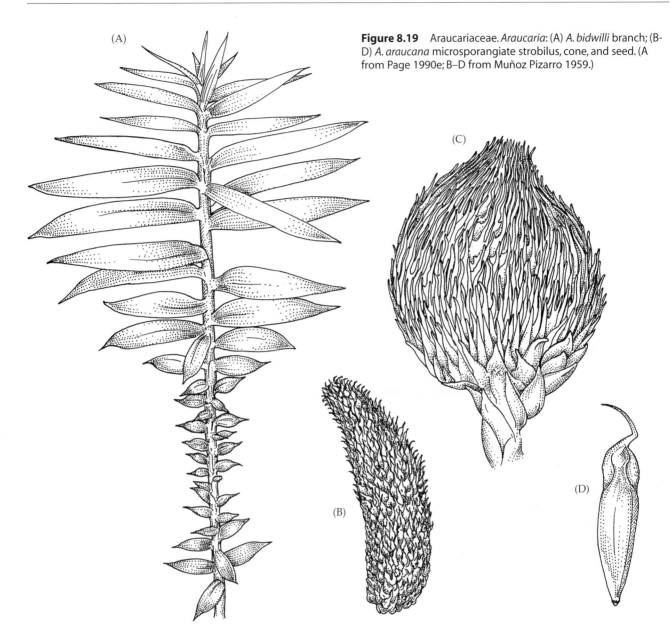

Figure 8.19 Araucariaceae. *Araucaria*: (A) *A. bidwilli* branch; (B–D) *A. araucana* microsporangiate strobilus, cone, and seed. (A from Page 1990e; B–D from Muñoz Pizarro 1959.)

extending back to 150 MYA, this is one of the world's rarest trees, with only about 43 adults in two populations 1.5 km apart. The trees, some of which are 500 to 1000 years old, have an unusual bark, described as "bubbling chocolate." Phylogenetic analyses of *rbcL* sequences show that this genus is the sister to the remainder of the family.

One of the largest and longest-lived trees in this family is *Agathis australis*, commonly called Kauri. One particular individual in northern New Zealand was 51.5 m tall and 13.8 m in circumference and about 2000 years old at the beginning of 2001. The Maori name for this individual is Tane Mahuta, which translates as "god of the forest."

References: Gilmore and Hill 1997; Page 1990e; Setogouchi et al. 1998; Stefanovic et al. 1998; http://www.rbgsyd.gov.au/html/wollemi.html.

Taxaceae Gray
(Yew Family)

Small to moderately sized trees or shrubs, usually not resinous or only slightly resinous; fragrant or not. Wood without resin canals. Leaves simple, persistent for several years, shed singly, spiral (opposite in one species), often twisted so as to appear 2-ranked, *linear, flattened, entire, acute at apex*, with 0–1 resin canals. Dioecious (rarely monoecious). Microsporangiate strobili with 6–14 microsporophylls; microsporangia 2–9 per microsporophyll, radially arranged around the microsporophyll or limited to its abaxial surface; pollen nonsaccate. **Ovules solitary** *and cones lacking*; **seeds with a hard outer layer, associated with a fleshy, usually brightly colored aril**; cotyledons 2 (occasionally 1 or 3) (Figure 8.20).

Distribution and ecology: Mostly Northern Hemisphere, extending south to Guatemala and Java, with one

Figure 8.20 Taxaceae. *Taxus floridana:* (A) leafy shoot with microsporangiate strobili at time of pollen release (× 0.7); (B) detail of abaxial surface of leaf (× 3.3); (C–E) microsporangiate strobili before, during, and after shedding of pollen (× 6.5); (F) detail of microsporophyll (× 27); (G) shoot with arillate ovule (× 2). (From Price 1990, *J. Arnold Arbor.* 71: pp. 69–91.)

endemic genus of New Caledonia. Taxaceae tend to grow in damp valley bottom sites where leaf litter accumulates.

Genera/species: 5/20. ***Major genera:*** *Taxus* (10 spp.) and *Torreya* (4).

Economic plants and products: *Taxus* is widely grown as an ornamental in North America and Europe. *Torreya* is less important as an ornamental, but its wood, edible seed, and seed oil are valued in Asia. *Taxus* contains taxol, one of several highly poisonous alkaloids in the leaves, stems, and seeds. Taxol's potent antimitotic activity makes it of potential use as an anticancer chemotherapeutic compound.

Discussion: Taxaceae are unique among the conifers in that their solitary seed is not associated with cone scales. The aril is an outgrowth of the axis below the seed. Some systematists have removed Taxaceae from the conifers because of this lack of a cone, but embryology, wood anatomy, chemistry, and leaf and pollen morphology unquestionably tie this family to other conifers. The cone is thought to be lost, and the solitary seed with an aril is therefore derived feature.

Diverse data from *rbcL* DNA sequences, morphology, anatomy, and alkaloid chemistry divide the family into two clades, one including *Taxus, Austrotaxus,* and *Pseudotaxus,* and the other comprising *Torreya* and *Amentotaxus.* The family is apparently most closely related to the Cephalotaxaceae, an East Asian monogeneric family characterized by paired ovules along a cone axis in association with a small outgrowth considered to be a reduced cone scale. Usually only one or two seeds mature, and these develop a juicy outer layer that resembles, but is not homologous with, the aril of Taxaceae. The solitary, drupelike seeds of many Podocarpaceae also resemble the arillate seeds of Taxaceae, but DNA sequence data indicate this fleshiness arose more than once (Stefanovic et al. 1998).

References: Hils 1993; Page 1990f; Price 1990; Stefanovic et al. 1998.

Gnetales

The Gnetales are of particular interest in plant evolution because they show features of conifers—seeds not enclosed in an ovary—and of angiosperms—vessels in the wood, somewhat flowerlike structures, and double fertilization. Together with angiosperms, Gnetales are sometimes referred to as anthophytes because of the presence of flowerlike structures or flowers, compound strobili with at least the rudiments of both megasporangia and microsporangia. Recent molecular phylogenetic studies (Bowe et al. 2000; Chaw et al. 2000) do not support anthophytes as a clade and instead link Gnetales with conifers, but some doubt the molecular resolution of relationships among these plants and continue to support a link with angiosperms (Doyle 1998) or at least do not rule out this relationship (Rydin et al. 2002). Gnetales show several possible synapomorphies: presence of enveloping bracts around the ovules and microsporangia and a micropylar projection of the integument that produces a pollination droplet. The Gnetales consist of three genera—*Gnetum, Welwitschia,* and *Ephedra*—each of which is morphologically highly distinct. *Gnetum* (Gnetaceae) contains about 40 species of mostly tropical, dioecious lianas (less often trees or shrubs) with opposite, simple, broad leaves and seeds enclosed in a fleshy, brightly colored envelope. There is one species of *Welwitschia* (Welwitschiaceae), a bizarre plant of southwestern African deserts with a massive and short stem and two enormous, straplike leaves that live for the entire life of the individual (up to 2000 years). *Ephedra* (Ephedraceae) is treated below.

References: Kubitzki 1990; Price 1996; Rydin et al. 2002.

Ephedraceae Dumortier
(Mormon Tea or Joint Fir Family)

Mostly shrubs, less often clambering vines, and rarely small trees; often spreading by rhizomes. Wood with vessels. *Branches numerous, whorled or clustered, longitudinally grooved; usually green and photosynthetic.* Leaves opposite or whorled, **scale-like, fused basally into a sheath, often shed soon after developing**; resin canals absent. Mostly dioecious. Pollen strobili in whorls of 1–10, each consisting of 2–8 series of opposite or whorled bracts, the apical bracts each subtending a stalk with 2–10(–15) microsporangia. Pollen furrowed, not winged. Ovulate strobili of 2–10 series of opposite or whorled bracts, those toward the apex subtending a pair of fused bracts forming a casing around the single ovule. Seeds 1–2(–3) per strobilus, yellow to dark brown; cotyledons 2 (Figure 8.21).

Distribution and ecology: Temperate regions worldwide except Australia. Ephedraceae often grow in dry, sunny habitats, such as deserts and steppes, and can occur as high as 4000 m in the Andes and Himalayas.

Genera/species: 1(*Ephedra*)/ca. 50.

Economic plants and products: *Ephedra* has long been used for a variety of medicinal purposes, such as cough and circulatory weakness. Its primary use today is for the alkaloid ephedrine, which functions as a vessel constrictant.

Discussion: The photosynthetic stem and small leaves make these plants superficially resemble equisetophytes. Pollination is by wind, less often by insects that are attracted by nectar produced by the ovulate strobili. Dispersal is either by wind, promoted by keeled wings on the bracts of the seed strobilus, or by birds, which are attracted to the bright yellow, orange, or red, juicy, outer bracts.

References: Kubitzki 1990; Price 1996; Stevenson 1993.

Figure 8.21 Ephedraceae. *Ephedra distachya*: (A) staminate branch.; (B) staminate strobilus; (C) staminate reproductive structure; (D) ovulate strobilus; (E) ovulate reproductive structure. (From *Flora Ibérica* 1986.)

Literature Cited and Suggested Readings

Items marked with an asterisk are especially recommended to those readers who are interested in further information on the topics discussed in this chapter.

Adams, R. P. 1993. *Juniperus*. In *Flora of North America*, vol. 2, 414–420.

Axsmith, B. J., T. N. Taylor and E. L. Taylor. 1998. Anatomically preserved leaves of the conifer *Notophytum krauselii* (Podocarpaceae) from the Triassic of Antarctica. *Am. J. Bot.* 85: 704–713.

Alston, A. H. G. 1932. Selaginellaceae. In The pteridophyta of Madagascar by C. Christiansen. *Dansk. Bot. Ark.* 7: 193–200.

*Beck, C. B. (ed.). 1988. *Origin and evolution of gymnosperms*. Columbia University Press, New York.

Billington, C. 1952. Ferns of Madagascar. *Bull. Cranbrook Inst. Sci.* 32: 114.

Bogler, D. J., A. Marraccini, and J. Francisco-Ortega. In press. Molecular systematic studies of cycads: evidence from trnL intron and ITS2 rDNA sequences. *Mem. New York Bot. Gard.*

Bowe, L. M. and C. W. dePamphilis. 1997. Conflict and congruence among three plant genetic compartments: Phylogenetic analysis of seed plants from gene sequences of *cox1*, *rbcL*, and 18S rDNA. *Am. J. Bot.* 84: 178–179.

Bowe, L. M., G. Coat and C. W. dePamphilis. 2000. Phylogeny of seed plants based on all three genomic compartments: Extant gymnosperms are monophyletic and Gnetales' closest relatives are conifers. *Proc. Nat. Acad. Sci. USA.* 97: 4092-4097.

Brownlie, G. 1977. *The pteridophyte flora of Fiji*. J. Cramer, Vaduz.

Brunsfeld, S. J., P. S. Soltis, D. E. Soltis, P. A. Gadek, C. J. Quinn, D. D. Strenge and T. A. Ranker. 1994. Phylogenetic relationships among the genera of Taxodiaceae and Cupressaceae: Evidence from *rbcL* sequences. *Syst. Bot.* 19: 253–262.

Chaw, S.-M., C. L. Parkinson, Y. Cheng, T. M. Vincent and J. D. Palmer. 2000. Seed plant phylogeny inferred from all three genomes: Monophyly of extant gymnosperms and origin of Gnetales from conifers. *Proc. Nat. Acad. Sci. USA.* 97: 4086-4091.

Conant, D. S., L. A. Raubeson, D. K. Attwood and D. B. Stein. 1995. The relationships of Papuasian Cyatheaceae to New World tree ferns. *Am. Fern J.* 85: 328–340.

Conant, D. S, and D. B. Stein. 2001. Phylogenetic and geographic relationships of the tree ferns (Cyatheaceae) on Mount Kinabalu. *Sabah Parks Nature Journal* 4: 25-43.

Crane, P. R. 1988. Major clades and relationships in the higher gymnosperms. In *Origins and evolution of gymnosperms*, C. B. Beck. (ed.), 218–272. Columbia University Press, New York.

Des Marais, D. L., K. M. Pryer, and A. R. Smith. 2001. Phylogeny, character evolution, and biogeography of extant horsetails (*Equisetum*). Abstract # 338 at 2001 annual meetings of the Botanical Society of America, the American Society of Plant Taxonomists, and the International Organization of Plant Systematists. http://www.botany2001.org/section11/abstracts/15.shtml.

*DiMichele, W. A. and J. E. Skog. 1992. The Lycopsida: A symposium. *Ann. Missouri Bot. Gard.* 79: 447–449.

Donoghue, M. J. 1994. Progress and prospects in reconstructing plant phylogeny. *Ann. Missouri Bot. Gard.* 81: 405–418.

Doyle, J. A. 1998. Phylogeny of vascular plants. *Ann. Rev. Ecol. Syst.* 29: 567-599.

Doyle, J. A., M. J. Donoghue and E. A. Zimmer. 1994. Integration of morphological and ribosomal RNA data on the origin of angiosperms. *Ann. Missouri Bot. Gard.* 81: 419–450.

Eckenwalder, J. E. 1976. Reevaluation of Cupressaceae and Taxodiaceae: A proposed merger. *Madrono* 23: 237–256.

Farjon, A. 1990. *Pinaceae: Drawings and descriptions of the genera Abies, Cedrus, Pseudolarix, Keteleeria, Nothotsuga, Tsuga, Cathaya, Pseudotsuga, Larix, and Picea.* Koeltz Scientific Books, Konigstein, Germany.

Farjon, A. and B. T. Styles. 1997. *Pinus* (Pinaceae). Monograph 75, Flora Neotropica. The New York Botanical Garden, New York.

Flora Ibérica. 1986. Jardín Botánico Real, Madrid.

*†Flora of North America Editorial Committee. 1993. *Flora of North America*, vol. 2, *Pteridophytes and gymnosperms*. Oxford University Press, New York.

Flora Tsinlingensis. 1974. Pteriodophyta. Academia Sinica, Beijing.

Friis, E. M., W. G. Falconer and P. R. Crane. 1987. *The origins of angiosperms and their biological consequences*. Cambridge University Press, Cambridge.

Gadek, P. A., D. L. Alpers, M. M. Helsewood, and C. J. Quinn. 2000. Relationships within Cupressaceae senso lato: A combined morphological and molecular approach. *Am. J. Bot.* 87: 1044-1057.

*Gifford, E. M. and A. S. Foster. 1988. *Morphology and evolution of vascular plants*, 3rd ed. W. H. Freeman, New York.

Gilmore, S. and K. D. Hill. 1977. Relationships of Wollemi Pine (*Wollemia nobilis*) and a molecualr phylogeny of the Araucariaceae. *Telopea* 7: 275-291.

Goremykin, V., V. Bobrova, J. Pahnke, A. Toitsky, A. Antonov and W. Martin. 1996. Noncoding sequences from the slowly evolving chloroplast inverted repeat in addition to *rbcL* data do not support gnetalean affinities of angiosperms. *Mol. Biol. Evol.* 13: 383–396.

Haider, K. 1954. Zur Morphologie und Physiologie der Sporangien leptosporangiate Farne. *Planta* 44: 370–411.

Hart, J. A. 1987. A cladistic analysis of conifers: Preliminary results. *J. Arnold Arbor.* 68: 269–307.

Hart, J. A. and R. A. Price. 1990. The genera of Cupressaceae (including Taxodiaceae) in the southeastern United States. *J. Arnold Arbor.* 71: 275–322.

*Hasebe, M., T. Omori, N. Nakazawa, T. Sano, M. Kato and K. Iwatsuki. 1994. *rbcL* gene sequences provide evidence for the evolutionary lineages of leptosporangiate ferns. *Proc. Nat. Acad. Sci., USA* 91: 7530–7534.

Hasebe, M., P. G. Wolf, K. M. Pryer, K. Ueda, M. Ito, R. Sano, G. J. Gastony, J. Yokoyama, J. R. Manhart, N. Murakami, E. H. Crane, C. H. Haufler and W. D. Hauk. 1995. Fern phylogeny based on *rbcL* nucleotide sequences. *Am. Fern J.* 85: 134.

Hauke, R. L. 1990. Equisetaceae. In *Families and genera of vascular plants*, vol. 1, K. U. Kramer and P. S. Green (volume eds.), 46–48.

Hauke, R. L. 1993. Equisetaceae. In *Flora of North America*, vol. 2, 76–84.

Hewitson, W. 1962. Comparative morphology of Osmundaceae. *Ann. Missouri Bot. Gard.* 49: 57–93.

Hils, M. H. 1993. Taxaceae. In *Flora of North America*, vol. 2, 423–427.

Holttum, R. E. 1963. *Flora Malesiana II*, vol. 1, Pteridophyta: Morphology, key, Gleicheniaceae, Schizaeaceae. 1–61.

Jermy, A. C. 1990a. Isoetaceae. In *Families and genera of vascular plants*, vol. 1, K. U. Kramer and P. S. Green (volume eds.), 26–31.

Jermy, A. C. 1990b. Selaginellaceae. In *Families and genera of vascular plants*, vol. 1, K. U. Kramer and P. S. Green (volume eds.), 39–45.

Johnson, D. M. 1993a. Marsileaceae. In *Flora of North America*, vol. 2, 331–335.

Johnson, D. M. 1993b. *Matteuccia; Onoclea*. In *Flora of North America*, vol. 2, 249–251.

Johnson, L. A. S. and K. L. Wilson. 1990. Cycadales. In *Families and genera of vascular plants*, vol. 1, K. U. Kramer and P. S. Green (volume eds.), 363–377. Springer-Verlag, Berlin.

Jones, D. L. 1993. *Cycads of the world*. Smithsonian Institution Press, Washington, DC.

Kelch, D. G. 1997. The phylogeny of Podocarpaceae based on morphological evidence. *Syst. Bot.* 22: 113–131.

Kelch, D. G. 1998. Phylogeny of Podocarpaceae: Comparison of evidence from morphology and 18S rDNA. *Am. J. Bot.* 85: 975–985.

Kenrick, P. 2000. The relationships of vascular plants. *Phil. Trans. R. Soc. Lond. B.* 355: 847-855.

Kenrick, P. and P. R. Crane. 1997. *The origin and early diversification of land plants. A cladistic study*. Smithsonian Institution Press, Washington.

Knobloch, I. W. and D. S. Correll. 1962. *Ferns and fern allies of Chihuahua, Mexico*. Renner, TX.

Korall P. and P. Kenrick. 2002. Phylogenetic relationships in Selaginellaceae based on *rbcL* sequences. *Am. J. Bot.* 89: 506-517.

Kramer, K. U. 1990a. Cyatheaceae. In *Families and genera of vascular plants*, vol. 1, K. U. Kramer and P. S. Green (volume eds.), 69–74.

Kramer, K. U. 1990b. Marsileaceae. In *Families and genera of vascular plants*, vol. 1,

---

†Articles from this flora are cited throughout this bibliography. They are listed by author and cited as *Flora of North America*, vol. 2, with appropriate page ranges.

K. U. Kramer and P. S. Green (volume eds.), 180–183.

Kramer, K. U. 1990c. Osmundaceae. In *Families and genera of vascular plants*, vol. 1, K. U. Kramer and P. S. Green (volume eds.), 197–200.

Kramer, K. U. 1990d. Psilotaceae. In *Families and genera of vascular plants*, vol. 1, K. U. Kramer and P. S. Green (volume eds.), 22–25.

*§Kramer, K. U. and P. S. Green (volume eds.). 1990. *Pteridophytes and gymnosperms*. Vol. 1 in *Families and genera of vascular plants*, K. Kubitzki (ed.). Springer-Verlag, Berlin.

Kubitzki, K. 1990. Gnetatae. In *Families and genera of vascular plants*, vol. 1, K. U. Kramer and P. S. Green (volume eds.), 378–391.

Kusumi, J., Y. Tsumura, H. Yoshimaru, and H. Tachida. 2000. Phylogenetic relationships in Taxodiaceae sensu stricto based on *matK* gene, *chlL* gene, *trnL-trnF* IGS region, and *trnL* intron sequences. *Am. J. Bot.* 87: 1480-1488.

Landry, G. P. 1993. Zamiaceae. In *Flora of North America*, vol. 2, 347–349.

*Lellinger, D. B. 1985. *A field manual of the ferns & fern-allies of the United States & Canada*. Smithsonian Institution Press, Washington, DC.

Lupia, R., H. Schneider, G. M. Moeser, K. M. Pryer and P. R. Crane. 2000. Marsileaceae sporocarps and spores from the late Cretaceous of Georgia, U.S.A. *Int. J. Plantg Sci.* 161: 975-988.

Madalski, J. 1954. *Atlas flory Polskiej i ziem oscennych 1*. Panstowowe Wydawnictwo Naukowe, Warsaw.

*Manhart, J. R. 1995. Chloroplast 16S rDNA sequences and phylogenetic relationships of fern allies and ferns. *Am. Fern J.* 85: 182–192.

Mishler, B. D., L. A. Lewis, M. A. Buchheim, K. S. Renzaglia, D. J. Garbary, C. F. Delwiche, F. W. Zechman, T. S. Kantz and R. L. Chapman. 1994. Phylogenetic relationships of the "green algae" and "bryophytes." *Ann. Missouri Bot. Gard.* 81: 451–483.

Muñoz Pizarro, C. 1959. Sinopsis de la Floa Chilena. Universidad de Chile, Santiago.

Nickrent, D. L., C. L. Parkinson, J. D. Palmer, and R. J. Duff. 2000. Multigene phylogeny of land plants with special reference to bryophytes and the earliest land plants. *Mol. Biol. Evol.* 17: 1885-1895.

*Nimsch, H. 1995. *A reference guide to the gymnosperms of the world*. Koeltz Scientific Books, Champaign, IL.

Nixon, K. C., W. L. Crepet, D. Stevenson and E. M. Friis. 1994. A reevaluation of seed plant phylogeny. *Ann. Missouri Bot. Gard.* 81: 484–553.

Norstog, K. J. and P. K. S. Fawcett. 1989. Insect-cycad symbiosis and its relations to the pollination of *Zamia fufuracea* (Zamiaceae) by *Rhopalotria mollis* (Curculionidae). *Am. J. Bot.* 76: 1380–1394.

*Norstog, K. J. and T. J. Nichols. 1997. *The biology of cycads*. Cornell University Press, Ithaca, NY.

Øllgaard, B. 1990. Lycopodiaceae. In *Families and genera of vascular plants*, vol. 1, K. U. Kramer and P. S. Green (volume eds.), 31–39.

Øllgaard, B. 1992. Neotropical Lycopodiaceae: An overview. *Ann. Missouri Bot. Gard.* 79: 687–717.

Page, C. N. 1990a. Cupressaceae. In *Families and genera of vascular plants*, vol. 1, K. U. Kramer and P. S. Green (volume eds.), 302–318.

Page, C. N. 1990b. Ginkgoaceae. In *Families and genera of vascular plants*, vol. 1, K. U. Kramer and P. S. Green (volume eds.), 284–289.

Page, C. N. 1990c. Pinaceae. In *Families and genera of vascular plants*, vol. 1, K. U. Kramer and P. S. Green (volume eds.), 319–331.

Page, C. N. 1990d. Podocarpaceae. In *Families and genera of vascular plants*, vol. 1, K. U. Kramer and P. S. Green (volume eds.), 332–346.

Page, C. N. 1990e. Araucariaceae. In *Families and genera of vascular plants*, vol. 1, K. U. Kramer and P. S. Green (volume eds.), 294-299.

Page, C. N. 1990f. Taxaceae. In *Families and genera of vascular plants*, vol. 1, K. U. Kramer and P. S. Green (volume eds.), 348–353.

Page, C. N. 1990f. Taxodiaceae. In *Families and genera of vascular plants*, vol. 1, K. U. Kramer and P. S. Green (volume eds.), 353–361.

Paris, C. A. 1993. *Adiantum*. In *Flora of North America*, vol. 2, 125–130.

Pérez Arbeláez, E. 1928. Die natürlich Gruppe der Davalliaceen (Sm.). Kaulf. *Bot. Abh. Goebel* 14: 1–96.

Phipps, C. J., T. N. Taylor, E. L. Taylor, N. R. Cuneo, L. D. Boucher and X. Yao. 1998. *Osmunda* (Osmundaceae) from the Triassic of Antarctica: An example of evolutionary stasis. *Am. J. Bot.* 85: 888–895.

*Price, R. A. 1989. The genera of Pinaceae in the southeastern United States. *J. Arnold Arbor.* 70: 247–305.

*Price, R. A. 1990. The genera of Taxaceae in the southeastern United States. *J. Arnold Arbor.* 71: 69–91.

*Price, R. A. 1996. Systematics of Gnetales: A review of morphological and molecular evidence. *Intl. J. Plant Sci.* 157: S40–S49.

Price, R. A. and J. M. Lowenstein. 1989. An immunological comparison of the Sciadopytiaceae, Taxodiaceae, and Cupressaceae. *Syst. Bot.* 14: 141–149.

Price, R. A., J. Olsen-Stojkovich and J. M. Lowenstein. 1987. Relationships among the genera of Pinaceae: An immunological comparison. *Syst. Bot.* 12: 91–97.

Pryer, K. M. 1999. Phylogeny of marsileaceous ferns and relationships of the fossil *Hydropteris pinnata* reconsidered. *Int. J. Plant Sci.* 160:931-954.

*Pryer, K. M., A. R. Smith and J. E. Skog. 1995. Phylogenetic relationships of extant ferns based on evidence from morphology and *rbcL* sequences. *Am. Fern J.* 85: 203–282.

*Pryer, K. M., H. Schneider, A. R. Smith, R. Cranfill, P. G. Wolf, J. S. Hunt, and S. D. Sipes. 2001. Horsetails and ferns are a monophyletic group and the closest relatives of seed plants. *Nature* 409: 618-622.

Rai, H. S., H. O'Brien, R. G. Olmstead, and S. W. Graham. 2001. Phylogeny of cycads and their placement in the seed plants, as

inferred from a large chloroplast data set. Abstract # 536 at 2001 annual meetings of the Botanical Society of America, the American Society of Plant Taxonomists, and the International Organization of Plant Systematists. http://www.botany25001.org/section12/abstracts/15.shtml.

Raubeson, L. A. and R. K. Jansen. 1992. Chloroplast DNA evidence of the ancient evolutionary split in vascular land plants. *Science* 255: 1697-1699.

Richardson, D. M. (ed.). 1998. *Ecology and biogeography of Pinus*. Cambridge University Press, Cambridge.

Renzaglia, K. S. R. J. Duff, D. L. Nickrent, and D. J. Garbary. 2000. Vegetative and reproductive innovations of early land plants: Implications for a unified phylogeny. *Phil. Trans. R. Soc. Lond. B.* 355: 769-793.

Renzaglia, K. S., T. H. Johnson, H. D. Gates, and D. P. Whittier. 2001. Architecture of the sperm cell of *Psilotum. Am. J. Bot.* 88: 1151-1163.

Rothwell, G. W. 1999. Fossils and ferns in the resolution of land plant phylogeny. *Bot. Rev.* 65: 188-218.

Rydin, C., M. Kallersjo, & E. M. Friis. 2002. Seed plant relationships and the systematic position of Gnetales based on nuclear and chloroplast DNA: conflicting data, rooting problems, and the monophyly of conifers. *Int. J. Plant Sci.* 163: 197-214.

Setogouchi, H. T. A. Osawa, J.-C. Pintaud, T. Jaffre, and J.-M. Veillon. 1998. Phylogenetic relationships within Araucariaceae based on *rbcL* gene sequences. *Am. J. Bot.* 85: 1507-1516.

Singer, S. R. 1997. Plant life cycles and angiosperm development. In *Embryology: Constructing the organism*, S. F. Gilbert and A. M. Raunio (eds.), 493–514. Sinauer Associates, Sunderland, MA.

Singh, H. 1978. *Embryology of gymnosperms*. Gebruder Borntraeger, Berlin.

Smith, A. R. 1995. Non-molecular phylogenetic hypotheses for ferns. *Am. Fern J.* 85: 104–122.

Smith, A. R. and R. B. Cranfill. 2002. Intrafamilial relationships of the thelypteroid ferns (Thelypteridaceae). Am. Fern J. In press.

Soltis, D. E. and P. S. Soltis. 1988. Are lycopods with high chromosome numbers ancient polyploids? *Am. J. Bot.* 75: 238–247.

*Sporne, K. R. 1970. *The morphology of pteridophytes*, 3rd ed. Hutchinson University Library, London.

*Sporne, K. R. 1974. *The morphology of gymnosperms*, 2nd ed. Hutchinson University Library, London.

Stefanovic, S., M. Jager, J. Deutsch, J. Broutin and M. Masselot. 1998. Phylogenetic relationships of conifers inferred from partial 28S rRNA gene sequences. *Am. J. Bot.* 85: 688–697.

Stein, D. B., D. S. Conant, M. E. Ahearn, E. T. Jordan, S. A. Kirch, M. Hasebe, K. Iwatsuki, M. K. Tan and J. A. Thomson. 1992. Structural rearrangements of the chloroplast genome provide an important phylogenetic link in ferns. *Proc. Nat. Acad. Sci. USA* 89: 1856–1860.

*Stevenson, D. W. (ed.). 1990. *The biology, structure, and systematics of the Cycadales*. Memoirs of the New York Botanical Garden, Bronx, NY.

---

§Articles from this reference are cited throughout this bibliography. They are listed by author and cited as *Families and genera of vascular plants*, vol. 1, K. U. Kramer and P. S. Green (volume eds.), with appropriate page ranges.

Stevenson, D. W. 1991. The Zamiaceae in the southeastern United States. *J. Arnold Arbor.*, Suppl. Series 1: 367–384.

Stevenson, D. W. 1992. A formal classification of the extant cycads. *Brittonia* 44: 220–223.

Stevenson, D. W. 1993. Ephedraceae. In *Flora of North America*, vol. 2, 428–434.

Stewart, W. N. and G. W. Rothwell. 1993. *Paleobotany and the evolution of plants*, 2nd ed. Cambridge University Press, Cambridge.

Taylor, T. N. and E. L. Taylor. 1993. *The biology and evolution of fossil plants*. Prentice Hall, Englewood Cliffs, NJ.

Taylor, W. C. 1984. *Arkansas ferns and fern allies*. Milwaukee.

Taylor, W. C., N. T. Luebke, D. M. Britton, R. J. Hickey and D. F. Brunton. 1993. Isoetaceae. In *Flora of North America*, vol. 2, 64–75.

Thieret, J. W. 1993a. Pinaceae. In *Flora of North America*, vol. 2, 352–398.

Thieret, J. W. 1993b. Psilotaceae. In *Flora of North America*, vol. 2, 16–17.

Tomlinson, P. B. 1992. Apsects of cone morphology and development in Podocarpaceae (Coniferales). *Int. J. Pl. Sci.* 153: 572-588.

Tryon, A. F. and R. C. Moran. 1997. *The ferns and fern allies of New England*. Massachusetts Audubon Society, Lincoln, MA.

*Tryon, R. M. and A. F. Tryon. 1982. *Ferns and allied plants*. Springer-Verlag, New York.

Tsumura, Y., K. Yoshimura, N. Tomaru, and K. Ohba. 1995. Molecular phylogeny of conifers using RFLP analysis of PCR-amplified specific chloroplast genes. *Theor. Appl. Genet.* 94: 1222-1236.

Valdespino, I. A. 1993. Selaginellaceae. In *Flora of North America*, vol. 2, 39–63.

Vangerow, S., T. Teerkorn, and V. Knoop. 1999. Phylogenetic information in the mitochondrial nad5 gene of pteridophytes: RNA editing and intron sequences. *Plant. Biol.* 1:235-243.

Vining, T. F. 1999. Molecular phylogenetics of Pinaceae. Ph.D. dissertation, University of Maine, Orono, Maine.

Wagner, W. H., Jr. 1990. Ophioglossaceae. In *Families and genera of vascular plants*, vol. 1, K. U. Kramer and P. S. Green (volume eds.), 193–197.

Wagner, W. H., Jr. and J. M. Beitel. 1992. Generic classification of modern North American Lycopodiaceae. *Ann. Missouri Bot. Gard.* 79: 676–686.

Wagner, W. H., Jr. and J. M. Beitel. 1993. Lycopodiaceae. In *Flora of North America*, vol. 2, 18–37.

Wagner, W. H., Jr. and A. R. Smith. 1993. Pteridophytes. In *Flora of North America*, vol. 2, 247–266.

Wagner, W. H., Jr. and F. S. Wagner 1993. Ophioglossaceae. In *Flora of North America*, vol. 2, 85-109.

Wang, X.-Q., D. C. Tank, and T. Sang. 2000. Phylogeny and divergence times in Pinaceae: Evidence from three genomes. *Mol. Phyl. Evol.* 17: 773-781.

Watson, F. D. and J. E. Eckenwalder. 1993. Cupressaceae. In *Flora of North America*, vol. 1, 399–422.

Webster, T. R. 1992. Developmental problems in *Selaginella* (Selaginellaceae) in an evolutionary context. *Ann. Missouri Bot. Gard.* 79: 632–647.

Whetstone, R. D. 1993. Ginkgoaceae. In *Flora of North America*, vol. 2, 350–351.

Whetstone, R. D. and T. A. Atkinson. 1993. Osmundaceae. In *Flora of North America*, vol. 2, 107–109.

Wikström, N. 1999. *Evolution of Lycopodiaceae (Lycopsida) – Relationships and patterns of diversification*. Akademitryck, Edsbruk, Sweden.

Wikström, N. and P. Kenrick. 1997. Phylogeny of Lycopodiaceae (Lycopsida) and the relationships of *Phylloglossum drummondii* Kunze based on *rbcL* sequences. *Int. J. Plant Sci.* 156: 862-871.

Wikström, N. and P. Kenrick. 2000. Relationships of *Lycopodium* and *Lycopodiella* based on combined plastid *rbcL* gene and *trnL* intron sequence data. *Syst. Bot.* 25: 495-510.

Wolf, P. G. 1997. Evaluation of *atpB* nucleotide sequences for phylogenetic studies of ferns and other pteridophytes. *Am. J. Bot.* 84: 1429–1440.

Wolf, P. G., K. M. Pryer, A. R. Smith, and M. Hasebe. 1998. Phylogenetic studies of extant pteridophytes. In *Molecular systematics of plants of Plants II: DNA sequencing*, D. E. Soltis, P. S. Soltis, and J. J. Doyle (eds.), 541-556. Kluwer Academic Publishers, Boston.

Wolf, P. G., S. D. Sipes, M. R. White, M. L. Martines, K. M. Pryer, A. R. Smith and K. Ueda. 1999. Phylogenetic relationships of the enigmatic fern families Hymenophyllopsidaceae and Lophosoriaceae: evidence from *rbcL* nucleotide sequences. *Pl. Syst. Evol.* 219: 263-270.

9 *Phylogenetic Relationships of Angiosperms*

The **angiosperms,** or flowering plants, are the dominant group of land plants. The monophyly of this group is strongly supported, as discussed in Chapter 7, and angiosperms are possibly sister to a group including all or many other extant seed plants (Goremykin et al. 1996; Chaw et al. 1997; Bowe et al. 2000), although some phylogenetic analyses support a close relationship with gnetophytes (Crane 1985; Donoghue and Doyle 1989; Chase et al. 1993; Doyle et al. 1994; Doyle 1996, 1998). The angiosperms have a long fossil record, going back to the earliest Cretaceous (ca. 135 million years ago); their fossils increase in abundance as one moves through the Cretaceous (Beck 1976; Crane et al. 1995). The group possibly originated during the Jurassic, more than 140 million years ago, but the lineage that gave rise to angiosperms may have been separate from other extant seed plants for twice that long.

Angiosperms have traditionally been divided into **monocots** (plants with a single cotyledon, radicle aborting early in growth with the root system adventitious, stems with scattered vascular bundles and usually lacking secondary growth, leaves with parallel venation, flowers 3-merous, and pollen grains usually monosulcate) and **dicots** (plants with two cotyledons, radicle not aborting and giving rise to mature root system, stems with vascular bundles in a ring and often showing secondary growth, leaves with a network of veins forming a pinnate to palmate pattern, flowers 4- or 5-merous, and pollen grains predominantly tricolpate or modifications thereof). But cladistic analyses based on nuclear, chloroplast, and mitochondrial DNA sequences and morphology do not support this dichotomy (Chase et al. 1993; Doyle 1996, 1998; Doyle et al. 1994; Donoghue and Doyle 1989; Graham and Olmstead 2000; Mathews and Donoghue 1999; Savolainen et al. 2000a; Soltis et al. 2000; Zimmer et al. 2000). In all published cladistic analyses the "dicots" form a paraphyletic complex, and features such as two cotyledons, a persistent radicle, stems with vascular bundles in a ring, secondary growth, and leaves with net venation are plesiomorphic within angiosperms; that is, these features evolved earlier in the phylogenetic history of tracheophytes.

In contrast, the monophyly of the monocots is supported by many synapomorphies, including leaves with parallel venation, embryos with a single cotyledon, sieve cell plastids with several cuneate protein crystals, stems with scattered vascular bundles, and an adventitious root system (although several of these characters are homoplasious; see page 22). The monophyly of monocots is also supported by 18S rDNA, *atpB*, *rbcL*, and *matK* nucleotide sequences (Chase et al. 1993, 1995a, 2000; Bharathan and Zimmer 1995; Källersjö et al. 1998; Soltis et al. 1997, 1998, 2000; Hilu et al. 2001).

Although the "dicots" apparently are not monophyletic, a large number of species traditionally considered within this group do constitute a well-supported clade: the **eudicots**, or **tricolpates** (Donoghue and Doyle 1989; Chase et al. 1993; Doyle et al. 1994; Soltis et al. 1997, 2000; Källersjö et al. 1998; Hoot et al. 1999; Savolainen et al. 2000a, b; Hilu et al. 2001). Synapomorphies of this monophyletic group include tricolpate pollen (or modifications of this basic pollen type), plus *rbcL*, *atpB*, *matK*, and 18S rDNA nucleotide sequences.

Phylogenetic relationships among the basal angiosperms—those that are neither monocots nor eudicots—are only now becoming clear. These families have often been considered to represent a paraphyletic, primitive angiosperm complex, and are treated as the superorder Annonanae by Thorne (1992) and as the subclass Magnoliidae by Cronquist (1968, 1981, 1988) and Takhtajan (1980, 1997). Most have ethereal oils in scattered spherical cells within the parenchymatous tissues and elongate vessel elements with slanted, scalariform perforation plates. Their flowers usually are radially symmetrical; the carpels and stamens are usually distinct and free, and the former have a poorly developed style and an elongate stigmatic region. The pollen is usually monosulcate, and the seeds have a tiny embryo with copious endosperm. Various families within this complex have been proposed as representing the "most primitive" extant angiosperm group, based either on subjective arguments (Cronquist 1968, 1981, 1988; Thorne 1974, 1992; Takhtajan 1969, 1980, 1997; Dahlgren 1977, 1983) or the results of phylogenetic analyses of morphological or single-gene character sets (see Donoghue and Doyle 1989; Zimmer et al. 1989; Loconte and Stevenson 1991; Hamby and Zimmer 1992; Chase et al. 1993; Doyle et al. 1994; Loconte and Stevenson 1991; Soltis et al. 1997; Doyle and Endress 2000). As can be seen by comparing Figures 9.1 through 9.4, hypothesized relationships among these families have varied dramatically, depending on the characters used in the analysis.

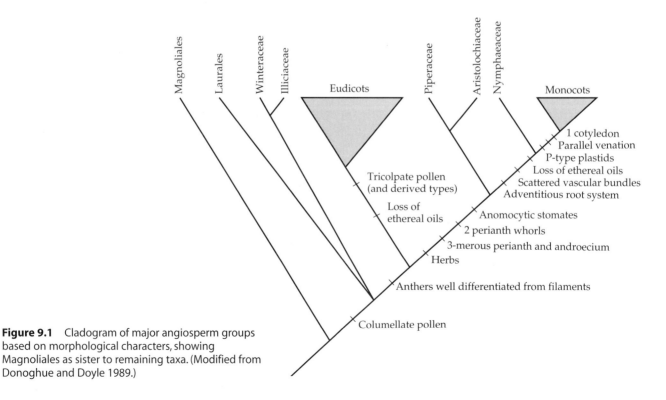

Figure 9.1 Cladogram of major angiosperm groups based on morphological characters, showing Magnoliales as sister to remaining taxa. (Modified from Donoghue and Doyle 1989.)

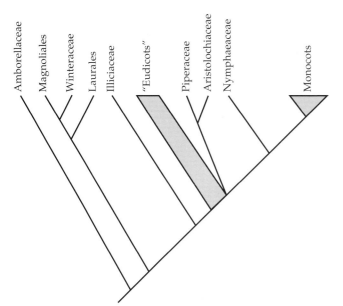

Figure 9.2 Cladogram of major angiosperm groups based on morphological characters, showing a nonmonophyletic eudicots. (Modified from Doyle and Endress 2000.)

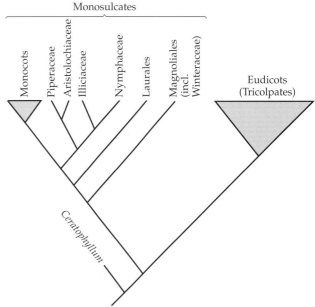

Figure 9.4 Cladogram of major angiosperm groups based on both morphology and nuclear ribosomal RNA. (Modified from Doyle et al. 1994.)

Fortunately, phylogenetic relationships among the nonmonocot, nontricolpate angiosperms have been clarified by numerous recent phylogenetic analyses that have been based on combined data sets. These analyses have used characters from two to several different genes (Graham and Olmstead 2000; Qiu et al. 2000; Savolainen et al. 2000a; Soltis et al. 1998, 2000), and sometimes also morphology (Doyle et al. 1994, Doyle 1998; Doyle and Endress 2000; Nandi et al. 1998), in order to clarify

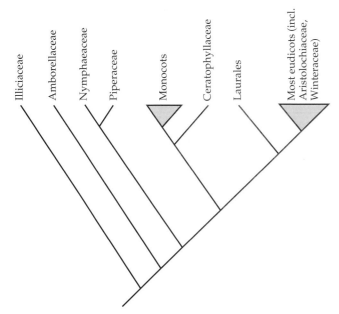

Figure 9.3 Cladogram of major angiosperm groups based on nuclear ribosomal DNA nucleotide sequences. (Modified from Soltis et al. 1997.)

groupings within these angiosperms. The cladograms in Figure 9.5 and 9.6 summarize the relationships among major clades as presented by Soltis et al. (1998, 2000), Graham and Olmstead (2000), Mathews and Donoghue (1999), and Qiu et al. (2000). At this point, most analyses support the position of Amborellaceae, Nymphaeaceae, and Illiciaceae (and relatives) as the three lineages that first diverged from the remaining angiosperm groups. These three "basal" lineages of angiosperms may be characterized by carpel margins usually sealed by a secretion, whereas the carpel margins of most other angiosperms are closed by postgenital fusion of the epidermal layers (Doyle 1998; Endress and Igersheim 2000).

These basal families have a distinctive type of carpel development in which a meristematic cross-zone develops between the arms of the U-shaped primordium and the carpel grows up like a tube; most other angiosperms have plicate carpels, i.e., the developing carpel is like a leaf folded down the middle with the ovules in the plicate zone (Doyle and Endress 2000). In addition, *Amborella* (the only genus in Amborellaceae) lacks vessels, and the genera of Nymphaeaceae either lack vessels or have vessels of an unusual, tracheid-like form (Schneider et al. 1995; Schneider and Carlquist 1996). Most other angiosperms have vessels, although these specialized water-conducting cells have been lost in a few groups (such as Winteraceae and some Chloranthaceae). Illiciaceae and relatives have ethereal oils in specialized spherical cells in the leaf mesophyll, so their leaves have pellucid dots; such dots are also present in the leaves of other families often regarded as primitive (e.g., Magnoliaceae, Annonaceae, Winteraceae, Lauraceae). But pellucid dots are absent from Nymphaea-

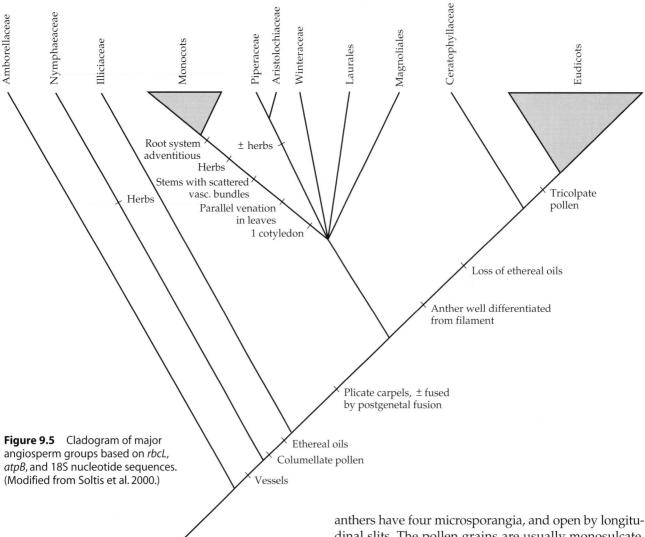

Figure 9.5 Cladogram of major angiosperm groups based on *rbcL*, *atpB*, and 18S nucleotide sequences. (Modified from Soltis et al. 2000.)

ceae and Amborellaceae; their lack, therefore, is likely to be the original condition within angiosperms.

It is probable that the Amborellaceae and Illiciaceae retain many plesiomorphic (ancestral) angiosperm characteristics, and many of these features have been retained by other nonmonocot, noneudicot families, such as Magnoliaceae, Annonaceae, Lauraceae, and Winteraceae. Possible ancestral vegetative characters include the perennial, evergreen, woody habit with primary vascular bundles in a ring; the production of vesselless wood; unilacunar nodes with two vascular traces; and alternate, simple, more or less coriaceous, pinnately veined leaves with entire or sparsely toothed margins, and lacking stipules.

As judged by common characteristics in these families, the flowers of early angiosperms are likely to have been insect-pollinated, radially symmetrical, perfect and hypogynous, with the parts several to numerous, spirally arranged, and free and distinct. The perianth is characteristically of tepals (i.e., not differentiated into a calyx and corolla). In these families, stamens are generally poorly differentiated into an anther and filament;

anthers have four microsporangia, and open by longitudinal slits. The pollen grains are usually monosulcate, and well-developed columellae are often lacking. Their carpels have an elongated stigma and a style that is not well differentiated from the ovary; the ovules are attached to the locule walls (often lateral) and are typically anatropous, with two integuments and a thick megasporangium wall (nucellus). Fruits of ancestral angiosperms may have been indehiscent (and fleshy) or dehiscent (by a single slit, i.e., follicles), with the seeds large, with abundant endosperm, a minute embryo, with two cotyledons. In contrast, the Nymphaeaceae possess many specialized characters that likely evolved in association with their aquatic habit (e.g., herbaceousness, stems with air canals and scattered vascular bundles, leaves with mucilage, elongate petioles, and dissected, peltate or nearly peltate blades, and flowers on long pedicels).

Most of the remaining nonmonocot, noneudicot angiosperms probably constitute a monophyletic group, which is here called the magnoliid clade (or simply the **magnoliids**; Figure 9.6). Some have divided the magnoliid families into two groups: the woody magnoliids (including the Magnoliales, Laurales, and Canellales) and the Piperales (including Aristolochiaceae, Piper-

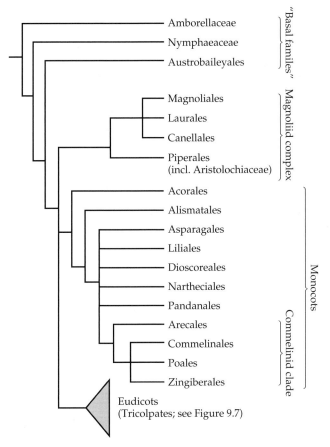

Figure 9.6 Major angiosperm clades based on the classification of the Angiosperm Phylogeny Group (APG) and updated by recent phylogenetic analyses based on morphology and numerous molecular data sets. (Modified from APG 1998, 2002, Doyle and Endress 2000, Graham and Olmstead 2000, Qiu et al. 2000, and Soltis et al. 2000.)

aceae, and relatives). The woody magnoliids are trees, shrubs, or lianas with alternate or opposite, usually pinnately veined, coriaceous leaves, and paracytic stomates. Their stems usually lack swollen nodes, and their wood often lacks broad rays. Flowers typically have several to numerous parts. Perianth parts are arranged spirally or in whorls of three, and stamens are often laminar. The filament is poorly differentiated from the anther, and the connective tissue is often well developed. The Piperales are usually herbaceous (but sometimes secondarily soft-woody) plants. Their stems generally have swollen nodes and distinct vascular bundles (separated by parenchyma tissue, or, when wood develops, the rays are very broad). Leaves are alternate, often more or less palmately veined, thin-textured, and with anomocytic stomates. Flowers have numerous to few parts; those of the perianth and androecium usually are in whorls of three. The filament is well differentiated from the anther, and the connective is usually inconspicuous.

Note that the Nymphaeaceae and monocots share some of the above-listed characters of Piperales, including herbaceous habit, a tendency for 3-merous flowers,

and filaments well differentiated from anthers. Given the cladogram shown in Figure 9.5, it is simplest to assume that these characters evolved independently in these three monophyletic groups. It is also noteworthy that stems with scattered vascular bundles have evolved independently in Nymphaeaceae, Piperaceae, and monocots. It is thus not surprising that some have considered Piperales and/or Nymphaeaceae to be closely related to monocots (Donoghue and Doyle 1989; Chase et al. 1993; Doyle et al. 1994; Doyle and Endress 2000).

Magnoliids, along with the morphologically similar but not closely related Amborellaceae, Nymphaeaceae, Illiciaceae, and relatives, traditionally have been considered to retain the greatest number of plesiomorphic features within the angiosperms (Cronquist 1968, 1981, 1988; Takhtajan 1969, 1980; Thorne 1974, 1992; Dahlgren 1977, 1983) This viewpoint has received some support from morphology-based cladistic analyses (Donoghue and Doyle 1989; Loconte and Stevenson 1991; Doyle et al. 1994), and receives at least partial support in recent molecular analyses because of the "basal" placement of Amborellaceae, Nymphaeaceae, and Illiciaceae (see, e.g., Mathews and Donoghue 1999, 2000; Graham and Olmstead 2000; Soltis et al. 2000). Note, however, that magnoliids are more closely related cladistically to either the monocots or eudicots than they are to the three "basal" angiosperm clades (Figure 9.5).

Although the angiosperm cladogram is clearly rooted by the "basal" clades Amborellaceae, Nymphaeaceae, Illiciaceae, and relatives, the placement of a few groups is still equivocal. Ceratophyllaceae, especially, is variously placed in the cladistic analyses cited above; it may be sister to the monocots (Graham and Olmstead 2000) or sister to the eudicots (Soltis et al. 2000). It is noteworthy that the analysis of *rbcL* sequences of Chase et al. (1993) placed this family as sister to all other angiosperms (see Figure 9.4). The Chloranthaceae are another family of problematic placement. In addition, it is currently unclear whether the magnoliids are more closely related to the monocots or to the eudicots. The identity of the early divergent angiosperm clades will continue to generate interest and arguments. It is anticipated, however, that progress will be made most rapidly through analyses combining data from several sources—morphological as well as molecular (Soltis et al. 1998).

Although rapid progress has been made in the last decade, we do not yet have a complete understanding of phylogenetic relationships among angiosperms. It is obvious, however, that a simplistic division of the angiosperms into monocots and dicots does not accurately reflect phylogenetic history. A cladogram presenting a conservative estimate of our knowledge of phylogenetic relationships among angiosperms is presented in Figures 9.6, 9.7, and 9.8; this tree is modified from that of the Angiosperm Phylogeny Group (1998, 2002). Hierarchial characterizations of all APG families may be found at http://www.mobot.org/MOBOT/research/Apweb/.

Figure 9.7 Major clades within the eudicot (tricolpate) clade. (Modified from APG 1998 and Soltis et al. 2000.)

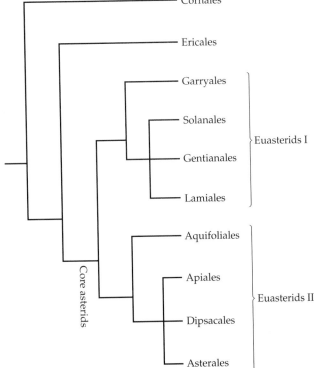

Figure 9.8 Major clades within the asterids. (Modified from APG 1998 and Soltis et al. 2000.)

The estimated 257,400 angiosperm species in 13,678 genera (Thorne 1992, 2001) occur in an extremely wide array of terrestrial habitats and exhibit an amazing diversity of morphological, anatomical, biochemical, and physiological characters. These plants are usually divided into about 450 families—Thorne (1992, 2001) recognized 490, Cronquist (1988) 387, Dahlgren (1983) 463, Takhtajan (1980) 589, and the Angiosperm Phylogeny Group (1998) 462.

This text gives detailed treatments of 142 angiosperm families, and an additional 100 families are briefly characterized (Table 9.1). As in the presentation of nonangiosperm tracheophytes in Chapter 8, each family treatment includes a **description** (in which useful identifying characters are indicated in *italic print* and presumed synapomorphies in **boldface**); a brief summary of **distribution** (with indication of ecology when the family occurs in only a limited array of plant communities or ecological conditions); a statement regarding estimated **number of genera and species** (including a listing of major genera); a listing of major **economic plants and products**, and a **discussion**, which includes information regarding characters supporting the group's monophyly, a brief overview of phylogenetic relationships within the family, information regarding pollination biology and fruit dispersal, and notes on other matters of biological interest; and a list of **references** that are useful sources of additional information. In addition to these elements, each of the angiosperm family presentations includes a **floral formula**—a graphic formula that summarizes floral symmetry and the number, fusion, and insertion of floral parts.

The family descriptions presented here are necessarily somewhat generalized, and exceptional conditions usually are not indicated. In these descriptions, anthers are assumed to be 2-locular and opening by longitudinal slits, and ovules are assumed to be anatropous, have two integuments, and a thick megasporangium, unless otherwise indicated. Endosperm is considered to be present in the seeds, and the embryo is considered to be straight, unless otherwise noted. The stem of any nonmonocot family is assumed to contain a ring of vascular bundles (eustele), while that of monocots is considered to have scattered bundles; therefore, only divergent conditions are described. Likewise, the embryos of any nonmonocot family are assumed to have two cotyledons, while those of monocots have only a single cotyledon, unless stated otherwise.

The drawings that illustrate many families were prepared in connection with the Generic Flora of the Southeastern United States Project unless otherwise indicated. Closely related families are treated within orders (i.e., names ending in -ales; see Appendix 1). **Ordinal treatments** include an outline of characters supporting the group's monophyly (ordinal synapomorphies) and a brief discussion of phylogenetic relationships of the families within the group. A **key** to all (or at least the most important) families constituting each order is included. Families with formal treatments in this text are indicated in **bold** in these keys. Circumscription of some families (and orders) has been altered from that traditionally used in order to render these groups monophyletic. Related orders are grouped into informal higher-level taxonomic groups. We largely follow the classification of the Angiosperm Phylogeny Group (APG 1998, 2002) because this classification is based on published cladistic analyses.

Families have been chosen for formal treatment on the basis of their number of genera and species, floristic dominance (especially in North America), economic importance, and phylogenetic interest. Table 9.1 lists the angiosperm families in an arrangement reflecting our understanding of their phylogenetic relationships, based on recent cladistic analyses and the classification proposed by the Angiosperm Phylogeny Group. Families receiving full coverage in the text are indicated in **boldface**, while those characterized only briefly are shown in *italics*. Because the phylogenetic relationships of several clades are still unclear, the sequence of orders employed in this chapter is somewhat arbitrary. Indeed any linear sequence is necessarily arbitrary because it cannot reflect the branching pattern expressed in a cladogram.

"BASAL FAMILIES"

Amborellales

Amborellaceae Pinchon
(Amborella Family)

Shrubs or small trees, lacking ethereal oils and pellucid dots; wood lacking vessels. Hairs simple. *Leaves alternate and 2-ranked, simple,* undulate, *obscurely serrate* (the teeth chloranthoid), with pinnate venation; stipules absent. Inflorescences determinate, axillary. **Flowers unisexual (plants dioecious),** radial. *Tepals 7-11, ± distinct,* imbricate. *Stamens numerous,* poorly differentiated into an anther and filament, the latter short, represented by staminodes in carpellate flowers; **pollen grains monoaperturate (ulcerate) at distal pole, with poorly defined aperture margin.** *Carpels 5 or 6,* **borne on a slightly concave receptacle that tears with age;** ovaries superior, with lateral placentation; stigma extending down style on adaxial surface, with 2 flanges. Ovule 1. *Fruit an aggregate of* **drupes,** the pit with a pitted surface; seed with a minute embryo.

Floral formula:

Staminate: * –9–11– , ∞ , 0

Carpellate:* –7– 8– , 1–2•̱ , <u>5–6</u> , drupes

Distribution and ecology: Endemic to New Caledonia, in moist, shaded understory of montane forests.

Genera/species: 1/1. Represented by a single genus and species, *Amborella trichopoda.*

Economic plants and products: None.

Discussion: *Amborella* has traditionally been placed in the Laurales because its flowers have a more or less concave receptacle, and its fruits are drupaceous. It has long been considered to be a primitive angiosperm, as suggested by its vessel-less wood, flowers with distinct tepals, numerous stamens with the anther and filaments more or less undifferentiated, several distinct carpels with stigmatic ridges unfused (and closed only by a secretion layer), and seeds with abundant endosperm and a minute embryo. The stomates are paracytic to anomocytic. These characters, along with the persistent, alternate, simple leaves, had suggested a placement within the magnoliids. Very recent DNA-based phylogenetic analyses (Graham and Olmstead 2000; Mathews and Donoghue 1999, 2000; Qiu et al. 2000; Soltis et al. 1999, 2000), however, distinguish *Amborella* from other "primitive angiosperms" such as Magnoliales, Canellales, and Laurales, and suggest that it may be the sister group of all other angiosperms (see Figures 9.5 and 9.6). However, this hypothesis requires additional study as some analyses have placed *Amborella* in a

TABLE 9.1   Major families of angiosperms as classified in this book.[a]

"BASAL FAMILIES"
Amborellales
Amborellaceae (p. 223)
Nymphaeales
Nymphaeaceae (p. 226) (includes
Barclayaceae, Cabombaceae)
Austrobaileyales
Illiciaceae (p. 228)
Schisandraceae (p. 228)

Chloranthaceae (p. 235)
(placement uncertain)

MAGNOLIID COMPLEX
Magnoliales
Magnoliaceae (p. 230)
Annonaceae (p. 232)
Myristicaceae (p. 233)
Degeneriaceae (p. 230)
Laurales
Lauraceae (p. 234)
Calycanthaceae (p. 234)
Hernandiaceae (p. 234)
Monimiaceae (p. 234)
Siparunaceae (p. 234)
Canellales
Winteraceae (p. 236)
Canellaceae (p. 236)
Piperales
Piperaceae (p. 237)
Aristolochiaceae (p. 238)
Lactoridaceae (p. 237)
Saururaceae (p. 237)
Hydnoraceae (p 237)

Ceratophyllaceae (p. 241)
(placement uncertain)

MONOCOTS
Acorales
Acoraceae (p. 244)
Alismatales
Araceae (p. 242) (includes
Lemnaceae)
Alismataceae (p. 245) (includes
Limnocharitaceae)
Hydrocharitaceae (p. 245) (includes
Najadaceae)
Potamogetonaceae (p. 247)
Butomaceae (p. 242)
Cymodoceaceae (p. 242)
Posidoniaceae (p. 242)
Ruppiaceae (p. 242)
Tofieldiaceae (p. 253)
Zannichelliaceae (p. 242)
Zosteraceae (p. 242)
Liliales
Liliaceae (p. 249)

"*Uvulariaceae*" (p. 250)
Colchicaceae (p. 250)
Smilacaceae (p. 251)
Melanthiaceae (p. 251)
Trilliaceae (p. 253)
Alstroemeriaceae (p. 251)
Calochortaceae (p. 250)
Nartheciales
Nartheciaceae (p. 253)
Asparagales
Asparagaceae (p. 255)
Ruscaceae (p. 257) (includes
Convallariaceae, Nolinaceae,
Dracaenaceae)
Agavaceae (p. 259)
Hyacinthaceae (p. 260)
Alliaceae (p. 260)
Amaryllidaceae (p. 261)
Asphodelaceae (p. 262)
Iridaceae (p. 263)
Orchidaceae (p. 264)
Agapanthaceae (p. 261)
Hemerocallidaceae (p. 254) (includes
Phormiaceae, Johnsoniaceae)
Hypoxidaceae (p. 256)
Themidaceae (p. 261)
Xanthorrhoeaceae (p. 254)
Dioscoreales
Dioscoreaceae (p. 267)
Burmanniaceae (p. 269)
Taccaceae (p. 269)

COMMELINOID CLADE
Arecales
Arecaceae (p. 269)
Commelinales
Commelinaceae (p. 272)
Haemodoraceae (p. 273)
Pontederiaceae (p. 275)
Philydraceae (p. 272)
Poales
Bromeliaceae (p. 278)
Typhaceae (p. 280) (includes
Sparganiaceae)
Eriocaulaceae (p. 282)
Xyridaceae (p. 282)
Juncaceae (p. 284)
Cyperaceae (p. 284)
Restionaceae (p. 286)
Poaceae (p. 287)
Flagellariaceae (p. 277)
Joinvilleaceae (p. 277)
Mayacaceae (p. 277)
Zingiberales
Zingiberaceae (p. 293)
Marantaceae (p. 295)
Cannaceae (p. 297)

Costaceae (p. 292)
Heliconiaceae (p. 292)
Musaceae (p. 292)
Strelitziaceae (p. 292)

EUDICOTS (TRICOLPATES)
"BASAL TRICOLPATES"
Ranunculales
Menispermaceae (p. 298)
Ranunculaceae (p. 300)
Berberidaceae (p. 302)
Papaveraceae (p. 304)
(includes Fumariaceae)
Lardizabalaceae (p. 299)
Proteales[b] and other "basal tricolpates"
Platanaceae (p. 306)
Proteaceae (p. 307)
Nelumbonaceae (p. 306)
Trochodendraceae (p. 306) (includes
Tetracentraceae)
Buxaceae (p. 306)

CORE EUDICOTS (CORE TRICOLPATES)

***Caryophyllid clade (includes
Caryophyllales and Polygonales)***
Caryophyllales
Caryophyllaceae (p. 310)
Phytolaccaceae (p. 310)
Nyctaginaceae (p. 313)
Amaranthaceae (p. 313) (includes
Chenopodiaceae)
Aizoaceae (p. 316)
"Portulacaceae" (p. 317)
Cactaceae (p. 317)
Petiveriaceae (p. 309)
Simmondsiaceae (p. 309)
Polygonales
Droseraceae (p. 321)
Polygonaceae (p. 323)
Plumbaginaceae (p. 321)
Nepenthaceae (p. 321)
Santalales
Loranthaceae (p. 325)
Viscaceae (p. 326)
"*Olacaceae*" (p. 325)
"*Santalaceae*" (p. 325)
Misodendraceae (p. 325)
Opiliaceae (p. 325)
Schoepfiaceae (p. 325)

ROSID CLADE
Saxifragales
Saxifragaceae (p. 329)
Crassulaceae (p. 330)
Hamamelidaceae (p. 332)
Altingiaceae (p. 334)
Cercidiphyllaceae (p. 328)

---

[a] Families receiving full coverage in the text are indicated in boldface, while those only briefly characterized are in *italics* and listed following those in boldface; thus the list does not necessarily place related families side by side. Page numbers (in parentheses) indicate the discussion of the family in this chapter.

[b] Proteales include Platanaceae, Proteaceae, and Nelumbionaceae.

TABLE 9.1   *(continued)*

Grossulariaceae (p. 328)
Haloragaceae (p. 328)
Iteaceae (p. 328)
Paeoniaceae (p. 328)
Vitales
Vitaceae (p. 334)
Leeaceae (p. 336)
Geraniales
Geraniaceae (p. 336)

Eurosids I
Zygophyllales
Zygophyllaceae (p. 338)
Krameriaceae (p. 338)
Oxalidales
Oxalidaceae (p. 338)
Cephalotaceae (p. 339)
Cunoniaceae (p. 339)
Celastrales
Celastraceae (p. 339) (includes
Hippocrateaceae)
Parnassiaceae (p. 340)
Malpighiales
Malpighiaceae (p. 341)
Euphorbiaceae (p. 343)
Phyllanthaceae (p. 347)
Clusiaceae (p. 347)
Rhizophoraceae (p. 349)
Violaceae (p. 351)
Passifloraceae (p. 351)
Salicaceae (p. 354; includes
Flacourtiaceae, in part)
Achariaceae (p. 341)
Chrysobalanaceae (p. 341)
Picrodendraceae (p. 341)
Podostemaceae (p. 341)
Putranjivaceae (p. 341)
Fabales
Fabaceae (p. 356)
Polygalaceae (p. 361)
Surianaceae (p. 356)
Rosales
Rosaceae (p. 365)
Rhamnaceae (p. 372)
Ulmaceae (p. 373)
Celtidaceae (p. 375)
Moraceae (p. 375)
Urticaceae (p. 378)
Cannabaceae (p. 363)
Cecropiaceae (p. 363)
Cucurbitales
Cucurbitaceae (p. 380)
Begoniaceae (p. 382)
Datiscaceae (p. 380)
Fagales
Fagaceae (p. 383)
Betulaceae (p. 386)
Casuarinaceae (p. 387)
Myricaceae (p. 389)
Juglandaceae (p. 391)
Nothofagaceae (p. 382)
Rhoipteleaceae (p. 383)

Ticodendraceae (p. 387)
Myrtales (placement uncertain; either in
Eurosids I or II)
Lythraceae (p. 394) (includes
Sonneratiaceae, Trapaceae,
Punicaceae)
Onagraceae (p. 396)
Combretaceae (p. 397)
Myrtaceae (p. 398)
Melastomataceae (p. 400)
Memecylaceae (p. 394)
Vochysiaceae (p. 393)

Eurosids II
Brassicales
Brassicaceae (p. 402) (includes
Capparaceae)
Bataceae (p. 404)
Caricaceae (p. 404)
Moringaceae (p. 404)
Resedaceae (p. 404)
Malvales
Malvaceae (p. 405) (includes Tiliaceae,
Sterculiaceae, Bombacaceae)
Cistaceae (p. 409)
Dipterocarpaceae (p. 410)
Thymelaeaceae (p. 405)
Sapindales
Rutaceae (p. 412)
Meliaceae (p. 412)
Simaroubaceae (p. 414)
Anacardiaceae (p. 416) (includes
Julianaceae)
Burseraceae (p. 416)
Sapindaceae (p. 418) (includes
Aceraceae, Hippocastanaceae)

ASTERID CLADE (= SYMPETALAE)
Cornales
Hydrangeaceae (p. 421)
Cornaceae (p. 422) (includes
Nyssaceae)
Loasaceae (p. 421)
Ericales
Sapotaceae (p. 426)
Ebenaceae (p. 428)
Primulaceae (p. 428)
Myrsinaceae (p. 430)
Theaceae (p. 430)
Ericaceae (p. 431) (includes
Pyrolaceae, Monotropaceae,
Empetraceae, Epacridaceae**)**
Sarraceniaceae (p. 433)
Lecythidaceae (p. 435)
Polemoniaceae (p. 436)
Actinidiaceae (p. 426)
Balsaminaceae (p. 426)
Clethraceae (p. 426)
Cyrillaceae (p. 426)
Fouquieriaceae (p. 426)
Styracaceae (p. 426)
Symplocaceae (p. 426)
Ternstroemiaceae (p. 426)
Theophrastaceae (p. 426)

Euasterids I
Garryales
Garryaceae (p. 421)
Solanales
Solanaceae (p. 437) (includes
Nolanaceae)
Convolvulaceae (p. 440) (includes
Cuscutaceae)
Boraginaceae (p. 442; includes
Hydrophyllaceae, in part; Lenno-
aceae) [placement uncertain]
Hydroleaceae (p. 438f)
Gentianales
Rubiaceae (p. 445)
Gentianaceae (p. 447)
Apocynaceae (p. 448) (includes
Asclepiadaceae)
Gelsemiaceae (p. 444)
Loganiaceae (p. 444)
Lamiales
Oleaceae (p. 454)
Gesneriaceae (p. 455)
Plantaginaceae (p. 456) (includes
Callitrichaceae, Scrophulariaceae in
part)
Scrophulariaceae (p. 458)
Orobanchaceae (p. 459) (includes
Scrophulariaceae, parasitic species)
Bignoniaceae (p. 459)
Acanthaceae (p. 462) (includes
Mendonciaceae)
Lentibulariaceae (p. 464)
Verbenaceae (p. 464)
Lamiaceae (p. 466) (includes many
genera typically treated as
Verbenaceae)
Avicenniaceae (p. 454)
Buddlejaceae (p. 454)
Calceolariaceae (p. 454)
Myoporaceae (p. 454)
Phrymaceae (p. 459)

Euasterids II
Aquifoliales
Aquifoliaceae (p. 468)
Helwingiaceae (p. 470)
Apiales
Apiaceae (p. 470) (includes
Araliaceae, Hydrocotylaceae)
Pittosporaceae (p. 471)
Dipsacales
Caprifoliaceae (p. 474) (includes
Dipsacaceae, Valerianaceae,
Diervillaceae, Linnaeaceae)
Adoxaceae (p. 476) (includes
Sambucus, Viburnum)
Asterales
Campanulaceae (p. 478) (includes
Lobeliaceae)
Asteraceae (p. 480)
Calyceraceae (p. 477)
Goodeniaceae (p. 477)
Menyanthaceae (p. 477)
Stylidiaceae (p. 477)

clade also containing Nymphaeaceae (Barkman et al. 2000) or Nymphaeaceae and Illiciaceae, with these families, together, comprising the sister to all other angiosperms (Savolainen et al. 2000a). Especially distinctive features of Amborellaceae include the vesselless wood, lack of ethereal oils, unisexual flowers with a shallow, cup-like receptacle, nearly inaperturate pollen, and aggregate of red drupes.

References: Bailey and Swamy 1948; Barkman et al. 2000; Endress 2001; Endress and Ingersheim 2000; Feild et al. 2001; Graham and Olmstead 2000; Hesse 2001; Mathews and Donoghue 1999; Philipson 1993; Qiu et al. 2000; Sampson 1993; Soltis et al 1999, 2000; Young 1982.

Nymphaeales

Nymphaeaceae Salisbury
(Water Lily Family)

Aquatic, **rhizomatous herbs**; *stem with vascular bundles usually scattered, with conspicuous air canals* and usually also **laticifers**; usually with distinct, stellate-branched sclereids projecting into the air canals; often with alkaloids (but not of the benzyl-isoquinoline type). Hairs simple, **usually producing mucilage (slime)**. *Leaves* alternate and spiral, opposite, or occasionally whorled, simple, entire to toothed or dissected, short- to *long-petiolate, with blade submerged, floating, or emergent*, with palmate to pinnate venation; stipules present or absent. Inflorescences of solitary flowers. **Flowers** bisexual, radial, **with a long pedicel** *and usually floating or raised above the surface of the water*, **with girdling vascular bundles in receptacle**. Tepals 4–12, distinct to connate, imbricate, often petal-like. *Petals (petal-like staminodes) lacking or 8 to numerous*, inconspicuous to showy, often intergrading with stamens. *Stamens 3 to numerous*, the innermost sometimes represented by staminodes; filaments distinct, free or adnate to petaloid staminodes, slender and well differentiated from anthers to laminar and poorly differentiated from anthers; pollen grains usually monosulcate or lacking apertures. Carpels 3 to numerous, distinct or connate; ovary/ovaries superior to inferior, *if connate then with several locules and* **placentation parietal (the ovules scattered on the partitions)**; *stigmas often elongate and radiating on an expanded circular to marginally lobed and grooved disk*, often surrounding the receptacle, which is visible as a knob or circular bump. Ovule 1 to numerous, anatropous to orthotropous. Nectaries lacking or sometimes present on the staminodes, although a sweet fluid may also be secreted by the stigma. *Fruit an aggregate of nuts or few-seeded indehiscent pods, a berry, or sometimes an irregularly dehiscent fleshy capsule*; seeds **usually operculate** (opening by a cap), often arillate; endosperm ± lacking, but with abundant perisperm (Figure 9.9).

Floral formula:

*, $\widehat{4\text{--}12}$, $\widehat{0\text{--}\infty}$, 3–∞, $\widehat{1\text{--}\infty}$; berry, nuts

Distribution and ecology: Widespread from tropical to cold temperate regions, occurring in rivers, ponds, lakes, and other freshwater wetlands.

Genera/species: 8/70. **Major genera:** *Nymphaea* (40 spp.) and *Nuphar* (15). These genera, along with *Cabomba* and *Brasenia*, occur in the continental United States and Canada.

Economic plants and products: Species of *Nymphaea* (water lily), *Nuphar* (yellow water lily, spatterdock), and *Victoria* (Amazon water lily) provide ornamental plants for pools. *Cabomba* is a popular aquarium plant.

Discussion: The family comprises two subfamilies: Cabomboideae, including *Cabomba* and *Brasenia*, and Nymphaeoideae, with the remaining genera; both are often recognized as families (Les et al. 1999). Cabomboideae are united by the presence of free-floating stems in addition to the rhizomes. They also have distinct carpels, nutlike fruits, and six-tepaled flowers that lack petal-like staminodes. Nymphaeoideae are separated from Cabomboideae by numerous synapomorphies: the presence of star-shaped sclerids, four or more sepals, spirally inserted stamens with laminar filaments, syncarpous fleshy fruits (berries), and laminar placentation (Les et al. 1999; Moseley et al. 1993). *Nuphar* likely is sister to the remaining genera of Nymphaeoideae. It has a superior ovary with separate stigmas and inconspicuous staminodes, while *Barclaya, Ondinea, Nymphaea, Victoria,* and *Euryale* have perigynous/epigynous flowers with a continuous stigmatic surface, showy staminodes, and zonosulcate pollen, all of which are likely synapomorphic (Les et al. 1999). Within this clade *Victoria* and *Euryale* share the apomorphy of prickles on the petiole and abaxial leaf surface; additionally both are acaulescent, short-lived herbs. Floral anatomy indicates that *Nymphaea* may be the sister group of this *Victoria* + *Euryale* clade. Within Nymphaeoideae the highest number of staminodial petals, stamens, and carpels occur in *Nymphaea* and *Victoria*, but even the flowers of the more primitive *Nuphar* possess relatively numerous parts. The flowers of *Cabomba* have relatively few parts, and the numerous floral parts seen in Nymphaeoideae probably are the result of secondary increase. Nymphaeaceae, as here circumscribed, clearly are monophyletic (Donoghue and Doyle 1989; Zimmer et al. 1989; Hamby and Zimmer 1992; Qiu et al. 1993; Chase et al. 1993; Doyle et al. 1994; Les et al. 1999) and there is little reason to subdivide the group.

Molecular evidence strongly supports the placement of Nymphaeaceae as the second of three "basal" angiosperm clades (see Figure 9.5, and Mathews and Donoghue 1999, 2000; Qiu et al. 2000; Soltis et al. 2000). The family has sometimes been considered closely related to monocots on the basis of its scattered vascular bundles, lack of a vascular cambium, adventitious root

Figure 9.9 Nymphaeaceae. *Nymphaea odorata:* (A) habit (× 0.3); (B) petiole, in cross section (× 5); (C) flower (× 1); (D) flower, in longitudinal section (× 1.5); (E) petal (× 1.5); (F–H) petal-like stamens (× 1.5); (I–K) inner stamens (× 1.5); (L) gynoecium showing numerous carpels (× 2); (M) gynoecium in cross section (× 1.5); (N) fruit, with coiled floral peduncle (× 0.5); (O) fruit (× 1); (P) dehiscing fruit (× 1.5); (Q) seed, with aril (× 15); (R) seed (× 30); (S) seed coat (greatly magnified); (T) longitudinal seciton of seed, with embryo, endosperm, and perisperm. (× 30). (Original art prepared for the Generic Flora of the Southeast U.S. Project. Used with permission.)

system, and frequently 3-merous flowers (Doyle and Endress 2000). Nelumbonaceae have often been placed in Nymphaeaceae, but all recent evidence places them within the tricolpate clade (Donoghue and Doyle 1989; Moseley et al. 1993; Chase et al. 1993; Qiu et al. 1993, 2000; Savolainen et al. 2000b; Soltis et al. 2000). Although superficially similar to Nymphaeaceae, Nelumbonaceae have numerous carpels that are sunken in pits within an enlarged, funnel-shaped, spongy receptacle.

The fragrant, showy flowers of Nymphaeaceae attract various insects (beetles, flies, and bees), which gather pollen or, less commonly, nectar. However, flowers of *Brasenia* lack nectar glands and are wind-pollinated, having numerous easily shaken anthers. Outcrossing is favored by protogyny. Flowers of *Victoria* and some species of *Nymphaea* attract beetles by providing food bodies (starch-filled carpel appendages) and producing heat along with a strong fruity odor. The flowers open and close daily and trap the beetles. In other species of *Nymphaea*, flies and small bees gather pollen from 2- or 3-day-old flowers. They are then attracted to a pool of sweet stigmatic fluid in young (1-day-old) flowers, in which they frequently drown. Pollen on their bodies becomes suspended in the stigmatic fluid and eventually germinates. The fleshy fruits of many Nymphaeaceae mature underwater and rupture irregularly due to swelling of the mucilaginous aril surrounding the seeds, which are water dispersed. In *Nuphar* the carpel segments separate and float away. Vegetative reproduction commonly occurs through production of rhizomes or specialized shoots/tubers.

References: Chase et al. 1993; Dahlgren and Clifford 1982; Donoghue and Doyle 1989; Doyle et al. 1994; Endress 2001; Hamby and Zimmer 1992; Ito 1986, 1987; Les et al. 1991; Orgaard 1991; Osborn et al. 1991; Osborn and Schneider 1988; Qiu et al. 1993, 2000; Les et al. 1999; Mathews and Donoghue 1999, 2000; Meeuse and Schneider 1979; Moseley et al. 1993; Padgett et al. 1999; Schneider and Carlquist 1995; Schneider and Jeter 1982; Schneider and Williamson 1993; Schneider et al. 1995; Savolainen et al. 2000b; Soltis et al. 2000; Taylor and Hickey 1996; Thorne 1974, 1992; Wiersema 1988; Wood 1959a; Zimmer et al. 1989.

Austrobaileyales

Illiciaceae A. C. Smith
(Star Anise Family)

Trees or shrubs; nodes unilacunar; with scattered spherical cells containing ethereal oils (aromatic terpenoids) and branched sclereids; often with tannins. Hairs simple. *Leaves alternate and spiral,* often clustered at the tips of the shoots, simple, *entire,* with pinnate venation, *blade with pellucid dots; stipules lacking. Inflorescences determinate but reduced to 1–3 flowers, axillary.* Flowers bisexual, radial,

with convex to shortly conical receptacle. *Tepals usually numerous,* distinct, the outer usually sepal-like, and the innermost sometimes minute, imbricate. *Stamens usually numerous,* distinct; *filaments short and thick, poorly differentiated from the anthers;* anthers with connective tissue extending between and beyond the apex of the pollen sacs; pollen grains tricolpate, but colpus position different from that of the eudicots. *Carpels usually 7 to numerous,* distinct, in a single whorl; ovaries superior, with ± basal placentation; stigma extending down style on adaxial surface. **Ovules 1 per carpel.** Nectar produced at base of stamens. **Fruit a starlike aggregate of 1-seeded follicles; seeds with smooth, hard coat**; embryo minute; *endosperm homogeneous* (Figure 9.10).

Floral formula: *, -7–∞- , 7–∞ , 7–∞; follicles

Distribution: Southeastern Asia, southeastern United States, Cuba, Hispaniola, and Mexico; mainly in moist forests.

Genus/species: 1/37. **Genus:** *Illicium.*

Economic plants and products: Anise oil is extracted from *Illicium verum* (star anise). Some species are used medicinally, and a few are used as ornamentals.

Discussion: Illiciaceae belongs within Austrobaileyales, an order containing 4 families and about 100 species. The monophyly of Austrobaileyales is supported by DNA characters (Renner 1999; Qiu et al. 2000; Soltis et al. 2000) and possibly also by hardened mesotestal cells. Austrobaileyaceae and Trimeniaceae are early divergent clades, and the more specialized Illiciaceae and Schisandraceae (a small family of vines with unisexual flowers) are additionally united by the apomorphic characters of pollen grains with three (or six) furrows (colpi), a feature that is otherwise limited to the eudicot clade, and unilacunar nodes. Austrobaileyales are among the most basal lineages of angiosperms (see Figures 9.5, 9.6, and the introduction to this chapter).

The genus *Illicium* is divided into two subgroups: section *Illicium,* which has lax, elongate inner tepals, and section *Cymbostemon,* which has ± erectly held, ovate to suborbicular inner tepals. The former probably is paraphyletic.

Flowers of *Illicium* are pollinated by a wide variety of small insects, particularly flies. The plants are self-incompatible. The follicles dehisce elastically, shooting out the smooth seeds. Both pollination and seed dispersal appear to be quite local. Most species also form extensive clones due to the production of rhizomes.

References: Endress 2001; Keng 1993; Qiu et al. 2000; Renner 1999; Roberts and Haynes 1983; Smith 1947; Soltis et al. 2000; Thien et al. 1983; Thorne 1974; White and Thien 1985; Wood 1958.

Figure 9.10 Illiciaceae. (A–I) *Illicium floridanum:* (A) fruiting branch (× 0.5); (B) opening flower bud with receptive carpels (× 4); (C) flower, later stage at shedding of pollen (× 1.5); (D) stamens, inner, outer, and an unusual subtepaloid form (× 7); (E) two carpels on receptacle (× 4); (F) carpel in cross section; note single ovule (× 15); (G) mature fruit, with single seed, endosperm stippled (× 3); (H) mature fruit in longitudinal section (× 3); (I) seed (× 3). (J) *I. parviflorum:* stamens (× 3). (From Wood 1958, *J. Arnold Arb.* 39: p. 317.)

MAGNOLIID COMPLEX

Magnoliales

Magnoliales are considered monophyletic (see Donoghue and Doyle 1989; Qiu et al. 1993) on the basis of *rbcL, atpB,* and 18S sequence characters, along with their 2-ranked leaves, at least outer whorl of perianth 3-merous, seeds with a fleshy seed coat or aril, trilacunar to multilacunar nodes, stratified phloem, an adaxial plate of vascular tissue in the leaf midrib, and P-type sieve elements that contain one to several protein crystals plus starch grains (or sometimes only starch grains). Some of these are homoplasious; possible additional synapomorphies are septate pith, ruminate endosperm, and boat-shaped pollen. The group has often been considered part of the ancestral angiosperm complex (Takhtajan 1969, 1980, 1997; Thorne 1974, 1992; Cronquist 1968, 1981, 1988; Dahlgren 1983) or as the basal clade within angiosperms

(Donoghue and Doyle 1989). If this is actually the case, then it is noteworthy that the group has retained granular monosulcate pollen (Walker and Walker 1984) and primitive vessels with scalariform perforations. However, DNA sequences (Zimmer et al. 1989; Qiu et al. 1993, 2000; Soltis et al. 1997, 2000; Mathews and Donoghue 1999, 2000; Savolainen et al. 2000a), morphology, and combined evidence from DNA and morphology (Doyle et al. 1994; Nandi et al. 1998) suggest that Magnoliales may not be basal within angiosperms (see Figs. 9.5 and 9.6). These families have retained numerous apparently plesiomorphic floral characters such as distinct and often numerous, spirally arranged stamens and carpels, superior ovaries, and seeds with a minute embryo and copious endosperm. The order consists of 6 families and about 2840 species; noteworthy families include **Annonaceae**, **Magnoliaceae**, **Myristicaceae**, and Degeneriaceae.

Key to Major Families of Magnoliales

1. Stamens monadelphous, their filaments connate into a tube or column; seeds arillate ..... **Myristicaceae**

1. Stamens distinct; seeds not arillate 2

2. Stipules present, sheathing the stem, and enveloping the apical bud; nodes multilacunar; endosperm homogeneous; perianth usually of tepals; receptacle elongate .... **Magnoliaceae**

2. Stipules lacking; nodes 3- to 5-lacunar; endosperm ruminate; perianth of sepals and petals; receptacle short to ± hemispherical 3

3. Carpel 1, the stigma running nearly its entire length; stamens laminar, with 3 veins, not packed into a tight ball; embryo with 3 or 4 cotyledons .... Degeneriaceae

3. Carpels more than 1, often numerous, the stigma ± elongate, but restricted to apical portion of each carpel; stamens short and stout, with an expanded connective, with 1 vein, and usually packed into a ball-like configuration; embryo with 2 cotyledons .... **Annonaceae**

References: Canright 1952, 1960; Cronquist 1981, 1988; Dahlgren 1983; Donoghue and Doyle 1989; Doyle and Endress 2000; Doyle et al. 1994; Endress 1986, 1994; Igersheim and Endress 1997; Loconte and Stevenson 1991; Mathews and Donoghue 1999, 2000; Nandi et al. 1998; Qiu et al. 1993, 2000; Savolainen et al. 2000a; Soltis et al. 1997, 2000; Takhtajan 1969, 1980, 1997; Taylor and Hickey 1992; Thorne 1974, 1992; Walker and Walker 1984; Weberling 1988; Wood 1958; Zimmer et al. 1989.

Magnoliaceae A. L. de Jussieu
(Magnolia Family)

Trees or shrubs; **nodes multilacunar**; *with spherical cells containing ethereal oils (aromatic terpenoids)*; with alkaloids, usually of the benzyl-isoquinoline type. Hairs simple to stellate. *Leaves alternate*, spiral to 2-ranked, simple, sometimes lobed, *entire*, with pinnate venation, *blade with pellucid dots*; **stipules present, surrounding the terminal bud. Inflorescence a solitary, terminal flower**, but sometimes appearing axillary (on a short shoot). **Flowers** usually bisexual, radial, **with an elongate receptacle**. *Tepals 6 to numerous*, distinct, occasionally the outer 3 differentiated from the others and ± sepal-like, imbricate. *Stamens numerous*, distinct, often with 3 veins; *filaments short and thick, poorly differentiated from the anthers*; anthers with connective tissue often extending beyond the apex of the pollen sacs; pollen grains monosulcate. *Carpels* usually numerous, distinct, *on an elongate receptacle*; ovaries superior, with lateral placentation; stigma usually extending down style on adaxial surface, but sometimes reduced and terminal. Ovules usually 2 per carpel, sometimes several. Nectaries lacking. **Fruit an aggregate of follicles, which usually become closely appressed as they mature and open along the carpel midvein (i.e., abaxial surface)**, sometimes fleshy, with individual fruits becoming fused to one another as they mature, forming an indehiscent or irregularly dehiscent berrylike structure, or an aggregate of samaras; *seed with red to orange, fleshy coat* (except *Liriodendron*), usually dangling from a slender thread; embryo minute; *endosperm homogeneous* (Figure 9.11 and Figure 4.47A).

Floral formula: *, -6-∞-, ∞, ⚭; follicles, samaras

Distribution and ecology: Temperate to tropical regions of eastern North America and eastern Asia, and tropical South America; mainly in moist forests.

Genera/species: 2/220. **Genera:** *Magnolia* (218 spp.) and *Liriodendron* (2).

Economic plants and products: *Liriodendron tulipifera* (tulip tree, tulip-poplar) and several species of *Magnolia* are important ornamentals. Species of both genera are also used for timber.

Discussion: Cladistic analyses of *rbcL* sequences (Qiu et al. 1993) and morphological characters support the monophyly of Magnoliaceae. The *rbcL* and *ndhF* genes (Qiu et al. 1993, 1995; Kim et al. 2001) indicate that the family is composed of two clades: represented by *Liriodendron* and by *Magnolia* s. l. Recognition of *Talauma*, *Michelia*, and *Manglietia* leads to a paraphyletic *Magnolia*, and therefore, we circumscribe this genus broadly. The monophyly of *Liriodendron* is supported by lobed leaves, carpels with a restricted stigma, samaroid fruits, and seeds with a thin, more or less dry coat. The monophyly of *Magnolia* is supported by the follicles opening along the abaxial (or outer) surface. These can be described as backward-opening follicles because in most families with follicular fruits the opening occurs along the adaxial surface.

The showy flowers of Magnoliaceae are mainly pollinated by beetles, which may be trapped in the flower for a period of time, and often eat pollen and/or various floral tissues. *Liriodendron*, however, is bee-pollinated. Pro-

Figure 9.11 Magnoliaceae. (A–I) *Magnolia virginiana:* (A) branch with terminal flower (× 0.3); (B) androecium (part removed) and gynoecium, on elongate receptacle (× 3.5); (C) stamens, adaxial surface (× 4); (D) stamens, in cross section (× 6); (E) gynoecium in longitudinal section, note two ovules in each carpel (× 5); (F) nearly mature fruit (× 0.75); (G) mature fruit with pendulous seeds (× 0.75); (H) seed in longitudinal section, note copious endosperm and minute embryo (× 2.5); (I) seed, with fleshy outer seed coat removed (× 2.5). (J–L) *M. grandiflora:* (J) flower bud (× 0.75); (K) floral receptacle with androecium (half of stamens removed) and gynoecium (× 2); (L) stamens, adaxial surface (× 4). (M) *M. tripetala:* stamen, adaxial surface (× 4). (N–Q) *M. acuminata:* (N) branch with terminal flower (× 0.3); (O) opening flower bud (× 0.3); (P) stamen, adaxial surface (× 4); (Q) anther in cross section (× 6). (From Wood 1974, *A student's atlas of flowering plants*, p. 36.)

togyny and self-incompatibility lead to outcrossing. The seeds of *Magnolia*, with bright red, pink, or orange, fleshy seed coats, hang on thin springy threads – actually vessels with spiral thickenings – when the follicle opens and are dispersed by birds. The fleshy syncarps of some of the tropical species are also colorful and bird-dispersed. The samaras of *Liriodendron* are dispersed by wind.

References: Agababian 1972; Canright 1952, 1953, 1960; Endress 1994b; Kim et al. 2001; Qiu et al. 1993, 1995; Gottsberger 1977, 1988; Nooteboom 1993; Thien 1974; Weberling 1988b; Wood 1958.

Annonaceae Adanson
(Pawpaw or Annona Family)

Trees, shrubs, or lianas, with conspicuously fibrous bark; *nodes trilacunar;* **vessel elements with simple perforations,** and wood with broad rays; *with scattered spherical cells containing ethereal oils (aromatic terpenoids), and often with scattered sclereids;* usually with alkaloids of the benzyl-isoquinoline type; often with tannins. Hairs simple, sometimes stellate, or peltate scales. *Leaves alternate, 2-ranked,* simple, entire, *often short petioled,* with pinnate venation, blade *with pellucid dots; stipules lacking.* Inflorescences determinate, sometimes reduced to a single flower, terminal or axillary. *Flowers* usually bisexual, radial, opening and then usually **gradually increasing in size before flowering**, *with a short, flat to ± hemispherical receptacle. Sepals usually 3,* distinct or slightly connate, valvate or imbricate. *Petals usually 6,* distinct, *the outer 3 often larger and differentiated from the inner,* imbricate and/or valvate. *Stamens usually numerous, appearing peltate and packed into a ball-like or disklike configuration,* distinct, with 1 vein; *filaments short and thick, poorly differentiated from the anthers;* anthers with connective tissue extending beyond the apex of anther sacs; pollen grains various, but often monosulcate, sometimes in tetrads or polyads. *Carpels (1-) 3 to numerous,* usually distinct, usually spirally arranged; ovaries superior, with lateral placentation; stigma extending down style on adaxial surface, or ± terminal. Ovules 1 to numerous per carpel. Nectaries or food tissues sometimes on inner petals. *Fruit an aggregate of berries,* these sometimes becoming connate as they develop; seeds **with raphe curving around the top**, often arillate; embryo minute; *endosperm ruminate* (Figure 9.12).

Floral formula: *, 3, 6, ∞, ⟨3–∞⟩; berries

Distribution and ecology: Widely distributed in tropical and subtropical regions and very characteristic of lowland wet forests.

Genera/species: 128/2300. **Major genera:** *Guatteria* (250 spp.), *Xylopia* (150), *Uvaria* (110), *Annona* (110), *Polyalthia* (100), *Artabotrys* (100), and *Rollinia* (65). The family is represented in the continental United States by *Asimina, Deeringothamnus,* and *Annona.*

Economic plants and products: Several species of *Annona* and *Rollinia* (cherimoya, guanabana, soursop, sugar apple, sweetsop) produce edible fruits. Berries of *Asimina triloba* (pawpaw) are also edible. Flowers of *Cananga odorata* (ylang-ylang) are used in perfumes, and fruits of *Monodora myristica* are used as a substitute for *Myristica fragrans* (nutmeg; Myristicaceae). *Annona, Cananga,* and *Polyalthia* are grown as ornamentals.

Discussion: The monophyly of the family is supported by DNA sequences (Qiu et al. 1993, 2000) and several morphological features (Doyle and Le Thomas 1994). A base chromosome number of eight probably is also synapomorphic (but reductions to $n = 7$ and increases to $n = 9$ have occurred). Tribes and genera have been distinguished by features such as indumentum, structure of the bud, sepal and petal aestivation, shape and texture of anthers, number and shape of carpels, extent of fruit connation, and number of seeds. Fruits in which the carpels become fused as they develop, forming a fleshy syncarp, as in *Annona* or *Rollinia,* are clearly derived in relation to persistently distinct berries, as in *Asimina* or *Cananga.* Connate carpels, as in *Monodora,* are also derived. *Anaxagorea* has retained follicle fruits, but most genera are united by the apomorphy of fleshy, indehiscent fruits (berries). *Anaxagorea,* along with a few other basal genera, also has laminar and elongated stamens, endosperm with irregular ruminations, while most genera form a clade based on more or less peltate stamens (packed into a characteristic ball-like or disk-like configuration) and spinose or thin-flat endosperm ruminations (extensions of seed coat into endosperm). A large group, which includes *Xylopia, Monodora, Asimina, Annona,* and *Artabotrys,* are characterized by pollen grains that lack germination apertures (Doyle and Le Thomas 1996).

Flowers of most Annonaceae show various specializations leading to pollination by beetles, including closed flowers, fruity odors, feeding tissues, thick, fleshy petals, and structural protection of the reproductive organs. Some species of *Annona* produce heat; beetles often stay in the flowers overnight, mate there, and frequently eat floral tissues. Many species of Annonaceae produce a sticky stigmatic fluid, which protects the carpels; the often hardened and more or less peltate anther connectives reduce consumption of pollen. Outcrossing in this family is favored by protogyny. The fleshy fruits are dispersed by birds, mammals, and turtles.

References: Doyle and Le Thomas 1994, 1996; Endress 1994b; Gottsberger 1988; van Heusden 1992; Kessler 1993; Norman and Clayton 1986; Qiu et al. 1993, 2000; Thorne 1974; Vander Wyk and Canright 1956; Walker 1971; Wood 1958.

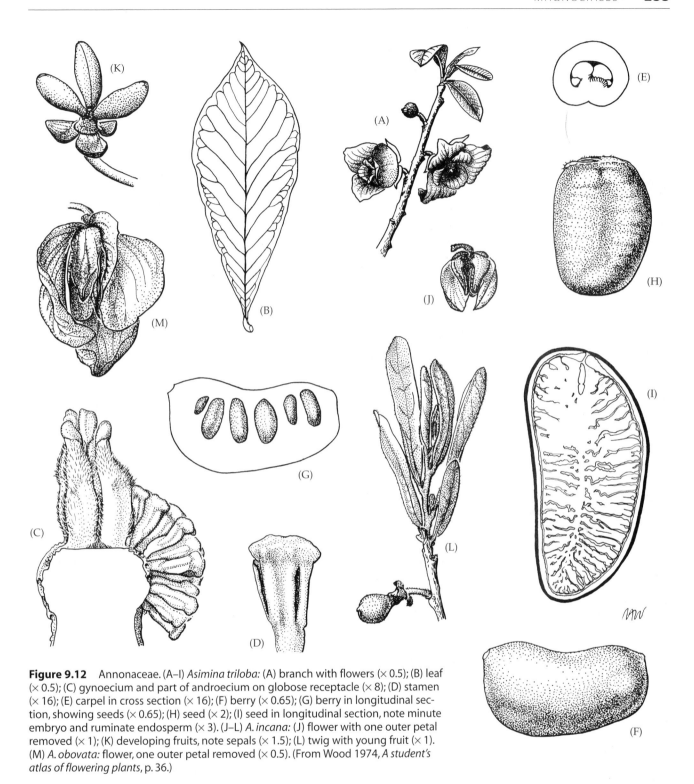

Figure 9.12 Annonaceae. (A–I) *Asimina triloba:* (A) branch with flowers (× 0.5); (B) leaf (× 0.5); (C) gynoecium and part of androecium on globose receptacle (× 8); (D) stamen (× 16); (E) carpel in cross section (× 16); (F) berry (× 0.65); (G) berry in longitudinal section, showing seeds (× 0.65); (H) seed (× 2); (I) seed in longitudinal section, note minute embryo and ruminate endosperm (× 3). (J–L) *A. incana:* (J) flower with one outer petal removed (× 1); (K) developing fruits, note sepals (× 1.5); (L) twig with young fruit (× 1). (M) *A. obovata:* flower, one outer petal removed (× 0.5). (From Wood 1974, *A student's atlas of flowering plants*, p. 36.)

Myristicaceae R. Brown
(Nutmeg Family)

Trees, **with bark exuding a reddish sap when slashed,** *and usually with distinctive growth form, the trunk erect, the lateral branches appearing whorled and ± horizontal; with scattered spherical cells containing ethereal oils* (aromatic terpenoids); *with hallucinogenic phenolic compounds* (e.g., myristicin), *tannins, and sometimes indole alkaloids.*

Hairs simple, T-shaped, branched, or stellate. Leaves alternate, usually 2-ranked, simple, entire, with pinnate venation, *blade with pellucid dots; stipules absent.* Inflorescences determinate, sometimes reduced and forming fascicles or heads, axillary. *Flowers inconspicuous, fragrant,* **unisexual (plants usually dioecious),** radial. **Tepals** usually 3, **connate,** valvate. *Stamens 2-numerous;* **filaments connate into a ± solid column**; anthers distinct or also connate; pollen grains monosulcate to inaperturate. **Carpel 1;**

ovary superior, *with ± basal placentation;* stigma 1, small, often bilobed. **Ovule 1.** Nectaries lacking. **Fruit ± leathery follicle, also dehiscing abaxially, exposing a single large seed and often colorful aril**; embryo minute; *endosperm ruminate.*

Floral formula:

Staminate: * –③– , 2–∞ , 0

Carpellate: * –③– , 0 , <u>1</u> , fleshy capsule

Distribution and ecology: Widely distributed in tropical regions and very characteristic of lowland wet forests.

Genera/species: 17/370. **Major genera:** *Horsfieldia* (100), *Myristica* (70), and *Virola* (50). No species occur in the continental United States or Canada.

Economic plants and products: The seeds of *Myristica fragrans* (nutmeg tree) provide the spice nutmeg, while the aril is the source of mace. Several species of *Virola* are the source of hallucinogenic snuff.

Discussion: The monophyly of Myristicaceae is supported by molecular (Soltis et al. 2000) as well as morphological features (see description). Phylogenetic relationships within the family are poorly understood.

The flowers are insect pollinated, especially by beetles and thrips, and often are night blooming. The distinctive fruits are dispersed by birds and mammals (primates) that are attracted by the arillate seeds (with red to orange aril contrasting with the black seed coat) and the often strikingly colored fruit wall (inner surface can be red, pink, or white).

References: Armstrong and Drummond 1986; Howe and Vande Kerckhove 1980; Kuhn and Kubitzki 1993; Smith and Wodehouse 1938; Wilson and Maculans 1967.

Laurales

Lauraceae A. L. de Jussieu

(Laurel Family)

Trees or shrubs, or twining, parasitic vine (*Cassytha*); *nodes unilacunar; with scattered spherical cells containing ethereal oils (aromatic terpenoids);* usually with tannins; usually with benzyl-isoquinoline and/or aporphine alkaloids. **Leaves alternate and spiral**, occasionally opposite, *but never 2-ranked*, simple, rarely lobed, **entire**, with usually pinnate venation, or sometimes the lowermost pair of secondary veins more prominent and arching toward apex and venation ± palmate, and all veins clearly visible, connected to adaxial and abaxial leaf surfaces by lignified tissue, blade *with pellucid dots; stipules lacking.* Inflorescences determinate to seemingly indeterminate, axillary. *Flowers* bisexual or unisexual (plants then ± dioecious), radial, *with distinctly concave receptacle*, usually small, pale green, white, or yellow. *Tepals usually 6*,

distinct or slightly connate, imbricate. *Stamens usually 3–12; filaments often with pairs of basal-lateral nectar- or odor-producing appendages* (staminodes), the 3 innermost stamens often also reduced to nectar- or odor-producing staminodes; *anthers opening by 2 or 4 flaps that curl from the base upward and pull out the sticky pollen*, often dimorphic; pollen grains without apertures, exine reduced to tiny spines. **Carpel 1**; *ovary superior, with ± apical placentation;* stigma 1, capitate, truncate, lobed, or elongate. *Ovule 1.* **Fruit a drupe** or occasionally a 1-seeded berry, *often associated with the persistent fleshy to woody receptacle (and sometimes also tepals) that often contrast in color with the fruit (i.e., fruit with a cupule);* **embryo large, with fleshy cotyledons; endosperm lacking** (Figure 9.13).

Floral formula:

*, ⟨6⟩, 3–12 + paired glands, <u>1</u>; drupe

Distribution and ecology: Widespread in tropical and subtropical regions and especially diverse in Southeast Asia and northern South America; characteristic of tropical wet forests.

Genera/species: 50/2500. **Major genera:** *Litsea* (400 spp.), *Ocotea* (350), *Cinnamomum* (350), *Cryptocarya* (250), *Persea* (200), *Beilschmiedia* (150), *Nectandra* (120), *Phoebe* (100), and *Lindera* (100). The following occur in the continental United States and/or Canada: *Cassytha*, *Cinnamomum*, *Licaria*, *Lindera*, *Litsea*, *Nectandra*, *Persea*, *Sassafras*, and *Umbellularia*.

Economic plants and products: The family contains spice plants such as *Laurus nobilis* (bay leaves), *Cinnamomum verum* (cinnamon), *C. camphora* (camphor), and *Sassafras albidum* (sassafras). *Persea americana* (avocado) is an important tropical fruit tree. *Beilschmiedia, Ocotea, Litsea,* and a few other genera contain species used for timber.

Discussion: Lauraceae belong to the large order Laurales, which consists of 7 families and about 3400 species; major families include Calycanthaceae, **Lauraceae**, Monimiaceae, Siparunaceae, and Hernandiaceae. Laurales are clearly monophyletic; synapomorphies include their unilacunar nodes, opposite leaves, cup-shaped receptacle, and pollen with sculptured apertures; some of these have been lost in many species. The monophyly of the order is supported in cladistic analyses based on both DNA sequences and morphology (Qiu et al. 1993, 2000; Renner 1999; Soltis et al. 2000; Doyle and Endress 2000). Calycanthaceae are probably basal within Laurales (Doyle and Endress 2000; Renner and Chanderbali 2000). The more derived families are united by the additional apomorphies of inaperturate pollen grains with a thin exine and spinules, stamens with paired appendages, anthers opening by flaps, and a single ovule per carpel (Donoghue and Doyle 1989; Renner 1999; Doyle and Endress 2000). Within this distinctive

Figure 9.13 Lauraceae. *Sassafras albidum:* (A) fruiting branch (× 0.5); (B) staminate flower, androecium with nine stamens in three whorls of three (× 5); (C) carpellate flower, note staminodes (× 5); (D) two stamens of third whorl, each with two glands, note anthers opening by introrsely and laterally opening valves (× 6); (E) stamen of first whorl, note introrsely opening valves (two removed) (× 6); (F) ovary in longitudinal section, with single apical ovule (× 10); (G) mature drupe and cupule (× 3); (H) fruit and cupule in longitudinal section, note embryo with large, fleshy cotyledons (one cotyledon removed to show plumule and radicle) (× 4). (From Wood 1974, *A student's atlas of flowering plants*, p. 37.)

clade, Siparunaceae (and close relatives) form a clade distinct from a Lauraceae + Hernandiaceae + Monimiaceae clade. Lauraceae are likely sister to Hernandiaceae (Doyle and Endress 2000), although molecular studies (Renner 1999; Renner and Chanderbali 2000) possibly support a closer relationship with Monimiaceae. Lauraceae clearly are monophyletic (see description, and Renner and Chanderbali 2000). Lauraceae differ from Monimiaceae in their usually alternate and spiral, entire leaves (vs. opposite, toothed to entire leaves) and single carpel (vs. numerous carpels). Lauraceae differ from Hernandiaceae in having a superior ovary (vs. inferior) and drupaceous fruits (vs. fruits nuts, associated with accessory structures). Siparunaceae, which often have been included within Monimiaceae, are distinctive— they have the several unusual apomorphic characters, e.g., bisporangiate anthers that open by a single slit, ovules with only a single integument, and flowers closed by a "roof" due to the extreme curvature of the cup-shaped receptacle (Renner et al. 1997).

Chloranthaceae have sometimes been placed in Laurales (Donoghue and Doyle 1989), but this placement is not supported by DNA-based cladistic analyses. Chloranthaceae are distinctive among basal angiosperm lineages because they have reduced flowers, opposite and often toothed leaves, and swollen nodes with sheathing stipules. They are likely not related to any of the orders of the Magnoliid group and are variable placed in recent phylogenetic analyses (but perhaps a lineage diverging from remaining angiosperms after *Amborella*, Nymphaeaceae, and Austrobaileyales, or possibly sister to monocots).

Lauraceae traditionally have been divided into two subfamilies. *Cassytha* is placed in the monotypic subfamily Cassythoideae on the basis of numerous specializations relating to its parasitic and viny habit, while all remaining genera are placed in the heterogeneous, and probably paraphyletic "Lauroideae." Recent molecular studies (Rohwer 2000) suggest that *Cassytha*, along with *Hypodaphnis*, an aberrant genus with a truly inferior ovary, may have diverged early in the evolution of Lauraceae. In another study (Chanderbali 2001) *Cassytha* is placed with *Neocinnamomum*. In either position "Lauroideae" is rendered paraphyletic.

Characters such as wood anatomy, stamen number, arrangement and number of anther locules, tepal persistence and form, fruit and cupule morphology, and inflorescence structure have been stressed in taxonomic groupings within "Lauroideae" (Burger 1988; Rohwer et al. 1991; van der Werff 1991; Rohwer 1993a, 1994; van der Werff and Richter 1996). Three tribes currently are recognized (see van der Werff and Richter 1996). Laureae have apparently racemose to umbellate inflorescences with involucral bracts and introrse anthers of the third whorl, and include *Litsea, Lindera, Laurus, Sassafras,* and *Umbellularia.* Perseeae, such as *Ocotea, Nectandra, Licaria, Persea, Phoebe,* and *Cinnamomum,* have cymose inflorescences lacking involucral bracts, and extrorse anthers of the third whorl. Recent molecular studies cast doubt on these tribal distinctions. Finally, Cryptocaryeae, such as *Beilschmiedia* and *Cryptocarya,* include plants similar to Perseeae, but their inflorescences have the lateral flowers of the three-flowered cymose units not quite opposite. The Cryptocaryeae are likely sister to the rest of the family (Rohwer 2000; Chanderbali 2001). Identification of genera and species is extremely difficult without flowering and fruiting material. Generic delimitation, which is often problematic, is discussed by van der Werff (1991) and Rohwer et al. (1991).

Flowers of Lauraceae are insect-pollinated, with flies and bees being the most common visitors. Modified staminodes that are paired at the base of some of the stamens produce odor and/or separate the stamens of the inner whorl and outer whorls; they sometimes also secrete nectar. In *Persea americana* outcrossing is enforced by a complicated system that involves two floral types, A and B flowers. The flowers have two periods of opening on successive days. In A flowers stigmas are receptive in the morning of the first day, while anthers open in the afternoon of the second day. In B flowers stigmas are receptive in the afternoon of the first day, and anthers shed their pollen during the morning of the second day. All the flowers open on any particular tree will be in the same stage. Thus, trees of both A and B types are necessary for cross-pollination. Many other species of *Persea* (and some other genera) have similar systems. The drupes are dispersed mainly by birds, but dispersal by mammals also occurs. The color of the drupes often contrasts with that of the cupule, increasing the attractiveness of the fruits.

References: Boyle 1980; Burger 1988; Chanderbali 2001; Cronquist 1981, 1988; Dahlgren 1983; Donoghue and Doyle 1989; Doyle and Endress 2000; Doyle et al. 1994; Endress and Igersheim 1997; Kubitzki and Kurz 1984; Qiu et al. 1993, 2000; Renner 1999; Renner and Chanderbali 2000; Renner et al. 1997; Rohwer 1993a, 1994, 2000; Rohwer et al. 1991; Soltis et al. 2000; Takhtajan 1969, 1980; Thorne 1974, 1992; van der Werff 1991; Weberling 1988b; Wood 1958.

Canellales

Winteraceae Lindley
(Winter's Bark Family)

Trees or shrubs; nodes trilacunar; **vessels lacking,** with elongate, slender tracheids only; *with scattered spherical cells containing ethereal oils (aromatic terpenoids).* Hairs usually lacking. *Leaves alternate and spiral,* simple, *entire,* with pinnate venation, blade *with pellucid dots,* abaxial surface of stomates usually plugged by wax deposits; stipules lacking. Inflorescences determinate, sometimes reduced to a single flower, terminal or axillary. Flowers usually bisexual, radial, with short receptacle. *Sepals usually 2–4,* distinct to connate, valvate, sometimes falling as a cap. *Petals usually 5 to many,* distinct, imbricate. *Stamens numerous,* distinct; *filaments ± flattened to laminar, usually poorly differentiated from the anthers;* anthers with connective tissue sometimes extending beyond the apex of the pollen sacs; **pollen grains uniporate, usually released in tetrads.** *Carpels 1 to numerous,* usually distinct, in a single whorl; ovaries superior, with lateral placentation; stigma extending down adaxial surface of style or ± capitate. Ovules 1 to several. Nectaries usually lacking. *Fruit a cluster of follicles or berries,* sometimes becoming connate as they mature; embryo minute; *endosperm homogeneous.*

Floral formula: ∗, (2–4), 5–∞, ∞, 1–∞; berries, follicles

Distribution and ecology: New Guinea, Australia, New Caledonia (and other islands of the southwestern Pacific), Madagascar, South America, and Mexico.

Genera/species: 5/90. **Major genera:** *Tasmannia* (40 spp.) and *Bubbia* (30). The family is not represented in the continental United States or Canada, but *Drimys* occurs in Mexico. Most are understory species of moist forests, cool montane forests, or restricted to swampy habitats.

Economic plants and products: The bark of *Drimys winteri* (Winter's bark) has been used medicinally.

Discussion: Winteraceae are most closely related to Canellaceae, and together, these two families constitute the Canellales. Canellaceae differ from Winteraceae in that their flowers have three sepals, five to 12 petals, six to numerous stamens that are fused into a tube, and two to six, connate carpels. Potential synapomorphies of Canellales include leaves with branched sclereids and flowers with a cortical vascular system; the group's monophyly is supported in both molecular (Chase et al. 1993; Qiu et al. 2000; Soltis et al. 2000) and morphological investigations (Doyle and Endress 2000). The group is characterized by aporphine alkaloids, trilacunar nodes, and sieve tube plastids with starch and protein crystalloids and/or fibers. In some phylogenetic analyses the Canellales have been hypothesized to be most closely related to Magnoliales, while in others they are sister to

Piperales (Doyle and Endress 2000; Graham and Olmstead 2000; Mathews and Donoghue 2000; Qiu et al. 2000); the latter is currently thought to be the more likely position.

Phylogenetic relationships within Winteraceae have been assessed through cladistic analyses of morphology (Vink 1988) and nuclear DNA sequences (Suh et al. 1993; Karol et al. 2000). Morphological and DNA characters agree on the close association of *Zygogynum, Exospermum, Belliolum,* and *Bubbia*. Vink (1988) treats this group as *Zygogynum* s.l. Suh et al. (1993), however, maintain *Bubbia*, which they consider to be sister to the *Zygogynum-Exospermum-Belliolum* clade. *Drimys* possesses the unusual autapomorphy of a reduced, more or less capitate stigma. *Tasmannia* may be sister to the above-mentioned genera, a hypothesis supported by ribosomal DNA sequences and by its low chromosome number. Finally, *Takhtajania* likely is sister to all other members of the family.

The lack of vessels in Winteraceae has often been considered a retained ancestral condition (see discussion in Bailey and Nast 1945; Thorne 1974; Cronquist 1981, 1988), a conclusion that is quite unparsimonious (Young 1981; Doyle and Endress 2000).

Flowers of Winteraceae, which are small to medium-sized, usually with a delicate, whitish corolla, are pollinated by a variety of insects, especially small beetles, thrips, primitive moths, and flies. Pollen is the major pollinator reward, but in some species fluids produced by the stigmas or staminal glands provide additional rewards. Some species of *Tasmannia* are wind pollinated. Many species show self-incompatibility. Vertebrate dispersal characterizes the berry-fruited species.

Winteraceae are often claimed to be the angiosperm family that has retained more ancestral features than any other (see discussion in Cronquist 1969, 1981, 1988 and Thorne 1974, 1992), but this claim depends upon particular assumptions about early angiosperm phylogeny.

References: Bailey and Nast 1945; Cronquist 1969, 1981, 1988; Doyle and Endress 2000; Doyle et al. 1990, 1994; Feild et al. 2000; Gottsberger 1988; Gottsberger et al. 1980; Karol et al. 2000; Keng 1993; Suh et al. 1993; Thien 1980; Thorne 1974, 1992; Vink 1988, 1993; Young 1981.

Piperales

Piperales are here circumscribed broadly, including 5 families, two of which, **Piperaceae** and **Aristolochiaceae,** are treated here. The monophyly of this group is strongly supported by phylogenetic analyses based on DNA sequence characters (Chase et al. 1993; Graham and Olmstead 2000; Qiu et al. 2000; Mathews and Donoghue 2000; Soltis et al. 2000). There is also support from morphology (Doyle and Endress 2000) and possible morphological synapomorphies include two-ranked leaves, sheathing leaf base, nuclear endosperm, a single adaxial prophyll. These plants are also characterized by

stems with swollen nodes, distinct vascular bundles, often forming wood with broad rays, and vessel elements with simple perforations. They are herbs or soft-woody plants, with benzyl-isoquinoline and aporphine alkaloids, often with palmately-veined leaves and more or less 3-merous flowers. Piperales often have been associated with the woody magnoliids (Cronquist 1981, 1988; Thorne 1992), and this view has received support from molecular-based phylogenetic analyses (Graham and Olmstead 2000; Qiu et al. 2000; Mathews and Donoghue 2000; Soltis et al. 2000). However, their 3-merous flowers and the monocot-like sieve tube plastids found in some Aristolochiaceae may suggest an affinity to the monocots (Dahlgren and Clifford 1982; Donoghue and Doyle 1989; Doyle and Endress 2000). Ribosomal RNA sequences and combined molecular/morphological cladistic analyses also indicate that Piperales may be the sister group to the monocots (Zimmer et al. 1989; Doyle et al. 1994; Doyle and Endress 2000). However, the order is here tentatively retained within the magnoliid clade.

Within Piperales, it is evident that Piperaceae are most closely related to Saururaceae (a small family of 4 genera and 6 species). The two families form a clade on the basis of tetracytic stomata, indeterminate, terminal, spicate inflorescences, bilateral floral symmetry, minute pollen grains, orthotropous ovules, perisperm as nutritive tissue in the seed, and anatomical or developmental details of the flowers and seeds (Donoghue and Doyle 1989; Tucker et al. 1993; Doyle et al. 1994; Igersheim and Endress 1998; Doyle and Endress 2000). The Piperaceae + Saururaceae clade is also strongly supported by cladistic analyses based on DNA sequences (Zimmer et al. 1989; Doyle et al. 1994; Qiu et al. 1993, 2000; Soltis et al. 1997, 2000; Mathews and Donoghue 2000). Aristolochiaceae are most closely related to Lactoridaceae and Hydnoraceae (see discussion under Aristolochiaceae).

References: Chase et al. 1993; Cronquist 1981, 1988; Dahlgren and Clifford 1982; Donoghue and Doyle 1989; Doyle and Endress 2000; Doyle et al. 1994; González and Rudall 2001; Graham and Olmstead 2000; Igersheim and Endress 1998; Qiu et al. 1993, 2000; Mathews and Donoghue 2000; Soltis et al. 1997, 2000; Thorne 1992; Tucker et al. 1993; Zimmer et al. 1989.

Piperaceae C. A. Agardh
(Pepper Family)

Herbs to small trees, sometimes epiphytic; *nodes often ± swollen or jointed*; vessel elements with usually simple perforations; **stem with vascular bundles of more than 1 ring or ± scattered;** *with spherical cells containing ethereal oils*; often with alkaloids. Hairs simple. Leaves usually alternate and spiral, simple, entire, with palmate to pinnate venation, *with pellucid dots*; stipules lacking or adnate to petiole (and petiole sometimes sheathing stem). Inflorescences indeterminate, *of thick spikes,* **densely covered**

Key to Major Families of Piperales

1. Flowers minute, lacking a perianth, ± densely clustered in thick spikes; ovary superior, the placentation basal, with a single ovule; fruit a drupe; stems with primary vascular bundles in more than 1 ring, or ± scattered .... **Piperaceae**

1. Flowers conspicuous, with perianth of 3 connate, showy sepals, and sometimes also with 3 petals, not in spikes; ovary inferior or half-inferior, the placentation axile or parietal with intruded placentas; fruit a septicidal capsule or cluster of follicles; stem with primary vascular bundles in a single ring .... **Aristolochiaceae**

---

with minute flowers, terminal or axillary, often displaced to a position opposite the leaf due to development of axillary shoot. **Flowers** bisexual or unisexual (plants monoecious or dioecious), appearing to be radial, **inconspicuous, each with a broadly triangular to peltate bract.** *Perianth lacking.* Stamens 1–10, often 6; filaments usually distinct; pollen grains monosulcate or lacking apertures. **Carpels 1–4, connate**; ovary superior, *with basal placentation*; stigmas 1–4, capitate, lobed, or brushlike. **Ovule 1 per gynoecium**, orthotropous, with 1 or 2 integuments. Nectaries lacking. **Fruit usually a drupe**; endosperm scanty, supplemented by perisperm (Figure 9.14).

Floral formula: $*, -0-, 1–10, \underline{\textcircled{1–4}}$; drupe

Distribution and ecology: Widely distributed in tropical and subtropical regions. Species of *Peperomia* are commonly epiphytes in moist broad-leaved forests.

Genera/species: 6/2020. *Major genera: Peperomia* (1000 spp.) and *Piper* (1000). Both occur within the continental United States.

Economic plants and products: Fruits of *Piper nigrum* provide black and white pepper, one of the oldest and most important spices. Leaves of *Piper betle* (betel pepper) are chewed (along with various spices and fruits of *Areca catechu*, betel nut) and have a mildly stimulating effect. A few species of *Piper* are used medicinally. Several species of *Peperomia* are grown as ornamentals because of their attractive foliage.

Discussion: Piperaceae are clearly monophyletic (Tucker et al. 1993). Members of the related family Saururaceae can be easily separated from Piperaceae by their two to ten ovules per carpel and distinct styles (and often distinct carpels, both plesiomorphies), although Saururaceae probably are monophyletic (Tucker et al. 1993).

Within Piperaceae, *Zippelia* is the sister group to the remaining species, which form a clade on the basis of a reduction to three carpels and capitate stigmas (Tucker et al. 1993). Relationships among the remaining genera are problematic. *Piper* (circumscribed broadly, including

Macropiper, Pothomorphe) is extremely diverse, varying in inflorescence position (terminal or axillary), structure (solitary or clustered spikes or racemes), sexuality (bisexual or unisexual, and the plants then dioecious), and number of stamens. Recent cladistic analysis based on nrDNA ITS sequences support the recognition of three major clades within the genus: South Pacific species, Asian species, and Neotropical species (Jaramillos and Manos 2001). *Piper* may be most closely related to *Peperomia* on the basis of an unusual development of the female gametophyte. *Peperomia* is the most derived member of this family, and shows numerous apomorphies, such as a single carpel, two unilocular stamens, a 16-nucleate female gametophyte (vs. 8 nuclei in other genera), ovules with a single integument, unusually small pollen grains that lack apertures, herbaceous habit, and succulent leaves.

The tiny flowers of Piperaceae probably are insect-pollinated, but more work on the pollination biology of the family is needed. Outcrossing may predominate due to protogyny. The drupes of *Piper* are mainly dispersed by birds and bats, while those of *Peperomia* are often sticky and may be externally transported by animals.

References: Borstein 1991; Jaramillos and Manos 2001; Semple 1974; Tebbs 1993; Thorne 1974, 1992; Tucker et al. 1993; Wood 1971.

Aristolochiaceae Adanson
(Dutchman's-Pipe Family)

Herbs, lianas, or occasionally shrubs; with spherical cells containing ethereal oils; **with aristolochic acids (bitter, yellow, nitrogenous compounds)** or alkaloids. Hairs simple. *Leaves alternate and spiral,* simple, sometimes lobed, *entire, with palmate venation, with pellucid dots*; stipules usually lacking. Inflorescences various. Flowers bisexual, radial to bilateral. *Sepals 3, connate, often bilateral, tubular, and S-shaped or pipe-shaped, with spreading 3-lobed to 1-lobed limb*, **showy**, dull red and mottled, valvate, and deciduous (distinct and green in *Saruma*). Petals usually lacking or vestigial, but in *Saruma* present and well developed, distinct, yellow, and imbricate. Stamens usually 6–12; filaments distinct, often adnate to

Figure 9.14 Piperaceae. (A) *Peperomia glabella:* leaf (× 0.75). (B–G) *P. humilis:* (B) flowering shoot (× 0.15); (C) portion of inflorescence (× 27); (D) flower with bract, showing gynoecium and two stamens (× 27); (E) gynoecium (× 27); (F) portion of spike with a mature drupe (× 18); (G) endocarp of drupe (× 27). (H–L) *P. obtusifolia:* (H) flowering shoot (× 0.75); (I) flower with bract, showing gynoecium and two stamens (× 27); (J) bract (× 27); (K) tip of spike with partly mature fruits, three removed to show immersion of base of fruit in tissue of axis, apex of spike with undeveloped flowers shown only in outline (× 9); (L) mature drupe, showing hooked apex and position of stigma (× 1991). (From Borstein 1991, *J. Arnold Arbor. Suppl. Ser.* 1, p. 359.)

style; pollen grains usually lacking apertures (monosulcate in *Saruma*). Carpels 4–6, connate (distinct in *Saruma*), often twisting in development; **ovary/ovaries half-inferior**, *inferior in the more derived taxa,* with axile placentation, or parietal with intruded placentas; stigmas 4–6, often lobed and spreading. Ovules numerous. Nectaries often of patches of glandular hairs on calyx tube. *Fruit a septicidal capsule* (cluster of follicles in *Saruma*), often pen-

dulous and opening from the base; seeds flattened, winged, or associated with fleshy tissue (Figure 9.15).

Floral formula:

* or X, ③, 3 or 0, 6–12, $\overline{(4-6)}$; capsule

Distribution and ecology: Widespread in tropical and temperate regions.

Figure 9.15   Aristolochiaceae. *Aristolochia tomentosa:* (A) branch with flower and bud 0.5; (B) flower (× 1); (C) flower with half of perianth and ovary removed, and detail of ovules in a single locule (× 1); (D) ovary in cross section (× 3); (E) receptive stigmas, at opening of flower (× 4); (F) stigmas at time of shedding of pollen (but adnate anthers shown as closed) now folded inward (× 4); (G) fruit (× 1); (H) seed, upper surface (× 1.5); (I) seed, lower surface, with corky funiculus (× 1.5); (J) seed, lower surface, corky funiculus removed (× 1.5). (From Wood 1974, *A student's atlas of flowering plants,* p. 21.)

Genera/species: 7/460. ***Major genera:*** *Aristolochia* (370 spp.), *Asarum* (70). *Asarum* and *Aristolochia* occur in the continental United States.

Economic plants and products: Many species of *Asarum* (wild ginger, incl. *Hexastylis*) and *Aristolochia* (Dutchman's-pipe; incl. *Isotrema, Pararistolchia*) are cultivated as ornamentals because of their unusual flowers or variegated leaves. Some species of *Aristolochia* are used medicinally.

Discussion: Some studies using DNA sequences indicate that the Aristolochiaceae may be paraphyletic, with *Lactoris* (Lactoridaceae) placed within it (see Qiu et al. 2000; Soltis et al. 2000), but the family here is considered monophyletic on the basis of its usually more or less inferior ovary and distinctive chemistry. The achlorophyllous root-parasitic Hydnoraceae, which have syntepalous flowers that arise endogenously from roots of the host, may also be closely related to the Aristolochiaceae (D. Nickrent, pers. com.).

Thorne (1992) and Huber (1993) divided Aristolochiaceae into Asaroideae, including *Asarum* and *Saruma,*

which are perennial herbs with radial flowers in which the stamens are free or only slightly adnate to the gynoecium, and Aristolochioideae, including *Aristolochia,* which are subshrubs to lianas with usually bilateral flowers in which the stamens are strongly adnate to the style.

Flowers of Aristolochiaceae are mainly fly-pollinated. The pollination syndrome is quite specialized in *Aristolochia,* which has elaborate trap flowers with a highly modified calyx. Flies are attracted to these flowers by their dull red and often mottled coloration and odor (which may be fruity to fetid). Nectar, produced by glandular hairs on the calyx tube, functions as a pollinator reward. Pollen-bearing flies become trapped within an inflated portion of the calyx tube, which may be entered only through the constricted opening, which often is provided with downward-pointing hairs. During the first phase of the flower's life, pollen is deposited on the receptive stigmas that crown the gynoecium, projecting over the still unopened anthers. After pollination, the stigmas wither and become erect, thus exposing the dehiscing anthers. The flies become covered with pollen, and are allowed to leave when the downward-pointing hairs and/or the calyx tube wither. The seeds of *Aristolochia*, which are typically flattened, are dispersed by wind from the hanging, parachute-like capsules, although some species are water-dispersed, or have sticky seeds, leading to external transport by animals; a few have fleshy fruits and are vertebrate dispersed. Dispersal of the seeds by ants is common in herbaceous taxa (e.g., *Asarum*); these seeds possess an aril-like structure.

References: Cronquist 1981; Faegri and van der Pijl 1980; Huber 1993; Kelly 1997; Qiu et al. 2000; Soltis et al. 2000; Thorne 1974, 1992.

A CLADE OF UNCERTAIN POSITION

Ceratophyllales

Ceratophyllaceae S. F. Gray
(Hornwort Family)

Submersed, aquatic herbs; roots lacking, but often with colorless rootlike branches anchoring the plant; stems with a single vascular strand with central air canal surrounded by elongate starch-containing cells; with tannins. Leaves whorled, simple, often **dichotomously dissected,** entire to serrate, **lacking stomates and a cuticle;** stipules lacking. **Inflorescences of solitary, axillary flowers. Flowers unisexual (plants monoecious),** *radial, inconspicuous,* **with a whorl of 7 to numerous bracts** (possibly tepals). *Stamens 10 to numerous,* distinct; filaments not clearly differentiated from anthers; anthers with connective prolonged beyond pollen sacs, and forming 2 prominent teeth; **pollen grains lacking apertures, with reduced exine, forming branched pollen tubes. Carpel 1;** ovary superior, **with ± apical placentation;** stigma elongated and extending along one side of style. Ovule 1 per carpel, **orthotropous, with 1 integument.** Nectaries lacking. **Fruit an achene, often with 2 or more projections, along with persistent style;** seed with testa ± obliterated; endosperm lacking**.

Floral formula: Staminate: *, -7–∞-, 10–∞, 0
Carpellate: *, -7–∞-, 0, <u>1</u>; achene

Distribution and ecology: Cosmopolitan; forming floating masses in freshwater habitats.

Genus/species: 1/6 **Genus:** *Ceratophyllum.*

Economic plants and products: *Ceratophyllum* is ecologically important in that it provides protection from predation for newly hatched fish; the foliage and fruits are important food items for migratory waterfowl; sometimes it is weedy, choking waterways.

Discussion: This family shows numerous adaptations to life as a submerged aquatic herb, has a fossil record extending back to the Early Cretaceous, and is an ancient and highly modified taxon of uncertain affinity. In the Chase et al. (1993) cladistic analysis based on *rbcL* nucleotide sequences the Ceratophyllaceae is the sister group of all other angiosperms. However, in more recent molecular analyses the family is sister to the monocots (Graham and Olmstead 2000; Qiu et al. 2000). In an analysis based on 18S, *rbcL*, and *atpB* sequences, however, it is weakly placed as the sister group of the eudicots (Soltis et al. 2000).

The species of *Ceratophyllum* are highly variable and taxonomically difficult. Relationships within the family are based on variation in the leaves and fruits.

The inconspicuous flowers of Ceratophyllaceae are submerged, and pollen is dispersed by water currents, as are the achenes, although dispersal by birds also occurs. The persistent style and variously developed appendages attach the small fruits to vegetation or sediments. Vegetative reproduction by fragmentation is common.

References: Chase et al. 1993; Cronquist 1981; Dahlgren 1989; Dahlgren 1983; Dilcher 1989; Endress 1994a; Graham and Olmstead 2000; Les 1988, 1989, 1993; Les et al. 1991; Qiu et al. 1993, 2000; Soltis et al. 2000; Takhtajan 1980; Thorne 1974, 1992; Wood 1959a.

MONOCOTS

The monocots are considered monophyletic based on their parallel-veined leaves, embryo with a single cotyledon, sieve cell plastids with several cuneate protein crystals, stems with scattered vascular bundles, and adventitious root system. Sieve cell plastids with several protein crystals also occur in some Aristolochiaceae (*Saruma* and *Asarum*), while scattered vascular bundles and adventitious roots also occur in the Nymphaeaceae and some Piperaceae. Several monocots have pinnate to palmate leaves with obviously reticulate venation patterns (see Dahlgren et al. 1985; Chase et al. 1995b), but these are probably reversals associated with life in shaded forest understory habitats. In addition, the leaves of most monocots, even those with a well-developed blade and petiole, are formed almost entirely from the basal end of the leaf primordium, while the leaves of nonmonocots are mainly derived from the apical end of the primordium. Several features often considered to characterize monocots, such as 3-merous flowers with two perianth whorls and herbaceous habit, also occur in Nymphaeaceae and some magnoliids. Monocots typically have monosulcate pollen, probably a retention of an ancestral angiosperm feature. It is noteworthy that monocots never have glandular-toothed leaves, any teeth being more or less spiny (as in Ceratophyllaceae).

The monophyly of the monocots is supported by DNA sequences and morphology (Chase et al. 1993, 1995a,b, 2000; Bharathan and Zimmer 1995; Stevenson and Loconte 1995; Soltis et al. 1997, 2000; Savolainen et al. 2000a). The taxonomic diversity of monocots is presented in detail by Kubitzki (1998a,b).

Alismatales

Cladistic analyses of both nuclear and chloroplast DNA sequences (Chase et al. 1993, 1995b, 2000; Duvall et al. 1993; Källersjö et al. 1998; Soltis et al. 2000) support the monophyly of the Alismatales, as do the possible morphological synapomorphies of stems with small scales or glandular hairs within the sheathing leaf bases at the nodes, extrorse anthers, and a large embryo (Dahlgren and Rasmussen 1983; Dahlgren et al. 1985; Stevenson and Loconte 1995). The Araceae are sister to the remaining families of the order, which constitute a subclade based on their seeds lacking endosperm and root hair cells shorter than other epidermal cells. All members of this subclade occur in wetland or aquatic habitats. Two major clades are recognized within this aquatic clade. The first contains Alismataceae, Hydrocharitaceae, and Butomaceae, and is supported by the apomorphies of perianth differentiated into sepals and petals, stamens more than six and/or carpels more than three (a secondary increase), and ovules scattered over the inner surface of the locules. Major families constituting the second clade are Potamogetonaceae, Ruppiaceae, Zosteraceae, Posidoniaceae, Zannichelli-aceae, and Cymodoceaceae. This group is diagnosed on the basis of pollen that usually lacks apertures and more or less lacks an exine (Dahlgren and Rasmussen 1983; Cox and Humphries 1993). Plants of marine habitats evolved within Hydrocharitaceae, and more than twice in the sea grass families (Zosteraceae, Cymodoceaceae, Ruppiaceae, and Posidoniaceae) (Les et al. 1997a). The order probably diverged very early in the evolution of monocots. Alismatales contains about 14 families and about 3320 species; major families include **Araceae**, **Alismataceae**, **Hydrocharitaceae**, Butomaceae, **Potamogetonaceae**, Ruppiaceae, Zosteraceae, Posidoniaceae, Zannichelliaceae, and Cymodoceaceae.

References: Chase et al. 1993, 1995b; Cox and Humphries 1993; Dahlgren and Rasmussen 1983; Dahlgren et al. 1985; Duvall et al. 1993; Källersjö et al. 1998; Les et al. 1997; Soltis et al. 2000; Stevenson and Loconte 1995.

Araceae Adanson
(Arum Family)

Terrestrial to aquatic herbs, often with rhizomes or corms, vines with aerial roots, epiphytes, or floating aquatics, the latter often very reduced with ± thalloid vegetative body; **grooved raphide crystals** *of calcium oxalate present, and asso-*

Key to Major Families of Alismatales

1. Flowers on a thick axis, the spadix, surrounded by or associated with a leaflike bract, the spathe; plants of various habitats . . . . **Araceae**

1. Flowers not on a spadix; plants aquatic or of wetlands 2

2. Perianth of sepals and petals 3

2. Perianth of tepals or lacking 5

3. Ovary inferior; carpels connate; fruits berries . . . . **Hydrocharitaceae**

3. Ovary superior; carpels ± distinct; fruits follicles or achenes 4

4. Pollen grains monosulcate; laticifers lacking and sap watery; fruits follicles; embryo straight . . . . Butomaceae

4. Pollen grains usually 4-multiporate or occasionally lacking apertures; laticifers present and sap milky; fruits achenes or follicles; embryo curved . . . . **Alismataceae**

5. Pollen grains globose or ellipsoid; plants of fresh, alkaline, or brackish water 6

5. Pollen grains threadlike; plants of marine environments 8

6. Flowers bisexual 7

6. Flowers unisexual . . . . Zannichelliaceae

7. Stamens 2, each with a small appendage; carpels on long stalks . . . . Ruppiaceae

7. Stamens 4, each with a large appendage; carpels sessile . . . . **Potamogetonaceae**

8. Flowers bisexual; stamens 3; carpel 1 . . . . Posidoniaceae

8. Flowers unisexual; stamens 1 or 2; carpels 2 9

9. Carpels distinct; stamens 2, connate . . . . Cymodoceaceae

9. Carpels connate; stamen 1 . . . . Zosteraceae

ciated chemicals causing irritation of mouth and throat if eaten; cyanogenic compounds often present, and sometimes with alkaloids; often with laticifers, mucilage canals, or resin canals, and latex watery to milky. Hairs simple, but often lacking. *Leaves alternate and spiral or 2-ranked,* sometimes basal, usually simple, *blade often well developed,* sometimes strongly lobed, pinnately to palmately compound, usually entire, with parallel, pinnate, or palmate venation, sheathing at base; stipules lacking, but glandular hairs or small scales sometimes present at the node inside the leaf sheath. **Inflorescences indeterminate,** usually terminal, **forming a spike of numerous small flowers packed onto a fleshy axis (a spadix), which may be lacking flowers** **toward its apex, which is subtended by a large leaflike to petal-like bract (a spathe),** but reduced in floating aquatic taxa. Flowers bisexual to unisexual (plants usually monoecious), radial, **lacking individual bracts.** Tepals usually 4–6 or lacking, distinct to connate, inconspicuous and often fleshy, valvate or imbricate. Stamens 1–6 (–12); filaments distinct to connate; anthers sometimes opening by pores, distinct to connate; pollen grains various. Carpels usually 2–3, connate; ovary usually superior, placentation various; stigma 1, punctate or capitate. Ovules 1 to numerous, anatropous to orthotropous. Nectaries lacking. *Fruit usually a berry,* but occasionally an utricle, drupe, or nutlike; endosperm sometimes lacking (Figure 9.16).

Figure 9.16 Araceae. (A–J) *Arisaema triphyllum:* (A) habit (× 0.2); (B) spathe, apical view (× 0.75); (C) staminate inflorescence, spathe removed (× 1.5); (D) portion of staminate spadix (× 9.75); (E) staminate flower (× 11.5); (F) carpellate inflorescence, spathe removed (× 1.5); (G) carpellate flower, longitudinal section, showing ovules (× 11.5); (H) ovule (× 30); (I) fruiting spadix (× 0.75); (J) fruit in longitudinal section, showing seed (× 3). (K–M) *A. triphyllum* var. *stewardsonii:* (K) inflorescence, lateral view (× 0.75); (L) inflorescence, back view (× 0.75); (M) spathe, apical view (× 0.75). (N) *A. dracontium:* habit (× 0.2). (From Wilson 1960a, *J. Arnold Arbor.* 41: p. 59.)

Floral formula: *, -4-6- or -0-, 1–6,(1–3); berry, utricle

Distribution and ecology: Cosmopolitan, but best developed in tropical and subtropical regions; very common in tropical forests and wetlands.

Genera/species: 109/2830. ***Major genera:*** *Anthurium* (900 spp.), *Philodendron* (500), *Arisaema* (150), *Homalomena* (140), *Amorphophallus* (100), *Schismatoglottis* (100), *Spathiphyllum* (60), *Monstera* (50), *Pothos* (50), *Xanthosoma* (40), *Dieffenbachia* (40), and *Syngonium* (30). Noteworthy genera of the continental United States and/or Canada are *Arisaema, Landoltia, Lemna, Lysichiton, Orontium, Peltandra, Pistia, Spirodela, Symplocarpus,* and *Wolffia.*

Economic plants and products: The starchy corms of *Alocasia, Colocasia* (taro), and *Xanthosoma* (yautia) are eaten (after proper treatment to remove the irritating chemicals). The berries of *Monstera* occasionally are eaten. The family contains numerous ornamentals, including *Philodendron, Zantedeschia* (calla lily), *Anthurium, Caladium, Colocasia, Dieffenbachia* (dumbcane), *Epipremnum, Monstera, Spathiphyllum, Syngonium, Aglaonema, Xanthosoma* (elephant's-ear), *Scindapsus, Spathicarpa,* and *Zamioculca.*

Discussion: Araceae are considered monophyletic based on morphology (Grayum 1990; Mayo et al. 1995) and cpDNA sequences (Chase et al. 1993; French et al. 1995). The family may be a fairly early divergent lineage within monocots and probably is sister to the remaining families of Alismatales (Dahlgren and Rasmussen 1983; Dahlgren et al. 1985; Chase et al. 1995b; Stevenson and Loconte 1995).

Araceae have been divided into several subfamilies based on variation in habit, leaf arrangement and morphology, inflorescence structure, floral morphology, pollen structure, anatomy, and chromosome number (Grayum 1990). Phylogenetic relationships have also been assessed within Araceae through the use of *rbcL* sequences (French et al. 1995) and morphology (Mayo et al. 1997). A few genera of very reduced floating aquatics, including *Spirodela, Landoltia, Lemna, Wolffia,* and *Wolfiella,* are often segregated as Lemnaceae (see den Hartog 1975; Landolt 1980, 1986; Landolt and Kandeler 1987; Cronquist 1981; Dahlgren et al. 1985). In *Lemna, Landoltia,* and *Spirodela* the spathe is represented by a membranous sheath, while it is completely lacking in *Wolffia* and *Wolfiella.* It is reasonably clear that these genera have been derived from less modified members of the subfamily Aroideae (French et al. 1995; Mayo et al. 1995; Stockey et al. 1997). Although *rbcL* sequences also place the much larger floating aquatic aroid *Pistia* in the Aroideae (see below), these genera may not be sister taxa (French et al. 1995).

Gymnostachyos, Orontium, and relatives have condensed non-cormlike, thickened stems and are sister to a large clade, the core aroids, which are united by their conspicuous spathe, leaves with expanded blades, major internode of the inflorescence between the spathe and next leaf below, the formation of a continuation shoot in the axil of the penultimate leaf below the spathe, and more or less basal placentation. Within the core aroids, the monoecious taxa comprise a large subclade, the Aroideae (incl. 74 genera, such as *Zamioculca, Dieffenbachia, Spathicarpa, Philodendron, Caladium, Syngonium, Aglaonema, Zantedeschia, Amorphophallus, Peltandra, Asarum, Arum, Arisaema, Alocasia, Colocasia,* and *Pistia* (Mayo et al. 1997). A second major clade within the core aroids, the Monsteroideae, is delimited on the basis of a undifferentiated spathe (i.e., lacking a tubular portion) that is soon deciduous or marcescent, with a distinct basal abscission zone. This tribe includes genera such as *Monstera, Scindapsus,* and *Epipremnum; Spathiphyllum* and relatives may be related (French et al. 1995). *Pothos* and relatives, and probably also *Anthurium,* form the Pothoideae, another core aroid subgroup; these plants are characterized by fine leaf venation with the secondary and tertiary veins forming cross veins to the primaries (Mayo et al. 1997).

Acorus has been removed from the Araceae and placed in its own family, the Acoraceae (Grayum 1987, 1990; Bogner and Nicolson 1991; Thorne 1992; Bogner and Mayo 1998), as suggested by morphology, secondary compounds, and DNA sequence characters. *Acorus* is a wetland herb with narrow, equitant leaves, small bisexual flowers with six tepals, six stamens, and two or three fused carpels, which are borne on a spadix, and berry fruits. Unlike Araceae, Acoraceae contain ethereal oils in specialized spherical cells and lack raphide crystals. Cladistic analyses based on *rbcL* sequences (Chase et al. 1993, 1995b) suggest that *Acorus* is the sister group to the remaining monocots.

Inflorescences of Araceae are pollinated by several groups of insects, especially beetles, flies, and bees. The inflorescence usually produces a strong odor (sweet to noxious) and often heat. The gynoecium matures before the androecium, and when the flowers are unisexual, the carpellate flowers mature before the staminate, leading to outcrossing. In *Arisaema,* small (generally young) plants are staminate and larger (older) plants are carpellate, again leading to outcrossing. Dispersal of the green to brightly colored berries is presumably by birds or mammals. The utricles of *Lemna* and relatives are water-dispersed.

References: Bogner and Mayo 1998; Bogner and Nicolson 1991; Chase et al. 1993, 1995b; Croat 1980; Cronquist 1981; Dahlgren and Rasmussen 1983; Dahlgren et al. 1995; den Hartog 1975; French et al. 1995; Grayum 1987, 1990; Igersheim et al. 2001; Landolt 1980, 1986; Landolt and Kandeler 1987; Les and Crawford 1999; Les et al. 1997b, 2002; Maheshwari 1958; Mayo et al. 1995, 1998; Prychid and Rudall 2000; Ray 1987a,b; Stevenson and Loconte 1995; Stockey et al. 1997; Wilson 1960a.

Alismataceae Ventenat
(Water Plantain Family)

Aquatic or wetland, rhizomatous herbs; **laticifers present, the latex white**; tissues ± aerenchymatous. Hairs usually lacking. Leaves alternate, spiral or 2-ranked, usually ± basal, simple, entire, usually with a well-developed blade, with parallel or palmate venation, sheathing at base; sometimes polymorphic, with narrower submerged, and broader floating and/or emergent blades; stipules lacking; small scales present at the node inside the leaf sheath. Inflorescences determinate, but often appearing indeterminate, with branches or flowers often ± whorled, terminal, borne at the apex of a scape. Flowers bisexual or unisexual (plants then monoecious), radial, *with perianth differentiated into a calyx and corolla. Sepals 3, distinct, imbricate. Petals 3, distinct, imbricate and crumpled, usually white or pink. Stamens usually 6 to numerous;* filaments distinct; **pollen grains usually 2- to polyporate**. *Carpels (3–) 6 to numerous,* distinct, ovaries superior, with ± basal placentation; stigma 1, minute. Ovules few to more commonly 1 per carpel. Nectaries at base of carpels, stamens, or perianth parts. *Fruit a cluster of achenes* (rarely follicles); **embryo strongly curved**; endosperm lacking (Figure 9.17).

Floral formula: *, 3, 3, 6–∞, <u>6–∞</u>; achenes

Distribution and ecology: Widely distributed; plants of freshwater marshes, swamps, lakes, rivers, and streams.

Genera/species: 16/100. **Major genera:** *Echinodorus* (45 spp.) and *Sagittaria* (35). The family is represented in the continental United States and/or Canada by the above and *Alisma, Damasonium,* and *Limnocharis*.

Economic plants and products: *Sagittaria* (arrowhead), *Alisma* (water plantain), *Echinodorus* (burr-heads), and *Hydrocleis* (water poppy) provide pond and/or aquarium ornamentals. The rhizomes of *Sagittaria* may be eaten.

Discussion: Alismataceae are defined broadly (including the Limnocharitaceae; see Pichon 1946; Thorne 1992) and considered monophyletic on the basis of morphological characters (Dahlgren et al. 1985). The genera with achenes and only a single basal ovule per carpel (e.g., *Alisma, Sagittaria,* and *Echinodorus*) may form a monophyletic subgroup (Chase et al. 1993, 1995b).

Species are often difficult to identify due to extensive variation in leaf morphology, which correlates with environmental parameters such as light intensity, water depth, water chemistry, and rate of flow (Adams and Godfrey 1961). Submerged leaves are usually linear, while floating or emergent leaves are petiolate with elliptic to ovate blade with an acute to sagittate base. Several different leaf forms may occur on the same plant.

Alismataceae (and Butomaceae) have often been considered to represent primitive monocots (Cronquist 1981; Hutchinson 1973) due to their numerous, distinct stamens and carpels, the result of bearing flowers that are superficially rather like those of Ranunculaceae. However, developmental and anatomical studies have indicated that these numerous stamens are actually due to secondary increase from an ancestral condition of two whorls of three parts.

The showy flowers of Alismataceae are pollinated by various nectar-gathering insects (often bees and flies). In *Alisma* and *Echinodorus* the flowers are bisexual, while they are usually unisexual in *Sagittaria*. The achenes are often dispersed by water; they float due to the presence of spongy tissue and are resinous on the outer surface. They are also eaten (and dispersed) by waterfowl.

References: Adams and Godfrey 1961; Chase et al. 1993, 1995b; Cronquist 1981; Dahlgren et al. 1985; Haynes et al. 1998; Hutchinson 1973; Pichon 1946; Rogers 1983; Thorne 1992; Tomlinson 1982.

Hydrocharitaceae A. L. de Jussieu
(Frog's-Bit or Tape Grass Family)

Aquatic herbs, completely submerged to partly emergent, and rooted in the substrate or floating and unattached, in freshwater or marine habitats, often rhizomatous; tissues ± aerenchymatous. Hairs unicellular, thick-walled, prickle-like along leaf margins and/or veins. Leaves alternate and spiral, opposite, or whorled, along stem or in a basal rosette, simple, entire or serrate, sometimes with a well-developed blade, with parallel or palmate venation, or only midvein evident, sheathing at base; stipules lacking; small scales present at the node inside the leaf sheath. Inflorescences determinate, sometimes reduced to a solitary flower, axillary, subtended by 2 often connate bracts. Flowers bisexual or unisexual (plants then monoecious or dioecious), usually radial, *with perianth differentiated into calyx and corolla. Sepals 3, distinct, valvate. Petals 3, distinct,* usually white, imbricate, *sometimes lacking.* Stamens 2 or 3 to numerous; filaments distinct to connate; pollen grains usually lacking apertures, in *Thalassia* and *Halophila* united into threadlike chains. *Carpels usually 3–6, connate;* **ovary inferior**, with ovules scattered over surface of locules, the placentae often ± deeply intruded; styles often divided, appearing twice the number of carpels; stigmas elongate and papillose. Ovules numerous. Nectar often secreted from staminodes. **Fruit fleshy,** *either a berry or irregularly to valvately opening capsule;* embryo sometimes curved; endosperm lacking.

Floral formula:

Staminate: *, 3, 3, 2–∞, 0

Carpellate: *, 3, 3, 0, $\overline{(3–6)}$; berry, fleshy capsule

Distribution and ecology: Widely distributed, but most common in tropical and subtropical regions, in freshwa-

Figure 9.17 Alismataceae. (A–M) *Sagittaria latifolia:* (A) flowering plant with rhizomes (× 0.4); (B) staminate flower, bracts and pedicels of two other flowers below (× 3); (C) central part of staminate flower in longitudinal section, note sterile carpels in center (× 4.5); (D) stamen with anther dehiscing (× 9); (E) two stamens and sterile carpel (× 9); (F) carpellate flower in longitudinal section, immediately after fall of petals (× 4.5); (G) carpel from carpellate flower (× 18); (H) ovule (× 38); (I) fruiting "head," an aggregate of achenes (× 1.5); (J, K) opposite sides of mature achene, note resin ducts and crest on side in K (× 9); (L) seed (× 18); (M) embryo (× 18). (N–P) *S. australis:* (N, O) opposite sides of achene (× 9); (P) seed (× 18). (Q) *S. lancifolia:* achene with one resin duct (× 18). (R) *S. graminea:* achene with one resin duct above crest and one below it (× 18). (S) *S. engelmanniana:* achene with crest and several resin ducts (× 9). (From Rogers 1983, *J. Arnold Arbor.* 64: p. 410.)

ter (most genera) and marine habitats (*Enhalus, Halophila, Thalassia*).

Genera/species: 17/100. **Major genera:** *Ottelia* (40 spp.) and *Elodea* (15). *Egeria, Elodea, Halophila, Hydrilla, Hydrocharis, Limnobium, Najas, Ottelia, Thalassia,* and *Vallisneria* occur in the continental United States and/or Canada.

Economic plants and products: Several genera, including *Hydrilla, Egeria,* and *Elodea* (waterweeds), *Vallisneria* (tape grass), and *Limnobium* (frog's-bit), are used as aquarium plants. Species of *Elodea, Hydrilla,* and *Lagarosiphon* are pernicious aquatic weeds.

Discussion: Hydrocharitaceae, although monophyletic (Dahlgren and Rasmussen 1983), are morphologically heterogeneous, and have been divided into three to five subfamilies (Dahlgren et al. 1985). *Najas* has very reduced flowers with a basal, erect ovule, but its placement within Hydrocharitaceae is supported by seed-coat anatomy and DNA sequences (Les et al. 1993; Les and Haynes 1995).

The family shows an interesting array of pollination mechanisms. Several species in *Egeria, Limnobium, Stratiotes,* and *Blyxa* have showy flowers that are held above the surface of the water and pollinated by various, usually nectar-gathering insects. In *Vallisneria, Enhalus,* and *Lagarosiphon,* the staminate flowers become detached and float on the water's surface, where they come into contact with the carpellate flowers (see Chapter 4). In *Elodea* the staminate flowers either detach or remain attached; their anthers explode, scattering pollen grains on the water's surface. In *Hydrilla* pollen transport may occur by wind or water. Finally, in *Thalassia* and *Halophila* pollination takes place underwater. Both outcrossing or selfing can occur. The fleshy fruits ripen below the water surface; fruits and/or seeds are either water- or animal-dispersed. Vegetative reproduction by fragmentation of rhizomes is common.

References: Cook 1982, 1998; Cox and Humphries 1993; Cox and Tomlinson 1988; Dahlgren et al. 1985; Dahlgren and Rasmussen 1983; Haynes 1988; Haynes and Holm-Nielsen 2001; Kaul 1968, 1970; Les and Haynes 1995; Les et al. 1993; Lowden 1982; Tomlinson 1969b.

Potamogetonaceae B. C. J. Dumortier
(Pondweed Family)

Aquatic, rhizomatous herbs. Stems with reduced vascular bundles often in a ring, with air cavities; tannins often present. Hairs lacking. Leaves alternate and spiral, or opposite, blade sometimes well developed, simple, entire, with parallel venation or with only a single midvein, sheathing at base, the sheath open, **and ± separated from blade so that it appears to be a stipule**, the leaves sometimes heteromorphic, with submerged and floating forms; 2 to several small scales present at the node, inside the leaf sheaths. *Inflorescences* indeterminate, terminal and axillary, *spikelike* and elevated above or lying on water's surface. Flowers bisexual, radial, *not associated with bracts* (at maturity). Tepals lacking. **Stamens 4, with well-developed appendages at base of anther that form what appears to be a ± fleshy perianth**; pollen grains without functional aperture, globose to ellipsoid. **Carpels usually 4**, distinct; *ovaries superior, with ± basal to apical placentation*; stigma 1, truncate to capitate. *Ovule 1, ± anatropous to orthotropous.* Nectaries lacking. *Fruit a cluster of achenes or drupes*; endosperm lacking.

Floral formula:

*, -0-, 4 (appendaged), 4; achenes, drupes

Distribution and ecology: Cosmopolitan; herbs of lakes, rivers, and other wetland habitats.

Genera/species: 4/100. **Major genera:** *Potamogeton* (90 spp.) and *Stuckenia* (6); both occur in North America.

Economic plants and products: Although the family is of little direct economic importance, many species provide wildlife food.

Discussion: *Ruppia,* a genus of alkaline, brackish, or occasionally salt water, is often placed here, but its inclusion makes the family biphyletic (Les et al. 1997a). Ruppiaceae are characterized by flowers with two stamens that have minute appendages, pollen slightly elongated, and carpels with long stalks.

In *Potamogeton* and *Stuckenia* the flowers are raised above the surface of the water and wind-pollinated, while flowers of *Ruppia* are held at the water surface and are water-pollinated. The fruits are animal- or water-dispersed.

References: Haynes 1978; Haynes et al. 1998b; Les et al. 1997a.

Petaloid Monocots

The Liliales, Asparagales, and Dioscoreales together constitute the petaloid monocots (Dahlgren et al. 1985; Thorne 1992; Chase et al. 1995b), a group sometimes referred to as the superorder Lilianae. These plants traditionally have been considered to be related, and the group is characterized by flowers with showy tepals (or petals) and endosperm lacking starch. The monophyly of Lilianae has been questioned (Goldblatt 1995; Stevenson and Loconte 1995; Chase et al. 1995a), and although it has received some support in analyses based on morphology and *rbcL* sequences (Chase et al. 1995b) more recent, multi-gene analyses suggest that the assemblage is not monophyletic (Chase et al. 2000; Soltis et al. 2000).

Liliales

The monophyly of Liliales is supported by cladistic analyses based on morphology and *rbcL, atpB,* and 18S sequences (Chase et al. 1995a,b; Stevenson and Loconte 1995; Goldblatt 1995; Källersjö et al. 1998; Chase et al. 2000; Soltis et al. 2000). The order is here circumscribed on the basis of the phylogenetic analyses of Chase and associates (Chase et al. 1995b, 2000). Synapomorphies supporting this group include nectaries mostly on the base of tepals or filaments, septal nectaries nearly always lacking, extrorse anthers and the frequent presence of spots on the tepals. The outer epidermis of the seed coat has a cellular structure and lacks phytomelan (a black crust); the inner part of the seed coat also has cellular structure, both plesiomorphies (see also Stevenson et al. 2000; Rudall et al. 2000a). Putative phylogenetic relationships within the order are shown in Figure 9.18.

It is important to note that the order Liliales and family Liliaceae here are quite narrowly delimited, following Dahlgren et al. (1985) and recent cladistic analyses (see references cited above). Many of the families here treated within Dioscoreales, Asparagales, and Liliales were formerly considered within a more broadly cir-cumscribed Liliales (Cronquist 1981; Thorne 1992). Cronquist (1981) placed most petaloid monocots with six-stamened flowers into a very broadly circumscribed—and we now know clearly polyphyletic—Liliaceae. Others have divided the petaloid monocots with six stamens into Liliaceae, including species with a superior ovary, and Amaryllidaceae, including species with an inferior ovary (Lawrence 1951). This separation is also artificial, separating clearly related genera such as *Agave* and *Yucca* (Agavaceae), *Crinum* (Amaryllidaceae) and *Allium* (Alliaceae), as is discussed in the family treatments (see Asparagales). Although our knowledge of relationships within Liliales is now much improved, the delimitation of some families is still problematic (see especially discussion under Liliaceae, Melanthiaceae, and "Uvulariaceae"). The Liliales include around 10 families and ca. 1300 species; major families include **Liliaceae, "Uvulariaceae," Colchicaceae, Smilacaceae, Trilliaceae**, and **Melanthiaceae**.

References: Chase et al. 1995a,b, 2000; Cronquist 1981; Dahlgren et al. 1985; Goldblatt 1995; Källersjö 1998; Lawrence 1951; Rudall et al. 2000a; Soltis et al. 2000;

Key to Major Families of Liliales

1. Vines, climbing by paired stipular tendrils at the base of the petiole ....**Smilacaceae**

1. Herbs, not climbing, and lacking tendrils 2

2. Ovary inferior; leaves usually twisted at base...Alstroemeriaceae

2. Ovary superior; leaves not twisted at base 3

3. Bulbs present..4

3. Rhizomes or corms present 6

4. Capsule loculicidal; megagametophyte of *Fritillaria* type (egg, synergids, and one polar nucleus haploid, antipodals and second polar nucleus triploid)....................................**Liliaceae**

4. Capsule septicidal or ventricidal; megagametophyte of *Polygonum* type (all cells haploid)..............5

5. Capsules with carpel units splitting apart and their ventral margins separating; flowers usually small, tepals not spotted; anther locules confluent..........................**Melanthiaceae**

5. Capsules septicidal; flowers large, tepals often with spots or lines; anther locules not confluent...Calochortaceae

6. Corms present..**Colchicaceae**

6. Rhizomes present...7

7. Leaves whorled, with parallel to palmate venation and pinnate secondary veins; perianth of sepals and petals...**Trilliaceae**

7. Leaves alternate, with ± parallel venation; perianth of tepals 8

8. Inflorescences determinate, usually few-flowered and paniculate, with cymose branching, or reduced to paired or solitary flowers; flowers minute to conspicuous; style 1, but sometimes apically 3-branched ....**Uvulariaceae**

8. Inflorescences indeterminate, simple or compound racemes or spikes; flowers ± minute; styles 3 ....**Melanthiaceae**

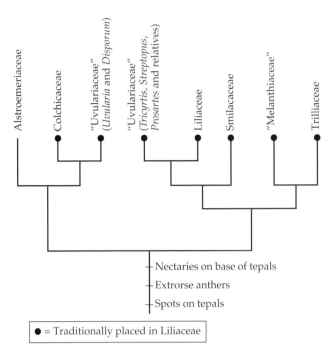

Figure 9.18 Cladogram showing hypothesized relationships within Liliales. (Modified from Chase et al. 2000, Rudall et al. 2000, and Vinnersten and Bremer 2001.)

others triploid. *Nectar produced at base of tepals. Fruit a loculicidal capsule*, occasionally a berry; seeds usually flat and disk-shaped, *seed coat not black*; endosperm oily, **its cells pentaploid** (Figure 9.19).

Floral formula: *, -6-, 6, ③; capsule, berry

Distribution and ecology: Widely distributed in temperate regions of the Northern Hemisphere; mainly spring-blooming plants of prairies, mountain meadows, and other open communities.

Genera/species: 12/500. ***Major genera:*** *Fritillaria* (100 spp.), *Gagea* (90), *Tulipa* (80), and *Lilium* (80). *Clintonia, Erythronium, Fritillaria, Medeola,* and *Lilium* occur in the continental United States and/or Canada.

Economic plants and products: *Tulipa* (tulips), *Fritillaria* (fritillary), *Lilium* (lilies), and *Erythronium* (trout lilies, adder's-tongue) are important ornamentals.

Discussion: Liliaceae, as here defined (Dahlgren et al. 1985), are probably monophyletic (Chase et al. 1995a,b). *Medeola* and *Clintonia* are often placed in

Stevenson and Loconte 1995; Stevenson et al. 2000; Thorne 1992.

Liliaceae Adanson
(Lily Family)

Herbs usually with **bulbs and contractile roots**; steroidal saponins often present. Hairs simple. *Leaves alternate and spiral, or whorled, along stem or in a basal rosette*, simple, entire, *with parallel venation*, often sheathing at base; stipules lacking. *Inflorescence* usually determinate, sometimes reduced to a single flower, *terminal. Flowers* bisexual, radial to slightly bilateral, *conspicuous. Tepals 6, distinct*, imbricate, *petaloid, often with spots or lines. Stamens 6*; filaments distinct; pollen grains usually monosulcate. *Carpels 3, connate; ovary superior, with axile placentation*; stigma 1, 3-lobed, or 3, ± elongated and extending along inner face of style branches. Ovules numerous, usually with 1 integument and a ± thin megasporangium; **megagametophyte developing from 4 megaspores** (*Fritillaria* type), with some cells haploid and

Figure 9.19 Liliaceae. *Lilium lancifolium:* (A) leaf (× 1); (B) gynoecium (× 1); (C) ovary, in cross-section (× 5); (D) flowering plant (× 0.3). (From Hutchinson 1973, *The families of flowering plants*, 3rd ed., p. 755.)

Uvulariaceae or Trilliaceae (Dahlgren et al. 1985). *Calochortus* (mariposa lily) is often placed here and surely is closely related (Chase et al. 2000), but this genus has septicidal capsules and the typical (or *Polygonum*) type of megagametophyte. Analyses based on DNA sequences (Chase et al. 1995a) support a placement of *Calochortus* either within Liliaceae (along with some members of the "Uvulariaceae") or in the Calochortaceae (Dahlgren et al. 1985).

The showy flowers of this family are insect-pollinated, especially by bees, wasps, butterflies, and moths; nectar and/or pollen are employed as pollinator rewards. The seeds are dispersed by wind or water; a few have arillike structures and are dispersed by ants.

References: Chase et al. 1995a,b, 2000; Dahlgren et al. 1985; Tamura 1998.

"Uvulariaceae" C. S. Kunth
(Bellflower Family)

Herbs with creeping rhizomes. Hairs simple. *Leaves alternate and spiral, along stem or in a basal rosette,* simple, entire, *with parallel venation,* and in *Prosartes* and *Tricyrtis* with clearly reticulate venation between the primary veins, sheathing at base; stipules lacking. *Inflorescences* determinate, but sometimes appearing indeterminate, rarely reduced to a single flower, *terminal. Flowers* bisexual, radial, *usually not conspicuous. Tepals 6, distinct,* imbricate, *petaloid, sometimes spotted.* Stamens 6; filaments distinct to connate; pollen grains monosulcate. *Carpels 3, connate; ovary superior, with axile placentation;* stigmas 3, ± elongated and extending along inner face of style branches. Ovules ± numerous, with usually 1 integument and a ± thin megasporangium. *Nectar produced at base of tepals. Fruit a septicidal or loculicidal capsule,* or sometimes a berry; seeds flat and disklike to globose, *not black.*

Floral formula: *, -6-,⟨6⟩,③; capsule, berry

Distribution: Widely distributed, mostly in temperate regions of the Northern Hemisphere.

Genera/species: 8/45. **Major genera:** *Disporum* (15 spp.), *Tricyrtis* (11), *Prosartes* (5), *Uvularia* (5), and *Streptopus* (4). The family is represented in the continental United States and/or Canada by *Prosartes, Scoliopus, Streptopus,* and *Uvularia.*

Economic plants and products: Members of genera such as *Tricyrtis* and *Uvularia* are used as ornamentals.

Discussion: "Uvulariaceae," as delimited by Dahlgren et al. (1985), are undoubtedly biphyletic (Chase et al. 1995a, 2000; Rudall et al. 2000a; Vinnersten and Bremer 2001). More study is needed before the group can be definitively restructured, but it is clear that *Uvularia*

and *Disporum* are not very closely related to *Prosartes, Scoliopus, Streptopus,* and *Tricyrtis* (Shinwari et al. 1994). The former are closely allied to Colchicaceae, while the latter are related to Liliaceae (Chase et al. 1995a, 2000; Rudall et al. 2000). *Scoliopus* has sometimes been placed in the Trilliaceae. It may eventually be necessary to segregate the taxa with ascending ovules and three-parted stigmas as the Tricyrtidaceae, or to expand the circumscriptions of Liliaceae and Colchicaceae (as in APG2 in prep.).

Flowers of "Uvulariaceae" are insect-pollinated, with nectar often providing the pollinator reward.

References: Chase et al. 1995a, 2000; Dahlgren et al. 1985; Rudall et al. 2000a; Shinwari et al. 1994; Vinnersten and Bremer 2001.

Colchicaceae A. P. de Candolle
(Colchicum Family)

Herbs with corms or elongated, corm-like structures; various alkaloids, including those of the colchicine-type (with a tropolone ring), present, and steroidal saponins absent. Hairs simple. *Leaves alternate and spiral, along stem or in a basal rosette,* simple, entire, *with parallel venation,* usually sheathing at the base, occasionally ending in a tendril; stipules lacking. Inflorescence determinate or indeterminate, sometimes reduced to a single flower, terminal or axillary. *Flowers* bisexual, radial, conspicuous. *Tepals 6,* distinct to connate, imbricate, petaloid, *often variegated, spotted, or with basal portion of a different color than the upper part.* Stamens 6; filaments distinct, sometimes adnate to tepals; pollen grains monosulcate to 2- or 4-sulcate or –foraminate. *Carpels 3, connate; ovary superior, with axile placentation;* stigmas 3, truncate to elongate. Ovules numerous, with 1 or 2 integuments and a ± thin megasporangium. *Nectar produced at base of tepals or stamens. Fruit a loculicidal or septicidal capsule;* seeds ± globose, sometimes arillate, the seed coat not black; endosperm oily.

Floral formula: *, ⟨-6-⟩, 6 , ③ , capsule

Distribution and ecology: Widespread in Africa, Europe, and Asia, to Australia and New Zealand, and especially characteristic of winter-rainfall ("Mediterranean") climates.

Genera/species: 13/190. **Major genera:** *Colchicum* (90 spp.), *Wurmbea* (40), *Androcymbium* (40), and *Iphigenia* (15). *Colchicum* and *Gloriosa* are naturalized (or persist after cultivation) in the continental United States.

Economic plants and products: *Colchicum* (autumn crocus) and *Gloriosa* (glory lily, flame lily) are important ornamentals. Several genera are used medicinally because of their highly toxic alkaloids. The alkaloid colchicine is well known for its antimitotic effect. It induces chromosome doubling and is thus much used in plant breeding to induce polyploidy.

Discussion: The monophyly of Colchicaceae, as here delimited, is supported by nucleotide sequences (Chase et al. 1995a, 2000; Rudall et al. 1996, 2000a) and morphology (especially the corm rootstock and unusual alkaloids). Colchicaceae are most closely related to certain genera of "Uvulariaceae" (*Uvularia* and *Disporum*), and these genera are sometimes included (Nordenstam 1998). Until the phylogenetic relationships within Liliales are better understood, we follow the circumscription of Dahlgren et al. (1985). These genera may form a clade sister to the Alstroemeriaceae.

Members of Colchicaceae are distinguished from Liliaceae and "Uvulariaceae" by their distinctive corms or corm-like rootstocks and alkaloid chemistry. In addition, they differ from Liliaceae in having the normal, *Polygonum* type megagametophyte development.

The showy flowers of members of Colchicaceae are pollinated by various insects (bees, wasps, flies, butterflies, and moths); nectar and/or pollen are employed as pollinator rewards. The seeds often show no obvious dispersal adaptations; some have arils or aril-like structures and may be dispersed by ants or birds.

References: Chase et al. 1995a, 2000; Dahlgren et al. 1985; Nordenstam 1998; Rudall et al. 1996, 2000a; Wildman and Pursey 1968.

Smilacaceae Ventenat
(Catbrier Family)

Vines or rarely erect herbs, *often with thick, tuberlike rhizomes*; steroidal saponins present. Hairs simple; prickles often present. **Leaves** alternate and spiral, simple, entire to spinose-serrate, **differentiated into a petiole and blade**, **with palmate venation, with primary veins converging, and clearly connected by a reticulum of higher-order veins**, *with a pair of tendrils near petiole base. Inflorescences* determinate, *usually umbellate,* terminal or axillary. *Flowers usually unisexual (plants dioecious),* radial, and inconspicuous. *Tepals 6,* distinct to slightly connate, imbricate. Stamens usually 6; filaments distinct to slightly connate; *anthers usually unilocular* due to confluence of two locules; pollen grains monosulcate or ± lacking apertures. Carpels 3, connate; ovary superior, usually with axile placentation; stigmas 3, ± elongate. **Ovules 1 or 2 in each locule**, anatropous to orthotropous. Nectar produced at base of tepals and stamens. **Fruit a 1–3-seeded berry**; seeds ± globose, not black (Figure 9.20).

Floral formula: Staminate: *, (-6-), 6 , 0

  Carpellate: *, (-6-), 0, ③; berry

Distribution: Widespread in tropical to temperate regions.

Genera/species: 2/317. **Genera:** *Smilax* (310 spp.) and *Rhipogonum* (7). The family is represented in the continental United States and Canada only by *Smilax*.

Economic plants and products: Several species of *Smilax* are used medicinally; the genus was also the source of the flavoring sarsaparilla. Young stems, berries, and tubers are occasionally eaten.

Discussion: Although the monophyly of Smilacaceae is supported by morphological data and *rbcL* sequences (Chase et al. 1995a; Judd 1998), an analysis using *rbcL*, *trnL-F*, and morphology suggests that it may be paraphyletic since *Philesia* usually is excluded, as Philesiaceae (Rudall et al. 2000). The family has often been considered to be related to Dioscoreaceae (Dahlgren et al. 1985), which are also vines with net-veined leaves. However, morphology (Conran 1989) and DNA sequences (Chase et al. 1993, 1995a, 2000; Soltis et al. 2000) support a placement in the Liliales. The family is often more broadly defined (Cronquist 1981), including genera such as *Luzuriaga* and *Petermannia*, but the inclusion of these genera renders the family polyphyletic (Chase et al. 1995a; Rudall et al. 2000a). In contrast, the family is restricted to *Smilax* by Thorne (1992).

The large genus *Smilax* is clearly monophyletic, as evidenced by the synapomorphies of paired tendrils arising from the petiole, imperfect flowers, anthers with confluent locules, and umbellate inflorescences.

The small flowers of Smilacaceae are pollinated by insects (bees, flies). The fruits are bird-dispersed.

References: Chase et al. 1993, 1995a, 2000; Conran 1989, 1998; Cronquist 1981; Dahlgren et al. 1985; Judd 1998; Rudall et al. 2000a; Soltis et al. 2000; Thorne 1992.

Melanthiaceae Batsch
(Death Camas Family)

Herbs **from short to elongate, usually bulblike rhizomes**; steroidal saponins and **various toxic alkaloids present**. Hairs simple. *Leaves* alternate and spiral, along stem or in a basal rosette, simple, entire, *with parallel venation*, sheathing at base; stipules lacking. *Inflorescences indeterminate*, terminal. *Flowers* usually bisexual, usually radial, *small. Tepals 6*, distinct to slightly connate, imbricate. Stamens 6; filaments distinct; anthers only 2-locular, **the locules usually confluent, opening by a single slit, resulting in a peltate appearance**; pollen grains usually monosulcate. Carpels 3, connate; ovary superior to slightly inferior, with axile placentation; *styles 3*; stigmas 3. Ovules 2 to numerous in each locule. Nectaries at base of tepals. **Fruit usually a ventricidal capsule** (carpels splitting apart and their ventral

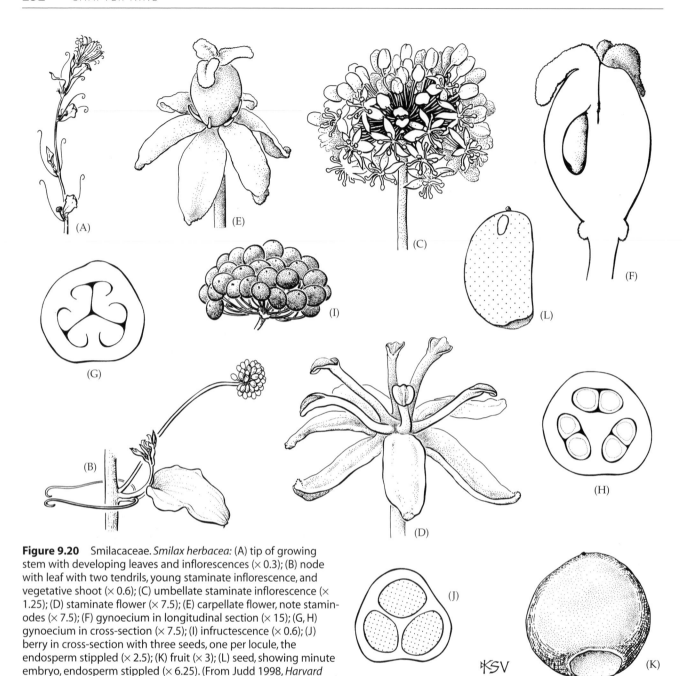

Figure 9.20 Smilacaceae. *Smilax herbacea:* (A) tip of growing stem with developing leaves and inflorescences (× 0.3); (B) node with leaf with two tendrils, young staminate inflorescence, and vegetative shoot (× 0.6); (C) umbellate staminate inflorescence (× 1.25); (D) staminate flower (× 7.5); (E) carpellate flower, note staminodes (× 7.5); (F) gynoecium in longitudinal section (× 15); (G, H) gynoecium in cross-section (× 7.5); (I) infructescence (× 0.6); (J) berry in cross-section with three seeds, one per locule, the endosperm stippled (× 2.5); (K) fruit (× 3); (L) seed, showing minute embryo, endosperm stippled (× 6.25). (From Judd 1998, *Harvard Pap. Bot.* 3(2): p. 162.)

margins separating to release seeds); seeds flattened, often winged or appendaged, not black.

Floral formula: *, ⌒6̅⌒, 6, ③; capsule

Distribution and ecology: Widely distributed in temperate and/or montane habitats; typically of herb-dominated communities.

Genera/species: 11/100. **Major genera:** *Veratrum* (45 spp.), *Schoenocaulon* (24), *Toxicoscordion* (10), and *Anticlea* (10). All of the above genera, along with *Amianthium*, *Stenanthium*, *Xerophyllum*, and *Zigadenus*, occur in the continental United States and/or Canada.

Economic plants and products: A few genera, especially *Veratrum*, are used as ornamentals; several genera have various medicinal or insecticidal uses (due to their poisonous alkaloids). The leaves of *Xerophyllum* are used for making baskets.

Discussion: The monophyly of Melanthiaceae is problematic; as usually circumscribed (Dahlgren et al. 1985) the group is obviously polyphyletic, as indicated by DNA sequences (Chase et al. 1993, 1995a,b, 2000; Rudall et al. 2000a) and morphology (Goldblatt 1995). Phenetic studies confirm the heterogeneity of this group (Ambrose 1980). Thus, the family here excludes the tribes Narthecieae and Tofieldieae, which are not at all closely related to the remaining species (Zomlefer 1997a,b,c). Tofieldiaceae (e.g., *Tofieldia* and *Harperocallis*) are distinct from Melanthiaceae (as here circumscribed) in their 2-ranked, equitant leaves, 2-sulcate pollen, presence of druses in parenchymatous tissues, and stalked ovary. Nartheciaceae (e.g., *Aletris*, *Narthecium*, *Lophiola*) have unusual roots with air spaces in the cortex, a single style, and loculicidal capsules; some also have 2-ranked, equitant leaves. These two families also lack the *Veratrum* alkaloids and flattened to winged seeds characteristic of Melanthiaceae as defined here. *Chionographis*, *Chamaelirium*, and *Xerophyllum* are tentatively included within Melanthiaceae; their inclusion may render the family nonmonophyletic. It is also evident (Chase et al. 1995a, 2000; Rudall et al. 2000a; Zomlefer et al. 2001) that *Trillium* is closely related to Melanthiaceae, and its treatment as Trilliaceae may make Melanthiaceae paraphyletic.

Most members of the family belong to the clearly monophyletic Melanthieae, i.e., *Amianthium*, *Anticlea*, *Toxicoscordion*, *Schoenocaulon*, *Stenanthium*, *Veratrum*, and *Zigadenus*, a group characterized by inflorescences with bisexual or bisexual and staminate flowers, nectaries near the base of the tepals, anther locules apically confluent, opening by a single slit and resulting in a peltate appearance, a ventricidal capsule, and veratrum-alkaloids (Zomlefer et al. 2001). The characters indicated in boldface in the family description apply only to this tribe.

The small flowers of Melanthiaceae are pollinated by bees and flies insects (and possibly also wind). The small seeds are probably wind dispersed.

References: Ambrose 1980; Dahlgren et al. 1985; Chase et al. 1993, 1995a,b, 2000; Goldblatt 1995; Rudall et al. 2000a; Tamura 1998c; Zomlefer 1997a,b,c; Zomlefer et al. 2001.

Trilliaceae Chevallier
(Trillium Family)

Herbs with slender to thick and tuberlike rhizomes; vascular bundles of stem usually in 3 rings; roots contractile; steroidal saponins present. Hairs simple. **Leaves whorled (and usually the same number as the sepals)**, simple, entire, sessile, **with ± palmate venation, with primary veins converging, secondary veins pinnate, and higher-order veins forming a distinct reticulum**; stipules lacking. **Inflorescences reduced to a single flower**, terminal. Flowers bisexual, radial, conspicuous, **the perianth differentiated into a calyx and corolla**. Sepals and petals usually 3 or 4, distinct, imbricate. Stamens 6 or 8 (rarely numerous); filaments distinct; pollen grains usually lacking apertures. Carpels 3–10, connate; *ovary superior, usually with axile placentation*; stigmas 3, ± elongate. Ovules numerous in each locule; megagametophyte formed from two megaspore nuclei (*Allium* type). Nectar lacking, produced by base of perianth parts, or in septa of ovary. **Fruit a fleshy capsule or berry; seeds usually arillate**, not black (Figure 9.21).

Floral formula:

*, 3–8, 3–8, 6–16, ③–⑧; fleshy capsule, berry

Distribution and ecology: Widely distributed in temperate regions of the Northern Hemisphere; mainly herbs of shaded forest habitats.

Genera/species: 1/50; ***Genus:*** *Trillium*.

Economic plants and products: A few species of *Trillium* are used medicinally or as ornamentals.

Discussion: The haploid chromosome set of Trilliaceae consists of five large chromosomes, and this condition may be an additional synapomorphy of the family. The characters that have been used to distinguish *Paris* (and a couple of other segregate genera) from *Trillium* are inconsistent, and their recognition probably would make *Trillium* paraphyletic (Kato et al. 1995b; Zomlefer 1996). Two large subgenera traditionally recognized within *Trillium* are subg. *Trillium*, with pedicellate flowers, and subg. *Phyllantherum*, with sessile flowers. The

Figure 9.21 Trilliaceae. *Paris polyphylla:* note whorled and net-veined leaves and terminal flower with sepals (leaflike structures) and petals (elongate structures) (× 0.75). (From Engler 1889, in Engler and Prantl, *Die natürlichen Pflanzenfamilien*, II(5): p. 83.)

pedicellate condition is plesiomorphic, calling the monophyly of subg. *Trillium* into question. Cladistic analyses based on cpDNA restriction sites (Kato et al. 1995a) and morphology (Kawano and Kato 1995) also indicate that subg. *Trillium* is paraphyletic and subg. *Phyllantherum* is monophyletic.

The showy flowers are pollinated by various insects (especially flies, beetles, and bees). Floral visitors to some species, however, are infrequent. Both selfing and agamospermy have been documented in the group. Dispersal by ants is characteristic of species producing capsules with arillate seeds, while bird dispersal is likely in those with colorful berries.

References: Dahlgren et al. 1985; Kato et al. 1995a,b; Kawano and Kato 1995; Samejima and Samejima 1987; Tamura 1998c; Zomlefer 1996.

Asparagales

The monophyly of Asparagales is supported by cladistic analyses based on morphology, 18S, and several chloroplast DNA sequences (Chase et al. 1995a,b, 2000; Conran 1989; Rudall et al. 1997; Källersjö et al. 1998; Fay et al. 2000; Soltis et al. 2000; Stevenson et al. 2000; McPherson and Graham 2001; but contrast with Stevenson and Loconte 1995). Supporting characters include their characteristic seeds, which have the outer epidermis of the coat obliterated (in most fleshy-fruited species) or present and with a black and carbonaceous phytomelan crust in many dry-fruited species. The inner part of the seed coat is usually completely collapsed. In contrast, seeds of Liliales always have a well-developed outer epidermis, lack phytomelan, and usually retain a cellular structure in the inner portion of the coat. Asparagales also can be distinguished from Liliales by their usually nonspotted tepals, nectaries in the septa of the ovary (instead of at base of tepals or stamens), and sometimes anomalous secondary growth (vs. lack of secondary growth).

The order consists of about 25 families and about 26,800 species; major families include **Orchidaceae**, Hypoxidaceae, **Iridaceae**, **Amaryllidaceae**, **Alliaceae**, **Hyacinthaceae**, Laxmanniaceae, **Agavaceae**, **Asparagaceae**, **Convallariaceae**, **Asphodelaceae**, and Hemerocallidaceae.

Relationships within the order have been investigated by Dahlgren et al. (1985), Rudall and Cutler (1995), Chase et al. (1995a,b, 1996), Stevenson and Loconte (1995), Rudall et al. (1997), Fay et al. (2000), and McPherson and Graham (2001). A paraphyletic basal assemblage, called the "lower" Asparagales, including families such as Orchidaceae, Hypoxidaceae, Iridaceae, Asphodelaceae, and Hemerocallidaceae, is characterized by simultaneous microsporogenesis (the four microspores separate all at once from one another after both meiotic divisions have been completed). In contrast, the so-called "higher" Asparagales, including

Alliaceae, Amaryllidaceae, Hyacinthaceae, Agavaceae, Asparagaceae, and Ruscaceae, has successive microsporogenesis (Figure 9.22). In these families a cell plate is laid down immediately after the first meiotic division and another in each of the daughter cells after the second meiotic division (Dahlgren and Clifford 1982). If the current phylogeny holds, we have an almost paradoxical – and evolutionarily fascinating – situation. Many of the "lower" Asparagales have a microsporogenesis pattern that is apomorphic (within the monocots), and they also have an inferior ovary, a feature that is also apomorphic. The microsporogenesis pattern and ovary position of most "higher" Asparagales are like those of more basal monocots!

Orchidaceae and Hypoxidaceae are among the earliest branching clades in Asparagales. Hemerocallidaceae are close to Asphodelaceae and Xanthorrhoeaceae, and all three families have anthraquinones. Within "higher" Asparagales, Alliaceae, Amaryllidaceae, and Agapanthaceae may form a clade, based on their bulbous rootstock, scapose umbellate inflorescences, and also cpDNA sequences (Fay et al. 2000). Ruscaceae and Asparagaceae may be related, as evidenced by their indehiscent, usually fleshy fruits; this relationship receives preliminary support from some analyses of DNA sequences (Chase et al. 1995a,b, 1996, 2000; Fay et al. 2000) but not others (Rudall et al. 1997). Finally, Themidaceae, Hyacinthaceae, Anthericaceae, and Agavaceae may form a clade (see Fay et al. 2000), although morphological support for this group is unclear. This clade may be sister to the Ruscaceae + Asparagaceae clade.

Some familial limits are currently unclear, and certain botanists have proposed that several of the families recognized here be combined. Because criteria for determining absolute rank (see Chapter 3 and Apppendix 1) do not exist, this is really an arbitrary taxonomic decision, although factors such as ease of recognition and level of support for the group's monophyly should be considered. The latter, of course, is vital. The recognition of nested monophyletic groups, and the determination of their diagnostic features, is the most important consideration—exactly which of these groups are positioned at the rank of "family," which are recognized as "subfamilies," etc., is much less significant. Thus, some botanists treat the Alliaceae, Amaryllidaceae, and Agapanthaceae, which together constitute a well-supported clade, as a single family (Alliaceae s.l.). As discussed above (and see also the family treatments), the fact that these three groups are closely related is beyond doubt, and it is an arbitrary decision whether they are considered as three closely related families or as three clades within a more broadly circumscribed family. Alliaceae s.l. are diagnosed by their scapose umbellate inflorescences with spathelike bracts. Likewise, some combine Xanthorrhoeaceae, Asphodelaceae, and Hemerocallidaceae, recognizing these three clades within a broadly circumscribed Xanth-

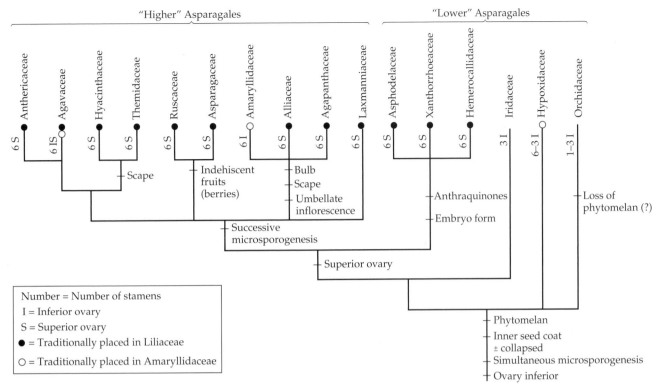

Figure 9.22 Cladogram showing hypothesized relationships within Asparagales. (Modified from Chase et al. 2000, Fay et al. 2000, Soltis et al. 2000, and McPherson and Graham 2001.)

orrhoeaceae. The presence of anthra-quinones and a seedling with a nonphotosynthetic cotyledon support the monophyly of Xanthorrhoeaceae s.l. Finally, it has been suggested (see Angiosperm Phylogeny Group 2002) that Agavaceae, Asparagaceae, Hyacinthaceae, Ruscaceae, Themidaceae, and a few small families can be combined; the resulting broadly circumscribed family would be called Asparagaceae.

In using the "family" treatments provided in this text, remember that these named groups represent clades. In your systematics class you may be called upon to know the name and diagnostic characters of each of these groups, only some of them, or of groups within them, or your instructor may focus on more inclusive clades, such as the groups discussed above. The clades within Asparagales considered in this textbook (see key) may be treated as (1) Asparagaceae, Ruscaceae, Asphodelaceae, Hemerocallidaceae, Agavaceae, Hyacinthaceae, Alliaceae, Amaryllidaceae, Iridaceae, Hypoxidaceae, and Orchidaceae (for a total of 11 families), or as (2) Asparagaceae s.l., Alliaceae s.l., Xanthorrhoeaceae s.l., Iridaceae, Hypoxidaceae, and Orchidaceae (for a total of only 5 families). A similar taxonomic situation—resulting from a current lack of consensus in the community of plant systematists regarding the number of "significant clades," the clades generally discussed (and here recog-

nized at the rank of family)—occurs in several other orders, especially Liliales, Malpighiales, Cornales, Lamiales, and Dipsacales.

References: APG 2002; Chase et al. 1995a,b, 1996, 2000; Conran 1989; Dahlgren et al. 1995; Dahlgren and Clifford 1982; Fay et al. 2000; Judd 2000; Källersjö et al. 1998; McPherson and Graham 2001; Rudall et al. 1996; Rudall and Cutler 1995; Soltis et al. 2000; Stevenson and Loconte 1995; Stevenson et al. 2000; Tomlinson and Zimmerman 1969; Zomlefer 1998.

Asparagaceae A. L. Jussieu
(Asparagus Family)

Rhizomatous herbs to shrubs, or scrambling to twining vines; *stems woody to withering annually, usually green, those associated with scale leaves forming leaflike phylloclades*, or stems reduced; steroidal saponins and ethereal oils present. Hairs simple. *Leaves alternate and spiral, simple, entire, usually reduced, ± scale-like, with a spiny basal spur,* the venation indistinct; stipules lacking. Inflorescences determinate, sometimes reduced to a single flower, axillary. Flowers bisexual or unisexual (plants monoecious or dioecious), radial, usually small. *Tepals 6, ± distinct, petaloid, not spotted,* imbricate. *Stamens usually 6,* repre-

Key to Major Families of Asparagales

1. Ovary inferior 2

1. Ovary superior 6

2. Stamens adnate to style, usually only 1 or 2; one perianth member highly modified, forming a lip; seeds lacking endosperm; placentation usually parietal .... **Orchidaceae**

2. Stamens not fused to style, usually 3 or 6; all perianth parts alike or 3 outer ± differentiated from 3 inner; seeds with endosperm; placentation usually axile 3

3. Stamens 3; leaves equitant; seed coat not black, with cellular structure .... **Iridaceae**

3. Stamens 6; leaves not equitant; seed coat with outer epidermis obliterated, or present and with black crust 4

4. Inflorescences scapose, umbellate (or reduced to a solitary flower); plants from a bulb, with contractile roots .... **Amaryllidaceae**

4. Inflorescences paniculate, racemose, fasciculate, or reduced to a solitary flower, usually not scapose; plants from a rhizome or corm, the roots contractile or not 5

5. Plants with rosettes of fleshy fibrous leaves; karyotype dimorphic, of 5 large and 25 small chromosomes; nectaries in the septa of the ovary; roots not contractile .... **Agavaceae** (Agavoideae)

5. Plants nonsucculent, leaves thin-textured and not strongly fibrous; karyotype not dimorphic; nectaries lacking; roots usually contractile .... Hypoxidaceae

6. Fruit fleshy, a berry 7

6. Fruit dry, hard or leathery, usually a capsule, or triangular and nutlike 8

7. Leaves usually rudimentary, forming small scales and plants with cylindrical to flattened green cladodes; seeds black .... **Asparagaceae**

7. Leaves usually ± large and photosynthetic; stems cylindrical, green to brown, but not the major photosynthetic organ of the plant; seeds not black .... **Ruscaceae**
(various herbaceous genera, along with ± woody plants, i.e., Dracaeneae)

8. Seeds not black; guard cells rich in oil; fruits dry, triangular, and nutlike .... **Ruscaceae** (Nolineae)

8. Seeds black (with phytomelan); guard cells lacking oil; fruits various, but usually capsules and never as above 9

9. Plants from a bulb; inflorescence scapose (atop an elongate internode) 10

9. Plants from a rhizome; inflorescence usually not scapose 11

10. Inflorescence umbellate; plants usually with an onion or garlic smell .... **Alliaceae**

10. Inflorescence with an axis; plants without an onion or garlic smell .... **Hyacinthaceae**

11. Plants with woody trunk (or rhizome) and strongly fibrous leaves; anthers small in relation to filaments; chromosomes dimorphic (5 large and 25 small) .... **Agavaceae** (Yuccoideae)

11. Plants with or without a woody trunk; leaves not strongly fibrous; anthers not small; chromosomes more uniform in size 12

12. Pollen monosulcate; inflorescences indeterminate, forming simple or compound racemes or spikes; seeds usually arillate; leaves usually succulent, often having conspicuous gelatinous central zone, and often with colored sap produced by specialized cells associated with vascular bundles of leaves; filaments free from tepals .... **Asphodelaceae**

12. Pollen trichotomosulcate, with Y-shaped aperture; or if monosulcate then inflorescences determinate, forming scorpioid cymes; seeds not arillate; leaves thin, lacking gelatinous zone and colored sap; filaments ± adnate to tepals .... Hemerocallidaceae
[incl. Phormiaceae, Johnsoniaceae]

sented by staminodes in carpellate flowers; filaments distinct to connate, adnate to tepals; pollen grains monosulcate. *Carpels 3, connate, ovary superior, with axile placentation;* stigma usually 1, capitate to 3-lobed. Nectaries in septa of ovary. Ovules 1 to several in each locule, anatropous to ± orthotropous; **megagametophyte curved-asymmetric.** *Fruit usually a few-seeded berry;* seeds angular to ± globose, *seed coat with phytomelan* (black crust) and inner layers collapsed.

Floral formula: *, – 6 – , ⑥ , ③ , capsule

Distribution and ecology: Widely distributed in the Old World from Europe and Africa to eastern Asia, Malesia, and Australia (*Asparagus*), and disjunct in Mexico (*Hemiphylacus*). The family is characteristic of regions of arid to Mediterranean climates.

Genera/species: 2/305. **Major genus:** *Asparagus* (300 species). The family is represented in the continental United States and Canada only by four naturalized species of *Asparagus*.

Economic plants and products: The young shoots of *Asparagus officinalis* are eaten as a vegetable. Numerous species of *Asparagus* are used medicinally; several are popular ornamentals, e.g., *A. aethiopicus* and *A. setaceus*.

Discussion: Monophyly of Asparagaceae is supported by a few embryological features, i.e., the megagametophyte is curved-asymmetric and becomes even more so during development, and the ovule has a persistent nucellar epidermis of enlarged, cytoplasm-rich cells (Rudall et al. 1998). Most species belong to the large genus *Asparagus*, a clade diagnosed by several distinctive morphological specializations (leaves reduced to non-photosynthetic scales that bear in their axils green, solitary to clustered, terete to flattened phylloclades, and fruits berries). Recently published DNA-based phylogenetic analyses (Chase et al. 1995a, 1996; Fay et al.2000; Rudall et al. 1997) also support the monophyly of Asparagaceae (including both *Asparagus* and *Hemiphylacus*).

The flattened photosynthetic organs of *Asparagus* have been the subject of much study and almost endless controversy (see Judd 2002; Kubitzki and Rudall 1998), however, most authors consider them to be flattened stems, a view accepted here. It is curious that similar photosynthetic branches have evolved independently in *Ruscus* and relatives (Ruscaceae). Certainly, Ruscaceae and Asparagaceae are close relatives.

The flowers of *Asparagus* are pollinated by various bees and beetles; the usually red to blue or black fruits are dispersed by birds.

References: Chase et al. 1995a, 1996; Fay et al. 2000; Judd 2002; Kubitzki and Rudall 1998; Malcomber and Sebsebe 1993; Rudall et al. 1997, 1998.

Ruscaceae R. Roemer
(Butcher's broom Family)

Rhizomatous herbs to trees; stems sometimes with anomalous secondary growth, sometimes with resin canals, occasionally flattened and photosynthetic (phylloclades); steroidal saponins present. Hairs simple. *Leaves* usually alternate and spiral, along stem or in a basal rosette, simple, entire, *with parallel venation,* occasionally petiolate, sheathing at base; stipules lacking. Inflorescences determinate, sometimes reduced to a single flower, terminal or axillary. *Flowers* bisexual, radial, *usually small.* Tepals (4) 6, *distinct, or more commonly connate and perianth then urn-shaped to bell-shaped, or wheel-shaped,* petaloid, *not spotted,* imbricate. Stamens (4) 6; filaments distinct or occasionally connate, often adnate to tepals; pollen grains monosulcate or lacking an aperture. *Carpels (2) 3, connate; ovary usually superior, with axile placentation;* stigma 1, capitate to 3-lobed. *Nectaries in septa of ovary.* Ovules 2 to few in each locule, anatropous to orthotropous. *Fruit usually a few-seeded berry;* seeds ± globose, **seed coat** with outer epidermis lacking cellular structure, and **lacking phytomelan** (black crust), and inner layers collapsed (Figure 9.23).

Floral formula:

*, ⑥ , 6 , ③ ; berry, or dry 3-angled and nutlike

Distribution and ecology: Widely distributed from temperate to tropical regions; herbaceous species often understory herbs of moist forests and woody species often of arid regions.

Genera/species: 28/500. **Major genera:** *Dracaena* (80 spp.), *Polygonatum* (50), *Sansevieria* (50), *Maianthemum* (33), *Ophiopogon* (30), and *Nolina* (25). The family is represented in the continental United States and/or Canada by *Convallaria, Maianthemum, Polygonatum, Nolina, Dasylirion,* and *Sansevieria*.

Economic plants and products: Several genera, including *Aspidistra* (cast-iron plant), *Convallaria* (lily of the valley), *Dracaena* (dragon tree), *Liriope* (border grass), *Maianthemum* (false Solomon's seal), *Polygonatum* (Solomon's seal), *Ophiopogon* (mondo grass), and *Sansevieria* (mother-in-law's tongue) are used as ornamentals.

Discussion: Ruscaceae are broadly circumscribed, including Convallariaceae, Nolinaceae, and Dracaenaceae, as suggested by Chase et al. (1995a, 1996, 2000), Rudall et al. (1997, 2000b), Fay et al. (2000), and Judd (2003). The monophyly of the family is supported by DNA sequences (Chase et al. 1995a, 2000; Bogler and Simpson 1995, 1996; Rudall et al. 1997, 2000b; Yamashita and Tamura 2000; Fay et al. 2000) and the lack of phytomelan in their seeds. Ruscaceae s. l. may be sister to Aspara-

Figure 9.23 Ruscaceae. (A–D) *Sansevieria cylindrica:* (A) flowering plant (× 0.3); (B) flower (× 1); (C) gynoecium (× 1.25); (D) ovary in cross-section (× 2.5). (E–H) *Dasylirion acrotrichum:* (E) flowering plant (× 0.05); (F) flower (× 1); (G) tepal and stamen (× 2); (H) cross-section of ovary (× 1). (I–P) *Dracaena draco:* (I) portion of inflorescence (× 1); (J) flower (× 4); (K) stamen (× 7); (L) ovary in cross-section (× 16); (M) gynoecium (× 7); (N) berry (× 1); (O) berry in cross-section (× 1); (P) seed (× 1). (From Engler 1889, in Engler and Prantl, *Die natürlichen Pflanzenfamilien* II(5): pp. 72, 73, 84.)

gaceae (Chase et al. 1995a, 1996, 2000; Bogler and Simpson 1996; Fay et al. 2000).

Herbaceous members of Ruscaceae form a paraphyletic complex (Conran 1989; Bogler and Simpson 1996; Rudall et al. 2000b; Yamashita and Tamura 2000) that includes to two woody clades, the Nolineae (*Nolina*, *Dasylirion*, *Calibanus*, and *Beaucarnea*) and the Dracaeneae (*Dracaena* and *Sansevieria*). The monophyly of the *Nolina* group (Nolineae) is supported by their dry, three-angled, and nutlike fruits, and by leaves with minute longitudinal ridges and guard cells containing large amounts of oil. The monophyly of *Dracaena* and *Sansevieria* (Dracaeneae) is supported by resin canals in their leaves and bark, which thus frequently stain dark red or orange-red, hence the name "dragon's blood," used for the sap in some species of *Dracaena*.

Among the herbaceous genera, *Ruscus*, *Semele*, and *Danaë* (the Rusceae) are distinctive in that they have flattened photosynthetic stems, with the leaves reduced to scales. They are often confused with members of the Asparagaceae, but lack phytomelan in the testa of their seeds. Traditionally, the Ruscaceae has been restricted to just these three small herbaceous genera.

The members of the Polygonateae, such as *Maianthemum* and *Polygonatum*, are easily recognized by their distinctive (and synapomorphic) habit. These plants have broad leaves borne along aerial stems, which are connected by a sympodial rhizome system. *Liriope*, *Ophiopogon*, and relatives (the Ophiopogoneae) are noteworthy because their fruits rupture early to expose seeds with a fleshy seed coat. These genera have spreading sympodial rhizomes with tufted clusters of often grasslike leaves; aerial stems are lacking. These leafy herbs, along with others, such as *Convallaria* and *Aspidistra*, have often been placed in the "Convallariaceae."

The small flowers of most Ruscaceae are insect-pollinated, especially by bees and wasps, which gather nectar or pollen. Their colorful berries usually are dispersed by birds. The dry, angular fruits of *Nolina* and relatives are wind-dispersed.

References: Bogler 1998; Bogler and Simpson 1995, 1996; Bos 1998; Chase et al. 1995a, 1996, 2000; Conran 1989; Conran and Tamura 1998; Dahlgren et al. 1995; Fay et al. 2000; Judd 2003; Rudall et al. 1996, 2000b; Soltis et al. 2000; Yamashita and Tamura 2000.

Agavaceae B. C. J. Dumortier
(Agave Family)

Usually *large rosette herbs, trees, or shrubs,* rhizomatous; stems with *anomalous secondary growth;* raphidelike crystals of calcium oxalate present; steroidal saponins present. Hairs simple. *Leaves* alternate and spiral, *in rosettes at base or ends of branches,* simple, usually succulent, entire to spinose-serrate, and *usually with a sharp spine at apex,* with parallel venation, *the vascular bundles associated with*

thick and tough fibers, sheathing at base; stipules lacking. Inflorescences determinate, *usually paniculate,* terminal. Flowers usually bisexual, radial to slightly bilateral, often showy. *Tepals 6, distinct to connate,* and perianth then tubular to bell-shaped, imbricate, *petaloid, not spotted* (*and usually white to yellow*). Stamens 6; filaments distinct, sometimes adnate to the perianth; pollen grains monosulcate. *Carpels 3, connate; ovary superior or inferior,* with axile placentation; stigma minute, capitate to 3-lobed. Ovules ± numerous in each locule. *Nectaries in septa of ovary. Fruit a loculicidal capsule,* but sometimes fleshy and berrylike; *seeds flat, the seed coat with black crust* (*phytomelan*) and inner layers ± collapsed; **karyotype of 5 large and 25 small chromosomes** (see Figure 4.23).

Floral formula: * or X, ⁶̣͜, 6, ③̲; capsule

Distribution and ecology: Widely distributed in warm temperate to tropical regions of the New World, and especially diverse in Mexico, but introduced in Old World; characteristic of arid and semiarid habitats.

Genera/species: 13/300. **Major genera:** *Agave* (300 spp.), *Yucca* (40), and *Hosta* (40). The family is represented in the continental United States and/or Canada by *Agave*, *Camassia*, *Chlorogalum*, *Furcraea*, *Hastingsia*, *Hesperoaloe*, *Hesperoyucca*, *Manfreda*, *Schoenolirion*, and *Yucca*.

Economic plants and products: Several species of *Agave* (sisal hemp), *Furcraea*, and *Yucca* are used as fiber sources, and a few species of *Agave* are fermented to produce tequila and mescal. Both *Agave* and *Yucca* are used in the manufacture of oral contraceptives (due to their steroidal saponins). Finally, several genera, including *Agave* (century plant), *Hosta*, *Manfreda*, *Polianthes*, and *Yucca*, are used as ornamentals.

Discussion: Agavaceae are related to Hyacinthaceae and Themidaceae, and the phylogenetic position of some genera is problematic. However, both phenotypic and DNA characters support the family's monophyly (Bogler and Simpson 1995, 1996). *Camassia*, *Hastingsia*, and *Chlorogalum* (usually treated in Hyacinthaceae, as subfam. Chlorogaloideae) and *Hosta* (usually placed in Hostaceae) are here included (see Chase et al. 1995a; Bogler and Simpson 1995, 1996; Pfosser and Speta 1999). *Hosta* is a genus of rhizomatous herbs with broad-bladed leaves with prominent parallel veins, one-sided racemes of lilylike flowers, and black-seeded capsules. These four genera also have a distinctive bimodal karyotype.

Woody Agavaceae usually are divided into the Yuccoideae (e.g., *Yucca*, *Hesperoaloe*, *Hesperoyucca*), with a superior ovary and tiny anthers, and the Agavoideae (e.g., *Agave*, *Furcraea*, *Manfreda*, *Polianthes*), with an inferior ovary and elongate anthers (Dahlgren et al. 1985); both are monophyletic (Bogler and Simpson 1995, 1996). The family has often been more broadly circumscribed (Cronquist 1981), including genera that are here considered in Ruscaceae (*Nolina, Dasylirion, Beaucarnea, Dracaena,* and *Sansevieria*) and *Cordyline* (Laxmanni-aceae). Such a broadly defined family is morphological-ly heterogeneous, unified only by the woody habit, and clearly polyphyletic (Dahlgren et al. 1985; Chase et al. 1995a,b; Bolger and Simpson 1995, 1996; Rudall et al. 1997).

The showy flowers of *Yucca* and *Hesperoyucca* are vis-ited by small moths of the genus *Tegeticula* (see Figure 4.23). Other genera of Agavaceae are pollinated by birds (many species of *Beschornea*) or bats (several species of *Agave*). The black seeds are typically dispersed by wind, and fleshy-fruited species are dispersed by animals.

References: Baker 1986; Bogler and Simpson 1995; Chase et al. 1995a,b; Cronquist 1981; Dahlgren et al. 1985; McKelvey and Sax 1933; Pfosser and Speta 1999; Rudall et al. 1997; Verhoek 1998.

Hyacinthaceae Batsch
(Hyacinth family)

Herbs **with bulbs** *and contractile roots;* with steroidal saponins, *poisonous steroids* (bufodienolids and cardeno-lids), and mucilage cells or canals. Hairs simple. *Leaves alternate and spiral, in a basal rosette, simple, entire, with par-allel venation,* sheathing at base; stipules lacking. *Inflores-cence indeterminate (usually a raceme), on a scape.* Flowers bisexual, usually radial, often showy. *Tepals 6,* distinct to connate, and perianth then bell-shaped to tubular, imbri-cate, petaloid, not spotted. *Stamens 6;* filaments distinct to connate, sometimes adnate to tepals; pollen grains monosulcate. *Carpels 3, connate, ovary superior;* stigma 1, capitate to 3-lobed. Ovules 1 to numerous in each locule. Nectaries in septa of ovary. *Fruit a loculicidal capsule;* seeds globose to flattened, occasionally with aril-like structures, the seed coat usually with phytomelan and inner layers compressed or collapsed.

Floral formula: *, $\widehat{(6)}$, $\widehat{(6)}$, $\underline{(3)}$, capsule

Distribution: Widespread from Europe and Africa to Asia, in temperate to tropical habitats, but most diverse in areas with a Mediterranean climate (pronounced summer dry season).

Genera/species: 63/850. **Major genera:** *Albuca* (60 spp.), *Eliokarmos* (60), *Ornithogalum* (50), *Muscari* (50), *Urginea* (50), and *Scilla* (30). A few species of *Muscari, Ornitho-galum,* and *Scilla* are sparingly naturalized.

Economic plants and products: Several genera, e.g., *Urginea* and *Thuranthos,* are used medicinally because of their cardenolides. Species of genera such as *Scilla, Mus-cari, Hyacinthus, Puschkinia, Hyacinthoides, Eucomis, Galto-nia,* and *Eliokarmos* are grown as ornamentals.

Discussion: Monophyly of Hyacinthaceae is supported by molecular data, which also supports exclusion of *Camassia, Chlorogalum,* and relatives, i.e., subfamily Chlorogaloideae (Chase et al. 1995a; Speta 1998). These genera are here considered as herbaceous members of Agavaceae, and they are distinguished from Hyacin-thaceae s.s. by their bimodal karyotype, and the combi-nation of more or less globose seeds with a firmly adher-ing testa, multinerved tepals, and distinctly 3-lobed stigmas. The exact placement of Hyacinthaceae within Asparagales is unclear, but they may be related to Themidaceae, and these two families may be associated with either the Agavaceae or, less likely, the Alliaceae + Amaryllidaceae clade (Chase et al. 1995a, 2000; Rudall et al. 1997; Fay et al. 2000; Soltis et al. 2000). This uncertain-ty in placement makes determination of morphological synapomorphies problematic, although the presence of bulbs and poisonous steroids are noteworthy. The inde-terminate inflorescence (usually racemose or spicate) easily distinguishes the group from Themidaceae, Alli-aceae, and Amaryllidaceae, all of which have umbellate inflorescences. The herbaceous habit, with a bulb, distin-guishes the family from Agavaceae.

The showy flowers are pollinated by a wide range of insects (bees, wasps, flies, moths) as well as birds, and both nectar and pollen are offered as floral rewards. Asexual reproduction occasionally occurs (agamo-spermy or production of bulblets). Dispersal of seeds may occur by water, wind, or movement by ants.

References: Chase et al. 1995a, 2000; Dahlgren et al. 1985; Fay et al. 2000; Rudall et al. 1997; Soltis et al. 2000; Speta 1998.

Alliaceae J. G. Agardh
(Onion Family)

Herbs with a bulb and contractile roots (or in *Tulbaghia,* a rhi-zome); stems reduced; **vessel elements with simple per-forations; laticifers present (and latex ± clear);** with steroidal saponins; **with onion- or garlic-scented sulfur compounds such as allyl-sulphides, propionaldehyde, propionthiol, and vinyl disulphide.** Hairs simple. Leaves alternate, usually 2-ranked, ± basal, simple, terete, angular, or flat, entire, with parallel venation, sheathing at base; stipules lacking. *Inflorescences* determi-nate, composed of one or more contracted helicoid cymes and *appearing to be an umbel, subtended by a few membranous spathelike bracts, terminal, at the end of a long scape.* Flowers bisexual, radial or bilateral, often showy; **individual flowers not associated with bracts.** *Tepals 6, distinct to connate,* and perianth then bell-shaped to tubu-lar, imbricate, *petaloid, not spotted;* a corona (outgrowth of the perianth) sometimes present. *Stamens 6 (3);* filaments distinct to connate, sometimes adnate to tepals, some-times appendaged; pollen grains monosulcate. *Carpels 3, connate; ovary superior,* with axile placentation; stigma 1, capitate to 3-lobed. Ovules 2 to numerous in each locule, anatropous to campylotropous. *Nectaries in septa of ovary. Fruit a loculicidal capsule; seeds globose to angular, the seed coat with phytomelan* and inner layers compressed or collapsed; embryo ± curved.

Floral formula: * or X, (-6-), (6), (3); capsule

Distribution and ecology: Widely distributed in temperate to tropical regions; frequently in semiarid habitats.

Genera/species: 13/645. **Major genera:** *Allium* (550 spp.), *Ipheion* (25), and *Tulbaghia* (24). The family is represented in the continental United States and/or Canada by *Allium* and *Nothoscordum*.

Economic plants and products: Several species of *Allium* (garlic, onions, shallots, chives, leeks, ramps) are important vegetables or flavorings. Their sap is mildly antiseptic, and several are used medicinally. A few genera, including *Allium, Gilliesia, Ipheion,* and *Tulbaghia* (society garlic), are used as ornamentals.

Discussion: The monophyly of Alliaceae is supported by morphology, chemistry, and *rbcL* sequences (Fay and Chase 1996). Alliaceae are closely related to Amaryllidaceae and Agapanthaceae. All three families are bulbous herbs with terminal umbellate inflorescences, which are subtended by spathaceous bracts and borne on a conspicuous scape; all of these features are probably synapomorphic. Cladistic analyses support the close relationship of these families (Chase et al. 1995a,b; Fay and Chase 1996; Rudall et al. 1996; Fay et al. 2000). Alliaceae and Agapanthaceae are often included within a broadened Amaryllidaceae (Hutchinson 1934, 1973), a decision that has merit from a practical standpoint, since all these plants have many features in common.

The Themidaceae, which are herbs from corms and include *Dichelostema, Triteleia,* and *Brodiaea,* are usually placed within Alliaceae. Recent cladistic analyses have indicated that these genera are more closely related to Hyacinthaceae than to typical Alliaceae (Chase et al. 1995a; Fay and Chase 1996; Rudall et al. 1997; Fay et al. 2000). Embryological characters do not support a close relationship between Alliaceae and Themidaceae (Berg 1996). *Agapanthus* (Agapanthaceae) also has traditionally been included in Alliaceae, but it lacks the sulfur-containing compounds, and cpDNA sequences do not support its placement in Alliaceae (Chase et al. 1995a; Fay et al. 2000).

The showy flowers of Alliaceae are pollinated by a variety of insects (especially bees and wasps). The seeds are predominantly wind- or water-dispersed. A few species produce bulblets in the inflorescence.

References: Berg 1996; Chase et al. 1995a,b; Dahlgren et al. 1985; Fay and Chase 1996; Fay et al. 2000; Hutchinson 1934, 1973; Mann 1959; Rahn 1998; Rudall et al. 1997.

Amaryllidaceae J. St. Hilaire
(Amaryllis or Daffodil Family)

Herbs from a bulb with contractile roots; stems reduced; vessel elements with scalariform perforations; **characteristic "amaryllis" alkaloids present**. Hairs simple. Leaves alternate, usually 2-ranked, ± basal, simple, flat, entire, with parallel venation, sometimes differentiated into a blade and petiole, sheathing at base; stipules lacking. *Inflorescences determinate, composed of one or more contracted helicoid cymes, and appearing to be an umbel,* sometimes reduced to a single flower, *subtended by a few membranous spathelike bracts, terminal, on a long scape. Flowers* bisexual, radial to bilateral, *showy, each associated with a filiform bract. Tepals 6, distinct to connate,* imbricate, *petaloid, not spotted;* a corona (outgrowth of the perianth) sometimes present. *Stamens 6;* filaments distinct to connate, sometimes adnate to the perianth, sometimes appendaged (and forming a staminal corona); pollen grains monosulcate or bisulcate. *Carpels 3, connate;* **ovary inferior**, with axile placentation; stigma 1, minute to capitate or 3-lobed. Ovules ± numerous in each locule, sometimes with 1 integument. *Nectaries usually in septa of ovary. Fruit a loculicidal capsule* or occasionally a berry; seeds dry to fleshy, flattened to globose, and sometimes winged, *the seed coat usually with black or blue crust,* but phytomelan sometimes lacking and outer epidermis then lacking cellular structure, inner layers also ± collapsed; embryo sometimes curved (Figure 9.24).

Floral formula: * or X, (-6-), (6), (3̄); capsule

Distribution: Widely distributed from temperate to tropical regions, and especially diverse in South Africa, Andean South America, and the Mediterranean region.

Genera/species: 59/870. **Major genera:** *Crinum* (130 spp.), *Hippeastrum* (70), *Zephyranthes* (60), *Hymenocallis* (50), *Cyrtanthus* (50), *Haemanthus* (40), and *Narcissus* (30). Common genera of the continental United States and/or Canada include *Crinum, Hymenocallis, Narcissus,* and *Zephyranthes.*

Economic plants and products: The family includes numerous ornamental genera: *Crinum* (crinum lily), *Eucharis* (Amazon lily), *Galanthus* (snowdrops), *Haemanthus* (blood lily), *Hippeastrum* (amaryllis), *Hymenocallis* (spider lily), *Narcissus* (daffodil, jonquil), *Zephyranthes* (rain lily, zephyr lily), *Cyrtanthus* (kaffir lily), *Amaryllis* (Cape belladonna), and *Nerine* (Guernsey lily).

Discussion: The monophyly of Amaryllidaceae is supported by secondary chemistry (amaryllid alkaloids), the inferior ovary, and *rbcL* and *trnL-F,* and ITS sequences (Chase et al. 1995a; Meerow et al. 1999). Tribal characterizations within the family are provided by Meerow

(A)

(B)

Figure 9.24   Amaryllidaceae. *Crinum kirkii:* (A) habit (× 0.2); (B) flower (× 0.5). (From Pax and Hoffman 1930, in Engler and Prantl, *Die natürlichen Pflanzenfamilien,* 2nd ed., p. 408.)

(1995), Meerow et al. (1999, 2000a, b), Dahlgren et al. (1985), and Traub (1957, 1963). The Amaryllideae (*Amaryllis, Nerine, Crinum,* and relatives) are sister to the remaining genera; Amaryllideae are easily recognizable by the apomorphies: presence of a sclerenchyma sheath in the scape, bisulcate pollen with a spiny exine, ovules with a single integument, non-dormant, water-rich seeds that lack phytomelan and have a chlorophyllous embryo. Among the remaining genera, Haemantheae (e.g., *Haemanthus, Clivia*) are diagnosed by berry fruits, and occur in Africa, while the rest of the genera are American and Eurasian and form a clade. A large group of American genera (*Hymenocallis, Eucharis, Habranthus, Zephyranthes, Hippeastrum,* and *Griffinia*) probably form a monophyletic group. These may be recognized by their obvolute spathaceous bracts (i.e., two bracts overlapping in bud so that one half of each is external and the other half is internal). The Eurasian genera, such as *Lycoris, Narcissus, Leucojum,* and *Galanthus,* also form a clade, which lacks apparent morphological support (Meerow et al. 1999, 2000a, b).

The showy flowers of Amaryllidaceae are pollinated by bees, wasps, butterflies, moths, and birds; most are outcrossing, but selfing also occurs. The seeds are usually wind- or water-dispersed, but some taxa, such as *Eucharis* subg. *Eucharis,* have blue seeds that contrast with bright orange capsules, leading to dispersal by birds. Large, fleshy seeds lacking phytomelan have evolved several times within the family. Some are water dispersed.

References: Chase et al. 1995a; Dahlgren et al. 1985; Meerow 1995; Meerow and Suijman 1998; Meerow et al. 1999, 2000a, b; Traub 1957, 1963.

Asphodelaceae A. L. de Jussieu
(Aloe Family)

Rhizomatous herbs to trees or shrubs; stems often with anomalous secondary growth; *anthraquinones often present.* Hairs simple. *Leaves* alternate, spiral or 2– ranked, *in rosettes at base or ends of branches,* simple, *often succulent, entire to spinose-serrate,* with parallel venation; *often* with vascular bundles (as seen in cross-section) arranged in a ring around central mucilaginous parenchyma tissue, with a cap of aloin cells at the phloem pole of most vascular bundles, these containing colored secretions and generally accumulating anthraquinones, not fibrous, **with parenchymatous inner bundle sheath cells**; sheathing at base; stipules lacking. Inflorescences indeterminate, terminal, but sometimes appearing lateral. Flowers usually bisexual, radial to bilateral, often showy. *Tepals 6, distinct to strongly connate* (perianth then ± tubular), imbricate, *petaloid, not spotted. Stamens 6*; filaments distinct; pollen grains monosulcate. *Carpels 3, connate; ovary superior*, with axile placentation; stigma 1, ± minute to discoid or slightly 3-lobed. Ovules 2 to numerous in each locule, anatropous to nearly orthotropous. *Nectaries in septa of ovary. Fruit a loculicidal capsule*; seeds **with dry aril** (occasionally lost) **that arises as an annular invagination at the distal end of the funiculus**, often flattened or winged, *the seed coat with phytomelan (a black crust)* and inner layers ± collapsed.

Floral formula: * or X, (-6-), 6, ③; capsule

Distribution and ecology: Distributed in temperate to tropical regions of the Old World, and especially diverse in southern Africa; usually in arid habitats.

Genera/species: 15/750. **Major genera:** *Aloe* (380 spp.), *Haworthia* (70), *Kniphofia* (70), and *Bulbine* (60). The family is represented in the continental United States only by two introduced taxa of *Aloe* (in southern Florida and California).

Economic plants and products: Several species of *Aloe* are used medicinally or in cosmetics; members of several genera, including *Aloe, Haworthia, Gasteria, Kniphofia*, and *Bulbine*, are used as ornamentals.

Discussion: The monophyly of Asphodelaceae is supported by *rbcL* sequences (Chase et al. 1995a; de Bruijn et al. 1995; Smith and Van Wyk 1998), arillate seeds, foliar anatomical characters, and possibly the base chromosome number of seven. The family is closely related to Xanthorrhoeaceae and Hemerocallidaceae and all three clades have anthraquinone pigments; some botanists have suggested that these all be treated within Xanthorrhoeaceae s. l.

Asphodelaceae are usually divided into two subfamilies (Dahlgren et al. 1985): the surely paraphyletic "Asphodeloideae" (including genera such as *Bulbine, Kniphofia*, and *Asphodelus*) and the specialized and clearly monophyletic Alooideae (including genera such as *Aloe, Gasteria*, and *Haworthia*). Synapomorphies for Alooideae include the distinctive leaves with a central gelatinous zone surrounded by vascular bundles associated with aloin cells and distinctive dimorphic karyotype (Smith and Van Wyk 1991).

The colorful flowers of Asphodelaceae are pollinated by various insects as well as birds. The seeds are mainly dispersed by wind.

References: Chase et al. 1995a; Dahlgren et al. 1985; de Bruijn et al. 1995; Judd 1997a; Smith and Van Wyk 1991, 1998.

Iridaceae A. L. de Jussieu
(Iris Family)

Herbs with rhizomes, corms, or bulbs, rarely (achlorophyllous; **styloids (large prismatic crystals) of calcium oxalate present in sheaths of vascular bundles** (these occasionally lost); tannins and/or various terpenoids often present. Hairs simple. *Leaves alternate, 2-ranked*, **equitant (oriented edgewise to the stem), and with a unifacial blade**, along the stem to basal, simple, entire, with parallel venation, sheathing at base; stipules lacking. **Inflorescences determinate, a scorpioid cyme**, often highly modified, sometimes reduced to a single flower, terminal. Flowers bisexual, radial to bilateral, conspicuous, subtended individually by 1 or 2 bracts. *Tepals 6*, the outer sometimes differentiated from the inner, distinct or connate, imbricate, *petaloid, sometimes spotted*. **Stamens (2) 3**; filaments distinct or connate, sometimes adnate to the perianth; anthers sometimes sticking to style branches; pollen grains usually monosulcate. *Carpels 3, connate; ovary usually inferior, with axile placentation*; style branches sometimes expanded and petaloid; stigmas (2) 3, terminal or on abaxial surface of style branches. Ovules few to numerous in each locule, anatropous or campylotropous. Nectaries in septa of ovary, on the tepals, or lacking. *Fruit a loculicidal capsule*; seeds sometimes arillate or with fleshy *seed coat*, the coat usually with cellular structure and ± *brown (black crust lacking)* (Figure 9.25).

Floral formula: * or X, (-6-), ③, ⓪; capsule

Distribution: Widely distributed.

Genera/species: 78/1750. **Major genera:** *Gladiolus* (255 spp.), *Iris* (250), *Moraea* (125), *Sisyrinchium* (100), *Romulea* (90), *Crocus* (80), *Geissorhiza* (80), *Babiana* (65), and *Hesperantha* (65). Noteworthy genera occurring in the continental United States and/or Canada are *Alophia, Calydorea, Iris, Nemastylis*, and *Sisyrinchium*.

Economic plants and products: The stigmas of *Crocus sativus* are the source of the spice saffron. Numerous genera, including *Crocus, Tigridia* (tiger flower), *Freesia, Iris, Ixia* (corn lily), *Romulea, Neomarica, Moraea* (butterfly lily), *Nemastylis, Belamcanda, Sisyrinchium* (blue-eyed grass), *Gladiolus, Crocosmia*, and *Trimezia*, are used as ornamentals.

Discussion: Cladistic analyses based on morphology and cpDNA sequences support the monophyly of Iridaceae

Figure 9.25 Iridaceae. (A) *Romulea purpurascens:* flower, note inferior ovary and three stamens (× 2.5). (B) *Crocus sativus:* (B) habit (× 1); (C) plant in longitudinal section (× 1); (D) style and stigmas (× 2). (From Pax 1889, in Engler and Prantl, *Die natürlichen Pflanzenfamilien*, pp. 142–143.)

(Goldblatt 1990; Rudall 1994; Chase et al. 1995a; Reeves et al. 2001). The phylogenetic position of Iridaceae is somewhat uncertain. Morphological characters (Stevenson and Loconte 1995; Chase et al. 1995b) place the family within Liliales, while DNA sequences place it within Asparagales (Chase et al. 1995a, 2000; Fay et al. 2000). Chloroplast DNA and morphology together placed the family within Asparagales, and therefore we consider the family to be a fairly basal member of this order.

Several major clades (often recognized as subfamilies) are evident within Iridaceae (Goldblatt 1990; Goldblatt et al. 1998; Rudall 1994; Reeves et al. 2001). The Isophysidoideae, including only *Isophysis*, is distinguished by a superior ovary and probably is sister to the remaining groups. The Nivenioideae (including *Aristea* and relatives) share the apomorphy of blue flowers; some members of this subfamily are shrubby. Iridoideae have flowers that often last only a single day; they can be recognized by the presence of nectaries on the tepals and by their very long style branches, divided below the level of the anthers, with each branch stigmatic only apically. The group also contains the free amino acids meta-carboxyphenylalanine and glycine. Several subclades are evident within Iridoideae, including the Sisyrinchieae (style branches alternating with stamens vs. opposite them in other members of subfamily; *Sisyrinchium* and relatives) and a clade comprising Irideae, Mariceae, and Tigrideae (tepals differentiated into a limb and claw, upper apices of the style branches with petaloid appendages; e.g., *Iris, Belamcanda, Moraea, Nemastylis, Trimezia,* and *Tigridia*). Finally, Ixioideae (e.g., *Ixia, Crocosmia, Geissorhiza, Crocus, Romulea, Freesia, Gladiolus,* and *Hesperantha*) are hypothesized to be

monophyletic based on their connate tepals, corms, sessile flowers, operculate pollen with micropunctate exine, and closed leaf sheaths. This group retains the ancestral feature of septal nectaries.

The showy flowers of Iridaceae are mainly insect-pollinated (especially by beetles, bees and flies), but some species are pollinated by birds. Nectar and/or pollen are the floral rewards. The seeds usually are dispersed by wind or water, but biotic dispersal also occurs.

References: Chase et al. 1995a,b, 2000; Dahlgren et al. 1985; Fay et al. 2000; Goldblatt 1990; Goldblatt et al. 1998; Reeves et al. 2001; Rudall 1994; Stevenson and Loconte 1995.

Orchidaceae Adanson
(Orchid Family)

Terrestrial or epiphytic herbs, or occasionally vines, *with rhizomes, corms, or root-tubers*, occasionally mycoparasitic; *stems often basally thickened and forming pseudobulbs; roots strongly mycorrhizal, often with spongy, water-absorbing epidermis composed of dead cells (velamen).* Hairs various. Leaves usually alternate, spiral or 2-ranked, often plicate, basal or along the stem, sometimes reduced, simple, entire, with usually parallel venation, sheathing at base; stipules lacking. Inflorescences indeterminate, sometimes reduced to a single flower, terminal or axillary. Flowers usually bisexual, **bilateral, usually resupinate (twisted 180° during development)**, often conspicuous. Outer tepals 3, distinct to connate, usually petaloid, imbricate. Inner tepals 3, distinct, sometimes spotted or variously colored, **the median one clearly differentiated from the 2 laterals, forming a lip (labellum)**, often with fleshy bumps or ridges and of an unusual shape or color pattern. **Stamens 3 or fewer (usually only 1 or 2), adnate to style and stigma, forming a column**; pollen usually *grouped into soft or hard masses (pollinia).* Carpels 3, connate; *ovary inferior, usually with parietal placentation,* but occasionally axile; style and stigma highly modified, with portion of the latter usually nonreceptive (rostellum), a portion of which may form a sticky pad (viscidium) attached to the pollinia. Ovules numerous, with a thin megasporangium wall (Figure 9.26). Nectar produced in a lip-spur, by sepal apices, or in septal nectaries, but often lacking. *Fruit a capsule opening by (1–) 3 or 6 longitudinal slits; seeds minute,* the seed coat crustose or membranous, **lacking phytomelan**, with only the outer layer persisting, the inner tissues collapsed; **embryo very minute; endosperm lacking** (Figures 9.26 and 9.27).

Floral formula: X, 5+1, $\overbrace{1 \text{ or } 2, \text{③}}$; capsule

Distribution and ecology: Widely distributed, but most diverse in tropical regions (where frequently epiphytic).

Genera/species: 775/19,500. *Major genera: Pleurothallis* (1120 spp.), *Bulbophyllum* (1000), *Dendrobium* (900), *Epidendrum* (800), *Habenaria* (600), *Eria* (500), *Lepanthes* (460), *Maxillaria* (420), *Oncidium* (420), *Masdevallia* (380), *Stelis* (370), *Liparis* (350), *Malaxis* (300), *Oberonia* (300), *Encyclia* (235), *Eulophia* (200), *Angraecum* (200), *Taeniophyllum* (170), *Phreatia* (160), *Polystachya* (150), *Calanthe* (150), *Vanilla* (100), and *Catasetum* (100). The family is represented in the continental United States and/or Canada by numerous genera; some of the more noteworthy include *Bletia, Calopogon, Calypso, Cleistes, Corallorhiza, Cypripedium, Encyclia, Epidendrum, Goodyera, Habenaria, Harrisella, Hexalectris, Liparis, Listera, Malaxis, Oncidium, Orchis, Platanthera, Pogonia, Ponthieva, Spiranthes, Tipularia, Triphora,* and *Zeuxine.*

Economic plants and products: Vanilla flavoring is extracted from the fruits of *Vanilla planifolia.* The family is economically important because of its numerous ornamentals, including *Cattleya* (corsage orchid), *Dendrobium, Epidendrum, Paphiopedilum* (slipper orchid), *Phalaenopsis, Vanda, Brassia, Cymbidium, Laelia, Miltonia, Oncidium, Encyclia,* and *Coelogyne.*

Figure 9.26 Floral diagrams of major groups of Orchidaceae. (A) Apostasioideae. (B) Cypripedioideae. (C) Orchidoideae and Epidendroideae. (D) Vanilloideae. Gray crescents = sepals; black crescents = uppermost petals; bicolored crescents = labellum (or lip tepals); stamens are indicated by four-lobed structures, with dots indicating pollen not in pollinia and solid black indicating pollen shed in pollinia. Staminodes indicated by small crescent or elliptical structures near stamen(s). Gynoecium indicated by circle showing placentation type; adnation and connation shown by lines connecting structures. (Modified from Endress 1994 and Dahlgren et al. 1985.)

Discussion: The monophyly of Orchidaceae is supported by morphology and *rbcL* sequences (Dressler 1981, 1993; Dressler and Chase 1995; Dahlgren et al. 1985; Burns-Balogh and Funk 1986; Judd et al. 1993; Chase et al. 2000; Fay et al. 2000). Phylogenetic relationships within the family have been addressed by several cladistic analyses of morphology and DNA sequences (Burns-Balogh and Funk 1986; Dressler 1986, 1993; Judd et al. 1993; Dressler and Chase 1995; Cameron et. al. 1999; Freudenstein and Rasmussen 1999; Cameron and Chase 2000; Freudenstein et al. 2000; Molvray et al. 2000) although delimitation of some groups is still unclear. *Apostasia* and *Neuwiedia* (subfamily Apostasioideae) are considered to be sister to the remaining orchids (Dressler 1993; Judd et al. 1993; Dressler and Chase 1995; Neyland and Urbatsch 1996; Cameron et. al. 1999; Freudenstein and Rasmussen 1999) (Figure 9.28). The monophyly of Apostasioideae is supported by their vessel elements with simple perforation plates and distinctive seeds. These two genera, and especially *Neuwiedia,* have

Figure 9.27 Orchidaceae. *Encyclia cordigera:* (A) habit (× 0.75); (B) flower (× 2); (C) sepals and petals (× 1.5); (D) labellum (or lip) (× 4); (E) column, ventral view, note terminal anther and stigma (in darkened depression (× 4); (F) column, dorsal view, note terminal anther (× 4); (G) column, laterial view (× 4). (Original drawings by Robert Dressler, University of Florida, Gainesville.)

retained many ancestral features, such as flowers with two (*Apostasia*) or three (*Neuwiedia*) stamens that are only slightly adnate to the style, axile placentation, pollen released as single, non-sticky grains, a symmetrical stigma, and all petals initiated simultaneously in development. The remaining orchids all have sticky (or fused) pollen and a broad, asymmetrical stigma with all lobes facing the center of the flower. Within this sticky-pollen clade, Cypripedioideae and Vanilloideae are basal. Cypripedioideae (e.g., *Cypripedium* and *Paphiopedilum*) are clearly monophyletic, as supported by their saccate (slipperlike) lip petal and median anther modified into a shieldlike staminode (see Figure 9.26); they have two functional stamens and lack pollinia. Vanilloideae (i.e., *Vanilla, Pogonia, Cleistes*) form a clade based on *rbcL* sequence data (Dressler and Chase 1995). This group is distinctive in that its flowers have only a single functional stamen and lack pollinia. Many Vanilloideae are vines

with net-veined leaves; their ovaries are sometimes 3-locular. All other orchids have pollinia and a completely fused filament and style (Dahlgren et al. 1985; Burns-Balogh and Funk 1986; Judd et al. 1993; see Figure 9.26). This pollinia-forming clade, like the Vanilloideae, has flowers with only a single functional stamen (monandry; the two lateral stamens are represented by slender staminodes or are entirely lacking). Morphological analyses (Freudenstein and Rasmussen 1999) and some molecular analyses (Cameron et al. 1999; Molvray et al. 2000) support the hypothesis that the monandrous orchids are monophyletic, but other molecular analyses (Cameron and Chase 2000) support the view that the reduction to a single functional stamen has occurred twice within orchids. Among monandrous orchids with pollinia, two large subfamilies, the Epidendroideae and the Orchidoideae (incl. Spiranthoideae), are recognized here (see also Figure 9.28; compare with Dressler 1981, 1986, 1993).

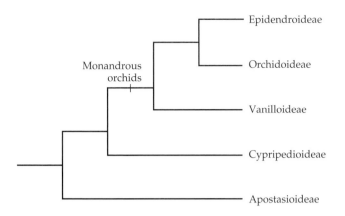

Figure 9.28 Phylogeny of Orchidaceae. (Modified from Cameron et al. 1999.)

Epidendroideae share the apomorphies of a beaked and incumbent anther (i.e., anther bent over apex of column), while Orchidoideae share the apomorphies of acute anther apex, soft stems, leaves convolute but not plicate, and lack of silica bodies (see also Stern et al. 1993; Dressler 1993). Epidendroideae contain numerous tropical epiphytes; representative genera include *Bulbophyllum, Catasetum, Dendrobium, Epidendrum, Encyclia, Maxillaria, Oncidium, Pleurothallis,* and *Vanda*. Generic delimitation in this group is notoriously problematic, and most large genera are nonmonophyletic. Representative members of the Orchidoideae include *Cynorkis, Diuris, Goodyera, Habenaria, Orchis, Platanthera, Spiranthes,* and *Zeuxine*.

Orchid flowers are extremely varied in form and attract a wide array of insects (bees, wasps, moths, butterflies, flies) as well as birds (see Chapter 4). Some attract generalist visitors, but many are quite specialized, attracting only one or a few species as pollinators. Pollen, nectar, or floral fragrances may be employed as pollinator rewards, although some flowers (e.g., *Cypripedium*) manipulate their pollinators and provide no reward, and some species of *Ophrys* and *Cryptostylis* mimic the form and scent of female bees, wasps, or flies, and are pollinated when the males attempt to mate with the flower (pseudocopulation, see Figure 6.12). Generally, the lip functions as a landing platform and provides visual or tactile cues orienting the pollinator. The pollinia become attached to the pollinator's body, and often are deposited in the stigma (usually a depression in the underside of the column) of the next flower visited. In some species pollination is a fairly uncommon event, and the flowers may remain functional and showy for many days, with wilting of the perianth occurring rapidly after fertilization. Most species are outcrossing, but selfing is known to occur. The tiny, dustlike seeds are dispersed by wind and require nutrients supplied by a mycorrhizal fungus for germination.

References: Burns-Balogh and Funk 1986; Cameron and Chase 2000; Cameron et. al. 1999; Chase et al. 2000; Cox et al. 1997; Dahlgren et al. 1985; Dressler 1961, 1981, 1986, 1993; Dressler and Chase 1995; Fay et al. 2000; Freudenstein and Rasmussen 1999; Freudenstein et al. 1999, 2000; Judd et al. 1993; Kocyan and Endress 2001; Molvray et al. 2000; Neyland and Urbatsch 1995; Pridgeon et al. 1999; Stern et al. 1993; van der Pijl and Dodson 1966.

Dioscoreales

Dioscoreaceae R. Brown
(Yam Family)

Twining vines with thick rhizomes or a large tuberlike swelling; *stem with vascular bundles in 1 or 2 rings;* steroidal sapogenins and alkaloids commonly present. Hairs simple to stellate; prickles sometimes present. *Leaves* usually alternate and spiral, simple, but sometimes palmately lobed or compound, entire, *differentiated into a petiole and blade, with palmate venation, the major veins converging and connected by a network of higher-order veins; the petiole usually with an upper and lower pulvinus,* sometimes with stipule-like flanges, not sheathing; bulbils sometimes present in leaf axils. Inflorescences determinate, but sometimes appearing indeterminate, axillary. *Flowers usually unisexual* (plants dioecious), radial. *Tepals 6,* distinct to slightly connate, imbricate. *Stamens 6* (3); filaments distinct to slightly connate, adnate to the base of the tepals; pollen grains monosulcate to variously porate. *Carpels 3, connate; ovary inferior, with axile placentation;* stigmas 3, minute to slightly bilobed. Ovules 2 to numerous in each locule. Nectaries in septa of ovary or base of tepals. *Fruit usually a triangular and 3-winged, loculicidal capsule,* but sometimes a berry or samara; **seeds usually flattened or winged,** the coat with yellow-brown to red pigments and crystals; embryo with 1 or occasionally 2 cotyledons (Figure 9.29).

Floral formula:

Staminate: *, -6-, 6, 0

Carpellate: *, -6-, 0,3; winged loculicidal capsule, samara, berry

Distribution: Widely distributed in the tropics and subtropics, with a few in temperate regions.

Genera/species: 3/404. ***Major genera:*** *Dioscorea* (400 spp.) and *Stenomeris* (2); both occur in the continental United States.

Economic plants and products: The starchy "tubers" of many species of *Dioscorea* (yams) are edible; these "tubers" should not to be confused with the roots of *Ipomoea batatas* (Convolvulaceae), which are also called yams. Other species are medicinally valuable due to

Figure 9.29 Dioscoreaceae. *Dioscorea quarternata:* (A) rhizome (× 0.7); (B) vine with fruits (× 0.7); (C) staminate flowers (× 10); (D) staminate flower, view from above (× 10); (E) stamens (× 50); (F) carpellate flowers (× 10); (G) carpellate flower, viewed from above (× 10); (H) ovary in cross-section (× 15); (I) ovule (greatly magnified); (J) capsule (× 1.5); (K) winged seed (× 3); (L) *D. villosa:* section of stem (× 7). (From Al-Shebaz and Schubert 1998, *J. Arnold Arbor.* 70: p. 72.)

the presence of alkaloids or steroidal sapogenins; the latter are used in anti-inflammatory medications and oral contraceptives.

Discussion: Dioscoreaceae are placed in Dioscoreales, an order with many vines with net-veined leaves. As circumscribed by Dahlgren et al. (1985), it also contained Trilliaceae, Stemonaceae, Taccaceae, Smilacaceae, and a few other families. Morphology (Conran 1989; Stevenson and Loconte 1995; Chase et al. 1995b) and DNA sequences suggest that the order, as so circumscribed, is not monophyletic. In this text, Trilliaceae and Smilacaceae are referred to the Liliales and Stemonaceae to the Pandanales, while the mycoparasitic Burmanniaceae are included here. Dioscoreales, as here circumscribed, are monophyletic (Chase et al. 1993, 1995a, 2000; Soltis et al. 2000; Caddick et al. 2000). Dioscoreales frequently have been considered to represent primitive monocots (Dahlgren et al. 1985; Stevenson and Loconte 1995), a position not supported by cladistic analyses based on DNA sequences (Chase et al. 1995a,b, 2000; Soltis et al. 2000).

Dioscoreaceae are easily differentiated from the closely related Taccaceae by their viny habit and usually unisexual flowers, axile (vs. parietal) placentation, and lack of filamentous bracts within the inflorescence. They may be distinguished from the phenetically similar Stemonaceae by 3-merous (vs. 2-merous) flowers and the consistently inferior ovary. Finally, Dioscoreaceae are easily separated from Burmanniaceae, a family of mycoparasitic herbs with scale-like leaves. Smilacaceae are also vines with net-veined leaves, but can easily be distinguished by their superior ovaries, few-seeded berries, and leaves with paired stipular tendrils.

The small genus *Stenomeris* (along with *Tacca*, Taccaceae) may be sister to the remaining taxa, which form a clade supported by the putative synapomorphies of unisexual flowers, underground root-tubers, three-winged fruits, and multi-layered seed coat with a crystalline inner layer. Phylogenetic relationships within the dioecious clade have been investigated on the basis of 18S nrDNA, *rbcL*, *atpB* sequences, and morphology (Caddick et al. 2000, 2002a) and it is evident that the recognition of genera such as *Tamus* and *Rajania* make *Dioscorea* paraphyletic. *Tacca* probably should be included in the family.

The inconspicuous flowers of Dioscoreaceae are insect-pollinated (mainly by flies). Dispersal is usually by wind, as indicated by the specialized fruits: three-winged capsules with flattened and/or winged seeds or samaras (as in species often segregated as *Rajania*).

References: Al-Shehbaz and Schubert 1989; Bouman 1995; Caddick et al. 2000, 2002a,b; Chase et al. 1995a,b, 2000; Conran 1989; Dahlgren et al. 1985; Soltis et al. 2000; Stevenson and Loconte 1995.

Commelinoid Monocots

The commelinoid monocots constitute a monophyletic group supported by *rbcL* sequences (Chase et al. 1993, 1995b; Duvall et al. 1993), *rbcL*, *atpB*, and 18S sequences (Chase et al. 2000; Soltis et al. 2000) and morphology (Dahlgren and Rasmussen 1983; Stevenson et al. 2000). Putative synapomorphies include *Strelitzia*-type epicuticular wax (Figure 4.34), endosperm with copious starch, starchy pollen, and UV-fluorescent compounds—ferulic and coumaric acids—in the cell walls (Dahlgren et al. 1985; Harley and Ferguson 1990; Barthlott and Fröhlich 1983; Harris and Hartley 1980; Zona 2001). Arecaceae (Palmae) have nonstarchy endosperm; Dahlgren et al. (1985) have suggested that this may represent an evolutionary loss. Analyses using DNA sequences (Chase et al. 1993, 1995b, 2000; Soltis et al. 2000) suggest that palms are the sister group to the remaining members of the commelinoid clade: Poales, Commelinales, and Zingiberales, so starchy endosperm may actually be a synapomorphy of the clade comprising these three orders.

Arecales

Arecaceae C. H. Schultz-Schultzenstein
(= Palmae Adanson)
(Palm Family)

Trees or shrubs with unbranched or rarely branched trunks, occasionally rhizomatous; **apex of stem with a large apical meristem**; tannins and polyphenols often present. Hairs various, and plants sometimes spiny due to modified leaf segments, exposed fibers, sharp-pointed roots, or petiole outgrowths. *Leaves* alternate and spiral, *often crowded in a terminal crown*, but sometimes well separated, simple and entire, *usually splitting in a pinnate to palmate fashion as the leaf expands, and at maturity appearing palmately lobed (with segments radiating from a single point), costapalmately lobed (± palmate segments diverging from a short central axis, or costa), pinnately lobed or compound (with well-developed central axis bearing pinnate segments)*, or rarely twice pinnately compound, differentiated into a petiole and blade, **the latter plicate**, and the segments either induplicate (V-shaped in cross-section) or reduplicate (Λ-shaped in cross-section; each segment with veins ± parallel to divergent, the petiole often with a flap (hastula), its base sheathing, with soft tissues often decaying to reveal various fiber patterns; stipules lacking. *Inflorescences* determinate and/or indeterminate, often *appearing compound-spicate*, axillary or terminal, with small to large, and deciduous to persistent bracts. *Flowers* bisexual or unisexual (and plants then monoecious to dioecious), radial, *usually sessile, with perianth usually differentiated into calyx and corolla*. Sepals 3, distinct to connate, usually imbricate. Petals usually 3, distinct to connate, imbricate to valvate. Stamens 3 or 6 to numerous; filaments distinct to connate, free or adnate to petals; pollen grains usually monosulcate. Carpels usually 3, but occasionally as many as 10, sometimes appearing to have a single carpel, distinct to connate; *ovary superior*, usually with axile placentation; stigmas various. **Ovules 1 in each locule**, anatropous to orthotropous. Nectaries in septa of ovary or lacking. **Fruit a drupe**, usually 1-seeded, often fibrous, or rarely a berry; endosperm with oils or carbohydrates, sometimes ruminate (Figure 9.30, 9.31).

Floral formula: *, ⌐3⌐, ⌐3⌐, ⌐6–∞⌐, ③; drupe, berry

Distribution: Widespread in tropical to warm temperate regions.

Genera/species: 200/2780. **Major genera:** *Calamus* (370 spp.), *Bactris* (200), *Daemonorops* (115), *Licuala* (100), and *Chamaedorea* (100). The family is represented in the continental United States by *Coccothrinax*, *Pseudophoenix*, *Rhaphidophyllum*, *Roystonia*, *Sabal*, *Serenoa*, *Thrinax*, *Washingtonia*, and a few naturalized genera.

Economic plants and products: Food plants come from *Areca* (betel nut), *Attalea* (American oil palm), *Bactris* (peach palm), *Cocos* (coconut), *Elaeis* (African oil palm), *Metroxylon* (sago palm), and *Phoenix* (date palm); a great many genera have an edible apical bud (palm cabbage). Other economically important palms include *Calamus* (rattan), *Copernicia* (carnuba wax), *Phytelephas* (vegetable

Figure 9.30 Arecaceae (Palmae, Coryphoideae). *Acoelorraphe wrightii*: (A) habit (greatly reduced); (B) junction of leaf blade and petiole, showing hastula (× 0.5); (C) portion of inflorescence (× 0.5); (D) portion of inflorescence axis with flowers(× 7); (E) calyx (× 14); (F) spread-out corolla and androecium (× 14); (G) petal; (H) gynoecium (× 23); (I) drupe (× 3.5); (J) seed (× 4.5). (From Zona 1997, *Harvard Pap. Bot.* 2(1): p. 97.)

Figure 9.31 Arecaceae (Palmae, Arecoideae). *Roystonia regia*: (A) habit (greatly reduced), note crownshaft and inflorescences; (B) portion of inflorescence axis, showing triads of flower buds, two staminate and one carpellate (× 4); (C) staminate flower bud with petal and three stamens removed (× 7); (D) open staminate flower, three stamens removed to show central pistillodium (× 7); (E, F) stamens, before and after dehiscence of anther (× 7); (G) nearly mature carpellate bud in partial section, showing nectariferous staminodial cup (hatched) and gynoecium (× 6.2); (H) corolla and staminoidal cup of carpellate flower (× 6.2); (I) drupe (× 2.8); (J) endocarp (× 2.8); (K) seed, showing hilar scar (× 2.3). (From Zona 1997, *Harvard Pap. Bot.* 2(1): p. 101.)

ivory), *Raphia* (raffia), and many genera that provide thatch. Finally, the family includes a large number of ornamentals: *Caryota* (fishtail palm), *Chamaerops* (European fan palm), *Livistona* (Chinese fan palm), *Roystonea* (royal palm), *Sabal* (cabbage palm), *Syagrus* (queen palm), *Washingtonia* (California fan palm), *Chamaedorea* (parlor palm), *Raphidophyllum* (needle palm), *Thrinax* (thatch palm), *Coccothrinax* (thatch palm), *Licuala*, *Veitchia*, *Acoelorraphe* (paurotis palm), *Butia* (jelly palm), *Copernicia*, *Dypsis*, and *Wodyetia* (foxtail palm).

Discussion: Arecaceae (or Palmae) are distinct, easily recognized, and monophyletic. Cladistic analysis of the family (Uhl et al. 1995) using morphology and cpDNA restriction sites supports placement of *Nypa*, a distinctive genus of Asian and western Pacific mangrove communities, as sister to the remaining palms. *Nypa* (Nypoideae) has a dichotomous-branching prostrate stem, and erect, pinnate, reduplicate leaves; fossils are known from Europe and America early in the Tertiary. A perianth of sepals and petals may be synapomorphic for the remaining palms.

However, analyses based on *rbcL, atpB, rps16, trnL-F*, and 18S sequences support a basal position for Calamoideae, another group with pinnate, reduplicate leaves, with *Nypa* possibly diverging next. Those genera with usually pinnate and reduplicate leaves may form a paraphyletic complex, while those with usually costapalmate or palmate and induplicate leaves—the Coryphoideae—form either a monophyletic (Uhl et al. 1995), paraphyletic group (Asmussen et al. 2000; Asmussen and Chase 2001), or variously resolved group (Hahn 2002).

Major clades of reduplicate-leaved palms (excluding *Nypa*) include the following: The Calamoideae have pinnate to palmate leaves and distinctive fruits that are covered with reflexed, imbricate scales (a distinctive synapomorphic feature). Noteworthy genera are *Raphia, Mauritia, Lepidocaryum, Metroxylon,* and *Calamus.* Hyophorbeae (e.g., *Chamaedorea, Hyophorbe*) have pinnate leaves and imperfect flowers in lines. Arecoideae have pinnate leaves and flowers in groups of three (triads), with one carpellate flower surrounded by two staminate flowers (a likely synapomorphy, but lost in some). Within the Arecoideae, a few clearly defined monophyletic groups are evident. Cocoeae have the inflorescence associated with a large, persistent, woody bract, and fruits with bony, three-pored endocarps, and include genera such as *Elaeis, Cocos, Syagrus, Attalea, Bactris, Desmoncus,* and *Jubaea.* Iriarteae (e.g., *Iriartea, Socratea*) have stilt roots, and leaf segments with blunt apices and diverging veins. Most Arecoideae are placed within a heterogeneous and definitely paraphyletic Areceae; representative genera include *Areca, Roystonea, Prestoea, Dypsis, Wodyetia, Veitchia, Ptychosperma,* and *Dictyosperma.* These palms sometimes have a crownshaft, a structure formed from a series of broad, overlapping leaf bases, which appears to be a vertical extension of the stem.

The Caryoteae (fishtail palms; e.g., *Caryota* and *Arenga*) form a distinctive clade that is usually placed within the Arecoideae because of their floral triads. This group has induplicate, blunt leaf segments with diverging veins. They probably are not closely related to other taxa with flowers in threes (triads) (Asmussen et al. 2000; Asmussen and Chase 2001; Hahn 2002)

Coryphoideae are traditionally divided into three tribes. The monogeneric Phoeniceae (*Phoenix*, the date palms) have distinctive pinnate leaves in which the basal segments are spinelike. Borasseae (e.g., *Latania, Borassus, Lodoicea,* and *Hyphaene*) have staminate flowers embedded in thickened inflorescence axes. Most genera of Coryphoideae are placed in a third tribe—the heterogeneous and clearly paraphyletic "Corypheae". Noteworthy genera include *Chamaerops, Rhaphis, Coccothrinax, Licuala, Copernicia, Corypha, Washingtonia, Sabal, Serenoa, Livistona, Thrinax, Rhapidophyllum,* and *Acoelorraphe.*

Flowers of palms usually are insect-pollinated (especially by beetles, bees, and flies); nectar is often employed as a pollinator reward (Henderson 1986). Palm fruits are usually fleshy and dispersed by a wide variety of mammals and birds, although some (e.g., *Nypa* and *Cocos*) are water dispersed and float in ocean currents (Zona and Henderson 1989).

References: Asmussen and Chase 2001; Asmussen et al. 2000; Dransfield 1986; Dransfield and Uhl 1998; Hahn 2002; Henderson 1986, 1995; Henderson et al. 1995; Moore 1973; Moore and Uhl 1982; Tomlinson 1990; Uhl and Dransfield 1987; Uhl et al. 1995; Zona 1997; Zona and Henderson 1989.

Commelinales

The Haemodoraceae, Pontederiaceae, Philydraceae, Commelinaceae, and possibly Hanguanaceae, form the Commelinales. These families have many-flowered helicoid cymes and an amoeboid tapetum—the innermost layer of the anther tapetum shows early breakdown of inner and radial cell walls, with the nuclei and cytoplasm moving into the anther cavity (Dahlgren and Clifford 1982), derived characters that they share with the likely related order Zingiberales. The monophyly of Commelinales is supported by DNA sequences (Chase et al. 1993, 1995a, 2000; Linder and Kellogg 1995; Soltis et al. 2000), but morphological synapomorphies are ambiguous. Pontederiaceae share with Haemodoraceae nontectate-columellate exine structure, while Philydraceae, like the Haemodoraceae, have harrow, unifacial leaves (Simpson 1990; Dahlgren et al. 1985). All three families have tannin cells in the perianth and sclereids in the placentae.

The placement of Commelinaceae has been disputed. Morphology places this family with Eriocaulaceae and Mayacaceae, in the group here called Poales (Stevenson and Loconte 1995), while DNA sequences place it with the Haemodoraceae, Pontederiaceae, and Philydraceae (Linder and Kellogg 1995; Chase et al. 2000). It is possible that flowers with sepals and petals and moniliform hairs, the characters that have been used to support a relationship with some members of the Poales, have evolved independently in Commelinaceae. As considered here, the Commelinales comprise 5 families and about 780 species.

References: Chase et al. 1993, 1995a, 2000; Duvall et al. 1993; Dahlgren and Clifford 1982; Dahlgren et al. 1985; Graham and Barrett 1995; Linder and Kellogg 1995; Simpson 1990; Soltis et al. 2000; Stevenson and Loconte 1995.

Commelinaceae Mirbel
(Spiderwort Family)

Herbs, sometimes succulent, with well-developed stems that are ± swollen at the nodes, or stems sometimes short; often with mucilage cells or canals containing raphides. Hairs simple, uniseriate or unicellular. Leaves alternate, 2-ranked or spiral, scattered along stem, simple, narrow or somewhat expanded, flat to sharply folded (V-shaped in cross-section) *and often with the opposite halves rolled separately against the midrib in bud*, entire, with parallel vena-

Key to Families of Commelinales

1. Perianth of sepals and petals; leaves with the opposite halves rolled separately against the midrib in bud; seed with a conical cap .... **Commelinaceae**
1. Perianth of tepals; leaves not as above; seeds lacking a conical cap... 2
2. Leaves with expanded, bifacial blade; petiole present....................................... **Pontederiaceae**
2. Leaves equitant, unifacial; petiole lacking 3
3. Stamen solitary; tepals 4 and distinct; flowers lacking septal nectaries; arylphenalenones absent, the roots never reddish.. Philydraceae
3. Stamens 3 or 6; tepals 6 and distinct to connate; flowers with septal nectaries, but these sometimes poorly developed; arylphenalenones (often reddish polyphenolics) present, often giving an orange-red to purple color to the rhizomes and roots........... **Haemodoraceae**

tion, **with closed basal sheath**; **stomata tetracytic**; stipules lacking. Inflorescences determinate, composed of few to several helicoid cymes, sometimes reduced to a solitary flower, terminal or axillary, often subtended by a folded, leafy bract. *Flowers* usually bisexual, radial to bilateral, *with perianth differentiated into a calyx and corolla. Sepals 3*, usually distinct, imbricate or with open aestivation. *Petals 3*, distinct and usually clawed to connate, and corolla then with a short to elongate tube and flaring lobes, *quickly self-digesting*, 1 petal sometimes differently colored and/or reduced, imbricate and crumpled in bud. Stamens 6, or 3 and then often with 3 staminodes; *filaments* slender, distinct to slightly connate, sometimes adnate to petals, *often with conspicuous moniliform (bead-like) hairs;* anthers occasionally with apical pores; pollen grains usually monosulcate. Carpels 3, connate; *ovary superior, with axile placentation;* stigma 1, capitate, fringed, or 3-lobed. Ovules 1 to several in each locule, anatropous to orthotropous. Nectaries lacking. *Fruit usually a loculicidal capsule* (occasionally a berry); **seeds with a conspicuous conical cap** (Figure 9.32.).

Floral formula: *, 3,⟨3⟩, 3 or 6,⟨3⟩; capsule

Distribution: Widespread in tropical to temperate regions.

Genera/species: 40/650. **Major genera:** *Commelina* (230 spp.), *Tradescantia* (60), *Aneilema* (60), *Murdannia* (45), and *Callisia* (20). *Callisia, Commelina, Gibasis, Murdannia,* and *Tradescantia* occur in the United States and/or Canada.

Discussion: The monophyly of Commelinaceae is supported by both morphological (see description) and molecular data (Linder and Kellogg 1995; Evans et al. 2000). The genus *Cartonema* may be sister to the remaining members of the family; it has yellow, radially symmetrical flowers, and lacks both the glandular microhairs and raphide canals present in other taxa (Faden 1998; Evans et al. 2000). Most genera of Commelinaceae belong to one of two large tribes (Faden and Hunt 1991): Trades-

cantieae (25 genera, e.g., *Callisia, Tradescantia,* and *Gibasis*) and Commelineae (13 genera, e.g., *Commelina, Murdannia,* and *Aneilema*). The former group is characterized by spineless pollen grains, medium to large chromosomes, radial flowers, and filament hairs (when present) moniliform; the latter has spiny pollen, small chromosomes, radial to bilateral flowers, and filament hairs (when present) usually not moniliform. There has been much convergence in floral characters within the family due to strong pollinator selection, and anatomical features (e.g., stomatal structure) may be quite useful in diagnosing major clades (Evans et al. 2000).

Flowers of Commelinaceae function for only a day at most. Pollination is usually accomplished by bees or wasps that gather pollen. Interestingly, the staminodes are often more conspicuous than the stamens, and bilateral flowers are held so that the odd sepal is adaxial. The moniliform hairs often present on the filaments of members of the Tradescantieae may delude bees into scraping them, as if gathering pollen. Self-pollination is common in some species.

References: Brenan 1966; Dahlgren et al. 1985; Evans et al. 2000; Faden 1998; Faden and Hunt 1991; Linder and Kellogg 1995; Tucker 1989.

Haemodoraceae R. Brown
(Bloodwort or Blood Lily Family)

Herbs, with rhizomes, corms, or bulbs, and roots often with orange-red pigment, **containing various phenalones, i.e., polyphenolic compounds**. *Hairs simple to dendritic or stellate, and usually densely covering inflorescence axes, bracts, and outer perianth parts. Leaves* alternate, *2-ranked, equitant, unifacial,* those of upper portion of stem reduced, simple, entire, with parallel venation, sheathing at base; stipules lacking. *Inflorescences determinate, consisting of a series of helicoid cymes,* but sometimes appearing indeterminate, terminal. Flowers bisexual, radial to bilateral. *Tepals 6,* showy, distinct to connate, the perianth

Figure 9.32 Commelinaceae. *Callisia cordifolia:* (A) stem with inflorescence (× 1.5); (B) inflorescence with buds and open capsules (× 9); (C) flower (× 22); (D) stamen (× 37); (E) longitudinal section of gynoecium (× 18); (F) open capsule (× 12); (G) seed, hilum below center (× 37); (H) seed from opposite side, conical cap at center (× 75). (From Tucker 1989, *J. Arnold Arbor.* 70: p. 119.)

tube (when present), sometimes with a slit along upper surface, imbricate or valvate. Stamens 3 or 6, sometimes reduced to 1, occasionally dimorphic; filaments free or adnate to the tepals; pollen grains monosulcate or 2–7-porate. *Carpels 3, connate;* **ovary** superior or **inferior,** with axile placentation; stigma 1, capitate to 3-lobed. Ovules 1 to numerous in each locule, anatropous to orthotropous. Nectaries in septa of ovary, but often poorly developed. *Fruit a loculicidal capsule;* seeds often winged.

Floral formula: * or X, -6-, 6 or 3 , ③; capsule

Distribution and ecology: Widespread in Australia, southern Africa, and northern South America, but a few in North America; often wetland plants.

Genera/species: 13 / 100. **Major genera:** *Conostylis* (30 spp.), *Haemodorum* (20), *Anigozanthos* (11). The family is represented in the United States by *Lachnanthes.*

Economic plants and products: Genera such as *Anigozanthos* (kangaroo paw), *Conostylis,* and *Lachnanthes* (blood lily, redroot) are used as ornamentals.

Discussion: Haemodoraceae are considered monophyletic due to the presence of arylphenalenones; they are the only vascular plants to possess these pigments, which produce the orange-red to purple coloration characteristic of the roots and rhizomes of many genera (Simpson 1990). Morphology supports recognition of two major clades: Haemodoreae and Conostylideae (Simpson 1990). The monophyly of Haemodoreae (e.g.,

Haemodorum, Lachnanthes, and *Xiphidium*) is supported by a reddish coloration in the roots and rhizomes, lack of sclereids in the placentas, and discoid seeds that are pubescent or marginally winged. Synapomorphies of Conostylideae (e.g., *Anigozanthos*) are multiseriate, branched hairs; six fertile stamens; pollen with rugulate (wrinkled) wall sculpturing and porate apertures; a nondeflexed style; and a base chromosome number of seven. The perianth of *Anigozanthos* has an adaxial slit; bilateral flowers have evolved within both tribes. Ovary position in the family has apparently undergone a shift from inferior to superior (Simpson 1990, 1993, 1998b).

The eastern North American *Lophiola* resembles Haemodoraceae in its unifacial leaves, inflorescence of helicoid cymes, and densely hairy flowers, but differs in pollen ultrastructure, hair morphology, and stem anatomy. *Lophiola* lacks the arylphenalenones so characteristic of Haemodoraceae, as well as the UV-fluorescent cell-wall-bound compounds that are present in Haemodoraceae (and all other Commelinoid monocots). *Lophiola* is now placed within Melanthiaceae (Ambrose 1985) or Nartheciaceae (Zomlefer 1997a,b,c and this text).

The variously colored flowers of Haemodoraceae are usually pollinated by insects (especially bees and butterflies), but bird pollination is characteristic of *Anigozanthos*. Nectar usually provides the pollinator reward, but *Xiphidium* is pollinated by pollen-gathering bees. The often small, flattened, hairy, or winged seeds of many species probably are dispersed by wind.

References: Ambrose 1985; Anderberg and Eldenäs 1991; Dahlgren et al. 1985; Robertson 1976; Simpson 1990, 1993, 1998a, b; Zomlefer 1997a,b,c.

Pontederiaceae Kunth
(Water Hyacinth Family)

Herbs, rhizomatous, floating to emergent aquatics; stems spongy. Hairs simple, only on reproductive parts. Leaves usually alternate and spiral, along stem or ± basal, ± differentiated into a petiole and blade, simple, entire, with parallel to palmate venation, sheathing at base; stipules lacking. Inflorescences determinate, but often appearing to be racemes or spikes, sometimes reduced to a single flower, terminal, but often appearing lateral, subtended by a bract. *Flowers bisexual, radial to bilateral, often tristylous. Tepals 6, showy,* variably **connate**, imbricate, *adaxial tepal of inner whorl often differentiated. Stamens usually 6*; **filaments adnate to the perianth tube**, *often unequal in length*; anthers opening by slits or pores; pollen grains with 1 or 2 furrows. Carpels 3, connate; ovary superior, with axile to occasionally intruded parietal placentation, sometimes with 2 locules sterile; stigma 1, capitate to 3-lobed. Ovules numerous to 1 in each locule. Nectaries often present in septa of ovary. *Fruit a loculicidal capsule or nut,* **surrounded by persistent basal portion of perianth tube** (Figure 9.33).

Floral formula: * or X, ⟨-6-, 6, ③⟩; capsule, nut

Distribution and ecology: Widespread in tropical and subtropical regions, with a few temperate species; plants of aquatic and wetland habitats.

Genera/species: 7/35. *Genera: Heteranthera* (12 spp.), *Eichhornia* (7), *Monochoria* (7), and *Pontederia* (6). The family is represented in the continental United States and/or Canada by *Eichhornia, Heteranthera,* and *Pontederia.*

Economic plants and products: Pontederia (pickerel weed) and *Eichhornia* (water hyacinth) are used as aquatic ornamentals; the latter is a very serious weed of still or slowly-flowing waters in the tropics and subtropics.

Discussion: The monophyly of Pontederiaceae has been supported by morphology (Eckenwalder and Barrett 1986) and molecular characters (Graham and Barrett 1995; Graham et al. 1998). Morphology supports the recognition of two major monophyletic groups. *Pontederia* and *Eichhornia* form a clade on the basis of a long-lived, perennial habit, a curved inflorescence axis, and bilateral, tristylous flowers, while *Heteranthera* and possibly *Monochoria* form a clade diagnosed by dimorphic stamens (fertile stamens and stamens providing food for insects) with basifixed anthers (Eckenwalder and Barrett 1986). However, molecular data suggest that *Monochoria* belongs to the *Pontederia + Eichhornia* clade. *Pontederia* is distinct due to the apomorphies of a gynoecium with two locules aborting and the third containing only a single ovule, indehiscent fruits (nuts) surrounded by a ribbed to toothed persistent basal portion of the perianth, and relatively large seeds. *Eichhornia* is not monophyletic.

Pontederiaceae is the only monocot family showing tristyly, which probably evolved just once within the group and then was lost several times. Only Oxalidaceae and Lythraceae also have tristylous species (Vuilleumier 1967).

The showy flowers of Pontederiaceae are pollinated by various insects, especially bees, flies, and butterflies. Outcrossing is characteristic of heterostylous species, but self-pollination, as in *Heteranthera*, is also common. The warty nuts of *Pontederia* are water dispersed, as are the small seeds of capsular fruited species. Asexual reproduction by fragmentation of the rhizome system is common in some species.

References: Barrett and Graham 1997; Cook 1998; Dahlgren et al. 1985; Eckenwalder and Barrett 1986; Graham and Barrett 1995; Graham et al. 1998; Lowden 1973; Ornduff 1966; Price and Barrett 1982; Rosatti 1987; Vuilleumier 1967.

Figure 9.33   Pontederiaceae. *Pontederia cordata:* (A) leaf blade and portion of petiole behind flowering stem with leaf and bract subtending inflorescence (× 0.4); (B) flower of long-styled form, with style and three mid-length stamens exserted (× 4); (C) flower of short-styled form, in semidiagrammatic longitudinal section (e.g., hairs not shown) showing two of three adaxial, mid-length stamens and two of three abaxial, long stamens (× 4); (D) flower of mid-styled form showing two of three adaxial, short stamens and two of three abaxial, long stamens (× 4); (E) flower of long-styled form, showing two of three adaxial, short stamens and two of three abaxial, mid-length stamens (× 4); (F) glandular hairs of staminal filaments (× 72); (G) ovary in longitudinal section, note fertile locule with a single pendulous ovule, dashed line shows position of cross-section in (H) (× 23); (H) ovary in cross-section, showing two aborted locules and one fertile locule (× 23); (I) terminal part of stem, with developing fruits (× 0.2); (J) fruit (× 4.5); (K) fruit enclosed in fleshy accrescent base of perianth (× 4.5). (From Rosatti 1987, *J. Arnold Arbor.* 68: p. 64.)

Poales

The monophyly of Poales, as delimited in this text, is well supported by DNA sequence data (Chase et al. 1995a, 2000; Soltis et al. 2000). Possible morphological characters supporting this grouping of families include silica bodies in the epidermis, styles separate or connate but strongly branched, the loss of raphide crystals, and the loss of septal nectaries. Wind pollination has evolved several times within Poales and is characteristic of the Typhaceae, Jun-caceae, Cyperaceae, Restionaceae, and Poaceae. The order is often delimited very narrowly (including only Poaceae, Restoniaceae, and close relatives, see Figure 9.34) but is here circumscribed broadly, including 18 families and about 19,500 species; major families include **Typhaceae, Bromeliaceae, Eriocaulaceae, Xyridaceae, Juncaceae, Cyperaceae, Restionaceae**, and **Poaceae**.

Major clades within the Poales are illustrated in Figure 9.34. Typhaceae and Bromeliaceae are isolated and likely

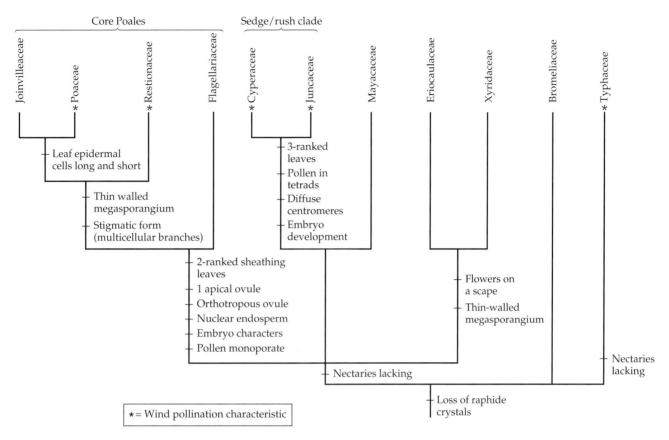

Figure 9.34 Cladogram showing hypothesized relationships within Poales. (Modified from Kellogg and Linder 1995, Chase et al. 2000, and Soltis et al. 2000.)

represent early divergent clades within the order. Eriocaulaceae and Xyridaceae are linked by their distinctive habit (i.e., rosette plants with dense cluster of flowers on a scape), perianth of a calyx and corolla, and ovules with thin-walled megasporangia (Dahlgren et al. 1985; Linder and Kellogg 1995). The remaining families mentioned above belong to either the sedge/rush clade or the core Poales.

Cyperaceae and Juncaceae (along with Thurniaceae) constitute the sedge/rush clade, which often has been recognized formally as the Juncales. This group is certainly monophyletic, based on morphology and *rbcL* sequences (Plunkett et al. 1995; Simpson 1995). Morphological synapomorphies include solid stems, 3-ranked leaves, loss of calcium oxalate raphides, pollen in tetrads (three of the grains reduced in Cyperaceae), chromosomes with diffuse centromeres, and details of embryo and pollen development. The genus *Prionium* (sometimes placed in Juncaceae, but here considered in Thurniaceae) may be sister to the Cyperaceae/Juncaceae clade (Munro and Linder 1998). All evidence points to Cyperaceae being monophyletic and sister to *Oxychloe* (Juncaceae) based on *rbcL* sequences (Muasya et al. 1998). Even if *Oxychloe* and *Prionium* are excluded from Juncaceae, its monophyly is not clear; both morphological and *rbcL* sequence data are ambiguous.

Members of the sedge/rush clade are wind-pollinated, superficially grasslike, and often confused with grasses. A common mnemonic is the following rhyme (origin uncertain): "Sedges have edges, and rushes are round, and grasses are hollow right to the ground." This rhyme refers to the sharply triangular stems of some (but not all) sedges (Cyperaceae), and the hollow stems of some (but not all) grasses (Poaceae). Rushes (Juncaceae) are indeed round-stemmed and solid, but the rhyme's characterization of sedges and grasses is an oversimplification.

The core Poales (or graminoid clade) includes Flagellariaceae, Restionaceae, Joinvilleaceae, and Poaceae, and a few other small families. Other than the Poaceae, which are cosmopolitan, these families are herbs of the Southern Hemisphere and especially the Pacific Ocean region. The largest family other than Poaceae in core Poales is Restionaceae, which grow mostly in South Africa and Australia, and the closest extant relative of Poaceae is Joinvilleaceae of the Pacific region. The monophyly of core Poales is supported by both morphological and DNA characters (Dahlgren and Rasmussen 1983; Dahlgren et al. 1985; Kellogg and Linder 1995; Linder and Kellogg 1995; Briggs et al. 2000; Chase et al. 2000; Soltis et al. 2000; Stevenson et al. 2000); morphological synapomorphies include two-ranked leaves, each with a sheath around the stem, stomates with dumbbell-shaped guard cells, small flowers with one apical orthotropous ovule per carpel, pollen monoporate with rim around pore, nuclear endosperm development, and embryo characters (Kellogg and Linder 1995; Soreng and Davis 1998). The clade also shows deletions in the ORF 2280 region of the chloroplast genome. Evolutionary relationships within core Poales are shown in Figure 9.34.

References: Briggs et al. 2000; Munro and Linder 1998; Plunkett et al. 1995; Simpson 1995; Chase et al. 1993, 1995b, 2000; Dahlgren et al. 1985; Dahlgren and Rasmussen 1983; Kellogg and Linder 1995; Linder and Kellogg 1995; Muasya et al. 1998; Soltis et al. 2000; Soring and Davis 1998; Stevenson and Loconte 1995; Stevenson et al. 2000.

Bromeliaceae A. L. de Jussieu
(Bromeliad Family)

Herbs, usually epiphytic; **solitary silica bodies** usually associated with epidermal cells. **Hairs water-absorbing peltate scales, occasionally merely stellate**. *Leaves* alternate and spiral, *often forming a tank-like basal rosette that holds water*, simple, entire to sharply serrate, with parallel venation, **containing water storage tissue** and air canals (often with stellate cells); sheathing at base; stipules lacking. Inflorescences indeterminate, terminal. *Flowers* usually bisexual, radial, **with perianth differentiated into a calyx and corolla**, *borne in the axils of often brightly colored bracts*. *Sepals 3*, distinct to connate, imbricate. *Petals 3*, distinct to connate, often with a pair of appendages at the base, imbricate. *Stamens 6*; filaments distinct to connate, sometimes adnate to petals; pollen grains monosul-

---

Key to Major Families of Poales

1. Leaves with water-absorbing peltate scales, occasionally merely stellate .... **Bromeliaceae**
1. Leaves lacking such hairs 2
2. Perianth differentiated into a calyx and usually colorful corolla 3
2. Perianth of tepals, or reduced to lacking 5
3. Leaves evenly distributed along stem; flowers solitary .... Mayacaceae
3. Leaves ± basal; flowers in heads or conelike spikes borne at apex of long scape 4
4. Flowers unisexual, in involucrate heads with numerous flowers open at one time; ovules 1 per locule .... **Eriocaulaceae**
4. Flowers bisexual, in conelike spikes, with only 1 or 2 flowers open at one time; ovules numerous on each placenta .... **Xyridaceae**
5. Leaves 3-ranked 6
5. Leaves 2-ranked 7
6. Perianth of 6 tepals; ovules 3–numerous; fruit a capsule .... **Juncaceae**
6. Perianth lacking or reduced to scales, bristles, or hairs; ovule 1; fruit an achene (nutlet) .... **Cyperaceae**
7. Inflorescence with numerous densely clustered flowers and appearing to be elongate/ cylindrical spikes or globose clusters, the staminate flowers positioned above the carpellate .... **Typhaceae**
7. Inflorescence not as above 8
8. Leaves circinately inrolled in bud, with coiled, tendril-apex at maturity .... Flagellariaceae
8. Leaves not circinately inrolled at tip 9
9. Leaves usually with the blade very reduced, consisting merely of a sheath, a ligule usually lacking; anthers with 1 locule, opening by a single slit .... **Restionaceae**
9. Leaves with a ± well developed blade, a ligule usually present; anthers with 2 locules, opening by 2 longitudinal slits 10
10. Perianth of 6 tepals; flowers not in spikelets; gynoecium 3-carpellate; fruit a drupe, 1-3 seeded; embryo undifferentiated .... Joinvilleaceae
10. Perianth reduced, usually represented by 2 or 3 lodicules; flowers in spikelets; gynoecium 2- or 3-carpellate; fruit usually a caryopsis, 1 seeded; embryo grass-like .... **Poaceae**

Figure 9.35 Bromeliaceae. (A, B) *Tillandsia recurvata:* (A) fruiting plant (× 0.4); (B) single stem with open capsule and seeds (× 0.75). (C–L) *T. usneoides:* (C) stem with flower and open fruit (× 0.75); (D) flower (× 2); (E) flower with two sepals, two petals, and five stamens removed to show gynoecium (× 3); (F) cross-section of ovary (× 22); (G) gynoecium in longitudinal section (× 22); (H) placenta and ovules (× 30); (I) open capsule with seeds (× 3); (J) seed with basal appendage of hairs (× 3); (K) seedling (× 6); (L) scale hair from leaf (× 75). (From Smith and Wood 1975, *J. Arnold Arbor.* 56: p. 386.)

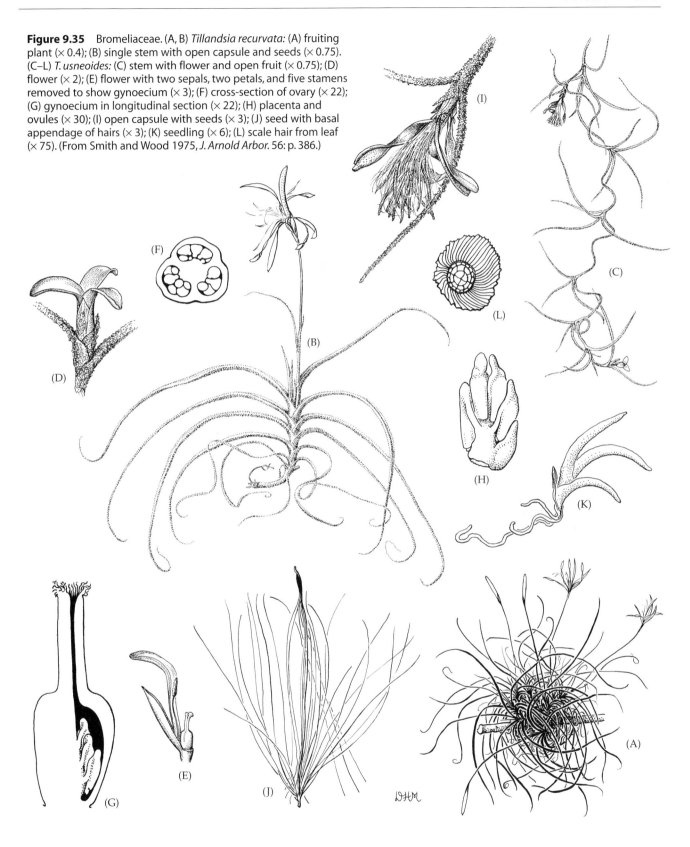

cate or disulcate, or with 2 to several pores. *Carpels 3, connate*; ovary superior to inferior, with axile placentation; stigmas 3, **usually spirally twisted**. Ovules numerous. Nectaries usually in septa of ovary. *Fruit a septicidal capsule or berry; seeds often winged or with tuft(s) of hairs* (Figure 9.35).

Floral formula: *, ⨪3⨪, ⨪3⨪, ⨪6⨪, $\overline{③}$; capsule, berry

Distribution and ecology: Tropical to warm temperate regions of the Americas (but one species of *Pitcairnia* in tropical Africa). An important group of epiphytes of moist montane forests; also occurring in xerophytic habitats.

Genera/species: 51/1520. ***Major genera:*** *Tillandsia* (450 spp.), *Pitcairnia* (250), *Vriesia* (200), *Aechmea* (150), *Puya* (150), and *Guzmania* (120). The family is represented in the continental United States by *Tillandsia, Catopsis,* and *Guzmania.*

Economic plants and products: The fruits of *Ananas comosus* (pineapple), a multiple fruit of fused berries associated with a fleshy, if somewhat fibrous, inflorescence axis, are important as an edible fruit. The dried stems and leaves of *Tillandsia usneoides* (Spanish moss) are used as stuffing material. Several genera, including *Aechmea, Billbergia, Bromelia, Guzmania, Neoregelia, Pitcairnia, Tillandsia,* and *Vriesia,* provide ornamentals.

Discussion: The monophyly of Bromeliaceae is supported by cpDNA restriction sites and *rbcL* sequence data (Ranker et al. 1990; Chase et al. 1993, 1995a), morphology, and possibly the base chromosome number of 25 (Dahlgren et al. 1985; Gilmartin and Brown 1987; Varadarajan and Gilmartin 1988). The family probably represents an early divergent clade within the superorder (Dahlgren et al. 1985; Linder and Kellogg 1995; Chase et al. 1993, 1995b); if basal, septal nectaries may be a plesiomorphy.

Bromeliaceae are divided into three subfamilies: Pitcairnioideae, Bromelioideae, and Tillandsioideae. Tillandsioideae are considered monophyletic on the basis of their entire leaf margins, hair-tufted seeds, and distinctive peltate scales that have several rings of isodiametric cells in the center of the scale and a fringe of 32 or 64 radiating cells. This clade has usually superior ovaries and capsular fruits. Bromelioideae are considered monophyletic due to their inferior ovaries and berry fruits; the subfamily has serrate leaves, perhaps a plesiomorphy. The monophyly of Pitcairnioideae is questionable; this group has serrate leaf margins, superior ovaries, and capsular fruits with winged seeds (all probable plesiomorphies). Analyses of *ndhF* sequences support the monophyly of Tillandsioideae and Bromelioideae, and the paraphyly of Pitcairnioideae (Terry et al. 1997a,b). The epiphytic habit evolved separately in Tillandsioideae and Bromelioideae.

Bromeliaceae show several adaptations to xerophytic and epiphytic conditions. The elongate, more or less concave leaves are typically clustered at the base of the plant, and their expanded bases form a water-retaining tank. The leaf surface is covered with water-absorbing peltate scales; each scale has a uniseriate stalk (of living cells), while the radiating cells of the scale are dead at maturity. The dead cells expand when wetted, drawing water into (and under) the scale, where it is osmotically taken into the leaf by the stalk cells. Water loss is reduced by location of stomates in furrows and a thick cuticle. Some hold water in a central "tank" while others directly absorb water held by elaborate scales. *Tillandsia usneoides* and *T. recurvata* grow happily on telephone wires. The plant's adventitious roots function mainly in holding it in place.

The showy flowers are pollinated by various insects, birds, or occasionally bats. The berries of members of the Bromelioideae are dispersed by birds or mammals, while the winged seeds of Pitcairnioideae and hair-tufted seeds of Tillandsioideae are wind dispersed.

References: Benzing 1980; Benzing et al. 1978; Brown and Gilmartin 1989; Chase et al. 1993, 1995b; Dahlgren et al. 1985; Gilmartin and Brown 1987; Linder and Kellogg 1995; Ranker et al. 1990; Smith and Till 1998; Smith and Wood 1975; Terry et al. 1997a,b; Varadarajan and Gilmartin 1988.

Typhaceae A. L. de Jussieu
(Cattail Family)

Rhizomatous herbs, aquatic or wetland, with leaves and stems distally floating or emergent. Hairs simple. *Leaves* alternate, *2-ranked*, simple, *linear*, entire, with parallel venation, *often spongy, with air canals and partitions containing stellate cells*, sheathing at base; stipules lacking. *Inflorescences* determinate, terminal, highly modified *with numerous densely clustered flowers and appearing to be elongate/cylindrical spikes or globose clusters, the staminate flowers positioned above the carpellate*, often subtended by a linear bract. **Flowers unisexual (plants monoecious)**, radial. **Tepals 1–6 and reduced,** *bractlike, numerous and bristlelike, or scale-like*, distinct. Stamens 1–8; filaments distinct or basally connate; anthers with connective sometimes expanded; pollen grains uniporate, in monads or tetrads. **Carpels 3,** connate, **usually only 1 functional**; ovary superior, with **apical placentation (and usually 1 locule)**, often on a stalk; stigma 1, extending along one side of style. **Ovule 1.** Nectaries lacking. *Fruit a drupe with dry-spongy "flesh" or an achene-like follicle*; seed or pit containing a pore, through which the embryo emerges (Figure 9.36).

Floral formula:

Staminate: *, (1–∞), 1–8, 0

Carpellate: *, 3–∞, 0, 1 ; drupe, achene-like follicle

Distribution and ecology: Widely distributed, especially in the Northern Hemisphere; characteristic of aquatic and wetland habitats, especially marshes.

Genera/species: 2/28. ***Genera:*** *Sparganium* (15 spp.) and *Typha* (13). Both occur in the continental United States and Canada.

Economic plants and products: *Typha* (cattails) and *Sparganium* (bur reeds) are occasionally used as ornamentals; the leaves of *Typha* are used as weaving material, and the

Figure 9.36 Typhaceae. (A–J) *Typha latifolia:* (A) base of plant with rhizome and sheathing leaf bases (× 0.3); (B) inflorescence and apex of stem, staminate portion above, carpellate below (× 0.3); (C) three staminate flowers, the three filaments of each flower connate (× 15); (D) dehiscing anther, with tetrads of pollen grains (× 31); (E) tetrad of pollen grains (× 31); (F) cluster of four carpellate flowers (× 31); (G) cluster of three carpellate flowers with many trichomes omitted to show stalked ovaries (× 15); (H) ovary in longitudinal section to show single apical ovule (× 31); (I) infructescence shedding fruits (× 0.3); (J) mature fruit, a stalked, achenelike follicle, with numerous hairs on the stalk (× 3.75). (K–M) *T. angustifolia:* (K) inflorescence, showing gap between staminate and carpellate portions (× 0.3); (L) pollen grain (× 1250); (M) portion of carpellate inflorescence to show three flowers, simple and glandular hairs, and spatulate sterile flower (× 15). (From Thieret and Luken 1996, *Harvard Pap. Bot.* 1(8): p. 32.)

starchy rhizomes, young staminate inflorescences, and pollen of both genera can be eaten.

Discussion: Cladistic analyses using morphology (Dahlgren et al. 1985; Linder and Kellogg 1995; Stevenson and Loconte 1995; Stevenson et al. 2000) strongly support the monophyly of Typhaceae (including Sparganiaceae). Analyses based DNA sequences are more equivocal, with some supporting (Chase et al. 1993; Duvall et al. 1993) and others not supporting (Chase et al. 2000) the monophyly of the group. An analysis of *rbcL* sequences and morphology (Chase et al. 1995b) also supported the monophyly of Typhaceae (incl. Sparganiaceae). Typhaceae here are circumscribed broadly (including Sparganiaceae) based on morphological characters (see description).

Typhaceae are wind-pollinated. The persistent perianth bristles of *Typha* help in the wind dispersal of the achenelike fruits (which after dispersal split open, releasing the single seed). The dry-spongy drupes of *Sparganium* may be dispersed by birds, mammals, or water (floating due to spongy outer layer of fruit).

References: Chase et al. 1993, 1995b, 2000; Dahlgren et al. 1985; Duvall et al. 1993; Linder and Kellogg 1995; Stevenson and Loconte 1995; Stevenson et al. 2000; Thieret 1982; Thieret and Luken 1996.

Eriocaulaceae Martynov
(Pipewort Family)

Herbs with shortened, cormlike stems or rhizomes; stems with vascular bundles in 1 or 2 rings. Hairs simple and uniseriate, or T-shaped. *Leaves* alternate and usually spiral, *usually in basal rosettes, or in tufts along branching stems*, simple, *narrow and grasslike*, entire, with parallel venation, sheathing at base; stipules lacking. **Inflorescences** indeterminate, **forming a head subtended by an involucre of stiff papery bracts**, *terminal, on a long scape*; scapes 1 to several, basally enclosed by a sheathing bract. **Flowers unisexual** (plants usually monoecious), radial to bilateral, *individually inconspicuous, with the perianth differentiated into a calyx and corolla, and often fringed with hairs*, usually in the axil of a papery bract. *Sepals* 2 or 3, distinct or connate, usually valvate. *Petals* 2 or 3, distinct (in carpellate flowers) or *connate (in staminate flowers)*, sometimes with nectar-producing glands near the apex, sometimes reduced (in carpellate flowers), usually valvate. Stamens 2–6, often unequal; filaments can be distinct or connate, **adnate to the petals**, sometimes arising from a stalk (formed by fused petals and filaments); anthers 1- or 2-locular; **pollen grains with an elongated and spiral germination furrow** (or, in some taxa, convoluted). Carpels 2 or 3, connate; *ovary superior, usually borne on a stalk, with axile placentation*; stigmas 2 or 3, minute. Ovules 1 in each locule, orthotropous, with a thin megasporangium. *Fruit a loculicidal capsule*; seed with testa ± longitudinally striated, reticulate, or "hairy."

Floral formula:

Staminate: * or X, (2–3), (2–3), 2–6, 0

Carpellate: *, (2–3), 2–3, 0, (2–3); capsule

Distribution and ecology: Widespread in tropical and subtropical regions, with a few extending into temperate habitats; usually in wetland habitats.

Genera/species: 9/1175. **Major genera:** *Paepalanthus* (500 spp.), *Eriocaulon* (400), *Syngonanthus* (200), and *Leiothrix* (65). The family is represented in the continental United States and/or Canada by *Eriocaulon*, *Lachnocaulon*, and *Syngonanthus*.

Economic plants and products: Dried inflorescences of *Syngonanthus* and *Eriocaulon* (pipeworts) are used in floral arrangements.

Discussion: Eriocaulaceae are distinctive, clearly monophyletic, and easily recognized by their involucrate heads of minute flowers. They are therefore often referred to as the "Compositae of the monocots." Two subfamilies are typically recognized: the Eriocauloideae (e.g., *Eriocaulon*), which have twice as many stamens as petals and an apical nectar gland on the petals, and the Paepalanthoideae (e.g., *Paepalanthus*, *Leiothrix*, *Syngonanthus*, and *Lachnocaulon*), which have as many stamens as petals and lack nectar glands. The "Paepalanthoideae" are not monophyletic, and *Paepalanthus* may represent a paraphyletic complex from which the other genera have evolved (Giulietti et al. 2000).

Flowers of Eriocaulaceae, which have anthers and styles clearly exserted, may be wind-pollinated, although nectaries on the flowers of *Eriocaulon* suggest that insect pollination also occurs. Floral visitors appear to be infrequent, and self-pollination is probably common. The seeds are presumably dispersed by wind or water.

References: Dahlgren et al. 1985; Giulietti et al. 2000; Kral 1966a, 1989; Stützel 1998.

Xyridaceae C. A. Agardh
(Yellow-Eyed Grass Family)

Herbs, with cormlike or bulblike stems, or occasionally rhizomes. Hairs simple or branched. *Leaves* alternate, *usually 2-ranked, often equitant and unifacial, flat to cylindrical, those of the upper portion of stem reduced*, simple, entire, with parallel venation, sheathing at base; stipules lacking. *Inflorescences* usually indeterminate and *forming a conelike head or spike, with spirally arranged, imbricate, persistent bracts, terminal, on a long scape*; scapes 1 to several, arising from axils of bracts or inner leaves. *Flowers* bisexual, **slightly bilateral**, *with perianth differentiated into a calyx and corolla, each in the axil of a stiff, papery to leathery bract*. Sepals 3, distinct and **dimorphic, the inner** one membranous and wrapped around the corolla, falling as the flower opens, and the 2 lateral ones sub-opposite, stiff and papery, usually keeled, and persistent. *Petals 3, distinct and clawed to connate, and then forming corolla with a narrow tube and distinctly 3-lobed, flaring limb*, imbricate, usually yellow or white, quickly wilting. **Stamens 3, opposite the petals, usually alternating with 3 staminodes**; filaments short and adnate to the petals; **staminodes apically 3-branched**, *densely covered*

with moniliform hairs; pollen grains monosulcate or lacking an aperture. Carpels 3, connate; *ovary superior, with parietal to free-central or axile placentation*; stigmas 3, ± capitate. Ovules usually numerous on each placenta, anatropous to orthotropous, with a thick to thin megasporangium. Nectaries lacking. *Fruit usually a loculicidal capsule*, surrounded by the persistent, dried corolla tube and clasped by the 2 lateral sepals; seeds minute, usually longitudinally ridged (Figure 9.37).

Figure 9.37 Xyridaceae. (A–N) *Xyris fimbriata:* (A) habit (× 0.2); (B) inflorescence (× 3); (C) inner sepal (× 9); (D) flower, showing the two persistent fimbriate lateral sepals and the subtending bract (behind flower) (× 7); (E) petal with filament of fertile anther adnate to claw, and staminode with free filament and much branched tip (× 9); (F) anther (× 9); (G) tip of staminode (× 18); (H) moniliform hairs of staminode (× 37); (I) gynoecium (× 7.5); (J) tip of three-parted style with stigmas (× 18); (K) ovary in cross-section, showing numerous ovules (most sectioned) (× 30); (L) adaxial side of placenta with ovules (× 18); (M) dehiscing capsule (× 9); (N) seed (× 60). (O, P) *X. brevifolia:* (O) habit (× 0.75); (P) seed (× 60). (From Kral 1983, *J. Arnold Arbor.* 64: p. 425.)

Floral formula: X, 1+2,③, 3+3•,③; capsule

Distribution and ecology: Widespread in tropical and subtropical regions, with a few extending into temperate habitats; characteristic of wetlands.

Genera/species: 5/300. **Major genus:** *Xyris* (260 spp.). The family is represented in the continental United States and Canada only by *Xyris*.

Economic plants and products: A few species of *Xyris* (yellow-eyed grass) are cultivated as ornamentals, especially in aquaria.

Discussion: The morphology of the boat-shaped sepals provides important diagnostic characters for species, most of which belong to *Xyris*.

The showy flowers of *Xyris* are ephemeral, and the corollas usually are expanded for no more than a few hours. Usually only one or two flowers per head are open at once. Flowers of sympatric species often open at different times of day. Nectaries are consistently lacking, and pollination may be predominantly accomplished by pollen-gathering bees. The staminodes, with their tufts of moniliform hairs, may facilitate pollination by gathering pollen and presenting it to bees, or may delude bees into scraping them as if gathering pollen. The minute seeds are dispersed by wind and/or water.

References: Dahlgren et al. 1985; Kral 1966b, 1983, 1992, 1998.

Juncaceae A. L. de Jussieu
(Rush Family)

Herbs, often with rhizomes; silica bodies absent; *stems round and solid. Leaves* alternate, *3-ranked*, basal or along lower portion of stem, *composed of a sheath and blade, the sheath usually open*; the blade simple, entire, with parallel venation, linear, flat or cylindrical; ligule and stipules lacking. Inflorescences basically determinate, terminal, highly branched, but often condensed and headlike. *Flowers usually bisexual*, but occasionally unisexual (plants dioecious), radial, inconspicuous. *Tepals 6, distinct*, imbricate, generally dull colored (green, red-brown, black), but sometimes white or yellowish. *Stamens (3–)6*; filaments distinct; pollen monoporate, in obvious tetrads. *Carpels 3, connate; ovary superior*, with axile or parietal (occasionally basal) placentation; ovules numerous (rarely 3); stigmas 3, usually elongate. Nectaries lacking. *Fruit a loculicidal capsule*, with 3 to many seeds.

Floral formula: *, -6-, (3–)6,③; capsule

Distribution and ecology: Worldwide, mostly temperate and/or montane. Often in damp habitats, but with notable exceptions, such as the weedy *Juncus trifidus*.

Genera/species: 6/400. **Major genera:** *Juncus* (300 spp.), *Luzula* (80). Both occur in the United States and Canada.

Economic plants and products: *Juncus effusus* (soft rush) and *J. squarrosus* (heath rush) are used to produce split rushes for baskets and chair bottoms. A few *Juncus* and *Luzula* species are used as ornamentals.

Discussion: Juncaceae may be nonmonophyletic as currently circumscribed (see discussion of order). It is not clear which, if any, of the family characteristics are synapomorphic; most are generalized features of monocotyledons.

Many members of this family look superficially like grasses, but the 3-ranked leaves, flowers with obvious tepals, and capsular fruit all make the distinction clear. In some species of *Juncus*, the large bract subtending the inflorescence is borne upright so that it looks like a continuation of the stem, and the inflorescence appears lateral.

The inconspicuous flowers of Juncaceae are predominantly wind pollinated; outcrossing is commonly brought about by protogyny, but some species are selfing. Dispersal of the small seeds occurs by wind, water, and external transport on animals.

References: Balslev 1998; Dahlgren et al. 1985.

Cyperaceae A. L. de Jussieu
(Sedge Family)

Herbs, generally rhizomatous; **stems usually ± triangular in cross-section**, often leafless above the base. *Leaves alternate, 3-ranked*, **with conical silica bodies**; *composed of a sheath and blade*, **the sheath closed**, the blade simple, entire to minutely serrate, with parallel venation, linear, flattened; stipules lacking; *ligule generally lacking. Inflorescence a complex arrangement of small spikes (spikelets)*, often subtended by bracts. *Flowers* bisexual or unisexual (plants then usually monoecious), *each subtended by a bract. Tepals lacking or reduced to usually 3–6 scales, bristles, or hairs.* **Stamens 1–3** (–6); filaments distinct; anthers not sagittate; **pollen** usually uniporate, **in pseudomonads (3 microspores degenerate and form part of the pollen wall)**. Carpels 2–3, connate; ovary superior, **with basal placentation; ovule 1**; styles 3, elongated. Nectaries lacking. **Fruit an achene (nutlet)**, often associated with persistent perianth bristles (Figure 9.38).

Floral formula: *, -0-6-, 1–3 (–6),③; achene

Distribution and ecology: Worldwide; often, but not exclusively, in damp sites.

Genera/species: 104/4500 **Major genera:** *Carex* (2000 spp.), *Cyperus* (600), *Fimbristylis* (300), *Scirpus* (300),

Figure 9.38 Cyperaceae. (A–H) *Scirpus (Schoenoplectus) tabernaemontani (S. validus)*: (A) underwater rhizome (× 0.65); (B) apex of stem with inflorescence (× 1.35); (C) single spikelet with lower flowers past anthesis, upper ones with anthers visible and styles exserted (× 16); (D) flower and subtending bract removed from spikelet, view of adaxial surface, stigmas exserted, anthers still included, note barbed bristles (× 27); (E) flower showing different maturation of stamens (× 16); (F) spikelet at stage later than that in C, immature achenes below and flowers with receptive stigmas above (× 16); (G) mature achene with persistent bristles (× 16); (H) same in vertical section, fruit wall hatched, endosperm stippled, embryo unshaded, the seed coat too thin to show (× 16). (I–J) *Trichophorum cespitosum (Scirpus cespitosus)*: (I) spikelet (× 11); (J) achene with bristles (× 16). (K–L) *Scirpus koilolepis*: (K) solitary spikelet, subtended by keeled bracts (× 16); (L) achene (× 16). (M–O) *S. erismaniae*: (M) basal flower in leaf axil (× 11); (N) achene from basal flower (× 16); (O) achene from spikelet borne on stem (× 16). (P) *S. cyperinus*: achene with elongate bristles (× 16). (From Tucker 1987, *J. Arnold Arbor.* 68: p. 372.)

Rhynchospora (200), *Scleria* (200), and *Eleocharis* (200). All of the above occur in North America; other common genera include *Cladium*, *Bulbostylis*, *Bulboschoenus*, *Eriophorum*, *Fuirena*, *Kyllinga*, *Schoenoplectus*, and *Trichophorum*.

Economic plants and products: *Cyperus papyrus* was used for making paper by ancient Egyptians and is also commonly planted as an ornamental. *Cyperus rotundus* (nut sedge) is an agricultural weed. *Cyperus esculentus* (nut sedge, chufa, tigernut) has edible underground storage organs, as do *Mariscus umbellatus*, *Scirpus tuberosus*, and *Eleocharis dulcis;* the last provides commercial "water chestnuts." Weaving materials are provided by stems and leaves of a few species of *Cyperus*, *Carex*, *Eleocharis*, *Lepironia*, and *Scirpus*. Roots of *Cyperus longus* (galingale) and *C. articulatus* are sweet-scented and used in perfumery. Roots of *Scirpus grossus* and *S. articulatus* are used in Hindu medicine. Various species of *Carex* are used as packing materials, hay, or straw.

Discussion: Cyperaceae contain unique conical silica bodies, which distinguish them from all other monocots. The family is apparently monophyletic (Muasya et al. 1998). In a comprehensive morphological study of the family, Bruhl (1995) recognized two subfamilies and ten tribes. The tribes represent groups that appear in both phenetic and cladistic analyses of the family, although only four of the ten have clear synapomorphies; phylogenetic relationships have also been assessed by a study of *rbcL* sequences (Muasya et al. 1998). The most common tribe in North America is the Cariceae, in which the spikelet prophyll forms a sac (the perigynium) enclosing the flower.

Like Juncaceae, Cyperaceae are often mistaken for grasses. They are distinguished by the more or less triangular stems, 3-ranked leaves, usual absence of a ligule, and closed sheaths. (The latter two characteristics occur in a few grasses, but not together.) Cyperaceae flowers are subtended by a single bract or, in *Carex*, bract plus prophyll, whereas most grass flowers are associated with two bracts (lemma and palea).

Most Cyperaceae are wind pollinated, but insect pollination has evolved several times, e.g., *Hypolytrum*, some species of *Ascolepis* and *Rhynchospora*. Dispersal of the fruits is by water (due to corky fruit wall or associated structures, or perianth bristles holding air bubbles, e.g., *Cladium*, *Remirea*, *Eleocharis*, *Fuirena*), external transport (by means of a hooked inflorescence axis, i.e., *Uncinia*, or muddy fruits sticking to feathers, fur, or clothing, e.g., *Fimbristylis*), birds (aril imitation, in *Scleria*), or wind (by action of elongate bristles, as in *Eriophorum* or *Scirpus*). The perigynium of *Carex* often assists in water dispersal.

References: Bruhl 1995; Goetghebeur 1998; Muasya et al. 1998; Tucker 1987.

Restionaceae R. Brown
(Restio Family)

Herbs, with rhizomes; *stems round, elliptic, or polyhedral in cross-section, and solid or hollow (in internodal region),* **anatomically distinctive, with "protective" cells lining substomatal chambers and 1 or 2 layers of chlorenchyma cells inside the epidermis, the chlorenchyma separated from the cortex by parenchymatous and sclerenchymatous rings;** silica bodies present. Hairs simple, sometimes flattened or peltate. *Leaves alternate, 2-ranked, usually with the blade very reduced and leaf consisting merely of an open sheath;* ligule usually lacking; stipules lacking. Inflorescences with flowers variously arranged, usually in spikelets. *Flowers usually unisexual (plants dioecious), often in dimorphic inflorescences,* radial, inconspicuous, each in the axil of a bract. *Tepals 3-6,* usually distinct, imbricate, generally dull colored or greenish. *Stamens (1-) 3,* sometimes represented by staminodes in carpellate flowers; filaments distinct to connate; *anthers with 1 locule, opening by a single slit;* pollen monoporate. Carpels 3, connate, ovary superior, with apical-axile placentation, and in some with only a single fertile locule; stigmas 3, elongate, often plumose. Ovules 1 per locule, orthotropous, with thin sporangium wall. Nectaries lacking. *Fruit a capsule, achene or nut*, often associated with a persistent perianth.

Floral formula:

Staminate: * – 6 – , ③ , 0

Carpellate:* – 6 – , 0 , ③ , capsule, achene, nut

Distribution and ecology: Distributed in the Southern Hemisphere and most diverse in Australia and South Africa; plants of low nutrient soils, and frequently of seasonally arid and/or fire-prone habitats.

Genera/species: 53/485. ***Major genera:*** *Restio* (90 spp.), *Ischyrolepis* (50), and *Thamnochortus* (35). This clade does not occur in North America.

Economic plants and products: A few genera are of horticultural interest, e.g., *Chondropetalus* and *Elegia*. *Thamnochortus* is used for thatch, and several genera provide important pasture plants.

Discussion: The monophyly of the family is supported by anatomical (Linder et al. 2000b) and molecular (Briggs et al. 2000) characters. Restionaceae are closely related to the Poaceae, Flagellariaceae, and especially Centrolepidaceae, as evidenced by their monoporate, spheroidal pollen grains and presence of silica bodies in parenchymatous tissues. The latter family (which has much more reduced flowers) may be nested within Restionaceae (Linder et al. 2000b), but is treated here as their sister group, as supported by cp-DNA sequence data (Briggs et al. 2000). The Restionaceae + Centrolepidaceae clade share a couple of remarkable embryological apomorphies (sporangium wall with elongated cells, polar nuclei surrounded by large starch bodies) and both have 1-loculed anthers. Finally, molecular data supports the exclusion of two small genera, *Lyginia* and *Hopkinsia*, which are distinctive in having 2-loculate anthers, a plesiomorphic feature. They belong to the Australian Anarthriaceae.

Members of Restionaceae belong to one of two large clades, the first containing the African genera, and the second the Australian genera (Briggs et al. 2000; Eldenäs and Linder 2000; Linder 2000; Linder et al. 2000a,b). Members of the African clade are characterized by pollen grains with raised or swollen aperture margins and stems with cavities beneath the stomata lined with thick-walled ("protective") cells; Australian genera have the pollen exine margins attenuate towards the germination pore, and most have lost "protective" cells. Long-pubescent rhizomes support the monophyly of the Australian clade.

At one time only a few large genera were recognized within Restionaceae, e.g., *Restio*, which included both African and Australian species. At this time many, much smaller genera are delimited, and *Restio* is considered to be restricted to Africa (Briggs and Johnson 1998a; Linder 1984, 1985). Recent studies have shown that additional generic recircumscriptions are required.

The inconspicuous flowers of Restoniaceae are wind-pollinated, although some species are selfing or apomictic. Wind dispersal of the small seeds (or achenes or nutlets, in genera with indehiscent fruits) is common; some have modified tepals (wings, awns, hairs) that presumably aid dispersal. A few genera (e.g., *Willdenowia* and relatives) have fruits associated with fleshy structures, and may attract ants. *Alexgeorgea* is unusual because its carpellate flowers are borne underground (but with the long style/stigma lying on the surface); pollination of these species is by wind, but the mechanism of dispersal of the nutlets is unknown.

References: Briggs and Johnson 1998a, b; Briggs et al. 2000; Carlquist 1976; Eldenäs and Linder 2000; Linder 1984, 1985, 2000; Linder et al. 2000a, b; Rudall and Linder 1988.

Poaceae Barnhart
(= Gramineae Adanson)
(Grass Family)

Herbs, often rhizomatous, but trees in tropical bamboos; *stems jointed, round to elliptical in cross-section, solid to hollow;* with silica bodies. *Leaves* alternate, *2-ranked, consisting of sheath, ligule, and blade; sheaths tightly encircling the stem, the margins overlapping but not fused* or, occasionally, united to form a tube; ligule a membranous flange or fringe of hairs at adaxial apex of sheath; blades simple, usually linear, usually with parallel venation, flat or sometimes rolled into a tube, continuous with the sheath or petiolate. Inflorescence a spike, panicle, cyme, or raceme of *spikelets. Spikelet composed of an axis bearing 2-ranked and* *closely overlapping basal bracts (glumes) and florets;* breaking up above the glumes or remaining intact at maturity; compressed parallel or perpendicular to the plane of arrangement of glumes and florets. *Glumes* usually 2, equal in size or unequal. **Florets** 1 to numerous per spikelet, made up of a bract (the *lemma*) subtending a flower and another bract (the *palea,* a prophyll) lying between the flower and the spikelet axis. Lemmas sometimes with 1 or more needle-like, straight or bent awns. Palea often translucent, smaller than, and partially enclosed by the lemma, often morphologically similar to a prophyll. Flowers small, bisexual or unisexual (plants then monoecious or dioecious), usually wind-pollinated, greatly reduced in size and number of floral parts. Lodicules (= perianth parts) mostly 2, translucent. Stamens (1–) 3 (–6, or numerous); *anthers usually sagittate;* pollen monoporate. Carpels 3, but often appearing as 2, connate; stigmas 2 (–3), plumose, papillae multicellular; ovary superior, with 1 locule and 1, subapical to nearly basal amphitropous or hemi-anatropous ovule, with a thin to thick megasporangium wall. **Fruit a single-seeded caryopsis (grain)** with fruit wall fused to the seed (or less often fruit wall free from the seed, and an achene, utricle, or berry); often associated with parts of the spikelet for dispersal. **Embryo with a highly modified cotyledon (scutellum), lateral in position** (Figures 9.39 and 9.40).

Floral formula: *, -2-, (1–)3(–6), (2–3); caryopsis

(Note: Useful in identification are features of the spikelet, including size, plane of compression, presence or absence of glumes, number of florets, presence of sterile or incomplete florets, number of veins on glumes and lemmas, presence or absence of awns, and aggregation of spikelets in secondary inflorescences.)

Distribution and ecology: Cosmopolitan; in desert to freshwater and marine habitats, and at all but the highest elevations. Native grasslands develop where there are periodic droughts, level to gently rolling topography, frequent fires, and in some instances grazing and certain soil conditions. Communities dominated by grasses, such as the North American prairie and plains, South American pampas, African veldt, and Eurasian steppes, account for about 24% of the Earth's vegetation. Woody bamboos play important roles in forest ecology in tropical and temperate Asia.

Genera/species: ca. 650/9700. Important genera are listed below under the major subgroups.

Economic plants and products: The economic importance of grasses lies in their paramount role as food: about 70% of the world's farmland is planted in crop grasses, and over 50% of humanity's calories come from grasses. People have cultivated cereals for at least 10,000 years. From the beginning of their domestication, wheat

Figure 9.39 Poaceae (Gramineae) subfamily Pooideae. *Poa pratensis:* (A) flowering plant with rhizomes at base (× 0.75); (B) apex of leaf sheath, ligule, and base of blade (× 4.5); (C) spikelet (× 17); (D) glumes (× 29); (E) floret before opening (× 17); (F) spikelet with lower flower open and stigmas receptive, the second floret open, anthers dehisced (× 17); (G) floret, lemma to left, palea to right (× 17); (H) dehisced anther (× 17); (I) lodicules and gynoecium (× 21); (J) portion of inflorescence with mature fruits, two fruits falling from spikelets (× 9); (K) floret in fruit (× 17); (L) caryopsis, lemma removed, palea to left (× 17); (M) caryopsis in longitudinal section, endosperm stippled, embryo unshaded (× 17); (N) embryo in diagrammatic longitudinal section, scutellum to left, coleoptile and coleorhiza to right, vascular tissue in black; (O) diagrammatic cross-section of embryo through scutellum, coleoptile, and first embryonic leaf at level of arrow in (N). (From Campbell 1985, *J. Arnold Arbor.* 66: p. 155.)

(*Triticum aestivum*), barley (*Hordeum vulgare*), and oats (*Avena sativa*) in the Near East, sorghum (*Sorghum bicolor*) and pearl millet (*Pennisetum americanum*) in Africa, rice (*Oryza sativa*) in southeastern Asia, and maize or corn (*Zea mays*) in Meso-America have made possible the rise of civilization. In terms of global production, the four most important crops are grasses: sugarcane (*Saccharum*

officinale), wheat, rice, and maize. Barley and sorghum are in the top twelve. Grasses are also used for livestock food, erosion control, turf production, and as a sugar source for the fermentation of alcoholic beverages, such as beer and whiskey. Bamboos are economically important in many tropical areas for their edible young shoots, fiber for paper, pulp for rayon, and strong stems for construction.

Figure 9.40   Poaceae (Gramineae) subfamily Panicoideae. (A–M) *Panicum (Dichanthelium) clandestinum*: (A) part of winter rosette (× 0.7); (B) inflorescence of chasmogamous spikelets (× 0.7); (C) upper part of plant, chasmogamous spikelets in fruit or shed from inflorescence, inflorescence of cleistogamous spikelets below (× 0.7); (D) upper part of leaf sheath, base of blade, and ligule (× 8); (E) chasmogamous spikelet at anthesis (× 8); (F) small first and larger second glume (× 14); (G) sterile lemma (pubescent) and sterile palea (× 14); (H) flower of cleistogamous spikelet (× 27); (I) fertile lemma (behind) and palea enclosing mature caryopsis (× 14); (J) mature caryopsis (× 14); (K) longitudinal section of caryopsis, embryo to left, endosperm stippled (× 16); (L) embryo in diagrammatic longitudinal section, scutellum to left, coleoptile and coleorhiza to right, vascular tissue in black, note mesocotyl above trace to scutellum; (M) diagrammatic cross-section of embryo through scutellum, coleoptile, and first embryonic leaf, at level of arrow in (L). (N) *P. anceps*: spikelet in fruit (× 14). (From Campbell 1985, *J. Arnold Arbor.* 66: p. 172.)

Subfamilial phylogenetic relationships Grasses are easily recognized, and their monophyly has been supported by morphological and DNA characters. Recent molecular systematic studies (Grass Phylogeny Working Group, GPWG 2001) support the recognition of 12 subfamilies. Anomochlooideae (native to Brazil), Pharoideae (Old and New World tropics), and Puelioideae (West Africa) are the three earliest diverging lineages, and together include only about 25 of the almost 10,000 species of the family. The remaining species fall into two large groups. One group (the BEP clade) includes Bambusoideae s.s., Ehrhartoideae, and Pooideae, and the other (the PACCAD clade) includes Panicoideae, Arundinoideae s.s., Chloridoideae, Centothecoideae, Aristidoideae, and Danthonioideae (Figure 9.41; Clark et al. 1995; Soreng and Davis 1998; Hilu et al. 1999; GPWG 2001). All subfamilies recognized by GPWG (2001) are monophyletic, although only a few have morphological synapomorphies that characterize all members. More often they are supported by sets of character states that characterize large subgroups.

For each of the five major subfamilies—Bambusoideae s.s., Ehrhartoideae, Chloridoideae, Panicoideae, and Pooideae—geographic distribution, internal phylogenetic and systematic structure, and important genera are presented. Characters distinguishing the major subfamilies are given in the accompanying key; note that the best structural characters distinguishing subfamilies are anatomical, but morphological characters are used whenever possible. Phylogenetic relationships among these major subfamilies (along with some smaller clades) are shown in Figure 9.41 (GPWG 2001).

Anomochlooideae (including *Anomochloa* and *Streptochaeta*) was originally included in Bambusoideae, but is clearly only distantly related. The species are herbaceous and have morphologically unique inflorescences that are difficult to interpret, and certainly do not resemble the spikelets that are characteristic of other grasses. They appear to be the sister group of the remaining members of the family (Figure 9.41; Clark et al. 1985; Soreng and Davis 1998), thus suggesting that the grass spikelet probably arose after the Anomochlooideae originated.

Pharoideae and Puelioideae were, like Anomochlooideae, also traditionally included in Bambusoideae. They have conventional spikelets that are one-flowered in Pharoideae and multi-flowered in Puelioideae. The shift from three to two stigmas occurred after the origin of Pharoideae but before the origin of Puelioideae. The leaf blades of Pharoideae are turned up-side-down, with the abaxial surface uppermost.

Bambusoideae s.s. includes both woody and herbaceous plants, and are almost exclusively tropical in dis-

---

Key to Major Clades of Poaceae

1. Specialized cells, called arm cells, within the leaves of most members; stamens often more than 3 2
1. Arm and fusoid cells absent; stamens 3 or fewer..3
2. Stigmas 3; mostly trees ....Bambusoideae
2. Stigmas 2; herbs...Ehrhartoideae
3. Spikelets compressed perpendicular to plane of arrangement of glumes and florets, not breaking at maturity into separate florets but falling as a unit, with 1 caryopsis-bearing floret..Panicoideae
3. Spikelets not compressed, or compressed parallel to plane of arrangement of glumes and florets, breaking up at maturity above the glumes, mostly with more than 1 caryopsis-bearing floret..4
4. Veins in leaves separated by more than 4 mesophyll cells; bundle sheaths with few chloroplasts, appearing clear in cross section (C_3 leaf anatomy); bicellular microhairs present or lacking in leaf epidermis..5
4. Veins in leaves separated by 2–4 cells; bundle sheaths with numerous chloroplasts, appearing distinctly green in cross section (C_4 anatomy); bicellular microhairs present in leaf epidermis 6
5. Bicellular microhairs absent in leaf epidermis; subsidiary cells parallel-sided; plants variable in habit, but mostly less than 1 m tall...Pooideae
5. Bicellular microhairs present in leaf epidermis; subsidiary cells dome-shaped, plants generally over 1 m tall ....Arundinoideae s.s.
6. Bicellular microhairs generally bulbous; awns, if present, unbranched ....Chloridoideae
6. Bicellular microhairs ± linear, awns divided into 3 parts.......................................Aristidoideae

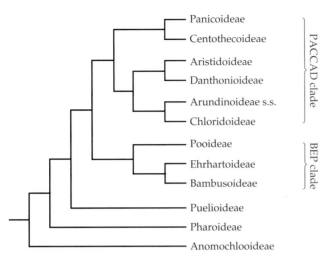

Figure 9.41 A phylogeny of Poaceae. (Modified from Clark et al. 1995 and Grass Phylogeny Working Group 2000.)

tribution. The woody bamboos form a monophyletic group that is sister to the clade including the herbaceous species. Woody bamboos, with stems up to 40 m in height, certainly do not resemble the turf in lawns. Flowering in many woody bamboos is also unusual, occurring in cycles of up to 120 years. Even though individual stems live for only one or a few decades, some form of "clock" directs stems to flower all at once throughout the range of a species. Important genera of woody bamboos are *Bambusa* (120 spp.), *Chusquea* (100), *Arundinaria* (50), *Sasa* (50), and *Phyllostachys* (45).

Ehrhartoideae include the Southern Hemisphere Ehrharteae, as well as the widespread Oryzeae. The latter are aquatic or wetland herbs. The most widely known Oryzeae are the commercially important Asian rice (*Oryza sativa*) and North American wild rice (*Zizania aquatica*).

Pooideae are largely temperate in distribution, especially in the Northern Hemisphere. Important genera include several cereals (wheat, barley, oats; see above under economic plants and products) as well as rye (*Secale cereale*), turf grasses (e.g., bluegrasses, *Poa*, 500 spp.), fescues (*Festuca*, 450), pasture grasses (e.g., *Phleum, Dactylis*), and some weeds (e.g., *Agrostis*, 220, and *Poa*). Other important grasses of this subfamily are *Stipa* (300), *Calamagrostis* (270), *Bromus* (150), and *Elymus* (150).

Chloridoideae bear distinctive bicellular hairs on the leaf epidermis; these may be synapomorphic for a subset of the group. All but two chloridoids show C_4 photosynthesis. The subfamily is best developed in arid and semiarid tropical regions, where its C_4 photosynthesis is presumably advantageous. Centers of distribution in Africa and Australia suggest a Southern Hemisphere origin. Some important genera are *Eragrostis* (350 spp.), *Muhlenbergia* (160), *Sporobolus* (160), *Chloris* (55), *Spartina* (15), and *Eustachys* (10); the first three of these appear to be polyphyletic.

Panicoideae have long been recognized taxonomically because of their distinctive spikelets (see key to Poaceae). The subfamily is primarily tropical and contains two large tribes, Andropogoneae and Paniceae, along with a number of smaller groups. Andropogoneae are relatively easy to recognize because the spikelets are often paired and grouped into a linear inflorescence. Paniceae are not as homogeneous as Andropogoneae. Important genera include *Panicum* (470 spp., demonstrably polyphyletic), *Paspalum* (330), *Andropogon* (100), *Setaria* (100), *Sorghum* (20), and *Zea* (4).

The other subfamilies in the PACCAD clade are structurally and genetically diverse. They range from small desert species (*Aristida*, 250 spp., the largest genus in the Aristidoideae) to giant wetland reeds (*Phragmites*, in the Arundinoideae s.s.).

Discussion: Poaceae rank behind Asteraceae, Orchidaceae, and Fabaceae in number of species, but are first in global economic importance and unsurpassed among angiosperms in land surface area dominated. The family's monophyly is supported by phenotypic characters (bracteate inflorescence, reduced perianth, fruit a caryopsis, and embryo and pollen wall features) as well as *rbcL* and *ndhF* sequences. Similarities to sedges (Cyperaceae) in habit and spikelets represent convergent evolution. Sedges are more closely related to rushes (Juncaceae), and Poaceae belong to core Poales (see Figure 9.34).

Grasses have been successful ecologically and have diversified extensively due to several key adaptations. The grass spikelet protects the flowers while permitting pollination when the lodicules open the spikelet. Spikelets have various adaptations for fruit dispersal. Versatility in breeding systems, including inbreeding and agamospermy, helps make some grasses successful colonizers. C_3 and C_4 leaf anatomy adapt grasses to a wide range of habitats. Meristems are located at the bases of the internodes and sheaths. As a result, grasses tolerate fire and grazing better than many other plants. Development of grasslands during the Miocene epoch (from around 25 to 5 million years ago) may have fostered the evolution of large herbivores, an important food source and stimulus for the evolution of *Homo sapiens*.

The economic and ecological importance of the family has motivated considerable systematic study. Early in the nineteenth century, differences between the spikelets of pooids and panicoids led to division of the whole family into these two groups by Robert Brown. Early in the twentieth century, leaf epidermal characters and chromosome number led to separation of chloridoid grasses from the pooids. In the middle of the twentieth century, internal leaf anatomy and embryological features set up the recognition of five to eight subfamilies. Differences in leaf anatomy are associated with different photosynthetic pathways. The C_3 pathway is more efficient in regions of cool and cold climate, whereas C_4 photosynthesis is advantageous in regions of high temperatures and low soil moisture. C$_3$ anatomy is the plesiomorphic state in the family; all C$_4$ species are in the PACCAD clade, although many C$_3$ taxa are included there as well.

Phylogenetic studies based on *rbcL, ndhF,* rpoC2, ITS, granule bound starch synthase I, and phytochrome B sequences agree with many phylogenetic relationships inferred from structural data. The PACCAD clade is supported by the embryological feature of a long mesocotyl internode (see Figure 9.40L), and this clade emerges strongly in all molecular analyses. The general picture of grass phylogeny is that (1) the basalmost clades are broad-leaved tropical forest grasses, such as Anomochlooideae and Pharoideae, (2) there is support for four large subfamilies (Bambusoideae s. s., Chloridoideae, Panicoideae, and Pooideae), and (3) bambusoid relatives and arundinoid grasses are spread over the phylogenetic tree (see Figure 9.41).

References: Barker et al. 1995; Campbell 1985; Clark and Pohl 1996; Clark et al. 1995, 2000; Clark and Judziewicz 1996; Clayton and Renvoize 1986; Grass Phylogeny Working Group 2001; Hilu et al. 1999; Judziewicz et al. 1999; Kellogg and Linder 1995; Kellogg and Watson 1993; Soderstrom et al. 1987; Soreng and Davis 1998; Tucker 1996; Watson and Dallwitz 1992.

Zingiberales

The monophyly of Zingiberales is supported by morphology (Dahlgren and Rasmussen 1983; Kress 1990, 1995; Stevenson and Loconte 1995; Stevenson et al. 2000), *rbcL* sequences (Chase et al. 1993, 1995b; Duvall et al. 1993; Smith et al. 1993; Källersjö et al. 1998), 18S rDNA sequences (Soltis et al. 1997), and *rbcL, atpB,* and 18S sequences (Chase et al. 2000; Soltis et al. 2000). Putative synapomorphies include large herbs with vessels more or less limited to the roots; presence of silica cells in the bundle sheath; leaves clearly differentiated into a petiole and blade, with pinnate venation, often tearing between the secondary veins, with the blade rolled into a tube in bud, and the petiole (and midvein) with enlarged air canals, flowers bilateral (but, in more specialized taxa, lacking a plane of symmetry); pollen usually lacking an exine; ovary inferior; and the arillate seeds with perisperm (Figure 9.42A). The monophyly of the order also was supported by Tomlinson's (1962) morphological and anatomical study.

Phylogenetic relationships within core Zingiberales are fairly well understood due to careful morphological and cpDNA analyses (Dahlgren and Rasmussen 1983; Kress 1990, 1995; Kress et al. 2001; Tomlinson 1962, 1969a). Cannaceae, Marantaceae, Zingiberaceae, and Costaceae form a clade based on reduction of the androecium to only a single functional stamen, presence of showy staminodes, seeds with more perisperm than endosperm, lack of raphide crystals in vegetative tissues, and leaves that are not easily torn (Figure 9.42B).

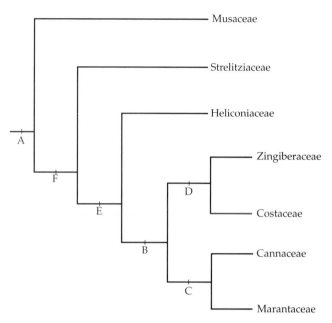

Figure 9.42 A phylogeny of Zingiberales, as discussed in the text. (Modified from Kress 1990, 1995, Kress et al. 2001, Soltis et al. 2000, Chase et al. 2000.)

Within this clade, Marantaceae and Cannaceae are hypothesized to be sister taxa, as evidenced by their flowers that lack a plane of symmetry, and androecium with only half of one stamen fertile, the other half of this stamen being expanded and staminodial (Figure 9.42C). Zingiberaceae and Costaceae constitute a clade that is supported by the unusual feature of the single functional stamen more or less grasping the style, a ligule at the apex of the leaf sheath, connate sepals, fused staminodes, and reduction of two of the three stigmas (Figure 9.42D). Heliconiaceae, Strelitziaceae, and Musaceae constitute a paraphyletic complex. The Heliconiaceae are hypothesized to be the sister taxon to the Cannaceae-Marantaceae-Zingiberaceae-Costaceae clade, as supported by the putative synapomorphies of a sterile outer median stamen, connate petals, and details of the root anatomy (Figure 9.42E). Strelitziaceae may be sister to the clade containing all of the above-listed families, all having 2-ranked leaves (Figure 9.42F). Musaceae (banana family) probably represent the sister group to a clade containing the remaining families of the order; this family has retained the plesiomorphic condition of spirally arranged leaves.

Zingiberales contain 8 families and about 1980 species; major families include **Cannaceae**, **Marantaceae**, **Zingiberaceae**, Costaceae, Heliconiaceae, Strelitziaceae, and Musaceae.

References: Chase et al. 1993, 1995b, 2000; Dahlgren and Rasmussen 1983; Duvall et al. 1993; Källersjö et al. 1998; Kress 1990, 1995; Kress et al. 2001; Smith et al. 1993; Soltis et al. 1997, 2000; Stevenson and Loconte 1995; Stevenson et al. 2000; Tomlinson 1962, 1969a.

Zingiberaceae Adanson
(Ginger Family)

Small to large, **spicy-aromatic herbs, scattered secretory cells containing ethereal oils, various terpenes, and phenyl-propanoid compounds.** Hairs simple. *Leaves* alternate, *2-ranked*, simple, entire, usually petiolate, *with a well-developed blade, pinnate venation,* sheathing base, *and a ligule*; petiole with air canals, these separated into segments by diaphragms composed of stellate-shaped cells; stipules lacking. Inflorescences determinate, but often appearing indeterminate, the cymose units in the axils of usually conspicuous bracts. *Flowers bisexual, bilateral,* usually lasting for only 1 day. *Sepals 3, connate,* imbricate. *Petals 3, connate, with one lobe often larger than the others,* imbricate. *Stamen 1, grooved, grasping the style;* staminodes usually 4, **2 large, connate, and forming a liplike structure (labellum),** *and 2 smaller, these distinct or connate with the 2 larger staminodes;* pollen grains monosulcate or lacking apertures, exine very reduced. Carpels 3, connate; *ovary inferior,* with usually axile placentation; *style* enveloped in groove between pollen sacs of the anther; stigma 1, funnel-shaped. Ovules ± numerous. **Nectaries 2, positioned atop the ovary.** *Fruit a dry to fleshy capsule or berry;* seeds usually arillate; endosperm and perisperm present (Figure 9.43).

Floral formula:

X,③,②+①,②•+2•+1,③; fleshy capsule, berry

Distribution and ecology: Widespread in tropical regions; chiefly in shaded to semi-shaded forest understory habitats; occasionally in wetlands. Asexual reproduction occurs in some species of *Globba*.

Genera/species: 50/1000. **Major genera:** *Alpinia* (150 spp.), *Amomum* (120), *Zingiber* (90), *Globba* (70), *Curcuma* (60), *Kaempferia* (60), and *Hedychium* (50). The family is represented in the continental United States by *Alpinia, Curcuma, Hedychium,* and *Zingiber* (all rarely naturalized).

Key to Major Families of Zingiberales

1. Functional stamens 5 or rarely 6 and staminodes lacking or inconspicuous; raphide crystals lacking; leaf blades typically tearing between the secondary veins..................................2

1. Functional stamens 1 or 1/2 and staminodes conspicuous, showy; raphide crystals present; leaf blades usually not tearing between the secondary veins...4

2. Leaves spirally arranged, the petiole (in cross-section) with 1 row of air canals; latex-producing cells present; fruits berries or fleshy capsules; perianth of tepals, 5 connate and 1 member of the inner whorl distinct...Musaceae

2. Leaves 2-ranked, the petiole with 2 rows of air canals; latex-producing cells lacking; fruits dry capsules, or fleshy schizocarps; perianth of sepals and petals, or tepals, but not as above............3

3. Ovules numerous in each locule; fruit a capsule; seeds with a colorful aril; flowers with a calyx and corolla, the sepals differing in color from the petals, which are dimorphic........Strelitziaceae

3. Ovules 1 in each locule; fruit a schizocarp, splitting into usually 3 drupelike segments; flowers with tepals, 5 connate and 1 member of the outer whorl distinct.....................Heliconiaceae

4. Androecium represented by a single functional stamen; flowers bilaterally symmetrical; sepals connate; leaf sheath associated with a ligule...5

4. Androecium represented by 1/2 functional stamen; flowers without a plane of symmetry; sepals distinct; leaf sheath lacking a ligule...6

5. Leaves 2-ranked, sheath usually open; plants with ethereal oils (spicy fragrant); at least 2 staminodes connate, forming a liplike structure; pollen exine very reduced..............**Zingiberaceae**

5. Leaves 1-ranked and spiral, sheath closed (at least initially); plants lacking ethereal oils; staminodes not connate; pollen with a well developed exine.......................................Costaceae

6. Ovule solitary in a single locule or in each of the 3 locules of the ovary; leaves petiolate, with an upper pulvinus; flowers in mirror-image pairs, the style held under pressure by modified and hooded staminode, released during pollination; fruit not warty..............**Marantaceae**

6. Ovules ± numerous in each of the 3 locules of the ovary; leaves ± lacking a petiole and upper pulvinus; flowers not in mirror-image pairs, the style not held under pressure or forcefully moving during pollination; fruit warty...................................**Cannaceae**

Figure 9.43   Zingiberaceae. *Hedychium coronarium:* (A) stem arising from rhizome (× 0.15); (B) portion of stem showing ligulate leaf bases (× 0.55); (C) stem, in cross-section (× 9); (D) leaf venation (× 9); (E) leaf sheath, in cross-section (× 9); (F) inflorescence (× 0.55); (G) flower, showing split, one-lobed calyx, narrow petals, two broad lateral staminodes, and liplike staminodes (with bifid apex) and single stamen grasping gynoecium (× 0.55); (H, I) anther, grasping style, adaxial and abaxial views (× 4.5); (J, K) stigma; (L) corolla tube, in cross-section above calyx (× 9); (M) cross-section of flower slightly above ovary, note two epigynous nectar glands and style (in groove in corolla tube) (× 9); (N) ovary in cross-section (× 9). (From Rogers 1984, *J. Arnold Arb.* 65: p. 24.)

Economic plants and products: The family contains several important spices, including *Zingiber* (ginger), *Curcuma* (turmeric), and *Amomum* and *Elettaria* (cardamom). The rhizomes of several species of *Curcuma* are used as a starch source. *Alpinia* (shell ginger), *Curcuma* (hidden lily), *Hedychium* (garland lily), *Globba*, *Nicolaia* (torch ginger), *Renealmia*, and *Zingiber* (ginger) contain ornamental species.

Discussion: The monophyly of the Zingiberaceae has been supported by DNA (Smith et al. 1993; Kress 1995; Wood et al. 2000) and morphology (Kress 1990). The family is closely related to the Costaceae, which often is included within Zingiberaceae as a subfamily (see Rogers 1984).

Flowers of Zingiberaceae are diverse in color and form and are mainly pollinated by bees, moths, butterflies, and birds. Many species are outcrossing, but self-pollination and vegetative reproduction also occur. Birds are the most common dispersal agents; the fleshy capsules are usually colorful and often contrast with the brightly colored arillate seeds.

References: Burtt and Smith 1972; Dahlgren et al. 1985; Rogers 1984; Kress 1990, 1995; Larsen et al. 1998; Smith et al. 1993; Wood et al. 2000.

Marantaceae Petersen
(Prayer Plant Family)

Herbs, with erect stem and short, tuberlike, starchy rhizomes. Hairs simple and surrounded by inflated epidermal cells. *Leaves* alternate, *usually 2-ranked*, simple, entire, petiolate, **with an upper pulvinus**, *a well-developed blade* **that folds upward at night**, *and pinnate venation* **with sigmoid secondary veins and evenly spaced cross-veins**, sheathing at base, lacking a ligule; petiole with air canals, these separated into segments by diaphragms composed of stellate-shaped cells; stipules lacking. Inflorescences determinate, but often appearing indeterminate, terminal. *Flowers* bisexual, *lacking a plane of symmetry* **but arranged in mirror-image pairs**. *Sepals 3, distinct*, imbricate. *Petals 3, connate*, imbricate. *Stamen 1, partly fertile and partly staminodial*; filament connate with staminodes and adnate to corolla; *anther unilocular (i.e., a half-anther, the other half expanded and sterile), depositing pollen onto the style before the flower opens; staminodes usually 3 or 4, petaloid and varying in shape*, basally connate and adnate to corolla, **1 from the inner androecial whorl forming a hooded structure with 1 or 2 small appendages (i.e., the hooded or cucullate staminode that holds the curved style under tension before the flower is triggered by a floral visitor), the second from the inner androecial whorl forming a callose-thickened structure, "the callose staminode," which often serves as a landing platform for insect visitors and helps to brace the hooded staminode**, and 1 or 2 from the outer androecial whorl ± petal-like; pollen grains lacking apertures, exine very reduced.

Carpels 3, connate; *ovary inferior*, with axile placentation, but 2 carpels often sterile and ± reduced; **style curved, held under tension by the hooded staminode that, when triggered by an insect, releases the style, which then elastically curves downward, scraping pollen from the insect's body and dusting it with pollen (held in a small cavity below the stigma)**; stigma 1, in depression between 3-lobed apex of style. **Ovules 1 in each locule, or solitary in the single functional carpel**. Nectaries in septa of ovary. *Fruit a loculicidal capsule or berry*; seeds usually arillate; embryo usually curved; endosperm and perisperm present.

Floral formula:

$$\$, 3, \underline{③}, 1\bullet \text{ or } 2\bullet + 2\bullet + \underline{(½ + ½}\bullet, \overline{③}); \text{ capsule, berry}$$

Distribution and ecology: Widely distributed in tropical and subtropical regions. Most occur in tropical rain forest margins and clearings or in wetlands.

Genera/species: 30/450. ***Major genus:*** *Calathea* (250 spp.). The family is represented in the continental United States only by *Thalia* (native) and *Maranta* (rarely naturalized).

Economic plants and products: West Indian arrowroot starch is obtained from the rhizomes of *Maranta arundinacea*. *Calathea, Ctenanthe, Maranta* (arrowroot, prayer plant), and *Thalia* are cultivated because of their decorative leaves.

Discussion: The monophyly of Marantaceae is supported by DNA and morphology (Kress 1990, 1995; Smith et al. 1993). Phylogenetic relationships within the family are poorly understood, but have recently been investigated based on morphology and *rps16* sequences (Andersson and Chase 2001). Traditionally, genera with only a single functional carpel (e.g., *Ischnosiphon, Maranta*, and *Thalia*) have been placed in the Maranteae, while genera with 3-locular ovaries (e.g., *Calathea* and *Marantochloa*) were placed in the Phrynieae. Both tribes, however, are polyphyletic.

The complex flowers of Marantaceae are mainly bee-pollinated and outcrossing; nectar provides the pollinator reward. The arils associated with the seeds of Marantaceae are often brightly colored and contain deposits of lipids, suggesting dispersal by birds or ants. Fruits of *Thalia* are water dispersed.

Some Marantaceae have deep red abaxial leaf surfaces, a possible adaptation for efficient use of light.

References: Andersson 1981, 1998; Andersson and Chase 2001; Classen-Bockhoff 1991; Dahlgren et al. 1985; Kennedy 2000; Kress 1990, 1995; Rogers 1984; Smith et al. 1993.

Figure 9.44   Cannaceae. *Canna flaccida:* (A) young shoot from rhizome (× 0.6); (B) blade and portion of petiole (× 0.6); (C) detail of upper part of sheathing petiole (× 1.15); (D) inflorescence, with open flower and flower bud, on open flower note sepals, reflexed petals, and five petaloid staminodes (× 0.75); (E) open flower, the fertile half-anther and tip of stigma visible in center (× 0.75); (F) detail of flower to show reflexed petals (× 0.75); (G) expanded staminode (× 0.75); (H) fertile stamen with half-anther attached to left margin (× 0.75); (I) detail of half anther after dehiscence (× 4.5); (J) longitudinal section near center of flower to show petals, staminodes, and style adnate to floral tube (× 1.5); (K) tip of style showing stigmatic line (near margin) (× 3); (L) ovary, in cross-section, the wall papillate (× 7.5); (M) capsule (× 0.75); (N) seed (× 4.5); (O) detail of seed coat (× 35); (P) seed in section, endosperm stippled (× 4.5); (Q) seedling (× 0.75). (From Rogers 1984, *J. Arnold Arb.* 65: p. 30.)

Cannaceae A. L. de Jussieu
(Canna Family)

Rhizomatous herbs; **mucilage canals present in rhizome and erect stem**. Plants glabrous. *Leaves* alternate, 2-ranked to spiral, simple, entire, ± *lacking a petiole, with a well-developed blade, the midvein of which possesses air canals, with pinnate venation,* sheathing at base, lacking a ligule, pulvinus, and stipules. Inflorescences determinate or indeterminate, terminal, *the main axis triangular in cross-section, with 3-ranked bracts,* each bract usually associated with reduced, 2- or 1-flowered cymes. *Flowers* bisexual, *lacking a plane of symmetry, often lasting only 1 day. Sepals 3, distinct,* imbricate. *Petals 3, connate,* imbricate. *Stamen 1;* filament connate with staminodes and adnate to corolla; *anther unilocular (i.e., a half-anther, the other half expanded and sterile); staminodes usually 3 or 4, petaloid, 1 larger than the others and recurved,* all basally connate and adnate to corolla; pollen grains lacking apertures, exine very reduced. Carpels 3, connate; *ovary inferior,* **externally papillate**, with axile placentation; **style flattened and ± petaloid**; stigma 1, extending along one edge of the style. Ovules ± numerous in each locule. Nectaries in septa of ovary. **Fruit a warty capsule**, usually splitting irregularly by disintegration of the fruit wall; *seeds spherical, black, associated with a tuft of hairs (modified aril);* endosperm and perisperm present (Figure 9.44).

Floral formula:

$, 3, ③, 1• or 2•+ 2• +(½ + ½•), ③; warty capsule

Distribution and ecology: Occurring in tropical and subtropical regions of the Americas; some species naturalized in the Old World; plants of moist openings in tropical forests, along rivers, or in wetlands.

Genera/species: 1/9. **Genus:** *Canna*. The family is represented in the continental United States by a few species of *Canna*.

Economic plants and products: The rhizomes of *Canna edulis* (Queensland arrowroot) are used as a starch source. Several species and various hybrids (e.g., *C. × generalis*) are used as ornamentals.

Discussion: The monophyly of the Cannaceae has been supported by DNA and morphology (Kress 1990, 1995; Smith et al. 1993).

The pollen is deposited on the style before the flower opens, either directly onto or somewhat below the stigma. Most species are selfing. Pollination biology has been poorly studied, but nectar-gathering bees, butterflies, moths, and birds may be the most frequent pollinators. The seeds are often water dispersed and long-lived.

References: Dahlgren et al. 1985; Kress 1990, 1995; Kubitzki 1998; Rogers 1984; Smith et al. 1993.

EUDICOTS (TRICOLPATES)

This large group is considered to be monophyletic on the basis of tricolpate pollen (or conditions derived from this pollen type) as well as *rbcL, atpB,* and 18S rDNA nucleotide sequence characters (Chase et al. 1993; Soltis et al. 1997, 1998, 2000; Hoot et al. 1999; Savolainen et al. 2000a,b). The clade also is characterized by cyclic flowers, that is, the parts arranged in whorls, with the members of individual whorls alternating. The staminal filaments are usually slender. Most members have plastids of sieve elements with starch grains (S-type). The major clades of tricolpates are shown in Figures 9.7 and 9.8.

"Basal Tricolpates"
Ranunculales

The Ranunculales are hypothesized to be monophyletic based on *rbcL, atpB,* and 18S rDNA sequences (Chase et al. 1993; Drinnan et al. 1994; Hoot and Crane 1995; Loconte et al. 1995; Soltis et al. 1997, 1998, 2000; Källersjö et al. 1998; Hoot et al. 2000; Savolainen et al. 2000a, b). The presence of the alkaloid berberine may also be synapomorphic. The order consists of 7 families and about 3490 species; major families include **Menispermaceae**, **Berberidaceae**, **Ranunculaceae**, and **Papaveraceae**. These families traditionally have been associated because of their predominantly herbaceous habit; toothed to lobed or even compound leaves; the presence of alkaloids, typically of the benzyl-isoquinoline type; hypogynous flowers with the parts usually distinct and free, and often with many stamens; and seeds with tiny embryos and copious endosperm (Cronquist 1981; Thorne 1974, 1992). The group has often been associated with the magnoliid complex (i.e., Magnoliales, Laurales, and Canellales: Cronquist 1969, 1981, 1988; Dahlgren 1983; Takhtajan 1969, 1980; Thorne 1974, 1992), but morphology (Donoghue and Doyle 1989; Doyle et al. 1994) and *rbcL, atpB,* and 18S rDNA sequences (Chase et al. 1993; Drinnan et al. 1994; Hoot and Crane 1995; Soltis et al. 2000; Savolainen et al. 2000a, b) indicate that Ranunculales are sister to the remaining members of the tricolpate clade.

Papaveraceae are monophyletic and probably sister to the remaining families within the order (Hoot and Crane 1995; Thorne 1974; Soltis et al. 2000) (Figure 9.45). This family differs from remaining members of the order in its syncarpous gynoecium, capsular fruits, quickly deciduous sepals, and presence of either laticifers and colored sap or specialized cells (with clear, mucilaginous sap). These characters, along with arillate seeds, are probably synapomorphic for Papaveraceae (or evolved very early in the evolution of this family). Morphological synapo-

Figure 9.45 Cladogram showing hypothesized relationships within Ranunculales. (Modified from Hoot et al. 1999 and Soltis et al. 2000.)

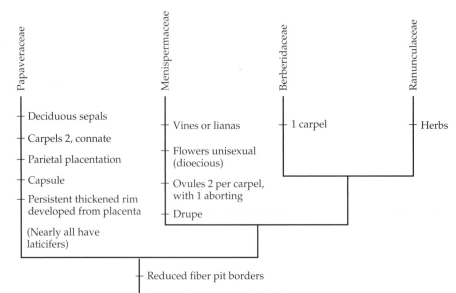

morphies of the clade comprising Berberidaceae, Menispermaceae, and Ranunculaceae are not readily apparent; the monophyly of this group is supported, however, by *rbcL, atpB,* and 18S sequences (Chase et al. 1993; Drinnan et al. 1994; Hoot and Crane 1995; Hoot et al. 1999; Soltis et al. 2000). Various relationships of these three families have been proposed (see Drinnan et al. 1994; Hoot and Crane 1995; Loconte and Estes 1989; Loconte and Stevenson 1991; Loconte et al. 1995). Berberidaceae apparently are most closely related to Ranunculaceae, as supported by nucleotide sequences (see Chase et al. 1993; Kim and Jansen 1995; Hoot and Crane 1995; Hoot et al. 1999) and similarities in floral form (Endress 1995).

References: Chase et al. 1993; Cronquist 1969, 1981, 1988; Dahlgren 1983; Donoghue and Doyle 1989; Doyle et al. 1994; Drinnan et al. 1994; Endress 1995; Endress and Igersheim 1999; Hoot and Crane 1995; Hoot et al. 1999; Källersjö et al. 1998; Kim and Jansen 1995; Loconte and Estes 1989; Loconte and Stevenson 1991; Loconte et al. 1995; Savolainen et al. 2000a, b; Soltis et al. 1997, 1998, 2000; Takhtajan 1969, 1980; Thorne 1974, 1992.

Menispermaceae A. L. de Jussieu
(Moonseed Family)

Twining vines or lianas, or less commonly shrubs or small trees; young stems with ring of vascular bundles separated by broad parenchymatous rays, *older stems often with anomalous secondary growth (successive rings of xylem and phloem) and flattened*; with various alkaloids (including the benzyl-isoquinoline type) and poisonous and very bitter sesquiterpenoids and diterpenoids. Hairs simple, sometimes glandular. *Leaves alternate and spiral, usually simple and entire, occasionally lobed or palmately compound, with usually palmate venation, often with an upper and lower pulvinus, often peltate or nearly so;* stipules lacking. Inflorescences usually determinate, axillary. **Flowers unisexual (the plants usually dioecious)**, radi-

al, *with 3-merous perianth.* Sepals usually 6, distinct, imbricate or valvate. Petals usually 6, distinct or connate, imbricate. Stamens (3-) 6-numerous, sometimes represented by staminodes in carpellate flowers; filaments distinct or connate; pollen grains usually tricolpate, tricolporate, or variously porate. *Carpels (1-) 3-6 (-numerous),* distinct, often on a gynophore; ovaries superior, with lateral placentation; stigma various, on very short style. **Ovules 2, but one aborting**, sometimes with a thin megasporangium wall. Nectaries lacking. **Fruit an aggregate of drupes**, the style sometimes nearly basal due to asymmetrical growth, *with endocarp ± curved, variously sculptured, usually laterally compressed; seeds curved; embryo usually curved;* endosperm sometimes ruminate, sometimes lacking. (Figure 9.46.)

Floral formula:

Staminate: * 6 , ⑥ , ⑥–∞ , 0

Carpellate:* 6 , ⑥ , ⑥–∞ , 3–6 , drupes

Distribution: Nearly pantropical, with a few genera extending into warm temperate regions; the group is characteristic of lowland tropical rain forests.

Genera/species: 71 / 450. **Major genera:** *Abuta* (35 spp.), *Cyclea* (30), *Stephania* (30), *Tinospora* (25), and *Cissampelos* (20). The family is represented in the continental United States and Canada by *Calycocarpum, Cocculus, Menispermum,* and *Cissampelos.*

Economic plants and products: Several species are used medicinally or as poisons, and the bark of members of

Figure 9.46 Menispermaceae. *Cocculus carolinus*: (A) pendent branch with fruit (× 0.6); (B) staminate flower, showing inner sepals, tow series of petals, and stamens, the short outer sepals not visible (× 18); (C) inner sepal, clasping petal, and stamen (× 18); (D) carpellate flower, outer sepal in foreground removed to show inner sepals, two series of petals, staminodia, and carpels (× 18); (E) inner sepal, clasping petal, and staminode (× 18); (F) gynoecium, the carpels with reflexed stigmas (× 18); (G) vertical section of carpel (× 12); (H) vertical section of fruit, showing coiled embryo surrounded by endosperm (white), bony endocarp (dark), and fleshy fruit wall (× 4.8); (I) endocarp (× 3.6). (From Ernst 1964, *J. Arnold Arbor.* 45: p. 31.)

several genera (including *Chondrodendron*) is used in the preparation of curare, which is an arrow poison of some South American Indian tribes, and medicinally as a muscle relaxant. A few species of *Cocculus* (coralbeads) and *Menispermum* (moonseed) are used as ornamentals.

Discussion: Monophyly of Menispermaceae is supported by several morphological characters. The family is most closely related to Ranunculaceae and Berberidaceae (Hoot and Crane 1995) but a close relationship to

Ranunculaceae and Papaveraceae has also been proposed (Loconte et al. 1995). Lardizabalaceae are also vines or lianas (e.g., *Akebia*, an occasional ornamental), but they differ from Menispermaceae in having compound leaves, several ovules per carpel, and fleshy follicles or berries. Dioecy has evolved independently in these two families.

Relationships within Menispermaceae are unclear, but the clade is usually divided into five to eight tribes on the basis of fruit and seed characters (endosperm present or absent, ruminate or homogeneous, embryo form, endocarp shape and sculpturing).

Pollination has not been studied but is probably by insects. The fleshy fruits are black, blue, purple, or red, small to quite large, and are presumably dispersed by birds and mammals.

References: Ernst 1964; Hoot and Crane 1995; Kessler 1993; Loconte et al. 1995; Mennega 1982; Thanikaimoni 1986.

Ranunculaceae Adanson
(Buttercup Family)

Herbs, shrubs, or occasionally vines; stems with vascular bundles often in several concentric rings or ± scattered; usually with alkaloids or ranunculin (a lactone glycoside); often with triterpenoid saponins. Hairs usually simple. *Leaves usually alternate and spiral*, occasionally opposite, *simple, sometimes lobed or dissected, to compound, usually serrate, dentate, or crenate*, with pinnate to occasionally palmate venation; *stipules usually lacking.* Inflorescences determinate, sometimes appearing indeterminate or reduced to a single flower, terminal. Flowers usually bisexual, radial to occasionally bilateral, with short to elongate receptacle. *Perianth parts usually not 3-merous. Tepals 4 to numerous, distinct*, and imbricate; *or perianth differentiated into calyx and corolla, then sepals usually 5, distinct, imbricate, and deciduous, and petals usually 5, distinct, imbricate, often with nectar-producing basal portion* or represented only by small nectar glands, probably derived from staminodes. *Stamens numerous;* filaments distinct; *anthers opening by longitudinal slits;* pollen grains tricolpate (or ± modified). *Carpels usually 5 to numerous,* occasionally reduced to 1, *usually distinct;* ovaries superior, with usually lateral placentation; stigmas punctate or

extending along one side of the style. Ovules 1 to numerous per carpel. *Fruit usually an aggregate of follicles or achenes*, occasionally a berry (Figure 9.47).

Floral formula:

* or X, -4–∞- or 5, 5, ∞, <u>1–∞</u> ; follicles, achenes, berries

Distribution: Widespread, but especially characteristic of temperate and boreal regions of the Northern Hemisphere.

Genera/species: 47/2000. **Major genera:** *Ranunculus* (400 spp.), *Aconitum* (250), *Clematis* (250), *Delphinium* (250), *Anemone* (150), and *Thalictrum* (100). Some of the numerous genera in the continental United States and/or Canada (in addition to those listed above) are *Actaea, Aquilegia, Caltha, Coptis, Hydrastis, Isopyrum, Myosurus, Trollius,* and *Xanthorhiza.*

Economic plants and products: The family is chiefly important for its numerous ornamental herbs, such as *Anemone* (windflower, including *Hepatica*), *Aconitum* (monkshood), *Actaea* (baneberry, including *Cimicifuga*), *Aquilegia* (columbine), *Caltha* (marsh marigold), *Clematis* (virgin's bower), *Delphinium* (larkspur), *Helleborus* (hellebore), *Ranunculus* (buttercup), *Thalictrum* (meadow rue), and *Trollius* (globeflower). A number of genera are highly poisonous.

Discussion: The monophyly of this group is supported by morphology and nucleotide sequences (Drinnan et al. 1994; Hoot and Crane 1995; Hoot 1995; Loconte and Estes 1989). *Hydrastis* is cladistically basal, having retained flowers with a 3-merous perianth, vessels with scalariform perforations, ovules with micropyle defined

---

Key to Major Families of Ranunculales

1. Gynoecium of 2 to many connate carpels, ovary with 1 locule and parietal placentas; fruits capsules; plants with clear, white, or colored sap ....**Papaveraceae**

1. Gynoecium of 1 to many distinct carpels, each ovary with lateral to ± basal placenta; fruits berries, follicles, or achenes; plants with clear sap (although wood may be yellow) or sap lacking 2

   2. Flowers usually bisexual; plants mostly herbs or shrubs, rarely lianas; fruits achenes, follicles, or berries 3

   2. Flowers usually unisexual; plants usually lianas; fruits drupes ....**Menispermaceae**

      3. Carpels solitary; stamens usually as many or twice as many as and opposite the petals; anthers usually opening by 2 flaps; perianth 3-merous; fruits usually fleshy, berries.......**Berberidaceae**

      3. Carpels usually numerous; stamens numerous, spirally arranged; anthers opening by 2 longitudinal slits; perianth usually not 3-merous; fruits usually dry, achenes or follicles ....**Ranunculaceae**

Figure 9.47 Ranunculaceae. *Caltha palustris:* (A) flowering plant (× 0.5); (B) flower (× 3); (C) stamen (× 12); (D) lateral view of carpel, with detail of nectar glands at base (× 10); (E) carpel in longitudinal section (× 10); (F) carpel in cross-section (× 15); (G) follicles from a five-carpellate flower (× 2); (H) seed (× 20); (I) seed in longitudinal section, note spongy seed coat, endosperm (stippled), and minute embryo (× 20). (From Wood 1974, *A student's atlas of flowering plants*, p. 29.)

by two integuments, and fleshy follicles (Loconte and Estes 1989); cpDNA restriction sites and sequence data also support the basal position of this genus, as well as that of *Glaucidium* (Hoot 1995; Johansson and Jansen 1993). The remaining genera of Ranunculaceae may form a clade based on vessels with simple perforations, 4- or 5-merous perianth, and dry fruits. A large clade (Ranunculoideae) is supported by the synapomorphies of large chromosomes, stomates longer than 35 μm, a chromosome number of 8, and nucleotide characters

(Hoot 1991, 1995). In addition, the alkaloid berberine has not been reported within this clade, and its loss may be synapomorphic. *Clematis, Ranunculus, Trautvetteria,* and *Anemone* probably form a subclade based on the synapomorphies of achenes, presence of ranunculin, and DNA sequence data (Johansson 1995; Johansson and Jansen 1993; Hoot 1991, 1995). A second subclade, including *Actaea, Caltha, Helleborus, Trollius, Nigella, Aconitum,* and *Delphinium,* may be diagnosed on the basis of petal-like tepals, but this grouping is not supported by nucleotide sequences (Hoot 1991, 1995). The remaining genera—*Xanthorhiza, Coptis, Aquilegia, Thalictrum,* and *Isopyrum,* constitute the "Thalictroideae," which are paraphyletic and basal within Ranunculaceae (along with *Hydrastis* and *Glaucidium*). These genera retain plesiomorphies—the presence of berberine, yellow creeping rhizomes, short, thin-walled hairs, and small chromosomes—that phenetically "link" them to Berberidaceae. Nucleotide sequences suggest that *Coptis* and *Xanthorhiza* form a clade that is sister to a clade containing the remaining genera of "Thalictroideae" + Ranunculoideae (Hoot 1995; Johansson and Jansen 1993).

Reduction in the number of ovules per carpel and the evolution of achenes have occurred several times within Ranunculaceae; petaloid staminodes also have evolved more than once, as suggested by their differing morphology, along with anatomical and developmental evidence (Hoot 1995). Woodiness probably is secondary within the family.

The wide range of floral structures within Ranunculaceae is associated with a diverse array of pollination syndromes. Most species are insect-pollinated, although some species of *Thalictrum* are wind-pollinated. *Anemone* and *Clematis* do not produce nectar and are pollinated by various pollen-gathering insects. In contrast, *Ranunculus, Delphinium,* and *Aquilegia* have modified nectar-secreting petals (sometimes spurred), and their flowers are visited by nectar-gathering insects (mainly bees) or hummingbirds. *Caltha* has nectar glands at the base of the carpels and is also bee-pollinated. Different maturation times of the androecium and gynoecium may promote outcrossing, although selfing is also common.

Dispersal mechanisms vary widely. Achenes of *Clematis* have persistent, long, hairy styles, and are dispersed by wind, while those of *Ranunculus* may contain tubercules or hooked spines, leading to external transport by animals. The small seeds of follicular species may be dispersed by wind or water, and some (e.g., *Helleborus*) are secondarily dispersed by ants. The berries of some species of *Actaea* are mainly bird-dispersed.

References: Cronquist 1981; Drinnan et al. 1994; Hoot 1991, 1995; Johansson 1995; Johansson and Jansen 1993; Jensen et al. 1995; Leppik 1964; Loconte and Estes 1989; Tamura 1963, 1965, 1993; Thorne 1974.

Berberidaceae A. L. de Jussieu
(Barberry Family)

Herbs or shrubs; stems with vascular bundles sometimes ± scattered; usually with alkaloids, *the wood usually colored yellow by berberine* (an isoquinoline alkaloid). Hairs simple. *Leaves usually alternate and spiral* (opposite in *Podophyllum*), *simple, sometimes lobed or dissected, to compound,* sometimes reduced and unifoliolate (some species of *Berberis*), entire to serrate or spinose-serrate, sometimes reduced to spines, with pinnate to palmate

venation; stipules lacking or present. Inflorescences various. Flowers bisexual, radial, with *perianth usually 3-merous.* Sepals usually 4 or 6, distinct, imbricate. *Outer petals 4 or 6, distinct, imbricate, lacking nectar glands,* occasionally lacking; *inner petals (probably petal-like staminodes) usually 6, nectar-producing,* showy or represented only by small scales, sometimes lacking. *Stamens 4 to numerous, but most often 6, usually opposite petals;* filaments usually distinct; *anthers opening by 2 flaps that open from the base* (longitudinal slits in *Nandina* and *Podophyllum*); pollen grains tricolpate (or modified). **Carpel 1**; ovary superior, with lateral to basal placentation; stigma capitate to 3-lobed. Ovules usually numerous, sometimes reduced to 1. *Fruit usually a berry,* sometimes ± dehiscent; seeds often arillate (Figure 9.48).

Floral formula: *, 4–6 , 4–6 , 4–6• + 4–∞, 1; berry

Distribution: Widespread, especially in temperate regions of the Northern Hemisphere and the Andes of South America.

Genera/species: 15/650. **Major genus:** *Berberis* (600 spp.). *Achlys, Berberis, Caulophyllum, Diphyllea, Jeffersonia, Nandina, Podophyllum,* and *Vancouveria* occur in the continental United States and/or Canada.

Economic plants and products: Genera such as *Berberis* (barberry, Oregon grape) and *Nandina* (sacred bamboo) contain valuable ornamentals. Many are extremely poisonous.

Discussion: The monophyly of Berberidaceae is supported by morphology and DNA-based analyses (Loconte 1993; Loconte and Estes 1989; Loconte et al. 1994; Hoot and Crane 1995). *Nandina* (in Nandinoideae) is sister to the remainder of the family (Berberidoideae), which is united by their herbaceous habit and anthers that open by two flaps. Several clades can be recognized within Berberidoideae (Meacham 1980; Loconte and Estes 1989; Loconte 1993; Kim and Jansen 1998a). *Leontice, Gymnospermum,* and *Caulophyllum* are characterized by flabelliform nectar-secreting petals (staminodes), pollen with reticulate sculpturing, basal placentation, and a chromosome base number of eight. The remaining genera of Berberidoideae form a clade on the basis of carpels with several ovules. Within this clade, a group containing *Ranzania* and *Berberis* may be diagnosed on the basis of inner petals with paired basal glands, sensitive stamens, and a base chromosome number of seven. *Berberis* has the apomorphies of secondary woodiness, spinose leaf teeth, irregular pollen apertures, basal placentation, and a large embryo. Some species of this genus have pinnately compound leaves (often segregated as *Mahonia*). *Podophyllum, Diphyllea,* and relatives form a clade based on stems with scattered vascular bundles and simple leaves with the veins ending in the teeth. Finally, *Epimedium, Vancouveria,*

Figure 9.48 Berberidaceae. *Podophyllum peltatum:* (A) young shoot (× 1); (B) flowering plant (× 0.33); (C) flower (× 1); (D) two stamens (× 5); (E) gynoecium (× 3.5); (F) gynoecium in longitudinal section (× 3.5); (G) ovary in cross-section (× 3.5); (H) berry (× 0.75); (I) fruit in cross-section, the seeds in a pulpy placenta (× 0.75); (J) seed (× 4). (From Wood 1974, *A student's atlas of flowering plants*, p. 33.)

Jeffersonia, and relatives may form a monophyletic group supported by leaves in a basal rosette and fruits that open by a horizontal slit. Chloroplast DNA restriction sites suggest, however, that the last group may be paraphyletic (Kim and Jansen 1995, 1998a).

The showy flowers of Berberidaceae are pollinated by various insects (mainly bees) that gather nectar or pollen. In *Berberis* the stamens are sensitive, and when touched by a bee as it probes for nectar, they spring inward, and the pollen-bearing flaps contact the insect's head. The often colorful berries of Berberidaceae typical-

ly are bird- or mammal-dispersed. Some have seeds with a hard, aril-like structure and are insect-dispersed. Fruits of *Vancouveria* are dispersed by wasps. *Caulophyllum* is distinct because the fruit ruptures early in development and the large, globose, fleshy blue seed develops in a completely exposed condition.

References: Cronquist 1981; Ernst 1964; Hoot and Crane 1995; Kim and Jansen 1995, 1998a; Loconte 1993; Loconte and Estes 1989; Loconte et al. 1995; Meacham 1980; Thorne 1974.

Papaveraceae Adanson
(Poppy Family)

Herbs to soft-wooded shrubs; stem with vascular bundles sometimes in several rings; *with laticifers present and plants with white, cream, yellow, orange, or red sap, or with specialized elongated secretory cells and sap then mucilaginous, clear*; sap with various alkaloids (including the benzyl-isoquinoline type). Hairs simple. *Leaves usually alternate and spiral, simple,* **but often lobed or dissected,** *entire to more commonly variously toothed, sometimes spinose,* with ± pinnate venation; *stipules lacking.* Inflorescences various. Flowers bisexual, radial (with numerous or only 2 main planes of symmetry) to bilateral. **Sepals usually 2 or 3,** usually distinct, imbricate, usually **quickly deciduous,** large and surrounding bud to small and bractlike. **Petals usually 4** *or 6,* sometimes numerous, distinct, imbricate *and often crumpled in bud and thus wrinkled when expanded; often the 2 (or 3) inner differentiated from the 2 (or 3) outer,* and sometimes with 1 or 2 of the outer petals with a prominent basal nectar spur or pouch and the 2 inner sticking together at apex, forming a cover over the stigmas. *Stamens numerous, to 6 that are ± connate in 2 groups of 3,* rarely reduced to 4; filaments distinct to connate; pollen grains tricolporate to polyporate. **Carpels 2** to numerous, **connate; ovary** superior, **with parietal placentation,** the placentas sometimes intruded; stigma(s) distinct to connate, 1 or equaling number of carpels, often discoid and lobed, sometimes capitate. Ovules usually numerous, but sometimes reduced to 1 or 2, often amphitropous to campylotropous. Nectaries lacking, or sometimes one or more of the filaments with a basal nectar gland. **Fruit a capsule,** *opening variously, but often by apical pores, valves, or longitudinal slits,* **often with a persistent thickened replum** (or replums, developed from the placenta, but often lost), occasionally a nut or lomentlike; **seeds sometimes arillate** (Figure 9.49).

Floral formula:

* or X, 2–3 , 4–6 (–∞), (4–∞) or ③ +③, (2–∞); capsule

Distribution: Widely distributed in mainly temperate regions; especially diverse in the Northern Hemisphere, but also in southern Africa and eastern Australia.

Genera/species: 40/770. **Major genera:** *Corydalis* (400 spp.), *Papaver* (100), *Fumaria* (50), and *Argemone* (30). Genera occurring in the continental United States and/or Canada include *Adlumia, Arctomecon, Argemone, Canbya, Chelidonium, Corydalis, Dendromecon, Dicentra, Eschscholzia, Fumaria, Hesperomecon, Meconella, Papaver, Platystemon, Romneya, Sanguinaria,* and *Stylophorum.* Delimitation of genera is often difficult (see Jork and Kadereit 1995).

Economic plants and products: *Papaver somniferum* (opium poppy) is the source of opium and derivatives

such as morphine, heroin, and codeine; the seeds of this species (which do not contain opium) are used as a spice. Many have showy flowers and are cultivated as ornamentals, such as species of *Argemone* (prickly poppy), *Eschscholzia* (California poppy), *Papaver* (poppy), *Macleaya* (plume poppy), *Corydalis* (harlequin), *Sanguinaria* (bloodroot), and *Dicentra* (Dutchman's-breeches, bleeding heart). Most species are highly poisonous.

Discussion: The monophyly of Papaveraceae is supported by morphology and nucleotide sequence data (Drinnan et al. 1994; Hoot and Crane 1995; Kadereit et al. 1994; Loconte et al. 1995). The family is broadly circumscribed here, including Fumariaceae (see below).

Two quite different hypotheses regarding phylogenetic relationships within the family have been proposed (compare Loconte et al. 1995 and Judd et al. 1994 with Blattner and Kadereit 1995, Kadereit et al. 1995, Hoot et al. 1997, and Schwarzbach and Kadereit 1995). The cladograms presented in these papers are quite similar in topology, but differ in how they are rooted.

Loconte et al. (1995) proposed (on the basis of morphological data) that the basal genera are in the subfamily Platystemonoideae; *Platystemon* and relatives show numerous and only slightly fused carpels (with free stigmas) that separate from one another at maturity. In contrast, Kadereit et al. (1995) and Hoot et al. (1997) proposed (on the basis of morphological and nucleotide sequence data) that *Pteridophyllum* is sister to the remaining genera; this genus lacks secretory cells, while the remaining genera are united by the possession of laticifers or idioblasts, a presumed synapomorphy.

Within the laticifer-idioblast clade, the Fumarioideae (e.g., *Dicentra, Corydalis,* and *Fumaria*) are sister to the remaining subfamilies, comprising genera such as *Papaver, Argemone, Meconopsis, Eschscholzia, Chelidonium,* and *Sanguinaria* (Kadereit et al. 1995; Hoot et al. 1997). The monophyly of the Fumarioideae is supported by stamens fused into two groups of three (diadelphous) in which the central stamen of each group has two locules and the two lateral stamens have only a single locule, staminal nectaries, and flowers in which the two outer petals look very different from the two inner petals (which are more or less connate over the stigma). In addition, one or both of the outer petals are saccate or spurred. Nonfumarioid members of the laticifer-idioblast clade form a subclade supported by numerous stamens. Within this group, the Chelidonioideae (e.g., *Chelidonium, Sanguinaria*) are possibly basal. Both Fumarioideae and Chelidonioideae have elongated, consistently two carpellate fruits with a replum (Figure 9.49F). Papaveroideae (e.g., *Papaver, Argemone,* and *Meconopsis*) are more derived and are diagnosed by their many-carpellate flowers and multicellular-multiseriate hairs. *Platystemon* and relatives are placed within Papaveroideae by Kadereit et al. (1995). *Eschscholzia*

Figure 9.49 Papaveraceae. *Chelidonium majus:* (A) flowering and fruiting branch (× 1); (B) opening flower bud, note two sepals (× 6.5); (C) flower (× 6.5); (D) stamen (× 22); (E) gynoecium (× 22); (F) two-valved capsule, note persistent rim (× 4.5); (G) seeds attached to rim (× 9); (H) arillate seed (× 44). (From Ernst 1962, *J. Arnold Arbor.* 43: p. 325.)

(Eschcholzioideae) are distinct due to their unicellular hairs, polycolpate pollen, and explosive fruits.

Fumaria, Dicentra, Corydalis, and relatives are often recognized (at least by many American systematists) as Fumariaceae (Cronquist 1981), although here they are included within Papaveraceae (as the subfamily Fumarioideae). Given the clear support for the monophyly of the clade including Fumarioideae, *Hypecoum, Pteridophyllum,* and remaining subfamilies (Papaveraceae s.s.), and the many features they have in common, we think it most useful to treat this clade as Papaveraceae s. l. (as in Thorne 1974, 1992; Takhtajan 1980).

Flowers of Papaveraceae attract bees, wasps, and flies as pollinators. Members of the Papaveroideae and Chelidonioideae, which have radially symmetrical flowers with an exposed androecium, do not produce nectar and are pollinated by pollen-gathering bees, flies, and beetles. The more specialized flowers of Fumarioideae have nectar spurs or sacs and are pollinated by bees. Bees probing for nectar depress the hooded inner petals, exposing the anthers, and the insect is dusted with pollen. *Bocconia* has inconspicuous flowers with pendulous stamens and is wind-pollinated. Both outcrossing and selfing may occur.

In many Papaveraceae the small seeds are shaken out of the capsules and thus passively dispersed (through the action of wind or contact with the plant), but in *Eschscholzia* the capsules dehisce explosively. The seeds are typically arillate, the arils (Figure 9.49H) being hard and oily, leading to dispersal by ants (several genera), or fleshy and brightly colored, leading to dispersal by birds (*Bocconia*). A fatty collar is produced around the base of the nuts of *Fumaria,* another adaptation for ant dispersal.

References: Blattner and Kadereit 1995; Chase et al. 1993; Cronquist 1981; Dahl 1990; Drinnan et al. 1994; Ernst

1962; Hoot and Crane 1995; Hoot et al. 1997; Jork and Kadereit 1995; Judd et al. 1994; Kadereit 1993; Kadereit and Sytsma 1992; Kadereit et al. 1994, 1995; Lidén 1986, 1993; Loconte et al. 1995; Ronse Decraene and Smets 1992; Schwarzbach and Kadereit 1995; Takhtajan 1980; Thorne 1974, 1992, 2001.

Proteales and Other "Basal Tricolpates"

Various families with reduced, wind-pollinated flowers that are often aggregated into dangling inflorescences (i.e., catkins or aments) have traditionally been considered to form a group of related families called the Hamamelidae (see Cronquist 1981, 1988; Takhtajan 1980; and the historical summary in Stern 1973). As shown by numerous lines of evidence, this assemblage clearly is not monophyletic (Thorne 1973a; Donoghue and Doyle 1989; Wolfe 1989; Crane and Blackmore 1989; Hufford and Crane 1989; Hufford 1992; Chase et al. 1993; Manos et al. 1993). Some families traditionally placed in the Hamamelidae, such as Platanaceae, Trochodendraceae, and Buxaceae, are early divergent lineages of the eudicots. Proteaceae and Nelumbonaceae are probably also related to these families. Morphology, *rbcL*, *atpB*, and 18S rDNA sequences support the near-basal placement of these families within the eudicots, with Proteaceae, Platanaceae, and Nelumbonaceae forming a clade, here called Proteales (Angiosperm Phylogeny Group 1998; Chase et al. 1993; Hufford 1992; Manos et al. 1993; Soltis et al. 1997, 2000). Proteales are diagnosed by gynoecia with 1 or 2 pendent, more or less orthotropous ovules per carpel. The Platanaceae + Proteaceae clade are supported by the putative synapomorphies of 4-merous flowers with stamens opposite the perianth parts and carpels with five vascular bundles, although the flowers of Platanaceae are highly reduced, making comparisons difficult. The families of Proteales, along with the phylogenetically adjacent Trochodendraceae and Buxaceae, are contrasted in the following key. Other hamamelid families with reduced flowers are placed in Saxifragales, Fagales or Malpighiales (see below).

Hamamelidaceae and Cercidiphyllaceae, in particular, sometimes have been considered to be related to Platanaceae (Hufford and Crane 1989; Schwarzwalder and Dilcher 1991; Hufford 1992), but this placement is not supported by DNA sequences (Chase et al. 1993; Savolainen et al. 2000a,b; Soltis et al. 1997, 2000). DNA evidence suggests that Hamamelidaceae and Cercidiphyllaceae belong near Saxifragaceae (and these two families are here referred to Saxifragales).

References: APG 1998; Chase et al. 1993; Crane and Blackmore 1989; Cronquist 1981, 1988; Donoghue and Doyle 1989; Hufford 1992; Hufford and Crane 1989; Manos et al. 1993; Savolainen et al. 2000a,b; Schwarzwalder and Dilcher 1991; Soltis et al. 1997, 2000; Stern 1973; Takhtajan 1980; Thorne 1973a; Wolfe 1989.

Platanaceae T. Lestiboudois
(Sycamore or Plane Tree Family)

Trees; **nodes multilacunar**; cyanides, *triterpenes*, and tannins present; *bark coming off in large, irregular sheets, leaving a patchy, light-colored, smooth surface.* **Hairs branched.** *Leaves alternate, 2-ranked or spiral, simple, usually palmately lobed and veined, ± coarsely toothed,* each tooth having a midvein that becomes attenuated toward a glandular

---

Key to Families of Proteales and Other "Basal Eudicots"

1. Plants aquatic with submerged rhizome; receptacle expanded with individual carpels sunken into its flattened apex; fruits nuts, embedded in an accrescent, spongy receptacle .... Nelumbonaceae

1. Plants of terrestrial habitats; receptacle not expanded; carpels free or connate; fruits achenes, follicles, or capsules 2

2. Vessels lacking; carpels with adaxial portion forming a nectar gland .... Trochodendraceae (incl. Tetracentraceae)

2. Vessels present; carpels lacking adaxial nectar gland 3

3. Carpels distinct, or carpel solitary; fruit not a capsule 4

3. Carpels at least partially connate; fruit a capsule .... Buxaceae

4. Perianth conspicuous, petaloid, and plants mainly animal pollinated; flowers usually bisexual; stamens 4, usually adnate to the perianth; leaves very often coriaceous; fruits usually follicles .... **Proteaceae**

4. Perianth minute, and plants wind pollinated; flowers unisexual; stamens 3–7, free from perianth; leaves not coriaceous; fruits achenes associated with hairlike bristles .... **Platanaceae**

apex, where it opens into a cavity; *petiole base expanded and enclosing axillary bud; stipules present, usually large and conspicuous.* **Inflorescences** indeterminate, a raceme of **globose heads**, but sometimes reduced to a single head, pendulous, axillary. *Flowers unisexual* (**and in unisexual heads, plants monoecious**), radial, very reduced and inconspicuous. Sepals 3–7, distinct or slightly connate, minute. Petals 3–7, distinct, minute, usually lacking in carpellate flowers. Stamens 3–7; *filaments very short; anthers with the connective prolonged into a peltate appendage;* pollen grains tricolporate. *Carpels usually 5–9, distinct; ovaries superior,* **placentation apical**; stigma elongated and extending along adaxial side of recurved style. Ovule 2 per carpel, **but 1 aborting**, orthotropous. Nectaries lacking. **Fruits ± linear achenes, subtended by long bristles, in dense, globose clusters**.

Floral formula: Staminate: *, (3–7), 3–7, 3–7, 0

Carpellate: *, (3–7), 0 , 0 , 5–9; achene

Distribution and ecology: Occurring in tropical to temperate regions of North America, south-central Europe, western Asia to Indochina; often along rivers or streams.

Genera/species: 1/9. **Genus:** *Platanus*.

Economic plants and products: Several species or hybrids of *Platanus* (plane tree, sycamore) are cultivated as ornamentals.

Discussion: The inconspicuous and unisexual flowers of *Platanus* are wind-pollinated; flowering takes place in the early spring, just as the leaves are emerging. The globose cluster of achenes breaks apart in the fall, and the individual fruits, with their hair tufts, are dispersed by wind (and sometimes secondarily by water).

References: Boothroyd 1930; Endress 1977; Endress et al. 1999; Ernst 1963b; Hufford and Crane 1989; Kubitzki 1993d.

Proteaceae A. L. de Jussieu
(Protea Family)

Trees or shrubs; nodes trilacunar; tannins commonly present, sometimes cyanogenic. Roots not forming mycorrhizae, *usually with clusters of specialized short lateral roots.* Hairs simple or forked, 3-celled, with basal cells short and terminal cell elongate, often thick-walled. Leaves usually alternate and spiral, simple or pinnately compound, sometimes lobed or deeply dissected, entire to serrate; stipules absent. Inflorescences determinate or indeterminate, terminal. Flowers usually bisexual, radial or bilateral, conspicuous. **Tepals 4**, distinct *or more commonly connate, often deeply cleft on one side,* or 3 tepals connate and 1 distinct, **valvate. Stamens 4;** *filaments usually adnate to tepals;* anthers usually with the connective pro-

longed and forming an appendage; pollen grains usually 3-porate or 3-colporate. **Carpel 1,** *often on a stalk;* ovary superior, with lateral placentation; stigma globose to elongated. Ovules 1 to numerous, anatropous to orthotropous. *Fruits follicles, nuts, achenes, drupes, or samaras;* seeds often winged; endosperm usually lacking.

Floral formula:

* or X, (4), 4 , 1; follicles, nut, achene, drupe

Distribution and ecology: Widespread in tropical and subtropical regions, especially southern Africa and Australia.

Genera/species: 80/1770. **Major genera:** *Grevillea* (200 spp.), *Hakea* (110), *Protea* (110), *Helicia* (80), *Leucadendron* (70), *Banksia* (50), and *Leucospermum* (40). The family is represented in North America only by a single introduced species of *Grevillea* (in Florida).

Economic plants and products: *Banksia, Embothrium, Grevillea, Hakea, Protea,* and *Telopea* are ornamentals. The seeds of *Macadamia integrifolia* (macadamia nut) are edible.

Discussion: The family usually is divided into two subfamilies, the Grevilleoideae (flowers in pairs) and the Proteoideae (flowers single); the former has been divided by Johnson and Briggs (1975) into four subfamilies. The monophyly of Proteoideae (including genera such as *Protea, Leucospermum, Leucadendron,* and *Conospermum*) is supported by small chromosomes, ovules reduced to one or two, and indehiscent, usually dry fruits. Grevilleoideae (e.g., *Embothrium, Grevillea, Macadamia, Banksia, Hakea, Helicia,* and *Telopea*) have their flowers in pairs, and their monophyly is supported by chloroplast DNA sequences (Hoot and Douglas 1998). Many have follicles and winged seeds, but indehiscent fruits with large, wingless seeds have evolved several times within the group. *Bellendena, Placospermum,* and *Toronia* may be sister to all other Proteaceae (Hoot and Douglas 1998), and Johnson and Briggs (1975) also noted their ancestral features.

The flowers of Proteaceae are pollinated by various insects, birds, and small mammals (marsupials and rodents). Species with follicles and winged seeds (e.g., *Embothrium, Telopea,* and *Grevillea*) are wind-dispersed, while those with dry to fleshy indehiscent fruits are dispersed by birds or mammals.

References: Hoot and Douglas 1998; Johnson and Briggs 1975; Purnell 1960; Rourke and Wiens 1977.

Core Eudicots (Core Tricolpates)

The monophyly of this large group (see Table 9.1 and Figure 9.7) is supported by *rbcL, atpB,* and 18S sequences (Chase et al. 1993; Soltis et al. 1998; Hoot et al. 1999; Savolainen et al. 2000a, b; Soltis et al. 2000). These plants have usually 4- or 5-merous flowers with the perianth differentiated into a calyx and corolla; placentation is usually axile.

Caryophyllid clade

The monophyly of the caryophyllid clade is supported by seed coat anatomy, anther wall development, vessel-elements with simple perforations, pollen shed in the 3-nucleate condition, and *rbcL, atpB, matK,* and 18S rDNA sequences. Their position, whether within the rosid complex, or sister to the asterid clade, is uncertain. The group includes two large orders (Caryophyllales and Polygonales), although a few families are of problematic place-

ment. The group is treated as Caryophyllales s. l. by the Angiosperm Phylogeny Group (1998, 2002).

Caryophyllales

The core Caryophyllales are clearly monophyletic, as evidenced by numerous distinctive synapomorphies, such as sieve tubes of phloem with plastids with peripheral ring of proteinaceous filaments and often a central protein crystal (whereas most members of the tricolpate clade have sieve tube plastids with starch grains); betalains forming red to yellow pigments (but anthocyanins in Caryophyllaceae); loss of the *rpl2* intron in cpDNA; a single whorl of tepals; placentation free-central to basal; embryo curved around the seed; and presence of

---

Key to Major Families of Caryophyllales

1. Gynoecium of 1 carpel, with a single ovule 2
1. Gynoecium of 2 to numerous carpels, with 1 to numerous ovules.................................3
2. Inflorescence indeterminate; flowers 4-merous, tepals 4, distinct; stamens 4 to numerous; perianth not differentiated ....Petiveriaceae
2. Inflorescence determinate; flowers usually 5-merous, tepals 5, connate; stamens usually 5 to numerous; perianth differentiated, the distal portion ± conspicuous and withering, the proximal portion accrescent, persistent, associated with the fruit in dispersal......... **Nyctaginaceae**
3. Inflorescence indeterminate; carpels ± distinct to connate and ovary with axile placentation, with 1 ovule in each locule; fruit a berry, developing from superior ovary...**Phytolaccaceae**
3. Inflorescence determinate; carpels connate and ovary with free-central, basal, parietal, or axile placentation, ovules 1 to numerous, and if axile, then ovules numerous in each locule; fruit a capsule, achene, utricle and developing from superior or inferior ovary, or a berry developing from ± inferior ovary 4
4. Flowers with hypanthium; ovary inferior to superior; stamens usually numerous 5
4. Flowers lacking a hypanthium; ovary superior; stamens few to numerous 6
5. Shoots ± equivalent; leaves succulent, never spiny, opposite or alternate; ovary superior to inferior, usually with axile placentation ....**Aizoaceae**
5. Shoots strongly dimorphic, the long shoots usually succulent and photosynthetic, the short shoots with a spine or cluster of spines (and sometimes also irritating hairs, glochids); stems succulent; leaves always alternate; ovary usually inferior, with basal or more commonly parietal placentation...**Cactaceae**
6. Flowers with 2 bracteoles that appear to form a calyx, tepals petaloid; stamens often numerous, sometimes as few as five ....**"Portulacaceae"**
6. Flowers lacking 2 sepal-like bracteoles, with tepals greenish to petaloid, or with 5 "petals" and 5 tepals appearing to be a calyx; stamens 5–10 7
7. Flowers usually with 5 sepals and 5 "petals"; leaves opposite, ovary with free-central to basal placentation, ovules usually numerous (sometimes only 1); fruit usually a capsule (occasionally achene); anthocyanins present; sieve cell plastids with central protein crystal ....**Caryophyllaceae**
7. Flowers usually with 5 tepals; leaves alternate or opposite and nodes not swollen; ovary with basal placentation, ovule usually 1; fruit an achene or utricle; betalains present; sieve cell plastids lacking central protein crystal ....**Amaranthaceae**

perisperm, with endosperm scanty or lacking (Behnke 1976, 1994; Bittrich 1993b; Chase et al. 1993; Downie and Palmer 1994a,b; Eckardt 1976; Mabry 1973, 1976; Manhart and Rettig 1994; Rettig et al. 1992; Rodman 1990, 1994; Rodman et al. 1984). The monophyly of the group has been strongly supported by *rbcL, atpB,* and 18S rDNA sequences (Chase et al. 1993; Soltis et al. 1997, 2000; Källersjö et al. 1998; Savolainen et al. 2000a,b). However, basal to the core Caryophyllales are three small families, including the monotypic Simmondsiaceae, from the southwestern United States and adjacent Mexico. Their morphological characters, so far as known, differ in part from those of the core families, but stems with successive cambia (concentric rings of xylem and phloem or concentric rings of vascular bundles) and unilacunar nodes may be synapomorphic for the order. These families also typically have separate styles or well-developed style branches with elongated stigmas. The Caryophyllales consists of 18 families and 8600 species; major families include **Aizoaceae, Caryophyllaceae, Phytolaccaceae,** Petiveriaceae, **Nyctaginaceae, Amaranthaceae** (including Chenopodiaceae), **Cactaceae,** and **Portulacaceae.** The order is most closely related to Polygonales, as indicated by morphology, chloroplast DNA (both structural rearrangements and *rbcL* sequences), and 18S rDNA sequences.

Phylogenetic relationships within core Caryophyllales have been much studied but are still poorly known (Downie et al. 1997; Downie and Palmer 1993a,b; Manhart and Rettig 1994; Rettig et al. 1992; Rodman et al. 1984; Rodman 1990, 1994) (Figure 9.50). Family limits are unclear, and many critical taxa are poorly known. Caryophyllaceae have often been considered the sister family to the remaining members of the order because they possess anthocyanins (as red or yellow pigments), while the other members of the order have betalains (Ehrendorfer 1976; Mabry 1973, 1976; Mabry et al. 1972). If this hypothesis is true, then betalains are not a synapomorphy of the Caryophyllales, and instead diagnose a clade containing all members of the order except Caryophyllaceae and Molluginaceae. Recent cladistic analyses (see references) cast doubt on this hypothesis, and suggest that Caryophyllaceae evolved from betalain-containing ancestors, and have lost betalains (along with the reacquisition of anthocyanins, the two changes possibly being biochemically interrelated; see Downie and Palmer 1994a,b). Phytolaccaceae, Petiveriaceae, Nyctaginaceae, and Aizoaceae may form one clade, supported by the presence of raphide crystals and characteristics of the chloroplast genome, and Portulacaceae and Cactaceae form another clade, based on the presence of

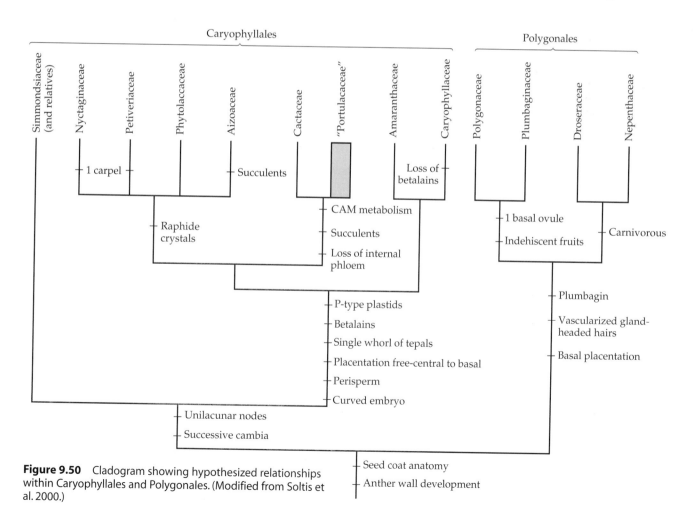

Figure 9.50 Cladogram showing hypothesized relationships within Caryophyllales and Polygonales. (Modified from Soltis et al. 2000.)

crassulacean acid metabolism (CAM), and the succulent habit. However, succulent leaves have also evolved in the Aizoaceae.

References: Behnke 1976, 1994; Behnke and Mabry 1994; Bittrich 1993b; Cronquist 1981, 1988; Downie and Palmer 1994a,b; Downie et al. 1997; Eckardt 1976; Ehrendorfer 1976; Källersjö et al. 1998; Mabry 1973, 1976; Mabry et al. 1972; Manhart and Rettig 1994; Rettig et al. 1992; Rodman 1990, 1994; Rodman et al. 1984; Savolainen et al. 2000a, b; Soltis et al. 1997, 2000.

Caryophyllaceae A. L. de Jussieu
(Carnation or Pink Family)

Usually herbs; stems sometimes with concentric rings of xylem and phloem; **anthocyanins present**; often with triterpenoid saponins. Hairs various. **Leaves opposite**, *simple, entire, often narrow, with pinnate venation, the secondary veins usually obscure and venation appearing ± parallel, the leaf pair often connected by a transverse nodal line, and nodes usually swollen;* stipules lacking or present. *Inflorescences determinate,* sometimes reduced to a single flower, *terminal.* Flowers usually bisexual, radial, sometimes with an androgynophore. *Tepals 4–5, distinct to connate,* imbricate, usually appearing to be sepals. *True petals lacking, but outer whorl of 4–5 stamens very often petal-like, here called "petals", these frequently bilobed, and sometimes differentiated into a long, thin, basal portion (claw) and an expanded apical portion (blade or limb) separated by appendaged joint.* Stamens 4–10; filaments distinct or slightly connate, sometimes adnate to "petals;" pollen grains tricolpate to polyporate. Carpels 2–5, connate; *ovary superior, with free-central* or occasionally basal *placentation;* stigmas minute to linear. Ovules usually numerous, occasionally few or only 1, ± campylotropous. Nectar produced by disk or staminal bases. *Fruit usually a loculicidal capsule, opening by valves or apical teeth,* but sometimes a utricle; embryo usually curved; endosperm ± lacking, replaced by perisperm (Figure 9.51).

Floral flormula:

$*, \widehat{4-5}, 4-5, \widehat{4-10}, \underline{\widehat{2-5}};$ capsule, utricle

Distribution and ecology: Widespread, but especially characteristic of temperate and warm temperate regions of the Northern Hemisphere, mostly of open habitats or disturbed sites.

Genera/species: 70/2200. **Major genera:** *Silene* (700 spp.), *Dianthus* (300), *Arenaria* (200), *Gypsophila* (150), *Minuartia* (150), *Stellaria* (150), *Paronychia* (110), and *Cerastium* (100). Numerous native and introduced genera occur in the continental United States and/or Canada; some of these, in addition to most of the above, include *Agrostemma, Drymaria, Geocarpon, Sagina, Saponaria, Spergula,* and *Stipulicida.*

Economic plants and products: The family is best known for ornamentals such as *Dianthus* (carnations, pinks), *Gypsophila* (baby's breath), *Saponaria* (soapwort), and *Silene* (catchfly, campion).

Discussion: The monophyly of Caryophyllaceae is supported by morphology, *rbcL,* and ORF2280 sequence characters. The family typically is divided into three subfamilies. The "Paronychioideae" (e.g., *Paronychia, Stipulicida, Spergula,* and *Spergularia*) probably are a paraphyletic, heterogeneous assemblage defined only by the presence of stipules (a likely plesiomorphy within Caryophyllaceae). Some members of this group have petals while others (e.g., *Paronychia*) lack them. The petaloid members of the "Paronychioideae" along with the "Alsinoideae" and Caryophylloideae probably constitute a monophyletic group, which may be diagnosed by the presence of often bilobed petals (Lüders 1907). Caryophylloideae and "Alsinoideae" also differ from most "Paronychioideae" in their embryo development and their basally connate leaves. Within this clade, the "Alsinoideae" (e.g., *Arenaria, Minuartia, Stellaria, Cerastium,* and *Sagina*) constitute a paraphyletic complex characterized by the symplesiomorphies of free sepals and nonjointed petals, while the Caryophylloideae (e.g., *Silene, Saponaria, Dianthus,* and *Gypsophila*) form a clade, based on their connate sepals and usually clawed and jointed petals.

Flowers of Caryophyllaceae are pollinated by various insects (flies, bees, butterflies, and moths) that gather nectar. Protandry leads to outcrossing in most species, but many of the weedy species have inconspicuous flowers, a reduced number of stamens, and are self-pollinated. The small or winged seeds of most species are shaken from the erect capsules by wind or passing animals. The dry utricles (associated with the persistent tepals) of *Paronychia* are probably also wind-dispersed. Sometimes the deciduous fruit cluster or the entire plant acts like a tumbleweed. Species such as *Sagina decumbens* have capsules that open widely only when wetted and have seeds that are dispersed by raindrops. Others are probably animal-dispersed, either by external transport or accidental ingestion as the entire plant is eaten.

References: Bittrich 1993a; Cronquist 1981; Lüders 1907; Thomson 1942.

Phytolaccaceae R. Brown
(Pokeweed Family)

Usually *herbs;* stem with concentric rings of vascular bundles or alternating concentric rings of xylem and phloem; betalains present; triterpenoid saponins present; raphide crystals (calcium oxalate) usually present. Hairs usually simple. *Leaves alternate and spiral, simple, entire, with pinnate venation;* stipules lacking. **Inflorescences indeterminate (racemes or spikes),** *terminal, but usually*

Figure 9.51 Caryophyllaceae. (A–J) *Silene virginica:* (A) flowering plant (× 0.33); (B) flower (× 3); (C) calyx lobes (× 3); (D) petal with stamen, note appendage at top of clawed portion of petal (× 4); (E) cross-section of flower through middle of ovary (× 5); (F) longitudinal section through base of flower, note hairy calyx, androgynophore, petals, stamen (on right), and free-central placenta (× 5); (G) ovule (× 25); (H) nearly mature fruit and calyx in longitudinal section, note seeds on free-central placenta (× 5); (I) seed (× 30); (J) seed in section, note curved embryo and perisperm (stippled) (× 20). (K) *S. caroliniana:* capsule (× 3). (L) *S. ovata:* petal (× 3). (M–N) *S. antirrhina:* (M) capsule, surrounded by dried calyx (× 3); (N) capsule in longitudinal section (× 3). (From Wood 1974, *A student's atlas of flowering plants*, p. 27.)

appearing lateral (and opposite the leaves). Flowers usually bisexual, radial. *Tepals usually 5 and distinct*, imbricate. *Stamens 10 to numerous*; filaments distinct to slightly connate; pollen grains tricolpate. *Carpels 3 to numerous, clearly to slightly connate*, or occasionally distinct; *ovary superior, usually with axile placentation; styles ± distinct*; stigmas ± linear. *Ovules 1 in each locule*, campylotropous. Nectar disk often present. **Fruit a berry**; embryo curved; endosperm lacking, replaced by perisperm (Figure 9.52.

Floral formula: *, -5- , ⟨10–∞⟩ , ⟨3–∞⟩ ; berry

Distribution and ecology: Widespread in tropical and warm temperate regions; characteristically early successional, sometimes with seeds that remain viable in the soil for several decades.

Genera/species: 4/30. **Major genus:** *Phytolacca* (25 spp.).

Economic plants and products: *Phytolacca* (pokeweed, pokeberry) is poisonous, containing a diverse array of chemicals, including single-chain to polymeric mitogens; ingestion or even contact with the sap should be avoided, but the young leaves are sometimes eaten after repeated boiling.

Discussion: Phytolaccaceae, as here circumscribed, are probably monophyletic. The family traditionally has been defined much more broadly, resulting in a polyphyletic assemblage that is difficult to characterize (Brown and Varadarajan 1985; Manhart and Rettig 1994; Rodman 1994). Taxa with four tepals, four (to numerous) stamens, a single carpel (with a single basal ovule), and

Figure 9.52 Phytolaccaceae. *Phytolacca americana* var. *americana*: (A) branch with flowers and immature fruits (× 0.7); (B) flower, stigmas not yet receptive (× 12); (C) gynoecium with receptive stigmas, in longitudinal section (× 20); (D) gynoecium in cross-section, with an ovule filling each locule (× 15); (E) ovule (× 1.5); (F) berry (× 4); (G) seed (× 7); (H) seed in section, note curved embryo and perisperm (netlike) (× 12). (From Rogers 1985, *J. Arnold Arbor.* 66: p. 12.)

drupes, achenes, or samaras, may be best segregated as Petiveriaceae (e.g., *Petiveria*, *Rivina*, *Trichostigma*). Petiveriaceae also differ from Phytolaccaceae in embryological and pollen characters. *Stegnosperma*, which has capsules, arillate seeds, and flowers with petaloid staminodes, is placed in its own family. Brown and Varadarajan (1985) placed *Gisekia* within Phytolaccaceae, as sister to the remaining genera, but *rbcL* sequences suggest that this genus is more closely related to Petiveriaceae.

Flowers of Phytolaccaceae attract bees, wasps, flies, and butterflies. The red to purple-black fruits often contrast with the bright red inflorescence axis and pedicels, and are dispersed by birds. The long-lived seeds probably allow species of *Phytolacca* to take advantage of temporally separated disturbance events at a single site.

References: Brown and Varadarajan 1985; Manhart and Rettig 1994; Rodman 1994; Rogers 1985; Rohwer 1993c.

Nyctaginaceae A. L. de Jussieu
(Four O'Clock Family)

Herbs, shrubs, or trees, usually with concentric rings of vascular bundles or alternating concentric rings of xylem and phloem; betalains present; raphide crystals (calcium oxalate) usually present. Hairs various. **Leaves usually opposite**, *simple, entire, with pinnate venation*; stipules lacking. *Inflorescences determinate*, terminal or axillary. *Flowers* usually bisexual and radial, *often associated with conspicuous bracts that may be sepal- or petal-like.* **Tepals** usually 5, **connate, forming a distinct tube**, induplicate-valvate to plicate or contorted, **the proximal portion of tube (persistent and green) differentiated from the distal (usually petal-like) portion**. Stamens usually 5; filaments distinct or slightly connate; pollen grains tricolpate to polyporate. *Carpel 1; ovary superior (but often appearing inferior because of its close association with the lower, often constricted, portion of the perianth tube)*, with basal placentation; stigma capitate. *Ovule 1*, usually campylotropous. Nectar disk present. **Fruit an achene or nut, enclosed in persistent, leathery to fleshy, basal portion of perianth tube,** *which may have 5 lines of sticky, gland-headed or hooked hairs, and appearing drupaceous*; embryo usually curved; endosperm ± lacking, replaced by perisperm (Figure 9.53).

Floral formula:

*, $\widehat{(5)}$, $\widehat{(5)}$, $\underline{1}$; achene, nut (with accrescent tepals)

Distribution: Widespread in tropical and subtropical regions.

Genera/species: 31 / 350. **Major genera:** *Neea* (80 spp.), *Guapira* (70), *Mirabilis* (50), *Pisonia* (40), *Abronia* (30), and *Boerhavia* (20). All but *Neea*, along with *Acleisanthes*, *Anulocaulis*, *Okenia*, and *Selinocarpus*, occur in the continental United States and/or Canada.

Economic plants and products: Several species of *Bougainvillea*, *Mirabilis* (four o'clock), and *Abronia* (sand verbena) are cultivated as ornamentals.

Discussion: Several tribes of Nyctaginaceae are recognized on the basis of variation in habit, leaf arrangement, bract formation, pubescence, stamen connation, stigma form, pollen morphology, and embryo shape.

Flowers of Nyctaginaceae attract a diversity of pollinators (bees, butterflies, moths, and birds); nectar is usually the pollinator reward. Species with fleshy, colorful, accrescent tepals usually are dispersed by birds. External transport occurs in species with sticky glands or hooked hairs on their accrescent tepals. Fruits of *Okenia* are pushed into the ground by rapid elongation of the pedicel. The colorful bracts of *Bougainvillea* become dry and papery in fruit, leading to wind dispersal of the associated achene.

References: Bittrich and Kuhn 1993; Bogle 1974; Cronquist 1981.

Amaranthaceae Adanson
(Amaranth Family)

Usually herbs or suffrutescent shrubs, sometimes succulent; usually with concentric rings of vascular bundles; betalains present; occasionally with C_4 photosynthesis; **plastids of the sieve elements with a ± peripheral ring of proteinaceous filaments, but without a central protein crystal**. Hairs simple to branched. *Leaves alternate and spiral, or opposite, simple*, usually entire or undulate, sometimes serrate or lobed, with pinnate venation, but veins often obscure, sometimes succulent; *stipules lacking*; nodes sometimes swollen. *Inflorescences determinate, terminal and axillary.* Flowers bisexual or, less commonly, unisexual (plants then monoecious to dioecious), radial, associated with fleshy to dry, papery bracts and/or bractlets, and often densely clustered. *Tepals usually 3–5*, distinct to slightly connate, *green and herbaceous or fleshy, to white (or reddish), dry, and papery*, imbricate. *Stamens 3–5, opposite tepals; filaments distinct, slightly to strongly connate*; anthers 2- or only 1-locular; **pollen grains 7-porate to polyporate, with pores scattered over the surface of the grain.** *Carpels usually 2 or 3, connate; ovary usually superior, with basal placentation*; stigmas 1 to 3, elongate to capitate. *Ovules 1 to few*, usually campylotropous. Nectar disk or glands often present. *Fruit usually an achene, utricle, or a circumscissile capsule (pyxis)*, usually associated with persistent, fleshy to dry, perianth and/or bractlets; embryo curved to spirally twisted; endosperm ± lacking, replaced by perisperm (Figures 9.54 and Figure 4.47E).

Floral formula:

*, $\widehat{3-5}$, $\widehat{(5)}$, $\underline{\widehat{2-3}}$; achene, utricle, 1-seeded capsule

Figure 9.53 Nyctaginaceae. (A–H) *Boerhavia diffusa:* (A) flowering stem (× 1); (B) two flowers (× 17); (C) flower in longitudinal section (× 22); (D) gynoecium and one of four stamens (× 36); (E) accrescent perianth base, enclosing fruit (× 17); (F) achene (× 17); (G) perianth base and fruit in longitudinal section, the embryo white and perisperm stippled (× 17); (H) same, in cross-section, the cotyledons above and hypocotyl below (× 22). (I) *B. erecta:* accrescent perianth enclosing fruit (× 17). (From Bogle 1974, *J. Arnold Arbor.* 55: p. 24.)

Distribution and ecology: Cosmopolitan and especially characteristic of disturbed, arid, or saline habitats.

Genera/species: 169/2360. ***Major genera:*** *Atriplex* (300 spp.), *Gomphrena* (120), *Salsola* (120), *Alternanthera* (100), *Chenopodium* (100), *Ptilotus* (100), *Suaeda* (100), *Iresine* (80), *Amaranthus* (60), *Corispermum* (60), and *Celosia* (50). Numerous genera have native or naturalized species within the continental United States and/or Canada; some of these include *Alternanthera, Amaranthus, Atriplex, Blutaparon, Celosia, Chenopodium, Froelichia, Iresine, Gomphrena, Grayia, Monolepis, Nitrophila, Salicornia, Salsola,* and *Suaeda.*

Economic plants and products: The leaves and/or roots of a few species, such as *Beta vulgaris* (beet, Swiss chard),

Spinacia oleracea (spinach), *Chenopodium* spp. (lamb's-quarters, goosefoot), and *Amaranthus* spp. (pigweed), are eaten. The seeds of several South American species of *Chenopodium* and *Amaranthus* are used to make flour. The family includes a few ornamentals, including *Celosia* (cockscomb), *Gomphrena* (globe amaranth), and *Iresine* (bloodleaf).

Discussion: Amaranthaceae are here broadly defined and include the Chenopodiaceae, which usually have been maintained as a separate family because of their usually distinct (vs. slightly to completely connate) stamens and greenish and membranous to fleshy (vs. white, white with green stripes, to pink or red, and dry/papery) tepals. The monophyly of Amaranthaceae, as broadly circumscribed, has been strongly supported by chloroplast

Figure 9.54 Amaranthaceae. (A–H) *Chenopodium album:* (A) tip of flowering shoot (× 0.5); (B) leaf (× 1); (C) tip of inflorescence branch (× 12); (D) opening flower, staminal filaments not yet elongated (× 25); (E) fruit, from above (× 10); (F) fruit, side view (× 10); (G) seed (× 10); (H) seed in section, note curved embryo surrounding perisperm (stippled) (× 10). (I–P) *C. ambrosioides:* (I) flowering shoot (× 0.5); (J) bisexual flower (× 25); (K) ovule (greatly magnified); (L) carpellate flower (× 15); (M) accrescent perianth enclosing fruit (× 10); (N) fruit (× 10); (O) seed, oriented as in fruit (× 10); (P) seed in section, perisperm stippled (× 10). (From Wood 1974, *A student's atlas of flowering plants*, p. 23.)

DNA restriction sites, *rbcL* sequences, ORF2280 sequences, and morphology (Rodman 1990, 1994; Rodman et al. 1984; Downie and Palmer 1994a,b; Manhart and Rettig 1994; Downie et al. 1997). Separation of Chenopodiaceae from Amaranthaceae likely is arbitrary and results in a paraphyletic Chenopodiaceae (Downie et al. 1997; Rodman 1990, 1994; Manhart and Rettig 1994).

Currently recognized subfamilies and tribes stress ovule number, embryo shape, number of anther locules, and perianth form. Many of these groups are probably not monophyletic. Genera with papery tepals and bractlets and monadelphous stamens—*Celosia, Iresine, Froelichia, Alternanthera, Blutaparon, Gomphrena,* and *Alternanthera*—may form a clade, which probably does not include *Amaranthus*. Within this group, genera with unilocular anthers—*Froelichia, Alternanthera, Gomphrena,* and *Iresine*—constitute a subclade. Greenish, often fleshy tepals or bracts and separate stamens characterize *Atriplex, Chenopodium, Salsola, Salicornia, Suaeda, Beta,* and *Spinacia*, a possibly paraphyletic complex. Spirally twisted embryos and reduced perisperm suggest that *Salsola, Suaeda,* and relatives form a monophyletic group. Hybridization and polyploidy are common in some genera, leading to taxonomic difficulties at the species level.

The usually small, densely clustered flowers of Amaranthaceae are pollinated by wind or by various insects; selfing as well as outcrossing may occur. The small, dry fruits or seeds, which are typically associated with the accrescent and sometimes hairy perianth, usually are dispersed by wind or water. *Salsola* includes tumbleweeds. A few species form burrlike inflorescence units that are externally transported by animals. In *Amaranthus* and *Celosia* the small seeds tend to fall from the parent plant but germinate only when the site is again disturbed. Many seeds are accidentally eaten and dispersed by browsing animals.

References: Blackwell 1977; Carolin 1983; Carolin et al. 1975; Cronquist 1981; Downie and Palmer 1994a,b; Downie et al. 1997; Judd and Ferguson 1999; Kühn et al. 1993; Manhart and Rettig 1994; Robertson 1981; Rodman 1990, 1994; Rodman et al. 1984; Townsend 1993.

Aizoaceae Martynov

(Stone Plant Family)

Succulent herbs; stem often with concentric rings of vascular bundles or alternating concentric rings of xylem and phloem; betalains present; alkaloids often present; raphide crystals (calcium oxalate) often present; usually with crassulacean acid metabolism (CAM) or C_4 photosynthesis; phytoferritin in phloem. Hairs various. *Leaves usually opposite, simple, usually entire and succulent (with clear cells in center of blade)*, with venation pinnate, but veins ± obscure, **epidermis (and stem) with many large, bladderlike cells**; stipules usually lacking. Inflorescences determinate, sometimes reduced to a single flower, terminal or axillary. Flowers usually bisexual, radial, **with a hypanthium**. *Tepals usually 5, ± connate, imbricate. Stamens usually 5 to numerous, the outer often elongate petal-like staminodes*; filaments distinct to slightly connate; pollen grains usually tricolpate. Carpels usually 2–5, connate; *ovary superior to inferior*, **with usually axile placentation**; stigmas ± linear. Ovules 1 to numerous in each locule, ± anatropous to campylotropous. Nectar disk usually present. *Fruit usually a loculicidal, septicidal, or circumscissile capsule*, sometimes fleshy; seeds sometimes arillate; embryo curved; endosperm lacking, replaced by perisperm.

Floral formula: *, ⑤ , 5–∞ + ∞•, ③–⑤ ; capsule

Distribution and ecology: Widespread in tropical and subtropical regions; primarily of coastal or arid habitats.

Genera/species: 127/2500. **Major genera:** *Conophytum* (290 spp.), *Delosperma* (150), *Lampranthus* (150), *Drosanthemum* (100), *Antimima* (60), *Lithops* (35), *Mesembryanthemum* (30), and *Carpobrotus* (30). Some genera occurring in the continental United States are *Carpobrotus, Cryophytum, Cypselea, Galenia, Sesuvium, Tetragonia,* and *Trianthema*.

Economic plants and products: The family contains numerous ornamentals, such as *Lampranthus, Dorotheanthus, Mesembryanthemum* (ice plant), *Ruschia,* and *Carpobrotus*. Some, such as *Lithops* (stone plants), are grown as curiosities. *Tetragonia* is used as a vegetable.

Discussion: The family appears to be monophyletic. Bittrich and Hartmann (1988) have investigated infrafamilial relationships. The species with numerous stamens, the outermost of which are petal-like staminodes (i.e., Mesembryanthemoideae and Ruschioideae), form a monophyletic group (Hartmann 1993), which has explosively diversified in South Africa and Australia. *Sesuvium* and relatives (i.e., Sesuvioideae) probably form a clade on the basis of their circumscissile capsules and arillate seeds. Generic limits have varied widely, with some systematists considering nearly all species within the large genus *Mesembryanthemum*.

Aizoaceae show several adaptations to extremely arid habitats. Bladderlike cells in the leaf (and stem) epidermis hold water, and the leaves themselves are succulent. Sometimes the plant is reduced to a single pair of opposite, nearly spherical leaves. In some genera (e.g., *Lithops*), only a clear "window" is exposed above the soil, with the chlorophyll-containing cells restricted to a thin layer along the sides or base of the nearly cylindrical leaves. These leaves are actually bracteoles, the adult plant evidently being a long-lived inflorescence. These and other bizarre features are probably adaptations to cope with intense sunlight.

The usually showy flowers attract bees, wasps, butterflies, flies, and beetles. The small seeds are usually dispersed by wind or water; in some genera the capsules open only when wetted.

References: Bittrich and Hartmann 1988; Bittrich and Struck 1989; Bogle 1970; Cronquist 1981; Hartmann 1993.

"Portulacaceae" Adanson
(Purslane Family)

Usually ± succulent herbs; mucilage cells common; betalains present; sometimes with crassulacean acid metabolism (CAM); phytoferritin present in phloem. Hairs usually simple. *Leaves opposite or alternate and spiral, simple, entire, with pinnate venation, veins ± obscure; stipules usually present, often scarious or tufts of short to elongate hairs.* Inflorescences determinate, sometimes appearing indeterminate or reduced to a single flower, terminal or axillary. Flowers usually bisexual and radial, *associated with 4 bracteoles, the inner 2 calyx-like. Tepals usually 4–6, occasionally numerous, petal-like,* distinct to slightly connate, imbricate. *Stamens usually 4–6, opposite tepals,* but sometimes fewer *or more numerous;* filaments distinct or slightly adnate to tepals; pollen grains tricolpate to polycolpate or polyporate. Carpels usually 2 or 3, connate; ovary superior to ± inferior, with ± free-central to basal placentation; stigmas usually linear. Ovules numerous to 1 per gynoecium, anatropous to campylotropous. Individual nectaries or a nectar disk. *Fruit usually a loculicidal or circumscissile capsule;* seeds sometimes arillate; embryo curved; endosperm lacking, replaced by perisperm (Figure 9.55).

Floral formula:

*, 2[bracts], (4–6 (–∞)), 4–∞ , (2–3) ; capsule

Distribution: Widely distributed in tropical and temperate regions; especially diverse in western North America and the Andes of South America.

Genera/species: 19/450. **Major genera:** *Calandrinia* (125 spp.), *Portulaca* (50), *Claytonia* (35), and *Talinum* (30). Several genera are common in the continental United States and/or Canada, including those listed above and *Calyptridium, Montia,* and *Lewisia.*

Economic plants and products: The leaves and young stems of *Portulaca oleracea* (purslane) are occasionally eaten. *Portulaca, Talinum* (flame flower), *Lewisia* (bitterroot), and *Calandrinia* (rock purslane) provide ornamentals.

Discussion: The monophyly of Portulacaceae has often been questioned, either in relation to the separation of two small families, Basellaceae and Didiereaceae (morphological data; Rodman 1990, 1994); the separation of the Cactaceae (ITS sequences; Hershkovitz and Zimmer 1997), or the divergent placement of *Portulaca* and *Clay-*

tonia (cpDNA restriction sites; Downie and Palmer 1994a,b). A recent cladistic analysis based on *ndhF* sequences suggests very strongly that the family is paraphyletic, although many details regarding relationships within it have not yet been clarified (Applequist and Wallace 2001). The Portulacaceae may eventually be separable into three families—an *Anacampseros* + *Portulaca* + *Talinum* clade (incl. Cactaceae, diagnosed by fruit wall strongly differentiated into two layers and most with axillary hairs), a *Portulacaria* + *Ceraria* clade (incl. Didiereaceae, mainly dioecious, woody plants with single-seeded, usually indehiscent fruits), and a *Cistanthe* + *Claytonia* + *Montia* + *Lewisia* + *Phemeranthus* + *Calandrinia* clade (herbs with clasping leaves). Basellaceae may represent yet another lineage. More study of this problematic group is required; the traditional circumscription of the family is retained here on the basis of flowers closely associated with a pair (or occasionally more) of sepal-like bracteoles.

Infrafamilial relationships have been investigated by Carolin (1987) in a morphology-based cladistic analysis. *Calandrinia* is probably paraphyletic. *Portulaca* and relatives may form a clade on the basis of their axillary hairs or scales (modified stipules), and the genus itself is noteworthy due to its half-inferior ovary.

Several genera (e.g., *Claytonia, Lewisia, Montia, Portulaca*) show a wide range of chromosome numbers, suggesting a complex evolutionary history through development of polyploidy followed by the formation of aneuploids (see Chapter 4).

Flowers of Portulacaceae usually open only in full sunlight, and then only for a short time; bees, flies, beetles, and butterflies visit the flowers to gather nectar. Time and duration of flowering often differ among related species (reproductive isolation). The tiny seeds are dispersed by wind and water. Those with arils may be ant-dispersed.

References: Applequist and Wallace 2001; Bogle 1969; Carolin 1987, 1993; Cronquist 1981; Downie and Palmer 1994a,b; Hershkovitz and Zimmer 1997; Manhart and Rettig 1994; Nyananyo 1990; Rodman 1990, 1994; Rodman et al. 1984.

Cactaceae A. L. de Jussieu
(Cactus Family)

Spiny *stem succulents (herbs to trees),* **shoots differentiated,** *with usually succulent (and cylindrical, conical, globose, or flattened, often ridged or jointed)* **long shoots producing photosynthetic leaves** *(although these are usually reduced and quickly falling), or leaves lacking,* **and short shoots (areoles) producing a spine or spine cluster,** *and often irritating hairs (glochids);* with crassulacean acid metabolism (CAM); phytoferritin present in phloem; often with alkaloids or triterpenoid saponins; betalains present. *Leaves of long shoots alternate and spiral, simple, entire, with pinnate or*

Figure 9.55 Portulacaceae. (A–H) *Portulaca oleracea:* (A) flowering and fruiting branch (× 0.75); (B) flower in longitudinal section (× 9); (C) nearly mature fruit, enclosed above by accrescent sepals (× 4.5); (D) same, with sepals removed (× 7); (E) base of circumscissile capsule after dehiscence, with basal funiculi (× 7); (F) upper part of fruit after dehiscence, with mature accrescent sepals (× 7); (G) seed (× 35); (H) curved embryo (× 35). (I–M) *P. pilosa:* (I) flowering and fruiting branch (× 0.5); (J) flower bud (× 12); (K) flower (× 6); (L) withered perianth adhering to upper part of circumscissile capsule (× 1.5); (M) base of capsule after dehiscence, with seed attached to placenta (× 1.5). (From Bogle 1969, *J. Arnold Arbor.* 50: p. 572.)

obscure venation, usually reduced to lacking; **leaves of short shoots modified into spines**; stipules lacking. *Inflorescences* determinate, but usually *reduced to a single flower, terminal, but with flowers sunken into the apex of a modified branch (and thus appearing to be axillary).* Flowers usually bisexual, radial to slightly bilateral, **with short to elongate hypanthium**, *and almost always developmentally modified, with ovary surrounded by apex of modified stem, thus outer portion of ovary and hypanthium with spine-bearing areoles.* **Tepals numerous and spirally arranged,** *usually distinct, all petal-like, or these often intergrading with outer, sepal-like tepals,* imbricate. **Stamens numerous**; pollen grains tricolpate to polycolpate or polyporate. Carpels 3 to numerous, connate; *ovary almost always inferior,* but in some species of *Pereskia* only half-inferior or even superior, *almost always with parietal placentation,* but in *Pereskia* ± basal; stigmas 3 to numerous, elongated and radiating. Ovules numerous, usually campylotropous. Nectary a ring on inner surface of hypanthium. **Fruit a berry,** *outer portion with nodes and internodes, and usually with spines and/or glochids at areoles;* seed sometimes covered by a bony aril; embryo usually curved; endosperm lacking, but perisperm sometimes present (Figure 9.56 and Figure 4.47F).

Floral formula: * or X, -∞-, ∞, ⎯③-∞⎯; berry

Distribution and ecology: Mainly in North and South America, but *Rhipsalis* occurs in tropical Africa, and several species of *Opuntia* have been introduced in Africa, Australia, and India. Plants typically of deserts and other arid habitats, but sometimes epiphytic in tropical forests.

Genera/species: 93/1400. ***Major genera:*** *Opuntia* (180 spp.), *Mammillaria* (170), *Echinopsis* (70), *Cleistocactus* (50), *Echinocereus* (50), *Rhipsalis* (50), and *Cereus* (40, or many more if various segregate genera are included). Several genera occur in the continental United States and/or Canada, including *Acanthocereus, Consolea, Cylindropuntia, Echinocactus, Echinocereus, Ferocactus, Harrisia, Lophophora, Mammillaria, Neolloydia, Opuntia, Pediocactus, Pilosocereus, Sclerocactus,* and *Thelocactus.*

Economic plants and products: The fruits of several species of *Opuntia* (prickly pear) are eaten. Nearly all genera are cultivated as ornamentals; some of the more common are *Opuntia, Carnegia* (giant saguaro), *Cereus* (hedge cactus, cereus), *Echinopsis* (sea-urchin cactus), *Epiphyllum* (orchid cactus), *Hylocereus* (night-blooming cereus), *Mammillaria* (pincushion cactus), *Melocactus* (Turk's-cap cactus), *Rhipsalis* (mistletoe cactus), and *Schlumbergera* (Christmas cactus). *Lophophora* contains mescaline alkaloids and is hallucinogenic.

Discussion: The family's monophyly is supported by numerous morphological characters and a 6 kb cpDNA inversion (R. Wallace, personal communication). It is likely that *Pereskia* is sister to all other Cactaceae. *Pereskia* retains numerous plesiomorphic character states: nonsucculent stems, well-developed and persistent leaves, cymose inflorescences, several styles, and in at least some species, superior ovaries with basal placentation. These character states are absent in all other members of the family. Most of the genera of Cactaceae are placed in Opuntioideae and Cactoideae, and these form a clade based on solitary flowers that are sunken into stem apices, inferior ovaries, and parietal placentation. The last three features are present in some species of *Pereskia*, and these species may actually be basal members of this clade; the monophyly of *Pereskia* is doubtful (Nyffeler 2002).

The Opuntioideae (i.e., *Opuntia, Consolea,* and relatives) are monophyletic based on the synapomorphies of the presence of glochids (spinelike hairs) on the areoles, the seed coat covered by a bony aril, and cpDNA characters. The monophyly of the Cactoideae, a large complex containing more than three-fourths of the species of Cactaceae, is supported by the extreme reduction or complete loss of leaves, the characteristic hilum region of the seeds, and a deletion of the *rpoC1* intron in the chloroplast genome (Nyffeler 2002; R. Wallace, pers. com.). Many species of Cactoideae have ribbed stems, another apomorphic feature. The family is taxonomically difficult, with problems in delimitation of both species and genera.

The showy flowers of Cactaceae are extremely variable in color and form and are visited by numerous different pollinators, including various insects (bees, flies, sphinx moths), birds, and bats, which may be attracted by either pollen or nectar. They are predominantly outcrossing. The berries are dispersed by birds or mammals, but several genera (e.g., *Cereus*) are at least partially dispersed by ants, which are attracted to the fleshy funiculi of the seeds. Fruits of some species are spiny, leading to external transport by mammals; others have jointed stems, which may easily break off when touched.

Cacti show numerous specializations for survival in dry habitats. They have a specialized habit, with dimorphic stems and leaves; some stems are photosynthetic and contain water storage tissue, and others are reduced and bear protective spine-leaves. Their CAM metabolism allows the stomates to open at night (when less water will be lost) in order to take in carbon dioxide, which is then stored as malic acid and used in photosynthesis during the next day. They have a wide-spreading and shallow root system, and sometimes also a deep taproot, as well as stems with a thick cuticle and an epidermis with sunken stomates.

References: Anderson 2001; Barthlott and Hunt 1993; Benson 1982; Boke 1964; Cronquist 1981; Gibson et al. 1986; Leins and Erbar 1994; Nyffeler 2002.

Figure 9.56 Cactaceae. *Opuntia pusilla:* (A) plant with immature fruits (× 0.5); (B) areole with spines and glochids (× 2); (C) single glochid (× 30); (D) flower (× 1); (E) underside of flower, sunken into apex of stem with areoles (× 1); (F) flower in longitudinal section (× 2); (G) stamen (× 14); (H) unexpanded stigmas (× 4); (I) ovary in cross-section (× 2); (J) ovule (greatly magnified); (K) berry (× 1); (L) seed, surrounded by bony aril (× 7); (M) seed in section, note curved embryo (× 7). (From Wood 1974, *A student's atlas of flowering plants*, p. 73.)

Polygonales

Polygonales are here defined broadly, including **Polygonaceae**, Plumbaginaceae, Nepenthaceae, **Droseraceae**, and six other families, and comprising about 2050 species. Evidence from cpDNA, 18S rDNA, cytochrome *c* amino acid sequences, fungal pathogens, and morphology has been used to suggest a relationship between the Caryophyllales and this group, especially Polygonaceae and Plumbaginaceae, and this order is included within Caryophyllales by the Angiosperm Phylogeny Group (1998). The monophyly of Polygonales is supported by *rbcL, atpB,* and 18S sequences (Williams et al. 1994; Soltis et al. 2000); supporting phenotypic features include scattered secretory cells containing plumbagin, a naphthaquinone (but these have been lost in several clades, such as Plumbaginaceae subfamily Staticoideae and Polygonaceae); an indumentum of stalked, vascularized, gland-headed hairs, which are often mucilage-producing (lost in a few clades, such as Polygonaceae); basal placentation (but parietal in some Droseraceae, and axile in Nepenthaceae); and starchy endosperm.

Polygonales may comprise two major clades (Figure 9.50). The first, comprising Polygonaceae and Plumbaginaceae, is supported by ovaries with a single basal ovule and usually indehiscent fruits (achenes or nutlets). The second, which includes Droseraceae and Nepenthaceae, is supported by molecular characters and probably the carnivorous habit as well. Most members of this clade also have circinate leaves and pollen grains in tetrads.

Droseraceae have been placed in the Rosales or "Violales," and some systematists have placed Nepenthaceae, Droseraceae, and Sarraceniaceae together on the basis of their carnivorous habit. The pitcherlike leaves of the Old World family Nepenthaceae are amazingly convergent with those of the New World Sarraceniaceae.

References: Behnke 1999; Chase et al. 1993; Cronquist 1981; Soltis et al. 1997, 2000; Williams et al. 1994.

Droseraceae Salisbury
(Sundew Family)

Insectivorous herbs, sometimes suffrutescent. Hairs stalked, gland-headed, producing mucilage and usually containing xylem. *Leaves* usually alternate and spiral, **adaxially circinate,** simple, with obscure venation, **the blade sensitive** *and forming a snap trap or covered with conspicuous, tentaclelike, mucilage-secreting hairs; insects caught in a trap or by sticky hairs are digested;* stipules present or lacking. Inflorescences determinate, sometimes reduced to a solitary flower, terminal. Flowers bisexual, radial. Sepals usually 5, slightly connate, imbricate. *Petals usually 5, distinct,* convolute. Stamens usually 5, occasionally numerous; filaments distinct or slightly connate; **pollen grains triporate to multiporate,** released in tetrads. *Carpels usually 3, connate; ovary superior, with basal or parietal placentation;* stigmas various. Ovules 3 to numerous. Fruit a loculicidal capsule (Figure 9.57).

Floral formula: *, ⑤, 5, ⑤, ③; capsule

Distribution and ecology: Widely distributed, commonly in wet, low-nutrient, acidic soils.

Genera/species: 4/110. **Major genus:** *Drosera* (80 spp.). *Drosera* and *Dionaea* occur in the continental United States and the former in Canada as well.

Economic plants and products: *Dionaea muscipula* (Venus's flytrap) and various species of *Drosera* (sundew) are occasionally cultivated as novelties.

Discussion: The phylogenetic systematics of Droseraceae has been investigated by Williams et al. (1994). The genus *Drosera* shows the derived condition of parietal placentation.

The white to purple flowers of Droseraceae are usually insect-pollinated; protandry often leads to outcross-

---

Key to Major Families of Polygonales

1. Plants carnivorous, with leaves highly modified and pitcher-like, a snap trap, or covered with sticky hairs 2
1. Plants not carnivorous 3
2. Leaves modified to form pitcher traps; flowers unisexual; filaments completely connate, forming a tube; placentation axile .... Nepenthaceae
2. Leaves modified to form snap traps or variously shaped and covered with sticky, insect-catching hairs; flowers bisexual; filaments distinct or only slightly connate; placentation basal or parietal .... **Droseraceae**
3. Stipules present, fused around stem at the node (i.e., an ocrea); carpels 2 or 3 .... **Polygonaceae**
3. Stipules lacking, or if present then not as above (i.e., lacking an ocrea); carpels 5 .... Plumbaginaceae

Figure 9.57 Droseraceae. (A–E) *Drosera tracyi:* (A) plant (× 0.5) (B) tip of developing leaf, note circinate verna-tion and glandular hairs (× 7); (C) flower (× 3); (D) gynoecium, three-carpellate (× 5); (E) seed (× 50). (F–J) *D. fili-formis:* (F) flower (× 3); (G) gynoecium (× 5); (H) ovary in cross-section (× 15); (I) infructescence (× 1); (J) capsule (× 5). (K–L) *D. capillaris:* (K) flowering plant (× 0.5); (L) seed (× 50). (M) *D. intermedia:* seed (× 50). (N) *D. rotundifo-lia:* seed (× 50). (O) *D. brevifolia:* seed (× 50). (From Wood 1974, *A student's atlas of flowering plants,* p. 45.)

Figure 9.59 Viscaceae. (A–L) *Phoradendron leucarpon:* (A) branch with fruits and flowers (× 1.5); (B) inflorescence (× 12); (C) detail of one fertile internode of staminate inflorescence with open flowers (× 18); (D) staminate flower in longitudinal section, one sessile anther at right, and sterile gynoecium in center (× 75); (E) one tepal with sessile anther (× 75); (F) same, from above, showing porose anther dehiscence (× 75); (G) carpellate inflorescence (× 12); (H) carpellate flower in longitudinal section to show large nectar disk and stigma, region of undifferentiated ovules indicated by circle of dots (× 75); (I) detail of infructescence with two berries (× 9); (J) berry in cross-section, endosperm stippled, embryo in center (× 18); (K) seed (× 18); (L) seed in section, endosperm stippled, embryo unshaded (× 18). (M) *P. rubrum:* tip of staminate plant (× 3). (From Kuijt 1982, *J. Arnold Arbor.* 63: p. 406.)

daceae, Cercidiphyllaceae, Paeoniaceae, **Hamamelidaceae**, and **Altingiaceae**, representing about 2470 species (Chase et al. 1993; Morgan and Soltis 1993; Soltis and Soltis 1997; Soltis et al. 1993, 1997). The group's monophyly is more or less supported by *rbcL, atpB,* and 18S rDNA sequences (Chase et al. 1993; Soltis and Soltis 1997; Soltis et al. 1997, 2000; Hoot et al. 1999; Savolainen et al. 2000b) and morphology (Hufford 1992). The order is clearly a member of the eudicot clade (Morgan and Soltis 1993; Soltis and Soltis 1997; Soltis et al. 1993, 1997, 2000), and may be sister to the remaining members of the rosid clade. DNA evidence suggests that Hamamelidaceae, Cercidiphyllaceae, and Altingiaceae should be included here (Soltis and Soltis 1997; Soltis et al. 2000), although these three families often were considered to be related to Platanaceae

(Schwarzwalder and Dilcher 1991; Hufford and Crane 1989). The order is morphologically similar to Rosales (and especially Rosaceae), but the major family, Saxifragaceae, can be readily separated from Rosaceae by the lack of stipules, fewer stamens, capsular fruits, and seeds possessing well-developed endosperm. The presence of a hypanthium, 2-carpellate gynoecia, and glandular leaf-teeth are possible synapomorphies of Saxifragales; members of this group have retained the plesiomorphies of 5-merous flowers with distinct parts. Many have striate pollen.

References: Chase et al. 1993; Hoot et al. 1999; Hufford 1992; Hufford and Crane 1989; Morgan and Soltis 1993; Savolainen et al. 2000b; Schwarzwalder and Dilcher 1991; Soltis and Soltis 1997; Soltis et al. 1993, 1997, 2000.

Key to Major Families of Saxifragales

1. Plants woody and nonsucculent 2

1. Plants herbaceous and/or succulents 6

2. Leaves of long shoots opposite; fruits follicles; plants dioecious . . . . Cercidiphyllaceae

2. Leaves always alternate; fruits capsules or berries; plants with bisexual flowers, or if unisexual, then plants monoecious 3

3. Resin ducts in stems and leaves, the latter with sweet, resinous odor when crushed; flowers unisexual, carpellate flowers in a globular head, with gynoecia of adjacent flowers maturing to form a globose multiple fruit; each capsule surrounded by numerous minute lobes or scales . . . . **Altingiaceae**

3. Resin ducts absent, leaves without sweet odor; flowers bisexual, or if imperfect, carpellate flowers not as above; fruits not multiple, sometimes associated with a persistent calyx but not surrounded by numerous lobes or scales 4

4. Styles 2; hypanthium ± absent; petals often elongate and coiled; anthers usually opening by flaps; hairs ± stellate . . . . **Hamamelidaceae**

4. Style 1; hypanthium present, short to elongate; petals various but never coiled; anthers opening by slits; hairs not stellate 5

5. Leaves pinnately to palmately veined; flowers with a conspicuous hypanthium extending well above the inferior ovary; placentation parietal; fruits berries . . . . Grossulariaceae

5. Leaves pinnately veined; flowers with a short hypanthium and superior ovary; placentation axile; fruits capsules . . . . Iteaceae

6. Usually submerged to emergent aquatic herbs; flowers inconspicuous, solitary and axillary or in terminal spikes; ovules 1 per carpel; fruits nutlike or drupaceous, sometimes schizocarpic . . . . Haloragaceae

6. Terrestrial herbs or succulents; flowers conspicuous, usually in various determinate or indeterminate inflorescences; ovules few to numerous per carpel; fruits capsules or aggregates of follicles 7

7. Plants succulent; carpels as many as the petals, distinct or united only at base; each carpel subtended by a scale-like nectar gland; hypanthium usually lacking . . . . **Crassulaceae**

7. Plants not succulent; carpels usually fewer than the petals, usually ± connate; each carpel not associated with a nectar gland, although nectar disk sometimes present; hypanthium variously developed . . . . **Saxifragaceae**

Saxifragaceae A. L. de Jussieu
(Saxifrage Family)

Herbs; vessel elements with simple perforations; often with tannins, sometimes cyanogenic. Hairs often simple. Leaves usually alternate and spiral, sometimes in a basal rosette, simple to pinnately or palmately compound, entire to serrate or dentate, with venation pinnate to palmate; **stipules lacking (**or represented by expanded margins of the petiole base). Inflorescences determinate to indeterminate, usually terminal. *Flowers* bisexual to unisexual (plants then monoecious to ± dioecious), radial to bilateral, *with a variously developed hypanthium.* Sepals usually 4 or 5, distinct to connate. *Petals usually 4 or 5, distinct, often clawed,* sometimes variously dissected, imbricate or convolute, sometimes reduced or lacking.

Stamens usually 3–10; pollen grains usually tricolpate or tricolporate. *Carpels 2 (–5),* ± connate or less commonly distinct; *ovary superior to inferior,* with axile or lateral placentation; stigmas separate, capitate. Ovules usually numerous on each placenta, with 1 or 2 integuments. Nectar disk often present around base of ovary. *Fruit a septicidal capsule or follicle* (Figure 9.60).

Floral formula:

; capsule, follicles

Distribution and ecology: Widely distributed in temperate and arctic regions, especially of the Northern Hemisphere, and often found in mountainous terrain.

Figure 9.60 Saxifragaceae. *Mitella diphylla:* (A) flowering plant (× 0.75); (B) detail of raceme (× 4); (C) flower (× 15); (D) flower in longitudinal section (× 17); (E) dehisced anther (× 35); (F) immature capsule (× 9); (G) top view of immature capsule (× 9) (H) floral cup and capsule in cross-section (× 9); (I) erect "splash cup" capsule after dehiscence (× 9); (J) seed (× 17). (From Spongberg 1972, *J. Arnold Arbor.* 53: p. 426.)

Genera/species: 30/550. **Major genera:** *Saxifraga* (325 spp.), *Heuchera* (55), *Chrysosplenium* (55), *Mitella* (20), and *Astilbe* (20). In addition to the above listed genera, noteworthy genera in the colder regions of the continental United States and Canada include *Boykinia*, *Leptarrhena*, *Sullivantia*, *Tellima*, *Tolmiea*, and *Tiarella*.

Economic plants and products: *Saxifraga*, *Astilbe*, and a few other genera are cultivated in rock gardens or perennial borders.

Discussion: Saxifragaceae, as here circumscribed, are considered monophyletic on the basis of cpDNA restriction sites, *rbcL*, *matK*, and 18S sequences (Chase et al. 1993; Johnson and Soltis 1994; Morgan and Soltis 1993; Soltis and Soltis 1997; Soltis et al. 1993, 1997) and morphology. In addition, members of the family share an *rpl2* intron deletion.

Saxifragaceae traditionally have been broadly circumscribed, including both woody and herbaceous taxa with opposite or alternate leaves, and have been impossible to characterize clearly (see Engler 1930; Schulze-Menz 1964). It is now clear that this broadly defined group is polyphyletic (Chase et al. 1993, 1995; Hufford 1992; Morgan and Soltis 1993; Soltis and Soltis 1997; Soltis et al. 1997). Some of the shrubby taxa previously associated with Saxifragaceae (e.g., *Itea* and *Ribes*) probably are closely related to the herbaceous saxifrages (Saxifragaceae, as here delimited), while others (e.g., *Hydrangea* and *Philadelphus*) are related to various groups within the asterid clade (see Hydrangeaceae). Even some herbaceous taxa are not closely related to Saxifragaceae, as recognized here; for example, *Parnassia* and *Lepuropetalon* (Parnassiaceae) are close to Celastraceae!

Until recently, relationships within Saxifragaceae s. s. have been poorly understood, probably as a result of the group's presumed rapid and recent diversification in cold temperate regions. Delimitation of some genera is problematic; many are monotypic, and some may not be monophyletic (e.g., *Saxifraga* and *Mitella*) (Soltis and Kuzoff 1995; Soltis et al. 1993, 1996, 2001). Hybridization sometimes causes taxonomic difficulties, and several intergeneric chloroplast capture events have been strongly supported (Soltis et al. 1991; Soltis and Kuzoff 1995). However, *matK*, *rbcL*, *trnL-F*, *psbA-trnH* nucleotide sequences and cpDNA restriction sites (Johnson and Soltis 1994; Soltis et al. 1991, 1995, 1996, 2001) support recognition of several monophyletic groups of genera within Saxifragaceae.

The flowers of Saxifragaceae are pollinated by various small, short-tongued insects (mainly flies and bees) that gather nectar and/or pollen. The numerous small seeds of *Saxifraga*, *Suksdorfia*, and *Boykinia* are borne on slender stems and probably scattered by wind or passing animals. In *Chrysosplenium* and *Mitella* seeds are bounced from the erect, "splash-cup" capsule by raindrops. Vegetative reproduction by bulbils, vivipary, stolons, and rhizomes also occurs.

References: Chase et al. 1993; Cronquist 1981; Engler 1930; Gornall and Bohm 1985; Hufford 1992; Johnson and Soltis 1994, 1995; Morgan and Soltis 1993; Savile 1975; Schulze-Menz 1964; Soltis and Hufford 2002; Soltis and Kuzoff 1995; Soltis and Soltis 1997; Soltis et al. 1991, 1993, 1995, 1996, 1997, 2001; Spongberg 1972; Thorne 1992.

Crassulaceae J. St. Hilaire
(Stonecrop Family)

Succulent *herbs* to shrubs; stem often with cortical or medullary vascular bundles; **with crassulacean acid metabolism (CAM)**; tannins present; often with alkaloids, sometimes cyanogenic. Hairs simple, but plants more commonly glabrous and glaucous. **Leaves** alternate and spiral, opposite, or whorled, sometimes in a basal rosette, simple or rarely pinnately compound, entire to crenate, dentate or serrate, **succulent**, with pinnate venation, but veins often obscure; *stipules lacking*. Inflorescences determinate, sometimes reduced to a solitary flower, terminal or axillary. Flowers usually bisexual, radial, *lacking a hypanthium*. Sepals usually 4 or 5, distinct to connate. Petals usually 4 or 5, distinct to connate (and then forming a ± tubular corolla), imbricate. *Stamens 4–10*; filaments distinct to slightly connate, free or adnate to corolla; anthers opening by terminal pores; pollen grains tricolporate. *Carpels usually 4 or 5*, distinct *to slightly connate at base; ovaries superior*, with lateral placentation (or axile at base, if carpels fused); stigmas minute. **Each carpel subtended by a scale-like nectar-producing gland**. Ovules few to numerous in each carpel; the megasporangium wall thin or thick. *Fruit an aggregate of follicles*, rarely a capsule (Figure 9.61).

Floral formula: *, (4–5), (4–5), (4–10), 4–5 ; follicles

Distribution and ecology: Widespread from tropical to boreal regions; plants very often of arid habitats.

Genera/species: 35/1500. **Major genera:** *Sedum* (450), *Crassula* (300), *Echeveria* (150), and *Kalanchoe* (125). These, along with *Diamorpha*, *Dudleya*, *Graptopetalum*, *Lenophyllum*, and *Villadia* occur in the continental United States and/or Canada.

Economic plants and products: *Sedum* (stonecrop), *Echeveria*, *Kalanchoe*, and *Sempervivum* (houseleek) are grown as ornamentals because of their distinctive succulent leaves.

Discussion: Analyses of *rbcL* and 18S rDNA sequences, along with morphology, support the monophyly of Crassulaceae (Chase et al. 1993; Morgan and Soltis 1993; Soltis and Soltis 1997; Soltis et al. 1997).

The family consists of two major clades: the Crassuloideae (i.e., *Tillaea* and *Crassula*), diagnosed by their flowers with a single whorl of stamens and ovules with a thin megasporangium, and the Sedoideae (e.g., *Aeonium*,

Figure 9.61 Crassulaceae. (A–D) *Sedum pulchellum:* (A) overwintering rosette (× 0.75); (B) flowering shoot (× 1.5); (C) flower (× 8); (D) cross-section through four carpels of gynoecium (× 30). (E–J) *S. pusillum:* (E) habit of mature plant (× 1.5); (F) immature follicles in longitudinal section, note nectaries (solid black) at base of carpels (× 9); (G) immature follicle (× 1.5); (H) mature, dehisced follicles (× 1.5); (I) seed (× 35); (J) embryo (× 35). (K) *S. glaucophyllum:* leafy shoot (× 1.5). (L) *S. ternatum,* leafy shoot (× 0.75). (M) *S. telephioides:* outline of leaf (× 0.75). (From Spongberg 1978, *J. Arnold Arbor.* 59: p. 206.)

Kalanchoe, Sedum, Echeveria, Villadia, and *Sempervivum)* diagnosed by a usually ridged seed coat. The Sedoideae have retained the plesiomorphic conditions of a thick megasporangium wall and stamens in two whorls. Within the Sedoideae, the *Kalanchoe* clade may represent a basal branch. The *Kalanchoe* group has sympetalous flowers and opposite leaves, while the remaining genera (i.e., the *Sedum* clade) usually have separate petals and alternate leaves. The monophyly of the *Sedum* clade may be supported by its alternate leaves, assuming that the opposite and decussate condition of Crassuloideae and the *Kalanchoe* clade represent the ancestral condition.

Restriction sites and *matK* sequences strongly support the monophyly of Crassuloideae and Sedoideae (van Ham 1994; van Ham and t'Hart 1998; Mort et al. 2001). Several currently recognized genera (especially *Sedum*) are probably polyphyletic (Mort et al. 1998). Hybridization is frequent, and even occurs between members of different genera, which is hardly surprising given the non-monophyly of *Sedum* and some other genera.

The family is of physiological interest because it exhibits crassulacean acid metabolism (CAM), an adaptation for growth in arid habitats. The stomates open primarily at night and close during the day, thus reducing water loss. Carbon fixation occurs in the leaves at night, leading to the formation of malic acid. During the day, when the stomates are closed, the fixed carbon is reduced to carbohydrate. Other adaptations to arid conditions are the succulent leaves (with abundant water storage tissue) and the frequent occurrence of a thick waxy coating covering the epidermis.

Flowers of most Crassulaceae are pollinated by a variety of insects, but some species of *Kalanchoe* have showy, sympetalous corollas and are bird-pollinated. The flowers are often protandrous, leading to outcrossing; but selfing may occur as well. The tiny seeds of Crassulaceae are probably dispersed by wind. Vegetative reproduction is common; some species of *Kalanchoe* produce numerous plantlets in adventitious buds associated with the leaf teeth.

References: Chase et al. 1992; Clausen 1975; Morgan and Soltis 1993; Mort et al. 1998, 2001; Soltis and Soltis 1997; Soltis et al. 1997; Spongberg 1978; van Ham 1994; van Ham and t'Hart 1998.

Hamamelidaceae R. Brown
(Witch Hazel Family)

Shrubs or trees; tannins often present. **Hairs stellate.** *Leaves alternate, often 2-ranked*, simple, entire to serrate, with pinnate or palmate venation; *stipules present, borne on the stem adjacent to the petiole base*. Inflorescences indeterminate, usually forming spikes, racemes, or heads; terminal or axillary. *Flowers bisexual or unisexual (and plants then monoecious), usually radial, showy to inconspicuous.* Sepals usually 4 or 5, distinct to connate, usually imbricate. *Petals usually 4 or 5, distinct, imbricate, valvate, or often circinately coiled in bud, sometimes lacking.* Stamens 4 or 5 and alternating with staminodes, or numerous; *anthers usually opening by 2 flaps*; pollen grains tricolpate to tricolporate. *Carpels 2, at least slightly connate*; **ovary half-inferior to inferior**, with axile placentation; *styles distinct, ± recurved, usually persistent*; stigmas 2, elongate along adaxial surface of style, or capitate. Ovules 1 to several per locule. Nectar sometimes produced by staminodes or inner basal surface of petals. *Fruit a loculicidal to septicidal, woody to leathery capsule*, **with woody exocarp and bony endocarp**; seeds with thick hard testa (Figure 9.62).

Floral formula: *, (4–5), 4–5 or 0 , 4–∞ , –②– ; capsule

Distribution: Scattered in tropical to temperate regions.

Genera/species: 25/80. **Major genera:** *Corylopsis* (20 spp.) and *Distylium* (15). *Hamamelis* and *Fothergilla* occur in the continental United States and/or Canada.

Economic plants and products: Several genera provide ornamental shrubs or trees, including *Hamamelis* (witch hazel), *Corylopsis* (winter hazel), *Distylium*, *Fothergilla* (witch alder), *Loropetalum*, and *Rhodoleia*. An extract of the bark of *Hamamelis* is used as an astringent.

Discussion: The family is variable, with many morphologically distinct genera. Its monophyly has sometimes been questioned (Chase et al. 1993; Schwarzwalder and Dilcher 1991; Soltis and Soltis 1996), and *Liquidambar* and relatives are here segregated as Altingiaceae. The placement of *Rhodoleia*, which has involucrate heads, is also problematic. Altingiaceae are characterized by stipules on the petiole base, resin canals, spiral leaves, unisexual flowers that lack perianth parts, monoecy, elongate anthers, capsules tightly arranged in a globose head, and distinctive pollen grains that have four or more apertures. Persistent floral appendages (possibly perianth parts) surround the fruits. They are easily distinguished from Hamamelidaceae, which have stipules borne on the stem, often 2-ranked leaves, usually bisexual flowers with evident perianth parts, shorter anthers, fruits in spikes, racemes, or nonglobose heads, and usually tricolpate or tricolporate pollen grains. The inclusion of Altingiaceae within Hamamelidaceae would result in a non-monophyletic group (Chase et al. 1993).

Several subfamilies are recognized within Hamamelidaceae, but most species belong to the Hamamelidoideae, a probably monophyletic group (Endress 1989a, 1993; Hufford and Crane 1989; Li et al. 1999). The Hamamelidoideae (e.g., *Corylopsis*, *Distylium*, *Fothergilla*, *Hamamelis*, and *Loropetalum*) may be united by the apomorphies of anthers with flaps, small stigmas, and leaves with pinnate venation with the secondary veins terminating in the teeth. Most also have 2-ranked leaves, carpels with only a single ovule, and ballistic seed dispersal. *Distylium*, *Fothergilla*, and relatives may form a clade characterized by loss of petals and aggregation of flowers into more or less capitate inflorescences (Endress 1989a). A possible clade containing *Hamamelis*, *Loropetalum*, and relatives has peculiar elongate petals that are circinately coiled in the bud (Figure 9.61H) (Endress 1989a), although it is not supported by ITS sequences (Li et al. 1999).

Both wind pollination and insect pollination (mainly by flies and bees) occur within the family (Endress 1977); either nectar or pollen is a pollinator reward. Flowers of the bird-pollinated *Rhodoleia* are aggregated and surrounded by showy inflorescence bracts. The reduced flowers of *Distylium* are wind-pollinated. *Fothergilla* has

Figure 9.62 Hamamelidaceae. (A–J) *Hamamelis vernalis:* (A) branch in spring with flowers
(× 1.5); (B) inflorescence of three flowers, petals removed to show sepals and bracts (× 6);
(C) flower (× 6); (D) flower in longitudinal section, stamen at right removed to show staminode
opposite petal (× 8); (E) two stamens, showing valves (× 22); (F) stamen, showing connective
(× 22); (G) staminodes (× 22); (H) branch in fall with fruits and flower buds (× 0.75); (I) capsule
(× 3); (J) seed (× 6). (K) *H. virginiana:* flower (× 8). (From Ernst 1963, *J. Arnold Arbor.* 44: p. 198.)

showy, inflated filaments and is probably secondarily insect-pollinated. Ballistic dispersal due to the ejection of the seeds is common. Many species grow along streams and are dispersed, at least in part, by water.

References: Bogle 1986; Chase et al. 1993; Endress 1977, 1989a,b, 1993; Ernst 1963b; Hufford and Crane 1989; Hufford and Endress 1989; Li et al. 1999; Manos et al. 1993; Schwarzwalder and Dilcher 1991; Soltis and Soltis 1997; Soltis et al. 1997; Tiffney 1986.

Altingiaceae Lindl.
(Sweet Gum Family)

Shrubs or trees; **secretory canals containing aromatic resinous compounds present in the bark, wood, and leaves**; iridoids present; tannins often present. Hairs simple. *Leaves alternate and spiral*, simple, *often palmately lobed*, entire to serrate, with pinnate or palmate venation; *stipules present*, **borne on the petiole base**. Inflorescences indeterminate, **the staminate in terminal racemes of globose stamen clusters, the carpellate in a globose head. Flowers unisexual (and plants monoecious)**, *radial, inconspicuous*. **Tepals (or floral appendages) only in carpellate flowers, then numerous**, represented by minute lobes or elongate scales. **Stamens numerous**; anthers elongate, opening by slits; **pollen grains polyporate**. *Carpels 2, slightly connate*; **ovary ± half-inferior**, with axile placentation; *styles distinct, ± recurved, persistent*; stigmas 2, elongate along adaxial surface of style. Ovules several per locule. Nectary absent. *Fruit a septicidal capsule*, **adjacent gynoecia combining to form a globose multiple fruit**; seeds often winged.

Floral formula:
  Staminate: *, -0- ,∞, 0
  Carpellate:*, -∞-, 0 , —②— ; capsule

Distribution: Temperate to tropical; Asia Minor, Southeast Asia, eastern North America south into Central America.

Genera/species: 3/12. *Major genera*: *Altingia* (7) and *Liquidambar* (4). Only *Liquidambar* occurs in the continental United States.

Economic plants and products: A fragrant gum is derived from several species of *Liquidambar* (sweet gum). Members of this genus are also used as ornamentals. Both *Altingia* and *Liquidambar* provide timber.

Discussion: The monophyly of Altingiaceae is supported by numerous morphological characters and *rbcL* sequences (Chase et al. 1993). The group is easily distinguished from Hamamelidaceae, in which it is often included (see discussion under that family). Persistent, short to elongate floral appendages of questionable homology surround the fruits of Altingiaceae (Bogle 1986).

Within *Liquidambar*, the western Asian species, *L. orientalis*, and the North American, *L. styraciflua*, form a

clade, which is the sister group of the southeastern Asian species *L. acalycina* and *L. formosana* (Li et al. 1997; Li and Donoghue 1999).

The flowers of Altingiaceae are pollinated by the wind and the seeds, which are often winged, are wind-dispersed.

References: Bogle 1986; Chase et al. 1993; Endress 1989a, 1993; Ernst 1963b; Li and Donoghue 1999; Li et al. 1997, 1999.

Vitales
Vitaceae A. L. de Jussieu
(Grape Family)

Usually lianas with leaf-opposed tendrils (modified inflorescences) that attach themselves by twining or by adhesive discs, but the lianous habit lost in a few, and these shrubs), the stem with ± swollen nodes; often with raphide sacs. Hairs various. *Leaves alternate, spiral or 2-ranked, simple to palmately or pinnately compound, palmately to pinnately veined*; stipules present. *Inflorescences* determinate, terminal, but *usually appearing leaf-opposed due to growth of the axillary branch from the opposing leaf axil*. Flowers bisexual or unisexual (plants then polygamodioecious or monoecious), radial. *Sepals usually 4–6, ± connate, small*, often represented only by an obscurely toothed or lobed rim. *Petals usually 4–6, distinct or distally falsely connate* (due to interlocked papillae, as in *Vitis*) and at flowering deciduous as a cap, valvate. *Stamens usually 4–6, opposite the petals*; pollen grains tricolporate. *Carpels 2, connate; ovary superior*, with axile placentation; stigma usually capitate. *Ovules 2 per locule*. Nectar disk prominent, usually forming a ring between the ovary and the stamens. *Fruit a berry*; **seeds 4, with a cordlike raphe on the adaxial surface, extending from the hilum to the seed apex and onto the convex abaxial side, where it joins a round to linear, depressed to somewhat elevated "chalazal knot," and also with a deep groove varying in shape and length flanking both sides of the raphe; endosperm 3-lobed** (Figure 9.63).

Figure 9.63 Vitaceae. (A–R) *Vitis rotundifolia:* (A) portion of flowering vine (× 0.5); (B) staminate flower, note that petals have dropped off (× 8); (C) inflorescence opposite petiole (× 0.33); (D) flower bud with falsely connate petals (× 10); (E) opening bud of bisexual flower, with petals forming a cap (× 10); (F) bisexual flower, note pseudoconnate petals falling as a cap (× 10); (G) gynoecium in longitudinal section (× 20); (H) ovary in cross-section, showing four ovules (× 20); (I) branch with tendril, opposite a leaf, and infructescence (× 0.5); (J) berry in cross-section (× 4); (K) seed (× 4); (L) seed, abaxial surface, note raphe and round, chalazal knot (× 4); (M, N) seeds, adaxial surface (× 4); (O) seed in cross-section, note three-lobed endosperm (stippled) (× 8); (P) seed in longitudinal section, endosperm stippled, embryo small, note chalazal knot (circular structure to left) (× 8); (Q, R) embryo (greatly magnified). (S–W) *V. vulpina:* (S) portion of vine with leaf and opposite infructescence (× 0.5); (T–V) seeds, abaxial surface (× 4); (W) seed, adaxial surface (× 4). (Original art prepared for the Generic Flora of the Southeastern U.S. Project. Used with permission.)

Floral formula: *,④⑤,④⑤, 4–5 ,②; berry

Distribution: Widely distributed, but most diverse in tropical and subtropical regions.

Genera/species: 12/700. **Major genera:** *Cissus* (300 spp.), *Vitis* (60), *Ampelopsis* (20), and *Parthenocissus* (15). These all occur in the continental United States and/or Canada.

Economic plants and products: Several species of *Vitis* (grapes) are of great commercial importance as sources of grapes, juice, wine, and raisins. A few species of *Parthenocissus* are used ornamentally.

Discussion: Vitaceae are easily recognized and are surely monophyletic. The family is sister to Leeaceae (Soltis et al. 2000), both having raphide sacs, specialized stalked, gland-headed hairs, a reduced calyx, stamens opposite the petals, and berries (Cronquist 1981). However, Leeaceae are shrubs or herbs that lack tendrils, and have flowers with connate filaments and secondarily divided ovary locules. Nucleotide sequences of *rbcL* also support the sister group relationship between Vitaceae and Leeaceae (Ingrouille et al. 2002).

Genera of Vitaceae are distinguished on the basis of structure of the nectar disk, configuration of endosperm in cross-section, length of the style, and number and connation of petals. Many are apparently non-monophyletic (Ingrouille et al. 2002), and there has been considerable parallel evolution in leaf shape and tendril and inflorescence form.

The small flowers, with their exposed nectaries, are visited by bees, wasps, flies, and beetles. Both selfing and outcrossing occur. The fleshy fruits are animal-dispersed.

References: Brizicky 1965b; Chase et al. 1993; Cronquist 1981; Gerrath et al. 2001; Ingrouille et al. 2002; Soltis et al. 2000; Suessenguth 1953.

Geraniales

Geraniaceae Adanson
(Geranium Family)

Usually herbs to subshrubs, *with stems usually jointed at the nodes.* Hairs simple, *often gland-headed, with aromatic oils. Leaves alternate and spiral, or opposite, simple and palmately lobed, dissected or compound,* ± serrate to entire, with ± palmate venation; stipules usually present. Inflorescences determinate, often umbel-like, sometimes reduced to a single flower, terminal or axillary. Flowers usually bisexual, radial or bilateral. *Sepals usually 5*, distinct or basally connate, occasionally the upper one prolonged backward into a nectar-producing spur that is an integral part of the pedicel. *Petals usually 5, distinct*, often emarginate, usually imbricate. Stamens (5–) 10–15; filaments distinct to slight-

ly connate; pollen grains often tricolporate. Carpels usually 5, connate; *ovary superior, ± lobed*, with axile placentation and usually *with a prominent, elongate, persistent, terminal sterile column; style 1*, stigmas 5, distinct, ± elongate. *Ovules 2 in each locule*, anatropous to campylotropous. Nectar glands usually alternate with the petals (or lacking). *Fruit a schizocarp with 5 one-seeded segments that separate elastically from the persistent central column, and often opening to release the seed*, or occasionally a schizocarp that lacks a central column (*Biebersteinia*), or a loculicidal capsule (*Hypseocharis*); embryo straight to curved; **endosperm scanty or lacking** (Figure 9.64).

Floral formula:

* or X,⑤, 5,⑩–⑮,⑤; schizocarp, capsule

Distribution: Widespread, especially in temperate and subtropical regions.

Genera/species: 7/750. **Major genera:** *Geranium* (300 spp.), *Pelargonium* (250), and *Erodium* (75). All major genera occur in the continental United States and/or Canada.

Economic plants and products: Several species of *Pelargonium* (geranium), *Geranium* (cranesbill), and *Erodium* (storksbill) are grown as ornamentals. Geranium oil, used in perfumes, is distilled from the leaves and shoots of several species of *Pelargonium*.

Discussion: The family usually has been considered closely related to the Oxalidaceae, but DNA-based cladistic analyses (Chase et al. 1993; Price and Palmer 1993; Savolainen et al. 2000b; Soltis et al. 2000) suggest that it is related to the Crossosomataceae, Staphyleaceae, and several other small families. The order is more narrowly circumscribed here, including only 5 other small families (see APG 1998). Morphological synapomorphies of Geraniales may include stamens opposite the petals, the floral nectary positioned outside the androecium, and glandular-toothed leaves. In addition, their vessel elements have simple perforations. Traditionally, Geraniales have been delimited much differently, to include families such as Zygophyllaceae, Oxalidaceae, Polygalaceae, Krameriaceae, and Malpighiaceae (e.g., Thorne 1992). None of these families are currently placed within the order.

Geraniaceae are a well-defined monophyletic group based on *rbcL* sequences and loss of the intron in the plastid gene *rpl16* (Price and Palmer 1993). *Hypseocharis* (usually placed in the Oxalidaceae) is sister to the remaining genera. This genus retains the plesiomorphic feature of capsular fruits. The remaining genera all have schizocarpic fruits, usually with a central stylar column (both apomorphies). Bilateral flowers and a spurred sepal are probably autapomorphies of *Pelargonium*, which is sister to a clade containing *Erodium*, *Geranium*, and relatives

Figure 9.64 Geraniaceae. (A–J) *Geranium maculatum:* (A) flowering plant (× 0.5); (B) flower (× 2.25); (C) stamen of outer whorl (× 6); (D) gynoecium with nectar glands on receptacle (below pubescent ovary) (× 6); (E) gynoecium in longitudinal section (× 9); (F) ovary in cross-section (× 12); (G) dehisced fruit, showing segments attached to recurved hygroscopic awns (× 3.0); (H) seed (× 7.5); (I) embryo (× 7.5); (J) embryo in cross-section to show folding of cotyledons (× 7.5). (K–M) *G. carolinianum:* (K) branch with flowers and fruits (× 0.5); (L) flower with receptive stigmas, anthers mostly fallen (× 4.5); (M) mature fruit before dehiscence (× 3). (From Robertson 1972, *J. Arnold Arbor.* 53: p. 190.)

(Price and Palmer 1993). Note that the spur can be recognized only as an indistinct line and bump on the pedicel.

The showy flowers of Geraniaceae are pollinated by a wide variety of insects; nectar guides are often present, and nectar is the reward. Many species are protandrous and outcrossing, but some are self-pollinating and weedy. The schizocarps of most species of *Geranium* dehisce explosively and may throw the seeds (or seeds and associated carpel wall) several meters. The sterile upper part of the ovary elongates greatly in fruit, cf. Figure 8.56D and 8.56M. The fruit opens by means of the segments separating from the central column, each segment being composed of a globose basal portion (one of the ovary lobes) and an awnlike upper portion (narrow outer layer of the column). The segments curve upward and inward (Figure 8.56G), and the seeds may be forcibly ejected from the globose portion of the segment due to the drying and contraction of the carpel wall. In *Erodium*, *Pelargonium*, and some species of *Geranium*, the schizocarpic segments retain the seeds and separate from the central column. The awnlike portion of the segment is hygroscopic, and varying weather conditions cause it to coil and uncoil repeatedly. This movement forces the seed into the soil.

References: Chase et al. 1993; Boesewinkel 1988; Price and Palmer 1993; Rama Devi 1991; Robertson 1972a; Savolainen et al. 2000b; Soltis et al. 2000; Yeo 1984.

Eurosids I

Zygophyllales

Zygophyllaceae R. Brown
(Creosote Bush Family)

Trees, shrubs, or herbs, with stems often sympodial and jointed at the nodes; **xylem with vessels, tracheids, and fibers arranged in horizontally aligned tiers;** usually producing steroidal or triterpenoid saponins, sesquiterpenes, and alkaloids. Hairs various. **Leaves usually opposite**, *usually 2-ranked,* **usually pinnately compound**, *without a terminal leaflet*, often strongly resinous, leaflets entire, with pinnate to palmate venation; *stipules usually present*. Inflorescences usually determinate, sometimes reduced to a solitary flower, terminal, but often appearing to be lateral. Flowers usually bisexual, radial. *Sepals usually 5, ± distinct. Petals usually 5, distinct, usually clawed*, imbricate or convolute. Stamens 10–15; filaments usually associated with basal glands or appendages; pollen grains often tricolporate. Carpels usually 5, sometimes reduced to 2, connate; *ovary superior, ridged or winged*, with axile placentation; stigma usually 1, capitate to lobed. Ovules 1 to several in each locule, anatropous to orthotropous. Nectar disk present at base of ovary. *Fruit usually a septicidal or loculicidal capsule or a schizocarp, sometimes spiny or winged*; seeds sometimes arillate; endosperm present.

Floral formula:

*, 5, 5, 10–15, ②–⑤; capsule, schizocarp

Distribution and ecology: Widespread in tropical and subtropical regions, especially in arid habitats.

Genera/species: 25/200. **Major genera:** *Zygophyllum* (80 spp.), *Fagonia* (40), *Balanites* (20), and *Tribulus* (20). Genera occurring in the continental United States include *Guaiacum*, *Kallstroemia*, *Larrea*, *Porlieria*, *Tribulus*, and *Zygophyllum*.

Economic plants and products: Some species of *Guaiacum* (lignum vitae) provide a strong, hard, heavy wood. The family also contains several ornamentals, including *Larrea* (creosote bush), *Guaiacum*, and *Tribulus* (caltrop).

Discussion: Zygophyllaceae are an isolated group, which are possibly sister to the remaining members of the Eurosids I clade. The family is considered to be monophyletic, based on morphological and DNA characters, after a few genera (e.g., *Nitraria*, *Malacocarpus*, and *Peganum*) are segregated as Peganaceae (Sapindales) (Dahlgren 1983; Gadek et al. 1996; Ronse Decraene et al. 1996). Analyses of *rbcL* sequence characters suggest that Zygophyllaceae are the sister group of Krameriaceae, a small family of root parasites with bilateral resupinate flowers, petaloid calyx, two highly modified and oil-producing petals, and globose, spiny fruits (Cronquist 1981; Källersjö et al. 1998; Savolainen et al. 2000b; Soltis et al. 2000).

Phylogenetic relationships within the Zygophyllaceae have been investigated based on *rbcL* and *trnL-F* sequences (Sheahan and Chase 2000); major clades include the Tribuloideae (e.g., *Tribulus*, *Kallstroemia*, *Balanites*), Larreoideae (e.g., *Bulnesia*, *Guaiacum*, *Larrea*), and Zygophylloideae (e.g., *Fagonia*, *Zygophyllum*).

The showy, nectar-producing flowers are pollinated by various insects. Capsules of *Guaiacum* open to expose colorful arillate seeds and are probably bird-dispersed. The winged schizocarps of *Bulnesia* are dispersed by wind, while the spiny schizocarps of *Tribulus* are externally transported by animals.

References: Cronquist 1981; Dahlgren 1983; Ronse Decraene et al. 1996; Gadek et al. 1996; Källersjö et al. 1998; Porter 1972; Sheahan and Chase 2000; Sheahan and Cutler 1993; Savolainen et al. 2000b; Soltis et al. 2000.

Oxalidales

Oxalidaceae R. Brown
(Wood Sorrel Family)

Herbs, often with bulblike tubers or fleshy rhizomes, to shrubs or trees; **with high levels of soluble and crystalline oxalates.** Hairs simple. *Leaves alternate and spiral, some-*

times forming a basal rosette, palmately to pinnately compound, or reduced and trifoliate or unifoliate, often with prominent pulvini and showing sleep movements, entire, often emarginate, with palmate to pinnate venation; stipules usually lacking. Inflorescences determinate, often umbel-like, sometimes reduced to a single flower, axillary. Flowers bisexual, radial, **heterostylous** (sometimes lost in weedy species). *Sepals 5, distinct. Petals 5, distinct* or very slightly connate, usually convolute. *Stamens* usually 10; filaments connate basally, *with outer filaments shorter than inner*; pollen grains usually tricolpate or tricolporate. Carpels usually 5, connate; ovary superior, ± lobed, with axile placentation; styles usually 5, distinct; stigmas usually capitate or punctate. Ovules usually several in each locule, **with a thin megasporangium wall**. Nectar produced by base of filaments or glands alternating with petals. *Fruit a loculicidal capsule* or berry, *often lobed or angled; seeds sometimes arillate, the outer part of testa often elastically turning inside out and ejecting the seed* (Figure 9.65).

Floral formula: *, 5, 5, ⑩, ⑤; capsule, berry

Distribution: Widespread, especially in tropical and subtropical regions.

Genera/species: 6/880. **Major genera:** *Oxalis* (800 spp.) and *Biophytum* (70). Only *Oxalis* occurs in the continental United States and Canada.

Economic plants and products: *Averrhoa carambola* (carambola, star fruit) has edible fruits; numerous cultivars exist that vary greatly in acidity (oxalate content). The tubers of *Oxalis tuberosa* (oca) are eaten in Andean South America.

Discussion: The monophyly of Oxalidaceae is supported by morphological and DNA characters (Price and Palmer 1993).

Oxalidaceae usually have been considered closely related to Geraniaceae on the basis of their actinomorphic, 5-merous flowers with ten stamens and lobed, syncarpous gynoecium. However, these similarities are symplesiomorphic. Separate styles nicely distinguish Oxalidaceae from Geraniaceae, but the feature is probably plesiomorphic. Phylogenetic studies based on *rbcL*, *atpB*, and 18S sequences indicate that Oxalidaceae are more closely related to Cunoniaceae and Cephalotaceae (and three other small families), here treated as the order Oxalidales, than to Geraniaceae (Chase et al. 1993; Price and Palmer 1993; Soltis et al. 2000). Cunoniaceae are tropical trees and shrubs with opposite, pinnately compound, stipulate leaves; Cephalotaceae are Australian insectivorous herbs with a rosette of pitcherlike leaves. Oxalidales are morphologically heterogeneous, but the order's monophyly is supported by recent molecular analyses. Members of Oxalidales have dry stigmas

(where known) and the seed exotegmen is fibrous with tracheids.

Heterostyly, both distyly and tristyly, is characteristic of Oxalidaceae. The family is characterized by outcrossing, but some weedy species are selfing. The showy, nectar-producing flowers of Oxalidaceae are pollinated by various insects. Most members of the family are self-dispersed by means of the explosive inversion of the smooth, turgid outer part of the testa.

References: Chase et al. 1993; Denton 1973; Ornduff 1972; Price and Palmer 1993; Robertson 1975; Soltis et al. 2000.

Celastrales

Celastraceae R. Brown

(Bittersweet Family)

Trees, shrubs, or lianas climbing by twining stems or hooklike branches; often with tannins. Hairs simple to branched. Leaves alternate and spiral or 2-ranked, or opposite, simple, entire to serrate, with pinnate venation; stipules present or lacking. Inflorescences usually determinate, terminal or axillary. Flowers bisexual or occasionally unisexual (plants monoecious to dioecious), radial, sometimes with a short hypanthium. *Sepals usually 4 or 5, distinct to slightly connate. Petals 4 or 5, distinct*, imbricate to occasionally valvate. **Stamens 3–5,** *alternating with petals*; filaments distinct to connate; pollen grains tricolporate or triporate, occasionally in tetrads or polyads. Carpels 2–5, connate; ovary superior to half-inferior, with axile placentation; stigma capitate to lobed. Ovules 2 to numerous per locule, sometimes with a thin megasporangium wall. **Conspicuous and usually intrastaminal nectar disk present,** sometimes adnate to ovary. *Fruit a loculicidal capsule (sometimes strongly 3-lobed), drupe, or berry; seeds often with an orange to red aril or winged*; endosperm sometimes lacking.

Floral formula:

*, (4–5), 4–5, (3–5), (2–5) ; capsule, drupe

Distribution: Widely distributed in tropical and subtropical regions, with a few genera extending into temperate areas.

Genera/species: 55/855. **Major genera:** *Maytenus* (200 spp.), *Euonymus* (200), *Salacia* (170), and *Hippocratea* (100). Noteworthy genera in the continental United States and/or Canada include *Celastrus, Crossopetalum, Euonymus, Hippocratea, Maytenus, Paxistima*, and *Schaefferia*.

Economic plants and products: *Celastrus* (bittersweet), *Euonymus*, and *Schaefferia* are used as ornamentals. The narcotic khat is derived from the leaves of *Catha*.

Figure 9.65   Oxalidaceae. (A–C) *Oxalis acetosella* ssp *montana:* (A) plant with chasmogamous flower and immature fruit from cleistogamous flower (× 0.7); (B) chasmogamous flower (× 3); (C) cleistogamous flower in off-center longitudinal section (× 18). (D–L) *O. stricta:* (D) habit (× 0.7); (E) mature fruit before dehiscence (× 4.5); (F) fruit in cross-section (× 1.8); (G) capsule during dehiscence, note seeds and arils (× 4.5); (H) seed enclosed by aril (× 18); (I, J) two views of testa splitting away from seed (× 18); (K) testa after eversion (× 18); (L) seed (× 18). (M, N) *O. dillenii:* (M) androecium and gynoecium from quasi-homostylous flower (× 15); (N) stigma (× 37). (O–Q) *O. grandis:* androecium and gynoecium from long-styled, midstyled, and short-styled flowers, respectively (× 9). (From Robertson 1975, *J. Arnold Arbor.* 56: p. 231.)

Discussion: Celastraceae are broadly circumscribed, including Hippocrateaceae, which are lianas with three stamens borne inside the disk, transverse anther dehiscence, berries or three-lobed capsules, and seeds that lack endosperm (Thorne 1992). The family is considered to be monophyletic on the basis of morphology and DNA sequences (Savolainen et al. 1994). Segregation of Hippocrateaceae would render Celastraceae paraphyletic (Clevinger and Panero 1998; Simmons and Hedin 1999; Simmons et al. 2000, 2001). Arils probably evolved three times in the family, in *Canotia*, in *Bhesia*, and in the com-

mon ancestor of all the remaining arillate genera (such as *Catha, Euonymus, Celastrus*, and *Maytenus*). In *Hippocratea* and relatives the aril has been modified into a basal wing, and in *Salacia* and relatives, into a mucilaginous pulp. The hippocrateoid group is likely monophyletic, as indicated by the loss of endosperm. The shrub *Brexia* (usually in Brexiaceae) is also included within Celastraceae.

The herbaceous genera *Parnassia* and *Lepuropetalon* (Parnassiaceae) are either closely related to Celastraceae, or nested within it, and possibly, therefore, should be included within it.

The often inconspicuous, green to white flowers are pollinated by bees, flies, and beetles; nectar is the reward. The fruits are often brightly colored capsules that open to expose arillate seeds that are dispersed by birds. The colorful drupes or berries of some species also are bird-dispersed. *Hippocratea* and relatives have winged seeds and are dispersed by wind.

References: Brizicky 1964a; Clevinger and Panero 1998; Hallé 1962; Källersjö et al. 1998; Savolainen et al. 1994; Simmons and Hedin 1999; Simmons et al. 2000, 2001; Thorne 1992.

Malpighiales

The monophyly of Malpighiales is indicated only by molecular data (*rbcL*, *atpB*, and 18S sequences) (Chase et al. 1993; Savolainen et al. 2000a,b; Soltis et al. 1998, 2000). The group is morphologically heterogeneous, but many have dry stigmas, a fibrous exotegmen, and trilacunar nodes. Several families are predominantly three-carpellate (e.g., Euphorbiaceae, Malpighiaceae, Passifloraceae, Violaceae). Groups such as Violaceae, Salicaceae, and Passifloraceae are distinct because of their parietal placentation (Cronquist 1981, 1988; Thorne 1992), and these families traditionally have been placed in the order "Violales."

Malpighiales contain about 35 families and about 13,100 species. Major families include Achariaceae, **Clusiaceae**, Chrysobalanaceae, **Euphorbiaceae**, **Malpighiaceae**, Ochnaceae, **Passifloraceae**, **Phyllanthaceae**, Picrodendraceae, Podostemonaceae, **Rhizophoraceae**, **Salicaceae**, and **Violaceae**.

Phylogenetic relationships within the order are problematic, but have recently been clarified by Chase et al. (2002). Salicaceae, Violaceae, Achariaceae, Turneraceae, and Passifloraceae are characterized by parietal placentation; most other Malpighiales have axile placentation. Achariaceae, Turneraceae, and Passifloraceae are related, and all share cyclopentenoid cyanogenic glucosides and cyclopentenyl fatty acids. Rhizophoraceae and Erythroxylaceae may be sister families; they share tropane and pyrrolidine alkaloids, opposite leaves, a terminal bud protected by stipules, and green embryos. Clusiaceae and Podostemaceae have similar xanthone pigments and secretory tissues/cells.

References: Chase et al. 1993, 2002; Cronquist 1981, 1988; Savolainen et al. 2000a,b; Soltis et al. 1998, 2000; Thorne 1992.

Malpighiaceae A. L. de Jussieu
(Barbados Cherry Family)

Shrubs, trees, lianas, or occasionally perennial herbs. **Hairs various, but always 1-celled, usually ± attached such that the hair has 2 arms, and is T-, V-, or Y-shaped,** *the stalk often short and the arms straight to twisted.* **Leaves usually opposite,** *simple,* usually entire, occasionally lobed, with usually pinnate venation, *often with 2 or more glands on petiole or abaxially on blade;* stipules usually present. Inflorescences determinate, but often appearing indeterminate, terminal or axillary. *Flowers usually bisexual, usually ± bilateral.* **Sepals 5,** distinct to connate basally, **with 2 conspicuous, abaxial, oil-producing glands on all 5 sepals or on the 4 lateral sepals,** oil glands vestigial or lacking in a few Neotropical species and most Paleotropical species. *Petals 5, distinct,* **usually clawed,** often with fringed or toothed margins, *the upper one usually slightly larger or smaller than the others and sometimes differentiated in color as well,* imbricate. Stamens usually 10; *filaments usually connate basally;* pollen grains usually 3–5-colporate or 4– to polyporate. *Carpels usually 3, connate;* ovary superior, with axile placentation; styles usually distinct; stigmas various. Ovules 1 in each locule; megagametophyte usually 16-nucleate. Nectaries lacking. *Fruit usually a samaroid schizocarp, schizocarp, drupe with usually 3-seeded, often ridged pit,* or nutlike; embryo straight to bent or coiled; endosperm ± lacking (Figure 9.66).

Floral formula:

X, ⑤, 1+4, ⑩, ③; samaroid schizocarp, drupe, berry, nutlike

Distribution: More or less pantropical, but especially diverse in South America.

Genera/species: 66/1200. **Major genera:** *Byrsonima* (150 spp.), *Heteropterys* (120), *Banisteriopsis* (92), *Tetrapterys* (90), *Stigmaphyllon* (90), and *Bunchosia* (75). *Aspicarpa, Byrsonima, Galphimia, Janusia,* and *Malpighia* occur in the continental United States.

Economic plants and products: The drupes of *Malpighia emarginata* (Barbados cherry) are edible and contain large amounts of vitamin C. *Malpighia, Stigmaphyllon, Galphimia,* and *Byrsonima* provide ornamentals. *Banisteriopsis caapi* contains narcotic alkaloids.

Discussion: The monophyly of Malpighiaceae is not in doubt, being supported by both morphological and cpDNA sequence characters (Chase et al. 1993; Soltis et al. 2000). Recognition of two traditional subfamilies on the basis of unwinged versus winged fruits is surely artificial (Anderson 1977). Fleshy fruits, as found in *Byrsonima* and *Malpighia*, have evolved several times. The form of the style and stigma, pollen structure, and chromosome number are of phylogenetic significance within the family (Anderson 1977). Two subfamilies are recognized based on DNA sequence data (Davis et al. 2001): Byrsonimoideae (including *Byrsonima* and a few other genera) have a base chromosome number of 6, while the remaining genera belong to the Malpighioideae and are characterized by x = 10.

In the Neotropics the family is principally pollinated by oil-gathering anthophorid bees, while most Old

Key to Major Families of Malpighiales

1. Plants aquatic, usually in rapidly flowing water; stems, roots, and leaves often not clearlydistinguished . . . . Podostemonaceae

1. Plants terrestrial or in mangrove communities; stems, roots, and leaves clearly distinguished 2

2. Plants with white to colored latex or other exudate in secretory canals, or clear to black resins in secretory cavities (pellucid or black dots) 3

2. Plants lacking resins or latex, without canals or cavities 4

3. Leaves usually opposite or whorled, lacking stipules, with clear, black, or ± colored resins or exudate in canals or cavities; flowers bisexual or unisexual, styles usually not divided and number of stigmas equaling or fewer than the carpels; fruit a capsule, berry, or drupe . . . . . . **Clusiaceae**

3. Leaves usually alternate, stipulate, with white (rarely colored) latex in laticifers; flowers always unisexual, styles forked to several times divided (and thus number of stigmas greater than number of carpels); fruit ± schizocarpic . . . . **Euphorbiaceae**

4. Style gynobasic; gynoecium apparently a single carpel or the ovary deeply lobed and thus appearing to be of distinct carpels 5

4. Style terminal; gynoecium obviously syncarpous 6

5. Gynoecium apparently a single carpel, with a lateral style, fruit a single drupe . . . . . . Chrysobalanaceae

5. Gynoecium syncarpous, the ovary deeply lobed, developing into several drupe-like units contrasting in color with an expanded receptacle . . . . Ochnaceae

6. Placentation axile 7

6. Placentation parietal 11

7. Flowers unisexual; styles not divided to several times divided (and thus more numerous than the carpels) 8

7. Flowers bisexual; styles not secondarily divided 9

8. Each locule with a single ovule; seeds often arillate . . . . **Euphorbiaceae**

8. Each locule with two ovules; seeds not arillate . . . . **Phyllanthaceae**

9. Hairs T-, V-, or Y-shaped, the stalk often short and the arms straight to twisted; sepals often with paired, abaxial, oil-producing glands; petals usually clawed . . . . **Malpighiaceae**

9. Hairs simple; sepals lacking glands; petals not clawed

10. Leaves opposite; stipules interpetiolar; petals usually fringed or hairy and individually enclosing a single stamen or a staminal group; often plants of mangrove communities . . . **Rhizophoraceae**

10. Leaves alternate; stipules lateral; petals not fringed or hairy and not enclosing a single stamen or group of stamens; plants not of mangrove habitats . . . . Ochnaceae

11. Flowers with a corona, consisting of 1 or more rows of filaments or scales, usually borne on a hypanthium; ovary usually on a short to long gynophore; often vines with tendrils; aril fleshy . . . . **Passifloraceae**

11. Flowers lacking a corona and usually a hypanthium as well; gynophore lacking; trees to herbs; always lacking tendrils; aril fleshy to hard and oily 12

12. Flowers usually bilateral; stamens usually 5, the filaments very short and stamens closely placed around style, connective prominent, some or all of the anthers with a glandlike or spurlike nectary on the back; style 1, often distally enlarged; stigma 1, sometimes lobed . . . . . **Violaceae**

12. Flowers radial; stamens usually 10 to numerous, sometimes few and then flowers in catkins, the filaments elongate and stamens not closely placed around style, connective not prominent, anthers lacking nectar-producing tissue; styles several to 1, not apically enlarged; stigmas 2 to several 13

13. Leaves without salicoid teeth; petals much more numerous than sepals; nectar disk lacking; anthers usually linear; cyanogenic glycosides present . . . . Achariaceae

13. Leaves with salicoid teeth; petals lacking, or if present than equal in number to the sepals, or sepals and petals both lacking; nectar disk or glands present, but these sometimes lacking if flowers in catkins; anthers usually globose or ovoid; cyanogenic compounds lacking . . . . . . **Salicaceae**

Figure 9.66 Malpighiaceae. *Byrsonima lucida:* (A) flowering branch (× 1.5); (B) flower (× 7.5) (C) calyx, showing glands on one sepal (glands on adjacent sepals removed) (× 11); (D) flower in longitudinal section (× 8); (E, F) stamens (× 22); (G) drupe (× 3); (H) pit (× 6); (I) pit in cross-section, showing three seeds and embryos (× 6); (J, K) opposite sides of the same embryo (× 9.5). (From Robertson 1972, *J. Arnold Arbor.* 53: p. 110.)

World species are pollinated by pollen-gathering bees. Most are outcrossing, but selfing also occurs; some have been shown to be agamospermous. Winged fruits, as in *Stigmaphyllon* or *Tetrapterys*, are wind-dispersed; fleshy fruits, as seen in *Malpighia*, *Byrsonima*, or *Bunchosia*, are bird- and/or mammal-dispersed.

References: Anderson 1977, 1979, 1990; Chase et al. 1993; Davis et al. 2001; Rao and Sarma 1992; Robertson 1972b; Soltis et al. 2000; Vogel 1990.

Euphorbiaceae A. L. de Jussieu
(Spurge Family)

Trees, shrubs, herbs, or vines, sometimes succulent and cactuslike; internal phloem sometimes present; chemically diverse, with alkaloids, di- or triterpenoids, tannins, and cyanogenic glucosides; *often with laticifers* containing milky or colored latex; *usually poisonous.* Hairs simple to branched, stellate, or peltate. *Leaves usually alternate and spiral or 2-ranked*, sometimes opposite, simple, sometimes palmately lobed or compound, entire to serrate, with pinnate to palmate venation; *sometimes with*

Figure 9.67 Euphorbiaceae. (A, B) *Croton linearis:* (A) branch with staminate flowers (× 0.75); (B) carpellate flower, two sepals removed to show nectar disk and staminodes (× 9). (C–M) *C. alabamensis:* (C) staminate flower (× 6); (D) branch with carpellate flowers (× 0.75); (E) scale from leaf (× 75); (F) carpellate flower (× 12); (G) same in longitudinal section (× 15); (H) branch with nearly mature fruit (× 1.5); (I) mature schizocarp (× 3); (J) same in cross-section to show three seeds, endosperm stippled, bony endocarps hatched (× 3); (K) persistent perianth and columella after dehiscence of fruit (× 3); (L) seed (× 4.5); (M) embryo (× 5.5). (N–O) *C. glandulosus:* (N) stellate hair from leaf (× 75); (O) seed (× 7.5). (P–Q) *C. punctatus:* stellate hairs from leaf (× 75). (R) *C. argyranthemus:* seed (× 7.5). (From Webster 1967, *J. Arnold Arbor.* 48: p. 356.)

paired nectar glands at base of blade or on petiole; stipules usually present. Inflorescences determinate, but *often highly modified, sometimes forming false flowers* as in the *cyathia* of *Euphorbia,* terminal or axillary. **Flowers unisexual** (plants dioecious or monoecious), usually radial, showy to inconspicuous. Sepals usually 2–6, distinct to slightly connate. Petals usually 0–5, distinct to slightly connate, valvate or imbricate, often lacking. Stamens 1 to numerous; filaments distinct to connate; pollen grains often tricolporate or polyporate. Carpels usually 3, connate; ovary superior, usually 3-lobed, with axile placentation; **styles** usually 3, **each usually bifid, or several times divided**; stigmas various. **Ovules 1 in each locule**. Nectar disk usually present. **Fruit usually a schizocarp, with segments elastically dehiscent from a persistent central column**; seeds often arillate; embryo straight to curved (Figures 9.67 and 9.68).

Floral formula:

Staminate: *, (2–6), (0–5), (1–∞), 0

Carpellate: *, (2–6), (0–5), 0, ③; schizocarp

Distribution: Widespread, but most diverse in tropical regions.

Genera/species: 320/6100. **Major genera:** *Euphorbia* (2000 spp.), *Croton* (750), *Acalypha* (400), *Macaranga* (250), *Manihot* (150), *Tragia* (150), *Jatropha* (150), *Mallotus* (120), *Sapium* (100), and *Dalechampia* (100). Numerous genera are represented in the continental United States and/or Canada; some of these include (in addition to most of the above) *Argythamnia, Bernardia, Cnidoscolus, Reverchonia, Sapium, Sebastiania,* and *Stillingia.*

Economic plants and products: *Hevea brasiliensis* (rubber tree) is the source of most natural rubber; *Aleurites moluccana* (candlenut tree) and *A. fordii* (tung tree) are sources of oils used in the manufacture of paints and varnishes; *Sapium sebiferum* (Chinese tallow tree) is a source of vegetable tallow and wax; *Euphorbia* spp. produce reduced hydrocarbon fuels. Many species are poisonous, and many have medicinal uses (see Rizk 1987). Species of *Euphorbia* and *Hippomane* have been used as arrow poisons, and members of several genera are used as fish poisons. Surprisingly, several species contain edible parts; the thick roots of *Manihot esculenta* (cassava, manioc, yuca) are an important starch source in tropical regions. The leaves of *Cnidoscolus chayamansa* (stinging nettle) are used as a vegetable, and the fruits of *Antidesma bunius* are eaten. Finally, *Euphorbia* (poinsettia, crown-of-thorns, various succulents), *Jatropha, Codiaeum* (croton), *Acalypha* (chenille plant), and other genera provide ornamentals.

Discussion: The family has often been considered close to Malvales, although molecular data suggest that it is

best placed in a broadly circumscribed Malpighiales (as in this text).

Euphorbiaceae are extremely diverse and have sometimes been divided into several families, or suggested to be polyphyletic. The monophyly of Euphorbiaceae, as broadly circumscribed, has not been supported in recent phylogenetic analyses based on DNA sequences (Chase et al. 2002; Savolainen et al. 2000b; Soltis et al. 2000; APG2 2002), and members of this "group" are here divided among four families: Euphorbiaceae s.s., Phyllanthaceae, Picrodendraceae, and Putranjivaceae. The mustard oil-containing *Drypetes* and *Putranjiva* are segregated as Putranjivaceae. The first three families are large, morphologically variable, and are discussed in more detail below. Evolutionary relationships within Euphorbiaceae, as broadly delimited, have been extensively studied by Webster (1967, 1987, 1994a,b), who recognized five subfamilies.

Phyllanthaceae (e.g., *Bischofia, Phyllanthus, Glochidion,* and *Antidesma*) are characterized by having two ovules per locule and nonarillate seeds (see treatment of this family). In contrast, Euphorbiaceae, as delimited in this text, have only a single ovule per locule and usually arillate seeds. Members of the Phyllanthaceae were considered as the subfamily Phyllanthoideae (of Euphorbiaceae) by Webster, who suggested that they represent the primitive group from which his other subfamilies were derived. Picrodendraceae (e.g., *Oldfieldia, Picrodendron,* and *Pseudanthus;* treated as Euphorbiaceae subfamily Oldfieldioideae by Webster) also have two ovules per locule. Spiny pollen supports the monophyly of Picrodendraceae; both Phyllanthaceae and Euphorbiaceae are characterized by nonspiny pollen. The monophyly of Euphorbiaceae, narrowly delimited, is supported by their having only one ovule per locule, unisexual flowers, ± divided styles, and schizocarpic fruits. Similar gynoecia with divided styles and schizocarpic fruits may have evolved independently in Phyllanthaceae. Picrodendraceae lacks divided styles and a nectar disk.

Euphorbiaceae s.s. has been divided into three subfamilies by Webster: the Acalyphoideae (*Acalypha, Alchornea, Tragia, Ricinus,* and relatives), which lack latex; and the Crotonoideae (*Croton, Manihot, Jatropha, Codiaeum, Aleurites, Cnidoscolus,* etc.) and Euphorbioideae (*Hippomane, Hura, Euphorbia, Gymnanthes, Stillingia, Sapium,* etc.), both of which have latex. The Crotonoideae have distinctive polyporate pollen, often stellate, peltate, or branched hairs, and colored to white, noncaustic sap, while the Euphorbioideae have tricolporate pollen, simple hairs, and white, often caustic sap. Euphorbioideae contain the large tribe Euphorbieae (mainly *Euphorbia,* which includes *Poinsettia, Chamaesyce,* etc.), which are considered monophyletic on the basis of their inflorescences, which are cyathia (Figure 9.67B). The carpellate flower is surrounded by numerous staminate flowers (each of which is reduced to a single stamen) within a cuplike structure formed from a highly reduced cymose

Figure 9.68 Euphorbiaceae (taxa with cyathia). (A–I) *Euphorbia corollata:* (A) flowering plant (× 0.75); (B) cyathium (× 12); (C) cyathium with walls removed to show small cymes of staminate flowers and bracteoles (× 12); (D) staminate flower with bracteole (× 12); (E) schizocarp (× 9); (F) segment of schizocarp, after dehiscence (× 9); (G) column of schizocarp (× 9); (H) seed (× 9); (I) embryo (× 6). (J) *E. innundata:* cyathium and carpellate flower with three-lobed perianth (× 6). (K–L) *E. commutata:* (K) cyathium, from above (× 12); (L) embryo (× 9). (M) E. dentata: cyathium (× 12). (N–Q) E. supina: (N) part of prostrate plant (× 0.75); (O) cyathium (× 24); (P, Q) seed, adaxial surface and above (× 24). (From Webster 1967, *J. Arnold Arbor.* 48: p. 396.)

inflorescence and associated bracts. One to five nectar glands, sometimes with petal-like appendages, are associated with the cuplike axis of each cyathium.

Most Euphorbiaceae are insect-pollinated (flies, bees, wasps, and butterflies), with nectar providing the floral attractant, yet some are probably pollinated by birds, bats, or other mammals. *Acalypha, Ricinus,* and *Alchornea,* among others, are wind-pollinated. Outcrossing is promoted by maturation of carpellate flowers before staminate ones. Most have elastic schizocarps. The large fruits of *Hura* or *Hevea* are able to explosively eject their seeds, shooting them several meters. Some are secondarily water-dispersed, while those with oily arils are sometimes secondarily dispersed by ants. Some taxa have fleshy arils (or indehiscent fleshy fruits) and are dispersed by birds.

References: Chase et al. 2002; Levin 1986; Park and Elisens 2000; Rao 1971; Rizk 1987; Savolainen et al. 2000b; Soltis et al. 2000; Sutter and Endress 1995; Webster 1967, 1987, 1994a,b.

Phyllanthaceae J. Agardh
(Phyllanthus Family)

Trees, shrubs, or herbs, sometimes with dimorphic (horizontal and erect) branches or phylloclades; chemically diverse, with alkaloids, triterpenoid saponins, and tannins; laticifers lacking; usually poisonous. Hairs simple. *Leaves alternate, often 2-ranked, usually simple,* entire to serrate, with pinnate venation; *stipules usually present.* Inflorescences determinate, axillary, occasionally reduced to a single flower. *Flowers unisexual (plants monoecious or dioecious), radial, ± inconspicuous.* Sepals usually 5, distinct or slightly connate. Petals usually 0–5, distinct to slightly connate, imbricate, often reduced or lacking. Stamens 3–8, occasionally fewer or more numerous; filaments distinct to connate; pollen grains tricolporate or polycolporate to polyporate, or inaperturate. *Carpels usually 3, connate; ovary superior,* **usually 3-lobed**, with axile placentation; *styles usually 3,* **each usually bifid**, occasionally entire; stigmas various. *Ovules 2 in each locule. Nectar disk usually present.* **Fruit usually a schizocarp, with segments elastically dehiscent from a persistent central column**, *sometimes a berry or drupe; seeds not arillate;* embryo straight to slightly curved.

Floral formula:

Staminate: *, ⑤, ⟨0–5⟩, ⟨3–∞⟩ , 0

Carpellate: *, ⑤, ⟨0–5⟩, 0 , ③ , schizocarp, drupe

Distribution: Widespread, but most diverse in tropical regions.

Genera/species: 55/1550. **Major genera:** *Phyllanthus* (600), *Glochidion* (300), *Antidesma* (150), and *Cleistanthus* (100). The group is represented in the continental United States and/or Canada by *Andrachne, Bischofia, Phyllanthus,* and *Savia.*

Economic plants and products: Several species are used medicinally, or have been used as fish-poisons. A few, such as *Bischofia* and *Phyllanthus,* provide ornamentals. A few species of *Antidesma* (bignay) and *Phyllanthus* (emblic, otaheite gooseberry) have edible drupes (high in vitamin C).

Discussion: Phyllanthaceae can easily be distinguished from Euphorbiaceae, in which they are often included, by their gynoecia that have 2-ovulate locules (versus 1-ovulate locules). They also the lack the laticifers (with white or colored latex) and arillate seeds so characteristic of many Euphorbiaceae. They can be distinguished from the small, and mainly southern Hemisphere family, Picrodendraceae (e.g., *Austrobuxus, Oldfieldia, Pseudanthus, Tetracoccus,* and *Picrodendron*), which also have 2-ovulate locules, by the presence of a nectar disk and frequently also bifid styles in their flowers, and in the lack of the spiny pollen characteristic of this clade. Recent molecular studies have shown that Phyllanthaceae, Picrodendraceae, and Euphorbiaceae represent three distinct lineages within Malpighiales, and these three clades likely are not each other's closest relatives (Chase et al. 2002; Savolainen et al. 2000b; Soltis et al. 2000). Phyllanthaceae are presumed to be monophyletic based on their frequently 3-lobed ovaries, bifid styles and schizocarpic fruits, all characters that evidently evolved in parallel in Euphorbiaceae. Phylogenetic relationships within the family are not well understood.

The small flowers of most Phyllanthaceae are pollinated by generalist insect visitors, mainly flies or small bees. Those with schizocarpic fruits have ballistic seed dispersal, while those with fleshy (and often colorful fruits) are mainly dispersed by birds.

References: Chase et al. 2002; Savolainen et al. 2000; Soltis et al. 2000; Webster 1955, 1956-1958, 1967, 1994a,b.

Clusiaceae Lindley
(= Guttiferae A. L. de Jussieu)
(Saint-John's-Wort Family)

Trees, shrubs, lianas, or herbs; **with clear, black, or colored resin or exudate in secretory canals or cavities**. Hairs simple, multicellular. **Leaves usually opposite or whorled**, *simple,* **entire**, with pinnate venation, **often with pellucid or black dots or canals**; stipules lacking, although paired glands may be present at the nodes, colleters common. Inflorescences determinate, sometimes reduced to a single flower, usually terminal. Flowers bisexual to unisexual (then plants usually dioecious), radial. *Sepals usually 2–5 and distinct. Petals usually 4–5, distinct, often asymmetrical,* imbricate or convolute. *Stamens usually numerous,* the innermost developing before the outermost, often fascicled; pollen grains usually tricolporate. *Carpels usually 2–5*, sometimes numerous, connate; *ovary superior*, with usually axile placentation, or sometimes parietal with deeply intruded placentas; stigmas peltate, lobed, punctate, or capitate. Ovules 2 to numerous per carpel, with a thin megasporangium. Nectaries usually lacking. *Fruit a variously dehiscent capsule*, berry, or drupe; seeds arillate or not; embryo straight, the cotyledons very large to small, and then the hypocotyl swollen; endosperm scanty or lacking (Figure 9.69).

Floral formula:

*, 2–10, 2–14, ⓧ, (3–5 (–∞)); capsule, berry, drupe

Distribution: Widespread; mainly tropical except for *Hypericum* and relatives.

Genera/species: 38/1100. **Major genera:** *Hypericum* (360 spp.), *Calophyllum* (200), *Garcinia* (200), *Clusia* (160), and *Mammea* (70). *Clusia, Hypericum*, and *Triadenum* occur in the continental United States and/or Canada.

Economic plants and products: The fruits of *Garcinia mangostana* (mangosteen) and *Mammea americana* (mamey apple) are highly prized. Some species of *Hypericum* are used medicinally, providing a popular treatment for depression. *Clusia* and *Hypericum* are used as ornamentals due to their showy flowers. Several genera provide timber.

Discussion: Clusiaceae are assumed to be monophyletic on the basis of anatomical and chemical synapomorphies. A cladistic analysis based on *rbcL* nucleotide sequences (Chase et al. 2002) suggests that Podostemaceae, an odd family of aquatics of rapidly flowing streams, may be nested within Clusiaceae. This hypothesis, however, is rather tentative, and the family is maintained here in its traditional circumscription. It is of interest that both Clusiaceae and Podostemaceae have canals, cavities, or individual cells with exudate and ovules with a thin sporangium wall. The traditional name for the family, Guttiferae, meaning "gum-bearing," refers to the clear to colored resinous sap characteristic of the group.

Figure 9.69 Clusiaceae (Guttiferae). (A–E) *Triadenum virginianum:* (A) plant with fruits (× 0.75); (B) flower, note stamens in threes with alternating staminodes (× 7.5); (C) ovary in cross-section (× 22); (D) capsule (× 5.5); (E) seed (× 37). (F) *T. walteri:* fascicle of stamens, note apical glands (× 18). (From Wood and Adams 1976, *J. Arnold Arbor.* 57: p. 89.)

Infrafamilial relationships have been considered by several workers, most recently Robson (1977) and Stevens (personal communication). Three subfamilies are recognized by Stevens (based on a morphology-based cladistic analysis). Kielmeyeroideae (e.g., *Calophyllum*, *Mammea*, and *Mesua*) are sister to the remaining genera and are woody plants with secretory canals and sometimes black-dotted leaves; an androecium that is not obviously fasciculate; short to elongate, usually connate styles; and an exposed apical bud. Clusioideae (e.g., *Garcinia* and *Clusia*) are woody plants with secretory canals and have usually short, connate styles, an embryo usually with a very large hypocotyl, and an apical bud often protected by the petiole bases. The Hypericoideae (e.g., *Hypericum*, *Triadenum*) are woody to herbaceous plants and lack exudate, but have black and/or clear dots in the leaves; they have elongate and often distinct styles, and an exposed apical bud. Hypericoideae are well developed in temperate regions, in contrast to the first two groups. Hypericoideae have often been recognized as a separate family (Hypericaceae). Recognition of Hypericaceae probably results in a paraphyletic Clusiaceae s.s. (Adams 1962; Robson 1977; Stevens, pers. com.).

Clusia is noteworthy for its broad range in habit. Some species begin growth as epiphytes, growing like a strangling fig; others have numerous adventitious roots that assist in support of the stems.

The showy flowers with conspicuous stamens characteristic of Clusiaceae are pollinated mainly by bees and wasps. Pollen is often the pollinator reward, but some taxa (e.g., *Clusia*) secrete terpenoid resins. Species with fleshy fruits (e.g., *Mammea*, *Calophyllum*, *Garcinia*) or capsules opening to expose colorful arillate seeds (e.g., *Clusia*) are usually dispersed by birds or mammals. Those with dry capsules and small seeds, as in *Hypericum*, are dispersed by wind and/or water.

References: Adams 1962; Chase et al. 2002; Maguire 1976; Ramirez and Gomez 1978; Robson 1977; Wood and Adams 1976.

Rhizophoraceae Persoon
(Red Mangrove Family)

Trees or shrubs, often with prop roots or pneumatophores; tannins usually present. Hairs usually simple. *Leaves opposite, with adjacent leaf pairs usually diverging from each other at an angle less than or greater than 90°*, simple, serrate or crenate to entire, with pinnate venation; **stipules interpetiolar, usually with colleters at the base of adaxial surface.** Inflorescences usually determinate, axillary. Flowers usually bisexual, radial, often with a hypanthium. *Sepals usually 4 or 5*, occasionally numerous, usually slightly connate, *thick, fleshy or leathery*, **valvate.** *Petals usually 4 or 5*, occasionally numerous, *distinct*, **usually fringed or hairy**, convolute or infolded, and **individually enclosing either a single stamen or a group of stamens in bud.** Stamens usually 8–10, occasionally numerous; filaments distinct or basally connate, sometimes short or lacking; *anthers sometimes multilocular and opening by a longitudinal flap*; pollen grains usually tricolporate. Carpels usually 2–6, connate; ovary superior to inferior, usually with axile placentation; stigma ± lobed. Ovules 2–8 in each locule. Nectar disk often present. *Fruit a septicidal capsule or several to 1-seeded berry*; seeds sometimes arillate; *embryo large, often with an elongated hypocotyl and germinating while still inside the fruit* (Figure 9.70).

Floral formula:

$$*, \widehat{(4-5)}, 4-5, \widehat{(8-10)}(-\infty), \overline{\widehat{(2-6)}}; \text{ capsule, berry}$$

Distribution and ecology: Pantropical; moist montane forests to tidally flooded wetlands (mangrove communities).

Genera/species: 12/84. **Major genera:** *Cassipourea* (55 spp.), *Rhizophora* (8), and *Bruguiera* (6). Only *Rhizophora* occurs in the continental United States.

Economic plants and products: *Rhizophora* is a source of tannin and charcoal. Mangrove swamps supply nutrition to marine communities and critical habitats for numerous marine organisms. They also stabilize shorelines and protect inland areas from wind and tides during tropical storms.

Discussion: The non-mangrove genera constitute a basal paraphyletic complex within Rhizophoraceae (Juncosa and Tomlinson 1988a,b; Toby and Raven 1988; Schwarzbach and Ricklefs 2000). In contrast, mangrove genera such as *Bruguiera* and *Rhizophora* form a well-supported clade, which can be diagnosed by their entire leaves, viviparous 1-seeded berries, and specializations relating to the structure and development of the seed or embryo. All mangrove species have anthers that open while still in the bud, depositing the pollen onto the hairy petals (another likely synapomorphy). Partially inferior ovaries, berries, prop roots, and lack of root hairs, typically considered to be characteristic of the mangrove taxa, actually evolved earlier in the evolution of the family, and also occur in tropical montane genera. *Rhizophora* is perhaps the most specialized genus in the family; it shows the unusual apomorphy of multiloculate anthers.

Most Rhizophoraceae produce nectar and are pollinated by butterflies, moths, various other insects, or birds. *Rhizophora* usually is wind-pollinated and does not produce nectar, although the flowers may be visited by pollen-gathering bees. Outcrossing is common. The mangrove species have fibrous "berries" that are one-seeded; the seed germinates while inside the fruit (viviparous), emerging from both seed coat and fruit up to 9 months

Figure 9.70 Rhizophoraceae. *Rhizophora mangle:* (A) fruiting branch, showing successive stages in development of hypocotyl (× 0.5); (B) inflorescence (× 1.5); (C) flower (× 3); (D) stamen, showing multiple pollen locules (× 22); (E) gynoecium and calyx in longitudinal section (× 5.5); (F) ovary in cross-section at position shown in E (× 5.5); (G–J) enlarging fruit in longitudinal section, showing progressive development from embryo to seedling, endosperm stippled (G–I × 3): 1, seed coat; 2, aborted ovule; 3, embryo; 4, cotyledons; 5, hypocotyl; (J) hypocotyl disarticulated from the cotyledonary tube, revealing plumule (× 1.4); (K) fruit with seedling, showing fully elongated cotyledonary tube and hypocotyl (× 0.7). (From Graham 1964, *J. Arnold Arbor.* 45: p. 288.)

before abscission. The cotyledons are fused into a tube; when the seedling is mature, the elongate hypocotyl (with its plumule) detaches from the cotyledons (and the berry) and falls from the tree. The seedling floats and survives in seawater. Eventually adventitious roots develop from the hypocotyl, which fix the seedling to the substrate. The initially horizontal seedlings become erected rapidly once rooted, such that the plumule is raised, snorkel-like, above the influence of tides.

The mangrove species have adapted to life in coastal estuary conditions by means of pneumatophores; prop roots; elongated, often curved and pointed, viviparous embryos; and the physiological blocking of salt uptake by the roots. Mangrove Rhizophoraceae have lost the ability to coppice (produce new sprouts from the old wood), which affects their ability to regenerate after major climatic or human-caused damage.

References: Dahlgren 1988; Graham 1964b; Juncosa and Tomlinson 1988a,b; Rabinowitz 1978; Schwarzbach and Ricklefs 2000; Toby and Raven 1988; Tomlinson 1986; Tomlinson and Cox 2000; Tomlinson et al. 1979.

Violaceae Batsch
(Violet Family)

Herbs to shrubs or trees; often with saponins and/or alkaloids. Hairs often simple. *Leaves alternate* and spiral or 2-ranked, occasionally opposite, sometimes forming a basal rosette, simple, sometimes lobed, entire to serrate, with pinnate to palmate venation; *stipules present.* Inflorescences indeterminate, sometimes reduced to a solitary flower, usually axillary. Flowers usually bisexual, *usually slightly to strongly bilateral. Sepals 5,* usually distinct. *Petals 5*, distinct, imbricate to convolute, sometimes the abaxial with a spur. *Stamens usually 5*, **their edges touching to form a ring around the gynoecium, filaments very short**, distinct to slightly connate, **the 2 abaxial anthers or all anthers with dorsal glandlike or spurlike nectaries, connective often with triangular, membranous, apical appendage,** *shedding pollen inward*, **which is picked up by the modified style**; pollen grains usually tricolporate. *Carpels usually 3, connate; ovary superior, with parietal placentation*; **style 1, usually curved or hooked and distally enlarged or modified; stigma usually expanded but with small receptive region**, sometimes lobed. Ovules 1 to many on each placenta. *Fruit usually a loculicidal capsule; seeds often arillate* (Figure 9.71).

Floral formula: X, 5, 5, ⑤, ③; capsule

Distribution: Widespread; mainly herbaceous in temperate regions.

Genera/species: 22/950. **Major genera:** *Viola* (500 spp.), *Rinorea* (300), and *Hybanthus* (100). *Viola* and *Hybanthus* occur within the continental United States and/or Canada.

Economic plants and products: Several species of *Viola* (violet, pansy) and *Hybanthus* (green-violet) are used as ornamentals. Rhizomes and roots of *Hybanthus ipecacuanha* have been used as a substitute for true ipecac (*Psychotria ipecacuanha*).

Discussion: Although Violaceae are clearly monophyletic, infrafamilial relationships are poorly understood. Most genera are placed in the Rinoreae (e.g., *Rinorea, Gloeospermum*, and *Rinoreocarpus*), a group characterized by radial to slightly bilateral flowers, or the Violeae (e.g., *Hybanthus* and *Viola*), which have strongly bilateral flowers. Analyses of *rbcL* sequences suggest that neither is monophyletic (Hodges, personal communication).

All genera are insect-pollinated. The flowers are visited by various flies, bees, wasps, and butterflies, and are typically outcrossing. Nectar guides are often present on the petals, and nectar, the pollinator reward, is stored in a modified petal-spur. Inconspicuous cleistogamous flowers may also be produced. The fairly small seeds may fall passively from the capsule valves or may be forcibly ejected. In some species the seeds are secondarily dispersed by ants, which are attracted by an oily aril.

References: Brizicky 1961b; Gates 1943.

Passifloraceae A. L. de Jussieu ex Roussel
(Passionflower Family)

Vines or lianas with axillary tendrils (modified inflorescences), usually with anomalous secondary growth, occasionally shrubs or trees lacking tendrils; usually with cyanogenic glucosides having a cyclopentenoid ring system, often also with alkaloids. Hairs various. *Leaves alternate and spiral, usually simple, often lobed*, entire to serrate, *venation usually ± palmate, usually with nectaries on the petiole (and lamina)*; stipules usually present. Inflorescences usually determinate, sometimes indeterminate or reduced to a single flower, axillary. Flowers usually bisexual, radial, usually with a cup-shaped to tubular hypanthium, often associated with conspicuous bracts. *Sepals usually 5*, distinct to slightly connate, *often petal-like. Petals usually 5 and distinct*, imbricate. **Complex corona borne on the apex and inner surface of the hypanthium,** *consisting of 1 to several rows of filaments, projections, or membranes. Stamens usually 5, often borne on a stalk along with gynoecium*, the filaments usually distinct; pollen grains tri- to 12-colporate. *Carpels usually 3, connate, ovary superior,* **and borne on a stalk** (often along with the androecium), *with parietal placentation*; stigmas usually 3, capitate to truncate. Ovules numerous on each placenta. Nectar disk at base of hypanthium. *Fruit a loculicidal capsule or berry; seeds usually flattened, with a fleshy aril* (Figure 9.72).

Floral formula: *, ⑤, 5, corona, 5, ③; capsule, berry

Figure 9.71 Violaceae. *Viola primulifolia:* (A) flowering plant (× 0.5); (B) flower (× 3); (C) flower, petals and two sepals removed, note sessile anthers closely placed around gynoecium (× 9); (D) lateral stamen, abaxial surface (× 16); (E) lateral stamen, adaxial surface (× 16); (F) lower stamen, abaxial surface, note nectar gland (× 16); (G) lower stamen, adaxial surface, note nectar gland, apical appendage, and open locule (× 16); (H) gynoecium, note unusual style with asymmetrically placed stigma (× 16); (I) ovary in cross-section (× 20); (J) nearly mature capsule (× 4); (K) capsule, seeds squeezed out by maturing valves (× 3); (L) seed (× 16); (M) cleistogamous flower (× 4); (N) cleistogamous flower, two sepals removed to expose reduced petals and two functional stamens (× 9). (From Wood 1974, *A student's atlas of flowering plants,* p. 71.)

Distribution: Widespread in tropical to warm temperate regions.

Genera/species: 18/630. **Major genera:** *Passiflora* (400 spp.) and *Adenia* (100). Only *Passiflora* occurs in the continental United States.

Economic plants and products: Several species of *Passiflora* (passionflower) have edible fruits; others are grown as ornamentals because of their showy flowers.

Discussion: The monophyly of Passifloraceae is supported most strongly by the presence of a corona in the flow-

Figure 9.72 Passifloraceae. *Passiflora incarnata:* (A) vine with tendrils and fruits (× 0.3); (B) apex of petiole with nectar glands (× 4); (C) bud with bracts (× 1.5); (D) bract with glands (× 7); (E) flower (× 1.5); (F) flower, central portion in partial longitudinal section, two stamens, one style removed (× 3); 1, sepal; 2, petal; 3, outer corona; 4, inner corona; 5, operculum; (G) ovary in cross-section (× 8.5); (H) young seed with developing aril (× 4); (I) older seed with aril (× 4); (J) seed with fleshy portion of seed coat removed (× 5.5). (From Brizicky 1961, *J. Arnold Arbor.* 42: p. 213.)

ers (de Wilde 1971, 1974). The "Paropsieae," which contains shrubs and trees that lack tendrils, probably represent a paraphyletic basal complex within the family. Passifloreae, in contrast, are clearly monophyletic, as evidenced by their viny habit, axillary tendrils, and specialized flowers.

Flowers of Passifloraceae may be white, green, red, blue, or purple, with the corona and perianth parts variously oriented and developed, and employ nectar as a pollinator reward. They attract a wide range of pollinators, including bees, wasps, moths, butterflies, birds, and bats. They also are the food plants of larvae of the butterfly genus *Heliconias*. Self-incompatibility is characteristic. The capsules or berries, with fleshy, usually colorful arils surrounding the seeds, usually lead to bird dispersal.

Variation in corona structure is often taxonomically significant. The corolla includes all structures between the perianth and the androecium, and is composed of several series of short to elongate filamentous structures. The operculum, a delicate membrane, is positioned between the corona and the gynoecium, and covers the nectar disk. The corona and operculum are outgrowths of the hypanthium. The corona is usually colorful, attracting pollinators and guiding them to the nectar. The operculum helps to hold nectar within the hypanthium.

References: Brizicky 1961a; de Wilde 1971, 1974; Killip 1938; Puri 1948.

Salicaceae Mirbel
(Willow Family)

Trees or shrubs; often with phenolic heterosides (salicin, populin), but usually lacking cyanogenic glycosides, usually with tannins. Hairs various. *Leaves deciduous, alternate, spiral or 2-ranked, simple,* usually serrate to dentate, **the teeth salicoid (i.e., with vein expanding at the tooth apex and associated with spherical, glandular setae)** except in *Casearia*, with pinnate to palmate venation, occasionally with pellucid dots or lines; *stipules usually present*. Inflorescences determinate to indeterminate, various in form, *sometimes in erect to pendent catkins,* sometimes reduced to a solitary flower, terminal or axillary. *Flowers bisexual or unisexual (plants dioecious),* radial, *often reduced, subtended by a usually hairy bract in Salix and Populus.* Sepals usually 3—8, distinct to slightly connate, sometimes ± vestigial, forming a disk-shaped to cup-shaped structure in *Populus* or represented by usually 1 or 2, small, often fringed, nectar glands in *Salix*. Petals 3—8, distinct, or lacking. Stamens 2 to numerous; filaments distinct to connate; pollen grains tricolpate, tricolporate, or lacking apertures. *Carpels usually 2–4, connate; ovary superior to half-inferior, with parietal placentation, the placentae sometimes deeply intruded and ovary then appearing axile;* stigmas usually 2–4, ± capitate to expanded and irregularly lobed. Ovules 1 to numerous on each placenta, sometimes with only 1 integument. Nectary a disk or of separate glands, sometimes lacking. *Fruit a loculicidal capsule, berry, or drupe; seeds often arillate or with basal tuft of hairs;* endosperm scanty or lacking (Figure 9.73).

Floral formula:　Staminate: *, (3–8), 3–8 or 0, 2–∞, 0
Carpellate: *, (3–8), 3–8 or 0, 0, (2–4),
capsule, berry, drupe

Distribution and ecology: Widespread, occurring from tropical to north temperate and arctic regions.

Genera/species: 58/1210. **Major genera:** *Salix* (450 spp.), *Casearia* (180), *Homalium* (180), *Xylosma* (85), *Populus* (35), and *Banara* (31). *Salix, Populus, Xylosma,* and *Flacourtia* occur in the continental United States and/or Canada.

Economic plants and products: *Salix* (willow) and *Populus* (poplar, cottonwood, aspen) provide lumber, wood pulp, and ornamentals. The bark of *Salix* was used medicinally due to the presence of salicylic acid, which reduces swelling and fever. Certain species of *Flacourtia* and *Dovyalis* are cultivated for their edible, fleshy fruits. Several genera provide useful ornamentals, e.g., *Oncoba, Casearia, Idesia,* and *Samyda*.

Discussion: Salicaceae are here circumscribed broadly, based on a phylogenetic analysis of *rbcL* sequences (Savolainen et al. 2000b; Soltis et al. 2000; Chase et al. 2002), and include not only *Salix* and *Populus*, but the numerous noncyanogenic members of "Flacourtiaceae". The monophyly of Salicaceae s.l. is well supported by DNA sequences, but the only morphological synapomorphy may be the distinctive salicoid tooth, but this is absent in *Casearia*, which may be basal. Such teeth are characteristic of a most other genera, although entire leaves have occasionally evolved. The presence of salicin, imperfect apetalous flowers, and (of course) salicoid teeth in *Salix* and *Populus* are shared with *Idesia, Itoa, Poliothyrsis,* and several other genera of "Flacourtiaceae" (Chase et al. 2002; Judd 1997b; Judd et al. 1994; Meeuse 1975). Molecular data support an especially close phylogenetic relationship of *Salix* and *Populus* to these genera, and also to the numerous noncyanogenic taxa traditionally placed in the "Flacourtiaceae" (e.g., *Azara, Banara, Dovyalis, Flacourtia, Homalium, Scolopia,* and *Xylosma*). In contrast, the cyanogenic "Flacourtiaceae" such as *Gynocardia, Kigellaria, Hydnocarpus,* and *Pangium* form a separate clade (Chase et al. 2002) within Malpighiales. Therefore, these cyanogenic genera have been segregated as the family Achariaceae. Members of this family also differ from Salicaceae in having many more petals than sepals, their linear (vs. globose or ovoid) anthers, and the lack of a nectar disk.

Phylogenetic relationships within Salicaceae are still poorly understood, but *Casearia* may be sister to the remaining genera. In addition, it is clear that *Populus* and

Figure 9.73 Salicaceae. *Salix caroliniana:* (A) winter bud (× 5); (B) expanding branchlet with young catkin (× 4); (C) staminate catkin (× 2); (D) staminate flower (× 10); (E) bract and nectar gland of staminate flower, stamens removed (× 20); (F) carpellate catkin (× 2); (G) carpellate flower, note lobed nectar gland at base (× 10); (H) stigmas (greatly magnified); (I) gynoecium in longitudinal section (× 20); (J) ovary in cross-section (× 40); (K) partly mature infructescence (× 2); (L) open capsule with escaping seeds (× 7); (M) seed, basal hairs removed (× 14); (N) embryo (greatly magnified); (O) leaf from rapidly growing summer shoot, note prominent stipules (× 1). (From Wood 1974, *A student's atlas of flowering plants*, p.

Salix form a clade, which is well nested within an array of tropical genera with less reduced, and usually insect pollinated flowers.

Within the *Salix* + *Populus* clade, *Populus* shows many plesiomorphic characters, including several bud scales, lack of floral nectar glands (thus wind-pollinated), two-integumented ovules, and a tendency toward palmate venation. The monophyly of both *Populus* and *Salix* is supported by molecular data (Chase et al. 2002). *Salix* has a single bud scale and flowers with the calyx modified into nectar-producing structures, thus possibly being secondarily insect-pollinated.

Systematists of the Englerian tradition considered *Salix* and *Populus* (i.e., Salicaceae s.s.) to be primitive within angiosperms, and placed the family within the "Amentiferae," a group with inconspicuous flowers in pendulous catkins (aments). Amentiferae include wind-pollinated families such as Platanaceae, Fagaceae, Betulaceae, and Juglandaceae, and are polyphyletic. The flowers of *Salix* and *Populus* may seem simple, but they are not primitive, and actually are highly reduced. It is clear that wind pollination has evolved numerous times within the angiosperms (Chase et al. 1993; Thorne 1973a, 1992).

The flowers of Salicaceae vary from large and showy to tiny and inconspicuous. The common presence of a nectar disk, ring of nectar glands, or nectar-secreting petal-appendages and the exposed androecium and gynoecium suggest that various unspecialized insects are the predominant pollinators. However, some genera, such as *Populus*, have reduced unisexual flowers and are wind pollinated. The flowers of *Populus* are aggregated on dangling catkins produced in early spring, usually before the leaves emerge, and have a reduced perianth and expanded stigmas. *Salix* flowers have nectar glands and an odor, and they attract various insects, although wind pollination is probably important as well.

Fruit type varies greatly. Dispersal by birds or mammals is characteristic of taxa with drupes (*Flacourtia*), berries (*Banara, Xylosma, Idesia, Oncoba, Scolopia, Dovyalis*), or capsules opening to expose colorful arillate seeds (*Samyda, Casearia*). Some genera (e.g., *Itoa, Poliothyrsis*) have capsules that open to release winged seeds and are wind dispersed. *Salix* and *Populus* have capsules

that open, releasing tiny, hairy seeds that are dispersed by wind and/or water.

Species limits within *Salix* are often difficult due to hybridization.

References: Argus 1974, 1986; Brunsfeld et al. 1992; Chase et al. 1993, 2002; Fisher 1928; Judd 1997b; Judd et al. 1994; Meeuse 1975; Savolainen et al. 2000b; Soltis et al. 2000; Thorne 1973a, 1992; Tollsten and Kundsen 1992.

Fabales

The monophyly of the Fabales is supported by *rbcL*, *atpB*, and 18S sequences (Chase et al. 1993; Källersjö et al. 1998; Savolainen et al. 2000a,b; Soltis et al. 2000). Morphological synapomorphies may include vessel elements with single perforations, a large, green embryo, and the absence of ellagic acid. Fabales contain 4 families and about 18,860 species. Major families are **Fabaceae**, **Polygalaceae**, and Surianaceae.

Fabaceae Lindley
(= Leguminosae Adanson)
(Legume or Bean Family)

Herbs, shrubs, trees, or vines/lianas climbing by twining or tendrils; **with a high nitrogen metabolism and unusual amino acids,** *often with root nodules containing nitrogen-fixing bacteria (Rhizobium);* sometimes with secretory canals or cavities; tannins usually present; often with alkaloids; sometimes cyanogenic; sieve cell plastids with protein crystals and usually also starch grains. Hairs various. *Leaves usually alternate, spiral to 2-ranked, pinnately (or twice pinnately) compound, to palmately compound, trifoliolate, or unifoliolate; entire* to occasionally serrate, with pinnate venation, occasionally leaflets modified into tendrils; *pulvinus of leaf and individual leaflets well developed, and leaf axis and leaflets usually showing sleep movements; stipules present, inconspicuous to leaflike,* occasionally forming spines. **Inflorescences almost always indeterminate,** sometimes reduced to a single flower, terminal or axillary. Flowers usually bisexual, radial to bilateral, **with a short,** *usually cup-shaped* **hypanthium.** *Sepals usually 5,* distinct to more commonly connate. Petals usually

---

Key to Major Families of Fabales

1. Leaves usually compound, with stipules and well-developed pulvini; fruit often a legume.....**Fabaceae**
1. Leaves usually simple, lacking stipules, pulvini not prominent; fruit a capsule, samara, nut, drupe, or berry..2
2. Carpels 1–5, distinct, styles gynobasic; flowers often radial; stamens usually opening by longitudinal slits...Surianaceae
2. Carpels usually 2–3, connate, style terminal; flowers ± bilateral; stamens usually opening by terminal pore...**Polygalaceae**

5, distinct or connate, valvate or imbricate, *all alike, or the uppermost petal differentiated in size, shape, or coloration (i.e., forming a banner or standard), and positioned internally or externally in bud, the 2 lower petals often connate or sticking together and forming a keel, or widely flaring.* Stamens 1 to numerous, but *usually 10,* hidden by the perianth to long-exserted, and sometimes showy; filaments distinct to connate, then commonly monadelphous or diadelphous (with 9 connate and 1, the adaxial, ± distinct); pollen grains tricolporate, tricolpate, or triporate, usually borne in monads, but occasionally in tetrads or polyads. **Carpel 1** (rarely 2-16), **distinct, elongate** (occasionally shortened), **and with a short gynophore**; *ovary superior, with lateral placentation; style 1, arching upward,* sometimes hairy; *stigma 1, small. Ovules 1 to numerous per carpel, borne in 2 rows along an upper placenta, often campylotropous.* Nectar usually produced by inner surface of hypanthium or an intrastaminal disk. **Fruit a legume**, sometimes a samara, loment, follicle, indehiscent pod,

achene, drupe, or berry; seeds often with hard coat with hourglass-shaped cells, sometimes arillate, and sometimes with a U-shaped line (pleurogram); embryo usually curved; endosperm often lacking (Figures 9.74, 9.75, 9.76; Table 9.2).

Floral formula: X or *, (5), (5), (10–∞), 1 ; legume

Distribution and ecology: Nearly cosmopolitan; the third largest family of angiosperms; occurring in a wide range of habitats.

Genera/species: 630/18,000. **Major genera:** *Astragalus* (2000 spp.), *Acacia* (1000), *Indigofera* (700), *Crotalaria* (600), *Mimosa* (500), *Desmodium* (400), *Tephrosia* (400), *Trifolium* (300), *Chamaecrista* (260), *Senna* (250), *Inga* (250), *Bauhinia* (250), *Adesmia* (230), *Dalbergia* (200), *Lupinus* (200), *Rhynchosia* (200), *Pithecellobium* (170), *Dalea* (150), *Lathyrus* (150), *Calliandra* (150), *Aeschynomene* (150), *Vicia* (140),

TABLE 9.2 *Diagnostic features for subfamilies of the Fabaceae (= Leguminosae).*

	Mimosoideae	"Caesalpinioideae"	Faboideae (= Papilionoideae)
Genera/species	40/2500	150/2700	429/12,615
Representative genera	*Acacia, Albizia, Calliandra, Inga, Leucaena, Mimosa, Parkia, Pithecellobium, Prosopis*	*Bauhinia, Caesalpinia, Cassia, Chamaecrista, Cercis, Delonix, Gleditsia, Parkinsonia, Senna, Tamarindus*	*Arachis, Astragalus, Baptisia, Crotalaria, Desmodium, Glycine, Indigofera, Lupinus, Melilotus, Phaseolus, Pisum, Robinia, Tephrosia, Trifolium, Vicia, Wisteria*
Habit	Trees to shrubs; occasionally herbs	Trees to shrubs; occasionally herbs	Herbs, shrubs, or trees
Leaves	Usually twice pinnately compound	Usually pinnately or twice pinnately compound	Pinnately compound to trifoliolate; occasionally unifoliolate
Inflorescence	Dense flowers opening ± simultaneously	± Lax flowers opening sequentially	± Lax flowers opening sequentially
Corolla	Radial Valvate Individually not showy	Usually bilateral (some radial) Imbricate, with upper petal usually innermost Usually showy	Bilateral (most) Imbricate, with upper petal outermost, two basal petals connate or coherent at apex — Banner — Wing — Keel Showy
Stamens	(10 – ∞) Showy	(10 – 1) Usually not showy	(10) or (9) + 1 Not showy
Pollen	Monads, tetrads, polyads	Monads	Monads
U-shaped line in seed (pleurogram)	Present	Usually lacking	Lacking

Figure 9.74   Fabaceae, subfamily Mimosoideae. *Albizia julibrissin:* (A) branch with flowers (× 0.4); (B) bisexual central flower of an individual inflorescence, showing elongated corolla tube and exserted staminal tube (× 2); (C) nonapical flower of an inflorescence, often functionally staminate (× 2); (D) anther (× 44); (E) ovary and lower part of style (× 9); (F) upper part of style and stigma (× 37); (G) ovary in longitudinal section from below, showing two rows of ovules (× 30); (H) ovary in longitudinal section, lateral view (× 30); (I) mature fruit (× 0.75); (J) seed, showing pleurogram (× 6); (K) seed in longitudinal section, showing large embryo (× 6); (L) leaflet (× 4.5); (M) base of petiole showing pulvinus and nectary (× 3). (From Elias 1974, *J. Arnold Arbor.* 55: p. 110.)

Albizia (130), *Swartzia* (130), *Lonchocarpus* (130), *Caesalpinia* (120), *Lotus* (100), *Millettia* (100), and *Erythrina* (100). Over a hundred genera occur in Canada and/or the continental United States; some of these are listed in Table 9.2.

Economic plants and products: Fabaceae are second only to Poaceae in economic importance. Important food plants include *Arachis* (peanuts), *Cajanus* (pigeon peas), *Cicer* (chickpeas), *Glycine* (soybeans), *Inga* (ice-

Figure 9.75 Fabaceae, subfamily "Caesalpinioideae." (A–J) *Senna bahamensis:* (A) flowering branch with immature fruits (× 0.75); (B) flower bud (× 3); (C) flower (× 2); (D) upper staminode (× 6); (E) functional lateral stamens (× 6); (F) functional lower stamen (× 6); (G) lower staminode (× 6); (H) gynoecium (× 6); (I) portion of immature fruit with developing seeds (× 4.5); (J) seed (× 6). (K–L) *S. obtusifolia:* (K) legume (× 0.75); (L) seed (× 6). (M–O) *Chamaecrista fasciculata:* (M) pinnately compound leaf with stipules (× 0.75); (N) flower bud (× 3); (O) seed (× 6). (From Robertson and Lee 1976, *J. Arnold Arbor.* 57: p. 38.)

cream bean), *Lens* (lentils), *Phaseolus* (beans), *Pisum* (peas), and *Tamarindus* (tamarind). However, it should be noted that many genera, including *Abrus* (rosary pea) and *Astragalus*, are highly poisonous. Many genera provide important forage plants, such as *Medicago* (alfalfa), *Melilotus* (sweet clover), *Trifolium* (clover), and *Vicia* (vetch). Some species can be plowed into the soil, resulting in a large increase in nitrogen levels, and form the basis of crop rotation. Ornamental species occur in *Acacia*, *Albizia* (mimosa), *Bauhinia* (orchid tree), *Calliandra* (powder puff), *Cassia*, *Cercis* (redbud), *Cytisus* (broom), *Delonix* (royal poinciana), *Erythrina*, *Gleditsia* (honey locust), *Laburnum* (goldenrain), *Lathyrus* (sweet pea), *Lupinus* (lupine), *Mimosa* (sensitive plant), *Parkinsonia* (Jerusalem thorn), *Robinia* (locust), and *Wisteria*. Commercial gums and resins are extracted from species of *Acacia* and *Hymenaea*; *Indigofera* (indigo) is used as a source of blue dye.

Figure 9.76 Fabaceae, subfamily Faboideae. *Vicia ludoviciana:* (A) tip of vine with flowers and fruits (× 0.3); (B) side view of flower bud (× 5); (C) flower (× 5); (D) banner petal (× 5); (E) inner surface of wing petal (× 5); (F) inner surface of keel petal (× 5); (G) keel seen from front (× 5); (H) androecium with nine stamens fused and one ± free (× 7); (I) gynoecium of one carpel (× 7); (J) young fruit, with one valve removed to show ovules (× 1.5); (K) mature legume (× 3); (L) dehisced legume (× 2.5); (M, N) seeds, note hilum half encircling seed (× 6); (O) seed in cross-section, note hilar region (dashed lines), and large embryo, cotyledon, and curved axis (× 8); (P) seedling. (From Wood 1974, *A student's atlas of flowering plants*, p. 60.)

Discussion: The monophyly of Fabaceae (or Legumi-nosae) is supported by several morphological features and DNA sequence data (Chappill 1994; Chase et al. 1993; Doyle 1994; Kajita et al. 2001). Nitrogen fixation occurs in numerous legumes, but is lacking in many early-diverging lineages; it is homoplasious and not synapomorphic for the family (Doyle et al. 1997).

Morphology (Hufford 1992) and DNA sequences (Chase et al. 1993; Soltis et al. 2000; Kajita et al. 2001) clearly place the family within the rosid complex, close to the Polygalaceae and Surianaceae (in the Fabales, as here circumscribed; see Angiosperm Phylogeny Group

1998). Pinnately compound leaves, flowers with imbri-cate perianth parts and (at least sometimes) with a dis-tinct nectar disk, along with similarities in woody anato-my and embryology have been used to suggest a relationship with Sapindales (Dickison 1981; Thorne 1992). Some botanists have considered Fabaceae to be closely related to Rosaceae because of the presence in both of stipulate leaves and hypanthia.

Three subgroups are generally recognized within Fabaceae: "Caesalpinioideae," Mimosoideae, and Faboideae (=Papilionoideae). In most classifications (Polhill et al. 1981) these are considered subfamilies, but they are sometimes treated as separate families (Cronquist 1981; Dahlgren 1983). (Diagnostic features for each subfamily are outlined in Table 9.2.) Phylogenetic analyses of morphological characters (Chappill 1994; Tucker and Douglas 1994) and DNA sequences (Doyle 1983; Doyle et al. 1997; Bruneau et al. 2001; Kajita et al. 2001) show that "Caesalpinioideae" are paraphyletic, with some genera more closely related to Mimosoideae and others more closely related to Faboideae than they are to one another. *Cercis* and its immediate relatives, including *Bauhinia*, may be sister to the rest of the family. Within Faboideae, it is clear that the temperate herbaceous lines are more recent derivatives of tropical woody groups, although the number of origins of the herbaceous habit remains uncertain (Lavin, personal communication). Detailed cladistic analyses are now available for several tribes of Fabaceae. *Swartzia* and *Sophora* (and their relatives) probably represent basal clades of Faboideae.

Several genera of Fabaceae show interesting coevolutionary relationships with various species of ants. Extrafloral nectar glands are common in Mimosoideae and "Caesalpinioideae," and the modified stipules of some species of *Acacia* are inhabited by ants, which protect the plant from herbivory (McKey 1989).

The flowers of Fabaceae are extremely variable in size, form, coloration, and pollinators. Nectar-gathering pollinators include bees, wasps, ants, butterflies, flies, beetles, birds, and bats, although bee pollination is most characteristic, especially of Faboideae (Arroyo 1981). The specialized bilaterally symmetrical flowers of this large group have a conspicuous banner (or standard) petal, which functions as a visual attractant, and two wings, which serve as a landing platform for visiting insects. When the bee lands and probes for nectar, it depresses the keel petals, which enclose the stamens and carpel, causing the stamens and stigma to contact the insect's underside. Various other specialized pollen presentation mechanisms occur; the stigma and stamens of the flowers of *Genista* and *Medicago*, for example, are explosively presented. Outcrossing is favored by protandry.

Fabaceae show a diversity of dispersal mechanisms. A large number of species have elastically dehiscing legumes. The valves are hygroscopic, and pressure builds as the fruit dries; eventually the two valves become coiled and separate from each other, tossing out the seeds. Other species have strongly flattened legumes that twist in the wind as they open, passively tossing out the small seeds. *Cassia*, *Gleditsia*, and *Enterocarpum* have more or less hard, indehiscent fruits in which the seeds are surrounded by a sweet to sour pulp. These fruits usually fall to the ground under the tree and are dispersed by mammals. Fruits of *Tamarindus* and *Inga* are mostly fleshy and form vertebrate-dispersed berries. Many species have fruits that open to expose either arillate (as in *Pithecellobium* and *Acacia*) or colorful seeds (as in *Abrus* and *Erythrina*) and are mainly dispersed by birds—the latter pair by deceit as the seeds are not edi-

ble. Members of many genera have fruits adapted for wind dispersal; the fruits may be indehiscent, with one or a few seeds, and winged, as in *Hymenolobium*, *Tipuana*, *Peltophorum*, and *Pterocarpus*, or inflated and balloonlike, as in *Oxytropis* and *Crotalaria*. *Trifolium* and *Astragalus* have various accessory structures for wind dispersal. The fruits of *Piscidia* are loments that split into one to a few seeded, winged segments. The loments of *Desmodium* and *Aeschynomene*, which break transversely into one-seeded segments, are covered with hooked hairs, leading to external transport by various animals. A few species of *Acacia* have ant-dispersed seeds.

References: Arroyo 1981; Augspurger 1989; Bruneau et al. 2001; Chappill 1994; Crisp and Weston 1987; Cronquist 1981; Dahlgren 1983; Dickison 1981; Doyle 1987, 1994, 1995; Doyle et al. 1997; Elias 1974; Herendeen et al. 1992; Heywood 1971; Hufford 1992; Isley 1998; Judd et al. 1994; Kajita et al. 2001; Lavin 1987; McKey 1989; Oldeman 1989; Polhill 1981; Polhill et al. 1981; Robertson and Lee 1976; Schrire 1989; Soltis et al. 2000; Thorne 1992; Tucker and Douglas 1994; Verma and Standley 1989; Weberling 1989.

Polygalaceae Hoffmannsegg and Link
(Milkwort Family)

Herbs to trees, or lianas; often producing triterpenoid saponins and methyl salicylate. Hairs simple. *Leaves alternate, spiral or 2-ranked, simple*, entire, with pinnate venation; stipules lacking, or paired glands or spines present. Inflorescences indeterminate, racemose to paniculate, sometimes reduced to a single flower, terminal or axillary. **Flowers** bisexual, ± **bilateral**. *Sepals usually 5, distinct to variably connate, often only the 2 abaxial ones connate, often with the 2 lateral ones larger than the others and petaloid. Petals (5–) 3, with 2 adaxial and 1 abaxial, distinct, but often all adnate to staminal tube, the abaxial one often boat-shaped, sometimes also appendaged, imbricate. Stamens (4–) 8 (–10); filaments distinct or connate*, **adnate to petals**; *anthers opening by 1 or 2 ± apical pores, or longitudinal slits;* **pollen grains polycolporate**. Carpels 2 or 3 (–8), connate; ovary superior, with axile placentation, but sometimes pseudomonomerous; style often with one branch stigmatic and the other sterile, ending in a hair tuft; stigma capitate. Ovules 1 in each locule. Nectar disk sometimes present. *Fruit a loculicidal capsule, samara, drupe, berry, or nut; seeds often with stiff hairs, sometimes arillate; endosperm present to lacking (Figure 9.77).*

Floral formula: X, ⑤, 3(–5), ④–10, ②–3; capsule

Distribution: Widespread; tropical to temperate.

Genera/species: 17/850. **Major genera:** *Polygala* (550 spp.), *Monnina* (125), and *Muraltia* (115). The first two occur in the continental United States and/or Canada.

Figure 9.77 Polygalaceae. (A–L) *Polygala paucifolia:* (A) flowering plant, with two chasmogamous flowers and one cleistogamous flower (at arrow) (× 1); (B) chasmogamous flower (× 4.5); (C) chasmogamous flower in lateral view, calyx and half of corolla removed to show androecium and gynoecium (× 4.5); (D) same, style removed (× 4.5); (E) anther shedding pollen (× 30); (F) empty anther (× 30); (G) ovary in longitudinal section, one ovule per locule (× 12); (H) ovule and immature aril (× 22); (I) mature seed with aril (× 12); (J) seed in longitudinal section, endosperm stippled, inner layer of seed coat hatched, embryo unshaded (× 12); (K) young capsules developed from cleistogamous flowers (× 3); (L) cleistogamous flower (× 15). (M–P) *P. grandiflora:* (M) flower (× 9); (N) flower in longitudinal section, with sepal and portion of lowermost petal removed, note connate filaments, adnate to petals (× 9); (O) capsule (× 9); (P) arillate seed (× 15). (From Miller 1971, *J. Arnold Arbor.* 52: p. 272.)

Economic plants and products: A few species of *Polygala* and *Securidaca* are cultivated as ornamentals.

Discussion: Polygalaceae are hypothesized to be monophyletic on the basis of morphological and *trnL-F* characters. The group is composed of two major clades (Persson

2001). The first, including *Xanthophyllum*, is diagnosed by the accumulation of aluminum in the tissues and distinct staminal filaments. The second major clade, including most of the family, is supported by the boat-shaped lower petal, stamens opening by terminal pores, and curved or bent style. Most of the genera in this clade (*Polygala, Mon-*

nina, Muraltia, and *Securidaca*) have several additional synapomorphies (e.g., flowers with three obvious petals, two carpels, and a bifid style with only one functional stigma) and constitute the Polygaleae, within which the large genus *Polygala* is nonmonophyletic.

Polygalaceae usually are considered to be related to Malpighiaceae or Krameriaceae, and all three families have bilaterally symmetrical flowers (Cronquist 1981). Analyses of *rbcL* sequences, however, place the family as the sister group of Fabaceae, in Fabales (Chase et al. 1993), as it is treated here. The flowers of most species of Polygalaceae do look remarkably similar to those of faboid legumes, yet the structures involved are not homologous, even if their ultrastructure is remarkably similar. The "wings" of legume flowers are lateral petals, while in Polygalaceae they are lateral sepals. The "keel" in flowers of Faboideae is formed from two fused or closely associated petals, while the similar structure of Polygalaceae is a single petal.

The showy flowers of many Polygalaceae attract various bees and wasps and function similarly to the flowers of faboid legumes. Self-pollination is also well known, occurring by a curving of the style or by movement of the sterile, hair-tufted style branch. Some species produce cleistogamous flowers. The samaras of *Securidaca* are dispersed by wind. The loculicidal capsules of *Polygala* release seeds with lobed, aril-like structures that are dispersed short distances by ants. Fleshy-fruited species are vertebrate-dispersed.

References: Chase et al. 1993; Cronquist 1981; Eriksen 1993a,b; Miller 1971a; Persson 2001; Verkerke 1985.

Rosales

The monophyly of Rosales receives strong support from analyses of *rbcL, atpB,* and 18S sequences (Chase et al. 1993; Källersjö et al. 1998; Savolainen et al. 2000a,b; Soltis et al. 2000) and other molecular data (APG 1998). The order is quite heterogeneous morphologically, but a reduction (or lack) of endosperm may be synapomorphic for these families. The presence of a hypanthium may also be synapomorphic, and this structure is seen in the families Rosaceae, Rhamnaceae, and some Ulmaceae; it has probably been lost in the more advanced families such as Celtidaceae, Moraceae, Cecropiaceae, and Urticaceae, which have very reduced flowers (Figure 9.78). Rosales also includes Elaeagnaceae (Russian olive).

Phylogenetic relationships within the order are still somewhat unclear, but Ulmaceae, Celtidaceae, Cannabaceae, Urticaceae, Moraceae, and Cecropiaceae probably constitute a clade, which is diagnosed by globose to elongate cystoliths (concretions of calcium carbonate) within specialized cells (lithocysts), reduced, inconspicuous flowers with 5 or fewer stamens, and 2-carpellate, unilocular ovaries with a single, apical (to basal) ovule (Humphries and Blackmore 1989; Judd et al. 1994). The

presence of urticoid teeth (which are nonglandular and have a slender median vein and converging lateral veins) and some characters of wood anatomy may also be synapomorphic. These families are often treated as the order Urticales (Cronquist 1981, 1988; Thorne 1992). They have sometimes been placed in the hamamelid complex (i.e., Platanaceae, Hamamelidaceae, Fagaceae, Betulaceae, etc.; see Cronquist 1981), but many systematists have considered them to be derived from Malvalean ancestors (Berg 1977, 1989; Dahlgren 1983; Thorne 1992). Here they are considered to be closely related to Rosaceae and Rhamnaceae and therefore are placed within Rosales.

Ulmaceae are probably sister to the Celtidaceae + Moraceae + Cecropiaceae + Urticaceae clade. This Celtidaceae + Moraceae + Cecropiaceae + Urticaceae clade is diagnosed by consistently unisexual flowers and curved embryos. Celtidaceae are included within Ulmaceae by most systematists (Cronquist 1981), creating a paraphyletic assemblage (i.e., Ulmaceae s.l.), which has been characterized by asymmetrical leaf bases, 2-ranked leaves, and inconspicuous flowers. The first character may be of little phylogenetic significance; it is clearly homoplasious because asymmetrical bases are lacking in many tropical Celtidaceae and occur in some Urticaceae and Moraceae. Two-ranked leaves and inconspicuous flowers are probably symplesiomorphies at this level. Celtidaceae and Ulmaceae are distinguished by pattern of leaf venation and vernation; floral morphology, sexuality, and anatomy; fruit type; embryo shape; pollen form; wood anatomy; flavonoid chemistry; type of sieve cell plastids; and chromosome number (Grudzinskaja 1967; Omori and Terabayashi 1993; Terabayashi 1991). For most of these features the Celtidaceae are more similar to Urticaceae and Moraceae than Ulmaceae, some of these similarities probably representing additional synapomorphies of the Celtidaceae + Moraceae + Cecropiaceae + Urticaceae clade. The Urticaceae + Cecropiaceae + Moraceae clade is supported by *rbcL* sequences and the presence of laticifers; within this clade, Moraceae are sister to a Cecropiaceae + Urticaceae clade, the latter being supported by laticifers restricted to the bark, a mucilaginous latex, pseudomonomerous gynoecia, basal placentation, and DNA features.

Phylogenetic placement of *Cannabis* and *Humulus* (Cannabaceae) has been argued. These two genera possess laticifers, suggesting a relationship with Urticaceae, but molecular characters place them firmly within Celtidaceae. They are distinguished by the combination of an herbaceous habit, dioecy, and two-carpellate gynoecia.

Rhamnaceae may be sister to the Ulmaceae-to-Urticaceae clade, as both groups show a reduction in the number of stamens (to a single whorl or less). The small family Elaeagnaceae (with distinctive, gold to silver peltate scales and lacking petals) may be sister to Rhamnaceae; both have basal ovules. Rosaceae may be the sister group of the remaining members of Rosales.

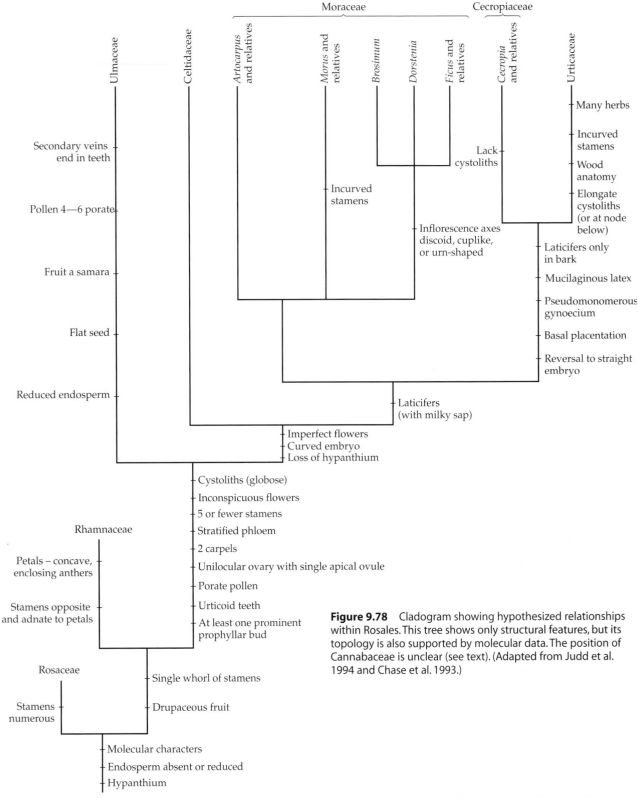

Figure 9.78 Cladogram showing hypothesized relationships within Rosales. This tree shows only structural features, but its topology is also supported by molecular data. The position of Cannabaceae is unclear (see text). (Adapted from Judd et al. 1994 and Chase et al. 1993.)

References: APG 1998; Baas et al. 2000; Berg 1977, 1989; Chase et al. 1993; Cronquist 1981, 1988; Dahlgren 1983; Grudzinskaja 1967; Humphries and Blackmore 1989; Judd et al. 1994; Källersjö et al. 1998; Miller 1970, 1971b; Omori and Terabayashi 1993; Savolainen et al. 2000a,b; Soltis et al. 2000; Sytsma et al. 1996; Terabayashi 1991; Thorne 1992; Wiegrefe et al. 1998; Zavada and Kim 1996.

The Rosales contain 11 families and about 6300 species; major families include **Rosaceae**, **Rhamnaceae**, **Ulmaceae**, **Celtidaceae**, Cannabaceae, **Urticaceae**, **Moraceae**, and Cecropiaceae.

Rosaceae A. L. de Jussieu

(Rose Family)

Herbs, shrubs, or trees (about three quarters of the genera contain woody plants), often rhizomatous, infrequently climbing; thorns sometimes present; often cyanogenic; **alkaloids absent**. Hairs typically simple, sometimes glandular, occasionally stellate; prickles sometimes present. Leaves usually alternate and spiral, simple to often palmately or pinnately compound, blade commonly with glandular-tipped teeth, with pinnate or palmate venation; *stipules usually present*. Inflorescences various. *Flowers often showy*, bisexual or infrequently unisexual (plants then dioecious or monoecious), usually radial, with a *hypanthium ranging from flat to cup-shaped or cylindrical and either free from or adnate to the carpels, often enlarging in fruit, with a nectar ring on the inside*. Sepals usually 5, sometimes alternating with epicalyx lobes. Petals usually 5, often clawed, imbricate, rarely absent. **Stamens usually numerous**, frequently 15 or more but sometimes 10 or fewer; filaments distinct or basally fused to nectar disk; pollen tricolporate. Carpels 1 to many, distinct or connate, sometimes adnate to hypanthium; ovary superior to inferior; styles same number as carpels; stigmas terminal; ovules 1, 2, or often more per carpel, basal, lateral, or apical (when carpels distinct) or ± axile placentation (when carpels connate). Fruit a follicle, achene (exposed or enclosed within the hypanthium, which is sometimes fleshy), pome, drupe, aggregate or accessory with drupelets or achenes, rarely a capsule. Endosperm usually absent (Figures 9.79–9.82).

Floral formula: $*, \underline{5}, \underline{5}, \underline{10-\infty}, \overline{\underline{(1-\infty)}}$; achene, druplet, follicles, drupe, pome

Distribution and ecology: The family is cosmopolitan and most abundant in the Northern Hemisphere. Some genera, such as *Lyonothamnus*, have very restricted distributions. In contrast, *Rubus*—the blackberries and raspberries—is native to six continents. Herbaceous members grow in temperate forest understory, salt marshes, freshwater marshes, arctic tundra, old fields, and roadsides. Woody members, such as *Rubus, Crataegus* (hawthorn), and some species of *Prunus* (cherries), are prominent in early stages of succession. Tree species, such as *Prunus serotina* (black cherry), may be compo-

---

Key to Major Families of Rosales

1. Indumentum of copious peltate scales and stellate hairs; flowers lacking petals and with a single carpel .... Elaeagnaceae
1. Indumentum various but usually not densely stellate; flowers with or without petals, the gynoecium with 1 to numerous, distinct to connate carpels 2

2. Flowers with nectar-producing hypanthium; carpels 1 to numerous, distinct to connate, and if connate, with axile placentation; stamens 4 to numerous 3
2. Flowers lacking a hypanthium, or if present, then not nectariferous; carpels 1 or 2, connate, with apical or basal placentation; stamens 1–5 4

3. Stamens 4 or 5, opposite the petals; petals concave or hooded, each enclosing a stamen at anthesis .... **Rhamnaceae**
3. Stamens 10 to numerous; petals flat to slightly cupped, not enclosing stamens at anthesis ...... **Rosaceae**

4. Secondary veins running directly into the teeth; flowers bisexual or unisexual; embryo straight; fruits dry, usually samaras or nutlike .... **Ulmaceae**
4. Secondary veins forming a series of loops; flowers consistently unisexual; embryo curved or straight; fruits drupes or achenes 5

5. Laticifers lacking; sap watery; fruits drupes, not multiple; plants woody .... **Celtidaceae**
5. Laticifers present, but sometimes reduced; sap milky to watery; fruits drupes or achenes, usually densely clustered and multiple and often also associated with accessory structures; plants woody to herbaceous 6

6. Latex throughout plant, white; cystoliths ± globose; gynoecium of usually 2 carpels, with apical placentation .... **Moraceae**
6. Latex restricted to bark or essentially lacking, milky to clear and mucilaginous; cystoliths ± elongate or lacking; gynoecium pseudomonomerous, with ± basal placentation 7

7. Cystoliths present and ± elongate; plants usually herbs; stamens incurved in bud .... **Urticaceae**
7. Cystoliths lacking; plants trees or shrubs; stamens ± straight in bud .... Cecropiaceae

Figure 9.79 Rosaceae, "spiraeoid group." *Aruncus dioicus:* (A) flowering stem of staminate plant (× 0.3); (B) portion of rhizome with stem base (× 0.75); (C) carpellate flower (× 17); (D) staminate flower (× 17); (E) staminate flower in longitudinal section, note nectar ring and rudimentary carpels (× 17); (F) bisexual flower (from "staminate" plant) in longitudinal section, petals removed (× 17); (G) fruits (× 7); (H) developing fruit in longitudinal section (× 17); (I) seed (× 30); (J) embryo (× 30); (K) same, side view (× 30). (From Robertson 1974, *J. Arnold Arbor.* 55: p. 326.)

Figure 9.80 Rosaceae, subfamily Rosoideae. (A–H) *Rosa carolina:* (A) branch with flower buds (× 0.75); (B) flower (× 1.5); (C) outer calyx lobe with two lateral appendages (× 3); (D) hairs on margin of calyx lobe (× 37); (E) flower in longitudinal section, note stamens, petals, and sepals attached at apex of cup-shaped hypanthium (× 4); (F) carpel (× 9); (G) achene (× 4.5); (H) achene in longitudinal section, note large embryo and lack of endosperm (× 7.5). (I– J) *R. setigera:* (I) immature fruit, with fleshy hypanthium (hip) surrounding achenes (× 4.5); (J) longitudinal section of fruit (× 4.5). (K) *R. eglanteria:* longitudinal section of fruit, note achenes attached to fleshy hypanthium at various levels (× 3). (From Robertson 1974, *J. Arnold Arbor.* 55: p. 613.)

nents of mature deciduous forests. Five genera, all in the Dryadeae as defined here—*Cercocarpus, Chamaebatiaria, Cowania, Dryas,* and *Purshia*—fix nitrogen through a symbiosis with the actinomycete *Frankia.*

Genera/species: 85/3000; the major genera are listed in the discussions that follow.

Economic plants and products: The family is important primarily for edible temperate zone fruits and ornamentals. Of first importance is *Malus domestica* (apple), a native of the Old World now with thousands of cultivars grown throughout temperate regions. *Prunus* has a wide variety of fruits: *P. dulcis* (almond), *P. armeniaca* (apricot), *P. avium* (sweet cherry), *P. cerasus* (sour cherry), *P. persica* (peach), and *P. domestica* (plum). Other edible fruits of commercial interest are *Pyrus* (pear), *Rubus* (blackberry, loganberry, raspberry), *Fragaria* (strawberry), *Cydonia* (quince), and *Eriobotrya* (loquat). Ornamental plants in the family include herbs, such as *Alchemilla* (lady's mantle), *Geum* (avens), *Filipendula* (meadowsweet), and *Potentilla* (cinquefoil), and many woody plants, notably *Amelanchier* (shadbush), *Chaenomeles* (flowering quince), *Cotoneaster*, *Crataegus* (hawthorn), *Exochorda* (pearlbush), *Kerria*, *Malus* (especially crab apples), *Photinia*, *Physocarpus* (ninebark), *Prunus* (especially Japanese flowering cherries), *Pyracantha* (firethorn), *Rhodotypos* (jetbead), *Rosa* (rose), *Sorbus* (mountain ash, rowan), and *Spiraea* (bridal wreath). Roses, perhaps the most popular and widely grown garden flower in the world, are complex hybrids developed from about nine wild species. *Prunus serotina* produces a wood that is valued for furniture and cabinetry; several genera supply timber.

Discussion: Even though Rosaceae display considerable diversity in anatomy, vegetative features, and fruit morphology, the family has long been considered monophyletic. Numerous stamens, absence of alkaloids, and sequences from *rbcL* sequences (Morgan et al. 1994), *ndhF* (Evans 1999), the granule-bound starch synthase gene (*GBSSI*; Evans et al. 2000) and other genes strongly support the monophyly of Rosaceae. Alternate leaves with stipules plus showy flowers with radial symmetry, numerous stamens, and a hypanthium aid in recognition of the family. Saxifragaceae, Fabaceae, Crassulaceae, and other groups have been proposed as potential relatives of the family, but several genes from the chloroplast and nucleus place Rosaceae as sister to remaining members of the Rosales.

Fruit type was the primary criterion in the traditional subdivision of Rosaceae into four subfamilies. For two of these subfamilies, Maloideae (Figure 9.81) and Amygdaloideae (Figure 9.82), fruit type was uniform. In pomes, the fruit type unique to the Maloideae, the hypanthium is fused to the ovary wall and often enlarges into an edible reward for fruit dispersers, as in apples and pears. Amygdaloideae (Prunoideae) bear drupes, and there is only one carpel in the flowers of the major genus, *Prunus*. Fruit type varies more in the other two subfamilies. Rosoideae (Figure 9.80) have achenes or drupelets. Achenes may be associated with a much-enlarged receptacle (*Fragaria*) or enclosed in a more or less urn-shaped or cylindrical receptacle in many genera. The fruit of *Rosa* is of this latter type, and the achenes together with the enclosing receptacle in this genus are called a hip. In *Rubus* the drupes are quite small, and therefore called drupelets, and aggregated. The fruit in the fourth traditional subfamily, "Spiraeoideae" (Figure 9.79), is a follicle or capsule.

Base chromosome number, various chemical constituents, distribution of rust fungal parasites, and DNA sequences from rDNA ITS 1 and 2, *matK*, *ndhF*, and *rbcL*

do not agree with the subfamilies delimited on the basis of fruit type. Rosoideae are defined more narrowly on the basis of molecular and chemical characters. Members of this more narrowly defined subfamily are herbs or shrubs with usually numerous carpels; a base chromosome number of seven (eight in one genus); the absence of the sugar alcohol sorbitol, cyanogenic glycosides, and flavones; and the presence of ellagic acid. The subfamily contains about 25 genera and 1200–1900 species, and the major genera are *Rubus* (400–740 spp.), *Potentilla* (200–500), *Alchemilla* (250), *Rosa* (250), *Geum* (40), and *Fragaria* (15). The first five of these genera present great systematic challenges due to the occurrence of hybridization, polyploidy, and/or agamospermy. *Rubus* is particularly challenging in Europe and North America and displays as much as any group the taxonomic complexity attending the interplay of hybridization, polyploidy, and agamospermy. *Potentilla*, whether broadly defined or after removal of many segregate genera, is not monophyletic. Several morphological characters in *Potentilla* and related genera in the Potentilleae, such as leaf type and whether the floral receptacle is swollen (as in strawberries) or not, are homoplasious with respect to molecular phylogenies. Molecular data place the Rosoideae as the sister to the remainder of the Rosaceae (see Figure 9.83).

The narrow definition of the Rosoideae requires removal and redefinition of three tribes—Adenostomateae, Dryadeae, and Kerrieae—that differ from this subfamily in their base chromosome number of 9 and chemical constituents. The Adenostomateae contain just *Adenostoma* (3 species of western North America), which is closely related to two genera from the tribe Sorbarieae of the "Spiraeoideae," *Chamaebatiaria* (2 species of the southwestern United States) and *Sorbaria* (15 species of Asia). Sorbarieae is therefore considered to contain *Adenostoma*. *Spiraeanthus*, a monotypic genus with a restricted distribution in Asia, nests within the Sorbarieae in *ndhF* trees. Another monotypic genus with a restricted distribution, *Lyonothamnus*, has been placed in the Sorbarieae by some workers. *Lyonothamnus floribundus* is native to just three islands off the southern Californian coast, and it has opposite, fernlike leaves and other morphological features that are quite unusual in the Rosaceae. This species has long been considered an isolated lineage, but molecular data place it as the sister group to the Dryadeae. Only part of the Dryadeae as traditionally defined by fruit type is removed from Rosoideae; the genera that are removed and make up this newly defined tribe are the five genera in the family with the capacity for nitrogen-fixing symbiosis (see "Distribution and ecology," above). These five genera all occur primarily in western North America; *Dryas* is also found in Europe. Three or four genera make up the Kerrieae as defined here, the best known being two Asian, monotypic genera used as ornamentals, *Kerria* and *Rhodotypos*.

Figure 9.81 Rosaceae, subfamily Maloideae. *Amelanchier laevis:* (A) flowering branch (× 0.75); (B) flower (× 4.5); (C) flower in longitudinal section, note nectar disk on inner surface of hypanthium, above carpels (× 9); (D) ovary in cross-section, at level of arrow in C (× 9); (E) fruiting branch (× 0.75); (F) pome (× 4.5); (G) pome in longitudinal section, note inferior ovary and seed (at left) (× 4.5); (H) embryo (× 9). (From Robertson 1974, *J. Arnold Arbor.* 55: p. 635.)

The traditional subfamily "Spiraeoideae" is clearly not monophyletic; its genera are widely separated on molecular trees. *Spiraea* (100 spp.), *Aruncus* (12), and about five other genera comprise the tribe Spiraeeae, which is monophyletic on the basis of molecular data and has the potential synapomorphy of loss of stipules. Other tribes

of this subfamily, such as Neillieae and Sorbarieae (which was discussed above), are not closely related to Spiraeeae. Neillieae contain three genera, *Neillia* (20 Asian species), *Physocarpus* (15 east Asian and North American species), and *Stephanandra* (5 Asian species). Many other genera that were part of the traditional "Spiraeoideae"

Figure 9.82 Rosaceae, subfamily Amygdaloideae. (A) *Prunus virginiana:* flowering branch (× 0.3).
(B–I) *P. serotina:* (B) fruiting branch (× 0.3); (C) tip of raceme (× 4.5); (D) flower (petals removed), in longitudinal section, note hypanthium, solitary carpel with two ovules (× 15); (E) ovary in longitudinal section, at
right angle to that of D (× 30); (F) pit from drupe (× 4.5); (G) pit in cross-section, the wall hatched, cotyledons unshaded (× 4.5); (H) seed (× 4.5); (I) embryo (× 4.5). (J–L) *P. pensylvanica:* (J) tip of branch with inflorescences (× 0.75); (K) shoot apex, showing leaf bases with glands and stipules (× 4.5); (L) flower (× 4.5).
(M) *P. caroliniana:* spinulose-serrate leaf (× 0.75). (From Robertson 1974, *J. Arnold Arbor.* 55: p. 658.)

align with various clades in the Rosaceae molecular phylogeny (Figure 9.83). For example, *Exochorda* groups with some genera of the Amygdaloideae (see below), *Spiraeanthus* with the Sorbarieae (see above), and four genera—*Kageneckia*, *Lindleya*, *Gillenia*, and *Vauquelinia*—with the Maloideae (see below).

Amygdaloideae have traditionally comprised one or a few genera with drupaceous fruits and a base chromosome number of eight. The subfamily contains one major genus, *Prunus*, with about 200 species mostly of the Northern Hemisphere. DNA sequences and ovule morphology show that the capsule-fruited *Exochorda* belongs in a clade that also includes two small genera with drupes and base chromosome number of eight, *Oemleria* and *Prinsepia*. Molecular data available at this time do not support a broadly defined subfamily of these three genera plus *Prunus*, and additional studies are required to resolve the relationships of these groups the within the family.

Ovule morphology and sequences from several genes confirm that four genera from the traditional "Spiraeoideae" with capsular or follicular fruits—*Kageneckia*, *Lindleya*, *Gillenia*, and *Vauquelinia*—are closely allied to the Maloideae. Ovules in the Maloideae plus these four genera are paired in each carpel, basally inserted, and anatropous. In addition, rust fungi in the genus *Gymnosporangium* infect only plants with pomes (i.e., the Maloideae as traditionally defined) plus *Gillenia* and *Vauquelinia*. *Kageneckia* and *Lindleya* have a haploid chromosome number of 17 (the same base number as the

Maloideae), and *Vauquelinia* has $x = 15$, which could have arisen from 17 by aneuploid reduction. Because of these diverse sources of evidence, a broader definition of the Maloideae to include *Kageneckia*, *Lindleya*, and *Vauquelinia* is used here. While the pome is no longer an autapomorphy for Maloideae s.l., the subfamily is still defined by the potential autapomorphy of a base chromosome number of 17. In this broad interpretation, the subfamily contains about 30 genera and 1000 species.

Maloideae are remarkable for the role that hybridization has played in their evolution. The base chromosome number of 17 hints that the subfamily could have arisen through allopolyploidization, and genetic studies confirm that members of the subfamily are allopolyploids. Since about 1930, a favored hypothesis for origin of the subfamily has been an ancient hybridization between an amygdaloid ($x = 8$) and a spiraeoid ($x = 9$). Recent studies of GBSSI falsify this hypothesis and identify instead ancestors of the subfamily closely related to *Gillenia*, a genus of herbaceous plants of the southeastern United States with $x = 9$. Aneuploidy could account for the Maloideae base number of 17 from 18.

Intergeneric hybridization, which is rather unusual in flowering plants, is relatively common in Maloideae and has contributed to uncertainty about the generic limits of the subfamily. There is the possibility that past gene flow is partly responsible for the lack of evolutionary divergence between the genera that has been observed in numerous genes and in wood anatomy. This lack of divergence is hard to explain given that several of the genera (e.g., *Amelanchier*, *Crataegus*, and *Photinia*) are relatively old, with fossils from 48 to 50 million years ago. Interspecific hybridization and agamospermy frequently occur in the larger genera, including *Crataegus* (265 spp.),

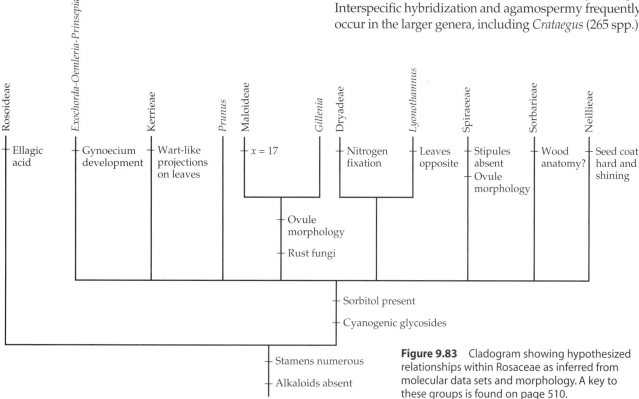

Figure 9.83 Cladogram showing hypothesized relationships within Rosaceae as inferred from molecular data sets and morphology. A key to these groups is found on page 510.

Cotoneaster (260), *Sorbus* (258), *Malus* (55), and *Amelanchier* (33), making recognition of species difficult. Other major genera are *Pyrus* (76) and *Photinia* (54).

The current status of Rosaceae phylogeny is that there are two large clades, Rosoideae s.s. and Maloideae s.l., and several small clades of two to seven genera. Other than the basal position of Rosoideae, there is uncertainty about the relationships among the other clades (Figure 9.83). The small clades include some tribes that may or may not have the traditional membership (e.g., Dryadeae, Kerrieae, Neillieae, Sorbarieae, and Spiraeeae) and some clusters of genera that have not been part of traditional classifications of the family (e.g., the *Exochorda-Oemleria-Prinsepia* clade). Additional work is required to resolve further relationships in this family.

Flowers of Rosaceae are unspecialized, with radial symmetry and flat or shallowly cup-shaped corollas, and are adapted primarily for generalist pollinators. Smaller flowers are visited by flies and short-tongued bees, while species with larger flowers are pollinated by long-tongued bees, wasps, butterflies, moths, and beetles. There has been considerable diversification for fruit dispersal, with adaptations for wind dispersal (winged seeds, plumose styles, and winged hypanthia that enclose the seeds), animal attachment (hooked prickles or barbs on the outside of the dispersal unit, hooked styles), and ingestion (small oil bodies for ant dispersal, and fleshy fruit walls and/or hypanthia for consumption by birds, mammals, and reptiles).

References: Alice et al. 2001; Bortiri et al. 2001; Campbell et al. 1995; Eriksson et al. 1998; Evans 1999; Evans and Dickinson, 1999 a,b; Lee and Wen. 2001; Morgan et al. 1994; Phipps et al. 1990; Robertson 1974; Robertson et al. 1991, 1992; Rohrer et al. 1991, 1994; Wolfe and Wehr 1988; Zhang 1992.

Rhamnaceae A. L. de Jussieu
(Buckthorn Family)

Trees, shrubs, often with thorns, or lianas with tendrils or twining stems, the axillary branches departing somewhat laterally; sometimes with nitrogen-fixing bacteria (*Frankia*) in root nodules; often with tannins. Hairs usually simple. *Leaves alternate and spiral*, or less commonly opposite, simple, entire to serrate, with strongly pinnate or palmate venation, tertiary veins often strongly ladderlike; *stipules present*, sometimes spinose. Inflorescences determinate, sometimes reduced to a single flower; axillary or terminal. *Flowers* usually bisexual, radial, *small, with disklike to cylindrical hypanthium.* Sepals usually 4 or 5, distinct, valvate. *Petals usually 4 or 5, distinct,* **usually ± concave or hooded and enclosing the anther at flowering,** *usually clawed.* Stamens 4 or 5, distinct, **opposite the petals; filaments adnate to base of petals**; pollen grains tricolporate. Carpels usually 2 or 3, connate; ovary superior to inferior, with axile placentation and usually 1

ovule attached at the base of each locule; stigmas usually ± capitate. Nectar-producing tissue on inner surface of hypanthium. **Fruit a dehiscent to indehiscent drupe (then ± schizocarpic) with 1 to several pits**, often with a conspicuous subbasal rim, rarely a samaroid schizocarp; endosperm present or lacking (Figure 9.84).

Floral formula:

*, 4–5, 4–5, 4–5, (2–3); drupe, sometimes dehiscent

Distribution and ecology: Nearly cosmopolitan, but especially diverse in tropical regions; characteristic of limestone soils. *Ceanothus* and some other genera fix nitrogen.

Genera/species: 45/850. **Major genera:** *Rhamnus* (150 spp.), *Phylica* (150), *Zizyphus* (100), *Gouania* (60), and *Ceanothus* (50). Noteworthy genera occurring in the continental United States and Canada are *Adolphia, Berchemia, Ceanothus, Colubrina, Gouania, Krugiodendron, Reynosia, Rhamnus, Sageretia,* and *Ziziphus.*

Economic plants and products: Fruits of *Ziziphus jujuba* (jujube) and the pedicels of *Hovenia dulcis* (raisin tree) are eaten. *Ceanothus, Colletia, Pomaderris, Rhamnus,* and *Phylica* provide ornamentals.

Discussion: The monophyly of Rhamnaceae is supported by morphology plus *rbcL* and *trnL-F* sequences (Richardson et al. 2000a). The family may be most closely related to Rosaceae, as supported by the presence of a well-developed hypanthium with a nectar-producing inner surface and stipulate leaves. However, DNA sequences (Chase et al. 1993; Soltis et al. 2000) suggest a closer relationship with Ulmaceae, Moraceae, and relatives.

Infrafamilial relationships have been studied by Richardson et al. (1997, 2000a,b); generic delimitation is often difficult. Extensive homoplasy is evident in ovary position, fruit type, and habit. Rhamnaceae may be composed of a few well-supported clades, for which no obvious morphological apomorphies are known. Major clades are the rhamnoids (e.g., *Rhamnus, Krugiodendron, Berchemia,* and *Reynosia*) and the ziziphoids (e.g., *Hovenia, Ziziphus, Ceanothus, Pomaderris, Phylica, Adolphia, Colubrina, Gouania,* and *Colletia*). Many rhamnoids have superior ovaries and drupaceous fruits, while ziziphoids usually have semi-inferior to inferior ovaries and schizocarpic fruits.

The usually inconspicuous flowers are visited by flies, bees, wasps, and beetles, and may be outcrossing or selfing. Seeds of fleshy-fruited genera such as *Rhamnus, Berchemia, Reynosia,* and *Krugiodendron* are usually dispersed by birds and/or mammals. Those with dehiscent, more or less schizocarpic drupes (e.g., *Ceanothus* and *Colubrina*) typically eject their seeds. The samaroid schizocarps of *Gouania* are wind-dispersed.

Figure 9.84 Rhamnaceae. (A–F) *Colubrina arborescens:* (A) branch with flowers and fruits (× 0.5); (B) flower, from above, note nectar disk and cup-shaped petals opposite the stamens (× 10); (C) flower in longitudinal section (× 10); (D) nearly mature fruit with very thin exocarp (× 6); (E) opening fruit, showing explosive endocarp and seeds (× 6); (F) persistent seeds after dehiscence of pericarp (× 6). (G–I) *C. elliptica:* (G) opening fruit, one segment of endocarp (to right) removed (× 6); (H) dehisced segment of endocarp (× 6); (I) seed (× 6). (Original art prepared for the Generic Flora of the Southeast U.S. Project. Used with permission.)

References: Brizicky 1964b; Chase et al. 1993; Cronquist 1981; Richardson et al. 1997, 2000a,b; Soltis et al. 2000.

Ulmaceae Mirbel
(Elm Family)

Trees; growing by a central plagiotropic axis, which secondarily becomes erect, and the lateral branches similar, thus with a spreading aspect; often with tannins; cystoliths present; laticifers absent. *Hairs simple, often with mineralized cell walls. Leaves alternate and 2-ranked, simple, simply or doubly serrate,* **with pinnate venation, secondary veins ending in the teeth,** *blade with asymmetrical base;* stipules present. Inflorescences determinate, forming fascicles, axillary. Flowers bisexual or unisexual (plants then monoecious, dioecious, or polygamous), radial, *inconspicuous,* with hypanthium. Tepals 4–9, distinct to connate, usually imbricate. Stamens 4–9, opposite the tepals; *filaments distinct, erect in bud;* **pollen grains 4- to 6-porate.** *Carpels 2, connate; ovary superior, with apical placentation and usually*

1 locule; stigmas 2, extending along adaxial side of styles. Ovule 1. **Fruit a samara or nutlet; seeds flat;** *embryo straight;* **endosperm of a single layer of cells and appearing absent** (Figure 9.85).

Floral formula: *, (4–9), 4–9, ②; samara, nut

Distribution: Widely distributed, but most diverse in temperate regions of the Northern Hemisphere.

Genera/species: 6/40. **Major genus:** *Ulmus* (25). *Ulmus* and *Planera* occur in the continental United States and/or Canada.

Economic plants and products: *Ulmus* (elm) and *Zelkova* provide timber and important ornamentals.

Discussion: Ulmaceae are monophyletic based on numerous morphological synapomorphies (Zavada and Kim 1996) and easily distinguished from Celtidaceae by

Figure 9.85 Ulmaceae. (A–H) *Ulmus americana:* (A) leaf (× 0.75); (B) branch with flowers (× 0.75); (C) flower at time of shedding of pollen (× 7); (D) flower with receptive stigmas (× 7); (E) gynoecium, in partial longitudinal section to show apical ovule (× 7); (F) samara (× 1.5); (G) seed (× 6); (H) embryo (× 6). (I–K) *U. rubra:* (I) leaf (× 0.75); (J) branch with samaras (× 0.75); (K) samara (× 1.5). (L) *U. crassifolia:* leaf (× 1.5). (M) *U. alata:* leaf (× 0.75). (From Elias 1970, *J. Arnold Arbor.* 51: p. 22.)

their leaves with secondary veins running directly into the teeth (vs. veins forming a series of loops); dry fruits, usually samaras (vs. fleshy fruits, i.e., drupes); 4–6-porate pollen with roughened exine (vs. 2–3-porate pollen with smooth exine); chemical makeup, with lignans, sesquiterpenes, and flavonols (vs. quebrachitol and glycoflavones); base chromosome number of 14 (vs. 10, 11); flowers bisexual or unisexual (vs. flowers consistently unisexual); seeds flat, with straight embryo (vs. seeds globular, with curved embryo); and styles with three vascular bundles (vs. a single bundle). In addition, *Ulmus* has plastids with protein crystals, while those of *Celtis* contain starch. The monophyly of Ulmaceae is also supported by cpDNA restriction sites (Wiegrefe et al.

1998). Within Ulmaceae, *Ulmus, Planera, Hemiptelea,* and *Zelkova* may form a clade on the basis of their unusual leaf vernation, the blade being folded and held to one side of the axis.

Ampelocera is the sister group of the remaining genera. *Ulmus, Planera,* and *Zelkova* form a clade; all have stipitate ovaries (Manchester and Tiffney 2001).

Two major clades are recognized within *Ulmus:* section *Oreoptelea,* which is diagnosed by elongate and articulated pedicels and ciliate-margined samaras, and section *Ulmus,* which has short, usually not obviously articulate pedicels and usually nonciliate-margined samaras. The monophyly of these two groups is strongly supported by cpDNA restriction sites (Wiegrefe et al. 1994).

The reduced flowers are wind-pollinated. The winged fruits of most species also are dispersed by wind. The warty, nutlike fruits of *Planera* are water-dispersed.

References: Elias 1970; Grudzinskaja 1967; Judd et al. 1994; Manchester 1989; Manchester and Tiffney 2001; Omori and Terabayashi 1993; Terabayashi 1991; Todzia 1993; Ueda et al. 1997; Wiegrefe et al. 1994, 1998; Zavada and Kim 1996.

Celtidaceae Link
(Hackberry or Sugarberry Family)

Trees or shrubs; cystoliths present; *laticifers absent.* Hairs simple, often with mineralized cell walls. *Leaves alternate and usually 2-ranked, simple, entire to serrate, with venation intermediate between pinnate and palmate,* usually with 3 main veins from the base, blade with symmetrical to asymmetrical base; *stipules present.* Inflorescences determinate, sometimes fasciculate, racemelike, or reduced to a solitary flower, axillary. *Flowers unisexual* (plants monoecious), radial, *inconspicuous.* Tepals usually 4 or 5, distinct to slightly connate, imbricate. Stamens 4 or 5 and opposite the tepals; filaments distinct, free to slightly adnate to tepals, erect or incurved in bud; pollen grains 2–3-porate. *Carpels 2, connate; ovary superior, with apical placentation* and 1 locule; stigmas elongate and extending along one side of style, sometimes divided. Ovule 1. *Fruit a drupe;* seeds globose; *embryo curved;* endosperm usually ± scanty (Figure 9.86).

Floral formula: Staminate: *, (4–5), 4–5, 0

Carpellate: *, (4–5), 0, ②; drupe

Distribution: Widely distributed in tropical to temperate regions.

Genera/species: 9/180. **Major genera:** *Celtis* (100 spp.) and *Trema* (55). Both occur in the continental United States and/or Canada.

Economic plants and products: *Celtis* (hackberry, sugarberry) provides timber and ornamental trees. The fruits are occasionally eaten.

Discussion: Celtidaceae are often included within Ulmaceae; see discussions under Ulmaceae and Rosales for an outline of the characters differentiating them. Some species of *Aphananthe* have secondary veins terminating in the teeth (as in Ulmaceae); the remaining features of this genus, however, are typical of Celtidaceae, and this venation condition apparently evolved independently from that of Ulmaceae. The genus is not a "link" between Celtidaceae and Ulmaceae. The probable inclusion of *Cannabis* and *Humulus* in the Celtidaceae will not affect its potential synapomorphies, since they are derived within it.

The inconspicuous flowers of Celtidaceae are wind-pollinated. The colorful drupaceous fruits have a sweet flesh and are adapted for dispersal by birds. Morphological synapomorphies of the group are unclear.

References: Elias 1970; Grudzinskaja 1967; Judd et al. 1994; Omori and Terabayashi 1993; Terabayashi 1991; Ueda et al. 1997.

Moraceae Link
(Mulberry or Fig Family)

Trees, shrubs, lianas, or rarely herbs; **with laticifers and milky sap, distributed in all parenchymatous tissues**; cystoliths present, usually globose; often with tannins. Hairs often simple, often with mineralized cell walls. *Leaves alternate, often 2-ranked but sometimes spiral, or opposite, usually simple,* sometimes lobed, *entire to serrate, with pinnate to palmate venation,* blade sometimes with cordate or asymmetrical base; *stipules usually present,* small to enlarged and leaving a circular scar on twig. *Inflorescences determinate, but sometimes appearing indeterminate,* axillary, *individual flowers usually congested and inflorescence axis often thickened and variously modified. Flowers unisexual* (plants monoecious), usually radial, *inconspicuous.* Tepals (0–) 4 or 5 (–8), distinct to connate, imbricate or valvate, often becoming fleshy and associated with mature fruits. *Stamens usually 1–5, opposite the tepals; filaments* distinct, *straight to incurved in bud;* anthers 2- or 1-locular; pollen grains usually 2–4- to multiporate. *Carpels 2, connate, sometimes with 1 carpel reduced;* ovary usually superior, *with apical placentation* and usually 1 locule; stigmas 2, extending along adaxial side of style, to capitate. Ovule 1, anatropous to campylotropous. *Fruit a drupe, dehiscent drupe, or achene, often closely clustered together and forming a multiple fruit; embryo curved to less commonly straight;* endosperm often lacking (Figure 9.87).

Floral formula:

Staminate: *, (4–5), 1–5, 0

Carpellate: *, (4–5), 0, ②; drupe; achene

Distribution: Widespread from tropical to temperate regions.

Genera/species: 53/1500. **Major genera:** *Ficus* (800 spp.) and *Dorstenia* (110). The family is represented in the continental United States and/or Canada by *Broussonetia, Fatoua, Ficus, Maclura,* and *Morus.*

Economic plants and products: Important edible fruits come from *Ficus* (figs), *Morus* (mulberries), *Artocarpus* (jackfruit, breadfruit), and *Brosimum* (breadnut). Several genera provide useful timber. The leaves of a few species of *Morus* are used as food for silkworms. Finally, genera such as *Ficus, Maclura* (Osage orange), and *Dorstenia* provide important ornamentals.

Figure 9.86   Celtidaceae. (A–I) *Trema micrantha:* (A) branch with staminate flowers (× 0.75); (B) same, leaves removed (× 3); (C) staminate flower with rudimentary gynoecium (× 7.5); (D) node with carpellate flowers (× 3); (E) carpellate flower (× 15); (F) same, in longitudinal section, showing solitary ovule (× 15); (G) drupe (× 7.5); (H) seed (× 15); (I) seed in partial longitudinal section to show embryo (× 15). (J) *T. lamarckiana:* leaf (× 0.75). (From Elias 1970, *J. Arnold Arbor.* 51: p. 38.)

Discussion: Moraceae are circumscribed narrowly, including woody species with conspicuous milky latex distributed throughout the plant, gynoecia with usually two evident carpels and a more or less apical ovule, and usually curved embryos. *Cecropia* and relatives (Cecropiaceae), which are phenetically intermediate between Moraceae and Urticaceae and are important early successional trees of tropical regions, are excluded. They are related to Urticaceae, as indicated by the restriction of laticifers to the bark, pseudomonomerous gynoecia, basal ovules, straight embryos, and cpDNA sequences (see Figure 9.78). Moraceae, as here circumscribed, are considered monophyletic on the basis of *rbcL* sequences (Sytsma et al. 1996). No morphological synapomorphies are known with certainty, although the presence of laticifers throughout all parts of the plant could be interpreted as

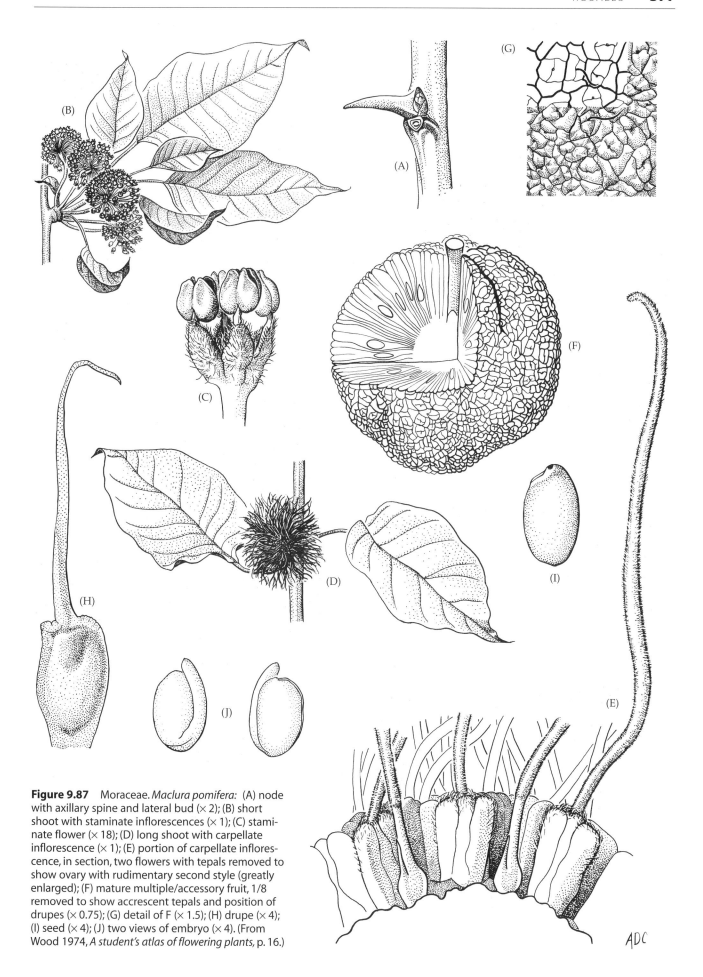

Figure 9.87 Moraceae. *Maclura pomifera:* (A) node with axillary spine and lateral bud (× 2); (B) short shoot with staminate inflorescences (× 1); (C) staminate flower (× 18); (D) long shoot with carpellate inflorescence (× 1); (E) portion of carpellate inflorescence, in section, two flowers with tepals removed to show ovary with rudimentary second style (greatly enlarged); (F) mature multiple/accessory fruit, 1/8 removed to show accrescent tepals and position of drupes (× 0.75); (G) detail of F (× 1.5); (H) drupe (× 4); (I) seed (× 4); (J) two views of embryo (× 4). (From Wood 1974, *A student's atlas of flowering plants,* p. 16.)

synapomorphic. Prior to the molecular study of Sytsma et al. (1996) the Moraceae had been considered metaphyletic (Judd et al. 1994).

Gynoecium reduction has occurred in several clades. It is common for the two styles to be slightly to strongly unequal in *Artocarpus*, *Dorstenia*, *Ficus*, and *Fatoua*. A complete loss of one of the two styles probably occurred in the ancestor of the Cecropiaceae + Urticaceae clade.

Ficus shows an amazing array of growth forms. Some species begin life as epiphytes, but eventually become ground-rooted and strangle the tree on which they were initially perched. Others have a broad spreading habit, with numerous adventitious roots functioning as supporting columns for the horizontally spreading branches.

The tiny flowers of Moraceae are often wind-pollinated, as in *Morus* and *Broussonetia*. The flowers of *Ficus*, in contrast, are pollinated by wasps that enter the syconium to lay eggs in the ovaries of the carpellate flowers. The wasp larvae feed on the ovary tissues of specialized flowers. Development of the larvae is timed with that of the staminate flowers, which open just as male and female wasps are emerging from their pupae. The wasps mate within the fig, and the females, carrying pollen, exit the developing inflorescence, and repeat their reproductive cycle. As the inflorescence continues to develop, the mature achenes are surrounded by the fleshy inflorescence axis. The evolution of figs apparently is well correlated with that of their fig-wasp pollinators (Weiblen 2000)

Moraceae exhibit a great diversity of inflorescence structures, although all based on a cymose theme, and of multiple fruits. In *Ficus* the tiny flowers are hidden on the inside of a hollow and cuplike inflorescence axis, with the achenes surrounded by a colorful, fleshy cup (syconium). In *Brosimum* the axis is also cuplike, but the staminate flowers are densely clustered on the outside, with a single carpellate flower in the center of the cup. The inflorescence axis of *Dorstenia* is discoid and flat-topped. In contrast, the inflorescence axes of *Artocarpus*, *Maclura*, and *Morus* are often elongate and covered by the densely packed flowers. The drupes of *Artocarpus* are densely aggregated. The massive infructescences of *Artocarpus* weigh up to 40 kg, and like those of *Morus* are associated with persistent, fleshy, accrescent tepals. The achenes of *Maclura* are also aggregated and associated with persistent fleshy tepals. Adaptive radiation, involving changes in the structure of the multiple fruits linked to changes in vertebrate dispersers (birds or mammals), appears to be characteristic of the more or less tropical, woody members of the family. The pits of *Dorstenia* are forcibly ejected from the infructescence; the fruits are explosive drupes!

References: Bechtel 1921; Berg 1978; Bonsen and ter Welle 1983; Chase et al. 1993; Cronquist 1981, 1988; Engler 1889; Faegri and van der Pijl 1980; Humphries and Blackmore 1989; Judd et al. 1994; Proctor and Yeo 1972; Sytsma et al. 1996; Thorne 1976, 1983, 1992; Wieblen 2000.

Urticaceae A. L. de Jussieu
(Nettle Family)

Herbs to rarely trees, shrubs, or vines; with laticifers restricted to the bark and producing milky sap, or reduced and sap clear, mucilaginous; **cystoliths** present, **± elongate**; sometimes with tannins. Hairs often simple, often with mineralized cell walls, sometimes stinging. *Leaves alternate and spiral or 2-ranked, or opposite, usually simple, entire to serrate, with pinnate to palmate venation*, blade sometimes with cordate or asymmetrical base; *stipules usually present. Inflorescences determinate,* axillary, *the individual flowers often congested,* sometimes reduced to a single flower. *Flowers unisexual* (plants monoecious to dioecious), usually radial, *inconspicuous. Tepals (3–) 4 (–6), distinct to connate, imbricate or valvate. Stamens usually 4 or 5, opposite the tepals; filaments* distinct, **incurved in bud, and elastically reflexed at anthesis**; anthers 2-locular; pollen grains usually 2- or 3- to multiporate. *Carpels apparently 1, but actually 2 with 1 extremely reduced (pseudomonomerous);* ovary superior, *with basal placentation* and 1 locule; stigmas 1 or 2, extending along adaxial side of style, to capitate or punctate. Ovule 1, orthotropous. *Fruit usually an achene; embryo straight;* endosperm sometimes lacking (Figure 9.88).

Floral formula: Staminate: *, ⊕, 4–5, 0

Carpellate: *, ⊕, 0, ①; drupe; achene

Distribution: Widespread from tropical to temperate regions.

Genera/species: 40/900. **Major genera:** *Pilea* (400), *Elatostema* (200), and *Boehmeria* (80). The family is represented in the continental United States and/or Canada by *Boehmeria*, *Hesperocnide*, *Laportea*, *Parietaria*, and *Pilea*.

Economic plants and products: Fiber is extracted from *Boehmeria nivea* (ramie) and *Urtica dioica* (stinging nettle). *Pilea* (artillery plant) and *Soleirolia* (baby's tears) provide important ornamentals.

Discussion: Urticaceae are circumscribed narrowly, including mainly herbaceous species with more or less elongate cystoliths; laticifers restricted to the bark, or very reduced, usually producing a clear, mucilaginous sap; pseudomonomerous gynoecia with a more or less basal ovule; and straight embryos. *Poikilospermum*, which has often been included in Cecropiaceae, is included here because it has dimorphic wood fibers and unlignified vessel elements (both synapomorphies of Urticaceae: Judd et al. 1994). The Urticaceae, as here circumscribed, are considered monophyletic on the basis of elongate cystoliths, incurved stamens, wood anatomy (Berg 1977, 1989; Cronquist 1981; Friis 1989, 1993; Humphries and Blackmore 1989; Judd et al. 1994), and *rbcL* sequences (Sytsma et al. 1996). In contrast to Moraceae, the family is mainly herbaceous.

Figure 9.88 Urticaceae. (A–K) *Urtica chamaedryoides:* (A) habit of young plant (× 0.75); (B) portion of adaxial surface of leaf, showing cystoliths (× 18); (C) stinging hair from petiole (× 18); (D) inflorescence with staminate flowers and young fruits (× 9); (E) staminate flower, after pollen has been shed (× 18); (F) carpellate flower (× 60); (G) ovary in longitudinal section, showing single, orthotropous ovule (× 74); (H) achene and accrescent perianth (× 22); (I) achene (× 30); (J) achene in longitudinal section, showing embryo and endosperm (stippled) (× 30); (K) wet achene, mucilage halo dotted (× 30). (L) *U. dioica:* achene (× 30). (From Miller 1971, J. Arnold Arbor. 52: p. 48.)

Aborted vascular bundles in gynoecia of *Laportea* and *Urtica* suggest that the unicarpellate ovary has been derived through abortion of a second carpel. The basal ovule of Urticaceae, likewise, has apparently been derived from an apical ovule, as in Moraceae. In *Boehmeria cylindrica* the vascular bundle supplying the ovule ascends the carpel wall for a short distance and then reverses direction to enter the ovule at the base of the ovary, suggesting a shift from apical to basal placentation in the common ancestor of Urticaceae and Cecropiaceae (see Figure 9.78).

Distinctive stinging hairs occur in genera such as *Urtica, Laportea,* and *Urera.* Each hair consists of a single, long, narrow, tapering stinging cell with a saclike base embedded in a multicellular outgrowth. The hair is closed at the tip by a small bulb that easily breaks off, producing an extremely sharp point. Upon contact with human skin, the hair punctures the surface, and compression of the base forces fluid containing histamines and acetylcholines into the wound. The process results in a reddening of the skin and painful itching and burning.

The tiny flowers of Urticaceae are usually wind-pollinated. In many species, the inflexed stamens extend elastically at anthesis, causing the anthers to release their pollen in a sudden burst. This facilitates the movement of pollen by air currents, even in species of forest understory habitats.

The achenes of Urticaceae often are associated with fleshy to dry tepals. The fruits may be eaten by birds, or dispersed by external transport on hair or feathers, by wind, or by ballistic methods.

References: Bechtel 1921; Berg 1977, 1978, 1989; Bonsen and ter Welle 1983; Chase et al. 1993; Cronquist 1981, 1988; Engler 1889; Friis 1989, 1993; Humphries and Blackmore 1989; Judd et al. 1994; Miller 1971b; Rohwer 1993b; Sytsma et al. 1996; Thorne 1976, 1983, 1992; Woodland 1989.

Cucurbitales

It is clear that Cucurbitaceae, Begoniaceae, and Datiscaceae are closely related, as they share the apomorphies of an inferior ovary, parietal placentae (often strongly intruded), unisexual flowers, the distinctive cucurbitoid tooth type, the presence of cucurbitacins (oxidized triterpenes), and separate vascular bundles in the stem. Analyses of DNA sequences and serological characters also support the monophyly of a group containing these three, along with four other small families (Källersjö et al. 1998; Savolainen et al. 2000b; Soltis et al. 2000). This clade is here recognized as the order Cucurbitales (following APG 1998).

References: Angiosperm Phylogeny Group 1998; Savolainen et al. 2000b; Soltis et al. 2000.

Cucurbitaceae A. L. de Jussieu
(Cucurbit Family)

Herbaceous or soft-woody vines, **usually with spirally coiled and often branched tendrils, borne ± laterally at the nodes** (possibly modified shoots); vascular bundles usually bicollateral, often in 2 concentric rings; alkaloids and bitter tetra- and pentacyclic triterpenoid saponins

usually present. Hairs simple, with calcified walls and a cystolith at the base. *Leaves alternate and spiral, usually simple, often palmately lobed, ± serrate, teeth cucurbitoid (i.e., with several veins entering tooth and ending at an expanded ± translucent glandular apex) with palmate venation;* **stipules lacking.** Inflorescences determinate, sometimes reduced to a single flower, axillary. *Flowers usually unisexual (plants monoecious or dioecious),* usually radial, **with short to elongate hypanthium,** often open for only a day. *Sepals usually 5,* usually connate, often reduced. **Petals** *usually 5,* **connate,** *bell-shaped, with a narrow tube and flaring lobes, or nearly flat, white, yellow to orange or red,* the lobes valvate or folded inward. **Stamens** 3-5, adnate to hypanthium, **variously connate and modified,** *usually appearing as 3 (or even appearing to be solitary due to complete connation of filaments and modification of anthers; filaments usually variously connate; anthers unilocular, but often appearing 2-locular or multilocular (due to connation),* **the locules usually bent to convoluted;** pollen grains various, with 3 to many furrows and/or pores. *Carpels usually 3, connate; ovary half-inferior to inferior, with parietal placentation, the placentae expanded and intruded; stigmas usually 3, each bilobed.* Ovules usually numerous on each placenta. Nectaries various. *Fruit a berry, the rind often leathery to hard (then a pepo), or occasionally a fleshy to dry, variously dehiscing capsule;* **seeds flattened, the seed coat with several layers,** the outermost sometimes fleshy; endosperm scanty or lacking (Figure 9.89 and Figure 4.47D).

Floral formula: Staminate: *,⑤,⑤,⑤, 0

Carpellate: *, 5, 5, 0,③; berry, capsule

Distribution: Widely distributed in the tropics and subtropics, with a few species occurring in temperate regions.

Genera/species: 118/825. ***Major genera:*** *Cayaponia* (60 spp.), *Momordica* (45), *Gurania* (40), and *Sicyos* (40). Noteworthy genera occurring in the continental United States and/or Canada are *Cayaponia, Cucumis, Cucurbita, Cyclanthera, Echinocystis, Ibervillea, Marah, Melothria, Momordica,* and *Sicyos.*

Economic plants and products: The family is very important for edible fruits and seeds, such as *Cucurbita* (pumpkins, winter and summer squashes, gourds), *Cucumis* (cantaloupe, muskmelon, honeydew melon, cucumber), *Citrullus* (watermelon), *Benincasa* (wax gourd), and *Sechium* (chayote). The dried fruits of *Lagenaria* (bottle gourd) are used as containers, and the dried fruits of *Luffa* (loofah) are used as a vegetable sponge. Some, such as *Momordica* (balsam apple), are used medicinally.

Discussion: Cucurbitaceae are easily recognized and monophyletic. Two subfamilies are recognized (Jeffrey

Figure 9.89 Cucurbitaceae. *Echinocystis lobatus:* (A) node of vine with leaf, flowers, fruit, and tendril (× 1); (B) staminate flower (× 5); (C) androecium (three stamens) (× 15); (D) carpellate flower (× 5); (E) ovary in longitudinal section, at position shown in F (× 10); (F) ovary in cross-section, note ovules embedded in fused parietal placentae (× 10); (G) mature fruit after dehiscence (× 1); (H) seed (× 8); (I) embryo, one cotyledon removed (× 8). (From Wood 1974, *A student's atlas of flowering plants,* p. 111.)

1967, 1980, 1990a,b). The "Zanonioideae," a small group characterized by separate styles, contains genera with numerous plesiomorphic features and probably is paraphyletic. The monophyly of Cucurbitoideae is supported by their completely connate styles and nrITS sequences (Jobst et al. 1998). This group is divided into several tribes based on characters such as ovule position and number, ornamentation of the pollen grains, shape of the hypanthium, fruit form, and characteristics of the androecium.

Flowers of Cucurbitaceae are showy and attract various insects, birds, and bats. Pollen and nectar are the rewards. Monoecy or dioecy leads to outcrossing. The highly modified androecium is similar in appearance to the style and stigma, and insects are fooled into visiting both staminate and carpellate flowers. The berries of most cucurbits are dispersed by various animals. *Echinocystis* has capsules that open explosively; in *Momordica*, the capsules open to expose brightly colored and fleshy seeds, which are dispersed by birds.

References: Chakravarty 1958; Jeffrey 1967, 1980, 1990a,b; Jobst et al. 1998; Robinson and Decker-Walters 1997.

Begoniaceae C. A. Agardh
(Begonia Family)

Herbs or soft-woody shrubs with jointed stems; tannins present; with large water-storage cells in leaf hypodermis. Hairs simple, often with a cystolith at the base. *Leaves alternate and 2-ranked, simple,* sometimes palmately lobed or palmate, usually ± *toothed, the teeth similar to those of Cucurbitaceae, with palmate venation, usually asymmetrical; stipules present, often large, persistent.* Inflorescences determinate, axillary. *Flowers unisexual (plants monoecious),* radial. **Perianth of petaloid tepals,** 2-10, but usually 4 (in 2 whorls) in staminate and 5 (in a single whorl) in carpellate flowers, distinct, imbricate or valvate, those of inner and outer whorls often ± differentiated. *Stamens 4-numerous,* the filaments distinct or basally connate; the connective expanded; pollen grains tricolporate. *Carpels usually 3, connate; ovary ± inferior,* **with axile placentation, each placenta expanded,** variably divided (derived from intruded parietal, and this condition present in a few), **usually with 3 prominent wings; stigmas 6, elongated, twisted, yellow, strongly papillose.** Ovules numerous in each locule. Nectaries lacking. *Fruit a loculicidal capsule,* **winged,** occasionally a berry. Seeds tiny, with ring of elongated testa cells that rupture, forming a cap (operculate); endosperm scanty or lacking.

Floral formula:

Staminate: * or X, 2 , 2 , ∞ , 0

Carpellate: * or X, – 5 – , 0 , ∞ , $\overline{③}$, capsule

Distribution: Widely distributed in the tropics and subtropics (excluding Australia), and especially characteristic of moist shaded understory habitats.

Genera/species: 2/920. **Major genus:** *Begonia* (919 spp.). The group is represented in the continental United States only by *Begonia* (naturalized). *Hillebrandia* is endemic to Hawaii.

Economic plants and products: The family contains numerous commercially important species and hybrids that are cultivated for their beautiful foliage and flowers.

Discussion: The Begoniaceae are fairly homogenous, with nearly all species belonging to the large genus *Begonia*, and the group's monophyly is supported by several morphological characters. *Hillebrandia sandwichensis* is sister to the remaining members of the family, which constitute the large and diverse genus *Begonia* (Swensen et al. 2001). Recognition of *Symbegonia* makes *Begonia* paraphyletic.

The showy staminate flowers are visited by pollen-gathering bees. The bright yellow twisted stigmas of the carpellate flowers mimic the anthers of staminate flowers, and thus also attract pollen-gathering bees (resulting in pollination by deceit; see also Proctor et al. 1996). The tiny seeds are presumably dispersed by wind or water (rain wash), and the prominent (often asymmetrical) wings may assist in shaking seeds out of the capsule. The cymes produce more staminate flowers than carpellate flowers and the staminate flowers also open sooner, leading to outcrossing. Vegetative propagation is common; plantlets develop from small tubers that often occur in the leaf axils or from adventitious buds that form on detached leaves.

References: Baranov 1981; de Lange and Bouman 1992; Proctor et el. 1996; Smith and Wasshausen 1986; Swensen et al. 2001.

Fagales

Fagales are monophyletic, based on their unisexual flowers with tepals very reduced or lacking, usually inferior ovary with one or two 1-integumented ovules per locule, pollen tube entering the ovule via the chalaza, absence of nectaries, and indehiscent 1-seeded fruits. They are trees or shrubs with tannins, alternate, stipulate leaves, typically wind-pollinated flowers aggregated in catkins, and seeds more or less lacking endosperm. They usually have gland headed and/or stellate hairs. The monophyly of Fagales is supported by chloroplast DNA restriction sites, *rbcL, atpB, matK,* and 18S sequences (Chase et al. 1993; Manos et al. 1993; Källersjö et al. 1998; Savolainen et al. 2000a,b; Soltis et al. 2000), and morphology (Hufford 1992). The order consists of 8 families and about 1115 species; major families include **Fagaceae,** Nothofagaceae, **Betulaceae, Casuarinaceae, Myricaceae,** and **Juglandaceae.** The group clearly belongs within the rosid complex (Chase et al. 1993; Hufford 1992; Thorne 1992). Fagales are not closely related to the Hamamelidaceae, Platanaceae, or other "basal" tricolpates, and the Hamamelidae (as traditionally defined; see Cronquist 1981, 1988; Stern 1973) are polyphyletic (Chase et al. 1993; Hufford 1992; Manos et al. 1993; Meurer-Grimes 1995; Thorne 1973a, 1992).

It is clear that either Fagaceae or, more probabably, Nothofagaceae are sister to the remaining families of Fagales (Hufford 1992; Manos et al. 1993, 2001; Manos and Steele 1997; Nixon 1989; Wolfe 1989). The Juglandaceae, Myricaceae, Casuarinaceae, and Betulaceae form a clade (core Fagales) on the basis of cpDNA features (restriction sites, *rbcL*, and *matK* nucleotide sequences) (Chase et al. 1993; Hufford 1992; Manos et al. 1993; Manos and Steele 1997; Nixon 1989; Wolfe 1989) (Figure 9.90). Members of core Fagales are characterized by more or less triporoporate pollen (pores with well-developed endoapertures) and ovules with multiple female gametophytes, and these may be an additional synapomorphies. This group of families represents the extant members of the Normapolles complex, well represented in the fossil record (Kedves 1989). Analyses based on *matK* and *rbcL* sequences (Manos and Steele 1997) support two major subclades within core Fagales, the first containing Betulaceae, Casuarinaceae, and Myricaceae, and the second Juglandaceae (and Rhoipteleaceae). Juglandaceae and Rhoipteleaceae also are united by the apomorphy of pinnately compound leaves, and Rhoipteleaceae are noteworthy in having the most Normapolles-like pollen of any living member of Fagales. Myricaceae shares with Juglandaceae peltate gland-headed hairs and a gynoecium with a single, orthotropous ovule, suggesting a closer relationship between these families than suggested by molecular data.

References: Chase et al. 1993; Cronquist 1981, 1988; Hufford 1992; Kedves 1989; Nixon 1989; Manos et al. 1993, 2001; Manos and Steele 1997; Meurer-Grimes 1995; Stern 1973; Thorne 1973a, 1992; Wolfe 1989.

Fagaceae Dumortier
(Beech or Oak Family)

Trees or shrubs; tannins present. *Hairs simple or stellate, often also glandular scales. Leaves usually alternate and spiral, simple, but often lobed, entire to serrate, with pinnate venation;* stipules present. *Inflorescences determinate, often erect and spicate, a dangling catkin, headlike cluster,* or even a solitary flower, terminal or axillary, the staminate and carpellate flowers on the same or different inflorescences. *Flowers unisexual (plants usually monoecious),* radial, ± inconspicuous, the staminate flowers in reduced cymes and associated with a bract, *the carpellate flowers usually in groups of 1–3 and* **associated with a scaly cupule.** Tepals usually 6, reduced and inconspicuous, distinct to slightly connate, imbricate. Stamens 4–numerous; filaments distinct; pollen grains tricolporate or tricolpate. *Carpels 3 (–12),* connate; *ovary inferior,* with axile placentation; stigmas separate, porose, or expanded along upper side of style. Ovules 2 in each locule, but all except 1 aborting. Nectaries usually lacking. **Fruit a nut, closely associated with a spiny to scaly, usually 4-valved or nonvalved cupule;** endosperm lacking (Figure 9.91).

Floral formula:

Staminate: *, ⦗6⦘, 4–∞, 0

Carpellate: *, ⦗6⦘, 0, ③; nut (with cupule)

Key to Major Families of Fagales

1. Fruits associated with a conspicuous cupule; carpels 2–6; pollen colpate or colporate 2
1. Fruits not associated with a cupule, although usually with variously developed bracts and bracteoles; carpels usually 2; pollen porate 3
2. Carpels 3 (–12); ovules with 2 integuments; stipules narrowly triangular, lacking colleters; leaves entire to serrate or lobed but never doubly serrate; plants of the Northern Hemisphere ....**Fagaceae**
2. Carpels 2–3; ovules with 1 integument; stipules peltate, with colleters; leaves entire to doubly serrate; plants of the Southern Hemisphere ....Nothofagaceae
3. Leaves pinnately compound ....**Juglandaceae**
3. Leaves simple 4
4. Leaves whorled, reduced to tiny scales; stems longitudinally ridged ....**Casuarinaceae**
4. Leaves alternate, with obvious blades; stems not ridged 5
5. Leaves entire to serrate; fruits achenes or drupes, usually with a covering of waxy papillae; plants dioecious or occasionally monoecious; ovule 1, basal, orthotropous ....**Myricaceae**
6. Leaves ± doubly serrate; fruits achenes, samaras, or nuts; plants monoecious; ovules 2 per locule at apex of incomplete septum, anatropous ....**Betulaceae**

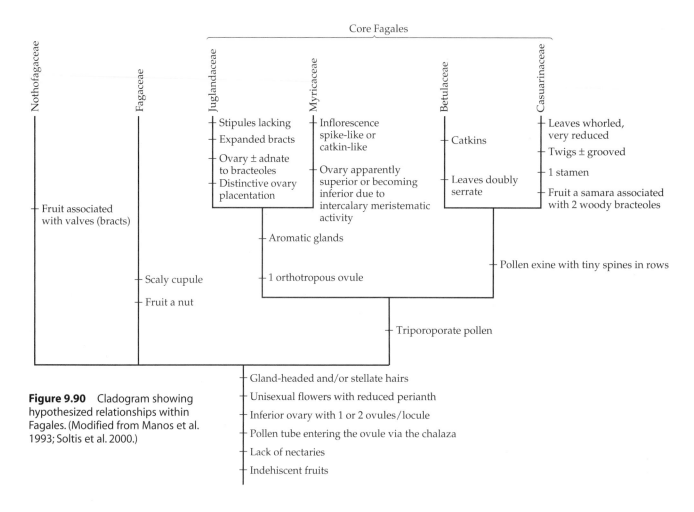

Figure 9.90 Cladogram showing hypothesized relationships within Fagales. (Modified from Manos et al. 1993; Soltis et al. 2000.)

Distribution: Widespread in tropical to temperate regions of the Northern Hemisphere.

Genera/species: 9/900. **Major genera:** *Quercus* (450 spp.), *Lithocarpus* (300), and *Castanopsis* (100). The family is represented in Canada and/or the continental United States by *Castanea, Chrysolepis, Fagus, Lithocarpus,* and *Quercus.*

Economic plants and products: The nuts of *Castanea* (chestnuts) are edible; those of *Quercus* (oak) and *Fagus* (beech) are also occasionally eaten. Cork is made from the bark of *Quercus suber. Quercus, Fagus,* and *Castanea* provide ornamental trees. The family is exceptionally important as a source of timber for construction, furniture, cabinetry, barrels, and other uses.

Discussion: The monophyly of Fagaceae is supported by morphology, cpDNA restriction sites (Manos et al. 1993), and *mat*K sequences (Manos and Steele 1997). *Fagus* (Fagoideae) is probably sister to the remaining genera (Manos et al. 1993), which constitute the Quercoideae. *Trigonobalanus,* a group that may have retained ancestral inflorescence conditions (see Fig. 9.92), probably is sister to the other genera of Quercoideae. *Castanea, Castanopsis, Lithocarpus,* and *Chrysolepis* may constitute a clade

(Nixon 1989; Manos et al. 2001) and have retained numerous plesiomorphic morphological characters: bisexual inflorescences, flowers with a less reduced perianth, exserted stamens, and tiny stigmas.

Quercus is considered to be monophyletic on the basis of its fruits: a single nut surrounded by a nonvalved cupule (forming an acorn); its monophyly has also been supported by DNA sequences (Manos and Steele 1997; Manos et al. 2001). The genus has separate carpellate and staminate inflorescences, the latter forming lax catkins. Two major phenetic groups can be recognized within *Quercus*: subgenus *Cyclobalanopsis* (those species with cyclically arranged cupule scales) and subgenus *Quercus* (those species with imbricate cupule scales). More study is needed to determine whether or not these represent clades. Subgenus *Quercus* is composed of three major groups. Red oaks (section *Lobatae*) are characterized by leaves often with bristle tips; flowers with elongate, linear-spatulate styles and usually retuse anthers; aborted ovules near the apex of the nut; and usually biennial fruits. White oaks (section *Quercus*) have leaves without bristle tips; flowers with short, abruptly dilated styles and usually apiculate anthers; aborted ovules at the base of the nut; and annual fruits. A third, smaller group, section *Protobalanus*, is similar to section *Quercus*, but differs in its biennial (vs. annual) fruits and clearly pubescent (vs. glabrous)

Figure 9.91 Fagaceae. (A–E) *Quercus ilicifolia:* (A) branch with staminate and carpellate flowers (× 1.5); (B) staminate flower (× 17.5); (C) carpellate flower (× 17.5); (D) carpellate flower in longitudinal section (× 30); (E) nut with scaly cupule (acorn) (× 2). (F–H) *Q. rubra:* (F) second-year carpellate flower in longitudinal section, showing ovule formation (× 12); (G) same in cross-section (× 37); (H) seed with aborted ovules (black dots toward apex of seed) (× 2). (I) *Q. imbricaria:* acorn (× 0.75). (J–K) *Q. alba:* (J) fruiting branch (× 0.5); (K) acorn (× 1.8). (From Elias 1971, *J. Arnold Arbor.* 52: pp. 182, 184.)

inner fruit wall (Nixon et al. 1995). Recent DNA analyses support the segregation of *Q. cerris, Q. suber,* and *Q. ilex,* and their relatives from section *Quercus,* and their recognition as section *Cerris* (Manos et al. 1999, 2001).

Variation in the cupule provides characters useful in generic recognition (Figure 9.92). The cupule is probably a cymose inflorescence in which the outer axes of the cyme are modified into cupule valves, which bear spines or scales (Brett 1964; Fey and Endress 1983; Nixon 1989). The valves may be more or less evident due to fusion. The cupules of *Nothofagus* (Nothofagaceae) probably are not homologous with those of Fagaceae. They may be composed of clustered bracts and stipules.

The inconspicuous flowers of *Fagus* and *Quercus* are borne in unisexual inflorescences that develop in the spring just before or with the leaves, and are wind-pollinated. *Castanea* and *Castanopsis* bear staminate flowers that are conspicuous, give off a strong odor, and are pollinated by flies, beetles, and bees. Wind pollination may have evolved three times within the family. The large nuts of Fagaceae are dispersed by birds and mammals (especially rodents).

Castanea has been devastated in North America by a disease caused by the fungus *Endothea parasitica.*

References: Abbe 1974; Brett 1964; Crepet and Nixon 1989; Elias 1971a; Endress 1977; Fey and Endress 1983; Jones 1986; Kaul and Abbe 1984; Kubitzki 1993b; Mac-Donald 1979b; Manos et al. 1993, 1999, 2001; Manos and Steele 1997; Nixon 1989; Nixon et al. 1995; Nixon and Crepet 1989.

Betulaceae S. F. Gray
(Birch Family)

Trees or shrubs; tannins present; bark smooth to scaly, sometimes exfoliating in thin layers, sometimes with prominent horizontal lenticels. *Hairs simple, gland-headed or peltate. Leaves alternate and spiral or 2-ranked, simple,* **doubly serrate**, *with pinnate venation, the secondary veins running into a serration; stipules present.* Inflorescences determinate, appearing spicate, **forming erect or pendu-**

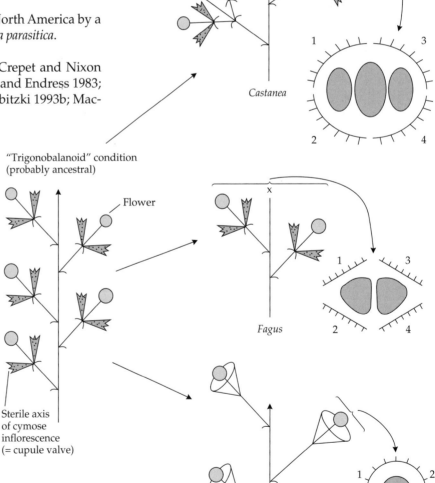

Figure 9.92 Hypothesis of evolution of various cupule forms in Fagaceae.

lous catkins (aments), terminal or axillary, sometimes exposed during the winter, solitary or in racemose clusters, with conspicuous bracts, *the staminate and carpellate flowers in separate inflorescences. Flowers unisexual (plants monoecious), radial, inconspicuous, usually 2 or 3, forming a cymose unit in the axil of each inflorescence bract, also usually associated with variously fused second- and third- order bracteoles. Tepals (0–) 1–4 (–6), reduced, ± distinct, sometimes lobed, slightly imbricate. Stamens usually (1–) 4 (–6),* sometimes appearing to be more due to the close association of the 3 flowers of the cymose unit; filaments short, distinct to basally connate, sometimes divided; pollen grains (2–) 3–7-(poro)porate. *Carpels usually 2, connate; ovary inferior,* with axile placentation (but incompletely 2-locular, and ovules borne at apex of incomplete septum); stigmas 2, running along adaxial surface of styles. Ovules usually 2 per locule, but all except 1 aborting, usually with 1 integument. Nectaries lacking. *Fruit an achene, nut, or 2-winged samara, associated with variously fused and developed bract-bracteole complex;* endosperm present or lacking (Figure 9.93).

Floral formula:

Staminate: *, -0–6-, ⌐1–4⌐, 0

Carpellate: *, -0–6-, 0,②; achene, samara, nut

Distribution and ecology: Widespread in temperate to boreal regions of the Northern Hemisphere, but *Alnus* extending into South America (Andes); species occur in early successional habitats or in wetlands, or as dominant forest trees. Nitrogen fixation occurs in specialized nodules (containing symbiotic bacteria) on the roots of *Alnus*.

Genera/species: 6/157. ***Major genera:*** *Betula* (60 spp.), *Alnus* (35), *Carpinus* (35), *Corylus* (15), and *Ostrya* (10). All of these occur in the continental United States and/or Canada.

Economic plants and products: The nuts of *Corylus* (filberts, hazelnuts) are edible. Several species of *Betula* (birch) and *Alnus* (alder) are important for timber or wood pulp; the latter is also important in land reclamation. *Betula, Corylus, Carpinus* (ironwood, hornbeam), and *Ostrya* (ironwood, hop hornbeam) provide valuable ornamentals.

Discussion: Betulaceae are considered monophyletic on the basis of having both their staminate and carpellate flowers in catkins and doubly serrate leaf margins (Crane 1989). Betulaceae and the Neotropical genus *Ticodendron* (Ticodendraceae) are sister taxa. *Ticodendron* has leaves with almost all the teeth directly vascularized by secondary veins, stipules encircling the twigs, and drupes. Betulaceae comprise two major monophyletic groups: Betuloideae (including *Alnus* and *Betula*) and Coryloideae (including *Carpinus, Corylus, Ostrya*, and

Ostryopsis) (Crane 1989; Chen et al. 1999). Betuloideae are hypothesized to form a clade on the basis of their flattened fruits, bracts and bracteoles fused into a scale, and carpellate flowers lacking a perianth. The monophyly of Coryloideae is supported by their staminate flowers lacking a perianth and carpellate flowers with the bracteoles connate and expanded. Nuclear ribosomal ITS and *rbcL* sequences also support these two subfamilies (Chen et al. 1999)

The development, anatomy, and morphology of the highly modified flowers and inflorescences of Betulaceae have received extensive study (Abbe 1935, 1974) and are important in generic identification (Figure 9.94). All members of the family have both staminate and carpellate catkins with flowers borne in cymose units consisting of one to three flowers and one to seven bract/bracteoles. The cymose units of the carpellate catkin of *Alnus* are reduced to two flowers, and each cyme is associated with a bract, two secondary bracteoles, and two tertiary bracteoles (all of which are connate, persistent, and woody). The cymose units of the carpellate catkin of *Betula* are usually composed of three flowers; these are associated with a bract and two fused bracteoles, forming a deciduous three-lobed "bract." In *Carpinus*, the two nutlets (of each cymose unit) are associated with a small bract, and each nutlet is adnate to an expanded secondary bracteole that is connate to the two adjacent tertiary bracteoles. In contrast, the achenes of *Ostrya* are completely surrounded by an expanded, bladderlike structure that develops from the fused secondary and two tertiary bracteoles. The nuts of *Corylus* are surrounded by the greatly expanded secondary bracteoles.

The inconspicuous flowers of Betulaceae are borne in usually erect carpellate catkins and pendulous staminate catkins. All members of the family bloom in the spring, before or just as the leaves emerge, and are wind-pollinated. The ovary and ovules are undeveloped at the time of pollination (as in Fagaceae). The small, two-winged samaras of *Betula* and *Alnus* are wind-dispersed. In some species of *Alnus* the wings are reduced, and these fruits, which float, are usually water-dispersed. The nutlets of *Carpinus* and *Ostrya* are associated with expanded and fused bracteoles, and both are wind- or water-dispersed. *Corylus* has large nuts that are dispersed by rodents.

References: Abbe 1935, 1974; Chen et al. 1999; Crane 1989; Dahlgren 1983; Furlow 1990; Kubitzki 1993a; Stone 1973; Tiffney 1986.

Casuarinaceae Lindley
(She-Oak Family)

Trees or shrubs with **slender, green, jointed, grooved twigs; roots with nodules containing nitrogen-fixing bacteria;** tannins present. Hairs simple or branched. **Leaves whorled, in groups of 4–20, simple, scale-like, ± connate, forming a toothed sheath at each node;** stip-

Figure 8.93 Betulaceae. (A–J) *Betula nigra:* (A) flowering branch, with two erect carpellate catkins and three pendulous staminate ones (× 0.75); (B) staminate cyme, view of adaxial side, showing stamens of three flowers, and tips of primary and secondary bracts (1, 2) (× 12); (C) same, side view, primary bract (1) and secondary bracteoles (2) at left (× 12); (D) staminate cyme as seen in B, anthers removed to show primary bract (1) and secondary bracteoles (2), single tepal (3) of each of three flowers, and partial filaments of each of six stamens (× 12); (E) two views of stamens, showing half-anthers (× 18); (F) carpellate cyme, view of adaxial side, showing three carpellate flowers, note that each flower is composed of two fused carpels (× 12); (G) carpellate cyme, abaxial view, showing bract complex with primary bract (1) fused to secondary bracteoles (2) (× 12); (H) branch with mature carpellate catkins (× 0.75); (I) abaxial side of primary bract complex (× 9); (J) fruit (× 9). (K–L) *B. lenta:* (K) branch with mature carpellate catkins (× 0.75); (L) abaxial side of bract complex of mature fruiting cyme (× 9). (From Furlow 1990, *J. Arnold Arbor.* 52: p. 71.)

Figure 9.94 Floral diagrams of the cymose units of the carpellate catkins of Betulaceae.

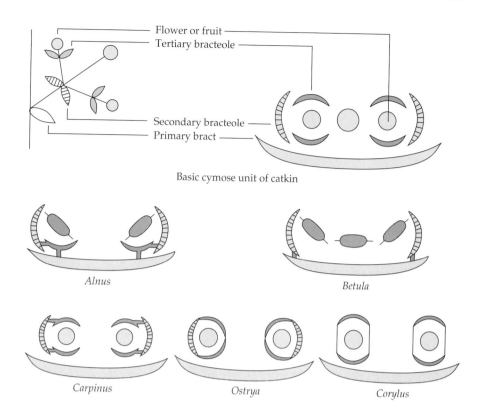

Flower or fruit

Tertiary bracteole

Secondary bracteole

Primary bract

Basic cymose unit of catkin

Alnus

Betula

Carpinus

Ostrya

Corylus

ules lacking. *Inflorescences indeterminate, terminal, forming catkins at the tips of lateral branches. Flowers unisexual (plants monoecious or dioecious), radial, inconspicuous, solitary in the axil of each inflorescence bract, and associated with 2 bracteoles.* Tepals lacking. **Stamen 1**; pollen grains usually triporoporate. *Carpels 2, connate; ovary presumably inferior,* with axile placentation; stigmas 2, running along inner surface of styles. Ovules 2 in each locule, or sometimes lacking in 1 of the locules, and all but 1 aborting, orthotropous, **with 2 integuments.** Nectaries lacking. **Fruit a samara, associated with 2 woody bracteoles, in conelike catkins**; endosperm lacking.

Floral formula: Staminate: *, -0-, 1, 0

Carpellate: *, -0-, 0,$\overline{②}$; samara

Distribution and ecology: Widely distributed in southeastern Asia, Australia, and the islands of the southwestern Pacific, but naturalized in coastal habitats in subtropical and tropical regions of Africa and the Americas. Often plants of xeric habitats.

Genera/species: 1/70; *Casuarina.* The family is represented in the continental United States (Florida) by three introduced species.

Economic plants and products: Several species are important timber trees or ornamentals. *Casuarina* is a problem weed tree in Florida.

Discussion: Casuarinaceae are easily recognized, and are considered monophyletic (see characters listed in bold

above). *Casuarina* has been divided into four genera based on characters such as nature of the stem grooves (shallow and open vs. deep and narrow), number of teeth per whorl, shape of the bracts in carpellate catkins, fruit color, and chromosome number (Johnson and Wilson 1989, 1993). However, a broad view of *Casuarina* is retained here.

The flowers of Casuarinaceae are wind-pollinated, and the fruits are wind-dispersed.

References: Johnson and Wilson 1989, 1993; Rogers 1982; Torrey and Berg 1988.

Myricaceae Blume
(Bayberry Family)

Aromatic trees or shrubs; triterpenes and sesquiterpenes present; tannins present; **roots usually with nodules that contain nitrogen-fixing bacteria.** *Peltate scales with a glandular, usually golden-yellow, swollen head, containing various aromatic oils and/or resins. Leaves alternate and spiral, simple* (deeply lobed in *Comptonia*), *entire to serrate, with pinnate venation; stipules absent, or present* (*Comptonia*). **Inflorescences** indeterminate, **often spikelike or catkinlike,** erect to ± pendulous, axillary, staminate and carpellate flowers usually in separate inflorescences. *Flowers unisexual* (*plants monoecious or dioecious*), *radial, inconspicuous, 1 in the axil of each inflorescence bract. Perianth lacking,* except in *Canacomyrica* where represented by 6 minute tepals at ovary apex, but flowers usually associated with bracts and bracteoles. *Stamens 2–9,* but appearing more numerous due to clustering of several flowers; pollen grains usually triporoporate. *Carpels 2, connate; ovary apparently superior (due to loss of perianth: Comptonia), becoming inferior due to intercalary meristematic activity around and/or beneath the gynoecium, forming a cuplike structure,* which raises the bracteoles up as part of the fruit wall (*Myrica gale* group), *or inferior even at the time of pollination, due to early intercalary activity* that forms a thick structure with (*Myrica,* most species) or without (*Canacomyrica*) papillae, with basal placentation; stigmas 2, elongated. Ovule 1 per gynoecium, orthotropous, with 1 integument. Nectaries lacking. *Fruit a drupe, covered either with waxy or fleshy papillae, or an achene,* not associated with conspicuous bracteoles (*Myrica,* most species; *Canacomyrica*), with 2 bracteoles fused to achene (*Myrica gale* group), or simply surrounding fruit (*Comptonia*); endosperm lacking, or nearly so (Figure 9.95).

Floral formula: Staminate: *, -0-, 1–9, 0

Carpellate: *, -0-, 0,$\underline{\overline{②}}$; drupe, achene

Distribution and ecology: Widespread in temperate to tropical regions; often early successional or in wetlands; plants associated with nitrogen-fixing, filamentous bacteria in root nodules.

Genera/species: 3/40. **Major genus:** *Myrica* (38 spp.).

Economic plants and products: Aromatic wax is extracted from the fruits of several species of *Myrica* (bayberry, wax myrtle, candleberry); a few species have edible fruits. Several species of *Myrica* are used as ornamental shrubs.

Figure 9.95 Myricaceae. (A–F) *Myrica pensylvanica:* (A) branch with staminate catkins (× 1.5); (B) staminate flower (× 14.5); (C) staminate flower, lateral view (× 14.5); (D) carpellate catkin (× 9); (E) carpellate flower with bracts (× 22); (F) carpellate flower in longitudinal section, showing basal ovule (× 30). (G–I) *M. cerifera:* (G) branch with fruits (× 0.75); (H) drupe (× 12); (I) fruit in longitudinal section, note waxy papillae, endocarp (indicated with numerous radiating lines), and embryo (× 12). (From Elias 1971, *J. Arnold Arbor.* 52: p. 310.)

Discussion: Myricaceae are tentatively considered to be monophyletic on the basis of their numerous specialized morphological features (MacDonald 1974, 1977, 1979a, 1989). *Comptonia* may be sister to a clade containing *Myrica* and *Canacomyrica*. *Comptonia* has retained stipules, diffuse porous wood, bracteoles that are free from the ovary, fruits that are small achenes, and does not show the intercalary development of a modified and cuplike portion of the inflorescence axis, which forms a large portion of the fruit wall in *Myrica* and *Canacomyrica*. Ovary/fruit development is most complex in *M. cerifera*, *M. pensylvanica*, *M. rubra*, and relatives; in this large group, waxy or fleshy papillae develop on the cuplike meristematic region that encircles the fruit. The result is a papillose drupe. In *Canacomyrica*, the cuplike meristematic region and resulting drupe are smooth. *Myrica gale* (and relatives) show the apomorphy of enlarged bracteoles that are strongly adnate to the fruit wall, and may be sister to a clade containing *Canacomyrica*, and the *M. cerifera* group. The recognition of *Canacomyrica* may make *Myrica* paraphyletic, and the distinctiveness of *M. cerifera* and relatives has led some to segregate these species as *Morella* (Wilbur 1994, 2001).

Abbe (1974) suggested that what appears to be the male flower (in the bract axil) is actually a cluster of staminate flowers, each of which consists of only a single stamen.

The reduced flowers of Myricaceae are wind-pollinated, and most species have separate staminate and carpellate catkins. The fleshy or waxy drupes of *Myrica cerifera* (and other similar species) are mainly bird-dispersed. The small fruits of *Myrica gale* are dispersed by water, the enlarged bracts acting as floats.

References: Abbe 1974; Elias 1971b; Kubitzki 1993c; MacDonald 1974, 1977, 1978, 1979a, 1989; Tiffney 1986; Wilbur 1994, 2001.

Juglandaceae A. P. de Candolle ex Perleb
(Walnut Family)

Aromatic trees; tannins present. Hairs various, often stellate, *and peltate scales with a glandular, swollen head containing various aromatic oils and/or resins. Leaves alternate and spiral*, occasionally opposite, **pinnately compound**, occasionally unifoliate, entire to serrate, with pinnate venation; **stipules lacking.** *Inflorescences indeterminate, erect to pendulous spikes or panicles, staminate and carpellate flowers often in separate inflorescences, and staminate inflorescence usually a catkin,* terminal or axillary. *Flowers unisexual (plants monoecious to less commonly dioecious),* ± radial, inconspicuous, 1 in the axil of each inflorescence bract and associated with 2 bracteoles, **the bracts** *sometimes 3-lobed,* **often expanded and forming a wing (or wings) associated with the fruit,** *or forming part of a cuplike husk that surrounds the fruit. Tepals 0–4,* inconspicuous, modified into a stigmatic disk in *Carya.* Stamens 3–numerous; filaments short; pollen grains triporoporate to polyporoporate.

Carpels usually 2, connate; ovary inferior, **and partially to completely adnate to the 2 bracteoles** *and often to the bract as well,* **ovary unilocular above and usually 2-locular below,** or apparently 4–8-locular due to false partitions, **with ovule borne at apex of incomplete septum**; stigmas usually 2, short to elongate and running along adaxial surface of style branches, often expanded. Ovule 1, orthotropous, with 1 integument. Nectaries lacking. *Fruit a nut or nutlet, often samara-like due to associated bracts and/or bracteoles, or a drupe, sometimes with outer husk splitting to release the bony pit;* embryo with *large cotyledons, these sometimes corrugated;* endosperm ± lacking (Figure 9.96).

Floral formula:

Staminate: *, -4–0-, 3–∞, 0

Carpellate: *, -4±0-, 0,$\overline{②}$; nut; nutlet; drupe (sometimes with ± dehiscent husk)

Distribution: Widespread from tropical to temperate regions.

Genera/species: 8/59. **Major genera:** *Carya* (25 spp.) and *Juglans* (20); both occur in the continental United States and/or Canada.

Economic plants and products: *Juglans regia* (walnut), *J. nigra* (black walnut), *Carya illinoensis* (pecan), and *C. ovata* (shagbark hickory) have edible nuts. *Juglans, Carya,* and *Engelhardia* are important as timber trees, and the first two genera, along with *Pterocarya* (wingnut), contain species that are used as ornamentals.

Discussion: The monophyly of Juglandaceae receives clear support from both morphology and *rbcL* sequences (Manchester 1987; Manning 1978; Smith and Doyle 1995). Phylogenetic relationships within Juglandaceae have been investigated in analyses based on morphology as well as molecular characters (Smith and Doyle 1995; Manos and Stone 2001), as well as more subjective considerations of morphological features (Manchester 1987; Manning 1978; Stone 1989, 1993). Engelhardioideae, including the tropical genera *Engelhardia, Alfaroa,* and *Oreomunnea,* are sister to the remaining genera. This group is characterized by three-lobed bracts and nuts with a fibrous shell. The remaining genera (i.e., *Platycarya, Carya, Cyclocarya, Pterocarya,* and *Juglans*—the Juglandoideae) form a well-supported clade; noteworthy synapomorphies of this group include buds with scales, odd-pinnately compound leaves with often serrate leaflets, bractlets that are completely fused to the ovary, isodiametric sclereids of the nut shell, wood with growth rings, and vessel elements with exclusively simple perforations. *Platycarya* is probably basal within Juglandoideae, with the remaining taxa linked by their drupaceous fruits, unisexual inflorescences, and elongate style branches. *Carya* is distinct due to its retention of highly modified, persistent tepals (form-

Figure 9.96 Juglandaceae. (A–M) *Carya ovata:* (A) flowering branch with staminate catkins and carpellate flowers (× 0.5); (B) staminate flower (× 14); (C) stamen (× 17); (D) bracts subtending staminate flower (× 14); (E) carpellate flower (× 8.5); (F) carpellate flower in longitudinal section, showing basal orthotropous ovule (× 8.5); (G) carpellate flower in cross-section, at level of lower arrow in F, showing 4-locular condition (× 8.5); (H) same, at level of upper arrow in F, showing ovule (× 8.5); (I) nut with dehiscent husk (× 1); (J) nut exposed by removal of two husk segments (× 1); (K) nut (× 1); (L) nut in cross-section, showing large embryo (unshaded) with corrugated cotyledons (× 1); (M) apical bud in winter condition (× 1.5). (N–S) *C. laciniosa:* (N) staminate flower (× 14); (O) stamen (× 17); (P) bracts subtending staminate flower (× 14); (Q) carpellate flower (× 8.5); (R) apical bud in winter condition (× 1.5); (S) seedling (× 0.25). (From Elias 1971, *J. Arnold Arbor.* 53: p. 37.)

ing a "stigmatic disk") in the carpellate flowers, and it has fruits with a dehiscent husk. It is sister to a clade containing *Juglans, Pterocarya,* and *Cyclocarya,* which are united by their stalked staminate catkins and twigs with chambered pith. *Juglans* has large drupes (with flesh derived from

adnate bract and bracteoles), while *Cyclocarya* and *Pterocarya* have retained small fruits with conspicuous winged bracteoles. Large fruits (with usually fleshy cotyledons) have evolved separately in *Alfaroa, Carya,* and *Juglans* (Stone 1989; Manos and Stone 2001).

The reduced flowers of Juglandaceae are clustered in dangling catkins and show numerous adaptations for wind dispersal. Flowering occurs as, or soon after, the leaves emerge; the ovules are not developed at the time of pollination. Fruits with a winglike bract or bracteoles such as *Englehardia*, *Cyclocarya*, and *Pterocarya*, are wind-dispersed. The large fruits of *Alfaroa* (nuts), *Juglans* (drupes), and *Carya* (nuts with a dehiscent outer husk) are adapted for dispersal by rodents.

References: Elias 1972; Manchester 1987; Manning 1938, 1940, 1948, 1978; Manos and Stone 2001; Smith and Doyle 1995; Stone 1973, 1989, 1993; Tiffney 1986.

Myrtales: *Incertae Sedis* within the Rosids

Myrtales are clearly monophyletic, as indicated by morphology, embryology, anatomy (Johnson and Briggs 1984), and *rbcL, atpB, ndhF,* and 18S sequences (Chase et al. 1993; Conti 1994; Conti et al. 1997; Källersjö et al. 1998;

Savolainen et al. 2000a,b; Soltis et al. 2000). Probable morphological synapomorphies include vessel elements with vestured (i.e., fringed) pits, stems with internal phloem, stipules absent or present as small lateral or axillary structures, flowers with short to elongate hypanthium, stamens incurved in the bud (but straight in Onagraceae), and a single style, the carpels being completely connate. In addition, these plants have simple, usually entire-margined, and often opposite leaves. Taxonomic placement of Myrtales is uncertain. The order consists of 14 families and about 9000 species; major families include **Lythraceae**, **Onagraceae**, **Myrtaceae**, **Melastomataceae**, Memecylaceae, and **Combretaceae**. Vochysiaceae, a South American family with one sepal usually larger than the others and swollen or spurred at the base, and usually only one fertile stamen, positioned opposite the spurred sepal, also belongs here, as indicated by the presence of internal phloem, a hypanthium, and *rbcL* and *ndhF* sequences (Conti 1994a,b). It is closely related to Myrtaceae.

Phylogenetic relationships of families within the order have been studied by Johnson and Briggs (1984),

---

Key to Major Families of Myrtales

1. Ovary unilocular, with apical placentation; fruit 1-seeded, a drupe, usually flattened, ridged, and/or winged; hairs long, straight, sharp-pointed, unicellular, and thick-walled ....**Combretaceae**

1. Ovary multilocular with axile placentation; fruit 1–many-seeded, usually a capsule or berry; hairs various, but not as above...2

2. Leaves with pellucid dots containing ethereal oils, thus aromatic when crushed; stamens usually numerous, the anther often with an apical secretory cavity.....................**Myrtaceae**

2. Leaves lacking pellucid dots, not aromatic; stamens equaling, half, or twice the number of petals to numerous, the anther lacking an apical secretory cavity...3

3. Leaves with 2–8 prominent secondary veins originating at or near base and converging toward apex, these usually connected by prominent tertiary veins oriented subperpendicular to midvein; anther connective, often with various appendages......**Melastomataceae**

3. Leaves usually with pinnate venation; anther lacking appendages, but sometimes with conspicuous gland 4

4. Leaves containing large branched sclerids; stamen connective thickened, with a conspicuous circular to depressed, terpenoid-producing gland; seeds often few, large, the often large embryo with thick or convolute cotyledons; trees or shrubs; calyx ± imbricate.......Memecylaceae

4. Leaves lacking large branched sclereids; stamens without gland on connective; seeds several to numerous, small, the minute embryo with short cotyledons; herbs to trees; calyx valvate...5

5. Pollen associated with viscin threads; stamens arising from rim of hypanthium, straight in bud; flowers usually 4- or 2-merous; petals smooth, not crumpled in bud; ovary inferior; hypanthium elongate with sepals ± reflexed (except *Ludwigia*).................**Onagraceae**

5. Pollen lacking viscin threads; stamens arising from inner surface of hypanthium, inflexed in bud; flowers usually 5-merous; petals wrinkled, crumpled in bud; ovary superior (except *Punica*); hypanthium short to elongate....................................**Lythraceae**

who employed structural characters, and Conti (1994a,b) and Conti et al. (1997), who used *rbcL* and *ndhF* sequences. Memecylaceae and Melastomataceae are sister taxa, whose pollen grains have well-developed poreless furrows (pseudocolpi) and lack floral nectaries. Onagraceae plus Lythraceae, with valvate calyx and grouped vessels in the wood, form another pair. Myrtaceae may be the sister group of the Melastomataceae + Memecylaceae clade. Relationships of Combretaceae are uncertain (Figure 9.97), but they may be associated with Onagraceae and Lythraceae.

References: Baas et al. 2000; Chase et al. 1993; Conti 1994a,b; Conti et al. 1997; Dahlgren and Thorne 1984; Johnson and Briggs 1984; Källersjö et al. 1998; Morley 1976; Patel et al. 1984; Renner 1993; Savolainen et al. 2000a,b; Soltis et al. 2000; Tobe 1989; Weberling 1988a.

Lythraceae J. St.-Hilaire
(Loosestrife Family)

Trees, shrubs, or herbs. Hairs various, sometimes silicified. *Leaves opposite,* less often whorled or alternate and spiral, simple, entire, *with pinnate venation;* stipules typically reduced, commonly appearing as a row of minute hairs. Inflorescences various. *Flowers bisexual, commonly di- or tristylous,* radial to occasionally bilateral, *with well-developed hypanthium,* frequently associated with an epicalyx.

Sepals usually 4–8, distinct or slightly connate, valvate, often rather thick. *Petals usually 4–8, distinct,* imbricate, **crumpled in bud and wrinkled at maturity,** occasionally lacking. *Stamens (4–) 8–16 (–numerous), usually attached well to slightly below summit of hypanthium, filaments unequal in length;* pollen usually tricolporate, sometimes with alternating poreless furrows. Carpels 2 to numerous, connate; *ovary superior* to rarely inferior, with axile placentation; stigma ± capitate. Ovules 2 to numerous in each locule. Nectaries often at base of hypanthium. *Fruit usually a dry, variously dehiscent capsule,* occasionally a berry; *seeds usually flattened and/or winged,* **seed coat with many-layered outer integument,** sometimes with epidermal hairs that expand and become mucilaginous upon wetting; endosperm ± absent (Figure 9.98).

Floral formula: * or X, 4–8, 4–8, 4–∞, 2–∞; capsule

Distribution and ecology: Widely distributed, but most species are tropical. Many occur in aquatic or semiaquatic habitats.

Genera/species: 30/600. *Major genera: Cuphea* (275 spp.), *Diplusodon* (72), *Lagerstroemia* (56), *Nesaea* (50), *Rotala* (45), and *Lythrum* (35). Genera occurring in the continental United States and/or Canada are *Ammannia, Cuphea, Decodon, Didiplis, Heimia, Lythrum, Nesaea, Peplis, Rotala,* and *Trapa.*

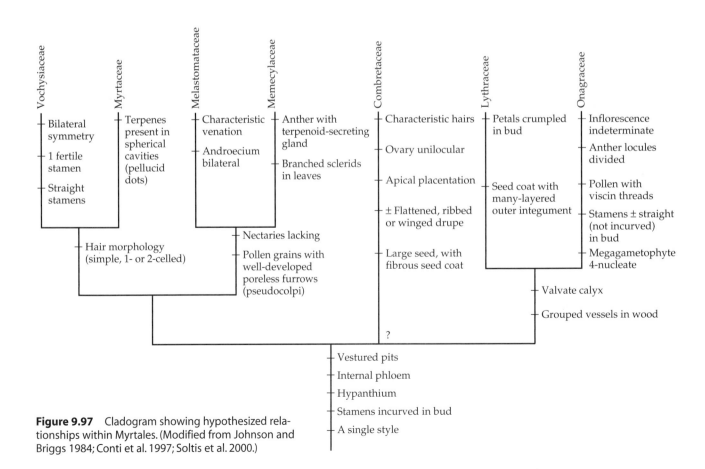

Figure 9.97 Cladogram showing hypothesized relationships within Myrtales. (Modified from Johnson and Briggs 1984; Conti et al. 1997; Soltis et al. 2000.)

Figure 9.89 Lythraceae. *Decodon verticillatus:*
(A) flowering branch (× 0.5); (B) stem in cross-section,
note spongy cortex (× 3); (C) inflorescence unit (× 4);
(D) flower in longitudinal section, note tristylous con-
dition (here a short style, medium and long stamens)
(× 4); (E) ovary in cross-section (× 8); (F) medium-
styled flower (× 4); (G) long-styled flower (× 4);
(H) capsule with persistent calyx (× 4); (I) seed (× 15);
(J) seed in section, endosperm stippled, embryo
unshaded (× 15); (K, L) embryo, two views (× 15).
(Original art prepared for the Generic Flora of the
Southeastern U.S. Project. Used with permission.)

Economic plants and products: Cuphea (Mexican heather), *Lagerstroemia* (crape myrtle), and *Lythrum* (loosestrife) provide ornamentals. The berries of *Punica* (pomegranates) contain numerous seeds with red, fleshy, and edible seed coats. *Lythrum salicaria* is a major exotic weed of wetlands in temperate North America.

Discussion: Lythraceae are easily recognized, and are considered monophyletic on the basis of morphological features (Graham et al. 1993a,b; Johnson and Briggs 1984). The family is here circumscribed broadly, including *Punica* (often placed in Punicaceae), *Sonneratia* and *Duabanga* (often placed in Sonneratiaceae), and *Trapa* (Trapaceae). This broad familial circumscription is supported by morphology (Graham et al. 1993b) and *rbcL* sequences (Conti 1994b). Graham et al. (1993b) noted that Lythraceae can be divided into two large clades. The first, containing *Sonneratia, Duabanga, Punica, Lagerstroemia,* and *Lawsonia,* is characterized by determinate inflorescences and wet stigmas, and the second, containing the remaining genera of the family, has indeterminate inflorescences, dry stigmas, and a reduced number of carpels. Within the latter clade, *Decodon* and *Pemphis* probably are basal, with the remaining genera delimited by the apomorphy of flattened seeds. Herbaceousness has evolved in several specialized, temperate genera.

Heterostyly is common. Most species of the family are pollinated by bees, beetles, and flies, although bird pollination occurs in *Cuphea,* in which the hypanthium is sometimes colorful and spurred, and *Sonneratia* is bat-pollinated. Nectar and pollen are employed as rewards. Cleistogamous or nearly cleistogamous flowers occur in *Peplis* and *Ammannia.* Seeds of Lythraceae are usually dispersed by wind or water. Some are buoyant due to spongy tissue in the outer seed coat.

References: Conti 1994b; Dahlgren and Thorne 1984; Duke and Jackes 1987; Ganders 1979; Graham 1964a; Graham et al. 1993a,b; Johnson and Briggs 1984; Redgrove 1929; Stubbs and Slabas 1982; Weberling 1988a.

Onagraceae Adanson
(Evening Primrose Family)

Herbs to shrubs or occasionally trees; raphides present. Hairs simple. *Leaves* alternate and spiral, opposite, or whorled, *simple,* entire to toothed, sometimes lobed, *with pinnate venation;* stipules small to lacking. **Inflorescences indeterminate**, terminal, or axillary and solitary. *Flowers* usually bisexual, radial or bilateral, *usually with well-developed hypanthium that is clearly prolonged above ovary* (except in *Ludwigia*). *Sepals* (2–) 4 (–7), distinct, valvate. *Petals* (2–) 4 (–7), distinct, sometimes clawed, occasionally lacking, imbricate, convolute, or valvate. *Stamens* (4–) 8, **not incurved in bud**, **anthers with septa dividing the sporogenous tissue within locules; pollen grains** in monads, tetrads, or polyads, usually triporate, occasion-

ally colpate, tricolporate, or biporate, **with unique paracrystalline beaded outer exine, and associated with viscin threads**. *Carpels usually 4,* connate; **ovary inferior**, usually with axile placentation; stigma capitate or clavate to 4-lobed or 4-branched. Ovules 1–numerous in each locule; **megagametophyte 4-nucleate** (i.e., *Oenothera*-type). Nectary usually near or at base of hypanthium. *Fruit a loculicidal capsule,* berry, or sometimes small, indehiscent, and nutlike; seeds sometimes winged or with a tuft of hairs; endosperm lacking (Figure 9.99).

Floral formula:

* or X, $\underbrace{4, 4,}$ 4 or 8, $\overline{\textcircled{4}}$; capsule, berry, nut

Distribution: Widely distributed and especially diverse in western North America and South America.

Genera/species: 16/650. *Major genera: Epilobium* (164 spp.), *Oenothera* (120), *Fuchsia* (110), *Ludwigia* (80), *Camissonia* (62), and *Clarkia* (45). *Chamaenerion* and *Circaea* also occur in North America.

Economic plants and products: Fuchsia, Oenothera (evening primrose), and *Clarkia* are ornamentals with showy flowers.

Discussion: The monophyly of Onagraceae is supported by morphology, rDNA, and *rbcL* sequence data (Hoch et al. 1993; Johnson and Briggs 1984; Sytsma and Smith 1988). Infrafamilial relationships have been investigated by numerous authors using morphology, rDNA, cpDNA restriction sites, *rbcL*, and ITS sequences (Bult and Zimmer 1993; Conti et al. 1993; Crisci et al. 1990; Hoch et al. 1993; Martin and Dowd 1986; Raven 1964, 1976, 1988; Sytsma and Smith 1992; Sytsma et al. 1998). All these workers found that *Ludwigia* is the sister taxon to the remaining genera. This genus lacks an elongate hypanthium and has a nectary around the base of the style. The rest of the family is united by the apomorphies of deciduous perianth, consistently 4-merous (or 2-merous) flowers, an elongate and deciduous hypanthium, and a reduction of carpel vascularization. Within this group, *Fuchsia, Circaea,* and possibly *Lopezia* form a clade; their outer integument develops in an unusual way. The remaining genera (e.g., *Epilobium, Chamaenerion, Camissonia, Gaura, Oenothera,* and *Clarkia*) form a clade diagnosed by loss of stipules and the occurrence of translocation heterozygosity (reciprocal interchange of parts of nonhomologous chromosomes; see Dietrich et al. 1977).

Common pollinators of Onagraceae include bees, moths, flies, and birds; nectar is the reward. Species may be self-pollinated or outcrossing. Dispersal in capsular-fruited genera is typically by wind or water. The berries of *Fuchsia* are bird-dispersed, and hooked hairs on the small, nutlike fruits of *Circaea* facilitate external transport. The seeds of *Epilobium* have a tuft of elongate hairs, while those of *Hauya* are winged; both are wind-dispersed.

Figure 9.99 Onagraceae. *Oenothera macrocarpa:* (A) part of flowering plant (× 0.5); (B) upper part of flower (× 1); (C) upper part of flower to show insertion of stamens at apex of hypanthium (× 1.5); (D) pollen grains connected by viscin threads (greatly magnified); (E) stigmas (× 30); (F) ovary in longitudinal section, with base of hypanthium and base of style, note ovules (× 3); (G) fruit (× 1); (H) fruit in cross-section (× 1); (I) seed (× 30); (J) seed in longitudinal section, note large embryo (× 30); (K) embryo (× 30). (From Wood 1974, *A student's atlas of flowering plants*, p. 77.)

References: Bult and Zimmer 1993; Carlquist 1975; Conti et al. 1993; Crisci et al. 1990; Dahlgren and Thorne 1984; Dietrich et al. 1997; Eyde 1982; Hoch et al. 1993; Johnson and Briggs 1984; Martin and Dowd 1986; Raven 1964, 1976, 1979, 1988; Skvarla et al. 1978; Sytsma et al. 1998a; Sytsma and Smith 1988, 1992; Tobe and Raven 1983.

Combretaceae R. Brown
(White Mangrove Family)

Trees, shrubs, or lianas, sometimes with erect, monopodial trunk supporting a series of horizontal, sympodial branches. Hairs various, **but some long, straight, sharp-pointed, unicellular, and very thick-walled, with a conical internal compartment at the base.** Leaves alternate and spiral, or opposite, **entire**, *with pinnate venation, often with domatia; petiole or base of blade often with 2 flask-shaped cavities, each containing a nectar gland;* stipules small or lacking. Inflorescences determinate, terminal or axillary. Flowers bisexual or unisexual (plants then monoecious, dioecious, or polygamous), usually radial, *with hypanthium slightly to conspicuously prolonged beyond the ovary. Sepals usually 4 or 5,* distinct or slightly connate, imbricate to valvate. *Petals usually 4 or 5*, distinct, imbricate or valvate, sometimes lacking. *Stamens 4–10*; filaments often long-exserted; pollen grains tricolporate or triporate, and often with poreless furrows. Carpels 2–5, connate; **ovary inferior and unilocular, with apical placentation**; stigma punctate to capitate. **Ovules few, pendulous on elongate funiculi from the tip of the locule.** Nectary usually a disk above ovary, often hairy. **Fruit a 1-seeded, ± flattened, ribbed and/or winged drupe; seed large, outer portion of seed coat fibrous**; embryo with usually folded or spirally twisted cotyledons; endosperm absent.

Floral formula: *, 4–5, 4–5 or 0, 4–10, (4–5); drupe
(usually ribbed or winged)

Distribution and ecology: Pantropical; *Laguncularia*, *Lumnitzera*, and *Conocarpus* are mangroves or mangrove associates; others occur in tropical broad-leaved forests or savannas.

Genera/species: 20/600. **Major genera:** *Combretum* (250 spp.) and *Terminalia* (200). The family is represented in the continental United States by *Bucida*, *Conocarpus*, *Laguncularia*, and *Terminalia*.

Economic plants and products: *Terminalia* (tropical almond) has edible fruits; other genera are ornamentals with showy flowers, such as *Combretum* and *Quisqualis* (Rangoon creeper), or distinctive habit, such as *Terminalia*, *Bucida* (black olive), and *Conocarpus* (buttonwood).

Discussion: Morphology (Dahlgren and Thorne 1984; Johnson and Briggs 1984) and *rbcL* sequence characters (Conti 1994b) both support the monophyly of Combretaceae.

The flowers typically produce nectar and attract insects, birds, or small mammals; the hypanthium varies from green to brightly colored. Many have bisexual flowers, and outcrossing is promoted by protogyny. *Conocarpus erectus*, however, is dioecious, and its flowers are densely packed in heads. In *Laguncularia racemosa*, some individuals bear staminate flowers, while others bear bisexual flowers that are very similar to the staminate ones. Flowers of *Terminalia* are also staminate or bisexual, but both occur on the same tree. The drupes of most genera are well adapted to dispersal by water due to their spongy mesocarp; some are distinctly winged and dispersed by wind, others are fleshy and bird or mammal-dispersed.

Domatia occur frequently in the family, usually associated with the veins on the abaxial leaf surface. These are usually inhabited by mites, as in *Terminalia* and *Conocarpus*, which probably protect the plant from fungi or tiny herbivores.

Laguncularia racemosa, the white mangrove, shows adaptations to daily flooding by salt water. Leaves contain

salt excretory glands, seeds germinate while still attached to the tree (vivipary; see Chapter 4), and pneumatophores (erectly growing roots containing aerenchyma tissue and functioning in gas exchange) may be present.

References: Conti 1994b; Dahlgren and Thorne 1984; Graham 1964b; Johnson and Briggs 1984; Stace 1965; Tomlinson 1980.

Myrtaceae Adanson
(Myrtle Family)

Trees or shrubs, often with flaky bark; **terpenes present. Hairs simple, unicellular or bicellular.** *Leaves opposite, or alternate and spiral*, rarely whorled, **entire**, *usually with pinnate venation*, **with scattered pellucid dots (i.e., spherical secretory cavities containing terpenoids and other aromatic, spicy-resinous compounds)**; stipules minute or lacking. Inflorescences determinate, but sometimes appearing indeterminate, terminal or axillary, sometimes reduced to a single flower. Flowers usually bisexual, radial, with well-developed hypanthium. *Sepals usually 4 or 5*, distinct to connate, imbricate, sometimes fused into a circumscissilely or irregularly rupturing cap. *Petals usually 4 or 5*, distinct to connate, imbricate, sometimes fused into a cap (and also adnate to the sepals), sometimes lacking. *Stamens usually numerous*, developing from the outside of the flower inward, *distinct or basally connate into 4 or 5 fascicles*; anthers often with connective with an apical secretory cavity; pollen grains usually tricolporate, with furrows fused. Carpels usually 2–5, connate; *ovary usually inferior to ± half-inferior*, with axile placentation, or less commonly parietal with intruded placentae; stigma usually capitate. Ovules 2 to numerous in each locule, anatropous to campylotropous. Nectariferous tissue atop ovary or lining inner surface of hypanthium. *Fruit usually a 1- to many-seeded berry or loculicidal capsule*, rarely a nut. Embryo with small to large cotyledons, these sometimes connate or folded or twisted together; endosperm scanty or lacking (Figure 9.100).

Floral formula:

*, (4–5), (4–5), ∞, (2–5); berry, capsule

Distribution: Pantropical, in a wide variety of habitats; also diverse in warm temperate Australia.

Genera/species: 144/3100. **Major genera:** *Eucalyptus* (500 spp.), *Eugenia* (400), *Myrcia* (300), *Syzygium* (300), *Melaleuca* (100), *Psidium* (100), and *Calyptranthes* (100). All of the above (except *Myrcia*), along with *Rhodomyrtus* and *Myrcianthes*, are native or naturalized in the continental United States.

Economic plants and products: *Eucalyptus* is an important source of timber. Many genera contain valuable

Figure 9.100 Myrtaceae. *Psidium longipes:* (A) flowering branch (× 0.5); (B) flower bud
(× 8); (C) flower (× 5); (D) stamen (× 25); (E) gynoecium and sepals (× 8); (F) stigma (× 25);
(G) ovary in longitudinal section, note ovules and resin cavities (× 8); (H, I) two ovaries in cross-
section, note resin cavities (× 12); (J) twig with berry (× 1.5); (K) berry (× 3); (L) berry in cross-
section (× 5); (M) seed (greatly enlarged); (N) seed in section, note large embryo. (Original art
prepared for the Generic Flora of the Southeastern U.S. Project. Used with permission.)

ornamentals due to their showy sepals, petals, and/or
stamens; these include *Myrtus* (myrtle), *Eucalyptus* (gum
tree), *Callistemon* (bottlebrush), *Melaleuca* (bottlebrush),
Leptospermum, and *Rhodomyrtus* (downy myrtle). Flower
buds of *Syzygium aromaticum* are cloves, and fruits of
Pimenta dioica are the source of allspice. Edible fruits are
common, including *Psidium guajava* (guava), *Syzygium*

jambos (rose apple), *S. malaccense* (Malay apple), *Myrcia-
ria cauliflora* (jaboticaba), *Eugenia uniflora* (Surinam cherry), and *Acca sellowiana* (pineapple guava). An oil, used
as a flavoring and antiseptic, is extracted from several
species of *Eucalyptus*. *Melaleuca* is an economically and
ecologically damaging weedy tree in the Everglades of
southern Florida.

Discussion: The infrafamilial relationships of this distinct and clearly monophyletic family have been investigated by cladistic analyses using both morphological features (Johnson and Briggs 1984), *rbc*L (Conti 1994), *mat*K (Wilson et al. 1996), *ndh*F sequences (Sytsma et al. 1998), and combined *mat*K and morphological data (Wilson et al. 2002). Two small genera, *Heteropyxis* and *Psiloxylon*, are cladistically basal, having retained perigynous flowers and stamens in only two whorls. Both genera also have stamens that are erect in the bud (a possible synapomorphy), and are sometimes segregated from Myrtaceae. The remaining genera of Myrtaceae—the "core" Myrtaceae—are united by the apomorphies of numerous stamens and at least partly inferior ovaries. These genera have usually been divided into the "Myrtoideae" (with berries and consistently opposite leaves) and "Leptospermoideae" (with capsules or nuts and alternate or opposite leaves). However, "Leptospermoideae," a mainly Australian complex including genera such as *Eucalyptus*, *Leptospermum*, *Metrosideros*, *Callistemon*, and *Melaleuca*, are paraphyletic and mainly basal within the "core" Myrtaceae, while the "Myrtoideae", including genera such as *Eugenia*, *Psidium*, *Calyptranthes*, *Syzygium*, and *Myrcianthes*, are polyphyletic because berry fruits have evolved independently in the myrtoid clade (e.g., *Eugenia*, *Psidium*, *Calyptranthes*, *Myrcianthes*, and *Myrtus*) and the *Acmena* group (e.g., *Acmena* and *Syzygium*).

Genera of "core" Myrtaceae can be divided among several well-supported clades, in addition to the Acmena and myrtoid groups mentioned above. *Callistemon* is closely related to *Melaleuca* (Melaleuca group). *Eucalyptus*, *Corymbia*, *Angophora*, and relatives constitute the *Eucalyptus* group, which is supported by wood anatomy. Most members of this clade have evolved various kinds of perianth caps (calyptras, formed by fusion of sepals and/or petals), but *Angophora* has separate petals. Cladistic analyses have shown that *Eucalyptus*, as broadly delimited, is not monophyletic, and segregates, such as *Corymbia*, could be recognized. Embryo characters often are stressed in generic delimitations.

The sweet-scented flowers of Myrtaceae are pollinated by various insects, birds, or mammals; nectar is the reward. In genera such as *Eucalyptus*, *Melaleuca*, and *Callistemon*, the stamens are more conspicuous than the petals, giving a bottlebrush effect. Fleshy-fruited species are dispersed by birds and mammals; capsular species frequently have small, and sometimes winged, seeds and are wind- or water-dispersed.

References: Conti 1994b; Conti et al. 1994; Dahlgren and Thorne 1984; Johnson 1976; Johnson and Briggs 1979, 1984; Schmid 1980; Sytsma et al. 1998b; Wilson 1960d; Wilson et al. 1996, 2002.

Melastomataceae A. L. de Jussieu
(Melastome or Meadow Beauty Family)

Trees, shrubs, vines, *or herbs*; cortical and/or pith vascular bundles usually present. Hairs various, often complex (e.g., stellate, dendritic, stalked glands, or peltate scales). *Leaves opposite,* entire to serrate, **with usually 2–8 subparallel secondary veins diverging from the base and converging toward the apex, these usually connected by prominent tertiary veins oriented ± perpendicular to midvein**; stipules lacking. Inflorescences determinate, terminal or axillary. Flowers usually bisexual, radial to bilateral, with well-developed hypanthium. *Sepals usually 3–5* and slightly connate, imbricate to valvate, sometimes fused into a circumscissilely or irregularly rupturing cap, sometimes associated with external projections. *Petals usually 3–5,* usually distinct and convolute. *Stamens usually 6–10,* sometimes dimorphic; *filaments bent,* **commonly twisted at anthesis, bringing the anthers to one side of the flower**; *anthers sometimes unilocular, opening by apical pores,* or sometimes longitudinal slits, *the connective often thickened or appendaged at base;* endothecium ephemeral; pollen grains usually tricolporate, with 3 alternating poreless furrows. Carpels usually 2–10, connate; ovary superior to inferior, usually with axile placentation; stigma capitate, punctate, or occasionally slightly lobed. Ovules usually numerous in each locule. *Nectaries usually lacking. Fruit a loculicidal capsule or berry; seeds usually numerous and small;* endosperm lacking (Figure 9.101).

Floral formula:

X or *, (3–5), 3–5, 6–10, (2–10); capsule, berry

Distribution and ecology: Pantropical; especially common in tropical montane habitats; often light-loving shrubs of early successional habitats.

Genera/species: 150 / 3000. ***Major genera:*** *Miconia* (1000 spp.), *Medinilla* (300), *Tibouchina* (240), *Leandra* (175), *Sonerila* (150), *Clidemia* (100), and *Microlicia* (100). The family is represented in the continental United States only by *Rhexia* and *Tetrazygia*.

Economic plants and products: Some genera contain ornamentals with showy flowers or striking leaves, including *Dissotis*, *Medinilla*, *Rhexia* (meadow beauty), and *Tibouchina* (princess flower).

Discussion: Morphological features (Renner 1993; Johnson and Briggs 1984) and DNA sequences (Conti 1994b; Sytsma et al. 1996; Clausing et al. 2000; Clausing and Renner 2001; Renner et al. 2001) support the monophyly of Melastomataceae. *Pternandra* probably is the sister group to the remaining genera (core Melastomataceae); this genus has retained an endothecium, while core Melastomataceae are united by the loss of an endotheci-

Figure 9.101 Melastomataceae. (A–J) *Rhexia virginica:* (A) tip of flowering plant (× 0.7); (B) flower (× 2.5); (C) same, side view with petals removed to show hypanthium and stamens (× 5.5); (D) hypanthium and gynoecium in longitudinal section (× 5.5); (E) ovary in cross-section (× 5.5); (F) ovule (× 65); (G) semidiagrammatic view of infructescence with leaves, bracts, and hairs omitted (× 0.7); (H) hypanthium enclosing mature capsule (× 5.5); (I) same, in longitudinal section, seeds removed (× 5.5); (J) seed (× 40). (K) *R. nashii:* flower bud in longitudinal section, petals removed to show inverted stamens (× 4). (L–M) *R. nuttallii:* (L) stamen from flower bud (× 4); (M) seed (× 40). (N) R. petiolata: stamen from open flower (× 16). (O–P) *Tetrazygia bicolor:* (O) leaf (× 0.7); (P) berry (× 16). (From Wurdack and Kral 1982, *J. Arnold Arbor.* 63: pp. 434, 438.)

um, and most of these genera also have anthers that open by apical pores (Clausing and Renner 2001). Members of core Melastomataceae are divided into several tribes. Most of these have capsular fruits, but berries are synapomorphic for the large tribe Miconieae (e.g., *Miconia, Clidemia, Leandra, Calycogonium, Conostegia, Tetrazygia, Henriettea, Tococa,* and *Mecranium*). Berries evolved independently in *Medinilla* (of the Dissochaeteae/Sonerileae complex) and Blakeeae (*Blakea, Topobea*). The Merianieae (e.g., *Merania*) and Rhexieae (*Rhexia*) are characterized by anthers with short connectives that are usually dorsally appendaged; the Rhexieae have the additional apomorphies of unilocular anthers and snail-shaped seeds. The Melastomateae (e.g., *Tibouchina, Melastoma, Monochaetum, Dissotis*) possess ovaries with a conic and often setose apex; this tribe also has snail-shaped seeds and may be closely related to *Rhexia*.

The flowers typically do not produce nectar, and are visited mainly by pollen-gathering bees, which vibrate or otherwise manipulate the anthers; the characteristic anther appendages may serve as a hold for the bee's legs. Most melastomes are facultative outcrossers, with the small stigma well separated from the anthers. The seeds of species with capsular fruits are usually wind- or raindrop-dispersed, while those with berries are dispersed by birds, mammals, turtles, or lizards; some seeds are secondarily ant-dispersed. Apomixis is common in species of disturbed habitats.

Some species (e.g., *Tococa*), have leaves with large pouches at the blade or petiole base that are inhabited by ants, while *Mecranium, Calycogonium,* and several other genera have special structures (domatia-forming hair tufts, pouches, etc.) in the axils of the secondary veins that are inhabited by predatory or fungus-eating mites.

References: Chase et al. 1993; Clausing and Renner 2001; Clausing et al. 2000; Cogniaux 1981; Judd 1986, 1989; Judd and Skean 1991; Renner 1989b, 1990, 1993; Stein and Tobe 1989; Vliet et al. 1981; Weberling 1988a; Whiffin 1972; Wilson 1950; Wurdack 1980, 1986; Wurdack and Kral 1982.

Eurosids II

Brassicales

Brassicaceae Burnett
(= Cruciferae Adanson)
(Crucifer, Mustard, or Caper Family)

Trees, *shrubs, or herbs; producing glucosinolates (mustard oil glucosides) and with myrosin cells;* often cyanogenic. Hairs diverse, simple to branched, stellate, or peltate. *Leaves usually alternate and spiral, sometimes in basal rosettes, simple, often pinnately dissected or lobed, or palmately or pinnately compound,* entire to serrate, with palmate or pinnate venation; stipules present or absent. *Inflorescences indeter-*

minate, occasionally reduced to a solitary flower, terminal or axillary. *Flowers usually bisexual, radial or bilateral, often lacking subtending bracts;* **receptacle prolonged, forming an elongate or shortened gynophore** (or androgynophore). *Sepals 4, distinct. Petals 4, distinct, often forming a cross,* often with an elongate claw and abruptly spreading limb, imbricate or convolute. *Stamens (2–) 6, or numerous, all ± the same length or the 2 outer shorter than the 4 inner (tetradynamous);* **filaments elongate** to rather short, distinct, or connate in pairs; pollen grains usually tricolporate or tricolpate. **Carpels usually 2,** connate; ovary superior, *with parietal placentation, frequently with the placentas forming a thick rim (replum) around the fruit and often connected by a false septum (a thin partition lacking vascular tissue) that divides the ovary into 2 chambers;* stigma capitate, sometimes bilobed. Ovules 1 to numerous on each placenta, anatropous to campylotropous. Nectar disk or gland usually present. *Fruit a berry or capsule, frequently with 2 valves breaking away from a replum and often additionally with a persistent septum (the fruit then a silique),* these short to elongate, globose to flattened; seeds with or without broad to narrow invagination, occasionally arillate; *embryo curved or folded;* endosperm scanty or absent (Figure 9.102).

Floral formula: * or X, , 4, 4, (2–) 6–∞ ,②; berry, capsule, silique-like capsule, silique

Distribution and ecology: Cosmopolitan, most diverse in the Mediterranean region, southwestern and central Asia, and western North America. Many species occur in early successional communities.

Genera/species: 419/4130. **Major genera:** *Capparis* (350 spp.), *Draba* (350), *Cleome* (200), *Erysimum* (180), *Cardamine* (170), *Lepidium* (170), *Arabis* (170), *Alyssum* (150), *Sisymbrium* (90), *Lesquerella* (90), *Heliophila* (70), *Thlaspi* (70), *Rorippa* (70), and *Hesperis* (60). Numerous genera occur in the continental United States and/or Canada; in addition to most of the above, noteworthy genera include *Barbarea, Brassica, Cakile, Caulanthus, Capsella, Cochlearia, Descurainia, Dimorphocarpa, Leavenworthia, Physaria, Platyspermum, Polanisia, Schoenocrambe, Stanleya, Streptanthus,* and *Warea*.

Economic plants and products: The family contains many important food plants, including both edible species, such as *Capparis spinosa* (capers), *Raphanus sativus* (radish), *Brassica oleracea* (cabbage, kale, broccoli, cauliflower, Brussels sprouts, kohlrabi), and *B. rapa* (Chinese cabbage, turnip), and sources of condiments, such as *Brassica juncea* (Chinese mustard), *B. nigra* (black mustard), *Sinapis alba* (white mustard), and *Armoracia rusticana* (horseradish). Table mustard is prepared from a mixture of the seeds of white mustard and either black mustard or Chinese mustard. Vegetable oil is extracted from the seeds of several species of *Brassica*, especially *B.*

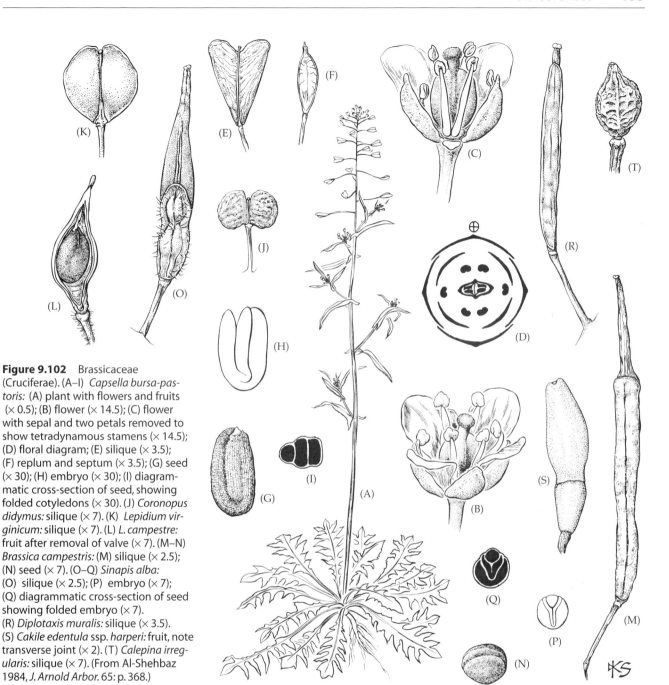

Figure 9.102 Brassicaceae (Cruciferae). (A–I) *Capsella bursa-pastoris:* (A) plant with flowers and fruits (× 0.5); (B) flower (× 14.5); (C) flower with sepal and two petals removed to show tetradynamous stamens (× 14.5); (D) floral diagram; (E) silique (× 3.5); (F) replum and septum (× 3.5); (G) seed (× 30); (H) embryo (× 30); (I) diagrammatic cross-section of seed, showing folded cotyledons (× 30). (J) *Coronopus didymus:* silique (× 7). (K) *Lepidium virginicum:* silique (× 7). (L) *L. campestre:* fruit after removal of valve (× 7). (M–N) *Brassica campestris:* (M) silique (× 2.5); (N) seed (× 7). (O–Q) *Sinapis alba:* (O) silique (× 2.5); (P) embryo (× 7); (Q) diagrammatic cross-section of seed showing folded embryo (× 7). (R) *Diplotaxis muralis:* silique (× 3.5). (S) *Cakile edentula* ssp. *harperi:* fruit, note transverse joint (× 2). (T) *Calepina irregularis:* silique (× 7). (From Al-Shehbaz 1984, *J. Arnold Arbor.* 65: p. 368.)

napus (canola, rapeseed oil). The family contains many ornamentals, such as *Cleome* (spider flower), *Hesperis* (rocket, dame's violet), *Erysimum* (wallflower), *Iberis* (candytuft), *Lunaria* (honesty, money plant), *Lobularia* (sweet alyssum), *Aurinia* (golden alyssum), and *Arabis* (rock cress). Weedy taxa are also common, e.g., *Capsella* (shepherd's purse), *Descurainia* (tansy mustard), *Lepidium* (peppergrass), and *Sisymbrium* (hedge mustard). *Arabidopsis thaliana* (thale or mouse-ear cress), a Eurasian weed, is the most widely used vascular plant in molecular and experimental biology.

Discussion: Brassicaceae may also be called Cruciferae, meaning "cross-bearing," in reference to the cruciform

arrangement of the petals. The family is considered to be monophyletic on the basis of the elongate gynophore and elongate, exserted stamens, although both of these are shortened in most derived members of the family (Rodman 1991b; Judd et al. 1994). Additional synapomorphies include the type of glucosinolates, the structure of the endoplasmic reticulum (Rodman 1981; Jorgensen 1981), and *rbcL* sequences (Rodman et al. 1993).

Brassicaceae are the largest family of the Brassicales, an order of 15 families characterized by the presence of glucosinolates, which contain sulfur. When these compounds react with myrosinase (contained in specialized spherical myrosin cells) they release hot, pungent mustard oils. The presence of glucosinolates (and myrosin cells) is synaptomorphic for members of Brassicales. The only other taxa that contains these compounds are *Drypetes* and *Putranjiva* (see Putranjivaceae, and discussion for Euphorbiaceae), and mustard oils are thus hypothesized to have evolved twice (Rodman et al. 1998). Phylogenetic relationships within the order have been investigated by cladistic analyses of morphological and DNA characters (Rodman 1991b; Rodman et al. 1993, 1996; Judd et al. 1994; Källersjö et al. 1998; Soltis et al. 1997, 2000; Karol et al. 1999; Savolainen et al. 2000b) (Figure 9.103). Brassicaceae are members of a morphologically distinct subclade of Brassicales, which may be diagnosed by 4-merous flowers, seeds with curved or folded embryos and lacking or nearly lacking endosperm, vessels with vestured pits, and protein-rich, unspecialized to vacuolar cisternae of the endoplasmic reticulum. Thus, these "cruciferous" features are also present in Resedaceae, Tovariaceae, Koeberliniaceae, and Bataceae. Relationships among basal families, such as Moringaceae + Caricaceae are well supported, and Tropaeolaceae are related to Akariaceae. These families have retained characters such as 5-merous flowers, ovaries with parietal placentation, straight embryos, and endosperm.

Morphology (Judd et al. 1994; Rodman 1991b) and *rbcL* sequences (Rodman et al. 1993; Savolainen et al. 2000b; Soltis et al. 2000) suggest that *Capparis* and relatives(i.e., "Capparoideae") form a basal paraphyletic complex within Brassicaceae. Members of the Cleomoideae (*Cleome* and relatives) and Brassicoideae (Temperate Zone mustards) form a monophyletic group based on the synapomorphies of an herbaceous habit, replum in the fruit, and *rbcL* sequences. The monophyly of the Cleomoideae is supported by palmately compound leaves and bilaterally symmetrical flowers, while the monophyly of Brassicoideae is indicated by the

frequent presence of a false septum in the ovary (and silique fruits) and folded embryos (with a loss of the seed invagination). Members of Brassicoideae also differ from Cleomoideae in their strongly reticulate and colpate (vs. smooth to shallowly reticulate and colporate) pollen. *Warea* and relatives (Thelypodieae) retain plesiomorphic features such as elongate, non- or only slightly tetradynamous stamens, and the presence of a distinct gynophore. Most members of Brassicoideae have short or obsolete gynophores and clearly tetradynamous stamens with short filaments. Members of the clade here referred to as Brassicoideae are, in earlier textbooks, recognized at the family level (as Brassicaceae), and are distributed predominantly in the temperate zone, whereas the predominantly tropical Cleomoideae and "Capparoideae" are placed together in a paraphylet-

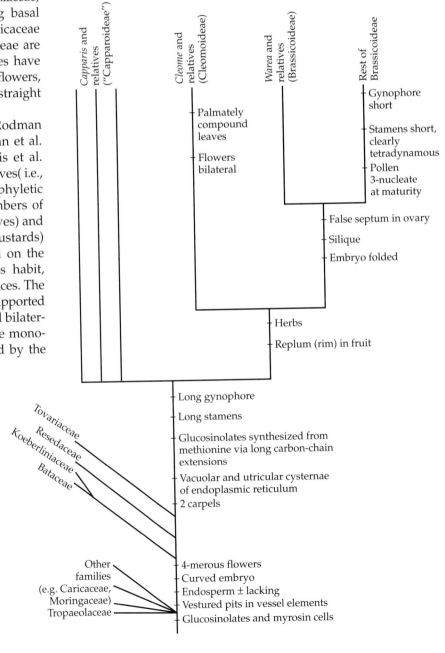

Figure 9.103 Cladogram showing hypothesized relationships within Brassicales (especially Brassicaceae). (Modified from Judd et al. 1994 and Soltis et al. 2000.)

ic "Capparaceae." A broadly defined Brassicaceae avoids recognition of arbitrarily delimited paraphyletic taxa (Judd et al. 1994).

Genera of Brassicoideae often are difficult to distinguish, and have been placed in some ten poorly defined tribes. Generic and tribal delimitation stress fruit morphology, calyx aestivation, flower color and symmetry, stigma form, number of seeds per locule, type of embryo folding, and indumentum (Al-Shehbaz 1984; Rollins 1993). It is, therefore, easier to identify fruiting specimens than ones in flower. Many commonly recognized genera, such as *Cleome, Brassica, Dentaria, Draba,* and *Arabidopsis,* are nonmonophyletic.

Flowers of Brassicaceae are frequently white, yellow, or pale to deep purple, and are pollinated by bees, flies, butterflies, moths, and beetles, which gather nectar. Pollination by birds or bats occurs in some tropical *Capparis.* Protogyny favors outcrossing, but many weedy species are selfing.

The small seeds may be explosively released from the siliques, as in *Cardamine,* but in most genera the valves of the silique (or capsule) merely fall away, exposing the seeds to the action of wind (or secondary dispersal by rain wash). In *Raphanus* and *Cakile* the corky fruits break transversely into one-seeded segments, which are then scattered by wind and/or water. Wings, bladders, or dust seeds facilitate dispersal by wind and have evolved several times within Brassicoideae. Fleshy fruits, as in *Capparis,* are dispersed by mammals or birds.

References: Al-Shehbaz 1984, 1985a,b, 1987, 1988a,b; Chase et al. 1993; Ernst 1963a; Jorgensen 1981; Judd et al. 1994; Källersjö et al. 2000; Karol et al. 1999; Rodman 1981; 1991a,b; Rodman et al. 1993, 1996, 1998; Rollins 1993; Savolainen et al. 2000b; Soltis et al. 2000; Sweeney and Price 2000; Vaughan et al. 1976; Warwick and Black 1993.

Malvales

Malvales are clearly monophyletic, as evidenced by their stratified phloem with fibrous and soft layers, wedge-shaped rays, mucilage (slime) canals and cavities, stellate hairs, connate sepals, malvoid leaf teeth (Judd and Manchester 1998), cyclopropenoid fatty acids, and *rbcL, atpB,* and 18S sequences (Alverson et al. 1998a; Fay et al. 1998; Källersjö et al. 1998; Savolainen et al. 2000a,b; Soltis et al. 1998, 2000). The complex vascular system that occurs in the petioles may also be synapomorphic. Stamens are frequently numerous and develop centrifugally from only a few trunk vascular bundles (evidence of a secondary increase from originally only two whorls). The order has been variously circumscribed, but probably consists of 9 families and 3560 species; major families include **Dipterocarpaceae**, **Cistaceae**, **Malvaceae**, and Thymelaeaceae. Phylogenetic analyses of DNA sequences clearly place the order within the Eurosid II clade, however, many systematists (e.g., Bessey 1915; Thorne 1992) had considered Malvales to be related to Urticales (here within Rosales, of Eurosid I) based on the common occurrence of bands of fibers in the phloem, alternate leaves with often palmate venation, and stipules.

References: Alverson et al. 1998a; Bessey 1915; Chase et al. 1993; Cronquist 1981; Fay et al. 1998; Judd and Manchester 1998; Källersjö et al. 1998; Savolainen et al. 2000a,b; Soltis et al. 1998, 2000; Thorne 1992.

Malvaceae Adanson
(Mallow Family)

Trees, shrubs, lianas, or herbs; mucilage canals (and often also mucilage cavities) present. Hairs various, but usually stellate or peltate scales. Leaves usually alternate, spiral or 2-ranked, simple, often palmately lobed, or palmately compound, entire

---

Key to Major Families of Malvales

1. Calyx and hypanthium colored and petaloid; internal phloem usually present .... Thymelaeaceae
1. Hypanthium absent and calyx usually not colorful; internal phloem lacking 2
  2. Placentation parietal; mucilage and/or resin canals lacking; sepals dimorphic (i.e., 3 large and 2 small), but not winglike .... **Cistaceae**
  2. Placentation axile; mucilage and/or resin canals present; sepals uniform or dimorphic but not as above, sometimes winglike 3
    3. Nectaries of densely packed glandular hairs (usually on calyx); anthers not appendaged; venation pinnate to palmate; mucilage canals present and resin canals lacking; calyx lobes usually not winglike; fruit a loculicidal capsule, schizocarp, indehiscent pod, nut, berry, drupe, or aggregate of follicles .... **Malvaceae**
    3. Nectaries not as above; anthers appendaged; venation usually pinnate; mucilage canals sometimes present and usually with resin canals; 2–5 calyx lobes becoming expanded and usually winglike in fruit; fruit usually a nut .... **Dipterocarpaceae**

to serrate, *the teeth malvoid* (i.e., with major vein unexpanded, and ending at the tooth apex), *with palmate or occasionally pinnate venation; stipules present.* Inflorescences indeterminate, mixed, or determinate, sometimes reduced to a single flower, axillary, **with determinate basic repeating unit and bearing 3 bracts, one of which is always sterile, while the others subtend lateral cymes or single flowers.** *Flowers* bisexual or unisexual, *usually radial, often associated with conspicuous bracts that form an epicalyx. Sepals usually 5, distinct to more commonly connate, valvate. Petals usually 5, distinct, imbricate,* convolute, or valvate, sometimes lacking. *Stamens 5 to numerous,* sometimes borne on a short to elongate androgynophore; *filaments distinct, basally connate and forming fascicles, but often strongly connate and forming a tube around the gynoecium (monadelphous); anthers 2-locular or unilocular (due to developmental splitting, with the apparent anther actually a half-anther),* usually unappendaged; staminodes sometimes present, these sometimes elongate and alternating with stamens or groups of stamens; pollen grains usually tricolporate, triporate, to multiporate, sometimes distinctly spiny. *Carpels 2 to many, con-*

Figure 9.104 Malvaceae (representative of complex with separate stamens). (A–G) *Triumfetta semitriloba:* (A) stem with flowers and fruits (× 1.5); (B) flower with two sepals and one petal removed (¥ 17); (C) flower in longitudinal section, with stamens removed (× 50); (D) fruit (× 8.5); (E) fruit prickle (× 17); (F) fruit in cross-section, showing seeds (× 11); (G) seed (× 22). (H) *T. pentandra:* fruit prickle (× 17). (From Brizicky 1965, *J. Arnold Arbor.* 46: p. 301.)

nate; ovary superior, placentation usually axile; stigma(s) capitate or lobed. Ovules 1 to numerous in each locule, anatropous to campylotropous. **Nectaries composed of densely packed, multicellular, glandular** *hairs on sepals and sometimes on petals or androgynophore. Fruit usually a loculicidal capsule, schizocarp, nut, indehiscent pod, aggregate of follicles,* drupe, or berry; seeds sometimes with hairs or arillate, occasionally winged; embryo straight to curved; endosperm present, often with cyclopropenoid fatty acids (Figures 9.104, 9.105).

Floral formula:

*, $\widehat{(5)}$, 5 or 0, $\widehat{(5-\infty)}$, $\widehat{(2-\infty)}$; capsule, schizocarp, nut, indehiscent pod, follicles

Distribution: Cosmopolitan.

Genera/species: 204/2330. **Major genera:** *Hibiscus* (300 spp.), *Sterculia* (250), *Dombeya* (250), *Sida* (200), *Pavonia* (200), *Grewia* (150), *Cola* (125), *Abutilon* (100), *Triumfetta* (100), *Bombax* (60), *Corchorus* (50), and *Tilia* (45). Noteworthy genera of the continental United States and Canada are *Abutilon, Callirhoe, Corchorus, Gossypium, Hibiscus, Kosteletzkya, Malva, Malvastrum, Malvaviscus, Modiola, Pavonia, Sida, Sidalcea, Sphaeralcea, Triumfetta, Tilia,* and *Urena*.

Economic plants and products: Important food plants include *Theobroma cacao* (chocolate, from seeds), *Cola nitida* and *C. acuminata* (cola seeds), *Durio zibethinus* (durian, fruit), and *Hibiscus esculentus* (okra, fruit). A few genera yield valuable timber; balsa wood is from *Ochroma pyramidale*. Hairs associated with the seeds are used as stuffing material (such as kapok, from species of *Ceiba* and

Figure 9.105 Malvaceae (representative of monadelphous clade). *Kosteletzkya virginica:* (A) flowering branch (× 0.5); (B) flower (× 1); (C) tip of staminal column with protruding styles, note half-stamens and teeth representing staminodes (× 10); (D) ovary in cross-section, each locule with one ovule (× 8); (E) capsule with calyx (× 4); (F) seed (× 7). (From Wood 1974, *A student's atlas of flowering plants,* p. 69.)

Bombax) or in fabrics (such as cotton, from species of *Gossypium*). The family contains numerous ornamentals, including *Tilia*, *Fremontodendron* (flannelbush), *Dombeya*, *Grewia*, *Firmiana* (Chinese parasol tree), *Ceiba*, *Abutilon*, *Althaea* (hollyhock), *Hibiscus* (hibiscus), *Pavonia*, *Malvaviscus* (turk's-cap), *Thespesia* (Portia tree), and *Malva* (mallow).

Discussion: The family is monophyletic and is circumscribed broadly; it usually has been divided into four families (i.e., Tiliaceae, Sterculiaceae, Bombacaceae, and Malvaceae s.s.). Traditional distinctions between these four families are arbitrary and inconsistent (Alverson et al. 1998, 1999 Baum et al. 1998; Bayer et al. 1999; Judd and Manchester 1998), and "Tiliaceae," "Sterculiaceae," and "Bombacaceae" are not monophyletic.

Infrafamilial relationships have been investigated by numerous systematists, and preliminary cladistic analyses based on morphology or DNA have been conducted by Judd and Manchester (1998), La Duke and Doebley (1995), Alverson et al. (1998, 1999), and Bayer et al.

(1999). General phylogenetic relationships are summarized in Figure 9.106. Note that the genera traditionally placed in "Tiliaceae" and "Sterculiaceae" (e.g., *Tilia*, *Grewia*, *Triumfetta*, *Berrya*, *Theobroma*, *Byttneria*, *Sterculia*, and *Walthera*) have retained often numerous stamens with 2-locular anthers. The stamens may be free or connate, and connation appears to have evolved more than once. Evolutionary relationships within these plants are still rather unclear but several monophyletic groups can be discerned (Figure 9.106)—some of which can not be diagnosed morphologically. The clade including *Grewia*, *Leuhea*, *Corchorus*, *Triumfetta*, *Apeiba*, and relatives (Grewioideae) has lost calyx fusion. *Theobroma*, *Byttneria*, *Guazuma* and relatives (Byttnerioideae) may be monophyletic due to the shared possession of five elongated staminodes that alternate with the unusually shaped petals (often clawed and basally hooded). Within this group, *Waltheria* and *Melochia* form a clade supported by the reduction of the androecium to only five stamens. *Dombeya* also has elongated staminodes that alternate with the petals, but it is more closely related to

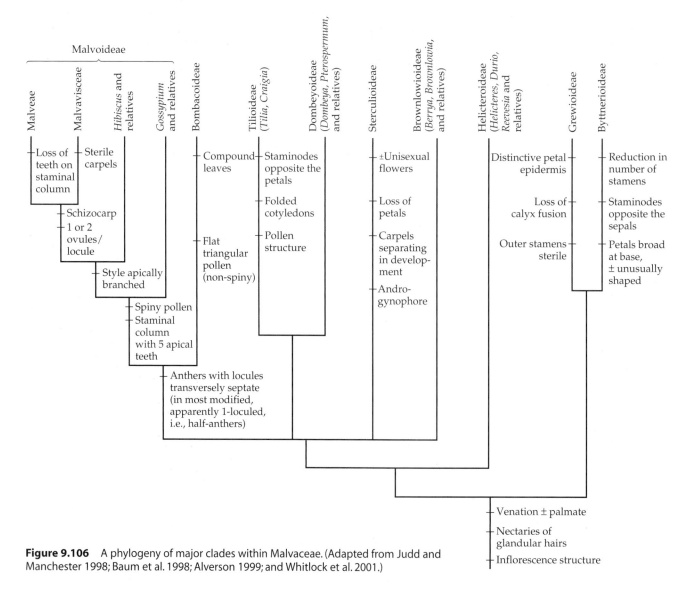

Figure 9.106 A phylogeny of major clades within Malvaceae. (Adapted from Judd and Manchester 1998; Baum et al. 1998; Alverson 1999; and Whitlock et al. 2001.)

Tilia. Dombeya is noteworthy in having spiny pollen. The clade containing *Sterculia, Cola, Firmiana,* and relatives (Sterculioideae) is supported by the distinctive unisexual, apetalous flowers with an elongate gynophore and carpels that separate at maturity, forming a cluster of follicles.

The remaining members of the family constitute a monophyletic group, supported by adnation of the androecium to the corolla and the presence of usually "unilocular" anthers. Anthers within this group may actually be half-anthers, although recent molecular evidence suggests that the underlying synapomorphy for this group may be transversely septate, 2-locular anthers that are strongly connate (Alverson et al. 1998b), which have become developmentally modified in many taxa so as to appear unilocular (opening by a single slit). The genera traditionally placed in "Bombacaceae" form a nonmonophyletic group, although a subgroup of "Bombacaceae"—*Adansonia, Pachira, Ceiba, Bombax, Pseudobombax,* and relatives (Bombacoideae)—may be monophyletic because all have palmately compound leaves and a distinctive pollen form (flattened, triangular, nonspiny). The largest group within the half-anther clade includes the group traditionally considered the Malvaceae (Malvoideae). The monophyly of Malvoideae is supported by globose spiny pollen, a staminal tube with five apical teeth, a well-developed epicalyx, and cpDNA restriction sites (La Duke and Doebley 1995). Molecular analyses associate several additional genera (e.g., *Quararibea, Matisia,* and *Fremontodendron*) with this clade. Within Malvoideae, the genera with loculicidal capsules and numerous seeds (e.g., *Hibiscus, Gossypium, Thespesia*) form a basal, paraphyletic complex. The more specialized malvoids are possibly united by the synapomorphies of schizocarpic fruits, carpels frequently more than five, and ovules only one or two per locule. Loss of apical teeth on the staminal column and cpDNA restriction sites diagnose Malveae (e.g., *Malva, Sida, Abutilon,* and *Sphaeralcea*). The monophyly of Malvavisceae (e.g., *Malvaviscus, Pavonia, Urena*) may be supported by sterile carpels alternating with the fertile ones, making the number of styles twice the number of ovary locules.

Flowers of Malvaceae are diverse, and attract bees, wasps, ants, flies, moths, birds, and bats. Nectar is the reward, and is usually produced on the inner surface of the connate sepals. Most species are outcrossing. Dispersal is also extremely varied. Capsular-fruited species have small seeds that are wind- or water-dispersed; they sometimes possess specialized dispersal structures such as wings or hairs. Many species with follicles (e.g., *Sterculia*) have seeds that contrast in color with the follicle wall, leading to dispersal by birds, but in *Firmiana* the follicle opens early in development and forms a dry, winglike structure for wind dispersal. Species with schizocarps are self-dispersed or adapted for external dispersal on birds or mammals. The large indehiscent pods of *Adansonia* contain an edible, somewhat sour, dry

flesh, and are dispersed by large mammals. The nuts of *Tilia* are borne on a cyme that is usually adnate to a conspicuous winglike bract; the entire infructescence and associated bract are dispersed by wind. Fleshy fruits or arillate seeds are usually dispersed by mammals or birds.

References: Alverson et al. 1998, 1999; Baum et al. 1998; Bayer 1998, 1999; Bayer et al. 1999; Edlin 1935; Brizicky 1965a, 1966a; Cronquist 1981; Chase et al. 1993; Fryxell 1988; Judd and Manchester 1998; La Duke and Doebley 1995; Robyns 1964; van Heel 1966; Whitlock et al. 2001.

Cistaceae Adanson
(Rockrose Family)

Shrubs or herbs, with tannins. *Hairs usually stellate* or peltate scales. Leaves opposite or alternate and spiral, simple, entire, usually pinnately veined, sometimes reduced, with a single vein; stipules present or absent. Inflorescences determinate, sometimes reduced to a single flower, terminal or axillary. Flowers bisexual, radial. *Sepals 5,* **the outer 2 distinctly narrower than the 3 inner**, or only 3, distinct to connate. *Petals 5 (–3), distinct,* imbricate, usually convolute. *Stamens usually numerous; filaments distinct*; anthers 2-locular; pollen grains usually tricolporate. *Carpels usually 3,* connate; ovary superior, *placentation parietal,* the placentae often intruded; stigma punctate to capitate, often 3-lobed. Ovules usually 4 to numerous on each placenta, usually orthotropous. Nectar disk present. *Fruit a loculicidal capsule*; embryo variously curved or folded.

Floral formula: $*, \overparen{(3+2)}, 5, \infty, \underline{\textcircled{3}}$; capsule

Distribution: Widely distributed in temperate regions, especially diverse in the Mediterranean region; mainly plants of open habitats and sandy or chalky soils.

Genera/species: 8/200. **Major genera:** *Helianthemum* (110 spp.) and *Lechea* (17). Both of these occur in the United States and/or Canada, along with *Hudsonia* and *Cistus*.

Economic plants and products: Several genera, including *Cistus* (rockrose), *Helianthemum* (sunrose, rockrose), and *Hudsonia* (beach heather), provide ornamentals.

Discussion: The monophyly of Cistaceae may be supported by the distinctive calyx. The family has often been considered related to Violaceae (or other members of Malpighiales with parietal placentation), but recent molecular analyses support a placement in Malvales and their floral development is very similar to that of Bixaceae. Phylogenetic relationships within Cistaceae are poorly understood, but *Lechea* may be an early divergent lineage and the large genus *Helianthemum* is very likely paraphyletic (including the specialized *Hudsonia*).

The flowers of Cistaceae may be showy (often bright yellow), attracting various bees, flies, and beetles, or inconspicuous and predominantly selfing. The flowers open during bright sunlight and usually remain open for only a few hours. The small seeds are dispersed by wind or rainwash.

References: Brizicky 1964c; Judd and Manchester 1998; Nandi 1998.

Dipterocarpaceae Blume
(Dipterocarp Family)

Trees; usually with branching resin canals in pith, wood, and bark, mucilage canals sometimes present in the cortex and pith; with tannins, *triterpenes and sesquiterpenes.* Hairs often fascicled, glandular, or peltate scales. *Leaves alternate, often 2-ranked, simple, entire, with pinnate venation, often with domatia; stipules present.* Inflorescences indeterminate or determinate, usually axillary. Flowers usually bisexual, radial, lacking an epicalyx. *Sepals 5, usually connate,* imbricate. *Petals 5, ± distinct,* imbricate and convolute. *Stamens (5-) 10-numerous,* occasionally on an androgynophore; filaments distinct or slightly connate; *anthers 2-locular,* **with terminal appendage developed from the connective;** pollen grains tricolpate or tricolporate. Carpels 2–4 (-5), connate; ovary superior, occasionally nearly inferior, with axile placentation; stigma capitate or lobed. Ovules 2–4 per locule. Nectary glandular. *Fruit usually a nut,* **the calyx persistent, with 2–5 calyx lobes becoming expanded** *and usually winglike.* Seeds lacking endosperm.

Floral formula:

* ⑤ , 5 , (10–∞) , (2–4) , nut (with winged calyx)

Distribution and ecology: Especially diverse in tropical Asia and Indomalaysia, where they often dominate monsoonal forests and rainforests, but also occurring in Africa and northern South America.

Genera/species: 17/550. **Major genera:** *Shorea* (195 spp.), *Hopea* (100), *Dipterocarpus* (70), and *Vatica* (75). The family does not occur in North America.

Economic plants and products: Several genera, especially *Shorea, Hopea, Dipterocarpus,* and *Vatica,* provide valuable hardwood timber.

Discussion: The monophyly of Dipterocarpaceae is supported by both morphological and molecular evidence (*rbcL* nucleotide sequences) (Dayanandan et al. 1999). The family is divided into three monophyletic groups, two of which (Monotoideae and Pakaraimoideae) are quite small (totaling about 35 species, taken together), and the third, the Dipterocarpoideae, containing the remaining species. Monotoideae (*Monotes, Marquesia, Pseudomonotes;* occurring in tropical Africa, Madagascar, and Colombia, South America) and Pakaraimoideae (*Pakaraimaea;* restricted to the Guayana highlands in northern South America) are cladistically basal, representing primitive elements in the family. They have stems lacking resin canals and have mucilage canals in the pith, tricolporate pollen, and equally expanded sepals. Monotoideae are distinguished from Pakaraimoideae in having petals longer (vs. shorter) than the sepals, flowers with an androgynophore (vs. this lacking), and differences in wood anatomy. The large and diverse Dipterocarpoideae are diagnosed by their branching resin canals, lack of mucilage canals in the pith, tricolpate pollen, and two or three calyx lobes conspicuously enlarged and wing-like. Phylogenetic relationships within the Dipterocarpoideae have been investigated by Dayanandan et al. (1999).

The flowers are insect pollinated, with thrips possibly being involved. The modified, wing-like calyx lobes allow the nuts to be dispersed only fairly short distances by wind, largely because the fruits are quite heavy.

References: Dayanandan et al. 1999; Londono et al. 1995; Maguire and Ashton 1977; Morton 1995.

Sapindales

Sapindales are clearly monophyletic, as indicated by the synapomorphies of pinnately compound leaves (occasionally becoming palmately compound, trifoliolate, or unifoliolate) and flowers with a distinct nectar disk. They are woody, with usually alternate and spiral exstipulate leaves and often rather small 4- or 5-merous flowers with imbricate perianth parts. The order's monophyly is strongly supported by analyses based on *rbcL, atpB,* and 18S sequence characters (Chase et al. 1993; Gadek et al. 1996; Källersjö et al. 1998; Savolainen et al. 2000a,b; Soltis et al. 1998, 2000). The order consists of 9 families and about 5800 species; major families include **Anacardiaceae, Burseraceae, Meliaceae, Rutaceae, Sapindaceae,** and **Simaroubaceae.**

Based on DNA sequences (Gadek et al. 1996; Soltis et al. 2000), Anacardiaceae and Burseraceae clearly form a clade, which is also supported by the presence of resin canals and biflavonoids in the leaves (Figure 9.107). A Meliaceae + Rutaceae + Simaroubaceae clade is supported by DNA sequences and the presence of bitter triterpenoids (Figure 9.107). DNA-based studies indicate that Simaroubaceae, as traditionally circumscribed, is polyphyletic, and this family is here narrowly defined. Picramniaceae (usually included within Simaroubaceae) does not belong in the order (Fernando et al. 1995; Fernando and Quinn 1995; Soltis et al. 1998, 2000), and instead is a basal rosid.

References: Chase et al. 1993, 1999; Fernando et al. 1995; Fernando and Quinn 1995; Gadek et al. 1996; Källersjö et al. 1998; Savolainen et al. 2000a,b; Soltis et al. 2000.

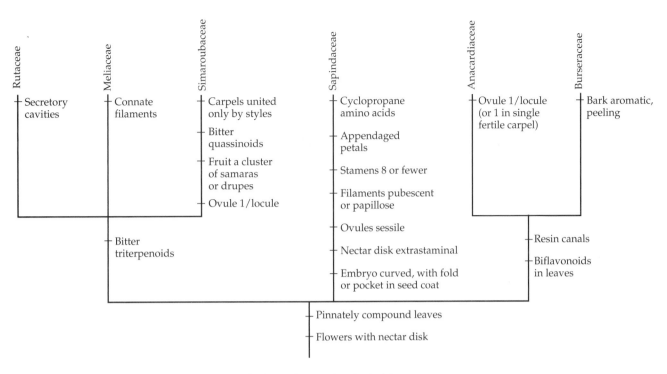

Figure 9.107 Cladogram showing hypothesized relationships within Sapindales. (Modified from Gadek et al. 1996 and Soltis et al. 2000.)

Key to Major Families of Sapindales

1. Stems, leaves, and fruits with pellucid dots containing aromatic ethereal oils . . . . **Rutaceae**

1. Stems, leaves, and fruits lacking pellucid dots 2

2. Plants strongly resinous, with vertical intercellular resin canals in the bark and associated with the phloem of major leaf veins 3

2. Plants lacking resin canals in bark and phloem of major leaf veins (although scattered secretory cells may be present) 4

3. Resins ± spicy-fragrant, not allergenic; bark smooth, often shredding; ovary with usually 2 ovules in each locule; placentation axile . . . . **Burseraceae**

3. Resins not spicy-fragrant, often allergenic; bark various, but not as above; ovary with only 1 ovule in each locule, and placentation axile, or more commonly only 1 locule fertile, and placentation apical . . . . **Anacardiaceae**

4. Stamens usually connate by their filaments, but if free, then fruit a capsule with winged seeds . . . . **Meliaceae**

4. Stamens distinct; fruit not a capsule with winged seeds 5

5. Nectar disk usually extrastaminal; ovules lacking a funiculus and broadly attached to an outgrowth of the placenta (obturator); stamens usually pubescent or papillose; leaves opposite or alternate; carpels persistently connate; bark not bitter . . . . **Sapindaceae**

5. Nectar disk intrastaminal; ovules with a funiculus; stamens glabrous; leaves alternate; carpels separating from each other after pollination; bark often very bitter . . . . **Simaroubaceae**

Rutaceae A. L. de Jussieu
(Citrus or Rue Family)

Usually trees or shrubs, sometimes with thorns, spines, or prickles; *usually with triterpenoid bitter substances*, alkaloids, and phenolic compounds; **with scattered secretory cavities (pellucid dots) containing aromatic ethereal oils**. Hairs various. *Leaves alternate and spiral*, or opposite, rarely whorled, *usually pinnately compound, or reduced and trifoliolate or unifoliolate*, occasionally palmately compound, *leaflets pellucid-dotted, especially near margin*, entire to crenate, with pinnate venation; stipules lacking. Inflorescences usually determinate, occasionally reduced to a single flower, terminal or axillary. Flowers bisexual or unisexual (plants then monoecious to dioecious), usually radial. *Sepals usually 4 or 5*, distinct to slightly connate basally. *Petals usually 4 or 5, distinct* or sometimes connate, usually imbricate. Stamens usually 8–10, sometimes numerous; filaments usually distinct but sometimes basally connate, glabrous or pubescent; pollen grains usually 3–6-colporate. *Carpels usually 4 or 5 to many, usually completely connate and with a single common style*, but occasionally with ovaries distinct and carpels only ± coherent by their styles; ovary superior, usually with axile placentation; stigmas various. Ovules 1 to several in each locule. *Nectar disk present, intrastaminal. Fruit a drupe, capsule, samara, cluster of follicles, or variously developed berry* (i.e., with mesocarp homogeneous or distinctly heterogeneous with a hard to leathery rind, with or without distinct partitions, and flesh or pulp derived from the ovary wall or from multicellular, juice-filled hairs); embryo straight to curved; endosperm present or lacking (Figure 9.108).

Floral formula: $*, \widehat{4\text{–}5}, \widehat{4\text{–}5}, \widehat{4\text{–}\infty}, \underline{\widehat{4\text{–}\infty}}$; berry, drupe, samara, cluster of follicles, capsule

Distribution: Nearly cosmopolitan, but mainly tropical and subtropical.

Genera/species: 155/930. **Major genera:** *Zanthoxylum* (200 spp.), *Agathosma* (180), and *Ruta* (60). Noteworthy genera in the continental United States and/or Canada include *Zanthoxylum*, *Amyris*, *Ptelea*, *Cneoridium*, *Poncirus*, and *Citrus*.

Economic plants and products: Several species of *Citrus* (oranges, tangerines, grapefruits, limes, lemons) are prized for their edible fruits. Fruits of *Fortunella* (kumquat) and *Casimiroa* (white zapote) are also eaten. *Ruta* (rue), *Zanthoxylum* (toothache tree), *Citrus*, and *Casimiroa* are used medicinally. The family contains many ornamentals, including *Murraya* (orange-jasmine), *Phellodendron* (cork tree), *Poncirus* (trifoliate orange), *Severinia* (boxthorn), and *Triphasia* (limeberry). The resinous compounds characteristic of Rutaceae are flammable, and thus the wood of a few genera, such as *Amyris* (torchwood), is used for fuel and torches.

Discussion: Rutaceae are broadly defined and their monophyly is evidenced by the presence of cavities containing ethereal oils, and appearing as pellucid dots in the mesophyll and other soft tissues, as well as *rbcL* and *atpB* sequences (Gadek et al. 1996; Morton et al. 1996; Chase et al. 1999). Traditionally recognized subfamilies are separated mainly by fruit type, carpel number, and extent of carpel connation. It is unlikely that any of these represent monophyletic groups (Morton et al. 1996; Chase et al. 1999) except for Aurantioideae (Citroideae, incl. *Citrus, Fortunella, Poncirus, Severinia, Atalantia, Aegle*, and relatives), a group characterized by globose berries and a base chromosome number of nine. Generic delimitations within this group are often problematic, especially those with berries with juice vesicles (e.g., *Citrus, Poncirus, Fortunella, Eremocitrus*, and *Microcitrus*). A recircumscribed Toddalioideae (incl. *Severinia, Zanthoxylum, Amyris, Phellodendron*, and relatives) may also be monophyletic, and these plants are characterized by fruits with a fibrous endocarp, usually forming follicles, capsules, or drupes with from one to several pits.

The family is notable for its variation in fruit types, which include a leathery-rinded berry ("hesperidium"; *Citrus, Poncirus*), hard-rinded berry (*Aegle*), typical berry with ± homogeneous flesh (*Casimiroa*), samara (*Ptelea*), drupe (*Phellodendron, Amyris*), cluster of follicles (*Zanthoxylum*), and capsule (*Ruta*).

Rutaceae are pollinated mainly by insects, especially bees and flies, which are attracted to the often showy, odor- and nectar-producing flowers. Most are outcrossing because the flowers are unisexual or, in bisexual flowers, because stigmas and anthers are separated physically or mature at different times. Selfing may occur as well. Asexual reproduction by agamospermy is frequent in *Zanthoxylum, Murraya, Poncirus*, and *Citrus*. Fleshy-fruited species are dispersed by mammals or birds. *Zanthoxylum*, with follicles of shiny black seeds that contrast in color with the green to red fruit wall, is also often bird-dispersed. The samaras of *Ptelea* are dispersed by wind.

References: Brizicky 1962b; Chase et al. 1999; de Silva et al. 1988; Fish and Waterman 1973; Gadek et al. 1996; Morton et al. 1996; Saunders 1934; Swingle 1967.

Meliaceae A. P. de Candolle
(Mahogany Family)

Trees or shrubs; commonly producing bitter triterpenoid compounds, usually with scattered secretory cells. Hairs usually simple, occasionally stellate or peltate scales. *Leaves usually alternate and spiral, once (or twice) pinnately compound*, occasionally trifoliolate or unifoliolate, leaflets usually entire, with pinnate venation; stipules lacking. Inflorescences usually determinate, axillary or less commonly terminal. *Flowers usually unisexual* (plants then

Figure 9.108 Rutaceae. *Poncirus trifoliata:* (A) flowering branch (× 0.75); (B) portion of twig, showing winged petiole with base of leaf blades, axillary thorn, and bud (× 1.5); (C) flower with two petals, one sepal, and several stamens removed, showing stamens, nectar disk, and gynoecium (× 6); (D) ovary in cross-section (× 9); (E) fruiting branch (× 0.75); (F) cross-section of mature berry, showing seeds embedded among pulp vesicles (× 1.5); (G) pulp vesicle with minute multicellular lateral appendages (× 4.5); (H) appendage of pulp vesicle (greatly magnified); (I) seed (× 4.5); (J, K) four of nine embryos from a single seed (resulting from agamospermy) (× 6). (From Brizicky 1962, *J. Arnold Arbor.* 43: p. 16.)

monoecious, dioecious, or polygamous), but often with well-developed staminodes or pistillodes, radial. *Sepals usually 4 or 5, distinct to ± connate. Petals usually 4 or 5,* distinct or slightly connate basally, imbricate, convolute, or valvate. *Stamens usually 4–10, occasionally more numerous;* **filaments connate, forming a tube, with or without apical appendages**, but separate filaments in *Cedrela* (a reversal), glabrous or pubescent; pollen grains 2- to 5-colporate. *Carpels usually 2–6, connate; ovary superior,* usually with axile placentation; stigma variously shaped, but usually capitate. Ovules usually 2 to numerous in each locule, anatropous to orthotropous. *Nectar disk present, intrastaminal. Fruit a loculicidal or septifragal capsule, drupe,* or berry; *seeds dry and winged* or with a fleshy seed coat; endosperm present or lacking (Figure 9.109).

Floral formula:

*, (4–5), 4–5, (4–10), (2–6); capsule, drupe

Distribution: Widespread in tropical and subtropical regions.

Genera/species: 51/550. *Major genera: Aglaia* (100 spp.), *Trichilia* (66), *Turraea* (65), *Dysoxylum* (61), and *Guarea* (35). *Melia* and *Swietenia* occur in the continental United States.

Economic plants and products: The family is important chiefly as a source of valuable timber, which comes from *Swietenia* (mahogany), *Cedrela* (West Indian cedar), and *Khaya* (African mahogany). *Azadirachta indica* (neem tree) is important medicinally and as a source of insecticides. *Melia azedarach* (chinaberry) and *A. indica* are ornamental trees.

Discussion: The monophyly of Meliaceae is supported by a preliminary analysis of *rbcL* sequences (Gadek et al. 1996) and morphology. Nearly all genera belong to either the Melioideae, with nonwinged seeds (in capsules, berries, or drupes), secondary xylem with one or two seriate rays, and naked buds; or to the Swietenioideae, with winged seeds (in capsules), secondary xylem with three to six seriate rays, and buds with scales (Pennington and Styles 1975). Swietenioideae, containing genera such as *Swietenia* and *Cedrela*, are probably monophyletic because of their distinctive flattened/winged seeds (in capsules) and scaly buds. *Cedrela* is distinct due to its separate filaments (a reversal) and erect petals. Melioideae, represented by genera such as *Trichilia, Guarea,* and *Melia,* are diverse morphologically and may be paraphyletic. *Melia* and *Azadirachta* share drupaceous fruits, a likely synapomorphy. The leaves of *Guarea* (and the related *Chisocheton*) are unusual in that the apex is meristematic and continues to produce new leaflets for several years.

Bees and moths are the most important pollinators of the small, nectar-producing flowers of Meliaceae. Outcrossing is enforced by unisexuality. Genera with drupes or capsular fruits and colorful fleshy seeds (e.g., *Trichilia*) are mainly dispersed by birds and various mammals (including bats). Winged seeds, as in *Swietenia* and *Cedrela*, are wind-dispersed.

References: Gadek et al. 1996; Miller 1990; Pennington and Styles 1975.

Simaroubaceae A. P. de Candolle
(Tree of Heaven Family)

Trees or shrubs, occasionally thorny; *scattered secretory cells often present in the leaves and bark, the pith conspicuous,* **with bitter triterpenoid compounds of the quassinoid type.** Hairs usually simple. *Leaves alternate and spiral, pinnately compound* to unifoliolate, leaflets entire to serrate, with pinnate venation; stipules usually lacking. Inflorescences determinate, terminal or axillary; catkins in *Leitneria. Flowers unisexual* (plants monoecious or rarely dioecious), but staminodes and pistillodes often well developed, radial. *Sepals 4 or 5,* but minute or lacking in *Leitneria,* distinct to slightly connate. *Petals usually 5,* distinct, rarely lacking (*Leitneria*), imbricate or valvate. Stamens usually 10, but reduced to 4 in *Leitneria,* in which they appear to be numerous due to clustering of 3 reduced flowers; filaments distinct, often basally appendaged; pollen grains tricolporate. *Carpels usually 5, but only 1 in Leitneria,* ± **united only by their styles;** *ovary superior, with axile placentation, but separating into individual carpels as the fruits develop;* stigma capitate to strongly lobed. **Ovules 1 per locule.** *Nectar disk present, intrastaminal,* but lost in *Leitneria.* **Fruit a cluster of samaras or ± dry to fleshy drupes;** endosperm ± lacking.

Floral formula:

Staminate: *, (4–5), 5, 10, (5) •
Carpellate: *, (4–5), 5, 10 •, (5); cluster of samaras
            or drupes

Distribution: Widespread in tropical or subtropical regions, with a few genera extending into temperate habitats.

Genera/species: 21/100. *Major genera: Simaba* (30 spp.), *Ailanthus* (15), and *Castela* (12). The family is represented in the continental United States by *Castela, Simarouba, Leitneria,* and *Ailanthus.*

Economic plants and products: Ailanthus (tree of heaven) and *Simarouba* (paradise tree) are often cultivated as ornamentals. Several genera are used medicinally.

Discussion: Simaroubaceae are considered monophyletic on the basis of morphology and *rbcL* sequences (Fernan-

Figure 9.109 Meliaceae. *Swietenia mahagoni:* (A) portion of inflorescence (× 1.2); (B) staminate flower (× 6); (C) same, with one sepal, petals, and part of staminal tube removed to show nectar disk and nonfunctional gynoecium (× 1.2); (D) ovary of staminate flower in cross-section, showing rudimentary ovules (× 14.5); (E) carpellate flower with petals and part of staminal tube removed, showing gynoecium, nectar disk, and rudimentary anthers (× 12); (F) ovary of carpellate flower in cross-section, showing ovules (× 14.5); (G) inflorescence with very young fruits (× 1.2); (H) branch with capsule (× 0.3); (I, J) stages in opening of capsule (× 0.6); (K) axis of fruit with seeds after fall of woody valves (× 0.6); (L) ridged fruit axis with seeds removed (× 0.6); (M) seed (× 1.2); (N) embryo (× 2.25). (From Miller 1990, *J. Arnold Arbor.* 71: p. 478.)

do et al. 1995; Gadek et al. 1996). Anatomical data are consistent with the hypothesis that the reduced, wind-pollinated flowers of *Leitneria* evolved from flowers similar to those of other Simaroubaceae, and the placement of this genus is also supported by serological data.

Kirkia, Picramnia, Suriana, and *Alvaradoa* were previously included within Simaroubaceae, resulting in a heterogeneous, clearly polyphyletic assemblage (Fernando et al. 1993, 1995; Gadek et al. 1996; Fernando and Quinn 1995; Savolainen et al. 2000b). These genera should be treated in the families Kirkiaceae (*Kirkia*)—Sapindales; Picramniaceae (*Alvaradoa, Picramnia*)—basal

rosid; and Surianaceae (*Suriana* and its relatives)—Fabales.

The flowers of Simaroubaceae are pollinated by various insects (especially bees) and birds. The flowers of *Leitneria* are wind-pollinated. Samaras (as in *Ailanthus*) are wind-dispersed, while the drupes of *Simarouba* are bird-dispersed.

References: Abbe and Earle 1940; Channell and Wood 1962; Cronquist 1944; Fernando et al. 1993, 1995; Fernando and Quinn 1995; Gadek et al. 1996; Petersen and Fairbrothers 1983; Savolainen et al. 2000b.

Anacardiaceae Lindley

(Sumac or Poison Ivy Family)

Trees, shrubs, or lianas; usually with tannins; *well-developed vertical resin canals in the bark and associated with the larger veins of the leaves, often also in other parenchymatous tissues, the resin clear when fresh but drying black, often causing dermatitis.* Hairs various. *Leaves usually alternate and spiral, pinnately compound, but sometimes trifoliolate or unifoliolate*, leaflets entire to serrate, with pinnate venation; stipules ± lacking. Inflorescences determinate, terminal or axillary. *Flowers almost always unisexual (plants usually dioecious)*, radial, small, often with well-developed staminodes or carpellodes. *Sepals usually 5*, distinct to slightly connate. *Petals usually 5*, distinct or slightly connate, ± imbricate. Stamens 5–10, occasionally more numerous or reduced to a single fertile stamen; filaments usually glabrous, usually distinct; pollen grains usually tricolporate or triporate. *Carpels typically 3*, sometimes 5, variously connate; ovary usually superior, sometimes all carpels fertile and gynoecium multilocular with axile placentation, *more commonly only 1 carpel fully developed and fertile (and others typically represented merely by their styles) and gynoecium ± asymmetrical and unilocular with apical placentation*; stigmas usually capitate. **Ovules 1 in each locule, or 1 in the single fertile carpel.** *Nectar disk present, intrastaminal. Fruit an often flattened asymmetrical drupe*; embryo curved to straight; endosperm scanty to lacking (Figure 9.110).

Floral formula:

Staminate: *, ⑤, 5, 5–10, ③(-5) •

Carpellate: *, ⑤, 5, 5–10 • ③(-5); drupe

Distribution: Mainly pantropical, with a few species in temperate regions.

Genera/species: 70/600. ***Major genera:*** *Rhus* (100 spp.), *Semecarpus* (50), *Lannea* (40), *Toxicodendron* (30), *Schinus* (30), and *Mangifera* (30). Noteworthy genera of the continental United States and/or Canada include *Cotinus*, *Metopium*, *Rhus*, *Schinus*, and *Toxicodendron*.

Economic plants and products: Fruits of *Mangifera indica* (mango) and *Spondias* (mombin, hog plum) are eaten, as are the roasted seeds of *Anacardium occidentale* (cashew) and *Pistacia vera* (pistachio). Fruits of several species of *Rhus* are used in drinks. A black lacquer is obtained from *Toxicodendron vernicifluum* (varnish tree). A few are ornamentals, including *Cotinus* (smoke tree), *Rhus* (sumac), and *Schinus* (Brazilian pepper). Finally, the group is of medical significance because so many of its taxa, particularly *Toxicodendron* (poison ivy, poison oak, poison sumac) and *Metopium* (poison wood), promote dermatitis in susceptible individuals due to the phenolic compound 3-n-pentadecycatechol in the resin. It is worth noting that mangoes and cashews, even though edible, may still cause an allergic reaction.

Discussion: Anacardiaceae and Burseraceae both have resin canals, biflavones, and clearly form a clade based on their DNA sequences (Gadek et al. 1996; Savolainen et al. 2000b; Soltis et al. 2000). Anacardiaceae are tentatively considered to be monophyletic on the basis of a reduced number of ovules, other morphological features, and *rbcL* sequences (Gadek et al. 1996; Terrazas and Chase 1996), although the recognition of Anacardiaceae may make Burseraceae paraphyletic.

The family is composed of two major subclades. Spondiadeae, which have retained many plesiomorphic features such as gynoecia with usually five carpels, multilocular ovaries, and fruits with a thick endocarp usually composed of lignified and irregularly oriented sclereids, may form a clade based on their septate fibers (Terrazas and Chase 1996). We note, however, that the group is often considered to be paraphyletic (Wannan and Quinn 1990, 1991). The remaining genera of the family form a large clade, clearly supported by gynoecia with three (or fewer) carpels, unilocular ovaries with apical placentation, and fruits with an endocarp that is composed of discrete and regularly arranged layers of cells.

Rhus and *Toxicodendron* have often been confused, and some botanists have united these two genera (and several others). Fruits of *Rhus* are glandular-pubescent and red, while those of *Toxicodendron* are glabrous and greenish to white. In addition, the resins of *Rhus* are not poisonous, while those of *Toxicodendron* cause a "poison ivy" rash. If combined, the resulting group would not be monophyletic.

The small, nectar-secreting flowers of Anacardiaceae are pollinated by various insects. Outcrossing is promoted by the more or less dioecious condition of members of this family. The large to small drupes are dispersed by various birds or mammals (including bats).

References: Brizicky 1962a; Cronquist 1981; Gadek et al. 1996; Gillis 1971; Savolainen et al. 2000b; Soltis et al. 2000; Terrazas and Chase 1996; Wannan and Quinn 1990, 1991.

Burseraceae Kunth

(Frankincense Family)

Trees or occasionally shrubs; **bark aromatic, smooth and often peeling;** *well-developed vertical resin canals in the bark and associated with the larger veins of the leaves, the resin clear, not allergenic.* Indumentum various. *Leaves usually alternate and spiral, usually pinnately compound*, but sometimes trifoliolate or unifoliolate, the leaflets entire to serrate, with pinnate venation; stipules ± lacking. Inflorescences usually determinate, axillary. *Flowers usually unisexual (and plants usually dioecious)*, radial, small, and often with well-developed staminodes or carpellodes.

Figure 9.110 Anacardiaceae. (A–D) *Rhus typhina:* (A) twig in winter condition (× 1.25); (B) staminate flower (× 10); (C) carpellate flower (× 10); (D) drupe (× 7). (E–N) *R. glabra:* (E) bud, breaking dormancy (× 1.25); (F) leaf (× 0.2); (G) inflorescence (× 0.5); (H) cymose cluster of carpellate flowers (× 7);
(I) carpellate flower in longitudinal section (× 10); (J) infructescence (× 0.8); (K) drupe (× 7); (L) glandular hair from fruit (greatly magnfied); (M) pit (× 10); (N) embryo (× 10). (Original art prepared for
the Generic Flora of the Southeastern U.S. Project. Used with permission.)

Sepals 4–5, usually slightly connate, usually imbricate, usually deciduous. *Petals 4–5*, distinct, usually imbricate. *Stamens in 2 or 1 whorls, twice as many as or equaling the petals; filaments usually distinct*, usually glabrous; anthers 2-loculed, opening by longitudinal slits; pollen grains usually tricolporate. Carpels 3–5, connate; ovary superior, with axile placentation; style 1; stigma 1, capitate to lobed. *Ovules 2 in each locule*, anatropous to campylotropous. *Nectar disk present, usually intrastaminal. Fruit a drupe with 1–5 pits, often with dehiscent valves*; embryo straight to curved; endosperm ± lacking.

Floral formula:

Staminate: * (4–5), 4–5 , 4–10 , (3–5)•

Carpellate:* (4–5), 4–5 , 4–10• , (3–5) ,

drupe, dehiscent drupe

Distribution: Mainly pantropical; especially diverse in tropical America and Africa.

Genera/species: 17/500. **Major genera:** *Bursera* (100), *Commiphora* (100), *Protium* (80), and *Canarium* (75). The family is represented in the continental United States only by *Bursera*.

Economic plants and products: Frankincense comes from *Boswella carteri* and myrrh from *Commiphora abyssinica* (and relatives). Some species of *Bursera* (gumbo-limbo) are used as ornamentals. Several genera provide timber species.

Discussion: The monophyly of Burseraceae is unclear, but has received preliminary support from a cladistic analysis of *rbcL* sequences (Gadek et al. 1995). The smooth bark is distinctive and may be synapomorphic. The small nectar-secreting flowers of Burseraceae are insect pollinated and the fleshy fruits are dispersed by birds.

References: Brizicky 1962; Gadek et al. 1995.

Sapindaceae A. L. de Jussieu
(Soapberry Family)

Trees, shrubs, or lianas with tendrils; often with tannins, usually with triterpenoid saponins in secretory cells, **with a diverse array of cyclopropane amino acids.** Hairs various. *Leaves alternate and spiral, or opposite, pinnately or palmately compound, trifoliolate, or unifoliolate, leaflets serrate or entire*, with pinnate or palmate venation; stipules lacking or present. Inflorescences determinate, terminal or axillary. *Flowers usually unisexual* (plants monoecious, ± dioecious, or polygamous), radial to bilateral. *Sepals usually 4 or 5*, distinct or sometimes basally connate. *Petals usually 4 or 5, sometimes lacking, distinct, often clawed*, **with ± basal appendages on adaxial surface** (these lost in a few), imbricate. **Stamens 8 or fewer**; *filaments distinct,*

usually pubescent or papillose; pollen grains usually tricolporate, the furrows sometimes fused with each other. **Carpels 2 or 3**, connate; ovary superior, usually with axile placentation; stigmas 2 or 3, minute to expanded. Ovules 1 or 2 in each locule, anatropous to orthotropous, **lacking a funiculus and broadly attached to a protruding portion of the placenta (the obturator). Nectar disk present, usually extrastaminal,** but sometimes with stamens borne upon it and ± intrastaminal. Fruit usually a loculicidal, septicidal, or septifragal capsule, arilloid berry, or schizocarp that splits into samara-like or sometimes drupelike segments, rarely a 1-seeded berry or drupe; *seeds often with an aril-like coat*; **embryo variously curved and with radicle separated from rest of embryo by deep fold or pocket in seed coat**; endosperm usually lacking (Figures 9.111, 9.112).

Floral formula:

Staminate: *, or X, (4–5), 4–5, 4–8, 0 or (2–3)•

Carpellate: *, or X, (4–5), 4–5, 4–8•, (2–3); capsule,

arilloid berry, drupaceous or samaroid schizocarp

Distribution: Mainly tropical and subtropical, with a few genera most diverse in temperate regions.

Genera/species: 147/2215. **Major genera:** *Serjania* (220 spp.), *Paullinia* (150), *Acer* (110), and *Allophylus* (100). Noteworthy genera occurring in the continental United States and/or Canada are *Acer, Aesculus, Cardiospermum, Cupania, Dodonaea, Exothea, Hypelate, Koelreuteria, Sapindus,* and *Serjania.*

Economic plants and products: The family contains several members that are sources of important tropical fruits, such as *Euphoria* (longan), *Litchi* (lychee), and *Nephelium* (rambutan), in which the flesh is derived from a large aril. The arils of *Blighia* (akee) are also eaten but are extremely poisonous if unripe. The seeds of *Meliccocus* (genip) have an edible fleshy coat, and those of *Paullinia cupana* (guarana) are used to make a caffeine-rich beverage. *Acer saccharum* (sugar maple) provides sugar and syrup. The fruits of species of *Sapindus* may be used as a natural soap due to the presence of saponins, and these poisonous compounds have led to the use of crushed branches and fruits of several genera as fish poisons. The group contains numerous ornamentals, including *Acer* (maple), *Aesculus* (horse chestnut), *Cardiospermum* (balloon vine), *Harpullia,* and *Koelreuteria* (goldenrain tree).

Discussion: The monophyly of Sapindaceae is supported by morphology and by DNA sequences (Gadek et al. 1996; Judd et al. 1994; Soltis et al. 2000). The presence of hypoglycin, an unusual and toxic non-protein amino acid, may be another synapomorphic feature of the group. The group is defined broadly, including both

Figure 9.111 Sapindaceae (representative of nonsamaroid taxa). *Exothea paniculata:* (A) branch with immature fruits (× 0.75); (B) staminate flower, showing nectar disk and pistillode (× 6); (C) bisexual flower, two stamens removed (× 6); (D) carpellate flower, showing staminodes (× 6); (E) gynoecium in longitudinal section, in plane marked by dashed line in F (× 9); (F) gynoecium in cross-section, in plane marked by dashed line in E (× 9); (G) fruit (× 3.5); (H) seed (× 3.5); (I) embryo (× 3.5). (From Brizicky 1963, *J. Arnold Arbor.* 44: p. 480.)

Aceraceae (maples and relatives) and Hippocastanaceae (horse chestnuts and relatives; see Judd et al. 1994). Morphological data suggest that if these are excluded, the remaining genera form a paraphyletic complex.

Infrafamilial phylogenetic relationships have been investigated using morphological data (Muller and Leenhouts 1976; Wolfe and Tanai 1987; Judd et al. 1994). Four well-supported clades can be recognized. The first is the hippocastanoid clade (i.e., *Aesculus* and relatives), which is united by opposite, palmately compound leaves, petals with marginal appendages, usually seven stamens, and large, leathery capsules that have sclerotic inclusions in the pericarp and open to release a single large seed. The other three clades are successively nest-ed. The second clade includes *Acer* and *Dipteronia* (traditional Aceraceae) and is diagnosed by opposite leaves, nonappendaged petals, more or less papillose stamens borne on a nectar disk, and two ovules per carpel (most reversals). The third is the samaroid clade, which includes *Acer, Dipteronia,* and numerous other tropical genera (e.g., *Thouinia* and *Serjania*). It is supported by the apomorphy of distinctive winged, schizocarpic fruits. The final, or sapindoid, clade is composed of the samaroid clade plus *Cupania, Euphoria, Sapindus, Blighia,* and relatives. These genera are united on the basis of their petals inserted at invaginations in the nectar disk, and ovules basally attached and reduced to one per carpel.

Figure 9.112   Sapindaceae (representatives of samaroid clade). (A–H) *Acer saccharum* ssp. *saccharum:*
(A) branch with flowers and expanding leaves (× 1.5); (B) leaf (× 0.5); (C) staminate flower (× 6); (D) staminate
flower in partial section, showing nectar disk and rudimentary gynoecium (× 9); (E) carpellate flower (× 6);
(F) carpellate flower in partial section, note nonfunctional stamens (× 9); (G) gynoecium in partial longitudinal
section (× 9); (H) samaroid schizocarp (× 1.5). (I–L) *A. saccharum* ssp. *floridanum:* (I) leaf (× 0.5); (J) samaroid
schizocarp (× 1.5); (K) seed (× 6); (L) embryo (× 6). (From Wood 1974, *A student's atlas of flowering plants*, p. 68.)

The flowers of Sapindaceae vary from small and radially symmetrical to fairly large, showy, and bilaterally symmetrical; they are pollinated by birds and a wide variety of insects, which are rewarded by floral nectar. *Dodonaea* and some species of *Acer* are wind-pollinated. The dioecious condition enforces outcrossing. Dispersal varies widely. Many tropical groups, such as *Blighia*, *Harpullia*, and *Cupania*, have capsules contrasting in color with the often arillate seeds and are dispersed by birds or mammals. *Litchi*, *Nephelium*, and *Euphoria* have indehiscent fruits, each containing a single seed associated with a fleshy aril, and are also animal-dispersed. Others, such as *Koelreuteria* and *Dodonaea*, have dry inflated or winged capsules that are wind-dispersed. Wind dispersal characterizes the samaroid clade, in which the winged mericarps spin like a propeller as they move through the air.

References: Brizicky 1963; Gadek et al. 1996; Hardin 1957; Judd et al. 1994; Muller and Leenhouts 1976; Umadevi and Daniel 1991; van der Pijl 1957; Wolfe and Tanai 1987.

Asterid Clade (Sympetalae)

This large and specialized subgroup of the tricolpate clade is hypothesized to be monophyletic on the basis of *rbcL, atpB, ndhF, matK,* and 18S rDNA sequence characters (Chase et al. 1993; Savolainen et al. 2000a,b; Soltis et al. 1997, 2000; Olmstead et al. 2000; Albach et al. 2001a) and probably ovules with only a single integument and a thin-walled megasporangium, although they are both homoplasious. Iridoids are widespread in the clade and could also be synapomorphic.

Cornales and Ericales may be successively sister to the rest of this clade. The core asterids, supported by the number of stamens equaling the number of petals, epipetalous stamens, an obviously sympetalous corolla, and DNA sequences, form the rest, although it should be noted that obviously sympetalous corollas and epipetalous stamens have also evolved in several members of the Ericales. The core asterids include two major clades, here called Euasterids I (Garryales, Gentianales, Lamiales, and Solanales) and Euasterids II (Aquifoliales, Apiales, Dipsacales, and Asterales) following the Angiosperm Phylogeny Group (1998) and recent phylogenetic analyses (Olmstead et al. 2000; Savolainen et al. 2000b; Soltis et al. 2000; Albach et al. 2001a,b).

Cornales

The monophyly of Cornales is strongly supported by *rbcL, ndhF, atpB, matK,* and 18S sequence characters (Xiang et al. 1993, 1998; Olmstead et al. 2000; Savolainen et al. 2000b; Soltis et al. 2000; Albach et al. 2001b), more or less inferior ovaries, usually reduced sepals, and an epigynous nectar disk. Many also have drupaceous fruits. The order, as here delimited, consists of perhaps three families, **Cornaceae** s.l. (including Nyssaceae and Alangiaceae), **Hydrangeaceae**, and Loasaceae, which together contain about 650 species (Hempel et al. 1995; Soltis et al. 1995; Xiang et al. 1998). Cornales may constitute the sister group to the rest of the asterid clade (Hempel et al. 1995; Olmstead et al. 1993; Xiang et al. 1993). Morphology also supports an asterid affinity, as these plants have ovules with a single integument, often a thin megasporangium, and iridoid compounds (Hufford 1992). Relationships within Cornales are still not well understood, and some botanists recognize Nyssaceae as distinct from Cornaceae.

Garryaceae and Vitaceae sometimes have been placed in Cornales. Garryaceae, a primarily western North American family of dioecious trees or shrubs with opposite, simple leaves, inconspicuous, 4-merous flowers borne in dangling catkins, and two-seeded berries, probably are more closely related to some families of the core asterid complex (Garryales, Euasterids I). Vitaceae are an early divergent lineage of the rosid clade.

References: Albach et al. 2001b; Downie and Palmer 1992; Hempel et al. 1995; Hufford 1992; Olmstead et al. 1993, 2000; Savolainen et al. 2000b; Soltis et al. 2000; Xiang et al. 1993, 1998.

Hydrangeaceae Dumortier
(Hydrangea Family)

Shrubs, small trees, lianas or herbs; usually with tannins; often with iridoids, aluminum, and raphide crystals. Hairs usually simple. *Leaves usually opposite, simple,* but

---

Key to Families of Cornales

1. Placentation parietal; leaves with hooked or sticky hairs..Loasaceae
1. Placentation axile or intrusive parietal; leaves with simple, Y- or T-shaped hairs..........................2
2. Fruit usually a septicidal or loculicidal capsule; ovules usually several to numerous in each locule..................................**Hydrangeaceae**
2. Fruit a drupe; ovules solitary in each locule...**Cornaceae**

sometimes lobed, entire to serrate or dentate, with pinnate or palmate venation; *stipules lacking.* Inflorescences determinate, terminal or axillary. Flowers bisexual, radial, those at margin of inflorescence sometimes sterile and with enlarged petal-like sepals. *Sepals usually 4 or 5, connate, the lobes often reduced. Petals usually 4 or 5 and distinct,* imbricate, convolute, or valvate. *Stamens 8 or 10 to numerous;* filaments distinct or slightly connate; pollen grains tricolpate or tricolporate. Carpels usually 2–5, connate; *ovary usually half-inferior to inferior,* **often ribbed,** with axile or deeply intruded parietal placentation; stigmas 2–5, usually elongate. Ovules usually several to numerous on each placenta, with 1 integument and a thin-walled megasporangium. *Nectar disk usually present atop ovary. Fruit usually a loculicidal or septicidal capsule;* seeds often winged.

Floral formula: *, (4–5), 4–5, (8–∞), (2–5) ; capsule

Distribution: Widespread, but especially characteristic of temperate to subtropical regions of the Northern Hemisphere.

Genera/species: 17/250. **Major genera:** *Philadelphus* (65 spp.), *Deutzia* (40), and *Hydrangea* (30). *Decumaria, Deutzia, Fendlera, Hydrangea,* and *Philadelphus* occur in the continental United States and/or Canada.

Economic plants and products: *Hydrangea* (hydrangea), *Decumaria* (climbing hydrangea), *Schizophragma* (climbing hydrangea), *Philadelphus* (mock orange), and *Deutzia* are frequently grown as ornamentals.

Discussion: The monophyly of Hydrangeaceae is supported by morphological and DNA characters (Albach et al. 2000b; Hempel et al. 1995; Hufford 1997; Morgan and Soltis 1993; Soltis et al. 1995; Xiang et al. 1993). This group has traditionally been included as a subfamily of Saxifragaceae (a basal tricolpate group), but cladistic analyses (Chase et al. 1993; Hempel et al. 1995; Hufford 1992; Morgan and Soltis 1993; Xiang et al. 1993) indicate that Hydrangeaceae are only distantly related to Saxifragaceae, and strongly support their placement near Cornaceae.

Fendlera and *Jamesia* (Jamesioideae) may be sister to the remaining genera of Hydrangeaceae (Hufford 1997; Hufford et al. 2001; here treated as Hydrangeoideae. The latter group is divided into two major clades: the Hydrangeae (*Hydrangea, Decumaria, Schizophragma,* and relatives) and the Philadelpheae (*Philadelphus, Deutzia, Fendlerella, Whipplea, Carpenteria,* and others) (Morgan and Soltis 1993; Soltis et al. 1995; Xiang et al. 1993; Hufford et al. 2001). Most members of the Hydrangeae have conspicuous sterile marginal flowers in the inflorescence, valvate petals, and more or less loculicidal capsules. Members of the Philadelpheae lack sterile flowers and usually have imbricate petals and septici-

dal capsules. The genus *Hydrangea* definitely is non-monophyletic.

Flowers of Hydrangeaceae may be large and showy, as in *Philadelphus,* or small (and then densely clustered and often associated with conspicuous sterile flowers), as in *Decumaria* or *Hydrangea.* The nectar produced in the epigynous floral disk rewards a variety of insect pollinators (butterflies, moths, flies, bees, wasps, and beetles). Outcrossing is promoted by protogyny, but self-pollination is also possible. The small seeds, which are often tailed or winged, are dispersed by wind.

References: Albach et al. 2001b; Chase et al. 1993; Hempel et al. 1995; Hufford 1992, 1997, 2001; Hufford et al. 2001; Morgan and Soltis 1993; Soltis et al. 1995; Spongberg 1972; Xiang et al. 1993.

Cornaceae Dumortier
(Dogwood Family)

Usually trees or shrubs; usually with iridoids. **Hairs often calcified, mesifixed, Y- or T-shaped.** *Leaves opposite, less commonly alternate and spiral, simple, usually entire,* but sometimes serrate, with pinnate to ± palmate venation, *secondary veins usually smoothly arching toward margin or forming a series of loops; stipules lacking.* Inflorescences determinate, terminal, sometimes associated with enlarged, showy bracts. Flowers bisexual or unisexual (plants then monoecious or dioecious), radial. *Sepals usually 4 or 5, distinct or connate, usually represented by small teeth,* sometimes lacking. *Petals usually 4 or 5,* ± distinct, imbricate or valvate. *Stamens 4–10;* filaments distinct; pollen grains usually tricolporate, **the apertures with an H-shaped thin region.** Carpels usually 2 or 3, connate, sometimes appearing to be a single carpel; ovary inferior, with axile placentation, **the axis lacking vascular bundles, and the ovules instead attached to vascular bundles that arch over the top of the each septum;** stigma usually capitate, lobed, or elongate. **Ovules 1 in each locule,** attached at apex, with 1 integument and a thin- to thick-walled megasporangium. *Nectar disk positioned atop ovary.* **Fruit a drupe, the pit 1- to few-seeded, ridged to winged, with thin regions** (i.e., germination valves, these sometimes obscure) (Figure 9.113).

Floral formula: *, (4–5), 4–5, 4–10, (2–3); drupe

Distribution: Widespread; especially common in north temperate regions.

Genera/species: 13/130. **Major genera:** *Cornus* (45 spp.), *Mastixia* (20), and *Nyssa* (10). *Cornus* and *Nyssa* occur in the continental United States and Canada.

Economic plants and products: Ornamental trees and shrubs occur in *Nyssa* (tupelo), *Davidia* (dove tree), and *Cornus* (dogwood).

ADC

Figure 9.113 Cornaceae. (A–E) *Cornus amomum:* (A) flowering branch (× 0.75);
(B) flower (× 9); (C) flower in longitudinal section, with petals and stamens removed (× 9);
(D) drupe (× 3); (E) pit in lateral view and from above (× 6). (F–H) *C. florida:* (F) flowering
branch (× 0.75); (G) flower (× 6); (H) pit from lateral view and from above (× 6). (I) *C.
alternifolia:* pit in lateral view and from above (× 6). (From Ferguson 1966, *J. Arnold Arbor.*
47: p. 111.)

Discussion: The monophyly of Cornaceae, as defined broadly (i.e., including Nyssaceae and Alangiaceae), is supported by morphology as well as *matK* and *rbcL* sequences (Xiang et al. 1998). Infrafamilial relationships have been investigated by Eyde (1988), Murrell (1993), Xiang et al. (1993, 1996, 1998), and Xiang and Murrell (1998). Two major clades can be recognized within Cornaceae: a nyssoid-mastixioid clade (*Nyssa, Camptotheca, Davidia, Mastixia, Diplopanax,* and probably *Curtisia*) with usually unisexual, 5-merous flowers; and a cornoid clade (*Cornus* and *Alangium*) with usually bisexual, 4-merous flowers.

Cornus has several specialized features, such as the number of stamens usually equaling that of the petals, ovules with the raphe positioned dorsally (i.e., apotropous), 4-merous flowers, and hairs coated with large crystals of calcium carbonate. Its monophyly is supported by morphology, cpDNA restriction sites, and *matK* and *rbcL* sequences. Combined evidence (morphology, restriction sites, *matK, rbcL,* 26S sequences: Xiang et al. 1993, 1996, 1998; Xiang and Murrell 1998; Fan and Xiang 2001) suggests that two major clades can be recognized within *Cornus:* blue-fruited dogwoods and red-fruited dogwoods. The latter contains the cornelian cherry clade (*C. mas* and relatives) and the big-bracted dogwood clade (*C. florida, C. kousa, C. nuttallii, C. canadensis,* and relatives).

The flowers of Cornaceae typically produce nectar and attract bees, flies, and beetles, but wind pollination may occur, as in *Davidia.* The white, blue, blue-black, purple, or red drupes, sometimes contrasting in color with the inflorescence axes, are dispersed by birds and mammals. The fruits of several species of *Nyssa* float well and are probably at least partly dispersed by water.

Key to Major Families of Ericales

1. Flowers resupinate, bilateral, with usually 3 sepals, the apparent lower one petaloid, often forming a nectar spur, with 5 petals, the seeming upper one distinct, concave and often sepal-like, the 4 others usually connate into 2 lateral pairs . . . . Balsaminaceae

1. Flowers not resupinate, radial or bilateral, with various numbers of distinct to connate sepals and petals, but not as above 2

2. Plants with laticifers and milky latex; hairs 2-branched, brownish, usually T-shaped; seeds with a large hilum . . . . **Sapotaceae**

2. Plants lacking laticifers; hairs not as above; seeds with a small hilum 3

3. Petals free 4

3. Petals connate 11

4. Plants carnivorous, with highly modified leaves that form pitcher-like traps . . . . **Sarraceniaceae**

4. Plants not carnivorous 5

5. Anthers inverting, or if apparently not inverting, then seed coat lacking 6

5. Anthers not inverting 9

6. Hairs usually stellate . . . . Clethraceae

6. Hairs not stellate 7

7. Stamens usually numerous; carpels usually incompletely fused, thus stigmas several; tissues with needle-like crystals of calcium oxalate . . . . Actinidiaceae

7. Stamens usually 2–10; carpels completely fused, thus a single stigma present; tissues lacking needle-like crystals 8

8. Seed coat present; pollen grains usually in tetrads; flowers pendulous or erect . . . . **Ericaceae**

8. Seed coat absent; pollen grains in monads; flowers ± erect . . . . Cyrillaceae

9. Ovary inferior or half-inferior; fruit usually opening by a circumscissile slit; stems with cortical vascular bundles . . . . **Lecythidaceae**

9. Ovary superior; fruit longitudinally dehiscent, or indehiscent; stems lacking cortical bundles 10

Species of *Cornus* can be identified in sterile condition by carefully tearing a leaf. The two halves will remain attached by delicate threads (unraveled spiral thickenings of the vessel elements), although some of the Hydrangeaceae also do this. The smoothly arcuate secondary veins are also diagnostic.

References: Eyde 1966, 1988; Eyde and Xiang 1990; Fan and Xiang 2001; Ferguson 1966a; Hufford 1992; Murrell 1993; Xiang 1990; Xiang et al. 1993, 1996, 1998; Xiang and Murrell 1998.

Ericales

The monophyly of Ericales has been strongly supported in analyses based on 18S, *rbcL*, *ndhF*, and *atpB* sequences (Chase et al. 1993; Kron and Chase 1993; Morton et al. 1997, 1998; Olmstead et al. 1993; Källersjö et al. 1998; Savolainen et al. 2000a,b; Soltis et al. 2000; Albach et al. 2001a,b). Ericales are here delimited broadly; these families usually have been divided into several smaller orders (e.g., Ebenales, Theales, Primulales, Polemoniales, Ericales s.s.). Morphological support for this group is weak, but a possible synapomorphy is the presence of theoid leaf teeth (i.e., a condition in which a single vein enters the tooth and ends in an opaque deciduous cap or gland; see Figure 4.13). Such teeth are found in at least some members of most of the included families (Hickey and Wolfe 1975). In the Ericaceae themselves, the condition is somewhat modified, each tooth being associated with a multicellular, usually gland-headed hair. Members of the Ericales usually can be distinguished from core asterids by their flowers with stamens usually twice the number of petals or numerous (vs. stamens equaling the number of petals or fewer). However, staminal reduction has

---

Key to Major Families of Ericales (continued)

10. Stamens with filaments only slightly longer than the anthers; fruit fleshy and indehiscent; embryo curved . . . . Ternstroemiaceae

10. Stamens with filaments several times longer than anthers; fruit dry, longitudinally dehiscent; embryo straight . . . . **Theaceae**

11. Placentation free-central 12

11. Placentation ± axile 14

12. Fruit a berry; flowers with conspicuous petaloid staminodes alternating with the corolla lobes; anthers extrorse; leaves lacking cavities containing resinous compounds and usually with elongate subepidermal fibers . . . . Theophrastaceae

12. Fruit a drupe or capsule; flowers lacking staminodes (or these very inconspicuous); anthers introrse; leaves often with elongate to spherical cavities containing resinous compounds, or these lacking, and lacking subepidermal fibers 13

13. Plants woody, chiefly tropical and subtropical; fruit a drupe, often 1-seeded . . . . **Myrsinaceae**

13. Plants herbaceous to suffrutescent, chiefly of temperate or cold regions; fruit dry, usually a capsule, usually with several to many seeds . . . . **"Primulaceae"**

14. Anthers inverting either early or late in development; pollen grains usually in tetrahedral tetrads . . . . **Ericaceae**

14. Anthers not inverting; pollen grains in monads 15

15. Indumentum of either stellate hairs or peltate scales . . . . Styracaceae

15. Indumentum various, but not as above 16

16. Plants with long and short shoots, the leaves of the long shoots quickly deciduous, but with a portion of the petiole hardened and persisting as a spine . . . . Fouquieriaceae

16. Shoots ± monomorphic, the leaves not forming a spine 17

17. Ovary inferior or half inferior, fruit a drupe . . . . Symplocaceae

17. Ovary superior, fruit a berry or capsule 18

18. Shrubs or trees, with black or dark-colored naphthaquinones; stamens usually 8 to numerous; fruit a berry associated with an expanded calyx . . . . **Ebenaceae**

18. Usually herbs, lacking naphthaquinones; stamens usually 5; fruit a capsule; calyx not expanded . . . . **Polemoniaceae**

occurred in Primulaceae and relatives, and increase in most Lecythidaceae, Actinidiaceae, and Theaceae.

Phylogenetic relationships within the order are not well understood, but recent work (Anderberg et al. 2002) suggests a number of groups. Sister to most of the order is a clade that includes Balsaminaceae. Fouquieriaceae plus Polemoniaceae may form a clade that is in turn sister to the remainder of the order. Relationships among many of the remaining families are problematic. It is clear, however, that Myrsinaceae, Primulaceae, and Theophrastaceae are part of a clade (often treated as Primulales), as evidenced by stamens equaling and opposite the corolla lobes; free-central placentation with a thick, more or less globose, central axis; and *rbcL* sequences (Anderberg and Ståhl 1995; Kron and Chase 1993; Olmstead et al. 1993). Relationships within this clade have been assessed by Anderberg and Ståhl (1995) and Anderberg et al. (1998). Theophrastaceae (probably including *Samolus*) are sister to "Primulaceae" + Myrsinaceae, and have retained the outer whorl of stamens as well-developed staminodes; many have pseudoverticillate leaves with subepidermal fibers. The "Primulaceae" + Myrsinaceae clade has the apomorphies of schizogenous secretory cavities with yellow to red to black resinous material (lost in some), introrse anthers, and loss of the outer staminal whorl.

Sapotaceae are somewhat isolated, but may be sister to Symplocaceae. Theaceae and Ternstroemiaceae may not be immediately related; both were previously included in Theaceae s.l.. The core Ericales include Actinidiaceae, Cyrillaceae, Clethraceae, Ericaceae, and Sarraceniaceae. They are clearly monophyletic (Anderberg 1992, 1993; Bayer et al. 1996; Judd and Kron 1993; Kron and Chase 1993); putative synapomorphies include anther inversion during development (so that the morphological base is apical), a usually hollow style that emerges from an apical depression in the ovary, and endosperm usually with haustoria at both ends (Anderberg 1992, 1993; Judd and Kron 1993; Kron 1996; Kron and Chase 1993). *Actinidia* and *Sarracenia* have many stamens, while *Actinidia* has incompletely fused carpels, so that the gynoecium has distinct styles. Cyrillaceae, Clethraceae, and Ericaceae may form a clade, which is supported by several embryological features. Morphological analyses support the sister group relationship of Clethraceae and Ericaceae (Anderberg 1993; Judd and Kron 1993), but molecular evidence suggests that Cyrillaceae may well be sister to Ericaceae (Anderberg et al. 2002). While some *rbcL* analyses (Kron and Chase 1993; Morton et al. 1997) suggest that Clethraceae are more closely related to Styracaceae (an ericalean family, but not a member of core Ericales) combined molecular data supports the inclusion of Clethraceae within core Ericales (Kron 1996).

Ericales includes 24 families and about 9450 species. Major families include Actinidiaceae, Balsaminaceae, Clethraceae, Cyrillaceae, **Ebenaceae**, **Ericaceae**, Fouquieriaceae, **Lecythidaceae**, **Myrsinaceae**, Polemoni-

aceae, **Primulaceae**, **Sarraceniaceae**, **Sapotaceae**, Styracaceae, Symplocaceae, Ternstroemiaceae, **Theaceae**, and Theophrastaceae.

References: Albach et al. 2001a,b; Anderberg 1992, 1993; Anderberg and Ståhl 1995; Anderberg et al. 1998, 2002; Bayer et al. 1996; Cronquist 1981; Hickey and Wolfe 1975; Judd and Kron 1993; Källersjö et al. 1998; Kron 1996; Kron and Chase 1993; Morton et al. 1997; Olmstead et al. 1993; Savolainen et al. 2000a,b; Soltis et al. 2000; Ståhl 1990; Stevens 1971.

Sapotaceae A. L. de Jussieu
(Sapodilla Family)

Trees or shrubs, sometimes with distinctly sympodial branches, or thorns; silica bodies often present; with tannins, often triterpenoids, and cyanogenic compounds; **with well-developed, elongate laticifers, the latex white. Hairs 2-branched, brownish, T-shaped, but one arm often ± reduced.** *Leaves alternate and spiral*, sometimes distinctly clustered at shoot apices, *simple, entire*, with pinnate venation; stipules present or absent. Inflorescences determinate, *usually fasciculate*, sometimes reduced to a single flower, axillary. Flowers bisexual, radial. *Sepals 4–8, sometimes dimorphic*, distinct or connate basally. *Petals 4–8, connate, sometimes with paired petaloid appendages (outgrowths from lower portion of corolla lobes)*, imbricate. *Stamens 8–16 and opposite the petals, usually alternating with staminodes*; filaments and staminodes adnate to corolla; pollen grains usually 3- or 4-colporate. *Carpels 2 to numerous, connate; ovary superior*, with axile placentation; stigma capitate to slightly lobed. Ovules 1 in each locule, with 1 integument and a thin-walled megasporangium. **Fruit a berry; seeds usually with a hard shiny testa and large hilum**; endosperm sometimes absent (Figure 9.114).

Floral formula: *, (4–8), (4–8), 4–16 + 4–8•, (2–∞); berry

Distribution and ecology: Pantropical, especially in wet lowland forests.

Genera/species: 53/1100. *Major genera:* *Pouteria* (325 spp.), *Palaquium* (110), *Planchonella* (100), *Madhuca* (100), *Sideroxylon* (75), and *Chrysophyllum* (70). *Chrysophyllum*, *Sideroxylon*, *Manilkara*, and *Pouteria* occur in the continental United States.

Economic plants and products: *Manilkara zapota* (sapodilla), *Pouteria mammosa* (mamey sapote), *Pouteria campechiana* (eggfruit), and *Chrysophyllum cainito* (star apple) provide delicious tropical fruits. Those of *Synsepalum dulcificum* affect the sense of taste; after eating even a portion of a single fruit, other food tastes sweet. Several genera are important sources of latex, such as *Palaquium* (gutta-percha) and *Manilkara zapota* (chicle, for chewing gum). Several genera provide economically important timber; others are useful

Figure 9.114 Sapotaceae. (A–H) *Manilkara jaimiqui* var. *emarginata:* (A) flowering and fruiting branch (× 0.75); (B) flower, after fall of corolla (× 4.5); (C) detached T-shaped hairs from calyx (× 75); (D) flower in longitudinal section, note staminodes alternating with stamens (× 7.5); (E) three corolla lobes from within, with appendages, stamens, and alternating staminodes (× 4.5); (F) three corolla lobes from without, to show dorsal appendages (× 4.5); (G) ovary in cross-section (× 6); (H) berry, the upper half removed to show single seed, locules with aborted ovules (× 1.5). (I–L) *M. zapota:* (I) portion of corolla from within, to show three corolla lobes, two petaloid staminodes alternating with stamens (× 4.5); (J) portion of corolla from without, to show three corolla lobes, tips of four staminodes (× 4.5); (K) ovary in cross-section (× 7.5); (L) seed, note elongate hilum (× 1.5). (From Wood and Channell 1960, *J. Arnold Arbor.* 41: p. 14.)

ornamentals, including *Chrysophyllum* (satinleaf), *Manilkara*, *Mimusops* (cherry mahogany), and *Sideroxylon* (buckthorn, ironwood, mastic).

Discussion: Sapotaceae are easily recognized and are hypothesized to be monophyletic (Morton et al. 1997;

Pennington 1991). The family usually has been considered closely related to Ebenaceae, a tropical family that also has alternate, entire, simple leaves, sympetalous flowers with epipetalous stamens, and superior ovaries developing into berries. However, Ebenaceae differ from Sapotaceae in their black or dark-colored naphthaquinones in the leaves, stems, and wood, absence of latex, dioecy, and lobed calyx that is persistent and expands as the fruit develops. Molecular data suggest that they may not form a clade.

Infrafamilial and generic relationships recently have been investigated by Pennington (1991), using characters such as position of the corolla lobes relative to the stamens and one another, position of the staminodes relative to the ovary, shape of the corolla, presence of petal appendages, size of the anthers, and position of the hilum scar. Various genera having twice as many stamens as corolla lobes form a heterogeneous complex, which is probably paraphyletic and basal. The remaining genera have a number of stamens equaling the number of the corolla lobes, and probably constitute a clade. *Mimusops*, *Manilkara*, and relatives may form a clade based on their calyx of two differentiated whorls. A single calyx whorl characterizes *Chrysophyllum*, *Pouteria*, *Synsepalum*, and *Sideroxylon* (incl. *Bumelia*, *Dipholis*, and *Masticodendron*).

The family is mainly insect-pollinated, although visitation by bats has been reported. Dispersal of the sweet berries is by various birds and mammals.

References: Kron and Chase 1993; Morton et al. 1997; Olmstead et al. 1993; Pennington 1991; Wood and Channell 1960.

Ebenaceae Gürke
(Ebony Family)

Trees or shrubs; **with black or dark-colored naphthaquinones (or related compounds) in nearly all tissues**; sometimes cyanogenic. Hairs simple, 2- or multibranched, sometimes glandular. *Leaves alternate, usually 2-ranked, simple, entire,* with pinnate venation, *usually with nectar glands on abaxial surface;* stipules absent. Inflorescences determinate, often reduced to a single flower, axillary. **Flowers unisexual (plants dioecious)**, radial. *Sepals 3–7, connate,* **usually persistent and variably enlarging as fruit develops**. *Petals 3–7, connate,* **± urn-shaped,** with imbricate or valvate lobes, often contorted. *Stamens (3-) 6-numerous; filaments usually adnate to corolla;* anthers occasionally opening by apical pores; replaced by staminodes in carpellate flowers; pollen grains tricolporate. *Carpels usually 3–8, connate; ovary superior, with axile placentation,* the locules often secondarily divided; stigmas 3–8, capitate to slightly elongated. Ovules 1 or 2 in each locule, with a thin-walled megasporangium. **Fruit a berry**, astringent until very ripe; seeds fairly large, with thin testa; endosperm sometimes ruminate.

Floral formula:

Staminate: *, ③–⑦ , ③–⑦ , 6–∞ , 0

Carpellate: *, ③–⑦ , ③–⑦ , 6–∞• , ③–⑧ , berry

Distribution: Pantropical, with a few species extending into temperate regions.

Genera/species: 3/500. *Major genera:* *Diospyros* (450), which is the only genus in the continental United States.

Economic plants and products: Several species of *Diospyros*, e.g., *D. digyna* (black sapote), *D. kaki* (Japanese persimmon), *D. lotus* (date plum), and *D. virginiana* (persimmon), have edible fruits; many species provide economically important timber (ebony).

Discussion: Ebenaceae are easily recognized and are hypothesized to be monophyletic based on their distinctive morphological characters. Their phylogenetic placement within Ericales is as yet unclear, and generic relationships within the group are also in need of investigation.

Ebenaceae are an especially important floristic element in the forests of Africa.

The family is insect pollinated and dispersal of the berries is by both birds and mammals.

References: Wood and Channel 1960.

"Primulaceae" Ventenat
(Primrose Family)

Herbs, sometimes suffrutescent; often with triterpenoid saponins and tannins; sometimes with secretory cavities containing yellow to red resinous materials. Hairs various, often elongate and septate. Leaves alternate and spiral, opposite, or whorled, often forming a basal rosette, simple, entire to serrate, sometimes lobed, with pinnate venation; stipules lacking. Inflorescences usually indeterminate, often highly modified, sometimes reduced to a single flower, terminal or axillary. *Flowers bisexual, usually radial, sometimes heterostylous.* Sepals usually 5, connate. *Petals usually 5, connate,* imbricate or twisted. *Stamens 5, opposite corolla lobes; filaments adnate to corolla;* anthers sometimes opening by apical pores; pollen grains tricolporate or 5–8 zonocolpate. *Carpels usually 5, connate; ovary superior* or occasionally half-inferior, *with free-central placentation (and ± globose central axis);* stigma capitate. *Ovules usually numerous,* anatropous to campylotropous, with 1 or 2 integuments and a thin-walled megasporangium. Nectaries usually lacking. *Fruit a capsule,* opening by valves or circumscissile, or occasionally indehiscent; seeds occasionally arillate (Figure 9.115).

Floral formula: *, ⑤, ⑤, 5, —⑤—; capsule

Distribution: Mainly temperate and cold regions of the Northern Hemisphere.

Genera/species: 20/1000. *Major genera:* *Primula* (500 spp.), *Lysimachia* (200), *Androsace* (100), and *Dodecatheon* (50). These, along with *Anagallis*, *Douglasia*, *Glaux*, *Hotto-*

Figure 9.115 Primulaceae. *Dodecatheon media:* (A) flowering plant (× 0.25); (B) flower bud (× 4); (C) flower, corolla lobes reflexed (× 4); (D) flower in longitudinal section (× 4); (E) androecial tube laid open, note that filaments are both connate and adnate to corolla tube (× 4); (F) stigma (greatly magnified); (G) ovary in cross-section, showing free-central placenta (× 13); (H) flower bud in cross-section (× 4); (I) infructescence (× 0.5); (J) capsule (× 4); (K) seed (× 15); (L) embryo (greatly magnified). (From Wood 1974, *A student's atlas of flowering plants*, p. 86.)

nia, Samolus, and *Trientalis,* occur in continental United States and/or Canada.

Economic plants and products: *Primula* (primrose), *Cyclamen* (sowbread), *Anagallis* (pimpernel), and *Dodecatheon* (shooting star) are cultivated as ornamentals.

Discussion: The monophyly of Primulaceae is supported by morphological characters (Anderberg and Ståhl 1995), but *rbcL* sequences strongly suggest that "Primulaceae" are paraphyletic, with Myrsinaceae and Theophrastaceae nested within their cladistic structure (Anderberg et al. 1998). Evidence from morphology, *ndhF, rbcL,* and *atpB* nucleotide sequences indicates that *Samolus* (flowers with staminodes, anthers opening toward center, ovary semi-inferior, fruit a capsule) must be transferred to Theophras-

taceae (fruit a berry, anthers opening outward). *Coris, Ardisiandra,* and genera of the Lysimachieae (e.g., *Anagallis, Cyclamen, Gaux, Lysimachia,* and *Trientalis*) should be placed within an expanded Myrsinaceae (Källersjö et al. 2000), and these often have resin cavities. These genera all have capsular fruits, unlike Myrsinaceae, as traditionally defined (see family description), which have drupaceous fruits. Once these genera are removed, Primulaceae s.s. are herbaceous plants with usually more or less campanulate corollas and capsular fruits; distinctive morphological synapomorphies of these plants, however, are not apparent. The family here is retained in its traditional circumscription, pending additional investigation. It is possible that "Primulaceae," Theophrastaceae, and Myrsinaceae, eventually will be combined, resulting in a clearly monophyletic *and* morphologically distinctive family.

The showy flowers of Primulaceae are pollinated by various insects. Heterostyly is common in *Primula,*but selfing also occurs. Recent work suggests that *Dodecatheon* is a buzz-pollinated lineage derived from within *Primula* (Mast et al. 2001; Trift et al. 2002). The small seeds are often wind- or water-dispersed, but some are dispersed by ants, which gather the oily arils.

References: Anderberg and Ståhl 1995; Anderberg et al. 1998; Channell and Wood 1959; Källersjö et al. 2000; Kron and Chase 1993; Mast et al. 2001; Olmstead et al. 1993; Ståhl 1990; Trift et al. 2002.

Myrsinaceae R. Brown
(Myrsine Family)

Trees, shrubs, or lianas, sometimes epiphytic or herbaceous; *scattered, elongate to spherical, resin cavities present, appearing as yellow-green, red, brown, or black dots to lines on leaves, stems, and floral parts*; with tannins, benzoquinones, and triterpenoid saponins. Hairs various, but often gland-headed. *Leaves alternate and spiral, simple, usually entire*, with pinnate venation; *stipules lacking*. Inflorescences determinate, sometimes reduced, terminal or axillary. Flowers bisexual or sometimes unisexual (then plants ± dioecious), radial. Sepals usually 4 or 5 and slightly connate. *Petals usually 4 or 5, connate, the corolla rotate or tubular with expanded lobes*, usually imbricate or convolute. *Stamens 4 or 5, opposite corolla lobes*; filaments distinct or connate, adnate to corolla; anthers sometimes opening by apical pores; pollen grains usually tricolporate. *Carpels usually 3–5, connate; ovary* superior, *with free-central placentation, the central placental axis thick, ± globose, completely filling the locule*; stigma usually capitate, sometimes lobed. **Ovules few to several**, anatropous to campylotropous, with thin megasporangium. Nectaries lacking. **Fruit usually a drupe with a single, 1- to few-seeded pit; seed with a depressed hilum**.

Floral formula: * ,④–5 ,④–5 , 4–5 ,③–5 ; drupe

Distribution: Widespread in tropical and subtropical regions, with a few warm temperate species.

Genera/species: 33/850. ***Major genera:*** *Ardisia* (300 spp.), *Myrsine* (200), and *Embelia* (130). *Myrsine* and *Ardisia* occur in the continental United States.

Economic plants and products: *Ardisia* (coralberry, marlberry), *Wallenia*, and *Myrsine* provide ornamentals.

Discussion: The monophyly of Myrsinaceae, as here circumscribed, is supported by morphology (Anderberg and Ståhl 1995). It is clear from morphology and *rbcL* sequences that *Maesa* (150 spp.), a distinctive mangrove genus with half-inferior ovaries, needs to be removed from the family (Anderberg and Ståhl 1995; Morton et al. 1997).

The small, pale green to white or pink flowers of Myrsinaceae are insect-pollinated; outcrossing is favored by protogyny or the more or less dioecious condition. The red to purple-black drupaceous fruits are dispersed by birds.

References: Anderberg and Ståhl 1995; Channell and Wood 1959.

Theaceae Mirbel
(Tea Family)

Trees or shrubs; usually with sclereids; tannins present. Hairs usually simple and unicellular. *Leaves alternate and spiral, sometimes 2-ranked, simple, toothed, the teeth theoid (i.e., with a deciduous glandular apex, see Figure 4.13), with pinnate venation; stipules lacking.* Inflorescences of solitary, axillary flowers. Flowers bisexual, radial, *the subtending bract(s) sometimes intergrading with calyx. Sepals usually 5, distinct or slightly connate basally, imbricate. Petals usually 5, distinct or very slightly connate basally*, imbricate, *slightly wrinkled along margin.* Stamens numerous, the ones nearest the gynoecium developing first, distinct or basally connate into a ring or 5 bundles that are opposite the petals; pollen grains tricolporate. *Carpels usually 3–5, connate;* ovary superior, with axile placentation; stigma 1 and lobed, to 3–5 and capitate. **Ovules 1 to few per locule,** with a thin-walled megasporangium. Nectariferous tissue at base of filaments or ovary. *Fruit a ± loculicidal capsule;* ± **few seeds, often flattened or winged; embryo large**; endosperm present or absent.

Floral formula: * ,⑤, 5,⑤ ,③–5 ; capsule

Distribution: Widespread in temperate to tropical regions.

Genera/species: 20/300. ***Major genera:*** *Camellia* (80 spp.), *Polyspora* (35), and *Laplacea* (30). *Franklinia, Gordonia,* and *Stewartia* occur in the continental United States.

Economic plants and products: Tea is made from the leaves of *Camellia sinensis. Camellia* (camellia), *Gordonia* (loblolly bay), *Stewartia,* and *Franklinia* (Franklin tree) provide ornamentals.

Discussion: Theaceae, as traditionally circumscribed, probably are not monophyletic, a conclusion supported by DNA-based cladistic analyses (Morton et al. 1997; Prince and Parks 2001). Here, the family is restricted to the genera traditionally placed in Theoideae. Genera of Ternstroemioideae (e.g., *Ternstroemia* and *Eurya*) are considered to represent a distinct family, the Ternstroemiaceae. The monophyly of both Theaceae and Ternstroemiaceae is supported by *rbcL* and *matK* sequences (Morton et al. 1997; Prince and Parks 2001). The morphological similarity between the flowers of some core Ericales, such as Actinidiaceae, and Theaceae is striking. Theaceae are composed of three monophyletic tribes: Theeae (*Camellia, Polyspora, Laplacea,* and relatives), Gordonieae (*Gordonia, Franklinia,* and *Schima*), and Stewartieae (*Stewartia*).

The numerous stamens of Theaceae (and some other Ericales) develop centrifugally and are served by a limited number of vascular bundles, suggesting that they are evolutionarily derived from few stamens (in two whorls). This interpretation is supported by cladistic analyses of morphological and DNA characters.

The showy flowers of Theaceae are pollinated by various insects. Their seeds are wind- or water-dispersed.

References: Chase et al. 1993; Kron and Chase 1993; Keng 1962; Morton et al. 1997; Olmstead et al. 1993; Wood 1959b.

Ericaceae A. L. de Jussieu
(Heath Family)

Trees, shrubs, lianas, sometimes epiphytic, occasionally mycoparasitic herbs lacking chlorophyll, strongly associated with mycorrhizal fungi. Hairs simple, usually multicellular and unicellular, sometimes dendritic, glandheaded, or peltate scales, but not stellate. *Leaves alternate and spiral, sometimes opposite or whorled, simple,* entire to serrate, sometimes revolute, with pinnate, ± parallel, or palmate venation, blade reduced in mycoparasites; stipules lacking. Inflorescences various. *Flowers usually bisexual,* rarely unisexual (then plants usually dioecious), radial to slightly bilateral, **usually ± pendulous**. Sepals usually 4 or 5, distinct to slightly connate. **Petals** *usually 4 or 5 and* **connate, often cylindrical to urn-shaped,** *with small to large, imbricate to valvate lobes, but sometimes ± bell-shaped or funnel-like,* occasionally distinct (a reversal). Perianth reduced to 2 or 3 sepals and petals, or 3 or 4 tepals in a few genera that are wind-pollinated. *Stamens 8–10,* but reduced to 2 or 3 in wind-pollinated species; filaments free or adnate to corolla, sometimes connate, **sometimes with paired projections (spurs) near or at junction with anther;** *anthers becoming inverted,* 2- or 1-locular, *usually opening by 2 apical pores,* **sometimes with 2 projections (awns)** *or with apex narrowed, forming a pair of tubules; pollen grains usually in tetrads,* usually tricolporate, sometimes associated with viscin threads. Carpels 2–10; ovary superior to inferior, usually with axile or deeply intruded parietal placentation; style 1, *hollow, internally fluted;* stigma capitate or slightly lobed. Ovules 1 to numerous per locule, with 1 integument and a thin-walled megasporangium. Nectariferous tissue around base or apex of ovary. *Fruit a septicidal or loculicidal capsule, berry, 1 or several-pitted drupe,* usually erectly held due to movement of pedicel; seed coat thin (Figure 9.116).

Floral formula: * or X, (4–5), (4–5), (4–10), (2–10);

capsule, berry, drupe

Distribution and ecology: Cosmopolitan, but especially common in tropical montane habitats, southern Africa, eastern North America, eastern Asia, and Australia; usually light-loving shrubs of acid soils.

Genera/species: 130/2700. **Major genera:** *Rhododendron* (800 spp.), *Erica* (600), *Vaccinium* (400), *Gaultheria* (150), *Leucopogon* (140), *Cavendishia* (100), and *Arctostaphylos* (50). Noteworthy genera (in addition to most of the above) in the continental United States and/or Canada are *Andromeda, Arbutus, Bejaria, Ceratiola, Chamaedaphne, Chimaphila, Corema, Empetrum, Gaylussacia, Kalmia, Leucothoe, Lyonia, Monotropa, Monotropsis, Oxydendrum, Pieris, Pterospora,* and *Pyrola.*

Economic plants and products: The edible fruits of *Vaccinium* (blueberries, cranberries) are economically important. The family contains many showy ornamentals, including *Arbutus* (madrone), *Calluna* (heather), *Erica* (heath), *Gaultheria* (wintergreen), *Kalmia* (mountain laurel), *Oxydendrum* (sourwood), *Pieris, Rhododendron* (azalea, rhododendron), and *Leucothoe* (fetterbush). *Gaultheria procumbens* is the original source of oil of wintergreen (methyl salicylate).

Discussion: Ericaceae are broadly circumscribed, including five families (Empetraceae, Epacridaceae, Monotropaceae, Pyrolaceae, and Vacciniaceae) that are sometimes recognized separately. Recognition of these segregates would render Ericaceae s.s. paraphyletic. As circumscribed here, the group is monophyletic, based on evidence from morphology, *rbcL, matK,* and 18S rDNA sequences (Anderberg 1993; Chase et al. 1993; Judd and Kron 1993; Kron 1996; Kron and Chase 1993; Kron et al. 2002; Soltis et al. 1997). The eastern Asian genus *Enkianthus* is sister to the remaining genera, which form a clade based on their pollen shed as tetrahedral tetrads, lack of a fibrous endothecium (innermost layer of anther locule), and seed lacking a vascular bundle in the raphe (Anderberg 1993, 1994; Judd and Kron 1993; Kron 1996; Kron and Chase 1993; Kron et al. 2002). Several subclades can be discerned within this clade (Figure 9.117).

The Ericoideae are characterized by their erect to more or less horizontal flowers, loss of stamen appendages, and usually septicidal capsules, but all these show homoplasy. The rhododendroid clade, including *Rhododendron* and *Menziesia,* is characterized by protective bracts at the inflorescence base, septicidal capsules, usually showy, ± bell-shaped bilateral flowers, viscin threads, and nonappendaged anthers. A morphologically similar clade, the phyllodocoids, includes general such as *Phyllodoce* and *Kalmia,* and also has septicidal capsules; their flowers are radially symmetrical. The ericoid clade (*Erica* and *Calluna*) is diagnosed by a persistent corolla and usually loculicidal capsules. All three groups are part of a clade that also includes the wind-pollinated empetroids. Empetroids—*Empetrum, Corema,* and *Ceratiola*—are clear-

Figure 9.116 Ericaceae. (A–G) Oxydendrum arboreum: (A) flowering branch (× 0.4); (B) flower (× 3); (C) flower in longitudinal section (× 9); (D) outer, inner, and lateral views of stamens (× 18); (E) portion of raceme with immature fruits (× 3); (F) opened capsule, one valve removed, note deeply immersed style (× 6); (G) seed (× 15). (From Wood 1961, J. Arnold Arbor. 42: p. 57.)

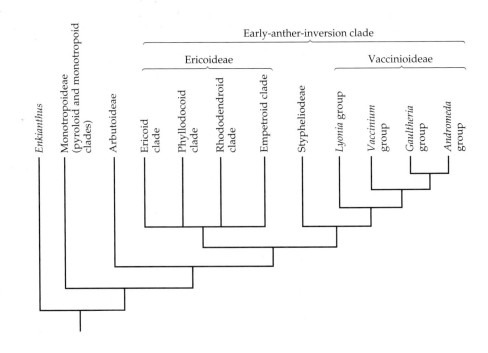

Figure 9.117 Cladogram showing hypothesized relationships within Ericaceae. (Adapted from Kron 1997 and Kron et al. 2002.)

ly monophyletic, based on their strongly lobed and expanded stigma, reduced corolla, and drupaceous fruit.

Another major group within Ericaceae is the Vaccinioideae, including *Vaccinium, Gaylussacia, Cavendishia,* and relatives with inferior ovaries, plus *Lyonia, Pieris, Chamaedaphne, Leucothoe, Gaultheria, Andromeda,* and others, with superior ovaries. The genera with inferior ovaries form a monophyletic subgroup that is especially diverse in the montane tropics. Vaccinioideae, characterized by a base chromosome number of 12, are sister to the Stypheliodeae (e.g., *Epacris, Styphelia,* and *Leucopogon*). The Stypheliodeae are monophyletic, as indicated by their monoloculate and epipetalous stamens in a single whorl, usually parallel-veined leaves, and usual lack of multicellular hairs.

The Ericoideae, Vaccinioideae, and Stypheliodeae all show early developmental inversion of their anthers, and DNA sequences strongly indicate that together they constitute a clade (Figure 9.117). In contrast, the Monotropoideae and Arbutoideae (see below) represent earlier divergent clades within the family. They, along with *Enkianthus,* have late-inverting anthers.

Monotropoideae are related on the basis of their more or less herbaceous habit and reduced embryos (Anderberg 1993). Within this group, the monotropoids (e.g., *Monotropa, Monotropsis,* and *Pterospora*) are easily recognized by their specialized mycoparasitic habit, associated with the loss of chlorophyll. These genera represent an extreme development of the fungal symbiotic relationship characteristic of the family. The fungi parasitized by various monotropoids are mycorrhizal, and are also attached to the roots of various forest trees. *Monotropa* (Indian pipe) thus indirectly parasitizes these trees, with nutrients passing from tree to fungus to monotropoid. The pyroloids (e.g., *Pyrola* and *Chimaphila*) are chlorophyllous herbs.

Arbutoideae (e.g., *Arctostaphylos* and *Arbutus*) are trees or shrubs with sympetalous, urceolate flowers, superior ovaries, and fleshy fruits (drupes, or berries with a fibrous inner fruit wall).

The pendulous, urn-shaped, cylindrical, or bell-shaped flowers of Ericaceae typically produce nectar and are visited by bees and wasps. The insect hangs onto the flower, probing for nectar, which is produced at the base of the corolla. In the process it brushes against the filaments or stamen appendages, causing pollen to fall onto its body. The stigma, which typically is centrally positioned at the narrowed corolla mouth, is situated so as to pick up pollen easily from visiting insects. In the montane tropics many species have tubular red flowers and are pollinated by birds. The viscin threads associated with the pollen of *Rhododendron* and related genera allow a large number of interconnected pollen tetrads to be simultaneously pulled out of an anther by pollinators. Capsular-fruited Ericaceae are largely wind-dispersed; most have small and/or winged seeds. The capsules are usually brought into an erect position before opening by movement of the pedicel.

Species with berries (e.g., *Vaccinium, Cavendishia*) or drupes (e.g., *Arctostaphylos, Gaylussacia, Empetrum,* and *Styphelia*) are usually bird-dispersed.

References: Anderberg 1992, 1993, 1994; Bidartondo and Bruns 2001; Crayn and Quinn 2000; Crayn et al. 1998; Cullings and Bruns 1992; Hermann and Palser 2000; Judd and Kron 1993; Kron and Chase 1993; Hileman et al. 2001; Kron 1997; Kron and Judd 1990; Kron and King 1996; Kron et al. 1999, 2002; Stevens 1971; Wallace 1975; Wood 1961.

Sarraceniaceae Dumortier
(Pitcher-Plant Family)

Carnivorous herbs or subshrubs; leaves alternate and spiral, **highly modified, forming pitcherlike traps,** *with ridge or laminar wing on adaxial side, and a flattened, relatively small, often hoodlike blade apically, inner surface often with retrorse hairs and glandular hairs,* sometimes in a basal rosette; stipules lacking. *Flowers large, ± pendulous, usually solitary on a scape,* bisexual, radial, *often associated with conspicuous bracts.* Sepals usually 5, distinct, often petaloid. Petals usually 5, distinct, imbricate. Stamens usually numerous; anthers sometimes inverting in development; pollen grains tricolporate to multicolporate. Carpels 3 or 5, connate; ovary superior, with axile or intruded parietal placentation; *style in Sarracenia expanded and peltate or umbrella-like with a small stigma under the tip of each of the 5 lobes;* stigmas truncate or minute. Ovules numerous, with 1 or 2 integuments and a thin-walled megasporangium. Nectaries absent. *Fruit a loculicidal capsule* (Figure 9.118).

Floral formula: *, 5, 5, ∞ ③-⑤; capsule

Distribution and ecology: Limited to North America and northern South America; in acidic habitats.

Genera/species: 3/15. **Genera:** *Sarracenia* (8 spp.), *Heliamphora* (6), and *Darlingtonia* (1).

Economic plants and products: *Darlingtonia* and *Sarracenia* (pitcher plants) are grown as novelties.

Discussion: Sarraceniaceae are superficially similar to Nepenthaceae and Droseraceae because of the shared carnivorous habit. However, morphology and DNA characters suggest that Sarraceniaceae are not closely related to these families, belonging instead to the asterid clade, within Ericales (Bayer et al. 1996; Hufford 1992; Kron and Chase 1993). The family may be the sister group of the carnivorous Roridulaceae. Morphology, *rbcL,* and ITS sequences (Bayer et al. 1996) support the monophyly of the family.

The tubular leaves of *Sarracenia* are variable, providing characters useful in specific delimitation. They may

ADC

Figure 9.118 Sarraceniaceae. (A–H) *Sarracenia oreophila:* (A) flowering plant (× 7.5); (B) flower (× 1); (C) androecium and gynoecium in longitudinal section (× 2); (D) upper part of ovary in cross-section (× 3); (E) lower part of ovary in cross-section (× 3); (F) capsule, note persistent and expanded style (× 1); (G) seed (× 12); (H) embryo (× 12). (I) *S. flava:* upper part of leaf (× 0.5). (J) *S. rubra:* upper part of leaf (× 1). (K, L) *S. purpurea:* (K) leaf (× 0.5); (L) leaf in longitudinal section, note apical and basal portions with own-ward-pointing hairs (× 0.5). (M) *S. psittacina:* leaves (× 1). (From Wood 1974, *A student's atlas of flowering plants*, p. 44.)

be erect or decumbent, and the apical hood usually prevents the entrance of rain. These remarkable leaves are often conspicuously colored, have strong odors, and are provided with nectar glands. The leaves of several species have translucent, windowlike spots. Insects attracted by the coloration or odor may fall or crawl into the fluid-filled pitchers, where they are often trapped by downward-pointing hairs. If unable to escape, they die and are digested by enzymes in the fluid, apparently providing nitrogen that is often limiting in the acidic habitats occupied by pitcher plants.

Flowers are generally visited by various pollen-gathering bees and wasps. Dispersal of the small seeds is probably by wind and water.

References: Bayer et al. 1996; de Buhr 1975; Hufford 1992; Kron and Chase 1993; McDaniel 1971; Wood 1960; Renner 1989a.

Lecythidaceae A. Richard
(Brazil-nut Family)

Trees, or occasionally shrubs or lianas; **stems with cortical vascular bundles**; with triterpenoid saponins, often tannins, and sometimes mucilage canals. Hairs usually simple, glandular or nonglandular. *Leaves alternate and spiral, often clustered at the apices of the branches, sometimes 2-ranked, simple, entire or toothed, the teeth theoid* (with a deciduous glandular apex, see Figure 4.13), with pinnate venation; stipules present or lacking. Inflorescences indeterminate, terminal or axillary, sometimes reduced to a solitary axillary flower. *Flowers bisexual, radial or bilateral (due to unusual development of the androecium).* Sepals usually 4-6, distinct or connate, imbricate or valvate. *Petals usually 4-6, distinct,* imbricate, occasionally lacking. *Stamens usually numerous, the ones nearest the gynoecium developing first, usually connate, and in the more specialized genera the fused portion asymmetrical, produced on one side of flower, forming a flat structure that may be curved over the ovary, with some of the stamens often reduced and staminodial;* pollen grains tricolporate, sometimes with the colpi fused together. *Carpels 2-8, connate;* **ovary inferior or half-inferior,** rarely superior, with axile placentation; style capitate or lobed. Ovules 1 to many per locule, with a thin-walled megasporangium. Nectaries lacking, or nectar produced by staminodes. *Fruit a capsule, often very large and bony, usually opening by a circumscissile slit (with a cap),* sometimes an indehiscent pod, drupe, or nut; seeds large, often with a fleshy aril, or aril flattened, forming a wing; embryo large, oily, often with a much-thickened hypocotyl; endosperm lacking or present (then usually ruminate).

Floral formula: X or *, (4-6), 4-6 , ∞ , 2-6

capsule (opening by a cap)

Distribution and ecology: Widespread in the tropics, and especially diverse in moist to wet forests of South America.

Genera/species: 30/440. **Major genera:** *Eschweilera* (120), *Gustavia* (50), *Lecythis* (50), and *Barringtonia* (40). The family does not occur in temperate North America; *Barringtonia* and *Couroupita* are occasionally cultivated in southern Florida.

Economic plants and products: *Bertholletia excelsa* provides edible seeds (Brazil nuts); the seeds of several species of *Lecythis* are also delicious. *Couroupita guianensis* (cannonball tree) and a few species of *Barringtonia* are used as ornamentals. Species of several genera provide useful timber.

Discussion: Both cpDNA nucleotide sequences and morphological characters (Morton et al. 1997a,b) support the monophyly of Lecythidaceae. The group consists of five major clades (which are here treated as subfamilies; see Morton et al. 1997b, 1998; Thorne 2001). *Napoleonaea* and *Crateranthus* (Napoleonaeoideae) are sister to the remaining members of the family, and form a clade diagnosed by extrorse anthers, an unusual androecium with an outer row of staminodes fused to form a radiating pseudocorolla, and the loss of petals. The next clade to diverge comprises *Asteranthos, Oubanguia, Scytopetalum,* and relatives (Scytopetaloideae); they are distinctive because they have seeds with ruminate endosperm. However, most genera and species belong to one of two large subfamilies that are sister taxa. Barringtonioideae (*Barringtonia, Planchonia,* and relatives) are an Old World clade characterized by pollen with fused colpi and a reduction to only one seed per fruit. Lecythidoideae are limited to the neotropics and constitute the largest subfamily (containing e.g., *Bertholletia, Couroupita, Eschweilera, Grias, Gustavia, Lecythis*). The monophyly of this group is supported by the haploid chromosome number of seventeen. Lecythidoideae contain all the genera having specialized bilaterally symmetrical flowers.

The large and showy flowers of Lecythidaceae usually are pollinated by various bees and wasps. The family shows an interesting evolutionary shift from species having radially symmetrical flowers with many stamens that use pollen as a pollinator reward to those with bilaterally symmetrical flowers with fewer fertile stamens (and many staminodes) that attract nectar-gathering euglossine bees. In some derived species the nectar is produced by modified staminodes. Bat pollination is also known (*Barringtonia,* and at least one species of *Lecythis*). The seeds of Lecythidaceae are dispersed by a wide variety of animals (mammals, including rodents, monkeys, bats; various birds and fish), which are attracted by the arillate seeds or inner parts of the fruit wall, which are eaten. Water dispersal is common in some taxa that occur along rivers (e.g., *Allantoma*); a few (e.g., species of *Cariniana* and *Couratari*) are wind dispersed, and these have the aril modified into a wing.

References: Mori and Prance 1990; Mori et al. 1978; Morton et al. 1995, 1997, 1998; Prance 1976; Prance and Mori 1978, 1979.

Polemoniaceae A. L. de Jussieu
(Phlox Family)

Herbs or occasionally shrubs, small trees, or lianas. Hairs various, often gland-headed. Leaves alternate and spiral, opposite, or whorled, simple, dissected, or pinnately compound, entire to serrate, with pinnate venation; stipules lacking. Inflorescences determinate, terminal, or sometimes solitary and axillary. *Flowers bisexual, radial, usually showy.* Sepals usually 5 and connate. *Petals usually 5, strongly connate, often forming a ± narrow tube, the lobes usually plicate and convolute. Stamens usually 5; filaments adnate to corolla tube;* pollen grains 4–many-colporate or porate. **Carpels 3**, connate; *ovary superior*, with axile placentation; stigmas on upper surface of style branches. Ovules 1 to numerous in each locule, with 1 integument and a thin-walled megasporangium. Nectar disk present. Fruit usually a loculicidal capsule; seed coat often mucilaginous when moistened (Figure 9.119).

Floral formula: *, ⑤, ⑤, 5, ③; capsule

Distribution: Widely distributed, but most diverse in temperate regions, especially western North America.

Genera/species: 26/380. **Major genera:** *Phlox* (70 spp.), *Linanthus* (40), *Gilia* (40), *Polemonium* (40), *Navarretia* (30), *Leptosiphon* (30), *Ipomopsis* (30). Additional important genera in the United States and/or Canada are *Aliciella*, *Alophyllum*, *Collomia*, *Eriastrum*, *Giliastrum*, *Leptodactylon*, *Loeselia*, and *Microsteris*.

Economic plants and products: The family is best known for its ornamentals, including, *Gilia*, *Phlox*, and *Polemonium*, with showy, variously colored flowers.

Discussion: Polemoniaceae have been placed in Solanales because of their radially symmetrical flowers with sympetalous, plicate corollas. DNA (Olmstead et al. 1993; Porter and Johnson 1998; Johnson et al. 1999) and morphology (Hufford 1992) indicate, however, that the family belongs within Ericales. Analyses of nucleotide sequences indicate that the woody tropical genera of Polemoniaceae form either a paraphyletic basal complex or a clade (Cobaeoideae), while the herbaceous and more temperate genera *Ipomopsis*, *Linanthus*, *Navarretia*, *Polemonium*, *Phlox*, and *Gilia* constitute a monophyletic group (Johnson et al. 1996; Porter and Johnson 1998, 2000; Prather et al. 2000; Steele and Vilgalys 1994).

Flowers of Polemoniaceae are variously colored (sometimes even within a species, as in *Phlox drummondii*; see Kelly 1920) and shaped, attracting bees, flies, beetles, butterflies, and moths, as well as birds and bats. Most are outcrossing due to protandry. Seed dispersal may be by external transport, facilitated by the mucilaginous seed coat, although movement of the small seeds by wind and/or water also occurs.

References: Grant 1959; Grant and Grant 1965; Hufford 1992; Johnson et al. 1996, 1999; Kelly 1920; Olmstead et al. 1993; Porter and Johnson 1998, 2000; Prather et al. 2000; Steele and Vilgalys 1994; Wilson 1960c.

Core Asterids

The remaining groups covered in this chapter form the clade of core asterids, which are divided into two subclades: Euasterids I and Euasterids II (see Figure 9.8). Core asterids have flowers in which the stamens are epipetalous and equaling (or less than) the corolla lobes in number, and a gynoecium usually of two fused carpels. Molecular data strongly supports the monophyly of this clade.

Euasterids I
Solanales

Solanales are tentatively taken as monophyletic, based on their characteristic radially symmetrical flowers with a plicate, sympetalous corolla. Members of this group have alternate, simple, exstipulate leaves and flowers in which the number of stamens equals the number of petals. They lack iridoids. The order consists of 6 families and about 7400 species; major families include **Solanaceae** (incl. Nolanaceae), **Convolvulaceae** (incl. Cuscutaceae), and perhaps **Boraginaceae** (incl. Hydrophyllaceae). Solanales clearly belong to the asterid clade, as evidenced by the single integument and thin megasporangium wall, and to the core asterid clade, as supported by their sympetalous flowers with epipetalous stamens and two fused carpels. Chloroplast DNA characters indicate that the order is most closely related to Gentianales and Lamiales (Downie and Palmer 1992; Olmstead et al. 1992, 1993, 2000; Soltis et al. 2000). Solanaceae and Convolvulaceae are considered to be sister families, based on the anatomical synapomorphy of internal phloem, similar alkaloid chemistry, and several cpDNA characters (Figure 9.120). Boraginaceae includes the traditionally recognized, "Hydrophyllaceae", and are characterized by their helicoid cymes and cpDNA features. Placement of the Boraginaceae within Solanales is somewhat problematic; Albach and colleagues (2001b) suggest that this family is the sister group of the Lamiales.

References: Albach et al. 2001b; Cronquist 1981, 1988; Dahlgren 1983; Downie and Palmer 1992; Olmstead et al.

Figure 9.119 Polemoniaceae. (A–F) *Phlox nivalis:* (A) flowering plant (× 1); (B) flower, view from above (× 0.75); (C) flower in longitudinal section, note epipetalous stamens (× 4); (D) capsule with persistent calyx (× 4); (E) seed (× 15); (F) embryo (× 15). (G–N) *P. divaricata* var. *laphamii:* (G) inflorescence (× 0.5); (H) flower bud with convolute corolla (× 4); (I) flower in longitudinal section (× 4); (J) glandular hairs (greatly magnified); (K) gynoecium, note three stigmas (× 15); (L) ovary in cross-section with single ovule per locule (× 30); (M) immature fruit (× 4); (N) immature fruit in longitudinal section (× 5.5). (O) *P. divaricata* var. *divaricata:* flower, view from above (× 0.75). (Unpublished original art prepared for the Generic Flora of the Southeast U.S. series; used with permission.)

1992a, 1993, 2000; Soltis et al. 2000; Takhtajan 1980, 1997; Thorne 1992.

Solanaceae Adanson

(Potato or Nightshade Fsamily)

Herbs, shrubs, trees, or vines; internal phloem usually present; various alkaloids present. Hairs diverse, but often stellate or branched, sometimes with prickles. *Leaves alternate and spiral*, often in pairs, the members of a pair both on the same side of the stem, *simple, sometimes deeply lobed* or even pinnately compound, entire to serrate, with pinnate venation; stipules lacking. Inflorescences determinate, sometimes reduced to a single flower, terminal but usually appearing lateral. *Flowers usually bisexual and radial. Sepals usually 5, connate*, persistent, sometimes enlarg-

Key to Major Families of Solanales

1. Ovary with 4 ovules 2
1. Ovary with more than 4 ovules (or if only 4, then placentation parietal)........................... 3
2. Stems lacking internal phloem and laticifers, sap not milky; inflorescence often distinctly scorpioid/helicoid; leaves with firm unicellular hairs that have a basal cystolith and often calcified or silicified walls and thus are rough to the touch; fruit a drupe with several pits or a schizocarp splitting into 4 nutlets .... **Boraginaceae**
2. Stems with internal phloem and usually also with laticifers (and sap milky); inflorescence and hairs not as above; fruit a capsule .... **Convolvulaceae**
3. Stems lacking internal phloem; inflorescence scorpioid/helicoid; placentation parietal; flowers lacking a nectar disk.. **Boraginaceae**
3. Stems with internal phloem; inflorescence usually not scorpioid/helicoid; placentation axile; flowers with a nectar disk... **Solanaceae**

ing as fruit develops. *Petals usually 5, connate, often strongly so, and forming a wheel-shaped, bell-shaped, funnel-shaped, or tubular corolla, distinctly plicate (with fold lines), with marginal portions of each corolla lobe often folded inward, and often convolute, sometimes imbricate or valvate. Stamens usually 5;* filaments epipetalous; anthers usually 2-locular, opening by longitudinal slits or terminal pores, sometimes sticking to each other; pollen grains usually 3- to 5-colpate or colporate. *Carpels usually 2 (–5),* **oriented obliquely to the median plane of the flower**, *connate; ovary superior, entire to deeply lobed, usually with axile placentation and 2 locules;* style terminal to gynobasic; stigma 2-lobed. *Ovules usually numerous in each locule,* but occasionally reduced to 1 per locule, with 1 integument and a thin-walled megasporangium. Nectar disk present or lacking. *Fruit usually a berry, septifragal capsule,* or schizocarp of nutlets; seeds often flattened (Figure 9.121).

Floral formula: *,⑤,⑤, 5, ②; berry, capsule

Distribution and ecology: Widespread, but most diverse in the Neotropics. Many species occur in disturbed habitats.

Genera/species: 147 / 2930. **Major genera:** *Solanum* (1400 spp.), *Lycianthes* (200), *Cestrum* (175), *Nicotiana* (100), *Physalis* (100), and *Lycium* (90). Numerous genera occur in the continental United States and/or Canada, some of which are *Capsicum, Datura, Solanum, Lycium, Nicotiana, Petunia,* and *Physalis.*

Economic plants and products: Most members of the family are poisonous due to the presence of tropane or steroid alkaloids. Solanaceae are the source of several pharmaceutical drugs, and some are powerful narcotics; some of these plants include *Nicotiana* (tobacco), *Atropa*

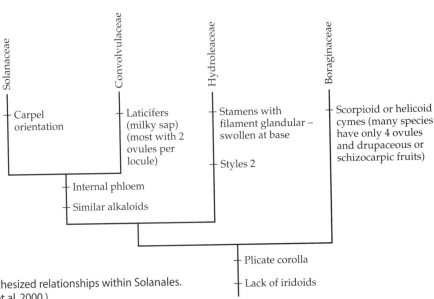

Figure 9.120 Cladogram showing hypothesized relationships within Solanales. (Modified from Ferguson 1999 and Soltis et al. 2000.)

Figure 9.121 Solanaceae. (A–J) *Physalis heterophylla:* (A) branch with flowers and fruits (× 0.35); (B) flower (× 2.5); (C) flower in longitudinal section (× 5.5); (D) stigma (greatly magnified); (E) ovary in cross-section (× 22); (F) mature accrescent calyx enclosing berry (× 4); (G) berry and calyx in longitudinal section (× 1.5); (H) berry and calyx in cross-section (× 1.5); (I) seed (× 12); (J) seed in section, endosperm stippled, embryo unshaded (× 12). (K–M) *P. walteri:* (K) branch with flowers and fruits (× 0.35); (L, M) branched hairs (greatly magnified). (From Wood 1974, *A student's atlas of flowering plants*, p. 98.)

(belladona), and *Datura* (jimsonweed). Surprisingly, the family also provides edible fruits, such as cayenne, red, and green peppers (*Capsicum* spp., which contain the alkaloid capsaicin, providing a "hot" flavor), tomatoes (*Solanum lycopersicum*), eggplants (*S. melongena*), tree tomatoes (*S. betacea*), and tomatillos (*Physalis ixocarpa*). The tubers of *Solanum tuberosum* (potatoes) are an important source of starch. Many genera provide ornamentals, including *Brunfelsia* (lady-of-the-night, yesterday-today-and-tomorrow), *Cestrum* (night-blooming jessamine), *Datura* (angel's trumpet), *Petunia*, *Physalis* (ground-cherry), and *Solanum* (nightshade).

Discussion: Solanaceae are considered monophyletic on the basis of morphological and cpDNA characters (Olmstead and Palmer 1991, 1992). Infrafamilial relationships have been investigated by D'Arcy (1979, 1991), using morphology, and by Olmstead and Palmer (1991, 1992, 1997), Olmstead and Sweere (1994), and Olmstead et al. (1995, 1999), based on a cladistic analysis of *rbcL* and *ndhF* sequences and cpDNA restriction site characters. The family has often been divided into two large subgroups, the Cestroideae (e.g., *Brunfelsia*, *Petunia*, *Cestrum*, and *Nicotiana*), defined by straight to slightly bent embryos and prismatic to subglobose seeds, and the Solanoideae (e.g., *Solanum*, *Capsicum*, *Lycianthes*, *Datura*, *Physalis*, *Lycium*, *Atropa*, and *Mandragora*), with curved embryos and flattened, discoidal seeds. Members of the Cestroideae typically have capsules, while both berries and capsules occur in Solanoideae. *Nolana* and relatives, which are especially distinct due to their gynobasic style and deeply lobed ovary, are included in Solanaceae (e.g., as the small subfamily Nolanoideae, D'Arcy 1979, 1991; Thorne 1992) or within Solanoideae (Olmstead et al. 1999), but systematists (e.g., Cronquist 1981) used to place them in their own family. The cladistic analyses of Olmstead and Palmer (1992) and Olmstead et al. (1995, 1999) demonstrated that the "Cestroideae" constitute a basal paraphyletic complex, which is divided into six subfamilies by Olmstead et al. (1995). Zygomorphic flowers have evolved several times within this complex. In contrast, monophyly of the Solanoideae is supported by cpDNA characters, discoidal seeds, and curved embryos. *Nolana* (Nolanoideae) nests within the Solanoideae; its recognition as family or subfamily has been due to over emphasis on its unusual multilobed ovary. The monophyly of the large genus *Solanum* is supported only if *Lycopersicon* and *Cyphomandra* are included (Bohs and Olmstead 1997; Olmstead and Palmer 1997; Spooner et al. 1993). *Solanum* s.l. is diagnosed by its distinctive wheel-shaped and deeply lobed corollas, connivent anthers usually opening by apical pores, and cpDNA restriction site, *rbcL*, and *ndhF* characters. *Solanum*, and many other members of Solanoideae, have berries (an apomorphic feature).

Flowers of Solanaceae are usually showy and attract various bees, wasps, flies, butterflies, and moths. *Solanum*

does not produce nectar and is pollinated by pollen-gathering bees and flies. Pollen is removed from the anthers by vibration or manipulation of the anthers. In contrast, *Cestrum* and *Datura* attract nectar-gathering insects, such as moths or butterflies. The brightly colored berries are dispersed by birds.

References: Cronquist 1981; D'Arcy 1979, 1991; Evans 1979; Johnston 1936; Olmstead et al. 1995, 1999; Olmstead and Palmer 1991, 1992; Olmstead and Sweere 1994; Roddick 1986, 1991; Spooner et al. 1993; Thorne 1992.

Convolvulaceae A. L. de Jussieu
(Morning Glory Family)

Usually twining and climbing herbs, often rhizomatous, occasionally with little or no chlorophyll and parasitic; the roots often storing carbohydrates; internal phloem usually present; **usually with laticifers, often with milky sap**; sometimes with alkaloids. Hairs various, often 2-branched or simple. *Leaves alternate and spiral, simple*, sometimes lobed or compound, usually entire, with pinnate or sometimes palmate venation, occasionally reduced; *stipules lacking*. Inflorescences determinate, sometimes reduced to a solitary flower, terminal or axillary. *Flowers usually bisexual, radial. Sepals usually 5, usually distinct or only very slightly connate*, persistent. *Petals usually 5, strongly connate and forming funnel-like corolla, distinctly plicate (with fold lines), with marginal portions of each corolla lobe folded inward and the middle portion of each lobe valvately arranged in bud, and often also convolute with a clockwise twist*, but occasionally merely imbricate. *Stamens usually 5; filaments epipetalous*, often of unequal lengths; pollen grains usually tricolpate to multiporate. *Carpels 2, connate; ovary superior*, entire to deeply 2- or 4-lobed, *usually with axile placentation*; style(s) terminal to gynobasic; stigma(s) 1 or 2, capitate, lobed, or linear. *Ovules usually 2 in each locule*, with 1 integument and usually a thin-walled megasporangium. Nectar disk usually lobed. *Fruit usually a septifragal, circumscissile, or irregularly dehiscing capsule*; embryo straight or curved, with folded cotyledons, or reduced (Figure 9.122 and Figure 4.47G,H).

Floral formula: *, 5, ⑤, 5, ②; capsule

Distribution: Widely distributed; most diverse in tropical and subtropical regions.

Genera/species: 55/1930. ***Major genera:*** *Ipomoea* (600 spp.), *Convolvulus* (250), *Cuscuta* (150), and *Jacquemontia* (120). Numerous genera occur in the continental United States and/or Canada; noteworthy genera other than those listed above are *Bonamia*, *Calystegia*, *Dichondra*, *Evolvulus*, *Merremia*, and *Stylisma*.

Economic plants and products: *Ipomoea batatas* (sweet potato) is important for its edible roots. Many species are

Figure 9.122 Convolvulaceae. (A–L) *Calystegia sepium:* (A) vine with flowers (× 0.5); (B) corolla opened out, note plications in corolla and epipetalous stamens (× 1); (C) flower in cross-section just above ovary, note two large bracteoles, free sepals, corolla tube with adnate filaments, and style (× 2); (D) gynoecium and nectar disk (× 2; enlargement × 8); (E) apex of style and stigmas (× 7); (F) ovary and irregular nectar disk in cross-section, note incomplete septum and four ovules (× 8); (G) four ovules, removed from ovary (× 8); (H) pollen grain, pantoporate (greatly magnified); (I) capsule with persistent bracts and sepals (× 2); (J) seed, adaxial surface (× 5); (K, L) two views of an embryo, note folded cotyledons (greatly enlarged). (M) *C. spithamaea:* flowering plant. (From Wood 1974, *A student's atlas of flowering plants*, p. 92.)

poisonous. *Ipomoea* (morning glory), *Jacquemontia*, *Porana* (Christmas vine), and *Dichondra* (ponyfoot) provide ornamentals.

Discussion: The monophyly of Convolvulaceae is supported by the presence of laticifers. *Humbertia* is probably sister to the remaining genera, which are united by a reduction to two ovules per carpel. The group is usually divided into two to four subfamilies, which are sometimes segregated as distinct families. Cuscutaceae are often segregated because of their numerous specializations associated with parasitism: reduction in chlorophyll content, scale-like leaves, haustoria, and reduced embryos. *Dichondra* and relatives are sometimes recognized as Dichondraceae due to their gynobasic styles. Recognition of either of these families makes Convolvulaceae s.s. paraphyletic (Neyland 2001).

The flowers of Convolvulaceae are usually showy, and attract various insects. They open for a day or less (often only a few hours), after which the corolla wilts. Some species of *Ipomoea* are pollinated by hummingbirds. Nectar is the reward. The seeds are relatively large, but are probably dispersed by wind, especially when retained in the rather papery and inflated capsule.

References: Austin 1979; Allard 1947; Neyland 2001; Wilson 1960b.

Boraginaceae Adanson
(Borage Family)

Herbs or shrubs to trees, occasionally lianas; internal phloem lacking; often with alkaloids. Hairs various, but *often unicellular, with a basal cystolith and often calcified or silicifed walls, and the plants rough to the touch. Leaves usually alternate and spiral, simple,* but sometimes deeply lobed or even compound, entire to serrate, with pinnate venation; *stipules lacking.* **Inflorescences determinate, usually forming helicoid or scorpioid cymes (i.e., axes coiled, bearing flowers along the upper side and straightening as the flowers mature),** usually terminal. *Flowers usually bisexual and radial. Sepals usually 5, distinct to connate. Petals usually 5, strongly connate and forming wheel-like, funnel-like, or tubular corolla, plicate (with fold lines),* imbricate or convolute. *Stamens usually 5; filaments epipetalous;* pollen grains tricolporate or triporate to polycolpate or polycolporate. *Carpels 2, connate; ovary superior,* spherical to deeply 4-lobed, with axile placentation *and 4 locules (actual locule of each carpel becomes divided by development of a false partition) or parietal placentation and usually numerous ovules;* style(s) terminal or gynobasic; stigma(s) 1 and 2-lobed, 2, or 4, capitate to truncate. Ovules 1 in each locule or numerous, with 1 integument and a thin-walled megasporangium. Nectar disk usually present around base of ovary. Fruit a drupe with 1 4-seeded, 2 2-seeded, or 4 1-seeded pit(s) a schizocarp of usually 4 1-seeded nutlets, or a loculicidal or irregularly

dehiscent capsule; embryo straight to curved; endosperm present to lacking (Figure 9.123).

Floral formula: *,⑤,⑤, 5 ,②; capsule, drupe, schizocarp of 2 or 4 nutlets

Distribution: Widely distributed in temperate and tropical regions.

Genera/species: 134/2650. ***Major genera:*** *Cordia* (320 spp.), *Heliotropium* (260), *Tournefortia* (150), *Onosma* (150), *Cryptantha* (150), *Phacelia* (150), *Myosotis* (100), *Cynoglossum* (75), and *Ehretia* (75). Numerous genera occur in the continental United States and/or Canada, including *Amsinckia, Bourreria, Cordia, Cryptantha, Cynoglossum, Draperia, Ellisia, Eriodictyon, Hackelia, Heliotropium, Hydrophyllum, Lappula, Lithospermum, Mertensia, Myosotis, Nama, Nemophila, Onosmodium, Phacelia, Plagiobothrys,* and *Romanzoffia.*

Economic plants and products: *Heliotropium* (heliotrope), *Mertensia* (Virginia bluebells), *Myosotis* (forget-me-not), *Cordia* (geiger tree and others), *Cynoglossum* (hound's-tongue), *Pulmonaria* (lungwort), and *Phacelia* provide ornamentals. Several species have been used as medicinal herbs, including *Borago officinalis* (borage), *Symphytum officinale* (comfrey), and *Lithospermum* spp. (puccoon); many species are poisonous.

Discussion: The family is circumscribed broadly (including Hydrophyllaceae, except *Hydrolea*; see Figure 9.120) and is considered to be monophyletic on the basis of its distinctive inflorescence form and DNA sequence data (Chase et al. 1993; Downie and Palmer 1992a; Ferguson 1999; Olmstead et al. 1992, 1993, 2000; Olmstead and Ferguson 2001). Schizocarpic fruits with four nutlets, 4-locular ovaries, and a gynobasic style evolved independently within Boraginaceae, Verbenaceae, and Lamiaceae (Cantino 1982).

Most systematists recognize four large subfamilies (Al-Shehbaz 1991). The tropical "Ehretioideae" (such as *Ehretia* and *Bourreria*) probably represent a paraphyletic complex characterized by the symplesiomorphies of a woody habit, terminal style divided into two branches, unlobed ovary, drupaceous fruit, flat cotyledons, and presence of endosperm. Cordioideae, which contains *Cordia* and two small genera, may constitute a related clade based on the four-branched style, plicate cotyledons, and lack of endosperm. This mainly tropical group has retained a woody habit, terminal styles, unlobed ovaries, and drupaceous fruits. The remaining two subfamilies both have schizocarpic fruits (with nutlets)—derived conditions that probably evolved independently. The monophyly of Heliotropioideae, which includes *Heliotropium, Tournefortia,* and *Argusia,* is supported by the often short, undivided style with a single stigma. Boraginoideae can be characterized by their gynobasic

Figure 9.123 Boraginaceae. *Mertensia virginica:* (A) flowering plant (× 0.35); (B) root with two stem bases (upper left) and rhizome (lower left) (× 0.75); (C) inflorescence (× 1.5); (D) flower (× 3); (E) flower, in longitudinal section (× 3); (F) flower after corolla has fallen, to show gynoecium (× 7.5); (G) gynoecium in longitudinal section, note gynobasic style (× 7.5); (H) nutlets and accrescent calyx (× 4.5); (I) abaxial surface of nutlet (× 9); (J) seed (× 9); (K) embryo (× 9). (From Al-Shehbaz 1991, *J. Arnold Arbor.* Suppl. Ser. 1: p. 91.)

style and deeply lobed ovary; this large subfamily (106 genera, including *Hackelia, Plagiobothrys, Lithospermum, Onosmodium, Cynoglossum, Symphytum, Borago, Mertensia,* and *Myosotis*) is well developed in temperate and subtropical regions and contains numerous herbaceous species. Members of all four subfamilies have gynoecia with only four ovules. Ehretioideae and Cordioideae often are segregated as "Ehretiaceae."

In addition, most genera traditionally placed in "Hydrophyllaceae" are closely related to the groups outlined above, and are here included within Boraginaceae. These plants have usually numerous ovules and capsular fruits (vs. gynoecia with only 4 ovules, developing into drupes or nutlets in the traditional Boraginaceae). Both numerous ovules and capsular fruits are likely ancestral within the Boraginaceae/Hydrophyllaceae clade. *Codon* may be related most closely to Boraginoideae, while *Phacelia, Nama,* and relatives probably are basal in a clade containing "Ehretioideae," Cordioideae, and Heliotropioideae. Nucleotide sequences of the *ndhF* chloroplast gene suggest that *Pholisma* and relatives (Lennoaceae), an unusual group of root parasites, may also belong to Boraginaceae, being within the ehretioid complex (Olmstead, personal communication).

Boraginaceae are usually pollinated by bees, wasps, butterflies, and flies, which gather nectar or pollen, but moths, flies, beetles, bats, and birds have also been recorded as pollinators. Both self- and cross-pollination occur; distyly has been documented in species of several genera. The flowers of *Mertensia, Myosotis,* and *Cryptantha* rapidly change color after pollination, as a signal to pollinators. The family shows diverse dispersal mechanisms. Species with drupaceous fruits are usually dispersed by birds or mammals. The drupes of some coastal species of *Cordia* are corky and will float in (and are dispersed by) water. The corky nutlets of *Argusia* and *Mertensia maritima* are also water-dispersed. In some genera with nutlets, the calyx is persistent and winglike, and dispersal is by wind. The nutlets of *Hackelia, Lappula,* and *Cynoglossum* have appendages and stick to fur or clothing, and *Myosotis* has hooks on its calyx. Many genera have an aril-like structure attached to the base of the nutlets that is hard and contains sugars, fats, and free amino acids; these are dispersed by ants. In *Heliotropium* and *Lithospermum* the nutlets are eaten by birds. Finally, the nutlets may be consumed and dispersed by browsing mammals. The small seeds of the capsular species are mainly wind- or water-dispersed. Again, some seeds have a hard, aril-like structure and may be at least partially dispersed by ants.

References: Al-Shehbaz 1991; Cantino 1982; Chase et al. 1993; Chuang and Constance 1992; Cronquist 1981; Downie and Palmer 1992; Ferguson 1999; Johnston 1950; Olmstead et al. 1992a, 1993, 2000; Olmstead and Ferguson 2001; Prior 1960; Wilson 1960c.

Gentianales

Gentianales are clearly monophyletic, based on the presence of vestured pits, stipules (sometimes reduced to a stipular line) and thick glandular hairs (colleters) on the adaxial surface of the stipules or base of the petiole (Bremer and Struwe 1992; Nicholas and Baijnath 1994; Struwe et al. 1994; Wagenitz 1959, 1992). The colleters produce mucilage and help to protect the shoot apex. Other possible synapomorphies include opposite leaves, a particular type of complex indole alkaloids, corollas that are convolute in bud (Bremer and Struwe 1992; Wagenitz 1992), *rbcL, matK, atpB, ndhF, 18S* sequences (Bremer et al. 1994; Chase et al. 1993; Olmstead et al. 1993; Endress et al. 1996; Källersjö et al. 1998; Backlund et al. 2000; Olmstead et al. 2000; Savolainen et al. 2000a,b; Soltis et al. 2000; Albach et al. 2001), and cpDNA restriction sites (Downie and Palmer 1992). The order consists of 5 families and about 14,200 species; families include **Gentianaceae**, **Rubiaceae**, and **Apocynaceae** (incl. Asclepiadaceae), Loganiaceae, and Gelsemiaceae. Gentianales are clearly within the asterid clade, as indicated by the single integument and a thin-walled megasporangium, and within the core asterid group because of their sympetaly and epipetalous stamens equaling the number of corolla lobes. Analyses of DNA sequences also strongly support the group's position within the asterid clade, and suggest that the order is most closely related to the Solanales. Most members of both orders have internal phloem and radially symmetrical flowers with the number of stamens equaling the number of corolla lobes. Gentianales are easily distinguished by their usually opposite leaves, stipules or at least a stipular line, and colleters.

"Loganiaceae," as traditionally circumscribed, constitute a nonmonophyletic undefinable assemblage. Thus, *Fagraea* and *Potalia* should be transferred to Gentianaceae, and *Buddleja* and relatives constitute the Buddlejaceae (of the Lamiales). *Gelsemium* and *Mostuea* are taxonomically isolated and are best treated as Gelsemiaceae (Figure 9.124), but within Gentianales (Bremer and

---

Key to Major Families of Gentianales

1. Ovary inferior; internal phloem lacking; stipules ± interpetiolar; placentation axile ....**Rubiaceae**

1. Ovary or ovaries superior; internal phloem usually present; stipules various (i.e., adjacent to the petiole base, ± interpetiolar, often reduced or lacking, and plant then usually with a nodal line); placentation axile, or parietal, or marginal..2

   2. Latex milky; style apically thickened, forming a stylar head; ovaries usually distinct, joined only by a common style or stylar head ....**Apocynaceae**

   2. Latex not milky; style not apically thickened and a stylar head lacking; ovaries usually connate.........3

      3. Stipules usually present, interpetiolar, these sometimes reduced to stipular lines at the node; placentation usually axile; plants usually woody ....Loganiaceae

      3. Stipules usually lacking; placentation parietal; plants usually herbaceous...................**Gentianaceae**

Figure 9.124 Cladogram showing hypothesized relationships within Gentianales. (Modified from Backlund et al. 2000 and Soltis et al. 2000.)

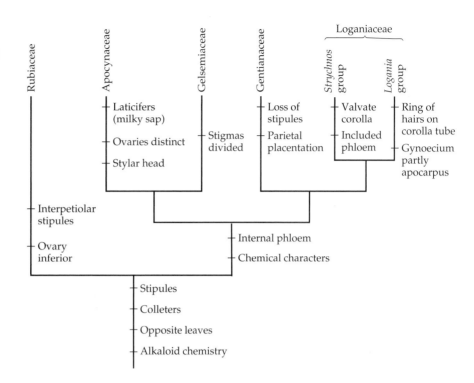

Struwe 1992; Endress et al. 1996; Struwe et al. 1994; Backlund et al. 2000; Soltis et al. 2000). Most genera of Loganiaceae, however, probably do constitute a monophyletic group, containing two major subclades: the *Strychnos* group (*Strychnos* and *Spigelia*) and the *Logania* group (*Geniostoma, Labordia, Logania, Mitreola,* and *Mitrasacme*). The former is characterized by a valvate corolla and included phloem, while the latter has a ring of hairs at the mouth of the corolla tube and a partly apocarpous gynoecium (Struwe et al. 1994). Loganiaceae have often been considered to form a basal complex within Gentianales, but recent analyses (Backlund et al. 2000; Olmstead et al. 2000) support the recognition of two major clades within the order, the first comprising Rubiaceae and the second including the remaining families (Figure 9.124). In addition to the characters shown in Figure 9.124, these two groups also differ in the biosynthesis of their iridoid and indole alkaloids (Jensen 1992).

References: Backlund et al. 2000; Bremer et al. 1994; Bremer and Struwe 1992; Chase et al. 1993; Downie and Palmer 1992; Endress et al. 1996; Jensen 1992; Källersjö et al. 1998; Nicholas and Baijnath 1994; Olmstead et al. 1993, 2000; Rogers 1986; Savolainen et al. 2000a,b; Soltis et al. 2000; Struwe et al. 1994; Wagenitz 1959, 1992; Wood 1983; Wood and Weaver 1982.

Rubiaceae A. L. de Jussieu
(Coffee or Madder Family)

Trees, shrubs, lianas, or herbs; lacking internal phloem; usually with iridoids, various alkaloids; raphide crystals common. Hairs various. *Leaves opposite or whorled, usual-*

ly entire, with pinnate venation; *stipules present,* **interpetiolar** *and usually connate,* occasionally leaflike, *with colleters on adaxial surface.* Inflorescences determinate, occasionally reduced to a single flower, terminal or axillary. *Flowers usually bisexual and radial, often heterostylous, frequently aggregated.* Sepals usually 4 or 5, connate, sometimes with colleters on adaxial surface. *Petals usually 4 or 5, connate,* forming a usually wheel-shaped to funnel-shaped corolla, adaxial surface often pubescent, the lobes valvate, imbricate, or contorted. *Stamens usually 4 or 5; filaments usually adnate to corolla* and positioned within corolla tube or at its mouth, sometimes basally connate; anthers 2-locular, opening by longitudinal slits; pollen grains usually tricolporate. *Carpels usually 2 (-5), connate;* **ovary inferior**, with usually axile placentation; stigma(s) 1 or 2, linear, capitate, or lobed. Ovules 1 to numerous in each locule, with 1 integument and a thin-walled megasporangium. Nectar disk usually present above ovary. Fruit a loculicidal to septicidal capsule, berry, drupe, schizocarp, or indehiscent pod; seeds sometimes winged; embryo straight to curved; endosperm present or lacking (Figure 9.125).

Floral formula: *, (4–5), (4–5), 4–5, (2–5); capsule, berry, drupe, schizocarp, indehiscent pod

Distribution: Cosmopolitan, but most diverse in tropical and subtropical regions.

Genera/species: 550/9000. *Major genera: Psychotria* (1500 spp.), *Galium* (400), *Ixora* (400), *Pavetta* (400), *Hedyotis* (400), *Tarenna* (370), *Randia* (250), *Gardenia* (250), *Palicourea* (250), *Mussaenda* (200), *Borreria* (150), and *Rondeletia* (125).

Figure 9.125 Rubiaceae. (A–E) *Diodia tetragona:* (A) flowering plant (× 0.4); (B) node with sessile axillary flower, note interpetiolar stipules (× 5); (C) gynoecium and calyx in longitudinal section, note inferior ovary (× 10); (D) nearly mature fruit (× 5); (E) drupaceous schizocarp in cross-section, endocarp hatched, endosperm stippled, embryo unshaded (× 10). (F–H) *D. teres:* (F) flower (× 5); (G) portion of schizocarp (× 5); (H) seed, adaxial surface (× 8). (From Wood 1974, *A student's atlas of flowering plants*, p. 106.)

Some of the numerous genera in the continental United States and/or Canada are *Casasia, Catesbaea, Cephalanthus, Chiococca, Diodia, Ernodia, Erithalis, Exostema, Galium, Guettarda, Hamelia, Hedyotis, Mitchella, Morinda, Pentodon, Pinckneya, Psychotria, Randia, Richardia,* and *Spermacoce.*

Economic plants and products: Coffee, a stimulating beverage containing caffeine, is made from the brewed seeds of *Coffea arabica* and *C. robusta*. Quinine, a drug used in treating malaria, comes from the bark of species of *Cinchona,* and ipecac, a drug used to induce vomiting, is derived from *Psychotria. Gardenia, Hamelia, Pentas, Randia, Rondeletia, Serissa, Hedyotis,* and *Ixora* provide ornamentals.

Discussion: Rubiaceae are an easily recognized monophyletic group. Phylogenetic relationships within the family have been investigated by cladistic analyses of morphological features (Bremer and Struwe 1992), molecular characters (Bremer and Jansen 1991; Bremer et al. 1995, 1999; Andersson and Rova 1999), and combined data (Andreasen and Bremer 2000).

Cinchonoideae, which include *Gardenia, Casasia, Ixora, Coffea, Erithalis, Chiococca, Exostema, Randia, Cinchona, Cephalanthus, Pinckneya, Mussaenda, Portlandia, Catesbaea,* and *Rondeletia,* have traditionally been defined either on the basis of ovaries with numerous ovules or by the presence of endosperm, lack of raphide crystals, and seeds with a pitted-ridged coat (all likely symplesiomorphies). The monophyly of this group has not been supported by cladistic studies employing morphological or cpDNA restriction site characters, but receives support from an analyses of *rbcL, ndhF,* or *rps16* intron sequences. However, a group of genera within Cinchonoideae, often segregated as Ixoroideae (e.g., *Ixora, Coffea, Garde-*

nia, Randia), are clearly monophyletic. This clade is characterized by flowers with the corolla contorted (with a left twist) and a specialized pollination presentation mechanism. The anthers shed their pollen before the flower opens, and pollen is presented to pollinators on the hairy upper portion of the style; only later does the stigma become receptive. This mechanism is similar to the plunger pollination system of the Asterales.

Rubioideae, comprising the majority of the species, are usually defined by the combination of the presence of raphides and valvate corolla lobes. Some members of this clade have numerous ovules, while others show a reduction to a single ovule per locule. This subfamily probably is monophyletic; potential synapomorphies include the presence of raphides, seeds with a smooth coat, the usually herbaceous habit (presumably reversed in *Psychotria* and *Palicourea*), and numerous cpDNA characters. Representative genera are *Pentas, Palicourea, Psychotria, Galium, Nertera, Hedyotis, Richardia, Diodia, Spermacoce, Pentodon, Morinda,* and *Mitchella*. Fleshy fruits and ovaries with a reduced number of ovules have evolved several times within Rubiaceae.

In *Galium* the stipules are expanded leaflike structures; the leaves, therefore, appear to be in whorls. The number of apparent leaves in a whorl varies depending upon the amount of fusion and splitting of the stipules.

The flowers of Rubiaceae are variable in color and shape, and may be pollinated by butterflies, moths, bees, flies, birds, or bats; wind pollination occurs in a few. Nectar is employed as the pollinator reward. Outcrossing is promoted by protandry, a specialized pollen presentation mechanism (see above), or heterostyly. The Rubiaceae contain more species with heterostyly than any other angiosperm family, and may contain more than all other angiosperms combined. Fleshy-fruited species are typically dispersed by birds. Seeds from capsular fruits may be small or winged and are often dispersed by wind. The fruits of some species of *Galium* have hooked hairs and are transported externally by animals.

References: Anderson 1973; Andersson 1992; Andersson and Rova 1999; Andreasen and Bremer 2000; Barrett and Richards 1990; Bremekamp 1966; Bremer and Jansen 1991; Bremer 1987, 1996; Bremer et al. 1995, 1999; Bremer and Struwe 1992; Kirkbride 1982; Lersten 1975; Manen et al. 1994; Nilsson et al. 1990; Robbrecht 1988; Rogers 1987; Verdcourt 1958; Vuilleumier 1967.

Gentianaceae A. L. de Jussieu
(Gentian Family)

Herbs, sometimes mycoparasites (with reduced leaves and lacking chlorophyll), to shrubs or small trees; stems often winged; usually with internal phloem; usually with iridoids. Hairs often simple. *Leaves usually opposite, simple, usually entire, often ± sessile,* with pinnate venation; **stipules usually lacking,** *but colleters often present at base of adaxial surface of petiole.* Inflorescences determinate, sometimes reduced to a single flower, axillary or terminal. *Flowers usually bisexual, radial.* Sepals usually 4 or 5 and connate, often with colleters on adaxial surface. *Petals usually 4 or 5, connate,* forming a wheel-shaped, funnel-shaped, or bell-shaped corolla, the lobes sometimes fringed, often with nectar glands and/or scales on adaxial surface of the tube, usually convolute, sometimes plicate at sinuses. *Stamens usually 4 or 5; filaments adnate to corolla;* anthers occasionally opening by terminal pores; pollen grains usually tricolporate or triporate. *Carpels 2, connate; ovary superior,* **with parietal placentation, the placentas sometimes deeply intruded and 2-lobed;** stigma ± capitate to strongly 2-lobed, the lobes sometimes spirally twisted. Ovules usually numerous on each placenta, with 1 integument and a thin-walled megasporangium. Nectar-producing disk or glands present. *Fruit usually a septicidal capsule.*

Floral formula: *, ④⑤, ④⑤, 4–5, ②; capsule

Distribution: Widely distributed, but most diverse in temperate and subtropical regions and in the montane tropics.

Genera/species: 84/970. **Major genera:** *Gentiana* (300 spp.), *Gentianella* (125), *Sebaea* (100), *Swertia* (100), and *Halenia* (70). These genera (except *Sebaea*) occur in the continental United States and/or Canada, along with *Bartonia, Centaurium, Eustoma, Frasera, Sabatia,* and *Voyria.*

Economic plants and products: *Gentiana* (gentian), *Centaurium* (centaury), *Eustoma* (prairie gentian), *Exacum* (German violet), and *Sabatia* (rose pink) provide ornamentals.

Discussion: Infrafamilial relationships have been investigated by Mészáros et al. (1996) and Thiv et al. (1999). It is clear that *Villarsia, Nymphoides, Lomatogonium, Obolaria,* and *Menyanthes* are only distantly related to Gentianaceae (Bremer et al. 1994; Chase et al. 1993; Downie and Palmer 1992; Michaels et al. 1993; Olmstead et al. 1993; Wood 1983; Wood and Weaver 1982). These genera are here placed in Menyanthaceae (Euasterid II), a family of wetland or aquatic herbs with alternate and spiral or 2-ranked leaves and an induplicate-valvately folded corolla. The two families also differ chemically and anatomically.

The colorful flowers of Gentianaceae are pollinated mainly by bees and butterflies. Nectar is the reward. Most are outcrossing due to protogyny. The small seeds are probably wind- or water-dispersed.

References: Bremer et al. 1994; Chase et al. 1993; Downie and Palmer 1992; Mészáros et al. 1996; Michaels et al. 1993; Olmstead et al. 1993; Struwe et al. 1994; Thiv et al. 1999; Wood and Weaver 1982.

Apocynaceae Adanson
(Milkweed Family)

Trees, shrubs, lianas, vines, and herbs, sometimes succulent and cactuslike; internal phloem present; **tissues with laticifers and sap usually milky**; with cardiac glycosides and various alkaloids; often with iridoids. Hairs various, often simple. *Leaves usually opposite*, sometimes alternate and spiral or 2-ranked or whorled, *entire*, with pinnate venation; *stipules reduced or lacking, colleters usually present at base of petiole*. Inflorescences determinate, but sometimes appearing indeterminate, occasionally reduced to a single flower, terminal or axillary, often not appearing strictly so. *Flowers usually radial*. Sepals usually 5, ± connate, sometimes reflexed, colleters often present at base of adaxial surface. *Petals usually 5, connate*, forming a wheel-shaped, bell-shaped, funnel-shaped, or tubular corolla, sometimes reflexed, often with coronal appendages or scales on inside or apex of tube, the lobes imbricate, valvate, or contorted. *Stamens usually 5*; filaments short, sometimes connate, always adnate to corolla; anthers often highly modified and 2-locular, *distinct or sticking together and forming a ring around the stylar head, free, sticking to the stylar head by intertwining trichomes and sticky secretions (viscin), or strongly adnate to stylar head by solid parenchymatous tissue*, the connective often with apical appendages; *staminal outgrowths (corona) often present and petal-like, hood-shaped and horn-shaped; pollen grains* usually tricolporate or di- or triporate, shed as monads, loosely sticking together by means of viscin, or as tetrads, also loosely sticking together, or strongly coherent and *forming hardened masses (pollinia)*, then translators (structures connecting pollinia from anther sacs of adjacent anthers) present, often with a sticky gland (corpusculum). *Carpels usually 2*, **connate by styles and/or stigmas only and ovaries distinct**, fully connate in *Allamanda, Carissa*, and *Thevetia; ovaries superior*, with usually lateral or axile placentation; **apical portion of style expanded and highly modified, forming a head, secreting viscin**, *usually abruptly enlarged and latitudinally differentiated into 3 zones* (i.e., specialized for pollen deposition, viscin secretion, and pollen reception), often 5-sided; stigmatic tissue often restricted to 5 small, lateral regions of the stylar head. Ovules 2 to numerous in each ovary, with 1 integument and a thin-walled (or secondarily thickened) megasporangium. Nectar glands or disk often present, or nectar secreted by stigmatic chambers (and held in the hollows of the staminal tube or coronal appendages). *Fruits often paired, each ovary usually developing into a fleshy or dry follicle, berry, or drupe*, the fruit surface smooth to covered with prickle-like outgrowths; *seeds flattened, often with a tuft of hairs;* embryo straight to bent (Figures 9.126, 9.127).

Floral formula: *,⑤,⑤,⑤,② ; 1 or 2 follicles, berry, drupe

Distribution: Widespread in tropical and subtropical regions, but with a few genera extending into temperate areas.

Genera/species: 355/3700. **Major genera:** *Asclepias* (230 spp.), *Tabernaemontana* (230), *Cynanchum* (200), *Ceropegia* (150), *Hoya* (150), *Matelea* (130), *Rauvolfia* (110), *Gonolobus* (100), *Secamone* (100), and *Mandevilla* (100). Noteworthy genera occurring in Canada and/or the continental United States include *Amsonia, Angadenia, Apocynum, Asclepias, Catharanthus, Cynanchum, Echites, Gonolobus, Macrosiphonia, Matelea, Morrenia, Pentalinon, Rhabdadenia, Sarcostemma,* and *Vallesia*.

Economic plants and products: Nearly all taxa are poisonous, and many have medicinal uses. *Catharanthus* (Madagascar periwinkle) provides antileukemia drugs, *Rauvolfia* provides a hypertension drug, and *Strophanthus* provides a drug used in treating heart ailments. Important ornamentals are *Allamanda, Amsonia* (bluestar), *Asclepias* (milkweed, butterfly weed), *Carissa* (Natal plum), *Catharanthus* (Madagascar periwinkle), *Hoya* (wax plant), *Nerium* (oleander), *Plumeria* (frangipani), *Stapelia* (carrion flower), *Trachelospermum* (confederate jasmine), and *Vinca* (periwinkle).

Discussion: Apocynaceae, as broadly circumscribed to include Asclepiadaceae, are clearly monophyletic, as indicated by their milky sap and highly modified gynoecium with separate ovaries and a differentiated stylar/stigmatic head, as well as *rbcL* and *matK* sequences (Chase et al. 1993; Civeyrel et al. 1998; Endress et al. 1996; Judd et al. 1994; Endress and Bruyns 2000; Endress and Stevens 2001; Potgieter and Albert 2001) (Figures 9.126 and 9.127). The syncarpous ovaries of *Carissa, Thevetia,* and *Allamanda* represent reversals. Apocynaceae exhibit a stepwise accumulation of androecial and gynoecial characters associated with increased efficiency and specialization of the pollination mechanism (Fallen 1986; Judd et al. 1994; Schick 1980, 1982). Included among these innovations are the enlarged, five-sided stylar head that is zonally differentiated, apical anther appendages, the differentiation of fertile and hardened sterile portions of the anthers, and adhesive hair pads on the style head and/or anthers.

One group of genera, which has often been segregated as Asclepiadaceae, but is treated here as the subfamily Asclepiadoideae, is further characterized by pollen agglutinated into pollinia and staminal appendages often associated with nectar storage. The pollinium of *Cryptostegia* and relatives (Periplocoideae) is composed of loosely agglutinated pollen tetrads resting on a cuplike translator. The remaining genera—*Asclepias, Calotropis, Cynanchum, Hoya, Matelea, Stapelia, Ceropegia, Gonolobus, Sarcostemma, Secamone,* and relatives—form a clade based on a specialized pollinium that develops by the agglutination of the pollen tetrads of a theca (i.e., pollen sac) into a hardened mass, and is attached to a yoke-shaped translator with a gland, as well as connation of the filaments. All but *Secamone* and relatives (Secamonoideae) have the additional synapomorphy of anthers with only two thecae; these genera constitute the large group called Asclepiadoideae (Fishbein 2001); see Figure 9.128. Recognition of the family Asclepiadaceae would make the restrictively circumscribed Apocynaceae paraphyletic (Civeyrel et al. 1998; Endress et al. 1996; Judd et al. 1994; Sennblad and Bremer 1996).

Apocynaceae show a great diversity of pollination mechanisms; frequent pollinators include various nectar-gathering insects (butterflies, moths, bees, flies). The stylar head plays an important role in pollination. In many moderately specialized genera it is differentiated into three discrete horizontal zones (see above). The receptive tissue of the stigma is restricted to the base of the stylar head, under a collarlike extension (pollen scraper, see Figure 9.127). Sticky fluid is produced in the middle zone, while the apex of the stylar head presents the pollen, which is deposited by the inwardly dehiscing anthers that surround the stylar head. As the pollinator probes for nectar, pollen from a previously visited flower is scraped off its mouthparts and deposited in the stigmatic region. The mouthparts become sticky as they contact the sticky fluid produced in the middle region of the stylar head, then pick up pollen from the apical region.

Figure 9.126 Apocynaceae (representative of taxa lacking pollinia). *Catharanthus roseus:* (A) stem with flowers and fruits (× 0.7); (B) flower bud, note contorted corolla and bulge at location of androecium and stylar head (× 4); (C) upper part of flower cut longitudinally, note anthers free from one another and from stylar head (× 8); (D) anther before dehiscence, side view (× 13.5); (E, F) anther before dehiscence, adaxial and abaxial views (× 13.5); (G) stylar head, lower ridge is pollen scraper (× 16); (H) base of flower after removal of two calyx lobes and corolla, note two nectaries (left and right) alternating with two ovaries (× 8); (I) ovary and nectaries in cross-section, note that ovaries are distinct (× 13.5); (J) pair of follicles (× 27); (K) follicle, after dehiscence (× 2.7); (L) seed (× 13.5); (M) seed in longitudinal section, endosperm stippled, embryo unshaded (× 13.5). (From Rosatti 1989, *J. Arnold Arbor.* 70: p. 366.)

Figure 9.127   Apocynaceae (representatives of pollinial clade). (A, B) *Asclepias syriaca:*
(A) very young fruit, cut longitudinally, after fall of corolla, androecium, and stylar head
(× 11); (B) nearly mature follicle (× 0.7). (C–F) *A. incarnata:* (C) distal part of stem with flowers and young fruits (× 0.5); (D) dehisced follicle, releasing seeds with apical tufts of hairs
(× 0.7); (E) seed, with apical tuft of hairs removed, note winged seed coat (× 4); (F) embryo
(× 5.5). (G–L) *A. syriaca:* (G) flower with corolla lobes reflexed, note conspicuous, hood-
shaped and horn-shaped staminal appendages (corona) (× 5); (H) flower from above, note
hood-shaped and horn-shaped staminal appendages opposite anther flaps over stylar head
and alternating with sticky glands (corpuscula) (× 5); (I) androecium (and, within, gynoecium) distal to level of adnation of corolla, with hood-shaped staminal appendages removed
(× 12); (J) anther, adaxial side, with two pollinia, each connected by a translator arm, a corpusculum, and another translator arm to one pollinium from adjacent anther (× 12);
(K) pair of pollinia (× 16); (L) gynoecium, the two carpels united only by stylar head (× 12).
(M) *A. connivens:* pair of pollinia (× 12). (N) *A. pedicellata:* side view of flower with two corolla lobes removed (× 5). (From Rosatti 1989, *J. Arnold Arbor.* 70: pp. 492, 494.)

In the most specialized genera (e.g., *Asclepias*) the androecium and gynoecium are adnate and the pollen is fused into pollinia. Nectar accumulates in the elaborate outgrowths of the stamens, appendages called hoods and horns. In the process of probing for nectar in the corona, the insect's leg may become lodged in the space between the anthers and make contact with the translator gland. When the insect pulls out its leg, it becomes stuck in the gland, pulling out the two pollinia. The pollinia are dislodged from the insect's leg when it slips into one of the stigmatic slits on the stylar head of another flower.

Most species of Apocynaceae have follicles with hair-tufted, flattened seeds that are dispersed by wind. In *Cameraria* the paired fruits are dry, distally winged, and dispersed by wind. The colorful (often blue or red) fleshy fruits of tropical genera, such as *Kopsia, Rauvolfia, Carissa,* and *Ochrosia,* are often mammal- or bird-dispersed. The seeds of *Tabernaemontana* are associated with a brightly colored, fleshy aril and dispersed by birds. Monkey dispersal has been reported for the arillate *Stemmadenia.*

References: Chase et al. 1993; Civeyrel et al. 1998; Endress and Bruyns 2000; Endress and Stevens 2001; Endress et al. 1983, 1996; Fishbein 2001; Judd et al. 1994; Fallen 1986; Kunze 1990, 1992; Potgieter and Albert 2001; Rosatti 1989; Safwat 1962; Schick 1980, 1982; Sennblad and Bremer 1996; Swarupanandan et al. 1996; Thomas and Dave 1991; Woodson 1954.

Lamiales

Lamiales (including Bignoniales or Scrophulariales) are clearly monophyletic. They are characterized by gland-headed hairs, oligosaccharides (which replace starch in carbohydrate storage), the frequent production of 6-oxygenated flavones, parenchyma tissue extending from anther connective into the locules, often diacytic stomates (i.e., stomates enclosed by one or more pairs of subsidiary cells whose common walls are at right angles to the guard cells), embryos usually of the onagrad type (*Veronica* variant), endosperm with a conspicuous micropylar haustorium, and protein inclusions in the nuclei of mesophyll cells (Dahlgren 1983; Judd et al. 1994; Wagenitz 1992; Yamazaki 1974), cpDNA restriction sites, and *rbcL, atpB, ndhF,* and 18S sequences (Lu 1990; Downie and Palmer 1992; Judd et al. 1994; Olmstead et al. 1992, 1993a,b, 2000, 2001; Olmstead and Reeves 1995; Källersjö et al, 1998; Wagstaff and Olmstead 1996; Soltis et al. 2000; Albach et al. 2001).

Oleaceae and the small families Tetrachondraceae and Plocospermataceae are probably basal within the order. Oleaceae are characterized by actinomorphic flowers (usually of four petals) with only two stamens, and flowers of Tetrachondraceae (incl. *Polypremum*) and Plocospermataceae are also radial. Most other members of the order have two-lipped, bilaterally symmetrical flowers forming a 2 + 3 pattern, that is, with two upper petals and three lower petals forming a corolla with two opposing lobes (i.e., bilabiate), but secondarily radial and four-petaled in Buddlejaceae (Figure 9.129). The remaining members of the order also have four stamens (two long

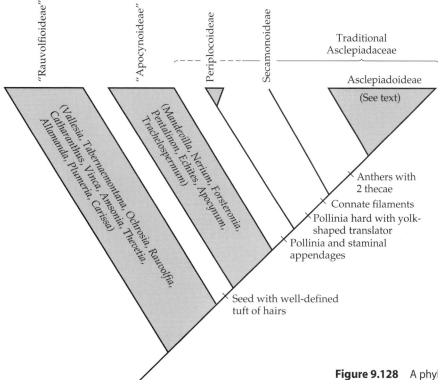

Figure 9.128 A phylogeny of Apocynaceae. (Modified from Judd et al. 1994 and Endress and Bruyns 2000.)

Key to Major Families of Lamiales

1. Perianth lacking; stamen usually 1; aquatic plants with inconspicuous water-pollinated flowers .... **Plantaginaceae** (in part, *Callitriche*)

1. Perianth present; stamens usually 2–4; terrestrial or aquatic plants with usually showy, animal- or wind-pollinated flowers 2

2. Mangrove trees or shrubs, with erect, pencil-like pneumatophores; leaves bearing subsessile glands that excrete a hypersaline solution that rapidly dries to form salt crystals; seed usually only 1 per fruit .... Avicenniaceae

2. Nonmangrove herbs to trees; lacking pneumatophores and salt-excreting glands; seeds (2–) 4 to numerous per fruit 3

3. Flowers radial, with 4 petals 4

3. Flowers bilateral, with 5 petals, the 2 upper forming one lip and the 3 lower forming an opposing lip 6

4. Stamens 2; ovules often 2 in each locule .... **Oleaceae**

4. Stamens 4; ovules ± numerous per locule 5

5. Plants herbaceous; leaves usually alternate and in basal rosettes, blade ± parallel-veined; flowers in spikes or heads, inconspicuous, wind pollinated; fruit a circumscissile capsule or achene .... **Plantaginaceae** (in part, *Plantago* and relatives)

5. Plants usually woody, leaves opposite, not in basal rosettes; blade pinnately veined; flowers in various cymose inflorescences, conspicuous and insect pollinated; fruit a variously dehiscing capsule, drupe, or samara .... Buddlejaceae

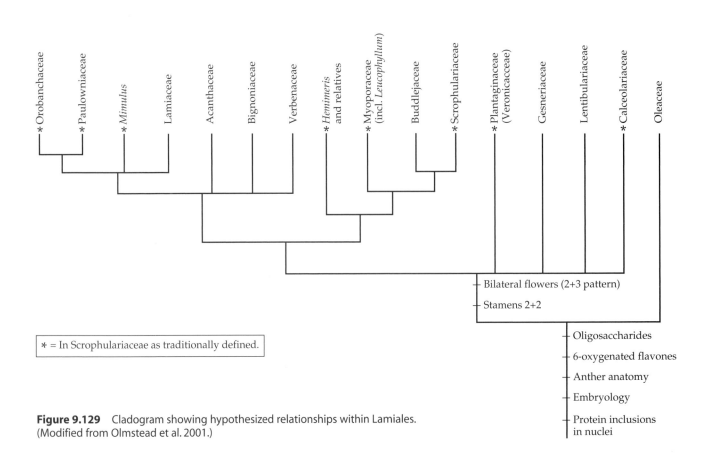

* = In Scrophulariaceae as traditionally defined.

Figure 9.129 Cladogram showing hypothesized relationships within Lamiales. (Modified from Olmstead et al. 2001.)

Key to Major Families of Lamiales (continued)

6. Ovules 2 per carpel, each locule divided by a false partition, thus with 1 ovule per chamber of ovary; fruits drupes (with 1–4 pits) or schizocarps of usually 4 nutlets; style terminal to gynobasic; leaves opposite..7

6. Ovules usually numerous, but if reduced to 2 per carpel, then ovary locules 2 (i.e., not secondarily divided by false partition); fruits usually capsules, sometimes berries or drupes; style always terminal; leaves alternate or opposite.........................8

7. Inflorescences indeterminate, the lateral units individual flowers or appearing to be individual flowers; ovules directly attached to margins of false partitions; style consistently terminal; stigma conspicuous, lobed...........................**Verbenaceae**

7. Inflorescences mixed, with ± indeterminate main axis and obviously determinate (cymosely branched) lateral units, these sometimes reduced and inflorescence then appearing pseudo-verticillate; ovules attached to false partitions very near the inrolled carpel margins; style terminal to gynobasic; stigma usually inconspicuous, at tips of stylar branches..........**Lamiaceae**

8. Plants insectivorous, with sticky, mucilaginous hairs, or bladders; placentation free-central, anthers 2...**Lentibulariaceae**

8. Plants not insectivorous; placentation usually axile or parietal, anthers usually 4..........................9

9. Ovary superior to inferior; anthers connate or sticking together in pairs; placentation parietal...**Gesneriaceae**

9. Ovary consistently superior; anthers usually distinct; placentation usually axile.........................10

10. Seeds with well-developed endosperm...11

10. Seeds with scanty or no endosperm...13

11. Plants hemiparasitic (and green) to holoparasitic (± lacking chlorophyll), forming haustorial connections to the roots of various host species; placentation axile or parietal...**Orobanchaceae**

11. Plants autotrophic (and green), lacking haustoria; placentation axile......................12

12. Anther locules confluent and anther opening by a single distal slit, the anther not sagittate; stamens 5, 4, or 2; flowers clearly to only slightly bilateral....................**Scrophulariaceae**

12. Anther locules distinct, opening by 2 distinct longitudinal slits, or apical portion of anther sacs adnate and opening by a single U- or V-shaped slit, the anther ± sagittate; stamens 4 or 2; flowers ± clearly bilateral...**Plantaginaceae**
(in part, numerous genera)

13. Fruits usually explosively dehiscent, seeds associated with an enlarged and specialized funiculus (retinaculum); often with cystoliths......................................**Acanthaceae**

13. Fruits dehiscent, but not explosively so, or indehiscent; lacking retinacula; cystoliths lacking..........14

14. Ovules ± numerous in each locule, borne on bilobed placentas; fruit usually an elongate capsule, never a drupe; leaves usually opposite, and compound, blades lacking pellucid dots; seeds usually flattened, winged...........................**Bignoniaceae**

14. Ovules usually 1 per locule, borne on unmodified placentas; fruit a drupe; leaves usually alternate and spiral, simple, blades usually with pellucid dots; seeds not flattened, unwinged...Myoporaceae

and two short), although sometimes these have been reduced to two. This clade is also characterized by endosperm with haustoria at both micropylar and chalazal ends, and lamellar protein inclusions in the nuclei.

Within the two-lipped corolla clade, "Scrophulariaceae," as traditionally delimited, are polyphyletic, unit-

ed only by symplesiomorphies (at this level of universality) such as flowers with two-lipped corollas, four stamens, and more or less globose capsules with small, endospermous seeds. Some members of this group are likely to be related more closely to members of various other families of Lamiales than they are to other scrophs

(Figure 9.129). The family here is divided into three major groups, each of which is apparently monophyletic: Scrophulariaceae s.s., Orobanchaceae, and Plantaginaceae. Verbenaceae and Lamiaceae are often considered sister taxa, and both have ovaries with four ovules, divided into four locules because of the development of a false septum, and aromatic, ethereal oils. Analyses of *rbcL* and *ndhF* sequences, however, support the hypothesis that these characters have evolved more than once within Lamiales (Olmstead and Reeves 1995; Wagstaff and Olmstead 1996; Oxelman et al. 1999). Analyses based on *rbcL, ndhF,* and *rps*2 nucleotide sequences (Olmstead et al, 2000b) suggest that Orobanchaceae are most closely related to *Paulownia* and Lamiaceae; Scrophulariaceae to Buddlejaceae and Myoporaceae, and Verbenaceae to Bignoniaceae. Plantaginaceae, along with Gesneriaceae and Calceolariaceae, may be early divergent lineages within the two-lipped corolla clade (Figure 9.129). Current family limits are uncertain and the families presented in this text are distinguished on very slight differences. Some systematists have even suggested that all the families with bilaterally symmetrical flowers (i.e., with a 2 + 3 corolla pattern and usually 4 stamens) be combined, forming a single large family.

Lamiales surely belong to the core asterid clade, as indicated by DNA sequences, sympetalous corollas, ovules with a single integument, and thin-walled megasporangium. The order is probably most closely related to the Solanales and Gentianales (Downie and Palmer 1992; Hufford 1992; Olmstead et al. 1992a,b, 1993).

The order consists of about 24 families and 17,800 species; major families include Avicenniaceae, **Oleaceae**, Buddlejaceae, **Bignoniaceae**, Myoporaceae, **Scrophulariaceae** s.s., **Orobanchaceae, Plantaginaceae, Gesneriaceae, Lentibulariaceae, Acanthaceae, Verbenaceae,** and **Lamiaceae** (Cantino 1982, 1990, 1992a,b; Cantino et al. 1992; Olmstead and Reeves 1995; Thorne 1992; Wagstaff and Olmstead 1996).

References: Albach et al. 2001; Armstrong 1985; Boeshore 1920; Cantino 1982, 1990, 1992a,b; Cantino et al. 1992; Cantino and Sanders 1986; Chase et al. 1993; Dahlgren 1983; Downie and Palmer 1992; Hufford 1992; Judd et al. 1994; Källersjö et al, 1998; Lu 1990; Olmstead et al. 1992a,b, 1993, 2000, 2001; Olmstead and Reeves 1995; Oxelman et al, 1999; Philbrick and Jansen 1991; Scotland et al. 1995; Soltis et al. 2000; Thorne 1992; Wagstaff and Olmstead 1996; Wagenitz 1992; Yamazaki 1974.

Oleaceae Hoffmannsegg and Link
(Olive Family)

Trees, shrubs, or lianas; sclereids often present; usually with phenolic glycosides; **buds 2 to several and superposed;** iridoids present. Hairs peltate scales. *Leaves usually opposite,* simple, pinnately compound, or trifoliolate, entire to serrate, with pinnate venation; *stipules lacking.*

Inflorescences determinate, sometimes reduced to a single flower, terminal or axillary. *Flowers bisexual or sometimes unisexual* (plants then polygamous to ± dioecious), *radial. Sepals usually 4, connate,* rarely lacking (in some *Fraxinus*). *Petals usually 4,* but sometimes more numerous, *connate,* imbricate or induplicate-valvate (i.e., folded inward with central portion of lobes meeting), rarely reduced or lacking (in some *Fraxinus*). **Stamens 2;** *filaments adnate to corolla;* pollen grains tricolpate or tricolporate. *Carpels 2, connate; ovary superior, with axile placentation;* stigma 2-lobed or capitate. *Ovules usually 2 in each locule,* with 1 integument and a thin-walled megasporangium. Nectar disk often present. *Fruit a loculicidal or circumscissile capsule, samara, berry, or drupe, often 1-seeded;* endosperm present to absent (Figure 9.130).

Floral formula: *, ④, ④, 2, ②; capsule, samara, drupe, berry

Distribution: Widely distributed in tropical to temperate regions.

Genera/species: 29 / 600. **Major genera:** *Jasminum* (230 spp.), *Chionanthus* (90), *Fraxinus* (60), *Ligustrum* (35), *Noronhia* (35), *Syringa* (30), *Menodora* (25), *Olea* (20), *Forestiera* (15), and *Osmanthus* (15). All of the above except *Olea* and *Noronhia* are represented in the continental United States and/or Canada.

Economic plants and products: *Olea europaea* (olive) produces edible fruits and is a source of cooking oil. *Fraxinus* (ash) contains commercially important timber species. *Forsythia* (golden bells), *Jasminum* (jasmine), *Ligustrum* (privet), *Chionanthus* (fringe tree), *Osmanthus* (fragrant olive, wild olive), *Syringa* (lilac), and *Noronhia* (Madagascar olive) provide useful ornamentals.

Discussion: Oleaceae are considered monophyletic on the basis of *rps16* and *trnL-F* sequences and several morphological synapomorphies (Wallander and Albert 2000). Because this family is phenetically quite divergent from most families of Lamiales, it sometimes has been placed in its own order (Oleales).

The family usually is divided into "Jasminoideae," characterized by ovules one to several, usually erect in each locule, and four to twelve petals, and Oleoideae, characterized by two pendulous ovules per locule and four petals. "Jasminoideae" are morphologically heterogeneous and paraphyletic (Kim and Jansen 1998b; Wallander and Albert 2000). Oleoideae are hypothesized to be monophyletic on the basis of the features cited above, the haploid chromosome number of 23, their flavone glycosides, and wood anatomy. Most Oleoideae have drupaceous fruits (e.g., *Chionanthus, Forestiera, Ligustrum, Noronhia, Olea,* and *Osmanthus*), but *Syringa* has loculicidal capsules, and *Fraxinus* has samaras. *Fraxinus* is also distinct due to its pinnately compound leaves (vs. simple

Figure 9.130 Oleaceae. (A–K) *Chionanthus virginicus:* (A) flowering branch (× 0.5); (B) staminate flower (× 3); (C) stamen from staminate flower (× 20); (D) nonfunctional gynoecium from staminate flower (× 3); (E) stamen from bisexual flower (× 20); (F) gynoecium from bisexual flower (× 15); (G) ovary in cross-section, with two ovules in each locule (× 30); (H) branch with dru- pes (× 0.5); (I) pit (× 2.5); (J) pit and seed in longitudinal section, endocarp wall hatched, endosperm stippled, embryo unshaded (× 4); (K) embryo (× 7.5). (L) *C. pygmaeus:* staminate flower, note two stamens (× 6). (From Wood 1974, *A student's atlas of flowering plants,* p. 87.)

in other Oleoideae) and reduced wind-pollinated flowers. Fruit type is variable within "Jasminoideae"; *Jasminum* has bilobed berries, *Fontanesia* has samaras, while *Forsythia* and *Menodora* have capsules.

The showy, variously colored, and often fragrant flowers of Oleaceae are pollinated by nectar-gathering bees, butterflies, and flies. The flowers of *Fraxinus* and *Forestiera* have a reduced perianth and are wind-pollinated. Both selfing and outcrossing occur. Fleshy-fruited taxa are dispersed by birds or mammals. The samaras of *Fraxinus* are wind-dispersed, as are the seeds of capsular species.

References: Bigazzi 1989; Johnson 1957; Kim and Jansen 1998b; Wallander and Albert 2000; Wilson and Wood 1959.

Gesneriaceae Richard and A. L. de Jussieu
(Gesneriad Family)

Herbs to shrubs, *often epiphytic;* often with phenolic glycosides; lacking iridoids. Hairs simple, often gland-headed. *Leaves usually opposite, usually simple,* entire to variously toothed, often softly hairy, with pinnate venation; *stipules usually lacking.* Inflorescences usually determinate, sometimes reduced to a single flower; axillary. Flowers bisexual, bilateral. Sepals 5, distinct to connate. *Petals 5, connate, the corolla usually 2-lipped,* the lobes imbricate. *Stamens 4, ± didynamous; filaments adnate to corolla;* **anthers sticking together in pairs or all together;** pollen grains usually tricolporate. *Carpels 2, connate;*

ovary superior to inferior, **with parietal placentation**; stigma 2-lobed. Ovules numerous on each placenta, with 1 integument and a thin-walled megasporangium. Nectar disk or glands usually present. *Fruit a loculicidal or septicidal capsule, sometimes fleshy, or a berry*; seeds small; endosperm present or absent.

Floral formula: X, ⑤, ②+③, ②+②, ②̄; capsule, berry

Distribution and ecology: Widely distributed in tropical regions.

Genera/species: 126/2850. **Major genera:** *Cyrtandra* (350 spp.), *Columnea* (250), *Besleria* (150), *Streptocarpus* (130), *Aeschynanthus* (100), *Chirita* (80), *Kohleria* (50), *Gesneria* (50), and *Sinningia* (40). No species occur naturally in the continental United States or Canada.

Economic plants and products: The family is notable chiefly for its numerous ornamental herbs, including *Episcia*, *Columnea*, *Sinningia* (gloxinia), *Streptocarpus* (Cape primrose, African violet), and *Aeschynanthus*.

Discussion: Gesneriaceae are considered monophyletic based on both morphology and *ndhF* sequences (Smith 1996; Smith et al. 1997). Most genera belong to one of two large subfamilies. Gesnerioideae are monophyletic, as evidenced by their half-inferior to completely inferior ovaries and large pollen grains. The monophyly of Cyrtandroideae, which has superior ovaries, is supported by an unusual embryonic development pattern, in which one cotyledon becomes larger than the other after germination. Chromosome number, secondary compounds, and *ndhF* sequences (Smith et al. 1997) also lend support to these two groups. Within Gesnerioideae, genera such as *Asteranthera*, *Besleria*, *Napeanthus*, and *Sinningia* are basal, while members of the Gloxinieae (*Kohleria*, *Gloxinia*, and relatives), Gesnerieae (*Gesneria*, *Sanango*, *Rhytidophyllum*) and Episcieae (e.g., *Codonanthe*, *Episcia*, *Alloplectus*, and *Columnea*) form a clade (Smith and Carroll 1997).

The showy flowers of Gesneriaceae show a diversity of pollination syndromes; major pollinators include bees, moths, butterflies, flies, bats, and birds. Nectar is used as a reward. Brightly colored calyces and even pigmented leaves sometimes assist in attracting pollinators. Species with fleshy fruits are usually bird-dispersed; the small seeds of capsular taxa are shaken from the capsule and may be dispersed by wind or water (rainwash).

References: Burtt 1977; Harrison et al. 1999; Möller and Cronk 1997; Smith 1996; Smith and Carroll 1997; Skog 1976; Smith et al. 1997; Wiehler 1983.

Plantaginaceae A. L. de Jussieu
(Snapdragon Family)

Herbs or less commonly shrubs, sometimes aquatic; autotrophic, lacking haustoria; often with phenolic gly-

cosides and triterpenoid saponins, and sometimes with cardiac glycosides. Hairs various, but usually simple, when glandular, the stalk elongate, *usually composed of 2 or more cells, and head ± globular to ellipsoidal*, **lacking vertical partitions**. *Leaves alternate and spiral, or opposite*, occasionally whorled, *simple*, entire to variously toothed, with pinnate venation, but ± parallel in *Plantago*; *stipules lacking*. Inflorescences various. *Flowers usually bisexual and bilateral, but ± radial in Plantago; reduced in Callitriche*. Sepals usually 4 or 5, connate. *Petals usually 5, or occasionally apparently 4 (due to fusion of 2 upper lobes), connate, the corolla 2-lipped*, sometimes with a basal nectar spur, the lower lip sometimes with a bulge obscuring the throat (personate), the lobes imbricate or valvate. *Stamens usually 4, didynamous*, sometimes reduced to 2, the fifth stamen sometimes present as a staminode (e.g., *Penstemon*); *filaments adnate to corolla; anthers 2-locular, locules distinct, opening by 2 longitudinal slits*, or apical portion of anther sacs sometimes adnate and opening by a single inverted, U- or V-shaped slit (i.e., sagittate), the pollen sacs divergent (anther sagittate); pollen grains often tricolporate. *Carpels 2, connate; ovary superior, with axile placentation, the placentas large, not divided*; stigma usually 2-lobed. *Ovules usually numerous in each locule*, with 1 integument and a thin-walled megasporangium. Nectar disk usually present, but lacking in *Plantago* and *Callitriche*. *Fruit usually a septicidal capsule, occasionally poricidal or circumscissile*; seeds angular or winged (Figure 9.131).

Floral formula:

X or *, ④–⑤, ②+③ or ④, 2+2 or 2, ②̲; capsule

Distribution: Nearly cosmopolitan, but most diverse in temperate areas.

Genera/species: 113/1800. **Major genera:** *Veronica* (400 spp.), *Penstemon* (250), *Plantago* (215), *Mimulus* (150), *Linaria* (120), *Lindernia* (100), *Nemesia* (50), *Antirrhinum* (40), *Limnophila* (35), *Gratiola* (20), *Scoparia* (20), and *Digitalis* (20). Noteworthy genera in the continental United States and/or Canada (in addition to most of those listed above) are *Bacopa*, *Callitriche*, *Chelone*, *Collinsia*, *Hippuris*, *Maurandya*, *Micranthemum*, *Scoparia*, and *Veronicastrum*.

Economic plants and products: The family is well known for its ornamentals, such as *Antirrhinum* (snapdragon), *Digitalis* (foxglove), *Mimulus* (monkey flower), *Penstemon* (beardtongue), *Russelia* (firecracker plant), and *Veronica* (speedwell). Species of *Digitalis* are used medicinally.

Discussion: Circumscription of this group has long been ambiguous, and various genera have been considered phenetically transitional to other families of Lamiales. Two major problems relate to the placement of *Scrophularia* and *Verbascum*, and to the disposition of parasitic genera such as *Pedicularis*, *Agalinis*, and *Castilleja*.

Figure 9.131 Plantaginaceae. *Penstemon canescens:* (A) flowering stem (× 1); (B) apex of flower bud (× 5); (C) flower (× 5); (D) flower in longitudinal section, note hairy staminode and stamens in two pairs (didynamous) (× 5); (E) anther (× 15); (F) ovary in longitudinal section (× 18); (G) ovary in cross-section, note numerous ovules (× 18); (H) capsule (× 4.5); (I) seed (× 36); (J) seed in longitudinal section, endosperm stippled, embryo unshaded (× 30). (From Wood 1974, *A student's atlas of flowering plants,* p. 99.)

Most of these genera have, until recently, been included in Scrophulariaceae, but *rbcL, ndhF,* and *rps2* sequences (Olmstead et al. 1992a, 1993, 2000, 2001; Olmstead and Reeves 1995) strongly support the transfer of *Scrophularia* and *Verbascum* to the *Selago* clade. Morphological characters also support this transfer, that is, *Scrophularia* and *Verbascum* share with *Selago* and relatives the apomorphies of confluent anther locules, anthers opening by a single distal slit, and club-shaped (nonsagittate) anthers (Weberling 1989). Also, they all have glandular hairs with a flattened, discoidal head of two to numerous cells that are separated by vertical partitions (a likely plesiomorphy). The rules of nomenclature (see Appendix 1) require that the *Selago*

clade be called Scrophulariaceae because this group contains *Scrophularia,* the type of the oldest available name. Members of the family formerly called Scrophulariaceae are here called Plantaginaceae, although the name Veronicaceae Durande has been used by Olmstead and associates (2001); see also Reveal et al. (1999).

Many parasitic herbs traditionally have been included in the Scrophulariaceae as traditionally defined. These plants maintain a physical connection to their host species by means of specially modified roots called haustoria. They range from hemiparasites, which are green and have chlorophyll, to holoparasites, which are white and lack chlorophyll. The parasitic genera are here

(

included in Orobanchaceae. Both morphology (i.e., the presence of root haustoria) and cpDNA sequences support this transfer (de Pamphilis et al. 1997).

Plantaginaceae, therefore, are circumscribed to exclude *Scrophularia*, *Verbascum*, and relatives, the parasitic genera, and the isolated genera *Paulownia* and *Schlegelia*. They include Callitrichaceae and Hippuridaceae, two families of specialized aquatic herbs with reduced flowers, and also a group of wind-pollinated herbs (Plantaginaceae s.s.; see the key, Cronquist 1981, and Rosatti 1984 for diagnostic features of *Plantago* and *Callitriche*). Recently published data suggests that a few additional autotrophic genera (e.g., *Calceolaria*, which has an extremely saccate corolla, *Leucophyllum*, which has secretory cavities in its leaves) should also be excluded (Olmstead et al. 2001). The placement of the large genus *Mimulus* is uncertain, but it is retained here pending additional study. The monophyly of Plantaginaceae is supported by cpDNA characters (Olmstead et al. 1993; Olmstead and Reeves 1995; Wagstaff and Olmstead 1996). The link of *Plantago* to insect-pollinated members of the family is especially evident chemically, particularly in the presence of aucubin (an iridoid glycoside), mannitol (a sugar alcohol), and linoleic and oleic acids (as major fatty acids). The four-lobed corolla of *Plantago* probably has been derived through the complete fusion of the upper two corolla lobes of an ancestral five-lobed corolla, as is also seen in some species of *Veronica*. Morphological synapomorphies of Plantaginaceae are not readily apparent, although the early development of the androecium relative to the corolla may support the group's monophyly, as may hair morphology.

Flowers of Plantaginaceae come in a great variety of shapes, colors, and sizes, and are pollinated mainly by nectar-gathering bees, flies, and birds. Nectar guides are frequent. The reduced flowers of *Callitriche* are water pollinated, while those of *Plantago* are wind-pollinated.

References: Armstrong 1985; Boeshore 1920; Cronquist 1981; de Pamphilis et al. 1997; Kamphy 1995; Miller 2001; Olmstead et al. 1992a,b, 1993, 2000, 2001; Olmstead and Reeves 1995; Pennell 1920, 1935; Philbrick and Jansen 1991; Rahn 1996; Reeves and Olmstead 1998; Rosatti 1984; Soekarjo 1992; Thieret 1967, 1971; Thorne 1992; Weberling 1989.

Scrophulariaceae A. L. de Jussieu
(Figwort Family)

Herbs or small shrubs; autotrophic; with iridoids. *Hairs usually simple, when glandular, with a short to elongate stalk cell and a flattened, discoidal head composed of 2 or 4 to numerous cells that are separated by vertical partitions*, sometimes stellate. Leaves alternate and spiral, or opposite, simple, entire to toothed, with pinnate venation; stipules lacking. Inflorescences indeterminate, terminal. *Flowers bisexual, bilateral to nearly radial.* Sepals usually 3–5, connate. *Petals 4 or 5, connate, the corolla often ± 2-lipped, or with narrow tube and broadly flaring, imbricate lobes. Stamens 5, 4, or 2*, the fifth stamen sometimes represented by a staminode; *filaments adnate to corolla; anthers 2-locular*, **but the anther sacs confluent and opening by a single distal slit oriented at right angles to the filament, anther base not sagittate;** pollen grains usually tricolporate. *Carpels 2, connate; ovary superior*, with axile placentation, the placentas undivided; stigma 2-lobed. Ovules numerous to 1 in each locule, with 1 integument and a **thick-walled megasporangium**. Nectar disk usually present. *Fruit a septicidal capsule or a schizocarp of 2 achenes.*

Floral formula: X, ④–5, ②+③, 5 or 2+2 or 2, ②;
capsule, schizocarp

Distribution: Widely distributed from temperate to tropical regions, but especially diverse in Africa.

Genera/species: 24/1200. **Major genera:** *Verbascum* (360 spp.), *Scrophularia* (250), *Selago* (160), *Sutera* (140), and *Manulea* (60). Only *Verbascum* and *Scrophularia* occur in the continental United States and Canada.

Economic plants and products: The family is of little economic importance; *Verbascum* is occasionally grown as an ornamental.

Discussion: The monophyly of Scrophulariaceae is clearly supported by morphology, *rbcL*, *ndhF*, and *rpo2* sequences (Olmstead et al. 1992a,b, 1993, 2001; Olmstead and Reeves 1995). *Verbascum* and *Scrophularia* are closely related, forming a clade on the basis of their distinctive seeds, hairy filaments, endosperm development, and cpDNA sequences (Olmstead and Reeves 1995; Thieret 1967; Olmstead et al. 2001). These genera probably form the sister clade to the remaining genera. *Selago* and relatives (Selagineae) form a clade based on their uniovulate locules and achenelike fruits. *Sutera* and relatives (Manuleeae) are phenetically intermediate between the *Verbascum* + *Scrophularia* clade and the Selagineae.

Members of Scrophulariaceae, as circumscribed here, probably are closely related to *Buddleja* and relatives (Buddlejaceae). *Buddleja* has radially symmetrical flowers with 4 corolla lobes and 4 stamens. The *Buddleja* + Scrophulariaceae clade is, in turn, likely sister to Myoporaceae (incl. *Leucophyllum*). Myoporaceae are characterized by leaf blades with pellucid dots and usually have only 1 ovule per locule and drupaceous fruits. *Leucophyllum* (likely sister to remaining members of this family) has densely hairy leaves with only 2 apical pellucid dots and has multiovulate locules and capsular fruits. Olmstead and associates (2001, but not APG 1998) treat Myoporaceae and Buddlejaceae within a re-circumscribed Scrophulariaceae. Pending additional study, Scrophulariaceae s.s., Buddlejaceae, and Myoporaceae are retained as three very closely related families (see key). *Phryma* (Phrymaceae) may also be related to these families.

Flowers of Scrophulariaceae are pollinated by a variety of nectar-gathering insects. The seeds or achenes are dispersed by wind.

References: Cronquist 1981; Hilliard 1994; Lersten and Beaman 1998; Olmstead and Reeves 1995; Olmstead et al. 1992a,b, 1993, 2001.

Orobanchaceae Vent.
(Broomrape Family)

Herbs; **hemiparasitic (**i.e., partial parasites, with chlorophyll**) to holoparasitic (**i.e., complete parasites, lacking chlorophyll**), with a single large or many small haustorial connections to the roots of host plants**; often with phenolic glycosides; usually with iridoids and **orobanchin, which causes the leaves (or the entire plant) to turn black in drying**. Hairs various, but usually simple, when glandular the stalks ± elongate, *usually composed of 2 or more cells, and the head ± globular to ellipsoidal,* **usually lacking vertical partitions.** *Leaves alternate and spiral, or opposite, simple, often pinnately lobed to dissected, sometimes reduced to scales,* entire to variously toothed, with pinnate venation; stipules lacking. **Inflorescences usually indeterminate**, sometimes reduced to a solitary flower, terminal or axillary. *Flowers bisexual, bilateral.* Sepals usually 5, connate. *Petals usually 5, connate, the corolla 2-lipped,* the lobes imbricate. *Stamens 4, didynamous,* the fifth stamen occasionally present as a staminode; *filaments adnate to corolla; anthers* 2-locular, *opening by 2 longitudinal slits,* 1 locule sometimes reduced or modified, pollen sacs divergent and anther sagittate; pollen grains often tricolporate. *Carpels 2, connate; ovary superior, with axile to parietal placentation, the placentas not divided (when placentation axile) or often divided (when placentation parietal);* stigma 2-lobed. Ovules usually numerous in each locule or on each placenta, with 1 integument and a thin-walled megasporangium. Nectar disk usually present around base of ovary. *Fruit a septicidal to loculicidal capsule;* seeds angular (Figure 9.132).

Floral formula: X,⑤,②+③, 2+2, ②; capsule

Distribution: Nearly cosmopolitan.

Genera/species: 61/1700. **Major genera:** *Pedicularis* (360 spp.), *Castilleja* (200), *Euphrasia* (200), *Orobanche* (100), *Buchnera* (100), *Agalinis* (60), *Striga* (50), and *Harveya* (40). Many genera occur in the continental United States and/or Canada; noteworthy genera (in addition to most of the above) include *Aureolaria, Conopholis, Epifagus, Melampyrum, Orthocarpus, Rhinanthus,* and *Seymeria.*

Economic plants and products: The family is of little economic importance, but a few taxa, such as *Striga* (witchweed), can severely damage crops.

Discussion: Orobanchaceae, as here circumscribed, are considered monophyletic on the basis of the hemi- to holoparasitic habit, hair morphology, and possibly racemose inflorescence. DNA-based cladistic analyses (de Pamphilis and Young 1995; de Pamphilis et al. 1997; Nickrent et al.1997; Young et al. 1999; Olmstead et al. 2001) also support the group's monophyly. The hemiparasitic genera, such as *Aureolaria, Agalinis, Pedicularis, Castilleja,* and *Buchnera,* traditionally have been placed within a broadly defined Scrophulariaceae (including both Plantaginaceae and Scrophulariaceae, as here delimited), based on numerous symplesiomorphic features. Such taxa have been placed within the Scrophulariaceae subfam. Rhinanthoideae, along with autotrophic genera such as *Veronica.* In most classifications (Cronquist 1981), the holoparasitic taxa (e.g., *Orobanche, Epifagus, Conopholis,* and *Boschniakia*) are placed within a narrowly circumscribed family, and parietal placentas are stressed as a defining character. However, the traditional distinction between Scrophulariaceae and Orobanchaceae is difficult to accept in light of genera, such as *Harveya,* that are holoparasitic but have axile placentation. Orobanchaceae encompass a continuum from chlorophyllous plants with numerous root haustoria to complete parasites that lack chlorophyll and have a single large haustorium. *Orobanche, Epifagus,* and *Conopholis,* for example, merely represent the end point of a continuum representing increasing specialization for a parasitic existence (Boeshore 1920). DNA-based analyses suggest that although haustoria have evolved only once, the holoparasitic habit has evolved several times within Orobanchaceae (de Pamphilis et al. 1997; Young et al. 1999).

The autotrophic genus *Lindenbergia* (usually treated as Scrophulariaceae) may be the sister group of these hemi- to holoparasitic genera. Morphological apomorphies linking it to them may include the abaxial lobes of the corolla outside the adaxial, details of hair morphology, and the indeterminate inflorescence. The genus perhaps should be included within Orobanchaceae.

Flowers of Orobanchaceae show a great diversity of shapes and colors and are pollinated by bees, wasps, flies, and birds. Bracts sometimes assist in pollinator attraction, as in *Castilleja.* Seeds are probably wind dispersed.

References: Boeshore 1920; de Pamphilis and Young 1995; de Pamphilis et al. 1997; Cronquist 1981; Thieret 1971; Olmstead et al. 2001; Young et al. 1999.

Bignoniaceae A. L. de Jussieu
(Bignonia or Trumpet Creeper Family)

Trees, shrubs, or lianas; the lianas with a characteristic kind of anomalous secondary growth that results in a 4- or many-lobed or furrowed xylem cylinder; usually with iridoids and phenolic glycosides. Hairs various, often

Figure 9.132 Orobanchaceae. *Conopholis americana:* (A) fruiting plant detached from root of Quercus rubra, which was to right, note scaly buds on "tuber (× 0.75);" (B) flower bud (× 6); (C, D) flowers (× 6); (E, F) stamens (× 12); (G) ovary in cross-section (× 9); (H) fruit in cross-section (× 3); (I, J) two seeds (× 18). (From Thieret 1971, *J. Arnold Arbor.* 52: p. 421.)

simple. *Leaves opposite or whorled,* occasionally alternate and spiral, **pinnately or palmately compound,** occasionally simple, entire to serrate, venation pinnate to palmate, the terminal and occasionally also some lateral leaflets sometimes modified into tendrils or hooks; *stipules lacking.* Inflorescences various. *Flowers bisexual, bilateral, usually large and showy.* Sepals 5, connate. *Petals 5, connate, the corolla ± 2-lipped,* the lobes usually imbricate. *Stamens usually 4, ± didynamous,* the fifth stamen sometimes represented by a small staminode, sometimes reduced to 2; *filaments adnate to corolla; anthers sagittate;* pollen grains diverse, sometimes in tetrads or polyads. *Carpels 2, connate; ovary superior, with usually axile placentation,* **the placentas divided (into 2 per locule);** *stigma strongly 2-lobed, each lobe sensitive* (i.e., closing after contact with pollinator). Ovules numerous, with 1 integument and a thin-walled megasporangium. Nectar disk

usually present. **Fruit a usually elongate,** *septicidal to loculicidal capsule,* occasionally a berry or indehiscent pod; **seeds usually flat, winged or fringed with hairs; endosperm lacking;** cotyledons deeply bilobed (Figure 9.133).

Floral formula: X, ⑤, (2+3), 2+2, ②; capsule

Distribution: Widely distributed in tropical and subtropical regions, with a few species in temperate climates; most diverse in northern South America.

Genera/species: 113/800. **Major genera:** *Tabebuia* (100 spp.), *Arrabidaea* (70), *Adenocalymma* (50), and *Jacaranda* (40). Only a few, including *Bignonia, Catalpa, Campsis, Macfadyena, Pithecoctenium,* and *Stenolobium,* occur in the continental United States.

Figure 9.133 Bignoniaceae. (A–H) *Catalpa bignonioides:* (A) branch of cymose inflorescence (× 0.5); (B) flower (× 2); (C) flower, most of corolla removed, exposing gynoecium, two fertile stamens, three staminodes (× 4); (D, E) staminodes (× 4); (F) capsule (× 0.5); (G) open capsule (× 0.5); (H) seed (× 3). (I–M) *C. speciosa:* (I) inflorescence branch (× 0.7); (J) staminodes (× 4); (K) ovary in cross-section, note divided placentas (× 12); (L) seed (× 12); (M) seedling. (From Wood 1974, *A student's atlas of flowering plants,* p. 102.)

Economic plants and products: Some species of *Catalpa* and *Tabebuia* are used for timber. *Spathodea* (African tulip tree), *Campsis* (trumpet creeper), *Bignonia* (cross vine), *Chilopsis* (flowering willow), *Clytostoma* (painted trumpet), *Crescentia* (calabash tree), *Jacaranda*, *Kigelia* (sausage tree), *Macfadyena* (cat-claw vine), *Pyrostegia* (flame vine), *Tabebuia*, *Stenolobium* (yellow bells), and *Tecomaria* (Cape honeysuckle) provide ornamentals.

Discussion: Bignoniaceae are easily recognized and probably monophyletic, as evidenced by *rbcL* and *ndhF* sequence data (Spangler and Olmstead 1999) and the morphological synapomorphies indicated in the family description. Evolutionary patterns within the family have been investigated by Gentry (1974, 1980, 1990), who recognized several tribes based on variation in habit, leaves, placentation, fruits, and seeds. Pinnately compound leaves are considered to be ancestral, but *Crescentia*, *Catalpa*, and *Tabebuia* contain species with reduced, trifoliolate, or even unifoliolate leaves. The lianous habit (with leaves modified for climbing) and anomalous growth are considered to be derived within Bignoniaceae and characterize a group of Neotropical genera (Bignonieae). *Crescentia*, *Parmentiera*, and *Kigelia* are distinct due to their indehiscent, often large, fruits.

Paulownia and *Schegelia* have been placed in Bignoniaceae or considered intermediate between this family and the Scrophulariaceae, as traditionally, but not here, defined. These genera lack the distinctive synapomorphies of Bignoniaceae (Armstrong 1985) and are here treated as members of small families (Spangler and Olmstead 1999). Their similarity to Plantaginaceae and Scrophulariaceae s.s. is probably symplesiomorphic.

Flowers of Bignoniaceae are variable in form, coloration, and time of anthesis. Pollinators are bees, wasps, butterflies, hawk moths, birds, and bats. Outcrossing is promoted by the sensitive stigma. The seeds of most species are dispersed by wind.

References: Armstrong 1985; Dobbins 1971; Gentry 1974, 1980, 1990; Manning 2000; Spangler and Olmstead 1999.

Acanthaceae A. L. de Jussieu
(Acanthus Family)

Herbs or occasionally vines or shrubs; often with anomalous secondary growth; usually with phenolic glycosides; often with iridoids, alkaloids, and diterpenoids; *cystoliths of various forms often present*. Hairs usually simple, but sometimes branched, dendritic, or stellate. *Leaves usually opposite*, but alternate and spiral in Nelsonioideae, *simple*, sometimes lobed, entire to serrate or dentate, with pinnate venation; *stipules lacking*. Inflorescences various. *Flowers bisexual, bilateral, often associated with large, often colorful bracts and bractlets.* Sepals usually 4 or 5, connate. *Petals 5, connate, the corolla ± 2-lipped*, the

lobes imbricate or convolute. *Stamens usually 4, didynamous, or sometimes only 2; filaments adnate to corolla;* anthers sometimes unilocular, often hairy, *often asymmetrical*, the pollen sacs sometimes widely separated on a modified connective; pollen grains various. *Carpels 2, connate; ovary superior, with usually axile placentation;* stigma funnel-shaped or 2-lobed, but 1 lobe sometimes reduced or lacking. *Ovules usually 2–10 in 2 rows in each locule*, rarely numerous (Nelsonioideae), anatropous to campylotropous, with 1 integument and a thin-walled megasporangium, *each ovule usually borne on a hook-shaped projection (retinaculum: i.e., a modified funiculus).* Nectar disk present. *Fruit a loculicidal capsule, almost always explosively dehiscent; seeds usually flat*, the coat sometimes mucilaginous; endosperm usually lacking (Figure 9.134).

Floral formula: X, ⑤, (2+3), 2+2 or 2, ②; capsule

Distribution: Widely distributed from tropical to warm temperate regions.

Genera/species: 256/2770. **Major genera:** *Justicia* (400 spp.), *Barleria* (250), *Strobilanthes* (250), *Ruellia* (200), *Thunbergia* (150), *Dicliptera* (150), and *Aphelandra* (150). Noteworthy genera in the continental United States include *Anisacanthus*, *Barleria*, *Carlowrightia*, *Dicliptera*, *Dyschoriste*, *Elytraria*, *Hygrophila*, *Justicia*, *Ruellia*, *Siphonoglossa*, and *Stenandrium*.

Economic plants and products: The family contains many ornamentals: *Aphelandra* (zebra plant), *Asystasia* (coromandel), *Barleria* (Philippine violet), *Justicia* (shrimp plant), *Eranthemum* (blue sage), *Fittonia* (mosaic plant), *Odontonema* (firespike), *Pachystachys* (yellow shrimp plant), *Sanchezia*, *Thunbergia* (clock vine, blue-sky vine), and *Acanthus* (bear's-breech).

Discussion: Most genera of Acanthaceae form a well-defined monophyletic group, but some aberrant genera (placed in Nelsonioideae and Thunbergioideae) have led some to doubt the group's naturalness (Bremekamp 1965). The family here is broadly circumscribed and is considered monophyletic on the basis of *ndhF, rbcL, trnL-F*, and ITS sequences (Hedrén et al. 1995; Scotland et al. 1995; McDade et al. 2000). Morphological synapomorphies for Acanthaceae are not known.

Nelsonioideae (e.g., *Nelsonia*, *Elytraria*) may represent a paraphyletic basal complex within the family; this group has retained many plesiomorphic features, such as alternate and spiral leaves, numerous ovules that lack retinacula, and seeds with endosperm. *Thunbergia* and relatives (Thunbergioideae) and the remaining genera (i.e., Acanthoideae) may form a clade supported by DNA sequence characters, loss of endosperm, reduction in ovule number and possibly opposite leaves. Most members of the family belong to the large and clearly

Figure 9.134 Acanthaceae. (A–G) *Justicia ovata* var. *angusta:* (A) flowering plant (× 0.7); (B) flower and flower buds (× 4); (C) anther (× 14); (D) nectar disk and ovary, with one side removed to show four ovules (× 17); (E) capsule, one seed removed, note retinacula (× 4); (F) moistened seed (× 8.5); (G) embryo (× 8.5). (H) *J. crassifolia:* opening capsule with two seeds, four retinacula (× 4). (I–L) *J. cooleyi:* (I) flower and bud (× 5.5); (J) corolla, opened out, with one stamen (× 5.5); (K) anther (× 14); (L) partly mature fruit and flower from which corolla has fallen (× 3). (M, N) *J. americana:* (M) corolla with stamens, from above (× 4); (N) anther (× 14). (From Long 1970, *J. Arnold Arbor.* 51: p. 304.)

monophyletic Acanthoideae (Scotland 1990; McDade et al. 2000). Synapomorphies of this clade include retinacula, explosive capsules, and a particular pattern of imbricate corolla lobes. Two monophyletic subgroups are represented within Acanthoideae: the larger, supported by the presence of cystoliths in the leaves, includes *Sanchezia, Hygrophila, Ruellia, Eranthemum, Dyschoriste, Strobilanthes, Barleria, Pachystachys, Dicliptera, Odontonema, Justicia,* and *Fittonia,* and the smaller, diagnosed by unilocular anthers, includes *Acanthus, Stenandrium,* and *Aphelandra* (Hedrén et al. 1995; Scotland et al. 1995; McDade and Moody 1999; McDade et al. 2000).

Interestingly, recent phylogenetic analyses based on molecular data support the inclusion of the mangrove family Avicenniaceae within Acanthaceae (Schwarzbach and McDade 2002), a placement also supported by articulated nodes, inflorescence structure, including flowers subtended by a bract and two bracteoles, a reduction in the number of ovules, and lack of endosperm.

Flowers of Acanthaceae, with their colorful corollas and often showy bracts, are pollinated by nectar-gathering bees, wasps, moths, butterflies, and birds. Protogyny leads to outcrossing, but some species of *Ruellia* produce cleistogamous flowers. The seeds are fairly large (compared to Plantaginaceae) and are flung some distance as the capsules dehisce explosively.

References: Bremekamp 1965; Hedrén et al. 1995; Long 1970; McDade and Moody 1999; McDade et al. 2000; Scotland 1990; Scotland et al. 1995.

Lentibulariaceae L. C. Richard
(Bladderwort Family)

Insectivorous herbs, aquatic or of wetlands, rooted in moist substrate to free-floating and lacking roots, sometimes epiphytic; vascular system often reduced; usually with iridoids. *Hairs sessile to stalked, gland-headed,* **some secreting mucilage and others digestive enzymes.** *Leaves alternate and spiral, or sometimes whorled, often in basal rosettes, simple, entire to finely divided,* **always highly modified,** *that is, flat and densely covered with sticky, mucilage-secreting and digestive hairs, and with margins inrolling in response to contact of glandular hairs with prey organism (Pinguicula), or tubular and spiraled, with-downward pointing and digestive hairs and a basal chamber (Genlisea), or not obviously foliaceous, highly dissected, bearing prey-catching bladders, each with 2 sensitive valves forming a trapdoor entrance, which opens inward in response to a stimulus conveyed by 4 sensory hairs and then immediately closes again, and lined on the inside with branched digestive hairs (Utricularia);* stipules lacking. **Inflorescences indeterminate,** on a scape, terminal, sometimes reduced to a single flower. Flowers bisexual, bilateral. Sepals 4 or 5, distinct to connate, often 2-lipped. *Petals 5, connate, the corolla 2-lipped, the lower lip usually with a nectar spur or sac,* and often with a bulge obscuring the throat, the lobes imbri-

cate. **Stamens 2**; *filaments adnate to corolla; anthers unilocular;* pollen grains tricolporate to multicolporate. *Carpels 2, connate; ovary superior, with free-central placentation;* **stigma unequally 2-lobed.** Ovules usually numerous, with 1 integument and a thin-walled megasporangium. **Nectar disk lacking,** *and nectar produced by corolla nectar spur. Fruit usually a circumscissile, loculicidal, or irregularly dehiscent capsule;* seeds small, **embryo ± undifferentiated;** endosperm lacking.

Floral formula: X, (4–5), (2+3), 2, ② ; capsule

Distribution and ecology: Widespread, occurring from tropical to boreal regions; insectivorous plants of wetlands, aquatic habitats, and moist forests. Some species of *Pinguicula* are epiphytes of moist tropical montane forests.

Genera/species: 3/180. **Major genera:** *Utricularia* (120 spp.) and *Pinguicula* (45). Both occur in the continental United States and Canada.

Economic plants and products: *Utricularia* (bladderwort) and *Pinguicula* (butterwort) are occasionally grown as novelty plants.

Discussion: Lentibulariaceae are clearly monophyletic, as evidenced by morphology and *rbcL* and *matK* sequences (Chase et al. 1993; Müller et al. 2000) although their exact phylogenetic placement within Lamiales is problematic. There are surprising differences in leaf form between *Pinguicula, Genlisea,* and *Utricularia.* However, the leaves of all three genera have glandular hairs that secrete both digestive enzymes and mucilage; a more or less unmodified leaf bearing gland-headed hairs may represent the ancestral condition. In *Utricularia* the hairs on the inside of the bladder also remove excess water from the lumen, which is then secreted by external glands. *Pinguicula* probably is the sister group of a clade containing *Genlisea* and *Utricularia* (Müller et al. 2000).

The showy (yellow, white, blue, purple) bilateral flowers are pollinated by nectar-gathering bees and wasps. In *Utricularia* the stigma is sometimes sensitive (closing when touched); outcrossing is characteristic.

References: Fineran 1985; Lloyd 1942; Müller et al. 2000; Taylor 1989; Wood and Godfrey 1957.

Verbenaceae Adanson
(Verbena Family)

Herbs, lianas, shrubs, or trees, sometimes with prickles or thorns; *stems usually square in cross-section;* usually with iridoids; often with phenolic glycosides. *Hairs simple, gland-headed, with ethereal oils (including terpenoids),* and *nonglandular, these, when present, usually unicellular, sometimes calcified or silicified. Leaves opposite* or occasionally whorled, simple, sometimes lobed, entire to serrate, with pinnate venation; *stipules lacking*. **Inflorescences indeterminate, the lateral units individual flowers or appearing as individual flowers, forming racemes, spikes, or heads**, terminal or axillary. Flowers bisexual, bilateral. *Sepals 5, connate*, the calyx tubular to bell-shaped, *persistent*, occasionally enlarged in fruit. *Petals 5 (but sometimes seemingly 4 by fusion of the upper pair), connate, the corolla weakly 2-lipped*, the lobes imbricate. *Stamens 4, didynamous; filaments adnate to corolla;* pollen grains usually tricolpate, **exine thickened near apertures**. *Carpels 2, connate; ovary superior, unlobed to ± 4-lobed, 2-locular but appearing 4-locular due to development of false septa,but sometimes 1 carpel suppressed and then* appearing only 2-locular, with axile placentation; style not apically divided, terminal; **stigma** usually 2-lobed, **conspicuous, with well-developed receptive tissue**. *Ovules 2 per carpel (i.e., usually 1 in each apparent locule), each marginally attached (attached directly to margins of false partitions), with 1 integument and a thin-walled megasporangium*. Nectar disk usually present. *Fruit a drupe with 2 or 4 pits (single and 2-lobed in Lantana), or a schizocarp splitting into 2 or 4 nutlets; endosperm lacking* (Figure 9.135).

Figure 9.135 Verbenaceae. (A–N) *Glandularia bipinnatifida:* (A) flowering plant (× 1.25); (B) corolla from bud (× 4); (C) flower with bract (× 4); (D) flower in longitudinal section, note epipetalous stamens (× 4); (E) stamen (× 24); (F) dehisced upper stamen, abaxial view, note gland (greatly enlarged); (G) tip of style and stigma, note expanded stigmatic tissue (greatly enlarged); (H) ovary with terminal style (× 20); (I) ovary in longitudinal section, note ovules (× 40); (J) infructescence (× 0.5); (K) nutlets within accrescent calyx (× 7); (L) nutlet, adaxial surface (× 15); (M) embryo (× 15); (N) calyx after fall of nutlets, note remains of nectar disk (× 2). (O) *Verbena stricta:* flowering plant, note indeterminate inflorescence (× 1). (From Wood 1974, *A student's atlas of flowering plants,* p. 95.)

Floral formula: X,⑤,(2+3), 2+2, ②; drupe, 4 nutlets

Distribution: Widely distributed from tropical to temperate regions.

Genera/species: 36/1035. ***Major genera:*** *Verbena* (200 spp.), *Lippia* (200), *Lantana* (150), *Citharexylum* (70), *Stachytarpheta* (70), *Glandularia* (60), and *Duranta* (30). All of the above occur in the continental United States and/or Canada; other noteworthy genera include *Aloysia*, *bouchea*, and *Priva*.

Economic plants and products: *Lippia* (lemon verbena) and *Priva* are used in herbal teas or yield essential oils. *Duranta* (golden dewdrop), *Lantana*, *Petraea* (queen's-wreath), *Stachytarpheta* (devil's-coachwhip), *Verbena*, and *Glandularia* provide ornamentals.

Discussion: Verbenaceae, as here circumscribed, are considered monophyletic on the basis of their morphology and *rbcL* sequences (Cantino 1992a,b; Chadwell et al. 1992; Judd et al. 1994; Olmstead et al. 1993; Wagstaff and Olmstead 1996). The family has been circumscribed much more broadly by most systematists (e.g., Cronquist 1981, 1988) and separated from Lamiaceae by the presence of a terminal (vs. gynobasic) style. Here we include only the traditional subfamily Verbenoideae (excluding the tribe Monochileae). As traditionally delimited, Verbenaceae are paraphyletic, while Lamiaceae are polyphyletic. In order to make Lamiaceae monophyletic, nearly two-thirds of the genera usually included within Verbenaceae (e.g., *Callicarpa*, *Clerodendrum*, *Vitex*, *Tectona*), are transferred to Lamiaceae (see Cantino 1992a,b; Cantino et al. 1992; Thorne 1992). As redefined, Verbenaceae can be distinguished from Lamiaceae by their indeterminate racemes, spikes, or heads (vs. inflorescences with an indeterminate main axis and cymosely branched lateral axes, these often reduced and forming pseudowhorls); ovules attached on the margins of false septa (vs. ovules attached on the sides of false septa); simple style with conspicuous, 2-lobed stigma (vs. usually apically forked style with inconspicuous stigmatic region at the tip of each style branch); pollen exine thickened near apertures (vs. not thickened); and nonglandular hairs exclusively unicellular (vs. multicellular, uniseriate). In addition, the flowers tend to be less strongly two-lipped. The style of Verbenaceae is exclusively terminal, while in Lamiaceae it varies from terminal to gynobasic.

The inclusion of *Petraea* within Verbenaceae is not supported by cpDNA data alone (Olmstead et al. 1993; Wagstaff and Olmstead 1996), but it is supported by a combination of morphology and cpDNA sequences (Cantino, pers. com.); its position should be considered doubtful.

The phenetically distinct *Phryma* (Phrymaceae), which has a pseudomonomerous gynoecium with a single basal ovule, may be related to Verbenaceae. It also has racemose inflorescences (Chadwell et al. 1992;

Thieret 1972). However, molecular evidence is suggestive of a relationship with Scrophulariaceae.

Avicennia is a specialized genus of about eight species characteristic of mangrove forests (Sanders 1997). It has often been included within a broadly defined Verbenaceae, but here is treated as distinct, Avicenniaceae (see key). The genus is likely is related to Acanthaceae (in which it probably should be placed).

The showy flowers of Verbenaceae are pollinated by nectar-gathering bees, wasps, and flies. Outcrossing is favored by protandry. The colorful drupaceous fruits are dispersed by birds. The nutlets are associated with a persistent calyx, and are released over time by the action of wind or mechanical contact. They may be secondarily dispersed by water (rainwash) or eaten by birds. A few are externally transported by animals.

References: Cronquist 1981, 1988; Thorne 1992; Cantino 1990, 1992a,b; Cantino et al. 1992; Chadwell et al. 1992; Judd et al. 1994; Olmstead et al. 1993; Sanders 1997, 2001; Thieret 1972; Tomlinson 1986; Turner 1999; Umber 1979; Wagstaff and Olmstead 1996.

Lamiaceae Martynov
(= Labiatae Adanson)
(Mint Family)

Herbs, shrubs, or trees; *stems often square in cross-section*; often with iridoids and phenolic glycosides. *Hairs gland-headed, with ethereal oils (including terpenoids), and simple, nonglandular, these, when present, usually multicellular and uniseriate or a mixture of multicellular and unicellular. Leaves usually opposite*, occasionally whorled, simple, sometimes lobed or dissected, or pinnately or palmately compound, entire to serrate; *stipules lacking. Inflorescences with indeterminate main axis and determinate (cymosely branched) lateral axes, often congested into pseudowhorls*, terminal or axillary. Flowers bisexual, usually bilateral. *Sepals usually 5, connate*, calyx radial to bilateral, ± tubular, bell-shaped, or wheel-shaped, persistent, occasionally enlarged in fruit. *Petals usually 5, connate, usually 2-lipped*, the lobes imbricate. *Stamens 4, didynamous to ± equal, sometimes reduced to 2; filaments adnate to corolla*; pollen grains tricolpate or hexacolpate. *Carpels 2, connate; ovary superior, unlobed to deeply 4-lobed, 2-locular but appearing 4-locular due to development of false septa*, with axile placentation; *style usually apically divided, terminal to gynobasic; stigmas 2, tiny and inconspicuous at apices of style branches. Ovules 2 per carpel (i.e., one in each apparent locule)*, **each laterally attached (attached on false septa very near the inrolled carpel margins)**, with 1 integument and a thin-walled megasporangium. Nectar disk often present. *Fruit a drupe with 1–4 pits, an indehiscent, 4-seeded pod, or a schizocarp splitting into 4 nutlets or 4 drupelets*; endosperm scanty or lacking (Figure 9.136).

Floral formula: X,⑤,(2+3), 2+2, ②; drupe, 4 nutlets

Figure 9.136 Lamiaceae (Labiatae). (A–J) *Salvia urticifolia:* (A) flowering plant, note reduced cymes forming pseudoverticillate inflorescence (× 0.5); (B) flower, side view (× 5); (C) flower, in longitudinal section (× 5); (D) stamen, side view, showing fertile anther half and sterile anther half (× 5); (E) stamens as seen in adaxial view (× 5); (F) stamens as seen from above and abaxially (× 5); (G) staminode (× 15); (H) gynoecium, note gynobasic style, lobed ovary and nectar disk (× 5); (I) nutlet (× 10); (J) embryo (× 10). (K–O) *S. lyrata:* (K) basal leaf (× 1); (L) flower in longitudinal section (× 5); (M) fertile stamen, both anther halves producing pollen (× 10); (N) tip of style with unequal style branches (greatly enlarged); (O) calyx in longitudinal section, with two (of four) nearly mature nutlets (× 10). (From Wood 1974, *A student's atlas of flowering plants,* p. 97.)

Distribution: Cosmopolitan.

Genera/species: 258/6970. **Major genera:** *Salvia* (800 spp.), *Hyptis* (400), *Clerodendrum* (400), *Thymus* (350), *Plectranthus* (300), *Scutellaria* (300), *Stachys* (300), *Nepeta* (250), *Vitex* (250), *Teucrium* (200), *Premna* (200), and *Callicarpa* (140). A large number of genera occur in the continental United States and/or Canada; noteworthy representatives (in addition to those listed above) include *Agastache, Ajuga, Collinsonia, Dicerandra, Dracocephalum, Glechoma, Hedeoma, Hyssopus, Lamium, Leonurus, Lycopus, Marrubium, Mentha, Monarda, Physostegia, Piloblephis, Prunella, Pycnanthemum, Pycnostachys, Satureja,* and *Trichostema.*

Economic plants and products: The family contains many species that are economically important either for their essential oils or for use as spices, including *Mentha* (peppermint, spearmint), *Lavandula* (lavender), *Marrubium* (horehound), *Nepeta* (catnip), *Ocimum* (basil), *Origanum* (oregano), *Rosmarinus* (rosemary), *Salvia* (sage), *Satureja* (savory), and *Thymus* (thyme). The tubers of a few species of *Stachys* are edible. *Tectona* (teak) is an important timber tree. Numerous genera provide ornamentals, including *Ajuga* (bugleweed), *Callicarpa* (beautyberry), *Clerodendrum*, *Plectranthus* (coleus), *Holmskioldia* (Chinese-hat plant), *Leonotis* (lion's ear), *Monarda* (bee balm), *Pycnanthemum* (mountain-mint), *Salvia*, *Scutellaria* (skullcap), and *Vitex* (monk's-pepper).

Discussion: Lamiaceae (or Labiatae), as here delimited, are considered monophyletic on the basis of laterally attached ovules, *rbcL,* and *ndhF* sequences (Cantino 1992a,b; Cantino et al. 1992; Junell 1934; Olmstead et al. 1993; Wagstaff and Olmstead 1996; Wagstaff et al. 1998). Phylogenetic relationships within Lamiales support a broad circumscription of the family, including many genera traditionally placed in Verbenaceae (see the discussion under that family for a listing of the most useful distinguishing features). The ovary of Lamiaceae varies from rounded to deeply four-lobed (vs. rounded to only moderately four-lobed in Verbenaceae), and the style is terminal to gynobasic (vs. consistently terminal in Verbenaceae). Most systematists have restricted Lamiaceae to those species with more or less gynobasic styles, a circumscription resulting in a polyphyletic assemblage because gynobasic styles have evolved more than once.

Several monophyletic groups can be distinguished within Lamiaceae (Cantino 1992a,b, Cantino et al. 1997; Wagstaff and Olmstead 1996; Wagstaff et al. 1998). Ajugoideae (or Teucrioideae) are considered monophyletic on the basis of their drupes with four pits, nonpersistent styles, and pollen grains with branched to granular columns in the exine. This clade includes *Clerodendrum, Teucrium, Trichostema, Ajuga,* and relatives. *Scutellaria, Holmskioldia,* and relatives constitute the Scutellarioideae, a group with rounded calyx lobes and nutlets with tuberculate or plumose outgrowths. Lamioideae and Nepetoideae together may constitute a monophyletic group, and both have completely gynobasic styles (Cantino 1992a). Analyses of *rbcL* and *ndhF* sequences, however, place Lamioideae in a clade with Scutellarioideae and Ajugoideae, and Nepetoideae as sister to either "Chloanthoideae" or *Callicarpa* (Wagstaff and Olmstead 1996; Wagstaff et al. 1998). The monophyly of Nepetoideae, which includes *Glechoma, Dicerandra, Hyptis, Lycopus, Melissa, Mentha, Monarda, Nepeta, Ocimum, Origanum, Plectranthus, Prunella, Pycnanthemum, Pycnostachys, Salvia, Satureja, Lavandula, Thymus,* and *Basilicum,* is supported by their hexacolpate pollen and DNA sequences; the other subfamilies have retained tricolpate pollen. Nepetoideae also lack endosperm and iridoids and have a high volatile terpenoid content. The mono-

phyly of Lamioideae, which includes *Lamium, Prasium, Galeopsis, Leonotis, Physostegia, Marrubium, Pogostemon,* and *Stachys,* is strongly supported by DNA data. *Gmelina, Callicarpa, Vitex, Premna, Cornutia,* and relatives are usually placed within the polyphyletic "Viticoideae;" *Tectona* and relatives have branched hairs, and have been treated as "Chloanthoideae" but are likely also polyphyletic.

The showy flowers of Lamiaceae are pollinated by bees, wasps, butterflies, moths, flies, beetles, and birds. The arching upper lip of the two-lipped corolla usually protects the stamens and stigma, while the lower lip provides a landing platform and is often showy. The pollinator is dusted with pollen on its back or head as it probes for nectar. In *Ocimum* and relatives, however, the stamens lie close to the lower lip and deposit pollen on the underside of the pollinator. The stamens of *Salvia* are highly modified, the connective is expanded, forming a lever arm. Most species are protandrous, and outcrossing is common. Some species of *Lamium* have cleistogamous flowers. Species with drupaceous fruits are usually dispersed by birds or mammals. The nutlets of many species are merely shaken from the persistent calyx by the action of wind or disturbance of the plant. Nutlets may also be eaten by birds or dispersed by water.

References: Abu-Asab and Cantino 1989; Cantino 1990, 1992a,b; Cantino et al. 1992; Cantino and Sanders 1986; Huck 1992; Judd et al. 1994; de Pamphilis and Young 1995; Junell 1934; Olmstead et al. 1993; Thorne 1992; Wagstaff and Olmstead 1996; Wagstaff et al. 1998.

Euasterids II

Aquifoliales

Aquifoliaceae de Candolle ex A. Richard
(Holly Family)

Trees or shrubs, bark ± smooth; sometimes with alkaloids, caffeine. Hairs simple. *Leaves alternate and spiral, simple,* entire to serrate, the teeth with deciduous glandular to spinose apex, with pinnate venation; *stipules minute.* Inflorescences determinate, axillary, but sometimes reduced to a solitary flower. *Flowers usually unisexual (plants dioecious), radial.* Sepals usually 4–6, slightly connate. *Petals usually 4–6, usually slightly connate,* imbricate. *Stamens usually 4–6 and slightly adnate to base of corolla;* pollen grains 3–4-colporate or porate; staminodes conspicuous in carpellate flowers. Carpels usually 4–6, connate; ovary superior, with axile placentation; **style very short or lacking**; *stigma capitate, discoid, or slightly lobed;* pistillode often conspicuous in staminate flowers. Ovules usually 1 in each locule, with 1 integument and a thin-walled megasporangium. Nectaries lacking, or nectar secreted by papillose swellings on adaxial petal surface. *Fruit a colorful* (red, orange-red, purple-black, or

Figure 9.137 Aquifoliaceae. (A–I) *Ilex glabra:* (A) branch with staminate flowers (× 0.5); (B) staminate flower (× 7); (C) staminate flower, in longitudinal section (× 7); (D) branch with carpellate flowers (× 0.5); (E) bud of carpellate flower (× 7); (F) carpellate flower, note staminodes (× 7); (G) carpellate flower in longitudinal section, note expanded stigma and very short style (× 7); (H) pit (× 8); (I) pit in longitudinal section, endocarp shown in radiating lines, endosperm stippled, embryo minute (× 8). (J–N) *I. opaca:* (J) fruiting branch (× 0.5); (K) drupe and tip of twig, note persistent stigma (× 2); (L) drupe with fleshy tissue removed from half, note four pits (× 3); (M) pit, abaxial surface (× 3); (N) pit, adaxial surface (× 3). (From Wood 1974, *A student's atlas of flowering plants*, p. 67.)

pink) *drupe* **with a broad stigmatic zone and usually 4–6 pits** (Figure 9.137).

Floral formula:

Staminate: *, (4–6), (4–6), 4–6, (4–6)•

Carpellate: *, (4–6), (4–6), 4–6•, (4–6); drupe (several pits)

Distribution and ecology: Widespread, but especially common in the montane tropics; characteristic of acidic soils.

Genera/species: 1/400. ***Genus:*** The single genus *Ilex.*

Economic plants and products: Numerous species of *Ilex* (holly) are cultivated for their ornamental foliage and colorful fruits. The leaves of *I. paraguariensis* are brewed to make the high-caffeine beverage *maté*. The leaves of *I. vomitoria*, also high in caffeine, were used by native Americans of the southeastern U.S. coastal plain to make the stimulating "black drink."

Discussion: Phylogenetic relationships within *Ilex* are obscure, although it is likely that recognition of *Nemopanthus* (on the basis of its slender, distinct petals) would create a paraphyletic *Ilex* (Powell et al. 2000). We follow Thorne (1992) in excluding two small tropical genera, *Phelline* and *Sphenostemon*. Aquifoliaceae probably are most closely related to Helwingiaceae (*Helwingia*), a group of shrubs with alternate and spiral, pinnately veined leaves and epiphyllous inflorescences (Soltis et al. 2000; Albach et al. 2001a,b).

Holly flowers are visited by various insects, especially bees. Outcrossing is characteristic due to dioecy. Some species may be at least partially wind pollinated. The drupes of hollies are dispersed by birds. The tiny embryo matures after being dispersed; germination is slow, often requiring 1 to 3 years.

References: Albach et al. 2001a,b; Brizicky 1964a; Olmstead et al. 1993; Powell et al. 2000; Soltis et al. 2000; Thorne 1992.

Apiales

Apiaceae Lindley
(= Umbelliferae A. L. de Jussieu)
(Carrot Family)

Herbs to lianas, shrubs, or trees, aromatic; *stems often hollow in internodal region; with secretory canals containing ethereal oils and resins, triterpenoid saponins, coumarins, falcarinone polyacetylenes, monoterpenes, and sesquiterpenes;* **with umbelliferose (a trisaccharide) as carbohydrate storage product**. Hairs various, sometimes with prickles. *Leaves alternate and spiral, pinnately or palmately compound to simple, then often deeply dissected or lobed*, entire to serrate, with pinnate to palmate venation; petioles ± sheathing; **stipules present** to absent. **Inflorescences** determinate, **forming simple umbels,** *these arranged in umbels, racemes, spikes, or panicles*, sometimes condensed into a head, often subtended by an involucre of bracts, terminal. Flowers usually bisexual but sometimes unisexual (plants then monoecious to dioecious), usually radial, small. *Sepals usually 5, distinct,* **very reduced**. *Petals usually 5, occasionally more, distinct,* but developing from a ring primordium, sometimes clearly connate, *often inflexed*, imbricate to valvate. **Stamens 5**, but occasionally numerous; filaments distinct; pollen grains usually tricolporate. *Carpels usually 2–5, occasionally numerous, connate;* **ovary inferior**, usually with axile placentation; **styles ± swollen at base to form a nectar-secreting structure (stylopodium) atop ovary**; stigmas usually 2–5, tiny, capitate to truncate, or elongate. Ovules 1 in each locule, with 1 integument and a thin-walled to less commonly

thick-walled megasporangium. *Fruit a drupe with 2–5 pits, or* **a schizocarp,** *the 2 dry segments (mericarps) often attached to an entire to deeply forked central stalk (carpophore); globular to elongated oil canals (vittae) often present in schizocarpic fruits;* fruit surface smooth or ribbed, sometimes covered with hairs, scales, or bristles, sometimes flattened or winged; **endosperm with petroselenic acid** (Figures 9.138, 9.139).

Floral formula: *, 5, ⑤, 5, (2–5); drupe, schizocarp

Distribution: Nearly cosmopolitan, diverse from tropical to temperate regions.

Genera/species: 460/4250. ***Major genera:*** *Schefflera* (600 spp.), *Eryngium* (230), *Polyscias* (200), *Ferula* (150), *Peucedanum* (150), *Pimpinella* (150), *Bupleurum* (100), *Oreopanax* (90), *Hydrocotyle* (80), *Lomatium* (60), *Heracleum* (60), *Angelica* (50), *Sanicula* (40), *Chaerophyllum* (40), and *Aralia* (30). Some of the numerous genera occurring in the continental United States and/or Canada are *Angelica, Apium, Aralia, Carum, Centella, Chaerophyllum, Cicuta, Conioselinum, Daucus, Eryngium, Hedera, Heracleum, Hydrocotyle, Ligusticum, Lomatium, Osmorhiza, Oxypolis, Panax, Pastinaca, Ptilimnium, Sanicula, Sium, Spermolepis, Thaspium, Torilis,* and *Zizia*.

Economic plants and products: Apiaceae contain many food and spice plants: *Anethum* (dill), *Apium* (celery), *Carum* (caraway), *Coriandrum* (coriander), *Cuminum* (cumin), *Daucus* (carrot), *Foeniculum* (fennel), *Pastinaca* (parsnip), *Petroselinum* (parsley), and *Pimpinella* (anise). However, many are extremely poisonous, such as *Conium* (hemlock, which Socrates is said to have taken for suicide) and *Cicuta* (water hemlock). *Panax quinquefolia* and *P. ginseng* (ginseng) and various species of *Aralia* (wild sarsaparilla) are important medicinally. A few genera contain useful ornamentals, including *Hedera* (English ivy), and *Schefflera* (umbrella tree).

Discussion: The monophyly of Apiaceae (or Umbelliferae) is supported by morphology, secondary chemistry, *rbcL*, and *matK* sequences (Downie and Palmer 1992; Hegnauer 1971; Judd et al. 1994; Olmstead et al. 1992, 1993; Plunkett et al. 1996a,b , 1997). Apiaceae are most closely related to Pittosporaceae, and these two, along with three or four small families, constitute Apiales (Dahlgren 1983; Savolainen et al. 2000b; Soltis et al. 2000; Albach et al. 2001a,b). Apomorphies of Apiaceae and Pittosporaceae include the distinctive resin/ethereal oil canals that are associated with the conducting tissues, a characteristic arrangement of lateral roots, the presence of falcarinone polyacetylenes, a tiny embryo, and reduced, bractlike leaves at the base of the shoots. Apiales clearly belong to the core asterid clade, as indicated by their ovules with a single integument and usually a thin megasporangium wall, sympetalous corollas (obvious in many Pittospo-

Figure 9.138 Apiaceae (or Umbelliferae), subfamily Aralioideae. (A–Q) *Aralia nudicaulis:* (A) flowering plant (× 0.3); (B) apex of rhizome (× 2); (C) staminate flower (× 8); (D) stamen (× 19); (E) staminate flower in longitudinal section (× 8); (F) rudimentary ovule (greatly magnified); (G) carpellate flower (× 8); (H) carpellate flower in longitudinal section (× 8); (I) functional ovule (greatly magnified); (J) staminate flower (× 8); (K) nonfunctional ovule (greatly magnified); (L) umbel of young fruits (× 1); (M) young fruit, note inferior ovary (× 5); (N) fruit in cross-section (× 5); (O) seed (× 8); (P) seed in section, endosperm stippled (× 8); (Q) embryo (greatly magnified). (R–T) *A. spinosa:* (R) portion of inflorescence (× 0.5); (S) flower (× 8); (T) drupe (× 3). (Unpublished original art prepared for the Generic Flora of the Southeast U.S. series; used with permission; treatment of Araliaceae in 1965 *J. Arnold Arbor.* 47: 126–136.)

Figure 9.139 Apiaceae (Umbelliferae), subfamily
Apioideae. (A–M) *Daucus carota:* (A) flowering plant (× 0.3);
(B) inflorescence in longitudinal section, a compound umbel
(× 0.5); (C) an umbellate unit from inflorescence (× 6); (D)
bisexual flower, note stylopodium, swollen region at base of
style (× 18); (E) bisexual flower after fall of stamens, in longitudinal section, note ovules (× 18); (F) bud, staminate flower
(× 7); (G) staminate flower, note stylopodium in center (× 18);
(H) staminate flower in longitudinal section (× 18); (I) sterile
central flower (× 11); (J) schizocarp (× 11); (K) dried schizocarp showing central carpophore (× 11); (L) fruit (schizocarp)
in cross-section (× 11); (M) embryo (greatly magnified).
(N) *D. pusillus:* schizocarp (× 11). (From Wood 1974, *A student's atlas of flowering plants,* p. 78.)

raceae), stamens in a single whorl (Judd et al. 1994; Judd 1996), *rbcL, ndhF*, and *atpB* sequences, and cpDNA restriction sites (Downie and Palmer 1992; Källersjö et al. 1998; Olmstead et al. 1992a, 1993, 2000; Plunkett et al. 1996a,b). The order probably is closely related to Dipsacales and Asterales (Olmstead et al. 1993, 2000). We note that all of these groups contain polyacetylenes.

Apiaceae here are circumscribed broadly, including both herbaceous species with two-carpellate, laterally flattened schizocarps with a well-developed carpophore and oil canals (vittae), which traditionally are placed within Apiaceae s.s. (Cronquist 1981); and woody species with two- to five-carpellate, usually globose drupes that lack oil cavities, which usually are segregated as a separate family, the Araliaceae (see Judd et al. 1994; Thorne 1973b). Although many members of Araliaceae and Apiaceae s.s. are distinct, the characters used to distinguish them are quite homoplasious (Plunkett et al. 1996a,b), and a few genera, such as *Myodocarpus, Delarbrea*, and *Mackinlaya*, are problematic. These taxa may represent remnants of a basal complex characterized by a woody habit, simple, stipulate leaves, and two-carpellate, schizocarpic fruits that gave rise to the core members of the aralioid and apioid clades. Herbaceousness and compound leaves undoubtedly have evolved several times within Apiaceae s.l.

Several clades can be discerned within Apiaceae s.l., although some are difficult to diagnose due to the lack of unambiguous morphological synapomorphies. Aralioideae (e.g., *Aralia, Dendropanax, Hedera, Oreopanax, Panax, Polyscias, Schefflera*, and *Tetrapanax*) are distinguished by usually globose, drupaceous fruits with several pits and by *matK* sequences. This clade is mainly tropical and woody; its leaves are stipulate and may be simple or compound. Saniculoideae (e.g., *Sanicula* and *Eryngium*) have a stylopodium separated from the style by a narrow groove. Apioideae (e.g., *Coriandrum, Angelica, Apium, Chaerophyllum, Cicuta, Daucus*, and *Spermolepis*) have compound umbels and schizocarpic fruits with a more or less forked carpophore (see Downie et al. 1998, 2000). Saniculoideae + Apioideae form a clade, often treated as Apiaceae s.s., that is supported by the herbaceous habit, elongated oil canals in the fruit, and *matK, rbcL, rpoC1*, and *rpl16* sequences. The presence of a carpophore may also be synapomorphic.

Several genera of Apiaceae s.l. (e.g., *Delarbrea* and *Myodocarpus*) cannot be placed with certainty within any of these three subfamilies. Genera traditionally placed in "Hydrocotyloideae" (such as *Centella* and *Hydrocotyle*) are scattered within the Apiaceae s.l., with some showing affinities to Apioideae and others to Aralioideae (Plunkett et al. 1996a,b, 1997). However, if Apiaceae and Araliaceae, in close to their traditional circumscriptions, were recognized, they would be poorly characterized morphologically, and certain genera would have no well-supported familial placement.

The usually small flowers of Apiaceae are typically densely aggregated and pollinated by a wide range of small nectar-gathering flies, beetles, bees, and moths. Cross-pollination is favored by protandry. Drupaceous fruits usually are dispersed by birds, while dry, schizocarpic fruits are often wind-dispersed. The fruit segments of genera such as *Daucus* are covered by bristles, leading to dispersal by external transport.

References: Albach et al. 2001a,b; Baumann 1946; Constance 1971; Cronquist 1981; Dahlgren 1983; Downie and Palmer 1992; Downie et al. 1998, 2000; Erbar 1991; Erbar and Leins 1988; Graham 1966; Hegnauer 1971; Jackson 1933; Judd et al. 1994; Källersjö et al. 1998; Olmstead et al. 1992a, 1993, 2000; Pimenov and Leonov 1993; Plunkett and Downie 1999; Plunkett et al. 1996a,b, 1997; Rodrigues 1957; Rogers 1950; Savolainen et al. 2000b; Soltis et al. 2000; Thorne 1973b, 1992; Wen et al. 2001.

Dipsacales

Dipsacales includes both Caprifoliaceae and Adoxaceae. The monophyly of this order is supported by opposite leaves and cellular endosperm development, *ndhF, rbcL, atpB*, and 18S nucleotide sequences (Donoghue 1983; Donoghue et al. 1992; Judd et al. 1994; Olmstead et al. 1993, 2000; Savolainen et al. 2000b; Soltis et al. 2000; Albach et al. 2001a,b). Phylogenetic relationships within Dipsacales are summarized in Figure 9.140.

References: Albach et al. 2001a,b; Donoghue 1983; Donoghue et al. 1992; Judd et al. 1994; Olmstead et al. 1993, 2000; Savolainen et al. 2000b; Soltis et al. 2000.

Key to Families of Dipsacales

1. Flowers ± bilateral, the style elongate with a capitate stigma; nectary composed of densely packed hairs on the inner surface of the corolla; pollen spiny ....**Caprifoliaceae**
1. Flowers radial, the style short with a lobed stigma; nectary glandular (atop ovary) or lacking; pollen non-spiny ....**Adoxaceae**

Caprifoliaceae Adanson
(Honeysuckle Family)

Herbs, shrubs, small trees, or lianas; often with phenolic glycosides, iridoids, and scattered secretory cells. Hairs various. *Leaves opposite, simple*, sometimes pinnately divided or compound, entire to serrate, with pinnate venation; stipules lacking. Inflorescences various. **Flowers** bisexual and **bilateral**. Sepals usually 5, connate. *Petals usually 5, connate, often with 2 upper lobes and 3 lower lobes, or a single lower lobe and 4 upper ones*, the lobes imbricate or valvate. *Stamens* (1–) *4 or 5*; filaments adnate to corolla; **pollen large, spiny**, usually tricolporate or triporate. *Carpels usually 2–5, connate*; ovary inferior, often elongate, with axile placentation, sometimes only 1 locule fertile; **style elongate; stigma capitate**. Ovules 1 to numerous in each locule, with 1 integument and a thin-walled megasporangium. **Nectar produced by closely packed glandular hairs on lower part of corolla tube**. Fruit a capsule, berry, drupe, or achene; endosperm present or lacking (Figure 9.141).

Floral formula: X, ⑤, ⑤, 4–5, (2–5); drupe,
berry, capsule, achene

Distribution: Widely distributed, especially in northern temperate regions.

Genera/species: 36/810. ***Major genera:*** *Valeriana* (200), *Lonicera* (150), *Scabiosa* (80), and *Valerianella* (50). Noteworthy genera of the continental United States and/or Canada include *Lonicera, Valeriana, Valerianella, Dipsacus, Linnaea, Symphoricarpos*, and *Diervilla*.

Economic plants and products: *Lonicera* (honeysuckle), *Abelia, Symphoricarpos* (snowberry), *Weigela*, and *Kolkwitzia* are used as ornamentals. *Dipsacus* (teasel) is a widespread weed.

Discussion: Caprifoliaceae are circumscribed broadly, including Dipsacaceae and Valerianaceae, but excluding *Sambucus* and *Viburnum*, which are referred to Adoxaceae. As here delimited, the family is monophyletic, as evidenced by morphology (Judd et al. 1994), *rbcL, atpB, atpB, ndhF* and.or 18 S sequences (Donoghue et al. 1992; Backlund and Bremer 1998; Backlund and Pyck 1998; Källersjö et al. 1998; Soltis et al. 2001; Albach et al. 2001a,b), and cpDNA restriction sites (Downie and Palmer 1992). The family is easily distinguished from Adoxaceae (including *Sambucus, Viburnum, Adoxa*, and relatives) by its bilateral (vs. radial) flowers with an elongate (vs. short) style, capitate (vs. lobed) stigma, spiny (vs. reticulate) pollen exine, and nectary composed of densely packed hairs on the inner surface of the lower portion of the corolla tube (vs. glandular nectary atop ovary or nectary lacking).

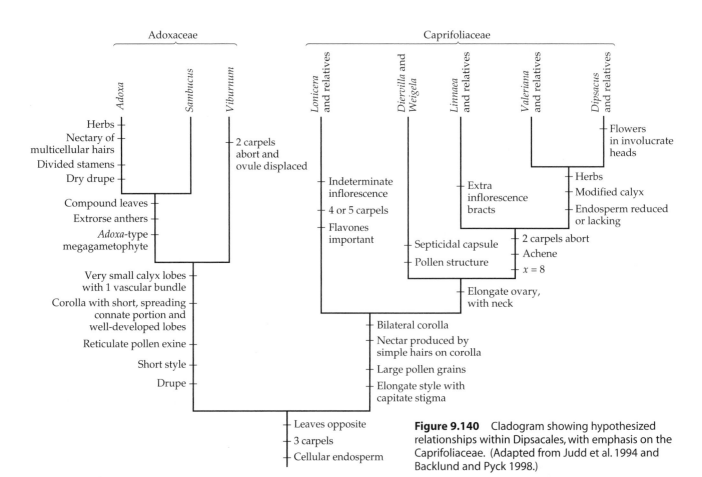

Figure 9.140 Cladogram showing hypothesized relationships within Dipsacales, with emphasis on the Caprifoliaceae. (Adapted from Judd et al. 1994 and Backlund and Pyck 1998.)

Figure 9.141 Caprifoliaceae. (A–H) *Lonicera sempervirens:* (A) fruiting branch (× 0.75); (B) flower (× 3); (C) corolla opened lengthwise to show attachment of stamens and distribution of hairs and nectar glands (× 2.3); (D) portion of inflorescence, corollas removed (× 12); (E) ovary in cross-section (× 15); (F) seed (× 9); (G) seed in cross-section, seed coat hatched, endosperm stippled, and embryo unshaded (× 9); (H) embryo (× 18). (I–J) *L. japonica:* (I) flowering branch (× 0.75); (J) flower (× 3). (From Ferguson 1966, *J. Arnold Arbor.* 47: p. 55.)

The best-supported clade within the Caprifoliaceae is the Linnaeeae, a tribe that includes *Linnaea, Dipelta, Abelia,* and *Kolkwitzia,* along with more specialized gen-

era such as *Valeriana* and *Dipsacus.* The monophyly of the Linnaeeae is supported by the reduction to one nectary (hair cluster), four or fewer stamens, a haploid chromosome number of eight, abortion of two of the three carpels (leaving a single-ovuled carpel occupying half of the ovary), and an achene fruit. A pappuslike calyx and lack of endosperm have evolved in *Valeriana, Dipsacus,* and relatives. We note that some systematists place *Valeriana* and relatives in Valerianaceae, *Dipsacus* and relatives in Dipsacaceae, and the remaining members of the

Linnaeeae in the segregate family Linnaeaceae (Backlund and Pyck 1998).

Diervilla and *Weigela* constitute a monophyletic group, which is supported by their septicidal capsules and tectate pollen with poorly developed columellae; this clade is occasionally segregated as the Diervillaceae (Backlund and Pyck 1998).

Lonicera, *Symphoricarpos*, and relatives probably form a clade, distinguished by determinate inflorescences and an often four- or five-carpellate gynoecium; Backlund and Pyck restrict the Caprifoliaceae to this group.

The showy flowers of Caprifoliaceae are pollinated by various nectar-gathering insects (mainly bees and wasps) and birds. The family shows a wide variety of dispersal syndromes.

References: Albach et al. 2001a,b; Backlund and Bremer, 1998; Backlund and Pyck 1998; Chase et al. 1993; Donoghue 1983; Donoghue et al. 1992; Downie and Palmer 1992; Eriksson and Donoghue 1997; Ferguson 1966b; Judd et al. 1994; Källersjö et al. 1998; Olmstead et al. 1992a, 2000; Soltis et al. 2000.

Adoxaceae E. Mey
(Moschatel or Elderberry Family)

Small trees, shrubs, or perennial herbs; with cyanogenic glycosides and iridoids. Hairs simple, stellate, or of peltate scales, glandular or nonglandular. *Leaves opposite, simple, trifoliolate, to pinnately compound*, entire or variously toothed, sometimes lobed, with pinnate to palmate venation; stipules present or lacking, sometimes glandular. Inflorescences determinate, often umbellate. *Flowers bisexual*, **radial**, a few species with sterile flowers around periphery of inflorescence. *Sepals 2–5, connate*, **reduced, with only a single vascular trace**. *Petals 4–5, connate, with usually short tube and well-developed*, imbricate or valvate *lobes*. *Stamens 5*, sometimes divided and appearing as 10; filaments adnate to base of corolla; pollen grains tricolpate or tricolporate, small to medium-sized, **with reticulate exine**. *Carpels 3–5, connate; ovary inferior or half-inferior*, with axile placentation; **style(s) short**; *stigma(s) capitate*. Ovules 1 in each locule, often only 1 functional, with 1 integument and a thin-walled megasporangium. *Nectar produced by glandular tissue atop ovary (Viburnum)*, by cushion-like group of multicellular hairs (*Adoxa*), or lacking (*Sambucus*). **Fruit a drupe**, *with 1–5 pits*. (Figure 9.142).

Floral formula: *, (2-5), (4-5), 5 , (3-5) , drupe

Distribution: Widespread in temperate regions of the Northern Hemisphere, but extending into the Africa, South America, Malesia, Australia, and New Zealand, especially in mountainous regions.

Genera/species: 5/245. **Major genera:** *Viburnum* (220 spp.) and *Sambucus* (20). The family is represented in the

continental United States and Canada by *Viburnum*, *Sambucus*, and *Adoxa*.

Economic plants and products: Numerous species of *Viburnum* (Arrow-wood) and *Sambucus* (Elder) are used as ornamentals, and the fruits are occasionally used for making jellies and wines. Several have medicinal uses.

Discussion: Monophyly of the Adoxaceae is supported by both morphological and DNA sequence (*rbcL* and ITS) characters (Donoghue et al. 1992; Eriksson and Donoghue; Judd et al. 1994). *Viburnum* (Opuloideae) is sister to *Sambucus* + *Adoxa* and relatives (Adoxoideae). The latter are linked based on their compound leaves, simple vessel perforations, extrorse anthers, and *Adoxa*-type megagametophyte development (i.e., developing from four megaspores, and eight-nucleate at maturity). Monophyly of the large and diverse genus *Viburnum* is supported by the unusual gynoecium development in which two carpels abort and the single functional ovule is displaced, developing in one of the sterile locules. *Adoxa*— and the closely related *Tetradoxa* and *Sinadoxa*— are especially distinctive, and easily distinguished from *Sambucus* by their herbaceous habit, nectary composed of multicellular hairs, divided stamens, and dry drupe (all synapomorphies). Morphological synapomorphies for *Sambucus* are uncertain, however, the small opening in the apical part of the endocarp is characteristic and may be synapomorphic; the production of sambunigrin (a cyanogenic glycoside) also may be a synapomorphy. Lack of nectaries in this genus has been suggested as a derived feature, but it could also represent the ancestral condition within Dipsacales.

It is noteworthy that rhizomatous herbs have evolved three times within Adoxaceae: in *Sambucus ebulus*, *S. adnata*, and relatives, in *S. gaudichaudiana*, and in the *Adoxa*, *Tetradoxa*, and *Sinadoxa* clade. The red-fruited species, e.g., *S. pubens* and *S. racemosa*, form a derived monophyletic group; red fruits have evolved independently in *Viburnum* and *Sambucus*.

The flowers, although small, are presented in showy clusters, and attract various insects, especially bees, wasps, and flies. The usually blue to red drupes are dispersed by birds.

References: Donoghue 1980, 1981, 1983a,b, 1985; Donoghue et al. 1992; Eriksson and Donoghue 1997; Ferguson 1966; Judd et al. 1994; Wilkinson 1948.

Asterales

The Asterales are monophyletic, as evidenced by their storage of carbohydrates as the oligosaccharide inulin, the presence of ellagic acid, and **plunger pollination**. The stamens are closely associated with one another (initially sticking together to completely connate) and form a tube around the style, with the anthers opening toward the inside. In plunger (or brush) pollination, pollen is pushed out of this tube, as out of a plunger, by specialized hairs on the style or by a specialized pollen-gathering cup. The style elongates to present pollen to floral visitors. Later the style branches diverge and the stigmas become receptive (Lammers 1992; Leins and Erbar 1990; Wagenitz 1977, 1992; Yeo 1993). The absence of a plunger mechanism in Menyanthaceae is likely a reversal. The monophyly of the order also is supported strongly by cpDNA restriction sites, *rbcL*, *atpB*, *ndhF*, and 18S sequences (Cosner et al. 1994; Downie and Palmer 1992; Källersjö et al. 1998; Michaels et al. 1993; Olmstead et al. 1992a, 1993, 2000; Savolainen et al. 2000b; Soltis et al. 2000; Albach et al. 2001a,b). The order consists of 12 families and about 24,900 species; major families include **Campanulaceae** (incl. Lobeliaceae), Menyanthaceae, Goodeniaceae, Calyceraceae, Stylidiaceae, and **Asteraceae**. Asterales clearly belong within the core group of the asterid clade (in Euasterid II), as evidenced by the

single integument, a thin megasporangium wall, sympetalous flowers with stamen number equaling corolla lobe number, and numerous cpDNA features (Anderberg 1992; Chase et al. 1993; Cosner et al. 1994; Cronquist 1981, 1988; Downie and Palmer 1992; Hufford 1992; Olmstead et al. 1992a, 1993; Wagenitz 1977, 1992). Analyses of DNA sequences suggest that the order is most closely related to Apiales and Dipsacales; all contain polyacetylenes.

Phylogenetic relationships of families within the order are rather unclear despite molecular studies (see above), and morphology (K. Bremer 1987, 1994). Campanulaceae (possibly along with a few small families) is the sister group to a clade containing Menyanthaceae, Goodeniaceae, Calyceraceae, and Asteraceae. These two clades are frequently recognized as a pair of related orders—Campanulales and Asterales. Various embryological and chemical features (Lammers 1992) seem to be most useful in diagnosing these two groups, and the

Figure 9.142 Adoxaceae. *Sambucus canadensis*: (A) flowering branch (× 0.4); (B) flower (× 9.6); (C) flower with corolla and stamens removed to show bracteoles and calyx lobes (× 9.6); (D) longitudinal section of flower with corolla and stamens removed to show solitary, pendulous ovule in each locule (× 16); (E) fruit (× 4.8); (F) pit (× 14.7). (From Ferguson 1966, *J. Arnold Arbor.* 47: p. 38.)

Key to Major Families of Asterales

1. Stamens 2, extrorse, adnate to style by filaments, forming a column..............................Stylidiaceae
1. Stamens 5, introrse, filaments not adnate to style, (although anthers often closely associated with style)...2

2. Ovary unilocular, with apical or basal ovule; flowers densely clustered in heads, which are surrounded by involucral bracts..3
2. Ovary multilocular, with axile placentation, or unilocular, with parietal placentation; flowers usually not in involucrate heads...4

3. Plants usually with resin canals or laticifers; ovary with basal ovule; anthers connate and filaments distinct; sepals forming a pappus of scales, bristles, or awns, or lacking.........**Asteraceae**
3. Plants with neither resin canals nor laticifers; ovary with apical ovule; anthers distinct and filaments connate; sepals not highly modified, of small lobes or teeth....................Calyceraceae

4. Ovary superior, with parietal placentation; flowers lacking a plunger or brush pollination mechanism, and style lacking pollen-gathering specializations.................Menyanthaceae
4. Ovary usually inferior or half-inferior, with axile placentation; flowers with a plunger pollination mechanism..5

5. Milky latex present; style with pollen-collecting hairs, these sometimes invaginating; flowers radial or bilateral; petals lacking marginal wings....................................**Campanulaceae**
5. Milky latex lacking; styles with a pollen-collecting cup; flowers bilateral; petals with marginal wings...Goodeniaceae

placement of a few families is questionable. The monophyletic group comprising Goodeniaceae, Calyceraceae, and Asteraceae seems well supported; Menyanthaceae have superior ovaries. Anatomical and developmental studies indicate that flowers with inferior ovaries may have evolved twice within Asterales. The inferior ovary of Campanulaceae evolved by adnation of the ovary to a hypanthium. In contrast, epigyny in Goodeniaceae, and possibly also Calyceraceae and Asteraceae, was brought about by the adnation of the perianth to the ovary. Morphological analyses suggest that Calyceraceae are sister to Asteraceae; both share an unusual kind of corolla venation (possibly synapomorphic). Both families also have flowers densely clustered in heads that are surrounded by an involucre of bracts and reduced, unilocular, and uniovulate ovaries. However, all these are considered to be parallelisms (see Lammers 1992). On the other hand, *rbcL* sequences support the sister group relationship of Goodeniaceae and Calyceraceae.

References: Albach et al. 2001a,b; Anderberg 1992; K. Bremer 1987, 1994; Brizicky 1966b; Chase et al. 1993; Cosner et al. 1994; Cronquist 1981, 1988; Downie and Palmer 1992; Källersjö et al. 1998; Lammers 1992; Leins and Erbar 1990; Michaels et al. 1993; Olmstead et al. 1992a, 1993, 2000; Savolainen et al. 2000b; Soltis et al. 2000; Wagenitz 1977, 1992; Yeo 1993.

Campanulaceae Adanson
(Bellflower or Lobelia Family)

Mostly herbs, but sometimes secondarily woody; plants storing carbohydrate as inulin (an oligosaccharide); **laticifers present with milky sap**; polyacetylenes present, but iridoids absent. Hairs usually simple, unicellular. *Leaves usually alternate and spiral, simple*, sometimes lobed, entire to serrate, with pinnate venation; stipules absent. Inflorescences various. *Flowers usually bisexual, radial to bilateral, with hypanthium, sometimes twisting 180¡ in development (resupinate). Sepals usually 5, connate. Petals usually 5, connate, forming a tubular or bell-shaped corolla, or 2-lipped to 1-lipped and then with a variously developed adaxial slit*, the lobes valvate. *Stamens usually 5; filaments distinct to distally connate, usually* **attached to disk at apex of ovary**; *anthers distinct but pressed together around the style or connate (syngenesious), forming a tube into which the pollen is shed, and the style then growing through this tube, picking up pollen with specialized, often invaginating hairs, or pushing it out, after which the stigmas become receptive (i.e., a plunger pollination mechanism)*; pollen grains with 3 to 12 apertures. *Carpels 2–5, connate*; **ovary usually inferior (or half-inferior)**, *with usually axile placentation; style with pollen-collecting hairs near the apex*; number of stigmas equaling number of carpels, globose to cylindrical. Ovules usually numerous, with 1 integument and a thin-walled megasporangium. Nectar disk present above ovary. *Fruit a loculicidal or poricidal capsule or a berry* (Figure 9.143).

Figure 9.143 Campanulaceae. (A–L) *Lobelia cardinalis:* (A) flowering stem (× 0.75); (B) flower (× 3); (C) flower with corolla removed, filaments monadelphous except at base, stigmas just beginning to expand (× 3); (D) anther tube formed by connation (× 6); (E) cross-section of anther tube after dehiscence of anthers, style in center (× 9); (F) tip of style with stylar brush and unexpanded stigmas at time of anther dehiscence (× 17.5); (G) expanded stigma, protruding from anther tube following dehiscence of anthers (× 6); (H) expanded stigmas, receptive to pollen (× 18); (I) ovary in cross-section (× 9); (J) ovary in longitudinal section, note half-inferior condition (× 4.5); (K) capsule (× 4.5); (L) seed (× 37). (M) *L. siphilitica:* flower (× 4.5). (From Rosatti 1986, *J. Arnold Arbor.* 67: p. 67.)

Floral formula: * or X,⑤,⑤,⑤, —②–5̄—; capsule, berry

Distribution: Widely distributed in temperate and subtropical regions and in the montane tropics.

Genera/species: 65/2200. *Major genera:* Lobelia (400 spp.), *Campanula* (450), *Centropogon* (200), *Siphocampylus* (225), and *Wahlenbergia* (270). Genera occurring in Canada and/or the continental United States include *Campanula*, *Downingia*, *Githopsis*, *Heterocodon*, *Howellia*, *Jasione*, *Legenere*, *Lobelia*, *Nemacladus*, *Parishella*, *Porterella*, *Triodanis*, and *Wahlenbergia*.

Economic plants and products: *Campanula* (bellflower, bluebell), *Lobelia* (cardinal flower, lobelia), and *Codonopsis* (bonnet bellflower) are used horticulturally.

Discussion: Monophyly of Campanulaceae, as here circumscribed, is supported by morphology and *rbcL* sequences (Cosner et al. 1994). Three subfamilies—Campanuloideae, "Cyphioideae," and Lobelioideae (Thorne 1992)—are often distinguished and may be recognized at the family level (Lammers 1992). The Campanuloideae, with radially symmetrical flowers and nonconnate anthers, are considered monophyletic on the basis of invaginating hairs on the upper portion of the style. The Lobelioideae constitute a clade based on their connate anthers, resupinate flowers, and one- to two-lipped corollas with a variously developed slit in the upper lobe (developmentally adaxial, but abaxial when flower is resupinate). Analysis of *rbcL* sequence variation also supports the monophyly of both subfamilies (Cosner et al. 1994).

Generic delimitations are often problematic; *Campanula*, *Centropogon*, and *Lobelia* are certainly not monophyletic. The showy flowers of Campanulaceae attract a wide range of floral visitors, especially bees and birds. Outcrossing is favored by the brush/plunger secondary pollen presentation mechanism. Dispersal of the small seeds of capsular species is by wind; species with berries are bird-dispersed.

References: Leins and Erbar 1990; Lammers 1992; Rosatti 1986; Shetler 1979; Cosner et al. 1994; Thorne 1992.

Asteraceae Martinov
(= Compositae Giseke)
(Aster or Composite Family)

Herbs, shrubs, or trees; storing carbohydrate as oligosaccharides, including inulin; *resin canals often present, laticifers often present, but one or the other of these sometimes lacking; polyacetylenes and terpenoid aromatic oils usually present;* **usually with sesquiterpene lactones (but lacking iridoids).** Hairs various. *Leaves alternate and spiral, opposite,* or whorled, *simple, but sometimes deeply lobed or dissected,* entire to variously toothed, with usually pin-nate or palmate venation; stipules lacking. **Flowers ± densely aggregated into indeterminate heads that are surrounded by an involucre of bracts (phyllaries);** the heads arranged in a determinate secondary inflorescence, terminal or axillary. *Flowers bisexual or unisexual, sometimes sterile, radial or bilateral.* **Sepals highly modified, forming a pappus** *composed of 2–many, persistent, sometimes connate, scales, awns, or bristles that are capillary, hairy, minutely barbed, or plumose,* or sometimes lacking. *Petals 5, connate, forming a radial and tubular corolla (disk flower),* forming a bilateral and 2-lipped corolla (i.e., with usually 2 petals in upper lip and 3 petals in lower lip), *or forming a bilateral and 1-lipped corolla with upper lip ± lacking and lower lip elongated, ± 3-lobed (ray flower), or forming a bilateral, elongated, tonguelike corolla ending in 5 small teeth (ligulate flower); the heads with only disk flowers (discoid), with disk flowers in center and ray flowers around the outside, the latter female or sterile (radiate), or with only ligulate flowers (ligulate), the corolla lobes valvate. Stamens usually 5; filaments distinct, adnate to corolla tube;* **anthers** *usually* **connate (syngenesious),** *often with apical or basal appendages, forming a tube around the style into which the pollen is shed, and the style then growing through this tube, pushing out or picking up pollen (with variously developed hairs) and presenting it to floral visitors, after which the stigmas become receptive (i.e., with a plunger pollination mechanism);* pollen grains usually tricolporate. *Carpels 2, connate; ovary inferior,* **with basal placentation;** *style branches 2, with stigmatic tissue covering the inner surface or in 2 marginal lines.* **Ovule 1 per ovary,** with 1 integument and a thin megasporangium. Nectary at apex of ovary. **Fruit an achene, crowned by a persistent pappus,** sometimes flattened, winged, or spiny; endosperm scanty or lacking (Figures 9.144–9.146).

Floral formula: * or X, ∞, ⑤,⑤,②̄; achene

Distribution and ecology: Cosmopolitan, found especially in temperate or tropical montane regions and open and/or dry habitats.

Genera/species: 1535/23,000. *Major genera:* *Senecio* (1500 spp.), *Vernonia* (1000), *Cousinia* (600), *Eupatorium* (600), *Centaurea* (600), *Hieracium* (500), *Helichrysum* (500), *Artemisia* (400), *Baccharis* (400), *Mikania* (400), *Saussurea* (300), *Cirsium* (270), *Aster* (250), *Jurinea* (250), *Bidens* (200), *Gnaphalium* (150), and *Solidago* (100). Generic limits are often problematic, and several of these large genera are frequently divided into numerous segregates (see Bremer 1994). Many genera occur in the continental United States and/or Canada (see various floras); especially important and/or widespread genera include *Ambrosia*, *Antennaria*, *Artemisia*, *Aster*, *Baccharis*, *Bidens*, *Coreopsis*, *Erigeron*, *Senecio*, *Centaurea*, *Chrysothamnus*, *Cirsium*, *Coreopsis*, *Eupatorium*, *Elephantopus*, *Gaillardia*, *Gamochaeta*, *Gnaphalium*, *Haplopappus*, *Helianthus*, *Hieracium*, *Liatris*, *Lactuca*, *Rudbeckia*, *Solidago*, *Verbesina*, and *Vernonia*.

Figure 9.144 Asteraceae (Compositae), subfamily "Cichorioideae" (with disk flowers). (A–I) *Cirsium horridulum* var. *vittatum:* (A) flowering plant (× 0.5); (B) flower, note inferior ovary, pappus bristles (× 3); (C) three of the five anthers from within, one the central in more detail (× 9); (D) base of anther from within (× 18); (E) detail of style, showing stylar brush at middle and lowermost part of united style branches with stigmatic line (greatly magnified); (F) achene with pappus (× 1.4); (G, H) basal, central, and upper portions of a single pappus filament (greatly magnified); (I) achene with pappus removed (× 10). (J) *C. horridulum* var. *horridulum:* flowering head (× 5). (K) *C. lecontei:* flowering head, the cobwebby hairs omitted (× 5). (From Scott 1990, *J. Arnold Arbor.* 71: p. 408.)

Economic plants and products: Food plants include *Cichorium* (endive, chicory), *Cynara* (artichoke), *Helianthus* (sunflower seeds and oil, Jerusalem artichoke), *Taraxacum* (dandelion greens), and *Lactuca* (let-tuce). *Artemisia* (wormwood, tarragon) contains spice plants. *Tanacetum* (pyrethrum) and *Pulicaria* (fleabane) contain species with insecticidal properties. *Ambrosia* (ragweed) is important as a major cause of hay fever, and several genera are serious weeds. Finally, ornamentals come from *Calendula* (marigold), *Dendranthema*, *Argyranthemum*, *Leucanthemum* (chrysanthemum),

Figure 9.145 Asteraceae (Compositae), subfamily "Cichorioideae" (with ligulate flowers). *Cichorium intybus:* (A) basal leaf (× 0.3); (B) upper part of flowering stem (× 0.6); (C) cluster of heads in axil of bract (× 2.5); (D) head of flowers (× 1); (E) ligulate flower, note five corolla teeth (× 5); (F) androecium, from bud just before opening, note apical and basal appendages (× 9.5); (G) flower bud in cross-section, showing corolla, five coherent anthers, and style with pollen-collecting hairs (× 24); (H) pollen grain in polar view (greatly magnified); (I) style branches with stigmas (× 18); (J) achene (× 14); (K) detail of pappus scales (× 24); (L) seed (× 9.5). (From Vuilleumier 1973, *J. Arnold Arbor.* 54: p. 49.)

Dahlia, Tagetes (French marigold), *Senecio, Spaghenticola, Gaillardia* (bandana daisy), *Helianthus* (sunflower), *Zinnia,* and many others.

Discussion: Asteraceae (or Compositae) form an easily recognized and obviously monophyletic group; both morphological and molecular synapomorphies are numerous (K. Bremer 1987, 1994, 1996; Jansen et al. 1991, 1992; Karis 1993; Karis et al. 1992; Keeley and Jansen 1991; Kim et al. 1992). The family is divided into several tribes (Table 9.3), which are often arranged into three subfamilies (K. Bremer 1987, 1994; Bremer and Jansen 1992). The Barnadesioideae, a small South American group of mainly trees and shrubs, is the sister group to the remaining genera (Figure 9.147). This group lacks the chloroplast DNA inversion that characterizes the remaining species (Jansen and Palmer 1987). The remaining tribes are more or less equally divided into the "Cichorioideae" and the Asteroideae (K. Bremer 1987, 1994; Carlquist 1976; Thorne 1992). The former is paraphyletic, but is retained here because phylogenetic relationships within the complex are still incompletely known; it is sometimes further divided (K. Bremer 1996). "Cichorioideae" are characterized by style branches with the inner surface stigmatic. Their heads are usually discoid, except in the distinct tribe Lactuceae, which has ligulate heads. Both resin canals and laticifers occur within this subfamily, and the latex system is especially well developed in Lactuceae (Table 9.3). Lactuceae are phenetically distinct and have sometimes been placed in their own subfamily (Cronquist 1955, 1977, 1981). The monophyletic Asteroideae can be diagnosed by the restriction of the stigmatic tissue to two marginal lines on each style branch; loss of laticifers; presence of ray flowers (and radiate heads, although these are lost in some genera); disk flowers with short lobes (but secondarily elongate in some), and cpDNA restriction site and *rbcL* sequence characters. Presumed synapomorphies of the currently recognized tribes are outlined in Table 9.3 (see also Bremer 1994; Solbrig 1963). Morphological features that are taxonomically important at the tribal level include characteristics of the style branches, (e.g., loca-

Figure 9.146 Asteraceae (Compositae), subfamily Asteroideae (with disk and ray flowers). (A–J) *Helianthus tuberosus:* (A) flowering plant (× 0.3); (B) head of flowers (× 0.5); (C) ray flower, sterile (× 2); (D) head, ray flowers removed, showing involucral bracts (phyllaries) (× 0.5); (E) disk flower and bract (× 10); (F) disk flower in longitudinal section, note anthers, pappus scales, and basal ovule (× 0.5); (G) style branches with their stigmatic lines (greatly magnified); (H) mature achene with pappus scales (× 10); (I) achene without pappus (× 8); (J) rhizome with overwintering tubers (× 1). (K–L) *H. annuus:* (K) achene (× 4.5); (L) embryo (× 4.5). (From Wood 1974, *A student's atlas of flowering plants*, p. 117.)

TABLE 9.3 *Characteristics and synapomorphies of tribes and subfamilies of Asteraceae (= Compositae).*

Tribe	Subfamily	Genera/species	Representative genera
1. Barnadesieae	Barnadesioideae	9/92	*Barnadesia, Chuquiraga, Dasyphyllum*
2. "Mutisieae"	"Cichorioideae"	76/970	*Chaptalia, Gerbera, Gochnatia, Mutisia, Trixis*
3. Cardueae (= Cynareae) (Thistles)	"Cichorioideae"	83/2500	*Carduus, Centaurea, Cirsium, Cynara, Echinops*
4. Lactuceae (= Cichorieae) (Dandelions and relatives)	"Cichorioideae"	98/1550	*Cichorium, Crepis, Hieracium, Krigia, Lactuca, Pyrrhopappus, Sonchus, Taraxacum, Tragopogon, Youngia*
5. Vernonieae (Ironweeds and relatives)	"Cichorioideae"	98/1300	*Elephantopus, Vernonia*
6. Liabeae	"Cichorioideae"	14/160	*Liabum*
7. Arctoteae (thistle-like plants)	"Cichorioideae"	16/200	*Arctotis, Berkheya*
8. Inuleae	Asteroideae	38/480	*Inula, Pulicaria, Telekia*
9. Plucheae	Asteroideae	28/220	*Pluchea, Sphaeranthus*
10. Gnaphalieae	Asteroideae	162/2000	*Anaphalis, Antennaria, Gamochaeta, Gnaphalium, Leontopodium*
11. Calenduleae	Asteroideae	8/110	*Calendula, Osteospermum*
12. Astereae	Asteroideae	174/2800	*Aster, Baccharis, Conyza, Erigeron, Haplopappus, Solidago*
13. Anthemideae	Asteroideae	109/1740	*Achillea, Anthemis, Argyranthemum, Artemisia, Chrysanthemum, Leucanthemum, Seriphidium, Tanacetum*
14. Senecioneae	Asteroideae	120/3200	*Erechtites, Senecio*
15. "Helenieae"	Asteroideae	110/830	*Arnica, Flaveria, Gaillardia, Helenium, Pectis, Tagetes*
16. Heliantheae	Asteroideae	189/2500	*Ambrosia, Bidens, Calea, Coreopsis, Cosmos, Dahlia, Helianthus, Iva, Rudbeckia, Verbesina, Viguiera, Zinnia*
17. Eupatorieae	Asteroideae	170/2400	*Ageratum, Carphephorus, Eupatorium, Garberia, Liatris, Mikania*

TABLE 9.3 *(continued)*

Flower types	Pappus types	Style branches	Major synapomorphies
Disk; 1+4 bilabiate	Usually bristles with long hairs	Inner surface stigmatic	Axillary spines; pubescence of long hairs on corolla, achene, pappus
Variable; 2+3 bilabiate, especially toward margin of head and/ or disk	Usually bristles	Inner surface stigmatic	None
Disk (deeply lobed)	Usually bristles or scales	Inner surface stigmatic (often with fused style branches)	Leaves dissected; leaves and involucral bracts spine-tipped; ring of hairs below style branches
Ligulate	Usually bristles	Inner surface stigmatic	Ligulate flowers; copious milky latex
Usually disk (deeply lobed)	Usually bristles	Inner suface stigmatic (near base)	Anthers with glandular apical appendages; style branches long, slender, pilose, acute at apex; anatomical details of anthers (endothecial thickenings)
Usually disk (deeply lobed) and ray	Bristles or scales	Inner surface stigmatic	Leaves opposite; trinerved ray flowers
Usually disk (deeply lobed) and ray	Usually scales or short cup	Inner surface stigmatic	Ring of hairs below style branches; anther not caudate; ray flowers; often spiny
Disk (lobes short) and ray	Scales or capillary bristles	Marginal stigmatic lines	Marginal flowers filiform; elongate crystals in epidermis of achene
Usually disk (lobes short)	Scales or capillary bristles	Marginal stigmatic lines	Marginal flowers filiform; style branches with apically rounded, sweeping hairs along the branches and on upper part of shaft
Usually disk (lobes short), rarely with rays	Usually capillary bristles	Marginal stigmatic lines	Pollen grains with thick basal layer regularly perforated (gnaphaloid); upper level of prominent columellae; $x = 7$
Disk (lobes short) and ray	Lacking	Marginal stigmatic lines	Loss of pappus
Usually disk (lobes short) and ray (rays sometimes lacking)	Usually bristles or scales	Marginal stigmatic lines	Epidermal cells of ray flower corolla with median thickening in outer wall; style branches with triangular-subulate sterile appendages, i.e., asteroid style branches adaxially glabrous
Usually disk (lobes short) and ray (but ray flowers sometimes lost)	Scales, short cup, or lacking	Marginal stigmatic lines	Leaves pinnately dissected; involucral bracts with scarious margins; scaly or reduced pappus; truncate style branches; ray epidermal cells papillose
Disk (lobes short) and ray (or only disk)	Usually capillary bristles	Marginal stigmatic lines	Involucre usually uniseriate (single whorl of bracts); chemistry of sesquiterpene lactones, pyrrolizidine alkaloids
Disk (lobes short) and ray (these sometimes lost)	Scales, short cup, bristles	Marginal stigmatic lines	None, due to segregation of following two tribes—based on opposite leaves, carbonized achene wall, endothecium of short cells (bracts lacking on receptacle)
Usually disk (lobes short) and ray (these sometimes lacking)	Awns, scales, bristles, or lacking	Marginal stigmatic lines	Bracts on receptacle; black anthers
Disk (lobes short, or occasionally long)	Bristles	Marginal stigmatic lines	Loss of ray flowers; hairy style bases and style branches with very long sterile appendages; style branches glandular between stigmatic lines; anatomy of anther endothecium

Figure 9.147 Cladogram showing hypothesized relationships within the Asteraceae. (Adapted from K. Bremer 1994.)

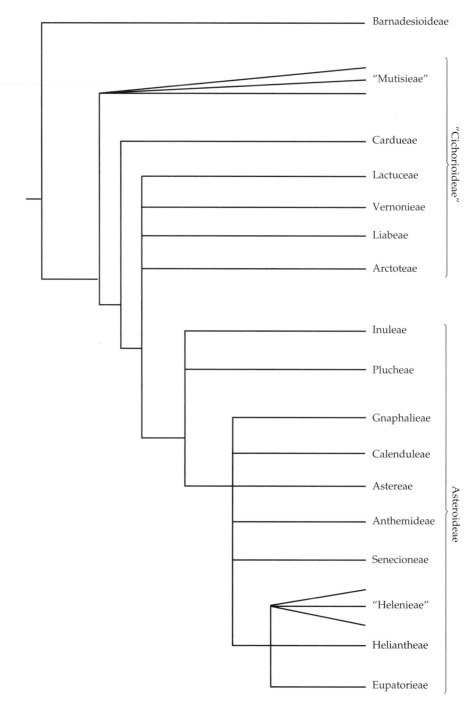

tion of stigmatic region, presence of hairs or sterile appendages, length and width, apex form); pappus form; corolla anatomy and shape; pollen morphology; anatomical and morphological features of the achenes; anatomy and form of the anthers; leaf arrangement; and presence or lack of nodal or marginal spines.

Crepis, Aster, Taraxacum, Tragopogon, Hieracium, and other genera are taxonomically difficult at the species level due to the combined action of hybridization, polyploidy, and agamospermy.

The tiny flowers of Asteraceae are not readily apparent; the involucrate heads usually function as—and at first glance may even appear to be—a single flower. In radiate heads the ray flowers serve to attract pollinators and the disk flowers mature centripetally. Pollinators usually land on the ray flowers and deposit pollen from other plants on the stigmas of older, marginal disk flowers. The filaments of many Asteraceae respond to touch by contracting abruptly to force pollen out of the plunger mechanism and onto the pollinator's body.

Corolla color is variable. Composite inflorescences are generally outcrossing and attract a wide array of generalist pollinators (butterflies, bees, flies, and beetles), but pollination by solitary bees is especially common. Some genera have reduced flowers that are wind-pollinated (e.g., *Ambrosia* and *Baccharis*), and some have heads reduced to a single flower, but these reduced heads are then aggregated into compound heads (e.g., *Echinops*).

The achenes of most members of Asteraceae are dispersed by wind, with the pappus bristles functioning as a parachute. The flattened and often winged fruit assists in wind dispersal. External transport on birds or mammals is facilitated by pappus modifications such as awns with retrorse barbs, fruit outgrowths, such as hooks or spines, or specialized involucral bracts.

References: Anderberg 1991a,b,c; Arriagada and Miller 1997; Brehm and Ownbey 1965; K. Bremer 1987, 1994, 1996; Bremer and Jansen 1992; Carlquist 1976; Cronquist 1955, 1977, 1980, 1981; Gustafsson 1996; Hansen 1992; Heywood et al. 1977; Jansen et al. 1991, 1992; Jansen and Palmer 1987; Jones 1982; Karis 1993; Karis et al. 1992, 2001; Keeley and Jansen 1991; Kim et al. 1992; Lane 1996; Leins and Erbar 1990; Michaels et al. 1993; Ownbey 1950; Rieseberg 1991; Scott 1990; Solbrig 1963; Soltis and Soltis 1989; Thorne 1992; Urtubey and Stuessy 2001; Vuilleumier 1969, 1973; Wagenitz 1992; Xiaoping and Bremer 1993.

Suggested Readings

In addition to the numerous books and scientific articles cited after each family, the following references will serve as additional sources of information on the morphological variation, geographic distribution, economic importance, and evolutionary relationships of angiosperm families.

Angiosperm Phylogeny Group. 1998. An ordinal classification for the families of flowering plants. *Ann. Missouri Bot. Gard.* 85: 531–553.

Angiosperm Phylogeny Group. 2002. An update of the Phylogeny Group classification for the orders and families of flowering plants: APGII. *Bot. J. Linnean Soc.* In press.

Bailey, L. H. and E. Z. Bailey. 1976. *Hortus third: A concise dictionary of plants cultivated in the United States and Canada.* Revised by the Bailey Hortorium. Macmillan, New York.

Benson, L. 1979. *Plant classification*, 2nd ed. D.C. Heath, Lexington, MA.

Cronquist, A. 1981. *An integrated system of classification of flowering plants.* Columbia University Press, New York.

Cronquist, A. 1988. *The evolution and classification of flowering plants*, 2nd ed. New York Botanical Garden, Bronx, NY.

Dahlgren, R. M. T. 1983. General aspects of angiosperm evolution and macrosystematics. *Nord. J. Bot.* 3: 119–149.

Dahlgren, R. M. T., H. T. Clifford and P. F. Yeo. 1985. *The families of the monocotyledons.* Springer-Verlag, Berlin.

Engler, A. and K. Prantl. 1887–1915. *Die natürlichen Pflanzenfamilien*, parts II-IV. G. Kreysing, Leipzig. [And ed. 2, incomplete, 1924–.]

Heywood, V. H. (ed.). 1978. *Flowering plants of the world.* Mayflower Books, New York.

Hickey, M. and C. J. King. 1988. *100 families of flowering plants of the world*, 2nd ed. Cambridge University Press, Cambridge.

Hutchinson, J. 1973. *The families of flowering plants*, 3rd ed. Oxford University Press, Oxford.

Jones, S. B. and L. E. Luchsinger. 1986. *Plant systematics.* McGraw-Hill, New York.

Kubitzki, K. 1990–1998. *The families and genera of vascular plants.* 4 vols. Springer–Verlag, Berlin. [Additional volumes are planned; detailed taxonomic treatments of families and genera with descriptions, information on anatomy, embryology, chromosomes, pollen structure, pollination, dispersal, and phylogeny; keys to genera are provided.]

Lawrence, G. H. M. 1951. *Taxonomy of vascular plants.* Macmillan, New York.

Maas, P. J. M. and L. Y. T. Westra. 1993. *Neotropical plant families.* Koeltz Scientific Books, Konigstein.

Mabberley, D. J. 1997. *The plant-book,* 2nd ed. Cambridge University Press, Cambridge.

Melchior, H. (ed.) 1964. *A. Engler's Syllabus der Pflanzenfamilien*, 12th ed. Vol. 2. Gebruder Borntraeger, Berlin.

Rendle, A. B. 1925. *The classification of flowering plants.* 2 vols. Cambridge University Press, Cambridge.

Stevens, P. F. (2001 onwards.) Angiosperm Phylogeny Website. Version 2 Aug 2001. http://www.mobot.org/MOBOT/research/Apweb/.

Takhtajan, A. 1980. Outline of the classification of flowering plants (Magnoliophyta). *Bot. Rev.* 46: 225–359.

Takhtajan, A. 1997. *Diversity and classification of flowering plants.* Columbia University Press, New York.

Thorne, R. F. 1992. Classification and geography of the flowering plants. *Bot. Rev.* 58: 225–348.

Thorne, R. F. 2000. The classification and geography of the monocotyledon subclasses Alismatidae, Liliidae, and Commelinidae. Pp. 75-124 *in* B.

Nordenstam, G. El-Ghazaly, M. Kassas, and T. C. Laurent (eds.), *Plant systematics for the 21st century.* Portland Press, London.

Thorne, R. F. 2001. The classification and geography of the flowering plants: dicotyledons of the class Angiospermae. *Bot. Rev.* 66: 441-647.

Van Balgooy, M. M. J. 1997. *Malesian seed plants. Vol. 1. Spot-characters. An aid for identification of families and genera.* Rijksherbarium/Hortus Botanicus, Leiden.

Van Balgooy, M. M. J. 1998. *Malesian seed plants. Vol. 2. Portraits of tree families.* Rijksherbarium/Hortus Botanicus, Leiden.

Van Balgooy, M. M. J. 2001. *Malesian seed plants. Vol. 3. Portraits of non-tree families.* Rijksherbarium/Hortus Botanicus, Leiden.

Watson, L. and M. J. Dallwitz. 1997. The families of flowering plants: Descriptions and illustrations. Website: <http://muse.bio.cornell.edu/delta/angio/www/index.htm>

Willis, J. C. 1973. *A dictionary of flowering plants and ferns*, 8th ed. Revised by H. K. Airy-Shaw. University Printing House, Cambridge.

Zomlefer, W. B. 1994. *Guide to flowering plant families.* The University of North Carolina Press, Chapel Hill.

Literature Cited

Abbe, E. C. 1935. Studies in the phylogeny of the Betulaceae. I. Floral and inflorescence anatomy and morphology. *Bot. Gaz.* 97: 1–67.

Abbe, E. C. 1974. Flowers and inflorescences of "Amentiferae." *Bot. Rev.* 40: 159–261.

Abbe, E. C. and T. T. Earle. 1940. Inflorescence, floral anatomy, and morphology of *Leitneria floridana. Bull. Torrey Bot. Club* 67: 173–193.

Abu-Asab, M. S. and P. D. Cantino. 1989. Pollen morphology of *Trichostema* (Labiatae) and its systematic implications. *Syst. Bot.* 14: 359–369.

Adams, P. 1962. Studies in the Guttiferae. II. Taxonomic and distributional observations on North American taxa. *Rhodora* 64: 231–242.

Adams, P. and R. K. Godfrey. 1961. Observations on the *Sagittaria subulata* complex. *Rhodora* 63: 247–266.

Agababian, V. S. 1972. Pollen morphology of the Magnoliaceae. *Grana* 12: 166–176.

Albach, D. C., P. S. Soltis, D. E. Soltis, and R. G. Olmstead. 2001a. Phylogenetic analysis of asterids based on sequences of four genes. *Ann. Missouri Bot. Gard.* 88: 162-212.

Albach, D. C., P. S. Soltis, D. E. Soltis. 2001b. Patterns of embryological and biochemical evolution in the asterids. *Syst. Bot.* 26: 242-262.

Albach, D. C., D. E. Soltis, M. W. Chase, and P. S. Soltis. 2001c. Phylogenetic placement of the enigmatic angiosperm *Hydrostachys. Taxon* 50: 781-805.

Al-Shehbaz, I. and B. G. Schubert. 1989. The Dioscoreaceae of the southeastern United States. *J. Arnold Arbor.* 70: 57–95.

Al-Shehbaz, I. A. 1984. The tribes of Cruciferae (Brassicaceae) in the southeastern United States. *J. Arnold Arbor.* 65: 343–373.

Al-Shehbaz, I. A. 1985a. The genera of Brassiceae (Cruciferae; Brassicaceae) in the southeastern United States. *J. Arnold Arbor.* 66: 279–351.

Al-Shehbaz, I. A. 1985b. The genera of Thelypodieae (Cruciferae; Brassicaceae) in the southeastern United States. *J. Arnold Arbor.* 66: 95–111.

Al-Shehbaz, I. A. 1987. The genera of Alysseae (Cruciferae; Brassicaceae) in the southeastern United States. *J. Arnold Arbor.* 68: 185–240.

Al-Shehbaz, I. A. 1988a. The genera of Anchonieae (Hesperideae) (Cruciferae; Brassicaceae) in the southeastern United States. *J. Arnold Arbor.* 69: 193–212.

Al-Shehbaz, I. A. 1988b. The genera of Arabideae (Cruciferae; Brassicaceae) in the southeastern United States. *J. Arnold Arbor.* 69: 85–166.

Al-Shehbaz, I. A. 1991. The genera of Boraginaceae in the southeastern United States. *J. Arnold Arbor.*, Suppl. 1: 1–169.

Albert, V. A., S. E. Williams, and M. W. Chase. 1992. Carnivorous plants: Phylogeny and structural evolution. *Science* 257: 1491–1495.

Alice, L. A., T. Eriksson, B. Eriksen, and C. S. Campbell. 2001. Hybridization and gene flow between distantly related species of *Rubus* (Rosaceae): Evidence from nuclear ribosomal DNA internal transcribed spacer region sequences. *Syst. Bot.* 26: 769–778.

Allard, H. A. 1947. The direction of twist of the corolla in the bud, and the twining of the stems in Convolvulaceae and Dioscoreaceae. *Castanea* 12: 88–94.

Alverson, W. S., K. G. Karol, D. A. Baum, M. W. Chase, S. M. Swensen, R. McCourt, and K. Sytsma. 1998a. Circumscription of the Malvales and relationships to other Rosidae: evidence from *rbcL* sequence data. *Amer. J. Bot.* 85: 876–887.

Alverson, W. S., B. A. Whitlock, R. Nyffeler, and C. Bayer. 1999. Phylogeny of core Malvales: Evidence from *ndhF* sequence data. *Amer. J. Bot.* 86: 1474-1486.

Ambrose, J. D. 1980. A re-evaluation of the Melanthioideae (Liliaceae) using numerical analyses. In *Petaloid monocotyledons*, Linnean Society Symposium Series no. 8, C. D. Brickell, D. F. Cutler and M. Gregory (eds.), 65–81. Academic Press, London.

Ambrose, J. D. 1985. *Lophiola*, familial affinity with the Liliaceae. *Taxon* 34: 140–150.

Anderberg, A. A. 1991a. Taxonomy and phylogeny of the tribe Plucheeae (Asteraceae). *Plant Syst. Evol.* 176: 145–177.

Anderberg, A. A. 1991b. Taxonomy and phylogeny of the tribe Inuleae (Asteraceae). *Plant Syst. Evol.* 176: 75–123.

Anderberg, A. A. 1991c. Taxonomy and phylogeny of the tribe Gnaphalieae (Asteraceae). *Opera Botanica* 104: 1–195.

Anderberg, A. A. 1992. The circumscription of the Ericales, and their cladistic relationships to other families of "higher" dicotyledons. *Syst. Bot.* 17: 660–675.

Anderberg, A. A. 1993. Cladistic relationships and major clades of the Ericales. *Plant Syst. Evol.* 184: 207–231.

Anderberg, A. A. 1994. Cladistic analysis of *Enkianthus* with notes on the early diversification of the Ericaceae. *Nordic J. Bot.* 14: 385–401.

Anderberg, A. A. and B. Ståhl. 1995. Phylogenetic interrelationships in the order Primulales, with special emphasis on the family circumscriptions. *Canad. J. Bot.* 73: 1699–1730.

Anderberg, A. A. and P. Eldenäs. 1991. A cladistic analysis of *Anigozanthos* and *Macropidia* (Haemodoraceae). *Austral. J. Bot.* 4: 655–664.

Anderberg, A. A., B. Ståhl, and M. Källersjö. 1998. Phylogenetic relationships in the Primulales inferred from *rbcL* sequence data. *Pl. Syst. Evol.* 211: 93-102.

Anderberg, A. A., B. Ståhl, and M. Källersjö. 2000. Maesaceae, a new primuloid family in the order Ericales s. l. *Taxon* 49: 183-187.

Anderberg, A. A., C. Rydin, and M. Källersjö. 2002. Phylogenetic relationships in the order Ericales s. l.: analyses of molecular data from five genes from the plastid and mitochondrial genomes. *Amer. J. Bot.* 89: 677–689.

Anderson, F. 2001. *The cactus family*. Timber Press, Portland, Oregon.

Anderson, W. R. 1973. A morphological hypothesis for the origin of heterostyly in the Rubiaceae. *Taxon* 22: 537–542.

Anderson, W. R. 1977. Byrsonimoideae, a new subfamily of the Malpighiaceae. *Leandra* 7: 5–18.

Anderson, W. R. 1979. Floral conservation in neotropical Malpighiaceae. *Biotropica* 11: 219–223.

Anderson, W. R. 1990. The origin of the Malpighiaceae: The evidence from morphology. *Mem. New York Bot. Gard.* 64: 210–224.

Andersson, L. 1981. The neotropical genera of Marantaceae: Circumscription and relationships. *Nordic J. Bot.* 1: 218–245.

Andersson, L. 1992. A provisional checklist of neotropical Rubiaceae. *Scripta Bot. Belg.* 1: 1–200.

Andersson, L. 1998. Marantaceae. In *The families and genera of vascular plants*. Vol. 4. Flowering plants—Monocotyledons, K. Kubitzki (ed.), 278-293.

Andersson, L. and M. W. Chase. 2001. Phylogeny and classification of Marantaceae. *Bot. J. Linn. Soc.* 135: 275-287.

Andersson, L. and H. E. Rova. 1999. The rps16 intron and the phylogeny of the Rubioideae (Rubiaceae). *Pl. Syst. Evol.* 214: 161–189.

Andreasen, K. and B. Bremer. 2000. Combined phylogenetic analysis in the Rubiaceae—Ixoroideae: morphology, nuclear and chloroplast DNA data. *Amer. J. Bot.* 87: 1731-1748.

Angiosperm Phylogeny Group. 1998. An ordinal classification for the families of flowering plants. *Ann. Missouri Bot. Gard.* 85: 531–553.

Angiosperm Phylogeny Group. 2002. An update of the Phylogeny Group classification for the orders and families of flowering plants: APGII. *Bot. J. Linnaean Soc.* In press.

Applequist, W. L. and R. S. Wallace. 2001. Phylogeny of the Portulacaceous cohort based on *ndhF* sequence data. *Syst. Bot.* 26: 406-419.

Argus, G. W. 1974. An experimental study of hybridization and pollination in *Salix* (willows). *Canad. J. Bot.* 52: 1613–1619.

Argus, G. W. 1986. The genus *Salix* (Salicaceae) in the southeastern United States. *Syst. Bot. Monogr.* 9: 1–170.

Armstrong, J. E. 1985. The delimitation of Bignoniaceae and Scrophulariaceae based on floral anatomy, and the placement of problem genera. *Amer. J. Bot.* 755–766.

Armstrong, J. E. and B. A. Drummond III. 1986. Floral biology of *Myristica fragrans* Houtt., the nutmeg of commerce. *Biotropica* 18: 32-38.

Arriagada, J. E. and N. G. Miller. 1997. The genera of Anthemideae (Compositae; Asteraceae) in the southeastern United States. *Harvard Pap. Bot.* No. 11: 1-46.

Arroyo, M. T. K. 1981. Breeding systems and pollination biology in Leguminosae. In *Advances in legume systematics*, part 2, R. M. Polhill and P. H. Raven (eds.), 723–769. Royal Botanic Gardens, Kew.

Asmussen, C. B., W. J. Baker, and J. Dransfield. 2000. Phylogeny of the palm family (Arecaceae) based on rps16 intron and *trnL-trnF* plastid DNA sequences. In *Monocots: Systematics and evolution*, K. L. Wilson and D. A. Morrison (eds.), 525-535. Collingwood, Australia.

Asmussen, C. B. and M. W. Chase. 2001. Coding and noncoding plastid DNA in palm systematics. *Amer. J. Bot.* 88: 1103-1117.

Augspurger, C. K. 1989. Morphology and aerodynamics of wind-dispersed legumes. *Monogr. Syst. Bot. Missouri Bot. Gard.* 29: 451–466.

Austin, D. F. 1979. Studies of the Florida Convolvulaceae I. Key to genera. *Florida Scientist* 42: 214–216.

Backlund, A. and B. Bremer. 1997. Phylogeny of the Asteridae s. str. based on rbcL sequences, with particular reference to the Dipsacales. *Pl. Syst. Evol.* 207: 225--254.

Backlund, A. and K. Bremer. 1998. To be or not to be: Principles of classification and monotypic plant families. *Taxon* 47: 391–401.

Backlund, A. and N. Pyck. 1998. Diervillaceae and Linnaeaceae, two new families of caprifolioids. *Taxon* 47: 657–661.

Backlund, A., B. Oxelman, and B. Bremer. 2000. Phylogenetic relationships within the Gentianales based on ndhF and rbcL sequences, with particular reference to the Loganiaceae. *Amer. J. Bot.* 87: 1029-1043.

Bailey, I. W. and C. G. Nast. 1945. The comparative morphology of the Winteraceae. VII. Summary and conclusions. *J. Arnold Arbor.* 26: 37–47.

Bailey, I. W. and B. G. L. Swamy. 1948. *Amborella trichopoda* Baill., a new morphological type of vesselless dicotyledon. *J. Arnold Arbor.* 29: 245-254.

Baker, H. G. 1986. Yuccas and yucca moths: A historical commentary. *Ann. Missouri Bot. Gard.* 73: 556–564.

Baranov, A. I. 1981. Studies in the Begoniaceae. *Phytologia Mem.* 4: 1-88.

Barker, N. P., H. P. Linder and E. H. Harley. 1995. Polyphyly of Arundinoideae (Poaceae): Evidence from rbcL sequence data. *Syst. Bot.* 20: 423–435.

Barkman, T. J., G. Chenery, J. R. McNeal, J. Luons-Weiler, W. J. Ellisens, G. Moore, A. D. Wolfe, and C. W. dePamphilis. 2000. Independent and combined analyses of sequences from all three genomic compartments converge on the root of flowering plant phylogeny. *Proc. Natl. Acad. Sci. U.S.A.* 97: 13166-13171.

Barrett, S. C. H. and S. W. Graham. 1997. Adaptive radiation in the aquatic plant family Pontederiaceae: insights from phylogenetic analyses. In *Molecular evolution and adaptive radiation*, T. J. Givnish and K. J. Sytsma (eds.), 225-238. Cambridge University Press, Cambridge.

Barrett, S. C. H. and J. H. Richards. 1990. Heterostyly in tropical plants. *Mem. New York Bot. Gard.* 55: 35–61.

Barthlott, W. and D. Fröhlich. 1983. Micromorphologie und Orientierungs muster epicuticularer Wachs-Kristalloide: ein neues systematisches Merkmal bei Monokotylen. *Plant Syst. Evol.* 142: 171–185.

Barthlott, W. and D. R. Hunt. 1993. Cactaceae. In *The families and genera of vascular plants.* vol. 2. Magnoliid, hamamelid and caryophyllid families, K. Kubitzki, J. G. Rohwer, and V. Bittrich (eds.), 161–197. Springer-Verlag, Berlin.

Balslev, H. 1998. Juncaceae. In *The families and genera of vascular plants,* vol. 4. Flowering plants—Monocotyledons, K. Kubitzki (ed.), 252-260. Springer-Verlag, Berlin.

Baum, D. A., W. S. Alverson and R. Nufeeler. 1998. A durian by any other

name: Taxonomy and nomenclature of the core Malvales. *Harvard Pap. Bot.* 3: 315–330.

Baumann, M. G. 1946. *Myodocarpus* und die Phylogenie der Umbelliferen-Frücht. *Ber. Schweiz. Bot. Ges.* 56: 13–112.

Bayer, C. 1998. Synflorescences of Malvaceae. *Nordic J. Bot.* 18: 335-338.

Bayer, C. 1999. The bicolor-unit—homology and transformation of an inflorescence structure unique to core Malvales. *Pl. Syst. Evol.* 214: 187-198.

Bayer, C., M. F. Fay, A. Y. de Bruijn, V. Savolainen, C. M. Morton, K. Kubitzki, W. S. Alverson, and M. W. Chase. 1999. Support for an expanded family concept of Malvaceae within a re-circumscribed order Malvales: combined analysis of plastid atpB and rbcL DNA sequences. *Bot. J. Linn. Soc.* 129: 267-303.

Bayer, R. J., L. Hufford, and D. E. Soltis. 1996. Phylogenetic relationships in Sarraceniaceae based on rbcL and ITS sequences. *Syst. Bot.* 21: 121–134.

Bechtel, A. R. 1921. The floral anatomy of the Urticales. *Amer. J. Bot.* 8: 386–410.

Beck, C. B. (ed.) 1976. *Origin and early evolution of angiosperms.* Columbia University Press, New York.

Behnke, H-D. 1976. Ultrastructure of sieve-element plastids in Caryophyllales (Centrospermae), evidence for the delimitation and classification of the order. *Plant Syst. Evol.* 126: 31–54.

Behnke, H-D. 1994. Sieve-element plastids: their significance for the evolution and systematics of the order. In *Caryophyllales: Evolution and systematics,* H.-D. Behnke and T. J. Mabry (eds.), 87–121. Springer-Verlag, Berlin.

Behnke, H.-D. 1999. P-type sieve-element plastids present in members of the tribes Triplareae and Coccolobeae (Polygonaceae) renew the links between the Polygonales and the Caryophyllales. *Pl. Syst. Evol.* 214: 15-27.

Behnke, H.-D. and W. Barthlott. 1983. New evidence from the ultrastructural and micromorphological fields in angiosperm classification. *Nordic J. Bot.* 3: 43–66.

Behnke, H.-D. and T. J. Mabry (eds.). 1994. *Caryophyllales: Evolution and systematics.* Springer-Verlag, Berlin.

Bensel, C. R. and B. F. Palser. 1975. Floral anatomy in the Saxifragaceae sensu lato. III. Kirengeshomoideae, Hydrangeoideae, and Escallonioideae. *Amer. J. Bot.* 62: 676–687.

Benson, N. L. 1982. *The cacti of the United States and Canada.* Stanford University Press, Stanford.

Benzing, D. H. 1980. *The biology of the bromeliads.* Mad River Press, Eureka, CA.

Benzing, D. H., J. Seemann and A. Renfrow. 1978. The foliar epidermis in Tillandsioideae (Bromeliaceae) and its role in habitat selection. *Amer. J. Bot.* 65: 359–365.

Berg, C. C. 1977. Urticales, their differentiation and systematic position. *Plant Syst. Evol.,* Suppl. 1: 349–374.

Berg, C. C. 1978. Cecropiaceae, a new family of the Urticales. *Taxon* 27: 39–44.

Berg, C. C. 1989. Systematics and phylogeny of the Urticales. In *Evolution, systematics, and fossil history of the Hamamelidae,* vol. 2, "Higher" Hamamelidae. Syst. Assoc. Special Vol. 40B, P. R. Crane and S. Blackmore (eds.), 193–200. Clarendon Press, Oxford.

Berg, R. Y. 1996. Development of ovule, embryo sac, and endosperm in *Dipterostemonx* and *Dichelostemma* (Alliaceae) relative to taxonomy. *Amer. J. Bot.* 83: 790–801.

Bessey, C. E. 1915. The phylogenetic taxonomy of flowering plants. *Ann. Missouri Bot. Gard.* 2: 109–164.

Bharatham, G. and E. A. Zimmer. 1995. Early branching events in monocotyledons: Partial 18S ribosomal DNA sequence analysis. In *Monocotyledons: Systematics and evolution,* P. J. Rudell, P. J. Cribb, D. F. Cutler and C. J. Humphries (eds.), 81–107. Royal Botanic Gardens, Kew.

Bidartondo, M. I. and T. D. Bruns. 2001. Extreme specificity in epiparasitic Monotropoideae (Ericaceae): widespread phylogenetic and geographical structure. *Mol. Ecol.* 10: 2285-2295.

Bigazzi, M. 1989. Ultrastructure of nuclear inclusions and the separation of Verbenaceae and Oleaceae (including *Nyctanthes*). *Plant Syst. Evol.* 163: 1–12.

Bittrich, V. 1993a. Caryophyllaceae. In *The families and genera of vascular plants,* vol. 2. Magnoliid, hamamelid and caryophyllid families, K. Kubitzki, J. G. Rohwer, and V. Bittrich (eds.), 206–236. Springer-Verlag, Berlin.

Bittrich, V. 1993b. Introduction to Centrospermae. In *The families and genera of vascular plants,* vol. 2, Magnoliid, hamamelid and caryophyllid families, K. Kubitzki, J. G. Rohwer, and V. Bittrich (eds.), 13–19. Springer-Verlag, Berlin.

Bittrich, V. and H. Hartmann. 1988. The Aizoaceae: A new approach. *Bot. J. Linn. Soc.* 97: 239–254.

Bittrich, V. and U. Kühn. 1993. Nyctaginaceae. In *The families and genera of vascular plants,* vol. 2, Magnoliid, hamamelid and caryophyllid families, K. Kubitzki, J. G. Rohwer, and V. Bittrich (eds.), 473–486. Springer-Verlag, Berlin.

Bittrich, V. and M. Struck. 1989. What is primitive in Mesembryanthemaceae? *S. Afr. J. Bot.* 55: 321–331.

Blackwell, W. H. 1977. The subfamilies of the Chenopodiaceae. *Taxon* 26: 395–397.

Blattner, F. R. and J. W. Kadereit. 1995. Three intercontinental disjunctions in Papaveraceae subfamily Chelidonioideae: evidence from chloroplast DNA. *Plant Syst. Evol.* Suppl. 9: 147–157.

Boeshore, I. 1920. The morphological continuity of Scrophulariaceae and Orobanchaceae. *Contr. Bot. Lab. Morris Arbor.* 5: 139–177.

Boesewinkel, F. D. 1988. The seed structure and taxonomic relationships of *Hypseocharis Remy. Acta Bot. Neerl.* 37: 111–120.

Bogle, A. L. 1969. The genera of Portulacaceae and Basellaceae in the southeastern United States. *J. Arnold Arbor.* 50: 566–598.

Bogle, A. L. 1970. The genera of Molluginaceae and Aizoaceae in the southeastern United States. *J. Arnold Arbor.* 51: 431–462.

Bogle, A. L. 1974. The genera of Nyctaginaceae in the southeastern United States. *J. Arnold Arbor.* 55: 1–37.

Bogle, A. L. 1986. The floral morphology and vascular anatomy of the Hamamelidaceae subfamily Liquidambaroideae. *Ann. Missouri Bot. Gard.* 73: 325–347.

Bogler, D. J. 1998. Nolinaceae. In *The families and genera of vascular plants,* vol 3. Flowering plants—Monocotyledons, K. Kubitzki (ed.), 392-397. Springer-Verlag, Berlin.

Bogner, J. and D. H. Nicolson. 1991. A revised classification of Araceae with dichotomous keys. *Willdenowia* 21: 35–50.

Bogler, D. J. and B. B. Simpson. 1995. A chloroplast DNA study of the Agavaceae. *Syst. Bot.* 20: 191–205.

Bogler, D. J. and B. B. Simpson. 1996. Phylogeny of Agavaceae based on ITS rDNA sequence variation. *Amer. J. Bot.* 83: 1225–1235.

Bogner, J. and S. J. Mayo. 1998. Acoraceae. In *The families and genera of vascular plants,* vol. 4. Flowering plants—Monocotyledons, K. Kubitzki (ed.), 7-11. Springer-Verlag, Berlin.

Bohs, L. and R. G. Olmstead. 1997. Phylogenetic relationships in *Solanum* (Solanaceae) based on *ndhF* sequences. *Syst. Bot.* 22: 5–17.

Boke, N. H. 1964. The cactus gynoecium: a new interpretation. *Amer. J. Bot.* 51: 598–610.

Bonsen, K. and B. J. H. ter Welle. 1983. Comparative wood and leaf anatomy of the Cecropiaceae (Urticales). *Bull. Mus. Nat. Hist. Nat. Paris,* Sér. 4, 5 (sect. B, Adansonia, No. 2): 151–177.

Boothroyd, L. E. 1930. The morphology and anatomy of the inflorescence and flower of the Platanaceae. *Amer. J. Bot.* 17: 678–693.

Borstein, A. J. 1991. The Piperaceae in the southeastern United States. *J. Arnold Arbor.,* Suppl. 1: 349–366.

Bortiri, E., S.-H. Oh, J. Jiang, S. Baggett, A. Granger, C. Weeks, M. Buckingham, D. Potter, and D. E. Parfitt. 2001. Phylogeny and systematics of *Prunus* (Rosaceae) as determined by sequence analysis of ITS and chloroplast *trnL-trnF* spacer DNA. *Syst. Bot.* 26: 797-807.

Bos, J. 1998. Dracaenaceae. In *The families and genera of vascular plants.* Vol. 3. Flowering plants—Monocotyledons, K. Kubitzki (ed.), 238-241. Springer-Verlag, Berlin.

Bouman, F. 1995. Seed structure and systematics in Dioscoreales. In *Monocotyledons: Systematics and evolution,* P. J. Rudall, P. J. Cribb, D. F. Cutler, and C. J. Humphries (eds.), 139–156. Royal Botanic Gardens, Kew.

Boyle, E. M. 1980. Vascular anatomy of the flower, seed and fruit of *Lindera benzoin.* *Bull. Torrey Bot. Club* 107: 409–417.

Brandbyge, J. 1993. Polygonaceae. In *The families and genera of vascular plants,* vol. 2, Magnoliid, hamamelid and caryophyllid families, K. Kubitzki, J. G. Rohwer, and V. Bittrich (eds.), 531–544. Springer-Verlag, Berlin.

Brehm, B. G. and M. Ownbey. 1965. Variation in chromatographic patterns in the *Tragopogon dubius-pratensis-porrifolius* complex (Compositae). *Amer. J. Bot.* 52: 811–818.

Bremekamp, C. E. B. 1965. Delimitation and subdivision of the Acanthaceae. *Bull. Bot. Surv. India* 7: 21–30.

Bremekamp, C. E. B. 1966. Remarks on the position, the delimitation and subdivision of the Rubiaceae. *Acta. Bot. Neerl.* 15: 1–33.

Bremer, B. 1987. The sister group of the paleotropical tribe Argostemmateae: a redefined neotropical tribe Hamelieae (Rubiaceae, Rubioideae). *Cladistics* 3: 35–51.

Bremer, B. 1996. Phylogenetic studies within Rubiaceae and relationships to other families based on molecular data. *Opera Bot. Belg.* 7: 33–50.

Bremer, B. and R. K. Jansen. 1991. Comparative restriction site mapping of chloroplast DNA implies new phylogenetic relationships within the Rubiaceae. *Acta Bot. Neerl.* 15: 1–33.

Bremer, B. and L. Struwe. 1992. Phylogeny of the Rubiaceae and Loganiaceae: Congruence or conflict between morphological and molecular data? *Amer. J. Bot.* 79: 1171–1184.

Bremer, B., K. Andreasen and D. Olsson. 1995. Subfamilial and tribal relationships in the Rubiaceae based on *rbcL* sequence data. *Ann. Missouri Bot. Gard.* 82: 383–397.

Bremer, B., R. G. Olmstead, L. Struwe, and J. A. Sweere. 1994. *rbcL* sequences support exclusion of *Retzia, Desfontainia,* and *Nicodemia* from the Gentianales. *Plant Syst. Evol.* 190: 213–230.

Bremer, B., R. K. Jansen, B. Oxelman, M. Backlund, H. Lautz, and K.-J. Kim. 1999. More characters or more taxa for a robust phylogeny—case study from the coffee family (Rubiaceae). *Syst. Biol.* 48: 413-435.

Bremer, K. 1987. Tribal interrelationships of the Asteraceae. *Cladistics* 3: 210–253

Bremer, K. 1994. *Asteraceae: Cladistics and classification.* Timber Press, Portland, OR.

Bremer, K. 1996. Major clades and grades of the Asteraceae. In *Compositae: Systematics.* Proceedings of the International Compositae Conference, Kew, D. J. N. Hind and H. J. Beentje (eds.), 1–7. Royal Botanic Gardens, Kew.

Bremer, K. and R. K. Jansen. 1992. A new subfamily of the Asteraceae. *Ann. Missouri Bot. Gard.* 79: 414–415.

Brenan, J. P. M. 1966. The classification of Commelinaceae. *J. Linn. Soc. Bot.* 59: 349–380.

Brett, D. W. 1964. The inflorescence of *Fagus* and *Castanea* and the evolution of the cupules of the Fagaceae. *New Phytol.* 63: 96–117.

Briggs, B. G. and L. A. S. Johnson. 1998a. New genera and species of Australian Restionaceae (Poales). *Telopea* 7: 345-373.

Briggs, B. G. and L. A. S. Johnson. 1998b. A guide to a new classification of Restionaceae and allied families. In *Australian rushes—biology, identification and conservation of Restionaceae and allied families,* K. A. Meney and J. S. Pate (eds.), 25-56. University of Western Australia Press, Perth.

Briggs, B. G., A. D. Marchant, S. Gilmore, and C. L. Porter. 2000. A molecular phylogeny of Restionaceae and allies. In *Monocots: Systematics and evolution,* K. L. Wilson and D. A. Morrison (eds.), 661-671. CSIRO Publ., Collingwood, Australia.

Brizicky, G. K. 1961a. The genera of Turneraceae and Passifloraceae in the southeastern United States. *J. Arnold Arbor.* 42: 204–218.

Brizicky, G. K. 1961b. The genera of Violaceae in the southeastern United States. *J. Arnold Arbor.* 42: 321–333.

Brizicky, G. K. 1962a. The genera of Anacardiaceae in the southeastern United States. *J. Arnold Arbor.* 43: 359–375.

Brizicky, G. K. 1962b. The genera of Rutaceae in the southeastern United States. *J. Arnold Arbor.* 43: 1–22.

Brizicky, G. K. 1962c. The genera of Simaroubaceae and Burseraceae in the southeastern United States. *J. Arnold Arbor.* 43: 173–186.

Brizicky, G. K. 1963. The genera of Sapindales in the southeastern United States. *J. Arnold Arbor.* 44: 462–501.

Brizicky, G. K. 1964a. The genera of Celastrales in the southeastern United States. *J. Arnold Arbor.* 45: 206–234.

Brizicky, G. K. 1964b. The genera of Rhamnaceae in the southeastern United States. *J. Arnold Arbor.* 45: 439–463.

Brizicky, G. K. 1964c. The genera of Cistaceae in the southeastern United States. *J. Arnold Arbor.* 45: 346–357.

Brizicky, G. K. 1965a. The genera of Tiliaceae and Elaeocarpaceae in the southeastern United States. *J. Arnold Arbor.* 46: 286–307.

Brizicky, G. K. 1965b. The genera of Vitaceae in the southeastern United States. *J. Arnold Arbor.* 46: 48–67.

Brizicky, G. K. 1966a. The genera of Sterculiaceae in the southeastern United States. *J. Arnold Arbor.* 47: 60–74.

Brizicky, G. K. 1966b. The Goodeniaceae in the southeastern United States. *J. Arnold Arbor.* 47: 293–300.

Brown, G. K. and G. S. Varadarajan. 1985. Studies in Caryophyllales I: Re-evaluation of classification of Phytolaccaceae s. l. *Syst. Bot.* 10: 49–63.

Brown, G. K. and A. J. Gilmartin. 1989. Stigma types in the Bromeliaceae—A systematic study. *Syst. Bot.* 14: 110–132.

Bruhl, J. J. 1995. Sedge genera of the world: Relationships and a new classification of the Cyperaceae. *Austr. Syst. Bot.* 8: 125–305.

Bruneau, A., F. Forest, P. S. Herendeen, B. B. Klitgaard, and G. P. Lewis. 2001. Phylogenetic relationships in the Caesalpinioideae (Leguminosae) as inferred from chloroplast *trnL* intron sequences. *Syst. Bot.* 26: 487-514.

Brunsfeld, S. J., D. E. Soltis and P. S. Soltis. 1992. Evolutionary patterns and processes in *Salix* sect. *Longifoliae:* Evidence from chloroplast DNA. *Syst. Bot.* 17: 239–256.

Bult, C. J. and E. A. Zimmer. 1993. Nuclear ribosomal RNA sequences for inferring tribal relationships within Onagraceae. *Syst. Bot.* 18: 48–63.

Burger, W. C. 1977. The Piperales and the monocots: Alternative hypothesis for the origin of the monocotyledon flower. *Bot. Rev.* 43: 345–393.

Burger, W. C. 1988. A new genus of Lauraceae from Costa Rica with comments on problems of generic and specific delimitation within the family. *Brittonia* 40: 275–282.

Burns-Balogh, P. and V. A. Funk. 1986. A phylogenetic analysis of the Orchidaceae. *Smithsonian Contr. Bot.* 61: 1–79.

Burtt, B. L. 1977. Classification above the genus, as exemplified by Gesneriaceae, with parallels from other groups. *Plant. Syst. Evol.,* Suppl. 1: 97–109.

Burtt, B. L. and R. M. Smith. 1992. Tentative keys to subfamilies, tribes, and genera of Zingiberaceae. *Notes Royal Bot. Gard. Edinburgh* 31: 171–176.

Caddick, L. R., P. J. Rudall, P. Wilkin, and M. W. Chase. 2000. Yams and their allies: systematics of Dioscoreales. In *Monocots: Systematics and evolution*, K. L. Wilson and D. A. Morrison (eds.), 475-487. Collingwood, Australia.

Caddick, L. R., J. R. Rudall, P. Wilkin, T. A. J. Hedderson, and M. W. Chase. 2002a. Phylogenetics of Dioscoreales based on combined analyses of morphological and molecular data. *Bot. J. Linn. Soc.* 138: 123-144.

Caddick, L. R., P. Wilkin, P. J. Rudall, T. A. J. Hedderson, and M. W. Chase. 2002b. Yams reclassified: a recircumscription of Dioscoreaceae and Dioscoreales. Taxon 51: 103–114.

Calder, D. M. and P. Bernhard. 1983. *The biology of mistletoes.* Academic Press, New York.

Cameron, K. M., M. W. Chase, W. M. Whitten, P. J. Kores, D. C. Jarrell, V. A. Albert, T. Tukawa, H. G. Hills and D. H. Goldman. 1999. A phylogenetic analysis of the Orchidaceae: Evidence from *rbcL* nucleotide sequences. *Amer. J. Bot.* 86: 208–224.

Cameron, K. M. and M. W. Chase. 2000. Nuclear 18S rDNA sequences of Orchidaceae confirm the subfamilial status and circumscription of Vanilloideae. In *Monocots: Systematics and evolution*, K. L. Wilson and D. A. Morrison (eds.), 457-464. Collingwood, Australia.

Campbell, C. S. 1985. The subfamilies and tribes of Gramineae (Poaceae) in the southeastern United States. *J. Arnold Arbor.* 66: 123–199.

Campbell, C. S., B. G. Baldwin, M. J. Donoghue and M. F. Wojciechowski. 1995. A phylogeny of the genera of Maloideae (Rosaceae): Evidence from internal transcribed spacers of nuclear ribosomal DNA sequences and congruence with morphology. *Amer. J. Bot.* 82: 903–918.

Canright, J. E. 1952. The comparative morphology and relationships of the Magnoliaceae. I. Trends in specialization in the stamens. *Amer. J. Bot.* 39: 484–497.

Canright, J. E. 1953. The comparative morphology and relationships of the Magnoliaceae. II. Significance of the pollen. *Phytomorphology* 3: 355–365.

Canright, J. E. 1960. The comparative morphology and relationships of the Magnoliaceae. III. Carpels. *Amer. J. Bot.* 47: 145–155.

Cantino, P. D. 1982. Affinities of the Lamiales: A cladistic analysis. *Syst. Bot.* 7: 237–248.

Cantino, P. D. 1990. The phylogenetic significance of stomata and trichomes in the Labiatae and Verbenaceae. *J. Arnold Arbor.* 71: 323–370.

Cantino, P. D. 1992a. Evidence for a polyphyletic origin of the Labiatae. *Ann. Missouri Bot. Gard.* 79: 361–379.

Cantino, P. D. 1992b. Toward a phylogenetic classification of the Labiatae. In *Advances in labiate science*, R. M. Harley and T. Reynolds (eds.), 27–32. Royal Botanic Gardens, Kew.

Cantino, P. D. and R. W. Sanders. 1986. Subfamilial classification of Labiatae. *Syst. Bot.* 11: 163–185.

Cantino, P. D., R. M. Harley, and S. J. Wagstaff. 1992. Genera of Labiatae: Status and classification. In *Advances in labiate science*, R. M. Harley and T.

Reynolds (eds.), 511–522. Royal Botanic Garden, Kew.

Carlquist, S. 1975. Wood anatomy of Onagraceae, with notes on alternative modes of photosynthate movement in dicotyledon woods. *Ann. Missouri Bot. Gard.* 62: 386–424.

Carlquist, S. 1976a. Tribal interrelationships and phylogeny of the Asteraceae. *Aliso* 8: 465–492.

Carlquist, S. 1976b. *Alexgeorgea*, a bizarre new genus of Restionaceae from Western Australia. *Austr. J. Bot.* 24: 281-295.

Carolin, R. C. 1987. A review of the family Portulacaceae. *Austr. J. Bot.* 35: 383–412.

Carolin, R. C. 1993. Portulacaceae. In *The families and genera of vascular plants*, vol. 2, Magnoliid, hamamelid and caryophyllid families, K. Kubitzki, J. G. Rohwer, and V. Bittrich (eds.), 554–555. Springer-Verlag, Berlin.

Carolin, R. C. 1983. The trichomes of the Chenopodiaceae and Amaranthaceae. *Bot. Jahrb. Syst.* 103: 451–466.

Carolin, R. C., S. W. L. Jacobs and M. Vesk. 1975. Leaf structure in Chenopodiaceae. *Bot. Jahrb. Syst.* 95: 226–255.

Chadwell, T. B., S. J. Wagstaff, and P. D. Cantino. 1992. Pollen morphology of *Phryma* and some putative relatives. *Syst. Bot.* 17: 210–219.

Chakravarty, H. L. 1958. Morphology of the staminate flowers in the Cucurbitaceae with special reference to the evolution of the stamens. *Lloydia* 21: 49–87.

Chanderbali, A. S., H. van der Werff, and S. S. Renner. 2001. Phylogeny and historical biogeography of Lauraceae: evidence from the chloroplast and nuclear genomes. *Ann. Missouri Bot. Gard.* 88: 104-134.

Channell, R. B. and C. E. Wood, Jr. 1959. The genera of the Primulales of the southeastern United States. *J. Arnold Arbor.* 40: 268–288.

Channell, R. B. and C. E. Wood, Jr. 1962. The Leitneriaceae in the southeastern United States. *J. Arnold Arbor.* 43: 435–438.

Chappill, J. A. 1994. Cladistic analysis of the Leguminosae: The development of an explicit hypothesis. In *Advances in legume systematics*, part 7, M. D. Crisp and J. J. Doyle (eds.), 1–9. Royal Botanic Gardens, Kew.

Chase, M. W. and 41 others. 1993. Phylogenetics of seed plants: An analysis of nucleotide sequences from the plastid gene *rbcL*. *Ann. Missouri Bot. Gard.* 80: 528–580.

Chase, M. W., M. R. Duvall, H. G. Hills, J. G. Conran, A. V. Cox, L. E. Eguiarte, J. Hartwell, M. F. Fay, L. R. Caddick, K. M. Cameron, and S. Hoot. 1995a. Molecular systematics of Lilianae. In *Monocotyledons: Systematics and evolution*, P. J. Rudall, P. J. Cribb, D. F. Cutler, and C. J. Humphries (eds.), 109–137. Royal Botanic Gardens, Kew.

Chase, M. W., D. W. Stevenson, P. Wilkin, and P. J. Rudall. 1995b. Monocot systematics: A combined analysis. In *Monocotyledons: Systematics and evolution*, P. J. Rudall, P. J. Cribb, D. F. Cutler and C. J. Humphries (eds.), 109–137. Royal Botanic Gardens, Kew.

Chase, M. W., P. J. Rudall, and J. G. Conran. 1996. New circumscriptions and a new family of asparagoid lilies: Genera formerly included in Anthericaceae. *Kew Bull.* 57: 667–680.

Chase, M. W., C. M. Morton, and J. A. Kallunki. 1999. Phylogenetic relationships of Rutaceae: a cladistic analysis of the subfamilies using evidence from *rbcL* and *atpB* sequence variation. *Amer. J. Bot.* 86: 1191-1199.

Chase, M. W., D. E. Soltis, P. S. Soltis, P. J. Rudall, M. F. Fay, W. H. Hahn, S. Sullivan, J. Joseph, M. Molvray, P. J. Kores, T. J. Givnish, K. J. Sytsma, and J. C. Pires. 2000. In *Monocots: Systematics and evolution*, K. L. Wilson and D. A. Morrison, (eds.), 3-16. CSIRO Publ., Collingwood, Australia.

Chase, M. W., S. Zmarty, M. D. Lledó, K. J. Wurdack, S. M. Swensen, and M. F.. Fay. 2002. When in doubt, put it in Flacourtiaceae: A molecular phylogenetic analysis based on plastid *rbcL* DNA sequences. *Kew Bull.* 57: 141–181.

Chen, Z.-O., S. R. Manchester, and H.-Y. Sun. 1999. Phylogeny and evolution of the Betulaceae as inferred from DNA sequences, morphology, and paleobotany. *Amer. J. Bot.* 86: 1168–1181.

Chuang, T. I. and L. Constance. 1992. Seeds and systematics in Hydrophyllaceae, tribe Hydrophylleae. *Amer. J. Bot.* 79: 257–264.

Civeyrel, L., A. LeThomas, K. Ferguson, and M. W. Chase. 1998. Critical reexamination of palynological characters used to delimit Asclepiadaceae in comparison to the molecular phylogeny obtained from plastid *matK* sequences. *Mol. Phylog. Evol.* 9: 517–527.

Clark, L. G., W. Zhang, and J. F. Wendel. 1995. A phylogeny of the grass family (Poaceae) based on *ndhF* sequence data. *Syst. Bot.* 20: 436–460.

Clark, L. G. and E. J. Judziewicz. 1996. The grass subfamilies Anomochlooideae and Pharoideae (Poaceae). *Taxon* 45: 641–645.

Clark, L. G. and R. W. Pohl. 1996. Agnes Chase's first book of grasses. Smithsonian Inst. Press, Washington, D.C.

Clark, L. G., M. Kobayashi, S. Mathews, R. E. Spangler, and E. A. Kellogg. 2000. The Puelioideae, a new subfamily of Poaceae. *Syst. Bot.* 25: 181-187.

Classen-Bockhoff, R. 1991. Untersuchungen zur Konstruktion des Bestäubungsapparates von *Thalia geniculata* (Marantaceen). *Bot. Acta* 104: 183–193.

Clausen, R. T. 1975. *Sedum of North America north of the Mexican plateau.* Cornell University Press, Ithaca, NY.

Clausing, G., K. Meyer, and S. S. Renner. 2000. Correlations among fruit traits and evolution of different fruits within Melastomataceae. *Bot. J. Linn. Soc.* 133: 303–326.

Clausing, G. and S. S. Renner. 2001. Molecular phylogenetics of Melastomataceae and Memecylaceae: implications for character evolution. *Amer. J. Bot.* 88: 486–498.

Clayton, W. D. and S. A. Renvoize. 1986. Genera Graminum. Grasses of the world. *Kew Bull. Add. Ser.* 13.

Clevinger, C. C. and J. L. Panero. 1998. Phylogenetic relationships of North American Celastraceae based on *ndhF* sequence data. *Amer. J. Bot.* 85(6) Suppl.: 120.

Cogniaux, A. 1891. Melastomaceae. In *Monographiae Phanerogamarum* 7, A. de Candolle and C. de Candolle (eds.), 1–1256. Masson, Paris.

Conran, J. G. 1989. Cladistic analyses of some net-veined Liliiflorae. *Plant Syst. Evol.* 168: 123–141.

Conran, J. G. 1998. Smilacaceae. In *Families and genera of vascular plants,* vol. 4. Flowering plants—Monocotyledons, K. Kubitzki (ed.), 417–422. Springer-Verlag, Berlin.

Conran, J. G. and M. N. Tamura. 1998. Convallariaceae. In *The families and genera of vascular plants* vol. 3. Flowering plants—Monocotyledons, K. Kubitzki (ed.), 186-198. Springer-Verlag, Berlin.

Constance, L. 1971. History of the classification of Umbelliferae (Apiaceae). *Bot. J. Linn. Soc.* 64, Suppl. 1: 1–11.

Conti, E. 1994. Phylogenetic relationships of Onagraceae and Myrtales: Evidence from *rbcL* sequence data. Ph.D. Dissertation, University of Wisconsin, Madison.

Conti, E., A. Fischback, and K. J. Sytsma. 1993. Tribal relationships in Onagraceae: Implications from *rbcL* data. *Ann. Missouri Bot. Gard.* 80: 672–685.

Conti, E., A. Litt, P. G. Wilson, S. A. Graham, B. G. Briggs, L. A. S. Johnson, and K. J. Sytsma. 1997. Interfamilial relationships in Myrtales: Molecular phylogeny and patterns of morphological evolution. *Syst. Bot.* 22: 629–647.

Cook, C. D. K. 1982. Pollination mechanisms in the Hydrocharitaceae. In Studies on aquatic vascular plants, J. J. Symoens, S. S. Hooper, and P. Compère (eds.), 1–5. Royal Botanical Society of Belgium, Brussels.

Cosner, M. E., R. K. Jansen, and T. G. Lammers. 1994. Phylogenetic relationships in the Campanulales based on *rbcL* sequences. *Plant Syst. Evol.* 190: 79–95.

Cox, A. V., A. M. Pridgeon, V. A. Albert, and M. W. Chase. 1997. Phylogenetics of the slipper orchids (Cypripedioideae, Orchidaceae): Nuclear rDNA ITS sequences. *Plant Syst. Evol.* 208: 197–223.

Cox, P. A. and P. B. Tomlinson. 1988. Pollination ecology of a seagrass, *Thalassia testudinum* (Hydrocharitaceae) in St. Croix. *Amer. J. Bot.* 75: 958-965.

Cox, P. A. and C. J. Humphries. 1993. Hydrophilous pollination and breeding system evolution in seagrasses: A phylogenetic approach to the evolutionary ecology of the Cymodoceaceae. *Bot. J. Linn. Soc.* 113: 217–226.

Crane, P. 1985. Phylogenetic analysis of seed plants and the origin of angiosperms. *Ann. Missouri Bot. Gard.* 72: 716–793.

Crane, P. 1989. Early fossil history and evolution of the Betulaceae. In *Evolution, systematics, and fossil history of the Hamamelidae,* vol. 2, "Higher" Hamamelidae. Syst. Assoc. Special Vol. 40B, P. R. Crane and S. Blackmore (eds.), 87–116. Clarendon Press, Oxford.

Crane, P. and S. Blackmore (eds.). 1989. *Evolution, systematics, and fossil history of the Hamamelidae,* vol. 1, Introduction and "lower" Hamamelidae, and vol. 2, "Higher" Hamamelidae. Syst. Assoc. Special Vols. 40A and B. Clarendon Press, Oxford.

Crane, P. F., E. M. Friis, and K. R. Pedersen. 1995. The origin and early diversification of angiosperms. *Nature* 374: 27–33.

Crayn, D. M., K. A. Kron, P. A. Gadek, and C. J. Quinn. 1998. Phylogenetics and evolution of epacrids: A molecular analysis using the plastid gene *rbcL* with a reap-

praisal of the position of *Lebetanthus. Austr. J. Bot.* 46: 187–200.

Crayn, D. M. and C. J. Quinn. 2000. The evolution of the *atpB-rbcL* intergenic spacer in the epacrids (Ericales) and its systematic and evolutionary implications. *Mol. Phylog. and Evol.* 16: 238–252.

Crepet, W. L. and K. C. Nixon. 1989. Earliest megafossil evidence of Fagaceae: Phylogenetic and biogeographic implications. *Amer. J. Bot.* 76: 842–855.

Crisci, J., E. A. Zimmer, P. C. Hoch, G. B. Johnson, C. Mudd, and N. S. Pan. 1990. Phylogenetic implications of ribosomal DNA restriction site variation in the plant family Onagraceae. *Ann. Missouri Bot. Gard.* 77: 523–538.

Crisp, M. D. and P. H. Weston. 1987. Cladistics and legume systematics, with an analysis of the Bossiaeeae, Brongniartieae, and Mirbelieae. In *Advances in legume systematics,* part 3, C. H. Stirton (ed.), 65–130. Royal Botanic Gardens, Kew.

Croat, T. B. 1980. Flowering behavior of the neotropical genus *Anthurium* (Araceae). *Amer. J. Bot.* 67: 888–904.

Cronquist, A. 1944. Studies in the Simaroubaceae. IV. Resume of the American genera. *Brittonia* 5: 128–147.

Cronquist, A. 1955. Phylogeny and taxonomy of the Compositae. *Amer. Midl. Nat.* 53: 478–511.

Cronquist, A. 1968. *The evolution and classification of flowering plants.* Houghton Mifflin, Boston.

Cronquist, A. 1977. The Compositae revisited. *Brittonia* 19: 137–153.

Cronquist, A. 1980. *Vascular flora of the southeastern United States,* vol. 1, Asteraceae. University of North Carolina Press, Chapel Hill.

Cronquist, A. 1981. *An integrated system of classification of flowering plants.* Columbia University Press, New York.

Cronquist, A. 1988. *The evolution and classification of flowering plants,* 2nd ed. New York Botanical Garden, Bronx.

Cullings, K. W. and T. D. Bruns. 1992. Phylogenetic origin of the Monotropoideae inferred from partial 28S ribosomal RNA gene sequences. *Canad. J. Bot.* 70: 1703–1708.

Dahl, A. 1990. Infrageneric division of the genus *Hypecoum* (Papaveraceae). *Nordic J. Bot.* 10: 129–140.

Dahlgren, G. 1989. The last Dalhgrenogram System of classification of the dicotyledons. In *Plant taxonomy, phytogeography, and related subjects,* K. Tan (ed.), 249–250. Edinburgh University Press, London.

Dahlgren, R. M. T. 1983. General aspects of angiosperm evolution and macrosystematics. *Nordic. J. Bot.* 3: 119–149.

Dahlgren, R. M. T. 1988. Rhizophoraceae and Anisophyllaceae: Summary statement, relationships. *Ann. Missouri Bot. Gard.* 75: 1259–1277.

Dahlgren, R. M. T. and H. T. Clifford. 1982. *The monocotyledons: A comparative study.* Academic Press, London.

Dahlgren, R. M. T. and F. N. Rasmussen. 1983. Monocotyledon evolution: Characters and phylogenetic estimation. *Evol. Biol.* 16: 255–395.

Dahlgren, R. M. T. and R. F. Thorne. 1984. The order Myrtales: Circumscription, variation, and relationships. *Ann. Missouri Bot. Gard.* 71: 633–699.

Dahlgren, R. M. T., H. T. Clifford, and P. F. Yeo. 1985. *The families of the monocotyledons.* Springer-Verlag, Berlin.

D'Arcy, W. 1979. The classification of the Solanaceae. In *The biology and taxonomy of the Solanaceae,* J. G. Hawkes, R. N. Lester, and A. D. Skelding (eds.), 3–48. Academic Press, London.

D'Arcy, W. 1991. The Solanaceae since 1976, with a review of its biogeography. In *Solanaceae 3: Taxonomy, chemistry, evolution,* J. G. Hawkes, R. N. Lester, M. Nee, and N. Estrada (eds.), 75–137. Royal Botanic Garden, Kew.

Davis, C. C., W. R. Anderson, and M. J. Donoghue. 2001. Phylogeny of Malpighiaceae: evidence from chloroplast *ndhF* and *trnL-F* nucleotide sequences. *Amer. J. Bot.* 88: 1830-1846.

Davis, J. I. 1995. A phylogenetic structure for the Monocotyledons, as inferred from chloroplast DNA restriction site variation, and a comparison of measures of clade support. *Syst. Bot.* 20: 503–527.

de Bruijn, A., V. A. Cox, and M. W. Chase. 1995. Molecular systematics of Asphodelaceae (Asparagales; Lilianae). *Amer. J. Bot.* 82 (6) Suppl.: 124.

de Buhr, L. E. 1975. Phylogenetic relationships of the Sarraceniaceae. *Taxon* 24: 297–306.

de Lange, A. and F. Bouman. 1992. Seed micromorphology of the genus *Begonia* in Africa: taxonomic and ecological implications. In *Studies in Begoniaceae III,* J. J. F. E. de Wilde (ed.), 1–82. Wageningen Agric. University Pap. Wageningen, The Netherlands.

de Pamphilis, C. W. and N. D. Young. 1995. Evolution of parasitic Scrophulariaceae/Orobanchaceae: Evidence from sequences of chloroplast ribososmal protein gene *rps2* and a comparision with traditional classification schemes. *Amer. J. Bot.* 82(6) Suppl.: 126.

de Pamphilis, C. W., N. D. Young, and A. D. Wolfe. 1997. Evolution of plastid gene *rps2* in a lineage of hemiparasitic and holoparasitic plants: Many losses of photosynthesis and complex patterns of rate variation. *Proc. Natl. Acad. Sci. USA* 94: 7362–7372.

de Silva, M. F., O. R. Gottlieb, and F. Ehrendorfer. 1988. Chemosystematics of the Rutales: Suggestions for a more natural taxonomy and evolutionary interpretation of the family. *Plant Syst. Evol.* 161: 97–134.

de Wilde, W. J. J. O. 1971. The systematic position of the tribe Paropsieae, in particular the genus *Ancistrothyrsus,* and a key to the genera of Passifloraceae. *Blumea* 19: 99–104.

de Wilde, W. J. J. O. 1974. The genera of the tribe Passifloreae (Passifloraceae) with special reference to flower morphology. *Blumea* 22: 37–50.

den Hartog, C. 1975. Thoughts about the taxonomical relationships within the Lemnaceae. *Aquatic Bot.* 1: 407–416.

Denton, M. E. 1973. A monograph of *Oxalis,* section *Ionoxalis* (Oxalidaceae) in North America. *Publ. Mus. Michigan State University, Biol. Ser.* 4: 455–615.

Dickison, W. C. 1981. The evolutionary relationships of the Leguminosae. In *Advances in legume systematics,* part 1, R. M. Polhill and P. H. Raven (eds.), 35–54. Royal Botanic Gardens, Kew.

Dietrich, W., W. L. Wagner and P. H. Raven. 1997. Systematics of *Oenothera* section

Oenothera subsection *Oenothera* (Onagraceae). *Syst. Bot. Monogr.* 50: 1–234.

Dilcher, D. L. 1989. The occurrence of fruits with affinities to Ceratophyllaceae in lower and mid-Cretaceous sediments. *Amer. J. Bot.* 76: 162.

Dobbins, D. R. 1971. Studies on the anomalous cambial activity in *Doxantha unguis-cati* (Bignoniaceae). II. A case of differential production of secondary tissues. *Amer. J. Bot.* 58: 697–705.

Donoghue, M. J. 1980. Flowering times in *Viburnum*. *Arnoldia* 40: 2-22.

Donoghue, M. J. 1981. Growth patterns in woody plants with examples from the genus *Viburnum*. *Arnoldia* 41: 2-23.

Donoghue, M. J. 1983a. A preliminary analysis of phylogenetic relationships in *Viburnum* (Caprifoliaceae s. l.). *Syst. Bot.* 8: 45-58.

Donoghue, M. J. 1983b. The phylogenetic relationships of *Viburnum*. In *Advances in cladistics*, vol. 2, N. I. Platnick and V. A. Funk (eds.), 143–166. Columbia University Press, New York.

Donoghue, M. J. 1985. Pollen diversity and exine evolution in *Viburnum* and the Caprifoliaceae sensu lato. *J. Arnold Arbor.* 66: 421–469.

Donoghue, M. J. and J. A. Doyle. 1989. Phylogenetic analysis of angiosperms and the relationships of Hamamelidae. In *Evolution, systematics, and fossil history of the Hamamelidae*, vol. 1, Introduction and "lower" Hamamelidae. Syst. Assoc. Special Vol. 40A, P. R. Crane and S. Blackmore (eds.), 17–45. Clarendon Press, Oxford.

Donoghue, M. J., R. G. Olmstead, J. F. Smith and J. D. Palmer. 1992. Phylogenetic relationships of Dipsacales based on *rbcL* sequences. *Ann. Missouri Bot. Gard.* 79: 333–345.

Downie, S. R. and J. D. Palmer. 1992. Restriction site mapping of the chloroplast DNA inverted repeat: A molecular phylogeny of the Asteridae. *Ann. Missouri Bot. Gard.* 79: 238–266.

Downie, S. R. and J. D. Palmer. 1994a. A chloroplast DNA phylogeny of the Caryophyllales based on structural and inverted repeat restriction site variation. *Syst. Bot.* 19: 236–252.

Downie, S. R. and J. D. Palmer. 1994b. Phylogenetic relationships using restriction site variation of the chloroplast DNA inverted repeat. In *Caryophyllales: Evolution and systematics*, H.-D. Behnke and T. J. Mabry (eds.), 223–233. Springer-Verlag, Berlin.

Downie, S. R., D. S. Katz-Downie and K.-J. Cho. 1997. Relationships in the Caryophyllales as suggested by phylogenetic analysis of partial chloroplast DNA *ORF2280* homolog sequences. *Amer. J. Bot.* 84: 253–273.

Downie, S. R., S. Ramanath, D. S. Katz-Downie and E. Llanas. 1998. Molecular systematics of Apiaceae subfamily Apioideae: Phylogenetic analysis of nuclear ribosomal DNA internal transcribed spacer and plastid *RPOC1* intron sequences. *Amer. J. Bot.* 85: 563–591.

Downie, S. R., D. S. Katz-Downie, and M. F. Watson. 2000. A phylogeny of the flowering plant family Apiaceae based on chloroplast DNA *rpl16* and *rpoC1* intron sequences: towards a suprageneric classification of subfamily Apioideae. *Amer. J. Bot.* 87: 273–292.

Doyle, J. A. 1996. Seed plant phylogeny and the relationships of the Gnetales. *Int. J. Plant Sci.* 157(6) Suppl.: 3–39.

Doyle, J. A. 1998. Phylogeny of vascular plants. *Annual Rev. Ecol. Syst.* 29: 567–599.

Doyle, J. A., C. L. Hotton and J. V. Ward. 1990. Early Cretaceous tetrads, zonasulculate pollen, and Winteraceae. II. Cladistic analysis and implications. *Amer. J. Bot.* 77: 1558–1568.

Doyle, J. A., M. J. Donoghue and E. A. Zimmer. 1994. Integration of morphological and ribosomal RNA data on the origin of angiosperms. *Ann. Missouri Bot. Gard.* 81: 419–450.

Doyle, J. A. and A. LeThomas. 1994. Cladistic analysis and pollen evolution in Annonaceae. *Acta Bot. Gall.* 141: 149–170

Doyle, J. A. and A. Le Thomas. 1996. Phylogenetic analysis and character evolution in Annonaceae. *Bull. Mus. Natl. Hist. Nat., Paris, 4ᵉ Sér., 18, 1996, sect. B, Adansonia, No. 3–4:* 279-334.

Doyle, J. A. and P. K. Endress. 2000. Morphological phylogenetic analyses of basal angiosperms: comparison and combination with molecular data. *Int. J. Pl. Sci.* 161 (Suppl.): S121-S153.

Doyle, J. J. 1987. Variation at the DNA level: Uses and potential in legume systematics. In *Advances in legume systematics*, part 3, C. H. Stirton (ed.), 1–30. Royal Botanic Gardens, Kew.

Doyle, J. J. 1994. Phylogeny of the legume family: An approach to understanding the origins of nodulation. *Annu. Rev. Ecol. Syst.* 25: 325–349.

Doyle, J. J. 1995. DNA data and legume phylogeny: A progress report. In *Advances in legume systematics*, part 7, M. D. Crisp and J. J. Doyle (eds.), 11–30. Royal Botanic Gardens, Kew.

Doyle, J. J., J. L. Doyle, J. A. Ballenger, E. E. Dickson, T. Kajita and H. Ohashi. 1997. A phylogeny of the chloroplast gene *rbcL* in the Leguminosae: Taxonomic correlations and insights into evolution of nodulation. *Amer. J. Bot.* 84: 541–554.

Dransfield, J. 1986. A guide to collecting palms. *Ann. Missouri Bot. Gard.* 73: 166–176.

Dransfield, J. and N. W. Uhl. 1998. Palmae. In *The families and genera of vascular plants.* Vol. 4. Flowering plants—Monocotyledons, K. Kubitzki (ed.), 306–389. Springer-Verlag, Berlin.

Dressler, R. L. 1968. The structure of the orchid flower. *Missouri Bot. Gard. Bull.* 49: 60–69.

Dressler, R. L. 1981. *The orchids: Natural history and classification.* Harvard University Press, Cambridge, MA.

Dressler, R. L. 1986. Recent advances in orchid phylogeny. *Lindleyana.* 1: 5–20.

Dressler, R. L. 1993. *Phylogeny and classification of the orchid family.* Dioscorides Press, Portland, OR.

Dressler, R. L. and M. W. Chase. 1995. Whence the orchids? In *Monocotyledons: Systematics and evolution*, P. J. Rudall, P. J. Cribb, D. F. Cutler, and C. J. Humphries (eds.), 217–226. Royal Botanic Gardens, Kew.

Drinnan, A. N., P. R. Crane, and S. B. Hoot. 1994. Patterns of floral evolution in the early diversification of non-magnoliid dicotyledons (eudicots). *Plant Syst. Evol.* Suppl. 8: 93–122.

Duke, N. C. and B. R. Jackes. 1987. A systematic revision of the mangrove genus *Sonneratia* (Sonneratiaceae) in Australasia. *Blumea* 32: 277–302.

Duvall, M. R., M. T. Clegg, M. W. Chase, W. D. Clark, W. J. Kress, L. E. Eguiarte, J. F. Smith, B. S. Gaut, E. A. Zimmer, and G. H. Learns, Jr. 1993. Phylogenetic hypotheses for the monocotyledons constructed from *rbcL* sequence data. *Ann. Missouri Bot. Gard.* 80: 607–619.

Eckardt, T. 1976. Classical morphological features of Centrospermous families. *Plant Syst. Evol.* 126: 5–25.

Eckenwalder, J. E. and S. C. H. Barrett. 1986. Phylogenetic systematics of Pontederiaceae. *Syst. Bot.* 11: 373–391.

Edlin, H. L. 1935. A critical revision of certain taxonomic groups of the Malvales. *New Phytol.* 34: 1–20, 122–143.

Ehrendorfer, F. 1976. Closing remarks: Systematics and evolution of Centrospermous families. *Plant Syst. Evol.* 126: 99–105.

Eldenäs, P. A. and H. P. Linder. 2000. Congruence and complementarity of morphological and *trnL-trnF* sequence data and the phylogeny of the African Restionaceae. *Syst. Bot.* 25: 692–707.

Elias, T. S. 1974. The genera of Mimosoideae (Leguminosae) in the southeastern United States. *J. Arnold Arbor.* 55: 67–118.

Elias, T. S. 1970. The genera of Ulmaceae in the southeastern United States. *J. Arnold Arbor.* 51: 18–40.

Elias, T. S. 1971a. The genera of Fagaceae in the southeastern United States. *J. Arnold Arbor.* 52: 159–195.

Elias, T. S. 1971b. The genera of Myricaceae in the southeastern United States. *J. Arnold Arbor.* 52: 305–318.

Elias, T. S. 1972. The genera of Juglandaceae in the southeastern United States. *J. Arnold Arbor.* 53: 26–51.

Endress, M. E. and P. V. Bruyns. 2000. A revised classification of the Apocynaceae s. l. *Bot. Rev.* 66: 1–56.

Endress, P. K. 1977. Evolutionary trends in the Hamamelidales-Fagales-group. *Plant Syst. Evol.*, Suppl. 1: 321–347.

Endress, P. K. 1986a. Floral structure, systematics, and phylogeny in Trochodendrales. *Ann. Missouri Bot. Gard.* 73: 297–324.

Endress, P. K. 1986b. Reproductive structures and phylogenetic significance of extant primitive angiosperms. *Plant Syst. Evol.* 152: 1–28.

Endress, P. K. 1989a. Phylogenetic relationships in the Hamamelidoideae. In *Evolution, systematics, and fossil history of the Hamamelidae*, vol. 1, Introduction and "lower" Hamamelidae. Syst. Assoc. Special Vol. 40A, P. R. Crane and S. Blackmore (eds.), 227–248. Clarendon Press, Oxford.

Endress, P. K. 1989b. A suprageneric taxonomic classification of the Hamamelidaceae. *Taxon* 38: 371–376.

Endress, P. K. 1993. Hamamelidaceae. In *The families and genera of vascular plants*, vol. 2, Magnoliid, hamamelid and caryophyllid families, K. Kubitzki, J. G. Rohwer and V. Bittrich (eds.), 322–331. Springer-Verlag, Berlin.

Endress, P. K. 1994a. Evolutionary aspects of the floral structure in *Ceratophyllum*. *Plant Syst. Evol.* 8: 175–183.

Endress, P. K. 1994b. Shapes, sizes and evolutionary trends in stamens of Magnoliidae. *Bot. Jahrb. Syst.* 115: 429–460.

Endress, P. K. 1994c. *Diversity and evolutionary biology of tropical flowers.* Cambridge University Press, Cambridge.

Endress, P. K. 1995. Floral structure and evolution in Ranunculanae. *Plant Syst. Evol.* Suppl. 9: 47–61.

Endress, P. K. 2001. The flowers in extant basal angiosperms and inferences on ancestral flowers. *Int. J. Plant Sci.* 162: 1111–1140.

Endress, P. K., M. Jenny, and M. E. Fallen. 1983. Convergent elaboration of apocarpous gynoecia in higher advanced dicotyledons (Sapindales, Malvales, Gentianales). *Nordic J. Bot.* 3: 293–300.

Endress, P. K. and A. Igersheim. 1997. Gynoecium diversity and systematics of the Laurales. *Bot. J. Linn. Soc.* 125: 93–168.

Endress, P. K. and A. Igersheim. 1999. Gynoecium diversity and systematics of the basal eudicots. *Bot. J. Linn. Soc.* 130: 305–393.

Endress, P. K. and A. Igersheim. 2000a. Gynoecium structure and evolution in basal angiosperms. *Int. J. Pl. Sci.* 161 (Suppl.): S211–S223.

Endress, P. K. and A. Igersheim. 2000b. The reproductive structures of the basal angiosperm *Amborella trichopoda* (Amborellaceae). *Int. J. Pl. Sci.* 161 (Suppl.): S237–S248.

Endress, M. E., B. Sennblad, S. Nilsson, L. Civeyrel, M. W. Chase, S. Huysmans, E. Grafsröm, and B. Bremer. 1996. A phylogenetic analysis of Apocynaceae s. str. and some related taxa in Gentianales: A multidisciplinary approach. *Opera Bot. Belg.* 7: 59–102.

Endress, M. E. and W. D. Stevens. 2001. The renaissance of the Apocynaceae s.l.: recent advances in systematics, phylogeny, and evolution: introduction. *Ann. Missouri Bot. Gard.* 88: 517–522.

Engler, A. 1888. Ulmaceae, Moraceae, and Urticaceae. *Die natürlichen Pflanzenfamilien*, Series III. Vol. 1: 59–119.

Engler, A. 1930. Saxifragaceae. In *Die natürlichen Pflanzenfamilien*, 2nd ed., vol. 18a, A. Engler and K. Prantl (eds.), 74–226. Engelmann, Leipzig.

Erbar, C. 1991. Sympetalae—a systematic character? *Bot. Jahrb. Syst.* 112: 417–451.

Erbar, C. and P. Leins. 1988. Blutenentwicklungs-geschichtliche Studies an *Aralia* und *Hedera* (Araliaceae). *Flora* 180: 391–406.

Eriksen, B. 1993a. Floral anatomy and morphology in the Polygalaceae. *Plant Syst. Evol.* 186: 17–32.

Eriksen, B. 1993b. Phylogeny of the Polygalaceae and its taxonomic implications. *Plant Syst. Evol.* 186: 33–55.

Eriksson, T. and M. J. Donoghue. 1997. Phylogenetic relationships of *Sambucus* and *Adoxa* (Adoxoideae, Adoxaceae) based on nuclear ribosomal ITS sequences and preliminary morphological data. *Syst. Bot.* 22: 555–573.

Ernst, W. R. 1962. The genera of Papaveraceae and Fumariaceae in the southeastern United States. *J. Arnold Arbor.* 43: 315–343.

Ernst, W. R. 1963a. The genera of Capparaceae and Moringaceae in the southeastern United States. *J. Arnold Arbor.* 44: 81–95.

Ernst, W. R. 1963b. The genera of Hamamelidaceae and Platanaceae in the southeastern United States. *J. Arnold Arbor.* 44: 193–210.

Ernst, W. R. 1964. The genera of Berberidaceae, Lardizabalaceae, and Menispermaceae in the southeastern United States. *J. Arnold Arbor.* 45: 1–35.

Evans, R. C. 1999. Molecular, morphological, and ontogenetic evaluation of relationships and evolution in the Rosaceae. Ph.D. dissertation, University of Toronto, Toronto.

Evans, R. C. and T. A. Dickinson. 1999a. Floral ontogeny and morphology in subfamily Amygdaloideae. *Int. J. Plant Sci.* 160: 955–979.

Evans, R. C. and T. A. Dickinson. 1999b. Floral ontogeny and morphology in subfamily Spiraeoideae. *Int. J. Plant Sci.* 160: 981–1012.

Evans, R. C., L. A. Alice, C. S. Campbell, T. A. Dickinson, and E. A. Kellogg. 2000. The granule-bound starch synthase (GBSSI) gene in the Rosaceae: multiple loci and phylogenetic utility. *Mol. Phyl.Evol.* 17: 388–400.

Evans, T. M., R. B. Faden, M. G. Simpson, and K. J. Sytsma. 2000. Phylogenetic relationships in the Commelinaceae. 1. A cladistic analysis of morphological data. *Syst. Bot.* 25: 668–691.

Evans, T. M., R. B. Faden, and K. J. Sytsma. 2000. Homoplasy in the Commelinaceae: a comparison of different classes of morphological characters. In *Monocots: Systematics and evolution*, K. L. Wilson and D. A. Morrison (eds.), 557–566. Collingwood, Australia.

Evans, W. C. 1979. Tropane alkaloids of the Solanaceae. In *The biology and taxonomy of the Solanaceae*, J. G. Hawkes, R. N. Lester and A. D. Skelding (eds.), 241–254. Academic Press, London.

Eyde, R. H. 1966. The Nyssaceae in the southeastern United States. *J. Arnold Arbor.* 47: 117–125.

Eyde, R. H. 1982. Evolution and systematics of the Onagraceae: Floral anatomy. *Ann. Missouri Bot. Gard.* 69: 735–747.

Eyde, R. H. 1988. Comprehending *Cornus*: Puzzles and progress in the systematics of the dogwoods. *Bot. Rev.* 54: 233–351.

Eyde, R. H. and Q. Xiang. 1990. Fossil mastixioid (Cornaceae) alive in eastern Asia. *Amer. J. Bot.* 77: 689–692.

Faden, R. B. 1998. Commelinaceae. *In The families and genera of vascular plants*, vol. 4. Flowering plants—Monocotyledons, K. Kubitzki (ed.), 109–128. Springer-Verlag, Berlin.

Faden, R. B. and D. R. Hunt. 1991. The classification of the Commelinaceae. *Taxon* 40: 19–31.

Faegri, K. and L. van der Pijl. 1980. *The principles of pollination ecology.* Pergamon Press, Oxford.

Fagerberg, W. R. and D. Allain. 1991. A quantitative study of tissue dynamics during closure in the traps of venus's flytrap *Dionaea muscipula* Ellis. *Amer. J. Bot.* 78: 647–657.

Fallen, M. E. 1986. Floral structure in the Apocynaceae: Morphological, functional, and evolutionary aspects. *Bot. Jahrb. Syst.* 106: 245–286.

Fan, C.-Z. and Q.-Y. Xiang. 2001. Phylogenetic relationships within *Cornus* (Cornaceae) based on 26S rDNA sequences. *Amer. J. Bot.* 88: 1131–1138.

Fay, M. F. and M. W. Chase. 1996. Resurrection of Themidaceae for the *Brodiaea* alliance, and recircumscription of Alliaceae, Amaryllidaceae and Agapanthoideae. *Taxon* 45: 441–451.

Fay, M. F., C. Bayer, W. S. Alverson, A. Y. De Bruijn, and M. W. Chase. 1998. Plastid *rbcL* sequence data indicate a close affinity between *Diegodendron* and *Bixa*. *Taxon* 47: 43–50.

Fay, M. F., P. J. Rudall, S. Sullivan, K. L. Stobart, A. Y. de Bruijn, G. Reeves, F. Qamaruz-Zaman, W.-P. Hong, J. Joseph, W. J. Hahn, J. G. Conran, and M. W. Chase. 2000. Phylogenetic studies of Asparagales based on four plastid DNA regions. In *Monocots: Systematics and evolution*, K. L. Wilson and D. A. Morrison (eds.), 360–371. Collingwood, Australia.

Feild, T. S., M. A. Zwieniecki, and N. M. Holbrook. 2000. Winteraceae evolution: an ecophysiological perspective. *Ann. Missouri Bot. Gard.* 87: 323–334.

Feild, T. S., T. Brodribb, T. Jaffré, and N. M. Holbrook. 2001. Acclimation of leaf anatomy, photosynthetic light use, and xylem hydraulics to light in *Amborella trichopoda* (Amborellaceae). *Int. J. Plant Sci.* 162: 999–1008.

Ferguson, D. M. 1999. Phylogenetic analysis and relationships in Hydrophyllaceae based on *ndhF* sequence data. *Syst. Bot.* 23: 253–268.

Ferguson, I. K. 1966a. The Cornaceae in the southeastern United States. *J. Arnold Arbor.* 47: 106–116.

Ferguson, I. K. 1966b. The genera of Caprifoliaceae in the southeastern United States. *J. Arnold Arbor.* 47: 33–59.

Fernando, E. S. and C. J. Quinn. 1995. Picramniaceae, a new family, and a recircumscription of Simaroubaceae. *Taxon* 44: 177–181.

Fernando, E. S., P. A. Gadek, and C. J. Quinn. 1995. Simaroubaceae, and artificial construct: Evidence from *rbcL* sequence variation. *Amer. J. Bot.* 82: 92–103.

Fernando, E. S., P. A. Gadek, D. M. Crayn, and C. J. Quinn. 1993. Rosid affinities of Surianaceae: Molecular evidence. *Mol. Phylog. Evol.* 2: 344–350.

Fey, B. S. and P. K. Endress. 1983. Development and morphological interpretation of the cupule in Fagaceae. *Flora, Morphol. Geobot. Oekophysiol.* 173: 451–468.

Fineran, B. A. 1985. Glandular trichomes in *Utricularia*: A review of their structure and function. *Israel J. Bot.* 34: 295–330.

Fish, F. and P. G. Waterman. 1973. Chemosystematics in the Rutaceae. II. The chemosystematics of the *Zanthoxylum/Fagara* complex. *Taxon* 22: 177–203.

Fishbein, M. 2001. Evolutionary innovation and diversification in the flowers of Asclepiadaceae. *Ann. Missouri Bot. Gard.* 88: 603–623.

Fisher, M. J. 1928. The morphology and anatomy of flowers of Salicaceae, I and II. *Amer. J. Bot.* 15: 307–326, 372–394.

French, J. C., M. G. Chung, and Y. K. Hur. 1995. Chloroplast DNA phylogeny of the Ariflorae. In *Monocotyledons: Systematics and evolution*, P. J. Rudall, P. J. Cribb, D. F. Cutler, and C. J. Humphries (eds.), 255–275. Royal Botanic Gardens, Kew.

Freudenstein, J. V. and F. N. Rasmussen. 1999. What does morphology tell us about orchid relationships? A cladistic analysis. *Amer. J. Bot.* 86: 225–248.

Freudenstein, J. V., D. M. Senyo, and M. W. Chase. 2000. Mitochondrial DNA and relationships in the Orchidaceae. In *Monocots: Systematics and evolution*, K. L. Wilson and D. A. Morrison (eds.), 421–429. Collingwood, Australia.

Friis, I. 1989. The Urticaceae: A systematic review. In *Evolution, systematics, and fossil history of the Hamamelidae*, vol. 2, "Higher" Hamamelidae. Syst. Assoc. Special Vol. 40B, P. R. Crane and S. Blackmore (eds.), 285–308. Clarendon Press, Oxford.

Friis, I. 1993. Urticaceae. In *The families and genera of vascular plants*, vol. 2, Magnoliid, hamamelid and caryophyllid families, K. Kubitzki, J. G. Rohwer, and V. Bittrich (eds.), 612–630. Springer-Verlag, Berlin.

Fryxell, P. A. 1988. Malvaceae of Mexico. *Syst. Bot. Monogr.* 25: 1–522.

Furlow, J. J. 1990. The genera of Betulaceae in the southeastern United States. *J. Arnold Arbor.* 71: 1–67.

Gadek, P. A., E. S. Fernando, C. J. Quinn, S. B. Hoot, T. Terrazas, M. C. Sheahan, and M. W. Chase. 1996. Sapindales: Molecular delimitation and infraordinal groups. *Amer. J. Bot.* 83: 802–811.

Ganders, F. R. 1979. The biology of heterostyly. *New Zealand J. Bot.* 17: 607–635.

Gates, B. N. 1943. Carunculate seed dissemination by ants. *Rhodora* 45: 438–445.

Gentry, A. H. 1974. Coevolutionary patterns in Central American Bignoniaceae. *Ann. Missouri Bot. Gard.* 61: 728–759.

Gentry, A. H. 1980. *Bignoniaceae.* Part 1 (Crescentieae and Tourrettieae). Fl. Neotropica Monogr. 25: 1–130.

Gentry, A. H. 1990. Evolutionary patterns in neotropical Bignoniaceae. *Mem. New York Bot. Gard.* 55: 118–129.

George, A. S. 1998. Proteus in Australia—an overview of the correct state of taxonomy of the Australian Proteaceae. *Aust. Syst. Bot.* 11: 257–266.

Gerrath, J. M., V. Posluszny, and N. G. Dengler. 2001. Primary vascular patterns in the Vitaceae. *Int. J. Pl. Sci.* 162: 729–745.

Gibson, A. C., K. C. Spencer, R. Bajai and, J. L. McLaughlin. 1986. The ever-changing landscape of cactus systematics. *Ann. Missouri Bot. Gard.* 73: 532–555.

Gillis, W. T. 1971. The systematics and ecology of poison-ivy and the poison-oaks (*Toxicodendron*, Anacardiaceae). *Rhodora* 73: 72–159, 161–237, 370–443, 465–540.

Gilmartin, A. J. and G. K. Brown. 1987. Bromeliales, related monocots and resolution of relationships among Bromeliaceae subfamilies. *Syst. Bot.* 12: 493–500.

Giulietti, A. M., V. L. Scatena, P. T. Sano, L. R. Parra, L. P. de Queiroz, R. M. Harley, N. L. Menezes, A. M. B. Ysepon, A. Salatino, M. L. Salatino, W. Vilegas, L. C. Santos, C. V. Ricci, M. C. B. Bonfim, and E. B. Miranda. 2000. Multidisciplinary studies on neotropical Eriocaulaceae. In *Monocots: Systematics and evolution*, K. L. Wilson and D. A. Morrison (eds.), 580–489. Collingwood, Australia.

Goetchebeur, P. 1998. Cyperaceae. In *The families and genera of vascular plants*, vol. 4. Flowering plants—Monocotyledons, K. Kubitzki (ed.), 141–190. Springer-Verlag, Berlin.

Goldblatt, P. 1990. Phylogeny and classification of Iridaceae. *Ann. Missouri Bot. Gard.* 77: 607–627.

Goldblatt, P. 1995. The status of R. Dahlgren's orders Liliales and Melanthiales. In *Monocotyledons: Systematics and evolution*, P. J. Rudall, P. J. Cribb, D. F. Cutler and C. J. Humphries (eds.), 181–200. Royal Botanic Gardens, Kew.

Goldblatt, P., J. C. Manning, and P. Rudall. 1998. Iridaceae. In *The families and genera of vascular plants*, vol. 3. Flowering plants—Monocotyledons, K. Kubitzki (ed.), 295–333. Springer-Verlag, Berlin.

González, F. and P. Rudall. 2001. The questionable affinities of *Lactoris*: evidence from branching pattern, inflorescence morphology, and stipule development. *Amer. J. Bot.* 88: 2143–2150.

Gornall, R. J. and B. A. Bohm. 1985. A monograph of *Boykinia, Peltoboykinia, Bolandra* and *Suksdorfia* (Saxifragaceae). *Bot. J. Linnean Soc.* 90: 1–71.

Gottsberger, G. 1977. Some aspects of beetle pollination in the evolution of flowering plants. *Plant Syst. Evol.*, 1: S211–S226.

Gottsberger, G. 1988. The reproductive biology of primitive angiosperms. *Taxon* 37: 630–643.

Gottsberger, G., I. Silberbauer-Gottsberger, and F. Ehrendorfer. 1980. Reproductive biology in the primitive relic angiosperm *Drimys brasiliensis* (Winteraceae). *Plant Syst. Evol.* 135: 11–39.

Graham, S. A. 1964a. The genera of Lythraceae in the southeastern United States. *J. Arnold Arbor.* 45: 235–250.

Graham, S. A. 1964b. The genera of Rhizophoraceae and Combretaceae in the southeastern United States. *J. Arnold Arbor.* 45: 285–301.

Graham, S. A. 1966. The genera of Araliaceae in the southeastern United States. *J. Arnold Arbor.* 47: 126–136.

Graham, S. A. and C. E. Wood, Jr. 1965. The genera of Polygonaceae in the southeastern United States. *J. Arnold Arbor.* 46: 91–121.

Graham, S. A., E. Conti, and K. Sytsma. 1993a. Phylogenetic analysis of the Lythraceae based on *rbcL* sequence divergence. *Amer. J. Bot. Suppl.* 80(6) Suppl.: 150.

Graham, S. A., J. V. Crisci, and P. C. Hoch. 1993b. Cladistic analysis of the Lythraceae sensu lato based on morphological characters. *Bot. J. Linnean Soc.* 113: 1–33.

Graham, S. W. and S. C. H. Barrett. 1995. Phylogenetic systematics of Pontederiales: Implications for breeding-system evolution. In *Monocotyledons: Systematics and evolution*, P. J. Rudall, P. J. Cribb, D. F. Cutler, and C. J. Humphries (eds.), 415–441. Royal Botanic Gardens, Kew.

Graham, S. W., J. R. Kohn, B. R. Morton, J. E. Eckenwalder, and S. C. H. Barrett. 1998. Phylogenetic congruence and discordance among one morphological and three molecular data sets from Pontederiaceae. *Syst. Biol.* 47: 545–567.

Graham, S. W. and R. G. Olmstead. 2000. Utility of 17 chloroplast genes for inferring the phylogeny of the basal angiosperms. *Amer. J. Bot.* 87: 1712–1730.

Grant, V. 1959. Natural history of the Phlox family. In *Systematic botany*, vol. 1. Martinus Nijhoff, The Hague.

Grant, V. and K. A. Grant. 1965. *Flower pollination in the phlox family.* Columbia University Press, New York.

Grass Phylogeny Working group. 2001. Phylogeny and subfamilial classification of the grasses (Poaceae). *Ann. Missouri Bot. Gard.* 88: 373–457.

Grayum, M. H. 1987. A summary of evidence and arguments supporting the removal of *Acorus* from Araceae. *Taxon* 36: 723–729.

Grayum, M. H. 1990. Evolution and phylogeny of the Araceae. *Ann. Missouri Bot. Gard.* 77: 628–697.

Grudzinskaja, I. A. 1967. Ulmaceae and reasons for distinguishing Celtidoideae as a separate family Celtidaceae Link. *Bot. Zhurn.* (Leningrad) 52: 1723–1749. [In Russian.]

Gustafsson, M. H. G. 1996. Phylogenetic hypotheses for Asteraceae relationships. In *Compositae: Systematics.* Proceedings of the International Compositae Conference, Kew, D. J. N. Hind and H. J. Beentje (eds.), 9–19. Royal Botanic Gardens, Kew.

Hahn, W. J. 2002. A molecular phylogenetic study of the Palmae (Arecaceae) based on *atpB, rbcL*, and 18S nrDNA sequences. *Syst. Biol.* 51: 92–112/

Hallé, N. 1962. Monographie des Hippocratéacées. *Mémoires de L'Inst. Francais D'Afrique Noire* 64: 1–245.

Hambey, R. K. and E. A. Zimmer. 1992. Ribosomal RNA as a phylogenetic tool in plant systematics. In *Molecular systematics of plants*, P. S. Soltis, D. E. Soltis and J. J. Doyle (eds.), 50–91. Chapman and Hall, New York.

Hansen, H. V. 1992. Studies in Calyceraceae with a discussion of its relationship to Compositae. *Nordic J. Bot.* 12: 63–75.

Hardin, J. W. 1957. A revision of the American Hippocastanaceae. *Brittonia* 9: 145–171.

Harley, M. M. and I. K. Ferguson. 1990. The role of the SEM in pollen morphology and plant systematics. In *Scanning electron microscope in taxonomy and functional morphology*. Syst. Assoc. Special Vol. 41, D. Claugher (ed.), 45–68. Clarendon Press, Oxford.

Harris, P. J. and R. D. Hartley. 1980. Phenolic constituents of the cell walls of monocotyledons. *Biochem. Syst. Ecol.* 8: 153–160.

Harrison, C. J., M. Möller, and C. B. Cronk. 1999. Evolution and development of floral diversity in *Streptocarpus* and *Saintpaulia*. *Ann. Bot.* 84: 49–60.

Hartmann, H. E. K. 1993. Aizoaceae. In *The families and genera of vascular plants*, vol. 2, Magnoliid, hamamelid and caryophyllid families, K. Kubitzki, J. G. Rohwer and V. Bittrich (eds.), 37–69. Springer-Verlag, Berlin.

Haynes, R. R. 1978. The Potamogetonaceae in the southeastern United States. *J. Arnold Arbor.* 59: 170–191.

Haynes, R. R. 1988. Reproductive biology of selected aquatic plants. *Ann. Missouri Bot. Gard.* 75: 805–810.

Haynes, R. R., D. H. Les, and L. B. Holm-Nielsen. 1998. Alismataceae. In *The families and genera of vascular plants*, vol. 4. Flowering plants—Monocotyledons, K. Kubitzki (ed.), 11–21. Springer-Verlag, Berlin.

Haynes, R. R., D. H. Les, and L. B. Holm-Nielsen. 1998. Potamogetonaceae. In *The families and genera of vascular plants*, vol. 4. Flowering plants—Monocotyledons, K. Kubitzki (ed.), 408–415. Springer-Verlag, Berlin.

Haynes, R. R. and L. B. Holm-Nielsen. 2001. The genera of Hydrocharitaceae in the southeastern United States. *Harvard Pap. Bot.* 5(2): 201–275.

Hedrén, M., M. W. Chase, and R. G. Olmstead. 1995. Relationships in the Acanthaceae and related families as suggested by cladistic analysis of *rbcL* nucleotide sequences. *Plant Syst. Evol.* 194: 93–109.

Hegnauer, R. 1971. Chemical patterns and relationships of Umbelliferae. In *The biology and chemistry of the Umbelliferae. Bot. J. Linn. Soc.* vol. 64, suppl. 1, V. H. Heywood (ed.), 267–277. Academic Press, London.

Hempel, A. C., P. A. Reeves, R. G. Olmstead, and R. K. Jansen. 1995. Implications of *rbcL* sequence data for higher order relationships of the Loasaceae and the anomalous aquatic plant *Hydrostachys* (Hydrostachyaceae). *Plant Syst. Evol.* 194: 25–37.

Henderson, A. 1986. A review of pollination studies in the palms. *Bot. Rev.* 52: 221–259.

Henderson, A. 1995. *The palms of the Amazon.* Oxford University Press, New York.

Henderson, A., G. Galeano, and R. Bernal. 1995. *Field guide to the palms of the Americas.* Princeton University Press, Princeton.

Herendeen, P. S., W. L. Crepet, and D. L. Dilcher. 1992. The fossil history of the Leguminosae: Phylogenetic and biogeographic implications. In *Advances in legume systematics*, part 4, P. S. Herendeen and D. L. Dilcher (eds.), 303–316. Royal Botanic Gardens, Kew.

Hermann, P. M. and B. F. Palser. 2000. Stamen development in the Ericaceae. I. Anther wall, microsporogenesis, inversion, and appendages. *Amer. J. Bot.* 87: 934–957.

Hershkovitz, M. A. and E. A. Zimmer. 1997. On the evolutionary origins of the cacti. *Taxon* 46: 217–232.

Hesse, M. 2001. Pollen characters of *Amborella trichopoda* (Amborellaceae): a reinvestigation. *Int. J. Pl. Sci.* 162: 201–208.

Heywood, V. H. 1971. The Leguminosae—A systematic purview. In *Chemotaxonomy of the Leguminosae*, J. B. Harborne, D. Boulter, and B. L. Turner (eds.), 1–29. Academic Press, London.

Heywood, V. H., J. B. Harborne, and B. L. Turner. 1977. An overture to the Compositae. In *The biology and chemistry of the Compositae*, vol. 1, V. H. Heywood, J. B. Harborne, and B. L. Turner (eds.), 1–20. Academic Press, London.

Hickey, L. J. and J. A. Wolfe. 1975. The bases of angiosperm phylogeny: Vegetative morphology. *Ann. Missouri Bot. Gard.* 62: 538–589.

Hileman, L. C., M. C. Vasey, and V. T. Parker. 2001. Phylogeny and biogeography of the Arbutoideae (Ericaceae): implications for the Madrean-Tethyan hypothesis. *Syst. Bot.* 26: 131–143.

Hilliard, O. M. 1994. *The Manuleae, a tribe of the Scrophulariaceae.* Edinburgh University Press, Edinburgh.

Hilu, K. W., L. A. Alice, and H. Liang. 1999. Phylogeny of Poaceae inferred from *matK* sequences. *Ann. Missouri Bot. Gard.* 86: 835–851.

Hilu, K. W., T. Borsch, V. Savolainen, P. S. Soltis, D. E. Soltis, M. W. Chase, K. Muller, T. Slotta, M. Powell, L. Chatrow, J. G. Rohwer, H. Sauquet, P. Cuenoud, C. Neinhuis, and L. A. Alice. 2001. Angiosperm phylogeny inferred from *matK* sequence data. P. 118. *Botany 2001—Abstracts.*

Hoch, P. C., J. V. Crisci, H. Tobe, and P. E. Berry. 1993. A cladistic analysis of the plant family Onagraceae. *Syst. Bot.* 18: 31–47.

Hoot, S. B. 1991. Phylogeny of the Ranunculaceae based on epidermal microcharacters and macromorphology. *Syst. Bot.* 16: 741–755.

Hoot, S. B. 1995. Phylogeny of the Ranunculaceae based on preliminary *atpB, rbcL* and 18S nuclear ribosomal DNA sequence data. *Plant Syst. Evol.* Suppl. 9: 241–251.

Hoot, S. B. and P. R. Crane. 1995. Interfamilial relationships in the Ranunculidae based on molecular systematics. *Plant Syst. Evol.* Suppl. 9: 119–131.

Hoot, S. B., A. W. Douglas. 1998. Phylogeny of the Proteaceae based on *atpB* and *atpB-rbcL* intergenic spacer region sequences. *Austr. Syst. Bot.* 11: 301–320.

Hoot, S. B., J. W. Kadereit, F. R. Blattner, K. B. Jork, A. E. Schwarzbach, and P. R. Crane. 1997. The phylogeny of the Papaveraceae s. l. based on four data sets: *atpB* and *rbcL* sequences, *trnK* restriction sites and morphological characters. *Syst. Bot.* 22: 575–590

Hoot, S. B., S. Magallón, and P. R. Crane. 1999. Phylogeny of basal eudicots based on three molecular data sets: *atpB, rbcL,* and 18S nuclear ribosomal DNA sequences. *Ann. Missouri Bot. Gard.* 86: 1–32.

Howe, H. F. and G. A. Vande Kerckhove. 1980. Nutmeg dispersal by tropical birds. *Science* 210: 925–927.

Huber, H. 1993. Aristolochiaceae. In *The families and genera of vascular plants*, vol. 2, Magnoliid, hamamelid and caryophyllid families, K. Kubitzki, J. G. Rohwer, and V. Bittrich (eds.), 129–137. Springer-Verlag, Berlin.

Huck, R. B. 1992. Overview of pollination biology in the Lamiaceae. In *Advances in labiate science*, R. M. Harley and T. Reynolds (eds.), 167–181. Royal Botanic Gardens, Kew.

Hufford, L. 1992. Rosidae and their relationships to other nonmagnoliid dicotyledons: A phylogenetic analysis using morphological and chemical data. *Ann. Missouri Bot. Gard.* 79: 218–248.

Hufford, L. 1997. A phylogenetic analysis of Hydrangeaceae based on morphological data. *Int. J. Plant Sci.* 158: 652–672.

Hufford, L. 2001. Ontogeny and morphology of the fertile flowers of *Hydrangea* and allied genera of tribe Hydrangeae. *Bot. J. Linn. Soc.* 137: 139–187.

Hufford, L. and P. R. Crane. 1989. A preliminary phylogenetic analysis of the "lower" Hamamelidae. In *Evolution, systematics and fossil history of the Hamamelidae*, vol. 1, Introduction and "lower" Hamamelidae. Syst. Assoc. Special Vol. 40A, P. R. Crane and S.

Blackmore (eds.), 175–192. Clarendon Press, Oxford.

Hufford, L. D. and P. K. Endress. 1989. The diversity of anther structures and dehiscence patterns among Hamamelididae. *Bot. J. Linnean Soc.* 99: 301–346.

Hufford, L., M. L. Moody, and D. E. Soltis. 2001. A phylogenetic analysis of Hydrangeaceae based on sequences of the plastid gene *matK* and their combination with *rbcL* and morphological data. *Int. J. Pl. Sci.* 162: 835–846.

Humphries, C. J. and S. Blackmore. 1989. A review of the classification of the Moraceae. In *Evolution, systematics and fossil history of the Hamamelidae*, vol. 2, "Higher" Hamamelidae. Syst. Assoc. Special Vol. 40B, P. R. Crane and S. Blackmore (eds.), 267–277. Clarendon Press, Oxford.

Hutchinson, J. 1934. *The families of flowering plants*, vol. 2, Monocotyledons. Macmillan, London.

Hutchinson, J. 1973. *The families of flowering plants*, 3rd ed. Clarendon Press, Oxford.

Igersheim, A., M. Buzgo, and P. K. Endress. 2001. Gynoecium diversity and systematics in basal monocots. *Bot. J. Linn. Soc.* 136: 1–65.

Igersheim, A. and P. K. Endress. 1997. Gynoecium diversity and systematics of the Magnolales and winteroids. *Bot. J. Linn. Soc.* 124: 213–271.

Igersheim, A. and P. K. Endress. 1998. Gynoecium diversity and systematics of the paleoherbs. *Bot. J. Linn. Soc.* 127: 289–370.

Iltis, H. H. 1999. Setchellanthaceae (Capparales), a new family for a relictual, glucosinolate-producing endemic of the Mexican deserts. *Taxon* 48: 257–275.

Ingrouille, M. J., M. W. Chase, M. F. Fay, D. Bowman, M. van der Bank, and A. D. E. Bruijn. 2002. Systematics of Vitaceae from the viewpoint of plastid *rbcL* DNA sequence data. *Bot. J. Linn. Soc.* 138: 421–432.

Isley, D. 1998. *Native and naturalized Leguminosae (Fabaceae) of the United States (exclusive of Alaska and Hawaii).* Brigham Young Univ., Provo, Utah.

Ito, M. 1986. Studies in the floral morphology and anatomy of Nymphaeales. III. Floral anatomy of *Brasenia schreberi* Gmel. and *Cabomba caroliniana* A. Gray. *Bot. Mag. Tokyo* 99: 169–184.

Ito, M. 1987. Phylogenetic systematics of the Nymphaeales. *Bot. Mag. Tokyo* 100: 17–36.

Jackson, G. A. 1933. A study of the carpophore of the Umbelliferae. *Amer. J. Bot.* 20: 121–144.

Jansen, R. K., H. J. Michaels and J. D. Palmer. 1991. Phylogeny and character evolution in the Asteraceae based on chloroplast DNA restriction site mapping. *Syst. Bot.* 16: 98–115.

Jansen, R. K., H. J. Michaels, R. S. Wallace, K-J. Kim, S. C. Keeley, L. E. Watson, and J. D. Palmer. 1992. Chloroplast DNA variation in the Asteraceae: Phylogenetic and evolutionary implications. In *Molecular systematics of plants*, P. S. Soltis, D. E. Soltis, and J. J. Doyle (eds.), 252–294. Chapman and Hall, New York.

Jansen, R. K. and J. D. Palmer. 1987. A chloroplast DNA inversion marks an ancient evolutionary split in the sunflower family, Asteraceae. *Proceed. Nat. Acad. Sci. U.S.A.* 84: 5818–5822.

Jaramillos, M. A. and P. S. Manos. 2001. Phylogeny and patterns of floral diversity in the genus *Piper* (Piperaceae*). Amer. J. Bot.* 88: 706–716.

Jeffrey, C. 1967. On the classification of the Cucurbitaceae. *Kew Bull.* 20: 417–426.

Jeffrey, C. 1980. A review of the Cucurbitaceae. *Bot. J. Linn. Soc.* 81: 233–247.

Jeffrey, C. 1990a. Appendix: An outline classification of the Cucurbitaceae. In *Biology and utilization of the Cucurbitaceae*, D. M. Bates, R. W. Robinson, and C. Jeffrey (eds.), 449–463. Cornell University Press, Ithaca, NY.

Jeffrey, C. 1990b. Systematics of the Cucurbitaceae—an overview. In *Biology and utilization of the Cucurbitaceae*, D. M. Bates, R. W. Robinson, and C. Jeffrey (eds.), 3–9. Cornell University Press, Ithaca, NY.

Jensen, S. R. 1992. Systematic implications of the distribution of iridoids and other chemical compounds in the Loganiaceae and other families of the Asteridae. *Ann. Missouri Bot. Gard.* 79: 284–302.

Jensen, U., S. B. Hoot, J. T. Johansson, and K. Kosuge. 1995. Systematics and phylogeny of the Ranunculaceae—a revised family concept on the basis of molecular data. *Pl. Syst. Evol. Suppl.* 9: 273–280.

Jobst, J., K. King and V. Hemleben. 1998. Molecular evolution of the Internal Transcribed Spacers (ITS1 and ITS2) and phylogenetic relationships among species of the family Cucurbitaceae. *Mol. Phylog. Evol.* 9: 204–219.

Johansson, J. T. 1995. A revised chloroplast DNA phylogeny of Ranunculaceae. *Plant Syst. Evol. Suppl.* 9: 253–261.

Johansson, J. T. and R. K. Jansen. 1993. Chloroplast DNA variation and phylogeny of the Ranunculaceae. *Plant Syst. Evol.* 187: 29–49.

Johnson, L. A. and D. E. Soltis. 1994. *matK* DNA sequences and phylogenetic reconstruction in Saxifragaceae s. str. *Syst. Bot.* 19: 143–156.

Johnson, L. A., J. L. Schultz, D. E. Soltis, and P. S. Soltis. 1996. Monophyly and generic relationships of Polemoniaceae based on *matK* sequences. *Amer. J. Bot.* 83: 1207–1224.

Johnson, L. A., D. E. Soltis, and P. S. Soltis. 1999. Phylogenetic relationships of Polemoniaceae inferred fgrom 18S ribosomal DNA sequences. *Pl. Syst. Evol.* 214: 65–89.

Johnson, L. A. S. 1957. A review of the family Oleaceae. *Contr. N. S. W. Nat. Herb.* 2: 397–418.

Johnson, L. A. S. 1976. Problems of species and genera in *Eucalyptus* (Myrtaceae). *Plant Syst. Evol.* 125: 155–167.

Johnson, L. A. S. and B. G. Briggs. 1975. On the Proteaceae: The evolution and classification of a Southern family. *J. Linnean Soc. Bot.* 70: 83–182.

Johnson, L. A. S. and B. G. Briggs. 1984. Myrtales and Myrtaceae: A phylogenetic analysis. *Ann. Missouri Bot. Gard.* 71: 700–756.

Johnson, L. A. S. and K. L. Wilson. 1989. Casuarinaceae: A synopsis. In *Evolution, systematics and fossil history of the Hamamelidae*, vol. 2, "Higher" Hamamelidae. Syst. Assoc. Special Vol. 40B, P. R. Crane and S. Blackmore (eds.), 167–188. Clarendon Press, Oxford.

Johnson, L. A. S. and K. L. Wilson. 1993. Casuarinaceae. In *The families and genera of vascular plants*, vol. 2, Magnoliid, hamamelid and caryophyllid families, K. Kubitzki, J. G. Rohwer and V. Bittrich (eds.), 237–242. Springer-Verlag, Berlin.

Johnston, I. M. 1936. A study of the Nolanaceae. *Contr. Gray Herb.* 113: 1–83.

Johnston, I. M. 1950. Studies in the Boraginaceae, XIX. *J. Arnold Arbor.* 31: 172–195.

Jones, J. H. 1986. Evolution of the Fagaceae: The implications of foliar features. *Ann. Missouri Bot. Gard.* 73: 228–275.

Jones, S. B. 1982. The genera of Vernonieae (Compositae) in the southeastern United States. *J. Arnold Arbor.* 63: 489–507.

Jorgensen, L. B. 1981. Myrosin cells and dilated cisternae of the endoplasmic reticulum in the order Capparales. *Nordic J. Bot.* 1: 433–445.

Jork, K. B. and J. W. Kadereit. 1995. Molecular phylogeny of the Old World representatives of Papaveraceae subfamily Papaveroidese with special emphasis on the genus *Meconopsis*. *Plant Syst. Evol. Suppl.* 9: 171–180.

Judd, W. S. 1986. Taxonomic studies in the Miconieae (Melastomataceae). I. Variation in inflorescence position. *Brittonia* 38: 238–242.

Judd, W. S. 1989. Taxonomic studies in the Miconieae (Melastomataceae). III. Cladistic analysis of axillary-flowered taxa. *Ann. Missouri Bot. Gard.* 76: 476–495.

Judd, W. S. 1996. The Pittosporaceae in the southeastern United States. *Harvard Pap. Bot.* No. 8: 15–26.

Judd, W. S. 1997a. The Asphodelaceae in the southeastern United States. *Harvard Pap. Bot.* No. 11: 109–123.

Judd, W. S. 1997b. The Flacourtiaceae in the southeastern United States. *Harvard Pap. Bot.* No. 10: 65–79.

Judd, W. S. 1998. The Smilacaceae in the southeastern United States. *Harvard Pap. Bot.* 3: 147–169.

Judd, W. S. 2000. The Hypoxidaceae in the southeastern United States. *Harvard Pap. Bot.* 5: 79–98.

Judd, W. S. 2001. The Asparagaceae in the southeastern United States. *Harvard Pap. Bot.* 6: 223–244.

Judd, W. S. 2003. The genera of Ruscaceae in the southeastern United States. *Harvard Pap. Bot.* 7: 93–149.

Judd, W. S. and J. D. Skean, Jr. 1991. Taxonomic studies in the Miconieae (Melastomataceae). IV. Generic realignments among terminal-flowered taxa. *Bull. Florida Mus. Nat. Hist., Biol. Sci.* 36: 25–84.

Judd, W. S. and K. A. Kron. 1993. Circumscription of Ericaceae (Ericales) as determined by preliminary cladistic analyses based on morphological, anatomical, and embryological features. *Brittonia* 45: 99–114.

Judd, W. S., W. L. Stern, and V. I. Cheadle. 1993. Phylogenetic position of *Apostasia* and *Neuwiedia* (Orchidaceae). *Bot. J. Linn. Soc.* 113: 87–94.

Judd, W. S., R. W. Sanders, and M. J. Donoghue. 1994. Angiosperm family pairs: Preliminary cladistic analyses. *Harvard Pap. Bot.* No. 5: 1–51.

Judd, W. S. and S. R. Manchester. 1998. Circumscription of Malvaceae (Malvales) as determined by a preliminary cladistic analysis employing morphological, paly-nological, and chemical characters. *Brittonia* 49: 384–405.

Judd, W. S. and T. K. Ferguson. 1999. The genera of Chenopodiaceae in the southeastern United States. *Harvard Pap. Bot.* 4: 365–416.

Judziewicz, E. J., L. G. Clark, X. Londoño, and M. J. Stern. 1999. *American bamboos*. Smithsonian Institution. Press, Washington, D. C.

Juncosa, A. M. and P. B. Tomlinson. 1988a. A historical and taxonomic synopsis of Rhizophoraceae and Anisophylleaceae. *Ann. Missouri Bot. Gard.* 75: 1278–1295.

Juncosa, A. M. and P. B. Tomlinson. 1988b. Systematic comparison and some biological characteristics of Rhizophoraceae and Anisophylleaceae. *Ann. Missouri Bot. Gard.* 75: 1296–1318.

Junell, S. 1934. Zur Gynäceummorphologie und Systematik der Verbenaceen und Labiaten. *Symb. Bot. Upsal.* 1(4): 1–219.

Kadereit, J. W. 1993. Papaveraceae. In *The families and genera of vascular plants*, vol. 2, Magnoliid, hamamelid and caryophyllid families, K. Kubitzki, J. G. Rohwer, and V. Bittrich (eds.), 494–506. Springer-Verlag, Berlin.

Kadereit, J. W. and K. J. Sytsma. 1992. Disassembling *Papaver*: A restriction site analysis of chloroplast DNA. *Nordic J. Bot.* 12: 205–217.

Kadereit, J. W., F. R. Blattner, K. B. Jork, and A. Schwarzbach. 1994. Phylogenetic analysis of the Papaveraceae s. l. (incl. Fumariaceae, Hypecoaceae and *Pteridophyllum*) based on morphological characters. *Bot. Jahrb. Syst.* 116: 361–390.

Kadereit, J. W., F. R. Blattner, K. B. Jork, and A. Schwarzback. 1995. The phylogeny of the Papaveraceae sensu lato: Morphological, geographical and ecological implications. *Plant Syst. Evol. Suppl.* 9: 133–145.

Kajita, T., H. Ohashi, Y. Tateishi, C. D. Bailey, and J. J. Doyle. 2001. *rbcL* and legume phylogeny with particular reference to Phaseoleae, Millettieae, and allies. *Syst. Bot.* 26: 515–536.

Kalkman, C. 1988. The phylogeny of the Rosaceae. *Bot. J. Linn. Soc.* 98: 37–59.

Källersjö. M., J. S. Farris, M. W. Chase, B. Bremer, M. F. Fay, C. J. Humphries, G. Petersen, O. Seberg, and K. Bremer. 1998. Simulatneous parsimony jackknife analysis of 2538 *rbcL* DNA sequences reveals support for major clades of green plants, land plants, and flowering plants. *Pl. Syst. Evol.* 213: 259–287.

Källersjö, M., G. Bergqvist, and A. A. Anderberg. 2000. Generic realignment in primuloid families of the Ericales s. l.: a phylogenetic analysis based on DNA sequences from three chloroplast genes and morphology. *Amer. J. Bot.* 87: 1325–1341.

Kampny, C. M. 1995. Pollination and floral diversity in Scrophulariaceae. *Bot. Rev.* 61: 350–366.

Karis, P. O. 1993. Morphological phylogenetics of the Asteraceae-Asteroideae, with notes on character evolution. *Plant Syst. Evol.* 186: 69–93.

Karis, P. O., M. Källersjö, and K. Bremer. 1992. Phylogenetic analysis of the Cichorioideae (Asteraceae), with emphasis on the Mutisieae. *Ann. Missouri Bot. Gard.* 79: 416–427.

Karis, P. O., P. Eldenäs, and M. Källersjö. 2001. New evidence for the systematic position of *Gundelia* L. with notes on delimitation of Arctoteae (Asteraceae). *Novon* 50: 105–114.

Karol, K. G., Y. Suh, G. E. Schutz, and E. A. Zimmer. 2000. Molecular evidence for the phylogenetic position of *Takhtajania* in the Winteraceae: inference from nuclear ribosomal and chloroplast gene spacer sequences. *Ann. Missouri Bot. Gard.* 87: 414–432.

Karol, K. G., J. E. Rodman, E. Conti, and K. J. Sytsma. 1999. Nucleotide sequence of *rbcL* and phylogenetic relationships of *Setchellanthus caeruleus* (Setchellanthaceae). *Taxon* 48: 303–315.

Kato, H., S. Kawano, R. Terauchi, M. Ohara and F. H. Utech. 1995a. Evolutionary biology of *Trillium* and related genera (Trilliaceae). I. Restriction site mapping and variation of chloroplast DNA and its systematic implications. *Pl. Species Biol.* 10: 17–29.

Kato, H., R. Terauchi, F. H. Utech, and S. Kawano. 1995b. Molecular systematics of the Trilliaceae sensu lato as inferred from *rbcL* sequence data. *Mol. Phylog. Evol.* 4: 184–193.

Kaul, R. B. 1968. Floral morphology and phylogeny in the Hydrocharitaceae. *Phytomorphology* 18: 13–35.

Kaul, R. B. 1970. Evolution and adaptation of the inflorescences in the Hydrocharitaceae. *Amer. J. Bot.* 57: 708–715.

Kaul, R. B. and E. C. Abbe. 1984. Inflorescence architecture and evolution in the Fagaceae. *J. Arnold Arbor.* 65: 375–401.

Kawano, S. and H. Kato. 1995. Evolutionary biology of *Trillium* and related genera (Trilliaceae) II. Cladistic analyses on gross morphological characters and phylogeny and evolution of the genus *Trillium*. *Plant Sp. Biol.* 10: 169–183.

Kedves, M. 1989. Evolution of the Normapolles complex. In *Evolution, systematics and fossil history of the Hamamelidae*, vol. 2, "Higher" Hamamelidae. Syst. Assoc. Special Vol. 40B, P. R. Crane and S. Blackmore (eds.), 1–7. Clarendon Press, Oxford.

Keeley, S. C. and R. K. Jansen. 1991. Evidence from chloroplast DNA for the recognition of a new tribe, Tarchonantheae and the tribal placement of *Pluchea* (Asteraceae). *Syst. Bot.* 16: 173–181.

Kellogg, E. A. and L. Watson. 1993. Phylogenetic studies of a large data set. I. Bambusoideae, Andropogonidae and Pooideae (Gramineae). *Bot. Rev.* 59: 273–343.

Kellogg, E. A. and H. P. Lindler. 1995. Phylogeny of Poales. In *Monocotyledons: Systematics and evolution*, P. J. Rudall, P. J. Cribb, D. F. Cutler, and C. J. Humphries (eds.), 511–542. Royal Botanic Gardens, Kew.

Kelly, J. P. 1920. A genetical study of flower form and flower color in *Phlox drummondii*. *Genetics* 5: 189–248.

Kelly, L. M. 1997. A cladistic analysis of *Asarum* (Aristolochiaceae) and implications for the evolution of herkogamy. *Amer. J. Bot.* 84: 1752–1765.

Keng, H. 1962. Comparative morphological studies in Theaceae. *Univ. Calif. Publ. Bot.* 33: 219–384.

Keng, H. 1993. Illiciaceae. In *The families and genera of vascular plants*, vol. 2, Magnoliid, hamamelid and caryophyllid

families, K. Kubitzki, J. G. Rohwer, and V. Bittrich (eds.), 344–347. Springer-Verlag, Berlin.

Kennedy, H. 2000. Diversification in pollination mechanisms in the Marantaceae. In *Monocots: Systematics and evolution*, K. L. Wilson and D. A. Morrison (eds.), 335–343. Collingwood, Australia.

Kessler, P. J. A. 1993a. Annonaceae. In *The families and genera of vascular plants*, vol. 2, Magnoliid, hamamelid and caryophyllid families, K. Kubitzki, J. G. Rohwer, and V. Bittrich (eds.), 93–129. Springer-Verlag, Berlin.

Kessler, P. J. A. 1993b. Menispermaceae. In *The families and genera of vascular plants*, vol. 2, Magnoliid, hamamelid and caryophyllid families, K. Kubitzki, J. G. Rohwer, and V. Bittrich (eds.), 402–418. Springer-Verlag, Berlin.

Killip, E. 1938. The American species of Passifloraceae. *Field Museum of Nat. Hist. Bot. Ser.* 19: 1–613.

Kim, J.-K. and R. K. Jansen. 1995. Phylogenetic implications of chloroplast DNA variation in the Berberidaceae. *Pl. Syst. Evol. Suppl.* 9: 341–349.

Kim, J.-K. and R. K. Jansen. 1998a. Chloroplast DNA restriction site variation and phylogeny of the Berberidaceae. *Amer. J. Bot.* 85: 1766–1778.

Kim, J.-K. and R. K. Jansen. 1998b. Paraphyly of Jasminoideae and monophyly of Oleoideae in Oleaceae. *Amer. J. Bot.* 85(6) Suppl.: 139.

Kim, J.-K., R. K. Jansen, R. S. Wallace, H. J. Michaels, and J. D. Palmer. 1992. Phylogenetic implications of *rbcL* sequence variation in the Asteraceae. *Ann. Missouri Bot. Gard.* 79: 428–445.

Kim, S., C.-W. Park, Y.-D. Kim, and Y. Suh. 2001. Phylogenetic relationships in family Magnoliaceae inferred from *ndhF* sequences. *Amer. J. Bot.* 88: 717–728.

Kim, Y.-D. and R. K. Jansen. 1998. Chloroplast DNA restriction site variation and phylogeny of the Berberidaceae. *Amer. J. Bot.* 85: 1766–1778.

Kirkbride, M. C. G. 1982. A preliminary phylogeny for the neotropical Rubiaceae. *Plant Syst. Evol.* 141: 115–121.

Kocyan, A. and P. K. Endress. 2001. Floral structure and development in *Apostasia* and *Neuwiedia* (Apostasioideae) and their relationship to other Orchidaceae. *Int. J. Pl. Sci.* 162: 847–867.

Kral, R. 1992. A treatment of American Xyridaceae exclusive of *Xyris*. *Ann. Missouri Bot. Gard.* 79: 819–885.

Kral, R. B. 1966a. Eriocaulaceae of continental North America north of Mexico. *Sida* 2: 285–332.

Kral, R. B. 1966b. *Xyris* (Xyridaceae) of the continental United States and Canada. *Sida* 2: 177–260.

Kral, R. B. 1983. The Xyridaceae in the southeastern United States. *J. Arnold Arbor.* 64: 421–429.

Kral, R. B. 1989. The genera of Eriocaulaceae in the southeastern United States. *J. Arnold Arbor.* 70: 131–142.

Kral, R. B. 1998. Xyridaceae. In The families and genera of vascular plants, vol. 4. Flowering plants—Monocotyledons, K. Kubitzki (ed.), 461–469. Springer-Verlag, Berlin.

Kress, W. J. 1990. The phylogeny and classification of the Zingiberales. *Ann. Missouri Bot. Gard.* 77: 698–721.

Kress, W. J. 1995. Phylogeny of the Zingiberanae: Morphology and mole-

cules. In *Monocotyledons: Systematics and evolution*, P. J. Rudall, P. J. Cribb, D. F. Cutler and C. J. Humphries (eds.), 443–460. Royal Botanic Gardens, Kew.

Kress, W. J., L. M. Prince, W. J. Hahn, and E. A. Zimmer. 2001. Unraveling the evolutionary radiation of the families of the Zingiberales using morphological and molecular evidence. *Syst. Biol.* 50: 926–944.

Kron, K. A. 1996. Phylogenetic relationships of Empetraceae, Epacridaceae, Ericaceae, Monotropaceae and Pyrolaceae: Evidence from nucleotide ribosomal 18S sequence data. *Ann. Bot.* 77: 293–303.

Kron, K. A. 1997. Phylogenetic relationships of Rhododendroideae (Ericaceae). *Amer. J. Bot.* 84: 973–980.

Kron, K. A. and W. S. Judd. 1990. Phylogenetic relationships within the Rhodoreae (Ericaceae) with specific comments on the placement of *Ledum*. *Syst. Bot.* 15: 57–68.

Kron, K. A. and M. W. Chase. 1993. Systematics of the Ericaceae, Empetraceae, Epacridaceae and related taxa based on *rbcL* sequence data. *Ann. Missouri Bot. Gard.* 80: 735–741.

Kron, K. A. and J. M. King. 1996. Cladistic relationships of *Kalmia, Leiophyllum* and *Loiseleuria* (Phyllodoceae, Ericaceae) based on nucleotide sequences from *rbcL* and nuclear ribosomal internal transcribed spacer regions (ITS). *Syst. Bot.* 21: 17–29.

Kron, K. A. and W. S. Judd. 1997. Systematics of the *Lyonia* group (Andromedeae, Ericaceae) and the use of species as terminals in higher-level cladistic analyses. *Syst. Bot.* 22: 479–492.

Kron, K. A., W. S. Judd, and D. M. Crayn. 1999. Phylogenetic analyses of Andromedeae (Ericaceae subfam. Vaccinioideae). *Amer. J. Bot.* 86: 1290–1300.

Kron, K. A., W. S. Judd, P. F. Stevens, D. M. Crayn, A. A. Anderberg, P. A. Gadek, C. J. Quinn, and J. L. Luteyn. 2002. A phylogenetic classification of Ericaceae: molecular and morphological evidence. *Bot. Rev.* 68: 335–423.

Kubitzki, K. 1993a. Betulaceae. In *The families and genera of vascular plants*, vol. 2, Magnoliid, hamamelid and caryophyllid families, K. Kubitzki, J. G. Rohwer, and V. Bittrich (eds.), 152–157. Springer-Verlag, Berlin.

Kubitzki, K. 1993b. Fagaceae. In *The families and genera of vascular plants*, vol. 2, Magnoliid, hamamelid and caryophyllid families, K. Kubitzki, J. G. Rohwer, and V. Bittrich (eds.), 301–309. Springer-Verlag, Berlin.

Kubitzki, K. 1993c. Myricaceae. In *The families and genera of vascular plants*, vol. 2, Magnoliid, hamamelid and caryophyllid families, K. Kubitzki, J. G. Rohwer, and V. Bittrich (eds.), 453–457. Springer-Verlag, Berlin.

Kubitzki, K. 1993d. Platanaceae. In *The families and genera of vascular plants*, vol. 2, Magnoliid, hamamelid and caryophyllid families, K. Kubitzki, J. G. Rohwer, and V. Bittrich (eds.), 521–522. Springer-Verlag, Berlin.

Kubitzki, K. (ed.) 1998a. *The families and genera of vascular plants*, vol. 3. Flowering plants–Monocotyledons: Lilianae (except Orchidaceae). Springer-Verlag, Berlin.

Kubitzki, K. (ed.) 1998b. *The families and genera of vascular plants*, vol. 4. Flowering plants–Monocotyledons: Alismatanae and Commelinanae (except Gramineae). Springer-Verlag, Berlin.

Kubitzki, K. 1998c. Cannaceae. In *The families and genera of vascular plants*, vol. 4. Flowering plants—Monocotyledons, K. Kubitzki (ed.), 103–106. Springer-Verlag, Berlin.

Kubitzki, K. 1998d. Typhaceae. In *The families and genera of vascular plants*, vol. 4. Flowering plants—Monocotyledons, K. Kubitzki (ed.), 457–461. Springer-Verlag, Berlin.

Kubitzki, K. and H. Kurz. 1984. Synchronized dichogamy and dioecy in neotropical Lauraceae. *Plant Syst. Evol.* 147: 253–266.

Kubitzki, K. and P. J. Rudall. 1998. Asparagaceae. In The families and genera of vascular plants, vol. 3. Flowering plants—Monocotyledons, K. Kubitzki (ed.), 125–129. Springer-Verlag, Berlin.

Kühn, U., V. Bittrich, R. Carolin, H. Freitag, I. C. Hedge, P. Uotila, and P. G. Wilson. 1993. Chenopodiaceae. In *The families and genera of vascular plants*, vol. 2, Magnoliid, hamamelid and caryophyllid families, K. Kubitzki, J. G. Rohwer, and V. Bittrich (eds.), 253–281. Springer-Verlag, Berlin.

Kühn, U. and K. Kubitzki. 1993. Myristicaceae. In *The families and genera of vascular plants*, vol. 2, Magnoliid, hamamelid and caryophyllid families, K. Kubitzki, J. G. Rohwer, and V. Bittrich (eds.), 457–467. Springer-Verlag, Berlin.

Kuijt, J. 1969. *The biology of parasitic flowering plants*. University of California Press, Berkeley.

Kuijt, J. 1981. Inflorescence morphology of Loranthaceae—an evolutionary synthesis. *Blumea* 27: 1–78.

Kuijt, J. 1982. The Viscaceae in the southeastern United States. *J. Arnold Arbor.* 63: 401–410.

Kunze, H. 1990. *Morphology and evolution of the corona in Asclepiadaceae and related families*. Akad. der Wiss. und der Literatur. Tropische und subtropische Pflanzenwelt 76. Steiner Verlag, Mainz-Stuttgart.

Kunze, H. 1992. Evolution of the translator in Periplocaceae and Asclepiadaceae. *Plant Syst. Evol.* 185: 99–122.

La Duke, J. C. and J. Doebley. 1995. A chloroplast DNA-based phylogeny of the Malvaceae. *Syst. Bot.* 20: 259–271.

Lammers, T. G. 1992. Circumscription and phylogeny of the Campanulales. *Ann. Missouri Bot. Gard.* 79: 388–413.

Landolt, E. 1980. *Biosystematic investigation in the family of duckweeds (Lemnaceae)*, vol. 1, Key to the determination of taxa within the family of Lemnaceae. Veröff. Geobot. Inst. ETH Stiftung Rübel Zürich 70: 13–21.

Landolt, E. 1986. *Biosystematic investigation in the family of duckweeds (Lemnaceae)*, vol. 2, The family of Lemnaceae, a monographic study, vol. 1. Veröff. Geobot. Inst. ETH Stiftung Rübel Zürich 71: 1–566.

Landolt, E. and R. Kandeler. 1987. *Biosystematic investigation in the family of duckweeds (Lemnaceae)*, vol. 4, The family of Lemnaceae, a monographic study, vol. 2. Veröff. Geobot. Inst. ETH Stiftung Rübel Zürich 95: 1–638

Lane, M. A. 1996. Pollination biology of Compositae. In *Compositae: Biology and utilization*. Proceedings of the International Compositae Conference, Kew, P. D. S. Caligari and D. J. N. Hind (eds.), 61–80. Royal Botanic Garden, Kew.

Larsen, K., J. M. Lock, H. Maas, and P. J. M. Maas. 1998. Zinbiberaceae. In *The families and genera of vascular plants*, vol. 4. Flowering plants—Monocotyledons, K. Kubitzki (ed.), 474–495. Springer-Verlag, Berlin.

Laubengayer, R. A. 1937. Studies in anatomy and morphology of the polygonaceous flower. *Amer. J. Bot.* 24: 329–343.

Lavin, M. 1987. A cladistic analysis of the tribe Robinieae (Papilionoideae, Leguminosae). In *Advances in legume systematics*, part 3, C. H. Stirton (ed.), 31–64. Royal Botanic Gardens, Kew.

Lawrence, G. H. M. 1951. *Taxonomy of vascular plants*. Macmillan, New York.

Lee, S. and J. Wen. 2001. A phylogenetic analysis of *Prunus* and the Amygdaloideae (Rosaceae) using ITS sequences of nuclear ribosomal DNA. *Amer. J. Bot.* 88: 150–160.

Leins, P. and C. Erbar. 1990. On the mechanism of secondary pollen presentation in the Campanulales-Asterales-complex. *Bot. Acta* 103: 87–92.

Leins, P. and C. Erbar. 1994. Putative origin and relationships of the order from the viewpoint of developmental flower morphology. In *Caryophyllales: Evolution and systematics*, H.-D. Behnke and T. J. Mabry (eds.), 303–316. Springer-Verlag, Berlin.

Leppik, E. E. 1964. Floral evolution in Ranunculaceae. *Iowa State Coll. Sci.* 39: 1–101.

Lersten, N. R. 1975. Colleter types in Rubiaceae, especially in relation to the bacterial leaf nodule symbiosis. *Bot. J. Linn. Soc.* 71: 311–319.

Lersten, N. R. and J. M. Beaman. 1998. First report of oil cavities in Scrophulariaceae and reinvestigation of air spaces in leaves of *Leucophyllum frutescens*. *Amer. J. Bot.* 85: 1646–1649.

Les, D. H. 1988. The origin and affinities of the Ceratophyllaceae. *Taxon* 37: 326–435.

Les, D. H. 1989. The evolution of achene morphology in *Ceratophyllum* (Ceratophyllaceae), IV. Summary of proposed relationships and evolutionary trends. *Syst. Bot.* 14: 254–262.

Les, D. H. 1993. Ceratophyllaceae. In *The families and genera of vascular plants*, vol. 2, Magnoliid, hamamelid and caryophyllid families, K. Kubitzki, J. G. Rohwer, and V. Bittrich (eds.), 246–250. Springer-Verlag, Berlin.

Les, D. H., D. K. Garvin, and C. F. Wimpee. 1991. Molecular evolutionary history of ancient aquatic angiosperms. *Proc. Natl. Acad. Sci. USA* 88: 10119–10123.

Les, D. H., M. A. Cleland, and M. Waycott. 1997a. Phylogenetic studies in Alismatidae, II: Evolution of marine angiosperms (seagrasses) and hydrophily. *Syst. Bot.* 22: 443–463.

Les, D. H., E. Landolt, and D. J. Crawford. 1997b. Systematics of the Lemnaceae (duckweeds): Inferences from micromolecular and morphological data. *Plant Syst. Evol.* 204: 161–177.

Les, D. H. and D. J. Crawford. 1999. *Landoltia* (Lemnaceae), a new genus of duckweeds. *Novon* 9: 530–533.

Les, D. H., E. L. Schneider, D. J. Padgett, P. S. Soltis, D. E. Soltis, and M. Zanis. 1999. Phylogeny, classification and floral evolution of water lilies (Nymphaeaceae; Nymphaeales): a synthesis of nonmolecular, *rbcL, matK*, and 18S rDNA data. *Syst. Bot.* 24: 28–46.

Les, D. H., D. J. Crawford, E. Landolt, J. D. Gabel, and R. T. Kimball. 2002. Phylogeny and systematics of Lamnaceae, the duckweed family. *Syst. Bot.* 27: 221–240.

Levin, G. A. 1986. Systematic foliar morphology of Phyllanthoideae (Euphorbiaceae). III. Cladistic analysis. *Syst. Bot.* 11: 515–530.

Li, J., A. L. Bogle, and A. S. Klein. 1997. Interspecific relationships and genetic divergence of the disjunct genus *Liquidambar* (Hamamelidaceae) inferred from DNA sequences of plastid gene *matK*. *Rhodora* 99: 299–240.

Li,, J., A. L. Bogle, and A. S. Klein. 1999. Phylogenetic relationships of the Hamamelidaceae inferred from sequences of internal transcribed spacers (ITS) of nuclear ribosomal DNA. *Amer. J. Bot.* 86: 1027–1037.

Li, J. and M. J. Donoghue. 1999. More molecular evidence for interspecific relationships in *Liquidambar* (Hamamelidaceae). *Rhodora* 101: 87–91.

Lidén, M. 1986. Synopsis of Fumarioideae (Papaveraceae) with a monograph of the tribe Fumarieae. *Opera Bot.* 88: 5–129.

Lidén, M. 1993. Fumariaceae. In *The families and genera of vascular plants*, vol. 2, Magnoliid, hamamelid and caryophyllid families, K. Kubitzki, J. G. Rohwer and V. Bittrich (eds.), 310–318. Springer-Verlag, Berlin.

Linder, H. P. 1985. Conspectus of the African species of Restionaceae. *Bothalia* 15: 387–503.

Linder, H. P. 2000. Vicariance, climate change, anatomy and phylogeny of Restionaceae. *Bot. J. Linn. Soc.* 134: 159–177.

Linder, H. P. and E. A. Kellogg. 1995. Phylogenetic patterns in the commelinid clade. In *Monocotyledons: Systematics and evolution*, P. J. Rudall, P. J. Cribb, D. F. Cutler, and C. J. Humphries (eds.), 473–496. Royal Botanic Gardens, Kew.

Linder, H. P., B. G. Briggs, and L. A. S. Johnson. 2000a. Restionaceae. In *The families and genera of vascular plants*, vol. 4, Flowering plants--Monocotyledons, K. Kubitzki (ed.), 425–445. Springer-Verlag, Berlin.

Linder, H. P., B. G. Briggs, and L. A. S. Johnson. 2000b. Restionaceae: a morphological phylogeny. In *Monocots: Systematics and evolution*, K. L. Wilson, and D. A. Morrison (eds.), 653–660. CSIRO Publ., Collingwood, Australia.

Lloyd, F. E. 1942. *The carnivorous plants.* Chronica Botanica, Waltham, MA.

Loconte, H. 1993. Berberidaceae. In *The families and genera of vascular plants*, vol. 2, Magnoliid, hamamelid and caryophyllid families, K. Kubitzki, J. G. Rohwer, and V. Bittrich (eds.), 147–152. Springer-Verlag, Berlin.

Loconte, H., L. M. Campbell, and D. W. Stevenson. 1995. Ordinal and familial relationships of ranunculid genera. *Plant Syst. Evol. Suppl.* 9: 99–118.

Loconte, H. and J. R. Estes. 1989. Phylogenetic systematics of Berberidaceae and Ranunculales (Magnoliidae). *Syst. Bot.* 14: 565–579.

Loconte, T. and D. W. Stevenson. 1991. Cladistics of the Magnoliidae. *Cladistics* 7: 267–296.

Long, R. W. 1970. The genera of Acanthaceae in the southeastern United States. *J. Arnold Arbor.* 51: 257–309.

Londoño, A. C., E. Alvarez, E. Forero, and C. M. Morton. 1995. A new genus and species of Dipterocarpaceae from the Neotropics. I. Introduction, taxonomy, ecology and distribution. *Brittonia* 47: 225–236.

Lowden, R. M. 1973. Revision of the genus *Pontederia* L. *Rhodora* 75: 426–483.

Lowden, R. M. 1982. An approach to the taxonomy of *Vallisneria* L. (Hydrocharitaceae). *Aquatic Bot.* 13: 269–298.

Lu, A.-M. 1990. A preliminary cladistic study of the families of the superorder Lamiiflorae. *Bot. J. Linnean Soc.* 103: 39–57.

Lüders. H. 1907. Systematische Untersuchungen über die Caryophyllaceen mit einfachem Diagramm. *Bot. Jahrb.* 40: 1–38.

Mabry, T. J. 1973. Is the order Centrospermae monophyletic? In *Chemistry in botanical classification*, G. Bendz and J. Santesson (eds.), 275–285. Academic Press, New York.

Mabry, T. J. 1976. Pigment dichotomy and DNA–DNA hybridization data for Centrospermous families. *Plant Syst. Evol.* 126: 79–94.

Mabry, T. J., L. Kilmer, and C. Chang. 1972. The betalains: Structure, function and biogenesis and the plant order Centrospermae. In *Recent advances in phytochemistry*, vol. 5, Structural and functional aspects of phytochemistry, V. C. Runeckles and T. C. Tso (eds.), 105–134. Academic Press, London.

MacDonald, A. D. 1974. Floral development in *Comptonia peregrina* (Myricaceae). *Canad. J. Bot.* 52: 2165–2169.

MacDonald, A. D. 1977. Myricaceae: Floral hypothesis for *Gale* and *Comptonia*. *Canad. J. Bot.* 55: 2636–2651.

MacDonald, A. D. 1978. Organogenesis of the male inflorescence and flowers of *Myrica esculenta. Canad. J. Bot.* 56: 2415–2423.

MacDonald, A. D. 1979a. Development of the female flower and gynecandrous partial inflorescence of *Myrica californica. Canad. J. Bot.* 57: 141–151.

MacDonald, A. D. 1979b. Inception of the cupule of *Quercus macrocarpa* and *Fagus grandifolia. Canad. J. Bot.* 57: 1777–1782.

MacDonald, A. D. 1989. The morphology and relationships of the Myricaceae. In *Evolution, systematics and fossil history of the Hamamelidae.* Syst. Assoc. Special Vol. 40B, P. R. Crane and S. Blackmore (eds.), 147–165. Clarendon Press, Oxford.

Maguire, B. 1976. Apomixis in the genus *Clusia* (Clusiaceae): A preliminary report. *Taxon* 25: 241–244.

Maguire, B. and P. S. Ashton. 1977. Pakaramoideae, Dipterocarpaceae of the Western Hemisphere. II. Systematic, geographic and phyletic considerations. Taxon 26: 359–368.

Maheshwari, S. C. 1958. *Spirodela polyrrhiza*: The link between the aroids and the duckweeds. *Nature* 181: 1745–1746.

Malcomber, S. T. and S. Sebsebe. 1993. The status of *Protasparagus* and *Myrsiphyllum* in the Asparagaceae. *Kew Bull.* 48: 63–78.

Manchester, S. R. 1987. The fossil history of the Juglandaceae. *Monogr. Syst. Bot. Missouri Bot. Gard.* 21: 1–137.

Manchester, S. R. 1989. Systematics and fossil history of the Ulmaceae. In *Evolution, systematics and fossil history of the Hamamelidae*, vol. 2, "Higher" Hamamelidae. Syst. Assoc. Special Vol. 40B, P. R. Crane and S. Blackmore (eds.), 221–251. Clarendon Press, Oxford.

Manchester, S. R. and B. H. Tiffney. 2001. Integration of paleobotanical and neobotanical data in the assessment of phytogeographic history of holarctic angiosperm clades. *Int. J. Plant Sci.* 162 (6, Suppl.): 519–527.

Manen, J. F., A. Natalil, and F. Ehrendorfer. 1994. Phylogeny of Rubiaceae–Rubieae inferred from the sequence of a cp-DNA intergene region. *Plant Syst. Evol.* 190: 195–211.

Manhart, J. R. and J. H. Rettig. 1994. Gene sequence data. In *Caryophyllales: Evolution and systematics*, H.-D. Behnke and T. J. Mabry (eds.), 235–246. Springer-Verlag, Berlin.

Mann, L. K. 1959. The *Allium* inflorescence: Some species of the section *Molium*. *Amer. J. Bot.* 46: 730–739.

Manning, S. D. 2000. The genera of Bignoniaceae in the southeastern United States. *Harvard Pap. Bot.* 5: 1–77.

Manning, W. E. 1938. The morphology of the flower of the Juglandaceae. I. The inflorescence. *Amer. J. Bot.* 25: 407–419.

Manning, W. E. 1940. The morphology of the flower of the Juglandaceae. II. The pistillate flowers and fruit. *Amer. J. Bot.* 27: 839–852.

Manning, W. E. 1948. The morphology of the flower of the Juglandaceae. III. The staminate flowers. *Amer. J. Bot.* 35: 606–621.

Manning, W. E. 1978. The classification within the Juglandaceae. *Ann. Missouri Bot. Gard.* 65: 1058–1087.

Manos, P. S. 1997. Systematics of *Nothofagus* (Nothofagaceae) based on rDNA spacer sequences: taxonomic congruence with morphology and plastid sequences. *Amer. J. Bot.* 84: 1137–1155.

Manos, P. S. and K. P. Steele. 1997. Phylogenetic analyses of "higher" hamamelididae based on plastid sequence data. *Syst. Bot.* 84: 1407–1419.

Manos, P. S., J. J. Doyle, and K. C. Nixon. 1999. Phylogeny, biogeography, and processes of molecular differentiation in *Quercus* subgenus *Quercus* (Fagaceae). *Mol. Phylog. and Evol.* 12: 333–349.

Manos, P. S., K. C. Nixon and J. J. Doyle. 1993. Cladistic analysis of restriction site variation within the chloroplast DNA inverted repeat region of selected Hamamelididae. *Syst. Bot.* 18: 551–562.

Manos, P. S. and D. E. Stone. 2001. Evolution, phylogeny and systematics of the Juglandaceae. *Ann. Missouri Bot. Gard.* 88: 231–269.

Manos, P. S., Z.-K. Zhou, and C. H. Cannon. 2001. Systematics of Fagaceae: Phylogenetic tests of reproductive trait evolution. *Int. J. Pl. Sci.* 162: 1361–1379.

Martin, P. G. and J. M. Dowd. 1986. Phylogenetic studies using protein sequences within the order Myrtales. *Ann. Missouri Bot. Gard.* 73: 441–448.

Mast, A. R. 1998. Molecular systematics of subtribe Banksiinae (Banksia and Dryandra; Proteaceae) based on cpDNA and nrDNA sequence data: implications for taxonomy and biogeography. *Austr. Syst. Bot.* 11: 321–342.

Mast, A. R., S. Kelso, A. J. Richards, D. L. Lang, D. M. S. Feller, and E. Conti. 2001. Phylogenetic relationships in *Primula* L. and related genera (Primulaceae) based on noncoding chloroplast DNA. *Int. J. Pl. Sci.* 162: 1138–1400.

Mathews, S. and M. J. Donoghue. 1999. The root of angiosperm phylogeny inferred from duplicate phytochrome genes. *Science* 286: 947–950.

Mathews, S. and M. J. Donoghue. 2000. Basal angiosperm phylogeny inferred from duplicate phytochromes A and C. *Int. J. Pl. Sci.* 161 (6 Suppl.): S41–S55.

Mayo, S. J., J. Bogner, and P. Boyce. 1995. The Arales. In *Monocotyledons: Systematics and evolution*, P. J. Rudall, P. J. Cribb, D. F. Cutler, and C. J. Humphries (eds.), 277–286. Royal Botanic Gardens, Kew.

Mayo, S. J., J. Bogner, and P. C. Boyce. 1998. Araceae. In *The families and genera of vascular plants*, vol. 4. Flowering plants—Monocotyledons, K. Kubitzki (ed.), 26–74. Springer-Verlag, Berlin.

McDade, L. A. and M. L. Moody. 1999. Phylogenetic relationships among Acanthaceae: evidence from noncoding *trnL-trnF* chloroplast DNA sequences. *Amer. J. Bot.* 86: 70–80.

McDade, L. A., S. E. Masta, M. L. Moody, and E. Waters. 2000. Phylogenetic relationships among Acanthaceae: evidence from two genomes. *Syst. Bot.* 25: 106–121.

McDaniel, S. T. 1971. The genus *Sarracenia* (Sarraceniaceae). *Bull. Tall Timbers Res. Sta.* 9: 1–36.

McKelvey, S. D. and K. Sax. 1933. Taxonomic and cytological relationships of *Yucca* and *Agave. J. Arnold Arbor.* 14: 76–80.

McKey, D. 1989. Interactions between ants and leguminous plants. *Monogr. Syst. Bot. Missouri Bot. Gard.* 29: 673–718.

McPherson, M. A. and S. W. Graham. 2001. Inference of Asparagales phylogeny using a large chloroplast data set. P. 126. *Botany 2001—Abstracts.*

Meacham, C. A. 1980. Phylogeny of the Berberidaceae with an evaluation of classifications. *Syst. Bot.* 5: 149–172.

Meerow, A. W. 1995. Towards a phylogeny of Amaryllidaceae. In *Monocotyledons: Systematics and evolution*, P. J. Rudall, P. J. Cribb, D. F. Cutler, and C. J. Humphries (eds.), 169–179. Royal Botanic Gardens, Kew.

Meerow, A. W. and D. A. Suijman. 1998. Amaryllidaceae In *The families and genera of vascular plants*, vol. 3. Flowering plants—Monocotyledons, K. Kubitzki (ed.), 83–110. Springer-Verlag, Berlin.

Meerow, A. W., M. F. Fay, C. L. Guy, Q.-B. Li, F. Q. Zaman, and M. W. Chase. 1999a. Systematics of Amaryllidaceae based on cladistic analysis of plastid *rbcL* and *trnL-F* sequence data. *Amer. J. Bot.* 86: 1325–1345.

Meerow, A. W., M. F. Fay, M. W. Chase, C. L. Guy, and Q.-B. Li. 1999b. The new phylogeny of the Amaryllidaceae. *Herbertia* 54: 180–203.

Meerow, A., M. F. Fay, M. W. Chase, C. l. Guy, Q.-B. Li, D. Snijman, and S.-L. Yang. 2000a. Phylogeny of Amaryllidaceae: molecules and morphology. In *Monocots: Systematics and evolution*, K. L. Wilson and D. A. Morrison (eds.), 372–386. Collingwood, Australia.

Meerow, A. W., C. L. Guy, Q.-B. Li, and S.-L. Yang. 2000b. Phylogeny of the American

Amaryllidaceae based on nrDNA ITS sequences. *Syst. Bot.* 25: 708–726.

Meeuse, A. D. J. 1975. Taxonomic relationships of Salicaceae and Flacourtiaceae. *Acta Bot. Neerl.* 24: 437–457.

Meeuse, B. J. D. and E. L. Schneider. 1979. *Nymphaea* revisited: A preliminary communication. *Israel J. Bot.* 28: 65–79.

Mennega, A. M. W. 1982. Stem structure of New World Menispermaceae. J. *Arnold Arbor.* 63: 145–171.

Mészáros, S., J. de Laet and E. Smets. 1996. Phylogeny of temperate Gentianaceae: A morphological approach. *Syst. Bot.* 21: 153–168.

Meurer-Grimes, B. 1995. New evidence for the systematic significance of acylated spermidines and flavonoids in pollen of higher Hamamelidae. *Brittonia* 47: 130–142.

Michaels, H. J., K. M. Scott, R. G. Olmstead, T. Szaro, R. K. Jansen, and J. D. Palmer. 1993. Interfamilial relationships of the Asteraceae: Insights from *rbcL* sequence variation. *Ann. Missouri Bot. Gard.* 80: 742–751.

Miller, N. G. 1970. The genera of Cannabaceae in the southeastern United States. *J. Arnold Arbor.* 51: 185–203.

Miller, N. G. 1971a. The genera of Polygalaceae in the southeastern United States. *J. Arnold Arbor.* 52: 267–284.

Miller, N. G. 1971b. The genera of Urticaceae in the southeastern United States. *J. Arnold Arbor.* 52: 40–68.

Miller, N. G. 1990. The genera of Meliaceae in the southeastern United States. *J. Arnold Arbor.* 71: 453–486.

Miller, N. G. 2001. The Callitrichaceae in the southeastern United States. *Harvard Pap. Bot.* 5(2): 277–301.

Möller, M. and Q. C. B. Cronk. 1997. Origin and relationships of *Saintpaulia* (Gesneriaceae) based on ribosomal DNA internal transcribed spacer (ITS) sequences. *Amer. J. Bot.* 84: 956–965.

Molvray, N., P. J. Kores, and M. W. Chase. 2000. Polyphyly of mycoheterotrophic orchids and functional influences on floral and molecular characters. In *Monocots: Systematics and evolution.* K. L. Wilson and D. A. Morrison (eds.), 441–448. Collingwood, Australia.

Moore, H. E. 1973. The major groups of palms and their distribution. *Gentes Herb.* 11: 27–141.

Moore, H. E. and N. W. Uhl. 1982. The major trends of evolution in palms. *Bot. Rev.* 48: 1–69.

Morgan, D. R. and D. E. Soltis. 1993. Phylogenetic relationships among members of Saxifragaceae sensu lato based on *rbcL* sequence data. *Ann. Missouri Bot. Gard.* 80: 631–660.

Morgan, D. R., D. E. Soltis and K. R. Robertson. 1994. Systematic and evolutionary implications of *rbcL* sequence variation in Rosaceae. *Amer. J. Bot.* 81: 890–903.

Mori, S. A., G. T. Prance, and A. B. Bolten. 1978. Additional notes on the floral biology of neotropical Lecythidaceae. *Brittonia* 30: 113–130.

Mori, S. A. and G. T. Prance. 1990. Lecythidaceae. Part II. The zygomorphic-flowered New World genera (*Couroupita, Corythophora, Bertholletia, Couratari, Eschweilera,* and *Lecythis*). *Flora Neotropica, Monograph* 21 (II): 1–376.

Morley, T. 1976. Memecyleae (Melastomataceae). *Flora Neotropica Monographs* 15: 1–295.

Mort, M. E., D. E. Soltis and P. S. Soltis. 1998. Molecular systematics of Crassulaceae based on *matK* sequence data. *Amer. J. Bot.* 85(6) Suppl.: 145–146.

Mort, M. E., D. E. Soltis, P. S. Soltis, J. Francisco-Ortega, and A. Santos-Guerra. 2001. Phylogenetic relationships and evolution of Crassulaceae inferred from *matK* sequence data. *Amer. J. Bot.* 88: 76–91.

Morton, C. M. 1995. A new genus and species of Dipterocarpaceae from the Neotropics. II. Stem anatomy. *Brittonia* 47: 237–247.

Morton, C.M., M. W. Chase, and J. Kallunki. 1996. Evaluation of the six subfamilies of Rutaceae using evidence from *rbcL* sequence variation. *Amer. J. Bot.* 83(6) Suppl.: 180–181.

Morton, C. M., K. A. Kron, and M. W. Chase. 1997a. A molecular evaluation of the monophyly of the order Ebenales based upon *rbcL* sequence data. *Syst. Bot.* 21: 567–586.

Morton, C. M., S. A. Mori, G. T. Prance, K. G. Karol, and M. W. Chase. 1997b. Phylogenetic relationships of Lecythidaceae: a cladistic analysis using *rbcL* sequence and morphological data. *Amer. J. Bot.* 84: 530–549.

Morton, C. M., G. T. Prance, S. M. Mori, and L. G. Thorburn. 1998. Re-circumscription of the Lecythidaceae. *Taxon* 47: 817–827.

Moseley, M. F., E. L. Schneider, and P. S. Williamson. 1993. Phylogenetic interpretations from selected floral vasculature characters in the Nymphaeaceae sensu lato. *Aquatic Bot.* 44: 325–342.

Muller, J. and P. W. Leenhouts. 1976. A general survey of pollen types in Sapindaceae in relation to taxonomy. In *The evolutionary significance of the exine.* Linn. Soc. Symp. Ser. No. 1, I. K. Ferguson and J. Muller (eds.), 407–455. Academic Press, London.

Müller, K., T. Borsch, L. Legendre, S. Porembski, and W. Barthlott. 2000. A phylogeny of Lentibulariaceae based on sequences of matK and adjacent noncoding regions. *Amer. J. Bot.* 87 (6, suppl.): 145–146.

Munro, S. L. and H. P. Linder. 1998. The phylogenetic position of *Prionium* (Juncaceae) within the order Juncales based on morphological and *rbcL* sequence data. *Syst. Bot.* 23: 43–55.

Murrell, Z. E. 1993. Phylogenetic relationships in *Cornus* (Cornaceae). *Syst. Bot.* 18: 469–495.

Musaya, A. M., D. A. Simpson, M. W. Chase, and A. Culham. 1998. An assessment of suprageneric phylogeny in Cyperaceae using rbcL DNA sequences. *Pl. Syst. Evol.* 211: 257–271.

Nandi, O. I. 1998. Floral development and systematics of Cistaceae. *Pl. Syst. Evol.* 212: 107–134.

Nandi, O. I., M. W. Chase, and P. L. Endress. 1998. A combined cladistic analysis of angiosperms using *rbcL* and non-molecular data sets. *Ann. Missouri Bot. Gard.* 85: 137–212.

Neyland, R. 2001. A phylogeny inferred from large ribosomal subunit (26S) rDNA sequences suggests that *Cuscuta* is a derived member of Convolvulaceae. *Brittonia* 53: 108–114.

Neyland, R. and L. E. Urbatsch. 1995. A terrestrial origin for the Orchidaceae suggested by a phylogeny inferred from *ndhF* chloroplast gene sequences. *Lindleyana* 10: 244–251.

Neyland, R. and L. E. Urbatsch. 1996. Phylogeny of subfamily Epidendroideae (Orchidaceae) inferred from *ndhF* chloroplast gene sequences. *Amer. J. Bot.* 83: 1195–1206.

Nicholas, A. and H. Baijnath. 1994. A consensus classification for the order Gentianales with additional details on the suborder Apocynineae. *Bot. Rev.* 60: 440–482.

Nickrent, D. L. 1996. Phylogenetic relationships of parasitic Santalales and Rafflesiales inferred from 18S rRNA sequences. *Amer J. Bot.* 83(6) Suppl.: 212.

Nickrent, D. L. and D. E. Soltis. 1995. A comparison of angiosperm phylogenies based upon complete 18S rDNA and *rbcL* sequences. *Ann. Missouri Bot. Gard.* 83: 208–234.

Nickrent, D. L. and R. J. Duff. 1996. Molecular studies of parasitic plants using ribosomal RNA. In *Advances in parasitic plant research,* M. T. Moreno, J. I. Cubero, D. Nerner, D. Joel, L. J. Musselman, and C. Parker (eds.), 28–52. Junta de Andalucia, Dirección General de Investigación Agraria, Córdoba, Spain.

Nickrent, D. L., R. J. Duff, A. E. Colwell, A. D. Wolfe, N. D. Young, K. E. Steinem, and C. W. de Pamphilis. 1997. Molecular phylogenetic and evolutionary studies of parasitic plants. In *Molecular systematics of plants,* vol. 2, DNA sequencing. D. Soltis, P. Soltis and J. Doyle (eds.), Chapter 8. Kluwer Academic, Boston.

Nilsson, L. A., E. Rabakonandrianina, B. Pettersson, and J. Ranaivo. 1990. "Ixoroid" secondary pollen presentation and pollination by small moths in the Malagasy treelet *Ixora platythyrsa* (Rubiaceae). *Plant Syst. Evol.* 170: 161–175.

Nixon, K. C. 1989. Origins of Fagaceae. In *Evolution, systematics and fossil history of the Hamamelidae,* vol. 2, "Higher" Hamamelidae. Syst. Assoc. Special Vol. 40B, P. R. Crane, and S. Blackmore (eds.), 23–43. Clarendon Press, Oxford.

Nixon, K. C. and W. L. Crepet. 1989. *Trigonobalanus* (Fagaceae): Taxonomic status and phylogenetic relationships. *Amer. J. Bot.* 76: 828–841.

Nixon, K. C., R. J. Jensen, P. S. Manos, and C. H. Muller. 1995, Fagaceae. In *Flora North America,* vol. 3, 436–506.

Nooteboom, H. P. 1993. Magnoliaceae. In *The families and genera of vascular plants,* vol. 2, Magnoliid, hamamelid and caryophyllid families, K. Kubitzki, J. G. Rohwer, and V. Bittrich (eds.), 391–401. Springer-Verlag, Berlin.

Nordenstam, B. 1998. Colchicaceae. In *The families and genera of vascular plants,* vol. 3. Flowering plants—Monocotyledons, K. Kubitzki (ed.), 175–185. Springer-Verlag, Berlin.

Norman, E. M. and D. Clayton. 1986. Reproductive biology of two Florida pawpaws: *Asimina obovata* and *A. pygmaea* (Annonaceae). *Bull. Torrey Bot. Club* 113: 16–22.

Nyananyo, B. L. 1990. Tribal and generic relationships in the Portulacaceae (Centrospermae). *Feddes Repert.* 101: 237–241.

Nyffeler, R. 2002. Phylogenetic relationships in the cactus family (Cactaceae) based on evidence from *trnK/makK* and *trnL-trnF* sequences. *Amer. J. Bot.* 89: 312–326.

Oldeman, R. A. A. 1989. Biological implications of leguminous tree architecture. *Monogr. Syst. Bot. Missouri Bot. Gard.* 29: 17–34.

Olmstead, R. and D. Ferguson. 2001. A molecular phylogeny of the Boraginaceae/Hydrophyllaceae. *Botany 2001 – Abstracts*: 131.

Olmstead, R. and J. D. Palmer. 1991. Chloroplast DNA and systematics of the Solanaceae. In *Solanaceae 3: Taxonomy, chemistry, evolution*, J. G. Hawkes, R. N. Lester, M. Nee, and N. Estrada (eds.), 161–168. Royal Botanic Gardens, Kew.

Olmstead, R. G. and J. D. Palmer. 1992. A chloroplast DNA phylogeny of the Solanaceae: Subfamilial relationships and character evolution. *Ann. Missouri Bot. Gard.* 79: 346–360.

Olmstead, R. G., H. J. Michaels, K. M. Scott, and J. D. Palmer. 1992a. Monophyly on the Asteridae and identification of their major lineages inferred from DNA sequences of *rbcL*. *Ann. Missouri Bot. Gard.* 79: 249–265.

Olmstead, R. G. and P. A. Reeves. 1995. Evidence for the polyphyly of the Scrophulariaceae based on chloroplast *rbcL* and *ndhF* sequences. *Ann. Missouri Bot. Gard.* 82: 176–193.

Olmstead, R. G., K. M. Scott, and J. D. Palmer. 1992b. A chloroplast DNA phylogeny for the Asteridae: Implication for the Lamiales. In *Advances in labiate science*, R. M. Harley and T. Reynolds (eds.), 19–25. Royal Botanic Gardens, Kew.

Olmstead, R. G., B. Bremer, K. M. Scott, and J. D. Palmer. 1993. A parsimony analysis of the Asteridae sensu lato based on *rbcL* sequences. *Ann. Missouri Bot. Gard.* 80: 700–722.

Olmstead, R. G. and J. A. Sweere. 1994. Combining data in phylogenetic systematics: An empirical approach using three molecular data sets in the Solanaceae. *Syst. Biol.* 43: 467–481.

Olmstead, R. G., J. A. Sweere, R. E. Spangler, L. Bohs, and J. D. Palmer. 1995. Phylogeny and provisional classification of the Solanaceae based on chloroplast DNA. In *Solanaceae IV*, M. Nee, D. E. Symon, R. N. Lester, and J. P. Jessop (eds.), 111–137. Royal Botanic Gardens, Kew.

Olmstead, R. G. and J. D. Palmer. 1997. Implications for the phylogeny, classification and biogeography of *Solanum* from cpDNA restriction site variation. *Syst. Bot.* 22: 19–29.

Olmstead, R. G., K.-J. Kim, R. K. Jansen, and S. J. Wagstaff. 2000. The phylogeny of the Asteridae sensu lato based on chloroplast *ndhF* gene sequences. *Mol. Phylogenetics and Evol.* 16: 96–112.

Olmstead, R. G., C. W. de Pamphilis, A. D. Wolfe, N. D. Young, W. J. Elisons, and P. A. Reeves. 2001. Disintegration of the Scrophulariaceae. *Amer. J. Bot.* 88: 348–361.

Omori, Y. and S. Terabayashi. 1993. Gynoecial vascular anatomy and its systematic implications in Celtidaceae and

Ulmaceae (Urticales). *J. Plant Res.* 106: 249–258.

Orgaard, M. 1991. The genus *Cabomba* (Cabombaceae)—a taxonomic study. *Nordic J. Bot.* 11: 179–203.

Orndurff, R. 1966. The breeding system of *Pontederia cordata* L. Bull. *Torrey Bot. Club* 93: 407–416.

Orndurff, R. 1972. The breakdown of trimorphic incompatibility in *Oxalis* section *corniculatae*. *Evolution* 26: 52–65.

Osborn, J. M. and E. L. Schneider. 1988. Morphological studies of the Nymphaeaceae *sensu lato*. XVI. The floral biology of *Brasenia schreberi*. *Ann. Missouri Bot. Gard.* 75: 778–794.

Osborn, J. M., T. N. Taylor, and E. L. Schneider. 1991. Pollen morphology and ultrastructure of the Cabombaceae: Correlations with pollination biology. *Amer. J. Bot.* 78: 1367–1378.

Ownbey, M. 1950. Natural hybridization and amphiploidy in the genus *Tragopogon*. *Amer. J. Bot.* 37: 487–499.

Oxelman, B., M. Backlund, and B. Bremer. 1999. Relationships of the Buddlejaceae s. l. investigated using parsimony Jackknife and Branch Support Analysis of chloroplast *ndhF* and *rbcL* sequence data. *Syst. Bot.* 24: 164–182.

Padgett, D. J., D. H. Les, and G. E. Crow. 1999. Phylogenetic relationships in *Nuphar* (Nymphaeaceae): evidence from morphology, chloroplast DNA, and nuclear ribosomal DNA. *Amer. J. Bot.* 86: 1316–1324.

Park, K. R. and W. S. Elisens. 2000. A phylogenetic study of tribe Euphorbieae (Euphorbiaceae). *Int. J. Plant Sci.* 161: 425–434.

Patel, V. C., J. J. Skvarla, and P. H. Raven. 1984. Pollen characters in relation to the delimitation of Myrtales. *Ann. Missouri Bot. Gard.* 71: 858–969.

Pennell, F. W. 1920. Scrophulariaceae of the southeastern United States. *Contr. New York Bot. Gard.* 221: 224–291.

Pennell, F. W. 1935. The Scrophulariaceae of eastern temperate North America. *Acad. Nat. Sci. Philadelphia Monogr.* 1: 1–650.

Pennington, T. D. 1991. *The genera of Sapotaceae*. Royal Botanic Gardens, Kew and New York Botanical Gardens, Bronx.

Pennington, T. D. and B. T. Styles. 1975. A generic monograph of the Meliaceae. *Blumea* 22: 419–540.

Persson, C. 2001. Phylogenetic relationships in Polygalaceae based on plastid DNA sequences from the *trnL-F* region. *Taxon* 50: 763–779.

Petersen, F. P. and D. E. Fairbrothers. 1983. A serotaxonomic appraisal of *Amphipterygium* and *Leitneria*: Two amentiferous taxa of Rutiflorae (Rosidae). *Syst. Bot.* 8: 134–148.

Pfosser, M. and F. Speta. 1999. Phylogenetics of Hyacinthaceae based on plastid DNA sequences. Ann. Missouri Bot. Gard. 86: 852–875.

Philbrick, C. T. and R. K. Jansen. 1991. Phylogenetic studies of North American *Callitriche* (Callitrichaceae) using chloroplast DNA restriction fragment analysis. *Syst. Bot.* 16: 478–491.

Philipson, W. R. 1993. Amborellaceae. In *The families and genera of vascular plants*, vol. 2, Magnoliid, hamamelid and caryophyllid families, K. Kubitzki, J. G. Rohwer,

and V. Bittrich (eds.), 92–93. Springer-Verlag, Berlin.

Phipps, J. B., K. R. Robertson, P. G. Smith, and J. R. Rohrer. 1990. A checklist of the subfamily Maloideae (Rosaceae). *Canad. J. Bot.* 68: 2209–2269.

Pichon, M. 1946. Sur les Alismatacées et les Butomacées. *Not. Syst.* 12: 170–183.

Pimenov, M. G. and M. V. Leonov. 1993. *The genera of the Umbelliferae*. Royal Botanic Gardens, Kew and Botanical Garden of Moscow University, Moscow.

Plunkett, G. M., D. E. Soltis, P. E. Soltis, and R. E. Brooks. 1995. Phylogenetic relationships between Juncaceae and Cyperaceae: Insights from *rbcL* sequence data. *Amer. J. Bot.* 82: 520–525.

Plunkett, G. M., D. E. Soltis, and P. S. Soltis. 1996a. Evolutionary patterns in Apiaceae: Inferences based on *matK* sequence data. *Syst. Bot.* 21: 477–495.

Plunkett, G. M., D. E. Soltis, and P. S. Soltis. 1996b. Higher level relationships of Apiales (Apiaceae and Araliaceae) based on phylogenetic analysis of *rbcL* sequences. *Amer. J. Bot.* 83: 499–515.

Plunkett, G. M., D. E. Soltis, and P. S. Soltis. 1997. Clarification of the relationship between Apiaceae and Araliaceae based on *matK* and *rbcL* sequence data. *Amer. J. Bot.* 84: 567–580.

Plunkett, G. M. and S. R. Downie. 1999. Major lineages within Apiaceae subfamily Apioideae: comparison of chloroplast restriction site and DNA sequence data. *Amer. J. Bot.* 86: 1014–1026.

Polhill, R. M. 1981. Papilionoideae. In *Advances in legume systematics*, part 1, R. M. Polhill and P. H. Raven (eds.), 191–208. Royal Botanic Gardens, Kew.

Polhill, R. M., P. H. Raven, and C. H. Stirton. 1981. Evolution and systematics of the Leguminosae. In *Advances in legume systematics*, part 1, R. M. Polhill, and P. H. Raven (eds.), 1–26. Royal Botanic Gardens, Kew.

Porter, D. M. 1972. The genera of Zygophyllaceae in the southeastern United States. *J. Arnold Arbor.* 53: 531–552.

Porter, J. M. and L. A. Johnson. 1998. Phylogenetic relationships of Polemoniaceae: Inferences from mitochrondrial *NAD1B* intron sequences. *Aliso* 17: 157–188.

Porter, J. M. and L. A. Johnson. 2000. A phylogenetic classification of Polemoniaceae. *Aliso* 19:55–91.

Powell, M., V. Savolainen, P. Cuénoud, J.-F. Manen, and S. Andrews. 2000. The mountain holly (*Nemopanthus mucronatus*: Aquifoliaceae) revisited with molecular data. KewBull. 55: 341–347.

Potgieter, K. and V. A. Albert. 2001. Phylogenetic relationships within Apocynaceae s.l. based on *trnL* intron and *trnL-F* spacer sequences and propagule characters. *Ann. Missouri Bot. Gard.* 88: 523–549.

Prance, G. T. 1976. The pollination and androphore structure of some Amazonian Lecythidaceae. *Biotropica* 8: 235–241.

Prance, G. T. and S. A. Mori. 1978. Observations on the fruits and seeds of neotropical Lecythidaceae. *Brittonia* 30: 21–33.

Prance, G. T. and S. A. Mori. 1979. Lecythidaceae. Part. I. The actinomorphic-flowered New World Lecythidaceae

(*Asteranthos, Gustavia, Grias, Allantoma,* and *Cariniana*). *Flora Neotropica, Monograph* 21: 1–270.

Prance, G. T. and V. Plana. 1998. The American Proteaceae. *Austr. Syst. Bot.* 11: 287–299.

Prather, L. A., C. J. Ferguson, and R. K. Jansen. Polemoniaceae phylogeny and classification: implications of sequence data from the chloroplast gene *ndhF. Amer. J. Bot.* 87: 1300–1308.

Price, R. A. and J. D. Palmer. 1993. Phylogenetic relationships of Geraniaceae and Geraniales from *rbcL* sequence comparisons. *Ann. Missouri Bot. Gard.* 80: 661–671.

Price, S. D. and S. C. H. Barrett. 1982. Tristyly in *Pontederia cordata* L. (Pontederiaceae). *Canad. J. Bot.* 60: 897–905.

Pridgeon, A. M., P. J. Cribb, M. W. Chase, and F. N. Rasmussen. 1999. *Genera Orchidacearum,* vol. 1. Oxford University Press, New York.

Prince, L. M. and C. R. Parks. 2001. Phylogenetic relationships of Theaceae inferred from chloroplast DNA sequence data. *Amer. J. Bot.* 88: 2309–2320.

Prior, P. V. 1960. Development of the helicoid and scorpioid cymes in *Myosotis laxa* Lehm. and *Mertensia virginica* L. *Proc. Iowa Acad. Sci.* 67: 76–81.

Proctor, M. and P. Yeo. 1972. *The pollination of flowers.* Taplinger, New York.

Proctor, M., P. Yeo, and A. Lack. 1996. *The natural history of pollination.* Timber Press, Portland, Oregon.

Prychid, C. J. and P. j. Rudall. 2000. Distribution of calcium oxalate crystals in monocotyledons. *In Monocots: Systematics and evolution,* K. L. Wilson and D. A. Morrison (eds.), 159–162. CSIRO Publ., Collingwood, Australia.

Puri, V. 1948. Studies in floral anatomy. V. On the structure and nature of the corona in certain species of the Passifloraceae. *J. Indian Bot. Soc.* 27: 130–149.

Purnell, H. M. 1960. Studies of the family Proteaceae. I. Anatomy and morphology of the roots of some Victorian species. *Austr. J. Bot.* 8: 38–50.

Qiu, Y.-L., M. W. Chase, D. H. Les, and C. R. Parks. 1993. Molecular phylogenetics of the Magnoliidae: Cladistic analyses of nucleotide sequences of the plastid gene *rbcL. Ann. Missouri Bot. Gard.* 80: 587–606.

Qiu, Y.-L., M. W. Chase, and C. R. Parks. 1995. A chloroplast DNA phylogenetic study of the eastern Asia-eastern North America disjunct section *Rhytidospermum* of *Magnolia* (Magnoliaceae). *Amer. J. Bot.* 82: 1582–1588.

Qiu, Y.-L., J. Les, F. Bernascond-Quadroni, D. E. Soltis, P. S. Soltis, M. Zanis, E. A. Zimmer, Z. Chen, V. Savolainen, and M. W. Chase. 1999. The earliest angiosperms: evidence from mitochondria, plastid and nuclear genomes. *Nature* 402: 404–407.

Qiu, Y. L., J. H. Lee, F. Bernasconi-Quadroni, D. E. Soltis, P. S. Soltis, M. Zanis, E. A. Zimmer, Z. D. Chen, V. Savolainen, and M. W. Chase. 2000. Phylogeny of basal angiosperms: Analyses of 5 genes from 3 genomes. *Int. J. Pl. Sci.* 161 (6 Suppl.) 161: S3–S27.

Rabinowitz, D. 1978. Dispersal properties of mangrove propagules. *Biotropica* 10: 47–57.

Rahn, K. 1996. A phylogenetic study of the Plantaginaceae. *Bot. J. Linn. Soc.* 120: 145–198.

Rahn, K. 1998. Alliaceae. In *The families and genera of vascular plants,* vol. 3. Flowering plants—Monocotyledons, K. Kubitzki (ed.), 70–78. Springer-Verlag, Berlin.

Rama Devi, D. 1991. Floral anatomy of *Hypseocharis* (Oxalidaceae) with a discussion on its systematic position. *Plant Syst. Evol.* 177: 161–164.

Ramivrez B., W. and L. D. Gomez P. 1978. Production of nectar and gums by flowers of *Monstera deliciosa* (Araceae) and some speices of *Clusia* (Guttiferae) collected by New World *Trigona* bees. *Brenesia* 14–15: 407–412.

Ranker, T. A., D. E. Soltis, and A. J. Gilmartin. 1990. Subfamilial phylogenetic relationships of the Bromeliaceae: Evidence from chloroplast DNA restriction site variation. *Syst. Bot.* 15: 425–434.

Rao, C. V. 1971. Anatomy of the inflorescence of some Euphorbiaceae with a discussion on the phylogeny and evolution of the inflorescnece including the cyathium. *Bot. Not.* 124: 39–64.

Rao, S. R. S. and V. Sarma. 1992. Morphology of 2-armed trichomes in relation to taxonomy: Malpighiales. *Feddes Repert.* 103: 559–565.

Raven, P. H. 1964. The generic subdivision of Onagraceae, tribe Onagreae. *Brittonia* 16: 276–288.

Raven, P. H. 1976. Generic and sectional delimitation in Onagraceae, tribe Epilobieae. *Ann. Missouri Bot. Gard.* 63: 326–340.

Raven, P. H. 1979. A survey of reproductive biology in Onagraceae. *New Zealand J. Bot.* 17: 575–593.

Raven, P. H. 1988. Onagraceae as a model of plant evolution. In *Plant evolutionary biology: A symposium honoring G. Ledyard Stebbins,* L. D. Gottlieb, and S. K. Jain (eds.), 85–107. Chapman and Hall, London.

Ray, T. S. 1987a. Diversity of shoot organization in the Araceae. *Amer. J. Bot.* 74: 1373–1387.

Ray, T. S. 1987b. Leaf types in the Araceae. *Amer. J. Bot.* 74: 1359–1372.

Redgrove, H. S. 1929. The cultivation and use of henna. *Gard. Chron. III* 85: 13–14.

Reeves, P. A. and R. G. Olmstead. 1998. Evolution of novel morphological and reproductive traits in a clade containing *Antirrhinum majus* (Scrophulariaceae). *Amer. J. Bot.* 85: 1047–1056.

Reeves, G. , M. W. Chase, P. Goldblatt, P. Rudall, M. F. Fay, A. V. Cox, B. Lejeune, and T. Souza-Chies. 2001. Molecular systematics of Iridaceae: evidence from four plastid DNA regions. *Amer. J. Bot.* 88: 2074–2087.

Renner, S. S. 1989a. Floral biological observations on *Heliamphora tatei* (Sarraceniaceae) and other plants from Cerro de la Neblina in Venezuela. *Plant Syst. Evol.* 163: 21–29.

Renner, S. S. 1989b. A survey of reproductive biology in neotropical Melastomataceae and Memecylaceae. *Ann. Missouri Bot. Gard.* 76: 496–518.

Renner, S. S. 1990. Reproduction and evolution in some genera of neotropical Melastomataceae. *Mem. New York Bot. Gard.* 55: 143–152.

Renner, S. S. 1993. Phylogeny and classification of the Melastomataceae and Memecylaceae. *Nordic J. Bot.* 13: 519–540.

Renner, S. S. 1999. Circumscription and phylogeny of the Laurales: evidence from molecular and morphological data. *Amer. J. Bot.* 86: 1301–1315.

Renner, S. S. and A. Chanderbali. 2000. What is the relationshp among Hernandiaceae, Lauraceae, and Monimiaceae, and why is this question so difficult to answer? *Int. J. Pl. Sci.* 161 (6 Suppl.) 161: S109–S119.

Renner, S. S., A. E. Schwarzbach, and L. Lohmann. 1997. Phylogenetic position and floral function of *Siparuna* (Siparunaceae: Laurales). *Int. J. Pl. Sci.* 158 (Suppl.): S89–S98.

Renner, S. S., G. Clausing, and K. Meyer. 2001. Historical biogeography of Melastomataceae: the roles of Tertiary migration and long-distance dispersal. *Amer. J. Bot.* 88: 1290–1300.

Rettig, J. H., H. D. Wilson, and J. R. Manhart. 1992. Phylogeny of the Caryophyllales: Gene sequence data. *Taxon* 41: 201–209.

Reveal, J. L., W. S. Judd, and R. Olmstead. 1999. (1405) Proposal to conserve the name Antirrhinaceae against Plantaginaceae (Magnoliophyta). *Taxon* 48: 182.

Richardson, J. E., M. F. Fay, Q. C. B. Cronk, and M. W. Chase. 2000. A revision of the tribal classification of Rhamnaceae. *Kew Bull.* 55: 311–340.

Richardson, J. E., M. E. Fay, Q. C. B. Cronk, D. Bowman, and M. W. Chase. 2000. A phylogenetic analysis of Rhamnaceae using *rbcL* and *trnL-F* plastid DNA sequences. *Amer. J. Bot.* 87: 1309–1324.

Rieseberg, L. H. 1991. Homoploid reticulate evolution in *Helianthus* (Asteraceae): Evidence from ribosomal genes. *Amer. J. Bot.* 78: 1218–1237.

Rizk, A.-F. M. 1987. The chemical constituents and economic plants of the Euphorbiaceae. *Bot. J. Linn. Soc.* 94: 293–326.

Robbrecht, E. 1988. Tropical woody Rubiaceae. *Opera Bot. Belg.* 1: 1–272.

Roberts, M. L. and R. R. Haynes. 1983. Ballistic seed dispersal in *Illicium* (Illiciaceae). *Plant Syst. Evol.* 143: 227–232.

Robertson, K. R. 1972a. The genera of Geraniaceae in the southeastern United States. *J. Arnold Arbor.* 53: 182–201.

Robertson, K. R. 1972b. The Malpighiaceae in the southeastern United States. *J. Arnold Arbor.* 53: 101–112.

Robertson, K. R. 1974. The genera of Rosaceae in the southeastern United States. *J. Arnold Arbor.* 55: 303–332, 344–401, 600–662.

Robertson, K. R. 1975. The genera of Oxalidaceae in the southeastern United States. *J. Arnold Arbor.* 56: 223–239.

Robertson, K. R. 1976. The genera of the Haemodoraceae in the southeastern United States. *J. Arnold Arbor.* 57: 205–216.

Robertson, K. R. 1981. The genera of Amaranthaceae in the southeastern United States. *J. Arnold Arbor.* 62: 267–313.

Robertson, K. R. and Y.-T. Lee. 1976. The genera of Caesalpinioideae (Leguminosae) in the southeastern United States. *J. Arnold Arbor.* 57: 1–53.

Robertson, K. R., J. B. Phipps, and J. R. Rohrer. 1992. Summary of leaves in the genera of Maloideae (Rosaceae). *Ann. Missouri Bot. Gard.* 79: 81–94.

Robertson, K. R., J. B. Phipps, J. R. Rohrer, and P. G. Smith. 1991. A synopsis of genera of the Maloideae (Rosaceae). *Syst. Bot.* 16: 376–394.

Robinson, R. W. and D. S. Decker-Walters. 1997. Cucurbits. *Crop Production Science in Horticulture # 6*. CAB International, Wallingford, U.K.

Robson, N. K. B. 1977. Studies in the genus *Hypericum* L. (Guttiferae). I. Infrageneric classification. *Bull. British Mus. (Nat. Hist.) Bot.* 5: 293–355.

Robyns, A. 1964. Bombacaceae. In *Flora of Panama VI*, R. E. Woodson, Jr. and R. W. Schery (eds.). *Ann. Missouri Bot. Gard.* 51: 37–68.

Roddick, J. G. 1986. Steroidal alkaloids of the Solanaceae. In *Solanaceae: Biology and systematics*, W. G. D'Arcy, (ed.), 201–222. Columbia University Press, New York.

Roddick, J. G. 1991. The importance of the Solanaceae in medicine and drug therapy. In *Solanaceae 3: Taxonomy, chemistry, evolution*, J. G. Hawkes, R. N. Lester, M. Nee and N. Estrada (eds.), 7–23. Royal Botanic Garden, Kew.

Rodman, J. E. 1981. Divergence, convergence, and parallelism in phytochemical characters: The glucosinolate–myrosinase system. In *Phytochemistry and angiosperm phylogeny*, D. A. Young and D. S. Seigler (eds.), 43–79. Praeger, New York.

Rodman, J. E. 1990. Centrospermae revisited, part 1. *Taxon* 39: 383–393.

Rodman, J. E. 1991a. A taxonomic analysis of glucosinolate-producing plants. I. Phenetics. *Syst. Bot.* 16: 598–618.

Rodman, J. E. 1991b. A taxonomic analysis of glucosinolate-producing plants. II. Cladistics. *Syst. Bot.* 16: 619–629.

Rodman, J. E. 1994. Cladistic and phenetic studies. In *Caryophyllales: Evolution and systematics*, H.-D. Behnke and T. J. Mabry (eds.), 279–301. Springer-Verlag, Berlin.

Rodman, J. E., M. K. Oliver, R. R. Nakamura, J. U. McClammer, Jr., and A. H. Bledsoe. 1984. A taxonomic analysis and revised classification of Centrospermae. *Syst. Bot.* 9: 297–323.

Rodman, J. E., R. A. Price, K. Karol, E. Conti, K. J. Sytsma, and J. D. Palmer. 1993. Nucleotide sequences of the *rbcL* gene indicate monophyly of mustard oil plants. *Ann. Missouri Bot. Gard.* 80: 686–699.

Rodman, J. E., K. G. Karol, R. A. Price, and K. J. Sytsma. 1996. Molecules, morphology and Dahlgren's expanded order Capparales. *Syst. Bot.* 21: 289–307.

Rodman, J. E., P. S. Soltis, D. E. Soltis, K. J. Sytsma, and K. G. Karol. 1998. Parallel evolution of glucosinolate biosynthesis inferred from congruent nuclear and plastid gene phylogenies. *Amer. J. Bot.* 85: 997–1006.

Rodrigues, R. L. 1957. Systematic anatomical studies of *Myrrhidendron* and other woody Umbellales. *University of California Publ. Bot.* 29: 145–318.

Rogers, C. L. 1950. The Umbelliferae of North Carolina and their distribution in the southeast. *J. Elisha Mitchell Sci. Soc.* 66: 195–266.

Rogers, G. K. 1982. The Casuarinaceae in the southeastern United States. *J. Arnold Arbor.* 63: 357–373.

Rogers, G. K. 1983. The genera of Alismataceae in the southeastern United States. *J. Arnold Arbor.* 64: 383–420.

Rogers, G. K. 1984. The Zingiberales (Cannaceae, Marantaceae, and Zingiberaceae) in the southeastern United States. *J. Arnold Arbor.* 65: 5–55.

Rogers, G. K. 1985. The genera of Phytolaccaceae in the southeastern United States. *J. Arnold Arbor.* 66: 1–37.

Rogers, G. K. 1986. The genera of Loganiaceae in the southeastern United States. *J. Arnold Arbor.* 67: 143–185.

Rogers, G. K. 1987. The genera of Cinchonoideae (Rubiaceae) in the southeastern United States. *J. Arnold Arbor.* 68: 137–183.

Rohrer, J. R., K. R. Robertson, and J. B. Phipps. 1991. Variation in structure among fruits of Maloideae (Rosaceae). *Amer. J. Bot.* 78: 1617–1635.

Rohrer, J. R., K. R. Robertson, and J. B. Phipps. 1994. Floral morphology of Maloideae (Rosaceae) and its systematic relevance. *Amer. J. Bot.* 81: 574–581.

Rohwer, J. G. 1993a. Lauraceae. In *The families and genera of vascular plants*, vol. 2, Magnoliid, hamamelid and caryophyllid families, K. Kubitzki, J. G. Rohwer, and V. Bittrich (eds.), 366–391. Springer-Verlag, Berlin.

Rohwer, J. G. 1993b. Moraceae. In *The families and genera of vascular plants*, vol. 2, Magnoliid, hamamelid and caryophyllid families, K. Kubitzki, J. G. Rohwer, and V. Bittrich (eds.), 438–453. Springer-Verlag, Berlin.

Rohwer, J. G. 1993c. Phytolaccaceae. In *The families and genera of vascular plants*, vol. 2, Magnoliid, hamamelid and caryophyllid families, K. Kubitzki, J. G. Rohwer, and V. Bittrich (eds.), 506–515. Springer-Verlag, Berlin.

Rohwer, J. G. 1994. A note on the evolution of the stamens in the Laurales, with emphasis on the Lauraceae. *Bot. Acta* 107: 103–110.

Rohwer, J. G. 2000. Toward a phylogenetic classification of the Lauraceae: evidence from *matK* sequences. *Syst. Bot.* 25: 60–71.

Rohwer, J. G., H. G. Richter, and H. van der Werff. 1991. Two new genera of neotropical Lauraceae and critical remarks on the generic delimitation. *Ann. Missouri Bot. Gard.* 78: 388–400.

Rollins, R. C. 1993. *The Cruciferae of continental North America*. Stanford University Press, Stanford.

Ronse Decraene, L-P. and J. R. Akeroyd. 1988. Generic limits in *Polygonum* and related genera (Polygonaceae) on the basis of floral characters. *Bot. J. Linn. Soc.* 98: 321–371.

Ronse Decraene, L. P. and E. Smets. 1992. An updated interpretation of the androecium of the Fumariaceae. *Canad. J. Bot.* 70: 1765–1776.

Ronse Decraene, L. P., J. de Laet, and E. F. Smets. 1996. Morphological studies in Zygophyllaceae. II. The floral development and vascular anatomy of *Peganum harmala*. *Amer. J. Bot.* 83: 201–215.

Rosatti, T. J. 1984. The Plantaginaceae in the southeastern United States. *J. Arnold Arbor.* 65: 533–562.

Rosatti, T. J. 1986. The genera of Sphenocleaceae and Campanulaceae in the southeastern United States. *J. Arnold Arbor.* 67: 1–64.

Rosatti, T. J. 1987. The genera of Pontederiaceae in the southeastern United States. *J. Arnold Arbor.* 68: 35–71.

Rosatti, T. J. 1989. The genera of suborder Apocynineae (Apocynaceae and Asclepiadaceae) in the southeastern United States. *J. Arnold Arbor.* 70: 307–401, 443–514.

Rourke, J. and D. Wiens. 1977. Convergent floral evolution in South African and Australian Proteaceae and its possible bearing on pollination by nonflying mammals. *Ann. Missouri Bot. Gard.* 64: 1–17.

Rudall, P. 1994. Anatomy and systematics of Iridaceae. *Bot. J. Linnean Soc.* 114: 1–21.

Rudall, P. and H. P. Linder. 1988. Megagametophyte and nucellus in Restionaceae and Flagellariaceae. *Amer. J. Bot.* 75: 1777–1786.

Rudall, P. and D. F. Cutler. 1995. Asparagales: A reappraisal. In *Monocotyledons: Systematics and evolution*, P. J. Rudall, P. J. Cribb, D. F. Cutler and C. J. Humphries (eds.), 157–168. Royal Botanic Gardens, Kew.

Rudall, P. A., C. A. Furness, M. W. Chase, and M. F. Fay. 1997. Microsporogenesis and pollen sulcus type in Asparagales (Lilianae). *Canad. J. Bot.* 75: 408–430.

Rudall, P. A., E. M. Engleman, L. Hanson, and M. W. Chase. 1998. Embryology, cytology and systematics of *Hemiphylacus*, *Asparagus* and *Anemarrhena* (Asparagales). *Pl. Syst. Evol.* 211: 181–199.

Rudall, P. A., K. L. Stobart, W.-P. Hong, J. G. Conran, C. A. Furness, G. C. Kite, and M. W. Chase. 2000a. Consider the lilies: systematics of Liliales. In *Monocots: Systematics and evolution*, K. L. Wilson and D. A. Morrison (eds.), 347–359. Collingwood, Australia.

Rudall, P. A., J. G. Conran, and M. W. Chase. 2000b. Systematics of Ruscaceae/Convallariaceae: a combined morphological and molecular investigation. *Bot. J. Linn. Soc.* 134: 73–92.

Safwat, F. M. 1962. The floral morphology of *Secamone* and the evolution of the pollinating apparatus of Asclepiadaceae. *Ann. Missouri Bot. Gard.* 49: 95–119.

Samejima, K. and J. Samejima. 1987. *Trillium genus illustrated*. Sapporo, Japan.

Sampson, F. B. 1993. Pollen morphology of the Amborellaceae and Hortoniaceae (Hoortonioideae: Monimiaceae). *Grana* 32: 154–162.

Sanders, R. W. 1997. The Avicenniaceae in the southeastern United States. *Harvard Pap. Bot.* No. 10: 81–92.

Sanders, R. W. 2001. The genera of Verbenaceae in the southeastern United States. *Harvard Pap. Bot.* 5(2): 303–358.

Saunders, E. R. 1934. On carpel polymorphism. IV. *Ann. Bot. II* 48: 643–692.

Savile, D. B. O. 1975. Evolution and biogeography of Saxifragaceae with guidance from their rust parasites. *Ann. Missouri Bot. Gard.* 62: 354–361.

Savolainen, V., J. F. Manen, E. Douzery, and R. Spichiger. 1994. Molecular phylogeny of families related to Celastrales based on *rbcL* 5' flanking regions. *Mol. Phylog. Evol.* 3: 27–37.

Savolainen, V., M. W. Chase, S. B. Hoot, C. M. Morton, D. E. Soltis, C. Bayer, M. F. Fay, A. Y. de Bruijn, S. Sullivan, and Y.-L. Qiu. 2000a. Phylogenetics of flowering plants based upon a combined analysis of plastid *atpB* and *rbcL* gene sequences. *Syst. Biol.* 49: 306–362.

Savolainen, V., M. F. Fay, D. C. Albach, A. Backlund, M. van der Bank, K. M. Cameron, S. A. Johnson, M. D. Lledó, J.-C. Pintaud, M. Powell, M. C. Sheahan, D. E. Soltis, P. S. Soltis, P. Weston, W. M. Whitten, K. J. Wurdack, and M. W. Chase. 2000b. Phylogeny of the eudicots: a nearly complete familial analysis based on *rbcL* gene sequences. *Kew Bull.* 55: 257–309.

Schick, B. 1980. Untersuchungen über Biotechnik der Apocynaceenblüte. I. Morphologie und Funktion des Narbenkopfes. *Flora, Morphol. Geobot. Oekophysiol.* 170: 394–432.

Schick, B. 1982. Untersuchungen über Biotechnik der Apocynaceenblüte. II. Bau und Funktion des Bestäubungsapparates. *Flora, Morphol. Geobot. Oekophysiol.* 172: 347–371.

Schmid, R. 1980. Comparative anatomy and morphology of *Psiloxylon* and *Heteropyxis* and the subfamilial and tribal classification of the Myrtaceae. *Taxon* 29: 559–595.

Schneider, E. L. and J. M. Jeter. 1982. Morphological studies of the Nymphaeaceae. XII. The floral biology of *Cabomba caroliniana. Amer. J. Bot.* 69: 1410–1419.

Schneider, E. L. and S. Carlquist. 1995. Vessels in the roots of *Barclaya rotundifolia. Amer. J. Bot.* 82: 1343–1349.

Schneider, E. L., S. Carlquist, K. Beamer and A. Kohn. 1995. Vessels in Nymphaeaceae: *Nuphar, Nymphaea* and *Ondinea. Int. J. Plant Sci.* 156: 857–862.

Schneider, E. L. and P. S. Williamson. 1993. Nymphaeaceae. In *The families and genera of vascular plants*, vol. 2, Magnoliid, hamamelid and caryophyllid families, K. Kubitzki, J. G. Rohwer and V. Bittrich (eds.), 486–493. Springer-Verlag, Berlin.

Schrire, B. D. 1989. A multidisciplinary approach to pollination biology in the Leguminosae. *Monogr. Syst. Bot. Missouri Bot. Gard.* 29: 183–242.

Schulze-Menz, G. K. 1964. Saxifragaceae. In *A. Engler's Syllabus der Pflanzenfamilien*, H. Melchior (ed.), 201–206. Gebruder Borntraeger, Berlin.

Schwarzbach, A. E. and J. W. Kadereit. 1995. Rapid radiation of North American desert genera of the Papaveraceae: Evidence from restriction site mapping of PCR-amplified chloroplast DNA fragments. *Plant Syst. Evol. Suppl.* 9: 159–170.

Schwarzbach, A. E. and R. E. Ricklefs. 2000. Systematic affinities of Rhizophoraceae and Anisophyllaceae and intergeneric relationships within Rhizophoraceae based on chloroplast DNA, nuclear ribosomal DNA, and morphology. *Amer. J. Bot.* 87: 547–564.

Schwarzbach, A. E. and L. A. McDade. 2002. Phylogenetic relationships of the mangrove family Avicenniaceae based on chloroplast and nuclear ribosomal DNA sequences. *Syst. Bot.* 27: 84–98.

Schwarzwalder, R. N. and D. L. Dilcher. 1991. Systematic placement of the Platanaceae in the Hamamelidae. *Ann. Missouri Bot. Gard.* 78: 962–969.

Scotland, R. W. 1990. Palynology and systematics of Acanthaceae. Ph.D. Thesis, University of Reading, England.

Scotland, R. W. J. A. Sweere, P. A. Reeves and R. G. Olmstead. 1995. Higher-level systematics of Acanthaceae determined by chloroplast DNA sequences. *Amer. J. Bot.* 82: 266–275.

Scott, R. W. 1990. The genera of Cardueae (Compositae; Asteraceae) in the south-eastern United States. *J. Arnold Arbor.* 71: 391–451.

Semple, K. S. 1974. Pollination in Piperaceae. *Ann. Missouri Bot. Gard.* 61: 868–871.

Sennblad, B. and B. Bremer. 1996. The familial and subfamilial relationships of Apocynaceae and Asclepiadaceae evaluated with *rbcL* data. *Plant Syst. Evol.* 202: 153–175.

Sheahan, M. C. and D. F. Cutler. 1993. Contribution of vegetative anatomy to the systematics of the Zygophyllaceae R. Br. *Bot. J. Linn. Soc.* 113: 227–262.

Sheahan, M. C. and M. W. Chase. 2000. Phylogenetic relationships within Zygophyllaceae based on DNA sequences of three plastid regions, with special emphasis on Zygophylloideae. *Syst. Bot.* 25: 371–384.

Shetler, S. G. 1979. Pollen-collecting hairs of *Campanula* (Campanulaceae). I. Historical review. *Taxon* 28: 205–215.

Shinwari, Z. K., R. Terauchi, F. H. Htech, and S. Kawano. 1994. Recognition of the New World *Disporum* section *Prosartes* (Liliaceae) based on the sequence data of the *rbcL* gene. *Taxon* 43: 353–366.

Sibaoka, T. 1991. Rapid plant movements triggered by action potentials. *Bot. Mag. Tokyo.* 104: 73–95.

Simmons, M. P. and J. P. Hedin. 1999. Relationships and morphological character change among genera of Celastraceae sensu lato (including Hippocrateaceae). *Ann. Missouri Bot. Gard.* 86: 723–757.

Simmons, M. P., V. Savolainen, C. C. Clevinger, R. H. Archer, S. Mathews, and J. I. Davis. 2000. Phylogeny of the Celastraceae inferred from morphology and nuclear and plastid loci. *Amer. J. Bot.* 87 (6, suppl.): 156–157.

Simmons, M. P., C. C. Clevinger, V. Savolainen, R. H. Archer, S. Mathews, and J. J. Doyle. 2001. Phylogeny of the Celastraceae inferred from phytochrome B gene sequence and morphology. *Amer. J. Bot.* 88: 313–325.

Simpson, D. 1995. Relationships within Cyperales. In *Monocotyledons: Systematics and evolution*, P. J. Rudall, P. J. Cribb, D. F. Cutler, and C. J. Humphries (eds.), 459–509. Royal Botanic Gardens, Kew.

Simpson, M. G. 1990. Phylogeny and classification of the Haemodoraceae. *Ann. Missouri Bot. Gard.* 77: 722–784.

Simpson, M. G. 1993. Septal nectary anatomy and phylogeny of the Haemodoraceae. *Syst. Bot.* 18: 593–613.

Simpson, M. G. 1998a. Haemodoraceae. In *The families and genera of vascular plants*, vol. 4. Flowering plants— Monocotyledons, K. Kubitzki (ed.), 212–222. Springer-Verlag, Berlin.

Simpson, M. G. 1998b. Reversal in ovary position from inferior to superior: evidence from floral ontogeny. *Int. J. Pl. Sci.* 159: 466–479.

Skog, L. E. 1976. A study of the tribe Gesnerieae, with a revision of *Gesneria* (Gesneriaceae: Gesnerioideae). *Smithsonian Contr. Bot.* 29: 1–182.

Skvarla, J. J., P. H. Raven, W. F. Chissoe, and M. Sharp. 1978. An ultrastructural study of viscin threads in Onagraceae pollen. *Pollen et Spores* 30: 5–143.

Smith, A. C. 1947. The families Illiciaceae and Schisandraceae. *Sargentia* 7: 1–224.

Smith, A. C. and R. P. Wodehouse. 1938. The American species of Myristicaceae. *Brittonia* 2: 393–510.

Smith, G. F. and B.-E. Van Wyk. 1991. Generic relationships in the Alooideae (Asphodelaceae). *Taxon* 40: 557–581.

Smith, J. F. 1996. Tribal relationships within Gesneriaceae: A cladistic analysis of morphological data. *Syst. Bot.* 21: 497–513.

Smith, J. F. and J. J. Doyle. 1995. A cladistic analysis of chloroplast DNA restriction site variation and morphology from the genera of the Juglandaceae. *Amer. J. Bot.* 82: 1163–1172.

Smith, J. F. and C. L. Carroll. 1997. A cladistic analysis of the tribe Episcieae (Gesneriaceae) based on *ndhF* sequences: Origin of morphological characters. *Syst. Bot.* 22: 713–724.

Smith, J. F., W. J. Kress, and E. A. Zimmer. 1993. Phylogenetic analysis of the Zingiberales based on *rbcL* sequences. *Ann. Missouri Bot. Gard.* 80: 620–630.

Smith, J. F., J. C. Wolfram, K. D. Brown, C. L. Carroll, and D. S. Denton. 1997. Tribal relationships in the Gesneriaceae: Evidence from DNA sequences of the chloroplast gene *ndhF*. *Ann. Missouri Bot. Gard.* 84: 50–66.

Smith, L. B. and C. E. Wood. 1975. The genera of Bromeliaceae in the southeastern United States. *J. Arnold Arbor.* 56: 375–397.

Smith, L. B. and D. C. Wasshausen. 1986. Begoniaceae, part I: illustrated key. *Smithsonian Contr. Bot.* 60: 1–129.

Smith, L. B. and W. Till. 1998. Bromeliaceae. In *The families and genera of vascular plants*, vol. 4. Flowering plants— Monocotyledons, K. Kubitzki (ed.), 74–99. Springer-Verlag, Berlin.

Soderstrom, T. R., K. W. Hilu, C. S. Campbell, and M. E. Barkworth (eds.). 1987. *Grass systematics and evolution*. Smithsonian Institution Press, Washington, DC.

Soekarjo, R. 1992. General morphology in *Plantago*. In *Plantago: A multidisciplinary study*, P. J. C. Kuiper and M. Bos (eds.), 6–12. Springer-Verlag, Berlin.

Solbrig, O. T. 1963. The tribes of the Compositae in the southeastern United States. *J. Arnold Arbor.* 44: 436–461.

Soltis, D. E. and P. S. Soltis. 1989. Allopolyploid speciation in *Tragopogon*: Insights from chloroplast DNA. *Amer. J. Bot.* 76: 1119–1124.

Soltis, D. E. and R. K. Kuzoff. 1995. Discordance between nuclear and chloroplast phylogenies in the *Huchera* group (Saxifragaceae). *Evolution* 49: 727–742.

Soltis, D. E. and P. S. Soltis. 1997. Phylogenetic relationships in Saxifragaceae sensu lato: A comparison of topologies based on 18S rDNA and *rbcL* sequences. *Amer. J. Bot.* 84: 504–522.

Soltis, D. E., D. R. Morgan, A. Grable, P. S. Soltis, and R. Kuzoff. 1993. Molecular systematics of Saxifragaceae sensu stricto. *Amer. J. Bot.* 80: 1056–1081.

Soltis, D. E., Q.-Y. Xiang, and L. Hufford. 1995. Relationships and evolution of Hydrangeaceae based on *rbcL* sequence data. *Amer. J. Bot.* 82: 504–514.

Soltis, D. E., R. K. Kuzoff, E. Conti, R. Gornall, and K. Ferguson. 1996. *mat*K and *rbcL* gene sequence data indicate that *Saxifraga* (Saxifragaceae) is polyphyletic. *Amer. J. Bot.* 83: 371–382.

Soltis, D. E., P. S. Soltis, T. G. Collier, and M. L. Edgerton. 1991. Chloroplast DNA vbariation within and among genera of the *Heuchera* group (Saxifragaceae): Evidence for chloroplast transfer and paraphyly. *Amer. J. Bot.* 78: 1091–1112.

Soltis, D. E., P. S. Soltis, D. L. Nickrent, L. A. Johnson, W. J. Hahn, S. B. Hoot, J. A. Sweere, R. K. Kuzoff, K. A. Kron, M. W. Chase, S. M. Swensen, E. Z. Zimmer, S.-M. Chaw, L. J. Gillespie, J. W. Kress, and K. J. Sytsma. 1997. Phylogenetic relationships among angiosperms inferred from 18S rDNA sequences. *Ann. Missouri Bot. Gard.* 84: 1–49.

Soltis, D. E., P. S. Soltis, M. E. Mort, M. W. Chase, V. Sarolainen, S. B. Hoot, and C. M. Morton. 1998. Inferring complex phylogenies using parsimony: An empirical approach using three large DNA data sets for angiosperms. *Syst. Biol.* 47: 32–42.

Soltis, D. E., P. S. Soltis, M. W. Chase, M. E. Mort, D. C. Albach, M. Zanis, V. Savolainen, W. H. Hahn, S. B. Hoot, M. F. Fay, M. Axtell, S. M. Swensen, L. M. Prince, W. J. Kress, K. C. Nixon, and J. S. Farris. 2000. Angiosperm phylogeny inferred from 18S rDNA, *rbcL*, and *atpB* sequences. *Bot. J. Linn. Soc.* 133: 381–461.

Soltis, D. E., R. K. Kuzoff, M. E. Mort, M. Zanis, M. Fishbein, L. Hufford, J. Koontz, and M. K. Arroyo. 2001. Elucidating deep-level phylogenetic relationships in Saxifragaceae using sequences for six chloroplast and nuclear DNA regions. *Ann. Missouri Bot. Gard.* 88: 669–693.

Soltis, D. E. and L. Hufford. 2002. Ovary position diversity in Saxifragaceae: clarifying the homology of epigyny. *Int. J. Plant Sci.* 163: 277–293.

Soreng, R. J. and J. I. Davis. 1998. Phylogenetics and character evolution in the grass family (Poaceae): Simultaneous analysis of morphological and chloroplast DNA restriction site character sets. *Bot. Rev.* 64: 1–85.

Spangler, R. E. and R. G. Olmstead. 1999. Phylogenetic analysis of Bignoniaceae based on the cpDNA gene sequences *rbcL* and *ndhF*. *Ann. Missouri Bot. Gard.* 86: 33–46.

Speta, F. 1998. Hyacinthaceae. In *The families and genera of vascular plants*, vol. 3. Flowering plants—Monocotyledons, K. Kubitzki (ed.), 261–285. Springer-Verlag, Berlin.

Spongberg, S. A. 1972. The genera of Saxifragaceae in the southeastern United States. *J. Arnold Arbor.* 53: 409–498.

Spongberg, S. A. 1978. The genera of Crassulaceae in the southeastern United States. *J. Arnold Arbor.* 59: 197–248.

Spooner, D. M., G. A. Anderson, and R. K. Jansen. 1993. Chloroplast DNA evidence for the interrelationships of tomatoes, potatoes, and pepinos (Solanaceae). *Amer. J. Bot.* 80: 676–688.

Stace, C. A. 1965. Cuticular studies an an aid to plant taxonomy. *Bull. Brit. Mus. (Nat. Hist.), Bot.* 4: 1–78.

Ståhl, B. 1990. *Taxonomic studies in the Theophrastaceae.* Dissertation. Department of Systematic Botany, University of Goteborg, Sweden.

Steele, L. P. and R. Vilgalys. 1994. Phylogenetic analyses of Polemoniaceae using nucleotide sequences of the plastid gene *matK. Syst. Bot.* 19: 126–142.

Stein, B. A. and H. Tobe. 1989. Floral nectaries in Melastomataceae and their systematic and evolutionary implications. *Ann. Missouri Bot. Gard.* 76: 519–531.

Stern, W. L. 1973. Development of the amentiferous concept. *Brittonia* 25: 316–333.

Stern, W. L., M. W. Morris, W. S. Judd, A. M. Pridgeon, and R. L. Dressler. 1993. Comparative vegetative anatomy and systematics of Spiranthoideae (Orchidaceae). *Bot. J. Linn. Soc.* 113: 161–197.

Stevens, P. F. 1971. A classification of the Ericaceae: Subfamilies and tribes. *Bot. J. Linn. Soc.* 64: 1–53.

Stevenson, D. W. and H. Loconte. 1995. Cladistic analysis of monocot families. In *Monocotyledons: Systematics and evolution*, P. J. Rudall, P. J. Cribb, D. F. Cutler, and C. J. Humphries (eds.), 543–578. Royal Botanic Gardens, Kew.

Stevenson, D. W., J. I. Davis, J. V. Freudenstein, C. R. Hardy, M. P. Simmons, and C. D. Specht. 2000. A phylogenetic analysis of the monocotyledons based on morphological and molecular character sets, with comments on the placement of *Acorus* and Hydatellaceae. In *Monocots: Systematics and evolution*, K. L. Wilson and D. A. Morrison (eds.), 17–24. CSIRO Publ., Collinswood, Australia.

Stockey, R. A., G. L. Hoffman, and G. W. Rothwell. 1997. The fossil monocot *Limnobiophyllum scutatum*: Resolving the phylogeny of Lemnaceae. *Amer. J. Bot.* 84: 355–368.

Stone, D. E. 1973. Patterns in the evolution of amentiferous fruits. *Brittonia* 25: 371–384.

Stone, D. E. 1989. Biology and evolution of temperate and tropical Juglandaceae. In *Evolution, systematics and fossil history of the Hamamelidae*, vol. 2, "Higher" Hamamelidae. Syst. Assoc. Special Vol. 40B, P. R. Crane and S. Blackmore (eds.), 117–145. Clarendon Press, Oxford.

Stone, D. E. 1993. Juglandaceae. In *The families and genera of vascular plants*, vol. 2, Magnoliid, hamamelid and caryophyllid families, K. Kubitzki, J. G. Rohwer, and V. Bittrich (eds.), 348–359. Springer-Verlag, Berlin.

Struwe, L., V. A. Albert and B. Bremer. 1994. Cladistics and family level classification of the Gentianales. *Cladistics* 10: 175–206.

Struwe, L., M. Thiu, J. W. Kadereit, A. S.-R. Pepper, T. J. Motley, P. J. White, J. H. E. Rova, K. Potgieter, and V. A. Albert. 1998. *Saccifolium* (Saccifoliaceae), an endemic of Sierra de la Neblina on the Brazilian-Venezuelan border, is related to a temperate-alpine lineage of Gentianaceae. *Harvard Pap. Bot.* 3: 199–214.

Stubbs, J. M. and A. R. Slabas. 1982. Ultrastructural and biochemical characterization of the epidermal hairs of the seeds of *Cuphea procumbens. Planta* 155: 392–399.

Stützel, T. 1998. Eriocaulaceae. In *The families and genera of vascular plants*, vol. 4. Flowering plants—Monocotyledons, K. Kubitzki (ed.), 197–207. Springer-Verlag, Berlin.

Suessenguth, K. 1953. Vitaceae. *Nat. Pflanzenfam.* 2, 20d: 174–371.

Suh, Y., L. B. Thien, H. E. Reeve, and E. A. Zimmer. 1993. Molecular evolution and phylogenetic implications of internal transcribed spacer sequences of ribosomal DNA in Winteraceae. *Amer. J. Bot.* 80: 1042–1055.

Sun, G., D. L. Dilcher, S. Zheng, and Z. Zhou. 1998. In search of the first flower: A Jurassic angiosperm, *Archaefructus*, from northeast China. *Science* 282: 1692–1695.

Sutter, D. and P. K. Endress. 1995. Aspects of gynoecial structure and macrosystematics in Euphorbiaceae. *Bot. Jahrb. Syst.* 116: 517–536.

Swarupanandan, K., J. K. Mangaly, T. K. Sonny, K. Kishorekumar, and S. Chand Basha. 1996. The subfamilial and tribal classification of the family Asclepiadaceae. *Bot. J. Linn. Soc.* 120: 327–369.

Sweeney, P. W. and R. A. Price. 2000. Polyphyly of the genus *Dentaria* (Brassicaceae): evidence from *trnL* intron and *ndhF* sequence data. *Syst. Bot.* 25: 468–478.

Swensen, S. M., W. L. Clement, L. L. Forrest, and M. C. Tebbit. 2001. *Hillebrandia sandwichensis*: evolutionary relationships and biogeography. *Botany 2001.* Abstracts. p. 95.

Swingle, W. T. 1967. The botany of *Citrus* and its wild relatives [revised by P. C. Reece]. In *The Citrus industry*, vol. 1, History, world distribution, botany and varieties, W. Reuther, H. J. Webber and L. D. Batchelor (eds.), 190–430. University of California Division of Agricultural Science, Berkeley.

Sytsma, K. J. and J. F. Smith. 1988. DNA and morphology: Comparisons in the Onagraceae. *Ann. Missouri Bot. Gard.* 75: 1217–1237.

Sytsma, K. J. and J. F. Smith. 1992. Molecular systematics of Onagraceae: Examples of *Clarkia* and *Fuchsia*. In *Molecular systematics of plants*, P. S. Soltis, D. E. Soltis and J. J. Doyle (eds.), 295–323. Chapman and Hall, New York.

Sytsma, K. J., E. Conti, M. Nepokroeff, C. Piers, Y. L. Qiu, and M. W. Chase. 1996. *rbcL* sequences clarify placement in Rosidae, composition and familial relationships. *Amer. J. Bot.* 83(6) Suppl.: 197.

Sytsma, K. J., D. A. Baum, A. Rodriguez, W. J. Hahn, L. Katinas, W. L. Wagner, and P. C. Hock. 1998a. An ITS phylogeny for Onagraceae: Congruence with three molecular data sets. *Amer. J. Bot.* 85(6) Suppl.: 160.

Sytsma, K. J., M. L. Zjhra, M. Nepokroeff, C. J. Quinn, and P. G. Wilson. 1998b. Phylogenetic relationships, morphological evolution and biogeography in Myrtaceae based on *ndhF* sequence analysis. *Amer. J. Bot.* 85(6) Suppl.: 161.

Takhtajan, A. 1969. *Flowering plants: Origin and dispersal.* Smithsonian Institution Press, Washington, DC.

Takhtajan, A. 1980. Outline of the classification of flowering plants (Magnoliophyta). *Bot. Rev.* 46: 225–359.

Takhtajan, A. 1997. *Diversity and classification of flowering plants.* Columbia University Press, New York.

Tamura, M. 1963. Morphology, ecology and phylogeny of the Ranunculaceae. II. Morphology of the Ranunculaceae. *Sci. Rep.* Osaka University 12: 141–156.

Tamura, M. 1965. Morphology, ecology and phylogeny of the Ranunculaceae. IV. Reproductive organs. *Sci. Rep.* Osaka University 14: 53–71.

Tamura, M. 1993. Ranunculaceae. In *The families and genera of vascular plants*, vol. 2, Magnoliid, hamamelid and caryophyllid families, K. Kubitzki, J. G. Rohwer and V. Bittrich (eds.), 563–583. Springer-Verlag, Berlin.

Tamura, M. N. 1998a. Liliaceae. In *The families and genera of vascular plants*, vol. 3.

Flowering plants—Monocotyledons, K. Kubitzki (ed.), 343–353. Springer-Verlag, Berlin.

Tamura, M. N. 1998b. Melanthiaceae. In *The families and genera of vascular plants,* vol 3. Flowering plants—Monocotyledons, K. Kubitzki (ed.), 369–380. Springer-Verlag, Berlin.

Tamura, M. N. 1998c. Trilliaceae. In *The families and genera of vascular plants*, vol 3. Flowering plants—Monocotyledons, K. Kubitzki (ed.), 444–452. Springer-Verlag, Berlin.

Taylor, D. W. and L. J. Hickey. 1992. Phylogenetic evidence for the herbaceous origin of angiosperms. *Plant Syst. Evol.* 180: 137–156.

Taylor, D. W. and L. J. Hickey. 1996. Evidence for and implications of an herbaceous origin for angiosperms. In *Flowering plant origin, evolution and phylogeny*, D. W. Taylor and L. J. Hickey (eds.), 232–266. Chapman and Hall, New York.

Taylor, P. 1989. The genus *Utricularia*: A taxonomic monograph. *Kew Bull.*, Add. Ser. 14: 1–724.

Tebbs, M. C. 1993. Piperaceae. In *The families and genera of vascular plants*, vol. 2, Magnoliid, hamamelid and caryophyllid families, K. Kubitzki, J. G. Rohwer, and V. Bittrich (eds.), 516–522. Springer-Verlag, Berlin.

Terabayashi, S. 1991. Vernation patterns in Celtidaceae and Ulmaceae (Urticales) and their evolutionary and systematic implications. *Bot. Mag. Tokyo* 104: 1–13.

Terrazas, T. and M. Chase. 1996. A phylogenetic analysis of Anacardiaceae based on morphology and *rbcL* sequence data. *Amer. J. Bot.* 83(6) Suppl.: 197.

Terry, R. G., G. K. Brown, and R. G. Olmstead. 1997a. Examination of subfamilial phylogeny in Bromeliaceae using comparative sequencing of the plastid locus *ndhF*. *Amer. J. Bot.* 84: 664–670.

Terry, R. G., G. K. Brown, and R. G. Olmstead. 1997b. Phylogenetic relationships in subfamily Tillandsioideae (Bromeliaceae) using *ndhF* sequences. *Syst. Bot.* 22: 333–345

Thanikaimoni, G. 1986. Evolution of Menispermaceae. *Canad. J. Bot.* 64: 3130–3133.

Thien, L. B. 1974. Floral biology of *Magnolia*. *Amer. J. Bot.* 61: 1037–1045.

Thien, L. B. 1980. Patterns of pollination in the primitive angiosperms. *Biotropica* 12: 1–13.

Thien, L. B., D. A. White, and L. Y. Yatsu. 1983. The reproductive biology of a relic: *Illicium floridanum* Ellis. *Amer. J. Bot.* 70: 719–727.

Thieret, J. W. 1972. The Phrymaceae in the southeastern United States. *J. Arnold Arbor.* 53: 226–233.

Thieret, J. W. 1982. The Sparganiaceae in the southeastern United States. *J. Arnold Arbor.* 63: 341–355.

Thieret, J. W. and J. O. Luken. 1996. The Typhaceae in the southeastern United States. *Harvard Pap. Bot.* No. 8: 27–56.

Thieret, W. 1967. Supraspecific classification in the Scrophulariaceae; a review. *Sida* 3: 87–106.

Thieret, W. 1971. The genera of Orobanchaceae in the southeastern United States. *J. Arnold Arbor.* 52: 404–434.

Thiv, M., L. Struwe, V. A. Albert, and J. W. Kadereit. 1999. The phylogenetic relationships of *Saccifolium bandeirae*

(Gentianaceae) reconsidered. *Harvard Pap. Bot.* 4: 519–526.

Thomas, V. and Y. Dave. 1991. Comparative and phylogenetic significance of the colleters in the family Apocynaceae. *Feddes Repert.* 102: 177–182.

Thomson, B. F. 1942. The floral morphology of the Caryophyllaceae. *Amer. J. Bot.* 29: 333–349.

Thorne, R. F. 1973a. The "Amentiferae" or Hamamelidae as an artificial group: A summary statement. *Brittonia* 25: 395–405.

Thorne, R. F. 1973b. Inclusion of the Apiaceae (Umbelliferae) in the Araliaceae. *Notes Roy. Bot. Gard. Edinburgh* 32: 161–165.

Thorne, R. F. 1974. A phylogenetic classification of the Annoniflorae. *Aliso* 8: 147–209.

Thorne, R. F. 1976. A phylogenetic classification of the Angiospermae. *Evol. Biol.* 9: 35–106.

Thorne, R. F. 1983. Proposed new realignments in the angiosperms. *Nordic J. Bot.* 3: 85–117.

Thorne, R. F. 1992. Classification and geography of the flowering plants. *Bot. Rev.* 58: 225–348.

Thorne, R. F. 2001. The classification and geography of the flowering plants: dicotyledons of the class Angiospermae. *Bot. Rev.* 66: 441–647.

Tiffney, B. H. 1986. Fruit and seed dispersal and the evolution of the Hamamelidae. *Ann. Missouri Bot. Gard.* 73: 394–416.

Tobe, H. 1989. The embryology of angiosperms: Its broad application to the systematic and evolutionary study. *Bot. Mag. Tokyo* 102: 351–367.

Tobe, H. and P. H. Raven. 1983. An embryological analysis of the Myrtales: Its definition and characteristics. *Ann. Missouri Bot. Gard.* 70: 71–94.

Tobe, H. and P. H. Raven. 1988. Seed morphology and anatomy of Rhizophoraceae, inter- and infrafamilial relationships. *Ann. Missouri Bot. Gard.* 75: 1319–1342.

Todzia, C. A. 1993. Ulmaceae. In *The families and genera of vascular plants*, vol. 2, Magnoliid, hamamelid and caryophyllid families, K. Kubitzki, J. G. Rohwer, and V. Bittrich (eds.), 603–611. Springer-Verlag, Berlin.

Tollsten, L. and J. T. Kundsen. 1992. Floral scent in dioecious *Salix* (Salicaceae)—a cue determining the pollination system? *Plant Syst. Evol.* 182: 229–237.

Tomlinson, P. B. 1962. Phylogeny of the Scitamineae: Morphological and anatomical considerations. *Evolution* 16: 192–213.

Tomlinson, P. B. 1969a. Commelinales—Zingiberales. In *Anatomy of the monocotyledons*, vol. 3, C. R. Metcalfe (ed.). Clarendon Press, Oxford.

Tomlinson, P. B. 1969b. On the morphology and anatomy of turtle grass, *Thalassia testudinum* (Hydrocharitaceae). *Bull. Mar. Sci.* 19: 286–305.

Tomlinson, P. B. 1980. *The biology of trees native to tropical Florida*. Published by the author, Petersham, MA.

Tomlinson, P. B. 1982. Helobiae (Alismatidae), vol. 7. In *Anatomy of the monocotyledons*, C. R. Metcalfe (ed.). Oxford University Press, Oxford.

Tomlinson, P. B. 1986. *The botany of mangroves*. Cambridge University Press, Cambridge.

Tomlinson, P. B. 1990. *The structural biology of palms*. Clarendon Press, Oxford.

Tomlinson, P. B. and M. H. Zimmerman. 1969. Vascular anatomy of monocots with secondary growth. *J. Arnold Arbor.* 50: 159–179.

Tomlinson, P. B., R. B. Primack, and J. S. Bunt. 1979. Preliminary observations on floral biology in mangrove Rhizophoraceae. *Biotropica* 11: 256–277.

Tomlinson, P. B. and P. A. Cox. 2000. Systematic and functional anatomy of seedlings in mangrove Rhizophoraceae: vivipary explained. *Bot. J. Linn. Soc.* 134: 215–231.

Torrey, J. G. and R. H. Berg. 1988. Some morphological features for generic characterization among the Casuarinaceae. *Amer. J. Bot.* 75: 684–874.

Townsend, C. C. 1993. Amaranthaceae. In *The families and genera of vascular plants*, vol. 2, Magnoliid, hamamelid and caryophyllid families, K. Kubitzki, J. G. Rohwer, and V. Bittrich (eds.), 70–91. Springer-Verlag, Berlin.

Traub, H. P. 1957. Classification of the Amaryllidaceae: Subfamilies, tribes and genera. *Plant Life* 13: 76–81.

Traub, H. P. 1963. *Genera of the Amaryllidaceae*. American Plant Life Society, La Jolla, CA.

Trift, I., M. Källersjö, and A. A. Anderberg. 2002. The monophyly of *Primula* (Primulaceae) evaluated by analysis of sequences from the chloroplast gene *rbcL*. *Syst. Bot.* 27: 396–407.

Tucker, G. C. 1987. The genera of Cyperaceae in the southeastern United States. *J. Arnold Arbor.* 68: 361–445.

Tucker, G. C. 1989. The genera of Commelinaceae in the southeastern United States. *J. Arnold Arbor.* 70: 97–130.

Tucker, G. C. 1996. The genera of Pooideae (Gramineae) in the southeastern United States. *Harvard Pap. Bot.* No. 9: 11–90.

Tucker, S. C., A. W. Douglas, and L. H.-X. Liang. 1993. Utility of ontogenetic and conventional characters in determining phylogenetic relationships of Saururaceae and Piperaceae (Piperales). *Syst. Bot.* 18: 614–641.

Tucker, S. C. and A. W. Douglas. 1994. Ontogenetic evidence and phylogenetic relationships among basal taxa of legumes. In *Advances in legume systematics*, part 6, I. K. Ferguson and S. Tucker (eds.), 11–32. Royal Botanic Gardens, Kew.

Turner, B. L. 1999. A reevaluation of the Mexican species of *Glandularia* (Verbenaceae). *Lundellia* 2: 55–71.

Ueda, K., K. Kosuge, and H. Tobe. 1997. A molecular phylogeny of Celtidaceae and Ulmaceae (Urticales) based on *rbcL* nucleotide sequences. *J. Plant Ses.* 110: 171–178.

Uhl, N. W. and J. Dransfield. 1987. *Genera palmarum*. L. H. Bailey Hortorium and International Palm Society, Ithaca, NY.

Uhl, N. W., J. Dransfield, J. I. Davis, M. A. Luckow, K. H. Hansen, and J. J. Doyle. 1995. Phylogenetic relationships among palms: Cladistic analyses of morphological and chloroplast DNA restriction site variation. In *Monocotyledons: Systematics and evolution*, P. J. Rudall, P. J. Cribb, D. F. Cutler, and C. J. Humphries (eds.), 623–661. Royal Botanic Gardens, Kew.

Umadevi, I. and M. Daniel. 1991. Chemosystematics of the Sapindaceae. *Feddes Repert.* 102: 607–612.

Umber, R. E. 1979. The genus *Glandularia* (Verbenaceae) in North America. *Syst. Bot.* 4: 72–102.

Urtuey, E. and T. F. Stuessy. 2001. New hypotheses of phylogenetic relationships in Barnadesioideae (Asteraceae) based on morphology. *Taxon* 50: 1043–1066.

van Heusden, E. C. H. 1992. Flowers of Annonaceae: Morphology, classification, and evolution. *Blumea*, Suppl. 7: 1–218.

van der Pijl, L. 1957. On the arilloids of *Nephelium, Euphoria, Litchi* and *Aesculus* and the seeds of Sapindaceae in general. *Acta Bot. Neerl.* 6: 618–641.

van der Pijl, L. and C. H. Dodson. 1966. *Orchid flowers, their pollination and evolution.* University of Miami Press, Coral Gables, FL.

van der Werff, H. 1991. A key to the genera of Lauraceae in the New World. *Ann. Missouri Bot. Gard.* 78: 377–387.

van der Werff, H. and H. G. Richter. 1996. Towards an improved classification of Lauraceae. *Ann. Missouri Bot. Gard.* 83: 409–418.

van Ham, R. C. H. 1994. *Phylogenetic implications of chloroplast DNA variation in the Crassulaceae.* Dissertation. University of Utrecht, Belgium.

van Ham, R. C. H. and H. T. t'Hart. 1998. Phylogenetic relationships in the Crassulaceae inferred from chloroplast DNA restriction-site variation. *Amer. J. Bot.* 85: 123–134.

van Heel, W. A. 1966. Morphology of the androecium in Malvales. *Blumea* 13: 177–394.

Vander Wyk, R. and J. E. Canright. 1956. The anatomy and relationships of the Annonaceae. *Trop. Woods* 104: 1–24.

Varadarajan, G. S. and A. J. Gilmartin. 1988. Phylogenetic relationships of groups of genera within the subfamily Pitcairnioideae (Bromeliaceae). *Syst. Bot.* 13: 283–293.

Vaughan, J. G., J. R. Pjelan, and K. E. Denford. 1976. Seed studies in the Cruciferae. In *The biology and chemistry of the Cruciferae,* J. G. Vaughan, A. J. MacLeod, and B. M. G. Jones (eds.), 119–144. Academic Press, London.

Verdcourt, B. 1958. Remarks on the classification of the Rubiaceae. *Bull. Jard. Bot. Etat* 28: 209–290.

Verhoek, S. 1998. Agavaceae. In *The families and genera of vascular plants,* vol. 3. Flowering plants—Monocotyledons, K. Kubitzki (ed.), 60–70. Springer-Verlag, Berlin.

Verkerke, W. 1985. Ovules and seeds of Polygalaceae. *J. Arnold Arbor.* 66: 353–394.

Verma, D-P. S. and J. Standley. 1989. The legume–*Rhizobium* equation: A coevolution of two genomes. *Monogr. Syst. Bot. Missouri Bot. Gard.* 29: 545–557.

Vink, W. 1988. Taxonomy in Winteraceae. *Taxon* 37: 691–698.

Vink, W. 1993. Winteraceae. In *The families and genera of vascular plants,* vol. 2, Magnoliid, hamamelid and caryophyllid families, K. Kubitzki, J. G. Rohwer, and V. Bittrich (eds.), 630–638. Springer-Verlag, Berlin.

Vinnersten, A. and K. Bremer. 2001. Age and biogeography of major clades in Liliales. *Amer. J. Bot.* 88: 1695–1703.

Vliet, G. J. C. M. van, J. Koek-Noorman, and B. J. H. ter Welle. 1981. Wood anatomy,

classification and phylogeny of the Melastomataceae. *Blumea* 27: 463–473.

Vogel, S. 1990. History of the Malpighiaceae in the light of pollination ecology. *Mem. New York Bot. Gard.* 55: 130–142.

Vuilleumier, B. S. 1967. The origin and evolutionary development of heterostyly in the angiosperms. *Evolution* 21: 210–226.

Vuilleumier, B. S. 1969. The genera of Senecioneae in the southeastern United States. *J. Arnold Arbor.* 50: 104–123.

Vuilleumier, B. S. 1973. The genera of Lactuceae (Compositae) in the southeastern United States. *J. Arnold Arbor.* 54: 42–93.

Wagenitz, G. 1959. Die systematische Stellung der Rubiaceae. *Bot. Jahr. Syst.* 79: 17–35.

Wagenitz, G. 1992. The Asteridae: Evolution of a concept and its present status. *Ann. Missouri Bot. Gard.* 79: 209–217.

Wagenitz, G. 1977. New aspects of the systematics of Asteridae. *Plant Syst. Evol.* Suppl. 1: 375–395.

Wagstaff, S. J. and R. G. Olmstead. 1996. Phylogeny of Labiatae and Verbenaceae inferred from *rbcL* sequences. *Syst. Bot.* 22: 165–179.

Wagstaff, S. J., L. Hickerson, R. Spangler, P. A. Reeves, and R. G. Olmstead. 1998. Phylogeny in Labiatae s. l., inferred from cpDNA sequences. *Pl. Syst. Evol.* 209: 265–274.

Walker, J. W. 1971. Pollen morphology, phytogeography and phylogeny of the Annonaceae. *Contr. Gray Herb.* 202: 1–131.

Walker, J. W. and A. G. Walker. 1984. Ultrastructure of Lower Cretaceous angiosperm pollen and the origin and early evolution of flowering plants. *Ann. Missouri Bot. Gard.* 71: 464–521.

Wallace, G. D. 1975. Interrelationships of the subfamilies of the Ericaceae and the derivation of the Monotropoideae. *Bot. Not.* 128: 286–298.

Wallander, E. and V. A. Albert. 2000. Phylogeny and classification of Oleaceae based on *rbs16* and *trnL-F* sequence data. *Amer. J. Bot.* 87: 1827–1841.

Wannan, B. S. and C. J. Quinn. 1990. Pericarp structure and generic affinities in the Anacardiaceae. *Bot. J. Linn. Soc.* 102: 225–252.

Wannan, B. S. and C. J. Quinn. 1991. Floral structure and evolution in the Anacardiaceae. *Bot. J. Linn. Soc.* 107: 349–385.

Watson, L. and M. J. Dallwitz. 1992. *The grass genera of the world.* C. A. B. International, Wallingford, U.K.

Warwick, S. I. and L. D. Black. 1993. Molecular relationships in subtribe Brassicinae (Cruciferae, tribe Brassiceae). *Canad. J. Bot.* 71: 906–918.

Weberling, F. 1988a. The architecture of inflorescences in the Myrtales. *Ann. Missouri Bot. Gard.* 75: 226–310.

Weberling, F. 1988b. Inflorescence structure in primitive angiosperms. *Taxon* 37: 657–690.

Weberling, F. 1989. Structure and evolutionary tendencies of inflorescences in the Leguminosae. *Monogr. Syst. Bot. Missouri Bot. Gard.* 29: 35–58.

Webster, G. L. 1955. Studies in the Euphorbiaceae, Phyllanthoideae. I. Taxonomic notes on the West Indian species of *Phyllanthus. Contr. Gray Herb.* 176: 45–63.

Webster, G. L. 1956–1958. Studies in the Euphorbiaceae, Phyllanthoideae. III. A monographic study of the West Indian species of *Phyllanthus. J. Arnold Arbor.* 37: 91–122, 217–268, 340–359; 38: 51–80, 170–198, 295–373; 39: 49–100, 111–212.

Webster, G. L. 1967. The genera of Euphorbiaceae in the southeastern United States. *J. Arnold Arbor.* 48: 303–430.

Webster, G. L. 1987. The saga of the spurges: A review of classification and relationships in the Euphorbiales. *Bot. J. Linn. Soc.* 94: 3–46.

Webster, G. L. 1994a. Classification of the Euphorbiaceae. *Ann. Missouri Bot. Gard.* 81: 3–32.

Webster, G. L. 1994b. Synopsis of the genera and suprageneric taxa of Euphorbiaceae. *Ann. Missouri Bot. Gard.* 81: 33–144.

Wen, J., G. M. Plunkett, A. D. Mitchell, and S. J. Wagstaff. 2001. The evolution of Araliaceae: a phylogenetic analysis based on ITS sequences of nuclear ribosomal DNA. *Syst. Bot.* 26: 144–167.

Whiffin, T. 1972. Observations on some upper Amazonian formicarial Melastomataceae. *Sida* 5: 32–41.

White, D. A. and L. B. Thien. 1985. The pollination of *Illicium parviflorum* (Illiciaceae). *J. Elisha Mitchell Sci. Soc.* 101: 15–18.

Whitlock, B. A., C. Bayer, and D. A. Baum. 2001. Phylogenetic relationships and floral evolution of the Byttnerioideae ("Sterculiaceae" or Malvaceae s. l.) based on sequences of the chloroplast gene, *ndhF. Syst. Bot.* 26: 420–437.

Weiblen, G. D. 2000. Phylogenetic relationships of functionally dioecious Ficus (Moraceae) based on ribosomal DNA sequences and morphology. *Amer. J. Bot.* 87: 1342–1357.

Wiegrefe, S. J., K. J. Sytsma, and R. P. Guries. 1994. Phylogeny of elm (*Ulmus,* Ulmaceae): Molecular evidence for a sectional classification. *Syst. Bot.* 19: 590–612.

Wiegrefe, S. J., K. J. Sytsma, and R. P. Gories. 1998. The Ulmaceae, one family or two? Evidence from chloroplast DNA restriction site mapping. *Pl. Syst. Evol.* 210: 249–270.

Wiehler, H. 1983. A synopsis of the neotropical Gesneriaceae. *Selbyana* 6: 1–219.

Wiens, D. and B. A. Barlow. 1971. The cytogeography and relationships of the Viscaceous and Eremolepidaceous mistletoes. *Taxon* 20: 313–332.

Wiersema, J. H. 1988. Reproductive biology of *Nymphaea* (Nymphaeaceae). *Ann. Missouri Bot. Gard.* 75: 795–804.

Wilbur, R. L. 1994. The Myricaceae of the United States and Canada: Genera, subgenera, and series. *Sida* 16: 93–107.

Wilbur, R. L. 2001. Five new combinations in the genus *Morella* (Myricaceae) for neotropical species. *Rhodora* 103: 120–122.

Wildinson, A. M. 1948. Floral anatomy and morphology of some species of the genus *Viburnum* of the Caprifoliaceae. *Amer. J. Bot.* 35: 455–465.

Wildman, W. C. and B. A. Pursey. 1968. Colchicine and related compounds. In *The alkaloids, chemistry and physiology 11,* R. H. F. Manske (ed.), 407–457. Academic Press, London.

Williams, S. E. 1976. Comparative sensory plysiology of the Droseraceae: Evolution of a plant sensory system. *Proc. Amer. Philos. Soc.* 120: 187–204.

Williams, S. E., V. A. Albert, and M. W. Chase. 1994. Relationships of Droseraceae: A cladistic analysis of *rbcL* sequence and morphological data. *Amer. J. Bot.* 81: 1027–1037.

Wilson, C. L. 1950. Vasculation of the stamen in the Melastomataceae, with some phyletic implications. *Amer. J. Bot.* 37: 431–444.

Wilson, K. A. 1960a. The genera of Arales in the southeastern United States. *J. Arnold Arbor.* 41: 47–72.

Wilson, K. A. 1960b. The genera of Convolvulaceae in the southeastern United States. *J. Arnold Arbor.* 41: 298–317.

Wilson, K. A. 1960c. The genera of Hydrophyllaceae and Polemoniaceae in the southeastern United States. *J. Arnold Arbor.* 41: 197–212.

Wilson, K. A. 1960d. The genera of Myrtaceae in the southeastern United States. *J. Arnold Arbor.* 41: 270–278.

Wilson, K. A. and C. E. Wood. 1959. The genera of Oleaceae in the southeastern United States. *J. Arnold Arbor.* 40: 369–384.

Wilson, P. G., P. A. Gadek, and C. J. Quinn. 1996. Phylogeny of Myrtaceae and its allies based on *matK* sequence data. *Amer. J. Bot.* 83(6) Suppl.: 202.

Wilson, P. G., M. M. O'Brien, P. A. Gadek, and C. J. Quinn. 2001. Myrtaceae revisited: A reassessment of infrafamilial groups. *Amer. J. Bot.* 88: 2013–2025.

Wilson, T. K. and L. M. Maculans. 1967. The morphology of the Myristicaceae. I. Flowers of *Myristica fragrans* and *M. malabarica. Amer. J. Bot.* 54: 214–220.

Wolfe, J. A. 1989. Leaf-architectural analysis of the Hamamelididae. In *Evolution, systematics and fossil history of the Hamamelidae*, vol. 1, Introduction and "lower" Hamamelidae. Syst. Assoc. Special Vol. 40A, P. R. Crane and S. Blackmore (eds.), 75–104. Clarendon Press, Oxford.

Wolfe, J. A. and T. Tanai. 1987. Systematics, phylogeny and distribution of *Acer* (maples) in the Cenozoic of North America. *J. Fac. Sci. Hokkaido Imperial University*, Ser. 4, Geology 22: 1–246.

Wolfe, J. A. and W. Wehr. 1988. Rosaceous *Chamaebatiaria*-like foliage from the Paleogene of western North America. *Aliso* 12: 177–200.

Wood, C. E. 1958. The genera of woody Ranales in the southeastern United States. *J. Arnold Arbor.* 39: 296–346.

Wood, C. E. 1959a. The genera of Nymphaeaceae and Ceratophyllaceae in the southeastern United States. *J. Arnold Arbor.* 40: 94–112.

Wood, C. E. 1959b. The genera of Theaceae of the southeastern United States. *J. Arnold Arbor.* 40: 413–419.

Wood, C. E. 1960. The genera of Sarraceniaceae and Droseraceae in the southeastern United States. *J. Arnold Arbor.* 41: 152–163.

Wood, C. E. 1961. The genera of Ericaceae in the southeastern United States. *J. Arnold Arbor.* 42: 10–80.

Wood, C. E. 1971. The Saururaceae in the southeastern United States. *J. Arnold Arbor.* 52: 479–485.

Wood, C. E. 1974. *A student's atlas of flowering plants: Some dicotyledons of eastern North America*. Harper & Row, New York. Prepared as part of the Generic Flora of the Southeastern U.S. Project.

Wood, C. E. 1983. The genera of Menyanthaceae in the southeastern United States. *J. Arnold Arbor.* 64: 431–445.

Wood, C. E. and P. Adams. 1976. The genera of Guttiferae (Clusiaceae) in the southeastern United States. *J. Arnold Arbor.* 57: 74–90.

Wood, C. E. and R. B. Channell. 1960. The genera of Ebenales in the southeastern United States. *J. Arnold Arbor.* 41: 1–35

Wood, C. E. and R. K. Godfrey. 1957. *Pinguicula* (Lentibulariaceae) in the Southeastern United States. *Rhodora* 59: 217–230.

Wood, C. E. and R. E. Weaver, Jr. 1982. The genera of Gentianaceae in the southeastern United States. *J. Arnold Arbor.* 63: 441–487.

Wood, T. H., W. M. Whitten, and N. H. Williams. 2000. Phylogeny of *Hedychium* and related genera (Zingiberaceae) based on ITS sequence data. *Edinb. J. Bot.* 57: 261–270.

Woodland, D. W. 1989. Biology of temperate Urticaceae (nettle) family. In *Evolution, systematics and fossil history of the Hamamelidae*, vol. 2, "Higher" Hamamelidae. Syst. Assoc. Special Vol. 40B, P. R. Crane and S. Blackmore (eds.), 309–318. Clarendon Press, Oxford.

Woodson, R. E. 1954. The North American species of *Asclepias* L. *Ann. Missouri Bot. Gard.* 41: 1–211.

Wurdack, J. J. 1980. Melastomataceae. In *Flora of Ecuador* No. 13, G. Harding and B. Sparre (eds.). University of Goteborg and Riksmuseum, Stockholm.

Wurdack, J. J. 1986. Atlas of hairs for neotropical Melastomataceae. *Smithsonian Contr. Bot.* 63: 1–80.

Wurdack, J. J. and R. Kral. 1982. The genera of Melastomataceae in the southeastern United States. *J. Arnold Arbor.* 63: 429–439.

Xiang, Q. Y., D. E. Soltis, D. R. Moran, and P. S. Soltis. 1993. Phylogenetic relationships of *Cornus* L. *sensu lato* and putative relatives inferred from *rbcL* sequence data. *Ann. Missouri Bot. Gard.* 80: 723–734.

Xiang, Q. Y., S. J. Brunsfeld, D. E. Soltis and P. S. Soltis. 1996. Phylogenetic relationships in *Cornus* based on chloroplast DNA restriction sites: Implications for biogeography and character evolution. *Syst. Bot.* 21: 515–534.

Xiang, Q. Y. and Z. Murrell. 1998. Relationships and biogeography of *Cornus* L. (Cornaceae) inferred from multiple molecular and morphological data sets. *Amer. J. Bot.* 85(6) Suppl.: 168.

Xiang, Q. Y., D. E. Soltis, and P. S. Soltis. 1998. Phylogenetic relationships of Cornaceae and close relatives inferred from *matK* and *rbcL* sequences. *Amer. J. Bot.* 85: 285–297.

Xiaoping, Z. and K. Bremer. 1993. A cladistic analysis of the tribe Astereae (Asteraceae) with notes on their evolution and subtribal classification. *Plant Syst. Evol.* 184: 259–283.

Yamashita, J. and M. N. Tamura. 2000. Molecular phylogeny of the Convallariaceae (Asparagales). *In Monocots: Systematics and evolution*, K. L. Wilson and D. A. Morrison (eds.), 387–400. CSIRO Publ., Collingwood, Australia.

Yamazaki, T. 1974. A system of Gamopetalae based on embryology. *J. Fac. Sci. University Tokyo*, Sect. 3, Bot. 11: 263–281.

Yeo, P. F. 1984. Fruit-discharge type in *Geranium* (Geraniaceae): Its use in classification and its evolutionary implications. *Bot. J. Linn. Soc.* 89: 1–36.

Yeo, P. F. 1993. Secondary pollen presentation. *Plant Syst. Evol.*, Suppl. 6: 1–268.

Young, D. A. 1981. Are the angiosperms primitively vesselless? *Syst. Bot.* 6: 313–330.

Young, D. A. 1982. Leaf flavonoids of *Amborella trichopoda. Biochem. Syst. Evol.* 10: 21–22.

Young, N. D., K. E. Steiner, and C. W. de Pamphilis. 1999. The evolution of parasitism in Scrophulariaceae/Orobanchaceae: plastid gene sequences refute an evolutionary transition series. *Ann. Missouri Bot. Gard.* 86: 876–893.

Zavada, M. S. and M. Kim. 1996. Phylogenetic analysis of Ulmaceae. *Plant Syst. Evol.* 200: 13–20.

Zhang, S.-H. 1992. Wood anatomy of the Rosaceae. Rijksherbarium/Hortus Botanicus, Leiden.

Zimmer, E. A., R. K. Hamby, M. L. Arnold, D. A. LeBlanc, and E. C. Theriot. 1989. Ribosomal RNA phylogenies and flowering plant evolution. In *The hierarchy of life*, B. Fernholm, K. Bremer, and H. Jornvall (eds.), 205–214. Elsevier, Amsterdam.

Zomlefer, W. B. 1996. The Trilliaceae in the southeastern United States. *Harvard Pap. Bot.* No. 1(9): 91–120.

Zomlefer, W. B. 1997a. The genera of Melanthiaceae in the southeastern United States. *Harvard Pap. Bot.* 2: 133–177.

Zomlefer, W. B. 1997b. The genera of Nartheciaceae in the southeastern United States. *Harvard Pap. Bot.* 2: 195–211.

Zomlefer, W. B. 1997c. The genera of Tofieldiaceae in the southeastern United States. *Harvard Pap. Bot.* 2: 179–194.

Zomlefer, W. B. 1998. The genera of Hemerocallidaceae in the southeastern United States. *Harvard Pap. Bot.* 3: 113–145.

Zomlefer, W. B., N. H. Williams, W. M. Whitten, and W. S. Judd. 2001. Generic circumscription and relationships in the tribe Melanthieae (Liliales, Melanthiaceae), with emphasis on *Zigadenus*: evidence from ITS and *trnL-F* sequence data. *Amer. J. Bot.* 88: 1657–1669.

Zona, S. 1997. The genera of Palmae (Arecaceae) in the southeastern United States. *Harvard Pap. Bot.* No. 11: 71–107.

Zona, S. 2001. Starchy pollen in commelinoid monocots. *Ann. Bot. II* 87: 109–116.

Zona, S. and A. Henderson. 1989. A review of animal-mediated seed dispersal in palms. *Selbyana* 11: 6–21.

Appendix: Groups of Rosaceae (see pp. 365–372)

Key to Major Groups of Rosaceae

1. Carpels usually numerous; fruit an achene or drupelet; $x = 7$ (rarely 8); sorbitol, cyanogenic glycosides, and flavones absent; ellagic acid present .... Rosoideae s. s.
1. Carpels usually 1–5; fruit a drupe, follicle, pome, or capsule; $x = 8, 9, 15, 16,$ or 17; sorbitol, cyanogenic glycosides, and flavones present; ellagic acid absent 2
2. Flowers usually with a single carpel (but rarely as many as 5); fruit a drupe (rarely a capsule); $x = 8$ .... *Prunus* and the *Exochorda-Oemleria-Prinsepia* clade
2. Flowers with 2–5 carpels (rarely only 1); fruit a pome, achene, follicle, capsule, or drupe; $x = 9, 15, 16, 17$ 3
3. Ovules paired, basally inserted, and anatropous 4
3. Ovules solitary, paired, or numerous in each carpel, variously inserted, and anatropous or otherwise but not paired, basally inserted, and anatropous 5
4. Shrubs and trees, $x = 17$ .... Maloideae s. l.
4. Herbs, $x = 9$ .... *Gillenia*
5. Stipules absent .... Spiraeeae
5. Stipules usually present although sometimes deciduous 6
6. Seed coat hard and shining .... Neillieae
6. Seed coat not hard and shining 7
7. Leaf epidermis with hard, wart-like projections .... Kerrieae
7. Leaf epidermis without hard, wart-like projections 8
8. Leaves opposite and fern-like .... *Lyonothamnus*
8. Leaves alternate and simple to compound but not fern-like 9
9. Nitrogen-fixing .... Dryadeae
9. Not nitrogen-fixing .... Sorbarieae

APPENDIX 1

Botanical Nomenclature

axonomic groups require names if we are to efficiently communicate (or gain access to) information regarding their identity, phylogenetic relationships, and other aspects of their biology. The naming of plants is called **botanical nomenclature**. The principles and rules of botanical nomenclature have been developed and adapted by a series of International Botanical Congresses and are listed in the International Code of Botanical Nomenclature (or ICBN; Greuter et al. 2000). The major goal of the ICBN is to provide one correct name for each taxonomic group (or taxon) within a stable system of names (a classification).

Scientific Names

Systematic botanists (as well as other scientists) use Latinized scientific names. Each taxon—for example, a species, genus, or family—has one such name that is used throughout the world. In contrast, common (or vernacular) names are often limited to a single language or to a particular geographic region. Sometimes the same common name (e.g., "bluebell") may be applied to several different taxa, and many species, especially if they are rare or of little economic importance, have no common names. Finally, common names are frequently misleading with respect to phylogenetic relationships; for example, poison oak, a species of *Toxicodendron* (Anacardiaceae), is not closely related to the true oaks of the genus *Quercus* (Fagaceae). The use of scientific names is therefore essential for the efficient, accurate communication of information about plants on a worldwide basis.

Scientific names of species are **binomials**; that is, they are composed of two words. The binomial system of nomenclature was first consistently used by Carolus Linnaeus in his *Species plantarum* (1753). The first word of a species name is a singular noun and is the name of the genus to which the plant is assigned. The second word is (1) an adjective modifying the generic name (and must agree in gender with the generic name), (2) a noun in apposition, or (3) a possessive noun; it is called the **specific epithet**.

Most specific epithets refer to a distinctive morphological, ecological, or chemical feature of the species (Table 1). Some epithets refer to the species'

511

TABLE 1 *Common specific epithets.*

Specific epithet	English meaning	Specific epithet	English meaning	Specific epithet	English meaning
acaulis	stemless	*brachycarpus*	short-fruited	*dentatus*	toothed
acicularis	needle-like	*brevipes*	short-footed	*didymus*	twinned, in pairs
aduncus	hooked	*brunneus*	deep brown	*digitatus*	fingered, palmate
aestivalis	summer	*bufonius*	of toads	*discolor*	having different colors
affinis	related	*caeruleus*	dark blue		
agrestis	of the field	*caesius*	bluish gray	*dulcis*	sweet
alatus	winged	*calvus*	bald, hairless	*dumosus*	bushy
albicans	whitish	*calycinus*	calyxlike	*echinatus*	prickly
albus	white	*campanulatus*	bell-shaped	*edulis*	edible
alpestris	alpine	*campestris*	of the fields	*effusus*	loose-spreading
alpinus	alpine	*candicans*	white, hoary	*elatior*	taller
alternans	alternating	*caninus*	of dogs	*elatus*	tall
altissimus	very tall	*capillaris*	hairlike	*elegans*	elegant
amabilis	lovely	*carinatus*	keeled	*ensifolius*	sword-leaved
amarus	bitter	*caudatus*	tailed	*eriocarpus*	woolly-fruited
ambigens	ambiguous	*cerifera*	wax-bearing	*esculentus*	edible
amoenus	charming	*cernuus*	drooping	*exiguus*	little, poor
amplexicaulis	clasping	*chloranthus*	green-flowered	*fallax*	deceptive
anceps	two-headed or two-edged	*chrysophyllus*	golden-leaved	*farinosus*	mealy
		chrysostomus	golden-mouthed	*fasciculatus*	fascicled
angustatus	narrow	*cinctus*	girdled	*fastigiatus*	erect and close together
angustifolius	narrow-leaved	*clandestinus*	concealed		
annotinus	year-old	*coarctatus*	pressed together, narrowed	*filipes*	threadlike stalks
annuus	annual			*fistulosus*	hollow, cylindrical
aphyllus	leafless	*coccineus*	scarlet	*flabellatus*	fanlike
apiculatus	tipped with a point	*comatus*	with hair	*flagellaris*	whiplike
appendiculatus	appendaged	*communis*	gregarious	*flavescens*	yellowish
applanatus	flattened	*commutatus*	changing	*flavus*	yellow
arcuatus	bowlike	*comosus*	bearded or tufted, long-haired	*flexilis*	flexible
arenarius	of sand			*floribundus*	free-flowering
areolatus	areolate, pitted	*concinnus*	neat	*floridus*	flowering
argenteus	silvery	*concolor*	colored similarly	*fluitans*	floating
argutus	sharp-toothed	*confertus*	crowded	*fluviatilis*	of a river
argyreus	silvery	*confinis*	bordered	*foetidus*	bad-smelling
aridus	arid	*conoideus*	conelike	*foliosus*	leafy
aristatus	awned	*contortus*	contorted	*formosus*	beautiful
arundinaceus	reedlike	*corniculatus*	horned	*frondosus*	leafy
arvensis	of cultivated fields	*cornutus*	horned	*fulgens*	shining
asper	rough	*coronarius*	used with garlands	*furcatus*	forked
atratus	blackened	*crassifolius*	thick-leaved	*geniculatus*	jointed
atropurpureus	dark purple	*crassipes*	thick-footed	*gracilis*	slender
atrosanguineus	dark blood-red	*crinitus*	hairy	*gramineus*	grassy
aureus	golden	*cristatus*	crested	*graveolens*	heavy-scented
australis	southern	*cuneiformis*	wedge-shaped	*hebecarpus*	pubescent-fruited
azureus	sky-blue	*dasycarpus*	shaggy-fruited	*helveolus*	pale-yellow
baccatus	berried	*dasystachys*	shaggy-spiked	*hirtus*	hairy
baculiformis	rod-shaped	*debilis*	weak	*holosericeus*	woolly, silky
bicolor	two-colored	*decapetalus*	ten-petaled	*humifusus*	sprawling
bidentata	two-toothed	*decipiens*	deceptive	*humilis*	dwarf
biennis	biennial	*decorus*	elegant	*hyemalis*	of winter
bifidus	twice-cut	*decumbens*	reclining	*hyperboreus*	far northern
biflorus	two-flowered	*deflexus*	bent downward	*hypogaeus*	underground
borealis	northern	*demissus*	low, weak	*hypoglaucus*	glaucous beneath

TABLE 1 *(continued)*

Specific epithet	English meaning	Specific epithet	English meaning	Specific epithet	English meaning
hystrix	bristly	*occidentale*	western	*sativus*	cultivated
incanus	hoary	*officinalis*	official	*saxatilis*	found on the rocks
inermis	unarmed	*oliganthus*	few-flowered	*scandens*	climbing
inodorus	without an odor	*oligocarpus*	few-fruited	*sclerophyllus*	hard-leaved
intumescens	swollen	*oligospermus*	few-seeded	*scoparius*	broomlike
junceus	rushlike	*operculatus*	with a lid	*sensibilis*	sensitive
lactatus	milky	*orientalis*	eastern	*septentrionalis*	northern
lacustris	of lakes	*ornatus*	adorned	*serotinus*	late-flowering
laevigatus	smooth	*orthocarpus*	straight-fruited	*serpens*	creeping
lanceolatus	lanceolate	*ovatus*	ovate	*serpyllifolius*	thyme-leaved
lanuginosus	woolly	*oxycanthus*	sharp-spined	*setaceus*	bristlelike
latifolius	broad-leaved	*paludosus*	marsh-loving	*speciosus*	showy
leptocladus	thin-stemmed	*palustris*	of swamps	*spectabilis*	spectacular
leucanthus	white-flowered	*parviflorus*	small-flowered	*squarrosus*	with parts recurved
linearis	linear	*parvifolius*	small-leaved	*stans*	erect
littoralis	of the seashore	*parvulus*	very small	*stellatus*	star-shaped
longipes	long-footed	*patens*	spreading	*stenophyllus*	narrow-leaved
lucidus	bright, clear	*pauciflorus*	few-flowered	*strictus*	strict, upright
lupulinus	hoplike	*pectinatus*	comblike	*tenellus*	slender, soft, tender
luteolus	yellowish	*pedatus*	like a foot	*tenuis*	slender, thin
macilentus	lean	*pentandrus*	five stamens	*teres*	terete
macranthus	large-flowered	*peregrinus*	exotic	*ternatus*	in threes
macrocarpus	large-fruited	*perennans*	perennial	*tetrapterus*	four-winged
macrophyllus	large-leaved	*plantagineus*	plantain-like	*thyrsiflorus*	thyrse-flowered
maculatus	spotted	*platycarpus*	broad-fruited	*tinctorius*	of dyes
maritimus	of the sea	*platycladus*	broad-branched	*tricoccus*	three-lobed
medius	intermediate	*platyphyllus*	broad-leaved	*tridens*	three-toothed
megarrhizus	large-rooted	*polyanthus*	many-flowered	*trifidus*	three-parted
micranthus	small-flowered	*polystachyus*	many-spiked	*tripteris*	three-winged
millefolius	many-leaved	*praecox*	precocious	*tristis*	sad, dull
mirabilis	wonderful	*prasinus*	grass-green	*trivialis*	common, ordinary
modestus	modest	*procera*	tall	*umbrosus*	shade-loving
mollis	soft	*pulchellus*	pretty, beautiful	*uncinatus*	hooked
moniliformis	constricted at regular intervals	*pulcher*	pretty, beautiful	*undulatus*	undulate
		pumilis	dwarf	*uniflorus*	one-flowered
monocephalus	single-headed	*pungens*	piercing	*urceolatus*	urn-shaped
monostachys	single-spiked	*pycnanthus*	densely flowered	*urens*	burning, stingin
montanus	of mountains	*quadrifolius*	four-leaved	*ursinus*	of bears
mutabilis	variable	*quinquefolius*	five-leaved	*usitatissimus*	most useful
nanus	dwarf	*ramosus*	branched	*vaginatus*	sheathed
natans	floating	*repens*	creeping	*validus*	strong
nemoralis	of groves	*retroflexus*	reflexed	*velutinus*	velvety
nictitans	blinking, nodding	*riparius*	of river banks	*venosus*	veiny
nigricans	black	*rostratus*	beaked	*vernalis*	vernal
nitens	shining	*rubellus*	reddish	*vernus*	of spring
nitidus	shining	*rubiginosus*	rusty	*versicolor*	variously colored
nivalis	snowy	*rufus*	red	*vestitus*	covered
niveus	snowy	*rugosus*	wrinkled	*vimineus*	of wickerwork
novae-angliae	of New England	*rupestris*	rock-loving	*virens*	green
noveboracensis	of New York	*saccharinus*	sugary, sweet	*volubilis*	twining
nudicaulis	naked-stemmed	*sagittatus*	arrowhead-shaped	*vulgaris*	common
nutans	nodding	*salinus*	salty	*vulpinus*	of foxes
obovatus	obovate	*sanguineus*	blood-red	*xanthocarpus*	yellow-fruited

geographic range; others honor the individual who first collected the plant or a scientist who contributed to the botanical knowledge of a particular geographic region or taxonomic group. Knowing the meanings of specific epithets aids in remembering scientific names; Table 1 presents a checklist of common specific epithets. Both the generic name and the specific epithet are italicized or underlined; the generic name is capitalized, and the ICBN recommends that the specific epithet always be written in lowercase. Finally, the specific epithet may not exactly repeat the generic name, as in *Benzoin benzoin*; such names are called **tautonyms**.

The specific epithet is often followed by one or more **authorities**: the name (or names) of the person (or persons) who first described the species. These names may be abbreviated, and a list of approved abbreviations has been prepared (Brummitt and Powell 1992). For example, the scientific name, including authority, for the white oak is *Quercus alba* Linnaeus. The genus is *Quercus* (oak), the specific epithet is *alba* (an adjective meaning "white"), and the name of the authority is Linnaeus (which may be abbreviated "L."). Other examples include *Pinus ponderosa* Dougl. (ponderosa pine), *Acer rubrum* L. (red maple), *Quercus virginiana* Mill. (live oak), and *Vaccinium corymbosum* L. (highbush blueberry).

Sometimes increased knowledge about a species' phylogenetic relationships results in a change in its name; this is a necessary consequence of the binomial system. In one such case, Thomas Walter described *Andromeda ferruginea* in 1788. Later Thomas Nuttall (1818) decided that this species (and its relatives) actually belongs to the genus *Lyonia* because it has capsules with distinctively thickened sutures. (Nuttall made this decision on the basis of his subjective evaluation of the pattern of character variation within these genera, but recent cladistic studies of this group of Ericaceae also suggest that *Andromeda*, as currently circumscribed, is not closely related to *Lyonia*.) Thus the specific epithet *ferruginea* was transferred by Nuttall to the genus *Lyonia*. The resulting scientific name, with authorities, is *Lyonia ferruginea* (Walt.) Nutt., and that is the currently accepted name of the species. Note that the describing authority's name is placed in parentheses and is followed by the transferring authority's name.

In a second case, German botanist G. C. Oeder described *Ledum groenlandicum* in 1771; he placed the species in the genus *Ledum* (at least in part) because it has radially symmetrical flowers with more or less distinct petals. A recent cladistic analysis (Kron and Judd 1990) of *Rhododendron* and related genera of the tribe Rhodoreae indicated that *Ledum groenlandicum* (and its close relatives) nests within the cladistic structure of *Rhododendron*, a genus in which nearly all species have sympetalous and slightly bilaterally symmetrical flowers. Therefore, this species has been transferred to *Rhododendron* and given the name *Rhododendron groenlandicum* (Oeder) Kron & Judd. The new name reflects the close phylogenetic relationship of this species to others that also have multicellular peltate scales: the lepidote rhododendrons, such as *Rhododendron minus* or *R. lapponicum*.

Name changes such as these can lead to difficulties in the information retrieval function of names (and classifications), but such changes also provide a more accurate reflection of hypothesized phylogenetic relationships (allowing our system of plant names to be more predictive).

Sometimes authority names are separated by the prepositions *ex* or *in*. Names separated by *ex* mean that the second author published a name that had been proposed (but was never published) by the first; the word *in* means that the first author published the species in a book or article edited (or written, in part) by the second. Examples include *Gossypium tomentosum* Nutt. ex Seem., which may be shortened to *G. tomentosum* Seem., and *Viburnum ternatum* Rehder in Sargent, which may be shortened to *V. ternatum* Rehder. The code allows that the author before the *ex* or after the *in* does not have to be listed.

Infraspecific taxa—subspecies or varieties—are sometimes recognized within variable species. Geographic races are often treated at the varietal or subspecific level, and in these cases varietal or subspecific epithets are provided, such as *Lyonia ligustrina* (L.) DC. var. *ligustrina* and *L. ligustrina* (L.) DC. var. *foliosiflora* (Michx.) Fernald, or *Carpinus caroliniana* Walt. subsp. *caroliniana* and *C. caroliniana* Walt. subsp. *virginiana* (Marsh.) Furlow. The epithet of the variety or subspecies that contains the type specimen of the species (type specimens will be described a little later) repeats the species epithet but without naming any authority; this variety is often called **nominate**, or "typical."

Scientific names of higher taxa—genera and above—are **uninomials**; they are composed of a single word. The names of genera are singular nouns, and the names of taxa above the rank of genus are plural nouns (in Latin). The ending of the noun often represents the rank at which the taxon is placed. The ICBN recognizes seven major (or obligatory) ranks (kingdom, phylum/division, class, order, family, genus, species), but allows others to be intercalated; such intercalation is often accomplished by the addition of the prefixes *super-* or *sub-*. The most commonly used ranks (from high to low), with their standard suffixes, are listed in Table 2.

It is common in writing a monograph or flora treatment to assign every taxon to a higher-ranked taxon, but this is not required by the code. Thus, if some genera within a family are placed in subfamilies, then it is conventional to put all genera in subfamilies, even if the result is multiple monotypic subfamilies or requires placing genera about which we know very little. The desire to place all taxa in the next higher taxon has in the past led to the creation of "garbage can" taxa—groups created purely to hold miscellaneous taxa about which little is known.

TABLE 2 *Rank of taxonomic categories recognized by the ICBN.[a]*

Category	Standard suffix
Kingdom	-bionta
Phylum (or **Division**)	-phyta
Subphylum (or Subdivision)	-phytina
Class	-opsida
Subclass	-idae
Superorder	-anae
Order	-ales
Suborder	-ineae
Superfamily	-ariae
Family	-aceae
Subfamily	-oideae
Tribe	-eae
Subtribe	-inae
Genus	None; italicized, initial capital letter
Species	None; genus name plus specific epithet; italicized

[a]The seven major, or obligatory, ranks are indicated in **boldface**.

Although the practice of assigning everything to a rank is pleasantly tidy, it is not necessary. Thus a genus can be placed in a family, but left unplaced (*incertae sedis*) as to subfamily or tribe. In the APG (Angiosperm Phylogeny Group) classification of angiosperm families followed in this text, some families are unplaced with respect to order even if they are placed with respect to higher-level group. For example, Buxaceae and Trochodendraceae are placed in the eudicot clade but are not placed in an order.

Generic names do not have standardized endings. Generic names are capitalized and italicized (or underlined); names above the generic level are capitalized but usually not italicized (among American workers, although Europeans often italicize these as well). Family names are based on the name of the type genus for the family (which is explained later in the section "Nomenclatural Principles")—for example, *Rosa* (Rosaceae), *Aster* (Asteraceae), *Erica* (Ericaceae), or *Cyperus* (Cyperaceae).

The use of eight traditional family names, however, is specifically allowed by the ICBN. These are Compositae (= Asteraceae), Cruciferae (= Brassicaceae), Gramineae (= Poaceae), Guttiferae (= Clusiaceae), Labiatae (= Lamiaceae), Leguminosae (= Fabaceae), Palmae (= Arecaceae), and Umbelliferae (= Apiaceae). (The names in parentheses are the newer names.) The ICBN applies to the ranks of family and below, although its rules and recommendations may be applied to higher ranks. For this reason there is appreciable variation in taxon names at ranks above the level of family.

An example of the classification hierarchy as applied to *Acer rubrum* (red maple) is presented in Table 3. Keep in mind that taxa, particularly above the level of species, represent monophyletic groups and are, as such, considered to be the products of evolution. On the other hand, the ranks of the taxonomic hierarchy are human constructs, having relative but not absolute meaning (see Chapter 2). It is thus more important to remember, for example, that the angiosperms are hypothesized to be a monophyletic group, supported by numerous synapomorphies, than to worry about what rank is most appropriate for this higher taxon.

ARGUMENTS AGAINST RANKS IN CLASSIFICATION

As discussed in Chapter 2, one of the major insights of phylogenetic classification is the observation that rank is arbitrary. Although a named group must be monophyletic, there is no way to determine a priori whether it should be named as a genus, family, or order—or some rank in between. If rank is arbitrary, then one logical step is to eliminate ranks altogether. Taxa could be placed in named groups, but the groups would not be designated as genus, family, order, or other rank. However, the elimination of ranks is problematic among orders, families, and genera. Groups assigned to those ranks are familiar and in common use, and acceptance of an entirely new sort of nomenclature is unlikely to happen rapidly or without protest. Nonetheless, an alternative system of nomenclature, known as the PhyloCode, is being developed.

The PhyloCode is designed entirely outside the rules of the ICBN, which governs the use of Linnaean ranks and has long been used by all plant taxonomists. It is an alternative nomenclatural system rather than a revision of the existing one (see the PhyloCode Web site: www.ohiou.edu/phylocode). Under the PhyloCode, names of taxa refer only to monophyletic groups and are defined in one of three ways:

TABLE 3 *An application of the classification hierarchy for Acer rubrum (red maple).*

Category	Taxon
Kingdom	Viridiplantae (green plants)
Phylum (or Division)	Embryophyta (embryophytes)
Subphylum (or Subdivision)	Tracheophytina (tracheophytes)
Class	Angiospermopsida (angiosperms)
Order	Sapindales
Family	Sapindaceae (soapberry family)
Genus	*Acer* (maple)
Species	*Acer rubrum* (red maple)

1. A **node-based definition** specifies the meaning of a taxon name by associating that name with a clade stemming from the immediate common ancestor of two (or more) designated descendants (Figure 1A).
2. A **stem-based definition** specifies the meaning of a name by associating the name with a clade of all organisms that share a more recent common ancestor with one designated descendant than with another (Figure 1B).
3. An **apomorphy-based definition** specifies the meaning of a taxon name by associating the name with a clade derived from the ancestor in which a particular apomorphic character arose (Figure 1C) (de Queiroz and Gauthier 1992).

The issue of ranking is particularly problematic with regard to species, and a full discussion of this subject would not fit in an introductory text. The system of binomial nomenclature (as outlined in the ICBN and described earlier) requires that every species be placed in a genus, and the genus name becomes the first element of the species name—the binomial. For example, *Quercus virginiana*, live oak, is one member of the large genus *Quercus*. Such binomial constructions may appear to be philosophically problematic if Linnaean ranks, such as genus, are eliminated from the classification system; binomials, however, are a particular kind of noun + adjective combination, and as such they have no particular philosophical connotations (Stevens 2002).

The situation is even worse if the phylogenetic placement is uncertain; *incertae sedis* is not an option for a species. Currently, a systematist describing a new species does her best to assign it to the correct genus, but in some cases doing so may be overstating the evidence. A solution to these problems would be to name species with a single word (a uninomial) rather than a binomial. Some proposals in the PhyloCode (Cantino et al. 1999) are for the use of uninomials.

Use of uninomials would increase the stability of species names but also would have several undesirable consequences, such as the loss of information regarding phylogenetic placement in the species name, and the confusion that would result from the shift from one system of naming to another. Furthermore, uninomials of different species may be the same, necessitating their being distinguished by numerical attachments; for example, *Quercus virginiana* might become "virginiana 3254" or "virginiana 3," a confusing system in studies dealing with more than a single genus.

The success and acceptance of phylogenetic nomenclature may depend on the decision of whether or not to have species names reflect as closely as possible the present binomial system, and thus possess "historical" information and connotations readily grasped by most users of systematic information, including students (Sytsma and Pires 2001).

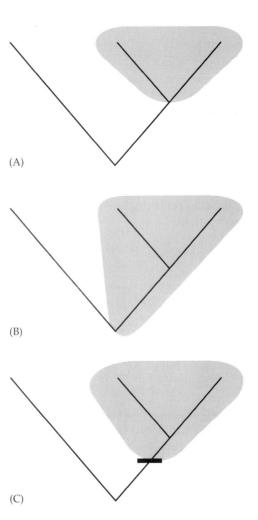

(A)

(B)

(C)

Figure 1   Ways to define taxon names. (A) Node-based definition. (B) Stem-based definition. (C) Apomorphy-based definition.

PRONUNCIATION OF SCIENTIFIC NAMES

Students are often hesitant to use scientific names because they seem difficult to pronounce. However, most are easier to pronounce than unknown English words. In addition, it is difficult to pronounce scientific names "incorrectly" because there is no universally agreed-upon system for pronouncing them!

Most North American botanists and horticulturists use "traditional English" pronunciation. In this system most letters of the alphabet are pronounced as they would be in English. All vowels may be either long or short, as in English. However, it is important to remember that a Latin word has as many syllables as it has vowels (or diphthongs—two vowels pronounced as one sound: *ea* as in meat, *eu* as in neutral). Every vowel or diphthong is pronounced, and there are no silent letters at the end of a word. However, when a word begins with *cn*, *gn*, *mn*, or *pt*, the first letter is silent.

Many Europeans pronounce Latin names using the "reformed academic" method, which attempts to approximate the pronunciation of educated Romans. The pronunciations of Latin American scientists are also quite divergent from those of the traditional English method; they are much more similar to the pronunciations of most Europeans.

NOMENCLATURAL PRINCIPLES

1. **Botanical nomenclature is independent of zoological nomenclature.** Although the codes of botanical and zoological nomenclature are similar in their basic principles, there are many differences in detail. One result of the independence of the two codes is that a "plant" and an "animal" may have the same scientific name; *Cecropia* (showy moths and weedy tropical trees of the Cecropiaceae) and *Pieris* (cabbage butterflies and shrubs of the Ericaceae) are just two examples, although creating more examples of such duplication is strongly discouraged.

Remember that the botanical code applies not only to the green plant clade but also to other eukaryotic clades—namely, stramenopiles (brown algae, golden algae, oomycete water molds), some alveolates (the dinoflagellates), red algae, true fungi (chytrids, zygomycetes, ascomycetes, basidiomycetes), and various "basal" eukaryote clades (euglenas, cellular slime molds, plasmodial slime molds). Some "protozoan" groups, such as euglenas and dinoflagellates, are sometimes considered "animals" and classified according to the nomenclatural rules of the zoological code. Therefore, a few groups of organisms may have two names—one under the botanical code, and a second under the zoological code.

2. **The application of names to taxonomic groups is determined by means of nomenclatural types.** When a new species or infraspecific taxon is described, the author must designate a particular specimen as the nomenclatural type. This specimen, deposited in a particular herbarium where it will be available for study, is the **holotype**, or name-bearing specimen. The name of the new species is therefore tied to this particular specimen, which illustrates what the author had in mind when he/she described the species. Duplicates of the holotype in other herbaria—that is, other portions of the same plant or other individuals of the same population, gathered at the same time and place as the holotype—are considered to be **isotypes**.

The holotype and isotypes may be consulted by systematists seeking to clarify the application of particular scientific names. For example, biological investigations may support the existence of two species where only one was previously recognized. To which of these two species should the original species name be applied? The type concept provides an unambiguous answer by requiring that the original species name be applied to the species to which the type specimen belongs.

The type specimen may not be at all "typical"; that is, it does not necessarily have the most common characters of the species to which it belongs. One should not disregard variation, or mistakenly think that all members of a species must possess certain "essential" or "key" features that are represented in the type specimen. The type is of no more importance than any other specimen in biological investigations, such as those determining specific delimitations or describing the pattern of variation; it is merely a nomenclatural aid. Once a systematist has determined, through systematic study, that recognition of certain species is justified (see Chapter 6 and Winston 1999), then type specimens are used in assigning names to those species.

The type for the name of a genus is the type for the name of a particular species within that genus. This type is cited by species' name; for example, *Lyonia ferruginea* is the type for the genus *Lyonia*. The type for the name of a family is the type for the name of a particular genus within that family (the type genus). As mentioned earlier, the name of the family is based on the name of this genus; for example, *Aster* is the type genus of Asteraceae, and *Erica* is the type genus of Ericaceae.

3. **The nomenclature of taxonomic groups is based on priority of publication.** The correct name for a taxon is the earliest name that is in accordance with the rules of nomenclature. Linnaeus's *Species plantarum* (published on 1 May 1753) is the starting point (for purposes of priority) for species names of vascular plants. Other plant groups may have other starting points. Later-published names for the same taxon are called **synonyms** and are not considered to be the correct name for the species. Names that duplicate names already in existence (for other species) are also not to be used and are called **homonyms**.

4. **Each taxon can bear only one correct name, the earliest that is in accordance with the rules, except in specified cases.** Certain widely used names are not actually the earliest published in accordance with the rules, but to avoid unnecessary name changes, many of these have been (or are being) conserved—that is, allowed to be considered the correct name of a taxon—through the special action of botanical congresses. In addition, the eight families singled out earlier may have more than one correct name.

5. **Scientific names are Latin or treated as Latin regardless of their derivation.** The use of Latin for scientific names originates from the use of Latin in medieval scholarship. Botanical publications were frequently written in Latin even as late as the middle of the nineteenth century. The use of Latinized names greatly facilitates communication among plant systematists, who belong to diverse cultural and language groups.

6. **The rules of nomenclature are retroactive unless expressly limited.** The ICBN is a system of rules and conventions to be followed, but these do not necessarily have a biological basis.

Requirements for Naming a New Species

The ICBN outlines the steps necessary in describing newly discovered taxa to ensure that the name is validly published; these steps are given here as they apply to species:

1. The species must be named. The name must be in Latin, in a binomial format, and it must not duplicate any name already in existence.
2. The rank of the name must be clearly indicated.
3. A type specimen must be designated.
4. The species must be described in Latin, or described in another language and accompanied by a Latin **diagnosis** (a brief statement of the characters of the species or a comparison with a similar species), or linked to a reference to a place where there is a Latin description. A useful reference for writing Latin descriptions and diagnoses is Stearn's *Botanical Latin* (1992).
5. All of this information must be **effectively published**; that is, it must be presented in a publication that is available to other botanists, such as a botanical journal or book. Publishing a new species in a seed catalogue, newspaper, e-mail message, or other ephemeral source does not qualify as effective publication.

If all of these guidelines are followed, the species name is considered to be **validly published**. However, just because a name is validly published does not mean that it is the correct name for a particular species. For example, the name could be a synonym of an earlier, validly published name. The rules of nomenclature regarding valid publication are relatively straightforward. It is much more difficult to justify the delimitation of a species (as discussed in Chapter 6; see also Winston 1999) than it is to fulfill the technical requirements of the ICBN.

CULTIVATED PLANTS

Plants produced in cultivation through hybridization, artificial selection, or other processes may receive cultivar names. The term **cultivar** is based on a combination of the words *cultivated* and *variety* (these entities were often called "varieties" in the earlier literature). Cultivars should not be confused with botanical varieties, which usually represent naturally occurring geographic races or morphologically distinct populations (see Chapter 6).

The term *cultivar* denotes an assemblage of cultivated plants that is clearly distinguished by particular character(s) (morphological, physiological, cytological, chemical, and so on) and that, following sexual or asexual reproduction, retains its distinguishing feature(s) (Brickell 1980). Cultivar names may be in any language *except*

Latin, and they are written with a capital initial letter. They are preceded by the abbreviation *cv.* or placed in single quotation marks. Cultivar names may be used after generic, specific, or common names. The following, for example, are equivalent designations of the same cultivar: *Citrullus* cv. Crimson Sweet; watermelon cv. Crimson Sweet; *Citrullus lanatus* cv. Crimson Sweet; *Citrullus lanatus* 'Crimson Sweet' (see Jeffrey 1977).

HYBRID NAMES

Hybrids between two species within the same genus may be designated by alphabetical listing of both species names, separated by the multiplication sign (×)—for example, *Verbascum lychnitis* × *V. nigrum*. Alternatively, a hybrid may be described and given its own epithet; for example, *Verbascum* ×*schiedeanum* (= *V. lychnitis* × *V. nigrum*). Hybrids between species in different genera also may be designated by the parental species names separated by the multiplication sign.

Alternatively, they may be represented by a condensed generic formula along with a specific epithet. A condensed generic formula is composed of elements of the generic names and is preceded by a multiplication sign—for example, ×*Dialaeliocattleya* (for an intergeneric hybrid of *Diacrium*, *Laelia*, and *Cattleya*). These conventions are usually not applied to species originating through natural hybridization and polyploidy (see Chapter 6), especially when they reproduce sexually.

Summary

The naming of plants is called botanical nomenclature, and the principles and rules for constructing and using scientific names are contained in the International Code of Botanical Nomenclature (ICBN). Scientific names are in Latin: Species names are composed of two words (that is, they are binomials), and the names of genera, families, and other higher taxa are uninomials (composed of only a single word).

The ICBN recognizes seven major ranks—kingdom, phylum/division, class, order, family, genus, and species—and the ranks above genus have standardized suffixes. The ranks of the taxonomic hierarchy are human constructs, having relative but not absolute meaning, whereas the taxa (monophyletic groups of organisms) placed in these ranks are the products of evolution and, as such, represent real entities. For this reason, some taxonomists have advocated the elimination of ranks.

Botanical nomenclature is independent of zoological nomenclature, although the two codes are similar in their basic principles. In both, the application of names to taxonomic groups is determined by means of nomenclatural types (name-bearing specimens). The correct name for a taxonomic group is the earliest name that is in accordance with the rules of nomenclature; this principle is referred to as priority of publication. Each taxonomic group can bear only one correct name (except in a few specified cases). Special rules govern the naming of hybrids and plants produced in cultivation.

The ICBN outlines the following necessary steps for describing a new species: (1) The species must be named with a Latin binomial. (2) The rank must be clearly indicated. (3) A type specimen must be designated. (4) The species must be described in Latin, or described in another language and provided with a diagnosis in Latin. (5)

All of this information must be effectively published. If these rules are followed, the species name is said to be validly published, and the name can compete for priority with other names. The earliest validly published name is the correct name for any particular species.

We note that it is much more difficult to understand the pattern of variation in nature, and thus to provide biological justification for the delimitation of a species, than it is to fulfill the technical requirements of the ICBN.

Literature Cited and Suggested Readings

Bailey, L. H. 1933. *How plants get their names.* Macmillan, New York. Reprinted 1963, Dover Publications, New York. [Historical development of the binomial system and the rules of nomenclature; information on the construction and pronunciation of scientific names; includes a list of common specific epithets.]

Brako, L., A. Y. Rossman and David F. Farr. 1995. *Scientific and common names of 7,000 vascular plants in the United States.* APS Press, American Phytopathological Society, St. Paul, MN. [Common name(s) for numerous vascular plants.]

Brickell, C. D. (ed.). 1980. *International code of nomenclature for cultivated plants—1980* (Regnum Vegetabile, Vol. 64). Bohn, Scheltema & Holkema, Utrecht, Netherlands.

Bridson, G. D. R. (ed.). 1991. *B-P-H/S: Botanico-periodicum-Huntianum/supplementum.* Hunt Institute for Botanical Documentation, Carnegie Mellon University, Pittsburgh, PA.

Brummitt, R. K. 1992. *Vascular plant families and genera.* Royal Botanic Gardens, Kew, UK. [A listing of genera of vascular plants of the world, both alphabetically and by family, as recognized in the Kew Herbarium.]

Brummitt, R. K. and C. E. Powell. 1992. *Authors of plant names.* Royal Botanic Gardens, Kew, UK. [A list of the personal names of nearly 30,000 botanical authorities, along with standardized abbreviations.]

Cantino, P. D., H. N. Bryant, K. de Queiroz, M. J. Donoghue, T. Eriksson, D. M. Hillis and M. S. Y. Lee. 1999. Species names in phylogenetic nomenclature. *Syst. Biol.* 48: 790–807. [Various methods of using uninomial names for species.]

De Queiroz, K. and J. Gauthier. 1992. Phylogenetic taxonomy. *Annu. Rev. Ecol. Syst.* 23: 449–480.

Farr, E. R., J. A. S. Leussink and F. A. Stafleu (eds.). 1979. *Index nominum genericorum (plantarum)*, 3 vols. (Regnum Vegetabile, Vols. 100–102]. Bohn, Scheltema & Holkema, Utrecht, Netherlands. [A listing of all genera of "plants" and fungi, giving for each the authority, bibliographic citation, type species, and taxonomic placement.]

Greuter, W. 1993. *Family names in current use for vascular plants, bryophytes, and fungi* (Regnum Vegetabile, Vol. 126). Published for the International Association for Plant Taxonomy by Koeltz Scientific Books, Königstein, Germany. [A listing of currently used family names, including authorities, literature citations, and type citations.]

Greuter, W., et al. (ed.). 2000. *International code of botanical nomenclature.* Koeltz Scientific Books, Königstein, Germany.

Greuter, W., R., R. K. Brummitt, E. Farr, N. Kilian, P. M. Koirk and P. C. Silva (eds.). 1993. *Names in current use for extant plant genera* (Regnum Vegetabile, Vol. 129). Published for the International Association for Plant Taxonomy by Koeltz Scientific Books, Königstein, Germany. [A listing of 28,041 generic names currently in use for extant algae, mosses, liverworts, vascular plants, and fungi; each listing includes the name, authority, literature citation, type citation, taxonomic disposition, and other nomenclatural information.]

Hooker, J. D. and B. D. Jackson. 1893–1990. *Index kewensis plantarum phanerogamarum*, 2 vols., 19 suppl. Oxford, UK. [A listing of all validly published names of vascular plants, including bibliographic citations.]

Jeffrey, C. 1977. *Biological nomenclature*, 2nd ed. Crane, Russak & Co., New York. [The legalistic complexities of the various nomenclatural codes presented in an easily understood manner; recommended for those interested in learning more about nomenclatural procedures.]

Kartesz, J. T. 1994. *A synonymized checklist of the vascular flora of the United States, Canada, and Greenland*, 2nd ed., 2 vols. Timber Press, Portland, OR. [A very useful listing of the vascular flora.]

Kartesz, J. T. and J. W. Thieret. 1991. Common names for vascular plants: Guidelines for use and application. *Sida* 14: 421–434. [Guidelines provided for the construction of common names for vascular plants.]

Kron, K. A. and W. S. Judd. 1990. Phylogenetic relationships within the Rhodoreae (Ericaceae) with specific comments on the placement of *Ledum. Syst. Bot.* 15: 57–68.

Lawrence, G. H. M., A. F. G. Buchheim, G. S. Daniels and H. Dolezal (eds.). 1968. *B-P-H: Botanico-periodicum-Huntianum.* Hunt Botanical Library, Pittsburgh, PA. [A listing of all periodical publications that regularly contain (or have contained) articles dealing with botany.]

Linnaeus, C. 1753. *Species plantarum*, 2 vols. Stockholm. [A listing of all species of plants then known, arranged according to Linnaeus's sexual system—that is, classes based on number and arrangement of the stamens; brief descriptions are provided; first consistent use of the binomial system and the starting point for priority for vascular plants.]

Mabberley, D. J. 1997. *The plant-book.* Cambridge University Press, Cambridge, UK. [A listing of all currently accepted generic and family names and commonly used English names of extant vascular plants; generic entries include authorities, family placement, approximate number of species, and geographic range; economically important plants and products are discussed briefly.]

McVaugh, R., R. Ross and F. A. Stafleu. 1968. *An annotated glossary of botanical nomenclature* (Regnum Vegetabile, Vol. 56). International Bureau for Plant Taxonomy and Nomenclature of the International Association for Plant Taxonomy, Utrecht, Netherlands. [Clear definitions of nomenclatural terms.]

Nuttall, T. 1818. *Genera of North American plants*, Vol. 1. Printed for the author by D. Heartt, Philadelphia.

Stafleu, F. A. and R. S. Cowan. 1976–1988. *Taxonomic literature*, 7 vols. (Regnum Vegetabile, Vols. 94, 98, 105, 110, 112, 115, 116). Bohn, Scheltema & Holkema, Utrecht, Netherlands. [An indispensable guide to botanical publications, both books and journal articles that were separately printed, with dates, biographies, and bibliographies of individual authors, and indications of where types of names they described are likely to be found.]

Stafleu, F. A. and E. A. Mennega. 1992–1995. *Taxonomic literature*, Suppl. 1–3 (Regnum Vegetabile, Vols. 125, 130, 132). Koeltz Scientific Books, Königstein, Germany.

Stearn, W. T. 1992. *Botanical Latin*, 4th ed. Timber Press, Portland, OR. [A comprehensive handbook of botanical Latin, summarizing grammar and syntax; contains an illustrated guide to descriptive terminology in both English and Latin, along with an extensive vocabulary.]

Stevens, P. F. 2002. Why do we name organisms? Some reminders from the past. *Taxon* 51: 11–26.

Sytsma, K. J. and J. C. Pires. 2001. Plant systematics in the next 50 years: Re-mapping the new frontier. *Taxon* 50: 713–732.

Willis, J. C. 1973. *A dictionary of the flowering plants and ferns*, 8th ed., revised by H. K. Airy Shaw. Cambridge University Press, London. [A listing of all extant genera of vascular plants, giving authorities, family placement, approximate number of species, and geographic range; brief family descriptions are also provided.]

Winston, J. E. 1999. *Describing species: Practical taxonomic procedure for biologists*. Columbia University Press, New York. [A practical manual for the process of describing new species, including nomenclature, recognizing species, species and subspecies concepts, taxonomic literature, species descriptions, using museum collections, etymology, and key construction.]

APPENDIX 2

Specimen Preparation and Identification

Acollection of dried plant specimens is called an **herbarium** (plural *herbaria*). Such collections are essential for systematic research. Herbarium specimens form the basis for most of our understanding of the pattern of variation in nature. They document the morphological variability of populations, species, and higher taxa, their geographic distributions, and their ecological characteristics, including blooming and fruiting times. In addition, small portions of the specimen can be removed (with permission) to study palynology, ultrastructure, micromorphology, anatomy, and (if the specimen is of sufficient quality) DNA.

Dried plant specimens also serve as **scientific vouchers** to document the presence of a species at a particular locality (in an environmental or floristic study), or the identity of a plant used in an experiment or from which a chromosome count, DNA sample, or chemical extract was obtained. Therefore, it is important to outline the steps involved in collecting, preserving, and identifying plants.

Collecting Plants

The plant collector must record certain facts while in the field. These include (1) locality: country or state, county, or other local governmental unit, distance from roads or cities, and latitude/longitude or section-township-range; (2) date of collection; (3) habitat type, with associated species; (4) elevation, especially in mountainous regions; and (5) any information concerning the plant that will not be evident in the pressed and dried specimen, such as flower color, fragrance of flowers and/or leaves, habit, plant size, presence of colored or milky sap, bark characters, frequency, and pollinators. The collector's name, along with the names of any individuals accompanying the collector, should be included with the field data. This information may be recorded in a field notebook (or collection book), with a tape recorder (for personal dictation), or on a portable computer, for eventual transfer to a specimen label.

The plant or plants collected should be representative of the variation seen in the population. Several specimens may be needed to document the observed pattern of variation adequately. Choose plants that are well developed and not

521

diseased. If possible, the entire plant, including underground parts, should be collected; never collect just one flower and/or leaf! Do not collect specimens lacking flowers and fruits; such material will be difficult to identify.

Large herbs may have to be sharply folded or cut into pieces when they are collected. If so, the parts should be sequentially labeled, and if the plant is too large to keep all of the parts, the collector may decide to keep only certain portions, such as pieces from the base, middle, and apex of the plant. In such cases care must be taken that no important information is lost. It is not necessary to preserve a portion of the root system when collecting trees or shrubs; the collector merely needs to cut off branches that document observed variation in vegetative and reproductive characters.

Certain kinds of plants require special treatment. Succulent plants or large, fleshy fruits can be cut longitudinally and/or in cross-section; preservation may be improved in these species if the plant material is initially killed by immersion in a preservative, such as ethyl alcohol, or in very hot water, or if the plant is first frozen. Floating or submerged aquatics can be floated in a tray of water and carefully lifted onto a piece of paper before further processing. Both staminate and carpellate flowers should be collected for dioecious or monoecious species.

Each plant collected should be assigned a number, which allows information concerning a particular specimen to be kept separate from observations on others. Most plant systematists simply begin with the number 1 and number consecutively throughout their careers, although other numbering systems are possible, such as using a new numbering sequence each year, or for each geographic region—for example, 97-1, 97-2 (for plants collected in 1997) or H-1, H-2 (for plants collected in Hispaniola).

Most systematic botanists consider the following items essential or at least very useful in the field: field notebook, field press or heavy-duty plastic bag, drying press, digging tools, pruning shears, newspapers, 10× hand lens, pocket knife, soft-leaded pencils, and maps. Other useful items include collecting jars or vials and liquid preservatives, camera and film, portable plant dryer, insect repellent, pole pruner, altimeter, compass, global positioning system unit, adhesive tape (for labeling specimens with numbers and for sealing specimen bags), and silica gel granules (to dry material for DNA studies). Figure 1 details the steps in creating an herbarium specimen. Figure 2 shows examples of herbarium specimen labels with pertinent information.

Pressing and Drying Plants

Once a plant has been collected and its field data recorded, the specimen needs to be pressed and dried with a **plant press**. It is important to press the plant before it wilts. A plant press, or drying press, consists of two 12" × 18" pieces of wood (the press panels); two press straps or ropes, which are used to tighten the press; blotters,

which absorb moisture from the plant specimens; and corrugated ventilators (cardboard or aluminum corrugates), which allow air to pass through the press.

The plant specimen is placed in a folded piece of newspaper (or newsprint) and arranged so that upper and lower leaf surfaces, flowers, and/or fruits are visible. The plant material should be sufficient to more or less fill the sheet of newspaper; for tiny plants, several individuals should be gathered, and large plants will need to be folded sharply or cut into several pieces. The specimen number is written on the newspaper or on a tag attached to the plant. This number allows reference to the information in the field notebook. Large, hard fruits may have to be removed from the plant, carefully labeled, and processed separately from the rest of the specimen.

The press is assembled as follows: press frame, corrugated ventilator, blotter, newspaper (with plant specimen), blotter, corrugated ventilator, blotter, newspaper (with plant specimen), blotter, corrugated ventilator, and so on, finishing with the second press frame. The press is tightened with two straps or ropes. The blotter on one side of a specimen with moderate-sized hard fruits, such as acorns or hickory nuts (in its folded piece of newspaper) may be replaced with a piece of foam rubber. The foam distributes the pressure evenly around the fruit, keeping leaves near the fruits from wrinkling. If foam is not available, wadded pieces of tissue paper placed around the fruits will also work. For some rigid plants it may be best to open the press after a day and rearrange the specimens because they will be somewhat more flaccid.

It may be impractical to carry a large drying press to remote collecting localities. Therefore, plant specimens are often carried in a **field press** (press frames with several corrugates and plenty of newspaper), a **vasculum** (plural *vascula*; a cylindrical metal container lined with moist newspaper), or a heavy-duty plastic bag to a place where they can be transferred to a drying press or processed in some other way.

Once in the plant press, the plants can be dried in several ways. The press may be placed in the sun or indoors in a dry place and the blotters changed every day, or it may be tied to the roof rack of a moving vehicle. However-

Figure 1 Preparing and storing herbarium specimens. ▶ (A) Steps in making an herbarium specimen. (1) Find a plant in nature that is well developed and not diseased; record the geographic locality, habitat, and associated plants in your field book. Collect portions of the plant that will be useful in identification. (2) Fold the plant to fit into the press; then fold newspaper over the plant. (3) Tighten the plant press straps. (4) Place the plant press in a dryer (if available). (5) Mount the dried specimen. (6) Identify the specimen using floras and monographs in the herbarium library. (7) For the finished herbarium specimen, transfer the information from the field book to the specimen label. (B) Prepared herbarium specimens are stored in special herbarium cases (left), which may form part of compactor-storage system (right). (Photos by W. S. Judd and R. E. Judd; interior photos taken in the University of Florida Herbarium.)

(A) (1)

(2)

(3)

(4)

(5)

(6)

(7)

(B)

JARDIN BOTANICO NACIONAL "DR. RAFAEL M. MOSCOSO"
SANTO DOMINGO, REPUBLICA DOMINICANA

44523   Melastomataceae

Miconia racemosa (Aubl.)DC

Arbusto 1.5m. de alto; fr. verde, pero
púrpura al madurar; margen del bosque.

República Dominicana: Cordillera Oriental
(cima), Prov. El Seibo: 16.5km. al Sur de
Miches, en la carretera a El Seibo y Pedro
Sánchez; bosque nublado, alterado por fuego
en los últimos 10 años, Clusia rosea,
Alchornea latifolia, Inga fagifolia, Ormosia
krugii, Buchenavia capitata, Sloanea
berteriana, Miconia spp, con manantial.
18°55'N 69°09'Oeste, elev 540-560m.
27 jun. 1990
T. Zanoni, F. Jiménez

**FLORA OF EGLIN AIR FORCE BASE
PLANTS OF FLORIDA**

Aristida condensata Chapm.
< Poaceae >

WALTON COUNTY: T2N, R20W, S29 Eglin Air Force
Base. Mossy Head 7.5' Quad. North of Range C-62 and W. and
E. of Little Alaqua Creek. From intersection of RR 381 and
RR 385, ca. 0.3 mile W. of RR 385. *Pinus palustris*
dominated sandhill community with thick understory of
Quercus laevis. Dominant herbaceous plants include
Andropogon ternarius and *Schizachyrium scoparium*. Tall,
robust, herb. Leaves and inflorescences are tinged with
dark purple. Abundant.

coll. Brenda Herring # 1289   Oct. 7, 1997
    with Jeff McAdoo, Nancy Gobris, and Jim Sheehan
det. B.J.H.
Gift to FLAS from the Nature Conservance / Brenda Herring, 23 Nov. 1998

Figure 2 Examples of labels from herbarium specimens.

Herbarium of the University of Florida
Gainesville, Florida U.S.A.
PLANTS OF DOMINICAN REPUBLIC

Miconia samanensis Urban

DISTRITO NACIONAL: Cordillera Central; Loma
Mariana Chica, ca. 6 km west of Villa Altagracia.
150-775 m elev. Disturbed moist montane forest at peak
(and ridge) of mountain, lower elevations with very
patchy and disturbed moist forest. Branches rigidly
ascending; new growth bright red. Shrub to 4 m tall;
occasional locally (but occurring just on peak and ridge,
ca. 750-775 m); petals pale pink, reflexed; stamens
white, actinomorphic; style white; fruits purple, turning
sky blue at maturity. Leaves with major veins impressed.

coll. Walter S. Judd # 6505   13 May 1992
    with James D. Skean, Jr.
det. W. S. Judd
Voucher-*Miconia* sect. *Chaenopleura* study.

Herbarium of the University of Florida
Gainesville, Florida, U.S.A.
**PLANTS OF: FLORIDA   COUNTY: ALACHUA
FLORA OF PAYNES PRAIRIE STATE PRESERVE**

Passiflora incarnata L.

S. of Gainesville. T10S, R20E, sec 25, SW ¼.
S. side of High Dike, ca. ½ mi W. of
pump station on Camp's Canal. Vine,
common. Sepals·green. Corolla white,
outer corona white and violet striped
in concentric circles,
inner corona violet.
10 May 1981

coll. Janet C. Easterday # 460

---

er, it is preferable to use an artificial heat source. **Plant
dryers** are boxlike or tablelike structures that suspend
the press over a source of heat, provided by metal heat
strips, variable-control hot plates, field stoves, light
bulbs, and/or fans. Too much heat (greater than 45°C)
can discolor the specimens and may even cause the
press and specimens to catch fire.

The press and dryer should be set up so as to allow
the warm air to flow through the plant press and remove
moisture. Large, hard fruits or cones should be removed
from the specimens and dried separately. Succulent or
delicate parts that may be best kept in a liquid preservative, such as alcohol or formalin, can be placed in a plastic container. Delicate flowers may be dipped in liquid
preservative and should be pressed in a folded piece of
tissue paper or nonabsorbent toilet paper. Bulky specimens, such as material with coriaceous leaves and/or

thick, succulent stems, may be placed directly against a
corrugate to speed drying. Aluminum corrugates dry
material more quickly than cardboard ventilators and do
not become dented with use. However, they are more
expensive and heavier.

The plant press should be examined carefully as it
dries, and tightened when necessary. It may also be
turned over if the heating is uneven. Most plant specimens will dry in 1 to 4 days, and the plant material usually will be rigid when dry.

When specimens are dry, they should be carefully
bundled to protect them from breakage. If the moisture
content of the air is high, specimens should be placed
(after cooling) into a tightly sealed plastic bag, to which
silica gel granules may be added.

If drying facilities are not immediately available, the
plant material can be temporarily preserved by the use
of alcohol (or other liquid preservatives). In this method,
a bundle of specimens in folded newspapers is compressed by hand and tied. The tied bundle is placed in a

heavy-duty plastic bag, and about half a liter of 60 to 70% alcohol is sprinkled inside. The bag is then sealed with tape. The alcohol vapors inside the bag will preserve the plant specimens for several weeks, and they can be pressed and dried later. Note, however, that specimens treated in this way will usually be unsuitable for subsequent chemical (including DNA) analysis.

A sealed bag of fresh plant specimens also can be safely kept in the refrigerator for a couple of days.

Conservation and the Law

Many plants that were once common are now rare, primarily due to habitat destruction, overexploitation by people, or competition with exotics (plants introduced from other parts of the world). Collectors should be conservation-minded. Rare or uncommon plants usually should not be collected; photographing the plant is a good alternative in such circumstances.

Many laws protect native plant species, and collecting should be done only with the proper permits (or permission of the landholder, if the plant is growing on private property). Permits are required for collecting in city, county, state, and national parks and forests. For collecting done in a national or state park or in a foreign country, a complete set of specimens should be given to the appropriate authorities. Many states also have laws protecting specific endangered species, and the federal Endangered Species Act protects endangered and threatened species on a national level. No listed species should be collected without the proper permits. And of course, one should never collect on private land without permission of the landowner.

Collecting in foreign countries often requires special governmental permission; local scientists also should be contacted wherever possible. International trade in plant material is covered in part by existing conservation legislation. The Convention on International Trade in Endangered Species of Wild Fauna and Flora (CITES) regulates the shipment of many plant groups, such as Orchidaceae, Cactaceae, Cycadaceae, Cyatheaceae, Nepenthaceae, Sarraceniaceae, and Zamiaceae. CITES-listed plants should not be collected without the proper authorization.

Plant Identification

Plants may be identified after they have been collected, pressed, and dried (as described earlier), but it is easier to identify fresh or liquid-preserved material. Dried material may be dissected more easily if it is first softened by being boiled in water containing a wetting agent, such as aerosol OT (dioctyl sodium sulfosuccinate), or a detergent. Before identifying a plant specimen, one should carefully observe certain features. A 10× hand lens or, even better, a good dissecting microscope will aid in observation, especially of floral features and hair types. Dissection is easier with a pair of dissect-ing needles, sharp-pointed forceps, and single-edged razor blades. A millimeter rule is essential for measuring. Critical observations include (1) habit; (2) leaf arrangement, shape, apex, base, margin, and venation; (3) hair characters; (4) floral features; (5) placentation and ovule number; and (6) fruit type.

KEYS

The systematist has several identification tools available, the most important of which are **dichotomous keys**. A dichotomous key presents the user with a series of choices between two mutually exclusive and parallel statements (leads of a **couplet**). If the user chooses correctly, he or she will be led to the name of the unknown object. The first dichotomous key to plants was published in 1778 by the French botanist Jean-Baptiste de Lamarck, and they have since become ubiquitous (Voss 1952). Dichotomous keys always have a flowchart structure, and they may be written in either an **indented** or a **bracketed** format (Figure 3). Bracketed keys are used throughout this book.

When using a dichotomous key, always read both leads of a couplet, do not guess on measurements, and look up any terms that you do not understand. Be sure to use a hand lens (or dissecting microscope) if necessary. Remember that living things are variable, so be sure to look at several leaves, flowers, or fruits in the process of observing the plant.

When constructing a key, keep in mind that the characters should be precisely defined, measurements should be used whenever possible (don't use terms like *large* and *small*), characters that are constant within a taxon are more useful than those that are variable, and features that are available throughout the growing season or are easily observed are preferred over those that are ephemeral or difficult to see.

Couplets should start with a noun followed by adjectives, and both leads of a couplet should begin with the same word (see Figure 3). The two leads should be parallel; for example, if leaf shape is given in the first, a contrasting condition needs to be presented in the second. Do not use unqualified negative statements; instead use positive statements wherever possible (e.g., "leaf apex acute" versus "leaf apex acuminate" is better than "leaf apex acute" versus "leaf apex not acute").

All parts of the key should be constructed dichotomously. Keys are most efficient when major couplets divide the entire set of included taxa into more or less equal-sized groups. A user is less likely to make an error if two or three characters are used per lead, but overly long leads lead to more mistakes.

A second kind of identification key is the **multi-access key**, or **polyclave**. This kind of key may be constructed through the use of punched cards, with one card per taxon and with the characters represented by holes around the margin of the card. If a taxon possesses a particular character, the hole for that character is notched.

Figure 3   Flowchart, bracketed, and indented keys to five imaginary plants.

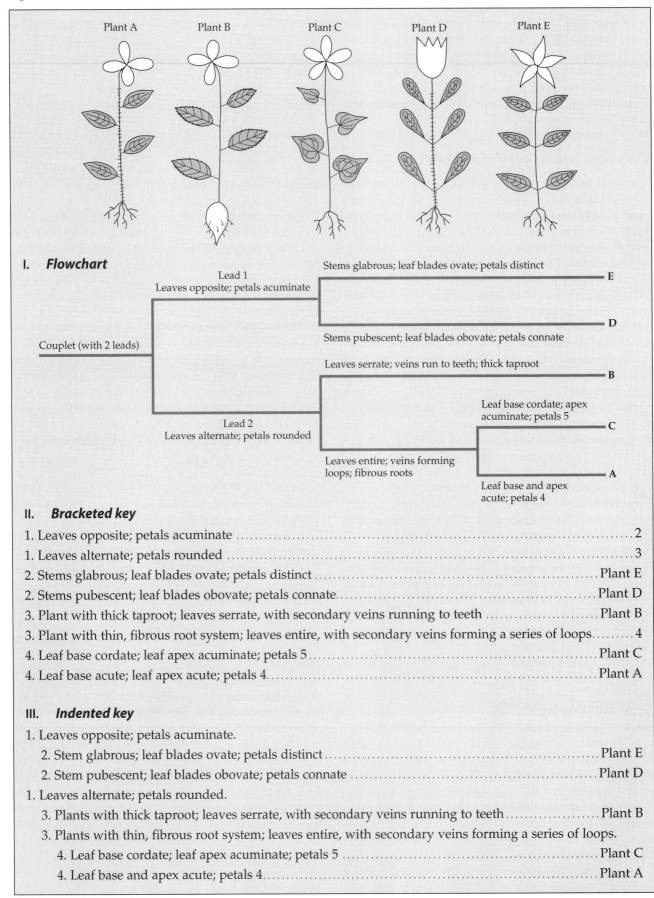

I.   Flowchart

Couplet (with 2 leads)

Lead 1
Leaves opposite; petals acuminate

Stems glabrous; leaf blades ovate; petals distinct — E

Stems pubescent; leaf blades obovate; petals connate — D

Lead 2
Leaves alternate; petals rounded

Leaves serrate; veins run to teeth; thick taproot — B

Leaf base cordate; apex acuminate; petals 5 — C

Leaves entire; veins forming loops; fibrous roots

Leaf base and apex acute; petals 4 — A

II.   Bracketed key

1. Leaves opposite; petals acuminate 2
1. Leaves alternate; petals rounded 3
2. Stems glabrous; leaf blades ovate; petals distinct ....Plant E
2. Stems pubescent; leaf blades obovate; petals connate........................Plant D
3. Plant with thick taproot; leaves serrate, with secondary veins running to teeth ....Plant B
3. Plant with thin, fibrous root system; leaves entire, with secondary veins forming a series of loops........4
4. Leaf base cordate; leaf apex acuminate; petals 5 ....Plant C
4. Leaf base acute; leaf apex acute; petals 4................................Plant A

III.   Indented key

1. Leaves opposite; petals acuminate.
    2. Stem glabrous; leaf blades ovate; petals distinct...........................Plant E
    2. Stem pubescent; leaf blades obovate; petals connate ....Plant D
1. Leaves alternate; petals rounded.
    3. Plants with thick taproot; leaves serrate, with secondary veins running to teeth ....Plant B
    3. Plants with thin, fibrous root system; leaves entire, with secondary veins forming a series of loops.
        4. Leaf base cordate; leaf apex acuminate; petals 5 ....Plant C
        4. Leaf base and apex acute; petals 4................................Plant A

The taxon cards are sorted by use of a large needle. For example, if the cards are stacked and the user runs the needle through the character "leaves alternate," all the cards notched at that spot, which represent species possessing this character, will fall out. Characters may be selected (and the cards sorted) in any sequence, and sorting may be repeated until only one card remains.

Alternatively, each card may represent a single character, with the taxa represented by punched holes. Each taxon possessing a particular feature will have its corresponding hole punched out on that character card. The user of such a key merely selects those cards representing characters evident on his or her specimen, then stacks them and holds them up to the light. When only one hole shows through the stack, that taxon represents the identity of the unknown plant.

Both kinds of punched-card keys operate by a step-by-step elimination process, as does a dichotomous key. However, any selection of characters may be used, and in any order, which is why these keys are called *multi-access*. They are quite advantageous when the plant material being keyed is incomplete. Even if the information on the specimen is not sufficient to identify a single taxon, the user may be able to obtain a short list of possible identities for the unknown plant. Thus, multi-access keys can be used as partial keys.

Polyclaves may also be published in written form. Each character is listed, followed by a list of all taxa (usually represented by a numerical designation) possessing that character (Figure 4). To use the key, one takes the intersection of the sets of taxa specified by a series of characters. Note that a multi-access key, of necessity, contains much more information than a traditional dichotomous key. It contains the data necessary to construct many different dichotomous keys—thus the appropriateness of the term *polyclave*.

It is relatively easy, given the proper computer software, to convert multi-access keys to a computer-interactive format. Computer-assisted identification programs such as the DELTA (Descriptive Language for Taxonomy) system (Pankhurst 1978, 1991; Watson and Dallwitz 1991, 1993; Dallwitz 1992; Askevold and O'Brien 1994) are sure to become more important in the future (Edwards and Morse 1995).

FLORAS AND MONOGRAPHS

Keys are usually presented in books together with plant descriptions, illustrations, distribution maps, and other biological information. It is important to read the appropriate taxon description after using a key; if a specimen has been properly keyed, the description of the identified species (or taxon) should match the plant in hand. The order of information in plant descriptions has been standardized as follows:

1. Habit (the plant as a whole)
2. Underground parts (roots, tubers, bulbs, and so on)
3. Vegetative buds

1. Leaves alternate (ABC)
1. Leaves opposite (DE)
2. Leaf blades ovate (ABCE)
2. Leaf blades obovate (D)
3. Leaves serrate (B)
3. Leaves entire (ACDE)
4. Stems pubescent (AD)
4. Stems glabrous (BCE)
5. Taproot present, thick (B)
5. Taproot lacking (ACDE)
6. Secondary veins forming loops (ACDE)
6. Secondary veins running to marginal teeth (B)
7. Leaf base cordate (C)
7. Leaf base acute or cuneate (ABDE)
8. Petals 4 (ABD)
8. Petals 5 (CE)
9. Petals distinct (ABCE)
9. Petals connate (D)
10. Petals acuminate (DE)
11. Petals apically rounded (ABC)

Figure 4 An example of a multi-access key for the imaginary plants shown in Figure 3.

4. Stems
5. Leaves (including arrangement, structure, petiole, shape of blade, base of blade, margin of blade, apex of blade, and venation)
6. Inflorescences
7. Flower (including symmetry, sexual condition, calyx, corolla, androecium, gynoecium, placentation, ovules, and nectaries)
8. Fruit (including dehiscence and fruit wall characteristics)
9. Seed (including coat, embryo, and endosperm)
10. Seedling
11. Chromosome number

A good general scheme for the description of any of the ten organs or plant parts listed here is the following: number and differentiation, place in relation to other organs, relative position, type of attachment, shape, size, color, surface texture, indumentum or detailed description of hairs, and anatomy (see Leenhouts 1968).

Illustrations, of course, are also useful in plant identification. Methods of botanical illustration have been outlined in several books (e.g., Zweifel 1961; West 1983; Holmgren and Angell 1986; Zomlefer 1994). In plant distribution maps, the range of the taxon is usually indicated by shading of the appropriate geographic region or by placement of dots at localities where the taxon has been collected; both approaches show the taxon's range at a glance (see Figures 4.43 and 6.27).

A **flora** is an account of the plants occurring in a particular area, including keys, descriptions, and illustrations. Floras may be local, applying to a relatively small region (e.g., Proctor 1984) or continental (e.g., Flora of North America Editorial Committee, 1993–2000). A selection of important North American floras is given at the end of this appendix.

In contrast, a **monograph** is a more comprehensive systematic study of a particular taxonomic group (taxon), also including keys, descriptions, and illustrations. A **taxonomic revision** is similar to a monograph, but is less comprehensive. Monographs and revisions usually are published in botanical journals such as those listed in Table 1. Keys, descriptions, and illustrations are now also available on numerous Internet sites.

PLANT SYSTEMATICS ON THE WORLD WIDE WEB

The World Wide Web is an ever growing resource for plant systematics, including a wealth of information about plants, access to floras and checklists, images of plants, detailed treatments of particular groups of plants, DNA sequences and other databases, as well as opportunities for further study and employment in the field. To get the same information or to reach the same Web site there are many starting points and even more paths to the destination. This section compiles leads to a few useful Web sites for plant systematics that can serve as starting points for further exploration.

To start, try the Internet Directory for Botany, or IDB (http://www.botany.net/IDB). The IDB organizes botani-

cal subjects into category lists, such as "Arboreta and Botanical Gardens"; "Botanical Societies, International Botanical Organizations"; "Botanical Museums, Herbaria, Natural History Museums"; "Checklists and Floras, Taxonomical Databases, Vegetation"; "Conservation and Threatened Plants"; and others. The "Checklists, Floras, Taxonomic Databases, Vegetation" subject page (http://www.botany.net/IDB/subject/botflor.html) leads to many other useful sites:

- A searchable database for plant chromosome numbers maintained by the Missouri Botanical Garden (http://mobot.mobot.org/W3T/Search/ipcn.html)
- A site for botanical glossaries (http://155.187.10.12/glossary/glossary.html)
- A database of aquatic, wetland, and invasive plants (http://aquat1.ifas.ufl.edu/database.html)
- An electronic version of the International Code of Botanical Nomenclature (the Tokyo Code of 1993) (http://www.bgbm.fuberlin.de/iapt/nomenclature/code/tokyo-e/)
- The International Plant Names Index (http://www.ipni.org), a database of the names and associated bibliographic details of all seed plants.
- The Missouri Botanical Garden databases (http://www.mobot.org/MOBOT/Research/alldb.shtml) with floras and parts of floras from various parts of the world

The large "global" section of the IDB "Checklists, Floras, Taxonomic Databases, Vegetation" subject page presents an annotated list of major works and projects, organized

---

TABLE 1 *Some important journals containing articles in systematic botany.*

Acta Botanica Mexicana	*Field Museum of Natural History, Botanical Series*
Aliso	*Harvard Papers in Botany*
American Fern Journal	*International Journal of Plant Science (= Botanical Gazette)*
American Journal of Botany	*Journal of Plant Research (= Botanical Magazine, Tokyo)*
Annals of the Missouri Botanical Garden	*Journal of the Arnold Arboretum*
Australian Journal of Botany	*Journal of the Linnaean Society—Botany*
Australian Systematic Botany	*Kew Bulletin*
Austrobaileya	*Madroño*
Blumea	*New Zealand Journal of Botany*
Bothalia	*Nordic Journal of Botany*
Botanical Review	*Novon*
Botanische Jahrbücher	*Phytologia*
Brittonia	*Plant Systematics and Evolution*
Bulletin of the Torrey Botanical Club	*Rhodora*
Canadian Journal of Botany	*Sida*
Castanea	*South African Journal of Botany*
Cladistics	*Systematic Biology*
Contributions from the U.S. National Herbarium	*Systematic Botany*
Darwiniana	*Taxon*
Edinburgh Journal of Botany (= Notes from the Royal Botanical Garden, Edinburgh)	

by continent. A sample of these sites, with one per continent, shows a wide range of plant systematic work:

- Conspectus of the Vascular Plants of Madagascar (http://www.mobot.org/MOBOT/Madagasc/welcome.html)
- Flora of China (http://flora.huh.harvard.edu/china/)
- Australian Flora & Vegetation Statistics (http://155.187.10.12/anbg/australian-flora-statistics.html) at the Australian National Botanic Gardens site
- Atlas Florae Europaeae (http://www.fmnh.helsinki.fi//map/afe/E_afe.htm?pageid=571&language=English)
- Flora of North America (http://hua.huh.harvard.edu/FNA/)
- Andean Botanical Information System (http://www.sacha.org/)

Major herbaria and botanical gardens have much of interest on their Web sites, including the history and nature of the institution, research at the institution, education (such as special courses and graduate study), special events, descriptions and in some cases searchable databases of collections, images, employment opportunities, and links to other sites. Here are some Web sites of several of the world's major herbaria and botanical gardens (sites for some other institutions are noted elsewhere in this section):

- Australian National Botanic Gardens, Canberra (http://www.anbg.gov.au/anbg/index.html)
- Fairchild Tropical Garden, Coral Gables, Florida (www.ftg.org/)
- Jardim Botânico do Rio de Janeiro, Brazil (http://www.jbrj.gov.br/)
- Missouri Botanical Garden, St. Louis (http://www.mobot.org/)
- Muséum National d'Histoire Naturelle, Paris (http://www.mnhn.fr/mnhn/pha/Collect.htm)
- New York Botanical Garden (http://www.nybg.org/)
- Royal Botanic Gardens, Kew, United Kingdom (http://www.rbgkew.org.uk/); a links page at this site (http://www/kew.org.uk/scihort/eblinks/botany.html) is very useful
- Royal Botanic Gardens Sydney, Australia (http://www.rbgsyd.gov.au/)
- Department of Systematic Biology—Botany, National Museum of Natural History, Smithsonian Institution, Washington, DC (http://www.nmnh.si.edu/botany/)

Sites representing floras, such as the one just listed for the Flora of North America (FNA), present a wealth of information on the history, purpose, and scope of floristic projects, as well as on the plants of the flora. The FNA site lists four volumes published at this time. One can search for the families that have been treated in these published volumes and access descriptions of families and genera, and keys to genera and species within the treated families. The site also lists the families that will be treated in the other 26 volumes that are in preparation for publication.

There are hundreds of regional or local floras in the world, and some of them have rich web sites. The Oregon Flora Project (http://www.oregonflora.org/) site, for example, is making the checklist of the plants of Oregon available as an online, searchable database. Currently the Asteraceae portion of the database is available. This site is also developing an Oregon Plant Atlas, with a distribution map of each plant species that has been recorded from the state. These maps will be combinable with different base maps of the state, showing features such as landforms, eco-regions, mean annual precipitation, and vegetation types. Another useful site is the Atlas of Florida Vascular Plants (http://www.plantatlas.usf.edu/about.asp). This site has nomenclatural data and county distribution maps for every species recorded from Florida.

Also useful are Web sites for professional societies, such as the American Society of Plant Taxonomists (ASPT; http://www.sysbot.org/), the Botanical Society of America (BSA; http://www.botany.org/), the International Association for Plant Taxonomy (IAPT; http://www.botanik.univie.ac.at/iapt/), the Society of Systematic Biologists (SSB; http://systbiol.org/), and the Willi Hennig Society (http://www.cladistics.org/). The BSA Web site has a page listing other Web sites of botanical interest (http://www.botany.org/bsa/www-bot.html), including the following categories: "Major Botany Links & Resources on the WWW"; "Jobs and Career Sites"; "How to find Universities, People and Discussion Groups on the Web"; "Funding Sources, Graduate Fellowships"; "Electronic Journals and Newsletters"; and "Internet Resource Guides."

The best source of DNA sequences is GenBank (http://www.ncbi.nlm.nih.gov/GenBank), supported by the National Institutes of Health as part of the National Center for Biotechnology Information (NCBI). This site supports databases that now amount to about 13 billion nucleotide bases from about 100,000 species and that are doubling about every 14 months. GenBank exchanges DNA sequence data on a daily basis with two other large databases: those of the European Molecular Biology Laboratory (EMBL) and the DNA Data Bank of Japan (DDBJ). One accesses sequence information at this site with the "Entrez" facility, a search-and-retrieval system that integrates information from databases at NCBI.

Journals routinely require that, prior to publishing a paper, authors submit DNA sequences (or amino acid sequences) included in a paper to GenBank. A unique alphanumeric sequence is assigned to each submission and is included in published papers using DNA sequences. Plant systematists routinely search GenBank to find out which genes have been sequenced for a set of taxa of interest and to add some previously published DNA sequences to data sets they are compiling. Students are encouraged to compile DNA sequence data for a group of plants that they can use to explore. If the data are the same as those or a portion of those in a published paper, comparisons can be made with the published results.

Some Web sites focus on phylogeny. For example, the Tree of Life Web Project (http://tolweb.org/tree/phylogeny.html) is a collection of about 2000 World Wide Web pages containing information about the diversity of life. These pages are authored by biologists from around the world. Each page contains information about one group, and the pages are linked one to another in the form of the evolutionary tree of organisms, with the pages branching off from a group's page giving information about the subgroups. TreeBASE (http://www.treebase.org/treebase/index.html), a relational database of phylogenetic information, stores phylogenetic trees and the data matrices used to generate them from published research papers.

For some groups of plants covered in this text, detailed systematic treatments are available on the Web. Sites focused on a particular group can have discussions of the work on the group, links to other Web sites with pertinent systematic information about the group, a list of collaborators, access to databases, and lists of publications. Here are three examples:

1. Phylogeny, character evolution, and diversification of extant ferns (http://www.fmnh.org/research_collections/botany/botany_sites/ferns/index.html)
2. Grass Phylogeny Working Group (http://www.ftg.fiu.edu/grass/gpwg/)
3. Neotropical Blueberries (http://www.nybg.org/bsci/res/lut2/), with information on Ericaceae

Many Web sites contain images of plants, which may be useful for those learning to identify plant families, genera, and species. The Texas A&M University Bioinformatics Working Group Vascular Plant Image Gallery (http://www.csdl.tamu.edu/FLORA/gallery.htm), for example, is a broad collection of angiosperms. The University of Wisconsin Virtual Foliage Home Page (http://www.wisc.edu/virtual.html) contains angiosperms, gymnosperms, ferns, and microscopy views. The Plant Systematics Teaching Collection at this site contains over 4000 images of vascular plants organized taxonomically. It is used as a reference in much the same way as an herbarium would be used, and it represents a subset of a larger collection available on CD-ROM. Many of the flora pages and sites already noted have images of some of the plants in the flora. Images are also available for many of the plants in the Missouri Botanical Garden's W^{3}TROPICOS database (http://mobot.mobot.org/W3T/Search/vast.html).

The Web goes far beyond this textbook in many directions, with searchable databases, thousands of images of plants, access to floras, and information about botanical gardens, herbaria, and employment opportunities. The Families of Flowering Plants homepage (http://www.biodiversity.uno.edu/delta/angio/index.html) includes descriptions and representative illustrations for all families of flowering plants. The Angiosperm Phylogeny Web site (http://www.mobot.org/MOBOT/Research/APweb/welcome.html) also holds information on all flowering plant orders and families, and it has the capacity for more frequent updates of information. The current classification of plant orders and families at this site is an elaboration of the Angiosperm Phylogeny Group classification (see Chapter 9). These sites are useful supplements to the information contained in this text.

HERBARIA, BOTANICAL GARDENS, AND TAXONOMIC EXPERTS

Our understanding of the flora of North America is not yet complete, and many tropical regions have yet more incomplete floras. In addition, more than 95% of plant genera have not received recent monographic treatment. Our knowledge of the world's flora clearly is not as advanced as many believe. Thus herbaria, in addition to the uses outlined previously, serve an important identification function.

An herbarium is like a "library" of pressed and dried plant specimens. An unknown specimen can be compared with identified herbarium specimens and identified by the process of comparison. This process can be time-consuming, but in many cases it is the only way a specimen can be identified. Many herbaria operate plant identification and information services. An herbarium is therefore an essential resource for any taxonomic program (involving public service, teaching, or research activities). Major herbaria are listed in Table 2.

Botanical gardens are also useful in plant identification, with the advantage over herbaria that plants can be observed in living condition. Finally, problematic collections may be sent to an expert on the systematics of a particular genus or family or on the flora of a particular region. (Lists of taxonomic experts may be found in Kiger et al. 1981 and Holmgren and Holmgren 1992.) Unfortunately, many genera and families have been studied by no, or too few, taxonomic experts.

Mounting and Processing Herbarium Specimens

Once plant material has been collected, pressed, dried, and identified, it is mounted (with glue, linen adhesive strips, and/or thread) on a standard-sized (in the United States, 11.5" × 16.5") sheet of archive-quality paper. Valuable loose plant parts, such as seeds, fruits, or flower parts, are placed in a small folded packet, which is glued to the sheet. The information from the field notebook is transferred to a typed or computer-generated specimen

TABLE 2 *Herbaria with 3 million or more specimens.*

Herbarium	Location	Number of specimens
Muséum National d'Histoire Naturelle	Paris, France	8,877,300
Royal Botanic Gardens	Kew, United Kingdom	6,000,000
Komarov Botanical Institute	St. Petersberg, Russia	5,770,000
Swedish Museum of Natural History	Stockholm, Sweden	5,600,000
New York Botanical Garden	New York, United States	5,300,000
Natural History Museum	London, United Kingdom	5,200,000
Conservatoire et Jardin Botaniques	Geneva, Switzerland	5,000,000
Harvard University	Cambridge, MA, United States	4,858,000
Smithsonian Institution	Washington, DC, United States	4,368,000
Institute de Botanique	Montpellier, France	4,000,000
Université de Lyon	Villeurbane, France	3,800,000
Naturhistorisches Museum	Vienna, Austria	3,750,000
Herbarium, Missouri Botanical Garden	St. Louis, MO, United States	3,700,000
Universitatis Florentinae,	Museo Botanico Firenze, Italy	3,600,000
Friedrich-Schiller-Universität	Jena, Germany	3,000,000
Rijksherbarium	Leiden, Netherlands	3,000,000

Source: Based on information from Holmgren et al. 1990.

label, giving the plant's name, the geographic locality at which it was collected, habitat, important information about the plant that is not apparent from the dried specimen, and date of collection, along with the collector's name and collection number (see Figure 2). The specimen label is glued to the sheet, usually in the lower right-hand corner. After the specimen has been mounted, the name of the herbarium is stamped on the sheet, documenting ownership, and an accession number is often added for accurate record keeping (see Figure 1A).

Specimens may be filed in the herbarium alphabetically by family, genus, and species, or according to a particular system of classification such as that of A. Engler or A. Cronquist (see Chapter 3 and the CD). Herbarium specimens will last indefinitely if handled with care. They are, however, susceptible to damage by various insects, fungi, and fire, and they should be stored in specialized cabinets (herbarium cases) in a climate-controlled environment (see Figure 1B). (Valuable information regarding the curation and operation of herbaria can be found in Fosberg and Sachet 1965, Bridson and Forman 1992, and Metsger and Byers 1999.)

Summary

A collection of dried plant specimens is called an herbarium; such collections are essential for systematic research and form the basis for our understanding of the pattern of variation in nature. Dried plant specimens serve as scientific vouchers to document the presence of a species at a particular locality, or the identity of a plant that was used in an experiment or from which a chromosome

count, chemical extract, or DNA, pollen, or anatomical sample was obtained.

The plant collector should record pertinent information regarding the geographic locality and important characteristics (e.g., plant height, flower or fruit color and odor, presence of latex, frequency of occurrence) not observable on the dried specimen. The names of the individuals making the collection and the date of collection should also be recorded. All of this information is included in a field notebook or collection book.

The plant collected should be representative of the variation seen in the population, and for herbaceous species generally the complete plant should be preserved. If a plant is too large to keep all of the parts, the collector may decide to keep only certain portions, and in such cases care must be taken that no important information is lost. In collecting trees and shrubs the collector merely needs to cut off branches that document the observed variation in vegetative and reproductive features.

Specimens are preserved by pressing and drying, and they are mounted on archive-quality paper in order to produce an herbarium specimen. A label containing the name of the plant and appropriate information from the field notebook also is affixed to the paper. The identified and mounted specimens are filed in an herbarium and serve as the database for both floristic and monographic (or revisionary) studies. Specimens may be filed in the herbarium alphabetically by family, genus, and species, or according to a particular system of classification. Herbarium specimens should last indefinitely if carefully handled, but they need to be protected from damage by fire, insects, and fungi. Usually they are stored in specialized cabinets (herbarium cases).

A preserved plant can be identified through the use of dichotomous keys, which present the user with a series of choices between two mutually exclusive and parallel statements, although multiple-access keys as employed in computer-assisted identification programs are increasingly popular. Keys, along with descriptions, illustrations, and distribution maps, are an important part of both floras and monographs.

A flora is an account of the plants occurring in a particular geographical area. A monograph is a more comprehensive systematic study of a particular taxonomic group; taxonomic revisions are similar to monographs but are less comprehensive. Both floras and monographs are very important in the process of plant identification. Plant specimens can also be identified by comparison with identified material housed in herbaria. Many herbaria operate identification and information services. Taxonomic experts working at the herbaria of universities, botanical institutes, or gardens are essential in the process of identification of plants on a worldwide basis.

Lists of major herbaria, botanical journals, floras, nomenclatural aids, and Web sites are included in this appendix.

Major North American Floras
(and other important references used in plant identification)

Abrams, L. 1923–1960. *Illustrated flora of the Pacific states: Washington, Oregon and California*, 4 vols. Stanford University Press, Stanford, CA. [Vol. 4 by L. Abrams and R. S. Ferris.]

Acevedo-Rodríguez, P. 1996. *Flora of St. John.* New York Botanical Garden, Bronx.

Adams, C. D. 1972. *Flowering plants of Jamaica*. University of the West Indies, Mona, Jamaica.

Allen, P. H. 1956. *The rain forests of Golfo Dulce*. Stanford University Press, Stanford, CA.

Bailey, L. H. 1949. *Manual of cultivated plants*, rev. ed. Macmillan, New York.

Batson, W. T. 1984. *A guide to the genera of the plants of eastern North America*, 3rd ed. University of South Carolina Press, Columbia.

Bisse, J. 1981. *Arboles de Cuba*. Ministerio de Cultura, Editorial Científico-Técnica, Havana, Cuba.

Calder, J. A. and R. L. Taylor. 1968. *Flora of the Queen Charlotte Islands*, 2 vols. (Canada Department of Agriculture, Research Branch, Monograph No. 4). Queen's Printer, Ottawa, Ontario.

Clewell, A. F. 1985. *Guide to the vascular plants of the Florida Panhandle*. University Presses of Florida, Tallahassee.

Correll, D. S. and H. B. Correll. 1972. *Aquatic and wetland plants of southwestern United States*. Environmental Protection Agency, Washington, DC.

Correll, D. S. and H. B. Correll. 1982. *Flora of the Bahama Archipelago*. J. Cramer, Vaduz.

Correll, D. S and M. C. Johnston. 1970. *Manual of the vascular plants of Texas*. Texas Research Foundation, Renner, TX.

Croat, T. B. 1978. *Flora of Barro Colorado Island*. Stanford University Press, Stanford, CA.

Cronquist, A. 1980. *Vascular flora of the southeastern United States*, Vol. 1: Asteraceae. University of North Carolina Press, Chapel Hill.

Cronquist, A., A. H. Holmgren, N. H. Holmgren, J. L. Reveal and P. K. Holmgren. 1972–1997. *Intermountain Flora*, Vols. 1, 3A, 3B, 4, 5, and 6. New York Botanical Gardens, Bronx.

Crow, G. E. and C. B. Hellquist. 2000. *Aquatic and wetland plants of northeastern North America*, 2 vols. University of Wisconsin Press, Madison.

Davis, P. H. and J. Cullen. 1979. *The identification of flowering plant families*, 2nd ed. Cambridge University Press, Cambridge, UK.

Diggs, G. M., Jr., B. L. Lipscomb and R. J. O'Kennon. 1999. *Shinners and Mahler's illustrated flora of north central Texas* (Sida, Botanical Miscellany, No. 16). Botanical Research Institute of Texas, Fort Worth.

Dorn, R. D. 1977. *Manual of vascular plants of Wyoming*, 2 vols. Garland, New York.

Duncan, W. H. and M. B. Duncan. 1988. *Trees of the southeastern United States*. University of Georgia Press, Athens.

Elias, T. S. 1980. *The complete trees of North America*. Van Nostrand Reinhold, New York.

Fassett, N. C. 1940. *A manual of aquatic plants.* McGraw-Hill, New York.

Fassett, N. C. 1976. *Spring flora of Wisconsin*, 4th ed. University of Wisconsin Press, Madison.

Fernald, M. L. 1950. *Gray's manual of botany*, 8th ed. American Book Co., New York.

Flora of North America Editorial Committee. 1993–2000. *Flora of North America north of Mexico*, Vols. 1–3, 22. Oxford University Press, New York.

Geesink, R., A. J. M. Leeuwenberg, C. E. Ridsdale and J. F. Veldkamp. 1981. *Thonner's analytical key to the families of flowering plants* (Leiden Botanical Series, Vol. 5). Leiden University Press, The Hague, Netherlands.

Gleason, H. A. 1952. *The new Britton and Brown illustrated flora of the northeastern United States and adjacent Canada*. New York Botanical Garden, New York.

Gleason, H. A. and A. Cronquist. 1991. *Manual of vascular plants of the northeastern United States and adjacent Canada*, 2nd ed. New York Botanical Garden, Bronx.

Godfrey, R. K. 1988. *Trees, shrubs, and woody vines of northern Florida and adjacent Georgia and Alabama*. University of Georgia Press, Athens.

Godfrey, R. K. and J. W. Wooten. 1979–1981. *Aquatic and wetland plants of the southeastern United States*, 2 vols. University of Georgia Press, Athens.

Gould, F. W. 1951. *Grasses of southwestern United States* (University of Arizona, Biological Science Bulletin, No. 7). University of Arizona, Tucson.

Great Plains Flora Association. 1986. *Flora of the Great Plains*. University Press of Kansas, Lawrence.

Haines, A. D. and T. F. Vining. 1998. *Flora of Maine*. V. F. Thomas, Bar Harbor, ME.

Hickman, J. C. (ed.). 1993. *The Jepson manual: Higher plants of California*. University of California Press, Berkeley.

Hinds, H. R. 1986. *Flora of New Brunswick*. Primrose Press, Fredericton, New Brunswick.

Hitchcock, C. L., A. Cronquist, M. Owenbey and J. W. Thompson. 1955–1969. *Vascular plants of the Pacific Northwest*, 5 vols. University of Washington Press, Seattle.

Holmgren, N. H. 1998. *The illustrated companion to Gleason and Cronquist's manual*. New York Botanical Garden, Bronx.

Howard, R. A. 1974–1989. *Flora of the Lesser Antilles*, 6 vols. Arnold Arboretum, Harvard University, Jamaica Plain, MA.

Hutchinson, J. 1968. *Key to the families of flowering plants of the world*. Clarendon Press, Oxford, UK.

Isley, D. 1990. *Vascular flora of the southeastern United States*, Vol. 3, Part 2: Leguminosae. University of North Carolina Press, Chapel Hill.

Isely, D. 1998. *Native and naturalized Leguminosae (Fabaceae) of the United States (exclusive of Alaska and Hawaii)*. Brigham Young University, Provo, UT.

Kearney, T. H. and R. H. Peebles. 1960. *Arizona flora*, 2nd ed. University of California Press, Berkeley.

Lellinger, D. B. 1985. *A field manual of the ferns and fern-allies of the United States and Canada*. Smithsonian Institution Press, Washington, DC.

Leon, Bro. and Bro. Alain. 1946–1962. *Flora de Cuba*, 5 vols. Cultural, Havana, Cuba.

Liogier, A. H. 1982–2000. *La flora de la Hispaniola*, 9 vols. University Central del

Este, San Pedro de Macorís, Dominican Republic.

Liogier, A. H. 1985–1995. *Descriptive flora of Puerto Rico and adjacent islands.* Universidad de Puerto Rico, Río Piedras.

Little, E. L., Jr., and F. H. Wadsworth. 1964. *Common trees of Puerto Rico and the Virgin Islands,* Vol. 1 (Agriculture Handbook, No. 249). U.S. Department of Agriculture, Washington, DC.

Little, E. L., Jr., R. O. Woodbury and F. H. Wadsworth. 1974. *Trees of Puerto Rico and the Virgin Islands,* Vol. 2 (Agriculture Handbook, No. 449). U.S. Department of Agriculture, Washington, DC.

Long, R. W. and O. Lakela. 1976. *A flora of tropical Florida.* Banyan Books, Miami, FL.

Looman, J. and K. F. Best. 1979. *Budd's flora of the Canadian Prairie Provinces* (Canada Department of Agriculture, Publication 1662). Research Branch, Agriculture Canada, Ottawa, Ontario.

Marie-Victorin, F. 1964. *Flore Laurentienne,* 2nd ed. Les Presses de l'Université de Montréal, Montreal, Quebec.

McVaugh, R. 1983–1993. *Flora Novo-Galiciana,* Vols. 5, 12–17. University of Michigan Herbarium, Ann Arbor.

Mohlenbrock, R. H. 1967–1990. *The illustrated flora of Illinois.* Southern Illinois University Press, Carbondale.

Moss, E. H. 1983. *Flora of Alberta,* 2nd ed. University of Toronto Press, Toronto, Ontario.

Munz, P. A. 1968. *Supplement to A California flora.* University California Press, Berkeley.

Munz, P. A. and D. D. Keck. 1959. *A California flora.* University of California Press, Berkeley.

Organization for Flora Neotropica. 1968–present. *Flora Neotropica.* New York Botanical Garden, Bronx.

Pease, A. S. 1964. *Flora of northern New Hampshire.* New England Botanical Club, Cambridge, MA.

Proctor, G. R. 1984. *Flora of the Cayman Islands* (Kew Bulletin Additional Series, 11). H.M.S.O., London.

Proctor, G. R. 1985. *Ferns of Jamaica.* British Museum (Natural History), London.

Radford, A. E., H. E. Ahles and C. R. Bell. 1964. *Manual of the vascular flora of the Carolinas.* University of North Carolina Press, Chapel Hill.

Rehder, A. 2001. *Manual of cultivated trees and shrubs: Hardy in North America* (reprint of 2nd ed.). Blackburn, Caldwell, NJ.

Roland, A. E. and E. C. Smith. 1966–1969. The flora of Nova Scotia (revised ed., 2 parts). *Proc. Nova Scotian Inst. Sci.* 26(2): 3–224; 26(4): 277–743.

Sargent, C. S. 1922. *Manual of the trees of North America,* 2 vols. Houghton Mifflin, New York.

Scoggan, H. J. 1957. *Flora of Manitoba* (National Museum of Canada Bulletin, No. 140). Department of Northern Affairs and National Resources, Ottawa, Ontario.

Scoggan, H. J. 1978. *The flora of Canada,* 4 parts. National Museum of Natural Sciences, National Museums of Canada, Ottawa.

Seymour, F. C. 1982. *Flora of New England,* 2nd ed. (Phytologia Memoirs, 5). Moldenke, Plainfield, NJ.

Shreve, F. and I. L. Wiggins. 1964. *Vegetation and flora of the Sonoran Desert,* 2 vols. Stanford University Press, Stanford, CA.

Simpson, D. R. and D. Janos. 1974. *Punch card key to the families of dicotyledons of the Western Hemisphere south of the United States.* Field Museum of Natural History, Chicago.

Small, J. K. 1933. *Manual of the southeastern flora.* Author, New York.

Sosa, V. (chief ed.). 1978–2000. *Flora de Veracruz,* 118 fasc. INIREB, Xalapa, Mexico.

Standley, P. C. 1920–1926. *Trees and shrubs of Mexico,* 5 vols. Government Printing Office, Washington, DC.

Standley, P. C. 1937–1939. *Flora of Costa Rica,* 2 vols. (Botanical Series, Field Museum of Natural History, Vol. 18). Chicago.

Standley, P. C., et al. 1946–1977. *Flora of Guatemala,* 13 parts (Chicago Natural History Museum, Fieldiana: Botany, Vol. 24). Chicago.

Stevens, W. D., C. U. Ulloa, A. Pool and O. M. Montiel (eds.). 2001. *Flora de Nicaragua,* 3 vols. Missouri Botanical Garden Press, St. Louis.

Steyermark, J. A. 1963. *Flora of Missouri.* Iowa State University Press, Ames.

Strausbaugh, P. D. S. and E. L. Core. 1971. *Flora of West Virginia,* 2nd ed. Seneca Books, Grantsville, WV.

Swink, F. 1990. *The key to the vascular flora of the northeastern United States and southeastern Canada.* Plantsmen's, Flossmoor, IL.

Tomlinson, P. B. 1980. *The biology of trees native to tropical Florida.* Author, Allston, MA.

Vines, R. A. 1960. *Trees, shrubs and woody vines of the Southwest.* University of Texas Press, Austin.

Voss, E. G. 1972–1997. *Michigan flora: A guide to the identification and occurrence of the native and naturalized seed plants of the state,* 3 vols. (Cranbrook Institute of Science Bulletin, 55, 59, 61). Cranbrook Institute of Science, Bloomfield Hills, MI.

Wagner, W. L., D. R. Herbst and S. H. Sohmer. 1990. *Manual of the flowering plants of Hawai'i,* 2 vols. University of Hawaii Press, Bishop Museum Press, Honolulu.

Weber, W. A. 1976. *Rocky Mountain flora,* 5th ed. Colorado Associated University Press. Boulder.

Welsh, S. L. 1974. *Anderson's flora of Alaska and adjacent parts of Canada.* Brigham Young University Press, Provo, UT.

Wiggins, I. L. 1980. *Flora of Baja California.* Stanford University Press, Stanford, CA.

Wood, C. E., Jr., et al. 1958–present. Generic flora of the southeastern United States. *J. Arnold Arbor.; Harvard Pap. Bot.*

Woodson, P. E., Jr., R. W. Schery, et al. 1943–1981. Flora of Panama. *Ann. Missouri Bot. Gard.* Vols. 30–67: *passim.*

Wunderlin, R. P. 1998. *Guide to the vascular plants of Florida.* University Press of Florida, Gainesville.

Wunderlin, R. P. and B. F. Hansen. 2000. *Flora of Florida,* Vol. 1. University Press of Florida, Gainesville.

Other Useful References

Askevold, I. S. and C. W. O'Brien. 1994. DELTA, an invaluable computer program for generation of taxonomic monographs. *Ann. Entomol. Soc. Am.* 87: 1–16.

Bridson, D. and L. Forman. 1992. *The herbarium handbook,* rev. ed. Royal Botanic Gardens, Kew, UK.

Dallwitz, M. J. 1974. A flexible computer program for generating diagnostic keys. *Syst. Zool.* 23: 50–57.

Dallwitz, M. J. 1992. DELTA and INTKEY. In *Proceedings of workshop on artificial intelligence and modern computer methods for systematic studies in biology,* R. Fortuner (ed.), University of California, Davis.

Dallwitz, M. J., T. A. Paine and E. J. Zurcher. 1993. *User's guide to the DELTA system: A general system for processing taxonomic descriptions,* 4th ed. CSIRO, Division of Entomology, Canberra, Australia.

DeWolf, G. P., Jr. 1968. Notes on making an herbarium. *Arnoldia* 28: 69–111.

Dransfield, J. 1986. A guide to collecting palms. *Ann. Missouri Bot. Gard.* 73: 166–176.

Edwards, M. and D. R. Morse. 1995. The potential for computer-aided identification in biodiversity research. *Trends Ecol. Evol.* 10: 152–158.

Fosberg, F. R. and M.-H. Sachet. 1965. *Manual for tropical herbaria* (Regnum Vegetabile, Vol. 39). International Bureau for Plant Taxonomy and Nomenclature, Utrecht, Netherlands.

Frodin, D. G. 2001. *Guide to standard floras of the world: An annotated, geographically arranged systematic bibliography of the principal floras, enumerations, checklists, and chorological atlases of different areas,* 2nd ed. Cambridge University Press, Cambridge, UK.

Haynes, R. R. 1984. Techniques for collecting aquatic and marsh plants. *Ann. Missouri Bot. Gard.* 71: 229–231.

Henderson, D. M. (ed.). 1983. *International directory of botanical gardens IV,* 4th ed. Koeltz Scientific Books, Königstein, Germany.

Holmgren, N. H. and B. Angell. 1986. *Botanical illustration: Preparation for publication.* New York Botanical Garden, Bronx.

Holmgren, P. K. and N. H. Holmgren. 1992. *Plant specialists index.* Koeltz Scientific Books, Königstein, Germany. [Index to specialists in the systematics of plants and fungi.]

Holmgren, P. K., N. H. Holmgren and L. C. Barnett. 1990. *Index Herbariorum,* Part I: *The herbaria of the world,* 8th ed. (Regnum Vegetabile, Vol. 120). New York Botanical Garden, Bronx.

Howard, R. A. 1969. The botanical garden—An unexploited source of information. *Boissiera* 14: 109–117.

Kiger, R. W., T. D. Jacobsen and R. M. Lilly. 1981. *International register of specialists and current research in plant systematics.* Hunt Institute for Botanical Documentation, Carnegie-Mellon University, Pittsburgh, PA.

Kudsen, J. W. 1972. *Collecting and preserving plants and animals.* Harper and Row, New York.

Lawrence, G. H. M. 1951. *Taxonomy of vascular plants*. Macmillan, New York. [Contains detailed chapters on plant identification, field and herbarium techniques, monographs and revisions, and floristics; see pp. 223–283.]

Leenhouts, P. W. 1968. *A guide to the practice of herbarium taxonomy* (Regnum Vegetabile, Vol. 58). International Bureau for Plant Taxonomy and Nomenclature, Utrecht, Netherlands.

Lot, A. and F. Chiang. 1986. *Manual de herbario*. Consejo Nacional de la Flora de México, Mexico.

Metsger, D. A. and S. C. Byers. 1999. *Managing the modern herbarium—An inter-disciplinary approach*. Society for the Preservation of Natural History Collections, Washington, DC.

Pankhurst, R. J. (ed.). 1975. *Biological identification with computers* (Systematics Association, Special Volume, No. 7). Academic Press, London.

Pankhurst, R. J. 1978. *Biological identification*. University Park Press, Baltimore, MD.

Pankhurst, R. J. 1991. *Practical taxonomic computing*. Cambridge University Press, Cambridge, UK.

Pohl, R. W. 1965. Dissecting equipment and materials for the study of minute plant structures. *Rhodora* 67: 95–96.

Robertson, K. R. 1980. *Observing, photographing, and collecting plants* (Circular 55). Illinois Natural History Survey, Urbana.

Smith, C. E., Jr. 1971. *Preparing herbarium specimens of vascular plants* (Agriculture Information Bulletin No. 348). Agricultural Research Service, U.S. Department of Agriculture, Washington, DC.

Soderstrom, T. R. and S. M. Young. 1983. A guide to collecting bamboos. *Ann. Missouri Bot. Gard.* 70: 128–136.

Stern, M. J. and T. Eriksson. 1996. Symbioses in herbaria: Recommendations for more positive interactions between plant systematists and ecologists. *Taxon* 45: 49–58.

Tilling, S. M. 1984. Keys in biological identification: Their role and construction. *J. Biol. Educ.* 18: 293–304.

Voss, E. G. 1952. The history of keys and phylogenetic trees in systematic biology. *J. Sci. Lab. Denison Univ.* 43: 1–25.

Watson, L. and M. J. Dallwitz. 1991. The families of angiosperms: Automated descriptions, with interactive identification and information retrieval. *Aust. Syst. Bot.* 4: 681–695.

Watson, L. and M. J. Dallwitz. 1992. *Grass genera of the world*. C.A.B. International, Wallingford, UK.

Watson, L. and M. J. Dallwitz. 1993. *The families of flowering plants: Interactive identifications and information retrieval* (CD-ROM version 1.0 for MS-DOS). CSIRO, Melbourne, Australia.

West, K. 1983. *How to draw plants*. Herbert Press, London.

Zomlefer, W. B. 1994. *Guide to flowering plant families*. University of North Carolina Press, Chapel Hill.

Zweifel, F. W. 1961. *A handbook of biological illustration*. University of Chicago Press, Chicago.

Glossary

Abaxial surface Surface facing away from the apex of the axis.

Acaulescent plant Plant that is apparently stemless; sometimes the stem is subterranean.

Accessory bud Extra bud (or buds) produced on either side of or above or below the axillary bud.

Accessory fruit Fruit (or a cluster of fruits) in which structures in addition to the matured gynoecium form a functional part of the fruit.

Achene Fairly small, indehiscent, dry fruit with a thin and close-fitting wall surrounding but free from the single seed.

Acorn Fruit of species belonging to the genus *Quercus*; a nut associated with a variously scaly cupule.

Actinomorphic *See* Radial symmetry.

Acuminate apex Apex with somewhat concave sides that taper to a sharp point (see Figure 4.10).

Acute apex Apex ending in a point of less than 90 degrees, with the sides of the tapered apex more or less straight to slightly convex (see Figure 4.10).

Acute base Base ending in a point of less than 90 degrees, with the sides of the tapered base slightly convex (see Figure 4.11).

Adaptation Any change in an organism resulting from natural selection.

Adaxial surface Surface facing toward the apex of the axis.

Adnate/adnation Fusion of unlike parts, as stamens with the corolla.

Adventitious embryony Development of the embryo directly from a somatic cell of the ovule without formation of a gametophyte.

Adventitious Developing from any plant part other than the normal one, e.g., adventitious roots develop from stems or leaves, not from the embryonic root (radicle).

Aerial root Root occurring above ground or water.

Agamic complex Group of species with two to several taxa that reproduce asexually.

Agamospecies Minimally differentiated individuals that reproduce asexually.

Agamospermy Asexual seed production; not involving fertilization of egg by sperm.

Aggregate fruit Fruit that develops from several separate carpels of a single flower.

Alkaloids Structurally diverse, physiologically active secondary compounds; derived from various amino acids or from mevalonic acid by various biosynthetic pathways (see pages 96–97).

Allele One of two or more molecular forms of a gene.

Allopatric Occurring in different geographical regions.

Allopatric speciation Divergence of geographically isolated populations, leading to evolution of new species.

Allopolyploid speciation Formation of a new species via hybridization followed by chromosome doubling in the hybrid.

Allopolyploidy Presence in a cell of two or more genomes from different species.

Alternate leaves Leaves borne one leaf per node along the stem (see Figure 4.4); such leaves may be spiral, 2-ranked, etc.

Ament *See* Catkin.

Amoeboid Having the form of an amoeba, i.e., with variable shape and various rounded extensions.

Amphiparacytic Stomate with guard cells surrounded by two pairs of subsidiary cells, each parallel to the guard cells (see Figure 4.35).

Anagenesis Evolutionary change within a single lineage over time.

Anatomy Study of the internal structure (e.g., cells, tissues).

Anatropous ovule   Ovule that is inverted and fused to the funiculus (stalk) so that the micropyle (opening) is situated next to the funiculus (see Figure 4.39).

Androecium   Collective term for all the stamens in a flower (see Figure 4.16).

Androgynophore   Stalk bearing the androecium (stamens) and gynoecium (carpels).

Aneuploidy   Having a chromosome number that is not an even multiple of the base number.

Anisocytic   Stomate surrounded by three subsidiary cells, of which one is distinctly smaller (see Figure 4.35).

Anisophylly   When the two leaves of a pair, or alternating leaves of the same flush, are diverse in shape or size.

Annual   Plant growing from seed to maturity within a single growing season.

Annulus   Band or cluster of thick-walled, hygroscopic cells in the sporangium wall of most ferns; facilitates release of spores.

Anomalocidal capsule   Capsule that opens irregularly.

Anomocytic   Stomate lacking differentiated subsidiary cells (see Figure 4.35).

Anther   Pollen-bearing part of the stamen, borne at the top of the filament (see Figure 4.16).

Antheridium   Male reproductive structure producing and protecting the sperm in embryophytes (land plants).

Antheridogens   Compounds released by archegonia of embryophytes that stimulate neighboring gametophytes to produce antheridia.

Anthocyanins   Violet, blue, red, and yellow flavonoid pigments not containing nitrogen; widespread among angiosperms.

Antibody   One of a diverse array of antigen-binding receptors.

Antigen   Molecule that white blood cells recognize as foreign and that triggers an immune response.

Antipodals   Three cells of the female gametophyte clustered at the end opposite the egg and synergids.

Aperture of pollen   Thin, variously shaped region of the pollen wall through which the pollen tube emerges during germination.

Apetalous   Lacking petals.

Apex (pl. apices)/apical   Tip of a structure.

Apical meristem   Group of dividing cells at the growing apex of a stem or root.

Apical placentation   Ovule or ovules attached at the apex of the ovary (see Figure 4.22).

Apocarpous   Carpels separate.

Apomixis   Asexual reproduction, usually used for asexual seed production (see agamospermy) but also applied to the production of plantlets by vegetative propagules (e.g., gemmae).

Apomorphy   Derived character state; also called an apomorphic character (or character state).

Apopetalous   Petals separate.

Aposepalous   Sepals separate.

Apotepalous   Tepals separate.

Arachnoid   Having a cobwebby appearance.

Archegonium (pl. archegonia)   Female reproductive structure producing and protecting the egg in embryophytes (land plants).

Areole   Spiny short-shoot characteristic of the Cactaceae.

Aril   A hard to juicy, often brightly colored outgrowth of the seed; usually developed from the funiculus or the outer integument.

Aromatic   Having a distinct, usually pleasant odor.

Articulate/articulation   Jointed.

Asexual reproduction   *See* Apomixis.

Asymmetrical flower   Flower lacking a plane of symmetry, i.e., neither radial nor bilateral.

Asymmetrical leaf   Leaf lacking a plane of symmetry; often wider on one side than the other, with widest point differently positioned on each side of the blade, and with an oblique base.

Attenuate apex   Apex tapering gradually to a narrow tip (see Figure 4.10).

Autapomorphy   Derived character state restricted to a single terminal taxon (in a cladistic analysis).

Authority. In nomenclature, the citation following the taxon name and consisting of the name or names of individuals who described the taxon or transferred it from one rank to another.

Autopolyploidy   Presence in a cell of three or more chromosome sets from the same species.

Awn   Narrow, bristlelike appendage, often at the tip of a structure such as a fruit or anther.

Axil   Region between the adaxial side of a leaf and the stem; literally, armpit.

Axile placentation   Ovules attached to the central axis of an ovary with two or more locules (see Figure 4.22).

Axillary branch   Branch that develops from an axillary bud.

Axillary bud   Bud located in the leaf axil (see Figure 4.2).

Backcrossing   Hybridization between F_1 hybrid individuals and members of either parental species.

Banner petal   Distinctive adaxial petal, usually the largest, of flowers, especially those of Fabaceae subfam. Faboideae; also called the **standard** or **flag petal**.

Bark   All tissues outside the vascular cambium in older trees, shrubs, or lianas.

Basal placentation   Ovule or ovules attached at the base of the ovary (see Figure 4.22).

Base number   Inferred ancestral haploid chromosome number of a taxon, usually prefixed by the letter "x", as in "$x =$".

Basifixed anther   Anther that is attached at its base to the filament.

Berry   Indehiscent, fleshy fruit with (one or) a few to many seeds; the flesh may be more or less homogeneous, or the outer part of the fruit may be firm, hard, or leathery.

Betalains   Nitrogenous red and yellow pigments; restricted to the Caryophyllales.

Bicollateral vascular bundle   Vascular bundle with phloem on two sides (abaxial and adaxial) of the xylem (see figure 4.38).

Biennial   Plant that lives for two years, growing vegetatively during the first and flowering/fruiting in the second.

Bifacial leaf   Leaf having anatomically distinct adaxial and abaxial surfaces.

Bijugate leaves   Opposite leaves in which those of adjacent nodes are rotated slightly less/more than 90 degrees.

Bilabiate   Two-lipped; typically applied to a calyx or corolla.

Bilateral symmetry   Divisible into equal halves by only one plane of symmetry; also called **zygomorphic** (see Figure 4.18).

Binary character   Character with only two character states.

Binomial nomenclature   System of naming developed by Linnaeus in which each species name is composed of two words: the genus name, a noun, and the specific epithet, an adjective or noun in apposition, e.g., *Acer rubrum.*

Biological nomenclature   Rules for naming living organisms as outlined in the various nomenclatural codes.

Biradial symmetry   Divisible into equal halves by only two planes of symmetry; compare with radial symmetry.

Biseriate   With two rows or whorls of structures, as a flower with two perianth whorls.

Biseriate hair   Hair composed of two rows of cells.

Bisexual flower   Flower with both androecium (stamens) and gynoecium (carpels); also called **perfect flower**.

Bivalents   Pairs of homologous chromosomes in a cell during meiosis.

Blade   Flat, photosynthetic portion of a leaf, also called the lamina (see Figure 4.3), or leaflets, in a compound leaf (see Compound leaf).

Bole   *See* Trunk.

Bootstrap analysis   Statistical technique that uses resampling and replication of characters to assess how well the data set support a given tree topology.

Botanical nomenclature   Rules for naming plants as outlined in the *International Code of Botanical Nomenclature.*

Bracketed key   Key in which the two leads of a couplet (pair of contrasting choices) are placed on adjacent lines so that the leads of a couplet are not separated by intervening lines (compare, Indented key).

Bract   A reduced leaf, in the axil of which arises a flower or an inflorescence branch.

Bracteole   Prophyll of a flower, usually very small, borne below the flower on a pedicel; also called a bractlet.

Bremer support   *See* Decay index.

Bud   Small embryonic shoots, whether floral or vegetative, often protected by modified leaves (bud scales), stipules, or hairs.

Bud-scale   Reduced, protective leaf associated with a bud.

Bud-scale scar   Scar left after a bud-scale falls from the plant.

Bulb   Short, erect, underground stem surrounded by thick, fleshy leaves or leaf bases.

C$_3$ plants   Plants having the photosynthetic pathway in which the first intermediate formed after fixation of carbon dioxide is the three-carbon compound phosphoglycerate.

C$_4$ plants   Plants having the photosynthetic pathway in which the first intermediate formed after fixation of carbon dioxide is the four-carbon compound oxaloacetate.

Calcium oxalate   Oxalic acid ($HO_2C \cdot CO_2H$) associated with calcium, often occurring in plant tissues as raphide or druse crystals (see Figure 4.37).

Callose   Glucose polymer, e.g., deposited during development in sieve areas of sieve-tube elements.

Calyculus   Reduced, rimlike calyx, as in members of Loranthaceae.

Calyx   Collective term for all sepals of a flower (see Figure 4.16).

Campylotropous ovule   Ovule that is curved so the micropyle is positioned near the funiculus (see Figure 4.39).

Canescent   Gray, with dense, short hairs.

Capitate   Enlarged at the tip, i.e., headlike.

Capitulum   *See* Head.

Capsule   Dry to rarely fleshy fruit from a two- to many-carpellate gynoecium that opens in various ways to release the seed or seeds.

Carbocyclic iridoids   Iridoids that have two ring systems, one composed entirely of carbon (see page 98.)

Cardenolides   Highly poisonous glucosides of a type of 23-carbon steroid.

Carnivorous   Capturing animals and digesting them, as applied to plants.

Carpel   Ovule-bearing unit(s) that make up the gynoecium.

Carpellate flower   Flower with a gynoecium (carpel or carpels) but no functional androecium (stamens).

Carpellode   Sterile carpel.

Carpophore   Stalk supporting each portion of the dehiscing schizocarp of many members of Apiaceae.

Caruncle   Rather firm outgrowth of a seed that develops from the outer integument (testa); in this text considered as a type of aril.

Caryopsis   Small, indehiscent, dry fruit with a thin wall surrounding and more or less fused to a single seed.

Catkin   Inflorescence consisting of a dense, elongated mass of inconspicuous, usually wind-pollinated flowers; also called an **ament**.

Caudate   With a tail-like appendage.

Caulescent plant   Plant with a distinct stem.

Cauliflory   Flowers or inflorescences borne on older stems and/or trunks.

Cellular endosperm Division of the primary endosperm nucleus, and all subsequent nuclear divisions, immediately followed by formation of cell walls.

Centrifugal development Developing first at the center and then gradually toward the periphery.

Centripetal development Developing first at the periphery and then gradually toward the center.

Centromere Small, constricted region of a chromosome having attachment sites for spindle fibers during nuclear division.

Chalaza Indistinct region of an ovule (or seed) where the integuments are connected to the nucellus, at the end opposite from the micropyle.

Character Any feature of the organism, often used to refer to those showing group-diagnosing variation.

Character displacement Divergence in two related species in the zone where they co-occur, resulting from natural selection against individuals with intermediate phenotypes.

Character state polarization Determination of the direction of character state change, i.e., determining what state is ancestral and what state (or states) are derived.

Character state One of the various conditions (or values) of a character observed across a group of taxa.

Chartaceous Of papery texture.

Chasmogamous Condition of a flower that opens and is usually cross-pollinated. Contrast with Cleistogamous.

Chlorophyll Light-sensitive green pigment that absorbs violet-to-blue and also red light, and is essential for photosynthesis.

Chloroplast capture Transfer of chloroplasts from one species to another via hybridization and subsequent introgression.

Chloroplast genome Circular DNA strand of the chloroplast, an endosymbiont within the eukaryotic host cell.

Chromosome Complexly folded DNA molecule of the nuclear genome in eukaryotes with many associated histone proteins.

Circinate Rolled in a coil from the top downward, with the apex nearest the center of the coil (see Figure 4.14).

Circumscissile Structure opening by a line around its circumference, with the top usually coming off as a lid, e.g., a circumscissile capsule.

Circumscription Delimitation ("drawing a line around") of any feature or features, as for instance determining the limits or boundaries of a taxon.

Clade Monophyletic group, made up of an ancestor and all of its descendents.

Cladistics Taxonomic philosophy/method in which relationships of taxa are based on recency of common ancestry as assessed by the pattern of shared derived character states (synapomorphies).

Cladogenesis Origination of new clades (monophyletic groups) through the splitting of evolutionary lineages.

Cladogram Branching diagram that shows hypothesized phylogenetic (sister-group) relationships of a group of organisms.

Class As a concept, a group of individuals that all possess a particular defining characteristic. As a taxonomic rank, the category between the higher rank of division (or phylum) and the lower rank of order; in phylogenetic classifications, a monophyletic group of orders.

Classification Delimitation, ordering, and ranking of taxa; or the system of internested groups resulting from this process.

Clawed With distinctly narrowed basal portion, as in the clawed petals of Malpighiaceae.

Cleistogamous Condition of a flower that never opens and is self-pollinated and self-fertilized. Contrast with Chasmogamous.

Cline Gradual change in a single character across a geographical or ecological gradient.

Coevolution Linked evolution of two or more closely interacting species.

Collateral vascular bundle Vascular bundle with phloem on one side and xylem on the other.

Collection number Number assigned to a preserved plant specimen, usually by its collector.

Colleter Multicellular "hair" that produces a sticky, mucilaginous, or resinous secretion, usually occurring on the adaxial surface of a stipule, petiole base, or calyx lobe of certain plants, especially members of Gentianales.

Colpate Pollen grain with long and grooved aperture(s), the colpus (colpi).

Columella "Little column" in the pollen exine, separating the tectum from the foot layer of the nexine (see Figure 4.48).

Column Structure formed by the fused style and stigma plus stamens of members of the Orchidaceae.

Combretaceous hair Thick-walled, unicellular hair with a sharp pointed apex and a bulbous base that appears bicellular because of a conical or concave cellulose membrane in the cell wall.

Common name Name of a plant in language of the culture or society where the plant occurs; also called the vernacular name.

Companion cell Parenchymatous cell with a nucleus and dense cytoplasm that is positioned next to a sieve tube element and is closely involved in its metabolism.

Complete flower Flower having a perianth, androecium, and gynoecium.

Compound leaf Leaf with two or more blades (leaflets).

Compound umbel Umbel of umbels, i.e., with the peduncle of each simple umbel arising from a common point.

Conducting tissues Xylem (water-conducting) and phloem (sucrose-conducting) tissues of a vascular plant.

Conduplicate Folded lengthwise with the adaxial surface within (see Figure 4.14).

Cone Ovulate reproductive structure of the conifers, consisting of reduced branches borne on a rather short axis.

Connate/connation Like parts fused, as the petals fused to form a flower with a sympetalous corolla.

Connective Portion of the stamen connecting the two pollen sacs of an anther (see Figure 4.16).

Consistency index Measure of the level of homoplasy of a character (or of all characters on a cladogram), equaling the minimum number of possible changes divided by the total number of actual changes on the tree (or the total tree length).

Contorted Parts imbricate and whorled, each in turn overlapped by that on one side and overlapping that on the other, the whorl tightly rolled, appearing twisted.

Convergent evolution Evolution of similar features from different ancestral conditions, resulting from similar selective pressures operating on distantly related, taxa, e.g., the evolution of stem succulence in both Cactaceae and Euphorbiaceae or carnivorous leaves in Sarraceniaceae and Nepenthaceae.

Convolute Rolled, with the margins overlapping (see Figure 4.14).

Coralloid root Short, thick, and much branched root, similar in appearance to some corals.

Cordate base Base heart-shaped, with rounded lobes (see Figure 4.11).

Coriaceous Thick, or of leathery texture.

Corm Short, erect, underground, more or less fleshy stem covered with thin, dry leaves (or leaf bases).

Corolla Collective term for all petals of a flower (see Figure 4.16).

Corona Usually showy outgrowths of the perianth parts, stamens, or receptacle.

Cortex Portion of a stem located between the pericycle (surrounding the primary vascular tissue) and the epidermis.

Cortical bundle Vascular bundle occurring in the cortex of a stem (see Figure 4.38).

Corymb Raceme with the pedicels of the lowermost flowers elongated, bringing all flowers to more or less the same level, i.e., a flat-topped raceme (see Figure 4.30).

Costapalmate More or less palmate leaf with an extension of the rachis, forming a midrib (or costa) in the blade, as in the leaves of many palms (Arecaceae).

Cotyledon Leaf, or one of two or occasionally more leaves developed at the first node of the embryo (see Figure 4.42).

Couplet A pair of contrasting choices in a dichotomous key.

Crenate margin Margin with rounded teeth (see Figure 4.12).

Crossing over Breakage and exchange of corresponding segments between a pair of homologous chromosomes during Prophase I of meiosis.

Cross-pollination Transfer of pollen from the anther of flowers of one individual to the stigma of flowers of a second individual.

Cryptic species Taxa that are reproductively isolated but are phenotypically very similar and hard to distinguish.

Crystal sand Granular mass composed of very tiny crystals of calcium oxalate.

Cucullate Hood-shaped.

Cultivar Variant/variety of a plant produced in cultivation through hybridization, artificial selection, or any other process.

Cuneate base Base that is wedge-shaped, triangular, and tapered to a point (see Figure 4.11).

Cupule Any cup-shaped structure, such as the scaly to spiny "cup" associated with the nut of members of Fagaceae, or swollen receptacle and perianth parts usually associated with the drupe of members of Lauraceae.

Cuticle Thin, waxy layer that covers the outer surface (epidermis) of most embryophytes.

Cyanogenesis Process by which hydrogen cyanide is released from cyanogenic compounds.

Cyanogenic glycosides Defensive compounds that are hydrolyzed to release hydrogen cyanide.

Cyathium Cuplike cymose inflorescence unit of species of *Euphorbia* that mimics a single complete flower and consists of one to few nectar glands, several staminate flowers, and a single carpellate flower.

Cycasins Poisonous compounds of cycads that cause neurological damage.

Cyclocytic Stomate surrounded by subsidiary cells forming one or two narrow rings around the guard cells; the number of cells in each ring usually is four or more (see Figure 4.35).

Cyme Determinate, compound inflorescence composed of repeating units of a pedicel bearing a terminal flower and below it, one or two bracteoles; each bracteole is associated with an axillary flower, and further bracteoles, and so on (see Figure 4.29).

Cystolith Stalked concretion composed of calcium carbonate that projects into the center of a specialized cell or lithocyst (see Figure 4.37).

Decay index Number of extra steps required to find trees (cladograms) that do not link a particular group of taxa.

Deciduous leaves Leaves that fall (are abscised) at the end of the growing season.

Decurrent base Lamina base tapering gradually to the petiole (see Figure 4.11), also used for leaf base that extends down the stem.

Decussate leaves Opposite leaves in which those of adjacent nodes are rotated 90 degrees from each other.

Dehiscence Method or process of opening of a structure, such as a fruit, anther, or sporangium.

Dehiscent drupe Fruit with a dry/fibrous to fleshy or leathery outer husk that eventually breaks apart or opens, exposing one or more nutlike pits, as in *Carya* (Juglandaceae).

Dendritic hair Hair with a treelike branching pattern (see Figure 4.15).

Dendrogram   Treelike diagram showing relationships, whether phenetic, cladistic, or evolutionary taxonomic.

Dentate margin   With coarse teeth that are perpendicular to the margin (see Figure 4.12).

Denticidal capsule   Capsule that opens by a series of apical teeth.

Determinate inflorescence   Inflorescence in which the axis is converted into a flower, resulting in the cessation of growth of that axis.

Diacytic   Stomate enclosed by a pair of subsidiary cells whose common wall is at right angles to the guard cells (see Figure 4.35).

Diadelphous   Stamens with their filaments fused to form two groups.

Diagnosis   Brief statement of the attributes that allow recognition of a taxon and its separation from one or more others.

Dichotomous branching   Branching by division of the apical meristem or apical cell, forming two equal or unequal branches.

Dichotomous key   Key composed of pairs of contrasting statements (couplets).

Didynamous stamens   Four stamens, two long and two short.

Dimorphic   With two distinct forms of the same structure.

Dioecious   With staminate and carpellate flowers borne on separate plants.

Diploid   With two full sets of chromosomes in each cell.

Disk flower   Radially symmetrical flower of many members of the Asteraceae.

Dissected   Deeply divided into many narrow segments.

Distichous   *See* Two-ranked.

Distyly   Flowers of different individuals of the same species having two different style lengths, the stamen lengths often varying inversely; an outcrossing mechanism.

Diterpenoids   Twenty-carbon terpenoid compounds (see page 98).

Division   Taxonomic category between the higher rank of kingdom and the lower rank of class; also called phylum (pl. phyla); in a phylogenetic classification a division is a monophyletic group of classes.

Dollo parsimony   Parsimony assumption that allows only one evolutionary gain of a character state but numerous losses.

Domatium (pl. domatia)   Small cavity, pouch, or tuft of hairs on a leaf or bract that functions as a tiny home for mites, ants, or other small organisms.

Double fertilization   In angiosperms, fusion of one sperm with the egg to form the zygote, and fusion of the second sperm with the polar nuclei to form the nutritive tissue, endosperm.

Doubly serrate margin   Margin with coarse teeth bearing smaller teeth on their margins, and leaf margin thus with both small and large teeth (see Figure 4.12).

Drupe   Indehiscent, fleshy fruit in which the outer part is more or less soft (to occasionally leathery or fibrous) and the center contains one or more hard pits or stones consisting of a bony endocarp surrounding a seed or seeds.

Drupelet   Drupe developing from a single carpel and, as traditionally used, one portion of an aggregate fruit, as in *Rubus* (Rosaceae).

Druse   Spherical, compound crystal of calcium oxalate in which the numerous component crystals protrude from the surface giving the cluster a star-shaped appearance (see Figure 4.37).

Ebracteate   Lacking bracts.

Ecotype   Distinct genetic entity of a species that is adapted to a particular set of ecological conditions.

Effective publication   Publication in a book or scientific journal as outlined in the *International Code of Botanical Nomenclature*.

Egg   Nonmotile, female gamete of embryophytes, often larger than the sperm, and formed in an archegonium or by a reduced female gametophyte.

Elaiosome   Hard, oily outgrowth (aril) of a seed that attracts ants.

Elater   Hygroscopic band attached to the spores of *Equisetum* that aid in dispersal.

Electrophoresis   Procedure using an electrical current to separate molecules (usually proteins) on the basis of their size and electrical charge.

Elliptic   Widest near the middle (see Figure 4.9).

Emarginate apex   Apex prominently notched (see Figure 4.10).

Embryo   Young sporophyte, from the first cell after fertilization (zygote) until germination of the seed (in seed plants) or emergence from protective gametophytic tissues (other embryophytes).

Embryo sac   Female gametophyte of the angiosperms (flowering plants).

Embryology   Study of the events leading to development of the male and female gametophytes, gamete development, syngamy, and early development of the embryonic sporophyte.

Endemic taxon   Taxon whose members are restricted to a particular geographic area.

Endexine   Lower part of the nexine, the innermost portion of the exine (see Figure 4.48).

Endocarp   Inner layer of the fruit wall or pericarp, as in the pit or stone of a peach or cherry (*Prunus*, Rosaceae).

Endosperm   Usually triploid nutritive tissue in the seed of most angiosperms, derived from the fusion of a sperm with the two polar nuclei of the female gametophyte.

Endothecium   Inner lining of the locule of an anther, usually fibrous.

Entire margin   With a smooth margin, lacking any teeth (see Figure 4.12).

Epicotyl   That portion of the embryonic stem above the cotyledons, and below the next leaves on the embryonic axis.

Epicuticular waxes   Wax deposits of variable form deposited on top of the cuticle on the surface of the plant.

Epidermis   Outermost layer of cells of the primary plant body, often with strongly thickened outer walls.

Epigynous   With perianth and stamens apparently borne upon the ovary (see Figure 4.21) and the ovary, therefore, inferior.

Epimatium   Modified ovuliferous scale that partly surrounds the ovule in members of the Podocarpaceae.

Epipetalous stamens   Stamens adnate (fused) to the corolla.

Epiphylly   Flower or inflorescence borne on a leaf.

Epiphyte   Plant growing upon another plant, which is used as a support.

Equitant leaves   Two-ranked leaves that are flattened in the plane of leaf insertion and with both surfaces morphologically and anatomically identical (i.e., unifacial), as in the leaves of *Iris* and relatives (Iridaceae).

Erose margin   Margin with irregular rounded to pointed teeth, as if gnawed (see Figure 4.12).

Essential oils   *See* Ethereal oils.

Ethereal oils   Aromatic, highly volatile, oily secondary plant products containing monoterpenoids and sesquiterpenoids (and other aromatic compounds), frequently in pellucid dots.

Eusporangium (pl. eusporangia)   Common sporangium form of embryophytes, sessile and with a wall composed of several cell layers.

Eustele   Vascular bundles of a stem forming a complete ring when viewed in cross-section (see Figure 4.38).

Even-pinnate leaf   Pinnately compound leaf with an even number of leaflets, i.e., lacking a terminal leaflet (see Figure 4.5).

Evergreen plant   Plant that is leafy throughout the year.

Evolution   Genetic change in a lineage (line of descent) over time.

Evolutionary taxonomy   Subjective approach to classification construction that attempts to reflect both overall similarity, ecologically significant features, and evolutionary relationships.

Evolutionary tree   Branching diagram representing phylogenetic relationships (evolutionary history) of a group of taxa; in this text considered basically equivalent to a cladogram, and frequently shortened to "tree."

Exalbuminous seeds   Mature seeds lacking endosperm.

Exine   Outer layer of the two-layered wall of a pollen grain (see Figure 4.48).

Exocarp   Outer layer of the fruit wall or pericarp.

Exserted floral structures   Sticking out, as in stamens extending beyond the corolla.

Extrafloral nectary   Nectar-producing structure positioned on a non-floral structure, such as a leaf, stipule, stem, bract, or inflorescence axis.

Extrastaminal nectar disk   Nectar disk positioned between the staminal whorl (or whorls) and the perianth parts.

Extrorse anther   Anther that opens outwards, toward the periphery of the flower.

F$_1$ generation   First filial generation; offspring resulting from the cross of individuals representing two different species or distinct genetic entities.

False indusium   In ferns, a revolute leaf margin or marginal lobe that protects developing sporangia in a sorus.

Family   Taxonomic category between the higher rank of order and the lower rank of genus; in a phylogenetic classification, a monophyletic group of genera.

Fascicle   Reduced, axillary inflorescence (either determinate or indeterminate) with flowers in a bundle or cluster and no obvious inflorescence axis (see Figure 4.30).

Felsenstein zone   A combination of very fast and very slow evolutionary rates that will make accurate phylogeny reconstruction impossible; taxa with faster rates (long branches) will appear to be closely related.

Fertilization   Fusion of the sperm nucleus and the egg nucleus.

Fiber   Any long, narrow, thick-walled, and lignified cell (see Figure 4.31).

Fibrous roots   Root system with roots of more or less equal thickness, and often well branched, the primary (tap) root absent or not obvious.

Field press   Device consisting of press frames, straps, corrugates, and newspapers used to press and transport plant specimens in the field.

Filament   Stalk of a stamen (see Figure 4.16).

Fissured xylem   Xylem broken up by the development of phloem or parenchyma tissue.

Fitch parsimony   Parsimony algorithm that treats all character states as unordered.

Fitness   An increase in adaptation to the environment, as brought about by genetic change; measured as reproductive success.

Flag petal   *See* Banner petal.

Flagellum (pl. flagella)   Tail-like motile structure of many eukaryotic cells, with a "9 plus 2" array of microtubules.

Flavonoids   Phenolic compounds that usually occur in a ring system derived through cyclization of an intermediate from a cinnamic acid derivative and three malonyl Co-A molecules (see page 99).

Fleshy root   Thick root with tissues for storing water or carbohydrates.

Flora   Enumeration of the plants occurring in a particular geographical area, usually with keys, descriptions, illustrations, and distribution maps; also the plants occurring within a designated region.

Floral cup   *See* Hypanthium.

Floral diagram   Diagram summarizing floral symmetry and the number, fusion, and insertion of the floral parts.

Floral formula Formula summarizing floral symmetry and the number, fusion, and insertion of floral parts.

Floral tube *See* Hypanthium.

Floret Very small flower, especially those of members of Asteraceae, Cyperaceae, and Poaceae; sometimes applied to the flowers plus associated bractlike structures.

Flower Reproductive structure of the angiosperms, consisting of a determinate, modified shoot, the floral axis or receptacle, bearing modified leaves, the perianth parts, stamens, and/or carpels.

Follicle Dry to rarely fleshy fruit derived from a single carpel that opens along a single (usually adaxial) longitudinal suture.

Foot layer (pollen) Outer layer of the nexine, separated from the tectum by the columellae (see Figure 4.48).

Free-central placentation Ovules attached to a free-standing column or central axis in the middle of a unilocular ovary (see Figure 4.22).

Frond Leaf of a fern, usually large and divided or deeply lobed.

Fruit Mature ovary with associated accessory parts, such as adnate hypanthium of *Malus* (apple, Rosaceae).

Funiculus (pl. funiculi) Stalk of an ovule; also called the **funicle** (see Figure 4.39).

Gametangium (pl. gametangia) Gamete-producing structure, e.g., archegonium or antheridium.

Gamete Haploid cell that fuses with another cell to form a zygote.

Gametophyte Haploid, gamete-producing generation of the plant reproductive cycle.

Gemma (pl. gemmae) Budlike structure or a cluster of cells that separate from the parent plant and grow, resulting in asexual propagation.

Gene Stretch of DNA controlling a heritable trait.

Gene flow Movement of alleles into and out of populations.

Gene trees Phylogenetic history of a gene, based on its DNA sequence.

Genealogy *See* phylogeny.

Genome All the DNA (genetic material) of the cells of a particular species; in photosynthetic eukaryotes comprised of the nuclear, chloroplast, and mitochondrial genomes.

Genotype Genetic constitution of an individual.

Genus (pl. genera) Taxonomic category between the higher rank of family and the lower rank of species; in a phylogenetic classification genera are monophyletic groups of species.

Geographical speciation *See* Allopatric speciation.

Glabrous Lacking hairs.

Glandular hair Hair bearing a knoblike secretory swelling (head) at its apex (see Figure 4.15).

Glaucous With a waxy covering, and thus often blue or white in appearance.

Globose More or less spherical.

Glucosinolates Mustard oil glucosides, hydrolyzed by myrosinases to yield pungent, hot-tasting mustard oils.

Glumes Pair of bracts at the base of a grass spikelet.

Glycosides Organic compounds that yield a sugar when hydrolyzed.

Grain *See* Caryopsis.

Gynobasic style Style that appears to be inserted at the base of the ovary.

Gynoecium Collective term for all of the carpels of a flower (see Figure 4.16).

Gynophore Stalk bearing a gynoecium.

Habit General appearance of a plant, e.g., whether a tree, shrub, vine, herb, etc.

Hair Epidermal outgrowth of diverse form, structure, and function (see Figure 4.15); often called a trichome.

Haploid With one full set made up of one member each of the pairs of chromosomes in the nucleus.

Haustorium (pl. haustoria) Specialized root of parasitic plants that penetrates another plant and absorbs water and nutrients.

Head Compact determinate or indeterminate inflorescence with a very short, often disklike axis and usually sessile flowers (See Figures 4.29, 4.30).

Helicoid cyme Coiled cyme in which the lateral branches all develop from the same side of the axis (see Figure 4.29).

Helobial endosperm Endosperm development in which the primary endosperm nucleus divides, forming a transverse wall, creating a small chalazal cell and a much larger micropylar cell, followed by a sequence of free nuclear divisions in the micropylar and sometimes also the chalazal cell.

Hemiparasite Parasitic plant that is green and able to produce at least some of its own carbohydrates but that has haustorial connections to other plants through which pass water as well as some carbohydrates.

Hemitropous ovule Ovule half-inverted, so the funicle is attached near the middle with the micropyle at right angles to it.

Herbaceous Not woody; the above-ground stem dying down at the end of the growing season.

Herbarium (pl. herbaria) Place where plant collections are stored, typically as pressed and dried plants mounted to sheets of paper, identified, and provided with locality and habitat data, so they can be further studied.

Herb Plant without a persistent above-ground woody stem, and either dying (in annuals) or dying back to the ground level (in herbaceous perennials) at the end of the growing season.

Heteromorphic Of two or more distinct forms.

Heterophylly Variation in form of the leaves produced during the life of the plant, e.g., sun/shade leaves or juvenile/adult leaves.

Heterosporous Producing two types of spores; small spores (microspores) that give rise to male gametophytes, and usually large spores (megaspores) that give rise to female gametophytes.

Heterostyly   Flowers in different individuals of the same species having two or three different style lengths, the stamen lengths often varying inversely; an outcrossing mechanism.

Heterozygosity   Presence of two or more alleles (molecular forms of a particular gene) within an individual.

Hexaploid   With six full sets of chromosomes in a cell.

Hilum   Scar on a seed that indicates the point of attachment of the funiculus, the stalk of the ovule.

Hirsute   Having long, often stiff hairs.

Hispid   Having stiff or rough hairs; bristly.

Holoparasite   Plant that lacks chlorophyll and obtains all of its water and nutrients from other individuals (hosts) through haustorial connections.

Holotype   Sole specimen of a species or infraspecific taxon used as the type by the author of a name; or the one specimen designated by the author as the type or holotype. The name-bearing specimen.

Homology   Similarity due to inheritance of a feature from a common ancestor.

Homonym   Name identical in spelling to the name of another plant.

Homoplasy (adj. homoplasious, homoplastic)   Similarity due to parallelism or reversal of character states.

Homosporous   Producing only one type of spore, all the same size and producing bisexual gametophytes.

Homozygosity   Presence of only one allele (molecular form of a particular gene) within an individual.

Hybrid   Offspring of two organisms belonging to different species.

Hybrid speciation   Production of a new species from two or more parental species via hybridization.

Hybrid swarm.   An array of individuals resulting from reproduction among F_1 hybrids and between hybrids and either parental species.

Hybridization   Process of hybrid formation.

Hygroscopic tissues   Tissues that swell, shrink, or change shape due to changes in moisture content.

Hypanthium   Flat, cuplike, or tubular structure on which the sepals, petals, and stamens are borne (see Figure 4.21); usually formed from the fused bases of the perianth parts and stamens or from a modified receptacle; also called a floral cup or floral tube.

Hypocotyl   Stem of an embryo below the cotyledon (or cotyledons).

Hypodermis (pl. hypodermes)   Differentiated layer or layers of cells beneath the epidermis.

Hypogynous   With perianth parts and stamens arising from below the ovary, and the ovary thus superior (see Figure 4.21).

Identification   Determination of the name of an unknown plant.

Imbricate   Overlapping, as of leaves or petals in bud, i.e., arranged like tiles or shingles on a roof, or scales of a fish.

Imperfect flower   Flower lacking either an androecium (stamens) or a gynoecium (carpels); also called a unisexual flower.

Included   Structure enclosed by others, as the stamens enclosed or hidden by the corolla, or scattered phloem strands surrounded by secondary xylem.

Incomplete flower   Flower lacking one or more of the floral parts, i.e., missing either the perianth, androecium, and/or gynoecium.

Indehiscent   Not opening.

Indehiscent pod   Fairly dry fruit with few to many seeds that fails to open.

Indented key   Key with each pair of contrasting choices (leads) that are equally indented and followed by the appropriate couplets, and thus the two leads are often widely separated in the key.

Indeterminate inflorescence   Inflorescence in which the main axis produces only lateral flowers, branches, or groups of flowers, so that the lowermost or outermost flowers usually open first, with the main axis (and lateral axes, if present) often elongating as the flowers develop, without the production of terminal flowers.

Indeterminate thyrse   Inflorescence with an indeterminate main axis and cymose lateral branches (see Figure 4.30).

Indumentum   General term for the hairs (trichomes) on the surface of a plant.

Indusium (pl. indusia)   Protective flap or outgrowth covering immature sori in ferns that develops from the abaxial surface of the leaf (or frond).

Inferior ovary   Ovary that is positioned beneath the point of attachment of the other floral parts, which appear, therefore, to arise from its apex (see Figure 4.21).

Infructescence   Mature inflorescence, with flowers replaced by fruits.

Insertion of floral parts   Pattern of attachment of floral parts to the receptacle.

Intectate   Exine of pollen grain lacking a tectum (outermost layer, separated from inner layers by small columns) (see Figure 4.48).

Integument   Protective outer layer (or layers) surrounding the sporangium of an ovule, which will develop into the seed coat (see Figure 4.39).

Internal phloem   Primary phloem in the form of strands or a continuous ring (as seen in cross-section) at the inner boundary of the xylem.

Internode   Stem between two adjacent nodes (see Figure 4.2).

Interpetiolar stipules   Stipules positioned at the node between the petiole bases of the opposing leaves.

Intine   Inner layer of the two-layered pollen wall (see Figure 4.48).

Intrastaminal nectar disk   Nectar disk positioned between the androecium (staminal whorl or whorls) and gynoecium.

Introgression   Permanent incorporation of genes and/or organelles of one species into another through extensive hybridization and back-crossing.

Intron   Non-coding portion of a gene transcript, excised before translation into protein.

Introrse anther   Anther that opens inwards, toward the floral axis.

Intruded placenta (pl. placentae)   Placenta that extends into the locule of the ovary (see Figure 4.22).

Involucre   Series of bracts surrounding a flower or inflorescence.

Inulin   Oligosaccharide containing about 30 fructose (and possibly also one or a few glucose) units.

Involute   With margins rolled inward toward the adaxial surface (see Figure 4.14).

Iridoids   Nine- or ten-carbon derivatives of a simple terpene unit (see page 98).

Isolating mechanism   Mechanism that prevents gene flow between closely related species.

Isotype   Duplicate specimen of the holotype, being a part of the same individual (or gathering) that forms the holotype.

Karyotype   Preparation of metaphase chromosomes sorted by length, centromere location, and other features.

Keel petals   Two abaxial petals of the flowers of Fabaceae subfam. Faboideae; sometimes also used of similar-shaped petals of other families, e.g., Polygalaceae.

Key   Series of contrasting choices used to identify an unknown organism by the process of elimination.

Kingdom   Highest (most inclusive) taxonomic category in the Linnaean hierarchial system.

Kranz anatomy   Foliar veins surrounded by conspicuously enlarged bundle sheath cells containing chloroplasts; associated with C4 photosynthesis.

Labellum   "Lip" of an orchid perianth; the adaxial member of the inner whorl of perianth parts (but due to rotation of the flower, at anthesis usually placed as the lowermost perianth part), and differentiated from the other perianth parts in size, form, and/or coloration; or two (or four) large, petaloid, and connate staminodes forming a lip-like structure in gingers and relatives.

Lanate   Woolly, with long intertwined, somewhat matted hairs.

Lanceolate   Narrowly ovate to narrowly elliptic; the term is ambiguous and so use is not recommended.

Laticifer   Cell or series of tubular cells, branched or unbranched, containing latex.

Latrorse anther   Anther that opens on the sides.

Leaf (pl. leaves)   Usually flat, determinate, photosynthetic part of a plant, borne on a branch or stem (axis) (see Figure 4.2), or more generally, in flowering plants, an appendicular structure of determinate growth usually subtending a bud or branch.

Leaf axil   Space in the angle between a leaf and the stem that bears it.

Leaf gap   Parenchymatous interruption left in the secondary vascular system of angiosperms by the departure of a vascular bundle(s) (or traces) to the leaves, hence, a trilacunar leaf gap, with three gaps (lacunae) left by the departure of three bundles to the leaf (see Figure 4.33).

Leaflet   One of the blades of a compound leaf.

Leaf scar   Remains of a point attachment of a leaf.

Legume   Dry, more or less elongated fruit derived from a single carpel that opens, often explosively, along two longitudinal sutures; the most common fruit type of members of Fabaceae.

Lemma   Outermost or lower of the two bracts that surround the grass flower (floret).

Lenticel   Wartlike protuberance on the stem surface involved in gas exchange.

Leptosporangium   Stalked sporangium with a wall of only one cell layer (and in most ferns with a cluster or row of specialized hygroscopic cells, an annulus).

Liana   Woody climbing plant.

Ligulate flower   Flower with a straplike, usually five-toothed, corolla limb characteristic of many Asteraceae.

Ligule   Adaxial projection from the top of the leaf sheath, as in gingers (Zingiberaceae) or grasses (Poaceae).

Lineage   Ancestor-descendent sequence of populations.

Linear   Long and very narrow, i.e., very narrowly oblong.

Lithocyst   Enlarged cell containing a cystolith (a large, crystal-like concretion).

Lobate base   Base with two well developed rounded lobes (see Figure 4.11).

Lobed   Having large rounded projections along margin (see Figure 4.12).

Local speciation   Divergence of small, geographically isolated populations at the periphery of the range of the parental species, usually resulting from a combination of natural selection and genetic drift, leading to formation of a new species.

Locule   Compartment, cavity, or chamber within an ovary or anther (see Figure 4.22).

Loculicidal capsule   Capsule that splits longitudinally between the septa (or placentae, if placentation is parietal) and into the locules (chambers) of the ovary.

Lodicules   Reduced perianth parts of members of Poaceae.

Loment   Dry, schizocarpic fruit derived from a single carpel that breaks transversely into one-seeded segments.

Long branch attraction   Lineages that evolve rapidly may contain large numbers of nucleotide similarities due to parallel evolution and may, therefore, group together in a cladistic analysis even though not each other's closest relatives.

Long shoot   Stem with long internodes; this term is applied only in plants in which internode length is clearly bimodal and both long and short shoots are present.

Mangrove Tropical tree that grows in tidally flooded coastal banks or estuaries, showing various adaptations such as salt tolerance or secretion, vivipary, or the development of prop-roots; or the plant community occurring in such regions.

Marcescent Withering, but persisting on the plant.

Marine Of or found in the sea (but not in fresh water).

Medullary bundle Vascular bundle occurring in the pith of a stem or within a cylinder of vascular tissue or ring of vascular bundles within a petiole (see Figure 4.38).

Megagametophyte Female gametophyte, e.g., the embryo sac, of angiosperms (see Figure 4.40).

Megaphyll Large leaf thought to have evolved by planation and webbing of a lateral branch system; characteristic of the ferns and seed plants. Forms leaf gaps in the central vascular system (or stele).

Megasporangium (pl. megasporangia) Multicellular, spore-producing structure (sporangium) that bears megaspores.

Megaspore Spore that germinates to produce a female gametophyte (megagametophyte).

Megasporocyte Cell that undergoes meiosis and gives rise to megaspores.

Meiosis Two-stage nuclear division process that reduces the chromosome number of a cell by half, to the haploid number, with each daughter nucleus receiving one of each type of chromosome, and followed by the production of spores.

Membranous. Thin, soft, and flexible, like a membrane; also called membranaceous.

Mericarp One or few-seeded segment of a schizocarp.

-merous. The numerical plan of a flower, as in a *Lilium* flower, which is 3-merous, or a *Geranium* flower, which is 5-merous.

Microgametophyte Male gametophyte.

Microhair Minute, two-celled hair found on the leaves of most Poaceae.

Micromorphology Study of the minute structures and surface features best seen with a scanning electron microscope.

Microphyll Small leaf with a single vein, and not associated with a leaf gap in the central vascular system (stele); characteristic of the Lycophytes.

Micropyle Passage through the integument or integuments of an ovule through which the pollen grain or pollen tube usually enters (see Figure 4.39).

Microspecies Minimally differentiated series of populations derived from uniparental reproduction (selfing or asexual means).

Microsporangium (pl. microsporangia) Sporangium that produces microspores.

Microspore Spore that germinates to produce a male gametophyte.

Microsporocyte Cell that undergoes meiosis and gives rise to microspores or pollen.

Midvein Central vein of a leaf or other organ (part), in a leaf often called the midrib.

Minimum spanning tree Shortest path in multivariate space connecting an array of taxa (objects).

Mitochondrial genome Genes contained on the circular DNA strand of the mitochondrion, an endosymbiont of the host eukaryotic cell.

Mitosis Nuclear division that maintains the parental chromosome number for daughter cells, and the basis for growth in size and asexual reproduction in plants.

Monad Generally, a single individual, free from others; specifically, a single pollen grain.

Monadelphous stamens Stamens fused by their filaments in a single group, usually forming a tube.

Monocolpate Pollen grain with a single, long, grooved aperture (see Figure 4.46).

Monoecious Staminate and carpellate flowers separate, but borne on a single plant.

Monograph Comprehensive systematic study of a particular taxonomic group, including keys, descriptions, illustrations, distribution maps, etc.

Monophyletic group Group composed of an ancestor and all of its descendants; diagnosed by synapomorphies (shared derived characters), a clade.

Monopodial shoot Shoot formed through the action of a single apical meristem.

Monoporate Pollen grain with a single porelike aperture (see Figure 4.46).

Monosulcate Pollen grain with a single, long and grooved aperture located at the pole (see Figure 4.46).

Monotelic *See* Determinate inflorescence.

Morphology Study of the form and structure (especially external) of living organisms.

Mucilage Slimy or sticky fluids, often polysaccharides.

Mucronate apex Apex terminated by a distinct, short and abrupt point or mucro (see Figure 4.10).

Multicellular hair Hair composed of several cells.

Multilacunar node Node with five or more leaf gaps (see Figure 4.33)

Multiple fruit Fruit produced by the gynoecia of several closely clustered flowers.

Multiseriate hair Hair composed of several rows of cells (see Figure 4.15).

Multistate character Character with three or more states.

Multivalent Grouping (synapsis) of three or more chromosomes in a cell during meiosis.

Mustard oils Hot, pungent compounds resulting from the hydrolyzation of glucosinolates in the mustards and their relatives (Brassicales).

Mutation A heritable change in the molecular structure of DNA.

Mycorrhiza Symbiotic association of various fungi with roots of vascular plants.

Myrosin cell Cell containing the enzyme myrosinase, which hydrolyzes glucosinolates to produce mustard oils.

Naked bud Bud not covered by scale leaves.

Native species Species that occur naturally in an area.

Natural   Of classification or relationships, agreeing with a particular author's idea of what is natural; a useless, if much used, term.

Naturalized species   Species brought from another region by human action that maintain themselves in their new region, reproducing and spreading, without human intervention.

Nectar disk   Disklike or ringlike floral nectary.

Nectar spur   Hollow, slender, saclike projection of a floral part, usually a petal or sepal.

Nectary   Nectar-producing gland, often forming projections, lobes, or disklike structures.

Neospecies   Recently evolved species.

Nesting   Property of hierarchies, whereby smaller, less inclusive, groups are subsets, that is, are completely included within, larger, more inclusive, groups.

Nexine   Inner portion of the exine of pollen, composed of the endexine and foot layer (see Figure 4.48).

Node   Region of the stem where the leaf and bud are borne (see Figure 4.2).

Nominate   Infraspecific taxon that contains the type of a species; often called the "typical" infraspecific taxon.

Nonglandular hair   Hair that does not produce a secretion.

Nucellus   In angiosperms, another name for the sporangium wall within the ovule.

Nuclear endosperm   Successive divisions of the primary endosperm nucleus unaccompanied by cell wall formation (free nuclear divisions), leading to a multinucleate mass of protoplasm; sometimes later converted into cellular tissue.

Nuclear genome   DNA in the chromosomes within the nucleus of a eukaryotic cell.

Nucleotide   Small organic compound with a five-carbon sugar, nitrogen-containing base, and phosphate group; the structural unit of nucleic acids (DNA and RNA).

Nut   Fairly large, indehiscent, dry fruit with a thick bony wall surrounding a single seed.

Oblique   Having unequal sides or an asymmetrical base.

Oblong   With the sides nearly or quite parallel for most of their length (see Figure 4.9).

Obovate   Widest near the apex, i.e., the terminal half broader than the basal (see Figure 4.9).

Obtuse apex   Apex blunt, having an angle greater than 90 degrees, with straight to slightly convex sides (see Figure 4.10).

Obtuse base   Base blunt, having an angle greater than 90 degrees, with slightly convex sides (see Figure 4.11).

Ocrea (pl. ocreas, ocreae)   Nodal sheath or tube formed by the stipule(s), e.g., characteristic of most members of Polygonaceae.

Odd-pinnate   Pinnately compound leaf with an odd number of leaflets, i.e., a terminal leaflet is present (see Figure 4.5).

Oil gland   Gland producing oils, i.e., long-chain hydrocarbons.

Operational Taxonomic Unit (OTU)   Terminal taxon used in an analysis of relationships, especially in phenetic studies.

Operculate   Opening by a lid or cap.

Opposite   Pair of leaves or other structures borne along a stem, the members of which are positioned on opposing sides of the stem (see Figure 4.4).

Order   Taxonomic category between the higher rank of class and the lower rank of family; in a phylogenetic classification, a monophyletic group of families.

Ordering   Determining the sequence in which three or more states of a character originated.

Orthotropic   Growing erect.

Orthotropous ovule   Straight ovule with the micropyle at the apex and the funiculus at the base (see Figure 4.39).

OTU   *See* Operational Taxonomic Unit.

Ovary   Ovule-bearing part of a carpel (or several fused carpels).

Ovate   Widest near the base; i.e., with the broader end nearer the point of attachment (see Figure 4.9).

Ovule   Structure in seed plants comprised of the female gametophyte, the sporangium, one or two, rarely three, integuments, and a funiculus (stalk); after fertilization, it develops into the seed (See Figures 4.16, 4.17, 4.39).

Palea   Innermost or upper bract, of the two bracts surrounding the grass flower (floret); possibly a bracteole/prophyll.

Palmate venation   Three or more primary veins (or well-developed secondary veins) arising from at or near the base of the blade (see Figure 4.8).

Palmately compound leaf   Leaf with more than three leaflets attached to a common point, like the fingers of a hand (see Figure 4.5).

Palynology   Study of the form and structure of pollen and spores.

Panicle   Indeterminate inflorescence with two or more orders of branching, each axis bearing flowers or higher order axes (see Figure 4.30).

Panicle-like cyme   Determinate inflorescence in which the lateral branches produce several internodes before ending in a terminal flower (see Figure 4.29).

Pantoporate   Pollen grain with numerous porelike apertures.

Papillate   Covered with short, rounded, nipple-like bumps or projections (papillae).

Pappus   Highly modified calyx consisting of awns, scales, hairs, or fine bristles, as in members of Asteraceae.

Paracytic   Stomate surrounded by two subsidiary cells parallel to the guard cells (see Figure 4.35).

Parallel venation   Several to many parallel veins running the length of the blade (see Figure 4.8).

Parallelism   Separate origination of the same character state in two (or more) organisms.

Paraphyletic group   Group containing a common ancestor and some, but not all, of its descendents; diagnosed by symplesiomorphies.

Paratetracytic   Stomate surrounded by four subsidiary cells, two of them being parallel to the guard cells and the other pair being polar and often smaller.

Parietal placentation   Ovules attached to the wall of the ovary (see Figure 4.22).

Parsimony   Acceptance of the simplest hypothesis (given the premises/assumptions) that explains the data.

Parthenogenesis   Development of an embryo from an unfertilized egg.

PCR   Polymerase chain reaction; a laboratory technique in which many copies of a DNA sequence can be replicated enzymatically.

Pedicel   Stalk of a single flower in an inflorescence (see Figure 4.16).

Peduncle   Stalk of a solitary, terminal flower or of an inflorescence (see Figure 4.28).

Pellucid dots   Translucent dots on the surface of a leaf, bract, petal, or other structure usually resulting from cavities or ethereal oil cells in leaf mesophyll or other parenchymatous tissues.

Peltate   Flat structures attached by their surfaces rather than the base or margin, e.g., like an umbrella.

Peltate scale   Shield-shaped or umbrella-like hair.

Pendulous   Hanging or drooping downward.

Peptide bonds   Bonds linking the amino acids in a protein.

Percurrent tertiary veins   Tertiary veins linking the secondaries, forming a ladderlike pattern (see Figure 4.6); often called scalariform.

Perennial   Plant that lives for three or more years and usually flowers and fruits repeatedly.

Perfect flower   Flower having both androecium (stamens) and gynoecium (carpels). Also known as **bisexual flower**.

Perforation plate   Area of the cell wall connecting two vessel elements (see Figure 4.31).

Perianth   Collective term for calyx and corolla, or all tepals (when calyx and corolla are not distinguished), of a flower (see Figure 4.16).

Pericarp   Mature ovary wall of the fruit.

Pericytic   Stomate surrounded by a single subsidiary cell (see Figure 4.35).

Perigynium   Urn-shaped or saclike prophyll surrounding the carpellate flower of some Cyperaceae, e.g., *Carex*.

Perigynous   With perianth parts and stamens borne on an hypanthium that surrounds, but is not fused to, the superior ovary (see Figure 4.21).

Perisperm   Diploid nutritive tissue in the seeds of some angiosperms derived from the sporangium wall (nucellus).

Petal   Member of the inner perianth whorl, usually colorful and assisting in attracting pollinators (see Figure 4.16).

Petaloid   Functioning like a petal, whether sepal, tepal, staminode, or pistil (lode).

Petiole   Stalk of a leaf (see Figure 4.3).

Petiolule   Stalk of a leaflet.

Phenetics   Method of classifying organisms based on overall similarity, i.e., the sum of similarities and differences, with uniform rates of change assumed.

Phenogram   Treelike diagram representing phenetic relationships, i.e., with taxa clustered on the basis of overall similarity.

Phenotype   Observable trait or traits of an individual.

Phloem   Sucrose-conducting tissue of vascular plants, composed of sieve elements or cells, parenchyma cells, fibers, and sclereids.

Photosynthesis   Biochemical reactions occurring within chloroplasts that trap energy from the sun and convert it to chemical energy, followed by fixation of carbon dioxide and synthesis of glucose phosphates that become converted to sucrose.

Phyletic trend   Gradual change in a character on an evolutionary tree (cladogram).

Phyllaries   Involucral bracts of the heads of members of the Asteraceae.

Phylloclade   Flattened, leaflike stem.

Phylogenetic classification   Classification accurately reflecting hypothesized phylogenetic relationships among a group of taxa.

Phylogeny   Evolutionary history of a group of organisms.

Phylum   Taxonomic category between the higher rank of kingdom and the lower rank of class; in plants usually called division.

Phytomelan   Opaque, black, carbonaceous substance that forms a crust on the testa of most Asparagales.

Pilose   Having scattered, long, slender, soft hairs.

Pin   Long-styled floral form in a heterostylous species.

Pinnate venation   Where secondary veins arise along the length of the single primary vein, like the teeth of a comb or the lateral units of a feather (see Figure 4.7).

Pinnately compound leaf   Compound leaf with leaflets more than three and attached along two sides of an axis (or rachis), i.e., featherlike (see Figure 4.5).

Pistil   Ovule-bearing part of a flower, formed from one or more carpels (see Figure 4.19).

Pistillode   Sterile pistil.

Pit   Central, hard portion of a drupe, composed of a seed or seeds surrounded by a bony endocarp (inner wall of the fruit); also called a stone or pyrene; also areas of the cell wall of vessels and tracheids.

Pith   Soft tissue in the center of a stem, usually consisting of more or less isodiametric cells.

Placenta (pl. placentae)   Place or part of the ovary to which the ovules are attached.

Placentation   Arrangement of ovules within the ovary.

Plagiotropic   Growing horizontally.

Plane   Flat, as used of leaf margins (see Figure 4.12).

Plant   Any member of the green plant clade, i.e., an organism with a cellulose cell wall, chlorophyll a and b, chloroplast surrounded by two membranes, starch as carbohydrate storage product, and often having cells with two anterior, whiplash flagella; other photosynthetic eukaryotes are not plants.

Plant dryer   Box- or table-like device used to suspend a plant press over a source of heat.

Plant press   Device consisting of press frames, straps, corrugates, and blotting paper; used to press and dry a plant specimen.

Plasmodesmata   Pore and associated strand of cytoplasm that crosses the primary cell wall and connects two adjacent cells.

Plesiomorphy   Ancestral character state; also called a plesiomorphic character state.

Pleurogram   Fine, U-shaped or more or less circular line or groove in the seed coat of some Fabaceae.

Plicate   Folded, as in a fan (see Figure 4.14).

Plunger pollination   Secondary pollen presentation mechanism in which the modified style picks up and presents to the pollinators pollen from introrsely dehiscent anthers, as in flowers of members of Campanulaceae, Asteraceae, and a few other families.

Pneumatophore   Specialized root involved in gas exchange in some mangrove or swamp species.

Point mutation   Change of a single nucleotide in a strand of DNA.

Polar nuclei   Two nuclei located in the middle of the female gametophyte (embryo sac) of angiosperms, which fuse with a sperm nucleus to form the primary endosperm nucleus, and eventually the endosperm (see Figure 4.40).

Pollen   Microspore containing a male gametophyte; in angiosperms germinating to form a pollen tube, which rapidly transports the sperm to the ovule (and egg).

Pollen sac   Chamber in the anther (of the stamen) that contains the pollen.

Pollen tube   Tube formed by the germinating pollen grain that carries the sperm to the ovule.

Pollination   Transference of pollen from the anther to the stigma of the same individual (selfing) or a different individual (outcrossing).

Pollination syndromes   Floral characteristics associated with pollination by various biotic or abiotic means, e.g., bird-pollination, bee-pollination, wind-pollination, etc.

Pollinium (pl. pollinia)   Mass of pollen grains transported as a unit, as in many Orchidaceae and Apocynaceae.

Polocytic   Stomate with guard cells partially surrounded by a single subsidiary cell (see Figure 4.35).

Polyacetylenes   Non-nitrogenous compounds formed from the linking of acetate units via fatty acids.

Polyad   Small cluster of pollen grains.

Polyclave   Multi-access key, as in a punched card key or computer interactive identification.

Polycolpate   Pollen grain with several long, grooved apertures (see Figure 4.46).

Polycolporate   Pollen grain with several long, grooved apertures, each with a central pore (see Figure 4.46).

Polygamous   With both bisexual and unisexual flowers on the same plant.

Polymerase chain reaction   *See* PCR.

Polypeptide   Polymer of amino acids.

Polyphyletic group   Group with two or more ancestors, but not including the true common ancestor of its members.

Polyploidy   With three or more complete sets of chromosomes in a cell.

Polyporate   Pollen grain with many porelike apertures.

Polysaccharide   Polymer of sugars.

Polytelic   *See* Indeterminate inflorescence.

Pome   Indehiscent, fleshy fruit in which the outer part is soft and the center contains papery to cartilaginous structures enclosing the seeds; characteristic fruit of apples, pears, quinces, and most other members of Rosaceae subfam. Maloideae.

Population   Individuals of a given locality that form a single interbreeding community.

Porate aperture   Round and porelike aperture.

Pore   Small opening, usually circular or elliptic.

Poricidal dehiscence   Opening by means of a pore(s) or flap(s), as in a poricidal capsule.

Prickle   Sharp-pointed hair, involving epidermal tissue, or emergence, involving epidermal and subepidermal tissues but not vascularized.

Primary growth   Growth of the plant body from activity of the stem and root apical meristems, or a primary thickening meristem (as in palms). Contrast with Secondary growth.

Primary metabolite   Molecules forming an essential part of the metabolic pathways of an organism.

Primary protein structure   Sequence of amino acids in a protein.

Primary vascular tissue   Xylem and phloem of the primary plant body, produced from and differentiated immediately behind the apical meristem.

Primary vein   Midvein of an appendicular structure (leaf, bract, petal, etc.) (venation pinnate), or one of several equally prominent veins arising from the base of an appendicular structure (venation palmate).

Primary xylem   Xylem of the primary plant body, produced from and differentiated immediately behind the apical meristem.

Primordium   Organ (plant part) or organized series of cells in their earliest stage of development.

Proleptic shoot   Axillary shoot that develops from a bud that shows a period of dormancy, often with bud scale scars at the base.

Prop roots   Adventitious roots usually arising from the basal portion of a trunk and assisting in support of the plant.

Prophyll   Basal leaf or leaves of axillary shoots, commonly two and lateral, but one and often two-keeled in monocots.

Protandry   Maturation of anthers, and shedding of pollen before stigmas become receptive (in a bisexual flower or monoecious individual).

Protein   Any of numerous organic molecules composed of amino acids linked together in chains.

Protogyny   Maturation of stigmas before pollen is shed (in a bisexual flower or monoecious individual).

Pseudobulb   Thickened, cormlike internode (or series of internodes) on the stem of many epiphytic orchids (Orchidaceae).

Pseudogene   A nonfunctional gene.

Pseudomonomerous   Falsely appearing to be comprised of a single carpel, due to extreme fusion and reduction of one or more carpels.

Pseudoterminal bud   Axillary bud that has taken over the function of a terminal bud, in a sympodial shoot.

P-type plastids   Plastids in sieve tube elements that accumulate proteins (or proteins and starch) in the form of crystalloids or filaments.

Ptyxis   Way in which an individual leaf or perianth part is folded in bud.

Puberulent   Having minute, short hairs.

Pubescent   Covered with short, soft hairs; or more generally, as in this text, having hairs of any type.

Pulvinus (pl. pulvini)   Swollen portion of the petiole (or petiolule) involved in movement, usually positioned at the petiole base, but sometimes at the apex (see Figure 4.3).

Pyrene   *See* Pit.

Pyxis   Circumscissile capsule.

Race   Geographically defined population or aggregate of populations that differ morphologically from other races of the species; often treated at the taxonomic rank of subspecies or variety.

Raceme   Simple, indeterminate inflorescence with a single axis bearing pedicellate flowers (see Figure 4.30).

Raceme-like cyme   Reduced cyme with a single axis terminated by a flower and bearing pedicellate flowers (see Figure 4.29).

Rachis   Main axis of a structure, such as a leaf or inflorescence.

Radial symmetry   Divisible into equal halves by two or more planes of symmetry (see Figure 4.18); also called actinomorphic.

Radicle   Embryonic root.

Rank   Level in a taxonomic hierarchy of a category and of the taxa of that category.

Ranking   Placement of a taxon in a category of the taxonomic hierarchy.

RAPD   Random amplified polymorphic DNA.

Raphe   Portion of the funiculus of an ovule that is fused to the integument; usually represented by a ridge.

Raphides   Needle-shaped crystals occurring within a closely packed bundle (see Figure 4.37).

Ray flower   Bilaterally symmetrical, straplike, apically often three-lobed, carpellate or sterile flower occurring around the periphery of the head of many members of Asteraceae.

Ray   *See* Wood ray.

Receptacle   Floral axis that bears the flower parts (see Figure 4.16).

Recurved   Curved backward.

Reduplicate   Folded downward with the abaxial surface within (see Figure 4.14).

Reflexed   Bent sharply backward or downward.

Regular symmetry   *See* Radial symmetry.

Remane's criterion (pl. criteria)   One of the criteria used in the determination of similarity (loosely called homology) between structures on different organisms.

Replum   Persistent, thickened rim (modified placenta) of the fruit that bears the seeds in many Brassicaceae and Papaveraceae.

Reproductive isolating mechanisms   Any heritable feature that prevents interbreeding between one or more genetically divergent populations.

Resins   Sticky aromatic hydrocarbons that harden when oxidized.

Restriction enzyme   One of a series of enzymes, derived from various bacteria, which cut DNA at a particular nucleotide sequence.

Resupinate   Twisted 180 degrees; turned up-side down.

Retention index   Measure applicable to either a single character or an entire set of characters on a cladogram, which reflects for a given tree topology how much of the variation displayed by the character reflects true synapomorphy; thus, a measure of phylogenetic informativeness of a character.

Reticulate tertiary veins   Tertiary veins forming a netlike pattern (see Figure 4.6).

Retinaculum (pl. retinacula)   Prominently thickened funiculus associated with a seed of most Acanthaceae, which functions in shooting out the seeds during capsule dehiscence.

Retuse apex   Slightly notched apex (see Figure 4.10).

Reversal   Derived character state changing back to the ancestral state.

Revolute   Rolled toward the abaxial side (see Figure 4.12).

Rhizome   Horizontal stem, often underground or on the surface of the ground, bearing scalelike leaves; often called a stolon (or runner) if above ground and having elongated internodes.

Ribosome   Complex of ribosomal RNA and proteins that synthesizes protein from a messenger RNA template.

Root   Portion of the plant axis lacking nodes and leaves, usually branching irregularly and found below ground.

Rooting (of evolutionary trees)   Determining the position of the root, i.e., converting a network into an evolutionary tree (or cladogram), and thus providing directionality to hypotheses of character state change.

Rounded   Round in outline; used to describe apices and bases (See Figures 4.10, 4.11).

Ruminate endosperm   Endosperm with regular to irregular ingrowths of the seed coat.

Saccate   Bag-shaped, bladder-like, or pouched.

Saccus (pl. sacci) Bladder- or wing-like appendage on the pollen grains of some conifers.

Sagittate base Arrowhead-shaped base (see Figure 4.11).

Samara Winged, indehiscent, dry fruit containing a single seed (or rarely two seeds).

Scabrous Rough.

Scalariform perforation plate A "plate" as in the cell wall of a vessel-element, with several to many ladderlike bars with openings in between.

Scalelike leaf Small, flat leaf.

Scape Erect, leafless stem bearing an inflorescence or flower at its apex; usually composed of a single, elongated internode.

Schizocarp Dry to rarely fleshy fruit breaking into one-seeded (or few-seeded) segments (mericarps).

Scientific name Latin name of a taxon, in accordance with the rules outlined in the *International Code of Botanical Nomenclature* (or other nomenclatural codes).

Sclereid Cell that is usually not strongly elongated in only one direction and that has a thick, usually lignified wall and is dead at maturity.

Scorpioid cyme Coiled cyme in which the lateral branches (and flowers) develop alternately on opposite sides of the axis (see Figure 4.29).

Scutellum (pl. scutella) Shield-shaped and specialized cotyledon of the embryo of members of Poaceae, to which the remaining portion of the embryo is laterally attached.

Seco-iridoids Iridoid compounds that lack a carbocyclic ring.

Secondary growth Growth resulting from the activity of the vascular cambium, producing xylem and phloem, and of the cork cambium, which produces bark, both of which result in an increase in stem diameter.

Secondary metabolites Various organic compounds of ecological significance.

Secondary phloem Phloem produced by the vascular cambium.

Secondary vascular tissue Xylem and phloem produced by the vascular cambium.

Secondary vein Vein that branches from a primary vein.

Secondary xylem Xylem produced by the vascular cambium.

Secretory structures Cavities, ducts, or other glandular structures that produce a secretion (nectar, oils, resins, volatile compounds, etc.)

Seed Product of the ovule after fertilization, comprising the embryo, its nutritive tissue, and the seed coat.

Seed coat Protective outer layer or layers of the seed, developed from the integument or integuments.

Self-compatibility Ability of an individual's own pollen to effect pollination and self-fertilization.

Self-fertilization Union of a sperm and egg from the same individual.

Self-incompatibility Failure of an individual's own pollen to effect self-fertilization, typically due to its failure to germinate on the stigma (sporophytic incompatibility) or failure to grow through the style (gametophytic incompatibility).

Self-pollination Transference of pollen from the anther to the stigma of the same individual.

Sensu lato In the broad sense; with a wide or general interpretation. Abbreviated s.l.

Sensu stricto In the strict or narrow sense; with a restricted interpretation. Abbreviated s.s.

Sepal One member of the outer perianth whorl, when floral whorls are differentiated, usually green and protecting the inner flower parts in bud (see Figure 4.16).

Separate floral parts Parts of the flower not fused.

Septal nectary Nectar producing tissue located in the septa of an ovary, as in many monocots.

Septifragal capsule Capsule with longitudinal valves formed by splitting across the ends of the septa (partitions), which remain attached to the center part of the fruit.

Septicidal capsule Capsule with longitudinal valves formed by splitting down the middle of the septa (partitions between the locules).

Septum (pl. septa) Partition or cross-wall, as in an ovary.

Sequencing convention Rule stating that taxa forming a pectinate portion of a cladogram may be placed at the same taxonomic rank and arranged in their order of branching from the base.

Sericeous Silky, with usually long, thin, appressed hairs.

Serology Technique that assesses evolutionary relationships using an animal's immunological antibody-antigen reaction when exposed to a variety of foreign, including plant, proteins.

Serrate margin Margin with saw-teeth, i.e., the teeth point forward (see Figure 4.12).

Sesquiterpene lactone Type of terpenoid molecule, known primarily from Asteraceae.

Sessile Lacking a stalk.

Sessile leaf Leaf lacking a petiole.

Sexine Outer portion of the exine of a pollen grain.

Shoot Stem.

Short shoot Stem with short internodes, other shoots having distinctly longer internodes; compare with Long shoot.

Shrub Woody plant that is usually shorter than a tree and produces several stems or trunks from the base.

Sibling species Taxa that are reproductively isolated without much phenotypic divergence; also called cryptic species.

Sieve cell Long, slender, enucleate, sucrose-conducting cell of the phloem, which does not form a portion of a sieve tube.

Sieve tube element Long sucrose conducting, enucleate cell of the phloem of flowering plants that forms part of the sieve tube.

Silica Dioxide form of silicon: SiO_2.

Silique   Fruit derived from a two-carpellate gynoecium in which the two halves of the fruit split away from a persistent partition (around the rim of which the seeds are attached); sometimes restricted only to such fruits that are more than twice as long as wide (with shorter fruits then called silicles).

Simple fruit   Fruit developing from a single carpel or several fused carpels of a single flower.

Simple hair   Unbranched hair.

Simple leaf   Leaf with a single blade.

Sister groups   Two taxa that are each other's closest relatives.

Sorus   Cluster of sporangia on the surface of a fern leaf; appearing as variously shaped brown patches on the abaxial surface.

Southern blot   Method of determining lengths of DNA fragments by transferring denatured fragments from an electrophoresis gel to a nylon membrane.

Spadix   Spike with a thickened, fleshy axis, as characteristic of members of the Araceae (see Figure 4.30).

Spathe   Large bract surrounding or subtending an inflorescence, as the often showy inflorescence bract of members of the Araceae (see Figure 4.30).

Speciation   Evolutionary process by which species are formed.

Species (pl. species)   A basic grouping of organisms. (See various definitions pertaining to particular species concepts on pages 144–147.)

Species tree   A tree (cladogram) illustrating a hypothesized evolutionary history of an array of species.

Specific epithet   The second word of a binomial species name; usually an adjective that modifies the generic name, as *alba* in the species name *Quercus alba* (see Table 1, pages 512–513).

Sperm   Motile, male gamete, often flagellated, formed in an antheridium or by a reduced male gametophyte.

Spike   Simple, indeterminate inflorescence with a single axis bearing sessile flowers (see Figure 4.30).

Spikelet   Small spike, as in the basic inflorescence units of members of Poaceae and Cyperaceae.

Spine   Reduced, sharp-pointed leaf or stipule, or sharp-pointed marginal tooth.

Sporangium (pl. sporangia)   Spore-bearing structure.

Spore   Reproductive cell resulting from meiotic cell divisions in a sporangium.

Sporocarp   Specialized reproductive leaves of members of Marsileaceae.

Sporophyll   Modified leaf that bears one or more sporangia.

Sporophyte   Diploid, spore-producing generation of the plant reproductive cycle.

Sporopollenin   Polymer with saturated and unsaturated hydrocarbons and phenolics that are cross-linked to each other; the most resistant biopolymer on Earth.

Spur   Variously shaped outgrowth of staminal filaments or connectives in Melastomataceae, Ericaceae, and a few other taxa; also used for tubular nectar-containing outgrowths of the calyx or corolla, the nectar spur.

Stamen   Pollen-bearing part of a flower, composed of a filament (stalk) and anther (pollen sacs, sporangia) (see Figure 4.16).

Staminate flower   Flower with an androecium (stamen or stamens) but not a functional gynoecium (carpel or carpels).

Staminode   Sterile stamen.

Standard petal   *See* Banner petal.

Starch   White, tasteless, solid carbohydrate composed of chains of glucose units; usually the carbohydrate storage compound of green plants.

Stellate hair   Hair with branches radiating outward like rays from a star.

Stem   The plant axis bearing leaves with axillary buds at the nodes separated by internodes; usually above ground.

Steroids   Hydrophobic molecules based on four connected carbon rings; often physiologically active.

Stigma   Part of the carpel (or several fused carpels) that receives and facilitates the germination of the pollen (see Figure 4.16).

Stipe   Petiole of a fern leaf; sometimes also used more generally to apply to any stalk, as in stipitate carpels.

Stipule scar   Mark left on the stem after a stipule has fallen off (abscised).

Stipule   One, usually of a pair of appendages located on either side of (or on) the petiole base; part of the leaf.

S-type plastid   Plastid in the sieve tube elements that accumulates starch.

Stolon   Rhizome having an elongated internode; see Rhizome.

Stoma (pl. stomata)   Controllable opening between two guard cells in the epidermis of the vascular plants (and mosses); also called a **stomate**.

Strigose   Having stiff hairs, all pointing in one direction.

Strobilus (pl. strobili)   Conelike cluster of modified, spore-bearing leaves inserted directly on an axis.

Style   More or less elongated part of the carpel (or of several fused carpels) between the stigma(s) and the ovary (see Figure 4.16), specialized for pollen tube growth.

Styloid   Large, elongated crystal with acute or square ends.

Stylopodium (pl. stylopodia)   Enlarged, nectariferous, basal portion of the style of the flowers of Apiaceae.

Subsidiary cell   Epidermal cell lying next to a guard cell and differing in size and/or shape from other epidermal cells.

Subspecies (pl. subspecies)   An infraspecific rank, usually applied to geographical races within a morphologically variable species.

Succulent   Fleshy and juicy.

Suffrutescent plant   Plant with woody base and herbaceous upper portion.

Sulcate   Pollen grain with a sulcus.

Sulcus   Long, grooved aperture positioned at the pole of a pollen grain.

Superior ovary   Ovary that arises above the point of insertion of the other flower parts (see Figure 4.21).

Superposed bud   Bud (or buds) located above or below the axillary bud.

Sylleptic shoot   Axillary shoot that elongates at the same time as the shoot on which it is borne, the branch thus often lacking basal bud-scale scars but with an elongated first internode.

Sympatric   Occurring together in the same geographical region.

Sympatric speciation   Speciation occurring without establishment of a geographical barrier; in plants, often involving hybridization and subsequent polyploidization.

Sympetalous flower   Flower with fused petals.

Symplesiomorphy   Shared ancestral character state.

Sympodial shoot   Shoot formed through activity of a series of axillary meristems.

Synandrous   Having the stamens fused.

Synapomorphy   Shared derived character state.

Syncarpous   Carpels fused, forming a compound pistil.

Syconium (pl. syconia)   Accessory and multiple fruit characteristic of figs (species of *Ficus,* Moraceae).

Syncolpate   Condition resulting from the fusion of the furrows (colpi) of a pollen grain.

Synergids   Two cells next to the egg in the female gametophyte of the angiosperms (see Figure 4.40).

Syngameon   Most inclusive unit of interbreeding in a group of hybridizing species.

Syngenesious stamens   Stamens fused by their anthers.

Synonym   One of two or more names applied to the same taxon.

Synsepalous flower   Flower with fused sepals.

Syntepalous flower   Flower with fused tepals.

Systematics   The science of organismal diversity; frequently used in a sense roughly equivalent to taxonomy.

Systematist   A person trained in the discipline of systematics.

Tannins   Yellow, astringent, bitter phenolic compounds.

Tapetum   Innermost layer of the anther wall, which produces enzymes, hormones, and nutritive materials used during pollen formation.

Taproot   Major root of seed plants; usually enlarged and growing downward, sometimes more or less aborted.

Tautonym   Name of a species in which the specific epithet exactly repeats the generic name; not permissible in the botanical code.

Taxon (pl. taxa)   Group of organisms at any level in the taxonomic hierarchy.

Taxonomic category   Rank of the taxonomic hierarchy, e.g., kingdom, division, class, order, family, genus, and species (see Table 2, page 515).

Taxonomic revision   Taxonomic study similar to a monograph, but less comprehensive, and usually including keys, descriptions, illustrations, and distribution maps for a particular taxon.

Taxonomist   Person trained in the discipline of taxonomy.

Taxonomy   Theory and practice of grouping individuals into species, arranging species into larger groups, and giving those groups names, thus producing a classification.

Tectate exine   Exine with the outer layer, or tectum, supported by numerous small columns (columellae) (see Figure 4.48).

Tegmen (pl. tegmenta)   Inner part of the seed coat, developing from the inner integument.

Tendril   Elongated and twining structure (modified from an inflorescence, leaf, or stem) assisting in climbing.

Tepal   One of a series of perianth parts that are not differentiated into calyx and corolla.

Terminal   Occurring at the tip or apex of a structure.

Terminal bud   Bud at the apex of a stem (in a monopodial shoot).

Terminal style   Style arising from the apex of an ovary.

Terpenoids   Structurally diverse compounds formed by the union of five-carbon isopentenoid pyrophosphate units formed in the mevalonic acid pathway (see pages 97–98.)

Tertiary vein   Vein that branches from a secondary vein, typically forming either a net-like (reticulate) or a ladder-like (percurrent, scalariform) pattern in a leaf.

Testa (pl. testae)   Outer part of the seed coat, developing from the outer integument.

Tetracytic   Stomate surrounded by four subsidiary cells.

Tetrad   Group of four; usually applied to pollen grains united in a group of four that have not separated after meiosis.

Tetradynamous stamens   With four long and two short stamens.

Tetraploid   With four complete sets of chromosomes in a cell.

Thorn   Reduced, sharp-pointed stem.

Three-ranked leaves   Leaves borne in three planes along the stem.

Thrum   Short-styled floral form in a heterostylous species.

Tomentose   Having densely matted soft hairs.

Topology   Pattern of branching of an evolutionary tree or cladogram.

Tracheid   Imperforate, elongate, thick-walled, water-conducting cell with bordered pits, occurring in xylem and dead at maturity.

Tree   A woody plant with a single main trunk (or bole). (Also used as an abbreviation for "evolutionary tree," or cladogram.)

Trichome   *See* Hair.

Tricolpate   Pollen grain with three long, grooved apertures (see Figure 4.46).

Tricolporate Pollen grain with three long, grooved apertures, each with a central pore (see Figure 4.46).

Trifoliolate leaf Compound leaf with three leaflets; often called trifoliate (see Figure 4.5).

Trigger flower Flower with one or more moveable parts that cause the pollen to be forcibly deposited on the pollinator.

Trilacunar node Node with three leaf gaps (see Figure 4.33).

Triploid With three full sets of chromosomes in a cell.

Triporate Pollen grain with three equatorial, porelike apertures (see Figure 4.47).

Tristichous *See* Three-ranked.

Tristyly Flowers of different individuals of the same species having three different style lengths, the stamen lengths often varying inversely; an outcrossing mechanism.

Triterpenes Thirty-carbon terpenoid compounds. (See page 98.)

Truncate Apex or base appearing as if cut off at the end, nearly straight across (see Figures 4.10, 4.11).

Trunk Main stem of a tree below the branches.

T-shaped hairs Branched hair in the shape of the letter T (see Figure 4.15).

Tuber Swollen, fleshy portion of a rhizome or root involved in water and/or carbohydrate storage.

Tubule Little tube, as on the anthers of species of *Vaccinium* (blueberries) and relatives (in the Ericaceae).

Twice-pinnately compound leaf Leaf with two orders of pinnate axes and leaflets borne on second order axes (see Figure 4.5).

Twining Spiraling around a support in order to climb.

Two-ranked leaves Leaves borne along just two sides of a stem, i.e., all leaves in the same plane.

Type specimen Name-bearing specimen of a species or infraspecific taxon.

Ultrastructure Structures best observed using a transmission electron microscope.

Umbel Determinate or indeterminate inflorescence in which all flowers have pedicels of equal or unequal length that arise from a single region at the apex of the inflorescence axis (see Figures 4.29, 4.30).

Unicellular hair Hair composed of only a single cell (see Figure 4.15).

Unifacial leaf Leaf that has both surfaces alike, i.e., not anatomically differentiated into an adaxial and abaxial surface.

Unifoliolate leaf Compound leaf that through evolutionary reduction has only a single leaflet; typically distinguished from a simple leaf by the presence of a distinct joint or a pulvinus at the blade-petiole junction, e.g., *Citrus*, many *Berberis*.

Unilacunar node Node with a single leaf gap (see Figure 4.33).

Uninformative character Character that is not useful in constructing hypotheses of phylogenetic relationship.

Uninomial Name consisting of one word, such as the name of a genus (e.g., *Erica* or *Acer*) or a family (Ericaceae or Sapindaceae).

Uniseriate With a single series or whorl of structures, as a single whorl of bracts, or perianth parts.

Uniseriate hair Hair composed of a single row of cells (see Figure 4.15).

Unisexual flower Flower lacking either an androecium (stamens) or a gynoecium (carpels); see Imperfect.

Univalent A solitary chromosome for which there is no homologue in the cell, so that during meiosis it does not pair with another chromosome.

Utricle Small, indehiscent, dry fruit with a thin wall (bladderlike) that is loose and free from the single seed.

Valid publication Publication of scientific names in accordance with criteria for valid publication outlined in the *International Code of Botanical Nomenclature* (see page 518).

Valvate Meeting at the edges without overlapping, as sepals or petals when in the bud.

Valve One of the segments of a dehiscent fruit, separating from other such units at maturity in order to release the seeds.

VNTR Variable number tandem repeats; DNA regions that are repeated a large number of times throughout the genome.

Variety An infraspecific rank below that of subspecies although used by some as equivalent to it; usually used for geographical races and well-marked ecotypes within morphologically variable species of plants.

Vascular cambium Cylinder of dividing cells that produces secondary xylem toward the center and secondary phloem toward the outside of a stem (see Figure 4.38).

Vascular tissues Conducting tissues, xylem and phloem.

Vasculum Cylindrical metal container used to carry freshly collected plant specimens.

Vein Vascular bundle, usually visible externally, as in a leaf.

Velutinous Velvety.

Venation Pattern of veins in a leaf or other plant part.

Venn diagram In systematics, a representation using a set of nested ovals, circles, or parentheses to describe nested monophyletic groups (see Figure 2.4A).

Vernation Way in which leaves or perianth parts are folded or arranged in the bud in relation to one another.

Versatile Structure, such as an anther, that is attached at its midpoint.

Vessel Series of cells that are joined to form a tubelike structure, the adjacent ends more or less breaking down, functioning in water transport in xylem.

Vessel element One of the cellular components of a vessel (see Figure 4.31).

Vestured pit   Pit in the xylem with the pit cavity wholly or partially lined with small projections from the cell wall.

Villous   Covered with long, fine, soft hairs.

Vine   Herbaceous climbing plant.

Viscid   Covered with a sticky substance.

Viscin   Elastic and/or somewhat sticky material often covering pollen grains.

Vitta (pl. vittae)   Aromatic oil or resin canal in the pericarp of the fruits of many Apiaceae.

Viviparous fruit   Fruit with the seeds germinating while fruit is still attached to the parent plant, as in many mangrove species.

Wagner parsimony   Type of parsimony used in a cladistic analysis in which all characters are treated as ordered.

Weighting   Assignment of differential importance to characters in a phenetic or cladistic analysis.

Whorled leaves   Three or more leaves at a single node (see Figure 4.4).

Wing   Thin, often membranous extension or outgrowth of a structure. Also used for the lateral pair of petals in the flowers of Fabaceae subfam. Faboideae and for the lateral petaloid sepals of many Polygalaceae.

Wood ray   Ribbonlike aggregate of cells extending radially in the secondary xylem or wood.

Woody   Hard in texture (containing secondary xylem).

Xerophyte   Plant adapted to dry conditions.

Xylem   Water-conducting tissue of the vascular plants, composed mainly of tracheids and/or vessel elements and parenchyma cells.

Zonate apertures   Ring-shaped or band-shaped apertures of a pollen grain.

Zygomorphic symmetry   *See* Bilateral symmetry.

Zygote   First cell of a sporophyte; formed by the fusion of the sperm and egg at fertilization.

Taxonomic Index

Boldface items refer to **family discussions**. Page numbers in *italics* refer to material in illustrations

Subject Index

Page numbers in *italics* refer to material in illustrations